GEOLOGISTS
AND
THE HISTORY OF GEOLOGY

GEOLOGISTS AND THE HISTORY OF GEOLOGY

An International Bibliography from the Origins to 1978

by

WILLIAM A.S. SARJEANT

Volume Four

Geologists Indexed by Country and Specialty

M

First published 1980 by

THE MACMILLAN PRESS LTD
London and Basingstoke

Companies and Representatives throughout the world

British Library Cataloguing in Publication Data
ISBN 0 333 29393 2

Manufactured in the United States of America

Section VIII

INDEX OF GEOLOGISTS BY NATIONALITY AND COUNTRY

Introduction

The index here provided is divided into two sections. In the first, geologists are indexed according to their nationality *at the time of their birth.* It would have been desirable to index geologists by adopted nationality also, but this proved too difficult to ascertain. Biographies or obituary notices do not usually make it clear whether a geologist, though he might have lived in a particular country for a number of years, ever became a naturalized citizen of that country; and many persons cling tenaciously to their nationality of birth, however long they might live in another land.

Since geologists are highly mobile people, the country of birth is not always, not even usually, the only country in which they worked. Many economic geologists, in particular, work during their lives in many different countries; other geologists pay visits of varying duration to other countries and, during that time, contribute to the elucidation of the geology of those countries. Yet others have worked at length on the rocks, minerals or fossils of lands they never visited. Some geologists never undertook geological work in their country of birth, for reasons of early emigration or of lack of employment opportunities. In contrast, in many countries (e.g. such African nations as Chad and Mali), all geological work has until recently been done by geologists coming from elsewhere. In the second index, therefore, geologists are indexed according to the countries in which they worked or undertook field work and research; or whose geology, as represented by specimens or fossils, they studied from afar.

In both sections, for convenience, a chronological subdivision is also adopted, into four periods:

I. Prior to 1800
II. 1800-1849
III. 1850-1899
IV. 1900 to present

In the nationality list, the period chosen refers to the date of birth of the subject; thus a geologist born in 85 B.C. or one born as late as 1799 would be shown as "Prior to 1800". In the list of countries worked in, I have tried to indicate the period in which the subject's work was done in that country or on that country's geology; but this is more difficult. Even when I was aware of the exact period when the geologist visited a country, I was not always so sure when his work on the geology of that country began or ended, for work done in preparation for a visit is always possible and, even after leaving that country and completing the resultant work, a continuing interest in its geology is probable. In consequence, a geologist's name will in many cases appear in the lists for two or, rarely, three successive periods.

Prospectors, diviners and mining engineers are treated in similar fashion;

however, since they are not truly geologists, their names are segregated into appendices at the end of these sections.

As noted in the Introduction to Section VI (Volume 2), countries are wherever possible identified by their present names, though familiar earlier names are cross-indexed. However, in some cases this has proved difficult, since biographical notices refer to work in French Equatorial Africa, Indochina, or some other geographic unit nowadays represented on the map by a cluster of nations; the geologist referred to may have worked in all these or only in one. For these reasons, such old names appear also here and are cross-indexed to their present component nations. There are instances also where biographical notices are tiresomely unspecific, referring to work in Central America, South East Asia or just Africa; since, though such references are vague, they might nevertheless be helpful in tracking down a particular geologist, they are indexed here as "Asia (unspecified)", "Central America (unspecified)" etc.

ALGERIA
1850-1899
GENTIL, Louis, (1868-1925)

ARGENTINA
Up to 1800
MUÑIZ, Francisco Javier, (1795-1871)

1800-1849
MERCERAT, Alcides, (-)

1850-1899
AMEGHINO, Florentino, (1854-1911)
CASTELLANOS, Alberto, (1896-1968)
JURADO, Martín Doello, (1884-1948)
KRAGLIEVICH, Lucas, (1886-1932)
MARENGO, Paulo, (1863-1929)
MORENO, Francisco Josué Pascasio, (1852-1919)
PASTORE, Franco, (1885-1958)
PEIRANO, Abel Antonio, (1896-1969)
POZZI, Santiago, (-1929)

1900 to present
BORRELLO, Angel V., (1918-1971)
CATTOI, Noemí Violeta, (1911-1965)
DOLGOPOL DE SÁEZ, Mathilde, (1901-1957)
HARRINGTON, Horacio Jaime, (1910-1973)
LEANZA, Armando F., (1919-1975)
MENÉNDEZ, Carlos Alberto, (1921-1976)
MORGAN, Eilir Evans, (?1911-1969)
MORTOLA, Edelmira Ines, (-1973)
SOBRAL, José María, (-1961)
TEZON, Roberto Vitelmo, (1921-1973)

ARMENIAN SSR (USSR)
1800-1849
ARZRUNI, Andreas, (1847-1898)

AUSTRALIA
1800-1849
HANNAN, Patrick, (1842-1925) [PROSPECTORS, DIVINERS AND MINING ENGINEERS]
PETTERD, William Frederick, (1849-1910)

1850-1899
ANDREWS, Ernest Clayton, (1870-1948)
BRAGG, William Lawrence, Sir, (1890-1971)
BRYAN, Walter Heywood, (1891-1966)
COMBE, A. D., (1893-1949)
COTTON, Leo Arthur, (1883-1963)
DART, Raymond Arthur, (1893-)
DUGGLEBY, Arthur Francis, (1893-1945)
DUNSTAN, Benjamin, (1864-1933)
ELLIOT SMITH, Grafton, Sir, (1871-1937)
HALL, Thomas Sergeant, (1858-1915)
JACK, Robert Lockhart, (1878-1964)

AUSTRALIA 1850-1899 Cont'd.

JUNNER, Norman Ross, (1891-1973)
LASSETER, Lewis Harold Bell "Das", (1883-1930) [PROSPECTORS, DIVINERS AND MINING ENGINEERS]
LEAHY, Michael, (-) [PROSPECTORS, DIVINERS AND MINING ENGINEERS]
MACDONALD, Alexander, (-) [PROSPECTORS, DIVINERS AND MINING ENGINEERS]
MADIGAN, Cecil Thomas, (1889-1946)
MENNELL, Frederick Philip, (1880-1966)
MORGAN, Peter (Percy) Gates, (1867-1927)
OSBORNE, George Davenport, (1899-1955)
RICHARDSON, Gwen, (-) [PROSPECTORS, DIVINERS AND MINING ENGINEERS]
SIMPSON, Edward Sydney, (1875-1939)
SINGLETON, F. A., (-1947)
SPENGLER, Erich, (1886-1962)
STILLWELL, Frank Leslie, (1888-1963)
TAYLOR, Griffith, (1881-1963)
TEALE, Edmund Oswald, Sir, (1874-1971)
THOMAS, David, (1880-1934)
TILLEY, Cecil Edgar, (1894-1973)
WARD, Leonard Keith, (1879-1964)
WILMOTH, Leslie James, (1881-1963)
WILSON, Reginald Charles, (1888-1943)
WOOLNOUGH, Walter George, (1876-1958)

1900 to present
ALDERMAN, Arthur Richard, (1901-)
ALEXANDER, Geoffrey Newman, (1908-1975)
CARROLL, Dorothy, (1907-1970)
CLUNE, Frank, (-) [PROSPECTORS, DIVINERS AND MINING ENGINEERS]
CRESPIN, Irene, (-)
DIPROSE, Kenward Thomas, (1949-1977)
EDWARDS, Austin Burton, (1909-1960)
FRASER, John A., (-) [PROSPECTORS, DIVINERS AND MINING ENGINEERS]
HILL, Dorothy, (-)
MILES, Keith Rodney, (1915-1972)
RAGGATT, Harold George, Sir, (1900-1968)
SULLIVAN, C. J., (1914?-)

Time unknown
DULHUNTY, John, (-)
RANDS, William Henry, (-)

AUSTRIA
Up to 1800
ANKER, Matthias Joseph, (1771-1843)
FICHTEL, Leopold von, (fl.1798)
HAIDINGER, Wilhelm Karl von, Ritter, (1795-1871)
PARTSCH, Paul, (1791-1856)
POPPELACK, Joseph, (?1780-1859)
RIEPL, Franz Xaver, (1790-1857)

AUSTRIA Up to 1800 Cont'd.

ROSTHORN, Franz von, Edler, (1795-1877)

1800-1849

ABERLE, Carl, (1818-1892)
AICHHORN, Sigmund, (1814-1892)
ALTH, Alois von, Edler, (1819-1886)
ANDRIAN-WERBURG, Ferdinand von, Freiherr, (1835-1914)
ASCHER, Hans, (1848-1953)
AUINGER, Mathias, (?1810-1890)
BRAUER, Friedrich Moritz, (1832-1904)
BREZINA, Maria Aristides, (1848-1909)
CANAVAL, Josef Leodegar, (1820-1898)
EHRLICH, Franz Carl, (1808-1886)
FALLAUX, Cornelius, (-1885)
FITZINGER, Leopold Joseph, (1802-1884)
FREYER, Heinrich, (1802-1865)
FUGGER, Friedrich Eberhard, (1842-1919)
GOBANZ, Alexander, (1826-1895)
GOBANZ, Josef, (-1899)
GRIESBACH, Karl Ludolf, (1847-1907)
HABERFELNER, Josef, (1830-1913)
HARPER, Lewis, (-)
HAUER, Franz von, Ritter, (1822-1899)
HOCHSTETTER, Ferdinand von, Ritter, (1829-1884)
HOERNES, Moritz, (1815-1868)
HOLLER, Anton, (1826-1909)
KARRER, Felix, (1825-1903)
KASTNER, Karl, (1847-1907)
KENNGOTT, Gustav Adolf, (1818-1897)
KNER, Rudolf, (1810-1869)
KOCH, Gustav Adolf, (?1846-1921)
KORNHUBER, Andreas, (1824-1905)
KRAŠAN, Franz, (1840-1907)
KREUTZ, Felix Franz Xaver, (1844-1910)
KUDERNATSCH, Johann, (1819-1856)
LANG, Victor von, Edler, (1838-1921)
LOEHR, August von, Ritter, (1847-1917)
LUDWIG, Ernst, (1842-1915)
MARSCHALL, August Friedrich, Count of Burgholtzhausen and Tromsdorf, (1805-1887)
MOJSISOVICS, Johann August Georg Edmund, Edler von Mojsvàr, (1839-1907)
MOLL, Johann Paul Karl von, (fl.1798)
PAYER, Julius von, Ritter, (1841-1912)
PICHLER, Adolf, (1819-1900)
PILAR, Georg, (1847-1893)
PRINZINGER, Heinrich, (1822-1908)
REDTENBACHER, Anton, (1841-1911)
REYER, Eduard, (1849-1914)
ROLLE, Friedrich, (1827-1887)
RUMPF, Johann, (1841-1923)
RUSSEGGER, Josef, (-)
SAUER, Julius, (1849-1911)
SCHENK, Joseph August, (1815-1891)

AUSTRIA 1800-1849 Cont'd.

SCHMIDT, Oskar, (1823-1886)
SCHRAUF, Albrecht, (1837-1897)
SEELAND, Ferdinand, (1822-1901)
SOHM, Bonifaz, Pater, (1847-1923)
STEINDACHNER, Franz, (1834-1919)
SUESS, Eduard, (1831-1914)
TAURER-GALLENSTEIN, Hans von, Ritter, (1846-1927)
TIETZE, Emil, (1845-1931)
TOULA, Franz, (1845-1920)
UNGER, Franz Joseph Andreas Nicolas, (1800-1870)
WEINBERGER, Isidor, (1838-1915)
WOLF, Heinrich, (1825-1882)
ZEPHAROVICH, Victor Leopold von, Ritter, (1830-1890)
ZUGMAYER, Heinrich, (1840-1917)
ZWANZIGER, Gustav Adolf, (1837-1893)

1850-1899

ADAMETZ, Lotte, (1879-1966)
AMPFERER, Otto, (1875-1947)
ARTHABER, Gustav von, Edler, (1864-1943)
BANDL, Erich, (1880-1960)
BAYER, Josef, (1882-1931)
BERWERTH, Friedrich Martin, (1850-1918)
BLASCHKE, Friedrich, (1883-1911)
BOBIES, Carl August, (1898-1958)
BÖHMERSHEIM, August Böhm von, Edler, (1858-1930)
BRUDER, George, (1856-1916)
CZERMAK, Friedrich, (1890-1960)
DIENER, Carl, (1862-1928)
DITTLER, Emil, (1882-1945)
DOELTER y CISTERLICH, Cornelio August, (1850-1930)
DREGER, Julius, (1861-1945)
DRUGČEVIČ, Ferdinand, (1856-1928)
FIRTSCH, Georg, (1859-1914)
FOLGNER, Raimund, (1882-1916)
FRAUSCHER, Karl, (1852-1914)
FRIEDL, Karl, (1898-1966)
FUSSENEGGER, Siegfried, (1894-1960)
GEYER, Georg, (1857-1938)
GÖTZINGER, Gustav, (1880-1969)
GRANIGG, Bartel, (1883-1951)
GRENGG, Roman, (1884-1972)
GROSSPIETSCH, Oskar, (-1920)
GÜLDNER, Max, (1872-1947)
GUTMANN, Max von, Ritter, (1857-1930)
HAITINGER, Max, (1868-1946)
HAMMER, Wilhelm, (1875-1942)
HANDLIRSCH, Anton, (1865-1935)
HATTEY, Hans, (1889-1955)
HAUSER, Alois, (1899-1955)
HEINSHEIMER, Georg J., (1887-1970)
HERITSCH, Franz, (1882-1945)
HILBER, Vincenz, (1853-1931)
HIMMELBAUER, Alfred, (1884-1943)

AUSTRIA 1850-1899 Cont'd.

HOERNES, Rudolf, (1850-1912)
HOFMANN, Elise, (1889-1955)
HUSSAK, Franz Eugen, (1856-1911)
IPPEN, Joseph Anton, (-?1918)
JÄGER, Robert, (1890-1915)
KAMPTNER, Erwin, (1889-1972)
KAUTSKY, Fritz, (1890-1963)
KITTL, Ernst Anton Leopold, (1854-1913)
KLEBELSBERG, Raimund von, (1886-1957)
KNETT, Josef, (1869-1946)
KOBER, Leopold, (1883-1970)
KOECHLIN, Rudolf Ignatz, (1862-1939)
KÖLBL, Leopold, (1895-1970)
KORNITZER, Louis, (1873-) [PROSPECTORS, DIVINERS AND MINING ENGINEERS]
KOSSMAT, Franz, (1871-1938)
KRASSER, Fridolin, (1863-1922)
KREJCI-GRAF, Karl, (-?1956)
KÜHN, Othmar, (1892-1969)
KYRLE, Georg, (1887-1937)
LECHNER, Karl, (1899-1958)
LICHTENECKER, Norbert, (1897-1938)
LÖWL, Ferdinand, (1856-1908)
MACHATSCHKI, Felix Karl Ludwig, (1895-1970)
MAY DE MADIIS, Alexis Freiherr von, (1852-1911)
MOHR, Hannes, (1882-1967)
MURR, Josef, (-1932)
NOTH, Rudolf, (1884-1954)
OSTADAL, Rudolf, (1897-1970)
PENCK, Walther, (1888-1923)
PETRASCHEK, Wilhelm, (1876-1967)
PIA, Julius von, (1887-1943)
RACHOY, Josef, (-1902)
REBEL, Hans, (1861-)
REISER, Karl August, (1853-1922)
RICHARZ, Stephan, Pater, (1874-1934)
RIEDEL, Josef, (1891-1966)
ROSENBURG, Georg, (1897-1969)
ROSIWAL, August Karl, (1860-1923)
ROTKY, Otto, (1870-1951)
SCHAFFER, Franz Xavier, (1876-1954)
SCHARF, Willi, (1896-1971)
SCHARIZER, Rudolf, (1859-1935)
SCHMIDT, Walter, (1885-1945)
SCHUSTER, Max, (1856-1887)
SCHWINNER, Robert Gangolf, (1878-1953)
SEEMAN, Fritz, (1884-1914)
SINGER, Max, (1874-1952)
SÖLCH, Johann, (1883-1951)
SOMMERMEIER, Leopold, (1885-1963)
SPITZ, Albrecht, (1883-1918)
SPREITZER, Hans, (1897-1973)
STINY, Josef, (1880-1958)
STREIFF-BECKER, Rudolf, (1873-1959)

AUSTRIA 1850-1899 Cont'd.

SUESS, Franz Eduard, (1867-1941)
TELLER, Friedrich, (1852-1913)
TERTSCH, Hermann, (1880-1963)
TRAUTH, Friedrich, (1883-1967)
TROLL-OBERGFELL, Oskar von, (1882-1972)
TSCHERNIG, Emil, (1896-1970)
VETTERS, Hermann, (1880-1941)
WAAGEN, Lukas, (1877-1959)
WÄHNER, Franz, (1856-1932)
WAITZ, Paul, (1876-1961)
WENGER, Marian, (1864-1939)
WETTSTEIN, Richard, (1863-1931)
WINKLER-HERMADEN, Artur, (1890-1963)

1900 to present

ABERER, Ferdinand, (1913-1965)
BESCHEL, Roland, (1928-1971)
BISTRITSCHAN, Karl, (1916-1957)
BÜRGL, Hans, (1907-1966)
CLAR, Eberhard, (1902-)
FRITSCH, Wolfgang, (1928-1970)
HASSINGER, Hugo, (-1952)
HOCHSTETTER, Harald Egbert von, Ritter, (1902-1972)
HOINKES, Herfried, (1916-1973)
JURASKY, Karl Alfons, (1903-1945)
JUST, Theodor Karl, (1904-1959)
KÜMEL, Friedrich, (1908-1955)
KÜPPER, Klaus, (1930-1957)
LINSBAUER, Alfred, (1943-1969)
OSBERGER, Rudolf A. J., (1924-1972)
PATTEISKY, Karl, (-?1967)
REITHOFER, Otto, (1902-1965)
SCHIENER, Alfred, (1906-1962)
SCHLAGER, Max, (1903-1972)
SCHUSTER, Rudolf, (1931-1968)
WEINBERGER, Ludwig, (1911-1966)
WEINFURTER, Emil, (1904-1968)
WICHE, Konrad, (1913-1969)

Time unknown

KERNER, Fritz, (-)

BANGLA DESH

1900 to present

RASHID, Muhammad Abdur, (1938-1969)

BELGIUM

Up to 1800

BOODTS, Anselmus Boëtius de, (1550-1634)
CRAHAY, Jacques Guillaume, (1789-1855)
ENGELSPACH, Auguste, (LARIVIÈRE, pseud.), (1799-1831)
HELMONT, Johann Baptista van, (1578-1644)
OMALIUS D'HALLOY, Jean-Baptist Julien d', (1783-1875)
SAUVEUR, Dieudonné Jean Joseph, (1797-1862)
SCHMERLING, Philippe Charles, (1791-1836)
VAN DER MAELEN, Philippe Marie Guillaume, (1795-1869)
VILLEFAGNE D'INGIHOUL, Hilarion Noel de, Baron, (-?1837)

1800-1849

BENEDEN, Edouard van, (1846-1910)
BENEDEN, Pierre Joseph van, (1809-1893)
BOSQUET, Joseph Augustin Hubert, (1814-1880)
BRIART, Alphonse, (1825-1898)
BUOHY, Victor, (1843-1887)
CAUCHY, F. P., (-?1842)
CHAPUIS, Félicien, (1824-1879)
COEMANS, Eugène H. L. G., (1825-1871)
CORNET, François Léopold, (1834-1887)
CRÉPIN, François, (1830-1903)
DE LA VALLÉE POUSSIN, Charles Louis Joseph Xavier, (1827-1903)
DELVAUX, Émile, (1837-1901)
DEWALQUE, Gilles Joseph Gustave, (1828-1905)
DORMAL, Victor, (-1900)
DUMONT, André Hubert, (1809-1857)
DUPONT, Édouard François, (1841-1911)
FIRKET, Adolphe, (1837-1905)
HABETS, Alfred, (1839-1908)
KONINCK, Laurent Guillaume de, (1809-1887)
LE HON, Henri Sebastian, (1809-1872)
MALAISE, Constantin H. L., (1834-1916)
MALHERBE, Renier, (-1891)
MORREN, Charles, (1807-1858)
MOURLON, Michel Félix, (1845-1915)
NYST, Pierre Henri, (1813-1880)
RENARD, Alphonse François, (1842-1903)
RUTOT, Aimé Louis, (1847-1933)
SMEYSTERS, Joseph, (1837-?1908)
SOREIL, Gustave Joseph, (1842-1907)
STORMS, Raymond, (1834-1897)
VAUX, Adolphe de, (-1899)
VINCENT, Gérard, (?1824-1899)
WAEL, Norbert Charles Louis de, (1817-1901)

1850-1899

BUTTGENBACH, Henri, (1874-1964)
CONRAD, Walter, (1888-1943)
CORNET, René Jules, (1865-1929)
DESTINEZ, Pierre, (-1911)

BELGIUM 1850-1899 Cont'd.

DORLODOT, Henry de, Chanoîne, (1855-1929)
DRUGMAN, Julien Léon Eugène, (1875-1950)
FORIR, Henri, (1856-1907)
FOURMARIER, Paul Frédéric Joseph Lucien Lauren, (1877-1970)
FRAIPONT, Julien, (1857-1910)
LAMBERT, Paul, (1869-1950)
LANCASTER, A., (-?1908)
LEDOUX, Auguste, (1887-1918)
LIBERT, Joseph, (1853-?1920)
LOHEST, Max, (1857-1926)
MEUNIER, Fernand, (1868-1926)
PAUW, L. F. de, (-)
PERGENS, Eduard Willem Gerard, (1862-1917)
RENIER, Armand Marie Vincent Joseph, (1876-1951)
SALÉE, Achille, Chanoine, (1883-1932)
SCHOEP, Alfred Ferdinandus Maria Ghislenus, (1881-1966)
SPRING, Walthère, (-)
STAINER, Xavier, (1865-)
STRAELEN, Victor Emile van, (1889-1964)

1900 to present
BILLIET, Valère Louis, (1903-1945)
DE DYCKER, Raymond, (1903-1947)
JAMOTTE, André, (1903-1951)
LECKWIJCK, Willem Peter van, (1902-1975)
LECOMPTE, Marius, (1902-1973)
TAZIEFF, Haroun, (1914-)
VANDENDRIESSCHE, Adrien, (1914-1940)
VERHOOGEN, Jean, (-)

BOLIVIA
1900 to present
REYES, Felix Celso, (1924-1972)

BRAZIL
Up to 1800
ANDRADA E SILVA, José Bonifácio de, (1762-1838)

1850-1899
COSTA SENA, Joaquim Candido da, (1852-1919)
DE ARAUJO FERRAZ, Jorge Palmiro, (1883-1926)
OLIVEIRA, Euzebio Paulo de, (1882-1939)

BULGARIA
1850-1899
ZLATARSKI, Georgi Nikolov, (1854-1909)

BURMA
1850-1899
CHHIBBER, Harbans Lal, (1899-1955)

CANADA

Up to 1800

GESNER, Abraham, (1797-1864)
LOGAN, William Edmond, Sir, (1798-1875)
ROY, Thomas, (-1842)

1800-1849

BAIN, Francis, (1842-1894)
BELL, Robert, (1841-1917)
BILLINGS, Elkanah, (1820-1876)
BILLINGS, Walter R., (?1849-1920)
BOWMAN, Amos, (1839-1894)
BROWN, Charles Barrington, (1839-1917)
BROWN, Henry Yorke Lyell "Stoney", (1844-1928)
DAWSON, George Mercer, (1849-1901)
DAWSON, John William, Sir, (1820-1899)
DOUGLAS, James, (1837-)
ELLS, Robert Wheelock, (1845-1911)
HARRINGTON, Bernard J., (1848-1907)
HARTT, Charles Frederic, (1840-1878)
ISBISTER, Alexander Kennedy, (1822-1883)
LAFLAMME, Joseph Clovis Kemner, Monseigneur, (1849-1910)
MATTHEW, George Frederic, (1837-1923)
MILLER, Willet Green, (1839-1925)
MURRAY, John, Sir, (1841-1914)
OGILVIE, William, (1846-1912)
ROTHWELL, Richard Pennefeather, (1836-1901)

1850-1899

ADAMS, Frank Dawson, (1859-1942)
ALCOCK, Frederick James, (1888-1972)
ALLAN, John Andrew, (1884-1955)
AMI, Henry Mark, (1858-1931)
BANNERMAN, Harold MacColl, (1897-1976)
BARLOW, Alfred Ernest, (1861-1914)
BATEMAN, Alan Mara, (1889-1971)
BELL, Alfred Hannam, (1895-1977)
BELL, James Macintosh, (1877-1934)
BENSLEY, Benjamin Arthur, (1875-1934)
BLACK, Davidson, (1884-1934)
BOWEN, Norman Levi, (1887-1956)
BOWLES, Oliver, (1877-1958)
BOYD, Walter Halcro, (1878-1960)
BROCK, Reginald Walter, (1874-1935)
BRUCE, Everend Lester "Louie", (1884-1949)
CAIRNES, Delorme Donaldson, (1879-1917)
CAMERON, A. E., (-)
CAMSELL, Charles, (1876-1958)
CHIPMAN, Kenneth G., (-)
COLEMAN, Arthur Philemon, (1852-1939)
COLLINS, William Henry, (1878-1937)
CONNELL, Frederick M., (1885-) [PROSPECTORS, DIVINERS AND MINING ENGINEERS]
CORNWALL, Ira Edmund, (1875-1962)
CROSS, Julian Gifford, (1889-)
DALY, Reginald Aldworth, (1871-1957)

CANADA 1850-1899 Cont'd.

DOUGLAS, George Vibert, (1892?-1958)
DOWLING, Donaldson Bogart, (1858-1925)
DRESSER, John Alexander, (1866-1954)
DRYSDALE, Charles Wales, (1885-1917)
DYER, William Spafford, (1894-1941)
ELLSWORTH, Hardy Vincent, (1889-1952)
EVANS, Charles Sparling, (1894-1962)
FERRIER, Walter F., (1865-1950)
GIROUX, N. J., (1859-1896)
HAWLEY, James Edwin, (1897-1965)
HAYES, Albert Orion, (1882-1965)
HUME, George Sherwood, (1893-1965)
LaBINE, Charles L., (1888-) [PROSPECTORS, DIVINERS AND MINING ENGINEERS]
LABINE, Gilbert Adelore, (1890-) [PROSPECTORS, DIVINERS AND MINING ENGINEERS]
LAMBE, Lawrence Morris, (1863-1919)
LOW, Albert Peter, (1861-1952)
MACKENZIE, Andrew Nicholls "Kelley", (1888-1951)
MACOUN, James Melville, (-1920)
MATTHEW, William Diller, (1871-1930)
McCONNELL, Richard George, (1857-1942)
McINNES, William, (1858-1925)
McLEARN, Frank Harris, (1885-1964)
MERRITT, Clifford A., (-)
MITCHELL, George M., (-) [PROSPECTORS, DIVINERS AND MINING ENGINEERS]
MOORE, Elwood S., (1879-1966)
MYLIUS, Louis Aubrey, (1890-1965)
PARKS, William Arthur, (1868-1936)
PENHALLOW, David Pearce, (1854-1910)
POOLE, Henry Skeffington, (1884-1917)
SANDERSON, James Owen Gresham "Pete", (1898-1963)
SANDS, John Melville, (1883-1962)
SOUTHWORTH, Charles, (1880-1968)
SPENCER, Joseph William Winthrop, (1851-1921)
STEWART, Grace Anne, (1893-1970)
STEWART, Irvine E., (1893-1965)
SUTTON, William John, (1859-1914)
TANTON, Thomas Leslie, (1890-1971)
THOMSON, Joseph Ellis, (1882-1944)
TYRRELL, Joseph Burr, (1858-1957)
UGLOW, William Lawrence, (1884-1926)
WALKER, Thomas Leonard, (1867-1942)
WARREN, Percival Sidney, (1890-1970)
WEEKS, Ludlow Jackson, (1899-1964)
WILLMOTT, Arthur Brown, (1866-1914)
WILSON, Alfred William Gunning, (1873-1954)
WILSON, Alice Evelyn, (1881-1964)
YOUNG, George Albert, (1898-1947)

1900 to present

AHO, Aaro E., (-1977)
AMBROSE, John Willis, (-1974)

CANADA 1900 to present Cont'd.

BAILEY, Ralph Fraser, (1920-1953)
BATEMAN, John Danvers, (1910-1965)
BEACH, Hugh Hamilton, (1909-1973)
BEAUCHEMIN, Pierre, (-) [PROSPECTORS, DIVINERS AND MINING ENGINEERS]
BOYLEN, Matthew James, (1911?-) [PROSPECTORS, DIVINERS AND MINING ENGINEERS]
BROCK, Byron Britton, (1904-1972)
BROWN, Eldon L., (-) [PROSPECTORS, DIVINERS AND MINING ENGINEERS]
BYRNE, Jerome Cotter, (1910-)
CALEY, John Fletcher, (1903-1971)
CRICKMAY, Geoffrey W., (1905-1971)
CURRIE, Philip John, (1949-)
FAIRBAIRN, Harold Williams, (1906-)
FRASER, Horace John "Dutch", (1905-1969)
GILLANDERS, Earl B., (-)
GOWER, John A., (1921-1972)
HARQUAIL, James A., (-)
HAYCOCK, Maurice, (1900-)
HELLENS, Alexander Daniel, (1911-) [PROSPECTORS, DIVINERS AND MINING ENGINEERS]
HEWITT, Robert Leigh, (1911-1956)
HUESTIS, Herman H., (1907-) [PROSPECTORS, DIVINERS AND MINING ENGINEERS]
HUTCHINS, Forbes M., (-)
JOHNSON, Hope, (-)
JOHNSTON, Ashton William, (1910-)
JOUBIN, Francis Renault, (1911-)
JOWSEY, Robert J., (1905-) [PROSPECTORS, DIVINERS AND MINING ENGINEERS]
KAY, George Marshall, (1904-1975)
LANDES, Robert William, (1910-1969)
LEDINGHAM, Glen Wallace, (1911-1978)
LINES, Frederic George, (1921-1963)
MACKENZIE, Graham Stewart, (1905-1960)
MALOUF, S. E., (1914-)
McBEAN, John Williams, (1913-)
MEEN, Victor Ben, (1910-1971)
MICHENER, Charles, (-)
MOORHOUSE, Walter Wilson, (1913-1969)
MOSHER, Alex, (1900-) [PROSPECTORS, DIVINERS AND MINING ENGINEERS]
MOSHER, Murdock C., (-) [PROSPECTORS, DIVINERS AND MINING ENGINEERS]
NEALE, Ernest Richard Ward, (1923-)
OLLERENSHAW, Neil, (-)
PERRY, Thomas Gregory, (1919-1972)
PHELPS, Frederick George, (1913-1964)
REDPATH, James B., (-) [PROSPECTORS, DIVINERS AND MINING ENGINEERS]
ROBERTSON, Andrew, (1903?-) [PROSPECTORS, DIVINERS AND MINING ENGINEERS]
RUSSELL, Loris Shano, (1904-)

CANADA 1900 to present Cont'd.

SHEARMAN, Harold Leigh, (1921-1974)
SMITH, James Robert, (1925-1972)
SPRATT, Joseph Grant, (1904-1978)
SPRINGER, Karl J., (-) [PROSPECTORS, DIVINERS AND MINING ENGINEERS]
SPROULE, John Campbell, (1905-1970)
STRANGWAY, David William, (1934-)
THOMPSON, Robert Mitchell, (1918-1967)
THOMSON, Robert, (1902-1970)
WARING, Waldo W., (1903-1955)
WARREN, Harry V., (?1904-)
WATTS, Murray Edmund, (1909-)
WATTS, Murray Edmund, (1909-) [PROSPECTORS, DIVINERS AND MINING ENGINEERS]
WEBB, John Benwell, (1901-1975)
WHITE, William Harrison, (1913-1972)
WILLIAMSON, John Thorburn, (1907-1958)
WILSON, John Tuzo, (1908-)

CHINA
Up to 1800
TU WAN, (fl.1115-1133)

1850-1899
CHAO, Yatseng T., (1898-1929)
TING, Ven Kiang, (?1887-1936)

1900 to present
YEE, Bing Q., (1924-1973)

COSTA RICA
1850-1899
TRISTÁN, José Fidel, (-1932)

CYPRUS
1900 to present
HADJISTAVRINOU, Yangos, (1934-1978)

CZECHOSLOVAKIA
Up to 1800
BORN, Ignaz von, Baron, (1742-1791)
ERCKER, Lazarus, (-1593)
KOMENSKÝ, Jan Ámos, (1592-1670)
LINDACKER, Jan Tadeáš, (1768-1816)
PRESL, Jan Svatopluk, (1791-1849)
PRESL, Karl Borivoj, (1794-1852)
REUSS, Franz Ambros, (1761-1830)
ŠTERNBERK, Kašpar Mariá, Hrabe, (1761-1838)
ZIPPE, František Xaver Maximilian, (1791-1863)
ZIPSER, Christian Andreas, (1783-1864)

1800-1849
BABÁNEK, František, (1836-1910)
BOŘICKÝ, Emanuel, (1840-1881)

CZECHOSLOVAKIA 1800-1849 Cont'd.

CORDA, August Josef, (1811-1849)
ČŽJŽEK, Johann Baptist Anton Karl, (1806-1855)
FEISTMANTEL, Karel, (1819-1885)
FEISTMANTEL, Otokar, (1848-1891)
FOETTERLE, František, (1823-1876)
FRIČ, Antonín, (1832-1913)
FRITSCH, Anton, (1832-1913)
HELMHACKER, Rudolf Václav, (1840-1915)
HÖFER-HEIMHALT, Hans, (1843-1924)
HOLUB, Emil, (1847-1902)
JOKÉLY, Johann, (1826-1862)
KOLENATÝ, Bedřich, (1812-1864)
KREJČÍ, Jan, (1825-1887)
KŘIŽ, Martin, (1841-1916)
KUSTA, Johann, (1845-1900)
LAUBE, Gustav C., (1839-1923)
MAKOWSKY, Alexander, (1833-1908)
MELION, Josef Vincenc, (1813-1905)
MENDEL, Johann Gregor, (1822-1884)
PAUL, Carl Maria, (1838-1900)
PETERS, Karl, (1825-1881)
PETTKO, Ján, (1812-1890)
PORTH, Emil, (1832-1858)
POŠEPNÝ, František, (1836-1895)
REUSS, August Emanuel von, Ritter, (1811-1873)
SCHMIDL, Adolph A., (1802-1863)
SIMONY, Friedrich, (1813-1896)
STOLICZKA, Ferdinand, (1838-1874)
TSCHERMAK VON SEYSENEGG, Gustav, (1836-1927)
VACEK, Michael, (1848-1925)
VRBA, Karel, (1845-1922)
WOLDŘICH, Jan Nepomuk, (1834-1906)

1850-1899

ABSOLON, Karel, (1877-1960)
BARVÍŘ, Jindřich Ladislav, (1863-1952)
BAUR, Georg Hermann Carl Ludwig, (1859-1893)
BAYER, Edvin Jaroslav, (1862-1927)
BAYER, František, (1854-1936)
BECKE, Friedrich Johann Karl, (1855-1931)
BOČEK, Antonín, (1880-1956)
BURKART, Eduard, (1865-1941)
ČAPEK, Václav, (1862-1926)
ČEPEK, Ladislav, (1899-1974)
ČERMÁK, Jiři, (1884-1955)
CHUDOBA, Karl F., (1898-1976)
CORNU, Felix, (1882-1909)
CVICIĆ, Jovan, (1865-1927)
DĚDINA, Václav, (1870-1956)
FENDRYCH, Otakar, (1860-1945)
GARTNER, Otokar, (1889-1961)
GRÄNZER, Josef, (1857-1934)
HANUŠ, František, (1860-1937)
HEIDE, Friedrich, (1891-)

GEOLOGISTS, etc., BY COUNTRY BORN

CZECHOSLOVAKIA 1850-1899 Cont'd.

HIBSCH, Josef Emanuel, (1852-1940)
HLAUSCHEK, Hans, (1896-1971)
HOFMANN, Adolf, (1853-1913)
HYNIE, Ota, (1899-1968)
JAHN, Jaroslav Jiljí, (1865-1934)
KAFKA, Josef, (1858-1929)
KAŠPAR, Jan, (1898-)
KATZER, Bedřich, (1861-1925)
KETTNER, Radim, (1891-1967)
KLOUČEK, Celda, (1855-1935)
KLVAŇA, Josef, (1857-1919)
KNIES, Jan, (1860-1925)
KODYM, Odolen, (1898-1963)
KOKTA, Jaroslav, (1894-)
KOVÁŘ, František, (1862-1941)
KRATOCHVÍL, Josef, (1878-1958)
KRYZ, Ferdinand, (1879-1958)
KUBIËNA, Walter L., (1897-1970)
KUČERA, Bruno, (1863-1934)
MAŠKA, Karel Jaroslov, (1851-1915)
MATĚJKA, Alois, (1898-)
MOSCHELESOVÁ, Julie, (1892-1956)
NOVÁK, Otomar Pravoslav, (1851-1892)
NOWACK, Ernst, (1891-1946)
OPPENHEIMER, Josef, (1883-1933)
PAĆAK, Oldřich, (1882-)
PALLA, Eduard, (1864-1922)
PERNER, Jaroslav, (1869-1947)
PETRBOK, Jaroslav, (1881-1960)
POČTA, Filip, (1859-1924)
PROCHÁZKA, Vladimír Josef, (1862-1913)
PROKOP, František, (1898-)
PURKYNĚ, Cyril, (1862-1937)
REDLICH, Karl A., (1869-1942)
REMEŠ, Mořic, (1867-1959)
RODLER, Alfred, (1861-1890)
ROSICKÝ, Vojtěch, (1880-1942)
RUDOLPH, Karl, (1881-1937)
RYBA, Frantisek, (1867-1918)
RYŠ, Josef, (1879-1960)
RZEHÁK, Anton, (1855-1923)
SCHUBERT, Richard Johann, (1876-1915)
ŠETLÍK, Jaromír, (1896-)
ŠIMEK, Antonín, (1887-1942)
SLAVÍK, František, (1876-1957)
SLAVÍKOVÁ, Ludmila, (1890-1943)
SMOLAŘ, Gotthard, (1863-1926)
SMOLÍK, Ladislav, (1895-1960)
SPACEK, Karel, (1866-1937)
SPIREK, Vincenz, (1852-1907)
ŠTĚRBA-BÖHM, Jan Stanislav, (1874-1938)
STOČES, Bohuslav, (1890-)
ŠUF, Jiří, (1891-1976)
SVOBODA, Josef, (1898-)

CZECHOSLOVAKIA 1850-1899 Cont'd.

UHLIG, Victor Karl, (1857-1911)
ULRICH, František, (1899-1941)
VELENOVSKÝ, Josef, (1858-1949)
VESELÝ, Václav, (1896-1956)
ZAHÁLKA, Čeněk, (1856-1938)
ZÁRUBA, Quido, (1899-)
ŽELÍZKO, Jan Vratislav, (1874-1938)

1900 to present
AMBROŽ, Vojtěch, (1905-1962)
AUGUSTA, Josef, (1903-1968)
BĚHOUNEK, Rudolf, (1902-1974)
BOUČEK, Bedřich, (1904-1975)
ČECH, Vladimír, (1901-)
ČECHOVIČ, Vsevolod, (1900-1961)
CEHÁK, Karl, (1903-1957)
CULEK, Antonin, (1904-1944)
DVOŘÁK, Arnošt, (1904-)
DVOŘÁK, Josef, (1904-)
FIALA, František, (1903-)
HEJTMAN, Bohuslav, (1901-)
HLÁVKA, Karel, (1905-)
HÓRAK, Lubomír, (1935-1962)
HROMADA, Karel, (1919-1974)
JELÍNEK, František, (1911-1966)
KALÁŠEK, Jan, (1903-)
KAMENICKÝ, Jakub, (1917-)
KETTNEROVÁ, Marja (Marie), (1900-1933)
KOUTEK, Jaromír, (1902-)
KRUTA, Tomáš, (1906-)
KUNSKÝ, Josef, (1903-)
LUCEK, Josef, (1930-1969)
NĚMEC, František, (1905-1972)
NĚMEJC, František, (1901-)
NETOLICKÝ, Jan, (1934-1963)
NOVÁČK, Radim, (1905-1942)
NOVÁK, Jiří, (?1902-1971)
NOVOTNÝ, Miroslav, (1921-1971)
PAUK, František, (1905-)
PLAS, Vladimír, (1905-)
PRADÁČOVÁ, Marcella, (1912-1972)
PŘIBYL, Alois, (1904-)
PROCHÁZKA, Miloš, (1923-1956)
ROST, Rudolf, (1902-)
SALÁT, Ján, (1919-1973)
SEKANINA, Josef, (1900-)
ŠPINAR, Zdeněk Vlastimil, (1906-)
STEINOCHER, Vladimír, (1902-1972)
STEJSKAL, Jan, (1902-1972)
SVOBODA, J. V., (1921-1972)
TUČEK, Karel, (1906-)
URBAN, Karel, (1905-)
URBÁNEK, Ladislav, (1901-)
VACHTL, Josef, (1907-)

CZECHOSLOVAKIA 1900 to present Cont'd.

VOSYKA, Stanislav, (1902-)
VRABKA, František, (1924-1957)
ZAPLETAL, Karel, (1903-1972)
ZÁZVORKA, Vlastislav, (1903-)
ŽEBERA, Karel, (1901-)
ZELENKA, Ladislav, (1900-1931)
ZOUBEK, Vladimír, (1903-)
ZPALETAL, Karel, (1903-1972)

DENMARK

Up to 1800

BARTHOLIN, Thomas, (1616-1680)
BARTHOLINUS, Erasmus, (1625-1698)
BRAHE, Tyge (Tycho), (1546-1601)
BRÜNNICH, Morten Thrane, (1737-1827)
ESMARCH, Jens, (1762-1839)
FORCHHAMMER, Johan Georg, (1794-1865)
MOLTKE, Adam Gottlob, Grev, (1710-1762)
PINGEL, Christian, (1793-1852)
STENO, Nicolaus, (1631-1687)
WAD, Gregers, (1755-1832)
WORM, Olle, (1588-1654)

1800-1849

COHEN, Emil Wilhelm, (1842-1905)
HARBOE, Edouard Georg, (1845-1919)
HOFF, Thomas Christian Frederik August, (1822-1882)
JESPERSEN, Johan Magnus Peder, (1833-1917)
JOHNSTRUP, Johannes Frederik, (1818-1894)
LUND, Peter Wilhelm, (1801-1880)
LÜTKEN, Christian Fredrik, (1827-1901)
PINGEL, Johann Victorinus, (1834-1919)
STEENSTRUP, Johannes Japetus Smith, (1813-1897)
TOPSØE, Haldor Frederik Axel, (1842-1935)
WALLICH, George Charles, (1815-1899)

1850-1899

BØGGILD, Ove Balthazar, (1872-1956)
BØGVAD, Richard, (1897-1952)
CHRISTENSEN, Carl Christian, (1860-1935)
CLAUSEN, Hans Carl Eisenreich, (1898-1971)
CLÉMENT, Charles Adolphe Antoine, (1860-1933)
FABER, Harald Nicolai, (1856-1943)
GRÖNWALL, Karl Anders Axel, (1869-1944)
HARTZ, Nikolaj Eeg Kruse, (1867-1937)
JESSEN, Axel Hans, (1868-1952)
KOCH, Lauge, (1892-1964)
KORNERUP, Andreas, (?1856-1880)
MADSEN, Marius Kristian, (1899-) [PROSPECTORS, DIVINERS AND MINING ENGINEERS]
MADSEN, Victor Christian, (1865-1947)
MALLING, Carl Peter, (1862-1927)
MILLER, Paul C., (1876-1945)
MILTHERS, Paul Christian Vilhelm Madsen, (1865-1962)
MORTENSEN, Theodor, (1868-1952)
NIELSEN, Kristian Brünnich, (1872-1942)
NORDMANN, Valdemar Johan Heinrich, (1872-1962)
NØRREGAARD, Erik Martin, (1880-1959)
POULSEN, Christian, (1896-1975)
RAVN, Jesper Peter Johansen, (1866-1951)
RØRDAM, Kristian, (1860-1939)
ROSENKJAER, H. N., (1851-1907)
RUTZEBECK, Hjalmar, (-) [PROSPECTORS, DIVINERS AND MINING ENGINEERS]
SCHWENNESEN, Alvin Theodore, (1883-1962)

DENMARK 1850-1899 Cont'd.

SORENSEN, Carl Christian, (1893-1965)
THORODDSEN, Thorvaldur, (1855-1921)
TUXEN, Christian Fr. A., (1850-1903)
USSING, Niels Viggo, (1864-1911)
WINGE, Herluf, (1857-1923)

1900 to present

BJERRUM, Laurits, (1919-1973)
LYSHEDE, Jens Martinus, (1911-1961)
MILTHERS, Keld, (1907-1960)
NØRVANG, Aksel, (1914-1968)
OLSEN, Gregers Vang, (1929-1960)
SKOU, Kristian, (1912-1967)
SORGENFREI, Theodor, (1915-1972)

EGYPT

1850-1899

HIGAZY, Riad Abdel-Megid, (1919-1967)

ENGLAND

Up to 1800

ADAM, William, (1794?-1873)
AIKEN, Arthur, (1773-1854)
ALLEN, William, (1770-1843)
ANNING, Mary, (1799-1847)
ARNOLD, Joseph, (1784-1818)
ARTIS, Edmund Tyrell, (1789-1847)
ASHMOLE, Elias, (1617-1692)
AULD, William, (fl.1810-1812)
AUSTIN, Thomas, Fort-Major, (1794-1881)
BABBAGE, Charles, (1792-1871)
BABINGTON, William, (1756-1833)
BAKER, William, (1787-1853)
BAKEWELL, Robert, (1768-1843)
BANKS, Sir Joseph, (1743-1820)
BARRATT, Joseph, (1796-1882)
BARROW, John, Sir, (1764-1848)
BEDDOES, Thomas, (1760-1808)
BEVIS (or BEVANS), John, (1693-1771)
BIGSBY, John Jeremiah, (1792-1881)
BOLTON, John, (1788-1873)
BOWERBANK, James Scott, (1797-1877)
BRADBURY, John, (1765-1823)
BRIGHT, Richard, (1789-1858)
BROOKE, Henry James, (1771-1857)
BROWN, E., (fl.1685)
BROWN, John, (1780-1859)
BRYDONE, Patrick, (1736-1818)
BUCKLAND, William, Reverend, (1784-1856)
BULLOCK, William, (1795-1840)
BURNET, Thomas, (1635-1715)
BUSHELL, Thomas, (?1594-1674) [PROSPECTORS, DIVINERS AND MINING ENGINEERS]
BUTLER, Samuel, (1774-1839)
CAVENDISH, Henry, (1731-1810)
CHILDREN, John George, (1777-1852)
CLARKE, Edward Daniel, (1769-1822)
CLARKE, William Branwhite, Reverend, (1798-1878)
CLIFT, William, (1775-1849)
COLBY, Thomas F., Major-General, (?1783-1852)
CONYBEARE, John Josiah, Reverend, (1779-1824)
CONYBEARE, William Daniel, Reverend (1787-1857)
COOPER, Thomas, (1759-1840)
CUMBERLAND, George, (1754-1848)
CURTIS, John, (1791-1862)
DALE, Samuel, (1659-1739)
DAMPIER, William, (1652-1715)
DARWIN, Erasmus, (1731-1802)
DARWIN, Francis Sacheverell, Sir, (1786-1859)

ENGLAND Up to 1800 Cont'd.

DAUBENY, Charles Giles Brindle, (1795-1867)
DAVY, Humphry, Sir, (1778-1829)
DE LA BECHE, Henry Thomas, Sir, (1796-1855)
DILLWYN, Lewis Weston, (1778-1855)
DIXON, Frederick, (1799-1849)
DRAYTON, Michael, (1563-1631)
EDWARDS, Frederic Erasmus, (1799-1875)
ENYS, John Samuel, (1797-1872)
ERLE, C., Reverend, (1790-1870?)
FAREY, John, (1766-1826)
FEATHERSTONHAUGH, George William, (1780-1866)
FOSTER, Westgarth, (1772-1835)
FOX, Robert Were, (1789-1877)
GILBERT, Davies, (1767-1839)
GILBERT, William, (1544-1603)
GRAHAM, Maria, Mrs., (1785-1842)
GREENOUGH, George Bellas, (1778-1855)
GREENWOOD, George, Colonel, (1799-1875)
GREGOR, William, Reverend, (1761-1817)
GREVILLE, Charles Francis, Right Honourable, (1749-1809)
HAILSTONE, John, Reverend, (1759-1847)
HALL, Elias, (1764-1853)
HALLEY, Edmond, (1656-1742)
HAMILTON, William, Sir (1730-1803)
HARIOT, Thomas, (1560-1621)
HARRIS, William, (1797-1877)
HEBERDEN, Thomas, (1703-1769)
HENNAH, Richard, Reverend, (?1765-1846)
HENSLOW, John Stevens, Reverend, (1796-1861)
HERSCHEL, John Frederick William, Sir, (1792-1871)
HESLOP, Mary Kingdon, (?1884-?1954)
HEULAND, John Henry, (1778-1856)
HIBBERT-WARE, Sameul, (1782-1848)
HILL, John, Sir, (1706-1775)
HOBBS, William, (fl.1700-1715)
HOLLAND, Henry, Sir, (1788-?1873)
HOME, Everard, Sir, (1756-1832)
HOOKE, Robert, (1635-1703)
HOOKER, William Jackson, Sir, (1785-1865)
HOPE, Frederick William, Reverend, (1797-1862)
HOPKINS, William, (1793-1866)
JENKIN, William, (1738-1820) [PROSPECTORS, DIVINERS AND MINING ENGINEERS]
JORDEN, Edward, (1569-1632)
JOSSELYN, John, (ca.1608-1675)
KEENE, William, (1798-1872)
KEILL, John, (1671-1721)
KEIR, James, (1735-1820)
KIDD, John, (1775-1851)
KIRBY, William, Reverend, (1759-1851?)
LARKIN, Nathaniel John, (1781-1855)
LETTSOM, John Coakley, (1744-1815)
LISTER, Martin, (1638-1712)
LONSDALE, William, (1794-1871)

ENGLAND Up to 1800 Cont'd.

MacENERY, John, Reverend, (1796-1841)
MACLEAY, William Sharp, (1792-1865)
MAJENDIE, Ashurst, (?1784-1867)
MALTHUS, Thomas Robert, (1766-1834)
MANTELL, Gideon Algernon, (1790-1852)
MARTIN, Peter John, (1786?-1860)
MARTIN, William, (1767-1810)
MATON, William George, (1774-1835)
MAWE, John, (1764-1829)
MERRET, Christopher, (1614-)
MICHELL, John, Reverend, (1724-1793)
MIERS, John, (1789-1879)
MILLINGTON, John, (1779-1886)
NEWCOMEN, Thomas, (1663-1739) [PROSPECTORS, DIVINERS AND MINING ENGINEERS]
NICOLS, Thomas, (fl. 1652)
NORTHAMPTON, Joshua Alwyne Spencer, Marquis of, (1790-1851)
NUTTALL, Thomas, (1786-1859)
OTLEY, Jonathan, (1766-1856)
PALEY, William, Archdeacon, (1743-1805)
PARISH, Woodbine, Sir, (1796-1882)
PARKINSON, James, (1755-1824)
PENNANT, Thomas, (1726-1798)
PHILLIPS, Richard, (1778?-1851)
PHILPOT, Elizabeth, (1780-1857)
PHILPOT, Mary, (1777-1838)
PLATT, Joshua, (1699-1773)
PLOT, Robert, (1640-1696)
PORTLOCK, Joseph Ellison, Major-General, (1794-1864)
PRATT, Samuel Peace, (1789-1863)
PRIESTLEY, Joseph, (1733-1804)
PYE SMITH, John, (1774-1851)
RASHLEIGH, Philip, (1729-1811)
RAY, John, (1627-1705)
RENNELL, James, Major, (1742-1830)
RICHARDSON, Benjamin, Reverend, (-1832)
ROSE, Caleb Burrell, (1790-1872)
ROYLE, John Forbes, (-1857?)
SANDERS, William, (1799-1875)
SCROPE, George Poulett, (né Thomson), (1797-1876)
SEDGWICK, Adam, Reverend, (1785-1873)
SHERLEY, Thomas, (1638-1678)
SIMPSON, Martin, (1798-1892)
SMITH, John, Captain, (1580-1631)
SMITH, William, (1769-1839)
SOWERBY, George Brettingham, I, (1788-1854)
SOWERBY, James, (1757-1822)
SOWERBY, James de Carle, (1787-1871)
STOKES, Charles, (1783-1853)
STRACHEY, John, (1671-1743)
STRANGE, John, (1732-1799)
STUKELEY, William, (1687-1765)
STUTCHBURY, Samuel, (1798-1859)
TAYLOR, John, (1780-1864)

ENGLAND Up to 1800 Cont'd.

TAYLOR, Richard Cowling, (1789-1851)
TOULMIN, George Hoggart, (ca.1750-1817)
TOWNSEND, Joseph, Reverend, (1739-1816)
TOWNSON, Robert, (1763-1827)
TRADESCANT, John, The Elder, (1587-1638)
TREVELYAN, Walter Calverley, Sir, (1797-1879)
TREVITHICK, Richard, (1771-1833) [PROSPECTORS, DIVINERS AND MINING ENGINEERS]
TRIMMER, Joshua, (1797-1857?)
WATERTON, Charles, (1782-1865)
WATSON, White, (1760-1835)
WEAVER, Thomas, (1773-1855)
WEBSTER, John, (1610-1682)
WEBSTER, Thomas, (1773-1844)
WELD, Isaac, Jr., (1774-1856)
WHEWELL, William, Reverend, (1794-1866)
WHISTON, William, (1666-1753)
WHITEHURST, John, (1713-1788)
WOLLASTON, William Hyde, (1766-1828)
WOOD, Nicholas, (1795-1865)
WOOD, Searles Valentine, (1798-1880)
WOODWARD, John, (1665-1728)
WOODWARD, Samuel, (1790-1838)
WOOLF, Arthur, (1766-1837) [PROSPECTORS, DIVINERS AND MINING ENGINEERS]

1800-1849

ABBOTT, George, (1844-1925)
ADAMS, John Couch, (1819-1892)
ALLPORT, Samuel, (1816-1897)
ANDERSON, Tempest, (1846-1913)
ANDREWS, W. S., (1847-1929)
ANSTED, David Thomas, (1814-1880)
ANSTIE, John, (-1900)
APLIN, Christopher d'Oyley Hale, (1819-1875)
ARCHER, Francis, (1839-1892)
ARGYLL, George Douglas, Duke of, (1823-1900)
ARMSTRONG, Henry Edward, (1848-1937)
ATHERSTONE, William Guybon, Honourable, (-1898)
ATTWOOD, George, (1845-1912)
ATTWOOD, Melville, (1812-1898)
AUSTIN, Charles Edward, (1819-1893)
AVELINE, William Talbot, (1822-1903)
BAILY, William Hellier, (1819-1888)
BAINBRIDGE, William, (1812-1860)
BAIRD SMITH, R., Colonel, (-)
BAKEWELL, Frederick Collier, (-1869)
BALSTON, William Edward, (1848-1918)
BARBER, Cecil Thomas, (-)
BARKAS, Thomas Pallister, (1819-1891)
BARLOW, Caleb, (1840-1908)
BARLOW, William, (1845-1934)
BARNES, William Henry, (1845-1935)
BARON, Richard, Reverend, (1847-1907)

ENGLAND 1800-1849 Cont'd.

BARRETT, Lucas, (1837-1862)
BARROW, John, Jr., (1808-1898)
BATES, Henry Walter, (1825-1892)
BAUERMAN, Hilary, (1833-1909)
BAYFIELD, Thomas Gabriel, (1817-1893)
BEALE, William Phipson, Sir, (1839-1922)
BEARDMORE, Nathaniel, (1816-1872)
BEASLEY, Henry Charles, (1836-1919)
BEESLEY, Thomas, (1818-1896)
BELL, Robert George, (1833-1888)
BELT, Thomas, (1832-1878)
BENNETT, Francis James, (1845-1920)
BENSTED, William Harding, (1802-1873)
BEST, Edward, (1824-1906)
BINNEY, Edward William, (1812-1881)
BIRD, Charles, (1843-1910)
BIRDS, James Adey, (1831-1894)
BLAKE, John Frederick, Reverend, (1839-1906)
BLAKE, John Hopwood, (1843-1901)
BLANFORD, Henry Francis, (1834-1893)
BLANFORD, William Thomas, (1832-1905)
BLOXAM, Thomas, (1836-?1872)
BONNEY, Thomas George, Reverend, (1833-1923)
BRADY, Antonio, Sir, (1811-1881)
BRADY, Henry Bowman, (1835-1891)
BRAYLEY, Edward William, (1802-1870)
BRINDLEY, William, (1832-1919)
BRISTOW, Henry William, (1817-1889)
BROCKBANK, Richard Bowman, (1824-1912)
BRODERIP, William John, (-1859)
BRODIE, Peter Bellinger, Reverend, (1815-1897)
BROWN, Alfred "Gogga", (1834-1920)
BROWN, John Allen, (1831-1903)
BROWN, Richard, (1805-1882)
BRYCE, James, (1806-1877)
BUCKLAND, Francis Trevelyan, (1826-1880)
BUCKMAN, James, (1814-1884)
BUNBURY, Charles James Fox, Sir, (1809-1886)
BUSK, George, (1807-1886)
BUTLER, Francis Henry, (1849-1935)
BUTLER, Samuel, (1835-1902)
CALLAWAY, Charles, (1838-1915)
CARPENTER, William Benjamin, (1813-1885)
CARTER, Henry John, (1813-1895)
CARTER, James, (1813-1895)
CASH, William, (1843-1914)
CAUTLEY, Proby T., Colonel Sir, (1802-1871)
CHADWICK, Samuel, (1845-1903)
CHAMPERKNOWNE, Arthur, (1839-1887)
CHAPMAN, Edward John, (1821-1904)
CHARLESWORTH, Edward, (1813-1893)
CHRISTY, Henry, (1810-1865)
CHURCH, Arthur Herbert, Sir, (1834-1915)
CHURCHILL, George Cheetham, (1822-1906)

ENGLAND 1800-1849 Cont'd.

CLARK, George Thomas, (1811-1898)
CLARK, John Willis, (1833-1910)
CLAYPOLE, Edward Waller, (1835-1901)
CLODD, Edward, (1840-1930)
CODRINGTON, Thomas, (1829-1918)
COLCHESTER, William, (1813-1898)
COLE, Edward Maule, Reverend, (1833-1911)
COLLINS, Joseph Henry, (1841-1916)
COOPER-KING, Charles, Lieutenant-Colonel, (1843-1898)
CORFIELD, William, (1843-1903)
COSSHAM, Handel, (1824-?1890)
CRANE, Edward, (1822-1901)
CROSS, John Edward, Reverend, (1821-1897)
CROSSKEY, H. W., Reverend, (1826-1893)
CUMMING, Joseph G., Reverend, (1812-1868)
CUNNINGTON, William, (1813-1906)
DAINTREE, Richard, (1831-1878)
DAKYNS, John Roche, (1836-1910)
DALLAS, William Sweetland, (1824-1890)
DALTON, William H., (?1848-1929)
DAMON, Robert, (1814-1889)
DARWIN, Charles Robert, (1809-1882)
DARWIN, Francis, (1848-1925)
DARWIN, George Howard, Sir, (1845-1912)
DAVIDSON, George, (1825-1911)
DAVIS, James William, (1846-1893)
DAWES, John Samuel, (1802-1878)
DAWKINS, William Boyd, Sir, (1837-1929)
DEANE, Henry, (1807-1874)
DE LA CONDAMINE, H. M., Reverend, (-1854)
DOVER, William Kinsey, (?1816-1891)
DOWKER, George, (1828-1899)
DRAPER, John William, (1811-1882)
DREW, Frederick, (1836-1891)
DUNCAN, Peter Martin, (1824-1891)
DUNN, Edward John, (1844-1937)
EGERTON, Philip de Malpas Grey, Sir, (1806-1881)
EGERTON, William Henry, Reverend, (1811-1910)
ELLIS, H., Right Honourable Sir, (-1855)
ENNISKILLEN, William Willoughby Cole, 3rd Earl of, (1807-1886)
ETHERIDGE, Robert, (1819-1903)
ETHERIDGE, Robert Jr., (1847-1920)
EVANS, Caleb, (1831-1886)
EVANS, John, Sir, (1823-1908)
FALCONER, Hugh, (1808-1865)
FEILDEN, Henry Wemyss, Colonel, (1838-1921)
FINCH, John, (fl.1823-1835)
FISHER, Osmond, Reverend, (1817-1914)
FITCH, Robert, (1802-1895)
FITZ-ROY, Robert, Vice-Admiral, (1805-1865)
FLETCHER, Hugh, (1848-1909)
FLETCHER, Thomas William, (1808-1893)
FLIGHT, Walter, (1841-1885)
FLOWER, John Wickham, (1807-1873)

ENGLAND 1800-1849 Cont'd.

FLOWER, William Henry, Sir, (1831-1899)
FOORD, Arthur Humphreys, (1845-1934)
FORBES, Edward, (1815-1854)
FOSTER, Clement Le Neve, Sir, (1841-1904)
FOX-STRANGWAYS, Charles Edward, (1844-1910)
FRANCIS, James, (1843-1925)
GALTON, Francis, Sir, (1822-1911)
GEIKIE, Archibald, Sir, (1835-1924)
GILBERT, Josiah, (1814-1892)
GLASS, Norman, Reverend, (1832-1895)
GODWIN-AUSTEN, Robert Alfred Cloyne, (1808-1884)
GOODCHILD, John George, (1844-1906)
GOULD, Charles, (-1895)
GREEN, Alexander Henry, (1832-1896)
GREEN, Upfield, (1834-1917)
GREEN, William Lowthian, (1819-1890)
GREENWELL, George Clementson, (1821-1900)
GREG, Robert Phillips, (1826-1906)
GREGORY, Augustus Charles, (1819-1905)
GREGORY, James Reynolds, (1832-1899)
GREGORY, Joshua William, (-1850)
GREY, George, Sir, (1812-1898)
GUNN, John, Reverend, (1801-1890)
GUNN, William, (1837-1902)
GUPPY, Robert John Lechmere, (1836-1916)
GURNEY, Henry Palin, Reverend, (1847-1904)
HALL, Marshall, Captain, (1831-1896)
HALL, Townshend Monckton, (1845-1899)
HANCOCK, Albany, (1806-1873)
HARGRAVES, Edward Hammond, (1816-1891)
HARKER, Allen, (1848-1894)
HARKNESS, Robert, (1816-1878)
HARMER, Frederic William, (1835-1923)
HARRISON, Benjamin, (1837-1921)
HARRISON, James, (1819-1868)
HARRISON, John Thornhill, (1815-1891)
HARRISON, William Jerome, (1845-1908)
HASTINGS, Barbara, Marchioness of, (1810-1858)
HAWKINS, Thomas, (1810-1889)
HAWKSHAW, John, Sir, (1811-1891)
HENWOOD, William Jory, (1805-1875)
HIGGINS, Henry Hugh, Reverend, (1814-1893)
HILL, William, (1849-1912)
HILLS, Richard Charles, (1848-1923)
HIND, Henry Youle, (1823-1908)
HINDE, George Jennings, (1839-1918)
HISLOP, Stephen, Reverend, (1817-1863)
HOFFMANN, George Christian, (1837-1917)
HOLGATE, Benjamin, (1838-1915)
HOLL, Harvey Buchanan, (1820-1886)
HOLMES, T. Vincent, (?1841-1923)
HONY, W. Edward, Archdeacon, (-)
HOOKER, Joseph Dalton, Sir, (1817-1911)
HOPKINSON, John, (1844-1919)

ENGLAND 1800-1849 Cont'd.

HORNE, William, (?1836-1928)
HOTBLACK, John Turner, (1848-1915)
HOW, Henry, (1828-1879)
HOWCHIN, Walter, Reverend, (1845-1937)
HOWELL, Henry Hyatt, (1834-1915)
HOWITT, Alfred William, (1830-1908)
HOWORTH, Henry Hoyle, Sir, (1842-1923)
HOWSE, Richard, (1821-1901)
HUDLESTON, Wilfrid Hudleston, (né Simpson), (1828-1909)
HUGGINS, William, Sir, (1824-1910)
HUGHES, Theodore W. H., (?1843-1907)
HUGHES, Thomas McKenny, (1832-1917)
HULKE, John Whitaker, (1830-1895)
HUNT, Arthur Roope, (1843-1914)
HURST, Thomas Grainge, (1824-1890)
HUTTON, Frederick Wollaston, (1836-1905)
HUTTON, William, (-1861?)
HUXLEY, Thomas Henry, (1825-1895)
IBBETSON, L. L. Boscawen, Captain, (-1869)
JACK, Robert Logan, (1845-1921)
JEBB, John Gladwyn, (1841-1893) [PROSPECTORS, DIVINERS AND MINING ENGINEERS]
JENKIN, Henry Charles Fleeming, (1833-1885)
JOHNSON, Henry, (1823-1885)
JONES, Thomas Rupert, (1819-1911)
JORDAN, James Brown, (1838-1915)
JUDD, John Wesley, (1840-1916)
JUKES, Joseph Beete, (1811-1869)
KAYE, Charles Turton, (1812-1846)
KEEPING, Henry, (1827-1924)
KINCH, Edward, (1848-1920)
KING, Samuel William, (1821-1868)
KING, William, (1809-1886)
KINGSLEY, Charles, Reverend, (1819-1875)
KITTO, Benedict, (1841-1928)
KOCH, Walter Edward, (1848-1916)
LAIDLAY, John Watson, (1808-1885)
LAING, Samuel, (1810-1897)
LAMBERT, Alan, (1837-1927)
LANKESTER, Edwin Ray, Sir, (1847-1929)
LAPWORTH, Charles, (1842-1920)
LA TOUCHE, James Digues, Reverend, (1824-1899)
LAW, Robert, (1840-1907)
LEBOUR, George Alexander Louis, (1847-1918)
LECKENBY, John, (1814-1877)
LEE, John Edward, (1808-1887)
LEEDS, Alfred Nicholson, (1847-1917)
LEEDS, Charles Edward, (1845-1912)
LIVEING, George Downing, (1827-1924)
LIVERSIDGE, Archibald, (1847-1927)
LLOYD, J. A., Colonel, (-1856?)
LLOYD, T. G. B., (1829-1876)
LOBLEY, James Logan, (1833-1913)
LOCKYER, Joseph Norman, Sir, (1836-1920)

ENGLAND 1800-1849 Cont'd.

LOFTUS, William Kennett, (1822-1859)
LOWRY, Joseph Wilson, (1803-1879)
LUBBOCK, John, Lord Avebury, (1834-1913)
LUCY, William Charles, (1822-1898)
LUDLAM, Henry, (1822-1880)
LYDEKKER, Richard, (1849-1915)
MACKESON, Henry Bean, (1811-1894)
MALLET, Robert, (1810-1881)
MANSEL-PLEYDELL, John Clavell, (1817-1902)
MANTELL, Walter Baldock Durant, Honourable, (1820-1895)
MARKHAM, Clements Robert, Sir, (1830-1916)
MARSHAM-TOWNSHEND, Robert Honourable, (1834-1915)
MARTEN, Henry John, (1827-1892)
MAW, George, (1832-1912)
MAWSON, Joseph, (1830-1927)
McMAHON, Charles Alexander, Lieutenant-General, (1830-1904)
MELLO, John Magens, Reverend, (1836-1914)
MEŸER, Charles John Adrian, (1832-1900)
MIALL, Louis Compton, (1842-1921)
MILLER, William Hallowes, (1801-1880)
MILLETT, Fortescue William, (1833-1915)
MILNE-EDWARDS, Henri, (1800-1885)
MITCHINSON, John, Right Reverend Bishop, (1833-1918)
MIVART, St. George, (1827-1900)
MÖCKLER, Frederick James, (?1845-1913)
MOLYNEUX, William, (1824-1882)
MOORE, Charles, (1814-1881)
MOORE, Henry Cecil, (1835-1908)
MORGAN, Edward Delmar, (1840-1909)
MORRIS, John, (1810-1886)
MORTIMER, John Robert, (1825-1911)
MORTON, George Highfield, (1826-1900)
MOSELEY, Henry Nottidge, (1845-1891)
MULLINS, John, (1838-) [PROSPECTORS, DIVINERS AND MINING ENGINEERS]
MURCHISON, Charles, (1830-1879)
MURRAY, John, III, (1808-1892)
MYLNE, Robert William, (1816-1890)
NESBIT, John Collis, (1819-1862)
NEWTON, Alfred, (1829-1907)
NEWTON, Edwin Tulley, (1840-1930)
NICHOLSON, Henry Alleyne, (1844-1899)
NORTON, Henry, (?1812-1892)
NORWOOD, Thomas Wilkinson, Reverend, (1829-1908)
ORMEROD, George Wareing, (1810-1891)
OWEN, Richard, Sir, (1804-1892)
PARKER, James, (1833-1912)
PARKER, William Kitchen, (1823-1890)
PARSONS, Henry Franklin, (1846-1913)
PATTISON, Samuel Rowles, (1809-1901)
PATTON, Andrew, (1821-1885)
PEACH, Benjamin Neeve, (1842-1926)
PEACH, Charles William, (1800-1886)
PEARCE, Chaning, (1811-1847)

ENGLAND 1800-1849 Cont'd.

PEARCE, Richard, (1837-1927)
PENGELLY, William, (1812-1894)
PENNING, William Henry, (1838-1902)
PERCEVAL, Spencer George, (1838-1922)
PERCY, John, (1817-1889)
PHILLIPS, John, (1800-1874)
PHILLIPS, John Arthur, (1822-1887)
PHILLIPS, William, (1800-1834)
PHILPOT, Margaret, (? -1845)
PIKE, John William, (1839-1874)
PIPER, George Harry, (1819-1897)
PLANT, Major John, (1819-?1892)
PLANT, Nathaniel, (1832-1875)
PLAYER, Jacob Hort, (1834-1908)
POLWHELE, Thomas Roxburgh, (1831-1909)
PORTER, Henry, (1832-1868)
POSTLETHWAITE, John, (1840-1925)
PRELLER, Charles du Riche, (1844-)
PRESTWICH, Joseph, Sir, (1812-1896)
QUEKETT, John Thomas, (1815-1861)
RANDALL, John, (1810-1910)
RANSON, John, (?1833-1866)
READE, Joseph Bancroft, Reverend, (1801-1870)
READE, Thomas Mellard, (1832-1909)
REDMAN, John Baldry, (1816-1899)
REDWOOD, Boverton, Sir, (1846-1919)
REED, William, (1810-1892)
REEVE, James, (1833-1920)
REEVE, Lovell, (1814-1865)
RICKARD, Francis Ignacio, (-)
RICKETTS, Charles, (1818-1904)
ROBERTS, George Edward, (1831-1865)
ROBINSON, Thomas, (1828-1906)
ROFE, John, (1801-1878)
RUDLER, Frederick William, (1840-1915)
RUSKIN, John, (1819-1900)
RUTLEY, Frank, (1842-1904)
SALMON, Henry Curwen, (1827-1873)
SALTER, John William, (1820-1869)
SALWEY, Humphry, (1803-1877)
SAWKINS, James Gay, (1806-1878)
SCOTT, Marcus Watson Tate, (1814-1875)
SCOULER, John, (-1874)
SEELEY, Harry Govier, (1839-1909)
SELWYN, Alfred Richard Cecil, (1824-1902)
SHARMAN, George, (1832-1914)
SHARP, Samuel, (1814-1882)
SHARPE, Daniel, (1806-1856)
SHIPMAN, James, (1848-1901)
SHORE, Thomas William, (1840-1905)
SHRUBSOLE, George William, (1827-1893)
SHRUBSOLE, William Hobbs, (1837-1927)
SLACK, Henry James, (1818-1896)
SLADE, Jeremiah, (1828-1905)

ENGLAND 1800-1849 Cont'd.

SLADEN, Walter Percy, (1849-1900)
SLATTER, Thomas James, (1834-1895)
SMITH, Augustus, (1804-1872)
SMITH, Edgar Albert, (1847-1916)
SMITH, Joshua Toulmin, (1816-1869)
SMITH, Reginald Allender, (?1847-1940)
SMITHE, Frederick, Reverend, (1822-1900)
SMYTH, Warington Wilkinson, Sir, (1817-1890)
SOLLAS, William Johnson, (1849-1936)
SOPWITH, Arthur, (1843-1920)
SOPWITH, Thomas, (1803-1879)
SORBY, Henry Clifton, (1826-1908)
SOWERBY, George Brettingham, II, (1812-1884)
SOWERBY, Henry, (1825-1891)
SPENCER, Herbert, (1820-1903)
SPIERS, William, Reverend, (-)
STANLEY, William Ford Robinson, (1828-1909)
STEPHENS, William John, (1829-1890)
STIFFE, Arthur William, Captain, (1849-1912)
STIRRUP, Mark, (1831-1907)
STOKES, Arthur Henry, (1842-1910)
STORY-MASKELYNE, Mervin Herbert Nevil, (1823-1911)
STRACHEY, Richard, General Sir, (1817-1908)
STRICKLAND, Hugh Edwin, (1811-1853)
SYMONDS, William Samuel, Reverend, (1818-1887)
TATE, Alexander Norman, (1837-1892)
TATE, George, (1805-1871)
TATE, Ralph, (1840-1901)
TAUNTON, John Hooke, (1820-1893)
TAWNEY, Edward Bernard, (1840-1882)
TAYLER, Joseph Walter, (-?1896)
TAYLOR, John Ellor, (1835-1895)
TEALL, Jethro Justinian Harris, Sir, (1849-1924)
TENISON-WOODS, Julian Edmund, Reverend, (1832-1889)
TENNANT, James, (1808-1881)
THACKERAY, Francis St. John, Reverend, (1832-1919)
THOMPSON, Beeby, (1848-1931)
THOMSON, John Turnbull, (1821-1884)
THORNTON, Richard, (1838-1863)
TIDDEMAN, Richard Hill, (1842-1917)
TOMLINSON, Charles, (1806-1897)
TOPLEY, William, (1841-1894)
TRENCH, Richard, (1836-1861)
TWELVETREES, William Harper, (1848-1919)
TYLDEN-WRIGHT, Charles, (1832-1900)
TYLER, Charles, Captain, (1826-1895)
TYLOR, Alfred, (1823-1884)
UDALL, John, (1841-1920)
USSHER, William Augustus Edmond, (1849-1920)
VALPY, Robert Harris, (1819-1904)
VICARY, William, (1811-1903)
VIVIAN, William, (1817-1879)
WALKER, John Francis, (1839-1907)
WALLACE, Alfred Russel, (1823-1913)

ENGLAND 1800-1849 Cont'd.

WARD, James Clifton, Reverend, (1843-1880)
WARD, John, (1837-1906)
WARDLE, Thomas, Sir, (1831-1909)
WATERHOUSE, John, (1806-1879)
WATSON, John, (1842-1918)
WESTON, Thomas Chesmer, (1832-1910)
WESTWOOD, John Obadiah, (1805-1893)
WETHERELL, Nathaniel Thomas, (1800-1875)
WHIDBORNE, George Ferris, Reverend, (1846-1910)
WHITAKER, William, (1836-1925)
WHITE, Walter, (-)
WHITEAVES, John Joseph Frederick, (1835-1909)
WHYMPER, Edward, (1840-1911)
WILBERFORCE, Samuel, Bishop, (1805-1873)
WILKINSON, Charles Smith, (1843-1891)
WILLIAMSON, William Crawford, (1816-1895)
WILSON, Edward, (?1849-1898)
WILTSHIRE, Thomas, Reverend, (1826-1902)
WINDOES, James, (1839-1897)
WINWOOD, Henry Hoyte, (1830-1920)
WITCHELL, Edwin, (1823-1887)
WOOD, Edward, (1808-1877)
WOOD, John George, Reverend, (1827-1889)
WOOD, Searles Valentine Jr., (1830-1884)
WOOD-MASON, James, (1846-1893)
WOODROW, James, (1828-1907)
WOODWARD, Charles Josiah, (1838-1932)
WOODWARD, Henry, (1832-1921)
WOODWARD, Horace Bolingbroke, (1848-1914)
WOODWARD, Samuel P., (1821-1865)
WORMS, Henry de, Baron, Lord Pirbright, (1840-1903)
WORTH, Richard Nicholas, (-1896)
WYNDHAM, T. Heathcote, (?1842-1876)
YOUNG, Alfred Prentice, (1841-1919)

1850-1899

ABBOTT, William James Lewis, (1853-1933)
ACLAND, Henry Dyke, (-1936)
ADAMSON, Samuel A., (-?1890)
ADYE, Ernest Howard, (1857-1918)
AIERS, Arthur Joseph, (1896-1965)
AIM, William Lennie, (?1872-1938)
AIMER, James David, (1899-1973)
ALLEN, Henry Atwool, (?1854-1934)
ANDRADE, Edward Neville da Costa, (1887-1971)
ANDREWS, Charles William, (1866-1924)
ANTHONY, Herbert Douglas, (1892-1968)
ASHCROFT, Frederick Noel, (1878-1949)
BADEN-POWELL, Donald Ferlys Wilson, (1897-1973)
BAILEY, Edward Battersby, Sir, (1881-1965)
BAILEY, Tom Esmond Geoffrey, Captain, (-1919)
BAKER, Benjamin Alfred, (1859-1939)
BAKER, Herbert Arthur, (1885-1954)
BALDWIN, Walter, (?1878-1956)

ENGLAND 1850-1899 Cont'd.

BALL, Francis Matthew, (1877-1956)
BALL, John, (-1941)
BANCROFT, John Bevis Beeston, (-1944)
BARCLAY, Francis Hubert, (1869-1935)
BARKER, Thomas Vipond, (1881-1931)
BARNES, Jonathan, (-1928)
BARNETT, Andrew Ketchen, (1852-1914)
BARRON, Thomas, (1867-1906)
BARROW, Geoffrey, (1899-1963)
BARROW, George, (1853-1932)
BARTLETT, Benjamin Pope, (-1928)
BATE, Dorothea Minola Alice, (1879-1951)
BATESON, William, (1861-1926)
BATHER, Francis Arthur, (1863-1934)
BAUER, Grace M., (1878-1965)
BAWDEN, Samuel Francis, (1854-1884)
BEADNELL, Hugh John Llewellyn, Major, (1874-1944)
BECHER, Harry Macdonald, (1855-1893)
BECKLES, Samuel H., (-1890)
BEEBY-THOMPSON, Arthur, (1873-1968)
BELINFANTE, Leo Baptiste Louis, (1861-1937)
BELL, Alfred Graham, (?1875-1964)
BEMROSE, Henry Howe, (1857-1939)
BENNETT, Arthur, (1873-1941)
BENNETT, Frederick William, (1860-1931)
BENNETT, Walter Hellyer, (1892-1971)
BENSON, William Noel, (1885-1957)
BION, Herbert Stanley, (1888-1915)
BIRLEY, Caroline, (1851-1907)
BISAT, William Sawney, (1886-1973)
BLACKETT, Patrick Maynard Stuart, Baron of Chelsea, (1897-1974)
BLACKMORE, Humphrey Purnell, (-1928)
BLAKE, George Stanfield, (-1940)
BLOOM, Edward Francis Dandridge, (-1942)
BOLTON, Herbert, (1863-1936)
BOSWELL, Katherine Cumming, (1888-1952)
BOSWELL, Percy George Hamnall, (1886-1960)
BOSWORTH, Thomas Owen, (1882-1928)
BOULTON, William Savage, (1867-ca.1954)
BOURCHIER, John Rickman, (1896-1927)
BOWER, Fredrik Orpen, (1855-1948)
BOWMAN, Herbert Lister, (1874-1942)
BRADY, Lionel Francis "Major", (1880-1963)
BRAGG, William Henry, Sir, (1862-1942)
BRAMMALL, Alfred, (?1879-1954)
BRODRICK, Harold, (1872-1946)
BROMEHEAD, Cyril Edward Nowill, (1885-1952)
BROUGHTON-EDGE, Arthur Broughton, (1895-1953)
BROWN, John Coggin, (1884-1962)
BROWNRIDGE, Charles, Colonel, (?1862-1951)
BRYDONE, Reginald Marr, (-1943)
BUCKLEY, Harold Eugene, (1897-1959)
BUCKMAN, Sidney Savory, (1860-1929)
BULL, Alfred Joseph, (1875-1950)

ENGLAND 1850-1899 Cont'd.

BULLEN, Robert Ashington, Reverend, (1850-1912)
BULLOWS, Wilfred Lowson, (?1869-1943)
BURROWS, Henry William, (?1860-1939)
BURTON, Ernest St. John, (?1874-1962)
BURTON, Richard Cooksey, (1890-1916)
BURY, Henry, (?1862-1958)
BUSK, Henry Gould, (?1891-1956)
BUTTERLEY, Archibald Donald, (1896-1957)
CANNON, Herbert Graham, (1897-1963)
CARPENTER, Philip Herbert, (1852-1891)
CARR, John Wesley, (1862-[1939])
CARRUTHERS, Robert George, (1880-1965)
CARTER, William Lower, (1855-1918)
CARUS-WILSON, C., (1857-1934)
CHAPMAN, Frederick, (1864-1943)
CHAPMAN, Sydney, (1888-1970)
CHARLESWORTH, John Kaye, (1889-)
CHUBB, Lawrence John, (?1888-1971)
CLARK, James Edmund, (1850-1944)
CLAYDEN, Arthur William, (1855-1944)
CLIFFORD, William Hugh, Right Honourable, (1858-1943)
CLOUGH, Charles Thomas, (1852-1916)
CLUTTERBUCK, Robert, (-1879)
COBBOLD, Edgar Sterling, (1851-1936)
COCKERELL, Theodore Dru Alison, (1866-1948)
COCKIN, George Marmaduke, (1852-1918)
COKE, George Elmsley, (? -1911)
COLE, Grenville Arthur James, (1859-1924)
COLENUTT, George William, (?1862-1944)
COLLINS, Arthur L., (1868-1902)
COLLINS, Henry Francis, (?1864-1924)
COLLINS, William Ion, (1897-1964)
CONNOR, James, (1894-1977)
CORNER, Frank, (1861-1939)
COX, Leslie Reginald, (1897-1965)
CRAMP, Alfred William, (?1873-1950)
CRICK, George Charles, (1856-1917)
CRICK, Walter Drawbridge, (1857-1903)
CROFTS, William Hastings, (1861-1935)
CROOK, Thomas, (1876-1937)
CROSFIELD, Margaret Chorley, (1859-1952)
CULLIS, Charles Gilbert, (1871-1941)
CUNNINGTON, C. H., Lieutenant, (1889-1918)
DAMPIER, William Cecil, Sir, (1867-1952)
DARELL, Robert Darell Smythe, (1851-1936)
DARWIN, Leonard, (1850-1943)
DAVIES, George Macdonald, (1885-1973)
DAVIES, Henry Nathaniel, (-1920)
DAVIES, Lewis Merson, Lieutenant-Colonel, (1882-1955)
DAVIS, Arthur George, (1892-1957)
DAVISON, Charles, (1858-1940)
DAVISON, Ernest Henry, (1877-1944)
DAWSON, Charles, (1864-1916)
DAY, Thomas Cuthbert, (1853-1935)

ENGLAND 1850-1899 Cont'd.

DEANE, Henry Jr., (-1924)
DEELEY, Richard Mountford, (1855-1944)
DELL, John Alexander, (1882-1964)
DENNIS, Frederick William, (1883-1939)
DE WAAL, Philip Henry Overbeck, Reverend, (-1942)
DEWEY, Henry, (1876-1965)
DEWHURST, Thomas, (1881-1973)
DIBLEY, George Edward, (1857-1943)
DIGBY, Bassett, (-)
DINES, Henry George, (1891-?1964)
DINHAM, Charles Hawker, (1883-1955)
DIXON, Ernest Edward Leslie, (1876-1963)
DIXON, Henry George Dacres, (1887-1947)
DOCWRA, Colin, (-1942)
DON, Archibald William Robertson, (1891-1916)
DOUGLAS, James Archibald, (1884-1978)
DUNCAN, Cecil Cooke, (1868-1948)
EARLE, Kenneth Wilson, (1891-1944)
EASTWOOD, Tom, (1889?-1970)
EDDINGTON, Arthur Stanley, Sir, (1882-1944)
EDMONDS, Charles, (1886-1964)
EDMUNDS, Francis Hereward, (1893-1960)
EDWARDS, Wilfred Norman, (1890-1956)
ELLES, Gertrude Lilian, (1872-1960)
ELLISON, John, (1859-1936)
ELSDEN, James Vincent, (?1856-1930)
EPPS, Frank James, (?1891-1974)
EVANS, Emile, (1883-1966)
EVANS, John William, (1857-1930)
EVANS, Percy, (1892-1974)
EYLES, Victor Ambrose, (1896-1978)
FAGG, Christopher Charles, (1883-1965)
FARMERY, Joseph Richard, (1872-1928)
FEARNSIDES, William George, (1879-1968)
FENNELL, Charles William, (?1863-1935)
FENNER, Edward George, (?1873-1941)
FERMOR, Lewis Leigh, Sir, (1880-1954)
FISHER, Ronald Aylmer, (1890-1962)
FLEET, William Frederick, (1890-1966)
FLETCHER, Lazarus, Sir, (1854-1921)
FLETCHER, Mark, Reverend, (1870-1956)
FLETT, John Smith, Sir, (1869-1947)
FOLEY, Mary Cecilia, (-1925)
FOOTE, R. Bruce, (-1912)
FORD, Hugh Alexander, (1885-1966)
FORSTER-COOPER, Clive, (1880-1947)
FOWLER, Joseph, (1872-1958)
FRANKS, Godfrey Firth, (1853-1917)
GARDINER, Charles Irving, (?1868-1940)
GARDINER, John Stanley, (1872-1946)
GARNETT, Cecil Stevenson, (1896-1951)
GARRETT, T. R. H., Major, (-1938)
GARWOOD, Edmund Johnston, (1864-1949)
GASTER, Christopher Theodore Augustine, (1878-1963)

ENGLAND 1850-1899 Cont'd.

GERRARD, John, (?1850-1920)
GIBBONS, Arthur, (1867-1938)
GIBSON, Walcot, (1864-1941)
GILBERT, Charles J., (-?1934)
GLAUERT, Ludwig, (1879-1963)
GOODE, R. H., Reverend, (1888-1969)
GOODRICH, Edwin Stephen, (1868-1946)
GOODYEAR, Edith, (1879-1959)
GORE, Charles Herbert Henry, (1866-1951)
GOSSLING, Frank Newbery, (1888-1946)
GOURLEY, Harold John Frederick, (1886-1956)
GRABHAM, George Walter, (1882-1955)
GRAHAM, William Vaux, (1859-1940)
GRANTHAM, Donald Robert, (1896-1969)
GRAY, Joseph William, (-1935)
GREEN, John Frederick Norman, (?1873-1949)
GREENE, Alfred Ernest, (1878-1940)
GREENLY, Edward, (1861-1951)
GREIG, Arthur, (1893-)
GROOM, Theodore Thomas, (1863-)
GWINNELL, Russell Frost, (1880-1913)
HALDANE, John Burdon Sanderson, (1892-1964)
HALL, Arthur Lewis, (1872-1955)
HALLIMOND, Arthur Francis, (1890-1968)
HALSE, George Whitfield, (1885-1968)
HAMLING, Joseph Green, (-?1935)
HARKER, Alfred, (1859-1939)
HARRIS, George Frederick, (1862-1906)
HARRISON, Edward Richard, Sir, (1872-1960)
HARRISON, Janet Mitchell Marr, (1893-1971)
HARRISON, John Burchmore, Sir, (1856-1926)
HARRISON, John Vernon, (1892-1972)
HARTLEY, Harold Brewer, Sir, (1878-1972)
HARTLEY, John Jerom, (1886-)
HATCH, Frederick Henry, (?1864-1932)
HAWARD, Frederick James Nairn, (1871-1953)
HAWKINS, Herbert Leader, (1887-1968)
HEARD, Albert, (1894-1944)
HERON-ALLEN, Edward, (1861-1943)
HERRIES, Robert Stansfield, (?1860-1941)
HESLOP, Septimus, (?1860-1945)
HIBBERT, Frank, Colonel, (?1889-1951)
HICKLING, Henry George Albert, (1883-1954)
HILL, James Bastian, (1861-1927)
HIMUS, Godfrey Wilfred, (1894-1964)
HIND, Wheelton, (1859-1920)
HINTON, Henry Arthur, (?1874-1937)
HINTON, Martin Alister Campbell, (1883-1961)
HITCHON, Hubert, (1899-1962)
HOBLEY, Charles William, (1867-1947)
HOBSON, Bernard, (1860-1933)
HOGBEN, George, (1853-1920)
HOLDEN, Henry Smith, (1887-1963)
HOLLAND, Thomas Henry, Sir, (1868-1947)

ENGLAND 1850-1899 Cont'd.

HOLLINGWORTH, Sydney Ewart, (1899-1966)
HOLLOWAY, Lionel E., (1898-1965)
HOLMES, Arthur, (1890-1965)
HONNOR, Frank Walter, (?1892-1976)
HOOLEY, Reginald Walter, (1866-1923)
HORNBY, John, (1880-1927) [PROSPECTORS, DIVINERS AND MINING ENGINEERS]
HOWES, George Bond, (1853-1905)
HUBBARD, George Hulme, (1899-1969)
HUDSON, Robert George Spencer, (1895-1965)
HUGHES, William John, (?1899-1961)
HUME, William Fraser, (1867-1949)
HUTCHINSON, Arthur, (1866-1937)
HUTCHINSON, Henry Neville, Reverend, (?1855-1927)
HUTTON, Harriet Mary, (1873-1937)
HUXLEY, Julian Sorell, Sir, (1887-1975)
HUXLEY, Leonard, (1860-1933)
HYLAND, John Shearson, (1866-1898)
ILLING, Vincent Charles, (1890-1969)
JAMES, W. Warwick, (?1874-1965)
JEFFREYS, Geoffrey, (1885-1953)
JENNINGS, Albert Vaughan, (1864-1903)
JESSON, Thomas, (?1850-1928)
JOHNSON, J. P., (1880-1918)
JOHNSON, Walter, (1867-1950)
JOHNSTON, Mary Sophia, (1875-1955)
JOHNSTON-LAVIS, Henry James, (1856-1914)
JONES, Ernest Lloyd, (?1863-1942)
JONES, Hubert Cecil, (1881-1963)
JONES, William Rupert, (1855-1915)
JOWETT, Albert, (1875-?1958)
JUKES-BROWNE, Alfred John, (1851-1914)
JULIAN, Henry Forbes, (1861-1912) [PROSPECTORS, DIVINERS AND MINING ENGINEERS]
JULIAN, Hester Forbes, Mrs., (1860-1934)
KEEPING, Walter, (1854-1888)
KELLOW, Moses, (-1943)
KENDALL, Percy Fry, (1856-1936)
KENNARD, Alfred Santer, (1870-1948)
KENNEDY, Horas Tristram, Lieutenant, (1889-1917)
KERR, John Graham, (1869-1957)
KIDNER, Henry, (?1851-1925)
KING, William Bernard Robinson, (1899-1963)
KITCHIN, Finlay Lorimer, (1870-1934)
KITSON, Albert Ernest, Sir, (1868-1937)
KNIPE, Henry Robert, (1855-1918)
KYNASTON, Herbert, (1868-1915)
LAKE, Philip, (1865-1949)
LAMPLUGH, George William, (1859-1926)
LANDON, Joseph, (-1902)
LANG, William Dickson, (1878-1966)
LANG, William Henry, (1874-1960)
LAPWORTH, Herbert, (1875-1933)
LARMAN, Mark Linington, (1899-1958)

ENGLAND 1850-1899 Cont'd.

LEECH, John George Chambré, (1890-1966)
LEES, George Martin, (1898-1955)
LE GROS CLARK, Wilfrid Edward, Sir, (1895-1971)
LEIGHTON, Thomas, (1858-1906)
LENOX-CONYNGHAM, Gerald Ponsonby, Colonel Sir, (1866-1956)
LEPPER, Gordon Willis, (1889-1950)
LEWIS, A. E., (-1906)
LIGHTFOOT, Ben, (1888-1966)
LINDON, Edward Byron, (1860-1891)
LOMAS, Joseph, (1860-1908)
LOMAX, James, (1857-?1934)
LONGBOTTOM, Arthur, (? -1910)
LONGDEN, Geoffrey Appleby, (1876-1964)
LONGSTAFF, Mary Jane, (née Donald), (1856-1935)
LOUIS, Henry, (1855-1939)
LYONS, Henry George, (1864-?1943)
MAIDWELL, Frederic Thomas, (1872-1921)
MANTLE, Harry G., (ca.1858-1937)
MARKS, Edward Seaborn, (?1875-1950)
MARR, Francis Alleyne, (1893-1942)
MARR, John Edward, (1857-1933)
MARSDEN, Ernest, (1889-1970)
MARSHALL, James Garth, (-1873)
MARSHALL, Patrick, (1869-1950)
MARSTON, Alvan Theophilus, (1889-1971)
MARTIN, Edgar Charles, (1884-1955)
MARTIN, Edward Alfred, (1864-1943)
MASLEN, Arthur John, (1871-1954)
MATLEY, Charles Alfred, (1866-1947)
MAUFE, Herbert Brantwood, (1879-1946)
MAWDSLEY, James Buckland, (1894-1964)
MAWSON, Douglas, Sir, (1882-1958)
MAYO, Herbert Theodore, (1889-1965)
McCABE, Joseph, (1867-1955)
McCALLUM, Robert Rowson, Lieutenant-Colonel, (1882-1934)
McCARTHY, Edward Thomas, (-) [PROSPECTORS, DIVINERS AND MINING ENGINEERS]
McCONNEL, J. C., (1860-1890)
McDONALD, William, (1876-1943)
McINTYRE, Peter, (1863-1940)
MEADOWS, Walter Hennesy, (?1864-1937)
MERRICKS, Frank, (?1863-1936)
MIDDLEMISS, Charles Stewart, (1859-1945)
MIERS, Henry Alexander, Sir, (1858-1942)
MIGEOD, Frederick William Hugh, (1872-1952)
MILNE, John, (1850-1913)
MILTON, John Herbert, (?1862-1925)
MITCHELL, Peter Chalmers, Sir, (1864-1945)
MOLYNEUX, Arthur John Charles, (?1866-1921)
MONCKTON, Horace Woolaston, (1857-1931)
MOORE, Edward William James, (1887-1970)
MOREY, Frank, (?1858-1925)
MORGAN, Conwy Lloyd, (?1852-1936)
MOYSEY, Lewis, (1869-1918)

ENGLAND 1850-1899 Cont'd.

MOY-THOMAS, James Alan, (-1944)
MUIR-WOOD, Helen Marguerite, (1896-1968)
NEWELL ARBER, Edward Alexander, (1870-1918)
NEWTON, Richard Bullen, (1850-1926)
NUTTALL, Winfred Laurence Falkiner, (1897-1969)
OATES, Frank, (1890-1945)
OCKENDEN, Maurice Albion, (1874-1954)
ODLING, Marmaduke, (?1888-1956)
OLDHAM, Richard Dixon, (1858-1936)
ORIEL, Benjamin, Reverend, (-1936)
OSMAN, Charles William, (1855-1932)
OSWALD, Felix, (1866-1958)
OVERY, Charles, Reverend, (1882-)
PALMER, L. S., (1890-1962)
PALMER, Rupert William, (1890-1922)
PARKER, Robert Luling, (1893-1973)
PARKER, Thomas Jeffrey, (1850-1897)
PARKER, William Albert, (1855-1918)
PARKINSON, Cyril, (1854?-1919)
PARKINSON, Donald, (1891-1974)
PARSONS, Ernest, (?1887-1963)
PARSONS, James, (1876-1908)
PARSONS, Leonard Miles, (1883-1956)
PEGG, E. J., (?1899-1948)
PENNINGTON, Rooke, (-?1887)
PENNY, Frederick Willoughby, (1887-1950)
PERCIVAL, Frederick George, (1887-1973)
PICKERING, William Henry, (1858-1912)
PILGRIM, Henry Guy Ellcock, (1875-1943)
PINFOLD, Ernest Sheppard, (1888-1969)
PLATNAUER, Henry Maurice, (1857-1939)
PLATT, William, (?1867-1943)
POCOCK, Roy Woodhouse, (1887-1954)
POLLARD, William, (1870-1946)
POLLOCK, Charles Murray, (1898-1969)
POULTON, Edward Bagnall, Sir, (1856-1943)
PRESTON, Henry, (1852-1940)
PRETYMAN, Frank Remington, (1891?-1917)
PRICHARD, Hesketh Vernon Hesketh, (1876-1922)
PRIEST, Simeon, (1876-1954)
PRIESTLEY, Raymond Edward, Sir, (1886-1974)
PRIOR, George Thurland, (1862-1936)
QUIRKE, Terence Thomas, (1886-1947)
RADLEY, Ernest George, (1880-1945)
RAISIN, Catherine Alice, (1855-1945)
RANSOME, Frederick Leslie, (1868-1935)
RAVEN, Charles, (1885-1964)
RAW, Frank, (?1875-1961)
READ, Herbert Harold, (1889-1970)
REED, Frederick Richard Cowper, (1869-1946)
REID, Clement, (1853-1916)
REID, Eleanor Mary, (née Wynne Edwards), (1860-1953)
REYNOLDS, Sidney Hugh, (1867-1949)
RHODES, James Eric Wynfield, (?1888-1976)

ENGLAND 1850-1899 Cont'd.

RICHARDSON, Linsdall, (1881-1967)
RICHARDSON, William Alfred, (1887-?1964)
RICKARD, Thomas Arthur, (1864-1953)
RIDLEY, Henry Nicholas, (1855-1956)
ROGERS, Arthur William, (1872-1946)
ROMANES, James, (1886-1942)
ROSS, Frank Stenhouse, (1890-1975)
ROWE, Arthur Walton, (1858-1926)
RUDMOSE BROWN, Robert Neal, (1879-1957)
RUFFORD, Philip James, (1852-1902)
RUSSELL, Arthur Edward Ian Montagu, Sir, (1878-1964)
RUSSELL, Frank Scott, (-1943)
RUTHERFORD, Hugo, Sir, (1887-1942)
RYDE, John Walter, (1898-1961)
SAINTY, James Edward, (?1882-1967)
SALTER, Alfred Edward, (?1863-1926)
SANDFORD, Kenneth Stuart, (1899-1971)
SANGER, Frederick Robert, (1874-1942)
SARGENT, Harry Lionel, (1871-1955)
SARGENT, Henry Crunden, (1851-1931)
SCHWARZ, Ernest Hubert Lewis, (1873-1928)
SCOTT, Dukinfield Henry, (1854-1934)
SEDGWICK, Adam, (1854-1913)
SELIGMAN, Gerald, (1886-1973)
SEWARD, Albert Charles, Sir, (1863-1941)
SEYLER, Clarence Arthur, (1866-1959)
SHAKESPEAR, Ethel Mary Reader, (née Wood), Dame, (1871-?1946)
SHANNON, William George St. John, (1889-1946)
SHERBORN, Charles Davies, (1861-1942)
SHERLOCK, R. L., (1875-1948)
SIBLY, Thomas Franklin, Sir, (1883-1948)
SIKES, Harold Lesley, (1881-1943)
SIMMONS, Eric Warr, Lieutenant, (1893-1915)
SIMMONS, William Charles, (1888-1951)
SIMPSON, George Clarke, Sir, (1878-1965)
SKEATS, Ernest Willington, (1875-1953)
SLATER, George, (-1956)
SLATER, Herbert Kelsall, (1875-1913)
SMALL, Evan William, (1859-1933)
SMITH, Bernard, (1881-1936)
SMITH, Herbert Gladstone, (-1950)
SMITH, Stanley, (1883-1955)
SMITH, Thomas Hodge, (1894-1945)
SOLLY, Richard Harrison, (1851-1925)
SOWERBY, Arthur de Carle, (1885-1954)
SPATH, Leonard Frank, (1882-1957)
SPENCER, Edmondson, (?1885-1970)
SPENCER, Leonard James, (1870-1959)
SPENCER, William Kingdon, (1878-1955)
STAMP, Laurence Dudley, Sir, (1898-1966)
STAPLES, Ernest Henry, (1883-1957)
STEBBING, William Pinchard Delane, (1873-1961)
STEWART, Philip Charteris Anstruther, (1874-1923)
STILES, Harold, Sir, (?1862-1946)

ENGLAND 1850-1899 Cont'd.

STOBBS, John Thomas, (?1866-1942)
STONELEY, Robert, (1894-1976)
STOPES, Henry, (1852-1902)
STOPES, Marie Charlotte Carmichael, (1880-1958)
STOW, George, (1876-?1955)
STRAHAN, Aubrey, Sir, (1852-1928)
STRAW, Stephen Henry, (?1892-1963)
SUTCLIFFE, William Henry, (1855-1913)
SUTTON, John Richard, (1864-1937)
SWEET, George, (-?1920)
SWEETING, George Scotland, (1889-1977)
SWINNERTON, Henry Hurd, (1875-1967)
TAYLOR, William P., (1859-) [PROSPECTORS, DIVINERS AND MINING ENGINEERS]
TEMPLEMAN, Andrew, (1887-1945)
THOMAS, Herbert Henry, (1876-1935)
THOMAS, Ronald Hawksby, Major, (1885-?1940)
THOROLD, William George Preston, (?1892-1943)
TILLYARD, Robin John, (1881-1937)
TIMMINS, Arthur, Major, (?1858-1940)
TIPPER, George Howlett, (1881-1947)
TOMBLESON, Frederick Binns, (?1872-1938)
TREACHER, Llewellyn, (1859-1943)
TRECHMANN, Charles Otto, (1851?-1917)
TRECHMANN, Charles Taylor, (1885-1964)
TRELAWNEY-ANSELL, E. C., (-) [PROSPECTORS, DIVINERS AND MINING ENGINEERS]
TREMLETT, Rex, (-) [PROSPECTORS, DIVINERS AND MINING ENGINEERS]
TROTTER, Frederick Murray, (1897-1968)
TRUEMAN, Arthur Elijah, Sir, (1894-1956)
TRUSCOTT, Samuel John, (?1870-1950)
TURTON, William Henry, (1862-1942)
TUTCHER, John William, (1858-1951)
TUTTON, Alfred Edwin Howard, (1864-1938)
TYRRELL, George Walter, (1883-1961)
VARTY SMITH, George, (-1924)
VASSALL, Henry, (1860-1926)
VAUGHAN, Arthur, (1868-1915)
VICKERS, Arthur Eric Jarvis, (1897-1958)
WADE, Arthur, (1878-1951)
WALKER, Edward Eaton, (-1903)
WARNES, Arthur Robert, (?1877-1942)
WARREN, Samuel Hazzledine, (1872-1958)
WASHINGTON-GRAY, Kenneth, (?1899-1963)
WATSON, David Meredith Seares, (1886-1973)
WATTISON, John Thomas, (1884-1974)
WATTS, William, (-1920)
WATTS, William Whitehead, (1860-1947)
WAY, Herbert W. L., (-) [PROSPECTORS, DIVINERS AND MINING ENGINEERS]
WELLS, Herbert George, (1866-1946)
WHEELER, Richard Vernon, (1883-1939)
WHITE, Henry Hopley, (-?1876)

ENGLAND 1850-1899 Cont'd.

WHITE, Stanley Battershill, (1899-1966)
WHITEHEAD, Talbot Haes, (1890-1966)
WILCOCKSON, William Howson, (1891-1976)
WILD, Robert Powley, (1882-1946)
WILKS, Mark, (?1862-1945)
WILLIAMS, Arthur Herbert, (?1865-1935)
WILLS, Robert Glover, (1878-1950)
WILSON, Charles Henry, (1851-1887)
WINGRAVE, Vitruvius Harold Wyatt, (1858-1938)
WINTER, William Percy, (?1870-1950)
WITHERS, Thomas Henry, (1883-1953)
WOOD, James George, (1888-1928)
WOODS, Ethel Gertrude, (-1939)
WOODS, Henry, (1868-1952)
WOODWARD, Arthur Smith, Sir, (1864-1944)
WOODWARD, Bernard Barham, (1853-1930)
WOODWARD, Harry Page, (1858-1917)
WOOLACOTT, David, (1872-1924)
WORDIE, James Mann, Sir, (?1890-1962)
WORLEY, W. F., (-1923)
WORSFORD, Frederick H., (1870-1954)
WRIGHT, William, (-1936)
WRIGLEY, Arthur George, (1885-1953)
YOUNG, Alfred Collett, (1862-1932)
YOUNG, George William, (1862-1929)
ZEALLEY, Arthur Edmund Victor, (1886-1918)

1900 to present

ALEXANDER, Frances Elizabeth Somerville, (1908-1958)
ANDREWS, Richard Wallis, (?1927-1967)
ARKELL, William Joscelyn, (1904-1958)
ASHBY, Donald Frederick, (?1914-1943)
BAKER, George, (1908-1975)
BARNETT, Peter, (1922-1946)
BATTLE, Walter Ravenhill Brown "Ben", (1919-1953)
BEASLEY, Cyril George, (1901-?1957)
BELL, John Arnold, (1939-1970)
BERNARD, Henry Meyners, (1953-1909)
BERRINGTON, Roger, (-) [PROSPECTORS, DIVINERS AND MINING ENGINEERS]
BISHOP, Walter William, (1931-1977)
BISHOPP, Douglas Wallace, (1900-1977)
BLACK, Maurice, (1904-1973)
BLOW, Walter Harry, (1924-1972)
BRAMWYCH, Michael, (1911-1974)
BROUGHTON, Derek George, (-1971)
BROWN, Peter Evans, (-)
BULLERWELL, William, (1916-1977)
BULMAN, Oliver Meredith Boone, (1902-1974)
BURNABY, Thomas Patrick, (1924-1968)
BURROWS, William Alfred, (1903-1970)
CAMPBELL, Keir Arthur, (-1955)
CARRINGTON, Richard, (-)
CASTELL, Cyril Philip, (1907-1972)

ENGLAND 1900 to present Cont'd.

CASTON, John E. F., (1925-1972)
CHALONER, William G., (-)
CRABTREE, John Edmund, (1940-1966)
CROFT, William Noble, (1915-1953)
CROWTHER, Arthur Frederick, (1929-1954)
DERRY, Duncan Ramsay, (?1908-)
DOBBIE, Charles Herbert, (1905-1976)
ENGLEHEART, Francis Henry Arnold, (-1963)
FALCON, Norman Leslie, (-)
FAULKNER, Joseph Stanislaus "Ginger", (1928-1963)
FORD, Trevor David, (1932-)
FOWLER, Gordon, Major, (-1953)
FUCHS, Vivian Ernest, Sir, (1908-)
GARNETT, Alice, (-)
GASKIN, Alan Robert James, (1927-1974)
GLEAVES, John T., (-1969)
GRAHAM, Roy, (1908-1939)
GREGORY, Hubert Harold, (-?1950)
GRESSWELL, Ronald Kay, (1905-1969)
GROVES, Arthur William, (1903-1956)
HAIG, David Salusbury, (?1913-1955)
HALL, Thomas Clifford Fitzwilliam, (1886-1956)
HARPER, John Childs, (1913-1977)
HARRY, William Trevelyan, (?1922-1964)
HARWOOD, Harold Benjamin, (-1956)
HAYWARD, Henry Arthur, (-1964)
HAYWARD, John, (-1958)
HENSON, Frederick Arthur, (1919-1957)
HEWITT, Henry Dixon, (-?1966)
HIDE, Raymond, (1929-)
HILL, Maurice Neville, (1919-1966)
HITCHEN, Charles Stansfield, (1904-1971)
HODGE, Matthew Burnop, (-1951)
HOLWILL, Frederick James Winsor, (1925-1966)
HUGHES, Harry, (1929-)
HURLEY, Patrick Mason, (1912-)
HUTCHINS, Philip Frank, (1927-1955)
INESON, Jack, (1917-1970)
JOYCE, John Raymond Foggan, (?1907-1963)
KENNERLEY, John Brian, (1937-1976)
KINGSBURY, Arthur William Gerald, (1906-1968)
KIRKPATRICK, Randolph, (-1950)
LEAKEY, Louis Seymour Bazett, (1903-1972)
LEAKEY, Mary Douglas (née Nicol), (1913-)
LEAKEY, Richard Erskine, (1944-)
LeLACHEUR, W. J., (-1923)
LINTON, David Leslie, (1906-1971)
LITTLE, Otway Henry, (-1956)
LOBJOIT, William Michael, (1932-1967)
LONSDALE, Kathleen, Dame, (née Yardley), (1903-1971)
LOWNDES, Ashley Gordon, (-1956)
MACDONALD, Gordon James Fraser, (1929-)
MACKAY, Robert Alexander, (?1904-1961)
MACMILLAN, William Edward Frank, (-1954)

ENGLAND 1900 to present Cont'd.

MARSH, Thomas William, (-1957)
MARSHALL, Alan Edward, (1938-1970)
McMATH, John Charles, (?1908-1954)
MILLER, Arthur Austin, (1900-1968)
MITCHELL, George Hoole "Mick", (1902-1976)
MOORE, Leslie Rowsell, (1914-)
MOTTRAM, Brian Hunter, (1922-1976)
NUTTALL, Ernest, (-?1953)
OAKLEY, Kenneth Page, (1911-)
OSBORNE WHITE, Harold J., (-1954)
PARGETER, Reginald Clive, (1919-1972)
PARRINGTON, Francis Rex, (-)
PEET, Geoffrey Alan, (1922-1959)
PENROSE, Evelyn, (-) [PROSPECTORS, DIVINERS AND MINING ENGINEERS]
POLLARD, Ernest Rex, (1922-1966)
PUEL, H. E., (-1964)
QUIRK, Roger Nathaniel, (?1909-1964)
READER, Thomas William, (-?1923)
REECE, Alan W., (1921-1960)
RICHARDSON, F. D. S. "Ricardo", (-1966)
ROE, Frank Walton "Peter", (?1914-1965)
RUTLEDGE, Harold, (1920-1954)
SAMPSON, David Noel, (1917-1971)
SANSOM, Charles Alan, (1902-1945)
SARJEANT, William Antony Swithin, (1935-)
SCHOFIELD, Eric, (1911-1944)
SHERBORNE, John Elliott, Sr., (1911-1974)
SMITH, Denis Raymond James, (1937-1964)
SMITH, Jeremy, (1929-1959)
SPIKE, Wilfred Ivan, (1925-1969)
STUDER, Joseph Paul, (1916-1951)
SUGDEN, Walter "Bill", (1923-1977)
SUTTON, John, (-)
SYLVESTER-BRADLEY, Peter Colley, (1914?-1978)
TAYLOR, Herbert, (-1963)
TAYLOR, James Haward, (1909-1968)
THOMAS, Henry Dighton, (1900-1966)
TIDY, Anthony John Lintott, (1928-1969)
TOLANSKY, Samuel, (?1908-1973)
TOZER, Charles Fallows, (1918-1966)
WAGER, Lawrence Rickard, (1904-1965)
WARBURG, Robert Oscar, (1915-1965)
WHEELER, Charles Douglas, (?1922-1944)
WHITTARD, Walter Frederick, (1902-1966)
WILSON, Vernon, (1907-1971)
WOOD, Frederick Wilfred, (?1904-1961)
WOOLDRIDGE, Sidney William, (1900-1963)
WYATT, Michael, (-1956)

Time unknown

BEER, Gavin R. de, Sir, (-)

ESTONIAN SSR (USSR)

Up to 1800

PANDER, Christian Heinrich, (1794-1865)

1800-1849

GREWINGK, Konstantin, (1819-1887)
HELMERSEN, Gregor von, (1803-1885)
SCHMIDT, Friedrich Bogdanovich "Bohdan", (1832-1908)

1850-1899

ANDRUSOV, Dimitri, (1897-1976)
BALK, Robert, (1899-1955)
BEKKER, Hendrik, (1891-1925)
POPOFF, Boris Anfirovich, (1871-1950)
TAMMANN, Gustav Heinrich Johann Apoll, (1861-1938)
THOMSON, Paul William, (1891-1957)
TOLL, Eduard von, Baron, (1858-1902)

FINLAND

Up to 1800

BONSDORFF, Pehr Adolf von, (1791-1839)
FORSIUS (HELSINGFORSIUS), Sigfrid Aronius, (ca.1550-1624)
GADD, Pehr Adrian, (1727-1797)
GADOLIN, Johan, (1760-1852)
LAXMAN, Erik, (1737-1796)

1800-1849

ARPPE, Adolf Edvard, (1818-1894)
CHYDENIUS, Johan Jakob "Hannes", (1836-1890)
FURUHJELM, Enoch Hjalmar, (1823-1886)
GADOLIN, Axel Wilhelm, General, (1828-1892)
MOBERG, Karl Adolf, (1840-1901)
NORDMANN, Alexander von, (1803-1866)
WIIK, Fredrik Johan, (1839-1909)

1850-1899

AF SCHULTÉN, August Benjamin, (1856-1913)
AILIO, Julius, (1872-1933)
BERGHELL, Hugo, (1864-1922)
BRENNER, Thord, (1892-1949)
ESKOLA, Pentti Eelis, (1883-1964)
FROSTERUS, Gustaf Benjamin, (1866-1931)
HACKMAN, Victor Axel, (1866-1941)
JULIN, Albert Lindsay von, (1876-1918)
KANERVA, John, (1882-)
MÄKINEN, Eero, (1886-1953)
MIKKOLA, Erkki, (-?1940)
NYHOLM, Emil Teodor, (1869-1908)
PETRA (PETANDER), August Herman, (1853-1928)
RAMSAY, Wilhelm, (1865-1928)
SAARINEN, Juho Hugo, (-?1908)
SEDERHOLM, Jakob Johannes, (1863-1934)
TANNER, Väinö, (1881-1948)
WILKMAN, Wanold Wrydon, (1872-?1938)

1900 to present

BRANDER, Gunnar, (1903-1940)
MARMO, Vladi, (1914-1969)

FRANCE

Up to 1800

ASTRUC, Jean, (1684-1766)
AUBUISSON DE VOISINS, Jean François d', (1769-1841)
AUDOUIN, Jean Victor, (1797-1841)
BARRANDE, Joachim, (1799-1883)
BAUME, Antoine, (1728-1804)
BEAUMONT, Jean Baptiste Armand Louis Léonce Élie de, (1798-1874)
BEAUSOLEIL, Jean du Chastelet, Baron de .... et d'Auffembach, (ca 1578-1620)
BEAUSOLEIL, Martine de Bertereau, Baronne de .... et d'Auffembach, (fl.1600-1630)
BERTRAND DE DOUE, Jacques Mathieu, (1776-1862)
BEUDANT, François-Sulpice, (1787-1850)
BLAINVILLE, Henri Marie Ducrotay de, (1777-1850)

FRANCE Up to 1800 Cont'd.

BOISDUVAL, Jean-Baptiste Alphonse, (1799-1879)
BONNARD, Augustin Henri de, (1781-1857)
BOUCHER DE CRÈVECOEUR DE PERTHES, Jacques, (1788-1868)
BOUGUER, Pierre, (1698-1758)
BOURGUET, Louis, (1678-1742)
BOURNON, Jacques Louis de, Comte, (1751-1825)
BRARD, Cyprien Prosper, (1788-1838)
BROCHANT DE VILLIERS, André Jean Marie, (1772-1840)
BRONGNIART, Alexandre, (1778-1847)
BRUGUIÈRES, Jean Guillaume, (1750-1799)
BUFFON, Georges Louis Leclerc, comte de, (1707-1788)
CHARPENTIER, Jean de, (1786-1855)
COLLOT, Georges Henri Victor, (1750-1805)
CORDIER, Pierre Louis Antoine, (1777-1861)
CUVIER, Jean Léopold Nicolas Fréderic, (1769-1832)
DAUBENTON, Louis Jean Marie, (1716-1799)
DEFRANCE, Jacques Louis Marie, (1758-1850)
DELAFOSSE, Gabriel, (1796-1878)
DESCARTES, René du Perrot, (1596-1650)
DESHAYES, Gérard Paul, (1796-1875)
DESMAREST, Nicolas, (1725-1815)
DES MOULINS, Charles, (1798-1875)
DEZALLIER D'ARGENVILLE, Antoine Joseph, (1680-1765)
DOLOMIEU, Dieudonné Sylvain Guy Tancrède, called Déodat, Gratet de, (1750-1801)
DROUIN, Jean Baptiste, (1752-1792)
DU FAY, Charles François, (1698-1739)
DUFRÉNOY, Pierre Armand, (1792-1857)
DUFRESNE, Louis, (1752-1832)
DUHAMEL, Jean Pierre François Guillaume, (1730-1816)
DUVERNOY, George Louis, (1777-1855)
EUDES-DESLONGCHAMPS, Jacques Amand, (1794-1867)
FAUJAS DE SAINT FOND, Barthélémy, (1741-1819)
FÉRUSSAC, André Étienne Just Pascal Joseph François d'Audebard, Baron de, (1786-1836)
FONTENELLE, Bernard Le Bovier de, (1657-1757)
FOURIER, Jean Baptiste Joseph, (1768-1830)
GAUTIER, Henri, (1660-1734)
GEOFFROY SAINT-HILAIRE, Étienne, (1772-1844)
GRATELOUP, Jean Pierre Sylvestre de, (1782-1861)
GRAVES, Louis, (1791-1857)
GUETTARD, Jean Étienne, (1715-1786)
GUEYMARD, François Émile, (1788-1870)
HACQUET, Belsazar, (1739-1815)
HAÜY, René-Just, Abbé, (1743-1822)
HUOT, Jean Jacques Nicolas, (1790-1845)
LA CONDAMINE, Charles-Marie de, (1701-1774)
LAMARCK, Jean-Baptiste Pierre Antoine de Monet, Chevalier de, (1744-1829)
LAMOUROUX, Jean Vincent Félix, (1779-1825)
LAURILLARD, Charles Léopold, (1783-1853)
LAVOISIER, Antoine Laurent, (1743-1794)
LEFEBVRE, Alexandre, (1798-1868)
LÉVY, Armand, (1795-1861)

FRANCE Up to 1800 Cont'd.

MAILLET, Benoît de, (1656-1738)
MÉTHERIE, Jean Claude de la, (1743-1817)
MICHELIN, Jean Louis Hardouin, (1786-1867)
MONNET, Antoine Grimoald, (1734-1817)
MONTLOSIER, Fr. Dominique Reynaud, comte de, (1755-1838)
NICOLLET, Joseph Nicolas, (1786-1843)
PALISSY, Bernard, (1510-1589)
PASSY, Antoine François, (1792-1873)
PERRAULT, Pierre, (1611-1680)
PRÉVOST, Louis Constant, (1787-1856)
RÉAUMUR, René Antoine Ferchault de, (1683-1757)
ROBINEAU-DESVOIDY, André Jean Baptiste, (1799-1857)
ROUELLE, Guillaume François, (1703-1770)
ROZET, Antoine, Commandant, (1798-1858)
SERRES, Marcel de, (1783-1862)
SOULAVIE, Jean Louis Giraud, Abbé, (1752-1813)
TAVERNIER, Jean Baptiste, baron d'Aubonne, (1605-1689)
TERQUEM, Olry, (1797-1887)
THIRRIA, Charles Edouard, (1796-1868)
VOLNEY, Constantin François Chasseboeuf, Comte de, (1757-1820)

1800-1849

ARCHIAC, Étienne Jules Adolphe Dexmier de Simon, Vicomte d', (1802-1868)
ARNAUD, Hilaire, (1827-1908)
BARBOT DE MARNY, N. P., (1832-1877)
BARET, Charles, (1831-1910)
BAYAN, Joseph Félix Ferdinand, (1845-1874)
BAYLE, Émile, (1819-1895)
BELGRAND, Eugène, (1810-1878)
BERTHELIN, Jean Georges, (1840-1897)
BERTRAND, Émile, (1844-1909)
BERTRAND, Marcel Alexandre, (1847-1907)
BINET, Auguste, (-1900)
BIOCHE, Alphonse, (1844-1918)
BLEICHER, Marie-Gustave, (1838-1901)
BONAPARTE, Charles Lucien Jules Laurent, Prince of Canino, (1803-1857)
BOURDOT, Jules Damase, (1837-1906)
BOURGEOIS, Louis Alexis, Abbé, (1819-1878)
BOURGUIGNAT, Jules René, (1829-?1919)
BOUSSINGAULT, Joseph Dieudonné Jean-Baptiste, (1807-1887)
BOYER, Georges, (-1892)
BRAVAIS, Auguste, (1811-1863)
BRONGNIART, Adolphe Théodore, (1801-1876)
BUCAILLE, Ernest-Lucien, (1835-1891)
BURAT, Amédée, (1809-1883)
BUREAU, Edouard, (1830-1918)
CANDOLLE, Alphonse de, (1806-1893)
CARNOT, Marie Adolphe, (1839-1920)
CAUMONT, Arcisse de, (1801-1872)
CAZIOT, Eugène, (1844-1931)
CHANCOURTOIS, A. E. Béguyer de, (-1886?)
CHANTRE, Ernest, (1843-)

FRANCE 1800-1849 Cont'd.

CHAPER, Maurice Armand, (1834-1896)
CHELLONNEIX, Émile, (-1885)
CHENU, Jean Charles, (1808-1879)
COLLENOT, Jean Jacques, (1814-1893)
COLLOMB, Edouard, (1801-1875)
COLLOT, Louis Marie François, (1846-1915)
CONTEJEAN, Charles Louis, (1824-1907)
COQUAND, Henri, (1813-1881)
COTTEAU, Gustave Honoré, (1818-1894)
CROIZET, Abbé, (-1863)
CROSSE, Joseph Charles Hippolyte, (1826-1898)
DALIMIER, Paul Louis Marie, (1835-1863)
DAMOUR, Alexis, (1808-1902)
DARCY, Henri Philibert Gaspard, (1803-1858)
DAUBRÉE, Gabriel Auguste, (1814-1896)
DAUTZENBERG, Philippe, (1849-1935)
DELESSE, Achille Ernest Oscar Joseph, (1817-1881)
DE RANCE, Charles Eugène, (1847-1906)
DES CLOIZEAUX, Alfred Louis Olivier Legrand, (1817-1897)
DESNOYERS, Jules, (1801-1887)
DOLLFUS, Auguste, (1840-1869)
DOLLOT, Auguste, (1841-1924)
DONNEZAN, Albert, (1846-1914)
DOUVILLÉ, Joseph Henri Ferdinand, (1846-1937)
DUJARDIN, Félix, (1801-1860)
DUMAS, Émilien, (1804-1870)
DUVAL-JOUVE, Joseph, (1810-1883)
ÉTALLON, Claude Auguste, (1826-1862)
EUDES-DESLONGCHAMPS, Eugène, (1830-1889)
FALSAN, Albert, (1833-1902)
FAYOL, Henri, (1841-1925)
FERRY, Henri de, (1826-1869)
FILHOL, Edouard, (1814-1883)
FILHOL, Henri, (1843-1902)
FISCHER, Paul Henri, (1835-1893)
FLICHE, Paul, (1836-1908)
FONDOUCE, Paul Cazalis de, (1838-1932)
FONTANNES, Charles François, (1839-1886)
FOUCAULT, Jean Bernard Léon, (1819-1868)
FOUQUÉ, Ferdinand André, (1828-1904)
FOURNEL, Henri, (-)
FOURNET, Joseph Jean Baptiste Xavier, (1801-1869)
FRIEDEL, Charles, (1832-1899)
FROSSARD, Émilien, (1802-1881)
FUCHS, Philippe Jacques Edmond, (1837-1889)
GALEOTTI, Henri Guillaume, (1814-1858)
GARRIGOU, Félix, (1835-1920)
GAUDRY, Jean Albert, (1827-1908)
GAUTHIER, Victor Auguste, (1837-1911)
GAUTIER, Emile Justin Armand, (1837-1920)
GEOFFROY SAINT-HILAIRE, Isidore, (1805-1861)
GERVAIS, François Louis Paul, (1816-1879)
GIARD, Alfred, (1846-1908)
GIRARDOT, Charles Joseph Albert Julien, (1848-1930)

FRANCE 1800-1849 Cont'd.

GOSSELET, Jules Auguste, (1832-1916)
GRAD, Charles, (1842-?1905)
GRAND'EURY, François Cyrille, (1839-1917)
GRANDIDIER, Alfred, (1836-1921)
GRAS, Albin, (-1854)
GROSSOUVRE, Albert de, (1849-1932)
GRUNER, Louis, (1809-1883)
GUÉBHARD, Adrien, (1849-1924)
GUILLIER, Albert, (1839-1885)
GUIRAND, Edmond, (1812-1888)
HAIME, Jules, (1824-1856)
HAMY, Jules Ernest Théodore, (1842-1908)
HAUTEFEUILLE, Paul Gabriel, (1836-1902)
HÉBERT, Edmond, (1812-1890)
HOLLANDE, Dieudonné, (1845-1921)
JACQUEMONT, Victor, (1801-1832)
JACQUOT, Eugène, (1817-1903)
JEANJEAN, Adrien, (1821-1897)
JULIEN, Pierre Alphonse, (1838-1905)
LACHAT, Hippolyte, (1829-1901)
LAMBERT, Jules Mathieu, (1848-1940)
LAPPARENT, Albert Auguste de, (1839-1908)
LARTET, Edouard Armand Isidore Hippolyte, (1801-1871)
LARTET, Louis, (1840-1899)
LECOQ, Henri, (1802-1871)
LEMOINE, Armand Victor, (1837-1897)
LEYMERIE, Alexander, (1801-1878)
LOCARD, Arnould, (1841-1904)
LORTET, Louis Charles, (1836-1909)
LORY, Charles, (1823-1889)
MALLARD, François Ernest, (1833-1894)
MARCOU, Jules, (1824-1898)
MARÈS, Paul, (1826-1900)
MARION, Antoine Fortuné, (1846-1900)
MARTIN, David, (1842-1918)
MATHERON, Pierre Émile Philippe, (1807-1899)
MAYER-EYMAR, Karl David Wilhelm, (1826-1907)
MEUNIER, Étienne Stanislas, (1843-1925)
MICHEL, Léopold Jean, (1846-1919)
MICHEL-LÉVY, Auguste, (1844-1911)
MIEG-KROH, Mathieu, (1849-1911)
MILNE-EDWARDS, Alphonse, (1835-1900)
MORIÈRE, Jules Pierre Gilles, (1817-1888)
MORTILLET, Gabriel de, (1820-1898)
MUNIER-CHALMAS, Ernest Philippe Auguste, (1843-1903)
NICOLET, Adolphe Célestin, (1803-1871)
NODOT, Léonhard, (1802-1859)
NOULET, Jean Baptiste, (1802-1890)
OEHLERT, Daniel Victor, (1849-1920)
OLRY, Albert, (1847-1913)
ORBIGNY, Alcide Charles Victor Dessalines d', (1802-1857)
ORTLIEB, Jean, (1839-1890)
OUSTALET, Émile, (1844-1905)
PAILLETTE, Antoine Adrien, (-1858)

FRANCE 1800-1849 Cont'd.

PARANDIER, A. N., (1804-1901)
PARRAN, Alphonse, (1827-1903)
PERON, Alphonse, (1834-1908)
PERREY, Alexis, (1807-1882)
PIDANCET, Pierre Marie Just, (1823-1871)
PIETTE, Louis Edouard Stanislas, (1827-1906)
PILLET, Louis, (1819-1894)
PISANI DE SERRES, Felix, (1831-1920)
PISSIS, Aimé, (1812-1889)
PLANTÉ, Gaston, (1834-1889)
POMEL, Nicolas Auguste, (1821-1898)
POTIER, Alfred, (1840-1905)
QUATREFAGES DE BRÉAU, Jean Louis Armand de, (1810-1892)
RAMÈS, Jean-Baptiste, (1832-1894)
RAULIN, Félix Victor, (1815-1905)
RENAULT, Bernard, (1836-1904)
REQUIEN, Esprit, (-1852)
RÉVIL, Joseph, (1849-1931)
RIGAUX, Edmond, (1839-1915)
RIVIÈRE, Auguste, (1809-1877)
RIVOT, Louis Edouard, (1820-1869)
ROCHE, Auguste, (1827-1905)
ROUAULT, Marie, (1813-1881)
ROUVILLE, Paul Gervais de, (1823-1907)
SAINTE-CLAIRE DEVILLE, Charles, (1814-1876)
SAINTE-CLAIRE DEVILLE, Henri, (-1881)
SAPORTA, Louis Charles Joseph Gaston de, Marquis, (1823-1895)
SARASIN, Edouard, (1843-1917)
SAUVAGE, François Clément, (1814-1872)
SAUVAGE, Henri Émile, (1842-1917)
SCHLUMBERGER, Charles, (1825-1904)
SCHRADER, François Jean Daniel "Franz", (1844-1924)
SELYS-LONGCHAMPS, Michel Edmond de, Baron, (1813-1900)
SEUNES, Jean, (1849-1920)
STAHL, Jean Benjamin, (1816-1893)
THIOLLIÈRE, Victor, (1801-1859)
THOMAS, Philippe, (1843-1910)
THURMANN, Jules, (1804-1855)
TOUCAS, Joseph Aristide, (1843-1911)
TOURNAL, Paul, (1805-1872)
TOURNOÜER, Jacques Raoul, (1822-1882)
TRIGER, Jacques Jules, (1801-1867)
TROUESSART, Edouard Louis, (1842-1927)
VAILLANT, Léon Louis, (1834-1914)
VERNE, Jules, (1828-1905)
VERNEUIL, Philippe Edouard Poulletier de, (1805-1873)
VIQUESNEL, Auguste, (1800-1867)
WATELET, Adolphe, (1811-1880)
WOHLGEMUTH, Jules, (-1893)
ZEILLER, Charles René, (1847-1915)

1850-1899

ABRARD, René, (1892-1973)
ALLOITEAU, James Henri Ernest, (1890-1969)

FRANCE 1850-1899 Cont'd.

ARAMBOURG, Camille, (1885-1969)
ASTRE, Gaston, (1896-)
AUTHELIN, Charles, (1871-1903)
AZÉMA, Léon, Colonel, (1852-1931)
BARRABÉ, Louis, (1895-1961)
BARROIS, Charles Eugène, (1851-1939)
BASSE DE MÉNORVAL, Éliane Marguerite, (1899-)
BEAUGÉ, Alfred, (1878-1935)
BECQUEREL, Henri, (1852-1908)
BERGERON, Jules, (1853-1919)
BERNARD, Félix, (1863-1898)
BERTRAND, C. Eg., (1851-1920)
BERTRAND, Léon, (1869-1947)
BERTRAND, Paul, (1879-1944)
BÉZIER, T., (-1926)
BIGOT, Alexandre, (1863-1953)
BIRAUD, François, (1899-1936)
BLAYAC, Joseph, (1865-1936)
BLONDEL, Fernand, (1894-1968)
BOUBÉE, Nérée, (1893-1967)
BOULE, Marcellin, (1861-1942)
BOURCART, Jacques, (1891-1965)
BOUSSAC, Jean, (1885-1917)
BRALY, Adrien, (1864-1927)
BRESSON, Arthur, (1873-1946)
BREUIL, Henri, Abbé, (1877-1961)
BROGLIE, Maurice de, (1875-1960)
BRONGNIART, Charles Jules Edme, (1859-1899)
BRUET, Edmond, (1890-)
CARPENTIER, Alfred, Chanoine, (1878-1952)
CASTERET, Norbert, (1897-)
CAYEUX, Lucien, (1864-1944)
CHAPUT, Jules Ernest, (1882-1943)
CHELOT, Émile, (1862-1930)
CHEVENARD, Pierre, (1888-1960)
CHUDEAU, René, (1864-1921)
COMMONT, Victor, (1866-1918)
COPAUX, Hippolyte, (1872-1934)
CORBIÈRE, Louis, (1850-1941)
COSSMANN, Maurice, (1850-1924)
COTTREAU, Jean, (1877-1945)
COURTY, Georges, (1875-1953)
CUVILLIER, Jean, (1899-1969)
DAGUIN, Fernand, (1889-1948)
DALLONI, Marius Gustave, (1880-1959)
DEFLANDRE, Georges Victor, (1897-1973)
DEHÉE, René, (1898-1928)
DELÉPINE, Gaston-Georges, Monseigneur, (1878-1963)
DEMAY, André, (1890-1964)
DEPÉRET, Charles Jean Julien, (1854-1929)
DOLLÉ, Louis, (1878-1965)
DOLLFUS, Gustave Fréderic, (1850-1931)
DOLLO, Louis Antoine Marie Joseph, (1857-1931)
DONCIEUX, Louis, (1874-1960)

FRANCE 1850-1899 Cont'd.

DOUVILLÉ, Robert, (1881-1914)
DOUXAMI, Henri, (1871-1913)
DUBOIS, Georges, (1890-1953)
DUFFOUR, Alexis, (1877-1948)
DUPARQUE, André, (1892-1960)
DUTERTRE, August Pierre, (1890-1940)
ERHART, Henri, (-)
FABRE, Georges, (-1911)
FALLOT, Paul, (1889-1960)
FERRONNIÈRE, Georges, (1875-1922)
FLEURY, Ernest, (1878-1958)
FORTAU, René, (1867-1920)
FOURNIER, Eugène, (1871-1941)
FRIEDEL, Edmond, (1895-1972)
FRIEDEL, Georges, (1865-1933)
FURON, Raymond, (1898-)
GAUBERT, Marie Benoit Paul, (1865-)
GAUDEFROY, Christophe Léon, Chanoine, (1878-1971)
GAURIER, Ludovic Charles, Abbé, (1874-1931)
GÉRARD, Charles, Colonel, (1866-1953)
GIGNOUX, Maurice Irénée Marie, (1881-1955)
GODON, Joseph, Chanoine, (-1932)
GRAMONT, Antoine Alfred Arnaud Xavier Louis de, Comte, (1861-1923)
GROTH, Jean, (1899-1916)
GUILLAUME, Louis, (1894-1952)
HARLÉ, Edouard, (1850-1922)
HAUG, Gustave Émile, (1861-1927)
HELBRONNER, Paul, (1871-1938)
HENRY, J., (-1924)
JACOB, Charles François Étienne, (1878-1962)
JANET, Léon, (1861-1909)
JANNETTAZ, Edouard, (-1900)
JODOT, Paul, (1879-1961)
JOLEAUD, Léonce, (1880-1938)
JOURDAIN, André, (1890-1967)
JUNG, Jean, (1896-)
KAISIN, Félix, Sr., (1879-1948)
KERFORNE, Fernand, (1864-1927)
KILIAN, Charles Constant Wilfrid, (1862-1925)
KILIAN, François Théodore Conrad, (1898-1950)
LACROIX, François Antoine Alfred, (1863-1948)
LALOY, Roger, (1851-1909)
LAMARE, Pierre, (1894-1967)
LAPPARENT, Jacques de, (-?1949)
LAUNAY, Louis Alphonse Auguste de, (1860-1938)
LECOINTRE, Georges, (1888-1972)
LE MESLE, George, (-1895)
LEMOINE, Paul, (1878-1940)
LENNIER, Gustave, (-1905)
LERICHE, Maurice, (1875-1948)
LONGCHAMBON, Michel, (1886-1916)
LORY, Pierre, (1866-1956)
LUGEON, Maurice, (1870-1953)
LUTAUD, Léon, (1883-1964)

FRANCE 1850-1899 Cont'd.

MAILLARD, Gustave, (1860-1891)
MAIRE, Julien Victor, (1856-1941)
MARGERIE, Emmanuel Marie Pierre Martin Jacquin de, (1862-1953)
MARTIN, Henri, (1873?-1936)
MARTONNE, Emmanuel Louis Eugène de, (1873-1955)
MAUGIN, Charles, (1878-1958)
MICHEL-LÉVY, Albert, (1877-1955)
MIGAUX, Léon, (1897-1914)
MOISSAN, Henri, (1852-1907)
MONTESSUS DE BALLORE, Fernand Jean Baptiste Marie Bernard, Comte de, (1851-1923)
MORELLET, Lucien, (1882-1945)
MORET, Léon, (1890-1972)
MORGAN, Jacques Jean Marie de, (1857-1924)
NICKLÈS, René, (1859-1917)
ORCEL, Jean, (1896-)
PAQUIER, Victor Lucien, (1870-1911)
PARDÉ, Maurice, (1893-1973)
PERRIN, René, (1893-1966)
PERVINQUIÈRE, Léon, (1873-1913)
PITTIER, Henry François, (1857-1950)
PIVETEAU, Jean, (1898-)
PONTIER, G., (-1933)
PRIEM, Fernand, (1857-1919)
PRUVOST, Pierre Eugéne Marie Joseph, (1890-1967)
RIAZ, Auguste de, (-1920)
ROMEU, Albert de, (1875-1915)
SAVIN, Léon Héli, Lieutenant-Colonel, (1851-1907)
SAYN, Gustave, (1862-1933)
SCHLUMBERGER, Conrad, (1878-1936)
SCHLUMBERGER, Marcel, (1884-1953)
TEILHARD DE CHARDIN, Pierre, Father, S.J., (1881-1955)
TERMIER, Henri François Emile, (1897-)
TERMIER, Pierre, (1859-1930)
THEVENIN, Armand, (1870-1918)
THIÉRY, Paul, (-1927)
TOURNOUËR, André, (-?1930)
TRONQUOY, René, (1884-1915)
UNGEMACH, Henri Léon, (1879-1936)
URBAIN, Georges, (-1938)
URBAIN, Pierre, (1895-1968)
VALMONT DE BOMARE, Jacques Christophe, (-)
VASSEUR, Gaston Casimir, (1855-1915)
VÉSIGNIÉ, Louis, Colonel, (1870-1954)
VIENNOT, Pierre, (1891-1931)
VINCIENNE, Henri, (1898-1965)
VIRET, Jean, (1894-1970)
WALLERANT, Frédéric Félix Auguste, (1858-1936)
WELSCH, Jules, (1858-1929)

1900 to present

ALIMEN, Henriette, (1900-)
AUBOUIN, Jean, (1928-)
BAUDOIN, René, (-)

FRANCE 1900 to present Cont'd.

BELLAIR, Pierre, (1910-)
BERTHELOT, Charles, (-) [PROSPECTORS, DIVINERS AND MINING ENGINEERS]
BONTE, Antoine Jean Marie Joseph, (1908-)
BOUREAU, Edouard Léon François, (1913-)
CAIRE, André, (1923-)
CAYEUX DE SENARPONT, André, called CAILLEUX, André, (1907-)
CHAUVE, Pierre, (1930-)
CHERVET, Jean, (1904-1962)
COMBAZ, Paul, Chanoine, (-1930)
DANGEARD, Louis, (-)
DEFLANDRE-RIGAUD, Marthe, (-)
DELEAU, Paul Camille, (1904-)
DENIZOT, Georges, (-)
DURAND DELGA, Michel, (1923-)
GILLARD, Pierre A., (1907-1947)
GLANGEAUD, Louis, (1903-)
GLORY, André, Abbé, (-)
GOGUEL, Jean, (1908-)
GRAMBAST, Louis, (-1976)
GUILLAUME, Solange, (1932-)
HOFFSTETTER, Robert, (1908-)
HOURCQ, Victor, (1901-1961)
HUPÉ, Pierre, (1907-)
JACQUET, Fernand, (1908-1937)
LACOSTE, Jean Pierre Victor, (1901-1942)
LANQUINE, Antonin, (-)
LAPPARENT, Albert Félix de, (1905-1975)
LE CALVEZ, Jean, (1908-1954)
LELUBRE, Maurice, (1916-)
LISSAJOUS, Marcel, (-1921)
LOUBIÈRE, A., (-)
MALAVOY, Jean, (1903-1945)
MEYENDORFF, André de, (1918-1942)
MILLOT, Georges, (1917-)
MONGIN, Denise, (-)
NICOLAS, Jean, (1925-)
NICOLAS, Jean Paul, (1921-1967)
PALM-LAZARD, C. B. B., (1933-1960)
RAVIER, Jean, (1922-)
RENAULT, Philippe, (-)
ROCH, Edouard, (1901-)
ROUBAULT, Marcel Edouard, (1905-1974)
SCHNEEGANS, Daniel, (1907-1949)
THORAL, Marcel, (1900-1956)

GERMANY

Up to 1800

AGRICOLA, Georgius, (1494-1555)
ALBERTI, Friedrich August von, (1795-1878)
ALBERTUS MAGNUS, (1193-1280)
ALBINUS, Petrus, (1543-1598)
ANDRÉ, Kristián Karel, (1763-1831)
ASSMAN, C. G., (-)
AYCKE, Johann Christian, (1766-1854)
BAER, Karl Ernst von, (1792-1876)
BATSCH, August Johann Georg Karl, (1761-1802)
BECHER, Johann Joachim, (1635-1682)
BERINGER, Johann Bartholomaeus Adam, (1667-1740)
BESSER, Wilibald Swibert Joseph Gottlieb von, (1784-1842)
BISCHOFF, Karl Gustav, (1792-1870)
BLUMENBACH, Johann Friedrich, (1752-1840)
BOJANUS, Ludwig Heinrich, (1776-1827)
BOUÉ, Ami, (1794-1881)
BREITHAUPT, Johann Friedrich August, (1791-1873)
BUCH, Christian Leopold von, Freiherr, (1774-1853)
BÜHLER, Georg Wilhelm Christian von, (1797-1860)
BURKART, Hermann Josef, (1798-1874)
CARUS, Carl Gustav, (1789-1869)
CHARPENTIER, Johann Friedrich Wilhelm Toussaint von, (1728-1805)
COTTA, Heinrich, (1763-1844)
EBEL, Johann Gottfried, (1764-1830)
EGEN, P. N. C., (1793-1849)
EHRENBERG, Christian Gottfried, (1795-1876)
EHRHART, Balthazar, (1700-1756)
EMMERLING, Ludwig August, (1765-1841)
ENGELHARDT, Otto Moritz Ludwig von, (1779-1842)
ESCHWEGE, Wilhelm Ludwig, Baron, (1777-1855)
ESER, Anton Friedrich, (1798-1873)
FABRICIUS, Georg, (1516-1571)
FISCHER VON WALDHEIM, Johann Gotthelf Friedrich, (1771-1853)
FLURL, Mathias von, (1756-1823)
FORSTER, Adolarius Jacob, (1739-1806)
FORSTER, Johann Georg Adam, (1754-1794)
FORSTER, Johann Reinhold, (1729-1798)
FREIESLEBEN, Johann Carl, (1774-1846)
FUCHS, Johann Nepomuk von, (1774-1856)
FUCHS, Leonhart, (1501-1566)
FÜCHSEL, Georg Christian, (1722-1773)
GAUSS, Carl Friedrich, (1777-1855)
GEMMING, Carl Emil von, (1794-1880)
GERMAR, Ernst Friedrich, (1786-1853)
GERSDORF, Adolf Traugott von, (1744-1807)
GIESECKE, Charles Lewis, Sir, (né Metzle, Karl Ludwig), (1761-1833)
GLOCKER, Ernst Friedrich, (1793-1858)
GMELIN, Karl Christian, (1762-1837)
GOETHE, Johann Wolfgang von, (1749-1832)
GOLDENBERG, Friedrich, (1799-1881)
GOLDFUSS, Georg August, (1782-1848)
GROSSE, Karl Friedrich August, (1768-1847)
GUTBIER, Christian August von, (1798-1866)

GERMANY Up to 1800 Cont'd.

HÄBERLEIN, Friedrich Karl, (1787-1871)
HAGENOW, Friedrich von, (1797-1865)
HARDENBERG, Friedrich Leopold von, Freiherr, (1772-1801)
HARTMANN, Ernst Gustav Friedrich von, (1767-1851)
HAUSMANN, Johann Friedrich Ludwig, (1782-1859)
HECKEL, Johann Jakob, (1790-1857)
HEHL, Johann Carl Ludwig, (1774-1853)
HELMS, Anton Zacharias, (1751-1803) [PROSPECTORS, DIVINERS AND MINING ENGINEERS]
HENCKEL, Johann Friedrich, (1679-1744)
HERDER, Siegmund August Wolfgang von, Freiherr, (1776-1838)
HERSCHEL, Frederick William, Sir, (1738-1822)
HOENINGHAUS, Friedrich Wilhelm, (1770-1854)
HOFF, Karl Ernst Adolf von, (1771-1837)
HOFFMANN, Friedrich, (1797-1836)
HUMBOLDT, Friedrich Wilhelm Heinrich Alexander von, Freiherr, (1769-1859)
JÄGER, Georg Friedrich, (1785-1866)
KANT, Immanuel, (1724-1804)
KARSTEN, Dietrich Ludwig Gustav, (1768-1810)
KEFERSTEIN, Christian, (1784-1866)
KENTMANN, Johannes, (1518-1574)
KEPLER, Johannes, (1571-1630)
KIELMEYER, Carl Heinrich, (1765-1844)
KIRCHER, Athanasius, (1601-1680)
KLEIN, Jakob Theodor, (1685-1759)
KONIG, Karl Dietrich Eberhardt, called Charles, (1774-1851)
KÜHN, Karl Amadeus, (1783-1848)
KURR, Johann Gottlob von, (1798-1870)
LEHMANN, Johann Christoph, (1675-1739)
LEHMANN, Johann Gottlob, (1719-1767)
LENZ, Johann Georg, (1748-1832)
LEONHARD, Karl Cäsar von, (1779-1862)
LESKE, Nathanael Gottfried, (1751-1786)
LESSER, Friedrich Christian, (1692-1754)
LILL VON LILIENBACH, Karl, (1798-1831)
MANDELSLOH, Friedrich von, Graf, (1795-1870)
MARTIUS, Karl Friedrich Philipp von, (1794-1868)
MATHESIUS, Johannes, (1504-1565)
MENGE, Johann, (-)
MICHAELIS, Christian Friedrich, (fl. 1766-1780)
MITSCHERLICH, Eilhard, (1794-1863)
MOHS, Carl Friedrich Christian, (1773-1839)
MORITZ, Karl Philipp, (1757-1793)
MÜLLER, Moritz, (1791-1858)
NAU, Bernhard Sebastian von, (1766-1845)
NAUMANN, Carl Friedrich, (1797-1873)
NEUMANN, Franz Ernest, (1798-1895)
NÖGGERATH, Johann Jakob, (1788-1877)
NOSSENI, Johann Maria, (1544-1620)
OEYNHAUSEN, Karl August Ludwig, (1795-1865)
OKEN, Lorenz, (1779-1851)
OTTO, Ernst von, (1799-1863)
PALLAS, Peter Simon, (1741-1811)

GERMANY Up to 1800 Cont'd.

PARROT, Georg Friedrich, (1767-1852)
PLIENINGER, Wilhelm Heinrich Theodor von, (1795-1879)
POTT, Johann Heinrich, (1692-1777)
PUSCH, Georg Gottlieb, (1790-1846)
RASPE, Rudolf Erich, (1737-1794)
RAUMER, Karl Georg von, (1783-1865)
REINECKE, Johann Christoph Matthias, (1769-1818)
ROSE, Gustav, (1798-1873)
RÜLEIN VON CALW, Ulrich, (ca.1467-1523)
RUMPF, Ludwig, (1793-1862)
RÜPPELL, Eduard, (1794-1884)
SARTORIUS, Georg Christian, (1774-1838)
SCHÄFFER, Jacob Christian, (1718-1790)
SCHEELE, Karl Wilhelm, (1742-1786)
SCHLAGINTWEIT, Adolf, (?1799-1857)
SCHLOTHEIM, Ernst Friedrich von, Freiherr, (1764-1832)
SCHMEISSER, Johann August, (1767-1837)
SCHOEPF, Johann David, (1752-1800)
SCHULZE, Christian Friedrich, (1730-1775)
STEININGER, Johann, (1794-1874)
STELLER, Georg Wilhelm, (1709-1746)
STIFFT, Christian Ernst, (1780-1855)
TREBRA, Friedrich Wilhelm Heinrich von, (1740-1819)
TROIL, Uno von, (1746-1803)
VALENTINI, Michael Bernhard, (1657-1729)
VOGT, Johann Georg Mauriz, (1669-1731)
VOIGT, Johann Karl Wilhelm, (1752-1821)
VOLCKMANN, Georg Anton, (1636-1721)
WAGNER, Johann Andreas, (1797-1861)
WEISMANN, Gottlieb, (1798-1859)
WEISS, Christian Samuel, (1780-1856)
WEISSENBACH, Karl Gustav Adalbert von, (1797-1848)
WERNER, Abraham Gottlob, (1749-1817)
ZIETEN, Carl Hartwig von, (1785-1846)
ZIMMERMANN, Christian, (1765-1853)

1800-1849

ABBE, Ernst, (1840-1905)
ABICH, Hermann Wilhelm, (1806-1886)
AHLBURG, Johannes Heinrich Wilhelm, (1833-1919)
ANDRAE, Carl Justus, (1816-1885)
BALDAUF, Richard Julius, (1848-1931)
BALTZER, Richard Armin, (1842-1913)
BAUER, Heinrich E., (1840-1896)
BAUER, Max Hermann, (1844-1917)
BAUMHAUER, Heinrich Adolf, (1848-1926)
BAUR, Carl Theodor von, (1836-1911)
BECKER, Ewald, (-1873)
BEHN, Wilhelm Friedrich Georg, (1808-1878)
BEISSEL, Ignaz, (1820-1887)
BENECKE, Ernst Wilhelm, (1838-1917)
BERENDT, Gottlieb Michael, (1836-1920)
BERNHARDI, Reinhard, (fl.1834-1841)
BERTKAU, Philipp, (1849-1895)

GERMANY 1800-1849 Cont'd.

BEUST, Friedrich Constantin von, Freiherr, (1806-)
BEYRICH, August Heinrich Ernst, (1815-1896)
BIBRA, Ernst von, Freiherr, (1806-1878)
BLUM, Johann Reinhard, (1802-1883)
BOLL, Ernst Friedrich August, (1817-1868)
BÖLSCHE, Wilhelm, (1843-1893)
BOXBERG, Ida von, (1806-1893)
BRANCA, Wilhelm von, (1844-1928)
BRANDT, Johann Friedrich von, (1802-1879)
BRAUN, Heinrich Alexander, (1805-1877)
BRAUN, Karl Friedrich Wilhelm, (1800-1864)
BRAUNS, David August, (1827-1893)
BRONN, Heinrich Georg, (1800-1862)
BUNSEN, Robert Wilhelm, (1811-1899)
BURBACH, Otto Carl Heinrich, (1838-1888)
BURMEISTER, Karl Hermann Konrad, (1807-1892)
CALKER, Friedrich Julius Peter von, (1841-1913)
CASPARY, Robert, (1818-1887)
CASTELLI, Albin Benedict, (1822-1892)
CLARAZ, Georges, (1832-1930)
COMPTNER, Gustav, (1831-1922)
COTTA, Carl Bernhard von, (1808-1879)
CREDNER, Carl Friedrich Heinrich, (1809-1876)
CREDNER, Carl Hermann Georg, (1841-1913)
DAMES, Wilhelm Barnim, (1843-1898)
DATHE, Johann Friedrich Ernst, (1845-1917)
DECHEN, Ernst Heinrich Carl von, (1800-1889)
DEFFNER, Carl, (1817-1878)
DIEFFENBACH, Ernst, (1811-1855)
DOHRN, Felix Anton, (1840-1909)
DUNKER, Wilhelm Bernhard Rudolph Hadrian, (1809-1885)
ECK, Heinrich von, (1837-1925)
EISEL, Robert, (1826-1917)
EMMRICH, Hermann Friedrich, (1815-1879)
ENGEL, Theodor, (1842-1933)
ENGELMANN, August Heinrich, (1831-1899)
ERMAN, Georg Adolf, (1806-1877)
EWALD, Julius Wilhelm, (1811-1891)
FACK, M. W., (1823-)
FISCHER, Carl Ernst, (1818-1886)
FISCHER, Leopold Heinrich, (1817-1886)
FRAAS, Oskar Friedrich von, (1824-1897)
FRENZEL, Friedrich August, (1842-1902)
FRESENIUS, Georg, (1808-1866)
FRITSCH, Karl von, Freiherr, (1838-1906)
FUCHS, Carl Wilhelm C., (1837-1886)
FUCHS, Wilhelm, (1802-1853)
FUHLROTT, Johann Carl, (1804-1877)
FÜRBRINGER, Max, (1846-1920)
GÄTZSCHMANN, Moritz Ferdinand, (1800-1895)
GEINITZ, Hanns Bruno, (1814-1900)
GENTH, Friedrick August Ludwig Karl Wilhelm, later called Frederick Augustus, (1820-1893)
GERLAND, Georg Karl Cornelius, (1833-1919)

GERMANY 1800-1849 Cont'd.

GEYLER, Hermann Theodor, (1834-1889)
GIEBEL, Christoph Gottfried Andreas, (1820-1881)
GIRARD, Carl Adolph, (1814-1878)
GÖPPERT, Heinrich Robert, (1801-1884)
GREBE, Heinrich, (1831-1903)
GRIEPENKERL, Otto, (1820-1888)
GRODDECK, Albrecht Ludwig von, (1837-1887)
GROTH, Paul Heinrich von, Ritter, (1843-1927)
GRUNOW, Albert, (1826-1914)
GÜMBEL, Carl Wilhelm von, Ritter, (1823-1898)
GÜNTHER, A. F., (?1805-1871)
GÜNTHER, Albert Charles Lewis Gotthilf, (1830-1914)
GURLDT, Adolf Friedrich, (1829-1902)
HAAST, John Francis Julius von, Sir, (1824-1887)
HÄBERLEIN, Ernst, (1819-1895)
HAECKEL, Ernst Heinrich Philipp August, (1834-1919)
HAGEN, Hermann August, (1817-1893)
HAGEN, Max, (1831-1893)
HALFAR, Anton, (1836-1893)
HASSE, Carl, (1841-1922)
HASSENCAMP, Ernst, (1824-1881)
HAUCHECORNE, Heinrich Lambert Wilhelm, (1828-1900)
HAUPT, Karl, (1829-1882)
HENRICH, Ferdinand, (1837-1915)
HENSEL, Reinhold Friederich, (1826-1881)
HEYDEN, Lucas von, (1838-1915)
HEYMANN, Hermann, (-1871)
HILGARD, Eugene Woldemar, (1833-1916)
HILGENDORF, Franz, (1839-1904)
HIRSCHWALD, Julius, (1845-1928)
HOFMEISTER, Wilhelm Friedrich Benedict, (1824-1877)
HOHENEGGER, Ludwig, (1807-1864)
HOSIUS, August, (1825-1896)
JUNGHUHN, Franz Wilhelm, (1808-1864)
KAPFF, Sixt Friedrich Jakob, (1809-1887)
KARSTEN, Gustav, (1820-1900)
KARSTEN, Hermann, (1809-1877)
KAUP, Johann Jakob, (1803-1873)
KAYSER, Friedrich Heinrich Emanuel, (1845-1927)
KIESOW, Johann, (1846-1901)
KINKELIN, Georg Friedrich, (1836-1913)
KLAASSEN, Hendericus Martinus, (1828-1910)
KLEIN, Johann Friedrich Carl, (1842-1907)
KLIPSTEIN, August Wilhelm, (1801-1894)
KLOCKE, Friedrich, (1847-1885)
KOBELL, Franz von, (1803-1882)
KOCH, Carl, (1827-1882)
KOCH, Eduard, (1838-1897)
KOCH, Friedrich Eduard, (1817-1894)
KOCH, Gottlieb Karl David von, (1849-1914)
KOENEN, Adolf von, (1837-1915)
KRAPF, Ludwig, (1810-1881)
KRAUS, Gregor, (1841-1915)
KRAUSS, Ferdinand, (1812-1890)

GERMANY 1800-1849 Cont'd.

KREFFT, Gerhard, (1830-1881)
KUNTH, Karl Ernst Albrecht, (1842-1871)
LANDOIS, Hermann, (1835-1905)
LASAULX, Arnold von, (1839-1886)
LEICHHARDT, Friedrich Wilhelm Ludwig, (1813-1848)
LEINER, Ludwig, (1830-1901)
LEMBERG, Johannes, (1842-1902)
LENZ, Oskar, (1848-1925)
LEONHARD, Gustav von, (1816-1878)
LEUBE, Gustav, (1808-1881)
LEUNIS, Johann, (1802-1873)
LEUZE, Alfred, (1845-1899)
LIEBE, Karl Leopold Theodor, (1828-1894)
LIENENKLAUS, Ernst, (1849-1905)
LORETZ, Hermann, (1836-1917)
LOSSEN, Karl August, (1841-1893)
LUDWIG, Rudolph August Birminhold Sebastian, (1812-1880)
MARSSON, Theodor, (1816-1892)
MAUCH, Carl Gottlieb, (1837-1875)
MENGE, Anton, (1808-1880)
MEYER, Christian Erich Hermann von, (1801-1869)
MEYN, Ludwig, (1820-1878)
MÖBIUS, Karl August, (1825-1908)
MÜLLER, Carl Hermann, (1823-1907)
MÜLLER, Ferdinand von, (1825-1896)
MÜLLER, Hugo Heinrich Wilhelm, (1833-1915)
NEHRING, Alfred, (1845-1904)
NETTELROTH, Henry, (1835-1887)
NEUGEBOREN, Johann Ludwig, (1806-1887)
NEUMAYR, Melchior, (1845-1890)
NIES, Friedrich, (1839-1895)
NÖLDEKE, Karl, (1815-1898)
OCHSENIUS, Carl Christian, (1830-)
OPPEL, Albert, (1831-1865)
ORTH, Albert, (1835-1915)
OTTMER, Eduard Julius Otto, (1846-1886)
PAGENSTECHER, Heinrich Alexander, (1825-1889)
PFAFF, Friedrich, (1825-1886)
PHILIPPI, Rudolph Amandus, (1808-1904)
PLATZ, Philipp, (1827-1900)
PROBST, Joseph, (1823-1905)
QUENSTEDT, Friedrich August von, (1809-1889)
RADDE, Gustav, (1831-1903)
RAMMELSBERG, Karl Friedrich, (1813-1899)
RATH, Gerhard vom, (1830-1888)
REHM, Heinrich, (1828-1916)
REINACH, Albert von, Baron, (1842-1905)
REINHART, Herman Francis, (1832-1889) [PROSPECTORS, DIVINERS AND MINING ENGINEERS]
REINSCH, Paul Friedrich, (1836-1914)
REMELÉ, Adolf Karl, (1839-1915)
RICHTER, Reinhard, (1813-1884)
RICHTHOFEN, Paul Wilhelm Ferdinand von, Freiherr, (1833-1905)
ROEDER, Charles, (1848-1911)

GERMANY 1800-1849 Cont'd.

ROGER, Otto, (1841-1915)
RÖMER, Carl Ferdinand von, (1818-1891)
RÖMER, Friedrich Adolph, (1809-1869)
RÖMER, Hermann, (1816-1894)
ROMINGER, Carl Ludwig, (1820-1907)
ROSENBUSCH, Karl Harry Ferdinand, (1836-1914)
ROSSMÄSSLER, Emil Adolf, (1806-1867)
ROTH, Justus Ludwig Adolph, (1818-1892)
RÜST, David, (1831-1916)
SADEBECK, Alexander, (1843-1879)
SAEMANN, Louis, (1821-1866)
SANDBERGER, Fridolin Carl Ludwig von, (1826-1898)
SARTORIUS VON WALTERSHAUSEN, Wolfgang, (1809-1876)
SAUER, Gustav Adolf, (1825-1932)
SCHAAFHAUSEN, Hermann, (1816-1893)
SCHACKO, Gustav, (1824-1917)
SCHAFHÄUTL, Karl Franz Emil, (1803-1890)
SCHALCH, Ferdinand, (1848-1918)
SCHAUROTH, Karl von, Freiherr, (1818-1893)
SCHENKENBERG, F. C. A., (-)
SCHERING, Ernst, (1833-1897)
SCHIEL, Jacob Heinrich, (1813-)
SCHIMPER, Karl Friedrich, (1803-1867)
SCHIMPER, Philipp Wilhelm, (1808-1880)
SCHLAGINTWEIT, Hermann, (-)
SCHLAGINTWEIT, Robert, (-)
SCHLECHTENDAL, Dietrich von, (1834-1916)
SCHLEGEL, Hermann, (1804-1884)
SCHLEIDEN, Matthias Jakob, (1804-1881)
SCHLOENBACH, Georg Justin Carl Urban, (1841-1870)
SCHLUETER, Clemens August, (1835-1906)
SCHMID, Ernst Erhard, (1815-1885)
SCHMIDT, Adolf, (1812-1899)
SCHMIDT, Johann Friedrich Julius, (1825-1884)
SCHOLZ, Max, (1832-1892)
SCHÖNN, Rudolf, (-1889)
SCHOTT, Heinrich, (-1865)
SCHWALBE, Gustav, (1844-1916)
SCHWEINFURTH, Georg, (1836-1925)
SCHWIPPEL, Karl, (1821-1911)
SEEBACH, Karl Albert Ludwig von, (1839-1880)
SEGER, Hermann, (-1893)
SELIGMANN, Gustav, (1849-1920)
SEMPER, Otto, (1830-1907)
SEUBERT, Moritz August, (1818-1878)
SOLMS-LAUBACH, Hermann zu, Graf, (1842-1915)
SPEYER, Oscar, (1827-1882)
SPITZWEG, Carl, (1808-1885)
STACHE, Guido, (1833-1921)
STAPFF, Friedrich Moritz, (1836-1895)
STEIN, Johannes Philip Emil Friedrich, (1814-1882)
STELZNER, Alfred Wilhelm, (1840-1895)
STENZEL, Karl Gustav, (1826-1905)
STERZEL, Johann Traugott, (1841-1914)

GERMANY 1800-1849 Cont'd.

STRASBURGER, Eduard, (1844-1912)
STRENG, August, (1830-1897)
STROMBECK, August von, (1808-1900)
STRUCKMANN, Carl Eberhard Friedrich, (1833-1898)
STRÜVER, Johannes Karl Theodor, called Giovanni, (1842-1915)
THEOBALD, Gottfried Ludwig, (1810-1869)
THIELE, Friedrich, (1814-1899)
TROSCHEL, Franz Hermann, (1810-1882)
TSCHUDI, Johan Jacob von, (1818-1889)
ULFFERS, Ernst, (1837-1919)
ULRICH, George Henry Frederick, (1830-1900)
VALET, August Friedrich, (1811-1889)
VOELKEL, Josef, (1828-1906)
VOGELSANG, Hermann, (1838-1874)
VOIGT, Friedrich Gustav, (1844-1886)
WAAGEN, Wilhelm Heinrich, (1841-1900)
WACHSMUTH, Charles, (1829-1896)
WAPPLER, Rudolph Bruno, (1828-1888)
WEBER, Wilhelm, (1804-1891)
WEBSKY, Christian Friedrich Martin, (1824-1886)
WEINKAUFF, H. C., (1817-1886)
WEISBACH, Albin, (1833-1901)
WEISE, Ernst, (1843-1941)
WEISS, Christian Ernst, (1833-1890)
WIECHMANN, Carl Michael, (1828-1883)
WINKLER, Clemens Alexander, (1838-1904)
WIRTGEN, Philipp Wilhelm, (1806-1870)
WÖHLER, Friedrich, (1800-1883)
WOLF, Theodor, (1841-1924)
WUNDT, Georg von, (1845-1929)
WÜNSCH, Edward A., (1822-1895)
ZADDACH, Ernst Gustav, (1817-1881)
ZIETZ, Amandus Christian Heinrich, (1839-1921)
ZIRKEL, Ferdinand, (1838-1912)
ZITTEL, Karl Alfred von, (1839-1904)

1850-1899

ABEL, Othenio, (1875-1946)
AMMON, Ludwig von, (1850-1922)
ANDRÉE, Karl, (1880-1959)
BÄRTLING, Richard, (1878-1936)
BAUHANS, Hans, (1888-1918)
BECK, Carl Richard, (1858-1919)
BECKENKAMP, Jakob, (1855-1931)
BEHR, Johannes, (1875-1973)
BEHREND, Paul, (-?1905)
BENTZ, Alfred, (1897-1964)
BERCKHEMER, Fritz, (1890-?1954)
BERGEAT, Alfred, Edmund, (1866-1924)
BERGT, Walther Adolf, (1864-1941)
BEUSHAUSEN, Hermann Emil Louis, (1863-1904)
BEUTELL, Albert, (1859-1921)
BEYSCHLAG, Franz, (-?1935)
BIRKMAIER, Anton, (-1926)

GERMANY 1850-1899 Cont'd.

BLANCKENHORN, Max Ludwig Paul, (1861-1947)
BLÜMCKE, Adolf, (1857-1914)
BODENBERGER, Wilhelm, (1857-1941)
BÖHM, Georg, (1854-1913)
BORCHERT, Gustav, (1863-1934)
BOSCH, Carl, (1874-1940)
BÖSE, Emil, (1868-1972)
BRANDES, Theodor, (-1916)
BRAUNS, Reinhard Anton, (1861-1937)
BROILI, Ferdinand, (1874-1946)
BRÖSAMLEN, Otto, (1885-1966)
BRÜCKNER, Eduard, (1862-1927)
BRUHNS, Franz Siegfried Willy, (1864-1929)
BÜCKING, Ferdinand Carl Bertram Hugo, (1851-1932)
BUSZ, Karl Heinrich Emil Georg, (1863-1930)
CHRISTA, Emanuel, (1874-1948)
CLOOS, Ernst, (1898-1974)
CLOOS, Hans, (1885-1951)
CONWENTZ, Hugo, (1855-1922)
CORNELIUS, Hans Peter, (1888-)
CRAMER, Rudolf, (1882-1949)
CREDNER, Wilhelm, (1892-1948)
DAHLGRÜN, Fritz, (1894-1954)
DANZIG, Emil, (1855-1943)
DEECKE, Johannes Ernst Wilhelm, (1862-1934)
DEICHMÜLLER, Johannes, (1854-)
DE LA SAUCE, Wilhelm, (1882-1955)
DELHAES, Wilhelm, (1883-191)
DENCKMANN, August, (1860-1925)
DENINGER, Karl, (1878-1917)
DEUBEL, Fritz, (1898-1966)
DIETLEN, Karl Rudolf, (1857-1931)
DIETRICH, Wilhelm Otto, (1881-1964)
DIETZ, Curt, (1882-1965)
DÖDERLEIN, Ludwig, (1855-1936)
DOSS, Karl Bruno, (1861-1919)
DREHER, Otto, (1888-1945)
DREVERMANN, Fritz, (1875-1932)
EBERT, Theodor, (1857-1899)
EDINGER, Tilly, (1897-1967)
EHRINGHAUS, Arthur, (-)
EISENACK, Alfred, (1891-)
ERDMANNSDÖFFER, Otto Heinrich, (1876-1955)
ETZOLD, Franz, (1859-1928)
FALB, Rudolf, (-)
FELIX, Johannes Paul, (1859-1941)
FIEBELKORN, Max, (1869-1912)
FINCKH, Alfred, (-1914)
FINCKH, Ludwig, (1871-1930)
FINSTERWALDER, Richard, (1899-1963)
FINSTERWALDER, Sebastian, (1862-?1931)
FISCHER, Ernst, (1888-1914)
FISCHER, Hugo, (1864-1926)
FLIEGEL, Gotthard, (1873-1947)

GERMANY 1850-1899 Cont'd.

FOLLMANN, Otto, (1856-1926)
FRAAS, Eberhard, (1862-1915)
FRANK, Wilhelm, (1891-1914)
FRANKE, Adolf, (1860-1942)
FRECH, Fritz, (1861-1917)
FRIEDRICH, Paul, (1856-1918)
FUCHS, Alexander, (1874-1935)
GÄBERT, Carl, (-1931)
GAGEL, Kurt, (1865-1927)
GALLWITZ, Hans, (1896-1958)
GATTNAR, Josef, (1854-1928)
GEINITZ, Eugen, (1854-1925)
GEYER, David, (1855-1932)
GOLDSCHMIDT, Victor, (1853-1933)
GOLDSCHMIDT, Victor Moritz, (1888-1947)
GÖLLER, August, (1878-1965)
GOSSNER, Balthasar, (1877-1937)
GOTHAN, Walther, (1879-1954)
GOTTSCHE, Carl Christian, (1855-1909)
GOTTSCHICK, Franz, (1865-1927)
GRIPP, Karl, (1891-)
GROEBER, Paul Friedrich Carl, (1885-1968)
GRÜNLING, Friedrich, (1857-1919)
GÜNZLER-SEIFFERT, Hans, (1883-1957)
GÜRICH, Georg, (1859-1938)
GUTENBERG, Beno, (1889-1960)
HAAS, Hippolyt Julius, (1855-1913)
HAASE, Erich, (1857-1894)
HÄBERLE, Daniel, (1864-1934)
HAHN, Friedrich Felix (Fritz), (1885-1914)
HANIEL, Curt Alfons, (1884-1914)
HARBORT, Erich, (1879-1929)
HAUFF, Bernhard, (1866-1950)
HAUTHAL, Rudolf, (1854-1928)
HEILAND, Carl August, (1899-1956)
HEMPEL, Walther Matthias, (1851-1916)
HENNIG, Edwin, (1882-)
HERRMANN, Fritz Heinrich Ulrich, (1883-1920)
HERZ, Otto, (1853-1905)
HERZENBERG, Robert, (1885-1955)
HESS, Hans, (1864-1940)
HILLER, Wilhelm, (1899-)
HINRICHS, Gustavus Detlef, (-1922)
HINTZE, Carl Adolf Ferdinand, (1851-1916)
HOFFMAN, Fr. Adolf, (1867-1898)
HOLZAPFEL, Eduard, (1853-1913)
HUENE, Friedrich von, Freiherr, (1875-1969)
JAECKEL, Otto, (1863-1929)
JANENSCH, Werner, (1878-1969)
JENTZSCH, Alfred, (1850-1925)
JOHNSEN, Arrien, (1877-1934)
KAISER, Friedrich Wilhelm Erich, (1871-1934)
KALKOWSKY, Ernst Louis, (1851-1938)
KARSCH, Ferdinand, (1853-1936)

GERMANY 1850-1899 Cont'd.

KEILHACK, Konrad, (1858-1944)
KESSLER, Paul, (?1882-1927)
KLAATSCH, Hermann, (1863-1916)
KLÄHN, Hans, (1884-1933)
KLAUTZSCH, Adolf, (1869-1927)
KLEBS, Richard, (1850-1911)
KLEMM, Gustav, (1858-1938)
KLOCKMANN, Friedrich Ferdinand Hermann, (1858-1937)
KOCKEL, Carl Walter, (1898-1966)
KOENIGSBERGER, Johann Georg, (1874-1946)
KOERT, Willi, (1875-1927)
KÖHLER, Alexander, (1893-1955)
KOKEN, Ernst Friedrich Rudolph Karl von, (1860-1912)
KOLBE, Hermann Julius, (1855-)
KOLBE, Robert Wilhelm, (1882-1960)
KOLBECK, Friedrich Ludwig Wilhelm, (1860-1943)
KONTHER, Fritz, (1895-1968)
KORN, Johannes, (1862-1927)
KRAFFT VON DELLMENSINGEN, Albrecht, (1871-1901)
KRANTZ, Friedrich Ludwig Robert, (1859-1926)
KRAUS, Ernst, (1889-1970)
KRÄUSEL, Richard, (1890-1966)
KRUSCH, Paul, (1869-1939)
KURTZ, Fritz, (1854-1920)
LACHMANN, Richard, (1885-1916)
LAUE, Max Theodor Felix von, (1879-1960)
LAUFER, Ernst, (1850-1893)
LEHMANN, Johannes Georg, (1851-1925)
LEHMANN, Otto, (1855-1922)
LEITZ, Ernst, Senior, (1871-1956)
LENK, Hans, (1863-1938)
LEONHARD, Richard, (1870-1916)
LEONHARDT, Johannes, (1893-1959)
LEPPLA, August, (1859-1924)
LEPSIUS, Karl Georg Richard, (1851-1915)
LEUZE, Johannes, (1883-1915)
LIEBISCH, Theodor, (1852-1922)
LINGELSHEIM, Alexander von, (1874-1937)
LINSTOW, Otto von, (1872-1929)
LOEWE, Fritz, (1895-1974)
LOOMIS, Theodor, (1875-1909)
LORENZ, Theodor, (1875-1909)
MAACK, Reinhard, (1892-1969)
MAAS, Günther Heinrich Julius Max, (1871-1905)
MAAS, Otto, (1867-1916)
MACHATSCHEK, Fritz, (-1957)
MARKGRAF, Richard, (1856-1916)
MARTIN, Johann Karl Ludwig, (1851-1942)
MAUCHER, Wilhelm, (1879-1930)
MAY, Walther, (1868-)
MEISTER, Ernst, (1887-1939)
MENZEL, Karl Gustav Johannes, (1875-1914)
MENZEL, Paul, (1864-1927)
MEYER, Erich, (1874-1915)

GERMANY 1850-1899 Cont'd.

MEYER, Hans, (1858-1929)
MICHAEL, Richard, (1869-1925)
MIELEITNER, Karl, (1890-1923)
MILCH, Ludwig, (1867-1928)
MÖRICKE, Wilhelm, (1861-1897)
MÜGGE, Johannes Otto Konrad, (1858-1932)
MÜLLER, Heinrich, (1887-1914)
MÜLLER, Karl Christian Gottfried Ludwig, (1862-1906)
MYLIUS, Hugo, (1876-1918)
NAUMANN, Edmund, (1854-1927)
NESSIG, Wilhelm Robert, (1861-1932)
OEBBEKE, Konrad, (1853-1932)
OERTEL, Walter, (1889-1924)
OPPENHEIM, Leo Paul, (1863-1934)
ORTMANN, Arnold Edward, (1863-1927)
OSANN, Carl Alfred, (1859-1923)
PAECKELMANN, Werner, (1890-1952)
PATRUNKY, H., (1854-1930)
PAULCKE, Wilhelm, (1873-1949)
PENCK, Friedrich Carl Albrecht, (1858-1945)
PETERSEN, Georg, (1898-   )
PFIZENMAYER, Eugen Wilhelm, (1869-   )
PHLEPS, Otto, (1868-1928)
PIETZCKER, Franz, (1885-1914)
PIETZSCH, Kurt, (1884-1964)
PLAGEMANN, A., (1858-1908)
POCKELS, Friedrich Carl Alwin, (1865-1913)
POMPECKJ, Josef Felix, (1867-1930)
POTONIÉ, Henry Edmond, (1857-1913)
POTONIÉ, Robert, (1889-1974)
PRATJE, Otto, (1890-1925)
PUFAHL, Otto Hermann August, (1855-1924)
QUITZOW, August Alfred Wilhelm, (1881-1914)
RAMDOHR, Paul, (1890-   )
RAU, Karl, (1870-1966)
REBEUR-PASCHWITZ, Ernst von, (1861-1895)
RECK, Hans, (1886-1937)
REICHERT, Friedrich, (1878-   )
REIS, Otto Maria, (1862-1934)
RICHTER, Emma, (1888-1956)
RICHTER, Ernst Julius, (-1868)
RICHTER, Rudolf, (1881-1957)
RIMANN, Eberhard, (1882-1944)
RINNE, Friedrich Wilhelm Berthold, (1863-1933)
RODOLPH, Emil, (1853-1915)
ROSE, Hermann, (1883-1976)
ROTHPLETZ, August, (1853-1918)
RUDOLPH, Emil, (1853-1915)
RUEDEMANN, Rudolf, (1864-1956)
SALOMON-CALVI, Wilhelm, (1868-1941)
SAPPER, Karl Theodor, (1866-1945)
SCHEIBE, Robert, (1859-1923)
SCHELLWIEN, Ernst, (1866-1906)
SCHEUMANN, Karl Hermann, (1881-1964)

GERMANY 1850-1899 Cont'd.

SCHINDEWOLF, Otto Heinrich, (1896-1972)
SCHLOSSER, Max, (1854-1933)
SCHLUNCK, Johannes, (1876-1915)
SCHMIDT, Hermann, (1892-1978)
SCHMIDT, Wilhelm Erich, (1882-1944)
SCHNEIDERHÖHN, Hans, (1887-1962)
SCHOENFLIES, Arthur Moritz, (1853-1928)
SCHOETENSACK, Otto, (1850-1912)
SCHOLZ, Erich, (1884-1914)
SCHÖNFELD, Georg, (1878-1926)
SCHOTTLER, Wilhelm, (1869-1932)
SCHREIBER, Hans, (1859-?1936)
SCHRÖDER, Hans Adolf Karl Eckart, (1899-1968)
SCHRÖDER, Richard, (1853-1916)
SCHROEDER, Henry Carl, (1859-1927)
SCHULZ, Karl, (1885-1928)
SCHUMACHER, Eugen, (1851-1922)
SCHUMACHER, Friedrich, (1884-)
SCHUMANN, Karl, (1851-1904)
SCHÜTZE, Ewald, (1873-1908)
SCHWERTSCHLAGER, Joseph, (1853-1924)
SIEGERT, Leo, (1872-1917)
SIMONS, Herbert Samuel, (1891-1956)
SOENDEROP, Fritz Hermann Gustav, (1876-1914)
SPANDEL, Erich, (1855-1909)
SPANGENBERG, Kurt, (1889-1957)
STAFF, Hans von, (1883-1915)
STARK, Peter, (1888-1932)
STEINMANN, Gustav, (1856-1929)
STEINMETZ, Hermann, (1879-1964)
STILLE, Hans, (1876-1966)
STOLLER, Jakob, (1873-1930)
STOLLEY, Ernst, (1869-1944)
STOLTZ, Karl, (?1858-1927)
STÜBEL, Alfons, (-?1904)
STUTZER, Otto, (1881-1936)
SÜSSMILCH, Carl Adolf, (?1874-1946)
TANNHÄUSER, Felix, (1874-1924)
TIETZE, Oskar Adolf Eduard, (1874-1920)
TODTMANN, Emmy Mercedes, (1888-1973)
TORNAU, Friedrich Karl August, (1877-1914)
TRIEBEL, Erich, (1894-1971)
TROLL, Carl, (1899-)
TRÜSTEDT, Otto, (1866-1929)
UHLIG, Carl Ludwig Gustav, (1872-?1938)
UHLIG, Johannes, (1883-1918)
UNTERMANN, Gerhard Ernest, Jr., (1898-1975)
VATER, Heinrich, (1859-1930)
VOIGT, Waldemar, (1850-1919)
VOLLRATH, Paul, (-1929)
WAGNER, Georg, (1885-1972)
WAHNSCHAFFE, Gustav Albert Bruno Felix, (1851-1914)
WALTHER, Johannes, (1860-1937)
WANNER, Johannes, (1878-1957)

GERMANY 1850-1899 Cont'd.

WEBER, Carl Albert, (1856-1931)
WEGENER, Alfred Lothar, (1880-1930)
WEGNER, Theodor, (1880-1934)
WEIDENREICH, Franz, (1873-1947)
WEIGELT, Johannes, (1890-1948)
WEILER, Wilhelm, (1890-1972)
WEINSCHENK, Ernst Heinrich Oskar Kasimir, (1865-1921)
WELTNER, Wilhelm, (1854-1917)
WEPFER, Emil, (1882-1930)
WETZEL, Otto Christian August, (1891-1971)
WETZEL, Walter, (1887-)
WEYLAND, Hermann, (1888-)
WICHMANN, Carl Ernst Arthur, (1851-1927)
WILCKENS, Otto Rudolf, (1876-1943)
WINDHAUSEN, Anselm, (1882-1932)
WOLDSTEDT, Paul, (1888-1973)
WOLFF, Wilhelm, (1872-1951)
WÜLFING, Ernst Anton, (1860-1930)
WÜST, Ewald, (1875-1934)
ZEISE, Oskar, (1860-1925)
ZIMMERMANN, Ernst, (1860-1944)

1900 to present

BREDDIN, Hans, (1900-1973)
BROTZEN, Fritz, (1902-1968)
DÜCKER, Alfred, (1908-1969)
EHRMANN, Martin L., (1903-1972)
ERTEL, Hans, (1904-1971)
FABIAN, Rudolf, (1912-1943)
FROHBERG, Max Hans, (1901-1970)
GEIST, Frank Otto William, (-1963)
GENTNER, Wolfgang, (1906-)
GIERTZ, Johannes, (1916-1943)
GOREAU, Thomas Fritz, (1924-1970)
GÜMBEL, Emil J., (-1966)
HANIEL, Klaus, (1914-1942)
HÄNTZSCHEL, Walter, (1904-1972)
HARRASSOWITZ, Hermann, (-1956)
HEINZELMANN, Karl, (1911-1966)
HOERNES, Dieter, (1912-1955)
HORN, Hans Lorenz, (1918-1941)
JUNG, Karl, (-?1972)
KARL, Franz, (1918-1972)
KAUENHOWEN, Walter, (1900-1942)
KAUFMAN, Rudolf, (1909-ca 1942)
KLEBER, Will, (-1970)
KLINGER, Fritz Erdmann, (-?1955)
KOENIGSWALD, Gustav Heinrich Ralph von, (1902-)
KUBELLA, Karl, (1922-1973)
LAUTERBACH, Robert, (1915-)
LEDERSTEGER, Karl, (-1973)
LEUTWEIN, Friedrich, (1911-1974)
LOTZE, Franz, (1903-1971)
MÄGDEFRAU, Karl, (1907-)

GERMANY 1900 to present Cont'd.

MARTINI, Hans Joachim, (1908-1969)
MARTINUSSEN, Eugen, (1916-1942)
MERTIN, Hans, (1902-1941)
MICHELS, Franz-Xaver, (1900-1973)
NÄGELE, Julius, (1906-1973)
NEUHAUS, Alfred, (1903-1975)
NEUMANN, Hans, (1910-1941)
OECHSLE, Richard, (1908-)
PFANNENSTIEL, Max, (1902-1976)
PLEININGER, Felix, (-?1954)
PLESSMANN, Werner, (1928-1972)
RATHSBURG, Alfred, (-)
REICHARDT, Wolfgang, (1909-1937)
SCHMIDT, Wolfgang, (1914-1972)
SCHMÖGER, Carl August, (-1970?)
SCHÜLLER, Arno, (1908-1968)
SCHWARZBACH, Martin, (-)
SOLOMONICA, Paul, (1908-1954)
SPRINGORUM, Gerd, (1915-1941)
STAESCHE, Karl, (1902-)
STROBEL, Ernst Gustav, (1903-1974)
TEICHERT, Curt, (-)
THIENHAUS, Rolf, (1913-1968)
TRÖGER, Walter Ehrenreich, (1901-1963)
WASMUND, Erich, (1902-1945)
WEYL, Richard, (-)
WICHER, Carl A., (1901-1957)
WIENHOLZ, Eva, (1930-1969)
WROOST, Volkmar, (1912-1936)
WUNDERLICH, Hans-Georg, (1928-1974)
ZEUNER, Frederick Everard, (1905-1963)
ZÖLLICH, Martin Sylvester, (1913-1940)

Time unknown
DACQUÉ, Edgar, (-)
KARSTEN, Carl Johann, (-)

GREECE
Up to 1800
AGATHARCHIDES of Cnidus, (170-100 B.C.)
ARISTOTLE, (384-322 B.C.)
PAUSANIAS, (ca.130-180 A.D.)
PYTHAGORAS, (fl.532 B.C.)
SICULUS, Diodorus, (ca.85-21 B.C.)
STRABO, (63B.C.-23 A.D.?)
THEOPHRASTUS, (371-ca.286 B.C.)

1850-1899
MITZOPOULOS, Maximos K., (1896-1968)

HUNGARY

Up to 1800

BENKÖ, Ferencz, (fl.1786)
FICHTEL, Johann Ehrenreich von, (1732-1795)
FRIDVALSKY, János, (1730-1784)
TELEKI, Domokos, Graf, (-1803)

1800-1849

BRUIMANN, Vilmos, (1819-?1896)
EÖTVÖS, Lorand, (1848-1919)
ESZTERHÁZY, Miklós, Herczeg, (1816-1894)
ETTINGSHAUSEN, Constantin von, Freiherr, (1826-1897)
FUCHS, Theodor, (1842-1925)
HANTKEN, Miksa von Prudnik, (1821-1893)
HANUSZ, István, (1840-1909)
HAZSLINSZKY, Hazslini Frigyes Ágost, (1818-1896)
HERBICH, Franz, (1821-1887)
HOFMANN, Károly, (1839-1891)
INKEY BÉLA, Palini, (1847-1921)
KOCH, Antal von Bodrog, (1843-1927)
KRENNER, József Sándor, (1839-1920)
LÓCZY, Lajos, (1849-1920)
NAGYSUR, János Böckh von, (1840-1909)
NENDTVICH, Károly, (1811-?1892)
PETHÖ, Gyula, (1848-1902)
PRIMICS, György, (1849-1893)
PRUGBERGER, József, (1818-1889)
RÓMER, Flóris, (1815-?1889)
RÓNAY, Jácint, Bishop, (1814-1889)
SAJÓHELYI, Frigyes, (1848-1940)
STAUB, Morić, (1842-1904)
STUR, Dénes, (1827-1893)
STÜRZENBAUM, József, (1845-1881)
SZABÓ, József, (1822-1894)
XÁNTUS, János, (1825-1894)
ZSIGMONDY, Vilmos, (1821-1888)

1850-1899

ADDA, Kálmán, (1862-1901)
ANDRÉANSZKY, Gábor, (1895-1967)
BALLENEGGER, Róbert, (1882-1969)
BALOGH, Ernö, (1882-1969)
BÁNYAI, János, (1886-1971)
BÖHM, Ferenc, (1881-1940)
DARÁNYI, Ignác, (-1925)
DARNAY-DORNYAY, Béla, (1887-1965)
DUDICH, Endre, (1895-1971)
EMSZT, Kálmán, (1873-1957)
FEJÉRVÁRY, Géza Gyula Imre, Baron von Komlóskeresztes, (1894-1932)
FRANZENAU, Agoston, (1856-1919)
GAÁL, István, (1877-1957)
GEDEON, Tihamér, (1898-1967)
GERBER, Frigyes, (1856-1897)
GÖRGEY VON GÖRGÖ, Rudolf, (1886-1915)
GÜLL, Vilmos, (1876-1909)
HALAVÁTS, Gyula, (1853-1926)

HUNGARY 1850-1899 Cont'd.

HEILPRIN, Angelo, (1853-1907)
HOLLENDONNER, Ferenc, (1882-1935)
HORUSITZKY, Henrik, (1870-1945)
ILOSVAY, Lajos, (1851-1937)
JABLONSKI, Eugene, (1892-1975)
KADIĆ, Ottokár, (1876-1957)
KALECSINSZKY, Sandór, (1857-1911)
KORMOS, Tivadar, (1881-1948)
KUTASSY, Endre, (1898-1938)
LAMBRECHT, Kálmán, (1889-1936)
LIEBUS, Adalbert, (1876-1945)
LIFFA, Aurél, (1872-1957)
LÖRENTHEY, Imre, (1867-1917)
LÖW, Márton, (1885-1943)
MAROS, Imre, (1882-1937)
MAURITZ, Béla, (1881-1971)
MELCZER, Gusztáv, (1869-1907)
MIHÁLTZ, István, (1897-1964)
NAGYSUR, Hugo Böckh von, (1874-1931)
NÁRAY-SZABÓ, István, (1899-1972)
NOPCSA VON FELSÖ-SZILVÁS, Ferenc, baron, (1877-1933)
NOSZKY, Jenö, Sr., (1880-1951)
PANYITY, Louis Samuel, (1890-1943)
PAPP, Károly, (1873-1963)
PÁVAI, Vajna Ferenc, (1886-1964)
POSEWITZ, Tivadar, (1850-1917)
RAKUSZ, Gyula, (1896-1932)
ROTH, Samu, (?1854-?1888)
ROZLOZSNIK, Pál, (1880-1940)
SCHAFARZIK, Ferenc, (1854-1927)
SCHERF, Emil, (1889-1967)
SCHMIDT, Sándor, (1855-?1905)
SCHRÉTER, Zóltan, (1882-1970)
SEMSEY UR SEMSEI, Andor, (-1923)
SIGMOND, Elek, (1873-1939)
SÜMEGHY, József, (1892-1955)
SZÁDECZKY, Gyula, Baron Szádeczky-Kardoss of Szádecsne and Kafdosfalva, (1860-1935)
SZONTAGH, Tamás, (1851-1937)
SZTERÉNYI, Hugo, (1857-1909)
TAEGER, Henrik, (1881-1939)
TAUSCH VON GLÖCKELSTHURN, Leopold, (1858-1899)
TELEGDI ROTH, Károly, (1886-?1956)
TOBORFFY, Zoltán, (1883-1927)
TOKODY, László, (1893-1964)
TÖRÖK, Zoltán, (1893-1963)
TRAXLER, Lászlo, (1868-1898)
TREITZ, Péter, (1866-1935)
VADÁSZ, Elemér, (1885-1970)
VENDEL, Aladár, (1886-1971)
VENDL, Maria, (1890-1945)
VÍGH, Gyula, (1889-1958)
VITÁLIS, István, (1871-1947)
VIZER, Vilmos, (1873-1941)

HUNGARY 1850-1899 Cont'd.

ZIMÁNYI, Károly, (1862-1941)

1900 to present

BEM-BÖHM, Boleszláv, (1900-1964)
CSAJÁGHY, Gábor, (1903-1972)
FÖLDVÁRI, Aladár, (1906-1973)
GROSSZ, Ádám, (1925-1973)
HERRMANN, Margit, (-?1957)
HORUSITZKY, Ferenc, (1901-1971)
KERTAI, György, (1912-1968)
KOCH, Sándor, (1906-)
KOLOSVÁRY, Gábor, (1901-1968)
KOZÁRY, Myron Theodore "Mirko", (1918-1966)
KULHAY, Gyula, (1910-1945)
MAJZON, László, (1904-1973)
NORDEN, John Alexander Erdelyi-Fazekas, (né Erdélyi-Fazekas), (1912-1977)
NOSZKY, Jenö Jr., (1909-1970)
PANTÓ, Gábor, (1917-1972)
PAPP, Ferenc, (1901-1969)
REICHERT, Robert, (1901-1937)
SCHEFFER, Viktor, (1900-1966)
SCHMIDT, Robert Eligius, (1902-1973)
SZABÓ, Pál Zóltán, (1901-1965)
TASNÁDI-KUBACSKA, András, (1902-1977)
VERTÉS, László, (1914-1968)
VIDACS, Aladár, (1909-?1965)

ICELAND

1850-1899

BÅRDHARSON, Gudhmundur G., (1880-1933)
EYTHORSSON, Jon, (1895-1968)
HANNESSON, Pálmi, (1898-1956)

1900 to present

PÅLSSON, Sveinn, (-)
TRYGGVASON, Tómas, (1907-1965)

INDIA

1800-1849

JERVIS, William Paget, (1831-1906)

1850-1899

CAMMIADE, Louis Aimé, (1872-?1951)
DAS-GUPTA, Hemchandra C., (1878-1933)
FOX, Cyril Sankey, Sir, (1886-1951)
KRISHNAN, Maharajapuram Sitaram, Padma Bhusan, (1898-1970)
MATHUR, Krishna Kumar, (-1936)
MUKERJI, Sushil Kumar, (1896-1934)
NARKE, Ganesh Govind, (-?1943)
RAMAN, Chandrasekhara Venkata, Sir, (1888-1970)
RAMA RAO, Bellu, (1890-1970)
RAMA RAO, Lakshmeswar, (1896-1974)
RAMA RAU, S. Sethu, Rao Bahadur, (1874-1929)
SAHA, Baidya Nath, Babu, (-1929)
SAHNI, Birbal, (1891-1949)
SAMBASIVA IYER, Valiathur S., (-?1943)
SEN GUPTA, Kiron Kumar, (1882-1943)
WADIA, Darashaw Nosherwan, (1883-1969)

1900 to present

BAJPAI, Mahesh Prasad, (1907-1938)
METRE, Waman Bapu, (1906-1970)
NAGAPPA, Yedatore, Shri, (1907-1960)
NAIDU, Purama Rangashayee Jagapathi, (1903-1970)
RAJNATH, Shri, (1904-1972)
SUBRAMANIAM, Anantharama Parameshwara, (1921-1973)
SWAMY, Kolagotla Bhimeswara, Sri, (-1950)

IRAQ

1900 to present

Al-NAQIB, Khorshid Mohammed, (?1927-1972)

IRELAND

Up to 1800

BOYLE, Robert, Honourable, (1627-1691)
FITTON, William Henry, (1780-1861)
GRIFFITH, Richard John, Sir, (1784-1878)
HAMILTON, William, Reverend, (1755-1797)
HUTTON, Robert, (1784-1870)
KIRWAN, Richard, (1733-1812)
McDONNELL, James, (?1763-1845)
MEADE, William, (1789-1862)
PENTLAND, Joseph Barclay, (1797-1873)

IRELAND Up to 1800 Cont'd.

PRIOR, Richard, (fl.1699)
SLOANE, Hans, Sir, (1660-1753)
USHER, James, Archbishop, (1581-1656)

1800-1849
ALLMAN, George James, (1812-1898)
BALL, John, (1818-1889)
BALL, Robert Stawell, Sir, (1840-1913)
BALL, Valentine, (1843-1895)
CLOSE, Maxwell Henry, Reverend, (1822-1903)
CONDON, Thomas, (1822-1907)
CROGHAN, George, (fl. 1765)
DU NOYER, George Victor, (-1869)
EGAN, Frederick William, (1836-1901)
FOOT, Frederick J., (-1867)
GANLY, Patrick, (1809-1899)
HAMILTON, Sir William R., (1805-1865)
HARDMAN, Edward Townley, (1845-1887)
HAUGHTON, Samuel, Reverend, (1823-1897)
KELVIN, William Thomson, Baron, (1824-1907)
KILROE, James Robinson, (1848-1927)
KINAHAN, George Henry, (1829-1908)
LEONARD, Hugh, (1841-1909)
MacADAM, James, (1801-1861)
MacHENRY, Alexander, (1843-1919)
MALLETT, Frederic Richard, (1841-1921)
McCOY, Frederick, Sir, (1823-1899)
McHENRY, Alexander, (1843-1919)
MEADE, Richard, (1827-1894)
MEDLICOTT, Henry Benedict, (1829-1905)
MOSS, Richard Jackson, (1847-1934)
NOLAN, Joseph, (1841-1902)
OGILBY, William, (1808-1873)
O'KELLY, Joseph, (1832-1883)
OLDHAM, Charles Aemilius, (1831?-1869)
OLDHAM, Thomas, (1816-1878)
O'REILLY, Joseph P., (1829-1905)
REYNOLDS, James Emerson, (1844-1920)
SCOTT, Robert Henry, (1833-1916)
STOKES, George Gabriel, Sir, (1819-1903)
SYMES, Richard Glascott, (1840-1906)
TUOMEY, Michael, (1805-1857)
TYNDALL, John, (1820-1893)
WHITLA, Francis, (fl.1830-1853)
WYATT-EDGELL, Henry Adrian, (1847-1866)
WYNNE, Arthur Beavor, (1835-1906)

1850-1899
BELL, Robert, (1864-1934)
CLELAND, Alexander M'Ivor, (1862-)
COTTER, Gerald de Purcell, (1881-1941)
CROOKSHANK, Henry, (1893-1972)
DWERRYHOUSE, Arthur Richard, (1867-1931)
GRAHAM, Richard Percival Devereux, (1880-1965)

IRELAND 1850-1899 Cont'd.

HINCH, John de Witt, (?1875-1931)
JOLY, John, (1857-1933)
MOON, Frederick William, (-1925)
RICHEY, James Ernest, (1886-1968)
VALENTINE, R. L., Lieutenant, (1890-1916)
WRIGHT, William Bourke, (1876-1939)

1900 to present
SMYTH, Louis Bouvier, (1933-?1953)

ISLE OF MAN (UNITED KINGDOM)
1800-1849
FORBES, David, (1828-1876)

ITALY
Up to 1800
ALBERTI, Leon Battista, (1404-1472)
ALDROVANDI, Ulisse, (1522-1605)
ARDUINO, Giovanni, (1714-1795)
ASCOLI, Caçco d', (1257-1327)
BASSI, Ferdinando, (1714-1774)
BECCARI, Jacopo Bartolommeo, (1682-1766)
BELGRADO, Jacopo, Padre, (fl.1738)
BIANCHI, Johannes-Janus Plancus, (1695-1775)
BIRINGUCCIO, Vannoccio, (1480-1538)
BOCCONE, Paolo, (1633-1704)
BREISLAK, Scipione, (1748-1826)
BROCCHI, Giovanni Battista (Giambattista), (1772-1826)
BRUNO, Filippo, called Giordano, (1548-1600)
CANALI, Luigi, (1759-1841)
CARDANO, Girolamo, (1501-1576)
CAROSI, Jan Filip, (1744-1799)
CATULLO, Tommaso Antonio, (1782-1869)
CIAMPINI, Giovanni Giustino, (1633-1698)
CIMARELLI, Vincenzo Maria, (-1660)
COLLINI, Cosimo Alessandro, (1727-1806)
COLONNA, Fabio, (1567-1650)
COSTA, Oronzio Gabriele, (1787-1867)
CURIONI, Giulio, (1796-1878)
DEGLI ALESSANDRI, Alessandro, (1461-1523)
FORTIS, Giovanni Battista Alberto, Abate, (1741-1803)
FRACASTORO, Girolamo Hieronymus, (1483-1553)
GALEAZZI, Domenico Mario Gusmano, (1686-ca.1775)
GEMMELLARO, Carlo, (1787-1866)
GRIMALDI, Francesco Antonio, (1740-1784)
GUALTIERI, Nicolaus, (1688-1744)
GUGLIELMINI, Domenico, (1655-1710)
GUIDONI, Gerolamo, (1794-1870)
IMPERATO, Ferrante, (1550-1631)
LUCRETIUS CARUS, Titus, (98-55 B.C.)
MARASCHINI, Pietro, (1724-1825)
MARSIGLI, Luigi Ferdinando, (ca.1330-1394)
MARZARI PENCATI, Giuseppe Conte, (1779-1836)
MONTI, Giuseppe, (1682-1760)

ITALY Up to 1800 Cont'd.

MONTICELLI, Teodoro, Abate, (1757-1845)
MORO, Anton Lazzaro, (1687-1740)
PAROLINI, Alberto Cav., (1788-1867)
PINI, Ermenegildo, (1739-1825)
PLINY THE ELDER (CAIUS PLINIUS SECUNDUS), (?20 A.D.-79 A.D.)
SAVI, Paolo, (1798-1871)
SCILLA, Agostino, (1639-1700)
SENECA, Lucius Annaeus, (ca.4 B.C. - 65 A.D.)
SOLDANI, Ambrogio, Pater, (1736-1808)
SPALLANZANI, Lazzaro, Abate, (1729-1799)
TARGIONI TOZZETTI, Giovanni, (1712-1783)
TENORE, Michele, (-)
TONDI, Matteo, (1762-1835)
VALLISNERI, Antonio, (1661-1730)
VANDELLI, Domenico, (1735-1816)
VINCI, Leonardo da, (1452-1519)
VIVENZIO, Giovanni, (-)

1800-1849

BALSAMO-CRIVELLI, Giuseppe, Marchese, (1800-1874)
BARETTI, Martino, (1841-1905)
BARGELLINI, Mariano, (1831-1906)
BARZANÒ, Gaetano, (-?1859)
BELLARDI, Luigi, (1818-1889)
BELLOTTI, Christoforo, (1823-1919)
BERTELLI, Petro Timoteo, (1826-1905)
BIANCONI, Gian Giuseppe, (1809-1878)
BOMBICCI PORTA, Luigi, (1833-1903)
BOTTI, Ulderigo, (1822-1905)
BRUNO, Carlo, (1831-1916)
BUCCHICH, Gregorio, (1828-1911)
CALCARA, Pietro, (1819-1854)
CAMPANI, Giovanni, (1820-1891)
CAPELLINI, Giovanni, (1833-1922)
CECCHI, Filippo, (1822-1887)
CESÀRO, Guiseppe Raimondo Pio, (1849-1939)
CIOFALO, Saverio, (1825-1925)
COCCHI, Igino, (1827-1913)
CORNALIA, Emilio, (1824-1882)
COSSA, Alfonso, (-)
CRISTOFORI, Joseph de, (1803-1837)
D'ACHIARDI, Antonio, (1839-1902)
DA RICO, Nicolò, Conte, (-?1845)
DEMARCHI, Lamberto, (1841-1907)
DE NICOLIS, Enrico, (1841-1908)
DEO GRATIAS, Pietro Perrando, Don, (?1846-1889)
DE ROSSI, Michele Stefano, (1834-1898)
DE VISIANI, Roberto, (1800-1878)
DE ZIGNO, Achille, Barone, (1813-1888)
DODERLEIN, Pietro, (1810-1895)
FINO, Vincenzo, (1848-1914)
FORESTI, Lodovico, (1829-1913)
GASTALDI, Bartolomeo, (1818-1879)
GEMMELLARO, Gaetano Giorgio, (1832-1904)

ITALY 1800-1849 Cont'd.

GIORDANO, Felice, (1825-1892)
GUISCARDI, Guglielmo, (1821-1885)
ISSEL, Arturo, (1842-1922)
LAIS, Giuseppe, Padre, (1845-1921)
LANINO, Giuseppe, (1832-1907)
LAWLEY, Robert, (1818-1881)
LIOY, Paolo, (1834-1911)
LOTTI, Bernardino, (1847-1933)
LOVISATO, Domenico, (1842-1916)
MARINELLI, Giovanni, (1846-1900)
MARINONI, Camillo, (1845-1883)
MASSALONGO, Abramo Bartolomeo, (1824-1860)
MENEGHINI, Giuseppe Giovanni Antonio, (1811-1889)
MENEGUZZO, Giovanni, (-)
MENIN, Aurelio, Padre, (-)
MICHELOTTI, Giovanni, (1814-1898)
MONTEROSATO, Tommaso di Maria, Marchese di, (1841-1927)
NEGRI, Arturo, (-1896)
NEGRI, Cristoforo, (-)
NICCOLI, Enrico, (1836-1915)
OMBONI, Giovanni, (1829-1910)
ORSINI, Antonio, (-1870)
PAGLIA, Enrico, (-1889)
PALMIERI, Luigi, (1807-1896)
PANEBIANCO, Ruggero, (1848-1930)
PANTANELLI, Dante, (1844-1913)
PARETO, Lorenzo, (1800-1865)
PASINI, Lodovico, (1804-1870)
PAULUCCI, Marianna, Marchesa, (1835-1919)
PELLATI, Nicola, (1835-1907)
PILLA, Leopoldo, (1805-1848)
PIRONA, Giulio Andrea, (1822-1895)
PISANI DE SERRES, Felix, (1831-1920)
PONZI, Giuseppe di Tomaso, (1805-1885)
POZZI, Giovanni, (-?1891)
RAGAZZONI, Giuseppe, (-1898)
RIVA-PALAZZI, Giovanni, (1838-1913)
ROVASENDA, Luigi di, Conte, (1826-1917)
SALMOJRAGHI, Francesco, (1837-?1910)
SCACCHI, Arcangelo, (1810-1893)
SCARABELLI GOMMI FLAMINI, Giuseppe, (1820-1905)
SECCO, Andrea, (-1889)
SEGUENZA, Giuseppe, (1833-1889)
SELLA, Quintino, (1827-1884)
SERPIERI, Alessandro, (1823-1885)
SILVESTRI, Orazio, (-1890)
SISMONDA, Angelo, (1807-1878)
SISMONDA, Eugenio, (1815-1870)
SORDELLI, Ferdinando, (?1837-1916)
SPEZIA, Giorgio, (1842-1912)
STABILE, Giuseppe, Abate, (1827-1869)
STATUTI, Augusto, (1829-1911)
STOPPANI, Antonio, (1824-1891)
STROBEL, Pellegrino, (1821-1895)

ITALY 1800-1849 Cont'd.

TACCHINI, Pietro, (1838-1905)
TARAMELLI, Torquato, (1845-1922)
TENORE, Gaetano, (1826-1903)
TERRIGI, Guglielmo, (1831-1892)
TRABUCCO, Giacomo, (1845-1925)
TRAVERSO, Giovanni Battista, (1843-1914)
UZIELLI, Gustavo, (1839-1911)
VERRI, Antonio, (1839-1925)
VILLA, Antonio, (-1885)
VILLA, Giovanni Battista, (-1887)
ZÀMARA, Giuseppe, (1837-1917)

1850-1899

ALOISI, Piero, (1881-1938)
ANELLI, Mario, (1882-1953)
ARTINI, Ettore, (1866-1928)
AZARI, Mario, (-?1932)
BALDACCI, Luigi, (1850-1927)
BARAFFAEL, Angelo, (1866-1928)
BARATTA, Mario, (1868-1935)
BASSANI, Francesco, (1853-1916)
BATTAGLIA, Rafaello, (1896-1958)
BELLINI, Raffaelo, (1874-1930)
BERTOLIO, Leopoldo Sollman, (-?1922)
BIAGI, Giuseppi, (1859-1905)
BIANCHI, Angelo, (1892-1970)
BIBOLINI, Aldo, (1876-1949)
BLANC, Gian Alberto, Barone, (1879-1966)
BOEGAN, Eugenio, (1875-1939)
BOERIS, Giovanni, (1867-1946)
BONARELLI, Guido, (1871-1951)
BONETTI, Filippo, (1854-1911)
BONOMINI, Celestino, Don, (1866-?1946)
BORSARI, Ferdinando, (1858-1891)
BRUGNATELLI, Luigi, (1859-1928)
BUCCA, Lorenzo, (1857-1930)
CACCIAMALI, Giambattista, (1856-1934)
CALCIATI, Cesare, (1885-1929)
CANAVARI, Mario, (1855-1928)
CAPACCI, Celso, (1854-1929)
CARAPEZZA, E., (-1915)
CARPANESE, Tullio, (-?1940)
CASTIGLIONI, Bruno, (1898-1945)
CERMENATI, Mario, (1868-1924)
CERULLI-IRELLI, Serafino, (1873-1946)
CHECCHIA-RISPOLI, Giuseppe, (1877-1947)
CIAMPI, Adolfo, (1876-1934)
CLERICI, Enrico, (1862-1938)
CONEDERA, Raimondo, (1859-1925)
CONTI, Piero Ginori, (1856-1939)
CORTESE, Emilio, (1856-1938)
CORTI, Benedetto, (1868-1907)
COZZAGLIO, Arturo, (1862-)
D'ACHIARDI, Giovanni, (1872-1944)

ITALY 1850-1899 Cont'd.

DAL PIAZ, Giorgio, (1872-1962)
DE AGOSTINI, Giovanni, (-)
DE ALESSANDRI, Giulio, (-?1920)
DE GREGORIO, Antonio, Marchese, (1855-1930)
DE LORENZO, Giuseppe, (1871-1957)
DE PRETTO, Olinto, (1857-1921)
D'ERASMO, Geremia, (1887-1962)
DE REGNY, Paolo Vinassa, (1871-1957)
DESIO, Ardito, (1897-)
DE STEFANI, Carlo, (1851-1924)
DI MARCHI, Luigi, (1857-1936)
DI STEFANO, Giovanni, (1856-1918)
D'OSSAT, Gioacchino De Angelis, (1865-1957)
DRAGO, Ciro, (1895-1960)
FABIANI, Ramiro, (1879-1954)
FAUSSONE, Luigi, (1880-1925)
FERRARIS, Erminio, (1851-1928)
FERUGLIO, Egidio, (1897-1954)
FOLCO, Carlo, (1866-1932)
FORMA, Ernesto, (1869-1911)
FORNASINI, Carlo, (1854-1931)
FOSSA-MANCINI, Enrico, (1884-1950)
FRANCHI, Secondo, (-?1939)
FRANCO, Pasquale, (1852-1907)
FRENGUELLI, Joaquín, (1883-1958)
FUCINI, Alberto, (1864-1940)
GEMMELLARO, Mariano, (1879-1921)
GIANNOTTI, Carlo, (1897-1929)
GIULIETTI, G. M., (-?1928)
GORTANI, Michele, (1883-1966)
GRILL, Emanuele, (1884-1961)
LINCIO, Gabrielle, (1874-1938)
MALLADRA, Alessandro, (1865-1945)
MANASSE, Ernesto, (1875-1922)
MARIANI, Ernesto, (-?1939)
MARINELLI, Olinto, (1874-1926)
MARTELLI, Alessandro, (1877-1934)
MARTINELLI, Cesare, (-1929)
MARTINOTTI, Anna, (-1931)
MATTEUCCI, Raffaele Vittorio, (1862-1909)
MATTIROLO, Ettore, (1853-1923)
MELI, Romolo, (1852-1920)
MELZI, Gilberto, Conte, (1868-1899)
MERCALLI, Giuseppe, (1850-1914)
MERCIAI, Giuseppe, (1877-1939)
MILLOSEVICH, Federico, (1875-1942)
MONETTI, Luigi, (1864-1926)
NIEVO, Ippolito, (1867-1916)
NOÉ, Emilio, (1881-1940)
NOVARESE, Vittorio, (1861-1948)
PALAZZO, Luigi, (1861-1933)
PARMA, Augusto, (1857-1926)
PARNISARI, Carlo, (-?1929)
PARONA, Carlo Fabrizio, (1855-1939)

ITALY 1850-1899 Cont'd.

PELLOUX, Alberto, (-)
PENTA, Francesco, (1899-1965)
PERRIER, Carlo, (1886-1948)
PETROCCHI, Carlo, (1898-1956)
PICCINELLI, Piero, Conte, (1865-1937)
PLATANIA, Gaetano, (1867-1938)
PORRIO, Cesare, (1865-1940)
PORTIS, Alessandro, (1853-1931)
QUERCIGH, Emanuele, (1885-1939)
REPOSSI, Emilio, (1876-1931)
RISTORI, Giuseppe, (1856-1905)
RIVA, Carlo, (1872-1902)
ROCCATI, Alessandro, (1872-1928)
RODIGHIERO, Andrea, (1892-1917)
ROSSI, Arturo, (1860-1892)
ROVERETO, Gaetano, (1870-1952)
SABATINI, Venturino, (1879-1922)
SACCO, Federico, (1864-1948)
SCACCHI, Eugenio, (1854-1929)
SCALIA, Salvatore, (1874-1923)
SCOTTI, Ariberto, (1885-1925)
SEGRÉ, Claudio, (1853-1928)
SEGUENZA, Luigi, (1873-1908)
SOMIGLIANA, Carlo, (?1859-1955)
STEFANINI, Giuseppe, (1882-1938)
STELLA, Augusto, (1863-1944)
TELLINI, Achille, (1866-1938)
TITTONI, Tommaso, (1855-1931)
TOMMASI, Annibale, (1858-1921)
TONINI, Lorenzo, ([1859]-1926)
TRENTANOVE, Giorgio Morando, (1874-1914)
TUCCIMEI, Giuseppe, (1851-1915)
UGOLINI, Riccardo, (1874-1936)
VACCHELLI, Nicola, Generale, (1870-1932)
VAGHI, Vittorio, (-1921)
VARDABASSO, Silvio, (1891-1966)
VERRI, Pietro, (-)
VIOLA, Carlo Maria, (1855-1925)
VISENTINI, Marco, (1885-1964)
ZAMBONINI, Ferruccio, (1880-1932)
ZENARI, Silvia, (1895-1956)
ZEZI, Pietro, (1884-1914)
ZUFFARDI, Pietro, (1885-1915)

1900 to present

AIROLDI, Marco, (1900-1937)
ANDREATTA, Benedettino Ciro, (1906-1960)
BLANC, Alberto Carlo, (1906-1960)
BOAGA, Giovanni, (1902-1961)
CONFORTO, Bruno, (1918-1963)
DAL PIAZ, Giambattista, (1927-)
FALINI, Filippo, (1917-1963)
FERASIN, Francesco, (1928-1958)
FIOCCO, Giorgio, (1931-)

ITALY 1900 to present Cont'd.

INEICHEN, Giovanni, (-1952)
LORENZI, Arrigo, (-1953)
MAGISTRETTI, Luigi, (-?1959)
MARCHESINI, Enrico, (1911-1968)
MORANDINI, Giuseppe, (1907-1969)
ONORATO, Ettore, (-)
PANICHI, Ugo, (-?1967)
PATRIZI, Saverio, (1902-1957)
PRINCIPI, Paolo, (-)
TONSO, Alessandro, (-)
TRENER, Giovanni Battista, (-?1954)
VERCELLI, Francesco, (-?1952)
ZACCAGNA, Domenico, (-?1940)

Time unknown

D'AREZZO, Restoro, (-)
D'ISENGARD, Luigi, Barone, (-)
LIPPI, Carmine Antonio, (-)

JAMAICA

1850-1899

LAKE, Horace Alexander, (1887-1957)

JAPAN

1850-1899

FUKUCHI, Nobuyo, (1877-1934)
HARADA, Toyokitsi, (-1894)
HIKI, Tadasu, (1866-1927)
ISHIMOTO, Mishio, (1893-1940)
JIMBŌ, Kotora, (1867-1924)
KIKUCHI, Dairoko, Baron, (1855-1917)
KOTÔ, Bundjirô, (1856-1935)
KUSAKABE, Shiroka, (1872-1924)
OMORI, Fusakichi, (1868-1923)
OTANI, Ryokichi, (1875-1934)
SEIKEI, Sekiya, (1855-1896)
SHIDA, Toshi, (1876-1936)
TERADA, Torahiko, (1878-1935)
TOKUNAGA, Shigeyasu, (1874-1940)
WADA, Tsunashirô, (1856-1920)
YABE, Hisakatsu, (1878-1969)
YAMASAKI, Naomasa, (1870-1929)

1900 to present

ASANO, Kiyoshi, (1910-)
KOBAYASHI, Teiichi, (1901-)
KUNO, Hisashi, (1910-1969)
NAGATA, Takesi, (1913-)
NAKAYA, Ukichiro, (1900-1962)
ŌCURA, Hirobumi, (1921-1969)
SHODA, Mikio, (1924-1974)

LATVIAN SSR (USSR)
Up to 1800
EICHWALD, Eduard Iwanowitsch von, (1795-1876)

1900 to present
GROSS, Walter Robert, (1903-1974)
MELLIS, Otto, (1906-1970)
ZANS, Verners Aleksandrs, (1904-1961)

LITHUANIAN SSR (USSR)
1850-1899
KAVECKIS, Mykolas, (1889-)
TOMKEIEFF, Sergei Ivanovich, (1892-1968)

1900 to present
HALICKA, Antonina, (1908-1973)

LUXEMBOURG
1850-1899
LUCIUS, Michel, (1876-1961)

MALAYSIA
1900 to present
HO, Robert, (1921-1972)

MEXICO
1800-1849
CASTILLO, Antonio del, Don, (1820-1895)

1850-1899
FLORES REYES, Teodoro, (1873-1955)
ORDOÑEZ, Ezequiel, (1867-1950)

1900 to present
DAVILA, Salvador Ortiz, (1902-1931)
GARFIAS, Valentin R., (-)
LÓPEZ VÁZQUEZ, Andrés, (1903-1954)
MALDONADO-KOERDELL, Manuel, (1908-1972)

NETHERLANDS

Up to 1800

BREDA, Jacob Gisbertus Samuel van, (1788-1867)
CAMPER, Pieter (Petrus), (1722-1789)
GODDING, Theodorus Joannes, Kanunnik, (1722-1797)
LEEUWENHOEK, Antoni van, (1632-1723)
MARUM, Martin van, (1750-1838)
PANHUYS, George Emile Auguste van, (1796-1871)
SPILBERGEN, Joris van, (1568-1620)
TROOST, Gerard, (1776-1850)

1800-1849

BEMMELEN, Jacob Maarten van, (1830-1911)
BINKHOORST TOT DEN BINKHOORST, Johan Theodoor, (1810-1876)
BLEEKRODE, S., (-)
BOGAERT, P. J. J., (1807-1894)
ELST, E. van de, (1826-1897)
FENNEMA, Reinder, (1849-?1897)
HOEVEN, Jan van der, (1801-1868)
RIEMSDIJK, A. W. G. van, (1803-1871)
STARING, W. C. H., (1808-1874)
THIERENS, F. F., (1811-)
UBAGHS, Johann Casimir, (1829-1894)
VERBEEK, Rogier Diederick Marius, (1845-1926)
WEYENBERGH, Hendrik, (1842-1885)
WINKLER, Tiberius Cornelius, (1822-1897)

1850-1899

BERNSEN, J. J. A., Pater, (1888-1932)
BOEKE, Hendrik Enno, (1881-1918)
BROUWER, Hendrik Albertus, (1886-1973)
COSIJN, Adrianus Johannes, (1890-1945)
DOORNINCK, Nico H. van, (1893-1956)
DUBOIS, Eugène, (1858-1940)
FLORSCHÜTZ, Frans, (1887-1965)
GROND, Gerard Joseph Alfons, (1890-1970)
GROOTHOFF, Christiaan Theodor, (1887-1969)
GRUTTERINK, Jan Adolf, (1879-1949)
HENNY, Gerhard, (1890-1944)
HENRY, Gerard, (1890-1943)
HOLST PELLEKAAN, William van, (1880-1949)
JONGMANS, Wilhelmus Josephus, (1878-1957)
KEMMERLING, Georges Laure Louis, (1888-1932)
KERCKHOFF, P. J. van, (-)
KRUITZINGA, Pieter, (1885-1967)
KUIPER, Willem Nicolaas, (1891-1952)
MARSMAN, J. H., (1892-1956)
MATTHES, François Emile, (1874-1948)
MEKEL, Johannes Antonius Alphonsus, (1891-1942)
MOLENGRAAFF, Gerard Johan Hendrik, (1887-1961)
MOLENGRAAFF, Gustaaf Adolf Frederik, (1860-1942)
NES, C. L. van, (?1878-1963)
PLESMAN, Albert, (1889-1953)
REINHOLD, Thomas, (1890-1955)
RHEDEN, J. J. Pannekoek van, (-1945)
RUTTEN, Louis Martin Robert, (1884-1946)

NETHERLANDS 1850-1899 Cont'd.

SCHERMERHORN, W., (1894-)
SCHOUTEN, C., (1891-1963)
SCHRUEDER, Antje, (1887-1952)
SMIT SIBINGA, Gerard Leonard, (1895-1963)
SONNEVELD, Jan, (1877-195)
STEENHUIS, Jacob Frederik, (1883-1958)
TESCH, P., (1879-1961)
UMBGROVE, Johannes Herman Frederik, (1899-1954)
VAN'T HOFF, Jacobus Hendricus, (1852-1911)
VENING MEINESZ, Felix Andries, (1887-1966)
VERSLUYS, Jan, (1880-1935)
VISSER, Philips Christiaan, (1882-1955)
VLERK, Isaak Martinus van der, (?1892-1974)
WATERSCHOOT VAN DER GRACHT, Willem Anton Josef Maria van, (1873-1943)
ZEIJLMANS VAN EMMICHOVEN, Cornelis Pieter Adriaan, (1897-1952)

1900 to present
ANDRAU, Evert Willem Karel, (1900-1951)
BAKELS, P. S., (-1960)
BAREN, F. H. van, (-1975)
BURGWAL, Rudolph Frans, (1917-1944)
EDELMANN, Cornelis Hendrik, (1901-1964)
ESCHER, Berend George, (-1967)
HEIMANS, E., (-?1937)
HUBRECHT, P. F., (-?1929)
ITERSON, F. K. Th. van, (-1957)
KLOMPÉ, Theodorus Hendricus Franciscus, (?1904-1963)
KUENEN, Phillip Henry, (1902-1976)
LINDEN, Bernard Hugo "Ben" van der, (-1958)
MUNCK, Victor C. E. A. de, (1920-1956)
PALM, Q. A., (1923-1960)
PET, Cees, (1900-1953)
POLDERVAART, Arie, (1918-1964)
RUTTEN, Martin Gerard, (1910-1970)
SATIJN, P. J. M., (?1907-1962)
SCHOLS, Hendrikus, (1901-1957)
TER MEULEN, J. A. C., (1902-1960)
WESTERVELD, J., (1905-1962)
ZIJLSTRA, Gerrit, (-1962)

NEW ZEALAND
1850-1899
ANDREW, Arthur Robert, (1881-1941)
BARRON, William Ernest, (1880-1935)
BARTRUM, John Arthur, (1885-1949)
COTTON, Charles Andrew, Sir, (1885-1970)
FINLAYSON, Alexander Moncreiff, (-1917)
MACLAREN, James Malcolm, (1873-1935)
MARWICK, John, (1891-1978)
THOMSON, James Allan, (1881-1928)

1900 to present
ALLAN, Robin Sutcliffe, (1900-1966)

NEW ZEALAND 1900 to present Cont'd.

BULLEN, Keith Edward, (1906-1976)
FINLAY, Harold John, (1901-1951)
HUTTON, Colin Osborne, (1910-1971)
PATTERSON, Arthur Lindo, (1902-1966)
THOMPSON, Hugh R., (1926-1959)
TOEBES, Cornelius, (?-1975)
WILLETT, Richard Wright, (1912-1974)

NORTHERN IRELAND

1800-1849

HULL, Edward, (1829-1917)
MACOUN, John, (1831-1920)
YOUNG, Robert, Right Honourable, (1822-1917)

1850-1899

CHRISTEN, Sydney Mary, (née THOMPSON), (-1923)
McLEAN, William, (?1877-1946)

NORWAY

Up to 1800

KEILHAU, Baltazar Mathias, (1797-1858)

1800-1849

BLYTT, Axel, (1843-1898)
COLLETT, Robert, (1842-1913)
HELLAND, Amund Theodor, (1846-1918)
HIORTDAHL, Thorstein Hallager, (1839-1925)
HOMAN, Christian Horrebow, (1826-1880)
KJERULF, Theodor, (1825-1888)
PETTERSEN, Karl, (1826-1890)
SARS, Michael, (1805-1869)

1850-1899

ANDERSEN, Olaf, (1884-1941)
BARTH, Thomas Frederik Weiby, (1899-1971)
BJERKNES, Vilhelm, (1862-1951)
BJØRLYKKE, Knut Olai, (1860-1946)
BRØGGER, Waldemar Christofer, (1851-1940)
DAMM, Carl Oluf Bernhard, (1869-1949)
FOSLIE, Steinar, (1887-1951)
GRØNLIE, Ole Tobias, (1871-1954)
HANSEN, Andreas M., (1857-1932)
HOLMSEN, Andreas, (1869-1955)
HOLMSEN, Gunnar, (1880-1976)
HOLTEDAHL, Olaf, (1885-1975)
HORNEMAN, Hans Henrik, (1878-1945)
KIAER, Johan Aschehong, (1869-1931)
KOLDERUP, Carl Fredrik, (1869-1942)
KOLDERUP, Niels-Henrik, (1898-1971)
MÜNSTER, Thomas Georg, (1855-1938)
OXAAL, John Thomas Fearnley, (1887-1942)
ØYEN, Peter Arnaeus, (1863-1932)
REUSCH, Hans Henrik, (1852-1922)
ROSENDAHL, Halvor, (1890-1963)

NORWAY 1850-1899 Cont'd.

SCHETELIG, Jakob Grubbe Cock, (1875-1935)
VOGT, Johan Herman Lie, (1858-1932)
VOGT, Thorolf, (1888-1958)

1900 to present
MUNSTERHJELM, Erik, (-) [PROSPECTORS, DIVINERS AND MINING ENGINEERS]
STRØM, Kaare, (1902-1967)
WERENSKIOLD, Werner, (-1962)

PAKISTAN
1900 to present
MUHAMMAD, Wali, (1935-1963)

PERU
1850-1899
BRAVO, José J., (1874-1927)
BROGGI, Jorge Albert, (1888-1966)
LISSON, Carlos I., (1868?-1957)

PHILIPPINE ISLANDS
1850-1899
ABARQUEZ, Ramon Fabella, (1891-1959)
ANDRES, Demetrio, (1889-1960)
FELICIANO, Jose Maria "Dako", (1887-1955)

1900 to present
HERRERA, Mariano B., (1918-1951)
TEVES, Juan S., "Steve", (1902-1958)

POLAND
Up to 1800
BORCH, Michael Jan, (1751-1810)
JAŚKIEWICZ, Jan D. P., (1749-1809)
JUNDZILL, Stanislaw Bonifacy, (1761-1847)
KLUK, Krzysztof, (1739-1796)
KOLLĄTAJ, Hugon, (1750-1812)
SAPIEHA, Aleksander, (1773-1812)
SENDIVOGIUS, Michael, (1556-1636 or 1646)
SOLTYK, Michal, (-)
STASZIC, Stanislaw, (1755-1826)
STRZELECKI, Pawel Edmund, Hrabia, (1797-1873)

1800-1849
BĄKOWSKI, Józef, (-1888)
BELZA, Józef, (1805-1888)
BERDAU, Feliks, (-?1888)
BOSNIASKI, Sigismund Grzymala, (1831-)
BRADETZKY, Samuel, (-)
CARNALL, Rudolf von, (1804-1874)
CHALUBIŃSKI, Tytus, (1820-1889)
CHALUPCZYŃSKI, Adam, (-1887)
CHOROSZEWSKI, Wincenty Ignacy Jędrzej, (1845-1901)
CZYRNIAŃSKI, Emilian, (1824-)
DOMEYKO, Ignacy, (1802-1889)
DUBICKI, Józef, (-?1890)
DYBOWSKI, Benedykt, (1833-1930)
DYBOWSKI, Wladyslaw, (1838-1910)
ESTREICHER, Alojzy, (-?1852)
GALLI, Edward Maryan de, (-?1893)
GIŽYCKI, Franciszek Ksawery, (-)
GLINOJECKI, Jan, (-)
HÜCKEL, Edward, (-?1896)
JASTRZĘBOWSKI, Wojciech, (-)
JELSKI, Konstanty, (1837-1896)
KONTKIEWICZ, Stanislaw, Sr., (1849-1924)

POLAND 1800-1849 Cont'd.

LABĘCKI, Hieronim, (1809-1862)
LOMNICKI, Marian Alojzy R., (1845-1915)
LUKASIEWICZ, Ignacy, (-)
NIEWIADOMSKI, Wincenty, (-?1892)
ŚLOŚARSKI, Antoni, (-?1897)
SULIMIRSKI, Filip, (-?1885)
SZTOLCMAN, J., (-)
TOROSIEWICZ, Teodor, (-)
TREJDOSIEWICZ, Jan, (-)
WAGA, Antoni Stanislaw Floryan, (-?1890)
WIŚNIOWSKI, Sygurd, (-)
ZARĘCZNY, Stanislaw Wawrzyniec, (1848-1909)
ZAWISZA, Jan, (-)
ZBORZEWSKI, Wojciech, (-)
ZDZITOWIECKI, Seweryn, (-?1881)
ŽEBROWSKI, Teofil, (-?1887)
ZEJSZNER, Ludwik, (1805-1871)
ZIELENIEWSKI, Michal, (-?1896)

1850-1899
ARCTOWSKI, Henryk Bronislaw, (1871-1958)
BOHDANOWICZ, Karol, (1864-1947)
BUJALSKI, Boleslaw, (1888-1945)
BUKOWSKI, Gejza, (1858-1937)
CZARNOCKI, Jan, (1889-1951)
DANYSZ-FLESZAROWA, Regina, (1888-1969)
DUNIN-BORKOWSKI, Stanislaw von, Graf, (1864-1934)
FRIEDBERG, Wilhelm, (1873-1941)
GĄSIOROWSKA, Zofia Halina, (1898-1957)
GOETEL, Walery, (1889-1972)
GOLDSCHLAG, Maurycy, (1891-1917)
GRZYBOWSKI, Józef, (1869-1922)
HEMPEL, Joachim, (1884-1955)
HIRSZBERG, Franciszek, (1863-1933)
HORWITZ, Ludwik, (1875-1943)
JANCZEWSKI, Edward Walery, (1887-1959)
JAROSZ, Jan, (1877-1944)
JOSSÉ, Kazimierz, (1892-1942)
KOKOSZYŃSKA, Bronislawa, (1897-1959)
KOMOROWICZ, Maurycy, (1881-1923)
KONTKIEWICZ, Stanislaw Jr., (1883-1951)
KOWALSKI, Ludwik, (1885-1943)
KOZLOWSKI, Roman, (1889-1977)
KRAJEWSKI, Stanislaw, (1890-1968)
KREUTZ, Stefan, (1883-1941)
KREUTZ, Szczęsny, (-?1887)
KROPACZEK, Boleslaw, (1886-1914)
KRYZYŻKIEWICZ, Józef, (1890-1955)
KUŹNIAR, Czeslaw Faustyn, (1889-1955)
KUŹNIAR, Wiktor, ([1871]-1935)
LENCEWICZ, Stanislaw, (1899-1944)
LEWIŃSKI, Jan Piotr, (1876-1939)
LILPOP, Jerzy, (1888-1945)
LIMANOWSKI, Mieczyslaw, (1876-1947)

POLAND 1850-1899 Cont'd.

LOZINSKI, Walery, (1880-1944)
LUBICZ-NIEZABITOWSKI, Edward, (1875-1946)
LUNIEWSKI, Adam, (1887-1945)
MALKOWSKI, Stanislaw, (1889-1962)
MAZUREK, Alojzy, (1885-1944)
MICHALSKI, Stanislaw, (1865-1949)
MOROZEWICZ, Józef Marian, (1865-1941)
NATURSKI, Jan, (-1936)
NECHAY, Wiktor, (1895-1940)
NOWAK, Jan, (1880-1940)
OPOLSKI, Zdzislaw, (1884-1941)
OSSOWSKI, Godfryd, (-?1897)
PACHUCKI, Czeslaw, (1898-1965)
PANOW, Eugeniusz, (1890-1958)
PAWLICA, Wladislaw, (1886-1919)
PAWLOWSKI, Stanislaw, (1882-1940)
PIECH, Kazimierz, (1893-1944)
PIOTROWSKI, Hugon Ludwik, (1897-1962)
PIWOWAR, Adam, (1874-1939)
PONIATOWSKI, Stanislaw Augusta, (1884-)
PREMIK, Józef, (1890-1963)
RABOWSKI, Ferdynand, (1884-1940)
RACIBORSKI, Maryan, (1863-1917)
REHMAN, Antoni, (-)
ROGALA, Wojciech, (1884-1947)
ROMER, Eugeniusz, (1871-1954)
ROSLONSKI, Romuald, (1880-1956)
ROZEN, Zygmunt, (-1936)
RUDZKI, Maurycy Pius, (-)
RUTKOWSKI, Feliks, (1878-1964)
RYCHLICKI, Jan, (-1942)
RYDZEWSKI, Bronislaw, (1884-1945)
RZACZYŃSKI, Gabriel, (-)
SAMSONOWICZ, Jan, (1888-1959)
SARJUSZ-MAKOWSKI, Arnold, (1876-1943)
SAWICKI, Ludomir, (-?1928)
SAWICKI, Ludwik, (1893-1972)
SIEMIRADZKI, Józef, (1858-1933)
SMOLEŃSKI, Jerzy, (1881-1940)
SMOLUCHOWSKI, Marian, (-)
STENZ, Edward, (1897-1956)
SUJKOWSKI, Zbigniew Leliwa, (1898-1954)
ŚWIDERSKI, Bohdan, (1892-1943)
SYNIEWSKA, Janina, (1893-1951)
SZAFER, Wladyslaw, (1886-1970)
SZAJNOCHA, Wladyslaw, (1857-1928)
TEISSEYRE, Wawrzyniec, (1860-1939)
THUGUTT, Stanislaw Józef, (1862-1956)
TOKARSKI, Julian, (1883-1961)
TOLWIŃSKI, Konstanty, (1876-1961)
WEIGNER, Stanislaw Wiktor, (1886-1935)
WEYBERG, Zygmunt, (1872-1944)
WINDIAKIEWICZ, Edward, (1858-1942)
WIŚNIOWSKI, Tadeusz, (1865-1933)

POLAND 1850-1899 Cont'd.

WOJNO, Tadeusz Jerzy, (1884-1971)
YACHEVSKY, Leonard Antonovich, (1858-1916)
ZAPALOWICZ, Hugo, (-)
ZUBER, Rudolf, (1858-1920)
ZUBER, Stanislaw, (1893-1947)
ŽUKOWSKI, Wladyslaw, (1873-1952)
ZWIERZYCKI, Józef, (1888-1961)

1900 to present
BADAK, Jerzy, (1932-1966)
BERES, Krzystztof, (-)
BOCHEŃSKI, Tadeusz Antoni, (1901-1958)
CHLEBOWSKI, Tadeusz, (1908-1961)
DRATH, Adam, (1904-1942)
FURMAŃSKI, Ignacy Jan, (1915-1972)
GUZIK, Kazimierz, (1911-1970)
JAROŃ, Bronislaw, (1905-1942)
KELUS, Aleksander, (1902-1946)
KIELAN-JAWOROWSKA, Zofia, (-)
KITAJEWSKI, Adam Maksymilian, (-)
KOBYLECKI, Mieczyslaw, (1909-1944)
KONGIEL, Roman, (1904-1960)
KWOKEK, Stefan, (1923-1970)
LUBICZ-SULIMIRSKI, Stefan, (1900-1953)
MORAWIECKI, Antoni, (1903-1971)
MOSSOCZY, Zbigniew, (1907-1963)
NIKLEWSKI, Jerzy, (1937-1971)
OBUCHOWICZ, Zbigniew Zygmunt, (1914-1965)
POBORSKI, Czeslaw, (1916-1964)
SKOCYLAS-CISZEWSKA, Kamila, (1902-1971)
ŚLIWA, Piotr, (1924-1959)
SMOLARSKA, Irena, (1928-1972)
ŚWIDZINSKI, Henryk, (1904-1969)
TEISSEYRE, Henryk, (1903-1975)
TRZASKA-JURKIEWICZOWA, Irena, (1912-1966)
WAWRYK, Wlodzimierz, (1902-1963)
WDOWIARZ, Jan, (1903-1956)
YAGODA, Herman, (1908-1964)
ZALESKI, Stanislaw, (-?1933)
ZAPART, Janusz Stanislaw, (1930-1961)
ŽYLA, Antoni, (1902-1966)

Time unknown
FONBERG, Ignacy, (-)
JURKIEWICZ, Karol, (-)
LASZKIEWICZ, Antoni, (-)
LUBOMĘSKI, Augusz, (-)
MALECKI, J., (-)
PISULEWSKI, Szymon, (-)
SAPALSKI, Jozef, (-)
SZAFARKIEWICZ, Józef, (-)

PORTUGAL

Up to 1800

DA COSTA, Emanuel Mendes, (1717-1791)

1800-1849

BAPTISTA, Isidoro Emilio, (1815-1863)
COSTA, Pereira da, (1809-1889)
GOMES, Jacinto Pedro, (1844-1910)
NERY DA ENCARNAÇAO DELGADO, Joaquim Filipe, (1835-1908)
PEREIRA DA COSTA, Francisco Antonio, (1809-1889)
RIBEIRO, Carlos, General, (1813-1882)
VASCONCELLOS DE PEREIRA CABRAL, Frederico Augusto de, (-)

1850-1899

BORGES, Alexandre, (1898-1955)
CARVALHO, Anselmo Ferraz de, (1878-1955)
DAMASIO, José Victorio, (-)
FREIRE DE ANDRADE, Carlos Benot, (1893-1956)
GONÇALVES GUIMARÃES, Antonio José, (1875-1919)
JESUS, Amílcar Mário de, (1895-1960)
MACHADE E COSTA, Alfredo Augusto d'Olveira, (1870-?1952)
PEREIRA DE LIMA, Wenceslau de Sousa, (1858-1920)
PEREIRA DE SAMPAIO FORJAZ PIMENTEL, António Augusto Álvares, (1893-)
SOUTO, Alberto, (1888-1961)
SOUZA BRANDÃO, Vicente de, (1863-1916)
VIANNA, António, (1887-1949)

1900 to present

DE SOUSA, Vasco Mendes, (1917-1962)
PILAR, Ludgero da Piedade, (1920-1967)
TORRE DE ASSUNÇÃO, Carlos Fernando, (1901-)

ROUMANIA

Up to 1800

BIELZ, Michael, (1787-1866)

1800-1849

BIELZ, Eduard Albert, (1827-1898)
COBÎLCESCU, Grigore, (1826-1892)
DRĂGHICEANU, Mathei M., (1844-1939)
PONI, Petru, (1841-1925)
ŞTEFANESCU, Gregoriu, (1838-1911)

1850-1899

CĂDERE, Dimitrie M., (1876-1941)
ISTRATI, Constantin I., (1850-1918)
JOHN VON JOHNESBERG, Conrad Heinrich, Edler, (1852-1918)
MACOVEI, Gheorghe, (1880-1969)
MRAZEC, Lodovic, (1867-1944)
MUNTEANI-MURGOCI, Gheorghe, (1872-1925)
PAPP, Simon, (1886-1970)
POPOVICI-HAŢEG, Valeriu, (1868-1929)
ROMAN, David, (1884-1927)
ŞTEFĂNESCU, Sabba, (1857-1931)
VÂLSAN, George, (1885-1935)

1900 to present

CERNESCU, Nicolae C., (1904-1967)
CONSTANTINESCU, Valeriu, (1921-1970)
GRIGORAŞ, Nicolae, (1913-1969)
MAXIM, Ion Al., (1902-1968)
MIRĂUŢĂ, Orest, (1931-1967)
POPOVICI, Viorica, (1918-1970)
SEMAKA, Alexandru, (1924-1970)

Time unknown

VOITEŞTI, Ion Popescu, (-)

SCOTLAND

Up to 1800

ALLAN, Thomas, (1777-1833)
BAIN, Andrew Geddes, (1797-1864)
BALD, Robert, (1776-1861)
BLACK, Joseph, (1728-1799)
BLACK, Samuel, (1785-1841)
BROWN, Robert, (1773-1858)
BROWN, Thomas, Captain, (1785-1862)
CLERK, John, Jr., Lord Eldin, (1757-1812)
CRICHTON, Alexander, Sir, (1763-1856)
DINWIDDIE, James, (-)
FLEMING, John, Reverend, (1785-1857)
GRANT, Robert Edmond, (1793-1874)
GUTHRIE, Matthew, (1743-1807)
HALL, James, Sir, (1761-1832)
HENDERSON, Ebenezer, (1784-1858)
HORNER, Leonard, (1785-1864)
HUNTER, John, (1728-1793)
HUTTON, James, (1726-1797)
JAMESON, Robert, (1774-1854)
LAUDER, Thomas Dick, Sir, (1784-1848)
LYELL, Charles, Sir, (1797-1875)
MACCULLOCH, John, (1773-1835)
MACGILLIVRAY, William, (1796-1852)
MACKENZIE, Alexander, Sir, (ca.1767-1830)
MACLAREN, Charles, (1782-1866)
MACLURE, William, (1763-1840)
MONBODDO, James Burnett, Lord, (1714-1799)
MURCHISON, Roderick Impey, Sir, (1792-1871)
NICOL, William, (ca.1768-1851)
PLAYFAIR, John, (1748-1819)
RICHARDSON, John, Sir, (1787-1865)
ROSE, Alexander, (1781-1860)
SIBBALD, Robert, Sir, (1641-1772)
SMITH, James, (1782-1867)
SOMERVILLE, Mary, (1780-1872)
THOMSON, Thomas, (1773-1852)
TRAILL, Thomas Stewart, (1781-1862)
URE, Andrew, (1778-1857)
WALKER, John, Reverend, (1731-1803)

1800-1849

ADAMS, Andrew Leith, (-1882)
ARMSTRONG, James, (1832-1892)
BALD, William, (-?1857)
BELL, Dugald, (1827-1892)
BENNIE, James, (1821-1901)
BRYSON, Alexander, (1816-1866)
BUCHANAN, John, (1819-1898)
CALVIN, Samuel, (1840-1911)
CAMPBELL, John Francis, (1821-1885)
CARRUTHERS, William, (1830-1922)
CHAMBERS, Robert, (1802-1871)
CRAIG, Robert, (?1821-1901)
CROLL, James, (1821-1890)

SCOTLAND 1800-1849 Cont'd.

DAVIDSON, Thomas, (1817-1885)
DICK, Allan Brugh, (1833-1926)
DICK, Robert, (1811-1866)
DUDGEON, Patrick, (-1895)
DUNLOP, Andrew, (1845-1915)
FERGUSSON, James, (1808-1885)
FORBES, James David, (1809-1868)
FRASER, James Pinkerton, (-)
GEIKIE, James, (1839-1915)
GILL, David, Sir, (1843-1914)
GLEN, David Corse, (1824-)
GORDON, George, Reverend, (1801-1893)
GORDON, John More, Reverend, (1849-1922)
HAMILTON, William John, (1805-1867)
HAY, Robert, (1835-1895)
HECTOR, James, Sir, (1834-1907)
HEDDLE, Matthew Forster, (1828-1897)
HOME, David Milne, (1805-1890)
HONEYMAN, David, Reverend, (1814-1889)
HORNE, John, (1848-1928)
HUNTER, Robert, Reverend, (1823-1897)
HUNTER-SELKIRK, John R. S., (1835-1898)
JAMIESON, Thomas Francis, (1829-1913)
KIRKBY, James Walker, (1834-1901)
KYLE, John Joseph Jolly, called Juan José, (1838-1922)
MACINTOSH, Charles, (1839-1922)
MACKINTOSH, Daniel, (1815-1891)
McKAY, Alexander, (1841-1917)
McMURTRIE, James, (1840-1914)
MILLER, Hugh, (1802-1856)
MITCHELL, Hugh, Reverend, (1822-1894)
MOORE, John Carrick, (1804-1898)
MUIR, John, (1838-1914)
MURDOCH, James Barclay, (1831-1906)
MURRAY, Alexander, (1810-1884)
MUSHET, David, (-?1859)
NICOL, James, (1810-1879)
OWEN, David Dale, (1807-1860)
OWEN, Richard Dale, (1810-1890)
PAGE, David, (1814-1879)
RAMSAY, Andrew Crombie, Sir, (1814-1891)
ROBERTSON, David, (1806-1896)
ROMANES, George John, (1848-1894)
SELKIRK, Dunbar James Douglas, Earl of, (1809-1885)
SLIMON, Robert, (1803-1882)
SMITH, John, (1845-1930)
SOMERVAIL, Alexander, (1841-1910)
STORRIE, John, (1844-1901)
THOMSON, James, (1823-1900)
TRAQUAIR, Ramsay Heatley, (1840-1912)
WATSON, Robert Boog, Reverend, (?1824-1910)
WRIGHT, Thomas, (1809-1884)
WYVILLE THOMSON, Charles, Sir, (1830-1882)
YOUNG, John, (1823-1900)

SCOTLAND 1800-1849 Cont'd.

YOUNG, John, (1835-1902)

1850-1899

ANDERSON, Charles, (1876-1944)
ANDERSON, William, (1860-1915)
BOYD, Alexander John, (1891-1969)
BOYLE, Robert, (1883-1917)
BREMNER, Alexander, (1863-1943)
BURNETT, George Alexander, (1892-1952)
CAIRNS, Robert, (1854-1911)
CAMPBELL, Robert, (1881-1957)
CHREE, Charles, (-)
CONAN DOYLE, Arthur, Sir, (1859-1930)
CORSTORPHINE, George Steuart, (1865-1919)
CRAIG, Robert Meldrum, (?1882-1956)
CURRIE, James, (1863-1930)
DRUMMOND, Henry, (1851-1897)
FALCONER, John Downie, (1876-1947)
GIBB, Alfred William, (1864-1937)
GIBB, Hugh, (1860-1932)
GORDON, William Thomas, (1884-1950)
GREGORY, John Walter, (1864-1932)
GUNN, Robert Marcus, (?1850-1909)
HANNAY, Harry Campbell, (1874-1961)
HERON, Alexander Macmillan, (1884-1971)
JOHNSTON, Robert Mackenzie, (-1918)
KEITH, Arthur, Sir, (1866-1955)
KENNEDY, William, (-1926)
KIDSTON, Robert, (1852-1924)
LAWRIE, William Henry, (?1884-1943)
LAWSON, Andrew Cowper, (1861-1952)
LEITCH, Patrick Arthur, (-1943)
MACCONOCHIE, Arthur, (1850-1922)
MACGREGOR, Murray, (1884-1966)
MACKIE, William, (1856-1932)
MANSON, William, (1891-1973)
MARIES, Alan C., (1890-1971)
McPHERSON, William, (-1928)
MILL, Hugh Robert, (1861-1950)
MILLER, Hugh Jr., (1850-1896)
MUIR, John Malcolm, (1885-1938)
OGILVIE, Francis Grant, Sir, (1858-1930)
OGILVIE GORDON, Maria Matilda, Dame, (1864-1939)
PARK, James, (1857-1946)
PEACOCK, Martin Alfred, (1898-1950)
PRINGLE, John, (1877-1948)
RAEBURN, Colin, (1894-1970)
SCOTT, Alexander, (1890-1951)
SHAND, Samuel James, (1882-1957)
STEWART, Thomas, (1857-1943)
TEMPLETON, James Clark, (1898-1950)
THOMSON, Alexander Gordon Milne, (1866-1919)
WALKER, Frederick, (1898-1968)
WALLACE, Robert Charles, (1881-1955)

SCOTLAND 1850-1899 Cont'd.

WILSON, J. S. Grant, (1855-1908)
WRIGHT, James, (1878-1957)

1900 to present
ANDERSON, Robert Beattie, (1904-1970)
BISSET, Charles Barron, (-?1966)
BROWN, James Macmillan, (1921-1970)
COCKBURN, Alexander Murray, (1901-1959)
CURRIE, Ethel Dobbie, (-1963)
CURRIE, Leslie D., (1904-1925)
DAVIDSON, Charles Findlay, (1911-1967)
GOODLET, George Anderson, (1910-1970)
HUTCHINSON, William W., (-)
HUTCHISON, Arthur Gilbertson, (1902-1949)
LAMONT, Archie, (-)
LEITCH, Duncan, (1904-1956)
MACDONALD, Simon, (1927-1971)
ROBBIE, James Andrew, (1910-1977)
SCOTT, James Frew, (1902-1956)
STENHOUSE, Andrew G., (-1950)
WALKER, George Forbes, (1916-1970)
WALTON, John, (-1971)

Time unknown
MACGREGOR, A. M., (-)

SOUTH AFRICA
1800-1849
BAIN, Thomas Charles John, (1829-1893)
DRAPER, David, (1849-1929)
EXTON, Hugh, (1833-1903)
RUBIDGE, Richard Nathaniel, (1821?-1869)
STOW, George B. William, (1822-1882)

1850-1899
BALL, Harry Standish, (?1888-1941)
BROOM, Robert, (1866-1951)
DRURY, James, (1879-)
FAIRBRIDGE, Kingsley, (1885-1924) [PROSPECTORS, DIVINERS AND MINING ENGINEERS]
HENDERSON, James Alexander Leo, (?1874-1940)
LESLIE, Thomas Nicholas, (-1942)
LEYDS, Louis Willem, (1887-1963)
MERENSKY, Hans, (1871-1952)
RAINIER, Peter W., (-) [PROSPECTORS, DIVINERS AND MINING ENGINEERS]
SMITH, James Leonard Brierley, (1897-)
TOIT, Alexander Logie du, (1878-1948)
WAGNER, Percy Albert, (1885-1929)

1900 to present
AHRENS, Louis Herman, (1918-)
BOARDMAN, Leslie Gray, (1913-1977)
BOOCOCK, Colin, (1925-1973)

SOUTH AFRICA 1900 to present Cont'd.

KITCHING, James, (-)
PARTRIDGE, Francis Chamberlain, (1903-1939)
SCHOLTZ, Douglas Leinster, (1902-1977)
WASSERSTEIN, Benno, (1906-1956)

SPAIN

Up to 1800

BORDA, José de la, Don, (1699-1778) [PROSPECTORS, DIVINERS AND MINING ENGINEERS]
DE PRADO, Casiano, Don, (1797-1866?)
ELHUYAR Y DE ZUBICE, Fausto de, (1755-1833)
GIMBERNAT, Carlos de, Don, (1765-1839)
RIO, Andrés Manuel del, (1764-1849)

1800-1849

AREITIO Y LARRINAGA, Alfonso de, (-1884)
BAUZÁ, Felipe, (1802-1875)
BOFILL Y POCH, Arturo, (1846-1929)
CASTRO Y SUERO, Manuel Fernandez de, (1825-1895)
MACPHERSON, José, Don, (1839-1902)
MALLADA, Lucas, (1841-1921)
ROSALES, Henry, (?1822-1916)
VIDAL I CARRERAS, Lluis Mariano, (1842-1922)
VILANOVA Y PIERA, Juan, (1822-1893)

1850-1899

ALMERA, Jaume, Canon, (1854-1919)
CALDERON Y ARANA, Salvador, (1851-1911)
GILLMAN, Clement, (?1883-1946)
HERNÁNDEZ-PACHECO, Francisco, (1899-1976)
PARDILLO VAQUER, Francisco, Don, (1884-1955)
ROYO Y GOMEZ, José, (1895-1958)
SELGA, Miguel, Reverend Father, S. J., (1879-1956)
VALLE DE LERSUNDI, Alfonso del, (1875-1944)

1900 to present

HERNÁNDEZ-PACHECO, Eduardo, (-1965)
RUIZ DE GAONA, Máximo, Padre, (-1971)

SRI LANKA

1800-1849

KELAART, E. F., (1818-1860)

1850-1899

COOMARASWAMY, Ananda K., (1877-1947)

SWEDEN

Up to 1800

BERGMAN, Torbern Olof, (1735-1784)
BERZELIUS, Jöns Jakob, (1779-1848)
CEDERBAUM, Peter Christopher, (1733-1795)
COLLING, Olof, (1677-1754)
CRONSTEDT, Axel Fredrik, (1722-1765)
ESPLING, Olof, (1751-1821)

SWEDEN Up to 1800 Cont'd.

FERBER, Johann Jacob, (1743-1790)
GYLLENBERG, Fredrik, (fl.1750)
GYLLENHAAL, Johan Gabriel, (1750-1788)
HERMELIN, Samuel Gustaf, (1744-1820)
HISINGER, Wilhelm, (1766-1852)
KALM, Pehr, (1716-1779)
LINNE, Carl von, (1707-1778)
NILSSON, Sven, (1787-1883)
NISSER, Pedro, (1799-1878) [PROSPECTORS, DIVINERS AND MINING ENGINEERS]
NORDENSKIÖLD, Nils Gustaf, (1792-1866)
SOLANDER, Daniel Carl, (1733-1782)
SWEDENBORG, Emanuel, (1688-1772)
TIDSTRÖM, Andreas Philip, (1723-1779)
TILAS, Daniel, (1712-1772)
VALLERIUS, Göran, (1683-1744)
VON ENGESTRÖM, Gustaf, (1738-1813)
WALLERIUS, Johan Gottschalk, (1709-1785)

1800-1849

ANGELIN, Nils Peter, (1805-1876)
ARRHENIUS, Johan, (1811-1889)
ATTERBERG, Albert, (1846-1916)
BLOMBERG, Karl Albert, (1844-1922)
BLOMSTRAND, Christian Wilhelm, (1826-1897)
DICKSON, Oscar, (1823-1897)
EGGERTZ, Victor, (1817-1889)
ERDMANN, Axel Joachim, (-)
ERDMANN, Edvard, (1840-1923)
FLINK, Gustaf, (1849-1931)
GELLERSTEDT, Göran, (1834-1913)
GUMAELIUS, Otto Joël, (1835-1899)
HAUAN, Kund Møller, (1840-1898)
IGELSTRÖM, Lars Johan, (1822-1897)
JÖNSSON, Jöns, (1848-1894)
KOLTHOFF, Gustaf, (1845-1914)
KURCK, Clas, (1849-1937)
LILJEVALL, Georg Gideon, (1848-1928)
LINDSTRÖM, Axel Fredrik, (1839-1911)
LINDSTRÖM, Jonas Gustaf Oscar, (1829-1901)
LINNARSSON, Jonas Gustaf Oscar, (1841-1881)
LOVÉN, Sven, (1809-1895)
LUNDGREN, Sven Anders Bernhard, (1843-1897)
NORDENSKIÖLD, Nils Adolf Erik, (1832-1901)
OLBERS, Elam Willgott, (1816-1895)
SANTESSON, Gustaf Henrik, (1847-1915)
SANTESSON, Karl Olof Birger, (1845-1893)
SAXE, Sjur Aasmundsen, (1808-1887?)
SIDENBLADH, Elias, (1836-1914)
SJÖGREN, Carl Anton Hjalmar, (1822-1893)
SOLITANDER, Carl Probus, (1843-1918)
STEENSTRUP, Knud Johannes Vogelius, (1842-1913)
SVEDMARK, Lennart Eugène, (1847-1922)
TIBERG, H. V., (1849-1913)

SWEDEN 1800-1849 Cont'd.

TORELL, Otto Martin, (1828-1900)
TÖRNEBOHM, Alfred Elis, (1838-1911)
TÖRNQUIST, Sven Leonhard, (1840-1920)
TROILIUS, Carl Oscar, (1813-[1899])
VON FEILITZEN, Carl Henrik Jobst, (1840?-1901)
VON POST, Hampus Adolf, (1822-1911)
WALMSTEDT, Lars Edward, (1819-1892)

1850-1899

AHLMANN, Hans Wilhelmsson, (1889-1974)
ALIN, Johan, (1879-1944)
AMINOFF, Gregori, (1883-1947)
ANDERSON, Gustavus Edwin, (1879-1940)
ANDERSSON, Axel R., (1858-1952)
ANDERSSON, Gunnar, (1865-1928)
ANDERSSON, Johann Gunnar, (1874-1960)
ANDERSSON, Oscar Frithiof, (1867-1911)
ARRHENIUS, Olof, (1895-)
ARRHENIUS, Svante August, (1859-1927)
BACKLUND, Erik, (1892-1938)
BACKLUND, Helge Götrik, (1878-1958)
BÄCKSTROM, Helge Mattias, (1865-1932)
BJAREBY, Alfred Gunnar, (1899-1967)
BONTHRON, John A., (-?1948)
BOOBERG, Gunnar, (1892-1944)
BORGSTRÖM, Leonard Henrik, (1876-?1954)
CARLHEIM-GYLLENSKÖLD, Vilhelm, (1859-1934)
CARLSON, Charles Gordon, (1895-1977)
CARSTENS, Carl Wilhelm, (1877-?1950)
DE GEER, Gerard Jacob, Baron, (1858-1943)
DE GEER, Sten, (1886-1933)
DUSÉN, Per, (1855-1926)
ENEROTH, Olof, (-1945)
ERDTMAN, Otto Gunnar Elias, (1897-1973)
ERIKSSON, J. V., (1883-1930)
FEGRAEUS, Torbern, (1853-1923)
FELLENIUS, Wolmar, (1876-?1957)
FINNEMAN, Karl Johan, (1880-1953)
FLODKVIST, Gunnar, (-?1946)
FLODKVIST, Herman, (1875-)
FUNKQUIST, Herman Per Anton, (1870-1952)
GAVELIN, Axel Olof, (1875-1947)
GEIJER, Per Adolf, (1886-1976)
GERTZ, Otto Daniel, (1878-1948)
GRANLUND, Erik, (1892-1938)
GRANÖ, Johannes Gabriel, (1882-1956)
GRANSTRÖM, Gustaf Abraham, (1851-1941)
GUSTAFSSON, Johan Peter, (-?1943)
HADDING, Assar, (1886-1962)
HAGLUND, Emil, (1877-1938)
HAMBERG, Axel, (1863-1933)
HEDSTROM, Herman Oskar, (1867-?1948)
HESSELMAN, Oscar August Henrik Wilhelm, (1874-1943)
HÖGBOM, Alvar, (1894-1939)

SWEDEN 1850-1899 Cont'd.

HÖGBOM, Arvid Gustaf, (1857-1940)
HÖGBOM, Ivar, (1892-1962)
HOLLENDER, Artur, (-?1946)
HOLM, Edvard Johann Gerhard, (1853-1926)
HOLMQUIST, Per Johan, (1866-1946)
HOLMSTRÖM, Leonard, (-?1922)
HÖRNER, Nils Gustaf, (1896-1950)
HUMMEL-GUMAELIUS, Torsten, (-?1949)
INBERG, Isak Johan, (1853-1893)
JOHANSSON, Harald Elias, (1880-1931)
JOHANSSON, Karl Fredrick, (1866-1933)
JOHANSSON, Simon, (1881-1944)
JUNGNER, Jonas Gustaf, (-?1942)
KAUDERN, Walter, (1881-1924)
KJELLMARK, Knut, (1866-?1944)
KJELLSTRÖM, Carl Johan Otto, (1855-1913)
LAGRELIUS, Axel, (1863-1944)
LINDGREN, Waldemar, (1860-1939)
LJUNGDAHL, Gustaf Samuel, (1882-1940)
LJUNGNER, Erik, (1892-1954)
LUNDBÖHM, Hjalmar, (1855-1926)
LUNDQVIST, Gösta, (1894-1967)
MAGNUSSON, Nils Harald, (1890-1976)
MAUZELIUS, Robert, (1864-1921)
MOBERG, Johan Christian, (1854-1915)
MÖLLER, Hjalmar August, (1866-1941)
MÖRTSELL, Sture, (-?1952)
MOSSBERG, Karl Elis, (-?1941)
MUNTHE, Henrik Vilhelm, (1860-1958)
NATHORST, Alfred Gabriel, (1850-1921)
NAUMANN, Einar Christian Leonard, (1891-1934)
NILSSON, Lars Albert, (1860-1906)
NORBERG, Hans L., (-?1918)
NORDENSKIÖLD, Nils Otto Gustaf, (1869-1928)
NORELIUS, Carl Oskar, (1861-1941)
ORTON, Bror Natanael, (-?1942)
OSVALD, Hugo, (1892-)
PETERSON, Olof August, (1865-1933)
PETERSSON, Walfrid, (1862-1933)
PHRAGMÉN, Gösta, (-?1944)
QUENSEL, Percy Dudgeon, (1881-1966)
RETZIUS, Emilia, (-?1904)
RICHERT, Johann Gustaf, (1857-1954)
RICHTER, Paul Bogoslav, (1854-1911)
RINDELL, Artur, (1852-1936)
ROSSBY, Carl-Gustaf Arvid, (1898-1957)
SAHLIN, Carl, (1861-1943)
SANDEGREN, Ragnar, (1887-1966)
SERNANDER, Rutger, (1866-1944)
SJÖGREN, Sten Anders Hjalmar, (1856-1922)
SUNDBERG, Karl, (1891-1938)
SUNDELIN, Uno, (-1926)
SUNDHOLM, Olof Herman, (-?1946)
TAMMEKANN, August, (1894-1959)

SWEDEN 1850-1899 Cont'd.

THOMASSON, Harald, (1893-1937)
THORNÉ, Sven Gustaf, (-?1950)
TIBERG, Björn Gustaf Georg, (1877-1948)
TROEDSSON, Gustaf T., (1891-1954)
TULLBERG, Sven Axel Theodor, (1852-1886)
UDDEN, Johan August, (1859-1932)
VON ECKERMANN, Claes Walther Harry, (1886-1969)
VON POST, Ernst Jakob Lennart, (1884-1951)
WADELL, Hakon, (1895-1962)
WALLÉN, Axel, (1877-1935)
WALLIN, Gustaf, (1871-1940)
WEIBULL, Kristian Oskar Mats, (1856-1923)
WESTERGÅRD, Anton Hilmer, (1880-1968)
WESTLUND, Erik Wiktor, (-?1954)
WIMAN, Carl Johan Josef Ernst, (1867-1944)
YNGSTRÖM, Lars, (1870-1944)
ZENZÉN, Nils, (1883-1959)

1900 to present

ASSARSSON, Gunnar, (-1963)
BENEDICKS, Carl, (-1958)
BERGSTRÖM, Gunnar, (-1954)
BJÖRNSSON, Sven, (1905-1950)
BJORULF, Hjalmar, (-?1956)
BODMAN, Gösta, (-1961)
CALDENIUS, Carl, (-1961)
DANIELLSON, Johan Patrik, (-?1954)
EKSTRÖM, Gunnar, (-1960)
FLORIN, Sten, (-)
FRÖDIN, John, (-1961)
GOLDKUTH, Algot, (-1957)
HALLGREN, Gunnar, (1907-)
HELLAND-HANSEN, Björn, (-1958?)
HÖGBOM, Bertil, (-1962)
HOLMBOE, Jens, (-?1943)
HORN AF RANTZIEN, Henning, (1922-1960)
JONSON, Per Abraham, (-?1957)
KIHLSTEDT, Folke H., (1901?-1956)
KROKSTRÖM, Torsten Hugo, (1905-1950)
LUNDBERG, Sven, (1900-?1948)
NORDENSKIÖLD, Karl Erik, (-?1954)
NORIN, Erik, (-)
ÖSTER, Johannes, (1902-1975)
RUTBERG, Karl, (-1961)
SAMUELSSON, Frans Gustaf, (-?1948)
SAMUELSSON, Gunnar, (-?1945)
SÄVE-SÖDERBERGH, Gunnar, (1910-1948)
SCHYTT, Valter, (1919-)
SIDENVALL, Karl Johan Ferdinand, (-?1957)
WALLROTH, Gunnar, (-?1949)
WESSLAU, Eric, (-1958)
WINGE, Knut Ingel, (-1958)
WRÅK, Walter Olof, (-?1933)

SWEDEN Cont'd.

Time unknown
ELZVIK, Erik, (-)
KAHLMETER, Henrik, (-)
TEGENGREN, Felix, (-)

SWITZERLAND
Up to 1800
BAUHIN, Jean, (1541-1613)
BERTRAND, Élie, (1712-1797)
BESSON, Henri, (fl.1786)
BONNET, Charles, (1720-1793)
COULON, Paul Louis, (1777-1855)
CROUSAZ, Jean-Pierre de, (1663-1750)
DUBOIS DE MONTPÉREUX, Frédéric, (1798-1850)
ESCHER VON DER LINTH, Hans Conrad, (1767-1823)
FONTAINE, Charles Aloyse, Chanoine, (1754-1834)
GAGNEBIN DE LA FERRIÈRE, Abraham, (1707-1800)
GESSNER, Conrad, (1516-1565)
GRUNER, Gottlieb Sigmund, (1717-1778)
HOFFMANN, Jan Leonard, (1710-1782)
HOTTINGER, Johann Heinrich, (1680-1756)
HUGI, Franz Joseph, (1796-1855)
KAPPELER, Moritz, (1685-1769)
KUHN, Bernhard Friedrich, (1762-1825)
LANG, Karl Nikolaus, (1670-1741)
LUC, Jean André de, (1727-1817)
MARTEL, Peter, (-)
MERIAN, Peter, (1795-1883)
MURALT, Johannes von, (1645-1733)
NECKER DE SAUSSURE, Louis Albert, (1786-1861)
PARACELSUS, (pseud. of HOHENHEIM, Philippus Theophrastus Bombastus von), (1493-1541)
PERRAUDIN, Jean-Pierre, (1767-1858)
ROUX, Jean Marc, (?1736-1805)
SAUSSURE, Horace Bénédict de, (1740-1799)
SAUSSURE, Nicolas Théodore de, (1765-1845)
SCHEUCHZER, Johann Gaspar, (1702-1729)
SCHEUCHZER, Johann Jakob, (1672-1733)
STUDER, Bernhard Rudolf, (1794-1887)
TREMBLEY, Abraham, (1710-1784)
VALLTRAVERS, Johann Rodolph von, (1723-)
VENETZ, Ignaz, (1788-1859)

1800-1849
AGASSIZ, Louis Jean Rodolphe, (1807-1873)
BALLY-PRIOR, Peter Eduard (né Bally), (1847-1926)
BIEDERMANN-IMHOOF, Adolf W. G., (1829-1900)
BOLL, Jacob, (1828-1880)
BRUNNER VON WATTENWYL, Karl, (1823-1914)
CAMPICHE, Gustave, (1809-1870)
CARTIER, Robert, (1810-1886)
CHOFFAT, Paul, (1849-1919)
COULON, Louis de, (1804-1894)
CRAMER, Carl Edward, (1831-1901)

SWITZERLAND 1800-1849 Cont'd.

DE LA HARPE, Philippe, (1830-1882)
DESOR, Pierre Jean Edouard, (1811-1882)
DUFOUR, Charles, (1827-1902)
EBRAY, Charles Henri Théophile, (1823-1879)
ERNI, Henri, (1822-1885)
ESCHER VON DER LINTH, Arnold, (1807-1872)
FAVRE, Ernest, (1845-1925)
FAVRE, Jean Alphonse, (1815-1890)
FISCHER-OOSTER, Carl von, (1807-1875)
FOREL, François Alphonse, (1841-1912)
FORSYTH MAJOR, Charles Immanuel, (1843-1923)
GAUDIN, Charles Théophile, (1822-1866)
GERLACH, Heinrich, (1822-1871)
GILLIÉRON, Victor, (1826-1890)
GREPPIN, Jean Baptiste, (-1881)
GRESSLY, Amanz, (1814-1865)
GUTZWILLER, Andreas, (1845-1917)
GUYOT, Arnold Henri, (1807-1884)
HAGENBACH-BISCHOFF, Eduard, (1833-1910)
HEER, Oswald, (1809-1883)
HEIM, Albert, (1849-1937)
HISELY, Charles, (1805-1873)
HUMBERT, Alois, (1829-1887)
JACCARD, Samuel Auguste, (1833-1895)
KAUFMANN, Franz Joseph, (-1892)
LANG, Franz Vinzenz, (1821-1899)
LESQUEREUX, Charles Leo, (1806-1889)
LORIOL LE FORT, Charles Louis Perceval de, (1828-1908)
MOESCH, Casimir, (1827-1898)
MONTMOLLIN, Auguste de, (1808-1898)
MORLOT, Adolf von, (1820-1867)
MÜHLBERG, Friedrich, (1840-1915)
MÜLLER, Albrecht, (1819-1890)
NÜESCH, Jakob, (1845-1915)
PICTET DE LA RIVE, François Jules, (1809-1872)
POURTALÈS, Louis François de, (1823-1880)
QUIQUEREZ, Auguste, (1801-1882)
RENEVIER, Eugène, (1831-1906)
RÜTIMEYER, Karl Ludwig, (1825-1895)
STUDER, Theophil, (1845-1922)
TRIBOLET, Georges de, (1830-1873)
VETTER, Benjamin, (1848-1893)
VOGT, Carl, (1817-1895)
VOLGER, Georg Heinrich Otto, (ca 1822-1897)
ZIEGLER, Jakob Melchior, (1801-1883)

1850-1899

AEBERHARDT, Berchtold, (1872-1912)
ARBENZ, Paul, (1880-1943)
ARGAND, Émile, (1879-1940)
BAUMBERGER, Ernst, (1866-1935)
BECK, Paul, (1882-1958)
BEDER, Robert, (1885-1930)
BERNOUILLI, Walter, (1885-1946)

SWITZERLAND 1850-1899 Cont'd.

BEUST, Fritz von, (1856-1908)
BRAUCHLI, R. W., (1895-1975)
BURCKHARDT, Carl, (1869-1935)
BURCKHARDT, Carl Rudolf, (1866-1908)
BUXTORF, August, (1877-)
CADISCH, Joos, (1895-)
COLLET, Léon William, (1880-1957)
DUBOIS, Auguste, (1862-1923)
DU BOIS, Georges C., (1874-)
DUPARC, Louis Claude, (1866-1932)
DU PASQUIER, Léon, (1864-1897)
ERB, Josef Theodor, (1874-1934)
ERNI, Arthur, (1885-1945)
FAESSLER, Carl, (1895-1957)
FISCHER, Otto, (1877-1943)
FISCHLI, Heinrich, (1857-1932)
FREI, Ernst, (1894-1974)
FREY, Alfred Paul, (1893-1963)
FRIEDLAENDER, Immanuel, (-?1918)
FRÜH, Jakob, (1852-)
GAGNEBIN, Élie, (1891-1949)
GERBER, Eduard, (1876-1956)
GLASER, Theodor, (1894-1922)
GOELDI, Emil August, (1859-1917)
GREPPIN, Eduard, (1856-1927)
GRUBENMANN, Ulrich, (1850-1924)
HANDSCHIN, Eduard, (1894-1962)
HEIM, Arnold, (?1882-1965)
HELBING, Hermann, (1880-1938)
HELBLING, Robert, (1874-1954)
HOTZ, Walter, (1884-1958)
HUGI, Emil, (1873-1937)
HUTTENLOCHER, Heinrich, (1890-1954)
JENNY, Fridolin, (1862-1933)
KISSLING, Ernst, (1865-1936)
KOBY, Fréderic Louis, (1852-1930)
LANG, Arnold, (1855-1914)
LEHNER, Ernst J., (1895-1968)
MERCANTON, Paul Louis, (1876-1963)
MEYER, Edouard, (1899-1974)
MONTAG, Emil, (1875-1960)
NIETHAMMER, Gottlob August, (1882-1915)
NIGGLI, Paul, (1888-1953)
OPPLIGER, Fritz, (1861-1932)
PETERHANS, Emil, (1899-1931)
PICCARD, Auguste, (1884-1962)
PREISWERK, Heinrich, (1876-1940)
REICHEL, Manfred, (1896-)
ROLLIER, Louis, (1859-1931)
ROMAN, Albert Fréderic, (1871-1943)
ROTH, Santiago, (1850-1924)
SARASIN, Paul, (1856-1929)
SCHARDT, W. Hans, (1858-1931)
SCHAUB, Samuel, (1882-1962)

SWITZERLAND 1850-1899 Cont'd.

SCHMIDT, Carl, (1862-1923)
SCHRÖTER, Carl, (1855-?1925)
SENN, Alfred, (1899-1949)
SORET, Charles, (-?1903)
STAUB, Rudolf, (1890-1961)
STEHLIN, Hans Georg, (1870-1942)
STEINER-BALZER, Alfred, (1879-1968)
STRÜBIN, Karl, (1876-1916)
TERNETZ, Carl, (1870-1928)
THALMANN, Hans E., (1899-1975)
TOBLER, August, (1872-1929)
TRIBOLET-HARDY, Frédéric Maurice de, (1852-1929)
TRÜMPY, Daniel, (1893-1971)
VALENTIN, Jean, (1877-1898)
WEBER, Julius, (1864-1924)
WETTSTEIN, Alexander, (1861-1887)
WIEDENMAYER, Carl, (1897-1951)
WILHELM, Oscar, (1897-1975)
ZYNDEL, Fortunat, (1882-1917)

1900 to present
CALAME, Jean Jacques, (1923-1973)
FICHTER, Hans Jakob, (1908-1977)
GUBLER, Jean, (1902-1940)
KLEIBER, Karl Ernst, (1912-1972)
LEUTENEGGER, Willy Otto, (1903-1972)
NICOL, Germain Louis, (1922-1973)
PICCARD, Jacques Ernst, (1922-)
RENAUD, André, (1904-1964)
RUTSCH, Rolf F., (1902-1975)
SCHAUB, Hans Peter, (1912-1965)
SCHMID, Kurt, (1906-1957)
STREIFF, Viktor, (1909-1973)
TRÜMPY, Rudolf D., (-)
WIRZ, Albert Eduard, (1910-1973)
WITZ, Albert Eduard, (1910-1973)

TURKEY

Up to 1800

CASSIUS, Dio Cocceianus, ([155-235] A.D.)
RAFINESQUE-SCHMALTZ, Constantine Samuel, (1783-1840)

1850-1899

FORTUNAT, Wincenty Karol, (1889-1967)
TASHMAN, Djavad Eyoub, (1893-1956)

1900 to present

TOKSÖZ, M. Nafi, (1934-)

UNION OF SOVIET SOCIALIST REPUBLICS

Up to 1800

ANDRZEJOWSKI, Antoni, (1785-1868)
AVICENNA, (980-1037)
BIRUNI, (pseud. of Abu-al-Rayán Muhammad ibn-Ahmad al-Biruni), (ca.972-1048)
LOMONOSOV, Mikhail Vasilyevich, (1711-1765)
NASIREDDIN, Mukhamed (called Tusi), (1201-1274)

1800-1849

BRANDT, Alexander von, (1844-1932)
DOROSZHIN, Peter P., (-)
HIRIAKOFF, Modeste, (1812-1894)
INOSTRANTZEV, Aleksandr Aleksandrovich, (1843-1920)
KARPINSKY, Alexander Petrovich, (1847-1936)
KEYSERLING, Alexander Andreevich, Count, (1815-1891)
KOKSHAROV, Nikola J. von, General, (1818-1893)
KOVALEVSKY, Alexander, (1840-1901)
KOVALEVSKY, Vladimir Onufrievich, (1843-1883)
KUPFFER, August Ernestovich, (1842-1918)
KUTORGA, Stepan Semenovich, (1805-1861)
MICKWITZ, Karl August von, (1849-1910)
SCHMALHAUSEN, Johannes Theodor, (1849-1894)
SCHRENCK, Leopold von, (1826-1894)
TRAUTSCHOLD, Hermann Adolfovich, (1817-1902)
VOLBORTH, Aleksandr Fedorovich, (1800-1876)
VOSKOBOINIKOV, N. A., (-)
WOEIKOF, Alexander, (1842-1916)
WYROUBOFF, Gr., (1843-1913)

1850-1899

ADAMS, Mikhail Ivanovich, (-)
AGAFONOFF, Valerian K., (1863-1955)
AMALITSKII, Vladimir Prochorovich, (1860-1917)
ANDRUSSOV, Nikolai Ivanovich, (1861-1924)
ARMASHEVSKY, Peter Jakovlevich, (-1919)
BARSONOV, Georgyi Pavlovich, (1897-)
BELYANKIN, Dmitri Stepanovich, (1876-1953)
BETECHTIN, Anatolii Georgievich, (1897-1962)
BODYLEVSKY, Vitaly Ivanovich, (1898-1968)
BOLDYREV, Anatolii Kapitonovich, (1883-1946)
BUBNOFF, Serge von, (1888-1957)
BYSTROV, Alexei Petrovich, (1899-1959)
CZARNOCKI, Stefan, (1878-1947)
DUBYANSKY, Viktor Viktorovich, (1880-1925)
FEDEROV, Evgraf Stepanovich, (1853-1919)
FERSMAN, Aleksander Yevgenyevich, (1883-1945)
GLINKA, Konstantin Dmitrievich, (1867-1927)
GLINKA, Sergei Fedorovich, (1855-1933)
GORIZDRO-KULCZYCKA, Zinaida, (1884-1949)
GORSKY, I. I., (1893-1975)
GOUBKIN, Ivan Mikhailovitch, (1871-1939)
GOUDKOFF, Paul Pavel, (1880-1955)
GRISHCHINSKY, Petr Iuustinovich, (-1920)
GROMOVA, Vera Isaacovna, (1891-1973)
GUBKIN, Ivan Mikhailovich, (1874?-)

UNION OF SOVIET SOCIALIST REPUBLICS 1850-1899 Cont'd.

HEINTZ, Anatol, (1898-1975)
HUSSAKOF, Louis, (1881-1965)
JÉRÉMINE, Elisabeth, (née Tschernaieff), (1879-1964)
KARANDYEEV, Vissarion Vissarionovich, (-1916)
KHARLICHKOV, Konstantin Vasil'evich, (-1921)
KHRUSHCHOV, Konstantin Dmitrievich, (? -1912)
KRECHTER, Vladimir, (?1897-1967)
KROTOV, Petr Ivanovich, (1852-1914)
KRYSHTOFOVICH, Afrikan Nikolaevich, (1885-1953)
KULIK, Leonid Alekseevich, (1883-1942)
KURNAKOV, Nikolai Semenovich, (1860-1941)
LOEWINSSON-LESSING, Franc Julyevich, (1859-1939)
LUKASZEWICZ, Józef, (1863-1928)
MICHALSKI, Aleksander Octavianovich, (1855-1904)
MUSHKETOV, Ivan Vassilievitch, (1850-1902)
NECHAEV, Aleksyei Vasil'evich, (1864-1915)
NEUBURG, Maria Feodorovna, (1894-1962)
NIKITIN, Vasily Vasiliivich, (1867-1942)
OBRUCHEV, Vladimir Afanas'evich, (1863-1956)
ORLOV, Nikolai Arsenevich, (-1919)
ORLOV, Uri Aleksandrovich, (1893-1966)
PAHLEN, Alexis von der Baron, (1850-1925)
PAPANIN, Ivan Dmitrievich "Vanya", (1895-)
POLANSKI, Jorge, (1892-1975)
RUDOLPH, Joseph, (1897-1953)
SAMOILOV, Yakov Vladimirovich, (1870-1925)
SMIRNOV, Sergei Sergei'evich, (1895-1947)
SOVINSKY, Viacheslav Nicholas, (1894-1962)
SUSCHKIN, Peter Petrovich, (1868-1928)
SZATSKIY, Mikolai Sergiechewich, (1895-1960)
TARASENKO, Vasily Efimovich, (1859-1926)
TCHERNYCHEV, Féodoss Nikolaievich, (1856-1914)
VELIKOVSKY, Immanuel, (1895-)
VERNADSKI, Vladimir Ivanovich, (1863-1945)
VIGILEV, Boris, (1883-1924)
VINOGRADOV, A. P., (1895-1975)
VULF, Georgii Viktorovich, (1863-1925)
YAKOVLEV, N. N., (1870-1966)
YEMTCHUKHNIKOV, G. A., (1885-1957)
ZAVOICO, Basil B., (1899-1975)
ZEMYATCHENSKY, Petr Andreevich, (1856-1942)
ZHEMCHUZHNYI, Sergei Fedorovitch, (1873-1929)

1900 to present

BELOUSSOV, Vladimir Vladimirovitch, (-)
BOGDANOV, Alexei Alexeievitch, (1906-1971)
DOLIVO-DOBROVOLSKY, Vadim Vladimirovich, (1904-1936)
EFREMOV, Ivan Antonovich, (1907-1972)
GORSHKOV, Georgii Stepanovich, (1921-1975)
ISLAMOV, Oskar Islamovich, (1910-1978)
KALININ, Gennady Pavlovich, (1916-1975)
KRYNINE, Paul Dimitri, (1902-1964)
MARKHININ, Y., (-)
MULLER, Siemon William, (1900-1970)

UNION OF SOVIET SOCIALIST REPUBLICS 1900 to present Cont'd.

OBRUCHEV, Dmitri, (1900-1970)
STRAKHOV, Nikolai Mikhailovich, (1900-)
TSCHIKHACHEV, Peter A., (-)
VLASOV, Kuz'ma Alekseevich, (1905-1964)

Time unknown
GRIGORIEV, Dmitri Pavlovitch, (-)
SEVASTIANOV, D. P., (-)
YANISHEVSKY, M. E., (-)

UNITED STATES OF AMERICA
Up to 1800
ANNAN, Robert, Reverend, (fl. 1782)
BARNES, Daniel Henry, (1785-1828)
BARTRAM, John, (1699-1777)
BARTRAM, William, (1739-1823)
BECK, Theodoric Romeyn, (1791-1855)
BRACKENRIDGE, George Weare, (1775-1856)
BURT, William Austin, (1792-1858)
CATLIN, George, (1796-1872)
CLARK, William, (1770-1828)
CLINTON, DeWitt, (1769-1828)
CORNELIUS, Elias C., (1794-1832)
COZZENS, Issachar, Jr., (1780-1865)
DANA, James Freeman, (1793-1827)
DARBY, William, (1775-1854)
DRAKE, Daniel, (1785-1852)
DUCATEL, Julius Timoleon, (1796-1849)
EATON, Amos, (1776-1842)
EATON, T.D., (-?1829)
EIGHTS, James, (1798-1882)
GIBBS, George, Colonel, (1776-1833)
GODMAN, John Davidson, (1794-1830)
GREEN, Jacob, (1790-1841)
HARE, Robert, (1781-1858)
HAYS, Isaac, (1796-1879)
HILDRETH, Samuel Prescott, (1783-1863)
HITCHCOCK, Edward, (1793-1864)
HUTCHINS, Thomas, (1730-1789)
JAMES, Edwin, (1797-1861)
JEFFERSON, Thomas, (1743-1826)
JOHNSTON, James Finlay Weir, (1796-1855)
KAIN, John H., (1759-1831)
KEATING, William Hypolitus, (1799-1840)
LEA, Isaac, (1792-1886)
LESUEUR, (1778-1846)
LINCECUM, Gideon, (1793-1874)
LOCKE, John, (1792-1856)
LONG, Stephen Harriman, (1784-1864)
MILLER, Samuel, (1769-1850)
MITCHELL, Elisha, Reverend, (1793-1857)
MITCHILL, Samuel Latham, (1764-1831)
MOORE, Nathaniel Fish, (1782-1872)
MORGAN, George, (fl.1766)
MORGAN, John, (fl.1765)

UNITED STATES OF AMERICA Up to 1800 Cont'd.

MORTON, Samuel George, (1799-1851)
OLMSTED, Denison, (1791-1859)
PARKER, Samuel, Reverend, (1779-1866)
PEALE, Charles Willson, (1741-1827)
PEALE, Titian Ramsay, (1799-1885)
PERCIVAL, James Gates, (1795-1856)
PERKINS, Benjamin Douglas, (1774-1810)
PETIVER, James, (1658-1718)
REDFIELD, William C., (1789-1857)
RENSSELAER, Jeremiah van, (1793-1871)
RHEA, Matthew, (1795-1879)
SAY, Thomas, (1787-1834)
SCHOOLCRAFT, Henry Rowe, (1793-1864)
SILLIMAN, Benjamin, Sr., (1779-1864)
SYMMES, John Cleves, (1780-1829)
THOMAS, David, (1776-1859)
THOMPSON, Zadock, (1796-1856)
TORREY, John, (1796-1873)
VANUXEM, Lardner, (1792-1848)
WAGNER, William, (1796-1885)
WAILES, Benjamin Leonard Couington, (1797-1862)
WARREN, John Collins, (1778-1856)
WATERHOUSE, Benjamin, (1753-1846)
WEBSTER, John White, (1793-1850)
WINTHROP, John, (1714-1779)
WISTAR, Caspar, (1761-1818)

1800-1849

ABBOT, Henry Larcom, (1831-1927)
ABERT, James W., Sr., (1820-1897)
ADAMS, Charles Baker, (1814-1853)
AGASSIZ, Alexander, (1835-1910)
ALDRICH, Truman Heminway, (1848-1932)
ALGER, Francis, (1807-1863)
ALLEN, Joel Asaph, (1838-1921)
BACHE, Alexander Dallas, (1806-1867)
BAILEY, Jacob Whitman, (1811-1857)
BAILEY, Loring Woart, (1839-1925)
BAIRD, Spencer Fullerton, (1823-1887)
BANNISTER, Henry Martyn, (-)
BARRIS, Willis Hervey, (1820-1901)
BECKER, George Ferdinand, (1847-1919)
BEMENT, Clarence S., (1843-1923)
BICKMORE, Albert Smith, (1839-1914)
BLAKE, John Marcus, (1838-1920)
BLAKE, William Phipps, (1826-1910)
BOLTON, Sarah Elizabeth Mary (née Knowles), (1841-1916)
BOUVÉ, Thomas Tracy, (1815-1896)
BOWEN, George Thomas, (1803-1828)
BRADLEY, Frank Howe, (1838-1879)
BREWER, William Henry, (1828-1910)
BROADHEAD, Garland Carr, (1827-1912)
BROOKS, Thomas Benton, (1836-1900)
BRUMBY, Richard T., (1804-1856)

UNITED STATES OF AMERICA 1800-1849 Cont'd.

BRUSH, George Jarvis, (1831-1912)
BUCKLEY, Samuel Botsford, (1809-1884)
BURROUGHS, John, (1837-1921)
CANFIELD, Frederick Alexander, (1849-1926)
CARPENTER, Franklin Reuben, (1848-1910)
CASWELL, John Henry, (1846-1909)
CHAMBERLIN, Thomas Chrowder, (1843-1928)
CHANDLER, Charles Frederick, (1836-1925)
CHAPIN, James Henry, (1832-1892)
CLARKE, Frank Wigglesworth, (1847-1931)
COMSTOCK, Theodore Bryant, (1849-1915)
CONRAD, Timothy Abbott, (1803-1877)
COOK, George Hammell, (1818-1889)
COOPER, James Graham, (1830-1902)
COPE, Edward Drinker, (1840-1897)
COUPER, James Graham, (1830-1902)
COUTHONY, Joseph Pitty, (1808-1864)
COX, Edward Travers, (1821-1907)
CREWS, George Curtis, (1837-1977)
DALL, William Healey, (1845-1927)
DANA, Edward Salisbury, (1849-1935)
DANA, James Dwight, (1813-1895)
DEANE, James, (1801-1858)
DEWEY, Chester, (1784-1867)
DONNELLY, Ignatius, (1831-1901)
DUTTON, Clarence Edward, Major, (1841-1912)
DWIGHT, William Buck, (1833-1906)
EGLESTON, Thomas, (1832-1900)
EMERSON, Benjamin Kendall, (1843-1932)
EMMONS, Ebenezer, (1800-1863)
EMMONS, Samuel Franklin, (1841-1911)
EMORY, William Hemsley, (1811-1887)
ENGELMANN, George, (1809-1884)
EVANS, John, (1812-1861)
FANNING, Tolbert, (1810-1874)
FERREL, William, (1817-1891)
FISKE, John, (1842-1901)
FONTAINE, William Morris, (1835-1913)
FOOTE, Albert E., (1846-1895)
FRAZER, Persifor, (1844-1909)
FRAZIER, Benjamin West, (1841-1905)
FRÉMONT, John Charles, (1813-1890)
FULLER, Homer Taylor, (1838-1910)
GABB, William More, (1839-1878)
GARDINER, Howard Calhoun, (1826-1917) [PROSPECTORS, DIVINERS AND MINING ENGINEERS]
GAY, Martin, (1803-1850)
GIBBES, Robert W., (1809-)
GILBERT, Grove Karl, (1843-1918)
GLENN, John, (-)
GRAY, Asa, (1810-1888)
GRINNELL, George Bird, (1849-1938)
GULICK, John Thomas, (1832-1923)
HAGUE, Arnold, (1840-1917)

UNITED STATES OF AMERICA 1800-1849 Cont'd.

HAGUE, James Duncan, (1836-1908)
HALL, Christopher Webber, (1845-1911)
HALL, James, (1811-1898)
HARGER, Oscar, (1843-1882)
HARRIS, Israel Hopkins, (1832-1897)
HAWN, Frederick, (1810-1898)
HAY, Oliver Perry, (1846-1930)
HAYDEN, Ferdinand Vandiveer, (1829-1887)
HILLIERS, John K., (-)
HITCHCOCK, Charles Henry, (1836-1919)
HOLBROOK, Levi, (1836-1922)
HOLDEN, Edward Singleton, (1846-1914)
HOLLAND, William Jacob, (1848-1932)
HOLMES, Francis S., (1815-1882)
HOLMES, William Henry, (1846-1933)
HOUGHTON, Douglass, (1809-1845)
HOVEY, Horace Carter, Reverend, (1833-1914)
HOWELL, Edwin Eugene, (1845-1911)
HUBBARD, Bela, (1814-1896)
HUBBARD, Lucius Lee, (1849-1933)
HUBBARD, Oliver Payson, (1809-1900)
HUNT, Thomas Sterry, (1826-1892)
HYATT, Alpheus, (1838-1902)
IRVING, Roland Duer, (1847-1888)
JACKSON, Charles Thomas, (1805-1880)
JACKSON, William Henry, (1843-1942)
JAMES, Uriah Pierson, (1811-1889)
JULIEN, Alexis Anastay, (1840-1919)
KANE, Elisha Kent, (1820-1857)
KENNICOTT, Robert, (1835-1866)
KERR, Washington Caruthers, (1827-1885)
KING, Clarence Rivers, (1842-1901)
LACOE, Ralph Dupuy, (1824-1901)
LAPHAM, Increase Allen, (1811-1875)
LE CONTE, Joseph, (1823-1901)
LEIDY, Joseph, (1823-1891)
LESLEY, J. Peter, (1819-1903)
LEVISON, Wallace Goold, (1846-1924)
LEWIS, Tayler, (1802-)
LIEBER, Oscar M., (-1862)
LITTLE, George, (1838-1924)
LOOMIS, Elias, (1811-1890)
LOUGHRIDGE, Robert Hills, (1843-1917)
LUTHER, Daniel Dana, (1840-1923)
LYMAN, Benjamin Smith, (1835-1920)
MACBRIDE, Thomas Huston, (1848-1934)
MACFARLANE, James, (1819-1885)
MACKAY, John William, (1831-1902) [PROSPECTORS, DIVINERS AND MINING ENGINEERS]
MARCY, Oliver, (1820-1899)
MARCY, Randolph Barnes, (1812-1887)
MARSH, Othniel Charles, (1831-1899)
MARTIN, Daniel Strobel, (1842-1925)
MARVINE, Archibald Robertson, (1848-1876)

UNITED STATES OF AMERICA 1800-1849 Cont'd.

MATHER, William Williams, (1804-1859)
MAURY, Matthew Fontaine, (1806-1873)
MEEK, Fielding Bradford, (1817-1876)
MENDENHALL, Thomas Corwin, (1841-1924)
MILLER, Samuel A., (1837-1897)
MILLS, James Ellison, (1834-1901)
MOORE, Francis, Jr., (1808-1864)
MORAN, Thomas, (1837-1926)
MUDGE, Benjamin Franklin, (1817-1879)
NASON, Henry Bradford, (1831-1895)
NEFF, Peter, (1827-1903)
NEWBERRY, John Strong, (1822-1892)
NILES, William Harmon, (1838-1910)
NORWOOD, Joseph Granville, (1807-1895)
ORTON, Edward Francis Baxter, (1829-1899)
O'SULLIVAN, T. H., (-1882)
PACKARD, Alpheus Spring, Jr., (1839-1905)
PERKINS, George Henry, (1844-1933)
PETTEE, William Henry, (1836-1904)
PLATT, Franklin, (1844-1900)
POWELL, John Wesley, (1834-1902)
PUMPELLY, Raphael, (1837-1923)
RAYMOND, Rossiter Worthington, (1840-1918)
REDFIELD, John Howard, (1815-1895)
RICE, William North, (1845-1928)
RIDDELL, John Leonard, (1807-1865)
ROCKWOOD, Charles Greene, (1843-1913)
ROEBLING, Washington Augustus, Colonel, (1837-1926)
ROGERS, Henry Darwin, (1808-1866)
ROGERS, William Barton, (1804-1882)
SAFFORD, James Merrill, (1822-1907)
SCUDDER, Samuel Hubbard, (1837-1911)
SEELY, Henry Martyn, (1828-1917)
SHALER, Nathaniel Southgate, (1841-1906)
SHUMARD, Benjamin Franklin, (1820-1869)
SHUMARD, George Getz, (1826-1867)
SILLIMAN, Benjamin Jr., (1816-1885)
SMITH, Eugene Allen, (1841-1927)
SMITH, Lawrence, (-1883)
SMOCK, John Conover, (1842-1925)
SNOW, Francis Huntington, (1840-1908)
SPRINGER, Frank, (1848-1924)
STEVENSON, John James, (1841-1924)
ST. JOHN, Orestes Hawley, (1841-1921)
ST. JOHN, Samuel, (1813-1876)
SWALLOW, George Clinton, (1817-1899)
TODD, James Edward, (1846-1922)
TRASK, John Boardman, (1824-1879)
VERRILL, Addison Emery, (1839-1926)
VOGDES, Anthony Wayne, General, (1843-1923)
WADSWORTH, Marshman Edward, (1847-1921)
WALL, George Parkes, (1832-1912)
WARD, Henry Augustus, (1834-1906)
WARD, Lester Frank, (1841-1913)

UNITED STATES OF AMERICA 1800-1849 Cont'd.

WHITE, Charles Abiathar, (1826-1910)
WHITE, Israel Charles, (1848-1927)
WHITFIELD, Robert Parr, (1828-1910)
WHITNEY, Josiah Dwight, (1819-1896)
WHITTLESEY, Charles, Colonel, (1808-1886)
WILLIAMS, Gardner Fred, (1842-1922)
WILLIAMS, Henry Shaler, (1847-1918)
WINCHELL, Alexander, (1824-1891)
WINCHELL, Newton Horace, (1829-1914)
WING, Augustus, Reverend, (1808-1876)
WOODWARD, Robert Simpson, (1849-1924)
WORTHEN, Amos Henry, (1812-1888)
WRIGHT, Albert Allen, (1846-1905)
WRIGHT, Charles E., (1843-1888)
WRIGHT, George Frederick, (1838-1921)
WYMAN, Jeffries, (1814-1874)

1850-1899

ABBE, Cleveland, Jr., (1872-1934)
ABERNATHY, George Elmer, (1888-1948)
ADAMS, George Irving, (1870-1932)
ADAMS, George Whitney, (1872-1930)
ADAMS, John Emery, (1899-1971)
ADAMS, Noah C., (1891-1927)
ADAMS, William Clyde, (1892-1957)
ADDINGTON, Arch Rombough, (1894-1972)
ADKINS, Walter Scott, (1890-1956)
AID, Harry, (1897-1927)
AINSWORTH, David, (1898-1967)
ALDREDGE, Robert Frank, (1895-1961)
ALEXANDER, Annie Montague, (1867-1950)
ALLEN, Donald Meredith, (1899-1971)
ALLEN, Walter J., (1890-1961)
ALLING, Harold Lattimore, (1888-1960)
AMBROSE, Arthur Warren, (1889-1952)
ANDERSON, Frank Marion, (1863-1945)
ANDERSON, George Harold, (1893-1956)
ANDREEN, Harry Mayo, (1891-1933)
ANDREWS, Roy Chapman, (1884-1960)
APPLIN, Esther Richards, (1895-1972)
ARGABRITE, William Greame, (1885-1963)
ARMSTRONG, Harold Kerr, (1893-1962)
ARNOLD, Ralph, (1875-1961)
ASHBURNER, Charles Albert, (1854-1889)
ASHLEY, George Hall, (1866-1951)
ATHY, Lawrence Ferdinand, (1898-1955)
ATKINSON, William H., (1898-1972)
AUGUR, Irving V., (1887-1930)
AURIN, Fritz Love, (1891-1969)
AVERY, Charles Dwight, (1877-1947)
BAGG, Rufus Mather, (1869-1946)
BAIN, Samuel McCutcheon, (-)
BAIRD, Chester Alvan, (1896-1959)
BAKER, Frank Collins, (1867-1942)

UNITED STATES OF AMERICA 1850-1899 Cont'd.

BALDWIN, David, (fl.1874-1881)
BALE, Hubert Eugene, (1898-1969)
BALL, John Rice, (1881-1953)
BALL, Max Waite, (1885-1954)
BALLOU, William Hosea, (1857-1937)
BARBER, Charles M, (1877-1954)
BARBOUR, Erwin Hinckley, (1856-1947)
BARBOUR, George Brown, (1891-1977)
BARBOUR, Thomas, (1884-1946)
BARNES, Roy Merrill, (1891-1960)
BARRELL, Joseph, (1869-1919)
BARRETT, Leslie Park, (1887-1972)
BARRINGER, Daniel Moreau, Sr., (1860-1929)
BARRINGTON, Thomas W., (-)
BARROW, Leonidas Theodore "Slim", (1895-1978)
BARTON, Donald Clinton, (1889-1939)
BARTON, George Hunt, (1852-1933)
BARTRAM, John Greer, (1893-1955)
BASCOM, Florence, (1862-1945)
BASSLER, Raymond Smith, (1878-1961)
BASTIN, Edson Sunderland, (1878-1953)
BATES, Mowry, (1875-1926)
BAUER, Lawson H., (1889-1954)
BAYLEY, William Shirley, (1861-1943)
BEACH, Floyd Kellogg, (1884-1964)
BEACH, John Osa, (1895-1964)
BEACHLER, Charles S., (1870-1894)
BECKER, Clyde M., (1882-1938)
BEECHER, Charles Emerson, (1856-1904)
BEEDE, Joshua William, (1871-1940)
BEEKLEY, Albert Leon, (1883-1952)
BELL, Olin Grogan, (1893-1970)
BELTZ, Edward W., (1891-1970)
BENNETT, Clyde M., (1881-1935)
BENNETT, Russell Hoadley, (1896-)
BENSON, Dale L., (1899-1968)
BENTZ, Ivan Vincent, (1896-1938)
BERKEY, Charles Peter, (1867-1955)
BEVIER, George Means, (1888-1972)
BEYER, Samuel Walker, (1865-1931)
BILLINGSLEY, Paul, (1887-1962)
BINNEY, Edwin, Jr., (1899-1928)
BLACKWELDER, Eliot, (1880-1969)
BLAIR, Charles Scofield, (1883-1971)
BLANCHARD, William Grant, Jr., (1894-1952)
BLANPIED, Bernerd William, (1897-1975)
BOLYARD, Garrett Lincoln, (1897-1963)
BOND, Lewis Alexander, (1896-1972)
BOND, Marshall, (1867-1941) [PROSPECTORS, DIVINERS AND MINING ENGINEERS]
BONTZ, Conrad K., (1894-1927)
BOUTWELL, John Mason, (1874-1968)
BOWEN, Charles Franklin, (1871-1953)
BOWEN, James P., (1889-1960)

UNITED STATES OF AMERICA 1850-1899 Cont'd.

BOWMAN, Isaiah, (1882-1950)
BOYD, Harold Edward, (1887-1952)
BOYLE, Walter J., (1897-1973)
BRACE, Orval L., (1893-1975)
BRADLEY, Everett L., (1897-1969)
BRADLEY, Wilmot Hyde, (1899-)
BRAMLETTE, Milton Nunn, (1896-1977)
BRANNER, John Casper, (1850-1922)
BRANSON, Edwin Bayer, (1877-1950)
BRANT, Ralph Allen, (1899-1958)
BREMNER, Carl St. John, (1895-1944)
BRIGHAM, Albert Perry, (1855-1932)
BRISCOE, Glenn Oliver, (1898-1974)
BRODERICK, Thomas Monteith, (1889-1965)
BROOKS, Alfred Hulse, (1871-1924)
BROWN, Amos Peaslee, (1864-1917)
BROWN, Barnum, (1873-1963)
BROWN, Otto E., (1896-1973)
BROWN, Thomas Clachar, (1882-1934)
BRYAN, William Jennings, (1860-1925)
BUCHER, Walter Herman, (1888-1965)
BUCKLEY, Ernest Robertson, (1872-1912)
BUEHLER, Henry Andrew, (1876-1944)
BURCHARD, Ernest Norman, (1875-1961)
BURFOOT, James Dabney "Dan", Jr., (1896-1966)
BURRESS, Walter March, (1893-1950)
BURTON, George Edgar, (1880-1953)
BURWELL, Edward Bouldin, Jr., (1894-1964)
BUTCHER, Cary Preston, (1898-1950)
BUTLER, Bertram Theodore, (1874-1958)
BUTLER, Gurdon Montague, (1881-1961)
BUTTERWORTH, Emerson McMillin, (1894-1961)
BUTTRAM, George Franklin, (1886-1966)
BUTTS, Charles, (1863-1946)
BUWALDA, John Peter, (1886-1954)
BYBEE, Halbert Pleasant, (1888-1957)
CADY, Gilbert Haven, (1882-1970)
CAHN, Lazard, (1865-1940)
CALL, Richard Ellsworth, (1856-1917)
CAMPBELL, Douglas Houghton, (1859-1953)
CAMPBELL, Guy, (1877-1970)
CAMPBELL, Harry A., (1893-1951)
CAMPBELL, Henry Donald, (1862-1934)
CAMPBELL, Marius Robinson, (1858-1940)
CAMPBELL, Robert Burns, (1892-1955)
CANNON, Robert Lee, (1899-1961)
CARLTON, Dave Pierce, (1890-1976)
CARMAN, Joel Ernest, (1882-1966)
CARNEY, Frank, (1868-1934)
CARPENTER, Everett, (1884-1968)
CARR, George Watson, (1899-1950)
CARSON, Carlton Mann "Kit", (1898-1972)
CARTER, Frank B., (1898-1968)
CASE, Ermine Cowles, (1871-1953)

UNITED STATES OF AMERICA 1850-1899 Cont'd.

CASSEL, Chester A. "Chet", (1897-1971)
CASTEEL, Sanford Henry, (1898-1952)
CASTER, Ermil Leslie, (1896-1955)
CATHCART, Stanley Holman, (1889-1953)
CHADWICK, George Halcott, (1876-1953)
CHAMBERLAIN, Arthur, (-)
CHAMBERLIN, Rollin Thomas, (1881-1948)
CHANEY, Ralph Works, (1890-1971)
CHENEY, Charles Albert, (1885-1937)
CHENEY, George Monroe, (1893-1952)
CHEVALIER, Jerome Archibald, (1877-1940)
CHISHOLM, William Francis, (1886-1944)
CHRISTMAN, Charles Gustave, (1873-1953)
CHURCH, James Edward, (1869-1959)
CLAPP, Frederick Gardner, (1879-1944)
CLARK, Bruce Lawrence, (1880-1945)
CLARK, Chester Charles, (1885-1953)
CLARK, Frank Rinker, (1881-1974)
CLARK, Howard, (1885-1972)
CLARK, Leslie Madison, (1898-1976)
CLARK, Robert Watson, (1884-1948)
CLARK, William Bullock, (1860-1917)
CLARKE, Carl W., (1896-1928)
CLARKE, John Mason, (1857-1925)
CLELAND, Herdman Fitzgerald, (1869-1935)
CLIFTON, Roland Leroy, (1884-1962)
CLINKSCALES, Albert Sidney, (1894-1969)
COFFIN, Reuben Clare, (1886-1972)
COLBURN, Burnham Standish, (1872-1959)
COLLINS, Melvin J., (1897-1936)
COLLOM, Roy Edward, (1881-1948)
COLONY, Roy Jed, (1870-1936)
CONDIT, Daniel Dale, (1886-1955)
CONKLING, Richard A., (1885-1952)
CONKLING, Russell C., (1897-1955)
CONNOLLY, Joseph Peter, (1890-1947)
COOK, Charles Wilford, (1882-1933)
COOK, Harold James, (1887-?1962)
COOK, James, (fl. 1870)
COOK, Margaret C., (1899-1968)
COOKE, Charles Wythe, (1887-1971)
COOKE, Harold Caswell, (1884-1956)
COOPER, Herschel H., (1894-1954)
CORTES, Henry Cornelius, (1892-1957)
CORYELL, Horace Noble, (1888-1965)
COTTINGHAM, Virgil Elisha, (1891-1958)
COX, Guy Henry, (1881-1922)
CRAIG, William Henry, (1897-1951)
CRAWFORD, David Johnston, (1899-1960)
CRAWFORD, Ralph Dixon, (1873-1950)
CRIDER, Albert Foster, (1873-1945)
CROOK, Alja Robinson, (1864-1930)
CROSBY, Irving Ballard, (1891-1959)
CROSBY, William Otis, (1850-1925)

UNITED STATES OF AMERICA 1850-1899 Cont'd.

CRUM, Harry Edwin, (1892-1955)
CULVER, Harold Eugene, (1883-1970)
CUMINGS, Willard Lawsen, (1874-1945)
CUNNINGHAM, George Magee, (1895-1963)
CURRIER, Louis Wade, (1890-1970)
CUSHING, Henry Platt, (1860-1921)
CUSHMAN, Joseph Augustine, (1881-1949)
DAKE, Charles Laurence, (1883-1934)
DALE, Nelson Clark, (1880-1962)
DAMON, Henry Gordon, (1883-1960)
DANIELS, James Ira, (1896-1967)
DARBY, Raymond Vern, (1896-1953)
DARROW, Clarence, (1857-1938)
DARTON, Nelson Horatio, (1865-1948)
DAVIS, Charles Albert, (1861-1916)
DAVIS, Elmer Fred, (1887-1974)
DAVIS, Leander S., (fl. 1871)
DAVIS, Ralph Emerson, (1884-1968)
DAVIS, Thornton, (1892-1969)
DAVIS, William Morris, (1850-1934)
DAWSON, Joseph Meadows, (1895-1946)
DAY, Arthur Louis, (1869-1960)
DEAN, Bashford, (1867-1928)
DECKER, Charles Elijah, (1868-1958)
DECKER, Charles LaVerne, (1898-1955)
DeGOLYER, Everette Lee, (1886-1956)
DE KOCH, William Gaspar, (1898-1954)
DENISON, Albert Rodger, (1897-1962)
DERBY, Orville Adelbert, (1851-1915)
DERDEN, Jesse Homer, (1890-1944)
DEUSSEN, Alexander, (1882-1959)
DeWOLF, Frank Walbridge, (1881-1957)
DICKERSON, Roy Ernest, (1877-1944)
DILLÉ, Glenn Scott, (1896-1963)
DITMARS, Raymond Lee, (1876-1942)
DOBBIN, Carroll Edward "Cy", (1892-1967)
DOBSON, Peter, (-)
DODSON, Floyd Cummings, (1895-1949)
DONOGHUE, David Patrick, (1891-1958)
DONOVAN, Percy Williams, (1879-1973)
DORSEY, George Edwin, (1892-1953)
DORSEY, Noah Ernest, (1873-1959)
DOUGLAS, James Mainwaring, (1887-1952)
DOUGLASS, Earl, (1862-1931)
DRAKE, Cecil, (1899-1949)
DRAKE, Noah Fields, (1864-1945)
DRESSER, Myron Amos, (1894-1973)
DUMBLE, Edwin Theodore, (1852-1927)
DUSTON, Arthur W., (1886-1938)
EAKLE, Arthur Starr, (1862-1931)
EASTMAN, Charles Rochester, (1868-1918)
EATON, Arthur, (1886-1923)
EATON, Joseph Edmund, (1895-1958)
EBY, James Brian, (1896-)

UNITED STATES OF AMERICA 1850-1899 Cont'd.

EDSON, Dwight J., (1896-1976)
EDSON, Fanny Carter, (1887-1952)
EDWARDS, Everett C., (1897-1972)
EDWARDS, Merwin Guy, (1889-1969)
EKBLAW, George Elbert, (1895-1972)
ELDRIDGE, George Homans, (1854-1905)
ELLEDGE, Emmett Russell, (1898-1960)
ELLIS, Brooks Fleming, (1897-1976)
ELLISON, Charles Walker, (1897-1953)
ELLISOR, Alva Christine, (1892-1964)
ELSON, William J., (1895-1965)
ELY, Fred Burchard, (1883-1946)
EMMONS, William Harvey, (1876-1948)
ENDLICH, Frederic Miller, (1851-1899)
ENGLISH, George Letchworth, (1864-1944)
ERDMANN, Charles Edgar, (1897-1971)
EYSSELL, Alfred Raab, (1898-1944)
FAIRCHILD, John Gifford, (1882-1965)
FARRINGTON, Oliver Cummings, (1864-1933)
FATH, Arthur Earl, (1887-1976)
FENNEMAN, Nevin Melancthan, (1865-1945)
FENNER, Clarence Norman, (1870-1949)
FERGUSON, Henry Gardiner, (1882-1966)
FERGUSON, John Logan, (1893-1973)
FERGUSON, Roy Noble, (1883-1961)
FETTKE, Charles Reinhard, (1888-1959)
FIELD, Richard Montgomery, (1885-1961)
FIGGINS, Jesse Dade, (1867-1944)
FINCH, Elmer Harrison, (1883-1953)
FISHER, Cassius Asa, (1872-1930)
FITTS, John, (1879-1942)
FLAHERTY, Robert J., (1884-1951)
FLEMING, John Adam, (1877-1956)
FOERSTE, August Frederick, (1862-1936)
FOHS, Ferdinand Julius, (1884-1965)
FOLGER, James Anthony, (1896-1956)
FOOTE, Harry Ward, (1875-1942)
FOOTE, Warren, (-1936)
FORD, William Ebenezer, (1878-1939)
FORT, Charles, (1874-1932)
FOSHAG, William Frederick, (1894-1956)
FOSTER, Margaret Dorothy, (1895-1970)
FOWLER, Charles Worth, Jr., (1899-1948)
FOWLER, George Malcolm, (1885-1972)
FUELLHART, Donald Egbert, (1899-1945)
FULLER, Richard Eugene, (1897-1976)
FURCRON, Aurelius Sydney, (1899-1971)
GALE, Hoyt Stoddard, (1876-1952)
GALLAGHER, William G., Jr., (1898-1932)
GALLOWAY, Jesse James, (1882-1962)
GALPIN, Sidney Longman, (1886-1962)
GARDINER, Thomas Moir, Jr., (1898-1930)
GARDNER, James Henry, (1883-1964)
GARDNER, Julia Anna, (1882-1960)

UNITED STATES OF AMERICA 1850-1899 Cont'd.

GARRETSON, Mary Welleck, (1896-1971)
GARRETT, Lovic Pierce, (1880-1943)
GARRETT, Melvin Miller "Dutch", (1894-1969)
GARRETT, Robert Ernest, (1888-1947)
GEIS, Wilfrid Hoy "Bill", (1892-1961)
GEORGE, William Owsley, (1892-1966)
GESTER, George Clark, (1884-1959)
GESTER, Stephen Howell, (1889-1967)
GETZENDANER, Frank Marshall, (1876-1957)
GIDLEY, James Williams, (1866-1931)
GILLES, Verner Arthur, (1886-1954)
GILLSON, Joseph Lincoln, (1895-1964)
GILLULY, James, (-)
GILMORE, Charles Whitney, (1874-1945)
GINTER, Roy Lamont, (1892-1958)
GIRTY, George Herbert, (1869-1939)
GLASS, Jewell Jeanette, (1888-1966)
GLENN, Leonidas Chalmers, (1871-1951)
GOLDMAN, Marcus Isaac, (1881-1965)
GOLDRING, Winifred, (1888-1971)
GOLDTHWAIT, James Walter, (1880-1947)
GOODE, George Brown, (1851-1896)
GOODSPEED, George Edward, (1887-1974)
GORDON, Charles Henry, (1857-1934)
GORDON, Dugald, (1886-1958)
GORDON, Samuel George, (1897-1952)
GOULD, Charles Newton, (1868-1949)
GRABAU, Amadeus William, (1870-1946)
GRAHAM, William Armstrong Patterson, (1899-1934)
GRANGER, Walter, (1872-1941)
GRANT, Ulysses Sherman, (1867-1932)
GRATACAP, Louis Pope, (1850-1917)
GRATON, Louis Caryl, (1880-1970)
GREEN, Guy Emmett "Squire", (1896-1965)
GREENE, Frank Cook, (1886-1975)
GREGORY, Herbert E., (1869-1952)
GREGORY, William King, (1876-1970)
GRIFFITH, Glenn Rodgers Vickery, (1891-1964)
GRIGGS, Robert F., (-)
GROUT, Frank Fitch, (1880-1958)
GUILD, Frank Nelson, (1870-1939)
GUNTER, Herman, (1885-1972)
GURLEY, William Frank Eugene, (1854-1943)
GWYNNE, Charles Sumner, (1885-1972)
HAGER, Lee, (1874-1944)
HAKE, Benjamin Franklin, (1895-1973)
HALL, Charles Monroe, (1870-1903)
HALL, Elwin B., (1891-1949)
HALL, George Martin, (1891-1941)
HAMILTON, Henry L., (1882-1925)
HAMMILL, Chester A., (1889-1941)
HAMMOND, John Hays, (1855-1936)
HANNA, G. Dallas, (1887-1970)
HANSON, Ernest Albin, (1899-1973)

UNITED STATES OF AMERICA 1850-1899 Cont'd.

HARDISON, Harvey, (1895-1949)
HARES, Charles Joseph, (1881-1970)
HARPER, O. C., (1897-1973)
HARRINGTON, George Leavitt, (1883-1972)
HARRIS, Gilbert Dennison, (1864-1952)
HARRISON, Thomas Samuel, (1881-1964)
HARTNAGEL, Chris Andrew, (1874-1962)
HASEMAN, William Peter, (1878-1932)
HATCHER, John Bell, (1861-1904)
HATFIELD, Willis Charles, "Luke", (1898-1970)
HAUPTMAN, Charles R., (1891-1971)
HAWKINS, Alfred Cary, (1887-1954)
HAWKINS, Glenn Dewayne, (1897-1970)
HAWORTH, Erasmus, (1855-1932)
HAYES, Charles Willard, (1858-1916)
HAYES, George Gibbons, (1889-)
HAZELTINE, Roy S., (1890-1972)
HAZZARD, Roy Thorpe, (1893-1976)
HEALD, Kenneth Conrad, (1888-1971)
HEATH, Daisy Winifred, (1875-1954)
HECK, Nicholas Hunter, (1882-1953)
HEDRICK, Oley Foster, (1891-1950)
HELLMAN, Milo, (1872-1947)
HENDERSON, Junius, (1865-1937)
HENDON, Bryan, (1896-1926)
HENDRICKSON, Victor James, (1892-1953)
HENLEY, Arthur Sidney, (1879-1936)
HENNEN, Ray Vernon, (1875-1958)
HERALD, Frank Edward, (1887-1965)
HERMANN, Adam, (-)
HEROLD, Stanley Carrollton, (1883-1958)
HEROY, William Bayard, Sr., (1883-1971)
HERRICK, Clarence Luther, (1858-1904)
HESNARD, Emile E., (1867-1929)
HESS, Frank L., (1871-1955)
HEWETT, Donnel Foster, (1881-1971)
HICE, Richard Robert, (1865-1925)
HIDDEN, William Earl, (1853-1918)
HIGGINS, Daniel Franklin, Jr., (1884-1930)
HIGGS, Morton Thomas, (1896-1956)
HILL, Frank A., (1858-1915)
HILL, Robert Thomas, (1858-1941)
HILLEBRAND, William Francis, (1853-1925)
HILLS, Thomas McDougall, (1881-1970)
HINDES, Earle P., (1888-1962)
HOBBS, William Herbert, (1864-1953)
HODGE, Edwin Thomas, (1887-1970)
HOGAN, Dana, (1890-1945)
HOLDEN, Roy Jay, (1870-1945)
HOLLICK, Arthur, (1857-1933)
HOLMAN, Eugene, (1895-1962)
HOLMES, Joseph Austin, (1859-1915)
HONESS, Arthur Pharaoh, (1887-1942)
HONESS, Charles William, (1885-1949)

UNITED STATES OF AMERICA 1850-1899 Cont'd.

HOOD, Forrest Wilson, (1897-1963)
HOPKINS, Edwin Butcher, (1882-1940)
HOPPER, Walter Everett, (1886-1971)
HORTON, Harold Merritt "Wog", (1897-1966)
HOSKINS, John Hobart, (1896-1957)
HOVEY, Edmund Otis, (1862-1924)
HOWARD, Edgar Billings, (1887-1943)
HOWE, Ernest, (1875-1932)
HOWE, Marshall Avery, (1867-1936)
HOWELL, Benjamin Franklin, (1890-1976)
HOWELL, Jesse V., (1891-1971)
HOWELL, William Frederick, (1895-1965)
HOYLE, Charles Robert, Sr., (1891-1954)
HUBBARD, Bela, (1890-1959)
HUBBARD, George David, (1871-1958)
HUDSON, Frank Samuel, (1890-1967)
HUDSON, George Henry, (1855-1934)
HUGHES, Donald Dudley, (1893-1946)
HUGHES, Richard, (1889-1969)
HULL, Joseph Poyer Deyo, (1889-1967)
HUMMEL, Carl Bernard, (1896-1923)
HUNT, Edwin Harold, (1897-1949)
HUNT, Walter Frederick, (1882-1975)
HUNTLEY, Charles Stirling, (1895-1926)
HUSSEY, Russell Claudius, (1888-1970)
HUTCHINSON, Frederick McAlpine, (1884-1952)
HUTCHINSON, Lon Lewis, (1877-1947)
HUTSON, Ezekiel Burney, (1897-1961)
HYDE, Clarence Edgar, (1894-1954)
ICKES, Eugene Law, (1885-1941)
IDDINGS, Arthur, (1891-1957)
IDDINGS, Joseph Paxson, (1857-1920)
INVILLIERS, Edward Vincent d', (1857-1928)
IRVING, John Duer, (1874-1918)
ISENBERGER, Nathan Perry, (1896-1956)
JACOBS, Elbridge Churchill, (1873-1957)
JAGGAR, Thomas Augustus, (1871-1953)
JAMES, Joseph Francis, (1857-1897)
JAMISON, Claude E., (1879-1939)
JEFFRIES, Charles Davis, (1896-1965)
JENNI, Clarence M., Colonel, (1896-)
JENNINGS, Charles I. "Al", (1894-1968)
JENNINGS, Philip H., (1899-1974)
JEWETT, John Mark, (1896-1970)
JILLSON, Willard Rouse, (1890-)
JOHANNSEN, Albert, (1871-1962)
JOHNSON, Bertrand Leroy, (1882-1962)
JOHNSON, Douglas Wilson, (1878-1944)
JOHNSTON, William Drumm, Jr., (1899-1972)
JONES, Daniel J., (1889-1965)
JONES, J. Claude, (1877-1932)
JONES, Ogden Sherman, (1891-1952)
JONES, William Francis, (1885-1941)
JORDAN, David Starr, (1851-1931)

UNITED STATES OF AMERICA 1850-1899 Cont'd.

JUDSON, Sidney Arthur, (1887-1971)
JUDSON, Wilber, (1880-1951)
KAY, Fred Hall, (1885-1943)
KAY, James LeFoy, (1892-1971)
KEELER, William Weaver, (1897-1943)
KEITH, Arthur, (1864-1944)
KELLOGG, Arthur Remington, (1892-1969)
KELLY, Donald, (1898-1949)
KEMP, James Furman, (1859-1926)
KENNEDY, Luther Eugene, (1891-1966)
KEPPLER, Leo George, (1890-1971)
KEW, William Stephen Webster, (1890-1961)
KEYES, Charles Rollin, (1864-1942)
KEYTE, Ivy Allen, (1878-1931)
KIMBALL, Kent K., (1892-1953)
KINDLE, Edward Martin, (1869-1940)
KINKEL, John Franklin, (1899-1936)
KINNEY, Harry Dill, (1886-1961)
KIRBY, Grady, (1889-1969)
KISLING, James William, Jr., (1898-1973)
KITSON, Howard Walde, (1883-1931)
KLEIN, A. Albert, (1889-1953)
KLEINPELL, William Darwin, (1898-1959)
KLINGER, Edgar Dewitt, (1899-1973)
KNEBEL, G. Moses, (1899-1974)
KNIGHT, Charles Robert, (1875-1953)
KNIGHT, Garold Louis "Jerry", (1894-1968)
KNIGHT, Samuel Howell, (1892-1975)
KNIGHT, Wilbur Clinton, (1858-1903)
KNOPF, Adolph, (1882-1966)
KNOWLTON, Frank Hall, (1860-1926)
KNOX, Thomas Knight, (1896-1949)
KOSCHMANN, Albert Herbert, (1896-1962)
KRAMER, William Baltser, III, (1896-1963)
KRAUS, Edward Henry, (1875-1973)
KRIDLER, George Matthew, (1889-1953)
KROENLEIN, George Alfred, (1898-1940)
KUMMEL, Henry Barnard, (1867-1945)
KUNZ, George Frederick, (1856-1932)
KURTZ, Robert Gourdin, (1898-1958)
LAHEE, Frederic Henry, (1884-1968)
LANDES, Henry, (1867-1936)
LANDWEHR, Walter Richard, (1893-1972)
LANE, Alfred Church, (1863-1948)
LANGTON, David W., Jr, (1864-?1908)
LARSEN, Esper Signius Jr., (1879-1961)
LARSH, Napoleon Bonaparte "Po", (1897-1967)
LARSH, Walter W., (1893-1951)
LEACH, Thomas Witt, (1896-1966)
LEATHERS, Walter S., (1874-)
LEDOUX, Albert Reid, (1852-1923) [PROSPECTORS, DIVINERS AND MINING ENGINEERS]
LEE, Henry E., (1875-1947)
LEE, Marvin, (1889-1953)

UNITED STATES OF AMERICA 1850-1899 Cont'd.

LEE, Olan Ivan, (1888-1952)
LEE, Willis Thomas, (1864-1926)
LEIGHTON, Henry, (1884-1963)
LEIGHTON, Morris Morgan, (1887-1971)
LEITH, Charles Kenneth, (1875-1956)
LEONARD, Arthur Gray, (1865-1932)
LEVERETT, Frank, (1859-1943)
LEVORSEN, Arville Irving, (1894-1965)
LEWIS, Henry Carvill, (1853-1888)
LEWIS, Joseph Volney, (1869-1969)
LIBBEY, William, (1855-1927)
LIDDLE, Ralph Alexander, (1896-1963)
LILLIGREN, Johnathan Martin, (1880-1956)
LINDSLEY, Thayer, (1882-) [PROSPECTORS, DIVINERS AND MINING ENGINEERS]
LLOYD, Edwin Russell, (1882-1955)
LOBECK, Armin Kahl, (1886-1958)
LONGYEAR, Edmund Joseph, (1864-1954) [PROSPECTORS, DIVINERS AND MINING ENGINEERS]
LONSDALE, John Tipton, (1895-1960)
LOOMIS, Frederic Brewster, (1873-1937)
LOUDERBACK, George Davis, (1874-1957)
LOUGHLIN, Gerald Francis, (1880-1946)
LOWELL, Percival, (1855-1916)
LOWMAN, Shepard Wetmore, (1899-1968)
LUCAS, Elmer Lawrence, (1898-1966)
LUCAS, Frederick Augustus, (1852-1929)
LUCKY, Maurice Cecil, (1895-1959)
LULL, Richard Swann, (1867-1957)
LUPTON, Charles Thomas, (1878-1935)
LUQUER, Lea McIlvaine, (1864-1930)
MacCLINTOCK, Paul, (1891-1970)
MacCURDY, George Grant, (1863-1947)
MACELWANE, James Bernard, Father, S.J., (1883-1956)
MACREADY, George Alexander, (1885-1955)
MANCHESTER, James Greenfield, (1871-1948)
MANSFIELD, Wendell Clay, (1874-1939)
MARBLE, John Putnam, (1897-1955)
MARBUT, Curtis Fletcher, (1863-1935)
MARKHAM, Edmund Olis, (1893-1960)
MARPLE, Mildred Fisher, (1898-1966)
MARSELL, Ray Everett, (1893-1971)
MARTIN, Francis Ivan "Shorty", (1896-1952)
MARTIN, George C., (1875-1943)
MARTIN, Handel Tongue, (1862-1931)
MASON, Shirley Lowell, (1893-1975)
MASSON, Charles E., (-) [PROSPECTORS, DIVINERS AND MINING ENGINEERS]
MATHER, Kirtley Fletcher, (1888-1978)
MATHEWS, Edward Bennett, (1869-1944)
MATSON, George Charlton, (1873-1940)
MATTHEWS, Edward Bennett, (1869-1944)
MAUCINI, Joseph John, (1893-1955)
MAY, Arthur Raymond, (1893-1948)

UNITED STATES OF AMERICA 1850-1899 Cont'd.

MAYFIELD, Samuel Martin, (1888?-1963)
McARTHUR, Donald, (1895-1960)
McCALLEY, Henry, (1852-1904)
McCALLIE, Samuel Washington, (1856-)
McCLAIN, Alan Hartman, (1899-1960)
McCLUNG, Clarence Erwin, (1870-1946)
McCOY, Alexander Watts, (1889-1944)
McFARLAND, Russell Scott, (1893-1968)
McFERRON, George Irwin, (1895-1969)
McGEE, William John, (1853-1912)
McGLOTHLIN, William Claude "Big Mack", (1895-1942)
McINTYRE, Paul Joseph, (1893-1963)
McKIM, John Walter, (1877-1927)
McKINSTRY, Hugh Exton, (1896-1961)
McNUTT, Vachel Harry, (1888-1936)
McVAY, Thomas Newkirk, (1891-1971)
MEAD, Roy Gibbons, (1881-1956)
MEAD, Warren Judson, (1883-1960)
MEHL, Maurice Goldsmith, (1887-1966)
MEINZER, Oscar Edward, (1876-1948)
MELL, Patrick Hues, (1850-1918)
MENDENHALL, Walter Curran, (1871-1957)
MERRIAM, John Campbell, (1869-1945)
MERRILL, George Perkins, (1854-1929)
MERRILL, Loren B., (?1853-1930)
MERRITT, John Wesley, (1885-1958)
METCALF, Roy J., (1889-1941)
MILLER, Benjamin Le Roy, (1874-1944)
MILLER, Carroll C., (1899-1974)
MILLER, Guy E., "Doc", (1893-1977)
MILLER, John Cummins, (1895-1951)
MILLER, Loye Holmes, (1874-1970)
MILLER, Wendell Zerbe, (1892-1958)
MILLER, Willard L., (1896-1973)
MILLER, William John, (1880-1965)
MILLS, Enos Abijah, (1870-1922)
MISER, Hugh Dinsmore, (1884-1969)
MIX, Charles Alexander, (1898-1954)
MONNETT, Victor Elvert, (1889-1972)
MONTGOMERY, Curtis Hall, (1897-1944)
MOODIE, Roy Lee, (1880-1934)
MOODY, Clarence Lemuel, (1888-1963)
MOODY, Graham Blair, (1889-1975)
MOORE, Andrew Charles, (1866-1928)
MOORE, Gilbert Parvin, (1898-1970)
MOORE, Prentiss D., (1895-1955)
MOORE, Raymond Cecil, (1892-1974)
MORAN, Robert Breck, (1879-1961)
MOREY, George Washington, (1888-1965)
MORGAN, Frank Albert, (1898-1976)
MORGAN, George Dillon, (1894-1950)
MORGAN, Glenn Beckley, (1876-1925)
MORLEY, Harold Thompson "Midge", (1897-1976)
MORRIS, Albert Ferd, (1898-1957)

UNITED STATES OF AMERICA 1850-1899 Cont'd.

MORRIS, Frederick Kuhne, (1885-1962)
MORSE, Paul Franklin, (1897-1929)
MORSE, William Clifford, (1874-1962)
MORTON, James F., (1870-1941)
MOSES, Alfred Joseph, (1859-1920)
MUNN, Malcolm John, (1874-1949)
MURDOCK, Joseph, (1890-1973)
MURRAY, Albert Nelson, (1894-1961)
MYERS, John Clark, (1894-1970)
MYERS, Julian Q., (1889-1950)
MYERS, Thurman H., (1890-1940)
MYERS, William Marsh, (1892-1951)
NARAMORE, Chester "Chet", (1879-1961)
NEDOM, Henry Arthur, (1895-1940)
NEEDHAM, Claude Ervin, (1894-1950)
NELSON, Lloyd Alvino, (1895-1964)
NELSON, Richard Newman, (1897-1964)
NELSON, Wilbur Armistead, (1890-1969)
NEWBY, Warner Wilson, (1898-1926)
NEWHOUSE, Walter Harry, (1897-1969)
NEWLAND, David Hale, (1872-1943)
NIGHTINGALE, William Thomas, (1897-1964)
NISBET, John McHenry, (1894-1962)
NOBLE, Earl Bart, (1894-1967)
NOBLE, Levi Fatzinger, (1882-1965)
NOÉ VON ARCHENEGG, Adolph Carl, (1873-1939)
NOLAN, Edward Dana, (1890-1926)
NOLAN, Philip Edward, (1887-1958)
NOMLAND, Jorgen O., (1885-1943)
NORTON, William Harmon, (-)
NORVILLE, Glen Smith, (1899-1963)
NORWOOD, Charles Joseph, (1853-1927)
NOTESTEIN, Frank B., (1885-1973)
NOWLAN, Harry Hackett, (1888-1958)
OAKES, Malcolm Christie, (1890-1977)
ODENBACH, Frederick L., Father, (1857-1933)
OFFICER, Herbert George, (1889-1937)
OGILVIE, Ida Helen, (1874-1963)
OHERN, Daniel Webster, (1870-1953)
OLCUTT, David Perry, (1896-1958)
OLDS, Thomas Hartman, (1883-1932)
ORCUTT, William Warren, (1869-1942)
OSBORN, Henry Fairfield, (1857-1935)
OWENS, Frith Cravens, (1898-1977)
PACKARD, Henry Jessup, (1883-1930)
PAINE, Paul, (1881-1957)
PALACHE, Charles, (1869-1954)
PALMER, Harold Schjoeth, (1890-1959)
PALMER, Robert Hastings, (1882-1948)
PARKER, Ralph D., (1898-) [PROSPECTORS, DIVINERS AND MINING ENGINEERS]
PARSONS, Arthur Leonard, (1873-1957)
PATTEN, William, (1861-1932)
PATTON, Leroy Thompson, (1880-1957)

UNITED STATES OF AMERICA 1850-1899 Cont'd.

PEABODY, Harlan William, (1881-1960)
PEATTIE, Roderick, (1891-)
PECK, Arthur Becker, (1892-1943)
PECK, Frederick Burritt, (1860-1925)
PEDRICK, Howard A. "Dad", (1863-?1930) [PROSPECTORS, DIVINERS AND MINING ENGINEERS]
PEGAU, Albert August, (1894-1963)
PEMBERTON, John Roy "Bill", (1884-1968)
PENFIELD, Samuel Lewis, (1856-1906)
PENROSE, Richard Alexander Fullerton, Jr., (1863-1931)
PEPPER, James Franklin, (1898-1963)
PEPPERBERG, Leon J., (1883-1937)
PERINI, Vincent Charles "Bud", Jr., (1895-1965)
PERRINE, Irving, (1884-1955)
PETERSON, Clarence J., (1885-1948)
PETSCH, Arthur Henry, (1895-1969)
PETTY, Dabney E., (1891-1963)
PETTY, Olive Scott, (1895-)
PEYTON, Garland, (1892-1964)
PHILBRICK, Eugene Patrick, (1897-1951)
PHILLIPS, Alexander Hamilton, (1866-1937)
PHILLIPS, Eldridge Douglas, (1894-1931)
PHILLIPS, William Battle, (1857-1918)
PILSBRY, Henry Augustus, (1862-1957)
PIRSSON, Louis Valentine, (1860-1919)
PLUMMER, Frederick Byron, (1886-1947)
PLUMMER, Helen Jeanne (née Skewes), (1891-1951)
POLLOK, Rudolf R., (1876-1929)
POND, Walter Franklin, (1885-1974)
POOR, Russell Spurgeon, (1899-1972)
PORCH, Edwin Leight, Jr., (1889-1948)
POTEET, James Harold, (1896-1965)
POTTER, Grover Cleveland, (1892-1975)
POTTER, Nelson Bryan, (1897-1965)
POWERS, Sidney, (1890-1932)
PRATLEY, Henry Hart, (1893-1941)
PRATT, Joseph Hyde, (1870-1942)
PRATT, Wallace Everette, (1885-)
PREECE, Rae, (1897-1961)
PRENTICE, Sydney Clark, (1870-1943)
PRETTYMAN, Thomas Mann, (1888-1940)
PRICE, Paul Holland, (1898-)
PROSSER, Charles Smith, (1860-1916)
PROUT, Franklin S. "Shorty", (1889-1940)
PURDUE, Albert Homer, (1861-1917)
PUTNAM, L. G., (1898-1936)
RAMSDELL, Lewis Stephen, (1895-1975)
RATH, Charles Martin, (1881-1959)
RATHBURN, Richard, (1852-1918)
RAVEN, Henry Cushier, (1889-1944)
REBER, Louis Ehrhart, Jr., (1889-1966)
REED, Ralph Daniel, (1889-1940)
REESIDE, John Bernard, Jr., (1889-1958)
RENAUD, Charles Louis, (1898-1964)

UNITED STATES OF AMERICA 1850-1899 Cont'd.

RENICK, Brink Coleman, (1897-1971)
REYNOLDS, Karl William, (1894-1970)
REYNOLDS, Roy A., (1889-1943)
RHOADES, Ralph Omer, (1895-1961)
RICH, John Lyon, (1884-1956)
RICHARDS, James Taylor "Dick", (1892-1953)
RICHARDSON, Charles Henry, (1862-1935)
RICHARDSON, George Burr, (1872-1949)
RIDGEWAY, Bertrand S., (1899-1942)
RIEBER, Frank, (1891-1948)
RIES, Heinrich, (1871-1951)
RIFE, Byron, (1892-1974)
RIGGS, Elmer Samuel, (1869-1963)
RIGGS, Robert Jennings, (1888-1965)
RITER, Samuel W., (1896-1941)
ROARK, Louis, (1890-1950)
ROBERTS, Hugh Marine, (1885-1965)
ROBERTS, Morgan Evan, (1894-1941)
ROBINSON, Bryan Franklin, (1897-1949)
ROBINSON, Ernest Guy, (1898-1973)
ROGERS, Austin Flint, (1887-1957)
ROGERS, Gaillard Sherburne, (1889-1919)
ROMER, Alfred Sherwood, (1894-1973)
ROSENHOLTZ, Joseph Leon, (1899-1963)
ROSS, Clarence Samuel, (1880-1975)
ROUNDY, Paul Vere, (1884-1937)
ROW, Charles Herbert, (1895-1970)
ROWE, Richard Burton, (1872-1902)
ROWLEY, Alden Bruce, (1888-1956)
RUBEY, William Walden, (1898-1974)
RUNNER, Joseph James, (1885-1970)
RUSS, Leon Fair, (1886-1953)
RUSSELL, Israel Cook, (1852-1906)
RUSSELL, Philip G., (1891-1971)
RUSSELL, Richard Joel, (1895-1971)
RUSSELL, William Low, (1897-1976)
RUSSOM, Vaughn Walter, (1891-1955)
RYAN, Russell Fossler, (1897-1935)
SALISBURY, Rollin D., (1858-1922)
SANDIDGE, John Roy, (1897-1976)
SARDESON, Frederick William, (1866-1959)
SAVILLE, Wilson Gordon, (1897-1954)
SAWYER, Roger W., (1895-1941)
SCHAEFFER, Charles Frederic August, (1860-1903)
SCHALLER, Waldemar Theodore, (1882-1967)
SCHENCK, Hubert Gregory, (1897-1960)
SCHILLING, Karl Howard, (1894-1945)
SCHLICHTEN, Otto Charles von, (1886-1950)
SCHMITT, Harrison Ashley, (1896-1966)
SCHMITTOU, Maurice B., (1899-1929)
SCHNEIDER, George William, (1899-1963)
SCHOEWE, Walter Henry, (1891-1966)
SCHOLL, Louis August, Jr., (1887-1953)
SCHRADER, Frank Charles, (1860-1944)

UNITED STATES OF AMERICA 1850-1899 Cont'd.

SCHRAMM, Eck Frank, (1883-1967)
SCHROYER, Charles Ross, (1887-1932)
SCHUCHERT, Charles, (1858-1942)
SCHWARZ, Melbert Edgar, (1898-1957)
SCOFIELD, Lloyd M., (1897-1969)
SCOTT, Gayle, (1894-1948)
SCOTT, Walter Winthrop, (1893-1939)
SCOTT, William Berryman, (1858-1947)
SEAMAN, Elizabeth Cochrane, (1865-1922)
SEARS, Julian Ducker, (1891-1970)
SEASHORE, Paul Theodore, (1898-1954)
SEE, Thomas Jefferson Jackson, (1866-)
SEGALL, Julius, (1884-1928)
SELLARDS, Elias Howard, (1875-1961)
SERVISS, Frederick LeVerne, (1895-1954)
SEVERY, Clarence Luther, (1885-1975)
SHALER, Millard King, (1880-1942)
SHATTUCK, George Burbank, (1869-)
SHAW, Eugene Wesley, (1881-1935)
SHAW, Everett Simeon, (1885-1955)
SHEARER, Harold Kurtz, (1889-1946)
SHEPARD, Edward Martin, (1854-1934)
SHERRILL, Richard Ellis, (1899-1953)
SHERWOOD, Theodore C., Jr., (1896-1928)
SHIDELER, William Henry, (1886-1958)
SHIMEK, Bohumil, (1861-1937)
SHIMER, Hervey Woodburn, (1872-1965)
SHORT, Maxwell Naylor, (1889-1952)
SHUFELDT, Robert Wilson, (1850-1934)
SHULER, Ellis William, (1881-1954)
SICKLER, John Miller, Jr., (1891-1970)
SINCLAIR, Ewart Gladstone, (1884-1927)
SINGEWALD, Joseph Theophilus, Jr., (1884-1963)
SISLER, James Donaldson, (1896-1935)
SLAWSON, Chester Baker, (1898-1964)
SLICHTER, Louis Byrne, (1896-)
SLOAN, Earle, (1858-1926)
SMITH, Aylwin L., (1898-1965)
SMITH, George Otis, (1871-1944)
SMITH, James Perrin, (1864-1931)
SMITH, Warren D., (1880-1950)
SMITH, William Richard Tangier, (1869-1972)
SNIDER, Luther Crocker, (1882-1947)
SNIVELEY, H. Norman, (1879-1959)
SNYDER, John Young, (1872-1939)
SOUTHER, John Barron, (1894-1964)
SPARKS, Dale Darrell Dolsun, (1894-1938)
SPOFFORD, Howard Nelson, (1881-1950)
STAINBROOK, Merrill Addison, (1897-1956)
STALDER, Walter "Cap", (1881-1949)
STANLEY, M. Wood, (1893-1929)
STANTON, Timothy William, (1860-1953)
STEBINGER, Eugene, (1883-1951)
STEIGER, George, (1869-1944)

UNITED STATES OF AMERICA 1850-1899 Cont'd.

STEINY, Homer Julius, (1892-1974)
STEPHENSON, Lloyd William, (1876-1962)
STERNBERG, Charles Hazelius, (1850-1943)
STERNBERG, Charles M., (1885-)
STERNBERG, George Fryer, (1883-1969)
STEUBING, William C., (1881-1943)
STEVENS, George D., (1898-1965)
STOCK, Chester, (1892-1950)
STOCKDALE, Paris Buell, (1896-1962)
STONE, Ralph Walter, (1876-1964)
STONER, Oscar E., Sr., (1892-1965)
SUMAN, John Robert, (1890-1972)
SUTTON, Arle Herbert, (1896-1972)
SUTTON, Frederick A., (1894-1950)
SWARTZ, Joel Howard, (1893-1971)
SWEET, Sidney R., (1878-1946)
SWIGER, Rual Bower, (1897-1950)
TABER, Stephen, (1882-1963)
TAFF, Joseph Alexander, (1862-1944)
TANDY, James Hiram "Spot", (1892-1962)
TARR, Ralph Stockman, (1864-1912)
TARR, William Arthur, (1881-1939)
TAYLOR, Charles Henry, (1877-1964)
TEAS, Livingstone Pierson, (1894-1970)
TEMPLETON, Eugene Crittenden, (1889-1932)
THIESSEN, Reinhardt, (1867-1938)
THOMAS, George Dewey, (1898-1975)
THOMAS, John Elmer, (1892-1949)
THOMAS, Norman Louis, (1897-1970)
THOMPSON, David Grosh, (1888-1943)
THOMPSON, E. G., (1898-1973)
THOMPSON, Jean Cleveland, (1887-1929)
THOMPSON, Robert Raymond, (1893-1954)
THOMPSON, Warren O., (1898-1968)
THORNBURG, Dwight Holbrook "Duke", (1895-1965)
THORPE, Malcolm Rutherford, (1891-1958)
TIEJE, Arthur Jerrold, (1891-1944)
TIGHT, William George, (1865-1910)
TILTON, John Littlefield, (1863-1930)
TOLMAN, Cyrus Fisher, (1873-1941)
TOMLINSON, Charles Weldon, (1892-1960)
TOOMEY, Charles Calvin, (1892-1957)
TOOTHAKER, Charles Robinson, (1873-1952)
TRAGER, Earl A., (1893-1978)
TRAINER, David Woolsey, Jr., (1899-1965)
TRASK, Parker Davies, (1899-1961)
TROXELL, Edward Leffingwell, (1884-1972)
TROXELL, John N., (1895-1970)
TWENHOFEL, William Henry, (1875-1957)
TWITCHELL, Mayville William, (1866-1927)
ULRICH, Edward Oscar, (1857-1944)
UPHAM, Warren, (1850-1934)
UREN, Lester Charles, (1888-1960)
UREY, Harold Clayton, (1893-)

UNITED STATES OF AMERICA 1850-1899 Cont'd.

VALERIUS, Claude N., (-)
VALERIUS, Michael Marcus, (1872-1947)
VAN AMRINGE, Edwin Verne, (1899-1956)
VAN COUVERING, Martin, (1888-1976)
VANDER LECK, Lawrence, (1890-1954)
VANDERLIP, Washington B., (-) [PROSPECTORS, DIVINERS AND MINING ENGINEERS]
VAN HISE, Charles Richard, (1857-1918)
VAN HORN, Frank Robertson, (1872-1933)
VAN INGEN, Gilbert, (1869-1928)
VAN TUYL, Francis M., (1887-1975)
VAUGHAN, Thomas Wayland, (1870-1952)
VAUX, George, Jr., (1863-1927)
VAUX, William S., Jr., (1872-1908)
VEATCH, Arthur Clifford, (1878-1938)
VERNON, Jesse, (1894-1972)
VER WIEBE, Walter August, (1887-1961)
VESTAL, Franklin Earl, (1884-1969)
VON ENGELN, Oskar Dietrich, (1880-1965)
VONSEN, Magnus, (1879-1954)
WAGENER, Charles Henry "Chick", (1897-1950)
WAGNER, Carroll M., (1893-1965)
WALCOTT, Charles Doolittle, (1850-1927)
WALCOTT, Mary Vaux, née Vaux, (-1940)
WALKER, Lucian Hickman, (1894-1970)
WALTMAN, William Dewitt, (1875-1955)
WANDKE, Alfred, (1887-1941)
WANLESS, Harold Rollin, (1898-1970)
WARD, Freeman, (1879-1943)
WARING, Gerald Ashley, (1883-1971)
WARREN, Charles Hyde, (1876-1950)
WASHBURNE, Chester Wesley, (1883-1971)
WASHINGTON, Henry Stephens, (1867-1934)
WASSON, Theron, (1887-1970)
WATSON, Thomas Leonard, (1871-1924)
WEAVER, Charles Edwin, (1880-1958)
WEAVER, Paul, (1888-1964)
WEDEL, Arthur Albert, (1898-1941)
WEEKS, Herbert John, (1897-1960)
WEIDMAN, Samuel, (1870-1945)
WELLER, James Marvin, (1899-1976)
WELLER, Stuart, (1870-1927)
WELLS, Francis Gerrit, (1898-1969)
WELLS, Horace Lemuel, (1855-1924)
WELLS, Roger Clark, (1877-1944)
WENTWORTH, Chester Keeler, (1891-1969)
WESTGATE, Lewis Gardner, (1868-1948)
WHEALTON, Rowland Gardiner "Peg", (1896-1956)
WHEELER, Orby Clinton, (1898-1970)
WHITE, Charles David, (1862-1935)
WHITE, Maynard Pressley, (1896-1961)
WHITE, Theodore Greely, (1872-1901)
WHITEHEAD, Robert Brooks, (1889-1936)
WHITEHEAD, Walter Lucius, (1891-1969)

UNITED STATES OF AMERICA 1850-1899 Cont'd.

WHITESIDE, Charles Edwin, (1897-1927)
WHITESIDE, Robert Massie, (1895-1936)
WHITLOCK, Herbert Percy, (1868-1948)
WHITTIER, William Harrison, (1889-1951)
WHITWELL, Earl Van Dorn, (1892-1963)
WIELAND, George Reber, (1865-1953)
WIGGLESWORTH, Edward, (1885-1945)
WILDER, Frank Alonzo, (1870-1930)
WILKERSON, Albert Samuel, (1897-1963)
WILLARD, Daniel Everett, (1862-1947)
WILLIAMS, Arthur James, (1877-1954)
WILLIAMS, George Huntington, (1856-1894)
WILLIAMS, George O. "Skipper", (1897-1967)
WILLIAMS, Ira Abraham, (1876-   )
WILLIAMS, John Francis, (1862-1891)
WILLIAMSON, Erskine Douglas, (1886-1923)
WILLIS, Bailey, (1857-1949)
WILLISTON, Samuel Wendell, (1852-1918)
WILSON, Eldred Dewey, (1898-1965)
WILSON, Malcolm Earl, Sr., (1885-1960)
WILSON, Roy Arthur, (1891-1970)
WILSON, Walter Byron "Shorty", (1885-1951)
WINCHELL, Alexander Newton, (1874-1958)
WINCHELL, Horace Vaughn, (1865-1923)
WINCHESTER, Dean Eddy, (1883-1936)
WING, Monta Eldo, (1897-1976)
WINTERER, Edward Virgil "Tubby", (1897-1944)
WINTON, Will McClain, (1885-1960)
WISSER, Edward Hollister, (1895-1970)
WOLF, Albert G., (1885-1964)
WOLF, Key, (1886-1960)
WOLFF, John Elict, (1857-1940)
WOOD, Robert Harrison, (1887-1956)
WOOD, Virgil O., (1892-1954)
WOODRUFF, Elmer Grant, (1872-1952)
WOODRUFF, John Grant, (1898-1971)
WOODWARD, Herbert Preston, (1899-1968)
WOODWORTH, Jay Backus, (1865-1925)
WORCESTER, Philip George, (1884-1970)
WORTMAN, Jacob Lawson, (1856-1926)
WRATHER, William Embry, (1883-1963)
WRIGHT, Andrew Clemmons, (1893-1968)
WRIGHT, Charles Will, (1879-   )
WRIGHT, Frank James, (1888-1954)
WRIGHT, Frederick Eugene, (1877-1953)
WRIGHT, Harry Favill, (1888-1944)
YEATES, William Smith, (1856-1908)
ZODAC, Peter, (1894-1967)
ZOLLER, Lawrence J., (1895-1973)

1900 to present

ADKINS, John Nathaniel, (1911-   )
AGOGINO, George, (-   )

UNITED STATES OF AMERICA 1900 to present Cont'd.

ALBERT, Herman W., (1882-) [PROSPECTORS, DIVINERS AND MINING ENGINEERS]
ALCORN, Avary Hunt, (1902-1967)
ANDERSON, Alfred Leonard, (1900-1964)
ANDERSON, Robert Lee, (1912-1959)
ANDERTON, Russell, (-) [PROSPECTORS, DIVINERS AND MINING ENGINEERS]
ANSBERRY, Leroy Eugene, (1923-1961)
ARMSTRONG, William Burrus, (1917-1954)
ASHMORE, Herman T., (1916-1976)
ASHMORE, Leland Wayne, (1916-1964)
BAIRD, Lucille Bailey, (1926-1963)
BALDWIN, William Felbert Jackson, Jr., (1927-1972)
BALLARD, James Logan, (1902-1974)
BALLARD, William Norval, (1906-1976)
BALLOU, Albert Lorenzo, Jr., (1917-1962)
BANDY, Mark Chance, (1900-1963)
BANDY, Orville Lee, (1917-1973)
BANKS, Thomas Russell, (1900-1948)
BARD, Richards J., (1900-1977)
BARLOW, Maerlant M. "Pete", (1906-1950)
BARTELL, Lawrence Denny, (1902-1955)
BARTLETT, Helen Blair, (1901-1969)
BASCOM, Willard, (1917-)
BASSINGER, William E., (1928-1977)
BATES, Robert Glenn, (1923-1972)
BECKER, Robert M., (1919-1975)
BECKER, Ted G., (1921-1975)
BECKMANN, Lawrence John, (1912-1976)
BECKWITH, Frank Asahel, (-1951)
BECKWITH, Radcliffe Harold, (1900-1964)
BEERY, Roy F., Jr., (1917-1964)
BERG, Ernest Lyle, (1910-1953)
BERMAN, Harry, (1902-1944)
BERNARD, Hugh Allen, (1915-1975)
BERRY, Edward Willard, (1900-1968)
BIEBER, Charles Leonard, (1901-1965)
BIGGS, Charles L. "Tib", (1919-1971)
BILLINGS, Katharine Stevens (neé Fowler-Lunn), (1902-)
BIRD, Roland T., (-1978)
BITTER, Francis, (1902-1967)
BLANKENSHIP, Joseph Croxton, (1930-1964)
BLEDSOE, Lucybelle, (1923-1966)
BLOODWORTH, Bill Lloyd, (1916-1971)
BLOOMER, Philip A., Jr., (1922-1967)
BLUMENTHAL, Milton Diez, (1922-1964)
BOCKELMAN, Frederick H., Jr., (1923-1968)
BOGARDUS, Robert Stephen "Bud", (1914-1953)
BOOTS, Paul Heath, (1909-1947)
BORN, Kendall Eugene, (1908-1947)
BOSS, Richard F., (1923-1976)
BOUCOT, Arthur James, (1924-)
BOWLING, Leslie, (1908-1977)
BRACE, William Francis, (1926-)

UNITED STATES OF AMERICA 1900 to present Cont'd.

BRADLEY, William Frank, (1908-1973)
BRAMKAMP, Richard Allen, (1910-1958)
BRANNOCK, Kent Combs "Casey", (1923-1973)
BRANSON, Carl Colton, (1906-1975)
BRANSON, Edwin Robert, (1908-1971)
BRAVINDER, Kenneth Mason, (1907-1946)
BREMER, B. E. "Boots", (1904-1974)
BREWER, Charles, Jr., (1900-1944)
BRILL, Virgil August, (1905-1972)
BRINKERHOFF, Ira A., (1904-1969)
BROWN, Alton Roe, (1930-1975)
BROWN, Bahngrell Walter "Bill", (1916-)
BROWN, John Albert, (1901-1954)
BROWN, Noel K., Jr., (1925-1968)
BROWN, Olin Teter, (1900-1962)
BRYAN, Joseph Jefferson, (1905-1951)
BRYAN, Roger Wesley, (1927-1971)
BUERGER, Martin Julian, (-)
BUMP, James Dye, (1903-1959)
BUNTE, Arnold S., (1902-1968)
BURDICK, Clifford, (-)
BURGER, Robert W. "Bush", (1906-1945)
BURNS, Russell Warren, (1912-1967)
BUTCHER, Walter William, (1916-1966)
BUTZ, Walter "Wally", (1905-1962)
BYRON, Clifford Winkler, (1901-1970)
CADY, Richard Carlysle, (1908-1943)
CAIN, James Walter, Jr., (1917-1941)
CAMP, John F. Jr., (1915-1976)
CAMP, Russell R., (1904-1967)
CAMPBELL, Arthur Byron, (1924-1971)
CAMPBELL, Ian, (1900-1978)
CANADA, William Richards, (1909-1962)
CANFIELD, Charles Reiter, (1906-1964)
CANTWELL, Thomas, Jr., (1927-)
CARR, William L., (1919-1975)
CARRITT, Dayton Ernest, (1915-)
CARROLL, Don Llewellyn, (1902-1954)
CARTER, Thomas Ray, (1927-1959)
CASEBEER, Charles Theodore, (1905-1941)
CASEY, John Hestwood, (1923-1967)
CHARLES, Homer Herbert, (1900-1963)
CHATMAN, Cecil La Mar, (1902-1938)
CHURCHILL, Frederick Holmes, (1901-1959)
CLARK, Paul Henry, (1909-1951)
CLINE, Lewis Manning, (1909-1971)
CLOUD, Preston Ercelle, Jr., (1912-)
CLOUGH, Kenneth H., (1903-1966)
COCHRAN, Kenneth L., (1910-1973)
COIL, Clarence, (-)
COLBERT, Edwin H., (1905-)
CONWAY, William Palen "Tony", Jr., (1914-1944)
COOMBS, Howard A., (-)
COOPER, Byron Nelson, (1912-1971)

UNITED STATES OF AMERICA 1900 to present Cont'd.

COOPER, Chalmer Lewis, (-)
COPASS, Jack Malcolm, (1906-1963)
COX, William Edgerton, (1912-1957)
CRANDALL, Richard Robbins, (1902-1949)
CRONEIS, Carey, (1901-1972)
CROSS, Rodman K., (1911-1970)
CROWN, Walter James, (1901-1951)
CUTTITTA, Frank, (1912-1974)
CUYLER, Robert Hamilton, (1908-1944)
DACEY, Reginald Lynn, (1908-1951)
DAMON, Fred T., (1912-1977)
DANE, Carle Hamilton, (1900-1968)
DAVIE, Thomas Early, (1906-1959)
DAWSON, Thomas Albert, (1916-1974)
DEAN, Lee Wallace, III, (1932-)
DEISS, Charles Frederick, (1903-1959)
DENNEN, William Henry, (1920-)
DENNING, Reynolds McConnell, (1916-1967)
DeRUDDER, Ronald D., (1932-1975)
DICKSON, Fielding Breeden, (1912-1972)
DIETZ, Robert Sinclair, (1914-)
DOELL, Richard Rayman, (1923-)
DONAVAN, Jack Hilder, (1921-1953)
DOUGLAS, John Gray, (1900-1974)
DOYLE, Robert, (1913-1963)
DUCOTE, James Wardy, (1919-1964)
DURHAM, Charles Albert, (1910-1954)
DUTY, Richard Emerson, (1908-1966)
EARDLEY, Armand John, (1901-1972)
EARGLE, Dolan Hoye, (1905-1973)
EARL, Eugene Leonard "Red", (1906-1956)
ELDER, Stanley Gordon, (1904-1957)
ELEY, Hugh Moore, (1902-1942)
ELLIOTT, Harold Charles, (1910-1972)
ELLIS, Charles W., Jr., (1934-1970)
ELLISON, Kenneth Alfred, (1902-1963)
ESAREY, Ralph Emerson, (1901-1975)
ESPACH, Ralph Homeward, (1900-1956)
EVANS, Vincent, (1907-1951)
EWING, William Maurice, (1906-1974)
FAHLE, Paul Bartlett, (1910-1954)
FANGUAY, Nolan Anthony, (1916-1942)
FANKUCHEN, Isidor, (1905-1964)
FARISH, Linn Markley, (1901-1944)
FARLEE, Richard Wilson, (1913-1961)
FERRERO, Clarence de Labretonne, (1903-1954)
FIEDLER, William M., (1914-1970)
FISHER, Bernard, (1918-1946)
FISK, Harold Norman, (1908-1964)
FITZGERALD, James, Jr., (1903-1951)
FITZGERALD, Paul Eugene, (1903-1961)
FITZGERALD, Thomas John, (1903-1973)
FLESCH, Gary Allen, (1949-1976)
FLETCHER, Corbin Drummond, (1900-1931)

UNITED STATES OF AMERICA 1900 to present Cont'd.

FLINT, Richard Foster, (1902-1976)
FLY, Paul J., (1907-1967)
FOLEY, Edward J., (1908-1971)
FORTINE, Dwight Hugh, (1904-1949)
FOSSHAGE, Ernest Willard, (1908-1955)
FOXWORTH, Wyckliff Riley, (1936-1965)
FRALICH, Charles Eugene, (1902-1964)
FRAZIER, David Edward, (1931-1976)
FREED, Richard, (1908-1967)
FREY, M. Gordon, (1913-)
FRIES, Carl, Jr., (1910-1965)
FRITTS, Crawford Ellsworth, (1927-1972)
FRIZZELL, Donald Leslie, (1906-1972)
FROST, Jack, (1900-1970)
FRY, Willis E., (1926-1973)
FRYXELL, Roald Hilding, (1934-1974)
FULTON, Loris John, (1906-1974)
GABRIEL, Alton, (1902-1954)
GAITHER, Benjamin Mason, (1919-1960)
GALLAGHER, David, (1906-1968)
GALLIE, John Francis, III, (1913-1951)
GANGI, Anthony Frank, (1929-)
GARDNER, Frank J., (1915-1975)
GARIEPY, Glen Bernard, (1904-1962)
GAULT, Hugh Richard, (1915-1961)
GEBELEIN, Conrad Dennis, (1945-1978)
GERMANN, John Christian, (1903-1954)
GETZENDANER, Avalette Edmond, (1902-1962)
GIERHART, Guy B. "Joe", (1903-1968)
GILL, James Rogers, (1922-1972)
GILLERMAN, Elliot "Bud", (1913-1974)
GLEASON, Charles D., (1908-1935)
GOERNER, Frederic Stephen, (1915-1969)
GOLDEN, John Muro, (1900-1945)
GOLLNICK, Robert Louis, (1918-1963)
GOODELL, Horace N. "Hank", (1915-1971)
GOODING, Ansel Miller, (1924-1976)
GOODSON, Joseph Arthur, (1910-1946)
GORMAN, Joseph M., (1910-1967)
GOYER, Sidney Carroll, (1934-1965)
GRAHAM, Daniel Willard, (1921-1954)
GRAHAM, David H., (1910-1944)
GRAHAM, John Warren, (1918-1971)
GRAHAM, Joseph John, (1909-1967)
GRANQUIST, William T., (1923-1974)
GRAWE, Oliver Rudolph, (1901-1965)
GRIMES, Russell Winters, (1900-1969)
GRIMM, Kenneth Edgar, (1912-1961)
GROSS, Arthur Bunert "Abie", (1904-1965)
GROSS, Henry, (-) [PROSPECTORS, DIVINERS AND MINING ENGINEERS]
GUILLOU, Robert Barton, (1923-1965)
GUINN, Delmar Richard, (1902-1955)
GUT, H. James, (1904-1969)

UNITED STATES OF AMERICA 1900 to present Cont'd.

HAHN, Emily, (1904-)
HALEPESKA, Dennis Leonard, (1920-1958)
HALL, Thomas O., (1906-1972)
HAM, William Eugene, (1916-1970)
HAMILTON, Peggy-Kay, (1922-1959)
HAMMOND, Weldon Woolf, (1908-1964)
HAMMONDS, G. Scott, (1910-1965)
HANDLEY, Howard Walter, (1905-1940)
HANSELL, James Myron, (1902-1970)
HANSON, Wallace W., (1921-1971)
HARRELL, Marshall Allen, (1900-1959)
HARRIS, Reginald W., (-)
HARRISON, George Haymond, (1903-1949)
HARRISON, John Wiley, (1916-1952)
HARTWIG, Albert Ernest, Jr., (1922-1966)
HAVARD, Charles Gentry, (1921-1950)
HAYWARD, Raymond Edward, (1932-1964)
HAZZARD, John Charles, (1903-1973)
HEADLEE, Larry A., (1938-1970)
HEEZEN, Bruce Charles, (1924-1977)
HENDERSON, Harold H. "Heinie", (1905-1949)
HERPERS, Henry F., Jr., (1915-1952)
HERRING, Donald Grant, Jr., (1918-1962)
HESS, Harry Hammond, (1906-1969)
HESSE, Curtis Julian, (1905-1945)
HIBBARD, Claude W., (1905-1973)
HILL, James Wilcott, (1920-1954)
HILL, Robert G., (1914-1973)
HILL, Robert Luce, (1937-1964)
HOFFMEISTER, William S., (-)
HOFMANN, Walter, (1919-1975)
HOLCOMB, Leroy D., (1919-1974)
HOLDEN, Edward Fuller, (1901-1925)
HOLLISTER, Joseph Steffens, (1903-1953)
HOLMES, George William, (1922-1970)
HOOKER, Marjorie, (1908-1976)
HOOKWAY, Lozell Charles, (1905-1938)
HOOVER, James Barr, (1902-1953)
HORBERG, Carl Leland, (1910-1955)
HORNER, Jack, (-)
HORNER, Seward Ellis, (1906-1954)
HOSKINS, Baker, Jr., (1900-1941)
HOUSER, Gertrude, (-)
HOUSTON, Samuel Henry, (1905-1947)
HOWER, John, Jr., (1927-)
HOYT, John Harger, (1928-1970)
HUBLEY, Richard Carleton, (1926-1957)
HUBMAN, Carl Wesley, (1905-1947)
HUDDLE, John Warfield, (1907-1975)
HUFFMAN, Edward Archibald, (1912-1954)
HUGHES, Richard John, Jr., (1915-1968)
HURNDALL, John P., (1902-1966)
HUSTON, Charles, (1908-) [PROSPECTORS, DIVINERS AND MINING ENGINEERS]

UNITED STATES OF AMERICA 1900 to present Cont'd.

IRELAND, Hubert Andrew, (1904-1973)
IRWIN, Marvin Leroy "Mel", (1915-1964)
JACOB, Charles Edward, (1914-1970)
JAMES, John William, (1921-1977)
JANES, Charles Arthur, (1928-1966)
JANOSKY, Robert A., (1923-1955)
JARVIS, Daniel, (1920-1973)
JENSEN, Christian "Cope", (1905-1961)
JEWELL, Willard B., (-)
JOESTING, Henry Rochambeau, (1903-1965)
JOHANSON, Donald C., (-)
JOHNS, Walter Wendell Schell, (1911-1963)
JOHNSON, Charles George, (1914-1969)
JOHNSON, Clifton Wood, (1905-1954)
JOHNSON, Earle Winston, (1902-1953)
JOHNSON, Joseph Wright, (1917-1957)
JOHNSON, Norman Lee, (1926-1970)
JOHNSON, Ralph Gordon, (1927-1976)
JOHNSON, W. Ray, (-1952)
JONES, Brynmor, (1902-1941)
JONES, Verner E., (1905-1977)
JORDAN, Louise, (1908-1966)
KAISEN, Ove C., (1902-1953)
KEYES, Wilson, (1901-1936)
KIDD, Major Gentry, (1904-1974)
KING, Jack Joseph, (1902-1929)
KING, Robert Bruce, (1923-1977)
KIRBY, Maurice E. "Bud", (1909-1963)
KIVI, Wilho John, (1918-1950)
KLINE, Virginia Harriett, (1910-1959)
KNOWLES, Gerald Elgin, (1923-1976)
KOCH, Harold Kenneth, (1927-1973)
KRALIS, Dan F., Jr., (1919-1962)
KREIDLER, William Lynn, (1911-1975)
KREPPS, William J., (1924-1976)
KRIEGER, Philip, (1900-1940)
LADNER, Albert L., (1905-1975)
LAKE, Merrill Evans, (1902-1938)
LAMMERS, Edward Chauncey Hinman, (1910-1971)
LAMSENS, O. Robert, (1922-1971)
LANCASTER, L. Kenneth, (1906-1971)
LAPHAM, Davis M., (1931-1974)
LARSEN, Esper Signius, III, (1912-1961)
LATIMER, Frank Parkes, (1903-1928)
LATTA, Albert Lloyd, (1907-1960)
LAW, John Melvin, (1914-1967)
LEATHEROCK, Otto, (1901-1941)
LEDBETTER, William Robert, (1921-1971)
LESTER, John Lawrence, (1914-1977)
LEWIS, Francis Leonard, Jr., (1921-1953)
LEWIS, G. Edward, (-)
LIBBY, Willard Frank, (1908-)
LIDZ, Louis, (1938-1967)
LILIENBORG, Burton Armand, (1900-1949)

UNITED STATES OF AMERICA 1900 to present Cont'd.

LIND, Charles McClelland, (1911-1957)
LINEHAN, Charles M., (1913-1973)
LLOYD, Edwin Russell "Rusty", Jr., (1924-1951)
LOPEZ, Abdon Joseph, (1924-1972)
LOSKAMP, Alvin Powell, (1900-1954)
LOW, Julian W., (1904-1968)
LUPTON, Benjamin Charles, (1908-1964)
LUTH, William Clair, (1934-)
MACDIARMID, Roy Angus, (1933-1966)
MacNAUGHTON, Lewis Winslow, (1902-1969)
MacROBERT OF DOUNESIDE AND CROMAR, Rachel, Lady (née Workman), (-1954)
MADDEN, Theodore Richard, (1925-)
MAEBIUS, Jed B., (1907-1969)
MALOTT, Clyde A., (-1950)
MANDARINO, Joseph Anthony, (-)
MANKIN, Charles John, (1932-)
MANN, Robert Scott, (1915-1956)
MARTI, Ernest John, (1913-1964)
MARTIN, Lewis, (1920-1973)
MARTYN, Phil F., (1903-1974)
MASON, Arnold Caverly, (1906-1961)
MAXEY, George Burke, (1917-1977)
MAXSON, John Haviland, (1906-1966)
MAYFIELD, William I., (1912-1972)
McCALL, John Gill, (?1923-1954)
McCAUGHEY, William John, (1882-1962)
McCAULEY, Dan Wesley, (1916-1967)
McCLOY, Walter Lowrie, Jr., (1912-1954)
McCLURE, Drexel Eugene, (1928-1977)
McCOLLOUGH, Edward Lamb, (1931-1955)
McCONNEL, Roger H., (1908-1971)
McDONALD, Stanley Matthews, (1915-1950)
McDOUGALL, John Edward, (1914-1964)
McFARLAND, Leland R., (1904-1973)
McGEHEE, Julius Rex, (1905-1963)
McGUINNESS, Charles Lee, (1914-1971)
McLAREN, Robert Lee, (1901-1953)
McLAUGHIN, Dean Benjamin, (1901-1965)
McLAUGHLIN, R. P., (-)
McMANNIS, William J., (1924-1973)
McMINN, Paul Meloy, (1919-1974)
McNEELY, Robert Barnes, (1902-1966)
McNEIL, Harold E., (1910-1965)
MEFFERD, Ralph L., (1909-1964)
MELTON, Frank A., (-)
MENCHER, Ely, (1913-)
MENKEN, Frederick Andrew, (1905-1965)
MERRICK, F. Kermit, (1915-1963)
MESSINA, Angelina Rose, (1910-1968)
MILLER, Buford, (1906-1962)
MILLER, Don John, (1919-1961)
MILLER, John Preston, (1923-1961)
MILLER, William D., (1930-1972)

UNITED STATES OF AMERICA 1900 to present Cont'd.

MILLER, William F., (1929-1976)
MILNER, Charles Albert, Jr., (1900-1971)
MITCHELL, James Garland, (1908-1972)
MITCHELL, Kingsley Camden, (1903-1931)
MOORE, Carl Allphin, (1911-1973)
MOORE, James Augustus, (1910-1955)
MOORE, Paul Brian, (1940-)
MORRILL, Junius Hughes, (1913-1961)
MORRIS, Harry Newton, Jr., (1920-1952)
MOSELEY, John Reed, (1900-1959)
MULFORD, John W., (1902-1964)
MULLEN, John Timothy, Jr., (1906-1959)
MURCHISON, Eugene Alexander, III, (1931-1953)
MURPHY, James Kenneth, (1900-1955)
MURRAY, George H., Jr., (1928-1971)
NANCE, James W., (1908-1977)
NASH, Carlton S., (1914-)
NEARHOOD, Clinton Elsworth, (1915-1957)
NEIGHBOURS, Glen Reed, (1917-1961)
NETTERSTROM, Paul Wagner, (1914-1960)
NEUMAN, Leonard Joseph, (1900-1953)
NININGER, Harvey Harlow, (-)
NININGER, Robert D., (-)
NYGREN, Walter Eric, (1907-1971))
O'BRIEN, Francis Eugene, (1916-1968)
O'CONNELL, Daniel Trugott, (1901-1971)
OLTMAN, Roy E., (-1977)
OWEN, William Joseph, (1929-1959)
OWENS, John Edward, (1912-1963)
PABST, Marie B., (1909-1963)
PAINE, William R., (-)
PALMER, Leland L., (1906-1978)
PARKER, Ben Hutchinson, (1902-1969)
PARKS, Roland Dane, (1900-1972)
PEABODY, Frank Elmer, (1914-1958)
PECORA, William Thomas, (1913-1972)
PENNEL, Harry David, (1907-1971)
PETERSON, Arthur Frederick, (1911-1967)
PETTY, Thomas Franklin, (1900-1962)
PICARD, Meredith Dane, (1927-)
PICKLESIMER, Harold Greeley, (1916-1952)
PILCHER, Benjamin Luther, Jr., (1906-1946)
PINSON, William Hamet, Jr., (1919-)
POSTEL, Albert Williams, (1909-1966)
POWELL, Ralph Sterling, (1901-1951)
PRESS, Frank, (1924-)
PRICE, Arthur Sebert, (1904-1948)
PUTNAM, William Clement, (1908-1963)
RABBITT, John Charles, (1907-1957)
RAINWATER, Edward Harriman "Rainey", (1909-1972)
RAMSEY, Thomas Yancey, (1910-1968)
RASMUSSEN, William Charles, (1917-1973)
RAY, Robert Hillyer, (1902-1968)
REA, Henry Carter, (1900-1963)

UNITED STATES OF AMERICA 1900 to present Cont'd.

REED, Eugene C., (1901-1976)
REED, Lyman Coleman, (1900-1970)
REESE, Richard Griffith, (1902-1964)
REICHERT, Stanley Orville, (1912-1970)
REID, William McCormick, (1911-1943)
REIFF, Allan Eugene, (1908-1933)
REISS, John W., (1920-1943)
REYNOLDS, Thomas Emmett, (-1972)
RHOADES, Roger F., (1905-1972)
RICE, Elmer Michael, (1900-1971)
RICH, Charles Irvin, (1918-1975)
RICHARDS, George L., Jr., (1908-1944)
RITTENHOUSE, Gordon, (1910-1974)
ROBERTS, Marion Stanton, (1909-1963)
ROBINSON, G. D., (-)
ROGATZ, Henry, (1902-1974)
ROGERS, Jesse Armstead, (1913-1962)
RONES, Morris, (1922-1965)
ROSEN, George D., (1914-1978)
RUKAS, Justin Matthew, (1910-1963)
SANFORD, Wendell Glenn, (1914-1958)
SAPPINGTON, Chester, (1905-1968)
SAUNDERS, Lowell Waller, (1901-1954)
SCHAIRER, John Frank, (1904-1970)
SCHLAIKJER, Erich Maren, (1905-1972)
SCHMIDT, Karl A. "Gus", (1900-1965)
SCHOLL, Guy Joseph, (1900-1968)
SCHWABENLAND, James Richard, (1926-1968)
SCOPES, John Thomas, (1900-)
SCOTT, Horace Lafayette, (1901-1976)
SCRUTON, Philip Challacombe, (1922-1962)
SHEPARD, Francis P., (-)
SHEPPS, Vincent Chester, (1928-1967)
SHERMAN, Roger Hiram, (1904-1954)
SHIPEK, Carl Joseph, (1917-1969)
SHROCK, Robert Rakes, (1904-)
SIMMONS, Jesse Elmore, (1907-1941)
SIMMONS, Kenneth Albert, (1906-1971)
SIMMONS, Marvin Gene, (1929-)
SIMPSON, George Gaylord, (1902-)
SIMPSON, Stephen Milton, Jr., (1929-)
SMITH, G. Wendell, (1917-1974)
SMITH, Gerald Julian, (1903-1976)
SMITH, Homer James, (1901-1954)
SMITH, Neal Johnstone, (1912-1971)
SMITH, Ned Myron, (1923-1976)
SMITH, Robert Hendee "Smokey", (1916-1946)
SPALDING, George M., (1930-1977)
SPEED, Carleton D., Jr., (1903-1970)
SPRAGUE, Harry Alvah, (1913-1950)
STEELE, Horace C., (1930-1977)
STEHR, Raymond Albert, (1903-1967)
STEIG, Maynard Herman, (1904-1965)
STELFOX, Reichie S., (1904-1932)

UNITED STATES OF AMERICA 1900 to present Cont'd.

STETSON, Henry Crosby, (1900-1955)
STEWART, Duncan, (1905-1969)
STIRTON, Ruben Arthur, (1901-1966)
STONE, A. John F., (1917-1969)
STONE, Robert L., (1912-1969)
STOW, Marcellus Henry, (1902-1957)
STRINGHAM, Bronson F., (1907-1968)
STUART, Duncan, VII, (1905-1969)
STUMM, Erwin Charles, (1908-1969)
SUFFIELD, Harry F., (1916-1966)
SULLINS, M. Joseph, (1930-1972)
SWAIN, James Fulton, (1903-1948)
SWANN, David H., (1915-1968)
SWEENEY, Henry Nunn, (1914-1972)
SZMUC, Eugene Joseph, (1927-1976)
TAGUE, Glenn Charles, (1905-1964)
TALLANT, Roy Leon, (1912-1961)
TANNER, Hugh Alexander, (1904-1954)
TAYLOR, Garvin Lawrence, (1908-1974)
TAYLOR, James Carlton, (1924-1976)
TAYLOR, W. Harlan, (1908-1972)
TETRICK, Paul Roderick, (1920-1956)
THAEMAR, James T., Sr., (1911-1972)
THIEL, Edward, (1928-1962)
THIES, Roland K., (1905-1973)
THOMAS, Harry George, (1929-1976)
THOMAS, Horace David "Bill", (1905-1967)
THOMAS, Leo A., (1917-1965)
THORNTON, Owen Fuller, (1916-1953)
THORSEN, Carl Peter, (1927-1964)
TILLOTSON, Allen Walter, (1905-1964)
TODD, Thomas W., (1927-1969)
TOLER, Henry N., (1901-1970)
TOMPKINS, Joseph D., (1905-1974)
TOWSE, Donald Frederick, (1924-)
TRICE, Edwin Leslie, Jr., (1929-1967)
TURNER, Joseph Harold, (1900-1950)
TYLER, Stanley Allen, (1906-1963)
UNTERMANN, Billie R., (1906-1973)
VANDER HOOF, Vertress Lawrence, (1904-1964)
VEIGEL, Edward C., (1920-1949)
VON ARX, William Stelling, (1916-)
WAGNER, Cary Richard, Jr., (1917-1962)
WAGNER, Walter Richard, (1928-1974)
WALDBAUM, David Robert, (1937-1974)
WALKER, Paul, (1934-1959)
WALLACE, Maurice Hudiburg, (1917-1961)
WALLS, James G., (1907-)
WALTERS, Mason G., (1903-1967)
WANGSNESS, Orrin Jerold, (1917-1967)
WAPPLER, Ed Dickinson, (1908-1937)
WARE, John McKee, (1905-1973)
WARE, Thomas Preston, (1909-1972)
WATERMAN, Willis D., (1925-1968)

UNITED STATES OF AMERICA 1900 to present Cont'd.

WEBB, David Knowlton, Jr., (1928-1973)
WEICHBRODT, Harold Tague, (1921-1966)
WEINZIERL, Laura Lee, (née Lane), (1900-1928)
WEISS, Eriol Joseph, (1918-1967)
WELLES, Samuel Paul, (-)
WESLEY, George Rutherford, (1905-1965)
WETMORE, Alexander, (1886-)
WEXLER, Harry, (1911-1962)
WHETTEN, John T., (-)
WHITAKER, George O., (-)
WHITAKER, Harvey, (1906-1964)
WHITCOMB, Bruce, (1900-1950)
WHITE, George Willard, (1903-)
WIER, Harvey James, Jr., (1924-1950)
WILBERT, Louis Joseph "Slick", Jr., (1919-1953)
WILDER, Newell Morris "Newt", (1908-1946)
WILKINSON, William Donald, (1901-1969)
WILSEY, James Arthur, Jr., (1921-1949)
WILSON, Bernard Wright, (1922-1976)
WILSON, Eugene Jack, (1921-1976)
WILSON, Harry H., (1902-1929)
WILSON, James W., (?1925-1964)
WILSON, Joseph Garland, (1902-1955)
WILSON, Leonard Richard, (1906-)
WILSON, Thomas Carroll, (1906-1969)
WILSON, William Jay, (1922-1971)
WIMBISH, Forrest E., (1900-1972)
WINCHESTER, John Widner, (1929-)
WINES, Donald Bradford, (1902-1958)
WINGETT, Raymond Oscar, (1909-1953)
WINTER, Claud Victor, Jr., (1925-1968)
WOODS, Earl Hazen, (1902-1950)
WRIGHT, Harold Douglas, (1921-1969)
WRIGHT, Jean Davies, (-)
WRIGHT, "Put", (-1962)
YEAGER, Lloyd Irvin, (1900-1938)
YOUNG, Karl Etienne "Happy", (1903-1936)
YOUNGQUIST, Walter, (-)
YULE, Frank Ingalls, (1926-1961)
ZAPP, Alfred Dexter, (1916-1962)
ZERFOSS, Samuel, (1912-1958)
ZIMMERMAN, Charles Collingwood, (1905-1961)

Time unknown

ANGELL, Cyrus D., (-)
JENSEN, David E., (-)
PERHAM, Frank, (-)
SCHORTMANN, Alvin, (-)
SCHORTMANN, Raymond, (-)
THOMPSON, Frederick Oliver, (-)
WOLFE, Caleb Wroe, (-)
YEDLIN, Neal, (-)

WALES

Up to 1800

EVANS, Lewis, (1700-1756)
LHUYD, Edward, (1660-1709)
OWEN, George, (1552-1613)
SMITH, Gilbert Nicholas, Reverend, (1796-1877)
WILLIAMS, Thomas, (1737-1802) [PROSPECTORS, DIVINERS AND MINING ENGINEERS]

1800-1849

DAVIES, David Christopher, (1827-1885)
DAVIES, Thomas, (1837-1892)
DAVIES, William, (1814-1891)
HICKS, Henry, (1837-1899)
HOMFRAY, David, (1822-1893)
HOPKINS, Evan, (?1810-1867)
JEFFREYS, John Gwyn, (1809-1885)
JENKINS, Henry Michael, (1841-1886)
LUXMOORE, Edward Bouverie, (1829-1893)

1850-1899

DAVID, Tannatt William Edgeworth, Sir, (1858-1934)
DAVIES, Arthur Morley, (1869-1959)
DAVIES, David, (1870-1931)
DAVIES, David Farr, (1871-1935)
DAVIES, John Henry, (1887-1971)
DODD, C., (-1928)
EDMONDS, Frederic Harrison, (1898-1965)
JACKSON, James Frederick, (1896-1966)
JEHU, Thomas John, (1871-?1943)
JOHNS, Cosmo, (?1866-1951)
JONES, Owen Thomas, (1878-1967)
JONES, William Richard, (1890-1970)
LEACH, Arthur Leonard, (1869-1957)
LEWIS, Herbert Price, (?1895-1947)
MORGAN, Joseph Bickerton, (1859-1894)
PHILLIPS, David Watkins, (1899-1962)
PUGH, William John, Sir, (1892-1974)
ROBERTS, Robert Davis, (1851-1911)
ROBERTS, Thomas, (1856-1892)
THOMAS, Abram Owen, (1876-1931)
THOMAS, Hugh Hamshaw, (1885-1962)

1900 to present

DIX, Emily, (1904-1973)
HOWARTH, W. E., (-1956)
HOWELLS, Jenkyn Owen, (-1950)
JONES, David Glynn, (1904-1974)
LEWIS, William Vaughan, (1907-1961)
OWENS, Eifion Wyn, (1951-1975)
THOMAS, Trevor Morgan, (?1919-1974)
WILLIAMS, Mervyn, (1934-1970)

YUGOSLAVIA

1800-1849

BRUSINA, Spiridion, (1845-1908)
LIPOLD, Markus Vincenc, (1816-1883)
PANČIČ, Joseph, (-1888)

1850-1899

ANIĆ, Dragutin, (1898-1961)
ATHANASIU, Sava, (-?1946)
BABIĆ, Bogdan, (1891-1961)
GORJANOVIĆ-KRAMBERGER, Dragutin, (1856-1937)
HINTERLECHNER, Karl, (1874-1932)
KIŠPATIĆ, Mišo, (1851-1926)
KOCH, Ferdo, (1874-?1936)
LUKOVIĆ, Milan, (1889-)
MARGETIČ, Marko, (1891-1955)
MARIĆ, Luka, (1899-)
MIHOLIĆ, Stanko, (1891-1960)
MOHOROVIČIĆ, Andrya, (1857-1936)
PETKOVIĆ, Vladamir K., (1873-1934)
PEVALEK, Ivo, (1893-)
POLJAK, Josip, (1882-1962)
RADIMSKY, Wenzel, (-1895)
RADOVANOVIĆ, Svetolik A., (1863-1928)
RAKOVEC, Ivan, (1899-)
SALOPEK, Marijan, (1883-1967)
TUČAN, Fran, (1878-1954)
ŽUKOVIĆ, Jovan M., (1856-1937)

1900 to present

ČELEBRIĆ, Dorde, (1929-1970)
GRČEV, Kočo, (1929-1963)
HERAK, Milan, (1917-)
JURKOVIĆ, Ivan, (1917-)
KOCHANSKY-DEVIDÉ, Vanda, (1915-)
PAVLOVIĆ, Stojan, (1903-)
PETKOVIĆ, Kosta, (1903-)
SIMONOVIĆ, Radivoj, (-1950)
TAJDER, Miroslav, (1909-)
TRIBUSSON, Boris, (1924-1957)
VOLOSHIN, Vadim "Charlie", (1934-1970)
ZIMERMAN, Sam, (1905-1972)

COUNTRY unspecified

1859-1899

PRICE, George McCready, (1870-)

INDEX OF GEOLOGISTS BY COUNTRY WORKED IN

ABYSSINIA

1800-1849

EHRENBERG, Christian Gottfried, (1795-1876)

1850-1899

BREUIL, Henri, Abbé, (1877-1961)
DE REGNY, Paolo Vinassa, (1871-1957)
DOUVILLÉ, Joseph Henri Ferdinand, (1846-1937)
EHRENBERG, Christian Gottfried, (1795-1876)
HATCH, Frederick Henry, (?1864-1932)
MANASSE, Ernesto, (1875-1922)
MARINELLI, Olinto, (1874-1926)
MARKHAM, Clements Robert, Sir, (1830-1916)
PALAZZO, Luigi, (1861-1933)
PORRIO, Cesare, (1865-1940)
PRIOR, George Thurland, (1862-1936)
SABATINI, Venturino, (1879-1922)

1900 to present

ARAMBOURG, Camille, (1885-1969)
BREUIL, Henri, Abbé, (1877-1961)
CALCIATI, Cesare, (1885-1929)
DE REGNY, Paolo Vinassa, (1871-1957)
DOUVILLÉ, Joseph Henri Ferdinand, (1846-1937)
HALL, Thomas Clifford Fitzwilliam, (1886-1956)
HATCH, Frederick Henry, (?1864-1932)
HOFMANN, Walter, (1919-1975)
JOHANSON, Donald C., (-)
LAMARE, Pierre, (1894-1967)
MANASSE, Ernesto, (1875-1922)
MARINELLI, Olinto, (1874-1926)
MARKHAM, Clements Robert, Sir, (1830-1916)
NOWACK, Ernst, (1891-1946)
PALAZZO, Luigi, (1861-1933)
PORRIO, Cesare, (1865-1940)
PRIOR, George Thurland, (1862-1936)
ROWLEY, Alden Bruce, (1888-1956)
SABATINI, Venturino, (1879-1922)
TEILHARD DE CHARDIN, Pierre, Father, S.J., (1881-1955)
TRAGER, Earl A., (1893-1978)
WING, Monta Eldo, (1897-1976)

ADEN <u>See</u> SOUTH YEMEN.

ADMIRALTY ISLANDS (PACIFIC OCEAN)

1850-1899

WICHMANN, Carl Ernst Arthur, (1851-1927)

1900 to present

WICHMANN, Carl Ernst Arthur, (1851-1927)

AFGHANISTAN

1850-1899

COLLINS, Arthur L., (1868-1902)
GRIESBACH, Karl Ludolf, (1847-1907)
GROEBER, Paul Friedrich Carl, (1885-1968)

AFGHANISTAN 1850-1899 Cont'd.

SEWARD, Albert Charles, Sir, (1863-1941)

1900 to present

CLAPP, Frederick Gardner, (1879-1944)
COLLINS, Arthur L., (1868-1902)
DRATH, Adam, (1904-1942)
DVOŘÁK, Arnošt, (1904-)
FRALICH, Charles Eugene, (1902-1964)
GRIESBACH, Karl Ludolf, (1847-1907)
GROEBER, Paul Friedrich Carl, (1885-1968)
LAPPARENT, Albert Félix de, (1905-1975)
LEUTWEIN, Friedrich, (1911-1974)
REA, Henry Carter, (1900-1963)
SEWARD, Albert Charles, Sir, (1863-1941)

AFRICA

1900 to present

CALAME, Jean Jacques, (1923-1973)
GOETEL, Walery, (1889-1972)

AFRICA (unspecified)

Up to 1800

RENNELL, James, Major, (1742-1830)

1800-1849

RENNELL, James, Major, (1742-1830)

1850-1899

ATTWOOD, George, (1845-1912)
BORSARI, Ferdinando, (1858-1891)
GREGORY, William King, (1876-1970)
KEYES, Charles Rollin, (1864-1942)
LAUNAY, Louis Alphonse Auguste de, (1860-1938)
MELZI, Gilberto, Conte, (1868-1899)
NEWTON, Richard Bullen, (1850-1926)
RAISIN, Catherine Alice, (1855-1945)
SCHWEINFURTH, Georg, (1836-1925)

1900 to present

ATTWOOD, George, (1845-1912)
BERNOUILLI, Walter, (1885-1946)
BIRAUD, Francois, (1899-1946)
CADY, Richard Carlysle, (1908-1943)
FURON, Raymond, (1898-)
GORDON, Samuel George, (1897-1952)
GREGORY, William King, (1876-1970)
HAZZARD, John Charles, (1903-1973)
HOURCQ, Victor, (1901-1961)
JABLONSKI, Eugene, (1892-1975)
JOHNSON, Norman Lee, (1926-1970)
JOUBIN, Francis Renault, (1911-)
KEYES, Charles Rollin, (1864-1942)
KOBER, Leopold, (1883-1970)
LAUNAY, Louis Alphonse Auguste de, (1860-1938)
MARCHESINI, Enrico, (1911-1968)

AFRICA (unspecified) 1900 to present Cont'd.

MIGAUX, Léon, (1897-1914)
NEWTON, Richard Bullen, (1850-1926)
RAISIN, Catherine Alice, (1855-1945)
RAMDOHR, Paul, (1890-)
SAPPINGTON, Chester, (1905-1968)
SCHWEINFURTH, Georg, (1836-1925)
TYRRELL, George Walter, (1883-1961)

AFRICA (unspecified) See also CENTRAL AFRICA; EAST AFRICA; NORTH AFRICA; WEST AFRICA.

ALBANIA

1800-1849

VIQUESNEL, Auguste, (1800-1867)

1850-1899

CVICIĆ, Jovan, (1865-1927)
DE REGNY, Paolo Vinassa, (1871-1957)
LIEBUS, Adalbert, (1876-1945)
MARTELLI, Alessandro, (1877-1934)
NAGYSUR, Hugo Böckh von, (1874-1931)
NOPCSA VON FELSŐ-SZILVÁS, Ferenc, baron, (1877-1933)
PORRIO, Cesare, (1865-1940)
VIQUESNEL, Auguste, (1800-1867)
WADE, Arthur, (1878-1951)

1900 to present

BOURCART, Jacques, (1891-1965)
CVICIĆ, Jovan, (1865-1927)
DE REGNY, Paolo Vinassa, (1871-1957)
HEINSHEIMER, Georg J., (1887-1970)
INEICHEN, Giovanni, (-1952)
LIEBUS, Adalbert, (1876-1945)
MARTELLI, Alessandro, (1877-1934)
MAYO, Herbert Theodore, (1889-1965)
NAGYSUR, Hugo Böckh von, (1874-1931)
NOPCSA VON FELSŐ-SZILVÁS, Ferenc, baron, (1877-1933)
NOWACK, Ernst, (1891-1946)
PAPP, Simon, (1886-1970)
PORRIO, Cesare, (1865-1940)
PREMIK, Józef, (1890-1963)
WADE, Arthur, (1878-1951)
WASHINGTON-GRAY, Kenneth, (?1899-1963)
ZUBER, Stanislaw, (1893-1947)

ALEUTIAN ISLANDS (NORTH PACIFIC OCEAN)

1800-1849

FURUHJELM, Enoch Hjalmar, (1823-1886)

1850-1899

EAKLE, Arthur Starr, (1862-1931)
FURUHJELM, Enoch Hjalmar, (1823-1886)

ALEUTIAN ISLANDS (NORTH PACIFIC OCEAN) Cont'd.

1900 to present
EAKLE, Arthur Starr, (1962-1931)
FISHER, Bernard, (1918-1946)

ALGERIA

Up to 1800
MAILLET, Benoît de, (1656-1738)

1800-1849
DESHAYES, Gérard Paul, (1796-1875)
MARÈS, Paul, (1826-1900)
POMEL, Nicolas Auguste, (1821-1898)

1850-1899
BIRLEY, Caroline, (1851-1907)
BLAYAC, Joseph, (1865-1936)
BRALY, Adrien, (1864-1927)
BREUIL, Henri, Abbé, (1877-1961)
CHUDEAU, René, (1864-1921)
DELÉPINE, Gaston-Georges, Monseigneur, (1878-1963)
DESHAYES, Gérard Paul, (1796-1875)
DOUVILLÉ, Joseph Henri Ferdinand, (1846-1937)
FOOTE, Albert E., (1846-1895)
FOURNEL, Henri, (-)
GENTIL, Louis, (1868-1925)
LOMAS, Joseph, (1860-1908)
MACLAREN, James Malcolm, (1873-1935)
MARÈS, Paul, (1826-1900)
MAW, George, (1832-1912)
PERON, Alphonse, (1834-1908)
POMEL, Nicolas Auguste, (1821-1898)
ROMAN, Albert Frédéric, (1871-1943)
SAUVAGE, Henri Émile, (1842-1917)
THOMAS, Philippe, (1843-1910)
WADE, Arthur, (1878-1951)

1900 to present
ARAMBOURG, Camille, (1885-1969)
BEEKLEY, Albert Leon, (1883-1952)
BIRAUD, François, (1899-1936)
BIRLEY, Caroline, (1851-1907)
BLAYAC, Joseph, (1865-1936)
BOURCART, Jacques, (1891-1965)
BOUREAU, Edouard Léon François, (1913-)
BRALY, Adrien, (1864-1927)
BRAMWYCH, Michael, (1911-1974)
BREUIL, Henri, Abbé, (1877-1961)
CAIRE, André, (1923-)
CHAUVE, Pierre, (1930-)
CHUDEAU, René, (1864-1921)
DALLONI, Marius Gustave, (1880-1959)
DELEAU, Paul Camille, (1904-)

ALGERIA 1900 to present Cont'd.

DELÉPINE, Gaston-Georges, Monseigneur, (1878-1963)
DOUVILLÉ, Joseph Henri Ferdinand, (1846-1937)
DURAND DELGA, Michel, (1923-)
FALLOT, Paul, (1889-1960)
GENTIL, Louis, (1868-1925)
GLANGEAUD, Louis, (1903-)
HOFFSTETTER, Robert, (1908-)
HUGHES, Richard, (1889-1969)
JÉRÉMINE, Elisabeth, (née Tschernaieff), (1879-1964)
JOLEAUD, Léonce, (1880-1938)
KILIAN, François Théodore Conrad, (1898-1950)
KOMOROWICZ, Maurycy, (1881-1923)
LAPPARENT, Albert Félix de, (1905-1975)
LEHNER, Ernst J., (1895-1968)
LEUTENEGGER, Willy Otto, (1903-1972)
LOMAS, Joseph, (1860-1908)
MACLAREN, James Malcolm, (1873-1935)
MARÈS, Paul, (1826-1900)
MAW, George, (1832-1912)
MEYENDORFF, André de, (1918-1942)
MIGAUX, Léon, (1897-1914)
MILLOT, Georges, (1917-)
PERON, Alphonse, (1834-1908)
PIVETEAU, Jean, (1898-)
RENAULT, Philippe, (-)
ROMAN, Albert Frédéric, (1871-1943)
ROUBAULT, Marcel Edouard, (1905-1974)
SARJEANT, William Antony Swithin, (1935-)
SAUVAGE, Henri Émile, (1842-1917)
SENN, Alfred, (1899-1949)
SMITH, G. Wendell, (1917-1974)
STREIFF, Viktor, (1909-1973)
THOMAS, Philippe, (1843-1910)
TRÜMPY, Daniel, (1893-1971)
TSCHIKHACHEV, Peter A., (-)
VINCIENNE, Henri, (1898-1965)
WADE, Arthur, (1878-1951)
WALLACE, Maurice Hudiburg, (1917-1961)
WARING, Gerald Ashley, (1883-1971)

ANDAMAN ISLANDS (INDIAN OCEAN)
1850-1899
OLDHAM, Richard Dixon, (1858-1936)
STOLICZKA, Ferdinand, (1838-1874)

1900 to present
OLDHAM, Richard Dixon, (1858-1936)

ANGOLA
1850-1899
BREUIL, Henri, Abbé, (1877-1961)
CHOFFAT, Paul, (1849-1919)
GREGORY, John Walter, (1864-1932)

ANGOLA Cont'd.

1900 to present
BORGES, Alexandre, (1898-1955)
BREUIL, Henri, Abbé, (1877-1961)
CHOFFAT, Paul, (1849-1919)
DAVIS, Thornton, (1892-1969)
FINCH, Elmer Harrison, (1883-1953)
FREIRE DE ANDRADE, Carlos Benot, (1893-1956)
GARRETT, Melvin Miller "Dutch", (1894-1969)
GREGORY, John Walter, (1864-1932)
HODGE, Edwin Thomas, (1887-1970)
HOLMES, Arthur, (1890-1965)
MASON, Shirley Lowell, (1893-1975)
NARAMORE, Chester "Chet", (1879-1961)
SANDIDGE, John Roy, (1897-1976)
TRÜMPY, Daniel, (1893-1971)
WALLACE, Maurice Hudiburg, (1917-1961)

ANGUILLA (WEST INDIES)
1850-1899
COTTEAU, Gustave Honoré, (1818-1894)

ANTARCTICA
1800-1849
EIGHTS, James, (1798-1882)
HOOKER, Joseph Dalton, Sir, (1817-1911)

1850-1899
ANDERSSON, Johann Gunnar, (1874-1960)
ARCTOWSKI, Henryk Bronislaw, (1871-1958)
DAVID, Tannatt William Edgeworth, Sir, (1858-1934)
EIGHTS, James, (1798-1882)
HOOKER, Joseph Dalton, Sir, (1817-1911)
PRIOR, George Thurland, (1862-1936)
SIMPSON, George Clarke, Sir, (1878-1965)
WIMAN, Carl Johan Josef Ernst, (1867-1944)

1900 to present
ANDERSSON, Johann Gunnar, (1874-1960)
ARCTOWSKI, Henryk Bronislaw, (1871-1958)
COTTON, Leo Arthur, (1883-1963)
COX, Leslie Reginald, (1897-1965)
CROFT, William Noble, (1915-1953)
DAVID, Tannatt William Edgeworth, Sir, (1858-1934)
DOUGLAS, George Vibert, (1892?-1958)
HEINSHEIMER, Georg J., (1887-1970)
HOLTEDAHL, Olaf, (1885-1975)
HOOKER, Joseph Dalton, Sir, (1817-1911)
JOHNSON, Charles George, (1914-1969)

ANTARCTICA 1900 to present Cont'd.

JOYCE, John Raymond Foggan, (?1907-1963)
KITCHING, James, (-)
MADIGAN, Cecil Thomas, (1889-1946)
MAWSON, Douglas, Sir, (1882-1958)
NAGATA, Takesi, (1913-)
ŌURA, Hirobumi (1921-1969)
PRIESTLEY, Raymond Edward, Sir, (1886-1974)
PRIOR, George Thurland, (1862-1936)
QUENSEL, Percy Dudgeon, (1881-1966)
REECE, Alan W., (1921-1960)
SCHYTT, Valter, (1919-)
SIMPSON, George Clarke, Sir, (1878-1965)
SIMPSON, George Gaylord, (1902-)
SMITH, Jeremy, (1929-1959)
SOBRAL, José María, (-1961)
SPATH, Leonard Frank, (1882-1957)
STILLWELL, Frank Leslie, (1888-1963)
STUART, Duncan, VII, (1905-1969)
TAYLOR, Griffith, (1881-1963)
THIEL, Edward, (1928-1962)
THOMSON, James Allan, (1881-1928)
TYRRELL, George Walter, (1883-1961)
WALKER, Paul, (1934-1959)
WEXLER, Harry, (1911-1962)
WIMAN, Carl Johan Josef Ernst, (1867-1944)
WORDIE, James Mann, Sir, (?1890-1962)

ANTIGUA (WEST INDIES)

1850-1899

BEEBY-THOMPSON, Arthur, (1873-1968)
GUPPY, Robert John Lechmere, (1836-1916)
THOMAS, Abram Owen, (1876-1931)

1900 to present

BEEBY-THOMPSON, Arthur, (1873-1968)
GUPPY, Robert John Lechmere, (1836-1916)
THOMAS, Abram Owen, (1876-1931)

ARABIA See SAUDI ARABIA.

ARCTIC (unspecified)

1800-1849

KANE, Elisha Kent, (1820-1857)
RICHARDSON, John, Sir, (1787-1865)
SALTER, John William, (1820-1869)

1850-1899

ANDERSSON, Johann Gunnar, (1874-1960)

ARCTIC (unspecified) 1850-1899 Cont'd.

CARPENTER, Philip Herbert, (1852-1891)
DE GEER, Gerard Jacob, Baron, (1858-1943)
EMERSON, Benjamin Kendall, (1843-1932)
HAMBERG, Axel, (1863-1933)
HOBBS, William Herbert, (1864-1953)
KANE, Elisha Kent, (1820-1857)
KITTL, Ernst Anton Leopold, (1854-1913)
KJELLSTRÖM, Carl Johan Otto, (1855-1913)
NORDENSKIÖLD, Nils Adolf Erik, (1832-1901)
RICHARDSON, John, Sir, (1787-1865)
SALTER, John William, (1820-1869)

1900 to present
ANDERSSON, Johann Gunnar, (1874-1960)
CARPENTER, Philip Herbert, (1852-1891)
CHARLESWORTH, John Kaye, (1889-)
DE GEER, Gerard Jacob, Baron, (1858-1943)
EMERSON, Benjamin Kendall, (1843-1932)
FOLEY, Edward J., (1908-1971)
HAMBERG, Axel, (1863-1933)
HOBBS, William Herbert, (1864-1953)
JABLONSKI, Eugene, (1892-1953)
KITTL, Ernst Anton Leopold, (1854-1913)
KJELLSTROM, Carl Johan Otto, (1855-1913)
LOEWE, Fritz, (1895-1974)
NODENSKIOLD, Nils Adolf Erik, (1832-1901)
PAPANIN, Ivan Dmitrievich "Vanya", (1895-)

ARGENTINA
Up to 1800
HELMS, Anton Zacharias, (1751-1803) [PROSPECTORS, DIVINERS AND MINING ENGINEERS]
MAWE, John, (1764-1829)

1800-1849
BURMEISTER, Karl Hermann Konrad, (1807-1892)
DARWIN, Charles Robert, (1809-1882)
FORBES, David, (1828-1876)
HELMS, Anton Zacharias, (1751-1803) [PROSPECTORS, DIVINERS AND MINING ENGINEERS]
MAWE, John, (1764-1829)
MIERS, John, (1789-1879)
MUÑIZ, Francisco Javier, (1795-1871)
NICOL, Germain Louis, (1922-1973)
ORBIGNY, Alcide Charles Victor Dessalines d', (1802-1857)
PARISH, Woodbine, Sir, (1796-1882)
RICKARD, Francis Ignacio, (-)

1850-1899
AMEGHINO, Florentino, (1854-1911)
BACKLUND, Helge Götrik, (1878-1958)
BODENBERGER, Wilhelm, (1857-1941)
BONARELLI, Guido, (1871-1951)
BURCKHARDT, Carl, (1869-1935)
BURMEISTER, Karl Hermann Konrad, (1807-1892)

ARGENTINA 1850-1899 Cont'd.

BUTTGENBACH, Henri, (1874-1964)
COURTY, Georges, (1875-1953)
DARWIN, Charles Robert, (1809-1882)
DUSÉN, Per, (1855-1926)
FORBES, David, (1828-1876)
GOTTSCHE, Carl Christian, (1855-1909)
HATCHER, John Bell, (1861-1904)
HAUTHAL, Rudolf, (1854-1928)
HELMS, Anton Zacharias, (1751-1803) [PROSPECTORS, DIVINERS AND MINING ENGINEERS]
HENNEN, Ray Vernon, (1875-1958)
HUENE, Friedrich von, Freiherr, (1875-1969)
KERR, John Graham, (1869-1957)
KLOCKMANN, Friedrich Ferdinand Hermann, (1858-1937)
KURTZ, Fritz, (1854-1920)
KYLE, John Joseph Jolly, called Juan José, (1838-1922)
LOOMIS, Frederic Brewster, (1873-1937)
LOVISATO, Domenico, (1842-1916)
MARTIN, Handel Tongue, (1862-1931)
MERCERAT, Alcides, (-)
MIERS, John, (1789-1879)
MORENO, Francisco Josué Pascasio, (1852-1919)
MUÑIZ, Francisco Javier, (1795-1871)
NICOL, Germain Louis, (1922-1973)
NORDENSKIÖLD, Nils Otto Gustaf, (1869-1928)
ORBIGNY, Alcide Charles Victor Dessalines d', (1802-1857)
PARISH, Woodbine, Sir, (1796-1882)
PAULCKE, Wilhelm, (1873-1949)
REICHERT, Friedrich, (1878-)
RICKARD, Francis Ignacio, (-)
ROTH, Santiago, (1850-1924)
SCOTT, William Berryman, (1858-1947)
SIEMIRADZKI, Józef, (1858-1933)
SONNEVELD, Jan, (1877-195)
STEINMANN, Gustav, (1856-1929)
STELZNER, Alfred Wilhelm, (1840-1895)
TANNHÄUSER, Felix, (1874-1924)
VALENTIN, Jean, (1877-1898)
WEYENBERGH, Hendrik, (1842-1885)
WIELAND, George Reber, (1865-1953)
WILLIS, Bailey, (1857-1949)
WINDHAUSEN, Anselm, (1882-1932)
WOOLNOUGH, Walter George, (1876-1958)

1900 to present

ALDREDGE, Robert Frank, (1895-1961)
AMEGHINO, Florentino, (1854-1911)
BACKLUND, Helge Götrik, (1878-1958)
BARD, Richards J., (1900-1977)
BEDER, Robert, (1885-1930)
BODENBERGER, Wilhelm, (1857-1941)
BONARELLI, Guido, (1871-1951)
BONTZ, Conrad K., (1894-1927)
BORRELLO, Angel V., (1918-1971)

ARGENTINA 1900 to present Cont'd.

BURCKHARDT, Carl, (1869-1935)
BUTTGENBACH, Henri, (1874-1964)
CARR, William L., (1919-1975)
CASTELLANOS, Alberto, (1896-1968)
CATHCART, Stanley Holman, (1889-1953)
CATTOI, Noemí Violeta, (1911-1965)
COURTY, Georges, (1875-1953)
DIXON, Henry George Dacres, (1887-1947)
DOLGOPOL DE SÁEZ, Mathilde, (1901-1957)
DRESSER, Myron Amos, (1894-1973)
DURHAM, Charles Albert, (1910-1954)
DUSÉN, Per, (1855-1926)
FERUGLIO, Egidio, (1897-1954)
FOSSA-MANCINI, Enrico, (1884-1950)
FOWLER, Charles Worth, Jr., (1899-1948)
FRENGUELLI, Joaquín, (1883-1958)
GOTTSCHE, Carl Christian, (1855-1909)
HARRINGTON, George Leavitt, (1883-1972)
HARRINGTON, Horacio Jaime, (1910-1973)
HATCHER, John Bell, (1861-1904)
HAUTHAL, Rudolf, (1854-1928)
HAYES, Albert Orion, (1882-1965)
HEIM, Arnold, (?1882-1965)
HEINSHEIMER, Georg J., (1887-1970)
HENNEN, Ray Vernon, (1875-1958)
HEROLD, Stanley Carrollton, (1883-1958)
HOTZ, Walter, (1884-1958)
HUBBARD, Bela, (1890-1959)
HUBMAN, Carl Wesley, (1905-1947)
HUENE, Friedrich von, Freiherr, (1875-1969)
JONES, Verner E., (1905-1977)
JURADO, Martín Doello, (1884-1948)
KERR, John Graham, (1869-1957)
KLOCKMANN, Friedrich Ferdinand Hermann, (1858-1937)
KRAGLIEVICH, Lucas, (1886-1932)
KUIPER, Willem Nicolaas, (1891-1952)
KURTZ, Fritz, (1854-1920)
KYLE, John Joseph Jolly, called Juan José, (1838-1922)
LEANZA, Armando F., (1919-1975)
LJUNGNER, Erik, (1892-1954)
LOOMIS, Frederic Brewster, (1873-1937)
LOVISATO, Domenico, (1842-1916)
MACHATSCHEK, Fritz, (-1957)
MARTIN, Handel Tongue, (1862-1931)
MENÉNDEZ, Carlos Alberto, (1921-1976)
MEYER, Edouard, (1899-1974)
MOORE, Gilbert Parvin, (1898-1970)
MORENO, Francisco Josué Pascasio, (1852-1919)
MORGAN, Eilir Evans, (?1911-1969)
MORTOLA, Edelmira Ines, (-1973)
NORDENSKIÖLD, Nils Otto Gustaf, (1869-1928)
NOTESTEIN, Frank B., (1885-1973)
PARKER, Ben Hutchinson, (1902-1969)
PASTORE, Franco, (1885-1958)

ARGENTINA 1900 to present Cont'd.

PAULCKE, Wilhelm, (1873-1949)
PEIRANO, Abel Antonio, (1896-1969)
PEMBERTON, John Roy "Bill", (1884-1968)
PENCK, Walther, (1888-1923)
POLANSKI, Jorge, (1892-1975)
POZZI, Santiago, (-1929)
PRICHARD, Hesketh Vernon Hesketh, (1876-1922)
QUENSEL, Percy Dudgeon, (1881-1966)
REED, Lyman Coleman, (1900-1970)
REICHERT, Friedrich, (1878-)
REYES, Felix Celso, (1924-1972)
ROMER, Alfred Sherwood, (1894-1973)
ROTH, Santiago, (1850-1924)
SCHAUB, Hans Peter, (1912-1965)
SCHMITTOU, Maurice B., (1899-1929)
SCOTT, William Berryman, (1858-1947)
SHAW, Everett Simeon, (1885-1955)
SIEMIRADZKI, Józef, (1858-1933)
SIMPSON, George Gaylord, (1902-)
SOBRAL, José María, (-1961)
SONNEVELD, Jan, (1877-195)
STEBINGER, Eugene, (1883-1951)
STEINMANN, Gustav, (1856-1929)
SULLINS, M. Joseph, (1930-1972)
SUTTON, Frederick A., (1894-1950)
TANNHÄUSER, Felix, (1874-1924)
TEZON, Roberto Vitelmo, (1921-1973)
TOURNOUËR, André, (-?1930)
TRÜMPY, Daniel, (1893-1971)
WALTON, John, (-1971)
WHITEHEAD, Walter Lucius, (1891-1969)
WIELAND, George Reber, (1865-1953)
WILLIS, Bailey, (1857-1949)
WINDHAUSEN, Anselm, (1882-1932)
WOOLNOUGH, Walter George, (1876-1958)

ARMENIAN SSR (USSR)
1850-1899
ARZRUNI, Andreas, (1847-1898)
MORGAN, Jacques Jean Marie de, (1857-1924)
VALENTIN, Jean, (1877-1898)

1900 to present
MORGAN, Jacques Jean Marie de, (1857-1924)

ASCENSION ISLAND (SOUTH ATLANTIC OCEAN)
1800-1849
DARWIN, Charles Robert, (1809-1882)

ASCENSION ISLAND (SOUTH ATLANTIC OCEAN) Cont'd.

1850-1899
DARWIN, Charles Robert, (1809-1882)
GILL, David, Sir, (1843-1914)

1900 to present
GILL, David, Sir, (1843-1914)

ASIA <u>See also</u> EASTERN ASIA; MIDDLE EAST; NEAR EAST; SOUTHEAST ASIA.

ASIA (unspecified)
Up to 1800
RENNELL, James, Major, (1742-1830)

1800-1849
RENNELL, James, Major, (1742-1830)

1850-1899
KEYES, Charles Rollin, (1864-1942)
LAUNAY, Louis Alphonse Auguste de, (1860-1938)

1900 to present
CADY, Richard Carlysle, (1908-1943)
FURON, Raymond, (1898-)
KEYES, Charles Rollin, (1864-1942)
LAUNAY, Louis Alphonse Auguste de, (1860-1938)
PENCK, Walther, (1888-1923)

ASSAM <u>See</u> INDIA.

AUSTRALIA
NEW SOUTH WALES
1800-1849
CLARKE, William Branwhite, Reverend, (1798-1878)
KEENE, William, (1798-1872)
LEICHHARDT, Friedrich Wilhelm Ludwig, (1813-1848)
STUTCHBURY, Samuel, (1798-1859)
SÜSSMILCH, Carl Adolf, (?1874-1946)

1850-1899
ANDERSON, William, (1860-1915)
ANDREWS, Ernest Clayton, (1870-1948)
BROWN, Charles Barrington, (1839-1917)
BULLEN, Robert Ashington, Reverend, (1850-1912)
CLARKE, William Branwhite, Reverend, (1798-1878)
DAVID, Tannatt William Edgeworth, Sir, (1858-1934)
ETHERIDGE, Robert Jr., (1847-1920)
KEENE, William, (1798-1872)
MACLAREN, James Malcolm, (1873-1935)
MARKS, Edward Seaborn, (?1875-1950)
STRZELECKI, Pawel Edmund, Hrabia, (1797-1873)
STUTCHBURY, Samuel, (1798-1859)
SÜSSMILCH, Carl Adolf, (?1874-1946)
ULRICH, George Henry Frederick, (1830-1900)
WILKINSON, Charles Smith, (1843-1891)

AUSTRALIA (NEW SOUTH WALES) 1850-1899 Cont'd.

WOOLNOUGH, Walter George, (1876-1958)

1900 to present

ANDERSON, William, (1860-1915)
ANDREWS, Ernest Clayton, (1870-1948)
BROWN, Charles Barrington, (1839-1917)
BULLEN, Keith Edward, (1906-1976)
BULLEN, Robert Ashington, Reverend, (1850-1912)
CARROLL, Dorothy, (1907-1970)
DAVID, Tannatt William Edgeworth, Sir, (1858-1934)
ETHERIDGE, Robert Jr., (1847-1920)
MACLAREN, James Malcolm, (1873-1935)
MARKS, Edward Seaborn, (?1875-1950)
OSBORNE, George Davenport, (1899-1955)
PHILLIPS, David Watkins, (1899-1962)
RAGGATT, Harold George, Sir, (1900-1968)
RAMDOHR, Paul, (1890-)
TAYLOR, Griffith, (1881-1963)
THOMAS, David, (1880-1934)
TILLYARD, Robin John, (1881-1937)
WOOLNOUGH, Walter George, (1876-1958)

AUSTRALIA
NORTHERN TERRITORY
1850-1899

WOOLNOUGH, Walter George, (1876-1958)

1900 to present

MADIGAN, Cecil Thomas, (1889-1946)
TAYLOR, Griffith, (1881-1963)
WOOLNOUGH, Walter George, (1876-1958)

AUSTRALIA
QUEENSLAND
1800-1849

APLIN, Christopher d'Oyley Hale, (1819-1875)
DUNSTAN, Benjamin, (1864-1933)
LEICHHARDT, Friedrich Wilhelm Ludwig, (1813-1848)
SÜSSMILCH, Carl Adolf, (?1874-1946)

1850-1899

APLIN, Christopher d'Oyley Hale, (1819-1875)
DAINTREE, Richard, (1831-1878)
DUNSTAN, Benjamin, (1864-1933)
ETHERIDGE, Robert Jr., (1847-1920)
HERON-ALLEN, Edward, (1861-1943)
JACK, Robert Logan, (1845-1921)
LINDON, Edward Byron, (1860-1891)
MACLAREN, James Malcolm, (1873-1935)
SÜSSMILCH, Carl Adolf, (?1874-1946)
SWEET, George, (-?1920)

1900 to present

BRYAN, Walter Heywood, (1891-1966)

AUSTRALIA (QUEENSLAND) 1900 to present Cont'd.

ETHERIDGE, Robert Jr., (1847-1920)
HERON-ALLEN, Edward, (1861-1943)
HILL, Dorothy, (-)
JACK, Robert Logan, (1845-1921)
MACLAREN, James Malcolm, (1873-1935)
REESE, Richard Griffith, (1902-1964)
SUGDEN, Walter "Bill", (1923-1977)
SWEET, George, (-?1920)

Time unknown
RANDS, William Henry, (-)

AUSTRALIA
SOUTH AUSTRALIA
1800-1849
NICOL, Germain Louis, (1922-1973)
SELWYN, Alfred Richard Cecil, (1824-1902)

1850-1899
JACK, Robert Lockhart, (1878-1964)
NICOL, Germain Louis, (1922-1973)
SELWYN, Alfred Richard Cecil, (1824-1902)
TATE, Ralph, (1840-1901)
WOODWARD, Harry Page, (1858-1917)
WOOLNOUGH, Walter George, (1876-1958)

1900 to present
BULLEN, Keith Edward, (1906-1976)
EDWARDS, Austin Burton, (1909-1960)
JACK, Robert Lockhart, (1878-1964)
MADIGAN, Cecil Thomas, (1889-1946)
MAWSON, Douglas, Sir, (1882-1958)
MILES, Keith Rodney, (1915-1972)
SELWYN, Alfred Richard Cecil, (1824-1902)
STIRTON, Ruben Arthur, (1901-1966)
TATE, Ralph, (1840-1901)
TILLYARD, Robin John, (1881-1937)
WOODWARD, Harry Page, (1858-1917)
WOOLNOUGH, Walter George, (1876-1958)

AUSTRALIA
TASMANIA
1800-1849
CLARKE, William Branwhite, Reverend, (1798-1878)
SELWYN, Alfred Richard Cecil, (1824-1902)

1850-1899
CLARKE, William Branwhite, Reverend, (1798-1878)
GOULD, Charles, (-1895)
PETTERD, William Frederick, (1849-1910)
SELWYN, Alfred Richard Cecil, (1824-1902)
TWELVETREES, William Harper, (1848-1919)

AUSTRALIA (TASMANIA) Cont'd.

1900 to present
PETTERD, William Frederick, (1849-1910)
SELWYN, Alfred Richard Cecil, (1824-1902)
TAYLOR, Griffith, (1881-1963)
TWELVETREES, William Harper, (1848-1919)

AUSTRALIA
VICTORIA
1800-1849
APLIN, Christopher d'Oyley Hale, (1819-1875)
HOPKINS, Evan, (?1810-1867)
McCOY, Frederick, Sir, (1823-1899)
MÜLLER, Ferdinand von, (1825-1896)
SELWYN, Alfred Richard Cecil, (1824-1902)

1850-1899
APLIN, Christopher d'Oyley Hale, (1819-1875)
CHAPMAN, Frederick, (1864-1943)
DAINTREE, Richard, (1831-1878)
ETHERIDGE, Robert Jr., (1847-1920)
HOPKINS, Evan, (?1810-1867)
KITSON, Albert Ernest, Sir, (1868-1937)
LINDON, Edward Byron, (1860-1891)
McCOY, Frederick, Sir, (1823-1899)
MÜLLER, Ferdinand von, (1825-1896)
SELWYN, Alfred Richard Cecil, (1824-1902)
SKEATS, Ernest Willington, (1875-1953)
ULRICH, George Henry Frederick, (1830-1900)
WILKINSON, Charles Smith, (1843-1891)

1900 to present
BAKER, George, (1908-1975)
BULLEN, Keith Edward, (1906-1976)
CHAPMAN, Frederick, (1864-1943)
EDWARDS, Austin Burton, (1909-1960)
ETHERIDGE, Robert Jr., (1847-1920)
KITSON, Albert Ernest, Sir, (1868-1937)
LOEWE, Fritz, (1895-1974)
SELWYN, Alfred Richard Cecil, (1824-1902)
SINGLETON, F. A., (-1947)
SKEATS, Ernest Willington, (1875-1953)

AUSTRALIA
WESTERN AUSTRALIA
1800-1849
GREGORY, Augustus Charles, (1819-1905)
GREGORY, Joshua William, (-1850)

1850-1899
GREGORY, Augustus Charles, (1819-1905)
GREGORY, Joshua William, (-1850)
HARDMAN, Edward Townley, (1845-1887)

AUSTRALIA (WESTERN AUSTRALIA) 1800-1849 Cont'd.

MACLAREN, James Malcolm, (1873-1935)
NEWELL ARBER, Edward Alexander, (1870-1918)
SARGENT, Harry Lionel, (1871-1955)
SIMPSON, Edward Sydney, (1875-1939)
WARD, Leonard Keith, (1879-1964)
WOODWARD, Harry Page, (1858-1917)
WOOLNOUGH, Walter George, (1876-1958)

1900 to present
CARROLL, Dorothy, (1907-1970)
GLAUERT, Ludwig, (1879-1963)
MACLAREN, James Malcolm, (1873-1935)
McKINSTRY, Hugh Exton, (1896-1961)
MILES, Keith Rodney, (1915-1972)
NEWELL ARBER, Edward Alexander, (1870-1918)
SARGENT, Harry Lionel, (1871-1955)
SIMPSON, Edward Sydney, (1875-1939)
SPATH, Leonard Frank, (1882-1957)
THOMSON, James Allan, (1881-1928)
WARD, Leonard Keith, (1879-1964)
WILMOTH, Leslie James, (1881-1963)
WOODWARD, Harry Page, (1858-1917)
WOOLNOUGH, Walter George, (1876-1958)

AUSTRALIA
STATE unspecified
Up to 1800
NISSER, Pedro, (1799-1878) [PROSPECTORS, DIVINERS AND MINING ENGINEERS]

1800-1849
DANA, James Dwight, (1813-1895)
DARWIN, Charles Robert, (1809-1882)
DIEFFENBACH, Ernst, (1811-1855)
ETTINGSHAUSEN, Constantin von, Freiherr, (1826-1897)
GREY, George, Sir, (1812-1898)
HARGRAVES, Edward Hammond, (1816-1891)
HOOKER, Joseph Dalton, Sir, (1817-1911)
JUKES, Joseph Beete, (1811-1869)
MACLEAY, William Sharp, (1792-1865)
MENGE, Johann, (-)
NISSER, Pedro, (1799-1878) [PROSPECTORS, DIVINERS AND MINING ENGINEERS]
PARKER, William Kitchen, (1823-1890)
ROSALES, Henry, (?1822-1916)
SALTER, John William, (1820-1869)
STEPHENS, William John, (1829-1890)

1850-1899
ANDERSON, Charles, (1876-1944)

AUSTRALIA (STATE unspecified) 1850-1899 Cont'd.

BOWEN, Charles Franklin, (1871-1953)
BRAGG, William Henry, Sir, (1862-1942)
BROWN, Henry Yorke Lyell "Stoney", (1844-1928)
CLIFFORD, William Hugh, Right Honourable, (1858-1943)
COLLINS, Henry Francis, (?1864-1924)
COTTEAU, Gustave Honoré, (1818-1894)
CRICK, George Charles, (1856-1917)
DANA, James Dwight, (1813-1895)
DARWIN, Charles Robert, (1809-1882)
DAVIS, William Morris, (1850-1934)
DEANE, Henry Jr., (-1924)
DIEFFENBACH, Ernst, (1811-1855)
DRUMMOND, Henry, (1851-1897)
DUNN, Edward John, (1844-1937)
ETTINGSHAUSEN, Constantin von, Freiherr, (1826-1897)
FEISTMANTEL, Otokar, (1848-1891)
GREGORY, John Walter, (1864-1932)
GREY, George, Sir, (1812-1898)
GUPPY, Robert John Lechmere, (1836-1916)
HALL, Thomas Sergeant, (1858-1915)
HANNAN, Patrick, (1842-1925) [PROSPECTORS, DIVINERS AND MINING ENGINEERS]
HARGRAVES, Edward Hammond, (1816-1891)
HENDERSON, James Alexander Leo, (?1874-1940)
HOFFMANN, George Christian, (1837-1917)
HOOKER, Joseph Dalton, Sir, (1817-1911)
HOWCHIN, Walter, Reverend, (1845-1937)
HOWITT, Alfred William, (1830-1908)
JACK, Robert Lockhart, (1878-1964)
JENKINS, Henry Michael, (1841-1886)
JOHNSTON, Robert Mackenzie, (-1918)
JUKES, Joseph Beete, (1811-1869)
KORNITZER, Louis, (1873-) [PROSPECTORS, DIVINERS AND MINING ENGINEERS]
KREFFT, Gerhard, (1830-1881)
LIVERSIDGE, Archibald, (1847-1927)
MACDONALD, Alexander, (-) [PROSPECTORS, DIVINERS AND MINING ENGINEERS]
MACLEAY, William Sharp, (1792-1865)
McCARTHY, Edward Thomas, (-) [PROSPECTORS, DIVINERS AND MINING ENGINEERS]
MOON, Frederick William, (-1925)
PARKER, William Kitchen, (1823-1890)
PILSBRY, Henry Augustus, (1862-1957)
ROSALES, Henry, (?1822-1916)
SALTER, John William, (1820-1869)
STEPHENS, William John, (1829-1890)
TEALE, Edmund Oswald, Sir, (1874-1971)
TENISON-WOODS, Julian Edmund, Reverend, (1832-1889)
VEATCH, Arthur Clifford, (1878-1938)
WAY, Herbert W. L., (-) [PROSPECTORS, DIVINERS AND MINING ENGINEERS]
ZIETZ, Amandus Christian Heinrich, (1839-1921)

AUSTRALIA (STATE unspecified) Cont'd.

1900 to present

ALDERMAN, Arthur Richard, (1901-)
ALEXANDER, Geoffrey Newman, (1908-1975)
ANDERSON, Charles, (1876-1944)
BAKER, George, (1908-1975)
BELL, John Arnold, (1939-1970)
BELTZ, Edward W., (1891-1970)
BENSON, William Noel, (1885-1957)
BERRINGTON, Roger, (-) [PROSPECTORS, DIVINERS AND MINING ENGINEERS]
BIEBER, Charles Leonard, (1901-1965)
BOWEN, Charles Franklin, (1871-1953)
BOYLE, Robert, (1883-1917)
BRAGG, William Henry, Sir, (1862-1942)
BREMNER, Carl St. John, (1895-1944)
BROUGHTON-EDGE, Arthur Broughton, (1895-1953)
BROWN, Henry Yorke Lyell "Stoney", (1844-1928)
BUTTERWORTH, Emerson McMillin, (1894-1961)
CARR, William L., (1919-1975)
CLAPP, Frederick Gardner, (1879-1944)
CLIFFORD, William Hugh, Right Honourable, (1858-1943)
CLUNE, Frank, (-) [PROSPECTORS, DIVINERS AND MINING ENGINEERS]
COLBERT, Edwin H., (1905-)
COLLINS, Henry Francis, (?1864-1924)
CONDIT, Daniel Dale, (1886-1955)
COTTON, Leo Arthur, (1883-1963)
CRESPIN, Irene, (-)
CRICK, George Charles, (1856-1917)
CRICKMAY, Geoffrey W., (1905-1971)
DART, Raymond Arthur, (1893-)
DAVIS, William Morris, (1850-1934)
DEANE, Henry Jr., (-1924)
DERRY, Duncan Ramsay, (?1908-)
DIPROSE, Kenward Thomas, (1949-1977)
DUNN, Edward John, (1844-1937)
FRASER, John A., (-) [PROSPECTORS, DIVINERS AND MINING ENGINEERS]
FREY, M. Gordon, (1913-)
GOLDMAN, Marcus Isaac, (1881-1965)
GREGORY, John Walter, (1864-1932)
GUPPY, Robert John Lechmere, (1836-1916)
HALL, Thomas Sergeant, (1858-1915)
HANNAN, Patrick, (1842-1925) [PROSPECTORS, DIVINERS AND MINING ENGINEERS]
HENDERSON, James Alexander Leo, (?1874-1940)
HO, Robert, (1921-1972)
HOFFMANN, George Christian, (1837-1917)
HOOKER, Joseph Dalton, Sir, (1817-1911)
HOWCHIN, Walter, Reverend, (1845-1937)
HOWELL, Benjamin Franklin, (1890-1976)
HOWITT, Alfred William, (1830-1908)
HUTCHISON, Arthur Gilbertson, (1902-1949)
JABLONSKI, Eugene, (1892-1975)

AUSTRALIA (STATE unspecified) 1900 to present Cont'd.

JACK, Robert Lockhart, (1878-1964)
JOHNSTON, Robert Mackenzie, (-1918)
JUNNER, Norman Ross, (1891-1973)
KING, Robert Bruce, (1923-1977)
KINNEY, Harry Dill, (1886-1961)
KORNITZER, Louis, (1873-) [PROSPECTORS, DIVINERS AND MINING ENGINEERS]
LAMSENS, O. Robert, (1922-1971)
LASSETER, Lewis Harold Bell "Das", (1883-1930) [PROSPECTORS, DIVINERS AND MINING ENGINEERS]
LEHNER, Ernst J., (1895-1968)
LIVERSIDGE, Archibald, (1847-1927)
MACDONALD, Simon, (1927-1971)
McCARTHY, Edward Thomas, (-) [PROSPECTORS, DIVINERS AND MINING ENGINEERS]
McMATH, John Charles, (?1908-1954)
MOON, Frederick William, (-1925)
NELSON, Richard Newman, (1897-1964)
NININGER, Harvey Harlow, (-)
NOTESTEIN, Frank B., (1885-1973)
PAPP, Simon, (1886-1970)
PENROSE, Evelyn, (-) [PROSPECTORS, DIVINERS AND MINING ENGINEERS]
PILSBRY, Henry Augustus, (1862-1957)
REED, Lyman Coleman, (1900-1970)
RHOADES, Roger F., (1905-1972)
RICHARDSON, Gwen, (-) [PROSPECTORS, DIVINERS AND MINING ENGINEERS]
ROSALES, Henry, (?1822-1916)
RUTLEDGE, Harold, (1920-1954)
SAPPINGTON, Chester, (1905-1968)
SCOFIELD, Lloyd M., (1897-1969)
SIMPSON, George Gaylord, (1902-)
SMITH, Thomas Hodge, (1894-1945)
SPENGLER, Erich, (1886-1962)
STILLWELL, Frank Leslie, (1888-1963)
SULLIVAN, C. J., (1914?-)
TAYLOR, James Haward, (1909-1968)
TEALE, Edmund Oswald, Sir, (1874-1971)
TEICHERT, Curt, (-)
TRECHMANN, Charles Taylor, (1885-1964)
TRÜMPY, Daniel, (1893-1971)
VEATCH, Arthur Clifford, (1878-1938)
WASHBURNE, Chester Wesley, (1883-1971)
WHETTEN, John T., (-)
WHITEHEAD, Walter Lucius, (1891-1969)
WILSON, Thomas Carroll, (1906-1969)
ZIETZ, Amandus Christian Heinrich, (1839-1921)

Time unknown

DULHUNTY, John, (-)

AUSTRIA

Up to 1800

ANKER, Matthias Joseph, (1771-1843)
FICHTEL, Leopold von, (fl.1798)
HACQUET, Belsazar, (1739-1815)
MOHS, Carl Friedrich Christian, (1773-1839)

1800-1849

ABERLE, Carl, (1818-1892)
ABICH, Hermann Wilhelm, (1806-1886)
AICHHORN, Sigmund, (1814-1892)
ANKER, Matthias Joseph, (1771-1843)
AUINGER, Mathias, (?1810-1890)
CANAVAL, Josef Leodegar, (1820-1898)
CHURCHILL, George Cheetham, (1822-1906)
COTTA, Carl Bernhard von, (1808-1879)
CREDNER, Carl Friedrich Heinrich, (1809-1876)
EHRLICH, Franz Carl, (1808-1886)
ETTINGSHAUSEN, Constantin von, Freiherr, (1826-1897)
FICHTEL, Leopold von, (fl.1798)
FITZINGER, Leopold Joseph, (1802-1884)
FUCHS, Wilhelm, (1802-1853)
GILBERT, Josiah, (1814-1892)
GOBANZ, Alexander, (1826-1895)
GRUNOW, Albert, (1826-1914)
HACQUET, Belsazar, (1739-1815)
HAIDINGER, Wilhelm Karl von, Ritter, (1795-1871)
HAUER, Franz von, Ritter, (1822-1899)
HECKEL, Johann Jakob, (1790-1857)
HOCHSTETTER, Ferdinand von, Ritter, (1829-1884)
HOERNES, Moritz, (1815-1868)
HOLLER, Anton, (1826-1909)
KARRER, Felix, (1825-1903)
KENNGOTT, Gustav Adolf, (1818-1897)
KNER, Rudolf, (1810-1869)
KORNHUBER, Andreas, (1824-1905)
KUDERNATSCH, Johann, (1819-1856)
MARSCHALL, August Friedrich, Count of Burgholtzhausen and Tromsdorf, (1805-1887)
MOHS, Carl Friedrich Christian, (1773-1839)
MOLL, Johann Paul Karl von, (fl.1798)
MORLOT, Adolf von, (1820-1867)
PARKER, William Kitchen, (1823-1890)
PARTSCH, Paul, (1791-1856)
PETERS, Karl, (1825-1881)
PICHLER, Adolf, (1819-1900)
POPPELACK, Joseph, (?1780-1859)
PŎSEPNÝ, František, (1836-1895)
PRÉVOST, Louis Constant, (1787-1856)
PRINZINGER, Heinrich, (1822-1908)
RIEPL, Franz Xaver, (1790-1857)
ROLLE, Friedrich, (1827-1887)
ROSTHORN, Franz von, Edler, (1795-1877)
RUSSEGGER, Josef, (-)
SCHRAUF, Albrecht, (1837-1897)
SEELAND, Ferdinand, (1822-1901)

AUSTRIA 1800-1849 Cont'd.

SIMONY, Friedrich, (1813-1896)
SMYTH, Warington Wilkinson, Sir, (1817-1890)
SZABÓ, József, (1822-1894)
UNGER, Franz Joseph Andreas Nicolas, (1800-1870)
WOLF, Heinrich, (1825-1882)
ZITTEL, Karl Alfred von, (1839-1904)

1850-1899
ABEL, Othenio, (1875-1946)
ABERLE, Carl, (1818-1892)
ABICH, Hermann Wilhelm, (1806-1886)
AICHHORN, Sigmund, (1814-1892)
AMPFERER, Otto, (1875-1947)
ANDRIAN-WERBURG, Ferdinand von, Freiherr, (1835-1914)
ARTHABER, Gustav von, Edler, (1864-1943)
ASCHER, Hans, (1848-1953)
AUINGER, Mathias, (?1810-1890)
BARBOT DE MARNY, N. P., (1832-1877)
BECKE, Friedrich Johann Karl, (1855-1931)
BERWERTH, Friedrich Martin, (1850-1918)
BÖHMERSHEIM, August Böhm von, Edler, (1858-1930)
BÖSE, Emil, (1868-1972)
BRAUER, Friedrich Moritz, (1832-1904)
BREZINA, Maria Aristides, (1848-1909)
BRÜCKNER, Eduard, (1862-1927)
BRUDER, George, (1856-1916)
BUKOWSKI, Gejza, (1858-1937)
CANAVAL, Josef Leodegar, (1820-1898)
CAPELLINI, Giovanni, (1833-1922)
CHRISTA, Emanuel, (1874-1948)
CHURCHILL, George Cheetham, (1822-1906)
COTTA, Carl Bernhard von, (1808-1879)
CREDNER, Carl Friedrich Heinrich, (1809-1876)
CRICK, George Charles, (1856-1917)
DIENER, Carl, (1862-1928)
DOELTER y CISTERLICH, Cornelio August, (1850-1930)
DREGER, Julius, (1861-1945)
DRUGČEVIČ, Ferdinand, (1856-1928)
EHRLICH, Franz Carl, (1808-1886)
ETTINGSHAUSEN, Constantin von, Freiherr, (1826-1897)
FIRTSCH, Georg, (1859-1914)
FITZINGER, Leopold Joseph, (1802-1884)
FRAUSCHER, Karl, (1852-1914)
FUCHS, Philippe Jacques Edmond, (1837-1889)
FUCHS, Theodor, (1842-1925)
FUCHS, Wilhelm, (1802-1853)
FUGGER, Friedrich Eberhard, (1842-1919)
GATTNAR, Josef, (1854-1928)
GEYER, Georg, (1857-1938)
GILBERT, Josiah, (1814-1892)
GOBANZ, Alexander, (1826-1895)
GOBANZ, Josef, (-1899)
GROSSPIETSCH, Oskar, (-1920)
GRUNOW, Albert, (1826-1914)

AUSTRIA 1850-1899 Cont'd.

GÜLDNER, Max, (1872-1947)
GUTMANN, Max von, Ritter, (1857-1930)
HABERFELNER, Josef, (1830-1913)
HAIDINGER, Wilhelm Karl von, Ritter, (1795-1871)
HAMMER, Wilhelm, (1875-1942)
HANDLIRSCH, Anton, (1865-1935)
HARADA, Toyokitsi, (-1894)
HAUER, Franz von, Ritter, (1822-1899)
HECKEL, Johann Jakob, (1790-1857)
HILBER, Vincenz, (1853-1931)
HINTERLECHNER, Karl, (1874-1932)
HOCHSTETTER, Ferdinand von, Ritter, (1829-1884)
HOERNES, Moritz, (1815-1868)
HOERNES, Rudolf, (1850-1912)
HOFMANN, Adolf, (1853-1913)
HOLLER, Anton, (1826-1909)
HUSSAK, Franz Eugen, (1856-1911)
JOHN VON JOHNESBERG, Conrad Heinrich, Edler, (1852-1918)
KARRER, Felix, (1825-1903)
KASTNER, Karl, (1847-1907)
KENNGOTT, Gustav Adolf, (1818-1897)
KITTL, Ernst Anton Leopold, (1854-1913)
KNER, Rudolf, (1810-1869)
KNETT, Josef, (1869-1946)
KOCH, Gustav Adolf, (?1846-1921)
KOECHLIN, Rudolf Ignatz, (1862-1939)
KORNHUBER, Andreas, (1824-1905)
KOSSMAT, Franz, (1871-1938)
KRAŠAN, Franz, (1840-1907)
KRASSER, Fridolin, (1863-1922)
KREUTZ, Felix Franz Xaver, (1844-1910)
KUDERNATSCH, Johann, (1819-1856)
KYNASTON, Herbert, (1868-1915)
LANG, Victor von, Edler, (1838-1921)
LIEBUS, Adalbert, (1876-1945)
LOEHR, August von, Ritter, (1847-1917)
LÖWL, Ferdinand, (1856-1908)
LUDWIG, Ernst, (1842-1915)
MACLAREN, James Malcolm, (1873-1935)
MARSCHALL, August Friedrich, Count of Burgholtzhausen and Tromsdorf, (1805-1887)
MAY DE MADIIS, Alexis Freiherr von, (1852-1911)
MOJSISOVICS, Johann August Georg Edmund, Edler von Mojsvàr, (1839-1907)
MORLOT, Adolf von, (1820-1867)
NEUMAYR, Melchior, (1845-1890)
OGILVIE GORDON, Maria Matilda, Dame, (1864-1939)
PARKER, William Kitchen, (1823-1890)
PARTSCH, Paul, (1791-1856)
PAUL, Carl Maria, (1838-1900)
PENCK, Friedrich Carl Albrecht, (1858-1945)
PETERS, Karl, (1825-1881)
PETRASCHEK, Wilhelm, (1876-1967)
PICHLER, Adolf, (1819-1900)

AUSTRIA 1850-1899 Cont'd.

PILAR, Georg, (1847-1893)
POPPELACK, Joseph, (?1780-1859)
POŠEPNÝ, František, (1836-1895)
PRÉVOST, Louis Constant, (1787-1856)
PRINZINGER, Heinrich, (1822-1908)
RACHOY, Josef, (-1902)
REBEL, Hans, (1861-)
REDTENBACHER, Anton, (1841-1911)
REISER, Karl August, (1853-1922)
REYER, Eduard, (1849-1914)
RICHARZ, Stephan, Pater, (1874-1934)
RIEPL, Franz Xaver, (1790-1857)
RODLER, Alfred, (1861-1890)
ROLLE, Friedrich, (1827-1887)
ROSIWAL, August Karl, (1860-1923)
ROSTHORN, Franz von, Edler, (1795-1877)
ROTKY, Otto, (1870-1951)
RUMPF, Johann, (1841-1923)
RUSSEGGER, Josef, (-)
SAUER, Julius, (1849-1911)
SCHAFFER, Franz Xavier, (1876-1954)
SCHARIZER, Rudolf, (1859-1935)
SCHLOENBACH, Georg Justin Carl Urban, (1841-1870)
SCHMIDT, Oskar, (1823-1886)
SCHRAUF, Albrecht, (1837-1897)
SCHUSTER, Max, (1856-1887)
SCHWINNER, Robert Gangolf, (1878-1953)
SEELAND, Ferdinand, (1822-1901)
SINGER, Max, (1874-1952)
SMYTH, Warington Wilkinson, Sir, (1817-1890)
SOHM, Bonifaz, Pater, (1847-1923)
STACHE, Guido, (1833-1921)
STEINDACHNER, Franz, (1834-1919)
STOLICZKA, Ferdinand, (1838-1874)
STREIFF-BECKER, Rudolf, (1873-1959)
SUESS, Eduard, (1831-1914)
SUESS, Franz Eduard, (1867-1941)
SZABÓ, József, (1822-1894)
TAURER-GALLENSTEIN, Hans von, Ritter, (1846-1927)
TAUSCH VON GLÖCKELSTHURN, Leopold, (1858-1899)
TELLER, Friedrich, (1852-1913)
TIETZE, Emil, (1845-1931)
TOULA, Franz, (1845-1920)
UHLIG, Victor Karl, (1857-1911)
UNGER, Franz Joseph Andreas Nicolas, (1800-1870)
VACEK, Michael, (1848-1925)
WAAGEN, Lukas, (1877-1959)
WAAGEN, Wilhelm Heinrich, (1841-1900)
WÄHNER, Franz, (1856-1932)
WAITZ, Paul, (1876-1961)
WENGER, Marian, (1864-1939)
WETTSTEIN, Richard, (1863-1931)
WOLF, Heinrich, (1825-1882)
YOUNG, Alfred Prentice, (1841-1919)

AUSTRIA 1850-1899 Cont'd.

ZEPHAROVICH, Victor Leopold von, Ritter, (1830-1890)
ZIRKEL, Ferdinand, (1838-1912)
ZITTEL, Karl Alfred von, (1839-1904)
ZUGMAYER, Heinrich, (1840-1917)
ZWANZIGER, Gustav Adolf, (1837-1893)

1900 to present

ABEL, Othenio, (1875-1946)
ABERER, Ferdinand, (1913-1965)
ADAMETZ, Lotte, (1879-1966)
AMPFERER, Otto, (1875-1947)
ANDRIAN-WERBURG, Ferdinand von, Freiherr, (1835-1914)
ARTHABER, Gustav von, Edler, (1864-1943)
ASCHER, Hans, (1848-1953)
BANDL, Erich, (1880-1960)
BAYER, Josef, (1882-1931)
BECKE, Friedrich Johann Karl, (1855-1931)
BERWERTH, Friedrich Martin, (1850-1918)
BESCHEL, Roland, (1928-1971)
BISTRITSCHAN, Karl, (1916-1957)
BLASCHKE, Friedrich, (1883-1911)
BOBIES, Carl August, (1898-1958)
BÖHMERSHEIM, August Böhm von, Edler, (1858-1930)
BÖSE, Emil, (1868-1972)
BRAUER, Friedrich Moritz, (1832-1904)
BREZINA, Maria Aristides, (1848-1909)
BRÜCKNER, Eduard, (1862-1927)
BRUDER, George, (1856-1916)
BUKOWSKI, Gejza, (1858-1937)
BÜRGL, Hans, (1907-1966)
CAPELLINI, Giovanni, (1833-1922)
CHRISTA, Emanuel, (1874-1948)
CHUDOBA, Karl F., (1898-1976)
CHURCHILL, George Cheetham, (1822-1906)
CLAR, Eberhard, (1902-)
CORNELIUS, Hans Peter, (1888-)
CRICK, George Charles, (1856-1917)
CZERMAK, Friedrich, (1890-1960)
DIENER, Carl, (1862-1928)
DITTLER, Emil, (1882-1945)
DOELTER y CISTERLICH, Cornelio August, (1850-1930)
DREGER, Julius, (1861-1945)
DRUGČEVIČ, Ferdinand, (1856-1928)
EBY, James Brian, (1896-)
FATH, Arthur Earl, (1887-1976)
FIRTSCH, Georg, (1859-1914)
FOLGNER, Raimund, (1882-1916)
FRAUSCHER, Karl, (1852-1914)
FRIEDL, Karl, (1898-1966)
FRITSCH, Wolfgang, (1928-1970)
FUCHS, Theodor, (1842-1925)
FUGGER, Friedrich Eberhard, (1842-1919)
FUSSENEGGER, Siegfried, (1894-1960)
GATTNAR, Josef, (1854-1928)

AUSTRIA 1900 to present Cont'd.

GEYER, Georg, (1857-1938)
GOLDSCHLAG, Maurycy, (1891-1917)
GÖRGEY VON GÖRGÖ, Rudolf, (1886-1915)
GÖTZINGER, Gustav, (1880-1969)
GRANIGG, Bartel, (1883-1951)
GRENGG, Roman, (1884-1972)
GROSSPIETSCH, Oskar, (-1920)
GRUNOW, Albert, (1826-1914)
GÜLDNER, Max, (1872-1947)
GUTMANN, Max von, Ritter, (1857-1930)
HABERFELNER, Josef, (1830-1913)
HAMMER, Wilhelm, (1875-1942)
HANDLIRSCH, Anton, (1865-1935)
HASSINGER, Hugo, (-1952)
HATTEY, Hans, (1889-1955)
HAUSER, Alois, (1899-1955)
HERITSCH, Franz, (1882-1945)
HILBER, Vincenz, (1853-1931)
HIMMELBAUER, Alfred, (1884-1943)
HINTERLECHNER, Karl, (1874-1932)
HLAUSCHEK, Hans, (1896-1971)
HOERNES, Rudolf, (1850-1912)
HOFMANN, Adolf, (1853-1913)
HOFMANN, Elise, (1889-1955)
HOINKES, Herfried, (1916-1973)
HOLLER, Anton, (1826-1909)
HUSSAK, Franz Eugen, (1856-1911)
IPPEN, Joseph Anton, (-?1918)
JÄGER, Robert, (1890-1915)
JOHN VON JOHNESBERG, Conrad Heinrich, Edler, (1852-1918)
JURASKY, Karl Alfons, (1903-1945)
KAMPTNER, Erwin, (1889-1972)
KARRER, Felix, (1825-1903)
KASTNER, Karl, (1847-1907)
KAUTSKY, Fritz, (1890-1963)
KITTL, Ernst Anton Leopold, (1854-1913)
KLEBELSBERG, Raimund von, (1886-1957)
KNETT, Josef, (1869-1946)
KOBER, Leopold, (1883-1970)
KOCH, Gustav Adolf, (?1846-1921)
KOECHLIN, Rudolf Ignatz, (1862-1939)
KÖLBL, Leopold, (1895-1970)
KORNHUBER, Andreas, (1824-1905)
KOSSMAT, Franz, (1871-1938)
KRAŠAN, Franz, (1840-1907)
KRASSER, Fridolin, (1863-1922)
KREJCI-GRAF, Karl, (-?1956)
KREUTZ, Felix Franz Xaver, (1844-1910)
KÜHN, Othmar, (1892-1969)
KÜMEL, Friedrich, (1908-1955)
KÜPPER, Klaus, (1930-1957)
KYNASTON, Herbert, (1868-1915)
KYRLE, Georg, (1887-1937)
LANG, Victor von, Edler, (1838-1921)

AUSTRIA 1900 to present Cont'd.

LECHNER, Karl, (1899-1958)
LICHTENECKER, Norbert, (1897-1938)
LIEBUS, Adalbert, (1876-1945)
LINSBAUER, Alfred, (1943-1969)
LOEHR, August von, Ritter, (1847-1917)
LÖWL, Ferdinand, (1856-1908)
LUDWIG, Ernst, (1842-1915)
MACHATSCHEK, Fritz, (-1957)
MACHATSCHKI, Felix Karl Ludwig, (1895-1970)
MACLAREN, James Malcolm, (1873-1935)
MARTINI, Hans Joachim, (1908-1969)
MAY DE MADIIS, Alexis Freiherr von, (1852-1911)
MILLOT, Georges, (1917-)
MOHR, Hannes, (1882-1967)
MOJSISOVICS, Johann August Georg Edmund, Edler von Mojsvàr, (1839-1907)
MURR, Josef, (-1932)
NOTH, Rudolf, (1884-1954)
OGILVIE GORDON, Maria Matilda, Dame, (1864-1939)
OSTADAL, Rudolf, (1897-1970)
PAPP, Simon, (1886-1970)
PATTEISKY, Karl, (-?1967)
PENCK, Friedrich Carl Albrecht, (1858-1945)
PETRASCHEK, Wilhelm, (1876-1967)
PIA, Julius von, (1887-1943)
PRINZINGER, Heinrich, (1822-1908)
RACHOY, Josef, (-1902)
REBEL, Hans, (1861-)
REDTENBACHER, Anton, (1841-1911)
REICHARDT, Wolfgang, (1909-1937)
REISER, Karl August, (1853-1922)
REITHOFER, Otto, (1902-1965)
REYER, Eduard, (1849-1914)
RICHARZ, Stephan, Pater, (1874-1934)
RIEDEL, Josef, (1891-1966)
ROSENBURG, Georg, (1897-1969)
ROSIWAL, August Karl, (1860-1923)
ROTKY, Otto, (1870-1951)
RUMPF, Johann, (1841-1923)
SAUER, Julius, (1849-1911)
SCHAFFER, Franz Xavier, (1876-1954)
SCHARF, Willi, (1896-1971)
SCHARIZER, Rudolf, (1859-1935)
SCHIENER, Alfred, (1906-1962)
SCHLAGER, Max, (1903-1972)
SCHMIDT, Walter, (1885-1945)
SCHUSTER, Rudolf, (1931-1968)
SCHWINNER, Robert Gangolf, (1878-1953)
SEELAND, Ferdinand, (1822-1901)
SEEMAN, Fritz, (1884-1914)
SINGER, Max, (1874-1952)
SOHM, Bonifaz, Pater, (1847-1923)
SÖLCH, Johann, (1883-1951)
SOLOMONICA, Paul, (1908-1954)

AUSTRIA 1900 to present Cont'd.

SOMMERMEIER, Leopold, (1885-1963)
SPITZ, Albrecht, (1883-1918)
SPREITZER, Hans, (1897-1973)
STACHE, Guido, (1833-1921)
STEINDACHNER, Franz, (1834-1919)
STINY, Josef, (1880-1958)
STREIFF-BECKER, Rudolf, (1873-1959)
SUESS, Eduard, (1831-1914)
SUESS, Franz Eduard, (1867-1941)
TAURER-GALLENSTEIN, Hans von, Ritter, (1846-1927)
TELLER, Friedrich, (1852-1913)
TERTSCH, Hermann, (1880-1963)
TIETZE, Emil, (1845-1931)
TOULA, Franz, (1845-1920)
TRAUTH, Friedrich, (1883-1967)
TROLL-OBERGFELL, Oskar von, (1882-1972)
TSCHERNIG, Emil, (1896-1970)
UHLIG, Victor Karl, (1857-1911)
VACEK, Michael, (1848-1925)
VETTERS, Hermann, (1880-1941)
WAAGEN, Lukas, (1877-1959)
WÄHNER, Franz, (1856-1932)
WAITZ, Paul, (1876-1961)
WEGENER, Alfred Lothar, (1880-1930)
WEINBERGER, Ludwig, (1911-1966)
WEINFURTER, Emil, (1904-1968)
WENGER, Marian, (1864-1939)
WETTSTEIN, Richard, (1863-1931)
WICHE, Konrad, (1913-1969)
WINKLER-HERMADEN, Artur, (1890-1963)
YOUNG, Alfred Prentice, (1841-1919)
ZIRKEL, Ferdinand, (1838-1912)
ZITTEL, Karl Alfred von, (1839-1904)
ZUGMAYER, Heinrich, (1840-1917)

Time unknown
KERNER, Fritz, (-)

AZORES (ATLANTIC OCEAN)
1900 to present
TORRE DE ASSUNÇÃO, Carlos Fernando, (1901-)

BAHAMA ISLANDS (WEST INDIES)
1800-1849
POURTALÈS, Louis François de, (1823-1880)

1850-1899
DIXON, Ernest Edward Leslie, (1876-1963)
POURTALÈS, Louis François de, (1823-1880)
SHATTUCK, George Burbank, (1869-)

1900 to present
DIXON, Ernest Edward Leslie, (1876-1963)
FIELD, Richard Montgomery, (1885-1961)
GEBELEIN, Conrad Dennis, (1945-1978)
GOLDMAN, Marcus Isaac, (1881-1965)
MAY, Arthur Raymond, (1893-1948)
SHATTUCK, George Burbank, (1869-)
WILSON, Roy Arthur, (1891-1970)

BAHRAIN (PERSIAN GULF)
1850-1899
BEEBY-THOMPSON, Arthur, (1873-1968)
PILGRIM, Henry Guy Ellcock, (1875-1943)

1900 to present
BEEBY-THOMPSON, Arthur, (1873-1968)
HUDSON, Robert George Spencer, (1895-1965)
JAMES, John William, (1921-1977)
NELSON, Richard Newman, (1897-1964)
NOMLAND, Jorgen O., (1885-1943)
PILGRIM, Henry Guy Ellcock, (1875-1943)
RHOADES, Ralph Omer, (1895-1961)

BALEARIC ISLANDS (SPAIN)
1850-1899
JOHNSTON, Mary Sophia, (1875-1955)

1900 to present
ASTRE, Gaston, (1896-)
JOHNSTON, Mary Sophia, (1875-1955)

BANGLA DESH
Up to 1800
GREENOUGH, George Bellas, (1778-1855)

1800-1849
GREENOUGH, George Bellas, (1778-1855)

1850-1899
GREENOUGH, George Bellas, (1778-1855)
HUGHES, Theodore W. H., (?1843-1907)

1900 to present
DAVIS, Ralph Emerson, (1884-1968)
HUGHES, Theodore W. H., (?1843-1907)

BANGLADESH <u>See also</u> INDIA.

BARBADOS (WEST INDIES)

1850-1899

BEEBY-THOMPSON, Arthur, (1873-1968)
FRANKS, Godfrey Firth, (1853-1917)
HARRISON, John Burchmore, Sir, (1856-1926)
JUKES-BROWNE, Alfred John, (1851-1914)
MITCHINSON, John, Right Reverend Bishop, (1833-1918)

1900 to present

BEEBY-THOMPSON, Arthur, (1873-1968)
BOSWORTH, Thomas Owen, (1882-1928)
FRANKS, Godfrey Firth, (1853-1917)
HARRISON, John Burchmore, Sir, (1856-1926)
JUKES-BROWNE, Alfred John, (1851-1914)
MITCHINSON, John, Right Reverend Bishop, (1833-1918)
ROMANES, James, (1886-1942)
SENN, Alfred, (1899-1949)

BASUTOLAND See LESOTHO.

BECHUANALAND See BOTSWANA.

BELGIUM

Up to 1800

BOODTS, Anselmus Boëtius de, (1550-1634)
BRUNO, Filippo, called Giordano, (1548-1600)

1800-1849

BENEDEN, Pierre Joseph van, (1809-1893)
BOSQUET, Joseph Augustin Hubert, (1814-1880)
BRIART, Alphonse, (1825-1898)
CAUCHY, F. P., (-?1842)
CHAPUIS, Félicien, (1824-1879)
COEMANS, Eugène H. L. G., (1825-1871)
CRAHAY, Jacques Guillaume, (1789-1855)
CRÉPIN, François, (1830-1903)
DE LA VALLÉE POUSSIN, Charles Louis Joseph Xavier, (1827-1903)
DEWALQUE, Gilles Joseph Gustave, (1828-1905)
DUMONT, André Hubert, (1809-1857)
GALEOTTI, Henri Guillaume, (1814-1858)
GODWIN-AUSTEN, Robert Alfred Cloyne, (1808-1884)
HAMILTON, William John, (1805-1867)
KONINCK, Laurent Guillaume de, (1809-1887)
LE HON, Henri Sebastian, (1809-1872)
MORREN, Charles, (1807-1858)
NYST, Pierre Henri, (1813-1880)
OMALIUS D'HALLOY, Jean-Baptist Julien d', (1783-1875)
RÖMER, Hermann, (1816-1894)
SAUVEUR, Dieudonné Jean Joseph, (1797-1862)
SCHMERLING, Philippe Charles, (1791-1836)
SELYS-LONGCHAMPS, Michel Edmond de, Baron, (1813-1900)
STORMS, Raymond, (1834-1897)
SYMONDS, William Samuel, Reverend, (1818-1887)

BELGIUM 1800-1849 Cont'd.

VAN DER MAELEN, Philippe Marie Guillaume, (1795-1869)
VERNEUIL, Philippe Edouard Poulletier de, (1805-1873)
VILLEFAGNE D'INGIHOUL, Hilarion Noel de, Baron, (-?1837)
VINCENT, Gérard, (?1824-1899)
WAEL, Norbert Charles Louis de, (1817-1901)

1850-1899

BENEDEN, Edouard van, (1846-1910)
BENEDEN, Pierre Joseph van, (1809-1893)
BIRDS, James Adey, (1831-1894)
BOSQUET, Joseph Augustin Hubert, (1814-1880)
BRIART, Alphonse, (1825-1898)
BUOHY, Victor, (1843-1887)
BUTTGENBACH, Henri, (1874-1964)
CESÁRO, Guiseppe Raimondo Pio, (1849-1939)
CHAMPERKNOWNE, Arthur, (1839-1887)
CHAPUIS, Félicien, (1824-1879)
COEMANS, Eugène H.L.G., (1825-1871)
CORNET, Francois Léopold, (1834-1887)
CORNET, René Jules, (1865-1929)
COTTEAU, Gustave Honoré, (1818-1894)
CRAHAY, Jacques Guillaume, (1789-1855)
CRÉPIN, Francois, (1830-1903)
DE LA VALLÉE POUSSIN, Charles Louis Joseph Xavier, (1827-1903)
DELÉPINE, Gaston-Georges, Monseigneur, (1878-1963)
DELVAUX, Émile, (1837-1901)
DEWALQUE, Gilles Joseph Gustave, (1828-1905)
DOLLO, Louis Antoine Marie Joseph, (1857-1931)
DORMAL, Victor, (-1900)
DRUGMAN, Julien Léon Eugène, (1875-1950)
DUMONT, André Hubert, (1809-1857)
DUPONT, Edouard François, (1841-1911)
FIRKET, Adolphe, (1837-1905)
FORIR, Henri, (1856-1907)
FOURMARIER, Paul Frédéric Joseph Lucien Lauren, (1877-1970)
FRAIPONT, Julien, (1857-1910)
GALEOTTI, Henri Guillaume, (1814-1858)
GODWIN-AUSTEN, Robert Alfred Cloyne, (1808-1884)
HABETS, Alfred, (1839-1908)
HAMILTON, William John, (1805-1867)
KONINCK, Laurent Guillaume de, (1809-1887)
LAMBERT, Paul, (1869-1950)
LANCASTER, A., (-?1908)
LE HON, Henri Sebastian, (1809-1872)
LERICHE, Maurice, (1875-1948)
LIBERT, Joseph, (1853-?1920)
LOHEST, Max, (1857-1926)
MALAISE, Constantin H. L., (1834-1916)
MALHERBE, Renier, (-1891)
McINTYRE, Peter, (1863-1940)
MEUNIER, Fernand, (1868-1926)
MORREN, Charles, (1807-1858)
MOURLON, Michel Félix, (1845-1915)
NYST, Pierre Henri, (1813-1880)
OMALIUS D'HALLOY, Jean-Baptiste Julien d', (1783-1875)

BELGIUM 1850-1899 Cont'd.

ORTLIEB, Jean, (1839-1890)
PAUW, L. F. de, (-)
PERGENS, Eduard Willem Gerard, (1862-1917)
RAISIN, Catherine Alice, (1855-1945)
RENARD, Alphonse François, (1842-1903)
RENIER, Armand Marie Vincent Joseph, (1876-1951)
RÖMER, Hermann, (1816-1894)
RUTOT, Aimé Louis, (1847-1933)
SAUVEUR, Dieudonné Jean Joseph, (1797-1862)
SELYS-LONGCHAMPS, Michel Edmond de, Baron, (1813-1900)
SMEYSTERS, Joseph, (1837-?1908)
SOREIL, Gustave Joseph, (1842-1907)
SPRING, Walthère, (-)
STAINER, Xavier, (1865-)
STORMS, Raymond, (1834-1897)
SYMONDS, William Samuel, Reverend, (1818-1887)
VAN DER MAELEN, Philippe Marie Guillaume, (1795-1869)
VAUGHAN, Arthur, (1868-1915)
VAUX, Adolphe de, (-1899)
VERNEUIL, Philippe Edouard Poulletier de, (1805-1873)
VINCENT, Gérard, (?1824-1899)
WAEL, Norbert Charles Louis de, (1817-1901)

1900 to present

BENEDEN, Edouard van, (1846-1910)
BILLIET, Valère Louis, (1903-1945)
BUTTGENBACH, Henri, (1874-1964)
CESÀRO, Guiseppe Raimondo Pio, (1849-1939)
CONRAD, Walter, (1888-1943)
CORNET, René Jules, (1865-1929)
DE DYCKER, Raymond, (1903-1947)
DE LA VALLÉE POUSSIN, Charles Louis Joseph Xavier, (1827-1903)
DELÉPINE, Gaston-Georges, Monseigneur, (1878-1963)
DELVAUX, Émile, (1837-1901)
DESTINEZ, Pierre, (-1911)
DEWALQUE, Gilles Joseph Gustave, (1828-1905)
DOLLO, Louis Antoine Marie Joseph, (1857-1931)
DORLODOT, Henry de, Chanoîne, (1855-1929)
DORMAL, Victor, (-1900)
DRUGMAN, Julien Léon Eugène, (1875-1950)
DUPONT, Édouard François, (1841-1911)
EDELMANN, Cornelis Hendrik, (1901-1964)
FIRKET, Adolphe, (1837-1905)
FORIR, Henri, (1856-1907)
FOURMARIER, Paul Frédéric Joseph Lucien Lauren, (1877-1970)
FRAIPONT, Julien, (1857-1910)
HABETS, Alfred, (1839-1908)
JAMOTTE, André, (1903-1951)
LAMBERT, Paul, (1869-1950)
LANCASTER, A., (-?1908)
LECKWIJCK, Willem Peter van, (1902-1975)
LECOMPTE, Marius, (1902-1973)
LEDOUX, Auguste, (1887-1918)
LERICHE, Maurice, (1875-1948)

BELGIUM 1900 to present Cont'd.

LOHEST, Max, (1857-1926)
MALAISE, Constantin H. L., (1834-1916)
McINTYRE, Peter, (1863-1940)
MEUNIER, Fernand, (1868-1926)
MOURLON, Michel Félix, (1845-1915)
NICOLAS, Jean Paul, (1921-1967)
PAUW, L. F. de, (-)
PERGENS, Eduard Willem Gerard, (1862-1917)
PIVETEAU, Jean, (1898-)
RAISIN, Catherine Alice, (1855-1945)
RENARD, Alphonse François, (1842-1903)
RENIER, Armand Marie Vincent Joseph, (1876-1951)
RUTOT, Aimé Louis, (1847-1933)
SALÉE, Achille, Chanoine, (1883-1932)
SCHOEP, Alfred Ferdinandus Maria Ghislenus, (1881-1966)
SMEYSTERS, Joseph, (1837-?1908)
SMITH, G. Wendell, (1917-1974)
SOREIL, Gustave Joseph, (1842-1907)
STAINER, Xavier, (1865-)
STRAELEN, Victor Emile van, (1889-1964)
VANDENDRIESSCHE, Adrien, (1914-1940)
VAUGHAN, Arthur, (1868-1915)
WAEL, Norbert Charles Louis de, (1817-1901)

BELIZE

1850-1899

BEEBY-THOMPSON, Arthur, (1873-1968)
WILSON, Charles Henry, (1851-1887)

1900 to present

BEEBY-THOMPSON, Arthur, (1873-1968)

BENIN

1900 to present

FURON, Raymond, (1898-)
MILLOT, Georges, (1917-)

BERMUDA ISLANDS (NORTH ATLANTIC OCEAN)

1800-1849

HUNTER, Robert, Reverend, (1823-1897)

1850-1899

BULLEN, Robert Ashington, Reverend, (1850-1912)
GREGORY, William King, (1876-1970)
HUNTER, Robert, Reverend, (1823-1897)
JULIAN, Henry Forbes, (1861-1912) [PROSPECTORS, DIVINERS AND MINING ENGINEERS]

BERMUDA ISLANDS (NORTH ATLANTIC OCEAN) Cont'd.

1900 to present
BULLEN, Robert Ashington, Reverend, (1850-1912)
GEBELEIN, Conrad Dennis, (1945-1978)
GREGORY, William King, (1876-1970)
GROSS, Henry, (-) [PROSPECTORS, DIVINES AND MINING ENGINEERS]
JULIAN, Henry Forbes, (1861-1912) [PROSPECTORS, DIVINERS AND MINING ENGINEERS]
STETSON, Henry Crosby, (1900-1955)

BHUTAN
1850-1899
PILGRIM, Henry Guy Ellcock, (1875-1943)

1900 to present
PILGRIM, Henry Guy Ellcock, (1875-1943)

BOHEMIA <u>See</u> CZECHOSLOVAKIA.

BOLIVIA
Up to 1800
HELMS, Anton Zacharias, (1751-1803) [PROSPECTORS, DIVINERS AND MINING ENGINEERS]

1800-1849
FORBES, David, (1828-1876)
HELMS, Anton Zacharias, (1751-1803) [PROSPECTORS, DIVINERS AND MINING ENGINEERS]
LLOYD, J. A., Colonel, (-1856?)
ORBIGNY, Alcide Charles Victor Dessalines d', (1802-1857)
PENTLAND, Joseph Barclay, (1797-1873)
PISSIS, Aimé, (1812-1889)

1850-1899
ADAMS, George Irving, (1870-1932)
BOWMAN, Isaiah, (1882-1950)
COURTY, Georges, (1875-1953)
EVANS, John William, (1857-1930)
FORBES, David, (1828-1876)
HAUTHAL, Rudolf, (1854-1928)
HELMS, Anton Zacharias, (1751-1803) [PROSPECTORS, DIVINERS AND MINING ENGINEERS]
LINDGREN, Waldemar, (1860-1939)
LLOYD, J. A., Colonel, (-1856?)
MILLER, Benjamin Le Roy, (1874-1944)
OCHSENIUS, Carl Christian, (1830-)
ORBIGNY, Alcide Charles Victor Dessalines d', (1802-1857)
PENTLAND, Joseph Barclay, (1797-1873)
PISSIS, Aimé, (1812-1889)
STELZNER, Alfred Wilhelm, (1840-1895)
TERNETZ, Carl, (1870-1928)

BOLIVIA Cont'd.

1900 to present

ADAMS, George Irving, (1870-1932)
BAKELS, P. S., (-1960)
BANDY, Mark Chance, (1900-1963)
BOWMAN, Isaiah, (1882-1950)
BROWN, Olin Teter, (1900-1962)
CARROLL, Don Llewellyn, (1902-1954)
COURTY, Georges, (1875-1953)
DENNING, Reynolds McConnell, (1916-1967)
DOUGLAS, James Archibald, (1884-1978)
DRESSER, Myron Amos, (1894-1973)
EVANS, John William, (1857-1930)
GORDON, Samuel George, (1897-1952)
HAKE, Benjamin Franklin, (1895-1973)
HARRINGTON, George Leavitt, (1883-1972)
HAUTHAL, Rudolf, (1854-1928)
HAYES, Albert Orion, (1882-1965)
HEALD, Kenneth Conrad, (1888-1971)
HERZENBERG, Robert, (1885-1955)
ISENBERGER, Nathan Perry, (1896-1956)
JONES, David Glynn, (1904-1974)
KOZLOWSKI, Roman, (1889-1977)
LINDGREN, Waldemar, (1860-1939)
MATHER, Kirtley Fletcher, (1888-1978)
MILLER, Benjamin Le Roy, (1874-1944)
MOORE, Gilbert Parvin, (1898-1970)
REYES, Felix Celso, (1924-1972)
RIGGS, Robert Jennings, (1888-1965)
ROMANES, James, (1886-1942)
SCHMITTOU, Maurice B., (1899-1929)
SEVERY, Clarence Luther, (1885-1975)
SINGEWALD, Joseph Theophilus, Jr., (1884-1963)
STEBINGER, Eugene, (1883-1951)
SULLINS, M. Joseph, (1930-1972)
SUTTON, Frederick A., (1894-1950)
TERNETZ, Carl, (1870-1928)
THOMAS, Norman Louis, (1897-1970)
WHITEHEAD, Walter Lucius, (1891-1969)

BORNEO

1800-1849

APLIN, Christopher d'Oyley Hale, (1819-1875)
GEYLER, Hermann Theodor, (1834-1889)

1850-1899

APLIN, Christopher d'Oyley Hale, (1819-1875)
BAILEY, Tom Esmond Geoffrey, Captain, (-1919)
BECHER, Harry Macdonald, (1855-1893)
BOWEN, Charles Franklin, (1871-1953)
CHAPER, Maurice Armand, (1834-1896)
GEYLER, Hermann Theodor, (1834-1889)
KORNITZER, Louis, (1873-) [PROSPECTORS, DIVINERS AND MINING ENGINEERS]
MOON, Frederick William, (-1925)

BORNEO 1850-1899 Cont'd.

NORDENSKIÖLD, Nils Adolf Erik, (1832-1901)
VERBEEK, Rogier Diederick Marius, (1845-1926)

1900 to present

BAILEY, Tom Esmond Geoffrey, Captain, (-1919)
BELTZ, Edward W., (1891-1970)
BERNOUILLI, Walter, (1885-1946)
BOURCHIER, John Rickman, (1896-1927)
BOWEN, Charles Franklin, (1871-1953)
CONDIT, Daniel Dale, (1886-1955)
COX, Leslie Reginald, (1897-1965)
DOORNINCK, Nico H. van, (1893-1956)
GARRETT, T. R. H., Major, (-1938)
HARRISON, John Vernon, (1892-1972)
HOTZ, Walter, (1884-1958)
KORNITZER, Louis, (1873-) [PROSPECTORS, DIVINERS AND MINING ENGINEERS]
LEUTENEGGER, Willy Otto, (1903-1972)
LINDEN, Bernard Hugo "Ben" van der, (-1958)
MOON, Frederick William, (-1925)
NIETHAMMER, Gottlob August, (1882-1915)
NORDENSKIÖLD, Nils Adolf Erik, (1832-1901)
NUTTALL, Winfred Laurence Falkiner, (1897-1969)
PET, Cees, (1900-1953)
POLLOCK, Charles Murray, (1898-1969)
REA, Henry Carter, (1900-1963)
ROE, Frank Walton "Peter", (?1914-1965)
RUTTEN, Louis Martin Robert, (1884-1946)
SMIT SIBINGA, Gerard Leonard, (1895-1963)
VERBEEK, Rogier Diederick Marius, (1845-1926)
WILHELM, Oscar, (1897-1975)
ZEIJLMANS VAN EMMICHOVEN, Cornelis Pieter Adriaan, (1897-1952)

BORNEO <u>See also</u> BRUNEI; SARAWAK.

BOTSWANA

1850-1899

BAIN, Thomas Charles John, (1829-1893)
JULIAN, Henry Forbes, (1861-1912) [PROSPECTORS, DIVINERS AND MINING ENGINEERS]
TOIT, Alexander Logie du, (1878-1948)

1900 to present

ANDERSON, Robert Beattie, (1904-1970)
BOOCOCK, Colin, (1925-1973)
JULIAN, Henry Forbes, (1861-1912) [PROSPECTORS, DIVINERS AND MINING ENGINEERS]
POLDERVAART, Arie, (1918-1964)
TOIT, Alexander Logie du, (1878-1948)
WASSERSTEIN, Benno, (1906-1956)

BRAZIL

Up to 1800

ANDRADA E SILVA, José Bonifácio de, (1762-1838)

BRAZIL Up to 1800 Cont'd.

BARROW, John, Sir, (1764-1848)
ESCHWEGE, Wilhelm Ludwig, Baron, (1777-1855)
GREGOR, William, Reverend, (1761-1817)
HUMBOLDT, Friedrich Wilhelm Heinrich Alexander von, Freiherr, (1769-1859)
LA CONDAMINE, Charles-Marie de, (1701-1774)
MAWE, John, (1764-1829)

1800-1849
AGASSIZ, Louis Jean Rodolphe, (1807-1873)
ANDRADA E SILVA, José Bonifácio de, (1762-1838)
ATTWOOD, Melville, (1812-1898)
BARROW, John, Sir, (1764-1848)
BATES, Henry Walter, (1825-1892)
BURMEISTER, Karl Hermann Konrad, (1807-1892)
DARWIN, Charles Robert, (1809-1882)
ELLIS, H., Right Honourable Sir, (-1855)
ESCHWEGE, Wilhelm Ludwig, Baron, (1777-1855)
GREGOR, William, Reverend, (1761-1817)
HENWOOD, William Jory, (1805-1875)
HUMBOLDT, Friedrich Wilhelm Heinrich Alexander von, Freiherr, (1769-1859)
MAWE, John, (1764-1829)
ORBIGNY, Alcide Charles Victor Dessalines d', (1802-1857)
WALLACE, Alfred Russel, (1823-1913)
WATERTON, Charles, (1782-1865)
WHITE, Charles Abiathar, (1826-1910)

1850-1899
AGASSIZ, Louis Jean Rodolphe, (1807-1873)
ATTWOOD, Melville, (1812-1898)
BATES, Henry Walter, (1825-1892)
BAUER, Heinrich E., (1840-1896)
BAUERMAN, Hilary, (1833-1909)
BOWEN, Charles Franklin, (1871-1953)
BRANNER, John Casper, (1850-1922)
BROWN, Charles Barrington, (1839-1917)
BURMEISTER, Karl Hermann Konrad, (1807-1892)
CLARAZ, Georges, (1832-1930)
COSTA SENA, Joaquim Candido da, (1852-1919)
COURTY, Georges, (1875-1953)
CUMINGS, Willard Lawsen, (1874-1945)
DARWIN, Charles Robert, (1809-1882)
DERBY, Orville Adelbert, (1851-1915)
DITMARS, Raymond Lee, (1876-1942)
DRAPER, David, (1849-1929)
ELLIS, H., Right Honourable Sir, (-1855)
ESCHWEGE, Wilhelm Ludwig, Baron, (1777-1855)
EVANS, John William, (1857-1930)
HARTT, Charles Frederic, (1840-1878)
HENNEN, Ray Vernon, (1875-1958)
HENWOOD, William Jory, (1805-1875)
HOWE, Ernest, (1875-1932)
HUENE, Friedrich von, Freiherr, (1875-1969)

BRAZIL 1850-1899 Cont'd.

HUMBOLDT, Friedrich Wilhelm Heinrich Alexander von, Freiherr, (1769-1859)
HUSSAK, Franz Eugen, (1856-1911)
KATZER, Bedřich, (1861-1925)
MARSHAM-TOWNSHEND, Robert Honourable, (1834-1915)
MAWSON, Joseph, (1830-1927)
MILLER, Benjamin Le Roy, (1874-1944)
MILLS, James Ellison, (1834-1901)
ORBIGNY, Alcide Charles Victor Dessalines d', (1802-1857)
PLANT, Nathaniel, (1832-1875)
PRIOR, George Thurland, (1862-1936)
RATHBURN, Richard, (1852-1918)
ROCCATI, Alessandro, (1872-1928)
ROTH, Santiago, (1850-1924)
SIEMIRADZKI, Józef, (1858-1933)
ST. JOHN, Orestes Hawley, (1841-1921)
STREIFF-BECKER, Rudolf, (1873-1959)
TERNETZ, Carl, (1870-1928)
THOMPSON, Beeby, (1848-1931)
TOOTHAKER, Charles Robinson, (1873-1952)
WALLACE, Alfred Russel, (1823-1913)
WATERTON, Charles, (1782-1865)
WHITE, Charles Abiathar, (1826-1910)
WINGE, Herluf, (1857-1923)

1900 to present

BAUERMAN, Hilary, (1833-1909)
BLAKE, George Stanfield, (-1940)
BOWEN, Charles Franklin, (1871-1953)
BRANNER, John Casper, (1850-1922)
BROUWER, Hendrik Albertus, (1886-1973)
BROWN, Charles Barrington, (1839-1917)
CANFIELD, Charles Reiter, (1906-1964)
CHUDOBA, Karl F., (1898-1976)
CLARAZ, Georges, (1832-1930)
COSTA SENA, Joaquim Candido da, (1852-1919)
COURTY, Georges, (1875-1953)
CUMINGS, Willard Lawsen, (1874-1945)
DE ARAUJO FERRAZ, Jorge Palmiro, (1883-1926)
DERBY, Orville Adelbert, (1851-1915)
DERRY, Duncan Ramsay, (?1908-)
DITMARS, Raymond Lee, (1876-1942)
DRAPER, David, (1849-1929)
EVANS, John William, (1857-1930)
GRAHAM, David H., (1910-1944)
GRATON, Louis Caryl, (1880-1970)
HARBORT, Erich, (1879-1929)
HENNEN, Ray Vernon, (1875-1958)
HOTZ, Walter, (1884-1958)
HOWE, Ernest, (1875-1932)
HUBBARD, Bela, (1890-1959)
HUENE, Friedrich von, Freiherr, (1875-1969)
HUSSAK, Franz Eugen, (1856-1911)
JOHNSTON, William Drumm, Jr., (1899-1972)

BRAZIL 1900 to present Cont'd.

KATZER, Bedřich, (1861-1925)
KREIDLER, William Lynn, (1911-1975)
KUBIËNA, Walter L., (1897-1970)
MAACK, Reinhard, (1892-1969)
MARSHAM-TOWNSHEND, Robert Honourable, (1834-1915)
MAWSON, Joseph, (1830-1927)
MILLER, Benjamin Le Roy, (1874-1944)
MILLS, James Ellison, (1834-1901)
MURDOCK, Joseph, (1890-1973)
OLIVEIRA, Euzebio Paulo de, (1882-1939)
OLTMAN, Roy E., (-1977)
PARDÉ, Maurice, (1893-1973)
PECORA, William Thomas, (1913-1972)
POND, Walter Franklin, (1885-1974)
PRIOR, George Thurland, (1862-1936)
RATHBURN, Richard, (1852-1918)
RIMANN, Eberhard, (1882-1944)
ROBERTS, Hugh Marine, (1885-1965)
ROCCATI, Alessandro, (1872-1928)
ROSE, Hermann, (1883-1976)
ROTH, Santiago, (1850-1924)
SCHMITTOU, Maurice B., (1899-1929)
SHEARER, Harold Kurtz, (1889-1946)
SIEMIRADZKI, Józef, (1858-1933)
SIMPSON, George Gaylord, (1902-)
SINGEWALD, Joseph Theophilus, Jr., (1884-1963)
SLAWSON, Chester Baker, (1898-1964)
ST. JOHN, Orestes Hawley, (1841-1921)
STREIFF-BECKER, Rudolf, (1873-1959)
TERNETZ, Carl, (1870-1928)
THOMAS, Ronald Hawksby, Major, (1885-?1940)
THOMPSON, Beeby, (1848-1931)
TOOTHAKER, Charles Robinson, (1873-1952)
TRIEBEL, Erich, (1894-1971)
WALLACE, Alfred Russel, (1823-1913)
WARING, Gerald Ashley, (1883-1971)
WHITE, Charles Abiathar, (1826-1910)
WINGE, Herluf, (1857-1923)
YOUNGQUIST, Walter, (-)

BRITISH GUIANA <u>See</u> GUYANA.

BRITISH HONDURAS <u>See</u> BELIZE.

BRITISH SOMALILAND <u>See</u> SOMALIA.

BRUNEI (BORNEO)
1900 to present
ANDREW, Arthur Robert, (1881-1941)
WILHELM, Oscar, (1897-1975)

BULGARIA

1800-1849

LILL VON LILIENBACH, Karl, (1798-1831)

1850-1899

CVICIĆ, Jovan, (1865-1927)
DOUVILLÉ, Joseph Henri Ferdinand, (1846-1937)
KOSSMAT, Franz, (1871-1938)
ZLATARSKI, Georgi Nikolov, (1854-1909)

1900 to present

CVICIĆ, Jovan, (1865-1927)
DOUVILLÉ, Joseph Henri Ferdinand, (1846-1937)
KOSSMAT, Franz, (1871-1938)
LECKWIJCK, Willem Peter van, (1902-1975)
SOVINSKY, Viacheslav Nicholas, (1894-1962)
ZLATARSKI, Georgi Nikolov, (1854-1909)

BURMA

1850-1899

BAILEY, Tom Esmond Geoffrey, Captain, (-1919)
BALL, Francis Matthew, (1877-1956)
BEEBY-THOMPSON, Arthur, (1873-1968)
BLANFORD, William Thomas, (1832-1905)
BROWN, Charles Barrington, (1839-1917)
FEARNSIDES, William George, (1879-1968)
HUGHES, Theodore W. H., (?1843-1907)
MACLAREN, James Malcolm, (1873-1935)
MARKS, Edward Seaborn, (?1875-1950)
MIDDLEMISS, Charles Stewart, (1859-1945)
MOON, Frederick William, (-1925)
OLDHAM, Richard Dixon, (1858-1936)
PORRIO, Cesare, (1865-1940)
RAMA RAU, S. Sethu, Rao Bahadur, (1874-1929)
REED, Frederick Richard Cowper, (1869-1946)
RICHTHOFEN, Paul Wilhelm Ferdinand von, Freiherr, (1833-1905)
STOLICZKA, Ferdinand, (1838-1874)

1900 to present

AMBROSE, John Willis, (-1974)
BAILEY, Tom Esmond Geoffrey, Captain, (-1919)
BALL, Francis Matthew, (1877-1956)
BARBER, Cecil Thomas, (-)
BEASLEY, Cyril George, (1901-?1957)
BEEBY-THOMPSON, Arthur, (1873-1968)
BION, Herbert Stanley, (1888-1915)
BLANFORD, William Thomas, (1832-1905)
BROWN, Charles Barrington, (1839-1917)
BROWN, John Coggin, (1884-1962)
BUSK, Henry Gould, (?1891-1956)
CONDIT, Daniel Dale, (1886-1955)
COTTER, Gerald de Purcell, (1881-1941)
CURRIE, Leslie D., (1904-1925)
DEWHURST, Thomas, (1881-1973)
EVANS, Percy, (1892-1974)
FEARNSIDES, William George, (1879-1968)

BURMA 1900 to present Cont'd.

FERMOR, Lewis Leigh, Sir, (1880-1954)
HAWLEY, James Edwin, (1897-1965)
HERON, Alexander Macmillan, (1884-1971)
HOLMES, Arthur, (1890-1965)
HUBMAN, Carl Wesley, (1905-1947)
HUGHES, Theodore W. H., (?1843-1907)
JONES, Hubert Cecil, (1881-1963)
JONES, William Richard, (1890-1970)
JURKOVIĆ, Ivan, (1917-)
KINNEY, Harry Dill, (1886-1961)
LEHNER, Ernst J., (1895-1968)
LEPPER, Gordon Willis, (1889-1950)
MACLAREN, James Malcolm, (1873-1935)
MARKS, Edward Seaborn, (?1875-1950)
MARR, Francis Alleyne, (1893-1942)
MAYO, Herbert Theodore, (1889-1965)
METRE, Waman Bapu, (1906-1970)
MIDDLEMISS, Charles Stewart, (1859-1945)
MOON, Frederick William, (-1925)
NAGAPPA, Yedatore, Shri, (1907-1960)
NIGHTINGALE, William Thomas, (1897-1964)
OLDHAM, Richard Dixon, (1858-1936)
PINFOLD, Ernest Sheppard, (1888-1969)
PORRIO, Cesare, (1865-1940)
RAMA RAU, S. Sethu, Rao Bahadur, (1874-1929)
REED, Frederick Richard Cowper, (1869-1946)
RICHTHOFEN, Paul Wilhelm Ferdinand von, Freiherr, (1833-1905)
RUDMOSE BROWN, Robert Neal, (1879-1957)
SANSOM, Charles Alan, (1902-1945)
SEN GUPTA, Kiron Kumar, (1882-1943)
SPATH, Leonard Frank, (1882-1957)
STAMP, Laurence Dudley, Sir, (1898-1966)
TEILHARD DE CHARDIN, Pierre, Father, S.J., (1881-1955)
WILMOTH, Leslie James, (1881-1963)

BURMA See also MERGUI ARCHIPELAGO.

BURUNDI
1900 to present
DE DYCKER, Raymond, (1903-1947)
JAMOTTE, André, (1903-1951)

CAMBODIA
1850-1899
JACOB, Charles François Étienne, (1878-1962)

1900 to present
GUBLER, Jean, (1902-1940)
JACOB, Charles François Étienne, (1878-1962)
RASMUSSEN, William Charles, (1917-1973)

CAMBODIA <u>See also</u> INDOCHINA (unspecified).

CAMEROONS
1900 to present
BERNOUILLI, Walter, (1885-1946)
HENNIG, Edwin, (1882-)
HOURCQ, Victor, (1901-1961)
JÉRÉMINE, Elisabeth, (née Tschernaieff), (1879-1964)
MILLOT, Georges, (1917-)
NICOLAS, Jean Paul, (1921-1967)
RECK, Hans, (1886-1937)
ROCH, Edouard, (1901-)
WILSON, Reginald Charles, (1888-1943)

CANADA
ALBERTA
1850-1899
BEEBY-THOMPSON, Arthur, (1873-1968)
BELL, Robert, (1841-1917)
BENSLEY, Benjamin Arthur, (1875-1934)
BROCK, Reginald Walter, (1874-1935)
COLEMAN, Arthur Philemon, (1852-1939)
DALY, Reginald Aldworth, (1871-1957)
DAWSON, George Mercer, (1849-1901)
DOWLING, Donaldson Bogart, (1858-1925)
FERRIER, Walter F., (1865-1950)
HECTOR, James, Sir, (1834-1907)
HENDERSON, James Alexander Leo, (?1874-1940)
LAMBE, Lawrence Morris, (1863-1919)
McCARTHY, Edward Thomas, (-) [PROSPECTORS, DIVINERS AND MINING ENGINEERS]
McCONNELL, Richard George, (1857-1942)
OGILVIE, Ida Helen, (1874-1963)
RIGGS, Elmer Samuel, (1869-1963)
STERNBERG, Charles Hazelius, (1850-1943)
TYRRELL, Joseph Burr, (1858-1957)
VAUX, William S., Jr., (1872-1908)
WHITEAVES, John Joseph Frederick, (1835-1909)

1900 to present
ALCOCK, Frederick James, (1888-1972)
ALLAN, John Andrew, (1884-1955)
AMBROSE, John Willis, (-1974)
BAILEY, Ralph Fraser, (1920-1953)
BALL, Max Waite, (1885-1954)
BEACH, Floyd Kellogg, (1884-1964)
BEACH, Hugh Hamilton, (1909-1973)

CANADA (ALBERTA) 1900 to present Cont'd.

BEEBY-THOMPSON, Arthur, (1873-1968)
BELL, Robert, (1841-1917)
BELTZ, Edward W., (1891-1970)
BENSLEY, Benjamin Arthur, (1875-1934)
BROCK, Reginald Walter, (1874-1935)
BURNS, Russell Warren, (1912-1967)
CLARK, Leslie Madison, (1898-1976)
COLEMAN, Arthur Philemon, (1852-1939)
CONKLING, Russell C., (1897-1955)
CURRIE, Philip John, (1949-   )
DALY, Reginald Aldworth, (1871-1957)
DAWSON, George Mercer, (1849-1901)
DE KOCH, William Gaspar, (1898-1954)
DOWLING, Donaldson Bogart, (1858-1925)
DUSTON, Arthur W., (1886-1938)
FATH, Arthur Earl, (1887-1976)
FERRIER, Walter F., (1865-1950)
HAKE, Benjamin Franklin, (1895-1973)
HALSE, George Whitfield, (1885-1968)
HECTOR, James, Sir, (1834-1907)
HENDERSON, James Alexander Leo, (?1874-1940)
HOWELLS, Jenkyn Owen, (-1950)
HUME, George Sherwood, (1893-1965)
IRWIN, Marvin Leroy "Mel", (1915-1964)
JOHNSON, Hope, (-   )
KANERVA, John, (1882-   )
LAMBE, Lawrence Morris, (1863-1919)
LANDES, Robert William, (1910-1969)
LEWIS, Francis Leonard, Jr., (1921-1953)
LINES, Frederic George, (1921-1963)
McCARTHY, Edward Thomas, (-   ) [PROSPECTORS, DIVINERS AND MINING ENGINEERS]
McCONNELL, Richard George, (1857-1942)
MICHENER, Charles, (-   )
MITCHELL, James Garland, (1908-1972)
MOORE, Prentiss D., (1895-1955)
MORRIS, Albert Ferd, (1898-1957)
NARAMORE, Chester "Chet", (1879-1961)
OGILVIE, Ida Helen, (1874-1963)
OLLERENSHAW, Neil, (-   )
PŘIBYL, Alois, (1904-   )
ROBINSON, Ernest Guy, (1898-1973)
RUSSELL, Loris Shano, (1904-   )
SANDERSON, James Owen Gresham "Pete", (1898-1963)
SCHOLL, Louis August, Jr., (1887-1953)
SLATER, George, (-1956)
SPRATT, Joseph Grant, (1904-1978)
SPROULE, John Campbell, (1905-1970)
STERNBERG, Charles Hazelius, (1850-1943)
STERNBERG, Charles M., (1885-   )
STERNBERG, George Fryer, (1883-1969)
STEWART, Grace Anne, (1893-1970)
THOMAS, Norman Louis, (1897-1970)
TYRRELL, Joseph Burr, (1858-1957)

CANADA (ALBERTA) 1900 to present Cont'd.

VAUX, William S., Jr., (1872-1908)
WALLACE, Robert Charles, (1881-1955)
WARREN, Percival Sidney, (1890-1970)
WEBB, John Benwell, (1901-1975)
WHITEAVES, John Joseph Frederick, (1835-1909)

CANADA
BRITISH COLUMBIA
1850-1899
ATTWOOD, George, (1845-1912)
BAUERMAN, Hilary, (1833-1909)
BEADNELL, Hugh John Llewellyn, Major, (1874-1944)
BOND, Marshall, (1867-1941) [PROSPECTORS, DIVINERS AND MINING ENGINEERS]
BOWMAN, Amos, (1839-1894)
BROCK, Reginald Walter, (1874-1935)
CAMSELL, Charles, (1876-1958)
CORNWALL, Ira Edmund, (1875-1962)
DALY, Reginald Aldworth, (1871-1957)
DAWSON, George Mercer, (1849-1901)
ELLS, Robert Wheelock, (1845-1911)
LAWSON, Andrew Cowper, (1861-1952)
McCARTHY, Edward Thomas, (-) [PROSPECTORS, DIVINERS AND MINING ENGINEERS]
McCONNELL, Richard George, (1857-1942)
RICKARD, Thomas Arthur, (1864-1953)
SUTTON, William John, (1859-1914)
TRELAWNEY-ANSELL, E. C., (-) [PROSPECTORS, DIVINERS AND MINING ENGINEERS]
VAUX, William S., Jr., (1872-1908)
WARREN, Charles Hyde, (1876-1950)
WHITEAVES, John Joseph Frederick, (1835-1909)

1900 to present
ALLAN, John Andrew, (1884-1955)
ATTWOOD, George, (1845-1912)
BATEMAN, Alan Mara, (1889-1971)
BAUERMAN, Hilary, (1833-1909)
BEADNELL, Hugh John Llewellyn, Major, (1874-1944)
BELL, Alfred Hannam, (1895-1977)
BILLINGSLEY, Paul, (1887-1962)
BOND, Marshall, (1867-1941) [PROSPECTORS, DIVINERS AND MINING ENGINEERS]
BROCK, Byron Britton, (1904-1972)
BROCK, Reginald Walter, (1874-1935)
CAMSELL, Charles, (1876-1958)
CHIPMAN, Kenneth G., (-)
CLARK, Leslie Madison, (1898-1976)
COCKBURN, Alexander Murray, (1901-1959)
CORNWALL, Ira Edmund, (1875-1962)
CURRIE, Philip John, (1949-)
DALY, Reginald Aldworth, (1871-1957)
DAWSON, George Mercer, (1849-1901)
DERRY, Duncan Ramsay, (?1908-)

CANADA (BRITISH COLUMBIA) 1900 to present Cont'd.

DIXON, Henry George Dacres, (1887-1947)
DRYSDALE, Charles Wales, (1885-1917)
ELLS, Robert Wheelock, (1845-1911)
ELLSWORTH, Hardy Vincent, (1889-1952)
EVANS, Charles Sparling, (1894-1962)
GOWER, John A., (1921-1972)
GRAHAM, Richard Percival Devereux, (1880-1965)
GRAHAM, Roy, (1908-1939)
HAKE, Benjamin Franklin, (1895-1973)
HAYCOCK, Maurice, (1900-)
HAYES, Albert Orion, (1882-1965)
HODGE, Edwin Thomas, (1887-1970)
HOWELLS, Jenkyn Owen, (-1950)
HUESTIS, Herman H., (1907-) [PROSPECTORS, DIVINERS AND MINING ENGINEERS]
HUTCHINSON, William W., (-)
JOHNSTON, Ashton William, (1910-)
LAWSON, Andrew Cowper, (1861-1952)
LILLIGREN, Johnathan Martin, (1880-1956)
McCARTHY, Edward Thomas, (-) [PROSPECTORS, DIVINERS AND MINING ENGINEERS]
McCONNELL, Richard George, (1857-1942)
McKINSTRY, Hugh Exton, (1896-1961)
NYGREN, Walter Eric, (1907-1971))
PARKER, Ralph D., (1898-) [PROSPECTORS, DIVINERS AND MINING ENGINEERS]
PEACOCK, Martin Alfred, (1898-1950)
RICKARD, Thomas Arthur, (1864-1953)
ROBERTSON, Andrew, (1903?-) [PROSPECTORS, DIVINERS AND MINING ENGINEERS]
SULLIVAN, C. J., (1914?-)
SUTTON, William John, (1859-1914)
THOMPSON, Robert Mitchell, (1918-1967)
TRELAWNEY-ANSELL, E. C., (-) [PROSPECTORS, DIVINERS AND MINING ENGINEERS]
UGLOW, William Lawrence, (1884-1926)
VAUX, William S., Jr., (1872-1908)
WARREN, Charles Hyde, (1876-1950)
WARREN, Harry V., (?1904-)
WHITE, William Harrison, (1913-1972)
WHITEAVES, John Joseph Frederick, (1835-1909)
WHITTIER, William Harrison, (1889-1951)
WILCOCKSON, William Howson, (1891-1976)
WOLF, Albert G., (1885-1964)

CANADA
LABRADOR
Up to 1800
BANKS, Sir Joseph, (1743-1820)

1800-1849
BANKS, Sir Joseph, (1743-1820)

CANADA (LABRADOR) Cont'd.

1850-1899
LIEBER, Oscar M., (-1862)
LOW, Albert Peter, (1861-1952)
WICHMANN, Carl Ernst Arthur, (1851-1927)

1900 to present
DOUGLAS, George Vibert, (1892?-1958)
LOW, Albert Peter, (1861-1952)
WICHMANN, Carl Ernst Arthur, (1851-1927)

CANADA
MANITOBA
1800-1849
KEATING, William Hypolitus, (1799-1840)

1850-1899
DOWLING, Donaldson Bogart, (1858-1925)
FERRIER, Walter F., (1865-1950)
HECTOR, James, Sir, (1834-1907)
McINNES, William, (1858-1925)
TYRRELL, Joseph Burr, (1858-1957)
WHITEAVES, John Joseph Frederick, (1835-1909)

1900 to present
ALCOCK, Frederick James, (1888-1972)
BATEMAN, John Danvers, (1910-1965)
BROWN, Eldon L., (-) [PROSPECTORS, DIVINERS AND MINING ENGINEERS]
BRUCE, Everend Lester "Louie", (1884-1949)
DERRY, Duncan Ramsay, (?1908-)
DOWLING, Donaldson Bogart, (1858-1925)
FERRIER, Walter F., (1865-1950)
HAKE, Benjamin Franklin, (1895-1973)
HECTOR, James, Sir, (1834-1907)
JOHNSTON, Ashton William, (1910-)
JOWSEY, Robert J., (1905-) [PROSPECTORS, DIVINERS AND MINING ENGINEERS]
LaBINE, Charles L., (1888-) [PROSPECTORS, DIVINERS AND MINING ENGINEERS]
LABINE, Gilbert Adelore, (1890-) [PROSPECTORS, DIVINERS AND MINING ENGINEERS]
McINNES, William, (1858-1925)
MOSHER, Alex, (1900-) [PROSPECTORS, DIVINERS AND MINING ENGINEERS]
TANTON, Thomas Leslie, (1890-1971)
TYLER, Stanley Allen, (1906-1963)
TYRRELL, Joseph Burr, (1858-1957)
WALLACE, Robert Charles, (1881-1955)
WARING, Waldo W., (1903-1955)
WHITEAVES, John Joseph Frederick, (1835-1909)

CANADA
NEW BRUNSWICK
1800-1849
DAWSON, John William, Sir, (1820-1899)
GESNER, Abraham, (1797-1864)
HENWOOD, William Jory, (1805-1875)
LOGAN, William Edmond, Sir, (1798-1875)

1850-1899
BAILEY, Loring Woart, (1839-1925)
CAMSELL, Charles, (1876-1958)
DAWSON, John William, Sir, (1820-1899)
ELLS, Robert Wheelock, (1845-1911)
GESNER, Abraham, (1797-1864)
HARTT, Charles Frederic, (1840-1878)
HENDERSON, James Alexander Leo, (?1874-1940)
HENWOOD, William Jory, (1805-1875)
LOGAN, William Edmond, Sir, (1798-1875)
MATTHEW, George Frederic, (1837-1923)
McINNES, William, (1858-1925)

1900 to present
ALCOCK, Frederick James, (1888-1972)
BAILEY, Loring Woart, (1839-1925)
BOYLEN, Matthew James, (1911?-) [PROSPECTORS, DIVINERS AND MINING ENGINEERS]
CALEY, John Fletcher, (1903-1971)
CAMSELL, Charles, (1876-1958)
ELLS, Robert Wheelock, (1845-1911)
ELLSWORTH, Hardy Vincent, (1889-1952)
EVANS, Charles Sparling, (1894-1962)
GOLDTHWAIT, James Walter, (1880-1947)
HAYES, Albert Orion, (1882-1965)
HENDERSON, James Alexander Leo, (?1874-1940)
HOWELL, Benjamin Franklin, (1890-1976)
MACKENZIE, Graham Stewart, (1905-1960)
MATTHEW, George Frederic, (1837-1923)
McINNES, William, (1858-1925)
MOORE, Prentiss D., (1895-1955)
NEALE, Ernest Richard Ward, (1923-)
ROBERTS, Hugh Marine, (1885-1965)
STOPES, Marie Charlotte Carmichael, (1880-1958)
WASHBURNE, Chester Wesley, (1883-1971)

CANADA
NEWFOUNDLAND
Up to 1800
BANKS, Sir Joseph, (1743-1820)

1800-1849
BANKS, Sir Joseph, (1743-1820)
JUKES, Joseph Beete, (1811-1869)
MURRAY, Alexander, (1810-1884)

1850-1899
JUKES, Joseph Beete, (1811-1869)

CANADA (NEWFOUNDLAND) 1850-1899 Cont'd.

KINDLE, Edward Martin, (1869-1940)
MURRAY, Alexander, (1810-1884)
NEWELL ARBER, Edward Alexander, (1870-1918)

1900 to present
BAKER, Herbert Arthur, (1885-1954)
BANNERMAN, Harold MacColl, (1897-1976)
BOYLEN, Matthew James, (1911?-) [PROSPECTORS, DIVINERS AND MINING ENGINEERS]
CONNELL, Frederick M., (1885-) [PROSPECTORS, DIVINERS AND MINING ENGINEERS]
DALE, Nelson Clark, (1880-1962)
FAIRBAIRN, Harold Williams, (1906-)
FRITTS, Crawford Ellsworth, (1927-1972)
HAYES, Albert Orion, (1882-1965)
HOWELL, Benjamin Franklin, (1890-1976)
JOHNSTON, Ashton William, (1910-)
KAY, George Marshall, (1904-1975)
KINDLE, Edward Martin, (1869-1940)
NEALE, Ernest Richard Ward, (1923-)
NEWELL ARBER, Edward Alexander, (1870-1918)
NEWHOUSE, Walter Harry, (1897-1969)
SHROCK, Robert Rakes, (1904-)
SPATH, Leonard Frank, (1882-1957)
TAYLOR, James Carlton, (1924-1976)
WEEKS, Ludlow Jackson, (1899-1964)

CANADA
NORTHWEST TERRITORIES
1850-1899
BELL, Robert, (1841-1917)
CAMERON, A. E., (-)
CAMSELL, Charles, (1876-1958)
DAWSON, George Mercer, (1849-1901)
ELLS, Robert Wheelock, (1845-1911)
LOW, Albert Peter, (1861-1952)
TYRRELL, Joseph Burr, (1858-1957)

1900 to present
BATEMAN, John Danvers, (1910-1965)
BATTLE, Walter Ravenhill Brown "Ben", (1919-1953)
BELL, Alfred Hannam, (1895-1977)
BELL, Robert, (1841-1917)
BYRNE, Jerome Cotter, (1910-)
CAMERON, A. E., (-)
CAMSELL, Charles, (1876-1958)
CHIPMAN, Kenneth G., (-)
DAWSON, George Mercer, (1849-1901)
DERRY, Duncan Ramsay, (?1908-)
ELLS, Robert Wheelock, (1845-1911)
HARQUAIL, James A., (-)
HAYCOCK, Maurice, (1900-)
HOLTEDAHL, Olaf, (1885-1975)

CANADA (NORTHWEST TERRITORIES) 1900 to present Cont'd.

HORNBY, John, (1880-1927) [PROSPECTORS, DIVINERS AND MINING ENGINEERS]
HUME, George Sherwood, (1893-1965)
JOHNSTON, Ashton William, (1910-)
KRALIS, Dan F., Jr., (1919-1962)
LaBINE, Charles L., (1888-) [PROSPECTORS, DIVINERS AND MINING ENGINEERS]
LABINE, Gilbert Adelore, (1890-) [PROSPECTORS, DIVINERS AND MINING ENGINEERS]
LOW, Albert Peter, (1861-1952)
MALOUF, S. E., (1914-)
MOSHER, Alex, (1900-) [PROSPECTORS, DIVINERS AND MINING ENGINEERS]
MOSHER, Murdock C., (-) [PROSPECTORS, DIVINERS AND MINING ENGINEERS]
MUNSTERHJELM, Erik, (-) [PROSPECTORS, DIVINERS AND MINING ENGINEERS]
SPROULE, John Campbell, (1905-1970)
TYRRELL, Joseph Burr, (1858-1957)
WALKER, Paul, (1934-1959)
WATTS, Murray Edmund, (1909-) [PROSPECTORS, DIVINERS AND MINING ENGINEERS]
WEBB, John Benwell, (1901-1975)
WEEKS, Ludlow Jackson, (1899-1964)
YOUNGQUIST, Walter, (-)

CANADA
NOVA SCOTIA
1800-1849
ALGER, Francis, (1807-1863)
BROWN, Richard, (1805-1882)
CHAPMAN, Edward John, (1821-1904)
DAWSON, John William, Sir, (1820-1899)
HONEYMAN, David, Reverend, (1814-1889)
HOW, Henry, (1828-1879)
JACKSON, Charles Thomas, (1805-1880)
LESLEY, J. Peter, (1819-1903)
LOGAN, William Edmond, Sir, (1798-1875)
SALMON, Henry Curwen, (1827-1873)

1850-1899
ALGER, Francis, (1807-1863)
BROWN, Richard, (1805-1882)
CHAPMAN, Edward John, (1821-1904)
DAWSON, John William, Sir, (1820-1899)
ELLS, Robert Wheelock, (1845-1911)
FLETCHER, Hugh, (1848-1909)
HONEYMAN, David, Reverend, (1814-1889)
HOW, Henry, (1828-1879)
JACKSON, Charles Thomas, (1805-1880)
LESLEY, J. Peter, (1819-1903)
LOGAN, William Edmond, Sir, (1798-1875)
MARSH, Othniel Charles, (1831-1899)
SALMON, Henry Curwen, (1827-1873)

CANADA (NOVA SCOTIA) 1850-1899 Cont'd.

WARREN, Charles Hyde, (1876-1950)

1900 to present

CHAPMAN, Edward John, (1821-1904)
DALE, Nelson Clark, (1880-1962)
DENNEN, William Henry, (1920-)
DERRY, Duncan Ramsay, (?1908-)
DOUGLAS, George Vibert, (1892?-1958)
ELLS, Robert Wheelock, (1845-1911)
FAIRBAIRN, Harold Williams, (1906-)
FERGUSON, John Logan, (1893-1973)
FLETCHER, Hugh, (1848-1909)
GOLDTHWAIT, James Walter, (1880-1947)
HAYES, Albert Orion, (1882-1965)
LESLEY, J. Peter, (1819-1903)
MATHER, Kirtley Fletcher, (1888-1978)
MOORE, Prentiss D., (1895-1955)
NEALE, Ernest Richard Ward, (1923-)
NEWHOUSE, Walter Harry, (1897-1969)
POOLE, Henry Skeffington, (1884-1917)
ROMER, Alfred Sherwood, (1894-1973)
SPROULE, John Campbell, (1905-1970)
STERNBERG, Charles M., (1885-)
WARREN, Charles Hyde, (1876-1950)
WEEKS, Ludlow Jackson, (1899-1964)
WRIGHT, "Put", (-1962)

CANADA

ONTARIO

Up to 1800

KALM, Pehr, (1716-1779)

1800-1849

CHAPMAN, Edward John, (1821-1904)
HOW, Henry, (1828-1879)
HUNT, Thomas Sterry, (1826-1892)
LOGAN, William Edmond, Sir, (1798-1875)
RIVOT, Louis Edouard, (1820-1869)
ROY, Thomas, (-1842)

1850-1899

BARLOW, Alfred Ernest, (1861-1914)
BEEBY-THOMPSON, Arthur, (1873-1968)
BELL, James Macintosh, (1877-1934)
BELL, Robert, (1841-1917)
BENSLEY, Benjamin Arthur, (1875-1934)
BILLINGS, Walter R., (?1849-1920)
BOYD, Walter Halcro, (1878-1960)
BROCK, Reginald Walter, (1874-1935)
CHAPMAN, Edward John, (1821-1904)
COLEMAN, Arthur Philemon, (1852-1939)
DOWLING, Donaldson Bogart, (1858-1925)
FERRIER, Walter F., (1865-1950)
GORDON, Charles Henry, (1857-1934)

CANADA (ONTARIO) 1850-1899 Cont'd.

HARRINGTON, Bernard J., (1848-1907)
HINDE, George Jennings, (1839-1918)
HOW, Henry, (1828-1879)
HUNT, Thomas Sterry, (1826-1892)
LAWSON, Andrew Cowper, (1861-1952)
LOGAN, William Edmond, Sir, (1798-1875)
McINNES, William, (1858-1925)
MILLER, Willet Green, (1839-1925)
MOORE, Elwood S., (1879-1966)
MUIR, John, (1838-1914)
PARKS, William Arthur, (1868-1936)
PARSONS, Arthur Leonard, (1873-1957)
RIVOT, Louis Edouard, (1820-1869)
WALKER, Thomas Leonard, (1867-1942)
WILLIAMS, George Huntington, (1856-1894)
WILLMOTT, Arthur Brown, (1866-1914)
WILSON, Alfred William Gunning, (1873-1954)

1900 to present
AMBROSE, John Willis, (-1974)
BANNERMAN, Harold MacColl, (1897-1976)
BARLOW, Alfred Ernest, (1861-1914)
BATEMAN, Alan Mara, (1889-1971)
BATEMAN, John Danvers, (1910-1965)
BEEBY-THOMPSON, Arthur, (1873-1968)
BELL, Alfred Hannam, (1895-1977)
BELL, James Macintosh, (1877-1934)
BELL, Robert, (1841-1917)
BENSLEY, Benjamin Arthur, (1875-1934)
BILLINGS, Walter R., (?1849-1920)
BOWEN, Norman Levi, (1887-1956)
BOYD, Walter Halcro, (1878-1960)
BOYLEN, Matthew James, (1911?-) [PROSPECTORS, DIVINERS AND MINING ENGINEERS]
BROCK, Reginald Walter, (1874-1935)
BRUCE, Everend Lester "Louie", (1884-1949)
BRYAN, Joseph Jefferson, (1905-1951)
BYRNE, Jerome Cotter, (1910-)
CALEY, John Fletcher, (1903-1971)
CHAPMAN, Edward John, (1821-1904)
COLEMAN, Arthur Philemon, (1852-1939)
CONNELL, Frederick M., (1885-) [PROSPECTORS, DIVINERS AND MINING ENGINEERS]
COOKE, Harold Caswell, (1884-1956)
CROSS, Julian Gifford, (1889-)
DERRY, Duncan Ramsay, (?1908-)
DOWLING, Donaldson Bogart, (1858-1925)
ELLSWORTH, Hardy Vincent, (1889-1952)
FAIRBAIRN, Harold Williams, (1906-)
FERRIER, Walter F., (1865-1950)
FRASER, Horace John "Dutch", (1905-1969)
FRITTS, Crawford Ellsworth, (1927-1972)
GILLANDERS, Earl B., (-)
GOLDTHWAIT, James Walter, (1880-1947)

CANADA (ONTARIO) 1900 to present Cont'd.

GORDON, Charles Henry, (1857-1934)
GRAHAM, Richard Percival Devereux, (1880-1965)
GROUT, Frank Fitch, (1880-1958)
HARQUAIL, James A., (-)
HARRINGTON, Bernard J., (1848-1907)
HAWLEY, James Edwin, (1897-1965)
HINDE, George Jennings, (1839-1918)
HUME, George Sherwood, (1893-1965)
HUSTON, Charles, (1908-) [PROSPECTORS, DIVINERS AND MINING ENGINEERS]
JOHNSTON, Ashton William, (1910-)
JOUBIN, Francis Renault, (1911-)
JOWSEY, Robert J., (1905-) [PROSPECTORS, DIVINERS AND MINING ENGINEERS]
LaBINE, Charles L., (1888-) [PROSPECTORS, DIVINERS AND MINING ENGINEERS]
LABINE, Gilbert Adelore, (1890-) [PROSPECTORS, DIVINERS AND MINING ENGINEERS]
LAWSON, Andrew Cowper, (1861-1952)
LEDOUX, Auguste, (1887-1918)
MADSEN, Marius Kristian, (1899-) [PROSPECTORS, DIVINERS AND MINING ENGINEERS]
MANDARINO, Joseph Anthony, (-)
MAWDSLEY, James Buckland, (1894-1964)
McBEAN, John Williams, (1913-)
McINNES, William, (1858-1925)
McKINSTRY, Hugh Exton, (1896-1961)
McLAUGHIN, Dean Benjamin, (1901-1965)
MEEN, Victor Ben, (1910-1971)
MILLER, Willet Green, (1839-1925)
MOORE, Elwood S., (1879-1966)
MOORHOUSE, Walter Wilson, (1913-1969)
MOSHER, Alex, (1900-) [PROSPECTORS, DIVINERS AND MINING ENGINEERS]
MOSHER, Murdock C., (-) [PROSPECTORS, DIVINERS AND MINING ENGINEERS]
MUIR, John, (1838-1914)
PARKER, Ralph D., (1898-) [PROSPECTORS, DIVINERS AND MINING ENGINEERS]
PARKS, William Arthur, (1868-1936)
PARSONS, Arthur Leonard, (1873-1957)
PEACOCK, Martin Alfred, (1898-1950)
PERRY, Thomas Gregory, (1919-1972)
QUIRKE, Terence Thomas, (1886-1947)
REDPATH, James B., (-) [PROSPECTORS, DIVINERS AND MINING ENGINEERS]
RITTENHOUSE, Gordon, (1910-1974)
ROBERTS, Hugh Marine, (1885-1965)
ROBERTSON, Andrew, (1903?-) [PROSPECTORS, DIVINERS AND MINING ENGINEERS]
RUDOLPH, Joseph, (1897-1953)
RUSSELL, Loris Shano, (1904-)
SHEARMAN, Harold Leigh, (1921-1974)
SMITH, Stanley, (1883-1955)

CANADA (ONTARIO) 1900 to present Cont'd.

SOUTHWORTH, Charles, (1880-1968)
STERNBERG, Charles M., (1885-)
STEWART, Irvine E., (1893-1965)
STRANGWAY, David William, (1934-)
TANTON, Thomas Leslie, (1890-1971)
THOMPSON, Hugh R., (1926-1959)
THOMPSON, Robert Mitchell, (1918-1967)
THOMSON, Joseph Ellis, (1882-1944)
THOMSON, Robert, (1902-1970)
TYLER, Stanley Allen, (1906-1963)
WALKER, Thomas Leonard, (1867-1942)
WALLACE, Robert Charles, (1881-1955)
WEBB, John Benwell, (1901-1975)
WEEKS, Ludlow Jackson, (1899-1964)
WHEELER, Orby Clinton, (1898-1970)
WILLMOTT, Arthur Brown, (1866-1914)
WILSON, Alfred William Gunning, (1873-1954)
WILSON, Alice Evelyn, (1881-1964)

CANADA
PRINCE EDWARD ISLAND
1850-1899
BAIN, Francis, (1842-1894)
ELLS, Robert Wheelock, (1845-1911)

1900 to present
ELLS, Robert Wheelock, (1845-1911)
SNIDER, Luther Crocker, (1882-1947)

CANADA
QUEBEC
Up to 1800
KALM, Pehr, (1716-1779)

1800-1849
CHAPMAN, Edward John, (1821-1904)
DAWSON, John William, Sir, (1820-1899)
HUNT, Thomas Sterry, (1826-1892)
RIVOT, Louis Edouard, (1820-1869)
SELWYN, Alfred Richard Cecil, (1824-1902)

1850-1899
BELL, Robert, (1841-1917)
CHAPMAN, Edward John, (1821-1904)
DAWSON, John William, Sir, (1820-1899)
DOUGLAS, James, (1837-)
DRESSER, John Alexander, (1866-1954)
ELLS, Robert Wheelock, (1845-1911)
HARRINGTON, Bernard J., (1848-1907)
HUNT, Thomas Sterry, (1826-1892)
KEITH, Arthur, (1864-1944)
LAFLAMME, Joseph Clovis Kemner, Monseigneur, (1849-1910)
LOW, Albert Peter, (1861-1952)
RICHARDSON, Charles Henry, (1862-1935)

CANADA (QUEBEC) 1850-1899 Cont'd.

RIVOT, Louis Edouard, (1820-1869)
SELWYN, Alfred Richard Cecil, (1824-1902)
SHALER, Nathaniel Southgate, (1841-1906)
WARREN, Charles Hyde, (1876-1950)
WHITEAVES, John Joseph Frederick, (1835-1909)
WILSON, Alfred William Gunning, (1873-1954)

1900 to present

ALCOCK, Frederick James, (1888-1972)
BANNERMAN, Harold MacColl, (1897-1976)
BEAUCHEMIN, Pierre, (-) [PROSPECTORS, DIVINERS AND MINING ENGINEERS]
BELL, Robert, (1841-1917)
BOYLEN, Matthew James, (1911?-) [PROSPECTORS, DIVINERS AND MINING ENGINEERS]
BRUET, Edmond, (1890-)
CHAPMAN, Edward John, (1821-1904)
COOKE, Harold Caswell, (1884-1956)
CORYELL, Horace Noble, (1888-1965)
CRICKMAY, Geoffrey W., (1905-1971)
DERRY, Duncan Ramsay, (?1908-)
DRESSER, John Alexander, (1866-1954)
ELLS, Robert Wheelock, (1845-1911)
FAESSLER, Carl, (1895-1957)
FAIRBAIRN, Harold Williams, (1906-)
GILLSON, Joseph Lincoln, (1895-1964)
GOLDTHWAIT, James Walter, (1880-1947)
GRAHAM, Richard Percival Devereux, (1880-1965)
HARRINGTON, Bernard J., (1848-1907)
HUME, George Sherwood, (1893-1965)
HUTCHINS, Forbes M., (-)
KEITH, Arthur, (1864-1944)
LAFLAMME, Joseph Clovis Kemner, Monseigneur, (1849-1910)
LOW, Albert Peter, (1861-1952)
MACKENZIE, Graham Stewart, (1905-1960)
MACOUN, James Melville, (-1920)
MADSEN, Marius Kristian, (1899-) [PROSPECTORS, DIVINERS AND MINING ENGINEERS]
MAWDSLEY, James Buckland, (1894-1964)
MEEN, Victor Ben, (1910-1971)
MOSHER, Alex, (1900-) [PROSPECTORS, DIVINERS AND MINING ENGINEERS]
MOSHER, Murdock C., (-) [PROSPECTORS, DIVINERS AND MINING ENGINEERS]
NEALE, Ernest Richard Ward, (1923-)
PERRY, Thomas Gregory, (1919-1972)
REDPATH, James B., (-) [PROSPECTORS, DIVINERS AND MINING ENGINEERS]
RICHARDSON, Charles Henry, (1862-1935)
ROBERTS, Hugh Marine, (1885-1965)
ROBERTSON, Andrew, (1903?-) [PROSPECTORS, DIVINERS AND MINING ENGINEERS]
SELWYN, Alfred Richard Cecil, (1824-1902)
SHALER, Nathaniel Southgate, (1841-1906)

CANADA (QUEBEC) 1900 to present Cont'd.

SPRINGER, Karl J., (-) [PROSPECTORS, DIVINERS AND MINING ENGINEERS]
SPROULE, John Campbell, (1905-1970)
STEWART, Irvine E., (1893-1965)
WARREN, Charles Hyde, (1876-1950)
WATTS, Murray Edmund, (1909-)
WATTS, Murray Edmund, (1909-) [PROSPECTORS, DIVINERS AND MINING ENGINEERS]
WEBB, John Benwell, (1901-1975)
WEEKS, Ludlow Jackson, (1899-1964)
WHITEAVES, John Joseph Frederick, (1835-1909)
WILLIAMSON, John Thorburn, (1907-1958)
WILSON, Alfred William Gunning, (1873-1954)

CANADA
SASKATCHEWAN
1800-1849
RICHARDSON, John, Sir, (1787-1865)

1850-1899
DAWSON, George Mercer, (1849-1901)
HECTOR, James, Sir, (1834-1907)
McINNES, William, (1858-1925)
RICHARDSON, John, Sir, (1787-1865)
TYRRELL, Joseph Burr, (1858-1957)
WHITEAVES, John Joseph Frederick, (1835-1909)

1900 to present
ALCOCK, Frederick James, (1888-1972)
BAILEY, Ralph Fraser, (1920-1953)
BRUCE, Everend Lester "Louie", (1884-1949)
BYRNE, Jerome Cotter, (1910-)
DAWSON, George Mercer, (1849-1901)
DE KOCH, William Gaspar, (1898-1954)
EDMONDS, Frederic Harrison, (1898-1965)
ELLSWORTH, Hardy Vincent, (1889-1952)
GILLANDERS, Earl B., (-)
HAKE, Benjamin Franklin, (1895-1973)
HECTOR, James, Sir, (1834-1907)
JOUBIN, Francis Renault, (1911-)
KRALIS, Dan F., Jr., (1919-1962)
MAWDSLEY, James Buckland, (1894-1964)
McINNES, William, (1858-1925)
NININGER, Harvey Harlow, (-)
RUSSELL, Loris Shano, (1904-)
SANDERSON, James Owen Gresham "Pete", (1898-1963)
SARJEANT, William Antony Swithin, (1935-)
SPROULE, John Campbell, (1905-1970)
STERNBERG, Charles M., (1885-)
TANTON, Thomas Leslie, (1890-1971)
TYRRELL, Joseph Burr, (1858-1957)
WARING, Waldo W., (1903-1955)
WEBB, John Benwell, (1901-1975)
WEEKS, Ludlow Jackson, (1899-1964)

CANADA (SASKATCHEWAN) 1900 to present Cont'd.

WHITEAVES, John Joseph Frederick, (1835-1909)

CANADA
YUKON
1850-1899
BROCK, Reginald Walter, (1874-1935)
CAMSELL, Charles, (1876-1958)
MACDONALD, Alexander, (-) [PROSPECTORS, DIVINERS AND MINING ENGINEERS]
McCONNELL, Richard George, (1857-1942)
MITCHELL, George M., (-) [PROSPECTORS, DIVINERS AND MINING ENGINEERS]
OGILVIE, William, (1846-1912)
TRELAWNEY-ANSELL, E. C., (-) [PROSPECTORS, DIVINERS AND MINING ENGINEERS]
TYRRELL, Joseph Burr, (1858-1957)

1900 to present
AHO, Aaro E., (-1977)
BELL, Alfred Hannam, (1895-1977)
BROCK, Reginald Walter, (1874-1935)
BROWN, Bahngrell Walter "Bill", (1916-)
CAIRNES, Delorme Donaldson, (1879-1917)
CAMSELL, Charles, (1876-1958)
CHIPMAN, Kenneth G., (-)
DERRY, Duncan Ramsay, (?1908-)
EVANS, Charles Sparling, (1894-1962)
HUME, George Sherwood, (1893-1965)
JOHNSTON, Ashton William, (1910-)
MANDARINO, Joseph Anthony, (-)
MASSON, Charles E., (-) [PROSPECTORS, DIVINERS AND MINING ENGINEERS]
McCONNELL, Richard George, (1857-1942)
MITCHELL, George M., (-) [PROSPECTORS, DIVINERS AND MINING ENGINEERS]
OGILVIE, William, (1846-1912)
SCHLAIKJER, Erich Maren, (1905-1972)
TRELAWNEY-ANSELL, E. C., (-) [PROSPECTORS, DIVINERS AND MINING ENGINEERS]
TYRRELL, Joseph Burr, (1858-1957)
UGLOW, William Lawrence, (1884-1926)

CANADA
PROVINCE unspecified
Up to 1800
BARTRAM, John, (1699-1777)
CLARK, William, (1770-1838)
MACKENZIE, Alexander, Sir, (ca.1767-1830)
WELD, Isaac, Jr., (1774-1856)

1800-1849
AGASSIZ, Louis Jean Rodolphe, (1807-1873)
AULD, William, (fl.1810-1812)
BIGSBY, John Jeremiah, (1792-1881)

CANADA (PROVINCE unspecified) 1800-1849 Cont'd.

BILLINGS, Elkanah, (1820-1876)
BLACK, Samuel, (1785-1841)
CHRISTY, Henry, (1810-1865)
CLARK, William, (1770-1838)
FEATHERSTONHAUGH, George William, (1780-1866)
FINCH, John, (fl.1823-1835)
GESNER, Abraham, (1797-1864)
HIND, Henry Youle, (1823-1908)
ISBISTER, Alexander Kennedy, (1822-1883)
MACKENZIE, Alexander, Sir, (ca.1767-1830)
SALTER, John William, (1820-1869)
SAY, Thomas, (1787-1834)
WELD, Isaac, Jr., (1774-1856)
WINCHELL, Alexander, (1824-1891)
WOODROW, James, (1828-1907)

1850-1899

ADAMS, Andrew Leith, (-1882)
ADAMS, Frank Dawson, (1859-1942)
AGASSIZ, Louis Jean Rodolphe, (1807-1873)
AMI, Henry Mark, (1858-1931)
BELL, James Macintosh, (1877-1934)
BIGSBY, John Jeremiah, (1792-1881)
BILLINGS, Elkanah, (1820-1876)
BIRLEY, Caroline, (1851-1907)
BLAKE, John Frederick, Reverend, (1839-1906)
BROCK, Reginald Walter, (1874-1935)
BROWN, Barnum, (1873-1963)
BROWN, Henry Yorke Lyell "Stoney", (1844-1928)
CHRISTY, Henry, (1810-1865)
COLEMAN, Arthur Philemon, (1852-1939)
COLLINS, William Henry, (1878-1937)
CREDNER, Carl Hermann Georg, (1841-1913)
CUMINGS, Willard Lawsen, (1874-1945)
DAVIS, William Morris, (1850-1934)
DOWLING, Donaldson Bogart, (1858-1925)
DRESSER, John Alexander, (1866-1954)
FEATHERSTONHAUGH, George William, (1780-1866)
FENNER, Clarence Norman, (1870-1949)
FOX-STRANGWAYS, Charles Edward, (1844-1910)
GESNER, Abraham, (1797-1864)
GIROUX, N. J., (1859-1896)
HANNAY, Harry Campbell, (1874-1961)
HATCH, Frederick Henry, (?1864-1932)
HIND, Henry Youle, (1823-1908)
HOFFMANN, George Christian, (1837-1917)
ISBISTER, Alexander Kennedy, (1822-1883)
JULIAN, Henry Forbes, (1861-1912) [PROSPECTORS, DIVINERS AND MINING ENGINEERS]
KEITH, Arthur, (1864-1944)
LAMBE, Lawrence Morris, (1863-1919)
LEDOUX, Albert Reid, (1852-1923) [PROSPECTORS, DIVINERS AND MINING ENGINEERS]
LINDGREN, Waldemar, (1860-1939)

CANADA (PROVINCE unspecified) 1850-1899 Cont'd.

LLOYD, T. G. B., (1829-1876)
LYELL, Charles, Sir, (1797-1875)
MARGERIE, Emmanuel Marie Pierre Martin Jacquin de, (1862-1953)
MORSE, William Clifford, (1874-1962)
NICHOLSON, Henry Alleyne, (1844-1899)
OGILVIE, William, (1846-1912)
PENHALLOW, David Pearce, (1854-1910)
REED, Frederick Richard Cowper, (1869-1946)
RICHARDSON, Charles Henry, (1862-1935)
RIES, Heinrich, (1871-1951)
ROMEU, Albert de, (1875-1915)
ROTHWELL, Richard Pennefeather, (1836-1901)
SALTER, John William, (1820-1869)
SARGENT, Harry Lionel, (1871-1955)
SLATER, Herbert Kelsall, (1875-1913)
SPENCER, Joseph William Winthrop, (1851-1921)
STEINMANN, Gustav, (1856-1929)
WALKER, Thomas Leonard, (1867-1942)
WELD, Isaac, Jr., (1774-1856)
WESTON, Thomas Chesmer, (1832-1910)
WHYMPER, Edward, (1840-1911)
WINCHELL, Alexander, (1824-1891)
WINCHELL, Horace Vaughn, (1865-1923)
WOODROW, James, (1828-1907)
WOODRUFF, Elmer Grant, (1872-1952)
YOUNG, George Albert, (1898-1947)

1900 to present

ADAMS, Frank Dawson, (1859-1942)
AHO, Aaro E., (-1977)
ALCOCK, Frederick James, (1888-1972)
AMBROSE, Arthur Warren, (1889-1952)
AMI, Henry Mark, (1858-1931)
ANDREWS, Richard Wallis, (?1927-1967)
BALL, Harry Standish, (?1888-1941)
BEACH, Floyd Kellogg, (1884-1964)
BEACH, Hugh Hamilton, (1909-1973)
BECKER, Clyde M., (1882-1938)
BELL, James Macintosh, (1877-1934)
BERNOUILLI, Walter, (1885-1946)
BESCHEL, Roland, (1928-1971)
BIRLEY, Caroline, (1851-1907)
BLAKE, George Stanfield, (-1940)
BLAKE, John Frederick, Reverend, (1839-1906)
BOSWORTH, Thomas Owen, (1882-1928)
BROCK, Reginald Walter, (1874-1935)
BROWN, Barnum, (1873-1963)
BROWN, Henry Yorke Lyell "Stoney", (1844-1928)
BRUCE, Everend Lester "Louie", (1884-1949)
CAIRNES, Delorme Donaldson, (1879-1917)
CAMPBELL, Harry A., (1893-1951)
CHISHOLM, William Francis, (1886-1944)
CLAPP, Frederick Gardner, (1879-1944)
COLEMAN, Arthur Philemon, (1852-1939)

CANADA (PROVINCE unspecified) 1900 to present Cont'd.

COLLINS, William Henry, (1878-1937)
COPASS, Jack Malcolm, (1906-1963)
CREDNER, Carl Hermann Georg, (1841-1913)
CRICKMAY, Geoffrey W., (1905-1971)
CUMINGS, Willard Lawsen, (1874-1945)
DAVIS, William Morris, (1850-1934)
DOWLING, Donaldson Bogart, (1858-1925)
DRESSER, John Alexander, (1866-1954)
DYER, William Spafford, (1894-1941)
EDWARDS, Merwin Guy, (1889-1969)
ELY, Fred Burchard, (1883-1946)
ESKOLA, Pentti Eelis, (1883-1964)
EVANS, Charles Sparling, (1894-1962)
FAESSLER, Carl, (1895-1957)
FARISH, Linn Markley, (1901-1944)
FENNER, Clarence Norman, (1870-1949)
FIEDLER, William M., (1914-1970)
FITZGERALD, Paul Eugene, (1903-1961)
FLAHERTY, Robert J., (1884-1951)
FOX-STRANGWAYS, Charles Edward, (1844-1910)
FREY, M. Gordon, (1913-)
FROHBERG, Max Hans, (1901-1970)
GEIS, Wilfrid Hoy "Bill", (1892-1961)
GOLLNICK, Robert Louis, (1918-1963)
GOODELL, Horace N. "Hank", (1915-1971)
GOWER, John A., (1921-1972)
GRATON, Louis Caryl, (1880-1970)
HALL, Thomas Clifford Fitzwilliam, (1886-1956)
HALSE, George Whitfield, (1885-1968)
HAMMOND, Weldon Woolf, (1908-1964)
HANNAY, Harry Campbell, (1874-1961)
HARRISON, Thomas Samuel, (1881-1964)
HATCH, Frederick Henry, (?1864-1932)
HELLENS, Alexander Daniel, (1911-) [PROSPECTORS, DIVINERS AND MINING ENGINEERS]
HEWETT, Donnel Foster, (1881-1971)
HEWITT, Robert Leigh, (1911-1956)
HIND, Henry Youle, (1823-1908)
HINDES, Earle P., (1888-1962)
HITCHEN, Charles Stansfield, (1904-1971)
HOFFMANN, George Christian, (1837-1917)
HUBBARD, Bela, (1890-1959)
HURLEY, Patrick Mason, (1912-)
IDDINGS, Arthur, (1891-1957)
JÉRÉMINE, Elisabeth, (née Tschernaieff), (1879-1964)
JUDSON, Wilber, (1880-1951)
JULIAN, Henry Forbes, (1861-1912) [PROSPECTORS, DIVINERS AND MINING ENGINEERS]
KEITH, Arthur, (1864-1944)
KENNEDY, William, (-1926)
KRIEGER, Philip, (1900-1940)
LAMBE, Lawrence Morris, (1863-1919)
LAMSENS, O. Robert, (1922-1971)
LANDES, Robert William, (1910-1969)

CANADA (PROVINCE unspecified) 1900 to present Cont'd.

LECKWIJCK, Willem Peter van, (1902-1975)
LEDINGHAM, Glen Wallace, (1911-1978)
LEDOUX, Albert Reid, (1852-1923) [PROSPECTORS, DIVINERS AND MINING ENGINEERS]
LEHNER, Ernst J., (1895-1968)
LINDGREN, Waldemar, (1860-1939)
LINDSLEY, Thayer, (1882-) [PROSPECTORS, DIVINERS AND MINING ENGINEERS]
LYELL, Charles, Sir, (1797-1875)
MACKAY, Robert Alexander, (?1904-1961)
MACKENZIE, Andrew Nicholls "Kelley", (1888-1951)
MANN, Robert Scott, (1915-1956)
MARGERIE, Emmanuel Marie Pierre Martin Jacquin de, (1862-1953)
McLEARN, Frank Harris, (1885-1964)
McNEELY, Robert Barnes, (1902-1966)
MENKEN, Frederick Andrew, (1905-1965)
MILLER, Wendell Zerbe, (1892-1958)
MORRIS, Frederick Kuhne, (1885-1962)
MORSE, William Clifford, (1874-1962)
NOLAN, Edward Dana, (1890-1926)
OGILVIE, William, (1846-1912)
OWEN, William Joseph, (1929-1959)
PAPP, Simon, (1886-1970)
PATTERSON, Arthur Lindo, (1902-1966)
PENHALLOW, David Pearce, (1854-1910)
PENROSE, Evelyn, (-) [PROSPECTORS, DIVINERS AND MINING ENGINEERS]
PHELPS, Frederick George, (1913-1964)
PRETYMAN, Frank Remington, (1891?-1917)
RAMDOHR, Paul, (1890-)
REA, Henry Carter, (1900-1963)
REECE, Alan W., (1921-1960)
REED, Frederick Richard Cowper, (1869-1946)
RICHARDSON, Charles Henry, (1862-1935)
RIDGEWAY, Bertrand S., (1899-1942)
RIES, Heinrich, (1871-1951)
ROBERTS, Morgan Evan, (1894-1941)
ROMEU, Albert de, (1875-1915)
ROTHWELL, Richard Pennefeather, (1836-1901)
ROUBAULT, Marcel Edouard, (1905-1974)
SARGENT, Harry Lionel, (1871-1955)
SCHOLTZ, Douglas Leinster, (1902-1977)
SCOFIELD, Lloyd M., (1897-1969)
SLATER, Herbert Kelsall, (1875-1913)
SMITH, James Robert, (1925-1972)
SPENCER, Joseph William Winthrop, (1851-1921)
STEHR, Raymond Albert, (1903-1967)
STEINMANN, Gustav, (1856-1929)
STETSON, Henry Crosby, (1900-1955)
STRANGWAY, David William, (1934-)
STUTZER, Otto, (1881-1936)
SUJKOWSKI, Zbigniew Leliwa, (1898-1954)
TAYLOR, Griffith, (1881-1963)
TEMPLETON, Eugene Crittenden, (1889-1932)

CANADA (PROVINCE unspecified) 1900 to present Cont'd.

THOMSON, Joseph Ellis, (1882-1944)
TRAGER, Earl A., (1893-1978)
WALCOTT, Mary Vaux, née Vaux, (-1940)
WALKER, Thomas Leonard, (1867-1942)
WASSON, Theron, (1887-1970)
WESTON, Thomas Chesmer, (1832-1910)
WHITEHEAD, Walter Lucius, (1891-1969)
WHYMPER, Edward, (1840-1911)
WILSON, John Tuzo, (1908-)
WILSON, Roy Arthur, (1891-1970)
WINCHELL, Horace Vaughn, (1865-1923)
WOODROW, James, (1828-1907)
WOODRUFF, Elmer Grant, (1872-1952)
YOUNG, George Albert, (1898-1947)

CANARY ISLANDS (NORTH ATLANTIC OCEAN)

Up to 1800

BUCH, Christian Leopold von, Freiherr, (1774-1853)
HUMBOLDT, Friedrich Wilhelm Heinrich Alexander von, Freiherr, (1769-1859)

1800-1849

BUCH, Christian Leopold von, Freiherr, (1774-1853)
HUMBOLDT, Friedrich Wilhelm Heinrich Alexander von, Freiherr, (1769-1859)
SAINTE-CLAIRE DEVILLE, Charles, (1814-1876)

1850-1899

BUCH, Christian Leopold von, Freiherr, (1774-1853)
CHUDEAU, René, (1864-1921)
HAECKEL, Ernst Heinrich Philipp August, (1834-1919)
HUMBOLDT, Friedrich Wilhelm Heinrich Alexander von, Freiherr, (1769-1859)
SAINTE-CLAIRE DEVILLE, Charles, (1814-1876)
SAUER, Gustav Adolf, (1825-1932)

1900 to present

BOURCART, Jacques, (1891-1965)
CHUDEAU, René, (1864-1921)
DENIZOT, Georges, (-)
HAECKEL, Ernst Heinrich Philipp August, (1834-1919)
KOMOROWICZ, Maurycy, (1881-1923)
ROMANES, James, (1886-1942)
SAUER, Gustav Adolf, (1825-1932)

CAPE VERDE ISLANDS (NORTH ATLANTIC OCEAN)

1800-1849

DARWIN, Charles Robert, (1809-1882)

1850-1899

BERGT, Walther Adolf, (1864-1941)

CAPE VERDE ISLANDS (NORTH ATLANTIC OCEAN) 1850-1899 Cont'd.

DARWIN, Charles Robert, (1809-1882)

1900 to present
BERGT, Walther Adolf, (1864-1941)
TORRE DE ASSUNÇAO, Carlos Fernando, (1901-)

CAROLINE ISLANDS (PACIFIC OCEAN)
1900 to present
JOHNSON, Charles George, (1914-1969)

CAYENNE <u>See</u> FRENCH GUIANA.

CELEBES <u>See</u> INDONESIA.

CENTRAL AFRICA (unspecified)
1900 to present
ROUBAULT, Marcel Edouard, (1905-1974)

CENTRAL AFRICAN EMPIRE
1900 to present
BIRAUD, François, (1899-1936)
FAULKNER, Joseph Stanislaus "Ginger", (1928-1963)
HOURCQ, Victor, (1901-1961)
LEUTWEIN, Friedrich, (1911-1974)
SANDFORD, Kenneth Stuart, (1899-1971)

CENTRAL AMERICA (unspecified)
1800-1849
BURKART, Hermann Josef, (1798-1874)

1850-1899
BURKART, Hermann Josef, (1798-1874)
DICKERSON, Roy Ernest, (1877-1944)
ORDOÑEZ, Ezequiel, (1867-1950)
PILSBRY, Henry Augustus, (1862-1957)
SPENCER, Joseph William Winthrop, (1851-1921)
WINCHELL, Horace Vaughn, (1865-1923)
WOEIKOF, Alexander, (1842-1916)

1900 to present
BLANCHARD, William Grant, Jr., (1894-1952)
BRISCOE, Glenn Oliver, (1898-1974)
DECKER, Charles LaVerne, (1898-1955)
DICKERSON, Roy Ernest, (1877-1944)
JONES, William Francis, (1885-1941)
JOUBIN, Francis Renault, (1911-)
MARTYN, Phil F., (1903-1974)
ORDOÑEZ, Ezequiel, (1867-1950)
PILSBRY, Henry Augustus, (1862-1957)
SPENCER, Joseph William Winthrop, (1851-1921)
STIRTON, Ruben Arthur, (1901-1966)
WINCHELL, Horace Vaughn, (1865-1923)

CENTRAL AMERICA (unspecified) 1900 to present Cont'd.

WOEIKOF, Alexander, (1842-1916)

CENTRAL ASIA (unspecified)

1850-1899

BOHDANOWICZ, Karol, (1864-1947)
OBRUCHEV, Vladimir Afanas'evich, (1863-1956)
SCHLOSSER, Max, (1854-1933)

1900 to present

BOHDANOWICZ, Karol, (1864-1947)
CALCIATI, Cesare, (1885-1929)
KRECHTER, Vladimir, (?1897-1967)
MACHATSCHEK, Fritz, (-1957)
NORIN, Erik, (-)
OBRUCHEV, Vladimir Afanas'evich, (1863-1956)
SCHLOSSER, Max, (1854-1933)

CEYLON <u>See</u> SRI LANKA.

CHAD

1850-1899

CHUDEAU, René, (1864-1921)
DOUVILLÉ, Joseph Henri Ferdinand, (1846-1937)

1900 to present

CHUDEAU, René, (1864-1921)
DALLONI, Marius Gustave, (1880-1959)
DOUVILLÉ, Joseph Henri Ferdinand, (1846-1937)
JÉRÉMINE, Elisabeth, (née Tschernaieff), (1879-1964)
MILLOT, Georges, (1917-)
NICOLAS, Jean Paul, (1921-1967)
ROCH, Edouard, (1901-)

CHANNEL ISLANDS (UNITED KINGDOM)

1850-1899

BIRDS, James Adey, (1831-1894)
DAVIES, Thomas, (1837-1892)
DUNLOP, Andrew, (1845-1915)

1900 to present

DAVIES, Thomas, (1837-1892)
DUNLOP, Andrew, (1845-1915)
HENSON, Frederick Arthur, (1919-1957)
SMITH, Herbert Gladstone, (-1950)

CHATHAM ISLANDS (SOUTH PACIFIC OCEAN)

1800-1849

DIEFFENBACH, Ernst, (1811-1855)

1850-1899

DIEFFENBACH, Ernst, (1811-1855)

CHILE

Up to 1800

MAWE, John, (1764-1829)
TREVITHICK, Richard, (1771-1833) [PROSPECTORS, DIVINERS AND MINING ENGINEERS]

1800-1849

DARWIN, Charles Robert, (1809-1882)
DOMEYKO, Ignacy, (1802-1889)
FORBES, David, (1828-1876)
GRAHAM, Maria, Mrs., (1785-1842)
MAWE, John, (1764-1829)
MIERS, John, (1789-1879)
ORBIGNY, Alcide Charles Victor Dessalines d', (1802-1857)
PHILIPPI, Rudolph Amandus, (1808-1904)
PISSIS, Aimé, (1812-1889)
RICKARD, Francis Ignacio, (-)
TREVITHICK, Richard, (1771-1833) [PROSPECTORS, DIVINERS AND MINING ENGINEERS]

1850-1899

ADAMS, George Irving, (1870-1932)
BEUTELL, Albert, (1859-1921)
COURTY, Georges, (1875-1953)
CUMINGS, Willard Lawsen, (1874-1945)
DARWIN, Charles Robert, (1809-1882)
DOMEYKO, Ignacy, (1802-1889)
FORBES, David, (1828-1876)
LINDGREN, Waldemar, (1860-1939)
MALLARD, François Ernest, (1833-1894)
MIERS, John, (1789-1879)
MILLER, Benjamin Le Roy, (1874-1944)
MONTESSUS DE BALLORE, Fernand Jean Baptiste Marie Bernard, Comte de, (1851-1923)
MÖRICKE, Wilhelm, (1861-1897)
NORDENSKIÖLD, Nils Otto Gustaf, (1869-1928)
OCHSENIUS, Carl Christian, (1830-)
ORBIGNY, Alcide Charles Victor Dessalines d', (1802-1857)
PHILIPPI, Rudolph Amandus, (1808-1904)
PISSIS, Aimé, (1812-1889)
PLAGEMANN, A., (1858-1908)
RICKARD, Francis Ignacio, (-)
SOLMS-LAUBACH, Hermann zu, Graf, (1842-1915)
STEINMANN, Gustav, (1856-1929)
WILLIS, Bailey, (1857-1949)

1900 to present

ADAMS, George Irving, (1870-1932)
BANDY, Mark Chance, (1900-1963)
BEUTELL, Albert, (1859-1921)
COURTY, Georges, (1875-1953)
CRABTREE, John Edmund, (1940-1966)
CUMINGS, Willard Lawsen, (1874-1945)
DERRY, Duncan Ramsay, (?1908-)
GORDON, Samuel George, (1897-1952)
HERZENBERG, Robert, (1885-1955)

CHILE 1900 to present Cont'd.

HOLLINGWORTH, Sydney Ewart, (1899-1966)
HOLLISTER, Joseph Steffens, (1903-1953)
HUSTON, Charles, (1908-) [PROSPECTORS, DIVINERS AND MINING ENGINEERS]
INESON, Jack, (1917-1970)
ISENBERGER, Nathan Perry, (1896-1956)
LINDGREN, Waldemar, (1860-1939)
MILLER, Benjamin Le Roy, (1874-1944)
MONTESSUS DE BALLORE, Fernand Jean Baptiste Marie Bernard, Comte de, (1851-1923)
NES, C. L. van, (?1878-1963)
NOBLE, Earl Bart, (1894-1967)
NORDENSKIÖLD, Nils Otto Gustaf, (1869-1928)
PHILIPPI, Rudolph Amandus, (1808-1904)
PLAGEMANN, A., (1858-1908)
POLLARD, Ernest Rex, (1922-1966)
QUENSEL, Percy Dudgeon, (1881-1966)
RIGGS, Robert Jennings, (1888-1965)
SEVERY, Clarence Luther, (1885-1975)
SINGEWALD, Joseph Theophilus, Jr., (1884-1963)
SOLMS-LAUBACH, Hermann zu, Graf, (1842-1915)
STEINMANN, Gustav, (1856-1929)
TAZIEFF, Haroun, (1914-)
THALMANN, Hans E., (1899-1975)
TRASK, Parker Davies, (1899-1961)
UNGEMACH, Henri Léon, (1879-1936)
WEAVER, Charles Edwin, (1880-1958)
WETZEL, Walter, (1887-)
WHITEHEAD, Walter Lucius, (1891-1969)
WILLIS, Bailey, (1857-1949)
WING, Monta Eldo, (1897-1976)

CHILE _See also_ JUAN FERNANDEZ ISLAND.

CHINA

Up to 1800

BARROW, John, Sir, (1764-1848)
DINWIDDIE, James, (-)
GREGOR, William, Reverend, (1761-1817)
TAVERNIER, Jean Baptiste, baron d'Aubonne, (1605-1689)
TU WAN, (fl.1115-1133)

1800-1849

BARROW, John, Sir, (1764-1848)
BLAKE, William Phipps, (1826-1910)
DINWIDDIE, James, (-)
ELLIS, H., Right Honourable Sir, (-1855)
GREGOR, William, Reverend, (1761-1817)
LAING, Samuel, (1810-1897)
ROYLE, John Forbes, (-1857?)
SALTER, John William, (1820-1869)
SCHENK, Joseph August, (1815-1891)
SCHLAGINTWEIT, Adolf, (?1799-1857)
SCHLAGINTWEIT, Hermann, (-)

CHINA 1800-1849 Cont'd.

SCHLAGINTWEIT, Robert, (-)

1850-1899

ADAMS, George Irving, (1870-1932)
ANDERSSON, Johann Gunnar, (1874-1960)
BECHER, Harry Macdonald, (1855-1893)
BLAKE, William Phipps, (1826-1910)
BÖHM, Georg, (1854-1913)
BREUIL, Henri, Abbé, (1877-1961)
DRAKE, Noah Fields, (1864-1945)
DRUMMOND, Henry, (1851-1897)
DUDGEON, Patrick, (-1895)
ELLIS, H., Right Honourable Sir, (-1855)
FUKUCHI, Nobuyo, (1877-1934)
GIRTY, George Herbert, (1869-1939)
GRABAU, Amadeus William, (1870-1946)
GROEBER, Paul Friedrich Carl, (1885-1968)
HAGUE, Arnold, (1840-1917)
HANNAY, Harry Campbell, (1874-1961)
JACK, Robert Lockhart, (1878-1964)
JAECKEL, Otto, (1863-1929)
KORNITZER, Louis, (1873-) [PROSPECTORS, DIVINERS AND MINING ENGINEERS]
KOTÔ, Bundjirô, (1856-1935)
LAING, Samuel, (1810-1897)
LOUDERBACK, George Davis, (1874-1957)
McCARTHY, Edward Thomas, (-) [PROSPECTORS, DIVINERS AND MINING ENGINEERS]
MUSHKETOV, Ivan Vassilievitch, (1850-1902)
NORDENSKIÖLD, Nils Adolf Erik, (1832-1901)
OBRUCHEV, Vladimir Afanas'evich, (1863-1956)
PUMPELLY, Raphael, (1837-1923)
REED, Frederick Richard Cowper, (1869-1946)
RICHTHOFEN, Paul Wilhelm Ferdinand von, Freiherr, (1833-1905)
ROYLE, John Forbes, (-1857?)
SALTER, John William, (1820-1869)
SCHENK, Joseph August, (1815-1891)
SCHLAGINTWEIT, Hermann, (-)
SCHLAGINTWEIT, Robert, (-)
SCHLOSSER, Max, (1854-1933)
STOLICZKA, Ferdinand, (1838-1874)
TOKUNAGA, Shigeyasu, (1874-1940)
WALKER, Thomas Leonard, (1867-1942)
WAY, Herbert W. L., (-) [PROSPECTORS, DIVINERS AND MINING ENGINEERS]
WILLIS, Bailey, (1857-1949)
WIMAN, Carl Johan Josef Ernst, (1867-1944)
WOEIKOF, Alexander, (1842-1916)

1900 to present

ADAMS, George Irving, (1870-1932)
ANDERSON, Gustavus Edwin, (1879-1940)
ANDERSSON, Johann Gunnar, (1874-1960)
ANDREWS, Roy Chapman, (1884-1960)

CHINA 1900 to present Cont'd.

BARBOUR, George Brown, (1891-1977)
BECKWITH, Radcliffe Harold, (1900-1964)
BELTZ, Edward W., (1891-1970)
BLACK, Davidson, (1884-1934)
BLACKWELDER, Eliot, (1880-1969)
BLAKE, William Phipps, (1826-1910)
BÖHM, Georg, (1854-1913)
BREUIL, Henri, Abbé, (1877-1961)
CADY, Gilbert Haven, (1882-1970)
CHANEY, Ralph Works, (1890-1971)
CHAO, Yatseng T., (1898-1929)
CLAPP, Frederick Gardner, (1879-1944)
DRAKE, Noah Fields, (1864-1945)
DVOŘÁK, Arnošt, (1904-)
EATON, Arthur, (1886-1923)
FUKUCHI, Nobuyo, (1877-1934)
GIRTY, George Herbert, (1869-1939)
GRABAU, Amadeus William, (1870-1946)
GROEBER, Paul Friedrich Carl, (1885-1968)
HAGUE, Arnold, (1840-1917)
HANNAY, Harry Campbell, (1874-1961)
HEIM, Arnold, (?1882-1965)
HERALD, Frank Edward, (1887-1965)
HERITSCH, Franz, (1882-1945)
HOWELL, Benjamin Franklin, (1890-1976)
HUBMAN, Carl Wesley, (1905-1947)
HUDSON, Frank Samuel, (1890-1967)
HUTCHISON, Arthur Gilbertson, (1902-1949)
JACK, Robert Lockhart, (1878-1964)
JAECKEL, Otto, (1863-1929)
KINNEY, Harry Dill, (1886-1961)
KLOMPÉ, Theodorus Hendricus Franciscus, (?1904-1963)
KOBAYASHI, Teiichi, (1901-)
KOENIGSWALD, Gustav Heinrich Ralph von, (1902-)
KORNITZER, Louis, (1873-) [PROSPECTORS, DIVINERS AND MINING ENGINEERS]
KOTÔ, Bundjirô, (1856-1935)
KRYSHTOFOVICH, Afrikan Nikolaevich, (1885-1953)
LOUDERBACK, George Davis, (1874-1957)
McCARTHY, Edward Thomas, (-) [PROSPECTORS, DIVINERS AND MINING ENGINEERS]
MEAD, Warren Judson, (1883-1960)
MORRIS, Frederick Kuhne, (1885-1962)
MULLER, Siemon William, (1900-1970)
MUSHKETOV, Ivan Vassilievitch, (1850-1902)
NIGHTINGALE, William Thomas, (1897-1964)
NORDENSKIÖLD, Nils Adolf Erik, (1832-1901)
NORIN, Erik, (-)
NOTESTEIN, Frank B., (1885-1973)
OBRUCHEV, Vladimir Afanas'evich, (1863-1956)
PARDÉ, Maurice, (1893-1973)
PUMPELLY, Raphael, (1837-1923)
REED, Frederick Richard Cowper, (1869-1946)
RICH, John Lyon, (1884-1956)

CHINA 1900 to present Cont'd.

RICHTHOFEN, Paul Wilhelm Ferdinand von, Freiherr, (1833-1905)
SCHLOSSER, Max, (1854-1933)
SCHÜLLER, Arno, (1908-1968)
SOWERBY, Arthur de Carle, (1885-1954)
SUTTON, Frederick A., (1894-1950)
SUTTON, John, (-)
TAYLOR, Griffith, (1881-1963)
TEILHARD DE CHARDIN, Pierre, Father, S.J., (1881-1955)
TEMPLETON, Eugene Crittenden, (1889-1932)
THOMAS, David, (1880-1934)
TING, Ven Kiang, (?1887-1936)
TOKUNAGA, Shigeyasu, (1874-1940)
VISSER, Philips Christiaan, (1882-1955)
WALKER, Lucian Hickman, (1894-1970)
WALKER, Thomas Leonard, (1867-1942)
WARING, Gerald Ashley, (1883-1971)
WELLER, James Marvin, (1899-1976)
WILLIS, Bailey, (1857-1949)
WILSON, John Tuzo, (1908-)
WIMAN, Carl Johan Josef Ernst, (1867-1944)
WOEIKOF, Alexander, (1842-1916)

Time unknown
TEGENGREN, Felix, (-)

CHRISTMAS ISLAND (PACIFIC OCEAN)
1850-1899
ANDREWS, Charles William, (1866-1924)
CHAPMAN, Frederick, (1864-1943)
MURRAY, John, Sir, (1841-1914)

1900 to present
ANDREWS, Charles William, (1866-1924)
CHAPMAN, Frederick, (1864-1943)
MURRAY, John, Sir, (1841-1914)

COCOS ISLANDS (PACIFIC OCEAN) See KEELING ISLANDS.

COLOMBIA
Up to 1800
HUMBOLDT, Friedrich Wilhelm Heinrich Alexander von, Freiherr, (1769-1859)
TREVITHICK, Richard, (1771-1833) [PROSPECTORS, DIVINERS AND MINING ENGINEERS]

1800-1849
HOPKINS, Evan, (?1810-1867)
HUMBOLDT, Friedrich Wilhelm Heinrich Alexander von, Freiherr, (1769-1859)
TREVITHICK, Richard, (1771-1833) [PROSPECTORS, DIVINERS AND MINING ENGINEERS]

1850-1899
ANDERSON, Frank Marion, (1863-1945)

COLOMBIA 1850-1899 Cont'd.

BERGT, Walther Adolf, (1864-1941)
BRALY, Adrien, (1864-1927)
GALE, Hoyt Stoddard, (1876-1952)
HAMMOND, John Hays, (1855-1936)
HENNEN, Ray Vernon, (1875-1958)
HOPKINS, Evan, (?1810-1867)
HUMBOLDT, Friedrich Wilhelm Heinrich Alexander von, Freiherr, (1769-1859)
NAGYSUR, Hugo Böckh von, (1874-1931)
ORDOÑEZ, Ezequiel, (1867-1950)
THOMPSON, Beeby, (1848-1931)

1900 to present

ALDREDGE, Robert Frank, (1895-1961)
ANDERSON, Frank Marion, (1863-1945)
ANDERTON, Russell, (-) [PROSPECTORS, DIVINERS AND MINING ENGINEERS]
ANDREW, Arthur Robert, (1881-1941)
ARGABRITE, William Greame, (1885-1963)
BARROW, Geoffrey, (1899-1963)
BELTZ, Edward W., (1891-1970)
BERGT, Walther Adolf, (1864-1941)
BONTZ, Conrad K., (1894-1927)
BOOTS, Paul Heath, (1909-1947)
BOURCHIER, John Rickman, (1896-1927)
BRALY, Adrien, (1864-1927)
BÜRGL, Hans, (1907-1966)
BUTTERWORTH, Emerson McMillin, (1894-1961)
CARLSON, Charles Gordon, (1895-1977)
CARR, William L., (1919-1975)
CLARK, Howard, (1885-1972)
CONKLING, Russell C., (1897-1955)
COOKE, Charles Wythe, (1887-1971)
CRAWFORD, David Johnston, (1899-1960)
CRONEIS, Carey, (1901-1972)
DONOGHUE, David Patrick, (1891-1958)
FAHLE, Paul Bartlett, (1910-1954)
FICHTER, Hans Jakob, (1908-1977)
FREY, Alfred Paul, (1893-1963)
GALE, Hoyt Stoddard, (1876-1952)
GESTER, George Clark, (1884-1959)
GETZENDANER, Avalette Edmond, (1902-1962)
GILLES, Verner Arthur, (1886-1954)
HAMMOND, John Hays, (1855-1936)
HATFIELD, Willis Charles, "Luke", (1898-1970)
HAWKINS, Glenn Dewayne, (1897-1970)
HAZZARD, John Charles, (1903-1973)
HEDRICK, Oley Foster, (1891-1950)
HENNEN, Ray Vernon, (1875-1958)
HEROLD, Stanley Carrollton, (1883-1958)
HUBBARD, George Hulme, (1899-1969)
HUBMAN, Carl Wesley, (1905-1947)
HUNTLEY, Charles Stirling, (1895-1926)
HURLEY, Patrick Mason, (1912-)

COLOMBIA 1900 to present Cont'd.

HYDE, Clarence Edgar, (1894-1954)
ISENBERGER, Nathan Perry, (1896-1956)
JOLEAUD, Léonce, (1880-1938)
KLEIBER, Karl Ernst, (1912-1972)
KNOWLES, Gerald Elgin, (1923-1976)
KOCH, Harold Kenneth, (1927-1973)
LAMSENS, O. Robert, (1922-1971)
MACKENZIE, Andrew Nicholls "Kelley", (1888-1951)
MARTI, Ernest John, (1913-1964)
MARTIN, Lewis, (1920-1973)
McARTHUR, Donald, (1895-1960)
MILLER, Wendell Zerbe, (1892-1958)
MORGAN, George Dillon, (1894-1950)
NAGYSUR, Hugo Böckh von, (1874-1931)
NOMLAND, Jorgen O., (1885-1943)
NOTESTEIN, Frank B., (1885-1973)
NYGREN, Walter Eric, (1907-1971))
ORDOÑEZ, Ezequiel, (1867-1950)
PALMER, Robert Hastings, (1882-1948)
PECORA, William Thomas, (1913-1972)
POLLOCK, Charles Murray, (1898-1969)
RAINIER, Peter W., (-) [PROSPECTORS, DIVINERS AND MINING ENGINEERS]
RHOADES, Ralph Omer, (1895-1961)
RITER, Samuel W., (1896-1941)
ROGERS, Gaillard Sherburne, (1889-1919)
ROYO Y GOMEZ, José, (1895-1958)
RUSS, Leon Fair, (1886-1953)
SCHERMERHORN, W., (1894-)
SCHOLL, Louis August, Jr., (1887-1953)
SCHROYER, Charles Ross, (1887-1932)
SMITH, Homer James, (1901-1954)
SPROULE, John Campbell, (1905-1970)
STIRTON, Ruben Arthur, (1901-1966)
STREIFF, Viktor, (1909-1973)
STUTZER, Otto, (1881-1936)
SULLINS, M. Joseph, (1930-1972)
SUTTON, Frederick A., (1894-1950)
THOMPSON, Beeby, (1848-1931)
TRÜMPY, Daniel, (1893-1971)
WARING, Waldo W., (1903-1955)
WEAVER, Charles Edwin, (1880-1958)
WHEELER, Orby Clinton, (1898-1970)
WHITTIER, William Harrison, (1889-1951)
WIEDENMAYER, Carl, (1897-1951)
WILSON, Joseph Garland, (1902-1955)
WILSON, Thomas Carroll, (1906-1969)

COMMANDER ISLAND (NORTH PACIFIC OCEAN) See KOMANDORSKI ISLANDS (Arctic U.S.S.R.)

COMORO ISLANDS (INDIAN OCEAN)
1850-1899
LACROIX, François Antoine Alfred, (1863-1948)

COMORO ISLANDS (INDIAN OCEAN) Cont'd.

1900 to present
LACROIX, François Antoine Alfred, (1863-1948)

CONGO REPUBLIC
1900 to present
BIRAUD, François, (1899-1936)
FAULKNER, Joseph Stanislaus "Ginger", (1928-1963)
HOURCQ, Victor, (1901-1961)
MILLOT, Georges, (1917-)

CORSICA See FRANCE.

COSTA RICA
Up to 1800
TREVITHICK, Richard, (1771-1833) [PROSPECTORS, DIVINERS AND MINING ENGINEERS]

1800-1849
TREVITHICK, Richard, (1771-1833) [PROSPECTORS, DIVINERS AND MINING ENGINEERS]

1850-1899
GABB, William More, (1839-1878)
JAGGAR, Thomas Augustus, (1871-1953)
PITTIER, Henry François, (1857-1950)
SAPPER, Karl Theodor, (1866-1945)
SEEBACH, Karl Albert Ludwig von, (1839-1880)

1900 to present
BEEKLEY, Albert Leon, (1883-1952)
BEVIER, George Means, (1888-1972)
CARLSON, Charles Gordon, (1895-1977)
FOLGER, James Anthony, (1896-1956)
GILLERMAN, Elliot "Bud", (1913-1974)
HALSE, George Whitfield, (1885-1968)
JAGGAR, Thomas Augustus, (1871-1953)
MACKENZIE, Andrew Nicholls "Kelley", (1888-1951)
PITTIER, Henry François, (1857-1950)
POLLARD, Ernest Rex, (1922-1966)
ROMANES, James, (1886-1942)
SAPPER, Karl Theodor, (1866-1945)
SEARS, Julian Ducker, (1891-1970)
STEWART, Duncan, (1905-1969)
STUART, Duncan, VII, (1905-1969)
SULLINS, M. Joseph, (1930-1972)
TRISTÁN, José Fidel, (-1932)
WALLACE, Maurice Hudiburg, (1917-1961)
WEYL, Richard, (-)

CUBA
Up to 1800
HUMBOLDT, Friedrich Wilhelm Heinrich Alexander von, Freiherr, (1769-1859)

CUBA Cont'd.

1800-1849

HUMBOLDT, Friedrich Wilhelm Heinrich Alexander von, Freiherr, (1769-1859)
POURTALÈS, Louis François de, (1823-1880)

1850-1899

ABEL, Othenio, (1875-1946)
BEEBY-THOMPSON, Arthur, (1873-1968)
BROWN, Barnum, (1873-1963)
BURCHARD, Ernest Norman, (1875-1961)
CASTRO Y SUERO, Manuel Fernandez de, (1825-1895)
CHAPER, Maurice Armand, (1834-1896)
COTTEAU, Gustave Honoré, (1818-1894)
CROSBY, William Otis, (1850-1925)
DARTON, Nelson Horatio, (1865-1948)
DICKERSON, Roy Ernest, (1877-1944)
FLORES REYES, Teodoro, (1873-1955)
HENNEN, Ray Vernon, (1875-1958)
HUMBOLDT, Friedrich Wilhelm Heinrich Alexander von, Freiherr, (1769-1859)
INVILLIERS, Edward Vincent d', (1857-1928)
IRVING, John Duer, (1874-1918)
MEINZER, Oscar Edward, (1876-1948)
MILLER, Benjamin Le Roy, (1874-1944)
POURTALÈS, Louis François de, (1823-1880)
WILLIAMS, Henry Shaler, (1847-1918)
WOODRUFF, Elmer Grant, (1872-1952)

1900 to present

ABEL, Othenio, (1875-1946)
BARBOUR, Thomas, (1884-1946)
BEEBY-THOMPSON, Arthur, (1873-1968)
BROWN, Barnum, (1873-1963)
BROWN, Noel K., Jr., (1925-1968)
BURCHARD, Ernest Norman, (1875-1961)
CASTRO Y SUERO, Manuel Fernandez de, (1825-1895)
CLAPP, Frederick Gardner, (1879-1944)
CROSBY, William Otis, (1850-1925)
DARTON, Nelson Horatio, (1865-1948)
DICKERSON, Roy Ernest, (1877-1944)
DUGGLEBY, Arthur Francis, (1893-1945)
FINCH, Elmer Harrison, (1883-1953)
FLORES REYES, Teodoro, (1873-1955)
HARRISON, Thomas Samuel, (1881-1964)
HENNEN, Ray Vernon, (1875-1958)
INVILLIERS, Edward Vincent d', (1857-1928)
IRVING, John Duer, (1874-1918)
JOHNSTON, William Drumm, Jr., (1899-1972)
KOZÁRY, Myron Theodore "Mirko", (1918-1966)
MacNAUGHTON, Lewis Winslow, (1902-1969)
MEAD, Warren Judson, (1883-1960)
MEINZER, Oscar Edward, (1876-1948)
MILLER, Benjamin Le Roy, (1874-1944)
MYERS, Thurman H., (1890-1940)

CUBA 1900 to present Cont'd.

NĚMEC, František, (1905-1972)
NOMLAND, Jorgen O., (1885-1943)
PALMER, Robert Hastings, (1882-1948)
POND, Walter Franklin, (1885-1974)
RUTTEN, Louis Martin Robert, (1884-1946)
RUTTEN, Martin Gerard, (1910-1970)
SCHAUB, Hans Peter, (1912-1965)
TABER, Stephen, (1882-1963)
WALLACE, Maurice Hudiburg, (1917-1961)
WASHBURNE, Chester Wesley, (1883-1971)
WILLIAMS, Henry Shaler, (1847-1918)
WILSON, Thomas Carroll, (1906-1969)
WIRZ, Albert Eduard, (1910-1973)
WOODRUFF, Elmer Grant, (1872-1952)

CURAÇAO (DUTCH WEST INDIES)

1900 to present

GRUTTERINK, Jan Adolf, (1879-1949)
MOLENGRAAFF, Gerard Johan Hendrik, (1887-1961)
RUTTEN, Louis Martin Robert, (1884-1946)
RUTTEN, Martin Gerard, (1910-1970)
VERSLUYS, Jan, (1880-1935)

CYPRUS

1850-1899

BATE, Dorothea Minola Alice, (1879-1951)
BEEBY-THOMPSON, Arthur, (1873-1968)
CULLIS, Charles Gilbert, (1871-1941)
JUKES-BROWNE, Alfred John, (1851-1914)
REED, Frederick Richard Cowper, (1869-1946)

1900 to present

BATE, Dorothea Minola Alice, (1879-1951)
BEEBY-THOMPSON, Arthur, (1873-1968)
BISHOPP, Douglas Wallace, (1900-1977)
BROUGHTON-EDGE, Arthur Broughton, (1895-1953)
CULLIS, Charles Gilbert, (1871-1941)
HADJISTAVRINOU, Yangos, (1934-1978)
JUKES-BROWNE, Alfred John, (1851-1914)
LEHNER, Ernst J., (1895-1968)
MACKAY, Robert Alexander, (?1904-1961)
RAEBURN, Colin, (1894-1970)
REED, Frederick Richard Cowper, (1869-1946)
SAMPSON, David Noel, (1917-1971)

CZECHOSLOVAKIA

Up to 1800

ANDRÉ, Kristián Karel, (1763-1831)
BORN, Ignaz von, Baron, (1742-1791)
DUHAMEL, Jean Pierre François Guillaume, (1730-1816)
ERCKER, Lazarus, (-1593)
HACQUET, Belsazar, (1739-1815)
KEPLER, Johannes, (1571-1630)
KOMENSKÝ, Jan Ámos, (1592-1670)

CZECHOSLOVAKIA Up to 1800 Cont'd.

LINDACKER, Jan Tadeáš, (1768-1816)
PETTKO, Ján, (1812-1890)
REUSS, Franz Ambros, (1761-1830)
ŠTERNBERK, Kašpar Marià, Hrabe, (1761-1838)
TAVERNIER, Jean Baptiste, baron d'Aubonne, (1605-1689)
VOGT, Johann Georg Mauriz, (1669-1731)

1800-1849
ANDRÉ, Kristián Karel, (1763-1831)
BARRANDE, Joachim, (1799-1883)
CORDA, August Josef, (1811-1849)
ČŽJŽEK, Johann Baptist Anton Karl, (1806-1855)
DUHAMEL, Jean Pierre François Guillaume, (1730-1816)
ETTINGSHAUSEN, Constantin von, Freiherr, (1826-1897)
FEISTMANTEL, Karel, (1819-1885)
FOETTERLE, František, (1823-1876)
GLOCKER, Ernst Friedrich, (1793-1858)
HACQUET, Belsazar, (1739-1815)
HANUŠ, František, (1860-1937)
HOHENEGGER, Ludwig, (1807-1864)
JOKÉLY, Johann, (1826-1862)
KOLENATÝ, Bedřich, (1812-1864)
KREJČÍ, Jan, (1825-1887)
LILL VON LILIENBACH, Karl, (1798-1831)
LINDACKER, Jan Tadeáš, (1768-1816)
LIPOLD, Markus Vincenc, (1816-1883)
MELION, Josef Vincenc, (1813-1905)
MENDEL, Johann Gregor, (1822-1884)
MURCHISON, Roderick Impey, Sir, (1792-1871)
PETTKO, Ján, (1812-1890)
POŠEPNÝ, František, (1836-1895)
PRESL, Jan Svatopluk, (1791-1849)
PRESL, Karl Borivoj, (1794-1852)
RAUMER, Karl Georg von, (1783-1865)
REUSS, August Emanuel von, Ritter, (1811-1873)
REUSS, Franz Ambros, (1761-1830)
RIEPL, Franz Xaver, (1790-1857)
SIMONY, Friedrich, (1813-1896)
ŠTERNBERK, Kašpar Marià, Hrabe, (1761-1838)
ZIPPE, František Xaver Maximilian, (1791-1863)
ZIPSER, Christian Andreas, (1783-1864)

1850-1899
ABSOLON, Karel, (1877-1960)
ANDRIAN-WERBURG, Ferdinand von, Freiherr, (1835-1914)
BABÁNEK, František, (1836-1910)
BARRANDE, Joachim, (1799-1883)
BARVÍŘ, Jindřich Ladislav, (1863-1952)
BAYER, Edvin Jaroslav, (1862-1927)
BAYER, František, (1854-1936)
BECKE, Friedrich Johann Karl, (1855-1931)
BOŘICKÝ, Emanuel, (1840-1881)
BRUDER, George, (1856-1916)
BURKART, Eduard, (1865-1941)

CZECHOSLOVAKIA 1850-1899 Cont'd.

ČAPEK, Václav, (1862-1926)
CVICIĆ, Jovan, (1865-1927)
ČŽJŽEK, Johann Baptist Anton Karl, (1806-1855)
DATHE, Johann Friedrich Ernst, (1845-1917)
DĚDINA, Václav, (1870-1956)
ETTINGSHAUSEN, Constantin von, Freiherr, (1826-1897)
FEISTMANTEL, Karel, (1819-1885)
FENDRYCH, Otakar, (1860-1945)
FOETTERLE, František, (1823-1876)
FRIČ, Antonín, (1832-1913)
FRITSCH, Anton, (1832-1913)
FUCHS, Philippe Jacques Edmond, (1837-1889)
GLOCKER, Ernst Friedrich, (1793-1858)
GRÄNZER, Josef, (1857-1934)
HANUŠ, František, (1860-1937)
HELMHACKER, Rudolf Václav, (1840-1915)
HIBSCH, Josef Emanuel, (1852-1940)
HINTERLECHNER, Karl, (1874-1932)
HÖFER-HEIMHALT, Hans, (1843-1924)
HOFMANN, Adolf, (1853-1913)
HOHENEGGER, Ludwig, (1807-1864)
JAHN, Jaroslav Jiljí, (1865-1934)
JOKÉLY, Johann, (1826-1862)
KAFKA, Josef, (1858-1929)
KATZER, Bedřich, (1861-1925)
KLOUČEK, Celda, (1855-1935)
KLVAŇA, Josef, (1857-1919)
KNIES, Jan, (1860-1925)
KOECHLIN, Rudolf Ignatz, (1862-1939)
KOLENATÝ, Bedřich, (1812-1864)
KOVÁŘ, František, (1862-1941)
KRATOCHVÍL, Josef, (1878-1958)
KREJČÍ, Jan, (1825-1887)
KŘIŽ, Martin, (1841-1916)
KUČERA, Bruno, (1863-1934)
KUSTA, Johann, (1845-1900)
LAUBE, Gustav C., (1839-1923)
LENZ, Oskar, (1848-1925)
LIEBUS, Adalbert, (1876-1945)
LIENENKLAUS, Ernst, (1849-1905)
LINNARSSON, Jonas Gustaf Oscar, (1841-1881)
LIPOLD, Markus Vincenc, (1816-1883)
MAKOWSKY, Alexander, (1833-1908)
MARTONNE, Emmanuel Louis Eugène de, (1873-1955)
MAŠKA, Karel Jaroslov, (1851-1915)
MELION, Josef Vincenc, (1813-1905)
MENDEL, Johann Gregor, (1822-1884)
MURCHISON, Roderick Impey, Sir, (1792-1871)
NOVÁK, Otomar Pravoslav, (1851-1892)
PALLA, Eduard, (1864-1922)
PAUL, Carl Maria, (1838-1900)
PERNER, Jaroslav, (1869-1947)
POČTA, Filip, (1859-1924)
PORTH, Emil, (1832-1858)

CZECHOSLOVAKIA 1850-1899 Cont'd.

PŎSEPNÝ, František, (1836-1895)
PRESL, Karl Borivoj, (1794-1852)
PROCHÁZKA, Vladimír Josef, (1862-1913)
PURKYNĚ, Cyril, (1862-1937)
RAUMER, Karl Georg von, (1783-1865)
REDLICH, Karl A., (1869-1942)
REMEŠ, Mořic, (1867-1959)
REUSS, August Emanuel von, Ritter, (1811-1873)
RIEPL, Franz Xaver, (1790-1857)
RYBA, Frantisek, (1867-1918)
RZEHÁK, Anton, (1855-1923)
SCHLOENBACH, Georg Justin Carl Urban, (1841-1870)
SHIMEK, Bohumil, (1861-1937)
SLAVÍK, František, (1876-1957)
SMOLAŘ, Gotthard, (1863-1926)
SPACEK, Karel, (1866-1937)
ŠTĚRBA-BŎHM, Jan Stanislav, (1874-1938)
SUESS, Franz Eduard, (1867-1941)
TSCHERMAK VON SEYSENEGG, Gustav, (1836-1927)
UHLIG, Victor Karl, (1857-1911)
VACEK, Michael, (1848-1925)
VELENOVSKÝ, Josef, (1858-1949)
VRBA, Karel, (1845-1922)
WAAGEN, Wilhelm Heinrich, (1841-1900)
WĂHNER, Franz, (1856-1932)
WEINBERGER, Isidor, (1838-1915)
WOLDŘICH, Jan Nepomuk, (1834-1906)
ZAHÁLKA, Čeněk, (1856-1938)
ŽELÍZKO, Jan Vratislav, (1874-1938)
ZIPPE, František Xaver Maximilian, (1791-1863)
ZIPSER, Christian Andreas, (1783-1864)

1900 to present

ABSOLON, Karel, (1877-1960)
AMBROŽ, Vojtěch, (1905-1962)
ANDRIAN-WERBURG, Ferdinand von, Freiherr, (1835-1914)
ANDRUSOV, Dimitri, (1897-1976)
AUGUSTA, Josef, (1903-1968)
BABÁNEK, František, (1836-1910)
BARVÍŘ, Jindřich Ladislav, (1863-1952)
BAYER, Edvin Jaroslav, (1862-1927)
BAYER, František, (1854-1936)
BECKE, Friedrich Johann Karl, (1855-1931)
BĚHOUNEK, Rudolf, (1902-1974)
BŎCEK, Antonín, (1880-1956)
BOUČEK, Bedřich, (1904-1975)
BRUDER, George, (1856-1916)
BURKART, Eduard, (1865-1941)
ČAPEK, Václav, (1862-1926)
ČECH, Vladimír, (1901-)
ČECHOVIČ, Vsevolod, (1900-1961)
CEHÁK, Karl, (1903-1957)
ČEPEK, Ladislav, (1899-1974)
ČERMÁK, Jiři, (1884-1955)

CZECHOSLOVAKIA 1900 to present Cont'd.

CHAPMAN, Sydney, (1888-1970)
CORNU, Felix, (1882-1909)
CULEK, Antonin, (1904-1944)
CVICIĆ, Jovan, (1865-1927)
DATHE, Johann Friedrich Ernst, (1845-1917)
DĚDINA, Václav, (1870-1956)
DVOŘÁK, Arnošt, (1904-)
DVOŘÁK, Josef, (1904-)
FENDRYCH, Otakar, (1860-1945)
FIALA, František, (1903-)
FRIČ, Antonín, (1832-1913)
FRITSCH, Anton, (1832-1913)
GARTNER, Otokar, (1889-1961)
GRÄNZER, Josef, (1857-1934)
HEIDE, Friedrich, (1891-)
HEJTMAN, Bohuslav, (1901-)
HELMHACKER, Rudolf Václav, (1840-1915)
HERITSCH, Franz, (1882-1945)
HIBSCH, Josef Emanuel, (1852-1940)
HINTERLECHNER, Karl, (1874-1932)
HLÁVKA, Karel, (1905-)
HÖFER-HEIMHALT, Hans, (1843-1924)
HOFMANN, Adolf, (1853-1913)
HÓRAK, Lubomír, (1935-1962)
HROMADA, Karel, (1919-1974)
HYNIE, Ota, (1899-1968)
JAHN, Jaroslav Jiljí, (1865-1934)
JELÍNEK, František, (1911-1966)
KAFKA, Josef, (1858-1929)
KALÁŠEK, Jan, (1903-)
KAMENICKÝ, Jakub, (1917-)
KAŠPAR, Jan, (1898-)
KATZER, Bedřich, (1861-1925)
KETTNER, Radim, (1891-1967)
KETTNEROVÁ, Marja (Marie), (1900-1933)
KLOUČEK, Celda, (1855-1935)
KLVAŇA, Josef, (1857-1919)
KNIES, Jan, (1860-1925)
KODYM, Odolen, (1898-1963)
KOECHLIN, Rudolf Ignatz, (1862-1939)
KOKTA, Jaroslav, (1894-)
KONTKIEWICZ, Stanislaw Jr., (1883-1951)
KOUTEK, Jaromír, (1902-)
KOVÁŘ, František, (1862-1941)
KRATOCHVÍL, Josef, (1878-1958)
KŘIŽ, Martin, (1841-1916)
KRUTA, Tomáš, (1906-)
KRYZ, Ferdinand, (1879-1958)
KUBIËNA, Walter L., (1897-1970)
KUČERA, Bruno, (1863-1934)
KUNSKÝ, Josef, (1903-)
LAUBE, Gustav C., (1839-1923)
LENZ, Oskar, (1848-1925)
LIEBUS, Adalbert, (1876-1945)

CZECHOSLOVAKIA 1900 to present Cont'd.

LIENENKLAUS, Ernst, (1849-1905)
LUCEK, Josef, (1930-1969)
MACHATSCHEK, Fritz, (-1957)
MAKOWSKY, Alexander, (1833-1908)
MARTONNE, Emmanuel Louis Eugène de, (1873-1955)
MAŠKA, Karel Jaroslov, (1851-1915)
MATĚJKA, Alois, (1898-)
MELION, Josef Vincenc, (1813-1905)
MOSCHELESOVÁ, Julie, (1892-1956)
NĚMEC, František, (1905-1972)
NĚMEJC, František, (1901-)
NETOLICKÝ, Jan, (1934-1963)
NOVÁČK, Radim, (1905-1942)
NOVÁK, Jiří, (?1902-1971)
NOVOTNÝ, Miroslav, (1921-1971)
OPPENHEIMER, Josef, (1883-1933)
PAČAK, Oldřich, (1882-)
PALLA, Eduard, (1864-1922)
PAPP, Simon, (1886-1970)
PAUK, František, (1905-)
PERNER, Jaroslav, (1869-1947)
PETRBOK, Jaroslav, (1881-1960)
PLAS, Vladimír, (1905-)
POČTA, Filip, (1859-1924)
PRADÁČOVÁ, Marcella, (1912-1972)
PŘIBYL, Alois, (1904-)
PROCHÁZKA, Miloš, (1923-1956)
PROCHÁZKA, Vladimír Josef, (1862-1913)
PROKOP, František, (1898-)
PURKYNĚ, Cyril, (1862-1937)
REDLICH, Karl A., (1869-1942)
REMEŠ, Mořic, (1867-1959)
ROSICKÝ, Vojtěch, (1880-1942)
ROST, Rudolf, (1902-)
RUDOLPH, Karl, (1881-1937)
RYBA, Frantisek, (1867-1918)
RYŠ, Josef, (1879-1960)
RZEHÁK, Anton, (1855-1923)
SALÁT, Ján, (1919-1973)
SCHUBERT, Richard Johann, (1876-1915)
SEKANINA, Josef, (1900-)
ŠETLÍK, Jaromír, (1896-)
SHIMEK, Bohumil, (1861-1937)
ŠIMEK, Antonín, (1887-1942)
SLAVÍK, František, (1876-1957)
SLAVÍKOVÁ, Ludmila, (1890-1943)
SMOLAŘ, Gotthard, (1863-1926)
SMOLÍK, Ladislav, (1895-1960)
SPACEK, Karel, (1866-1937)
ŠPINAR, Zdeněk Vlastimil, (1906-)
STEINOCHER, Vladimír, (1902-1972)
STEJSKAL, Jan, (1902-1972)
ŠTĚRBA-BÖHM, Jan Stanislav, (1874-1938)
STOČES, Bohuslav, (1890-)

CZECHOSLOVAKIA 1900 to present Cont'd.

SUESS, Franz Eduard, (1867-1941)
ŠUF, Jiří, (1891-1976)
SVOBODA, J. V., (1921-1972)
SVOBODA, Josef, (1898-)
TSCHERMAK VON SEYSENEGG, Gustav, (1836-1927)
TUČEK, Karel, (1906-)
UHLIG, Victor Karl, (1857-1911)
ULRICH, František, (1899-1941)
URBAN, Karel, (1905-)
URBÁNEK, Ladislav, (1901-)
VACEK, Michael, (1848-1925)
VACHTL, Josef, (1907-)
VELENOVSKÝ, Josef, (1858-1949)
VESELÝ, Václav, (1896-1956)
VOSYKA, Stanislav, (1902-)
VRABKA, František, (1924-1957)
VRBA, Karel, (1845-1922)
WÄHNER, Franz, (1856-1932)
WEINBERGER, Isidor, (1838-1915)
WOLDŘICH, Jan Nepomuk, (1834-1906)
ZAHÁLKA, Čeněk, (1856-1938)
ZAPLETAL, Karel, (1903-1972)
ZÁRUBA, Quido, (1899-)
ZÁZVORKA, Vlastislav, (1903-)
ŽEBERA, Karel, (1901-)
ZELENKA, Ladislav, (1900-1931)
ŽELÍZKO, Jan Vratislav, (1874-1938)
ZOUBEK, Vladimír, (1903-)
ZPALETAL, Karel, (1903-1972)

DAHLAK ISLANDS (RED SEA)
1800-1849
EHRENBERG, Christian Gottfried, (1795-1876)

1850-1899
EHRENBERG, Christian Gottfried, (1795-1876)

DAHOMEY <u>See</u> BENIN.

DENMARK
Up to 1800
BARTHOLIN, Thomas, (1616-1680)
BARTHOLINUS, Erasmus, (1625-1698)
BRÜNNICH, Morten Thrane, (1737-1827)
LUC, Jean André de, (1727-1817)
MOLTKE, Adam Gottlob, Grev, (1710-1762)
PINGEL, Christian, (1793-1852)
STENO, Nicolaus, (1631-1687)
WAD, Gregers, (1755-1832)
WORM, Olle, (1588-1654)

1800-1849
BRÜNNICH, Morten Thrane, (1737-1827)
FORCHHAMMER, Johan Georg, (1794-1865)
HAIDINGER, Wilhelm Karl von, Ritter, (1795-1871)
HAUSMANN, Johann Friedrich Ludwig, (1782-1859)
HOFF, Thomas Christian Frederik August, (1822-1882)
JOHNSTRUP, Johannes Frederik, (1818-1894)
LUC, Jean André de, (1727-1817)
LUND, Peter Wilhelm, (1801-1880)
LÜTKEN, Christian Fredrik, (1827-1901)
MORLOT, Adolf von, (1820-1867)
PINGEL, Christian, (1793-1852)
STEENSTRUP, Johannes Japetus Smith, (1813-1897)
TREVELYAN, Walter Calverley, Sir, (1797-1879)
WAD, Gregers, (1755-1832)
WALLICH, George Charles, (1815-1899)

1850-1899
BIRLEY, Caroline, (1851-1907)
BØGGILD, Ove Balthazar, (1872-1956)
CHRISTENSEN, Carl Christian, (1860-1935)
CLÉMENT, Charles Adolphe Antoine, (1860-1933)
FABER, Harald Nicolai, (1856-1943)
FORCHHAMMER, Johan Georg, (1794-1865)
HAIDINGER, Wilhelm Karl von, Ritter, (1795-1871)
HARBOE, Edouard Georg, (1845-1919)
HAUSMANN, Johann Friedrich Ludwig, (1782-1859)
HOFF, Thomas Christian Frederik August, (1822-1882)
JESPERSEN, Johan Magnus Peder, (1833-1917)
JESSEN, Axel Hans, (1868-1952)
JOHNSTRUP, Johannes Frederik, (1818-1894)
LUND, Peter Wilhelm, (1801-1880)
LÜTKEN, Christian Fredrik, (1827-1901)
MADSEN, Victor Christian, (1865-1947)
MALLING, Carl Peter, (1862-1927)

DENMARK 1850-1899 Cont'd.

MILTHERS, Paul Christian Vilhelm Madsen, (1865-1962)
MORLOT, Adolf von, (1820-1867)
MORTENSEN, Theodor, (1868-1952)
NIELSEN, Kristian Brünnich, (1872-1942)
NORDMANN, Valdemar Johan Heinrich, (1872-1962)
PINGEL, Christian, (1793-1852)
PINGEL, Johann Victorinus, (1834-1919)
RAVN, Jesper Peter Johansen, (1866-1951)
RØRDAM, Kristian, (1860-1939)
ROSENKJAER, H. N., (1851-1907)
STEENSTRUP, Johannes Japetus Smith, (1813-1897)
TAYLER, Joseph Walter, (-?1896)
TOPSØE, Haldor Frederik Axel, (1842-1935)
TREVELYAN, Walter Calverley, Sir, (1797-1879)
TUXEN, Christian Fr. A., (1850-1903)
USSING, Niels Viggo, (1864-1911)
WALLICH, George Charles, (1815-1899)
WEIBULL, Kristian Oskar Mats, (1856-1923)

1900 to present

BAIRD, Chester Alvan, (1896-1959)
BIRLEY, Caroline, (1851-1907)
BJERRUM, Laurits, (1919-1973)
BØGGILD, Ove Balthazar, (1872-1956)
BØGVAD, Richard, (1897-1952)
CHRISTENSEN, Carl Christian, (1860-1935)
CLÉMENT, Charles Adolphe Antoine, (1860-1933)
FABER, Harald Nicolai, (1856-1943)
HARBOE, Edouard Georg, (1845-1919)
JESPERSEN, Johan Magnus Peder, (1833-1917)
JESSEN, Axel Hans, (1868-1952)
KOCH, Lauge, (1892-1964)
LÜTKEN, Christian Fredrik, (1827-1901)
LYSHEDE, Jens Martinus, (1911-1961)
MADSEN, Victor Christian, (1865-1947)
MALLING, Carl Peter, (1862-1927)
MILTHERS, Keld, (1907-1960)
MILTHERS, Paul Christian Vilhelm Madsen, (1865-1962)
MORTENSEN, Theodor, (1868-1952)
NIELSEN, Kristian Brünnich, (1872-1942)
NORDEN, John Alexander Erdelyi-Fazekas, (né Erdélyi-Fazekas), (1912-1977)
NORDMANN, Valdemar Johan Heinrich, (1872-1962)
NØRREGAARD, Erik Martin, (1880-1959)
NØRVANG, Aksel, (1914-1968)
OLSEN, Gregers Vang, (1929-1960)
PINGEL, Johann Victorinus, (1834-1919)
POULSEN, Christian, (1896-1975)
RAVN, Jesper Peter Johansen, (1866-1951)
RØRDAM, Kristian, (1860-1939)
ROSENKJAER, H. N., (1851-1907)
SKOU, Kristian, (1912-1967)
SLATER, George, (-1956)
SORGENFREI, Theodor, (1915-1972)

DENMARK 1900 to present Cont'd.

TEICHERT, Curt, (-)
TOPSØE, Haldor Frederik Axel, (1842-1935)
TUXEN, Christian Fr. A., (1850-1903)
USSING, Niels Viggo, (1864-1911)
WEIBULL, Kristian Oskar Mats, (1856-1923)
WETZEL, Otto Christian August, (1891-1971)

DJIBOUTI
1850-1899
BREUIL, Henri, Abbé, (1877-1961)

1900 to present
BREUIL, Henri, Abbé, (1877-1961)

DOMINICAN REPUBLIC
1850-1899
BERGT, Walther Adolf, (1864-1941)
DARTON, Nelson Horatio, (1865-1948)
GABB, William More, (1839-1878)
MARVINE, Archibald Robertson, (1848-1876)

1900 to present
BERGT, Walther Adolf, (1864-1941)
CLAPP, Frederick Gardner, (1879-1944)
COOKE, Charles Wythe, (1887-1971)
DARTON, Nelson Horatio, (1865-1948)
HIGGS, Morton Thomas, (1896-1956)
LEHNER, Ernst J., (1895-1968)
NOTESTEIN, Frank B., (1885-1973)
ROCH, Edouard, (1901-)
WALLACE, Maurice Hudiburg, (1917-1961)

DUTCH WEST INDIES
1850-1899
MOLENGRAAFF, Gustaaf Adolf Frederik, (1860-1942)

1900 to present
MOLENGRAAFF, Gustaaf Adolf Frederik, (1860-1942)

DUTCH WEST INDIES <u>See also</u> CURAÇAO.

EAST AFRICA (unspecified)
1850-1899
ANDREWS, Charles William, (1866-1924)
KITSON, Albert Ernest, Sir, (1868-1937)
WILLIS, Bailey, (1857-1949)

1900 to present
ANDREWS, Charles William, (1866-1924)
BADEN-POWELL, Donald Ferlys Wilson, (1897-1973)
KITSON, Albert Ernest, Sir, (1868-1937)
LEAKEY, Mary Douglas (née Nicol), (1913-)
PARRINGTON, Francis Rex, (-)
WILLIS, Bailey, (1857-1949)

EASTER ISLAND (SOUTH PACIFIC OCEAN)
1900 to present
BANDY, Mark Chance, (1900-1963)

EASTERN ASIA (unspecified)
1850-1899
LORENZ, Theodor, (1875-1909)

1900 to present
LORENZ, Theodor, (1875-1909)

ECUADOR
Up to 1800
HUMBOLDT, Friedrich Wilhelm Heinrich Alexander von, Freiherr, (1769-1859)

1800-1849
HUMBOLDT, Friedrich Wilhelm Heinrich Alexander von, Freiherr, (1769-1859)

1850-1899
GALE, Hoyt Stoddard, (1876-1952)
HUMBOLDT, Friedrich Wilhelm Heinrich Alexander von, Freiherr, (1769-1859)
MEYER, Hans, (1858-1929)
TANNHÄUSER, Felix, (1874-1924)
THOMPSON, Beeby, (1848-1931)
WHYMPER, Edward, (1840-1911)
WOLF, Theodor, (1841-1924)

1900 to present
BENTZ, Alfred, (1897-1964)
BOSWORTH, Thomas Owen, (1882-1928)
BOYD, Harold Edward, (1887-1952)
BREMNER, Carl St. John, (1895-1944)
BRUET, Edmond, (1890-)
BUSK, Henry Gould, (?1891-1956)

ECUADOR 1900 to present Cont'd.

CARR, William L., (1919-1975)
CASTON, John E. F., (1925-1972)
DURHAM, Charles Albert, (1910-1954)
FRIZZELL, Donald Leslie, (1906-1972)
GALE, Hoyt Stoddard, (1876-1952)
HALSE, George Whitfield, (1885-1968)
HATFIELD, Willis Charles, "Luke", (1898-1970)
HAYES, Albert Orion, (1882-1965)
HAZZARD, John Charles, (1903-1973)
HIGGS, Morton Thomas, (1896-1956)
HOFFSTETTER, Robert, (1908-)
HOTZ, Walter, (1884-1958)
ISENBERGER, Nathan Perry, (1896-1956)
KENNERLEY, John Brian, (1937-1976)
KREIDLER, William Lynn, (1911-1975)
LANDES, Robert William, (1910-1969)
LEWIS, G. Edward, (-)
LIDDLE, Ralph Alexander, (1896-1963)
MAY, Arthur Raymond, (1893-1948)
MEYER, Hans, (1858-1929)
NYGREN, Walter Eric, (1907-1971))
SCHOLL, Louis August, Jr., (1887-1953)
SPROULE, John Campbell, (1905-1970)
TANNHÄUSER, Felix, (1874-1924)
THALMANN, Hans E., (1899-1975)
THOMAS, Ronald Hawksby, Major, (1885-?1940)
THOMPSON, Beeby, (1848-1931)
WALLACE, Maurice Hudiburg, (1917-1961)
WARING, Waldo W., (1903-1955)
WASSON, Theron, (1887-1970)
WEAVER, Charles Edwin, (1880-1958)
WHEELER, Orby Clinton, (1898-1970)
WHYMPER, Edward, (1840-1911)
WOLF, Theodor, (1841-1924)

CUADOR See also GALÂPAGOS ISLANDS.

EGYPT

Up to 1800

BROCCHI, Giovanni Battista (Giambattista), (1772-1826)
CLARKE, Edward Daniel, (1769-1822)
CORDIER, Pierre Louis Antoine, (1777-1861)
DOLOMIEU, Dieudonné Sylvain Guy Tancrède, called Déodat, Gratet de, (1750-1801)
GEOFFROY SAINT-HILAIRE, Étienne, (1772-1844)
MAILLET, Benoît de, (1656-1738)
SICULUS, Diodorus, (ca.85-21 B.C.)
STRABO, (63B.C.-23 A.D.?)
VOLNEY, Constantin François Chasseboeuf, Comte de, (1757-1820)

1800-1849

BROCCHI, Giovanni Battista (Giambattista), (1772-1826)
CLARKE, Edward Daniel, (1769-1822)
CORDIER, Pierre Louis Antoine, (1777-1861)

EGYPT 1800-1849 Cont'd.

DOLOMIEU, Dieudonné Sylvain Guy Tancrède, called Déodat, Gratet de, (1750-1801)
GEOFFROY SAINT-HILAIRE, Étienne, (1772-1844)
RÖMER, Hermann, (1816-1894)
RUSSEGGER, Josef, (-)
SCHENK, Joseph August, (1815-1891)
VOLNEY, Constantin François Chasseboeuf, Comte de, (1757-1820)

1850-1899
ADAMS, Andrew Leith, (-1882)
ANDREWS, Charles William, (1866-1924)
BALL, John, (-1941)
BARRON, Thomas, (1867-1906)
BEADNELL, Hugh John Llewellyn, Major, (1874-1944)
BEEBY-THOMPSON, Arthur, (1873-1968)
BLANCKENHORN, Max Ludwig Paul, (1861-1947)
BRINDLEY, William, (1832-1919)
CHAPMAN, Frederick, (1864-1943)
CORDIER, Pierre Louis Antoine, (1777-1861)
CRICK, George Charles, (1856-1917)
DI STEFANO, Giovanni, (1856-1918)
DOUVILLÉ, Joseph Henri Ferdinand, (1846-1937)
ELLIOT SMITH, Grafton, Sir, (1871-1937)
ERB, Josef Theodor, (1874-1934)
EVANS, John William, (1857-1930)
FORTAU, René, (1867-1920)
FOSTER, Clement Le Neve, Sir, (1841-1904)
FRAAS, Eberhard, (1862-1915)
GILL, David, Sir, (1843-1914)
HULL, Edward, (1829-1917)
HUME, William Fraser, (1867-1949)
LOMAS, Joseph, (1860-1908)
LYONS, Henry George, (1864-?1943)
MARKGRAF, Richard, (1856-1916)
MOON, Frederick William, (-1925)
MORGAN, Jacques Jean Marie de, (1857-1924)
RÖMER, Hermann, (1816-1894)
RUSSEGGER, Josef, (-)
SCHENK, Joseph August, (1815-1891)
SCHWEINFURTH, Georg, (1836-1925)
SZÁDECZKY, Gyula, Baron Szádeczky-Kardoss of Szádecsne and Kafdosfalva, (1860-1935)
WADE, Arthur, (1878-1951)

1900 to present
ANDREW, Arthur Robert, (1881-1941)
ANDREWS, Charles William, (1866-1924)
BALL, John, (-1941)
BARRON, Thomas, (1867-1906)
BEADNELL, Hugh John Llewellyn, Major, (1874-1944)
BEEBY-THOMPSON, Arthur, (1873-1968)
BERNOUILLI, Walter, (1885-1946)
BLANCKENHORN, Max Ludwig Paul, (1861-1947)
BRINDLEY, William, (1832-1919)

EGYPT 1900 to present Cont'd.

BÜRGL, Hans, (1907-1966)
BUSK, Henry Gould, (?1891-1956)
BUTTERWORTH, Emerson McMillin, (1894-1961)
CAMPBELL, Keir Arthur, (-1955)
CHAPMAN, Frederick, (1864-1943)
CHAPMAN, Sydney, (1888-1970)
CLAPP, Frederick Gardner, (1879-1944)
CRICK, George Charles, (1856-1917)
DERRY, Duncan Ramsay, (?1908-)
DI STEFANO, Giovanni, (1856-1918)
DOUVILLÉ, Joseph Henri Ferdinand, (1846-1937)
ELLIOT SMITH, Grafton, Sir, (1871-1937)
ERB, Josef Theodor, (1874-1934)
EVANS, John William, (1857-1930)
FATH, Arthur Earl, (1887-1976)
FOLEY, Edward J., (1908-1971)
FORTAU, René, (1867-1920)
FOSTER, Clement Le Neve, Sir, (1841-1904)
FOX, Cyril Sankey, Sir, (1886-1951)
FRAAS, Eberhard, (1862-1915)
GESTER, George Clark, (1884-1959)
GILL, David, Sir, (1843-1914)
HALSE, George Whitfield, (1885-1968)
HAMMOND, Weldon Woolf, (1908-1964)
HIGAZY, Riad Abdel-Megid, (1919-1967)
HULL, Edward, (1829-1917)
HUME, William Fraser, (1867-1949)
HUXLEY, Julian Sorell, Sir, (1887-1975)
JABLONSKI, Eugene, (1892-1975)
JURKOVIČ, Ivan, (1917-)
KROKSTRÖM, Torsten Hugo, (1905-1950)
LITTLE, Otway Henry, (-1956)
LOMAS, Joseph, (1860-1908)
LYONS, Henry George, (1864-?1943)
MARKGRAF, Richard, (1856-1916)
MOON, Frederick William, (-1925)
MORGAN, Jacques Jean Marie de, (1857-1924)
MORRIS, Frederick Kuhne, (1885-1962)
NIGHTINGALE, William Thomas, (1897-1964)
NOTESTEIN, Frank B., (1885-1973)
PFANNENSTIEL, Max, (1902-1976)
POLLOCK, Charles Murray, (1898-1969)
RHOADES, Ralph Omer, (1895-1961)
RUSSELL, William Low, (1897-1976)
SANDFORD, Kenneth Stuart, (1899-1971)
SCHERMERHORN, W., (1894-)
SCHWEINFURTH, Georg, (1836-1925)
SCHWENNESEN, Alvin Theodore, (1883-1962)
SZÁDECZKY, Gyula, Baron Szádeczky-Kardoss of Szádecsne and Kafdosfalva, (1860-1935)
TEILHARD DE CHARDIN, Pierre, Father, S.J., (1881-1955)
TRASK, Parker Davies, (1899-1961)
WADE, Arthur, (1878-1951)
WARING, Gerald Ashley, (1883-1971)

EGYPT 1900 to present Cont'd.

ZIMERMAN, Sam, (1905-1972)

EL SALVADOR

1850-1899

DOLLFUS, Auguste, (1840-1869)
MONTESSUS DE BALLORE, Fernand Jean Baptiste Marie Bernard, Comte de, (1851-1923)
SAPPER, Karl Theodor, (1866-1945)

1900 to present

BLANCHARD, William Grant, Jr., (1894-1952)
BURWELL, Edward Bouldin, Jr., (1894-1964)
HYDE, Clarence Edgar, (1894-1954)
MONTESSUS DE BALLORE, Fernand Jean Baptiste Marie Bernard, Comte de, (1851-1923)
MUNCK, Victor C. E. A. de, (1920-1956)
SAPPER, Karl Theodor, (1866-1945)
STIRTON, Ruben Arthur, (1901-1966)
WASSERSTEIN, Benno, (1906-1956)

ELLICE ISLANDS (PACIFIC OCEAN)

1850-1899

CHAPMAN, Frederick, (1864-1943)
DAVID, Tannatt William Edgeworth, Sir, (1858-1934)
GARDINER, John Stanley, (1872-1946)

1900 to present

CHAPMAN, Frederick, (1864-1943)
DAVID, Tannatt William Edgeworth, Sir, (1858-1934)
GARDINER, John Stanley, (1872-1946)

ENGLAND

Up to 1800

AIKIN, Arthur, (1773-1854)
BABINGTON, William, (1756-1833)
BAKEWELL, Robert, (1768-1843)
BANKS, Sir Joseph, (1743-1820)
BARROW, John, Sir, (1764-1848)
BEDDOES, Thomas, (1760-1808)
BEVIS (or BEVANS), John, (1693-1771)
BRADBURY, John, (1765-1823)
BROOKE, Henry James, (1771-1857)
BROWN, E., (fl.1685)
BRUNO, Filippo, called Giordano, (1548-1600)
BRYDONE, Patrick, (1736-1818)
BURNET, Thomas, (1635-1715)
BUSHELL, Thomas, (?1594-1674) [PROSPECTORS, DIVINERS AND MINING ENGINEERS]
BUTLER, Samuel, (1774-1839)
CAVENDISH, Henry, (1731-1810)
CHILDREN, John George, (1777-1852)
CLARKE, Edward Daniel, (1769-1822)
CLIFT, William, (1775-1849)
COOPER, Thomas, (1759-1840)

ENGLAND Up to 1800 Cont'd.

CUMBERLAND, George, (1754-1848)
DALE, Samuel, (1659-1739)
DAMPIER, William, (1652-1715)
DARWIN, Erasmus, (1731-1802)
DAVY, Humphry, Sir, (1778-1829)
DILLWYN, Lewis Weston, (1778-1855)
DRAYTON, Michael, (1563-1631)
FAREY, John, (1766-1826)
FAUJAS DE SAINT FOND, Barthélémy, (1741-1819)
FERBER, Johann Jacob, (1743-1790)
FORSTER, Adolarius Jacob, (1739-1806)
FORSTER, Johann Georg Adam, (1754-1794)
FORSTER, Johann Reinhold, (1729-1798)
FORSTER, Westgarth, (1772-1835)
GEOFFROY SAINT-HILAIRE, Étienne, (1772-1844)
GILBERT, Davies, (1767-1839)
GILBERT, William, (1544-1603)
GREENOUGH, George Bellas, (1778-1855)
GREGOR, William, Reverend, (1761-1817)
GREVILLE, Charles Francis, Right Honourable, (1749-1809)
HAILSTONE, John, Reverend, (1759-1847)
HALL, Elias, (1764-1853)
HARIOT, Thomas, (1560-1621)
HENNAH, Richard, Reverend, (?1765-1846)
HERSCHEL, Frederick William, Sir, (1738-1822)
HEULAND, John Henry, (1778-1856)
HILL, John, Sir, (1706-1775)
HOME, Everard, Sir, (1756-1832)
HOOKE, Robert, (1635-1703)
HUNTER, John, (1728-1793)
HUTTON, James, (1726-1797)
JENKIN, William, (1738-1820) [PROSPECTORS, DIVINERS AND MINING ENGINEERS]
JORDEN, Edward, (1569-1632)
JOSSELYN, John, (ca.1608-1675)
KEILL, John, (1671-1721)
KEIR, James, (1735-1820)
KIDD, John, (1775-1851)
KIRBY, William, Reverend, (1759-1851?)
KONIG, Karl Dietrich Eberhardt, called Charles, (1774-1851)
LETTSOM, John Coakley, (1744-1815)
LISTER, Martin, (1638-1712)
LUC, Jean André de, (1727-1817)
MALTHUS, Thomas Robert, (1766-1834)
MATON, William George, (1774-1835)
MAWE, John, (1764-1829)
MERRET, Christopher, (1614-)
MICHELL, John, Reverend, (1724-1793)
MORITZ, Karl Philipp, (1757-1793)
NEWCOMEN, Thomas, (1663-1739) [PROSPECTORS, DIVINERS AND MINING ENGINEERS]
NICOLS, Thomas, (fl. 1652)
OTLEY, Jonathan, (1766-1856)
PARKINSON, James, (1755-1824)

ENGLAND Up to 1800 Cont'd.

PENNANT, Thomas, (1726-1798)
PHILPOT, Mary, (1777-1838)
PLATT, Joshua, (1699-1773)
PLOT, Robert, (1640-1696)
PRIESTLEY, Joseph, (1733-1804)
PYE SMITH, John, (1774-1851)
RASHLEIGH, Philip, (1729-1811)
RASPE, Rudolf Erich, (1737-1794)
RAY, John, (1627-1705)
SHERLEY, Thomas, (1638-1678)
SICULUS, Diodorus, (ca.85-21 B.C.)
SLOANE, Hans, Sir, (1660-1753)
SMITH, William, (1769-1839)
SOLANDER, Daniel Carl, (1733-1782)
SOWERBY, James, (1757-1822)
STRACHEY, John, (1671-1743)
STRANGE, John, (1732-1799)
STUKELEY, William, (1687-1765)
TOWNSEND, Joseph, Reverend, (1739-1816)
TOWNSON, Robert, (1763-1827)
TRADESCANT, John, The Elder, (1587-1638)
TREVITHICK, Richard, (1771-1833) [PROSPECTORS, DIVINERS AND MINING ENGINEERS]
VALLTRAVERS, Johann Rodolph von, (1723-)
WATSON, White, (1760-1835)
WEAVER, Thomas, (1773-1855)
WEBSTER, John, (1610-1682)
WEBSTER, Thomas, (1773-1844)
WHITEHURST, John, (1713-1788)
WOLLASTON, William Hyde, (1766-1828)
WOODWARD, John, (1665-1728)
WOOLF, Arthur, (1766-1837) [PROSPECTORS, DIVINERS AND MINING ENGINEERS]

1800-1849

ADAM, William, (1794?-1873)
AIKIN, Arthur, (1773-1854)
ALLPORT, Samuel, (1816-1897)
ANNING, Mary, (1799-1847)
ANSTED, David Thomas, (1814-1880)
APLIN, Christopher d'Oyley Hale, (1819-1875)
ARNOLD, Joseph, (1784-1818)
ARTIS, Edmund Tyrell, (1789-1847)
ATHERSTONE, William Guybon, Honourable, (-1898)
ATTWOOD, Melville, (1812-1898)
AUSTIN, Thomas, Fort-Major, (1794-1881)
AVELINE, William Talbot, (1822-1903)
BABINGTON, William, (1756-1833)
BAILY, William Hellier, (1819-1888)
BAINBRIDGE, William, (1812-1860)
BAKEWELL, Robert, (1768-1843)
BANKS, Sir Joseph, (1743-1820)
BARKAS, Thomas Pallister, (1819-1891)
BARROW, John, Sir, (1764-1848)

ENGLAND 1800-1849 Cont'd.

BATES, Henry Walter, (1825-1892)
BAYFIELD, Thomas Gabriel, (1817-1893)
BEARDMORE, Nathaniel, (1816-1872)
BEDDOES, Thomas, (1760-1808)
BEESLEY, Thomas, (1818-1896)
BENSTED, William Harding, (1802-1873)
BEST, Edward, (1824-1906)
BIGSBY, John Jeremiah, (1792-1881)
BINNEY, Edward William, (1812-1881)
BOLTON, John, (1788-1873)
BOWERBANK, James Scott, (1797-1877)
BRACKENRIDGE, George Weare, (1775-1856)
BRADBURY, John, (1765-1823)
BRADY, Antonio, Sir, (1811-1881)
BRAYLEY, Edward William, (1802-1870)
BRIGHT, Richard, (1789-1858)
BRISTOW, Henry William, (1817-1889)
BROCKBANK, Richard Bowman, (1824-1912)
BRODERIP, William John, (-1859)
BRODIE, Peter Bellinger, Reverend, (1815-1897)
BROOKE, Henry James, (1771-1857)
BROWN, John, (1780-1859)
BROWN, Thomas, Captain, (1785-1862)
BRYDONE, Patrick, (1736-1818)
BUCKLAND, Francis Trevelyan, (1826-1880)
BUCKLAND, William, Reverend, (1784-1856)
BUCKMAN, James, (1814-1884)
BUNBURY, Charles James Fox, Sir, (1809-1886)
BUSK, George, (1807-1886)
BUTLER, Samuel, (1774-1839)
CARPENTER, William Benjamin, (1813-1885)
CARTER, Henry John, (1813-1895)
CARTER, James, (1813-1895)
CARUS, Carl Gustav, (1789-1869)
CASH, William, (1843-1914)
CAUTLEY, Proby T., Colonel Sir, (1802-1871)
CAVENDISH, Henry, (1731-1810)
CHAMBERS, Robert, (1802-1871)
CHAPMAN, Edward John, (1821-1904)
CHILDREN, John George, (1777-1852)
CHRISTY, Henry, (1810-1865)
CHURCHILL, George Cheetham, (1822-1906)
CLARK, George Thomas, (1811-1898)
CLARKE, Edward Daniel, (1769-1822)
CLARKE, William Branwhite, Reverend, (1798-1878)
CLIFT, William, (1775-1849)
COLBY, Thomas F., Major-General, (?1783-1852)
COLCHESTER, William, (1813-1898)
CONYBEARE, John Josiah, Reverend, (1779-1824)
CONYBEARE, William Daniel, Reverend, (1787-1857)
COOPER, Thomas, (1759-1840)
COSSHAM, Handel, (1824-?1890)
CRANE, Edward, (1822-1901)
CROSS, John Edward, Reverend, (1821-1897)

ENGLAND 1800-1849 Cont'd.

CROSSKEY, H. W., Reverend, (1826-1893)
CUMBERLAND, George, (1754-1848)
CUMMING, Joseph G., Reverend, (1812-1868)
CUNNINGTON, William, (1813-1906)
CURTIS, John, (1791-1862)
DALLAS, William Sweetland, (1824-1890)
DAMON, Robert, (1814-1889)
DARWIN, Charles Robert, (1809-1882)
DARWIN, Erasmus, (1731-1802)
DARWIN, Francis Sacheverell, Sir, (1786-1859)
DAUBENY, Charles Giles Brindle, (1795-1867)
DAVIES, William, (1814-1891)
DAVY, Humphry, Sir, (1778-1829)
DAWES, John Samuel, (1802-1878)
DEANE, Henry, (1807-1874)
DE LA BECHE, Henry Thomas, Sir, (1796-1855)
DE LA CONDAMINE, H. M., Reverend, (-1854)
DILLWYN, Lewis Weston, (1778-1855)
DIXON, Frederick, (1799-1849)
DOVER, William Kinsey, (?1816-1891)
DOWKER, George, (1828-1899)
DRAPER, John William, (1811-1882)
DREW, Frederick, (1836-1891)
DUNCAN, Peter Martin, (1824-1891)
EDWARDS, Frederic Erasmus, (1799-1875)
EGERTON, Philip de Malpas Grey, Sir, (1806-1881)
EGERTON, William Henry, Reverend, (1811-1910)
ELLIS, H., Right Honourable Sir, (-1855)
ENNISKILLEN, William Willoughby Cole, 3rd Earl of, (1807-1886)
ENYS, John Samuel, (1797-1872)
ERLE, C., Reverend, (1790-1870?)
ETHERIDGE, Robert, (1819-1903)
EVANS, John, Sir, (1823-1908)
FALCONER, Hugh, (1808-1865)
FAREY, John, (1766-1826)
FAUJAS DE SAINT FOND, Barthélemy, (1741-1819)
FEATHERSTONHAUGH, George William, (1780-1866)
FINCH, John, (fl.1823-1835)
FISHER, Osmond, Reverend, (1817-1914)
FITCH, Robert, (1802-1895)
FITTON, William Henry, (1780-1861)
FITZ-ROY, Robert, Vice-Admiral, (1805-1865)
FLETCHER, Thomas William, (1808-1893)
FLOWER, John Wickham, (1807-1873)
FORBES, David, (1828-1876)
FORBES, Edward, (1815-1854)
FORSTER, Adolarius Jacob, (1739-1806)
FORSTER, Westgarth, (1772-1835)
FOX, Robert Were, (1789-1877)
GALTON, Francis, Sir, (1822-1911)
GEINITZ, Hanns Bruno, (1814-1900)
GEOFFROY SAINT-HILAIRE, Étienne, (1772-1844)
GILBERT, Davies, (1767-1839)
GODWIN-AUSTEN, Robert Alfred Cloyne, (1808-1884)

ENGLAND 1800-1849 Cont'd.

GREENOUGH, George Bellas, (1778-1855)
GREENWELL, George Clementson, (1821-1900)
GREENWOOD, George, Colonel, (1799-1875)
GREG, Robert Phillips, (1826-1906)
GREGOR, William, Reverend, (1761-1817)
GREVILLE, Charles Francis, Right Honourable, (1749-1809)
GUNN, John, Reverend, (1801-1890)
HAIDINGER, Wilhelm Karl von, Ritter, (1795-1871)
HAILSTONE, John, Reverend, (1759-1847)
HALL, Elias, (1764-1853)
HANCOCK, Albany, (1806-1873)
HARGRAVES, Edward Hammond, (1816-1891)
HARRIS, William, (1797-1877)
HARRISON, James, (1819-1868)
HARRISON, John Thornhill, (1815-1891)
HASTINGS, Barbara, Marchioness of, (1810-1858)
HAWKINS, Thomas, (1810-1889)
HAWKSHAW, John, Sir, (1811-1891)
HENNAH, Richard, Reverend, (?1765-1846)
HENSLOW, John Stevens, Reverend, (1796-1861)
HENWOOD, William Jory, (1805-1875)
HERSCHEL, Frederick William, Sir, (1738-1822)
HERSCHEL, John Frederick William, Sir, (1792-1871)
HEULAND, John Henry, (1778-1856)
HIGGINS, Henry Hugh, Reverend, (1814-1893)
HIND, Henry Youle, (1823-1908)
HOLL, Harvey Buchanan, (1820-1886)
HOLLAND, Henry, Sir, (1788-?1873)
HOME, Everard, Sir, (1756-1832)
HOOKER, Joseph Dalton, Sir, (1817-1911)
HOOKER, William Jackson, Sir, (1785-1865)
HOPE, Frederick William, Reverend, (1797-1862)
HOPKINS, William, (1793-1866)
HORNER, Leonard, (1785-1864)
HOW, Henry, (1828-1879)
HOWSE, Richard, (1821-1901)
HUDLESTON, Wilfrid Hudleston, (né Simpson), (1828-1909)
HUGGINS, William, Sir, (1824-1910)
HURST, Thomas Grainge, (1824-1890)
HUTTON, William, (-1861?)
HUXLEY, Thomas Henry, (1825-1895)
JENKIN, William, (1738-1820) [PROSPECTORS, DIVINERS AND MINING ENGINEERS]
JOHNSON, Henry, (1823-1885)
JOHNSTON, James Finlay Weir, (1796-1855)
JONES, Thomas Rupert, (1819-1911)
KAYE, Charles Turton, (1812-1846)
KEEPING, Henry, (1827-1924)
KEIR, James, (1735-1820)
KIDD, John, (1775-1851)
KING, Samuel William, (1821-1868)
KING, William, (1809-1886)
KINGSLEY, Charles, Reverend, (1819-1875)
KIRBY, William, Reverend, (1759-1851?)

ENGLAND 1800-1849 Cont'd.

KLAASSEN, Hendericus Martinus, (1828-1910)
KONIG, Karl Dietrich Eberhardt, called Charles, (1774-1851)
LAING, Samuel, (1810-1897)
LARKIN, Nathaniel John, (1781-1855)
LA TOUCHE, James Digues, Reverend, (1824-1899)
LECKENBY, John, (1814-1877)
LEE, John Edward, (1808-1887)
LETTSOM, John Coakley, (1744-1815)
LONSDALE, William, (1794-1871)
LUC, Jean André de, (1727-1817)
LUCY, William Charles, (1822-1898)
LUDLAM, Henry, (1822-1880)
MACKESON, Henry Bean, (1811-1894)
MACKINTOSH, Daniel, (1815-1891)
MACLEAY, William Sharp, (1792-1865)
MAJENDIE, Ashurst, (?1784-1867)
MALLET, Robert, (1810-1881)
MALTHUS, Thomas Robert, (1766-1834)
MANSEL-PLEYDELL, John Clavell, (1817-1902)
MANTELL, Gideon Algernon, (1790-1852)
MARTEN, Henry John, (1827-1892)
MARTIN, Peter John, (1786?-1860)
MATON, William George, (1774-1835)
MAWE, John, (1764-1829)
MEADE, Richard, (1827-1894)
MIERS, John, (1789-1879)
MILLER, Hugh, (1802-1856)
MILLER, William Hallowes, (1801-1880)
MOORE, Charles, (1814-1881)
MORRIS, John, (1810-1886)
MORTIMER, John Robert, (1825-1911)
MORTON, George Highfield, (1826-1900)
MURRAY, John, III, (1808-1892)
MYLNE, Robert William, (1816-1890)
NESBIT, John Collis, (1819-1862)
NORTON, Henry, (?1812-1892)
OEYNHAUSEN, Karl August Ludwig, (1795-1865)
OGILBY, William, (1808-1873)
ORMEROD, George Wareing, (1810-1891)
OTLEY, Jonathan, (1766-1856)
OWEN, Richard, Sir, (1804-1892)
PARKER, William Kitchen, (1823-1890)
PARKINSON, James, (1755-1824)
PATTISON, Samuel Rowles, (1809-1901)
PATTON, Andrew, (1821-1885)
PEACH, Charles William, (1800-1886)
PEARCE, Chaning, (1811-1847)
PENGELLY, William, (1812-1894)
PERCY, John, (1817-1889)
PHILLIPS, John, (1800-1874)
PHILLIPS, William, (1800-1834)
PHILPOT, Elizabeth, (1780-1857)
PHILPOT, Margaret, (? -1845)
PHILPOT, Mary, (1777-1838)

ENGLAND 1800-1849 Cont'd.

PIPER, George Harry, (1819-1897)
PLANT, Major John, (1819-?1892)
PRESTWICH, Joseph, Sir, (1812-1896)
PRIESTLEY, Joseph, (1733-1804)
PYE SMITH, John, (1774-1851)
QUEKETT, John Thomas, (1815-1861)
RAMSAY, Andrew Crombie, Sir, (1814-1891)
RANDALL, John, (1810-1910)
RASHLEIGH, Philip, (1729-1811)
READE, Joseph Bancroft, Reverend, (1801-1870)
REDMAN, John Baldry, (1816-1899)
REED, William, (1810-1892)
REEVE, James, (1833-1920)
REEVE, Lovell, (1814-1865)
RICHARDSON, Benjamin, Reverend, (-1832)
RICKARD, Francis Ignacio, (-)
RICKETTS, Charles, (1818-1904)
ROFE, John, (1801-1878)
RÖMER, Hermann, (1816-1894)
ROSE, Caleb Burrell, (1790-1872)
RUSKIN, John, (1819-1900)
SALMON, Henry Curwen, (1827-1873)
SALTER, John William, (1820-1869)
SALWEY, Humphry, (1803-1877)
SANDERS, William, (1799-1875)
SAWKINS, James Gay, (1806-1878)
SCOTT, Marcus Watson Tate, (1814-1875)
SCOULER, John, (-1874)
SEDGWICK, Adam, Reverend, (1785-1873)
SELWYN, Alfred Richard Cecil, (1824-1902)
SHARP, Samuel, (1814-1882)
SHARPE, Daniel, (1806-1856)
SHRUBSOLE, George William, (1827-1893)
SIMPSON, Martin, (1798-1892)
SLACK, Henry James, (1818-1896)
SLADE, Jeremiah, (1828-1905)
SMITH, Augustus, (1804-1872)
SMITH, Joshua Toulmin, (1816-1869)
SMITH, William, (1769-1839)
SMITHE, Frederick, Reverend, (1822-1900)
SMYTH, Warington Wilkinson, Sir, (1817-1890)
SOPWITH, Thomas, (1803-1879)
SORBY, Henry Clifton, (1826-1908)
SOWERBY, George Brettingham, I, (1788-1854)
SOWERBY, George Brettingham, II, (1812-1884)
SOWERBY, Henry, (1825-1891)
SOWERBY, James, (1757-1822)
SOWERBY, James de Carle, (1787-1871)
SPENCER, Herbert, (1820-1903)
STOKES, Charles, (1783-1853)
STORY-MASKELYNE, Mervin Herbert Nevil, (1823-1911)
STRICKLAND, Hugh Edwin, (1811-1853)
TATE, George, (1805-1871)
TAUNTON, John Hooke, (1820-1893)

ENGLAND 1800-1849 Cont'd.

TAYLOR, John, (1780-1864)
TAYLOR, Richard Cowling, (1789-1851)
TENNANT, James, (1808-1881)
TOMLINSON, Charles, (1806-1897)
TOPLEY, William, (1841-1894)
TORELL, Otto Martin, (1828-1900)
TOWNSEND, Joseph, Reverend, (1739-1816)
TOWNSON, Robert, (1763-1827)
TREVELYAN, Walter Calverley, Sir, (1797-1879)
TREVITHICK, Richard, (1771-1833) [PROSPECTORS, DIVINERS AND MINING ENGINEERS]
TRIMMER, Joshua, (1797-1857?)
TYLOR, Alfred, (1823-1884)
VALLTRAVERS, Johann Rodolph von, (1723-)
VALPY, Robert Harris, (1819-1904)
VICARY, William, (1811-1903)
VIVIAN, William, (1817-1879)
WALLACE, Alfred Russel, (1823-1913)
WALLICH, George Charles, (1815-1899)
WATERHOUSE, John, (1806-1879)
WATSON, White, (1760-1835)
WEAVER, Thomas, (1773-1855)
WEBSTER, Thomas, (1773-1844)
WESTWOOD, John Obadiah, (1805-1893)
WETHERELL, Nathaniel Thomas, (1800-1875)
WHEWELL, William, Reverend, (1794-1866)
WHITE, Walter, (-)
WILBERFORCE, Samuel, Bishop, (1805-1873)
WILLIAMS, Thomas, (1737-1802) [PROSPECTORS, DIVINERS AND MINING ENGINEERS]
WILLIAMSON, William Crawford, (1816-1895)
WILTSHIRE, Thomas, Reverend, (1826-1902)
WITCHELL, Edwin, (1823-1887)
WOLLASTON, William Hyde, (1766-1828)
WOOD, Edward, (1808-1877)
WOOD, John George, Reverend, (1827-1889)
WOOD, Nicholas, (1795-1865)
WOOD, Searles Valentine, (1798-1880)
WOODWARD, Samuel, (1790-1838)
WOODWARD, Samuel P., (1821-1865)
WOOLF, Arthur, (1766-1837) [PROSPECTORS, DIVINERS AND MINING ENGINEERS]
WRIGHT, Thomas, (1809-1884)

1850-1899

ABBOTT, George, (1844-1925)
ABBOTT, William James Lewis, (1853-1933)
ADAMSON, Samuel A., (-?1890)
ADYE, Ernest Howard, (1857-1918)
AIKIN, Arthur, (1773-1854)
AIM, William Lennie, (?1872-1938)
ALLEN, Henry Atwool, (?1854-1934)
ALLPORT, Samuel, (1816-1897)
ANDREWS, Charles William, (1866-1924)

ENGLAND 1850-1899 Cont'd.

ANSTED, David Thomas, (1814-1880)
ANSTIE, John, (-1900)
APLIN, Christopher d'Oyley Hale, (1819-1875)
ARCHER, Francis, (1839-1892)
ARMSTRONG, Henry Edward, (1848-1937)
ATHERSTONE, William Guybon, Honourable, (-1898)
ATTWOOD, Melville, (1812-1898)
AUSTIN, Thomas, Fort-Major, (1794-1881)
AVELINE, William Talbot, (1822-1903)
BAILY, William Hellier, (1819-1888)
BAINBRIDGE, William, (1812-1860)
BAKER, Benjamin Alfred, (1859-1939)
BAKEWELL, Frederick Collier, (-1869)
BALDWIN, Walter, (?1878-1956)
BALL, Francis Matthew, (1877-1956)
BALL, John, (-1941)
BALSTON, William Edward, (1848-1918)
BARCLAY, Francis Hubert, (1869-1935)
BARKAS, Thomas Pallister, (1819-1891)
BARLOW, Caleb, (1840-1908)
BARLOW, William, (1845-1934)
BARNES, William Henry, (1845-1935)
BARNETT, Andrew Ketchen, (1852-1914)
BARROIS, Charles Eugène, (1851-1939)
BARROW, George, (1853-1932)
BATE, Dorothea Minola Alice, (1879-1951)
BATES, Henry Walter, (1825-1892)
BATESON, William, (1861-1926)
BATHER, Francis Arthur, (1863-1934)
BAUER, Grace M., (1878-1965)
BAUERMAN, Hilary, (1833-1909)
BAWDEN, Samuel Francis, (1854-1884)
BAYFIELD, Thomas Gabriel, (1817-1893)
BEADNELL, Hugh John Llewellyn, Major, (1874-1944)
BEALE, William Phipson, Sir, (1839-1922)
BEARDMORE, Nathaniel, (1816-1872)
BEASLEY, Henry Charles, (1836-1919)
BECKLES, Samuel H., (-1890)
BEESLEY, Thomas, (1818-1896)
BELINFANTE, Leo Baptiste Louis, (1861-1937)
BELL, Alfred Graham, (?1875-1964)
BELL, Robert George, (1833-1888)
BEMROSE, Henry Howe, (1857-1939)
BENNETT, Arthur, (1873-1941)
BENNETT, Francis James, (1845-1920)
BENNETT, Frederick William, (1860-1931)
BENSTED, William Harding, (1802-1873)
BEST, Edward, (1824-1906)
BIGSBY, John Jeremiah, (1792-1881)
BINNEY, Edward William, (1812-1881)
BIRD, Charles, (1843-1910)
BIRDS, James Adey, (1831-1894)
BLACKMORE, Humphrey Purnell, (-1928)
BLAKE, John Frederick, Reverend, (1839-1906)
BLAKE, John Hopwood, (1843-1901)

ENGLAND 1850-1899 Cont'd.

BOLTON, Herbert, (1863-1936)
BOLTON, John, (1788-1873)
BONNEY, Thomas George, Reverend, (1833-1923)
BOULTON, William Savage, (1867-ca.1954)
BOWER, Fredrik Orpen, (1855-1948)
BOWERBANK, James Scott, (1797-1877)
BOWMAN, Herbert Lister, (1874-1942)
BRACKENRIDGE, George Weare, (1775-1856)
BRADY, Antonio, Sir, (1811-1881)
BRADY, Henry Bowman, (1835-1891)
BRAGG, William Henry, Sir, (1862-1942)
BRAYLEY, Edward William, (1802-1870)
BRIGHT, Richard, (1789-1858)
BRINDLEY, William, (1832-1919)
BRISTOW, Henry William, (1817-1889)
BROCKBANK, Richard Bowman, (1824-1912)
BRODERIP, William John, (-1859)
BRODIE, Peter Bellinger, Reverend, (1815-1897)
BRODRICK, Harold, (1872-1946)
BROWN, John, (1780-1859)
BROWN, John Allen, (1831-1903)
BROWN, Thomas, Captain, (1785-1862)
BROWNRIDGE, Charles, Colonel, (?1862-1951)
BUCKLAND, Francis Trevelyan, (1826-1880)
BUCKLAND, William, Reverend, (1784-1856)
BUCKMAN, James, (1814-1884)
BUCKMAN, Sidney Savory, (1860-1929)
BULL, Alfred Joseph, (1875-1950)
BULLEN, Robert Ashington, Reverend, (1850-1912)
BULLOCK, William, (1795-1840)
BUNBURY, Charles James Fox, Sir, (1809-1886)
BURROWS, Henry William, (?1860-1939)
BURY, Henry, (?1862-1958)
BUSK, George, (1807-1886)
BUTLER, Francis Henry, (1849-1935)
CAIRNS, Robert, (1854-1911)
CALLAWAY, Charles, (1838-1915)
CARPENTER, William Benjamin, (1813-1885)
CARR, John Wesley, (1862-[1939])
CARRUTHERS, William, (1830-1922)
CARTER, Henry John, (1813-1895)
CARTER, James, (1813-1895)
CARTER, William Lower, (1855-1918)
CARUS, Carl Gustav, (1789-1869)
CARUS-WILSON, C., (1857-1934)
CASH, William, (1843-1914)
CAUTLEY, Proby T., Colonel Sir, (1802-1871)
CHADWICK, Samuel, (1845-1903)
CHAMBERS, Robert, (1802-1871)
CHAMPERKNOWNE, Arthur, (1839-1887)
CHAPMAN, Edward John, (1821-1904)
CHAPMAN, Frederick, (1864-1943)
CHARLESWORTH, Edward, (1813-1893)
CHREE, Charles, (-)

ENGLAND 1850-1899 Cont'd.

CHRISTY, Henry, (1810-1865)
CHURCH, Arthur Herbert, Sir, (1834-1915)
CHURCHILL, George Cheetham, (1822-1906)
CLARK, George Thomas, (1811-1898)
CLARK, James Edmund, (1850-1944)
CLARKE, William Branwhite, Reverend, (1798-1878)
CLAYDEN, Arthur William, (1855-1944)
CLODD, Edward, (1840-1930)
CLOUGH, Charles Thomas, (1852-1916)
CLUTTERBUCK, Robert, (-1879)
COBBOLD, Edgar Sterling, (1851-1936)
COCKIN, George Marmaduke, (1852-1918)
CODRINGTON, Thomas, (1829-1918)
COKE, George Elmsley, (? -1911)
COLBY, Thomas F., Major-General, (?1783-1852)
COLCHESTER, William, (1813-1898)
COLE, Edward Maule, Reverend, (1833-1911)
COLE, Grenville Arthur James, (1859-1924)
COLENUTT, George William, (?1862-1944)
COLLINS, Arthur L., (1868-1902)
COLLINS, Henry Francis, (?1864-1924)
COLLINS, Joseph Henry, (1841-1916)
CONYBEARE, William Daniel, Reverend, (1787-1857)
COOMARASWAMY, Ananda K., (1877-1947)
COOPER-KING, Charles, Lieutenant-Colonel, (1843-1898)
CORFIELD, William, (1843-1903)
CORNER, Frank, (1861-1939)
COSSHAM, Handel, (1824-?1890)
CRAMP, Alfred William, (?1873-1950)
CRANE, Edward, (1822-1901)
CRICK, George Charles, (1856-1917)
CRICK, Walter Drawbridge, (1857-1903)
CROFTS, William Hastings, (1861-1935)
CROOK, Thomas, (1876-1937)
CROSFIELD, Margaret Chorley, (1859-1952)
CROSS, John Edward, Reverend, (1821-1897)
CROSSKEY, H. W., Reverend, (1826-1893)
CULLIS, Charles Gilbert, (1871-1941)
CUMMING, Joseph G., Reverend, (1812-1868)
CUNNINGTON, William, (1813-1906)
CURTIS, John, (1791-1862)
DAKYNS, John Roche, (1836-1910)
DALLAS, William Sweetland, (1824-1890)
DALTON, William H., (?1848-1929)
DAMON, Robert, (1814-1889)
DARELL, Robert Darell Smythe, (1851-1936)
DARWIN, Charles Robert, (1809-1882)
DARWIN, Francis, (1848-1925)
DARWIN, Francis Sacheverell, Sir, (1786-1859)
DARWIN, George Howard, Sir, (1845-1912)
DARWIN, Leonard, (1850-1943)
DAUBENY, Charles Giles Brindle, (1795-1867)
DAVIES, Arthur Morley, (1869-1959)
DAVIES, David Christopher, (1827-1885)

ENGLAND 1850-1899 Cont'd.

DAVIES, Henry Nathaniel, (-1920)
DAVIES, Thomas, (1837-1892)
DAVIES, William, (1814-1891)
DAVIS, James William, (1846-1893)
DAVISON, Charles, (1858-1940)
DAVISON, Ernest Henry, (1877-1944)
DAWES, John Samuel, (1802-1878)
DAWKINS, William Boyd, Sir, (1837-1929)
DAWSON, Charles, (1864-1916)
DEANE, Henry, (1807-1874)
DEANE, Henry Jr., (-1924)
DEELEY, Richard Mountford, (1855-1944)
DE LA BECHE, Henry Thomas, Sir, (1796-1855)
DE LA CONDAMINE, H. M., Reverend, (-1854)
DE RANCE, Charles Eugène, (1847-1906)
DEWEY, Henry, (1876-1965)
DIBLEY, George Edward, (1857-1943)
DICK, Allan Brugh, (1833-1926)
DILLWYN, Lewis Weston, (1778-1855)
DIXON, Ernest Edward Leslie, (1876-1963)
DOVER, William Kinsey, (?1816-1891)
DOWKER, George, (1828-1899)
DRAPER, John William, (1811-1882)
DREW, Frederick, (1836-1891)
DRUGMAN, Julien Léon Eugène, (1875-1950)
DUNCAN, Cecil Cooke, (1868-1948)
DUNCAN, Peter Martin, (1824-1891)
DWERRYHOUSE, Arthur Richard, (1867-1931)
EDWARDS, Frederic Erasmus, (1799-1875)
EGERTON, Philip de Malpas Grey, Sir, (1806-1881)
EGERTON, William Henry, Reverend, (1811-1910)
ELLES, Gertrude Lilian, (1872-1960)
ELLIOT SMITH, Grafton, Sir, (1871-1937)
ELLIS, H., Right Honourable Sir, (-1855)
ELSDEN, James Vincent, (?1856-1930)
ENNISKILLEN, William Willoughby Cole, 3rd Earl of, (1807-1886)
ENYS, John Samuel, (1797-1872)
ERLE, C., Reverend, (1790-1870?)
ETHERIDGE, Robert, (1819-1903)
EVANS, Caleb, (1831-1886)
EVANS, John, Sir, (1823-1908)
EVANS, John William, (1857-1930)
FALCONER, Hugh, (1808-1865)
FARMERY, Joseph Richard, (1872-1928)
FEARNSIDES, William George, (1879-1968)
FEATHERSTONHAUGH, George William, (1780-1866)
FENNELL, Charles William, (?1863-1935)
FENNER, Edward George, (?1873-1941)
FINLAYSON, Alexander Moncreiff, (-1917)
FISHER, Osmond, Reverend, (1817-1914)
FITCH, Robert, (1802-1895)
FITTON, William Henry, (1780-1861)
FITZ-ROY, Robert, Vice-Admiral, (1805-1865)
FLETCHER, Lazarus, Sir, (1854-1921)

ENGLAND 1850-1899 Cont'd.

FLETCHER, Mark, Reverend, (1870-1956)
FLETCHER, Thomas William, (1808-1893)
FLIGHT, Walter, (1841-1885)
FLOWER, John Wickham, (1807-1873)
FLOWER, William Henry, Sir, (1831-1899)
FOORD, Arthur Humphreys, (1845-1934)
FOOTE, Albert E., (1846-1895)
FOOTE, R. Bruce, (-1912)
FORBES, David, (1828-1876)
FORBES, Edward, (1815-1854)
FOSTER, Clement Le Neve, Sir, (1841-1904)
FOWLER, Joseph, (1872-1958)
FOX, Robert Were, (1789-1877)
FOX-STRANGWAYS, Charles Edward, (1844-1910)
FRANCIS, James, (1843-1925)
FRANKS, Godfrey Firth, (1853-1917)
GALTON, Francis, Sir, (1822-1911)
GARDINER, Charles Irving, (?1868-1940)
GARWOOD, Edmund Johnston, (1864-1949)
GASTER, Christopher Theodore Augustine, (1878-1963)
GEINITZ, Hanns Bruno, (1814-1900)
GERRARD, John, (?1850-1920)
GIBBONS, Arthur, (1867-1938)
GIBSON, Walcot, (1864-1941)
GLASS, Norman, Reverend, (1832-1895)
GODWIN-AUSTEN, Robert Alfred Cloyne, (1808-1884)
GOODCHILD, John George, (1844-1906)
GOODRICH, Edwin Stephen, (1868-1946)
GOODYEAR, Edith, (1879-1959)
GORE, Charles Herbert Henry, (1866-1951)
GOSSLING, Frank Newbery, (1888-1946)
GOULD, Charles, (-1895)
GRAHAM, William Vaux, (1859-1940)
GREEN, Alexander Henry, (1832-1896)
GREEN, John Frederick Norman, (?1873-1949)
GREEN, Upfield, (1834-1917)
GREENE, Alfred Ernest, (1878-1940)
GREENLY, Edward, (1861-1951)
GREENOUGH, George Bellas, (1778-1855)
GREENWELL, George Clementson, (1821-1900)
GREENWOOD, George, Colonel, (1799-1875)
GREG, Robert Phillips, (1826-1906)
GREGORY, James Reynolds, (1832-1899)
GROOM, Theodore Thomas, (1863-)
GUNN, John, Reverend, (1801-1890)
GUNN, William, (1837-1902)
GÜNTHER, Albert Charles Lewis Gotthilf, (1830-1914)
GUPPY, Robert John Lechmere, (1836-1916)
GURNEY, Henry Palin, Reverend, (1847-1904)
HAIDINGER, Wilhelm Karl von, Ritter, (1795-1871)
HALL, Elias, (1764-1853)
HALL, Marshall, Captain, (1831-1896)
HALL, Townshend Monckton, (1845-1899)
HANCOCK, Albany, (1806-1873)

ENGLAND 1850-1899 Cont'd.

HANNAY, Harry Campbell, (1874-1961)
HARGRAVES, Edward Hammond, (1816-1891)
HARKER, Alfred, (1859-1939)
HARKER, Allen, (1848-1894)
HARKNESS, Robert, (1816-1878)
HARMER, Frederic William, (1835-1923)
HARRIS, George Frederick, (1862-1906)
HARRIS, William, (1797-1877)
HARRISON, Benjamin, (1837-1921)
HARRISON, Edward Richard, Sir, (1872-1960)
HARRISON, James, (1819-1868)
HARRISON, John Thornhill, (1815-1891)
HARRISON, William Jerome, (1845-1908)
HARTLEY, Harold Brewer, Sir, (1878-1972)
HASTINGS, Barbara, Marchioness of, (1810-1858)
HATCH, Frederick Henry, (?1864-1932)
HAWARD, Frederick James Nairn, (1871-1953)
HAWKINS, Thomas, (1810-1889)
HAWKSHAW, John, Sir, (1811-1891)
HENSLOW, John Stevens, Reverend, (1796-1861)
HENWOOD, William Jory, (1805-1875)
HERON-ALLEN, Edward, (1861-1943)
HERRIES, Robert Stansfield, (?1860-1941)
HERSCHEL, John Frederick William, Sir, (1792-1871)
HESLOP, Septimus, (?1860-1945)
HICKS, Henry, (1837-1899)
HIGGINS, Henry Hugh, Reverend, (1814-1893)
HILL, James Bastian, (1861-1927)
HILL, William, (1849-1912)
HIND, Henry Youle, (1823-1908)
HIND, Wheelton, (1859-1920)
HINDE, George Jennings, (1839-1918)
HINTON, Henry Arthur, (?1874-1937)
HOBSON, Bernard, (1860-1933)
HOLGATE, Benjamin, (1838-1915)
HOLL, Harvey Buchanan, (1820-1886)
HOLLAND, Henry, Sir, (1788-?1873)
HOLLAND, Thomas Henry, Sir, (1868-1947)
HOLMES, T. Vincent, (?1841-1923)
HOOKER, Joseph Dalton, Sir, (1817-1911)
HOOKER, William Jackson, Sir, (1785-1865)
HOOLEY, Reginald Walter, (1866-1923)
HOPE, Frederick William, Reverend, (1797-1862)
HOPKINS, William, (1793-1866)
HOPKINSON, John, (1844-1919)
HORNE, William, (?1836-1928)
HORNER, Leonard, (1785-1864)
HOTBLACK, John Turner, (1848-1915)
HOW, Henry, (1828-1879)
HOWCHIN, Walter, Reverend, (1845-1937)
HOWELL, Henry Hyatt, (1834-1915)
HOWES, George Bond, (1853-1905)
HOWSE, Richard, (1821-1901)
HUDLESTON, Wilfrid Hudleston, (né Simpson), (1828-1909)

ENGLAND 1850-1899 Cont'd.

HUGGINS, William, Sir, (1824-1910)
HUGHES, Thomas McKenny, (1832-1917)
HULKE, John Whitaker, (1830-1895)
HULL, Edward, (1829-1917)
HUNT, Arthur Roope, (1843-1914)
HURST, Thomas Grainge, (1824-1890)
HUTCHINSON, Arthur, (1866-1937)
HUTCHINSON, Henry Neville, Reverend, (?1855-1927)
HUTTON, Harriet Mary, (1873-1937)
HUTTON, William, (-1861?)
HUXLEY, Leonard, (1860-1933)
HUXLEY, Thomas Henry, (1825-1895)
HYLAND, John Shearson, (1866-1898)
IBBETSON, L. L. Boscawen, Captain, (-1869)
JAMES, W. Warwick, (?1874-1965)
JEBB, John Gladwyn, (1841-1893) [PROSPECTORS, DIVINERS AND MINING ENGINEERS]
JENKIN, Henry Charles Fleeming, (1833-1885)
JENKINS, Henry Michael, (1841-1886)
JENNINGS, Albert Vaughan, (1864-1903)
JERVIS, William Paget, (1831-1906)
JESSON, Thomas, (?1850-1928)
JOHNSON, Henry, (1823-1885)
JOHNSON, Walter, (1867-1950)
JOHNSTON, James Finlay Weir, (1796-1855)
JONES, Ernest Lloyd, (?1863-1942)
JONES, Owen Thomas, (1878-1967)
JONES, Thomas Rupert, (1819-1911)
JONES, William Rupert, (1855-1915)
JORDAN, James Brown, (1838-1915)
JOWETT, Albert, (1875-?1958)
JUDD, John Wesley, (1840-1916)
JUKES-BROWNE, Alfred John, (1851-1914)
JULIAN, Henry Forbes, (1861-1912) [PROSPECTORS, DIVINERS AND MINING ENGINEERS]
JULIAN, Hester Forbes, Mrs., (1860-1934)
KAYE, Charles Turton, (1812-1846)
KAYSER, Friedrich Heinrich Emanuel, (1845-1927)
KEEPING, Henry, (1827-1924)
KEEPING, Walter, (1854-1888)
KEITH, Arthur, Sir, (1866-1955)
KENDALL, Percy Fry, (1856-1936)
KENNARD, Alfred Santer, (1870-1948)
KERR, John Graham, (1869-1957)
KIDNER, Henry, (?1851-1925)
KINCH, Edward, (1848-1920)
KING, Samuel William, (1821-1868)
KING, William, (1809-1886)
KINGSLEY, Charles, Reverend, (1819-1875)
KITCHIN, Finlay Lorimer, (1870-1934)
KITTO, Benedict, (1841-1928)
KLAASSEN, Hendericus Martinus, (1828-1910)
KNIPE, Henry Robert, (1855-1918)
KOCH, Walter Edward, (1848-1916)

ENGLAND 1850-1899 Cont'd.

KONIG, Karl Dietrich Eberhardt, called Charles, (1774-1851)
KORNITZER, Louis, (1873-) [PROSPECTORS, DIVINERS AND MINING ENGINEERS]
KYNASTON, Herbert, (1868-1915)
LAING, Samuel, (1810-1897)
LAKE, Philip, (1865-1949)
LAMBERT, Alan, (1837-1927)
LAMPLUGH, George William, (1859-1926)
LANDON, Joseph, (-1902)
LANG, Victor von, Edler, (1838-1921)
LANG, William Dickson, (1878-1966)
LANG, William Henry, (1874-1960)
LANKESTER, Edwin Ray, Sir, (1847-1929)
LAPWORTH, Charles, (1842-1920)
LAPWORTH, Herbert, (1875-1933)
LARKIN, Nathaniel John, (1781-1855)
LA TOUCHE, James Digues, Reverend, (1824-1899)
LAW, Robert, (1840-1907)
LEACH, Arthur Leonard, (1869-1957)
LEBOUR, George Alexander Louis, (1847-1918)
LECKENBY, John, (1814-1877)
LEE, John Edward, (1808-1887)
LEEDS, Alfred Nicholson, (1847-1917)
LEEDS, Charles Edward, (1845-1912)
LEIGHTON, Thomas, (1858-1906)
LENOX-CONYNGHAM, Gerald Ponsonby, Colonel Sir, (1866-1956)
LEWIS, A. E., (-1906)
LEWIS, Henry Carvill, (1853-1888)
LIVEING, George Downing, (1827-1924)
LLOYD, T. G. B., (1829-1876)
LOBLEY, James Logan, (1833-1913)
LOCKYER, Joseph Norman, Sir, (1836-1920)
LOMAX, James, (1857-?1934)
LONGBOTTOM, Arthur, (? -1910)
LONGDEN, Geoffrey Appleby, (1876-1964)
LONGSTAFF, Mary Jane, (née Donald), (1856-1935)
LONSDALE, William, (1794-1871)
LOUIS, Henry, (1855-1939)
LUCY, William Charles, (1822-1898)
LUDLAM, Henry, (1822-1880)
LYELL, Charles, Sir, (1797-1875)
LYONS, Henry George, (1864-?1943)
MACKESON, Henry Bean, (1811-1894)
MACKINTOSH, Daniel, (1815-1891)
MACLEAY, William Sharp, (1792-1865)
MAIDWELL, Frederic Thomas, (1872-1921)
MAJENDIE, Ashurst, (?1784-1867)
MALLET, Robert, (1810-1881)
MANSEL-PLEYDELL, John Clavell, (1817-1902)
MANTELL, Gideon Algernon, (1790-1852)
MANTLE, Harry G., (ca.1858-1937)
MARGERIE, Emmanuel Marie Pierre Martin Jacquin de, (1862-1953)
MARKS, Edward Seaborn, (?1875-1950)
MARR, John Edward, (1857-1933)

ENGLAND 1850-1899 Cont'd.

MARSH, Othniel Charles, (1831-1899)
MARSHALL, James Garth, (-1873)
MARSHAM-TOWNSHEND, Robert Honourable, (1834-1915)
MARTEN, Henry John, (1827-1892)
MARTIN, Edward Alfred, (1864-1943)
MARTIN, Peter John, (1786?-1860)
MASLEN, Arthur John, (1871-1954)
MATLEY, Charles Alfred, (1866-1947)
MAW, George, (1832-1912)
McCABE, Joseph, (1867-1955)
McDONALD, William, (1876-1943)
McINTYRE, Peter, (1863-1940)
McMAHON, Charles Alexander, Lieutenant-General, (1830-1904)
McMURTRIE, James, (1840-1914)
MEADE, Richard, (1827-1894)
MEADOWS, Walter Hennesy, (?1864-1937)
MELLO, John Magens, Reverend, (1836-1914)
MERRICKS, Frank, (?1863-1936)
MEŸER, Charles John Adrian, (1832-1900)
MIALL, Louis Compton, (1842-1921)
MIDDLEMISS, Charles Stewart, (1859-1945)
MIERS, Henry Alexander, Sir, (1858-1942)
MIERS, John, (1789-1879)
MIGEOD, Frederick William Hugh, (1872-1952)
MILLER, Hugh, (1802-1856)
MILLER, Hugh Jr., (1850-1896)
MILLER, William Hallowes, (1801-1880)
MILLETT, Fortescue William, (1833-1915)
MILNE, John, (1850-1913)
MILTON, John Herbert, (?1862-1925)
MITCHELL, Peter Chalmers, Sir, (1864-1945)
MITCHINSON, John, Right Reverend Bishop, (1833-1918)
MÖCKLER, Frederick James, (?1845-1913)
MONTAG, Emil, (1875-1960)
MOORE, Charles, (1814-1881)
MOORE, Henry Cecil, (1835-1908)
MOREY, Frank, (?1858-1925)
MORGAN, Conwy Lloyd, (?1852-1936)
MORRIS, John, (1810-1886)
MORTIMER, John Robert, (1825-1911)
MORTON, George Highfield, (1826-1900)
MOSELEY, Henry Nottidge, (1845-1891)
MOYSEY, Lewis, (1869-1918)
MÜLLER, Hugo Heinrich Wilhelm, (1833-1915)
MULLINS, John, (1838-) [PROSPECTORS, DIVINERS AND MINING ENGINEERS]
MURCHISON, Charles, (1830-1879)
MURRAY, John, III, (1808-1892)
MYLNE, Robert William, (1816-1890)
NESBIT, John Collis, (1819-1862)
NEWELL ARBER, Edward Alexander, (1870-1918)
NEWTON, Alfred, (1829-1907)
NEWTON, Edwin Tulley, (1840-1930)
NEWTON, Richard Bullen, (1850-1926)

ENGLAND 1850-1899 Cont'd.

NICHOLSON, Henry Alleyne, (1844-1899)
NORTON, Henry, (?1812-1892)
NORWOOD, Thomas Wilkinson, Reverend, (1829-1908)
OCKENDEN, Maurice Albion, (1874-1954)
OEYNHAUSEN, Karl August Ludwig, (1795-1865)
OGILBY, William, (1808-1873)
OGILVIE, Francis Grant, Sir, (1858-1930)
OPPEL, Albert, (1831-1865)
ORMEROD, George Wareing, (1810-1891)
OSMAN, Charles William, (1855-1932)
OSWALD, Felix, (1866-1958)
OTLEY, Jonathan, (1766-1856)
OWEN, Richard, Sir, (1804-1892)
PARKER, James, (1833-1912)
PARKER, Thomas Jeffrey, (1850-1897)
PARKER, William Albert, (1855-1918)
PARKER, William Kitchen, (1823-1890)
PARKINSON, Cyril, (1854?-1919)
PARSONS, Henry Franklin, (1846-1913)
PATTISON, Samuel Rowles, (1809-1901)
PATTON, Andrew, (1821-1885)
PEACH, Charles William, (1800-1886)
PENGELLY, William, (1812-1894)
PENNING, William Henry, (1838-1902)
PENNINGTON, Rooke, (-?1887)
PERCEVAL, Spencer George, (1838-1922)
PERCY, John, (1817-1889)
PHILLIPS, John, (1800-1874)
PHILLIPS, William, (1800-1834)
PHILPOT, Elizabeth, (1780-1857)
PICKERING, William Henry, (1858-1912)
PIKE, John William, (1839-1874)
PIPER, George Harry, (1819-1897)
PLANT, Major John, (1819-?1892)
PLATNAUER, Henry Maurice, (1857-1939)
PLATT, William, (?1867-1943)
PLAYER, Jacob Hort, (1834-1908)
POLLARD, William, (1870-1946)
POLWHELE, Thomas Roxburgh, (1831-1909)
PORTER, Henry, (1832-1868)
POSTLETHWAITE, John, (1840-1925)
POULTON, Edward Bagnall, Sir, (1856-1943)
PRESTON, Henry, (1852-1940)
PRESTWICH, Joseph, Sir, (1812-1896)
PRIEST, Simeon, (1876-1954)
PRINGLE, John, (1877-1948)
PRIOR, George Thurland, (1862-1936)
PYE SMITH, John, (1774-1851)
QUEKETT, John Thomas, (1815-1861)
RAISIN, Catherine Alice, (1855-1945)
RAMSAY, Andrew Crombie, Sir, (1814-1891)
RANDALL, John, (1810-1910)
RANSON, John, (?1833-1866)
RAW, Frank, (?1875-1961)
READE, Joseph Bancroft, Reverend, (1801-1870)

ENGLAND 1850-1899 Cont'd.

READE, Thomas Mellard, (1832-1909)
REDMAN, John Baldry, (1816-1899)
REDWOOD, Boverton, Sir, (1846-1919)
REED, Frederick Richard Cowper, (1869-1946)
REED, William, (1810-1892)
REEVE, James, (1833-1920)
REEVE, Lovell, (1814-1865)
REID, Clement, (1853-1916)
REID, Eleanor Mary, (née Wynne Edwards), (1860-1953)
REYNOLDS, Sidney Hugh, (1867-1949)
RICKARD, Francis Ignacio, (-)
RICKARD, Thomas Arthur, (1864-1953)
RICKETTS, Charles, (1818-1904)
RIDLEY, Henry Nicholas, (1855-1956)
ROBERTS, George Edward, (1831-1865)
ROBERTS, Thomas, (1856-1892)
ROBINSON, Thomas, (1828-1906)
ROEDER, Charles, (1848-1911)
ROFE, John, (1801-1878)
RÖMER, Hermann, (1816-1894)
ROSE, Caleb Burrell, (1790-1872)
ROWE, Arthur Walton, (1858-1926)
RUDLER, Frederick William, (1840-1915)
RUFFORD, Philip James, (1852-1902)
RUSKIN, John, (1819-1900)
RUSSELL, Arthur Edward Ian Montagu, Sir, (1878-1964)
RUTLEY, Frank, (1842-1904)
SALMON, Henry Curwen, (1827-1873)
SALTER, Alfred Edward, (?1863-1926)
SALTER, John William, (1820-1869)
SALWEY, Humphry, (1803-1877)
SANDERS, William, (1799-1875)
SANGER, Frederick Robert, (1874-1942)
SARGENT, Harry Lionel, (1871-1955)
SARGENT, Henry Crunden, (1851-1931)
SAWKINS, James Gay, (1806-1878)
SCOTT, Dukinfield Henry, (1854-1934)
SCOTT, Marcus Watson Tate, (1814-1875)
SCOULER, John, (-1874)
SEDGWICK, Adam, Reverend, (1785-1873)
SEDGWICK, Adam, (1854-1913)
SEELEY, Harry Govier, (1839-1909)
SELWYN, Alfred Richard Cecil, (1824-1902)
SEWARD, Albert Charles, Sir, (1863-1941)
SEYLER, Clarence Arthur, (1866-1959)
SHAKESPEAR, Ethel Mary Reader, (née Wood), Dame, (1871-?1946)
SHARMAN, George, (1832-1914)
SHARP, Samuel, (1814-1882)
SHARPE, Daniel, (1806-1856)
SHERBORN, Charles Davies, (1861-1942)
SHERLOCK, R. L., (1875-1948)
SHIPMAN, James, (1848-1901)
SHORE, Thomas William, (1840-1905)
SHRUBSOLE, George William, (1827-1893)

ENGLAND 1850-1899 Cont'd.

SHRUBSOLE, William Hobbs, (1837-1927)
SIMPSON, George Clarke, Sir, (1878-1965)
SIMPSON, Martin, (1798-1892)
SKEATS, Ernest Willington, (1875-1953)
SLACK, Henry James, (1818-1896)
SLADE, Jeremiah, (1828-1905)
SLADEN, Walter Percy, (1849-1900)
SLATTER, Thomas James, (1834-1895)
SMALL, Evan William, (1859-1933)
SMITH, Augustus, (1804-1872)
SMITH, Joshua Toulmin, (1816-1869)
SMITH, Reginald Allender, (?1847-1940)
SMITHE, Frederick, Reverend, (1822-1900)
SMYTH, Warington Wilkinson, Sir, (1817-1890)
SOLLAS, William Johnson, (1849-1936)
SOLLY, Richard Harrison, (1851-1925)
SOMERVAIL, Alexander, (1841-1910)
SOPWITH, Arthur, (1843-1920)
SOPWITH, Thomas, (1803-1879)
SORBY, Henry Clifton, (1826-1908)
SOWERBY, George Brettingham, I, (1788-1854)
SOWERBY, George Brettingham, II, (1812-1884)
SOWERBY, Henry, (1825-1891)
SOWERBY, James de Carle, (1787-1871)
SPENCER, Herbert, (1820-1903)
SPENCER, William Kingdon, (1878-1955)
SPIERS, William, Reverend, (-)
STANLEY, William Ford Robinson, (1828-1909)
STEBBING, William Pinchard Delane, (1873-1961)
STIFFE, Arthur William, Captain, (1849-1912)
STILES, Harold, Sir, (?1862-1946)
STOBBS, John Thomas, (?1866-1942)
STOKES, Arthur Henry, (1842-1910)
STOKES, Charles, (1783-1853)
STOPES, Henry, (1852-1902)
STORY-MASKELYNE, Mervin Herbert Nevil, (1823-1911)
STOW, George, (1876-?1955)
STRAHAN, Aubrey, Sir, (1852-1928)
STRICKLAND, Hugh Edwin, (1811-1853)
SUTCLIFFE, William Henry, (1855-1913)
SUTTON, John Richard, (1864-1937)
SWINNERTON, Henry Hurd, (1875-1967)
TATE, Alexander Norman, (1837-1892)
TATE, George, (1805-1871)
TAUNTON, John Hooke, (1820-1893)
TAWNEY, Edward Bernard, (1840-1882)
TAYLER, Joseph Walter, (-?1896)
TAYLOR, John, (1780-1864)
TAYLOR, John Ellor, (1835-1895)
TAYLOR, William P., (1859-) [PROSPECTORS, DIVINERS AND MINING ENGINEERS]
TEALL, Jethro Justinian Harris, Sir, (1849-1924)
TENISON-WOODS, Julian Edmund, Reverend, (1832-1889)
TENNANT, James, (1808-1881)

ENGLAND 1850-1899 Cont'd.

THACKERAY, Francis St. John, Reverend, (1832-1919)
THOMAS, Herbert Henry, (1876-1935)
THOMPSON, Beeby, (1848-1931)
TIDDEMAN, Richard Hill, (1842-1917)
TIMMINS, Arthur, Major, (?1858-1940)
TOMBLESON, Frederick Binns, (?1872-1938)
TOMLINSON, Charles, (1806-1897)
TOPLEY, William, (1841-1894)
TORELL, Otto Martin, (1828-1900)
TÖRNQUIST, Sven Leonhard, (1840-1920)
TREACHER, Llewellyn, (1859-1943)
TRECHMANN, Charles Otto, (1851?-1917)
TRELAWNEY-ANSELL, E. C., (-) [PROSPECTORS, DIVINERS AND MINING ENGINEERS]
TRENCH, Richard, (1836-1861)
TREVELYAN, Walter Calverley, Sir, (1797-1879)
TRIMMER, Joshua, (1797-1857?)
TRUSCOTT, Samuel John, (?1870-1950)
TURTON, William Henry, (1862-1942)
TUTCHER, John William, (1858-1951)
TUTTON, Alfred Edwin Howard, (1864-1938)
TWELVETREES, William Harper, (1848-1919)
TYLDEN-WRIGHT, Charles, (1832-1900)
TYLER, Charles, Captain, (1826-1895)
TYLOR, Alfred, (1823-1884)
UDALL, John, (1841-1920)
USSHER, William Augustus Edmond, (1849-1920)
VALPY, Robert Harris, (1819-1904)
VARTY SMITH, George, (-1924)
VASSALL, Henry, (1860-1926)
VAUGHAN, Arthur, (1868-1915)
VICARY, William, (1811-1903)
VIVIAN, William, (1817-1879)
WALKER, John Francis, (1839-1907)
WALLACE, Alfred Russel, (1823-1913)
WALLICH, George Charles, (1815-1899)
WARD, James Clifton, Reverend, (1843-1880)
WARD, John, (1837-1906)
WARDLE, Thomas, Sir, (1831-1909)
WARNES, Arthur Robert, (?1877-1942)
WARREN, Samuel Hazzledine, (1872-1958)
WATERHOUSE, John, (1806-1879)
WATSON, John, (1842-1918)
WATTS, William, (-1920)
WATTS, William Whitehead, (1860-1947)
WAY, Herbert W. L., (-) [PROSPECTORS, DIVINERS AND MINING ENGINEERS]
WELLS, Herbert George, (1866-1946)
WESTWOOD, John Obadiah, (1805-1893)
WETHERELL, Nathaniel Thomas, (1800-1875)
WHIDBORNE, George Ferris, Reverend, (1846-1910)
WHITAKER, William, (1836-1925)
WHITE, Henry Hopley, (-?1876)
WHITE, Walter, (-)

ENGLAND 1850-1899 Cont'd.

WHITEAVES, John Joseph Frederick, (1835-1909)
WILBERFORCE, Samuel, Bishop, (1805-1873)
WILKS, Mark, (?1862-1945)
WILLIAMS, Arthur Herbert, (?1865-1935)
WILLIAMSON, William Crawford, (1816-1895)
WILLS, Robert Glover, (1878-1950)
WILSON, Edward, (?1849-1898)
WILTSHIRE, Thomas, Reverend, (1826-1902)
WINDOES, James, (1839-1897)
WINGRAVE, Vitruvius Harold Wyatt, (1858-1938)
WINTER, William Percy, (?1870-1950)
WINWOOD, Henry Hoyte, (1830-1920)
WITCHELL, Edwin, (1823-1887)
WOOD, Edward, (1808-1877)
WOOD, John George, Reverend, (1827-1889)
WOOD, Nicholas, (1795-1865)
WOOD, Searles Valentine, (1798-1880)
WOOD, Searles Valentine Jr., (1830-1884)
WOOD-MASON, James, (1846-1893)
WOODS, Henry, (1868-1952)
WOODWARD, Arthur Smith, Sir, (1864-1944)
WOODWARD, Bernard Barham, (1853-1930)
WOODWARD, Charles Josiah, (1838-1932)
WOODWARD, Henry, (1832-1921)
WOODWARD, Horace Bolingbroke, (1848-1914)
WOODWARD, Samuel P., (1821-1865)
WOOLACOTT, David, (1872-1924)
WORMS, Henry de, Baron, Lord Pirbright, (1840-1903)
WORSFORD, Frederick H., (1870-1954)
WORTH, Richard Nicholas, (-1896)
WRIGHT, Thomas, (1809-1884)
WÜNSCH, Edward A., (1822-1895)
YOUNG, Alfred Collett, (1862-1932)
YOUNG, Alfred Prentice, (1841-1919)
YOUNG, George William, (1862-1929)

1900 to present

ABBOTT, George, (1844-1925)
ABBOTT, William James Lewis, (1853-1933)
ACLAND, Henry Dyke, (-1936)
ADYE, Ernest Howard, (1857-1918)
AHRENS, Louis Herman, (1918-)
AIERS, Arthur Joseph, (1896-1965)
AIM, William Lennie, (?1872-1938)
AIMER, James David, (1899-1973)
ALEXANDER, Frances Elizabeth Somerville, (1908-1958)
ALLEN, Henry Atwool, (?1854-1934)
ANDRADE, Edward Neville da Costa, (1887-1971)
ANDREWS, Charles William, (1866-1924)
ANDREWS, Richard Wallis, (?1927-1967)
ANTHONY, Herbert Douglas, (1892-1968)
ARKELL, William Joscelyn, (1904-1958)
ARMSTRONG, Henry Edward, (1848-1937)
ASHBY, Donald Frederick, (?1914-1943)

ENGLAND 1900 to present Cont'd.

AVELINE, William Talbot, (1822-1903)
BADEN-POWELL, Donald Ferlys Wilson, (1897-1973)
BAILEY, Edward Battersby, Sir, (1881-1965)
BAKER, Benjamin Alfred, (1859-1939)
BAKER, Herbert Arthur, (1885-1954)
BALDWIN, Walter, (?1878-1956)
BALL, Francis Matthew, (1877-1956)
BALL, Harry Standish, (?1888-1941)
BALL, John, (-1941)
BALSTON, William Edward, (1848-1918)
BANCROFT, John Bevis Beeston, (-1944)
BARBER, Cecil Thomas, (-)
BARCLAY, Francis Hubert, (1869-1935)
BARKER, Thomas Vipond, (1881-1931)
BARLOW, Caleb, (1840-1908)
BARLOW, William, (1845-1934)
BARNES, Jonathan, (-1928)
BARNES, William Henry, (1845-1935)
BARNETT, Andrew Ketchen, (1852-1914)
BARNETT, Peter, (1922-1946)
BARRETT, Lucas, (1837-1862)
BARROIS, Charles Eugène, (1851-1939)
BARROW, Geoffrey, (1899-1963)
BARROW, George, (1853-1932)
BARTLETT, Benjamin Pope, (-1928)
BATE, Dorothea Minola Alice, (1879-1951)
BATESON, William, (1861-1926)
BATHER, Francis Arthur, (1863-1934)
BATTLE, Walter Ravenhill Brown "Ben", (1919-1953)
BAUER, Grace M., (1878-1965)
BAUERMAN, Hilary, (1833-1909)
BEADNELL, Hugh John Llewellyn, Major, (1874-1944)
BEALE, William Phipson, Sir, (1839-1922)
BEASLEY, Cyril George, (1901-?1957)
BEASLEY, Henry Charles, (1836-1919)
BELINFANTE, Leo Baptiste Louis, (1861-1937)
BELL, Alfred Graham, (?1875-1964)
BELL, John Arnold, (1939-1970)
BEMROSE, Henry Howe, (1857-1939)
BENNETT, Arthur, (1873-1941)
BENNETT, Francis James, (1845-1920)
BENNETT, Frederick William, (1860-1931)
BENNETT, Walter Hellyer, (1892-1971)
BERNARD, Henry Meyners, (1953-1909)
BEST, Edward, (1824-1906)
BION, Herbert Stanley, (1888-1915)
BIRD, Charles, (1843-1910)
BISAT, William Sawney, (1886-1973)
BISHOP, Walter William, (1931-1977)
BISSET, Charles Barron, (-?1966)
BLACK, Maurice, (1904-1973)
BLACKETT, Patrick Maynard Stuart, Baron of Chelsea, (1897-1974)
BLACKMORE, Humphrey Purnell, (-1928)
BLAKE, George Stanfield, (-1940)

ENGLAND 1900 to present Cont'd.

BLAKE, John Frederick, Reverend, (1839-1906)
BLAKE, John Hopwood, (1843-1901)
BLOOM, Edward Francis Dandridge, (-1942)
BLOW, Walter Harry, (1924-1972)
BOLTON, Herbert, (1863-1936)
BONNEY, Thomas George, Reverend, (1833-1923)
BOOTS, Paul Heath, (1909-1947)
BOSWELL, Katherine Cumming, (1888-1952)
BOSWELL, Percy George Hamnall, (1886-1960)
BOULTON, William Savage, (1867-ca.1954)
BOWER, Fredrik Orpen, (1855-1948)
BOWMAN, Herbert Lister, (1874-1942)
BRADY, Lionel Francis "Major", (1880-1963)
BRAGG, William Henry, Sir, (1862-1942)
BRAGG, William Lawrence, Sir, (1890-1971)
BRAMMALL, Alfred, (?1879-1954)
BRINDLEY, William, (1832-1919)
BROCKBANK, Richard Bowman, (1824-1912)
BRODRICK, Harold, (1872-1946)
BROMEHEAD, Cyril Edward Nowill, (1885-1952)
BROUGHTON, Derek George, (-1971)
BROUGHTON-EDGE, Arthur Broughton, (1895-1953)
BROWN, John Allen, (1831-1903)
BROWN, John Coggin, (1884-1962)
BROWN, Peter Evans, (-)
BROWNRIDGE, Charles, Colonel, (?1862-1951)
BRYDONE, Reginald Marr, (-1943)
BUCKLEY, Harold Eugene, (1897-1959)
BUCKMAN, Sidney Savory, (1860-1929)
BULL, Alfred Joseph, (1875-1950)
BULLEN, Robert Ashington, Reverend, (1850-1912)
BULLERWELL, William, (1916-1977)
BULMAN, Oliver Meredith Boone, (1902-1974)
BURNABY, Thomas Patrick, (1924-1968)
BURROWS, Henry William, (?1860-1939)
BURROWS, William Alfred, (1903-1970)
BURTON, Ernest St. John, (?1874-1962)
BURY, Henry, (?1862-1958)
BUTLER, Francis Henry, (1849-1935)
BUTTERLEY, Archibald Donald, (1896-1957)
CAIRNS, Robert, (1854-1911)
CALLAWAY, Charles, (1838-1915)
CARR, John Wesley, (1862-[1939])
CARRINGTON, Richard, (-)
CARRUTHERS, Robert George, (1880-1965)
CARRUTHERS, William, (1830-1922)
CARTER, William Lower, (1855-1918)
CARUS-WILSON, C., (1857-1934)
CASTELL, Cyril Philip, (1907-1972)
CHADWICK, Samuel, (1845-1903)
CHALONER, William G., (-)
CHAPMAN, Edward John, (1821-1904)
CHAPMAN, Frederick, (1864-1943)
CHAPMAN, Sydney, (1888-1970)

ENGLAND 1900 to present Cont'd.

CHARLESWORTH, Edward, (1813-1893)
CHARLESWORTH, John Kaye, (1889-)
CHREE, Charles, (-)
CHUBB, Lawrence John, (?1888-1971)
CHURCH, Arthur Herbert, Sir, (1834-1915)
CHURCHILL, George Cheetham, (1822-1906)
CLARK, James Edmund, (1850-1944)
CLAYDEN, Arthur William, (1855-1944)
CLODD, Edward, (1840-1930)
CLOUGH, Charles Thomas, (1852-1916)
COBBOLD, Edgar Sterling, (1851-1936)
COCKIN, George Marmaduke, (1852-1918)
CODRINGTON, Thomas, (1829-1918)
COLE, Edward Maule, Reverend, (1833-1911)
COLE, Grenville Arthur James, (1859-1924)
COLENUTT, George William, (?1862-1944)
COLLINS, Arthur L., (1868-1902)
COLLINS, Henry Francis, (?1864-1924)
COLLINS, Joseph Henry, (1841-1916)
COLLINS, William Ion, (1897-1964)
CONNOR, James, (1894-1977)
COOMARASWAMY, Ananda K., (1877-1947)
CORFIELD, William, (1843-1903)
CORNER, Frank, (1861-1939)
COX, Leslie Reginald, (1897-1965)
CRAMP, Alfred William, (?1873-1950)
CRANE, Edward, (1822-1901)
CRICK, George Charles, (1856-1917)
CRICK, Walter Drawbridge, (1857-1903)
CROFT, William Noble, (1915-1953)
CROFTS, William Hastings, (1861-1935)
CROOK, Thomas, (1876-1937)
CROSFIELD, Margaret Chorley, (1859-1952)
CULLIS, Charles Gilbert, (1871-1941)
CUNNINGTON, C. H., Lieutenant, (1889-1918)
CUNNINGTON, William, (1813-1906)
DAKYNS, John Roche, (1836-1910)
DALTON, William H., (?1848-1929)
DARELL, Robert Darell Smythe, (1851-1936)
DARWIN, Francis, (1848-1925)
DARWIN, George Howard, Sir, (1845-1912)
DARWIN, Leonard, (1850-1943)
DAVIDSON, Charles Findlay, (1911-1967)
DAVIES, Arthur Morley, (1869-1959)
DAVIES, George Macdonald, (1885-1973)
DAVIES, Henry Nathaniel, (-1920)
DAVIES, Thomas, (1837-1892)
DAVIS, Arthur George, (1892-1957)
DAVISON, Charles, (1858-1940)
DAVISON, Ernest Henry, (1877-1944)
DAWKINS, William Boyd, Sir, (1837-1929)
DAWSON, Charles, (1864-1916)
DEANE, Henry Jr., (-1924)
DEELEY, Richard Mountford, (1855-1944)

ENGLAND 1900 to present Cont'd.

DELL, John Alexander, (1882-1964)
DENNIS, Frederick William, (1883-1939)
DE RANCE, Charles Eugène, (1847-1906)
DE WAAL, Philip Henry Overbeck, Reverend, (-1942)
DEWEY, Henry, (1876-1965)
DIBLEY, George Edward, (1857-1943)
DICK, Allan Brugh, (1833-1926)
DIGBY, Bassett, (-)
DINES, Henry George, (1891-?1964)
DINHAM, Charles Hawker, (1883-1955)
DIX, Emily, (1904-1973)
DIXON, Ernest Edward Leslie, (1876-1963)
DIXON, Henry George Dacres, (1887-1947)
DOBBIE, Charles Herbert, (1905-1976)
DOCWRA, Colin, (-1942)
DON, Archibald William Robertson, (1891-1916)
DOUGLAS, James Archibald, (1884-1978)
DRUGMAN, Julien Léon Eugène, (1875-1950)
DUNCAN, Cecil Cooke, (1868-1948)
DWERRYHOUSE, Arthur Richard, (1867-1931)
EARLE, Kenneth Wilson, (1891-1944)
EASTWOOD, Tom, (1889?-1970)
EDINGER, Tilly, (1897-1967)
EDMONDS, Charles, (1886-1964)
EDMUNDS, Francis Hereward, (1893-1960)
EDWARDS, Wilfred Norman, (1890-1956)
EGERTON, William Henry, Reverend, (1811-1910)
ELLES, Gertrude Lilian, (1872-1960)
ELLIOT SMITH, Grafton, Sir, (1871-1937)
ELSDEN, James Vincent, (?1856-1930)
ENGLEHEART, Francis Henry Arnold, (-1963)
EPPS, Frank James, (?1891-1974)
ETHERIDGE, Robert, (1819-1903)
EVANS, Emile, (1883-1966)
EVANS, John, Sir, (1823-1908)
EVANS, John William, (1857-1930)
EYLES, Victor Ambrose, (1896-1978)
FAGG, Christopher Charles, (1883-1965)
FALCON, Norman Leslie, (-)
FANKUCHEN, Isidor, (1905-1964)
FARMERY, Joseph Richard, (1872-1928)
FEARNSIDES, William George, (1879-1968)
FENNELL, Charles William, (?1863-1935)
FENNER, Edward George, (?1873-1941)
FINLAYSON, Alexander Moncreiff, (-1917)
FIOCCO, Giorgio, (1931-)
FISHER, Osmond, Reverend, (1817-1914)
FLEET, William Frederick, (1890-1966)
FLETCHER, Lazarus, Sir, (1854-1921)
FLETCHER, Mark, Reverend, (1870-1956)
FOLEY, Mary Cecilia, (-1925)
FOORD, Arthur Humphreys, (1845-1934)
FOOTE, R. Bruce, (-1912)
FORD, Hugh Alexander, (1885-1966)

ENGLAND 1900 to present Cont'd.

FORD, Trevor David, (1932-)
FOSTER, Clement Le Neve, Sir, (1841-1904)
FOWLER, Gordon, Major, (-1953)
FOWLER, Joseph, (1872-1958)
FOX, Cyril Sankey, Sir, (1886-1951)
FOX-STRANGWAYS, Charles Edward, (1844-1910)
FRANCIS, James, (1843-1925)
FRANKS, Godfrey Firth, (1853-1917)
GALTON, Francis, Sir, (1822-1911)
GARDINER, Charles Irving, (?1868-1940)
GARNETT, Alice, (-)
GARNETT, Cecil Stevenson, (1896-1951)
GARWOOD, Edmund Johnston, (1864-1949)
GASKIN, Alan Robert James, (1927-1974)
GASTER, Christopher Theodore Augustine, (1878-1963)
GEINITZ, Hanns Bruno, (1814-1900)
GERRARD, John, (?1850-1920)
GIBBONS, Arthur, (1867-1938)
GIBSON, Walcot, (1864-1941)
GILBERT, Charles J., (-?1934)
GLEAVES, John T., (-1969)
GOODCHILD, John George, (1844-1906)
GOODE, R. H., Reverend, (1888-1969)
GOODRICH, Edwin Stephen, (1868-1946)
GOODYEAR, Edith, (1879-1959)
GORDON, William Thomas, (1884-1950)
GORE, Charles Herbert Henry, (1866-1951)
GOSSLING, Frank Newbery, (1888-1946)
GOURLEY, Harold John Frederick, (1886-1956)
GRABHAM, George Walter, (1882-1955)
GRAHAM, William Vaux, (1859-1940)
GRANTHAM, Donald Robert, (1896-1969)
GRAY, Joseph William, (-1935)
GREEN, John Frederick Norman, (?1873-1949)
GREEN, Upfield, (1834-1917)
GREENE, Alfred Ernest, (1878-1940)
GREENLY, Edward, (1861-1951)
GREGORY, Hubert Harold, (-?1950)
GREIG, Arthur, (1893-)
GRESSWELL, Ronald Kay, (1905-1969)
GROOM, Theodore Thomas, (1863-)
GROVES, Arthur William, (1903-1956)
GUNN, William, (1837-1902)
GÜNTHER, Albert Charles Lewis Gotthilf, (1830-1914)
GUPPY, Robert John Lechmere, (1836-1916)
GURNEY, Henry Palin, Reverend, (1847-1904)
GWINNELL, Russell Frost, (1880-1913)
HAIG, David Salusbury, (?1913-1955)
HALDANE, John Burdon Sanderson, (1892-1964)
HALL, Thomas Clifford Fitzwilliam, (1886-1956)
HALLEY, Edmond, (1656-1742)
HALLIMOND, Arthur Francis, (1890-1968)
HAMLING, Joseph Green, (-?1935)
HANNAY, Harry Campbell, (1874-1961)

ENGLAND 1900 to present Cont'd.

HARKER, Alfred, (1859-1939)
HARMER, Frederic William, (1835-1923)
HARPER, John Childs, (1913-1977)
HARRIS, George Frederick, (1862-1906)
HARRISON, Benjamin, (1837-1921)
HARRISON, Edward Richard, Sir, (1872-1960)
HARRISON, John Vernon, (1892-1972)
HARRISON, William Jerome, (1845-1908)
HARRY, William Trevelyan, (?1922-1964)
HARTLEY, Harold Brewer, Sir, (1878-1972)
HARTLEY, John Jerom, (1886-)
HARWOOD, Harold Benjamin, (-1956)
HATCH, Frederick Henry, (?1864-1932)
HAWARD, Frederick James Nairn, (1871-1953)
HAWKINS, Herbert Leader, (1887-1968)
HAYWARD, Henry Arthur, (-1964)
HAYWARD, John, (-1958)
HENSON, Frederick Arthur, (1919-1957)
HERON-ALLEN, Edward, (1861-1943)
HERRIES, Robert Stansfield, (?1860-1941)
HESLOP, Mary Kingdon, (?1884-?1954)
HESLOP, Septimus, (?1860-1945)
HEWITT, Henry Dixon, (-?1966)
HIBBERT, Frank, Colonel, (?1889-1951)
HICKLING, Henry George Albert, (1883-1954)
HIDE, Raymond, (1929-)
HILL, James Bastian, (1861-1927)
HILL, Maurice Neville, (1919-1966)
HILL, William, (1849-1912)
HIMUS, Godfrey Wilfred, (1894-1964)
HIND, Henry Youle, (1823-1908)
HIND, Wheelton, (1859-1920)
HINDE, George Jennings, (1839-1918)
HINTON, Henry Arthur, (?1874-1937)
HINTON, Martin Alister Campbell, (1883-1961)
HITCHON, Hubert, (1899-1962)
HOBSON, Bernard, (1860-1933)
HODGE, Matthew Burnop, (-1951)
HOLDEN, Henry Smith, (1887-1963)
HOLGATE, Benjamin, (1838-1915)
HOLLAND, Thomas Henry, Sir, (1868-1947)
HOLLINGWORTH, Sydney Ewart, (1899-1966)
HOLLOWAY, Lionel E., (1898-1965)
HOLMES, Arthur, (1890-1965)
HOLMES, T. Vincent, (?1841-1923)
HOLWILL, Frederick James Winsor, (1925-1966)
HONNOR, Frank Walter, (?1892-1976)
HOOKER, Joseph Dalton, Sir, (1817-1911)
HOOLEY, Reginald Walter, (1866-1923)
HOPKINSON, John, (1844-1919)
HORNE, William, (?1836-1928)
HOTBLACK, John Turner, (1848-1915)
HOWCHIN, Walter, Reverend, (1845-1937)
HOWELL, Henry Hyatt, (1834-1915)

ENGLAND 1900 to present Cont'd.

HOWES, George Bond, (1853-1905)
HOWSE, Richard, (1821-1901)
HUDLESTON, Wilfrid Hudleston, (né Simpson), (1828-1909)
HUDSON, Robert George Spencer, (1895-1965)
HUGGINS, William, Sir, (1824-1910)
HUGHES, Thomas McKenny, (1832-1917)
HUGHES, William John, (?1899-1961)
HULL, Edward, (1829-1917)
HUNT, Arthur Roope, (1843-1914)
HUTCHINS, Philip Frank, (1927-1955)
HUTCHINSON, Arthur, (1866-1937)
HUTCHINSON, Henry Neville, Reverend, (?1855-1927)
HUTTON, Harriet Mary, (1873-1937)
HUXLEY, Leonard, (1860-1933)
ICKES, Eugene Law, (1885-1941)
ILLING, Vincent Charles, (1890-1969)
INESON, Jack, (1917-1970)
JACKSON, James Frederick, (1896-1966)
JAMES, W. Warwick, (?1874-1965)
JENNINGS, Albert Vaughan, (1864-1903)
JERVIS, William Paget, (1831-1906)
JESSON, Thomas, (?1850-1928)
JOHNSON, J. P., (1880-1918)
JOHNSON, Walter, (1867-1950)
JONES, Ernest Lloyd, (?1863-1942)
JONES, Owen Thomas, (1878-1967)
JONES, Thomas Rupert, (1819-1911)
JONES, William Richard, (1890-1970)
JONES, William Rupert, (1855-1915)
JORDAN, James Brown, (1838-1915)
JOWETT, Albert, (1875-?1958)
JOYCE, John Raymond Foggan, (?1907-1963)
JUDD, John Wesley, (1840-1916)
JUKES-BROWNE, Alfred John, (1851-1914)
JULIAN, Henry Forbes, (1861-1912) [PROSPECTORS, DIVINERS AND MINING ENGINEERS]
JULIAN, Hester Forbes, Mrs., (1860-1934)
KAYSER, Friedrich Heinrich Emanuel, (1845-1927)
KEEPING, Henry, (1827-1924)
KEITH, Arthur, Sir, (1866-1955)
KENDALL, Percy Fry, (1856-1936)
KENNARD, Alfred Santer, (1870-1948)
KERR, John Graham, (1869-1957)
KIDNER, Henry, (?1851-1925)
KINCH, Edward, (1848-1920)
KING, William Bernard Robinson, (1899-1963)
KINGSBURY, Arthur William Gerald, (1906-1968)
KINNEY, Harry Dill, (1886-1961)
KIRKPATRICK, Randolph, (-1950)
KITCHIN, Finlay Lorimer, (1870-1934)
KITTO, Benedict, (1841-1928)
KLAASSEN, Hendericus Martinus, (1828-1910)
KNIPE, Henry Robert, (1855-1918)
KOCH, Walter Edward, (1848-1916)

ENGLAND 1900 to present Cont'd.

KORNITZER, Louis, (1873-) [PROSPECTORS, DIVINERS AND MINING ENGINEERS]
KYNASTON, Herbert, (1868-1915)
LAKE, Philip, (1865-1949)
LAMBERT, Alan, (1837-1927)
LAMONT, Archie, (-)
LAMPLUGH, George William, (1859-1926)
LANDON, Joseph, (-1902)
LANG, Victor von, Edler, (1838-1921)
LANG, William Dickson, (1878-1966)
LANG, William Henry, (1874-1960)
LANKESTER, Edwin Ray, Sir, (1847-1929)
LAPWORTH, Charles, (1842-1920)
LAPWORTH, Herbert, (1875-1933)
LARMAN, Mark Linington, (1899-1958)
LAW, Robert, (1840-1907)
LAWRIE, William Henry, (?1884-1943)
LEACH, Arthur Leonard, (1869-1957)
LEBOUR, George Alexander Louis, (1847-1918)
LEDINGHAM, Glen Wallace, (1911-1978)
LEECH, John George Chambré, (1890-1966)
LEEDS, Alfred Nicholson, (1847-1917)
LEEDS, Charles Edward, (1845-1912)
LEES, George Martin, (1898-1955)
LEIGHTON, Thomas, (1858-1906)
LEITCH, Duncan, (1904-1956)
LeLACHEUR, W. J., (-1923)
LENOX-CONYNGHAM, Gerald Ponsonby, Colonel Sir, (1866-1956)
LEWIS, A. E., (-1906)
LEWIS, William Vaughan, (1907-1961)
LIGHTFOOT, Ben, (1888-1966)
LINTON, David Leslie, (1906-1971)
LITTLE, Otway Henry, (-1956)
LIVEING, George Downing, (1827-1924)
LOBJOIT, William Michael, (1932-1967)
LOBLEY, James Logan, (1833-1913)
LOCKYER, Joseph Norman, Sir, (1836-1920)
LOMAX, James, (1857-?1934)
LONGBOTTOM, Arthur, (? -1910)
LONGDEN, Geoffrey Appleby, (1876-1964)
LONGSTAFF, Mary Jane, (née Donald), (1856-1935)
LONSDALE, Kathleen, Dame, (née Yardley), (1903-1971)
LOUIS, Henry, (1855-1939)
LOWNDES, Ashley Gordon, (-1956)
LYELL, Charles, Sir, (1797-1875)
LYONS, Henry George, (1864-?1943)
MacENERY, John, Reverend, (1796-1841)
MACHATSCHKI, Felix Karl Ludwig, (1895-1970)
MACMILLAN, William Edward Frank, (-1954)
MAIDWELL, Frederic Thomas, (1872-1921)
MANSEL-PLEYDELL, John Clavell, (1817-1902)
MANTLE, Harry G., (ca.1858-1937)
MARGERIE, Emmanuel Marie Pierre Martin Jacquin de, (1862-1953)
MARKS, Edward Seaborn, (?1875-1950)

ENGLAND 1900 to present Cont'd.

MARR, John Edward, (1857-1933)
MARSDEN, Ernest, (1889-1970)
MARSH, Thomas William, (-1957)
MARSHALL, Alan Edward, (1938-1970)
MARSHAM-TOWNSHEND, Robert Honourable, (1834-1915)
MARSTON, Alvan Theophilus, (1889-1971)
MARTIN, Edgar Charles, (1884-1955)
MARTIN, Edward Alfred, (1864-1943)
MASLEN, Arthur John, (1871-1954)
MATLEY, Charles Alfred, (1866-1947)
MAUFE, Herbert Brantwood, (1879-1946)
MAW, George, (1832-1912)
McCALLUM, Robert Rowson, Lieutenant-Colonel, (1882-1934)
McDONALD, William, (1876-1943)
McINTYRE, Peter, (1863-1940)
McMAHON, Charles Alexander, Lieutenant-General, (1830-1904)
McMURTRIE, James, (1840-1914)
MEADOWS, Walter Hennesy, (?1864-1937)
MELLO, John Magens, Reverend, (1836-1914)
MERRICKS, Frank, (?1863-1936)
MIALL, Louis Compton, (1842-1921)
MIDDLEMISS, Charles Stewart, (1859-1945)
MIERS, Henry Alexander, Sir, (1858-1942)
MIGEOD, Frederick William Hugh, (1872-1952)
MILLER, Arthur Austin, (1900-1968)
MILLETT, Fortescue William, (1833-1915)
MILNE, John, (1850-1913)
MILTON, John Herbert, (?1862-1925)
MITCHELL, George Hoole "Mick", (1902-1976)
MITCHELL, Peter Chalmers, Sir, (1864-1945)
MITCHINSON, John, Right Reverend Bishop, (1833-1918)
MÖCKLER, Frederick James, (?1845-1913)
MONCKTON, Horace Woolaston, (1857-1931)
MONTAG, Emil, (1875-1960)
MOORE, Edward William James, (1887-1970)
MOORE, Henry Cecil, (1835-1908)
MOORE, Leslie Rowsell, (1914-)
MOREY, Frank, (?1858-1925)
MORGAN, Conwy Lloyd, (?1852-1936)
MORTIMER, John Robert, (1825-1911)
MORTON, George Highfield, (1826-1900)
MOTTRAM, Brian Hunter, (1922-1976)
MOYSEY, Lewis, (1869-1918)
MUIR-WOOD, Helen Marguerite, (1896-1968)
MÜLLER, Hugo Heinrich Wilhelm, (1833-1915)
NEWELL ARBER, Edward Alexander, (1870-1918)
NEWTON, Alfred, (1829-1907)
NEWTON, Edwin Tulley, (1840-1930)
NEWTON, Richard Bullen, (1850-1926)
NICOLAS, Jean Paul, (1921-1967)
NORWOOD, Thomas Wilkinson, Reverend, (1829-1908)
NUTTALL, Ernest, (-?1953)
OAKLEY, Kenneth Page, (1911-)
OATES, Frank, (1890-1945)

ENGLAND 1900 to present Cont'd.

OCKENDEN, Maurice Albion, (1874-1954)
ODLING, Marmaduke, (?1888-1956)
OGILVIE, Francis Grant, Sir, (1858-1930)
ORIEL, Benjamin, Reverend, (-1936)
OSBORNE WHITE, Harold J., (-1954)
OSMAN, Charles William, (1855-1932)
OSWALD, Felix, (1866-1958)
OVERY, Charles, Reverend, (1882-)
PALMER, L. S., (1890-1962)
PARKER, James, (1833-1912)
PARKER, William Albert, (1855-1918)
PARKINSON, Cyril, (1854?-1919)
PARKINSON, Donald, (1891-1974)
PARRINGTON, Francis Rex, (-)
PARSONS, Henry Franklin, (1846-1913)
PARSONS, Leonard Miles, (1883-1956)
PATTERSON, Arthur Lindo, (1902-1966)
PATTISON, Samuel Rowles, (1809-1901)
PEET, Geoffrey Alan, (1922-1959)
PEGG, E. J., (?1899-1948)
PENNING, William Henry, (1838-1902)
PERCEVAL, Spencer George, (1838-1922)
PERCIVAL, Frederick George, (1887-1973)
PHILLIPS, David Watkins, (1899-1962)
PICKERING, William Henry, (1858-1912)
PLATNAUER, Henry Maurice, (1857-1939)
PLATT, William, (?1867-1943)
PLAYER, Jacob Hort, (1834-1908)
POCOCK, Roy Woodhouse, (1887-1954)
POLLARD, William, (1870-1946)
POLWHELE, Thomas Roxburgh, (1831-1909)
POSTLETHWAITE, John, (1840-1925)
POULTON, Edward Bagnall, Sir, (1856-1943)
PRESTON, Henry, (1852-1940)
PRETYMAN, Frank Remington, (1891?-1917)
PRIEST, Simeon, (1876-1954)
PRINGLE, John, (1877-1948)
PRIOR, George Thurland, (1862-1936)
PUEL, H. E., (-1964)
QUIRK, Roger Nathaniel, (?1909-1964)
RADLEY, Ernest George, (1880-1945)
RAISIN, Catherine Alice, (1855-1945)
RANDALL, John, (1810-1910)
RANSON, John, (?1833-1866)
RAVEN, Charles, (1885-1964)
RAW, Frank, (?1875-1961)
READE, Thomas Mellard, (1832-1909)
READER, Thomas William, (-?1923)
REDWOOD, Boverton, Sir, (1846-1919)
REED, Frederick Richard Cowper, (1869-1946)
REID, Clement, (1853-1916)
REID, Eleanor Mary, (née Wynne Edwards), (1860-1953)
REYNOLDS, Sidney Hugh, (1867-1949)
RHODES, James Eric Wynfield, (?1888-1976)

ENGLAND 1900 to present Cont'd.

RICHARDSON, Linsdall, (1881-1967)
RICHARDSON, William Alfred, (1887-?1964)
RICKARD, Thomas Arthur, (1864-1953)
RICKETTS, Charles, (1818-1904)
RIDLEY, Henry Nicholas, (1855-1956)
ROBBIE, James Andrew, (1910-1977)
ROBERTS, Thomas, (1856-1892)
ROBINSON, Thomas, (1828-1906)
ROE, Frank Walton "Peter", (?1914-1965)
ROEDER, Charles, (1848-1911)
ROSS, Frank Stenhouse, (1890-1975)
ROWE, Arthur Walton, (1858-1926)
RUDLER, Frederick William, (1840-1915)
RUFFORD, Philip James, (1852-1902)
RUSSELL, Arthur Edward Ian Montagu, Sir, (1878-1964)
RUSSELL, Frank Scott, (-1943)
RUTHERFORD, Hugo, Sir, (1887-1942)
RUTLEDGE, Harold, (1920-1954)
RUTLEY, Frank, (1842-1904)
RYDE, John Walter, (1898-1961)
SAINTY, James Edward, (?1882-1967)
SALTER, Alfred Edward, (?1863-1926)
SAMPSON, David Noel, (1917-1971)
SANDFORD, Kenneth Stuart, (1899-1971)
SANGER, Frederick Robert, (1874-1942)
SARGENT, Harry Lionel, (1871-1955)
SARGENT, Henry Crunden, (1851-1931)
SARJEANT, William Antony Swithin, (1935-)
SCHOFIELD, Eric, (1911-1944)
SCOTT, Dukinfield Henry, (1854-1934)
SEDGWICK, Adam, (1854-1913)
SEELEY, Harry Govier, (1839-1909)
SELIGMAN, Gerald, (1886-1973)
SELWYN, Alfred Richard Cecil, (1824-1902)
SEWARD, Albert Charles, Sir, (1863-1941)
SEYLER, Clarence Arthur, (1866-1959)
SHAKESPEAR, Ethel Mary Reader, (née Wood), Dame, (1871-?1946)
SHANNON, William George St. John, (1889-1946)
SHARMAN, George, (1832-1914)
SHERBORN, Charles Davies, (1861-1942)
SHERLOCK, R. L., (1875-1948)
SHIPMAN, James, (1848-1901)
SHORE, Thomas William, (1840-1905)
SHRUBSOLE, William Hobbs, (1837-1927)
SIBLY, Thomas Franklin, Sir, (1883-1948)
SIMMONS, Eric Warr, Lieutenant, (1893-1915)
SIMMONS, William Charles, (1888-1951)
SIMONS, Herbert Samuel, (1891-1956)
SIMPSON, George Clarke, Sir, (1878-1965)
SKEATS, Ernest Willington, (1875-1953)
SLADE, Jeremiah, (1828-1905)
SLATER, George, (-1956)
SMALL, Evan William, (1859-1933)
SMITH, Bernard, (1881-1936)

ENGLAND 1900 to present Cont'd.

SMITH, Denis Raymond James, (1937-1964)
SMITH, Herbert Gladstone, (-1950)
SMITH, Reginald Allender, (?1847-1940)
SMITH, Stanley, (1883-1955)
SOLLAS, William Johnson, (1849-1936)
SOLLY, Richard Harrison, (1851-1925)
SOMERVAIL, Alexander, (1841-1910)
SOPWITH, Arthur, (1843-1920)
SORBY, Henry Clifton, (1826-1908)
SOWERBY, Arthur de Carle, (1885-1954)
SPATH, Leonard Frank, (1882-1957)
SPENCER, Herbert, (1820-1903)
SPENCER, William Kingdon, (1878-1955)
SPIKE, Wilfred Ivan, (1925-1969)
STAMP, Laurence Dudley, Sir, (1898-1966)
STANLEY, William Ford Robinson, (1828-1909)
STAPLES, Ernest Henry, (1883-1957)
STEBBING, William Pinchard Delane, (1873-1961)
STIFFE, Arthur William, Captain, (1849-1912)
STILES, Harold, Sir, (?1862-1946)
STOBBS, John Thomas, (?1866-1942)
STOKES, Arthur Henry, (1842-1910)
STONELEY, Robert, (1894-1976)
STOPES, Henry, (1852-1902)
STOPES, Marie Charlotte Carmichael, (1880-1958)
STORY-MASKELYNE, Mervin Herbert Nevil, (1823-1911)
STOW, George, (1876-?1955)
STRAHAN, Aubrey, Sir, (1852-1928)
STRAW, Stephen Henry, (?1892-1963)
STUTZER, Otto, (1881-1936)
SUJKOWSKI, Zbigniew Leliwa, (1898-1954)
SUTCLIFFE, William Henry, (1855-1913)
SUTTON, John, (-)
SUTTON, John Richard, (1864-1937)
SWEETING, George Scotland, (1889-1977)
SWINNERTON, Henry Hurd, (1875-1967)
SYLVESTER-BRADLEY, Peter Colley, (1914?-1978)
TAYLOR, Griffith, (1881-1963)
TAYLOR, Herbert, (-1963)
TAYLOR, James Haward, (1909-1968)
TAYLOR, William P., (1859-) [PROSPECTORS, DIVINERS AND MINING ENGINEERS]
TEALL, Jethro Justinian Harris, Sir, (1849-1924)
TEMPLEMAN, Andrew, (1887-1945)
THACKERAY, Francis St. John, Reverend, (1832-1919)
THOMAS, Henry Dighton, (1900-1966)
THOMAS, Herbert Henry, (1876-1935)
THOMAS, Hugh Hamshaw, (1885-1962)
THOMPSON, Beeby, (1848-1931)
THOMPSON, Hugh R., (1926-1959)
THOROLD, William George Preston, (?1892-1943)
TIDDEMAN, Richard Hill, (1842-1917)
TILLEY, Cecil Edgar, (1894-1973)
TIMMINS, Arthur, Major, (?1858-1940)

ENGLAND 1900 to present Cont'd.

TOMBLESON, Frederick Binns, (?1872-1938)
TOMKEIEFF, Sergei Ivanovich, (1892-1968)
TÖRNQUIST, Sven Leonhard, (1840-1920)
TOZER, Charles Fallows, (1918-1966)
TREACHER, Llewellyn, (1859-1943)
TRECHMANN, Charles Otto, (1851?-1917)
TRELAWNEY-ANSELL, E. C., (-) [PROSPECTORS, DIVINERS AND MINING ENGINEERS]
TREMLETT, Rex, (-) [PROSPECTORS, DIVINERS AND MINING ENGINEERS]
TROTTER, Frederick Murray, (1897-1968)
TRUEMAN, Arthur Elijah, Sir, (1894-1956)
TRUSCOTT, Samuel John, (?1870-1950)
TURTON, William Henry, (1862-1942)
TUTCHER, John William, (1858-1951)
TUTTON, Alfred Edwin Howard, (1864-1938)
TWELVETREES, William Harper, (1848-1919)
UDALL, John, (1841-1920)
USSHER, William Augustus Edmond, (1849-1920)
VALPY, Robert Harris, (1819-1904)
VARTY SMITH, George, (-1924)
VASSALL, Henry, (1860-1926)
VAUGHAN, Arthur, (1868-1915)
VICARY, William, (1811-1903)
VICKERS, Arthur Eric Jarvis, (1897-1958)
WAGER, Lawrence Rickard, (1904-1965)
WALKER, John Francis, (1839-1907)
WALLACE, Alfred Russel, (1823-1913)
WALTON, John, (-1971)
WARBURG, Robert Oscar, (1915-1965)
WARD, John, (1837-1906)
WARDLE, Thomas, Sir, (1831-1909)
WARNES, Arthur Robert, (?1877-1942)
WARREN, Samuel Hazzledine, (1872-1958)
WATSON, David Meredith Seares, (1886-1973)
WATSON, John, (1842-1918)
WATTISON, John Thomas, (1884-1974)
WATTS, William, (-1920)
WATTS, William Whitehead, (1860-1947)
WEAVER, Paul, (1888-1964)
WELLS, Herbert George, (1866-1946)
WHEELER, Richard Vernon, (1883-1939)
WHIDBORNE, George Ferris, Reverend, (1846-1910)
WHITEAVES, John Joseph Frederick, (1835-1909)
WHITEHEAD, Talbot Haes, (1890-1966)
WHITTARD, Walter Frederick, (1902-1966)
WILCOCKSON, William Howson, (1891-1976)
WILD, Robert Powley, (1882-1946)
WILKS, Mark, (?1862-1945)
WILLIAMS, Arthur Herbert, (?1865-1935)
WILLIAMS, Mervyn, (1934-1970)
WILLS, Robert Glover, (1878-1950)
WILSON, Vernon, (1907-1971)
WILTSHIRE, Thomas, Reverend, (1826-1902)

ENGLAND 1900 to present Cont'd.

WINGRAVE, Vitruvius Harold Wyatt, (1858-1938)
WINTER, William Percy, (?1870-1950)
WINWOOD, Henry Hoyte, (1830-1920)
WITHERS, Thomas Henry, (1883-1953)
WOOD, Frederick Wilfred, (?1904-1961)
WOOD, James George, (1888-1928)
WOODS, Ethel Gertrude, (-1939)
WOODS, Henry, (1868-1952)
WOODWARD, Arthur Smith, Sir, (1864-1944)
WOODWARD, Bernard Barham, (1853-1930)
WOODWARD, Charles Josiah, (1838-1932)
WOODWARD, Henry, (1832-1921)
WOODWARD, Horace Bolingbroke, (1848-1914)
WOOLACOTT, David, (1872-1924)
WOOLDRIDGE, Sidney William, (1900-1963)
WORDIE, James Mann, Sir, (?1890-1962)
WORMS, Henry de, Baron, Lord Pirbright, (1840-1903)
WORSFORD, Frederick H., (1870-1954)
WRIGHT, William, (-1936)
WRIGLEY, Arthur George, (1885-1953)
YOUNG, Alfred Collett, (1862-1932)
YOUNG, Alfred Prentice, (1841-1919)
YOUNG, George William, (1862-1929)
ZEALLEY, Arthur Edmund Victor, (1886-1918)
ZEUNER, Frederick Everard, (1905-1963)

Time unknown
BEER, Gavin R. de, Sir, (-)
BIRDS, James Adey, (1831-1894)
PARKER, Thomas Jeffrey, (1850-1897)

EQUATORIAL GUINEA
1900 to present
HERNÁNDEZ-PACHECO, Francisco, (1899-1976)

ERITREA See ABYSSINIA.

ESTONIAN SSR (USSR)
1800-1849
ABICH, Hermann Wilhelm, (1806-1886)
GREWINGK, Konstantin, (1819-1887)
PANDER, Christian Heinrich, (1794-1865)

1850-1899
ABICH, Hermann Wilhelm, (1806-1886)
GREWINGK, Konstantin, (1819-1887)
MACLAREN, James Malcolm, (1873-1935)
PANDER, Christian Heinrich, (1794-1865)
SCHMIDT, Friedrich Bogdanovich "Bohdan", (1832-1908)

ESTONIAN SSR (USSR) 1850-1899 Cont'd.

TAMMANN, Gustav Heinrich Johann Apoll, (1861-1938)

1900 to present
BALK, Robert, (1899-1955)
BEKKER, Hendrik, (1891-1925)
MACLAREN, James Malcolm, (1873-1935)
SCHMIDT, Friedrich Bogdanovich "Bohdan", (1832-1908)
TAMMANN, Gustav Heinrich Johann Apoll, (1861-1938)
THOMSON, Paul William, (1891-1957)

ETHIOPIA *See* ABYSSINIA.

EUROPE (unspecified)
Up to 1800
HAMILTON, William, Sir, (1730-1803)
HUMBOLDT, Friedrich Wilhelm Heinrich Alexander von, Freiherr, (1769-1859)
VALLISNERI, Antonio, (1661-1730)

1800-1849
BOUÉ, Ami, (1794-1881)
HAMILTON, William, Sir, (1730-1803)
HUMBOLDT, Friedrich Wilhelm Heinrich Alexander von, Freiherr, (1769-1859)

1850-1899
BALDAUF, Richard Julius, (1848-1931)
BOUÉ, Ami, (1794-1881)
CAPELLINI, Giovanni, (1833-1922)
FEISTMANTEL, Otokar, (1848-1891)
FRAAS, Eberhard, (1862-1915)
FRECH, Fritz, (1861-1917)
HENDERSON, James Alexander Leo, (?1874-1940)
HORWITZ, Ludwik, (1875-1943)
HUMBOLDT, Friedrich Wilhelm Heinrich Alexander von, Freiherr, (1769-1859)
JACK, Robert Lockhart, (1878-1964)
JUDD, John Wesley, (1840-1916)
KEYES, Charles Rollin, (1864-1942)
LYELL, Charles, Sir, (1797-1875)
MILTHERS, Paul Christian Vilhelm Madsen, (1865-1962)
NEUMAYR, Melchior, (1845-1890)
PARSONS, Arthur Leonard, (1873-1957)
SALOMON-CALVI, Wilhelm, (1868-1941)
WASHINGTON, Henry Stephens, (1867-1934)

1900 to present
BALDAUF, Richard Julius, (1848-1931)
BENTZ, Alfred, (1897-1964)
BISAT, William Sawney, (1886-1973)
BOGDANOV, Alexei Alexeievitch, (1906-1971)
BUBNOFF, Serge von, (1888-1957)
CAPELLINI, Giovanni, (1833-1922)
COLBERT, Edwin H., (1905-)

EUROPE (unspecified) 1900 to present Cont'd.

CONKLING, Richard A., (1885-1952)
DE KOCH, William Gaspar, (1898-1954)
FRAAS, Eberhard, (1862-1915)
FRECH, Fritz, (1861-1917)
HENDERSON, James Alexander Leo, (?1874-1940)
HOCHSTETTER, Harald Egbert von, Ritter, (1902-1972)
HORWITZ, Ludwik, (1875-1943)
JACK, Robert Lockhart, (1878-1964)
JOUBIN, Francis Renault, (1911-)
JUDD, John Wesley, (1840-1916)
KEYES, Charles Rollin, (1864-1942)
LECKWIJCK, Willem Peter van, (1902-1975)
LYELL, Charles, Sir, (1797-1875)
MILTHERS, Paul Christian Vilhelm Madsen, (1865-1962)
NEWHOUSE, Walter Harry, (1897-1969)
PARSONS, Arthur Leonard, (1873-1957)
ROUBAULT, Marcel Edouard, (1905-1974)
SALOMON-CALVI, Wilhelm, (1868-1941)
SAPPINGTON, Chester, (1905-1968)
SCHNEIDERHÖHN, Hans, (1887-1962)
SCHUMACHER, Friedrich, (1884-)
TRECHMANN, Charles Taylor, (1885-1964)
WASHINGTON, Henry Stephens, (1867-1934)
WAWRYK, Wlodzimierz, (1902-1963)
WRIGHT, Charles Will, (1879-)

FAEROE ISLANDS (NORTH ATLANTIC OCEAN)

1800-1849

TREVELYAN, Walter Calverley, Sir, (1797-1879)

1850-1899

BIRLEY, Caroline, (1851-1907)
DUDGEON, Patrick, (-1895)
STOKES, Arthur Henry, (1842-1910)
TREVELYAN, Walter Calverley, Sir, (1797-1879)
ZIRKEL, Ferdinand, (1838-1912)

1900 to present

BIRLEY, Caroline, (1851-1907)
GÖRGEY VON GÖRGÖ, Rudolf, (1886-1915)
PEACOCK, Martin Alfred, (1898-1950)
STOKES, Arthur Henry, (1842-1910)
ZIRKEL, Ferdinand, (1838-1912)

FALKLAND ISLANDS

1800-1849

DARWIN, Charles Robert, (1809-1882)

1850-1899

DARWIN, Charles Robert, (1809-1882)

FARASAN ISLANDS (RED SEA)

1800-1849

EHRENBERG, Christian Gottfried, (1795-1876)

1850-1899

EHRENBERG, Christian Gottfried, (1795-1876)

1900 to present

JONES, David Glynn, (1904-1974)

FEDERAL REPUBLIC OF GERMANY <u>See</u> GERMANY.

FERNANDO DE NORONHA (SOUTH ATLANTIC OCEAN)

1850-1899

WILLIAMS, George Huntington, (1856-1894)

FERNANDO PO <u>See</u> EQUATORIAL GUINEA.

FIJI ISLANDS (PACIFIC OCEAN)

1850-1899

BRADY, Henry Bowman, (1835-1891)
DAVID, Tannatt William Edgeworth, Sir, (1858-1934)

1900 to present

DAVID, Tannatt William Edgeworth, Sir, (1858-1934)
FRASER, John A., (-) [PROSPECTORS, DIVINERS AND MINING ENGINEERS]
PEGAU, Albert August, (1894-1963)
STILLWELL, Frank Leslie, (1888-1963)
TAYLOR, Griffith, (1881-1963)

FIJI ISLANDS (PACIFIC OCEAN) 1900 to present Cont'd.

WILLIAMS, Mervyn, (1934-1970)

FINLAND

Up to 1800

BUCH, Christian Leopold von, Freiherr, (1774-1853)
GADD, Pehr Adrian, (1727-1797)
GADOLIN, Johan, (1760-1852)
GROSSE, Karl Friedrich August, (1768-1847)
KALM, Pehr, (1716-1779)
LAXMAN, Erik, (1737-1796)
LINNE, Carl von, (1707-1778)

1800-1849

ARPPE, Adolf Edvard, (1818-1894)
BONSDORFF, Pehr Adolf von, (1791-1839)
BRAVAIS, Auguste, (1811-1863)
BUCH, Christian Leopold von, Freiherr, (1774-1853)
ENGELHARDT, Otto Moritz Ludwig von, (1779-1842)
FURUHJELM, Enoch Hjalmar, (1823-1886)
GADOLIN, Axel Wilhelm, General, (1828-1892)
GADOLIN, Johan, (1760-1852)
GROSSE, Karl Friedrich August, (1768-1847)
NORDENSKIÖLD, Nils Gustaf, (1792-1866)
NORDMANN, Alexander von, (1803-1866)

1850-1899

AF SCHULTÉN, August Benjamin, (1856-1913)
AILIO, Julius, (1872-1933)
ARPPE, Adolf Edvard, (1818-1894)
BERGHELL, Hugo, (1864-1922)
BRAVAIS, Auguste, (1811-1863)
BRÜCKNER, Eduard, (1862-1927)
BUCH, Christian Leopold von, Freiherr, (1774-1853)
CHYDENIUS, Johan Jakob "Hannes", (1836-1890)
FROSTERUS, Gustaf Benjamin, (1866-1931)
FUCHS, Philippe Jacques Edmond, (1837-1889)
FURUHJELM, Enoch Hjalmar, (1823-1886)
GADOLIN, Axel Wilhelm, General, (1828-1892)
GADOLIN, Johan, (1760-1852)
HACKMAN, Victor Axel, (1866-1941)
INBERG, Isak Johan, (1853-1893)
JULIN, Albert Lindsay von, (1876-1918)
MOBERG, Karl Adolf, (1840-1901)
NORDENSKIÖLD, Nils Gustaf, (1792-1866)
NORDMANN, Alexander von, (1803-1866)
NYHOLM, Emil Teodor, (1869-1908)
PETRA (PETANDER), August Herman, (1853-1928)
POPOFF, Boris Anfirovich, (1871-1950)
RAMSAY, Wilhelm, (1865-1928)
SAARINEN, Juho Hugo, (-?1908)
SEDERHOLM, Jakob Johannes, (1863-1934)
SOLITANDER, Carl Probus, (1843-1918)
TANNER, Väinö, (1881-1948)
TRÜSTEDT, Otto, (1866-1929)

FINLAND 1850-1899 Cont'd.

WIIK, Fredrik Johan, (1839-1909)
WILKMAN, Wanold Wrydon, (1872-?1938)

1900 to present
AF SCHULTÉN, August Benjamin, (1856-1913)
AILIO, Julius, (1872-1933)
ANDREWS, Richard Wallis, (?1927-1967)
BERGHELL, Hugo, (1864-1922)
BESCHEL, Roland, (1928-1971)
BRANDER, Gunnar, (1903-1940)
BRENNER, Thord, (1892-1949)
BRÜCKNER, Eduard, (1862-1927)
CHYDENIUS, Johan Jakob "Hannes", (1836-1890)
ESKOLA, Pentti Eelis, (1883-1964)
FROSTERUS, Gustaf Benjamin, (1866-1931)
HACKMAN, Victor Axel, (1866-1941)
JULIN, Albert Lindsay von, (1876-1918)
LECKWIJCK, Willem Peter van, (1902-1975)
MÄKINEN, Eero, (1886-1953)
MARMO, Vladi, (1914-1969)
MARTINUSSEN, Eugen, (1916-1942)
MIKKOLA, Erkki, (-?1940)
MOBERG, Karl Adolf, (1840-1901)
NYHOLM, Emil Teodor, (1869-1908)
PETRA (PETANDER), August Herman, (1853-1928)
POPOFF, Boris Anfirovich, (1871-1950)
RAMSAY, Wilhelm, (1865-1928)
SAARINEN, Juho Hugo, (-?1908)
SEDERHOLM, Jakob Johannes, (1863-1934)
SOLITANDER, Carl Probus, (1843-1918)
SPRINGORUM, Gerd, (1915-1941)
STUTZER, Otto, (1881-1936)
TANNER, Väinö, (1881-1948)
TRÜMPY, Daniel, (1893-1971)
TRÜSTEDT, Otto, (1866-1929)
WIIK, Fredrik Johan, (1839-1909)
WILKMAN, Wanold Wrydon, (1872-?1938)
ZANS, Verners Aleksandrs, (1904-1961)

FRANCE
Up to 1800
ASTRUC, Jean, (1684-1766)
AUBUISSON DE VOISINS, Jean François d', (1769-1841)
BAUME, Antoine, (1728-1804)
BEAUSOLEIL, Jean du Chastelet, Baron de .... et d'Auffembach, (ca 1578-1620)
BEAUSOLEIL, Martine de Bertereau, Baronne de .... et d'Auffembach, (fl.1600-1630)
BERTRAND DE DOUE, Jacques Mathieu, (1776-1862)
BLAINVILLE, Henri Marie Ducrotay de, (1777-1850)
BOUGUER, Pierre, (1698-1758)
BOURGUET, Louis, (1678-1742)
BOURNON, Jacques Louis de, Comte, (1751-1825)
BREISLAK, Scipione, (1748-1826)

FRANCE Up to 1800 Cont'd.

BROCHANT DE VILLIERS, André Jean Marie, (1772-1840)
BRONGNIART, Alexandre, (1778-1847)
BRUGUIÈRES, Jean Guillaume, (1750-1799)
BRUNO, Filippo, called Giordano, (1548-1600)
BUCH, Christian Leopold von, Freiherr, (1774-1853)
BUFFON, Georges Louis Leclerc, comte de, (1707-1788)
COLLOT, Georges Henri Victor, (1750-1805)
CORDIER, Pierre Louis Antoine, (1777-1861)
CUVIER, Jean Léopold Nicolas Frédéric, (1769-1832)
DAUBENTON, Louis Jean Marie, (1716-1799)
DEFRANCE, Jacques Louis Marie, (1758-1850)
DESCARTES, René du Perrot, (1596-1650)
DESMAREST, Nicolas, (1725-1815)
DEZALLIER D'ARGENVILLE, Antoine Joseph, (1680-1765)
DROUIN, Jean Baptiste, (1752-1792)
DU FAY, Charles François, (1698-1739)
DUFRESNE, Louis, (1752-1832)
DUHAMEL, Jean Pierre François Guillaume, (1730-1816)
DUVERNOY, George Louis, (1777-1855)
FAUJAS DE SAINT FOND, Barthélémy, (1741-1819)
FERBER, Johann Jacob, (1743-1790)
FONTENELLE, Bernard Le Bovier de, (1657-1757)
FORSTER, Adolarius Jacob, (1739-1806)
GAUTIER, Henri, (1660-1734)
GEOFFROY SAINT-HILAIRE, Étienne, (1772-1844)
GUETTARD, Jean Étienne, (1715-1786)
HAÜY, René-Just, Abbé, (1743-1822)
HUTTON, James, (1726-1797)
LA CONDAMINE, Charles-Marie de, (1701-1774)
LAMARCK, Jean-Baptiste Pierre Antoine de Monet, Chevalier de, (1744-1829)
LAVOISIER, Antoine Laurent, (1743-1794)
MÉTHERIE, Jean Claude de la, (1743-1817)
MONNET, Antoine Grimoald, (1734-1817)
MONTLOSIER, Fr. Dominique Reynaud, comte de, (1755-1838)
PALISSY, Bernard, (1510-1589)
PERRAULT, Pierre, (1611-1680)
RÉAUMUR, René Antoine Ferchault de, (1683-1757)
ROUELLE, Guillaume François, (1703-1770)
SERRES, Marcel de, (1783-1862)
SOULAVIE, Jean Louis Giraud, Abbé, (1752-1813)
WARREN, John Collins, (1778-1856)

1800-1849

ARCHIAC, Étienne Jules Adolphe Dexmier de Simon, Vicomte d', (1802-1868)
AUBUISSON DE VOISINS, Jean François d', (1769-1841)
AUDOUIN, Jean Victor, (1797-1841)
BARRANDE, Joachim, (1799-1883)
BAYLE, Émile, (1819-1895)
BEAUMONT, Jean Baptiste Armand Louis Léonce Élie de, (1798-1874)
BELGRAND, Eugène, (1810-1878)
BERTRAND DE DOUE, Jacques Mathieu, (1776-1862)
BLAINVILLE, Henri Marie Ducrotay de, (1777-1850)

FRANCE 1800-1849 Cont'd.

BOISDUVAL, Jean-Baptiste Alphonse, (1799-1879)
BONNARD, Augustin Henri de, (1781-1857)
BOUCHER DE CRÈVECOEUR DE PERTHES, Jacques, (1788-1868)
BOUÉ, Ami, (1794-1881)
BOURDOT, Jules Damase, (1837-1906)
BOURGEOIS, Louis Alexis, Abbé, (1819-1878)
BOURGUIGNAT, Jules René, (1829-?1919)
BOURNON, Jacques Louis de, Comte, (1751-1825)
BOUSSINGAULT, Joseph Dieudonné Jean-Baptiste, (1807-1887)
BRARD, Cyprien Prosper, (1788-1838)
BRAVAIS, Auguste, (1811-1863)
BREISLAK, Scipione, (1748-1826)
BROCHANT DE VILLIERS, André Jean Marie, (1772-1840)
BRONGNIART, Adolphe Théodore, (1801-1876)
BRONGNIART, Alexandre, (1778-1847)
BUCH, Christian Leopold von, Freiherr, (1774-1853)
BUCKLAND, William, Reverend, (1784-1856)
BURAT, Amédée, (1809-1883)
BUSK, George, (1807-1886)
CAMPICHE, Gustave, (1809-1870)
CANDOLLE, Alphonse de, (1806-1893)
CAUMONT, Arcisse de, (1801-1872)
CHARPENTIER, Jean de, (1786-1855)
CHENU, Jean Charles, (1808-1879)
CHRISTY, Henry, (1810-1865)
COLLENOT, Jean Jacques, (1814-1893)
COLLOMB, Edouard, (1801-1875)
COLLOT, Georges Henri Victor, (1750-1805)
CONTEJEAN, Charles Louis, (1824-1907)
COQUAND, Henri, (1813-1881)
CORDIER, Pierre Louis Antoine, (1777-1861)
CROIZET, Abbé, (-1863)
CUVIER, Jean Léopold Nicolas Fréderic, (1769-1832)
DAMOUR, Alexis, (1808-1902)
DARCY, Henri Philibert Gaspard, (1803-1858)
DAUBENY, Charles Giles Brindle, (1795-1867)
DAUBRÉE, Gabriel Auguste, (1814-1896)
DEFRANCE, Jacques Louis Marie, (1758-1850)
DELAFOSSE, Gabriel, (1796-1878)
DELESSE, Achille Ernest Oscar Joseph, (1817-1881)
DES CLOIZEAUX, Alfred Louis Olivier Legrand, (1817-1897)
DESHAYES, Gérard Paul, (1796-1875)
DESMAREST, Nicolas, (1725-1815)
DES MOULINS, Charles, (1798-1875)
DESNOYERS, Jules, (1801-1887)
DESOR, Pierre Jean Edouard, (1811-1882)
DOMEYKO, Ignacy, (1802-1889)
DUFRÉNOY, Pierre Armand, (1792-1857)
DUFRESNE, Louis, (1752-1832)
DUHAMEL, Jean Pierre François Guillaume, (1730-1816)
DUJARDIN, Félix, (1801-1860)
DUMAS, Émilien, (1804-1870)
DUVAL-JOUVE, Joseph, (1810-1883)
DUVERNOY, George Louis, (1777-1855)

FRANCE 1800-1849 Cont'd.

EBRAY, Charles Henri Théophile, (1823-1879)
ERLE, C., Reverend, (1790-1870?)
EUDES-DESLONGCHAMPS, Jacques Amand, (1794-1867)
FAUJAS DE SAINT FOND, Barthélémy, (1741-1819)
FEATHERSTONHAUGH, George William, (1780-1866)
FÉRUSSAC, André Étienne Just Pascal Joseph François d'Audebard, Baron de, (1786-1836)
FILHOL, Edouard, (1814-1883)
FORBES, James David, (1809-1868)
FORSTER, Adolarius Jacob, (1739-1806)
FOUCAULT, Jean Bernard Léon, (1819-1868)
FOUQUÉ, Ferdinand André, (1828-1904)
FOURNET, Joseph Jean Baptiste Xavier, (1801-1869)
FROSSARD, Émilien, (1802-1881)
GAUDRY, Jean Albert, (1827-1908)
GEOFFROY SAINT-HILAIRE, Étienne, (1772-1844)
GEOFFROY SAINT-HILAIRE, Isidore, (1805-1861)
GERLACH, Heinrich, (1822-1871)
GERVAIS, François Louis Paul, (1816-1879)
GRAS, Albin, (-1854)
GRATELOUP, Jean Pierre Sylvestre de, (1782-1861)
GRAVES, Louis, (1791-1857)
GRUNER, Louis, (1809-1883)
GUEYMARD, François Émile, (1788-1870)
GUIRAND, Edmond, (1812-1888)
HAIDINGER, Wilhelm Karl von, Ritter, (1795-1871)
HAIME, Jules, (1824-1856)
HAMILTON, William John, (1805-1867)
HAÜY, René-Just, Abbé, (1743-1822)
HÉBERT, Edmond, (1812-1890)
HECKEL, Johann Jakob, (1790-1857)
HUOT, Jean Jacques Nicolas, (1790-1845)
JACQUOT, Eugène, (1817-1903)
JEANJEAN, Adrien, (1821-1897)
KEENE, William, (1798-1872)
KING, Samuel William, (1821-1868)
LAMARCK, Jean-Baptiste Pierre Antoine de Monet, Chevalier de, (1744-1829)
LAMOUROUX, Jean Vincent Félix, (1779-1825)
LARTET, Edouard Armand Isidore Hippolyte, (1801-1871)
LAURILLARD, Charles Léopold, (1783-1853)
LECOQ, Henri, (1802-1871)
LEFEBVRE, Alexandre, (1798-1868)
LÉVY, Armand, (1795-1861)
LEYMERIE, Alexander, (1801-1878)
LORY, Charles, (1823-1889)
MARCOU, Jules, (1824-1898)
MATHERON, Pierre Émile Philippe, (1807-1899)
MÉTHERIE, Jean Claude de la, (1743-1817)
MICHELIN, Jean Louis Hardouin, (1786-1867)
MILNE-EDWARDS, Henri, (1800-1885)
MONNET, Antoine Grimoald, (1734-1817)
MONTLOSIER, Fr. Dominique Reynaud, comte de, (1755-1838)
MORIÈRE, Jules Pierre Gilles, (1817-1888)

FRANCE 1800-1849 Cont'd.

MORLOT, Adolf von, (1820-1867)
MORTILLET, Gabriel de, (1820-1898)
NESBIT, John Collis, (1819-1862)
NICOLET, Adolphe Célestin, (1803-1871)
NODOT, Léonhard, (1802-1859)
NOULET, Jean Baptiste, (1802-1890)
OMALIUS D'HALLOY, Jean-Baptist Julien d', (1783-1875)
ORBIGNY, Alcide Charles Victor Dessalines d', (1802-1857)
PAILLETTE, Antoine Adrien, (-1858)
PARANDIER, A. N., (1804-1901)
PARETO, Lorenzo, (1800-1865)
PARRAN, Alphonse, (1827-1903)
PASSY, Antoine François, (1792-1873)
PENTLAND, Joseph Barclay, (1797-1873)
PERREY, Alexis, (1807-1882)
PHILLIPS, John Arthur, (1822-1887)
PIDANCET, Pierre Marie Just, (1823-1871)
PIETTE, Louis Edouard Stanislas, (1827-1906)
PILLET, Louis, (1819-1894)
PLANCHON, Jules Émile, (1823-1888)
POMEL, Nicolas Auguste, (1821-1898)
POTIER, Alfred, (1840-1905)
PRATT, Samuel Peace, (1789-1863)
PRESTWICH, Joseph, Sir, (1812-1896)
PRÉVOST, Louis Constant, (1787-1856)
QUATREFAGES DE BRÉAU, Jean Louis Armand de, (1810-1892)
RAULIN, Félix Victor, (1815-1905)
REQUIEN, Esprit, (-1852)
RIVIÈRE, Auguste, (1809-1877)
ROBINEAU-DESVOIDY, André Jean Baptiste, (1799-1857)
ROCHE, Auguste, (1827-1905)
RÖMER, Hermann, (1816-1894)
ROSE, Alexander, (1781-1860)
ROUAULT, Marie, (1813-1881)
ROUVILLE, Paul Gervais de, (1823-1907)
ROZET, Antoine, Commandant, (1798-1858)
SAEMANN, Louis, (1821-1866)
SAINTE-CLAIRE DEVILLE, Charles, (1814-1876)
SAPORTA, Louis Charles Joseph Gaston de, Marquis, (1823-1895)
SAUVAGE, François Clément, (1814-1872)
SCHLUMBERGER, Charles, (1825-1904)
SCROPE, George Poulett, (né Thomson), (1797-1876)
SELYS-LONGCHAMPS, Michel Edmond de, Baron, (1813-1900)
SERRES, Marcel de, (1783-1862)
SHARPE, Daniel, (1806-1856)
SOULAVIE, Jean Louis Giraud, Abbé, (1752-1813)
STAHL, Jean Benjamin, (1816-1893)
SYMONDS, William Samuel, Reverend, (1818-1887)
TERQUEM, Olry, (1797-1887)
THIOLLIÈRE, Victor, (1801-1859)
THIRRIA, Charles Edouard, (1796-1868)
TOURNAL, Paul, (1805-1872)
TOURNOÜER, Jacques Raoul, (1822-1882)
TRIGER, Jacques Jules, (1801-1867)

FRANCE 1800-1849 Cont'd.

VERNE, Jules, (1828-1905)
WARREN, John Collins, (1778-1856)
WATELET, Adolphe, (1811-1880)

1850-1899

AGAFONOFF, Valerian K., (1863-1955)
ARCHIAC, Étienne Jules Adolphe Dexmier de Simon, Vicomte d', (1802-1868)
ARNAUD, Hilaire, (1827-1908)
AUTHELIN, Charles, (1871-1903)
AZÉMA, Léon, Colonel, (1852-1931)
BARET, Charles, (1831-1910)
BARRANDE, Joachim, (1799-1883)
BARROIS, Charles Eugène, (1851-1939)
BATE, Dorothea Minola Alice, (1879-1951)
BAYAN, Joseph Félix Ferdinand, (1845-1874)
BAYLE, Émile, (1819-1895)
BEAUMONT, Jean Baptiste Armand Louis Léonce Élie de, (1798-1874)
BECQUEREL, Henri, (1852-1908)
BELGRAND, Eugène, (1810-1878)
BENECKE, Ernst Wilhelm, (1838-1917)
BERGERON, Jules, (1853-1919)
BERNARD, Félix, (1863-1898)
BERTHELIN, Jean Georges, (1840-1897)
BERTRAND, C. Eg., (1851-1920)
BERTRAND, Émile, (1844-1909)
BERTRAND, Léon, (1869-1947)
BERTRAND, Marcel Alexandre, (1847-1907)
BERTRAND DE DOUE, Jacques Mathieu, (1776-1862)
BÉZIER, T., (-1926)
BIGOT, Alexandre, (1863-1953)
BINET, Auguste, (-1900)
BIOCHE, Alphonse, (1844-1918)
BLAYAC, Joseph, (1865-1936)
BLEICHER, Marie-Gustave, (1838-1901)
BOISDUVAL, Jean-Baptiste Alphonse, (1799-1879)
BONAPARTE, Charles Lucien Jules Laurent, Prince of Canino, (1803-1857)
BOUCHER DE CRÈVECOEUR DE PERTHES, Jacques, (1788-1868)
BOUÉ, Ami, (1794-1881)
BOULE, Marcellin, (1861-1942)
BOURDOT, Jules Damase, (1837-1906)
BOURGEOIS, Louis Alexis, Abbé, (1819-1878)
BOURGUIGNAT, Jules René, (1829-?1919)
BOUSSINGAULT, Joseph Dieudonné Jean-Baptiste, (1807-1887)
BOYER, Georges, (-1892)
BRADY, Henry Bowman, (1835-1891)
BRALY, Adrien, (1864-1927)
BRAVAIS, Auguste, (1811-1863)
BRESSON, Arthur, (1873-1946)
BREUIL, Henri, Abbé, (1877-1961)
BROGLIE, Maurice de, (1875-1960)
BRONGNIART, Adolphe Théodore, (1801-1876)
BRONGNIART, Charles Jules Edme, (1859-1899)

FRANCE 1850-1899 Cont'd.

BROOKS, Alfred Hulse, (1871-1924)
BRUHNS, Franz Siegfried Willy, (1864-1929)
BUCAILLE, Ernest-Lucien, (1835-1891)
BUCH, Christian Leopold von, Freiherr, (1774-1853)
BÜCKING, Ferdinand Carl Bertram Hugo, (1851-1932)
BUCKLAND, William, Reverend, (1784-1856)
BURAT, Amédée, (1809-1883)
BUREAU, Edouard, (1830-1918)
BURROWS, Henry William, (?1860-1939)
BUSK, George, (1807-1886)
CAMPICHE, Gustave, (1809-1870)
CANDOLLE, Alphonse de, (1806-1893)
CARNOT, Marie Adolphe, (1839-1920)
CARPENTIER, Alfred, Chanoine, (1878-1952)
CAUMONT, Arcisse de, (1801-1872)
CAYEUX, Lucien, (1864-1944)
CAZIOT, Eugène, (1844-1931)
CHANCOURTOIS, A. E. Béguyer de, (-1886?)
CHANTRE, Ernest, (1843-)
CHARPENTIER, Jean de, (1786-1855)
CHELLONNEIX, Émile, (-1885)
CHELOT, Émile, (1862-1930)
CHENU, Jean Charles, (1808-1879)
CHOFFAT, Paul, (1849-1919)
CHRISTY, Henry, (1810-1865)
COLLENOT, Jean Jacques, (1814-1893)
COLLOMB, Edouard, (1801-1875)
COLLOT, Louis Marie François, (1846-1915)
COMMONT, Victor, (1866-1918)
CONTEJEAN, Charles Louis, (1824-1907)
COOMARASWAMY, Ananda K., (1877-1947)
COPAUX, Hippolyte, (1872-1934)
COQUAND, Henri, (1813-1881)
CORBIÈRE, Louis, (1850-1941)
CORDIER, Pierre Louis Antoine, (1777-1861)
COSSMANN, Maurice, (1850-1924)
COTTEAU, Gustave Honoré, (1818-1894)
COTTREAU, Jean, (1877-1945)
COURTY, Georges, (1875-1953)
CROIZET, Abbé, (-1863)
DALIMIER, Paul Louis Marie, (1835-1863)
DAMOUR, Alexis, (1808-1902)
DARCY, Henri Philibert Gaspard, (1803-1858)
DAUBENY, Charles Giles Brindle, (1795-1867)
DAUBRÉE, Gabriel Auguste, (1814-1896)
DAUTZENBERG, Philippe, (1849-1935)
DAVIS, William Morris, (1850-1934)
DELAFOSSE, Gabriel, (1796-1878)
DELÉPINE, Gaston-Georges, Monseigneur, (1878-1963)
DELESSE, Achille Ernest Oscar Joseph, (1817-1881)
DEPÉRET, Charles Jean Julien, (1854-1929)
DES CLOIZEAUX, Alfred Louis Olivier Legrand, (1817-1897)
DESHAYES, Gérard Paul, (1796-1875)
DES MOULINS, Charles, (1798-1875)

FRANCE 1850-1899 Cont'd.

DESNOYERS, Jules, (1801-1887)
DESOR, Pierre Jean Edouard, (1811-1882)
DOLLÉ, Louis, (1878-1965)
DOLLFUS, Auguste, (1840-1869)
DOLLFUS, Gustave Fréderic, (1850-1931)
DOLLOT, Auguste, (1841-1924)
DOMEYKO, Ignacy, (1802-1889)
DONCIEUX, Louis, (1874-1960)
DONNEZAN, Albert, (1846-1914)
DOUVILLÉ, Joseph Henri Ferdinand, (1846-1937)
DOUXAMI, Henri, (1871-1913)
DUFFOUR, Alexis, (1877-1948)
DUFRÉNOY, Pierre Armand, (1792-1857)
DUJARDIN, Félix, (1801-1860)
DUMAS, Émilien, (1804-1870)
DUVAL-JOUVE, Joseph, (1810-1883)
DUVERNOY, George Louis, (1777-1855)
EBRAY, Charles Henri Théophile, (1823-1879)
ERLE, C., Reverend, (1790-1870?)
ÉTALLON, Claude Auguste, (1826-1862)
EUDES-DESLONGCHAMPS, Eugène, (1830-1889)
EUDES-DESLONGCHAMPS, Jacques Amand, (1794-1867)
FABRE, Georges, (-1911)
FALSAN, Albert, (1833-1902)
FAYOL, Henri, (1841-1925)
FEATHERSTONHAUGH, George William, (1780-1866)
FERRONNIÈRE, Georges, (1875-1922)
FERRY, Henri de, (1826-1869)
FILHOL, Edouard, (1814-1883)
FILHOL, Henri, (1843-1902)
FIRKET, Adolphe, (1837-1905)
FISCHER, Paul Henri, (1835-1893)
FLEURY, Ernest, (1878-1958)
FLICHE, Paul, (1836-1908)
FONDOUCE, Paul Cazalis de, (1838-1932)
FONTANNES, Charles François, (1839-1886)
FORBES, James David, (1809-1868)
FOUCAULT, Jean Bernard Léon, (1819-1868)
FOUQUÉ, Ferdinand André, (1828-1904)
FOURMARIER, Paul Fréderic Joseph Lucien Lauren, (1877-1970)
FOURNET, Joseph Jean Baptiste Xavier, (1801-1869)
FOURNIER, Eugène, (1871-1941)
FRIEDBERG, Wilhelm, (1873-1941)
FRIEDEL, Charles, (1832-1899)
FRIEDEL, Georges, (1865-1933)
FROSSARD, Émilien, (1802-1881)
FUCHS, Philippe Jacques Edmond, (1837-1889)
GARRIGOU, Félix, (1835-1920)
GAUBERT, Marie Benoit Paul, (1865-)
GAUDEFROY, Christophe Léon, Chanoine, (1878-1971)
GAUDRY, Jean Albert, (1827-1908)
GAURIER, Ludovic Charles, Abbé, (1874-1931)
GAUTHIER, Victor Auguste, (1837-1911)
GAUTIER, Emile Justin Armand, (1837-1920)

FRANCE 1850-1899 Cont'd.

GEOFFROY SAINT-HILAIRE, Isidore, (1805-1861)
GÉRARD, Charles, Colonel, (1866-1953)
GERLACH, Heinrich, (1822-1871)
GERVAIS, François Louis Paul, (1816-1879)
GIARD, Alfred, (1846-1908)
GIRARDOT, Charles Joseph Albert Julien, (1848-1930)
GOSSELET, Jules Auguste, (1832-1916)
GRAD, Charles, (1842-?1905)
GRAND'EURY, François Cyrille, (1839-1917)
GRANDIDIER, Alfred, (1836-1921)
GRAS, Albin, (-1854)
GRATELOUP, Jean Pierre Sylvestre de, (1782-1861)
GRAVES, Louis, (1791-1857)
GROSSOUVRE, Albert de, (1849-1932)
GRUNER, Louis, (1809-1883)
GRÜNLING, Friedrich, (1857-1919)
GUÉBHARD, Adrien, (1849-1924)
GUEYMARD, François Émile, (1788-1870)
GUILLIER, Albert, (1839-1885)
GUIRAND, Edmond, (1812-1888)
HAIDINGER, Wilhelm Karl von, Ritter, (1795-1871)
HAIME, Jules, (1824-1856)
HAMILTON, William John, (1805-1867)
HAMY, Jules Ernest Théodore, (1842-1908)
HARLÉ, Edouard, (1850-1922)
HARRIS, George Frederick, (1862-1906)
HAUG, Gustave Émile, (1861-1927)
HAUTEFEUILLE, Paul Gabriel, (1836-1902)
HÉBERT, Edmond, (1812-1890)
HECKEL, Johann Jakob, (1790-1857)
HELBRONNER, Paul, (1871-1938)
HOLLANDE, Dieudonné, (1845-1921)
HUOT, Jean Jacques Nicolas, (1790-1845)
JACOB, Charles François Étienne, (1878-1962)
JACQUOT, Eugène, (1817-1903)
JANET, Léon, (1861-1909)
JANNETTAZ, Edouard, (-1900)
JEANJEAN, Adrien, (1821-1897)
JOHNSTON, Mary Sophia, (1875-1955)
JULIEN, Pierre Alphonse, (1838-1905)
KAYSER, Friedrich Heinrich Emanuel, (1845-1927)
KEENE, William, (1798-1872)
KERFORNE, Fernand, (1864-1927)
KILIAN, Charles Constant Wilfrid, (1862-1925)
KING, Samuel William, (1821-1868)
KORNITZER, Louis, (1873-) [PROSPECTORS, DIVINERS AND MINING ENGINEERS]
LACROIX, François Antoine Alfred, (1863-1948)
LALOY, Roger, (1851-1909)
LAMBERT, Jules Mathieu, (1848-1940)
LAPPARENT, Albert Auguste de, (1839-1908)
LARTET, Edouard Armand Isidore Hippolyte, (1801-1871)
LAUNAY, Louis Alphonse Auguste de, (1860-1938)
LAURILLARD, Charles Léopold, (1783-1853)

FRANCE 1850-1899 Cont'd.

LECOQ, Henri, (1802-1871)
LEFEBVRE, Alexandre, (1798-1868)
LE MESLE, George, (-1895)
LEMOINE, Armand Victor, (1837-1897)
LEMOINE, Paul, (1878-1940)
LENNIER, Gustave, (-1905)
LERICHE, Maurice, (1875-1948)
LÉVY, Armand, (1795-1861)
LEYMERIE, Alexander, (1801-1878)
LIENENKLAUS, Ernst, (1849-1905)
LOCARD, Arnould, (1841-1904)
LORTET, Louis Charles, (1836-1909)
LORY, Charles, (1823-1889)
LORY, Pierre, (1866-1956)
LYELL, Charles, Sir, (1797-1875)
MACLAREN, James Malcolm, (1873-1935)
MAILLARD, Gustave, (1860-1891)
MAIRE, Julien Victor, (1856-1941)
MALLARD, François Ernest, (1833-1894)
MARCOU, Jules, (1824-1898)
MARGERIE, Emmanuel Marie Pierre Martin Jacquin de, (1862-1953)
MARION, Antoine Fortuné, (1846-1900)
MARTIN, David, (1842-1918)
MARTIN, Henri, (1873?-1936)
MARTONNE, Emmanuel Louis Eugène de, (1873-1955)
MATHERON, Pierre Émile Philippe, (1807-1899)
MAUGIN, Charles, (1878-1958)
McINTYRE, Peter, (1863-1940)
MEUNIER, Étienne Stanislas, (1843-1925)
MICHEL, Léopold Jean, (1846-1919)
MICHELIN, Jean Louis Hardouin, (1786-1867)
MICHEL-LÉVY, Albert, (1877-1955)
MICHEL-LÉVY, Auguste, (1844-1911)
MIEG-KROH, Mathieu, (1849-1911)
MILNE-EDWARDS, Alphonse, (1835-1900)
MILNE-EDWARDS, Henri, (1800-1885)
MOISSAN, Henri, (1852-1907)
MONTESSUS DE BALLORE, Fernand Jean Baptiste Marie Bernard, Comte de, (1851-1923)
MORGAN, Jacques Jean Marie de, (1857-1924)
MORIÈRE, Jules Pierre Gilles, (1817-1888)
MORLOT, Adolf von, (1820-1867)
MORTILLET, Gabriel de, (1820-1898)
MUNIER-CHALMAS, Ernest Philippe Auguste, (1843-1903)
NESBIT, John Collis, (1819-1862)
NICKLÈS, René, (1859-1917)
NICOLET, Adolphe Célestin, (1803-1871)
NODOT, Léonhard, (1802-1859)
NOULET, Jean Baptiste, (1802-1890)
OEHLERT, Daniel Victor, (1849-1920)
OLRY, Albert, (1847-1913)
OMALIUS D'HALLOY, Jean-Baptist Julien d', (1783-1875)
OPPEL, Albert, (1831-1865)
ORBIGNY, Alcide Charles Victor Dessalines d', (1802-1857)

FRANCE 1850-1899 Cont'd.

OUSTALET, Émile, (1844-1905)
PAILLETTE, Antoine Adrien, (-1858)
PAQUIER, Victor Lucien, (1870-1911)
PARANDIER, A. N., (1804-1901)
PARETO, Lorenzo, (1800-1865)
PARRAN, Alphonse, (1827-1903)
PASSY, Antoine François, (1792-1873)
PENTLAND, Joseph Barclay, (1797-1873)
PERGENS, Eduard Willem Gerard, (1862-1917)
PERON, Alphonse, (1834-1908)
PERREY, Alexis, (1807-1882)
PERVINQUIÈRE, Léon, (1873-1913)
PHILLIPS, John Arthur, (1822-1887)
PIDANCET, Pierre Marie Just, (1823-1871)
PIETTE, Louis Edouard Stanislas, (1827-1906)
PILLET, Louis, (1819-1894)
PISANI DE SERRES, Felix, (1831-1920)
PLANCHON, Jules Émile, (1823-1888)
PLANTÉ, Gaston, (1834-1889)
POMEL, Nicolas Auguste, (1821-1898)
PORRIO, Cesare, (1865-1940)
POTIER, Alfred, (1840-1905)
PRATT, Samuel Peace, (1789-1863)
PRESTWICH, Joseph, Sir, (1812-1896)
PRÉVOST, Louis Constant, (1787-1856)
PRIEM, Fernand, (1857-1919)
PRINGLE, John, (1877-1948)
PUMPELLY, Raphael, (1837-1923)
QUATREFAGES DE BRÉAU, Jean Louis Armand de, (1810-1892)
RAMÈS, Jean-Baptiste, (1832-1894)
RAULIN, Félix Victor, (1815-1905)
RENAULT, Bernard, (1836-1904)
RENEVIER, Eugène, (1831-1906)
REQUIEN, Esprit, (-1852)
RÉVIL, Joseph, (1849-1931)
RIAZ, Auguste de, (-1920)
RIGAUX, Edmond, (1839-1915)
RIVIÈRE, Auguste, (1809-1877)
ROBINEAU-DESVOIDY, André Jean Baptiste, (1799-1857)
ROCHE, Auguste, (1827-1905)
ROMAN, Albert Frédéric, (1871-1943)
RÖMER, Hermann, (1816-1894)
ROMEU, Albert de, (1875-1915)
ROSE, Alexander, (1781-1860)
ROSENBUSCH, Karl Harry Ferdinand, (1836-1914)
ROUAULT, Marie, (1813-1881)
ROUVILLE, Paul Gervais de, (1823-1907)
ROZET, Antoine, Commandant, (1798-1858)
SAEMANN, Louis, (1821-1866)
SAINTE-CLAIRE DEVILLE, Charles, (1814-1876)
SAINTE-CLAIRE DEVILLE, Henri, (-1881)
SAPORTA, Louis Charles Joseph Gaston de, Marquis, (1823-1895)
SARASIN, Edouard, (1843-1917)
SAUVAGE, François Clément, (1814-1872)

FRANCE 1850-1899 Cont'd.

SAUVAGE, Henri Émile, (1842-1917)
SAVIN, Léon Héli, Lieutenant-Colonel, (1851-1907)
SAYN, Gustave, (1862-1933)
SCHLOENBACH, Georg Justin Carl Urban, (1841-1870)
SCHLUMBERGER, Charles, (1825-1904)
SCHLUMBERGER, Conrad, (1878-1936)
SCHMIDT, Carl, (1862-1923)
SCHRADER, François Jean Daniel "Franz", (1844-1924)
SCHUMACHER, Eugen, (1851-1922)
SCHWALBE, Gustav, (1844-1916)
SCROPE, George Poulett, (né Thomson), (1797-1876)
SELYS-LONGCHAMPS, Michel Edmond de, Baron, (1813-1900)
SERRES, Marcel de, (1783-1862)
SEUNES, Jean, (1849-1920)
SHARPE, Daniel, (1806-1856)
STAHL, Jean Benjamin, (1816-1893)
SYMONDS, William Samuel, Reverend, (1818-1887)
TERMIER, Pierre, (1859-1930)
TERQUEM, Olry, (1797-1887)
THEVENIN, Armand, (1870-1918)
THIOLLIÈRE, Victor, (1801-1859)
THIRRIA, Charles Edouard, (1796-1868)
THURMANN, Jules, (1804-1855)
TOUCAS, Joseph Aristide, (1843-1911)
TOURNAL, Paul, (1805-1872)
TOURNOÜER, Jacques Raoul, (1822-1882)
TRIGER, Jacques Jules, (1801-1867)
TROUESSART, Edouard Louis, (1842-1927)
VAILLANT, Léon Louis, (1834-1914)
VALMONT DE BOMARE, Jacques Christophe, (-)
VASSEUR, Gaston Casimir, (1855-1915)
VERNE, Jules, (1828-1905)
VÉSIGNIÉ, Louis, Colonel, (1870-1954)
WALLERANT, Frédéric Félix Auguste, (1858-1936)
WARREN, John Collins, (1778-1856)
WATELET, Adolphe, (1811-1880)
WELSCH, Jules, (1858-1929)
WOHLGEMUTH, Jules, (-1893)
WOODWARD, Henry, (1832-1921)
ZEILLER, Charles René, (1847-1915)

1900 to present

ABRARD, René, (1892-1973)
AGAFONOFF, Valerian K., (1863-1955)
ALIMEN, Henriette, (1900-)
ALLOITEAU, James Henri Ernest, (1890-1969)
ARNAUD, Hilaire, (1827-1908)
ASTRE, Gaston, (1896-)
AUBOUIN, Jean, (1928-)
AUTHELIN, Charles, (1871-1903)
AZÉMA, Léon, Colonel, (1852-1931)
BARET, Charles, (1831-1910)
BARRABÉ, Louis, (1895-1961)
BARROIS, Charles Eugène, (1851-1939)

FRANCE 1900 to present Cont'd.

BASSE DE MÉNORVAL, Éliane Marguerite, (1899-)
BATE, Dorothea Minola Alice, (1879-1951)
BAUDOIN, René, (-)
BECQUEREL, Henri, (1852-1908)
BELLAIR, Pierre, (1910-)
BENECKE, Ernst Wilhelm, (1838-1917)
BERGERON, Jules, (1853-1919)
BERTHELOT, Charles, (-) [PROSPECTORS, DIVINERS AND MINING ENGINEERS]
BERTRAND, Émile, (1844-1909)
BERTRAND, Léon, (1869-1947)
BERTRAND, Marcel Alexandre, (1847-1907)
BERTRAND, Paul, (1879-1944)
BEUDANT, François-Sulpice, (1787-1850)
BÉZIER, T., (-1926)
BIGOT, Alexandre, (1863-1953)
BIOCHE, Alphonse, (1844-1918)
BIRAUD, François, (1899-1936)
BLAYAC, Joseph, (1865-1936)
BLEICHER, Marie-Gustave, (1838-1901)
BLONDEL, Fernand, (1894-1968)
BONAPARTE, Charles Lucien Jules Laurent, Prince of Canino, (1803-1857)
BONTE, Antoine Jean Marie Joseph, (1908-)
BOUBÉE, Nérée, (1893-1967)
BOULE, Marcellin, (1861-1942)
BOURCART, Jacques, (1891-1965)
BOUREAU, Edouard Léon François, (1913-)
BOURGUIGNAT, Jules René, (1829-?1919)
BOUSSAC, Jean, (1885-1917)
BRALY, Adrien, (1864-1927)
BRESSON, Arthur, (1873-1946)
BREUIL, Henri, Abbé, (1877-1961)
BROGLIE, Maurice de, (1875-1960)
BROOKS, Alfred Hulse, (1871-1924)
BROUGHTON-EDGE, Arthur Broughton, (1895-1953)
BRUET, Edmond, (1890-)
BRUHNS, Franz Siegfried Willy, (1864-1929)
BÜCKING, Ferdinand Carl Bertram Hugo, (1851-1932)
BUREAU, Edouard, (1830-1918)
BURROWS, Henry William, (?1860-1939)
CAIRE, André, (1923-)
CARLSON, Charles Gordon, (1895-1977)
CARNOT, Marie Adolphe, (1839-1920)
CARPENTIER, Alfred, Chanoine, (1878-1952)
CASTERET, Norbert, (1897-)
CAYEUX, Lucien, (1864-1944)
CAYEUX DE SENARPONT, André, called CAILLEUX, André, (1907-)
CAZIOT, Eugène, (1844-1931)
CHANTRE, Ernest, (1843-)
CHAPUT, Jules Ernest, (1882-1943)
CHAUVE, Pierre, (1930-)
CHELOT, Émile, (1862-1930)
CHERVET, Jean, (1904-1962)

FRANCE 1900 to present Cont'd.

CHEVENARD, Pierre, (1888-1960)
CHOFFAT, Paul, (1849-1919)
CLAPP, Frederick Gardner, (1879-1944)
COLLOT, Louis Marie François, (1846-1915)
COMBAZ, Paul, Chanoine, (-1930)
COMMONT, Victor, (1866-1918)
CONTEJEAN, Charles Louis, (1824-1907)
COOMARASWAMY, Ananda K., (1877-1947)
COPAUX, Hippolyte, (1872-1934)
COSSMANN, Maurice, (1850-1924)
COTTREAU, Jean, (1877-1945)
COURTY, Georges, (1875-1953)
CUVILLIER, Jean, (1899-1969)
DAGUIN, Fernand, (1889-1948)
DALLONI, Marius Gustave, (1880-1959)
DAMOUR, Alexis, (1808-1902)
DANGEARD, Louis, (-)
DAUTZENBERG, Philippe, (1849-1935)
DAVIS, William Morris, (1850-1934)
DEFLANDRE, Georges Victor, (1897-1973)
DEFLANDRE-RIGAUD, Marthe, (-)
DEHÉE, René, (1898-1928)
DELEAU, Paul Camille, (1904-)
DELÉPINE, Gaston-Georges, Monseigneur, (1878-1963)
DEMAY, André, (1890-1964)
DENIZOT, Georges, (-)
DEPÉRET, Charles Jean Julien, (1854-1929)
DOLLÉ, Louis, (1878-1965)
DOLLFUS, Gustave Frédéric, (1850-1931)
DOLLOT, Auguste, (1841-1924)
DONCIEUX, Louis, (1874-1960)
DONNEZAN, Albert, (1846-1914)
DOUVILLÉ, Joseph Henri Ferdinand, (1846-1937)
DOUVILLÉ, Robert, (1881-1914)
DOUXAMI, Henri, (1871-1913)
DREHER, Otto, (1888-1945)
DUBOIS, Georges, (1890-1953)
DUFFOUR, Alexis, (1877-1948)
DUPARQUE, André, (1892-1960)
DURAND DELGA, Michel, (1923-)
DUTERTRE, August Pierre, (1890-1940)
EDELMANN, Cornelis Hendrik, (1901-1964)
ERHART, Henri, (-)
FABRE, Georges, (-1911)
FALLOT, Paul, (1889-1960)
FALSAN, Albert, (1833-1902)
FAYOL, Henri, (1841-1925)
FERRONNIÈRE, Georges, (1875-1922)
FILHOL, Henri, (1843-1902)
FIRKET, Adolphe, (1837-1905)
FLEURY, Ernest, (1878-1958)
FLICHE, Paul, (1836-1908)
FONDOUCE, Paul Cazalis de, (1838-1932)
FOUQUÉ, Ferdinand André, (1828-1904)

FRANCE 1900 to present Cont'd.

FOURMARIER, Paul Frédéric Joseph Lucien Lauren, (1877-1970)
FOURNIER, Eugène, (1871-1941)
FRIEDBERG, Wilhelm, (1873-1941)
FRIEDEL, Edmond, (1895-1972)
FRIEDEL, Georges, (1865-1933)
FURON, Raymond, (1898-)
GARRIGOU, Félix, (1835-1920)
GAUDEFROY, Christophe Léon, Chanoine, (1878-1971)
GAUDRY, Jean Albert, (1827-1908)
GAURIER, Ludovic Charles, Abbé, (1874-1931)
GAUTHIER, Victor Auguste, (1837-1911)
GAUTIER, Emile Justin Armand, (1837-1920)
GÉRARD, Charles, Colonel, (1866-1953)
GIARD, Alfred, (1846-1908)
GIGNOUX, Maurice Irénée Marie, (1881-1955)
GILLARD, Pierre A., (1907-1947)
GIRARDOT, Charles Joseph Albert Julien, (1848-1930)
GLANGEAUD, Louis, (1903-)
GLORY, André, Abbé, (-)
GODON, Joseph, Chanoine, (-1932)
GOGUEL, Jean, (1908-)
GÖRGEY VON GÖRGÖ, Rudolf, (1886-1915)
GOSSELET, Jules Auguste, (1832-1916)
GRAD, Charles, (1842-?1905)
GRAMBAST, Louis, (-1976)
GRAMONT, Antoine Alfred Arnaud Xavier Louis de, Comte, (1861-1923)
GRAND'EURY, François Cyrille, (1839-1917)
GRANDIDIER, Alfred, (1836-1921)
GROSSOUVRE, Albert de, (1849-1932)
GRÜNLING, Friedrich, (1857-1919)
GUÉBHARD, Adrien, (1849-1924)
GUILLAUME, Louis, (1894-1952)
GUILLAUME, Solange, (1932-)
GÜMBEL, Emil J., (-1966)
HAMY, Jules Ernest Théodore, (1842-1908)
HARLÉ, Edouard, (1850-1922)
HARRIS, George Frederick, (1862-1906)
HAUG, Gustave Émile, (1861-1927)
HAUTEFEUILLE, Paul Gabriel, (1836-1902)
HELBRONNER, Paul, (1871-1938)
HENRY, J., (-1924)
HERALD, Frank Edward, (1887-1965)
HIBBERT-WARE, Samuel, (1782-1848)
HLAUSCHEK, Hans, (1896-1971)
HOFFSTETTER, Robert, (1908-)
HOLLANDE, Dieudonné, (1845-1921)
HOWELL, Benjamin Franklin, (1890-1976)
HUPÉ, Pierre, (1907-)
ICKES, Eugene Law, (1885-1941)
JACOB, Charles François Étienne, (1878-1962)
JACQUOT, Eugène, (1817-1903)
JANET, Léon, (1861-1909)
JANNETTAZ, Edouard, (-1900)
JÉRÉMINE, Elisabeth, (née Tschernaieff), (1879-1964)

FRANCE 1900 to present Cont'd.

JODOT, Paul, (1879-1961)
JOHNSTON, Mary Sophia, (1875-1955)
JOLEAUD, Léonce, (1880-1938)
JOURDAIN, André, (1890-1967)
JULIEN, Pierre Alphonse, (1838-1905)
JUNG, Jean, (1896-)
KAISIN, Félix, Sr., (1879-1948)
KAYSER, Friedrich Heinrich Emanuel, (1845-1927)
KERFORNE, Fernand, (1864-1927)
KILIAN, Charles Constant Wilfrid, (1862-1925)
KORNITZER, Louis, (1873-) [PROSPECTORS, DIVINERS AND MINING ENGINEERS]
LACROIX, François Antoine Alfred, (1863-1948)
LALOY, Roger, (1851-1909)
LAMARE, Pierre, (1894-1967)
LAMBERT, Jules Mathieu, (1848-1940)
LANQUINE, Antonin, (-)
LAPPARENT, Albert Auguste de, (1839-1908)
LAPPARENT, Albert Félix de, (1905-1975)
LAPPARENT, Jacques de, (-?1949)
LAUNAY, Louis Alphonse Auguste de, (1860-1938)
LE CALVEZ, Jean, (1908-1954)
LECOINTRE, Georges, (1888-1972)
LELUBRE, Maurice, (1916-)
LEMOINE, Paul, (1878-1940)
LENNIER, Gustave, (-1905)
LERICHE, Maurice, (1875-1948)
LEUTWEIN, Friedrich, (1911-1974)
LIENENKLAUS, Ernst, (1849-1905)
LISSAJOUS, Marcel, (-1921)
LOCARD, Arnould, (1841-1904)
LONGCHAMBON, Michel, (1886-1916)
LORTET, Louis Charles, (1836-1909)
LORY, Pierre, (1866-1956)
LOUBIÈRE, A., (-)
LUTAUD, Léon, (1883-1964)
LYELL, Charles, Sir, (1797-1875)
MACLAREN, James Malcolm, (1873-1935)
MAILLARD, Gustave, (1860-1891)
MAIRE, Julien Victor, (1856-1941)
MARGERIE, Emmanuel Marie Pierre Martin Jacquin de, (1862-1953)
MARSHALL, Alan Edward, (1938-1970)
MARTIN, David, (1842-1918)
MARTIN, Henri, (1873?-1936)
MARTONNE, Emmanuel Louis Eugène de, (1873-1955)
MAUGIN, Charles, (1878-1958)
McINTYRE, Peter, (1863-1940)
MEUNIER, Étienne Stanislas, (1843-1925)
MICHEL, Léopold Jean, (1846-1919)
MICHEL-LÉVY, Albert, (1877-1955)
MICHEL-LÉVY, Auguste, (1844-1911)
MIEG-KROH, Mathieu, (1849-1911)
MIGAUX, Léon, (1897-1914)
MILLOT, Georges, (1917-)

FRANCE 1900 to present Cont'd.

MILNE-EDWARDS, Alphonse, (1835-1900)
MOISSAN, Henri, (1852-1907)
MONGIN, Denise, (-)
MONTESSUS DE BALLORE, Fernand Jean Baptiste Marie Bernard, Comte de, (1851-1923)
MORELLET, Lucien, (1882-1945)
MORET, Léon, (1890-1972)
MORGAN, Jacques Jean Marie de, (1857-1924)
MUNIER-CHALMAS, Ernest Philippe Auguste, (1843-1903)
NICKLÈS, René, (1859-1917)
NICOLAS, Jean, (1925-)
OEHLERT, Daniel Victor, (1849-1920)
OLRY, Albert, (1847-1913)
ORCEL, Jean, (1896-)
OUSTALET, Émile, (1844-1905)
PALM-LAZARD, C. B. B., (1933-1960)
PAQUIER, Victor Lucien, (1870-1911)
PARANDIER, A. N., (1804-1901)
PARDÉ, Maurice, (1893-1973)
PARRAN, Alphonse, (1827-1903)
PERGENS, Eduard Willem Gerard, (1862-1917)
PERON, Alphonse, (1834-1908)
PERRIN, René, (1893-1966)
PERVINQUIÈRE, Léon, (1873-1913)
PIETTE, Louis Edouard Stanislas, (1827-1906)
PISANI DE SERRES, Felix, (1831-1920)
PIVETEAU, Jean, (1898-)
PONTIER, G., (-1933)
PORRIO, Cesare, (1865-1940)
PRIEM, Fernand, (1857-1919)
PRINGLE, John, (1877-1948)
PRUVOST, Pierre Eugéne Marie Joseph, (1890-1967)
PUMPELLY, Raphael, (1837-1923)
RAULIN, Félix Victor, (1815-1905)
RAVIER, Jean, (1922-)
RENAULT, Bernard, (1836-1904)
RENAULT, Philippe, (-)
RENEVIER, Eugène, (1831-1906)
RÉVIL, Joseph, (1849-1931)
RIAZ, Auguste de, (-1920)
RIGAUX, Edmond, (1839-1915)
ROCH, Edouard, (1901-)
ROCHE, Auguste, (1827-1905)
ROMAN, Albert Frédéric, (1871-1943)
ROMEU, Albert de, (1875-1915)
ROSENBUSCH, Karl Harry Ferdinand, (1836-1914)
ROUBAULT, Marcel Edouard, (1905-1974)
ROUVILLE, Paul Gervais de, (1823-1907)
RUTTEN, Martin Gerard, (1910-1970)
SARASIN, Edouard, (1843-1917)
SAUVAGE, Henri Émile, (1842-1917)
SAVIN, Léon Héli, Lieutenant-Colonel, (1851-1907)
SAYN, Gustave, (1862-1933)
SCHLUMBERGER, Charles, (1825-1904)

FRANCE 1900 to present Cont'd.

SCHLUMBERGER, Conrad, (1878-1936)
SCHLUMBERGER, Marcel, (1884-1953)
SCHMIDT, Carl, (1862-1923)
SCHNEEGANS, Daniel, (1907-1949)
SCHRADER, François Jean Daniel "Franz", (1844-1924)
SCHUMACHER, Eugen, (1851-1922)
SCHWALBE, Gustav, (1844-1916)
SEUNES, Jean, (1849-1920)
SMITH, G. Wendell, (1917-1974)
TEILHARD DE CHARDIN, Pierre, Father, S.J., (1881-1955)
TERMIER, Henri François Emile, (1897-)
TERMIER, Pierre, (1859-1930)
THEVENIN, Armand, (1870-1918)
THIÉRY, Paul, (-1927)
THORAL, Marcel, (1900-1956)
THURMANN, Jules, (1804-1855)
TOUCAS, Joseph Aristide, (1843-1911)
TRONQUOY, René, (1884-1915)
TROUESSART, Edouard Louis, (1842-1927)
TRÜMPY, Daniel, (1893-1971)
URBAIN, Georges, (-1938)
URBAIN, Pierre, (1895-1968)
VAILLANT, Léon Louis, (1834-1914)
VASSEUR, Gaston Casimir, (1855-1915)
VERNE, Jules, (1828-1905)
VÉSIGNIÉ, Louis, Colonel, (1870-1954)
VIENNOT, Pierre, (1891-1931)
VINCIENNE, Henri, (1898-1965)
VIRET, Jean, (1894-1970)
WALLERANT, Frédéric Félix Auguste, (1858-1936)
WALTON, John, (-1971)
WELSCH, Jules, (1858-1929)
WILMOTH, Leslie James, (1881-1963)
WIMBISH, Forrest E., (1900-1972)
WOODWARD, Henry, (1832-1921)
ZEILLER, Charles René, (1847-1915)

FRANZ JOSEF LAND (ARCTIC)
1900 to present
PAPANIN, Ivan Dmitrievich "Vanya", (1895-)

FRENCH CAMEROONS See CAMEROONS.

FRENCH EQUATORIAL AFRICA See CHAD; CENTRAL AFRICAN EMPIRE; CONGO REPUBLIC.

FRENCH GUIANA
1800-1849
WATERTON, Charles, (1782-1865)

1850-1899
TERNETZ, Carl, (1870-1928)
WATERTON, Charles, (1782-1865)

FRENCH GUIANA Cont'd.

1900 to present
TERNETZ, Carl, (1870-1928)

FRENCH GUINEA *See* GUINEA.

FRENCH SOUDAN *See* MALI.

FRENCH TERRITORY OF THE AFARS AND THE ASSAS *See* DJIBOUTI.

GABON

1900 to present

HOURCQ, Victor, (1901-1961)
MIGAUX, Léon, (1897-1914)
NICOLAS, Jean Paul, (1921-1967)
SCHAUB, Hans Peter, (1912-1965)
SCHNEEGANS, Daniel, (1907-1949)
TRÜMPY, Daniel, (1893-1971)
WASHBURNE, Chester Wesley, (1883-1971)

GALÁPAGOS ISLANDS (ECUADOR)

1800-1849

DARWIN, Charles Robert, (1809-1882)

1850-1899

DARWIN, Charles Robert, (1809-1882)
WOLF, Theodor, (1841-1924)

1900 to present

HOFFSTETTER, Robert, (1908-)
SIMPSON, George Gaylord, (1902-)
STRAELEN, Victor Emile van, (1889-1964)
TRIEBEL, Erich, (1894-1971)
WOLF, Theodor, (1841-1924)

GERMAN DEMOCRATIC REPUBLIC See GERMANY.

GERMANY

Up to 1800

AGRICOLA, Georgius, (1494-1555)
ALBERTUS MAGNUS, (1193-1280)
ALBINUS, Petrus, (1543-1598)
ASSMAN, C. G., (-)
AYCKE, Johann Christian, (1766-1854)
BATSCH, August Johann Georg Karl, (1761-1802)
BECHER, Johann Joachim, (1635-1682)
BERINGER, Johann Bartholomaeus Adam, (1667-1740)
BLUMENBACH, Johann Friedrich, (1752-1840)
BOJANUS, Ludwig Heinrich, (1776-1827)
BRUNO, Filippo, called Giordano, (1548-1600)
BUCH, Christian Leopold von, Freiherr, (1774-1853)
CHARPENTIER, Johann Friedrich Wilhelm Toussaint von, (1728-1805)
COLLINI, Cosimo Alessandro, (1727-1806)
COTTA, Heinrich, (1763-1844)
CRICHTON, Alexander, Sir, (1763-1856)
DUHAMEL, Jean Pierre François Guillaume, (1730-1816)
EHRHART, Balthazar, (1700-1756)
EMMERLING, Ludwig August, (1765-1841)
ESCHWEGE, Wilhelm Ludwig, Baron, (1777-1855)
FABRICIUS, Georg, (1516-1571)
FAUJAS DE SAINT FOND, Barthélemy, (1741-1819)
FERBER, Johann Jacob, (1743-1790)
FISCHER VON WALDHEIM, Johann Gotthelf Friedrich, (1771-1853)

GERMANY Up to 1800 Cont'd.

FLURL, Mathias von, (1756-1823)
FORSTER, Adolarius Jacob, (1739-1806)
FORSTER, Johann Georg Adam, (1754-1794)
FORSTER, Johann Reinhold, (1729-1798)
FREIESLEBEN, Johann Carl, (1774-1846)
FUCHS, Johann Nepomuk von, (1774-1856)
FUCHS, Leonhart, (1501-1566)
FÜCHSEL, Georg Christian, (1722-1773)
GAUSS, Carl Friedrich, (1777-1855)
GERSDORF, Adolf Traugott von, (1744-1807)
GMELIN, Karl Christian, (1762-1837)
GOETHE, Johann Wolfgang von, (1749-1832)
GREENOUGH, George Bellas, (1778-1855)
HAMILTON, William, Sir, (1730-1803)
HARDENBERG, Friedrich Leopold von, Freiherr, (1772-1801)
HARTMANN, Ernst Gustav Friedrich von, (1767-1851)
HEHL, Johann Carl Ludwig, (1774-1853)
HELMS, Anton Zacharias, (1751-1803) [PROSPECTORS, DIVINERS AND MINING ENGINEERS]
HENCKEL, Johann Friedrich, (1679-1744)
HERDER, Siegmund August Wolfgang von, Freiherr, (1776-1838)
HOENINGHAUS, Friedrich Wilhelm, (1770-1854)
HUMBOLDT, Friedrich Wilhelm Heinrich Alexander von, Freiherr, (1769-1859)
KANT, Immanuel, (1724-1804)
KARSTEN, Dietrich Ludwig Gustav, (1768-1810)
KENTMANN, Johannes, (1518-1574)
KEPLER, Johannes, (1571-1630)
KIELMEYER, Carl Heinrich, (1765-1844)
KIRCHER, Athanasius, (1601-1680)
KLEIN, Jakob Theodor, (1685-1759)
KONIG, Karl Dietrich Eberhardt, called Charles, (1774-1851)
LEHMANN, Johann Christoph, (1675-1739)
LEHMANN, Johann Gottlob, (1719-1767)
LENZ, Johann Georg, (1748-1832)
LESKE, Nathanael Gottfried, (1751-1786)
LUC, Jean André de, (1727-1817)
MARSIGLI, Luigi Ferdinando, (ca.1330-1394)
MATHESIUS, Johannes, (1504-1565)
MOHS, Carl Friedrich Christian, (1773-1839)
MORITZ, Karl Philipp, (1757-1793)
NAU, Bernhard Sebastian von, (1766-1845)
NOSSENI, Johann Maria, (1544-1620)
PALLAS, Peter Simon, (1741-1811)
PARROT, Georg Friedrich, (1767-1852)
POTT, Johann Heinrich, (1692-1777)
RASPE, Rudolf Erich, (1737-1794)
REINECKE, Johann Christoph Matthias, (1769-1818)
RÜLEIN VON CALW, Ulrich, (ca.1467-1523)
SARTORIUS, Georg Christian, (1774-1838)
SCHÄFFER, Jacob Christian, (1718-1790)
SCHEELE, Karl Wilhelm, (1742-1786)
SCHLOTHEIM, Ernst Friedrich von, Freiherr, (1764-1832)
SCHMEISSER, Johann August, (1767-1837)

GERMANY Up to 1800 Cont'd.

SCHOEPF, Johann David, (1752-1800)
SCHULZE, Christian Friedrich, (1730-1775)
STENO, Nicolaus, (1631-1687)
STIFFT, Christian Ernst, (1780-1855)
STRANGE, John, (1732-1799)
TELEKI, Domokos, Graf, (-1803)
TREBRA, Friedrich Wilhelm Heinrich von, (1740-1819)
VALENTINI, Michael Bernhard, (1657-1729)
VOIGT, Johann Karl Wilhelm, (1752-1821)
VOLCKMANN, Georg Anton, (1636-1721)
WEBER, Wilhelm, (1804-1891)
WEISMANN, Gottlieb, (1798-1859)
WERNER, Abraham Gottlob, (1749-1817)
ZIMMERMANN, Christian, (1765-1853)

1800-1849

ALBERTI, Friedrich August von, (1795-1878)
ANDRAE, Carl Justus, (1816-1885)
ASSMAN, C. G., (-)
AYCKE, Johann Christian, (1766-1854)
BATSCH, August Johann Georg Karl, (1761-1802)
BEHN, Wilhelm Friedrich Georg, (1808-1878)
BEISSEL, Ignaz, (1820-1887)
BESSER, Wilibald Swibert Joseph Gottlieb von, (1784-1842)
BEUST, Friedrich Constantin von, Freiherr, (1806-)
BEYRICH, August Heinrich Ernst, (1815-1896)
BIBRA, Ernst von, Freiherr, (1806-1878)
BISCHOFF, Karl Gustav, (1792-1870)
BLUM, Johann Reinhard, (1802-1883)
BLUMENBACH, Johann Friedrich, (1752-1840)
BOJANUS, Ludwig Heinrich, (1776-1827)
BOLL, Ernst Friedrich August, (1817-1868)
BOUÉ, Ami, (1794-1881)
BOXBERG, Ida von, (1806-1893)
BRANDT, Johann Friedrich von, (1802-1879)
BRAUN, Heinrich Alexander, (1805-1877)
BRAUN, Karl Friedrich Wilhelm, (1800-1864)
BRAUNS, David August, (1827-1893)
BREITHAUPT, Johann Friedrich August, (1791-1873)
BRONN, Heinrich Georg, (1800-1862)
BUCH, Christian Leopold von, Freiherr, (1774-1853)
BÜHLER, Georg Wilhelm Christian von, (1797-1860)
BURKART, Hermann Josef, (1798-1874)
CARUS, Carl Gustav, (1789-1869)
CASPARY, Robert, (1818-1887)
CASTELLI, Albin Benedict, (1822-1892)
CHARPENTIER, Jean de, (1786-1855)
CHARPENTIER, Johann Friedrich Wilhelm Toussaint von, (1728-1805)
COLLINI, Cosimo Alessandro, (1727-1806)
COTTA, Carl Bernhard von, (1808-1879)
COTTA, Heinrich, (1763-1844)
CREDNER, Carl Friedrich Heinrich, (1809-1876)
CRICHTON, Alexander, Sir, (1763-1856)
DAUBENY, Charles Giles Brindle, (1795-1867)

GERMANY 1800-1849 Cont'd.

DECHEN, Ernst Heinrich Carl von, (1800-1889)
DEFFNER, Carl, (1817-1878)
DIEFFENBACH, Ernst, (1811-1855)
DUHAMEL, Jean Pierre François Guillaume, (1730-1816)
DUMONT, André Hubert, (1809-1857)
DUNKER, Wilhelm Bernhard Rudolph Hadrian, (1809-1885)
EGEN, P. N. C., (1793-1849)
EHRENBERG, Christian Gottfried, (1795-1876)
EISEL, Robert, (1826-1917)
EMMERLING, Ludwig August, (1765-1841)
EMMRICH, Hermann Friedrich, (1815-1879)
ESCHWEGE, Wilhelm Ludwig, Baron, (1777-1855)
ESER, Anton Friedrich, (1798-1873)
EWALD, Julius Wilhelm, (1811-1891)
FACK, M. W., (1823-)
FAUJAS DE SAINT FOND, Barthélémy, (1741-1819)
FISCHER, Carl Ernst, (1818-1886)
FISCHER, Leopold Heinrich, (1817-1886)
FISCHER VON WALDHEIM, Johann Gotthelf Friedrich, (1771-1853)
FLURL, Mathias von, (1756-1823)
FORSTER, Adolarius Jacob, (1739-1806)
FRAAS, Oskar Friedrich von, (1824-1897)
FREIESLEBEN, Johann Carl, (1774-1846)
FRESENIUS, Georg, (1808-1866)
FUCHS, Johann Nepomuk von, (1774-1856)
FUHLROTT, Johann Carl, (1804-1877)
GÄTZSCHMANN, Moritz Ferdinand, (1800-1895)
GAUSS, Carl Friedrich, (1777-1855)
GEINITZ, Hanns Bruno, (1814-1900)
GEMMING, Carl Emil von, (1794-1880)
GENTH, Friedrick August Ludwig Karl Wilhelm, later called
Frederick Augustus, (1820-1893)
GERMAR, Ernst Friedrich, (1786-1853)
GERSDORF, Adolf Traugott von, (1744-1807)
GEYLER, Hermann Theodor, (1834-1889)
GIEBEL, Christoph Gottfried Andreas, (1820-1881)
GIRARD, Carl Adolph, (1814-1878)
GMELIN, Karl Christian, (1762-1837)
GOETHE, Johann Wolfgang von, (1749-1832)
GOLDENBERG, Friedrich, (1799-1881)
GOLDFUSS, Georg August, (1782-1848)
GREENOUGH, George Bellas, (1778-1855)
GRIEPENKERL, Otto, (1820-1888)
GRUNOW, Albert, (1826-1914)
GÜMBEL, Carl Wilhelm von, Ritter, (1823-1898)
GÜNTHER, A. F., (?1805-1871)
GUTBIER, Christian August von, (1798-1866)
HÄBERLEIN, Ernst, (1819-1895)
HÄBERLEIN, Friedrich Karl, (1787-1871)
HAGEN, Hermann August, (1817-1893)
HAGENOW, Friedrich von, (1797-1865)
HAIDINGER, Wilhelm Karl von, Ritter, (1795-1871)
HAMILTON, William, Sir, (1730-1803)
HAMILTON, William John, (1805-1867)

GERMANY 1800-1849 Cont'd.

HARDENBERG, Friedrich Leopold von, Freiherr, (1772-1801)
HARTMANN, Ernst Gustav Friedrich von, (1767-1851)
HASSENCAMP, Ernst, (1824-1881)
HAUCHECORNE, Heinrich Lambert Wilhelm, (1828-1900)
HAUPT, Karl, (1829-1882)
HAUSMANN, Johann Friedrich Ludwig, (1782-1859)
HEHL, Johann Carl Ludwig, (1774-1853)
HELMS, Anton Zacharias, (1751-1803) [PROSPECTORS, DIVINERS AND MINING ENGINEERS]
HENSEL, Reinhold Friederich, (1826-1881)
HERDER, Siegmund August Wolfgang von, Freiherr, (1776-1838)
HEYMANN, Hermann, (-1871)
HOENINGHAUS, Friedrich Wilhelm, (1770-1854)
HOFF, Karl Ernst Adolf von, (1771-1837)
HOFFMANN, Friedrich, (1797-1836)
HOFMEISTER, Wilhelm Friedrich Benedict, (1824-1877)
HOHENEGGER, Ludwig, (1807-1864)
HORNER, Leonard, (1785-1864)
HOSIUS, August, (1825-1896)
HUMBOLDT, Friedrich Wilhelm Heinrich Alexander von, Freiherr, (1769-1859)
JÄGER, Georg Friedrich, (1785-1866)
KANT, Immanuel, (1724-1804)
KAPFF, Sixt Friedrich Jakob, (1809-1887)
KARSTEN, Dietrich Ludwig Gustav, (1768-1810)
KARSTEN, Gustav, (1820-1900)
KARSTEN, Hermann, (1809-1877)
KAUP, Johann Jakob, (1803-1873)
KEFERSTEIN, Christian, (1784-1866)
KIELMEYER, Carl Heinrich, (1765-1844)
KLIPSTEIN, August Wilhelm, (1801-1894)
KOBELL, Franz von, (1803-1882)
KOCH, Carl, (1827-1882)
KOCH, Friedrich Eduard, (1817-1894)
KOENEN, Adolf von, (1837-1915)
KONIG, Karl Dietrich Eberhardt, called Charles, (1774-1851)
KRAUSS, Ferdinand, (1812-1890)
KÜHN, Karl Amadeus, (1783-1848)
KURR, Johann Gottlob von, (1798-1870)
LENZ, Johann Georg, (1748-1832)
LEONHARD, Gustav von, (1816-1878)
LEONHARD, Karl Cäsar von, (1779-1862)
LEUBE, Gustav, (1808-1881)
LEUNIS, Johann, (1802-1873)
LUC, Jean André de, (1727-1817)
LUDWIG, Rudolph August Birminhold Sebastian, (1812-1880)
MANDELSLOH, Friedrich von, Graf, (1795-1870)
MARSSON, Theodor, (1816-1892)
MARTIUS, Karl Friedrich Philipp von, (1794-1868)
MENGE, Anton, (1808-1880)
MENGE, Johann, (-)
MEYER, Christian Erich Hermann von, (1801-1869)
MEYN, Ludwig, (1820-1878)
MÖBIUS, Karl August, (1825-1908)

GERMANY 1800-1849 Cont'd.

MOHS, Carl Friedrich Christian, (1773-1839)
MORRIS, John, (1810-1886)
MÜLLER, Carl Hermann, (1823-1907)
MÜLLER, Moritz, (1791-1858)
NAU, Bernhard Sebastian von, (1766-1845)
NAUMANN, Carl Friedrich, (1797-1873)
NEUMANN, Franz Ernest, (1798-1895)
NÖGGERATH, Johann Jakob, (1788-1877)
NÖLDEKE, Karl, (1815-1898)
OEYNHAUSEN, Karl August Ludwig, (1795-1865)
OTTO, Ernst von, (1799-1863)
PAGENSTECHER, Heinrich Alexander, (1825-1889)
PALLAS, Peter Simon, (1741-1811)
PARROT, Georg Friedrich, (1767-1852)
PFAFF, Friedrich, (1825-1886)
PHILIPPI, Rudolph Amandus, (1808-1904)
PLATZ, Philipp, (1827-1900)
PLIENINGER, Wilhelm Heinrich Theodor von, (1795-1879)
PROBST, Joseph, (1823-1905)
PUSCH, Georg Gottlieb, (1790-1846)
QUENSTEDT, Friedrich August von, (1809-1889)
RAMMELSBERG, Karl Friedrich, (1813-1899)
RAUMER, Karl Georg von, (1783-1865)
REINECKE, Johann Christoph Matthias, (1769-1818)
RICHTER, Reinhard, (1813-1884)
RÖMER, Carl Ferdinand von, (1818-1891)
RÖMER, Friedrich Adolph, (1809-1869)
RÖMER, Hermann, (1816-1894)
ROSE, Gustav, (1798-1873)
ROSSMÄSSLER, Emil Adolf, (1806-1867)
ROTH, Justus Ludwig Adolph, (1818-1892)
RUMPF, Ludwig, (1793-1862)
RÜPPELL, Eduard, (1794-1884)
SAEMANN, Louis, (1821-1866)
SANDBERGER, Fridolin Carl Ludwig von, (1826-1898)
SARTORIUS, Georg Christian, (1774-1838)
SARTORIUS VON WALTERSHAUSEN, Wolfgang, (1809-1876)
SCHAAFHAUSEN, Hermann, (1816-1893)
SCHACKO, Gustav, (1824-1917)
SCHAFHÄUTL, Karl Franz Emil, (1803-1890)
SCHAUROTH, Karl von, Freiherr, (1818-1893)
SCHENK, Joseph August, (1815-1891)
SCHENKENBERG, F. C. A., (-)
SCHIMPER, Karl Friedrich, (1803-1867)
SCHIMPER, Philipp Wilhelm, (1808-1880)
SCHLEGEL, Hermann, (1804-1884)
SCHLEIDEN, Matthias Jakob, (1804-1881)
SCHLOTHEIM, Ernst Friedrich von, Freiherr, (1764-1832)
SCHMEISSER, Johann August, (1767-1837)
SCHMID, Ernst Erhard, (1815-1885)
SCHMIDT, Adolf, (1812-1899)
SCHMIDT, Johann Friedrich Julius, (1825-1884)
SCHOEPF, Johann David, (1752-1800)
SCHÖNN, Rudolf, (-1889)

GERMANY 1800-1849 Cont'd.

SCHOTT, Heinrich, (-1865)
SCHWIPPEL, Karl, (1821-1911)
SCROPE, George Poulett, (né Thomson), (1797-1876)
SMYTH, Warington Wilkinson, Sir, (1817-1890)
SPEYER, Oscar, (1827-1882)
STEIN, Johannes Philip Emil Friedrich, (1814-1882)
STEININGER, Johann, (1794-1874)
STENZEL, Karl Gustav, (1826-1905)
STIFFT, Christian Ernst, (1780-1855)
STROMBECK, August von, (1808-1900)
TELEKI, Domokos, Graf, (-1803)
THEOBALD, Gottfried Ludwig, (1810-1869)
THIELE, Friedrich, (1814-1899)
TORELL, Otto Martin, (1828-1900)
TREBRA, Friedrich Wilhelm Heinrich von, (1740-1819)
TROSCHEL, Franz Hermann, (1810-1882)
UNGER, Franz Joseph Andreas Nicolas, (1800-1870)
VALET, August Friedrich, (1811-1889)
VERNEUIL, Philippe Edouard Poulletier de, (1805-1873)
VOELKEL, Josef, (1828-1906)
VOIGT, Johann Karl Wilhelm, (1752-1821)
WAGNER, Johann Andreas, (1797-1861)
WAPPLER, Rudolph Bruno, (1828-1888)
WEBER, Wilhelm, (1804-1891)
WEBSKY, Christian Friedrich Martin, (1824-1886)
WEINKAUFF, H. C., (1817-1886)
WEISMANN, Gottlieb, (1798-1859)
WEISS, Christian Samuel, (1780-1856)
WEISSENBACH, Karl Gustav Adalbert von, (1797-1848)
WERNER, Abraham Gottlob, (1749-1817)
WIECHMANN, Carl Michael, (1828-1883)
WIRTGEN, Philipp Wilhelm, (1806-1870)
WÖHLER, Friedrich, (1800-1883)
ZIETEN, Carl Hartwig von, (1785-1846)
ZIMMERMANN, Christian, (1765-1853)
ZITTEL, Karl Alfred von, (1839-1904)

1850-1899

ABEL, Othenio, (1875-1946)
AHLBURG, Johannes Heinrich Wilhelm, (1833-1919)
ALBERTI, Friedrich August von, (1795-1878)
AMMON, Ludwig von, (1850-1922)
ANDRAE, Carl Justus, (1816-1885)
ARZRUNI, Andreas, (1847-1898)
AYCKE, Johann Christian, (1766-1854)
BALDAUF, Richard Julius, (1848-1931)
BALL, John, (-1941)
BALTZER, Richard Armin, (1842-1913)
BÄRTLING, Richard, (1878-1936)
BAUER, Max Hermann, (1844-1917)
BAUMHAUER, Heinrich Adolf, (1848-1926)
BAUR, Carl Theodor von, (1836-1911)
BECK, Carl Richard, (1858-1919)
BECKENKAMP, Jakob, (1855-1931)

GERMANY 1850-1899 Cont'd.

BECKER, Ewald, (-1873)
BEHN, Wilhelm Friedrich Georg, (1808-1878)
BEHR, Johannes, (1875-1973)
BEHREND, Paul, (-?1905)
BEISSEL, Ignaz, (1820-1887)
BERENDT, Gottlieb Michael, (1836-1920)
BERGEAT, Alfred, Edmund, (1866-1924)
BERGT, Walther Adolf, (1864-1941)
BERNHARDI, Reinhard, (fl.1834-1841)
BERTKAU, Philipp, (1849-1895)
BEUSHAUSEN, Hermann Emil Louis, (1863-1904)
BEUST, Friedrich Constantin von, Freiherr, (1806-)
BEUTELL, Albert, (1859-1921)
BEYRICH, August Heinrich Ernst, (1815-1896)
BIBRA, Ernst von, Freiherr, (1806-1878)
BIRKMAIER, Anton, (-1926)
BISCHOFF, Karl Gustav, (1792-1870)
BLANCKENHORN, Max Ludwig Paul, (1861-1947)
BLUM, Johann Reinhard, (1802-1883)
BLÜMCKE, Adolf, (1857-1914)
BÖHM, Georg, (1854-1913)
BOLL, Ernst Friedrich August, (1817-1868)
BÖLSCHE, Wilhelm, (1843-1893)
BORCHERT, Gustav, (1863-1934)
BOSCH, Carl, (1874-1940)
BÖSE, Emil, (1868-1972)
BOUÉ, Ami, (1794-1881)
BOXBERG, Ida von, (1806-1893)
BRANCA, Wilhelm von, (1844-1928)
BRANDES, Theodor, (-1916)
BRANDT, Johann Friedrich von, (1802-1879)
BRAUN, Heinrich Alexander, (1805-1877)
BRAUN, Karl Friedrich Wilhelm, (1800-1864)
BRAUNS, David August, (1827-1893)
BRAUNS, Reinhard Anton, (1861-1937)
BREITHAUPT, Johann Friedrich August, (1791-1873)
BROILI, Ferdinand, (1874-1946)
BRONN, Heinrich Georg, (1800-1862)
BRÜCKNER, Eduard, (1862-1927)
BRUDER, George, (1856-1916)
BRUHNS, Franz Siegfried Willy, (1864-1929)
BUCH, Christian Leopold von, Freiherr, (1774-1853)
BÜCKING, Ferdinand Carl Bertram Hugo, (1851-1932)
BÜHLER, Georg Wilhelm Christian von, (1797-1860)
BURBACH, Otto Carl Heinrich, (1838-1888)
BURKART, Hermann Josef, (1798-1874)
BUSZ, Karl Heinrich Emil Georg, (1863-1930)
CALKER, Friedrich Julius Peter von, (1841-1913)
CARUS, Carl Gustav, (1789-1869)
CASPARY, Robert, (1818-1887)
CASTELLI, Albin Benedict, (1822-1892)
CHAMPERKNOWNE, Arthur, (1839-1887)
CHARPENTIER, Jean de, (1786-1855)
CHRISTA, Emanuel, (1874-1948)

GERMANY 1850-1899 Cont'd.

CLARAZ, Georges, (1832-1930)
COHEN, Emil Wilhelm, (1842-1905)
COMPTNER, Gustav, (1831-1922)
CONWENTZ, Hugo, (1855-1922)
COTTA, Carl Bernhard von, (1808-1879)
CREDNER, Carl Friedrich Heinrich, (1809-1876)
CREDNER, Carl Hermann Georg, (1841-1913)
CRICHTON, Alexander, Sir, (1763-1856)
CRICK, George Charles, (1856-1917)
DAMES, Wilhelm Barnim, (1843-1898)
DANZIG, Emil, (1855-1943)
DATHE, Johann Friedrich Ernst, (1845-1917)
DAUBENY, Charles Giles Brindle, (1795-1867)
DAVIS, William Morris, (1850-1934)
DECHEN, Ernst Heinrich Carl von, (1800-1889)
DEECKE, Johannes Ernst Wilhelm, (1862-1934)
DEFFNER, Carl, (1817-1878)
DEICHMÜLLER, Johannes, (1854-)
DENCKMANN, August, (1860-1925)
DENINGER, Karl, (1878-1917)
DIEFFENBACH, Ernst, (1811-1855)
DIENER, Carl, (1862-1928)
DIETLEN, Karl Rudolf, (1857-1931)
DÖDERLEIN, Ludwig, (1855-1936)
DOSS, Karl Bruno, (1861-1919)
DREVERMANN, Fritz, (1875-1932)
DRUGMAN, Julien Léon Eugène, (1875-1950)
DUMONT, André Hubert, (1809-1857)
DUNKER, Wilhelm Bernhard Rudolph Hadrian, (1809-1885)
EBERT, Theodor, (1857-1899)
ECK, Heinrich von, (1837-1925)
EHRENBERG, Christian Gottfried, (1795-1876)
EISEL, Robert, (1826-1917)
EMMRICH, Hermann Friedrich, (1815-1879)
ENGEL, Theodor, (1842-1933)
ERDMANNSDÖFFER, Otto Heinrich, (1876-1955)
ESCHWEGE, Wilhelm Ludwig, Baron, (1777-1855)
ESER, Anton Friedrich, (1798-1873)
ETZOLD, Franz, (1859-1928)
EWALD, Julius Wilhelm, (1811-1891)
FACK, M. W., (1823-)
FALB, Rudolf, (-)
FELIX, Johannes Paul, (1859-1941)
FIEBELKORN, Max, (1869-1912)
FINCKH, Ludwig, (1871-1930)
FINSTERWALDER, Sebastian, (1862-?1931)
FISCHER, Carl Ernst, (1818-1886)
FISCHER, Hugo, (1864-1926)
FISCHER, Leopold Heinrich, (1817-1886)
FISCHER VON WALDHEIM, Johann Gotthelf Friedrich, (1771-1853)
FLIEGEL, Gotthard, (1873-1947)
FOLLMANN, Otto, (1856-1926)
FRAAS, Eberhard, (1862-1915)
FRAAS, Oskar Friedrich von, (1824-1897)

GERMANY 1850-1899 Cont'd.

FRANKE, Adolf, (1860-1942)
FRECH, Fritz, (1861-1917)
FRENZEL, Friedrich August, (1842-1902)
FRESENIUS, Georg, (1808-1866)
FRIEDRICH, Paul, (1856-1918)
FRITSCH, Karl von, Freiherr, (1838-1906)
FUCHS, Alexander, (1874-1935)
FUCHS, Carl Wilhelm C., (1837-1886)
FUCHS, Johann Nepomuk von, (1774-1856)
FUCHS, Philippe Jacques Edmond, (1837-1889)
FUHLROTT, Johann Carl, (1804-1877)
FÜRBRINGER, Max, (1846-1920)
GAGEL, Kurt, (1865-1927)
GATTNAR, Josef, (1854-1928)
GÄTZSCHMANN, Moritz Ferdinand, (1800-1895)
GAUSS, Carl Friedrich, (1777-1855)
GEINITZ, Eugen, (1854-1925)
GEINITZ, Hanns Bruno, (1814-1900)
GEMMING, Carl Emil von, (1794-1880)
GENTH, Friedrick August Ludwig Karl Wilhelm, later called Frederick Augustus, (1820-1893)
GERLAND, Georg Karl Cornelius, (1833-1919)
GERMAR, Ernst Friedrich, (1786-1853)
GEYER, David, (1855-1932)
GEYLER, Hermann Theodor, (1834-1889)
GIEBEL, Christoph Gottfried Andreas, (1820-1881)
GIRARD, Carl Adolph, (1814-1878)
GOLDENBERG, Friedrich, (1799-1881)
GOLDSCHMIDT, Victor, (1853-1933)
GÖLLER, August, (1878-1965)
GOSSNER, Balthasar, (1877-1937)
GOTTSCHE, Carl Christian, (1855-1909)
GOTTSCHICK, Franz, (1865-1927)
GREBE, Heinrich, (1831-1903)
GREENOUGH, George Bellas, (1778-1855)
GRIEPENKERL, Otto, (1820-1888)
GRODDECK, Albrecht Ludwig von, (1837-1887)
GROEBER, Paul Friedrich Carl, (1885-1968)
GROTH, Paul Heinrich von, Ritter, (1843-1927)
GRÜNLING, Friedrich, (1857-1919)
GRUNOW, Albert, (1826-1914)
GÜMBEL, Carl Wilhelm von, Ritter, (1823-1898)
GÜNTHER, A. F., (?1805-1871)
GÜRICH, Georg, (1859-1938)
GUTBIER, Christian August von, (1798-1866)
HAAS, Hippolyt Julius, (1855-1913)
HAASE, Erich, (1857-1894)
HÄBERLE, Daniel, (1864-1934)
HÄBERLEIN, Ernst, (1819-1895)
HÄBERLEIN, Friedrich Karl, (1787-1871)
HAECKEL, Ernst Heinrich Philipp August, (1834-1919)
HAGEN, Hermann August, (1817-1893)
HAGEN, Max, (1831-1893)
HAGENOW, Friedrich von, (1797-1865)

GERMANY 1850-1899 Cont'd.

HAIDINGER, Wilhelm Karl von, Ritter, (1795-1871)
HALFAR, Anton, (1836-1893)
HAMILTON, William John, (1805-1867)
HARTMANN, Ernst Gustav Friedrich von, (1767-1851)
HASSE, Carl, (1841-1922)
HASSENCAMP, Ernst, (1824-1881)
HAUCHECORNE, Heinrich Lambert Wilhelm, (1828-1900)
HAUFF, Bernhard, (1866-1950)
HAUPT, Karl, (1829-1882)
HAUSMANN, Johann Friedrich Ludwig, (1782-1859)
HAUTHAL, Rudolf, (1854-1928)
HELMS, Anton Zacharias, (1751-1803) [PROSPECTORS, DIVINERS AND MINING ENGINEERS]
HEMPEL, Walther Matthias, (1851-1916)
HENRICH, Ferdinand, (1837-1915)
HENSEL, Reinhold Friederich, (1826-1881)
HERZ, Otto, (1853-1905)
HESS, Hans, (1864-1940)
HEYDEN, Lucas von, (1838-1915)
HEYMANN, Hermann, (-1871)
HILGENDORF, Franz, (1839-1904)
HINTERLECHNER, Karl, (1874-1932)
HINTZE, Carl Adolf Ferdinand, (1851-1916)
HIRSCHWALD, Julius, (1845-1928)
HOFF, Karl Ernst Adolf von, (1771-1837)
HOFFMAN, Fr. Adolf, (1867-1898)
HOFMEISTER, Wilhelm Friedrich Benedict, (1824-1877)
HOHENEGGER, Ludwig, (1807-1864)
HOLZAPFEL, Eduard, (1853-1913)
HORNER, Leonard, (1785-1864)
HOSIUS, August, (1825-1896)
HUENE, Friedrich von, Freiherr, (1875-1969)
HUMBOLDT, Friedrich Wilhelm Heinrich Alexander von, Freiherr, (1769-1859)
HUSSAK, Franz Eugen, (1856-1911)
JAECKEL, Otto, (1863-1929)
JÄGER, Georg Friedrich, (1785-1866)
JENTZSCH, Alfred, (1850-1925)
JOHNSEN, Arrien, (1877-1934)
JULIAN, Henry Forbes, (1861-1912) [PROSPECTORS, DIVINERS AND MINING ENGINEERS]
KAISER, Friedrich Wilhelm Erich, (1871-1934)
KALKOWSKY, Ernst Louis, (1851-1938)
KAPFF, Sixt Friedrich Jakob, (1809-1887)
KARSCH, Ferdinand, (1853-1936)
KARSTEN, Gustav, (1820-1900)
KARSTEN, Hermann, (1809-1877)
KAUP, Johann Jakob, (1803-1873)
KAYSER, Friedrich Heinrich Emanuel, (1845-1927)
KEFERSTEIN, Christian, (1784-1866)
KEILHACK, Konrad, (1858-1944)
KESSLER, Paul, (?1882-1927)
KIESOW, Johann, (1846-1901)
KINKELIN, Georg Friedrich, (1836-1913)

GERMANY 1850-1899 Cont'd.

KISSLING, Ernst, (1865-1936)
KLAATSCH, Hermann, (1863-1916)
KLAUTZSCH, Adolf, (1869-1927)
KLEBS, Richard, (1850-1911)
KLEIN, Johann Friedrich Carl, (1842-1907)
KLEMM, Gustav, (1858-1938)
KLIPSTEIN, August Wilhelm, (1801-1894)
KLOCKE, Friedrich, (1847-1885)
KLOCKMANN, Friedrich Ferdinand Hermann, (1858-1937)
KOBELL, Franz von, (1803-1882)
KOCH, Carl, (1827-1882)
KOCH, Eduard, (1838-1897)
KOCH, Friedrich Eduard, (1817-1894)
KOCH, Gottlieb Karl David von, (1849-1914)
KOENEN, Adolf von, (1837-1915)
KOENIGSBERGER, Johann Georg, (1874-1946)
KOERT, Willi, (1875-1927)
KOKEN, Ernst Friedrich Rudolph Karl von, (1860-1912)
KOLBE, Hermann Julius, (1855-)
KOLBECK, Friedrich Ludwig Wilhelm, (1860-1943)
KONIG, Karl Dietrich Eberhardt, called Charles, (1774-1851)
KORN, Johannes, (1862-1927)
KRANTZ, Friedrich Ludwig Robert, (1859-1926)
KRAUS, Gregor, (1841-1915)
KRAUSS, Ferdinand, (1812-1890)
KRUSCH, Paul, (1869-1939)
KUNTH, Karl Ernst Albrecht, (1842-1871)
KURR, Johann Gottlob von, (1798-1870)
KURTZ, Fritz, (1854-1920)
LANDOIS, Hermann, (1835-1905)
LASAULX, Arnold von, (1839-1886)
LAUE, Max Theodor Felix von, (1879-1960)
LAUFER, Ernst, (1850-1893)
LEHMANN, Johannes Georg, (1851-1925)
LEHMANN, Otto, (1855-1922)
LEINER, Ludwig, (1830-1901)
LEITZ, Ernst, Senior, (1871-1956)
LEMBERG, Johannes, (1842-1902)
LENK, Hans, (1863-1938)
LENZ, Oskar, (1848-1925)
LEONHARD, Gustav von, (1816-1878)
LEONHARD, Karl Cäsar von, (1779-1862)
LEONHARD, Richard, (1870-1916)
LEPPLA, August, (1859-1924)
LEPSIUS, Karl Georg Richard, (1851-1915)
LEUBE, Gustav, (1808-1881)
LEUNIS, Johann, (1802-1873)
LEUZE, Alfred, (1845-1899)
LEWIS, Henry Carvill, (1853-1888)
LIEBE, Karl Leopold Theodor, (1828-1894)
LIEBISCH, Theodor, (1852-1922)
LIENENKLAUS, Ernst, (1849-1905)
LINCIO, Gabrielle, (1874-1938)
LINGELSHEIM, Alexander von, (1874-1937)

GERMANY 1850-1899 Cont'd.

LINSTOW, Otto von, (1872-1929)
LOOMIS, Theodor, (1875-1909)
LORENZ, Theodor, (1875-1909)
LORETZ, Hermann, (1836-1917)
LOSSEN, Karl August, (1841-1893)
LUDWIG, Rudolph August Birminhold Sebastian, (1812-1880)
MAAS, Günther Heinrich Julius Max, (1871-1905)
MAAS, Otto, (1867-1916)
MANDELSLOH, Friedrich von, Graf, (1795-1870)
MARGERIE, Emmanuel Marie Pierre Martin Jacquin de, (1862-1953)
MARSH, Othniel Charles, (1831-1899)
MARSSON, Theodor, (1816-1892)
MARTIUS, Karl Friedrich Philipp von, (1794-1868)
MAUCHER, Wilhelm, (1879-1930)
MAY, Walther, (1868-)
MENGE, Anton, (1808-1880)
MENZEL, Karl Gustav Johannes, (1875-1914)
MENZEL, Paul, (1864-1927)
MEUNIER, Fernand, (1868-1926)
MEYER, Christian Erich Hermann von, (1801-1869)
MEYER, Erich, (1874-1915)
MEYER, Hans, (1858-1929)
MEYN, Ludwig, (1820-1878)
MICHAEL, Richard, (1869-1925)
MIELEITNER, Karl, (1890-1923)
MILCH, Ludwig, (1867-1928)
MÖBIUS, Karl August, (1825-1908)
MORRIS, John, (1810-1886)
MÜGGE, Johannes Otto Konrad, (1858-1932)
MÜLLER, Carl Hermann, (1823-1907)
MÜLLER, Karl Christian Gottfried Ludwig, (1862-1906)
MÜLLER, Moritz, (1791-1858)
MYLIUS, Hugo, (1876-1918)
NAUMANN, Carl Friedrich, (1797-1873)
NEHRING, Alfred, (1845-1904)
NESSIG, Wilhelm Robert, (1861-1932)
NEUMANN, Franz Ernest, (1798-1895)
NEUMAYR, Melchior, (1845-1890)
NIES, Friedrich, (1839-1895)
NÖGGERATH, Johann Jakob, (1788-1877)
NÖLDEKE, Karl, (1815-1898)
OCHSENIUS, Carl Christian, (1830-)
OEBBEKE, Konrad, (1853-1932)
OEYNHAUSEN, Karl August Ludwig, (1795-1865)
OPPEL, Albert, (1831-1865)
OPPENHEIM, Leo Paul, (1863-1934)
ORTH, Albert, (1835-1915)
OSANN, Carl Alfred, (1859-1923)
OTTMER, Eduard Julius Otto, (1846-1886)
OTTO, Ernst von, (1799-1863)
PAGENSTECHER, Heinrich Alexander, (1825-1889)
PALAZZO, Luigi, (1861-1933)
PARROT, Georg Friedrich, (1767-1852)
PATRUNKY, H., (1854-1930)

GERMANY 1850-1899 Cont'd.

PAULCKE, Wilhelm, (1873-1949)
PENCK, Friedrich Carl Albrecht, (1858-1945)
PFAFF, Friedrich, (1825-1886)
PHILIPPI, Rudolph Amandus, (1808-1904)
PHLEPS, Otto, (1868-1928)
PLAGEMANN, A., (1858-1908)
PLATZ, Philipp, (1827-1900)
PLIENINGER, Wilhelm Heinrich Theodor von, (1795-1879)
POCKELS, Friedrich Carl Alwin, (1865-1913)
POMPECKJ, Josef Felix, (1867-1930)
PORRIO, Cesare, (1865-1940)
POTONIÉ, Henry Edmond, (1857-1913)
PROBST, Joseph, (1823-1905)
PUFAHL, Otto Hermann August, (1855-1924)
QUENSTEDT, Friedrich August von, (1809-1889)
RAMMELSBERG, Karl Friedrich, (1813-1899)
RATH, Gerhard vom, (1830-1888)
RAU, Karl, (1870-1966)
RAUMER, Karl Georg von, (1783-1865)
REBEUR-PASCHWITZ, Ernst von, (1861-1895)
REHM, Heinrich, (1828-1916)
REINACH, Albert von, Baron, (1842-1905)
REINSCH, Paul Friedrich, (1836-1914)
REIS, Otto Maria, (1862-1934)
REMELÉ, Adolf Karl, (1839-1915)
RICHTER, Ernst Julius, (-1868)
RICHTER, Reinhard, (1813-1884)
RICHTHOFEN, Paul Wilhelm Ferdinand von, Freiherr, (1833-1905)
RINNE, Friedrich Wilhelm Berthold, (1863-1933)
RODOLPH, Emil, (1853-1915)
ROGER, Otto, (1841-1915)
RÖMER, Carl Ferdinand von, (1818-1891)
ROMER, Eugeniusz, (1871-1954)
RÖMER, Friedrich Adolph, (1809-1869)
RÖMER, Hermann, (1816-1894)
ROSE, Gustav, (1798-1873)
ROSENBUSCH, Karl Harry Ferdinand, (1836-1914)
ROSSMÄSSLER, Emil Adolf, (1806-1867)
ROTH, Justus Ludwig Adolph, (1818-1892)
ROTHPLETZ, August, (1853-1918)
RUDOLPH, Emil, (1853-1915)
RUMPF, Ludwig, (1793-1862)
RÜPPELL, Eduard, (1794-1884)
RÜST, David, (1831-1916)
RUTLEY, Frank, (1842-1904)
SADEBECK, Alexander, (1843-1879)
SAEMANN, Louis, (1821-1866)
SANDBERGER, Fridolin Carl Ludwig von, (1826-1898)
SARTORIUS VON WALTERSHAUSEN, Wolfgang, (1809-1876)
SAUER, Gustav Adolf, (1825-1932)
SCHAAFHAUSEN, Hermann, (1816-1893)
SCHACKO, Gustav, (1824-1917)
SCHAFHÄUTL, Karl Franz Emil, (1803-1890)
SCHALCH, Ferdinand, (1848-1918)

GERMANY 1850-1899 Cont'd.

SCHAUROTH, Karl von, Freiherr, (1818-1893)
SCHENK, Joseph August, (1815-1891)
SCHENKENBERG, F. C. A., (-)
SCHERING, Ernst, (1833-1897)
SCHIMPER, Karl Friedrich, (1803-1867)
SCHIMPER, Philipp Wilhelm, (1808-1880)
SCHLECHTENDAL, Dietrich von, (1834-1916)
SCHLEGEL, Hermann, (1804-1884)
SCHLEIDEN, Matthias Jakob, (1804-1881)
SCHLOENBACH, Georg Justin Carl Urban, (1841-1870)
SCHLOSSER, Max, (1854-1933)
SCHLUETER, Clemens August, (1835-1906)
SCHLUNCK, Johannes, (1876-1915)
SCHMID, Ernst Erhard, (1815-1885)
SCHMIDT, Adolf, (1812-1899)
SCHMIDT, Johann Friedrich Julius, (1825-1884)
SCHOENFLIES, Arthur Moritz, (1853-1928)
SCHOETENSACK, Otto, (1850-1912)
SCHOLZ, Max, (1832-1892)
SCHÖNFELD, Georg, (1878-1926)
SCHÖNN, Rudolf, (-1889)
SCHOTT, Heinrich, (-1865)
SCHOTTLER, Wilhelm, (1869-1932)
SCHRÖDER, Richard, (1853-1916)
SCHROEDER, Henry Carl, (1859-1927)
SCHUMACHER, Eugen, (1851-1922)
SCHUMANN, Karl, (1851-1904)
SCHÜTZE, Ewald, (1873-1908)
SCHWERTSCHLAGER, Joseph, (1853-1924)
SCHWIPPEL, Karl, (1821-1911)
SCROPE, George Poulett, (né Thomson), (1797-1876)
SEEBACH, Karl Albert Ludwig von, (1839-1880)
SEGER, Hermann, (-1893)
SELIGMANN, Gustav, (1849-1920)
SEMPER, Otto, (1830-1907)
SIEGERT, Leo, (1872-1917)
SMYTH, Warington Wilkinson, Sir, (1817-1890)
SOENDEROP, Fritz Hermann Gustav, (1876-1914)
SOLMS-LAUBACH, Hermann zu, Graf, (1842-1915)
SPANDEL, Erich, (1855-1909)
SPEYER, Oscar, (1827-1882)
STAPFF, Friedrich Moritz, (1836-1895)
STEIN, Johannes Philip Emil Friedrich, (1814-1882)
STEININGER, Johann, (1794-1874)
STEINMANN, Gustav, (1856-1929)
STELZNER, Alfred Wilhelm, (1840-1895)
STENZEL, Karl Gustav, (1826-1905)
STERZEL, Johann Traugott, (1841-1914)
STIFFT, Christian Ernst, (1780-1855)
STOLLER, Jakob, (1873-1930)
STOLLEY, Ernst, (1869-1944)
STOLTZ, Karl, (?1858-1927)
STRASBURGER, Eduard, (1844-1912)
STRENG, August, (1830-1897)

GERMANY 1850-1899 Cont'd.

STROMBECK, August von, (1808-1900)
STRUCKMANN, Carl Eberhard Friedrich, (1833-1898)
TANNHÄUSER, Felix, (1874-1924)
THEOBALD, Gottfried Ludwig, (1810-1869)
THIELE, Friedrich, (1814-1899)
TIETZE, Oskar Adolf Eduard, (1874-1920)
TORELL, Otto Martin, (1828-1900)
TÖRNQUIST, Sven Leonhard, (1840-1920)
TROSCHEL, Franz Hermann, (1810-1882)
UHLIG, Carl Ludwig Gustav, (1872-?1938)
UNGER, Franz Joseph Andreas Nicolas, (1800-1870)
VALET, August Friedrich, (1811-1889)
VAN'T HOFF, Jacobus Hendricus, (1852-1911)
VATER, Heinrich, (1859-1930)
VERNEUIL, Philippe Edouard Poulletier de, (1805-1873)
VETTER, Benjamin, (1848-1893)
VIOLA, Carlo Maria, (1855-1925)
VOELKEL, Josef, (1828-1906)
VOGELSANG, Hermann, (1838-1874)
VOIGT, Friedrich Gustav, (1844-1886)
VOIGT, Waldemar, (1850-1919)
WAGNER, Johann Andreas, (1797-1861)
WAHNSCHAFFE, Gustav Albert Bruno Felix, (1851-1914)
WALTHER, Johannes, (1860-1937)
WANNER, Johannes, (1878-1957)
WAPPLER, Rudolph Bruno, (1828-1888)
WEBER, Carl Albert, (1856-1931)
WEBSKY, Christian Friedrich Martin, (1824-1886)
WEINKAUFF, H. C., (1817-1886)
WEINSCHENK, Ernst Heinrich Oskar Kasimir, (1865-1921)
WEISBACH, Albin, (1833-1901)
WEISE, Ernst, (1843-1941)
WEISMANN, Gottlieb, (1798-1859)
WEISS, Christian Ernst, (1833-1890)
WEISS, Christian Samuel, (1780-1856)
WELTNER, Wilhelm, (1854-1917)
WEYENBERGH, Hendrik, (1842-1885)
WIECHMANN, Carl Michael, (1828-1883)
WILCKENS, Otto Rudolf, (1876-1943)
WILLIAMS, John Francis, (1862-1891)
WINDHAUSEN, Anselm, (1882-1932)
WINKLER, Clemens Alexander, (1838-1904)
WIRTGEN, Philipp Wilhelm, (1806-1870)
WÖHLER, Friedrich, (1800-1883)
WOLF, Theodor, (1841-1924)
WOLFF, Wilhelm, (1872-1951)
WÜLFING, Ernst Anton, (1860-1930)
WUNDT, Georg von, (1845-1929)
WÜNSCH, Edward A., (1822-1895)
WÜST, Ewald, (1875-1934)
ZEISE, Oskar, (1860-1925)
ZIETZ, Amandus Christian Heinrich, (1839-1921)
ZIMMERMANN, Christian, (1765-1853)
ZIMMERMANN, Ernst, (1860-1944)

GERMANY 1850-1899 Cont'd.

ZIRKEL, Ferdinand, (1838-1912)
ZITTEL, Karl Alfred von, (1839-1904)

1900 to present

ABEL, Othenio, (1875-1946)
AHLBURG, Johannes Heinrich Wilhelm, (1833-1919)
AMMON, Ludwig von, (1850-1922)
ANDRÉE, Karl, (1880-1959)
BALDAUF, Richard Julius, (1848-1931)
BALL, John, (-1941)
BALTZER, Richard Armin, (1842-1913)
BARTH, Thomas Frederik Weiby, (1899-1971)
BÄRTLING, Richard, (1878-1936)
BAUER, Max Hermann, (1844-1917)
BAUHANS, Hans, (1888-1918)
BAUMHAUER, Heinrich Adolf, (1848-1926)
BAUR, Carl Theodor von, (1836-1911)
BECK, Carl Richard, (1858-1919)
BECKENKAMP, Jakob, (1855-1931)
BEHR, Johannes, (1875-1973)
BEHREND, Paul, (-?1905)
BENTZ, Alfred, (1897-1964)
BERCKHEMER, Fritz, (1890-?1954)
BERENDT, Gottlieb Michael, (1836-1920)
BERGEAT, Alfred, Edmund, (1866-1924)
BERGT, Walther Adolf, (1864-1941)
BEUSHAUSEN, Hermann Emil Louis, (1863-1904)
BEUTELL, Albert, (1859-1921)
BEYSCHLAG, Franz, (-?1935)
BIRKMAIER, Anton, (-1926)
BLANCKENHORN, Max Ludwig Paul, (1861-1947)
BLÜMCKE, Adolf, (1857-1914)
BOEKE, Hendrik Enno, (1881-1918)
BÖHM, Georg, (1854-1913)
BORCHERT, Gustav, (1863-1934)
BOSCH, Carl, (1874-1940)
BÖSE, Emil, (1868-1972)
BRANCA, Wilhelm von, (1844-1928)
BRANDES, Theodor, (-1916)
BRAUNS, Reinhard Anton, (1861-1937)
BREDDIN, Hans, (1900-1973)
BROILI, Ferdinand, (1874-1946)
BRÖSAMLEN, Otto, (1885-1966)
BROTZEN, Fritz, (1902-1968)
BRÜCKNER, Eduard, (1862-1927)
BRUDER, George, (1856-1916)
BRUHNS, Franz Siegfried Willy, (1864-1929)
BUBNOFF, Serge von, (1888-1957)
BÜCKING, Ferdinand Carl Bertram Hugo, (1851-1932)
BUSZ, Karl Heinrich Emil Georg, (1863-1930)
CALKER, Friedrich Julius Peter von, (1841-1913)
CHRISTA, Emanuel, (1874-1948)
CHUDOBA, Karl F., (1898-1976)
CLARAZ, Georges, (1832-1930)

GERMANY 1900 to present Cont'd.

CLOOS, Hans, (1885-1951)
COHEN, Emil Wilhelm, (1842-1905)
COMPTNER, Gustav, (1831-1922)
CONWENTZ, Hugo, (1855-1922)
CORNELIUS, Hans Peter, (1888-)
CRAMER, Rudolf, (1882-1949)
CREDNER, Carl Hermann Georg, (1841-1913)
CREDNER, Wilhelm, (1892-1948)
CRICK, George Charles, (1856-1917)
DAHLGRÜN, Fritz, (1894-1954)
DANZIG, Emil, (1855-1943)
DATHE, Johann Friedrich Ernst, (1845-1917)
DAVIS, William Morris, (1850-1934)
DEECKE, Johannes Ernst Wilhelm, (1862-1934)
DEICHMÜLLER, Johannes, (1854-)
DE LA SAUCE, Wilhelm, (1882-1955)
DELHAES, Wilhelm, (1883-191)
DENCKMANN, August, (1860-1925)
DENINGER, Karl, (1878-1917)
DEUBEL, Fritz, (1898-1966)
DIENER, Carl, (1862-1928)
DIETLEN, Karl Rudolf, (1857-1931)
DIETRICH, Wilhelm Otto, (1881-1964)
DIETZ, Curt, (1882-1965)
DÖDERLEIN, Ludwig, (1855-1936)
DOSS, Karl Bruno, (1861-1919)
DREHER, Otto, (1888-1945)
DREVERMANN, Fritz, (1875-1932)
DRUGMAN, Julien Léon Eugène, (1875-1950)
DÜCKER, Alfred, (1908-1969)
EBY, James Brian, (1896-)
ECK, Heinrich von, (1837-1925)
EDINGER, Tilly, (1897-1967)
EHRINGHAUS, Arthur, (-)
EISEL, Robert, (1826-1917)
EISENACK, Alfred, (1891-)
ENGEL, Theodor, (1842-1933)
ERDMANNSDÖFFER, Otto Heinrich, (1876-1955)
ERHART, Henri, (-)
ERTEL, Hans, (1904-1971)
ETZOLD, Franz, (1859-1928)
FABIAN, Rudolf, (1912-1943)
FATH, Arthur Earl, (1887-1976)
FELIX, Johannes Paul, (1859-1941)
FIEBELKORN, Max, (1869-1912)
FINCKH, Alfred, (-1914)
FINCKH, Ludwig, (1871-1930)
FINSTERWALDER, Richard, (1899-1963)
FINSTERWALDER, Sebastian, (1862-?1931)
FISCHER, Ernst, (1888-1914)
FISCHER, Hugo, (1864-1926)
FLIEGEL, Gotthard, (1873-1947)
FOLLMANN, Otto, (1856-1926)
FRAAS, Eberhard, (1862-1915)

GERMANY 1900 to present Cont'd.

FRANK, Wilhelm, (1891-1914)
FRANKE, Adolf, (1860-1942)
FRECH, Fritz, (1861-1917)
FRENZEL, Friedrich August, (1842-1902)
FRIEDRICH, Paul, (1856-1918)
FRITSCH, Karl von, Freiherr, (1838-1906)
FUCHS, Alexander, (1874-1935)
FÜRBRINGER, Max, (1846-1920)
GÄBERT, Carl, (-1931)
GAGEL, Kurt, (1865-1927)
GALLWITZ, Hans, (1896-1958)
GATTNAR, Josef, (1854-1928)
GEINITZ, Eugen, (1854-1925)
GEINITZ, Hanns Bruno, (1814-1900)
GEIST, Frank Otto William, (-1963)
GENTNER, Wolfgang, (1906-)
GERLAND, Georg Karl Cornelius, (1833-1919)
GEYER, David, (1855-1932)
GIERTZ, Johannes, (1916-1943)
GILLMAN, Clement, (?1883-1946)
GOLDSCHMIDT, Victor, (1853-1933)
GÖLLER, August, (1878-1965)
GOSSNER, Balthasar, (1877-1937)
GOTHAN, Walther, (1879-1954)
GOTTSCHE, Carl Christian, (1855-1909)
GOTTSCHICK, Franz, (1865-1927)
GREBE, Heinrich, (1831-1903)
GRIPP, Karl, (1891-)
GROEBER, Paul Friedrich Carl, (1885-1968)
GROSS, Walter Robert, (1903-1974)
GROTH, Paul Heinrich von, Ritter, (1843-1927)
GRÜNLING, Friedrich, (1857-1919)
GRUNOW, Albert, (1826-1914)
GÜMBEL, Emil J., (-1966)
GÜRICH, Georg, (1859-1938)
GUTENBERG, Beno, (1889-1960)
HAAS, Hippolyt Julius, (1855-1913)
HÄBERLE, Daniel, (1864-1934)
HAECKEL, Ernst Heinrich Philipp August, (1834-1919)
HAHN, Friedrich Felix (Fritz), (1885-1914)
HANIEL, Curt Alfons, (1884-1914)
HANIEL, Klaus, (1914-1942)
HÄNTZSCHEL, Walter, (1904-1972)
HARBORT, Erich, (1879-1929)
HARRASSOWITZ, Hermann, (-1956)
HASSE, Carl, (1841-1922)
HAUCHECORNE, Heinrich Lambert Wilhelm, (1828-1900)
HAUFF, Bernhard, (1866-1950)
HAUTHAL, Rudolf, (1854-1928)
HEINZELMANN, Karl, (1911-1966)
HEMPEL, Walther Matthias, (1851-1916)
HENNIG, Edwin, (1882-)
HENRICH, Ferdinand, (1837-1915)
HERITSCH, Franz, (1882-1945)

GERMANY 1900 to present Cont'd.

HERRMANN, Fritz Heinrich Ulrich, (1883-1920)
HERZ, Otto, (1853-1905)
HESS, Hans, (1864-1940)
HEYDEN, Lucas von, (1838-1915)
HIBBERT-WARE, Samuel, (1782-1848)
HILGENDORF, Franz, (1839-1904)
HILLER, Wilhelm, (1899-)
HINTERLECHNER, Karl, (1874-1932)
HINTZE, Carl Adolf Ferdinand, (1851-1916)
HIRSCHWALD, Julius, (1845-1928)
HLAUSCHEK, Hans, (1896-1971)
HOERNES, Dieter, (1912-1955)
HOLZAPFEL, Eduard, (1853-1913)
HORN, Hans Lorenz, (1918-1941)
HUENE, Friedrich von, Freiherr, (1875-1969)
HUSSAK, Franz Eugen, (1856-1911)
JABLONSKI, Eugene, (1892-1975)
JAECKEL, Otto, (1863-1929)
JANENSCH, Werner, (1878-1969)
JENTZSCH, Alfred, (1850-1925)
JOHNSEN, Arrien, (1877-1934)
JULIAN, Henry Forbes, (1861-1912) [PROSPECTORS, DIVINERS AND MINING ENGINEERS]
JUNG, Karl, (-?1972)
KAISER, Friedrich Wilhelm Erich, (1871-1934)
KALKOWSKY, Ernst Louis, (1851-1938)
KARL, Franz, (1918-1972)
KARSCH, Ferdinand, (1853-1936)
KAUENHOWEN, Walter, (1900-1942)
KAUFMAN, Rudolf, (1909-ca 1942)
KAYSER, Friedrich Heinrich Emanuel, (1845-1927)
KEILHACK, Konrad, (1858-1944)
KESSLER, Paul, (?1882-1927)
KIESOW, Johann, (1846-1901)
KINKELIN, Georg Friedrich, (1836-1913)
KISSLING, Ernst, (1865-1936)
KLAATSCH, Hermann, (1863-1916)
KLÄHN, Hans, (1884-1933)
KLAUTZSCH, Adolf, (1869-1927)
KLEBER, Will, (-1970)
KLEBS, Richard, (1850-1911)
KLEIN, Johann Friedrich Carl, (1842-1907)
KLEMM, Gustav, (1858-1938)
KLINGER, Fritz Erdmann, (-?1955)
KLOCKMANN, Friedrich Ferdinand Hermann, (1858-1937)
KOCH, Gottlieb Karl David von, (1849-1914)
KOCKEL, Carl Walter, (1898-1966)
KOENEN, Adolf von, (1837-1915)
KOENIGSBERGER, Johann Georg, (1874-1946)
KOENIGSWALD, Gustav Heinrich Ralph von, (1902-)
KOERT, Willi, (1875-1927)
KÖHLER, Alexander, (1893-1955)
KOKEN, Ernst Friedrich Rudolph Karl von, (1860-1912)
KOLBE, Hermann Julius, (1855-)

GERMANY 1900 to present Cont'd.

KOLBE, Robert Wilhelm, (1882-1960)
KOLBECK, Friedrich Ludwig Wilhelm, (1860-1943)
KONTHER, Fritz, (1895-1968)
KORN, Johannes, (1862-1927)
KOZÁRY, Myron Theodore "Mirko", (1918-1966)
KRANTZ, Friedrich Ludwig Robert, (1859-1926)
KRAUS, Ernst, (1889-1970)
KRAUS, Gregor, (1841-1915)
KRÄUSEL, Richard, (1890-1966)
KRUSCH, Paul, (1869-1939)
KUBELLA, Karl, (1922-1973)
KUBIËNA, Walter L., (1897-1970)
KURTZ, Fritz, (1854-1920)
LACHMANN, Richard, (1885-1916)
LANDOIS, Hermann, (1835-1905)
LAUE, Max Theodor Felix von, (1879-1960)
LAUTERBACH, Robert, (1915-)
LEDERSTEGER, Karl, (-1973)
LEHMANN, Johannes Georg, (1851-1925)
LEHMANN, Otto, (1855-1922)
LEINER, Ludwig, (1830-1901)
LEITZ, Ernst, Senior, (1871-1956)
LEMBERG, Johannes, (1842-1902)
LENK, Hans, (1863-1938)
LENZ, Oskar, (1848-1925)
LEONHARD, Richard, (1870-1916)
LEONHARDT, Johannes, (1893-1959)
LEPPLA, August, (1859-1924)
LEPSIUS, Karl Georg Richard, (1851-1915)
LEUTWEIN, Friedrich, (1911-1974)
LEUZE, Johannes, (1883-1915)
LIEBE, Karl Leopold Theodor, (1828-1894)
LIEBISCH, Theodor, (1852-1922)
LIENENKLAUS, Ernst, (1849-1905)
LINCIO, Gabrielle, (1874-1938)
LINGELSHEIM, Alexander von, (1874-1937)
LINSTOW, Otto von, (1872-1929)
LOOMIS, Theodor, (1875-1909)
LORENZ, Theodor, (1875-1909)
LORETZ, Hermann, (1836-1917)
LOTZE, Franz, (1903-1971)
MAACK, Reinhard, (1892-1969)
MAAS, Günther Heinrich Julius Max, (1871-1905)
MAAS, Otto, (1867-1916)
MACHATSCHEK, Fritz, (-1957)
MACHATSCHKI, Felix Karl Ludwig, (1895-1970)
MÄGDEFRAU, Karl, (1907-)
MARGERIE, Emmanuel Marie Pierre Martin Jacquin de, (1862-1953)
MARTINI, Hans Joachim, (1908-1969)
MAUCHER, Wilhelm, (1879-1930)
MEISTER, Ernst, (1887-1939)
MENZEL, Karl Gustav Johannes, (1875-1914)
MENZEL, Paul, (1864-1927)
MERTIN, Hans, (1902-1941)

GERMANY 1900 to present Cont'd.

MEUNIER, Fernand, (1868-1926)
MEYER, Erich, (1874-1915)
MEYER, Hans, (1858-1929)
MICHAEL, Richard, (1869-1925)
MICHELS, Franz-Xaver, (1900-1973)
MILCH, Ludwig, (1867-1928)
MITSCHERLICH, Eilhard, (1794-1863)
MÜGGE, Johannes Otto Konrad, (1858-1932)
MÜLLER, Carl Hermann, (1823-1907)
MÜLLER, Heinrich, (1887-1914)
MÜLLER, Karl Christian Gottfried Ludwig, (1862-1906)
MYLIUS, Hugo, (1876-1918)
NARAMORE, Chester "Chet", (1879-1961)
NEHRING, Alfred, (1845-1904)
NESSIG, Wilhelm Robert, (1861-1932)
NEUHAUS, Alfred, (1903-1975)
NEUMANN, Hans, (1910-1941)
OEBBEKE, Konrad, (1853-1932)
OECHSLE, Richard, (1908-)
OERTEL, Walter, (1889-1924)
OPPENHEIM, Leo Paul, (1863-1934)
ORTH, Albert, (1835-1915)
OSANN, Carl Alfred, (1859-1923)
PAECKELMANN, Werner, (1890-1952)
PALAZZO, Luigi, (1861-1933)
PAPP, Simon, (1886-1970)
PATRUNKY, H., (1854-1930)
PAULCKE, Wilhelm, (1873-1949)
PENCK, Friedrich Carl Albrecht, (1858-1945)
PFANNENSTIEL, Max, (1902-1976)
PHILIPPI, Rudolph Amandus, (1808-1904)
PHLEPS, Otto, (1868-1928)
PIETZCKER, Franz, (1885-1914)
PIETZSCH, Kurt, (1884-1964)
PLAGEMANN, A., (1858-1908)
PLEININGER, Felix, (-?1954)
PLESSMANN, Werner, (1928-1972)
POCKELS, Friedrich Carl Alwin, (1865-1913)
POMPECKJ, Josef Felix, (1867-1930)
PORRIO, Cesare, (1865-1940)
POTONIÉ, Henry Edmond, (1857-1913)
POTONIÉ, Robert, (1889-1974)
PRATJE, Otto, (1890-1925)
PROBST, Joseph, (1823-1905)
PUFAHL, Otto Hermann August, (1855-1924)
QUITZOW, August Alfred Wilhelm, (1881-1914)
RAMDOHR, Paul, (1890-)
RATHSBURG, Alfred, (-)
RAU, Karl, (1870-1966)
RECK, Hans, (1886-1937)
REED, Lyman Coleman, (1900-1970)
REHM, Heinrich, (1828-1916)
REICHARDT, Wolfgang, (1909-1937)
REINACH, Albert von, Baron, (1842-1905)

GERMANY 1900 to present Cont'd.

REINSCH, Paul Friedrich, (1836-1914)
REIS, Otto Maria, (1862-1934)
REMELÉ, Adolf Karl, (1839-1915)
RICHTER, Emma, (1888-1956)
RICHTER, Rudolf, (1881-1957)
RICHTHOFEN, Paul Wilhelm Ferdinand von, Freiherr, (1833-1905)
RIMANN, Eberhard, (1882-1944)
RINNE, Friedrich Wilhelm Berthold, (1863-1933)
RODOLPH, Emil, (1853-1915)
ROGER, Otto, (1841-1915)
ROMANES, James, (1886-1942)
ROMER, Eugeniusz, (1871-1954)
ROSE, Hermann, (1883-1976)
ROSENBUSCH, Karl Harry Ferdinand, (1836-1914)
ROTHPLETZ, August, (1853-1918)
RUDOLPH, Emil, (1853-1915)
RÜST, David, (1831-1916)
RUTLEY, Frank, (1842-1904)
SAUER, Gustav Adolf, (1825-1932)
SCHACKO, Gustav, (1824-1917)
SCHALCH, Ferdinand, (1848-1918)
SCHEIBE, Robert, (1859-1923)
SCHEUMANN, Karl Hermann, (1881-1964)
SCHINDEWOLF, Otto Heinrich, (1896-1972)
SCHLECHTENDAL, Dietrich von, (1834-1916)
SCHLOSSER, Max, (1854-1933)
SCHLUETER, Clemens August, (1835-1906)
SCHLUNCK, Johannes, (1876-1915)
SCHMIDT, Hermann, (1892-1978)
SCHMIDT, Wilhelm Erich, (1882-1944)
SCHMIDT, Wolfgang, (1914-1972)
SCHMÖGER, Carl August, (-1970?)
SCHOENFLIES, Arthur Moritz, (1853-1928)
SCHOETENSACK, Otto, (1850-1912)
SCHOLZ, Erich, (1884-1914)
SCHÖNFELD, Georg, (1878-1926)
SCHOTTLER, Wilhelm, (1869-1932)
SCHRÖDER, Hans Adolf Karl Eckart, (1899-1968)
SCHRÖDER, Richard, (1853-1916)
SCHROEDER, Henry Carl, (1859-1927)
SCHÜLLER, Arno, (1908-1968)
SCHULZ, Karl, (1885-1928)
SCHUMACHER, Eugen, (1851-1922)
SCHUMACHER, Friedrich, (1884-)
SCHUMANN, Karl, (1851-1904)
SCHÜTZE, Ewald, (1873-1908)
SCHWARZBACH, Martin, (-)
SCHWERTSCHLAGER, Joseph, (1853-1924)
SCHWIPPEL, Karl, (1821-1911)
SELIGMANN, Gustav, (1849-1920)
SEMPER, Otto, (1830-1907)
SIEGERT, Leo, (1872-1917)
SIMONS, Herbert Samuel, (1891-1956)
SLATER, George, (-1956)

GERMANY 1900 to present Cont'd.

SOENDEROP, Fritz Hermann Gustav, (1876-1914)
SOLMS-LAUBACH, Hermann zu, Graf, (1842-1915)
SPANDEL, Erich, (1855-1909)
SPANGENBERG, Kurt, (1889-1957)
STAESCHE, Karl, (1902-)
STARK, Peter, (1888-1932)
STEINMANN, Gustav, (1856-1929)
STEINMETZ, Hermann, (1879-1964)
STENZEL, Karl Gustav, (1826-1905)
STERZEL, Johann Traugott, (1841-1914)
STILLE, Hans, (1876-1966)
STOLLER, Jakob, (1873-1930)
STOLLEY, Ernst, (1869-1944)
STOLTZ, Karl, (?1858-1927)
STRASBURGER, Eduard, (1844-1912)
STROBEL, Ernst Gustav, (1903-1974)
STÜBEL, Alfons, (-?1904)
STUTZER, Otto, (1881-1936)
TANNHÄUSER, Felix, (1874-1924)
TEICHERT, Curt, (-)
THIENHAUS, Rolf, (1913-1968)
THOMSON, Paul William, (1891-1957)
TIETZE, Oskar Adolf Eduard, (1874-1920)
TÖRNQUIST, Sven Leonhard, (1840-1920)
TRIEBEL, Erich, (1894-1971)
TRÖGER, Walter Ehrenreich, (1901-1963)
TROLL, Carl, (1899-)
TRÜMPY, Rudolf D., (-)
UHLIG, Carl Ludwig Gustav, (1872-?1938)
UHLIG, Johannes, (1883-1918)
VAN'T HOFF, Jacobus Hendricus, (1852-1911)
VATER, Heinrich, (1859-1930)
VELIKOVSKY, Immanuel, (1895-)
VIOLA, Carlo Maria, (1855-1925)
VOELKEL, Josef, (1828-1906)
VOIGT, Waldemar, (1850-1919)
VOLLRATH, Paul, (-1929)
WAGNER, Georg, (1885-1972)
WAHNSCHAFFE, Gustav Albert Bruno Felix, (1851-1914)
WALTHER, Johannes, (1860-1937)
WANNER, Johannes, (1878-1957)
WASMUND, Erich, (1902-1945)
WEBER, Carl Albert, (1856-1931)
WEGENER, Alfred Lothar, (1880-1930)
WEGNER, Theodor, (1880-1934)
WEIGELT, Johannes, (1890-1948)
WEILER, Wilhelm, (1890-1972)
WEINSCHENK, Ernst Heinrich Oskar Kasimir, (1865-1921)
WEISBACH, Albin, (1833-1901)
WEISE, Ernst, (1843-1941)
WELTNER, Wilhelm, (1854-1917)
WEPFER, Emil, (1882-1930)
WETZEL, Otto Christian August, (1891-1971)
WETZEL, Walter, (1887-)

GERMANY 1900 to present Cont'd.

WEYLAND, Hermann, (1888-)
WICHER, Carl A., (1901-1957)
WIENHOLZ, Eva, (1930-1969)
WILCKENS, Otto Rudolf, (1876-1943)
WINDHAUSEN, Anselm, (1882-1932)
WINKLER, Clemens Alexander, (1838-1904)
WOLDSTEDT, Paul, (1888-1973)
WOLF, Theodor, (1841-1924)
WOLFF, Wilhelm, (1872-1951)
WROOST, Volkmar, (1912-1936)
WÜLFING, Ernst Anton, (1860-1930)
WUNDERLICH, Hans-Georg, (1928-1974)
WUNDT, Georg von, (1845-1929)
WÜST, Ewald, (1875-1934)
ZEISE, Oskar, (1860-1925)
ZIETZ, Amandus Christian Heinrich, (1839-1921)
ZIMMERMANN, Ernst, (1860-1944)
ZIRKEL, Ferdinand, (1838-1912)
ZITTEL, Karl Alfred von, (1839-1904)
ZÖLLICH, Martin Sylvester, (1913-1940)

Time unknown
DACQUÉ, Edgar, (-)
KARSTEN, Carl Johann, (-)

GHANA

1850-1900
BEEBY-THOMPSON, Arthur, (1873-1968)
MACLAREN, James Malcolm, (1873-1935)
MARKS, Edward Seaborn, (?1875-1950)
McCARTHY, Edward Thomas, (-) [PROSPECTORS, DIVINERS AND MINING ENGINEERS]
TEALE, Edmund Oswald, Sir, (1874-1971)
WAY, Herbert W. L., (-) [PROSPECTORS, DIVINERS AND MINING ENGINEERS]

1900 to present
BEEBY-THOMPSON, Arthur, (1873-1968)
BILLINGS, Katharine Stevens (nee Fowler-Lunn), (1902-)
BISHOPP, Douglas Wallace, (1900-1977)
FURON, Raymond, (1898-)
HALL, Thomas Clifford Fitzwilliam, (1886-1956)
HUGHES, William John, (?1899-1961)
JUNNER, Norman Ross, (1891-1973)
LOBJOIT, William Michael, (1932-1967)
MACLAREN, James Malcolm, (1873-1935)
MARKS, Edward Seaborn, (?1875-1950)
McCARTHY, Edward Thomas, (-) [PROSPECTORS, DIVINERS AND MINING ENGINEERS]
OATES, Frank, (1890-1945)
SLAWSON, Chester Baker, (1898-1964)
TEALE, Edmund Oswald, Sir, (1874-1971)
WILD, Robert Powley, (1882-1963)
WILMOTH, Leslie James, (1881-1963)

GIBRALTAR

1800-1849

SMITH, James, (1782-1867)

1850-1899

SMITH, James, (1782-1867)

1900 to present

ACLAND, Henry Dyke, (-1936)
BRAMWYCH, Michael, (1911-1974)

GOLD COAST *See* GHANA.

GREAT BRITAIN *See* CHANNEL ISLANDS; ENGLAND; ISLE OF MAN; SCOTLAND; WALES.

GREECE

Up to 1800

AGATHARCHIDES of Cnidus, (170-100 B.C.)
ARISTOTLE, (384-322 B.C.)
MORO, Anton Lazzaro, (1687-1740)
PAUSANIAS, (ca.130-180 A.D.)
PYTHAGORAS, (fl.532 B.C.)
SICULUS, Diodorus, (ca.85-21 B.C.)
STRABO, (63B.C.-23 A.D.?)
THEOPHRASTUS, (371-ca.286 B.C.)

1800-1849

DARWIN, Francis Sacheverell, Sir, (1786-1859)
DESHAYES, Gérard Paul, (1796-1875)
FOUQUÉ, Ferdinand André, (1828-1904)
GOBANZ, Alexander, (1826-1895)
HAMILTON, William John, (1805-1867)
PORTLOCK, Joseph Ellison, Major-General, (1794-1864)
RAULIN, Félix Victor, (1815-1905)
RÖMER, Hermann, (1816-1894)
STRICKLAND, Hugh Edwin, (1811-1853)
TREVELYAN, Walter Calverley, Sir, (1797-1879)
VIQUESNEL, Auguste, (1800-1867)

1850-1899

BATE, Dorothea Minola Alice, (1879-1951)
BEEBY-THOMPSON, Arthur, (1873-1968)
BROWN, Barnum, (1873-1963)
BUKOWSKI, Gejza, (1858-1937)
BULLEN, Robert Ashington, Reverend, (1850-1912)
CAYEUX, Lucien, (1864-1944)
COLLINS, Henry Francis, (?1864-1924)
CVICIĆ, Jovan, (1865-1927)
DARWIN, Francis Sacheverell, Sir, (1786-1859)
DESHAYES, Gérard Paul, (1796-1875)
FOOTE, Albert E., (1846-1895)
FOUQUÉ, Ferdinand André, (1828-1904)

GREECE 1850-1899 Cont'd.

FRIEDLAENDER, Immanuel, (-?1918)
FUCHS, Philippe Jacques Edmond, (1837-1889)
GOBANZ, Alexander, (1826-1895)
HAMILTON, William John, (1805-1867)
JENKINS, Henry Michael, (1841-1886)
KOSSMAT, Franz, (1871-1938)
MARTELLI, Alessandro, (1877-1934)
PORTLOCK, Joseph Ellison, Major-General, (1794-1864)
RAULIN, Félix Victor, (1815-1905)
RÖMER, Hermann, (1816-1894)
SCHLOSSER, Max, (1854-1933)
SEEBACH, Karl Albert Ludwig von, (1839-1880)
STRICKLAND, Hugh Edwin, (1811-1853)
TREVELYAN, Walter Calverley, Sir, (1797-1879)
VIQUESNEL, Auguste, (1800-1867)

1900 to present
AUBOUIN, Jean, (1928-)
BATE, Dorothea Minola Alice, (1879-1951)
BEEBY-THOMPSON, Arthur, (1873-1968)
BELTZ, Edward W., (1891-1970)
BIANCHI, Angelo, (1892-1970)
BOURCART, Jacques, (1891-1965)
BROWN, Barnum, (1873-1963)
BUKOWSKI, Gejza, (1858-1937)
BULLEN, Robert Ashington, Reverend, (1850-1912)
CARLSON, Charles Gordon, (1895-1977)
CAYEUX, Lucien, (1864-1944)
COLLINS, Henry Francis, (?1864-1924)
CVICIĆ, Jovan, (1865-1927)
DERRY, Duncan Ramsay, (?1908-)
DIXON, Henry George Dacres, (1887-1947)
FOUQUÉ, Ferdinand André, (1828-1904)
FRIEDLAENDER, Immanuel, (-?1918)
HUXLEY, Julian Sorell, Sir, (1887-1975)
JURKOVIĆ, Ivan, (1917-)
KOSSMAT, Franz, (1871-1938)
LAPPARENT, Jacques de, (-?1949)
MARTELLI, Alessandro, (1877-1934)
MITZOPOULOS, Maximos K., (1896-1968)
PFANNENSTIEL, Max, (1902-1976)
POND, Walter Franklin, (1885-1974)
RAULIN, Félix Victor, (1815-1905)
RECK, Hans, (1886-1937)
ROCH, Edouard, (1901-)
SCHLOSSER, Max, (1854-1933)
SMITH, G. Wendell, (1917-1974)
TRÜMPY, Daniel, (1893-1971)

GREENLAND
Up to 1800
GIESECKE, Charles Lewis, Sir, (né Metzle, Karl Ludwig), (1761-1833)
PINGEL, Christian, (1793-1852)

GREENLAND Cont'd.

1800-1849

GIESECKE, Charles Lewis, Sir, (né Metzle, Karl Ludwig), (1761-1833)
JOHNSTRUP, Johannes Frederik, (1818-1894)
KANE, Elisha Kent, (1820-1857)
PINGEL, Christian, (1793-1852)
TORELL, Otto Martin, (1828-1900)

1850-1899

BACKLUND, Helge Götrik, (1878-1958)
BALDAUF, Richard Julius, (1848-1931)
BEUST, Fritz von, (1856-1908)
CHAMBERLIN, Thomas Chrowder, (1843-1928)
CHANCOURTOIS, A. E. Béguyer de, (-1886?)
EMERSON, Benjamin Kendall, (1843-1932)
HARTZ, Nikolaj Eeg Kruse, (1867-1937)
HOBBS, William Herbert, (1864-1953)
JOHNSTRUP, Johannes Frederik, (1818-1894)
KANE, Elisha Kent, (1820-1857)
KINDLE, Edward Martin, (1869-1940)
KJELLSTRÖM, Carl Johan Otto, (1855-1913)
KORNERUP, Andreas, (?1856-1880)
NATHORST, Alfred Gabriel, (1850-1921)
NORDENSKIÖLD, Nils Otto Gustaf, (1869-1928)
PAYER, Julius von, Ritter, (1841-1912)
PINGEL, Christian, (1793-1852)
RAVN, Jesper Peter Johansen, (1866-1951)
SALISBURY, Rollin D., (1858-1922)
SEWARD, Albert Charles, Sir, (1863-1941)
STEENSTRUP, Knud Johannes Vogelius, (1842-1913)
SWINNERTON, Henry Hurd, (1875-1967)
TOOTHAKER, Charles Robinson, (1873-1952)
TORELL, Otto Martin, (1828-1900)
UPHAM, Warren, (1850-1934)
USSING, Niels Viggo, (1864-1911)
WHYMPER, Edward, (1840-1911)
WRIGHT, George Frederick, (1838-1921)

1900 to present

BACKLUND, Helge Götrik, (1878-1958)
BALDAUF, Richard Julius, (1848-1931)
BATEMAN, John Danvers, (1910-1965)
BATTLE, Walter Ravenhill Brown "Ben", (1919-1953)
BESCHEL, Roland, (1928-1971)
BEUST, Fritz von, (1856-1908)
BØGVAD, Richard, (1897-1952)
BROWN, Peter Evans, (-)
CHAMBERLIN, Thomas Chrowder, (1843-1928)
DERRY, Duncan Ramsay, (?1908-)
EDWARDS, Wilfred Norman, (1890-1956)
EMERSON, Benjamin Kendall, (1843-1932)
FUCHS, Vivian Ernest, Sir, (1908-)
GORDON, Samuel George, (1897-1952)
HARRY, William Trevelyan, (?1922-1964)

GREENLAND 1900 to present Cont'd.

HARTZ, Nikolaj Eeg Kruse, (1867-1937)
HOBBS, William Herbert, (1864-1953)
KINDLE, Edward Martin, (1869-1940)
KJELLSTRÖM, Carl Johan Otto, (1855-1913)
KLEIBER, Karl Ernst, (1912-1972)
KOCH, Lauge, (1892-1964)
LOEWE, Fritz, (1895-1974)
MADSEN, Marius Kristian, (1899-) [PROSPECTORS, DIVINERS AND MINING ENGINEERS]
MILTHERS, Keld, (1907-1960)
NATHORST, Alfred Gabriel, (1850-1921)
NORDENSKIÖLD, Nils Otto Gustaf, (1869-1928)
NØRVANG, Aksel, (1914-1968)
PARGETER, Reginald Clive, (1919-1972)
PAYER, Julius von, Ritter, (1841-1912)
POULSEN, Christian, (1896-1975)
RAVIER, Jean, (1922-)
RAVN, Jesper Peter Johansen, (1866-1951)
REECE, Alan W., (1921-1960)
RENAUD, André, (1904-1964)
ROSSBY, Carl-Gustaf Arvid, (1898-1957)
SALISBURY, Rollin D., (1858-1922)
SÄVE-SÖDERBERGH, Gunnar, (1910-1948)
SCHAUB, Hans Peter, (1912-1965)
SCHYTT, Valter, (1919-)
SEWARD, Albert Charles, Sir, (1863-1941)
SPATH, Leonard Frank, (1882-1957)
STEENSTRUP, Knud Johannes Vogelius, (1842-1913)
STEWART, Duncan, (1905-1969)
SWINNERTON, Henry Hurd, (1875-1967)
TOOTHAKER, Charles Robinson, (1873-1952)
TRÜMPY, Rudolf D., (-)
UPHAM, Warren, (1850-1934)
USSING, Niels Viggo, (1864-1911)
VOGT, Thorolf, (1888-1958)
WAGER, Lawrence Rickard, (1904-1965)
WALKER, Paul, (1934-1959)
WEGENER, Alfred Lothar, (1880-1930)
WHYMPER, Edward, (1840-1911)
WOODS, Ethel Gertrude, (-1939)
WORDIE, James Mann, Sir, (?1890-1962)
WRIGHT, George Frederick, (1838-1921)

GUADELOUPE (WEST INDIES)
1850-1899
HOVEY, Edmund Otis, (1862-1924)

1900 to present
HOVEY, Edmund Otis, (1862-1924)

GUAM (PACIFIC OCEAN)
1900 to present
PICCARD, Jacques Ernst, (1922-)

GUATEMALA

1850-1899

BERGEAT, Alfred, Edmund, (1866-1924)
BROWN, Barnum, (1873-1963)
DOLLFUS, Auguste, (1840-1869)
HAGUE, Arnold, (1840-1917)
HAMMOND, John Hays, (1855-1936)
NAGYSUR, Hugo Böckh von, (1874-1931)
ORDOÑEZ, Ezequiel, (1867-1950)
SAPPER, Karl Theodor, (1866-1945)

1900 to present

ANDERSON, Robert Lee, (1912-1959)
BERGEAT, Alfred, Edmund, (1866-1924)
BOURCHIER, John Rickman, (1896-1927)
BROWN, Barnum, (1873-1963)
CARR, William L., (1919-1975)
CLARK, Chester Charles, (1885-1953)
COOKE, Charles Wythe, (1887-1971)
CRAWFORD, David Johnston, (1899-1960)
CRONEIS, Carey, (1901-1972)
DECKER, Charles LaVerne, (1898-1955)
HAGUE, Arnold, (1840-1917)
HAMMOND, John Hays, (1855-1936)
NAGYSUR, Hugo Böckh von, (1874-1931)
ORDOÑEZ, Ezequiel, (1867-1950)
PETSCH, Arthur Henry, (1895-1969)
SAPPER, Karl Theodor, (1866-1945)
SCHOLL, Guy Joseph, (1900-1968)
WADELL, Hakon, (1895-1962)

GUERNSEY See CHANNEL ISLANDS.

GUINEA

1850-1899

CHUDEAU, René, (1864-1921)
LACROIX, François Antoine Alfred, (1863-1948)

1900 to present

BASSINGER, William E., (1928-1977)
CHUDEAU, René, (1864-1921)
LACROIX, François Antoine Alfred, (1863-1948)
LAPPARENT, Jacques de, (-?1949)
MALAVOY, Jean, (1903-1945)
MILLOT, Georges, (1917-)
PERCIVAL, Frederick George, (1887-1973)

GUYANA

1800-1849

WATERTON, Charles, (1782-1865)

1850-1899

ARZRUNI, Andreas, (1847-1898)
BEEBY-THOMPSON, Arthur, (1873-1968)
BROWN, Charles Barrington, (1839-1917)
FRANKS, Godfrey Firth, (1853-1917)

GUYANA 1850-1899 Cont'd.

HARRISON, John Burchmore, Sir, (1856-1926)
HEILPRIN, Angelo, (1853-1907)
TATE, Ralph, (1840-1901)
WATERTON, Charles, (1782-1865)

1900 to present

BEEBY-THOMPSON, Arthur, (1873-1968)
BISHOPP, Douglas Wallace, (1900-1977)
BROWN, Charles Barrington, (1839-1917)
CARLSON, Charles Gordon, (1895-1977)
DERRY, Duncan Ramsay, (?1908-)
FRANKS, Godfrey Firth, (1853-1917)
GRANTHAM, Donald Robert, (1896-1969)
HARRISON, John Burchmore, Sir, (1856-1926)
HEILPRIN, Angelo, (1853-1907)
LEHNER, Ernst J., (1895-1968)
POLLARD, Ernest Rex, (1922-1966)
RICHARDSON, Gwen, (-) [PROSPECTORS, DIVINERS AND MINING ENGINEERS]
ROMANES, James, (1886-1942)
TATE, Ralph, (1840-1901)

HAITI

1850-1899

BERGT, Walther Adolf, (1864-1941)
DICKERSON, Roy Ernest, (1877-1944)
GUPPY, Robert John Lechmere, (1836-1916)

1900 to present

BEEKLEY, Albert Leon, (1883-1952)
BERGT, Walther Adolf, (1864-1941)
CLAPP, Frederick Gardner, (1879-1944)
DICKERSON, Roy Ernest, (1877-1944)
GARRETSON, Mary Welleck, (1896-1971)
GUPPY, Robert John Lechmere, (1836-1916)
JONES, William Francis, (1885-1941)
LEHNER, Ernst J., (1895-1968)
ROCH, Edouard, (1901-)

HAWAII (NORTH PACIFIC OCEAN) See UNITED STATES OF AMERICA (HAWAII).

HOLLAND See NETHERLANDS.

HONDURAS

1850-1899

DITMARS, Raymond Lee, (1876-1942)

1900 to present

DECKER, Charles LaVerne, (1898-1955)
DITMARS, Raymond Lee, (1876-1942)
HARRISON, John Vernon, (1892-1972)
MACKENZIE, Andrew Nicholls "Kelley", (1888-1951)
SOUTHER, John Barron, (1894-1964)
TEMPLETON, Eugene Crittenden, (1889-1932)
WALLACE, Maurice Hudiburg, (1917-1961)

HONG KONG

1850-1899

HANNAY, Harry Campbell, (1874-1961)
KORNITZER, Louis, (1873-) [PROSPECTORS, DIVINERS AND MINING ENGINEERS]

1900 to present

BROCK, Byron Britton, (1904-1972)
HANNAY, Harry Campbell, (1874-1961)
KORNITZER, Louis, (1873-) [PROSPECTORS, DIVINERS AND MINING ENGINEERS]
UGLOW, William Lawrence, (1884-1926)
WALKER, Lucian Hickman, (1894-1970)

HUNGARY

Up to 1800

BENKÖ, Ferencz, (fl.1786)
BORN, Ignaz von, Baron, (1742-1791)
DUHAMEL, Jean Pierre François Guillaume, (1730-1816)
ESMARCH, Jens, (1762-1839)
FERBER, Johann Jacob, (1743-1790)
FICHTEL, Johann Ehrenreich von, (1732-1795)

HUNGARY Up to 1800 Cont'd.

FRIDVALSKY, János, (1730-1784)
STASZIC, Stanislaw, (1755-1826)
TAVERNIER, Jean Baptiste, baron d'Aubonne, (1605-1689)
TELEKI, Domokos, Graf, (-1803)

1800-1849

BRUIMANN, Vilmos, (1819-?1896)
COTTA, Carl Bernhard von, (1808-1879)
DAUBENY, Charles Giles Brindle, (1795-1867)
DUHAMEL, Jean Pierre François Guillaume, (1730-1816)
ESMARCH, Jens, (1762-1839)
ESZTERHÁZY, Miklós, Herczeg, (1816-1894)
ETTINGSHAUSEN, Constantin von, Freiherr, (1826-1897)
GRUNOW, Albert, (1826-1914)
HANTKEN, Miksa von Prudnik, (1821-1893)
HAUER, Franz von, Ritter, (1822-1899)
HAZSLINSZKY, Hazslini Frigyes Ágost, (1818-1896)
HECKEL, Johann Jakob, (1790-1857)
HERBICH, Franz, (1821-1887)
KUDERNATSCH, Johann, (1819-1856)
NENDTVICH, Károly, (1811-?1892)
NEUGEBOREN, Johann Ludwig, (1806-1887)
PARTSCH, Paul, (1791-1856)
PETERS, Karl, (1825-1881)
POŠEPNÝ, František, (1836-1895)
PRUGBERGER, József, (1818-1889)
RÓMER, Flóris, (1815-?1889)
RÓNAY, Jácint, Bishop, (1814-1889)
SCHMIDL, Adolph A., (1802-1863)
STASZIC, Stanislaw, (1755-1826)
STUR, Dénes, (1827-1893)
TELEKI, Domokos, Graf, (-1803)
XÁNTUS, János, (1825-1894)
ZIPSER, Christian Andreas, (1783-1864)
ZSIGMONDY, Vilmos, (1821-1888)

1850-1899

ADDA, Kálmán, (1862-1901)
BRUIMANN, Vilmos, (1819-?1896)
BUKOWSKI, Gejza, (1858-1937)
COTTA, Carl Bernhard von, (1808-1879)
DAUBENY, Charles Giles Brindle, (1795-1867)
DU BOIS, Georges C., (1874-)
EMSZT, Kálmán, (1873-1957)
EÖTVÖS, Lorand, (1848-1919)
ESZTERHÁZY, Miklós, Herczeg, (1816-1894)
ETTINGSHAUSEN, Constantin von, Freiherr, (1826-1897)
FRANZENAU, Agoston, (1856-1919)
FUCHS, Philippe Jacques Edmond, (1837-1889)
FUCHS, Theodor, (1842-1925)
GAÁL, István, (1877-1957)
GERBER, Frigyes, (1856-1897)
GRUNOW, Albert, (1826-1914)
GÜLL, Vilmos, (1876-1909)

HUNGARY 1850-1899 Cont'd.

HALAVÁTS, Gyula, (1853-1926)
HANTKEN, Miksa von Prudnik, (1821-1893)
HANUSZ, István, (1840-1909)
HAUER, Franz von, Ritter, (1822-1899)
HAZSLINSZKY, Hazslini Frigyes Ágost, (1818-1896)
HECKEL, Johann Jakob, (1790-1857)
HERBICH, Franz, (1821-1887)
HOFMANN, Károly, (1839-1891)
HOLLENDONNER, Ferenc, (1882-1935)
HORUSITZKY, Henrik, (1870-1945)
ILOSVAY, Lajos, (1851-1937)
INKEY BÉLA, Palini, (1847-1921)
KADIČ, Ottokár, (1876-1957)
KALECSINSZKY, Sandór, (1857-1911)
KOCH, Antal von Bodrog, (1843-1927)
KOECHLIN, Rudolf Ignatz, (1862-1939)
KRENNER, József Sándor, (1839-1920)
KUDERNATSCH, Johann, (1819-1856)
LIFFA, Aurél, (1872-1957)
LINGELSHEIM, Alexander von, (1874-1937)
LÓCZY, Lajos, (1849-1920)
LŐRENTHEY, Imre, (1867-1917)
MELCZER, Gusztáv, (1869-1907)
NAGYSUR, Hugo Böckh von, (1874-1931)
NAGYSUR, János Böckh von, (1840-1909)
NENDTVICH, Károly, (1811-?1892)
NEUGEBOREN, Johann Ludwig, (1806-1887)
NOPCSA VON FELSŐ-SZILVÁS, Ferenc, baron, (1877-1933)
PAPP, Károly, (1873-1963)
PARTSCH, Paul, (1791-1856)
PAUL, Carl Maria, (1838-1900)
PERGENS, Eduard Willem Gerard, (1862-1917)
PETERS, Karl, (1825-1881)
PETHŐ, Gyula, (1848-1902)
PŐSEPNÝ, František, (1836-1895)
POSEWITZ, Tivadar, (1850-1917)
PRIMICS, György, (1849-1893)
PRUGBERGER, József, (1818-1889)
RÓMER, Flóris, (1815-?1889)
RÓNAY, Jácint, Bishop, (1814-1889)
ROTH, Samu, (?1854-?1888)
SAJÓHELYI, Frigyes, (1848-1940)
SCHAFARZIK, Ferenc, (1854-1927)
SCHMIDL, Adolph A., (1802-1863)
SCHMIDT, Sándor, (1855-?1905)
SEMSEY UR SEMSEI, Andor, (-1923)
SIGMOND, Elek, (1873-1939)
STAUB, Moric, (1842-1904)
STUR, Dénes, (1827-1893)
STŰRZENBAUM, József, (1845-1881)
SZÁDECZKY, Gyula, Baron Szádeczky-Kardoss of Szádecsne and Kafdosfalva, (1860-1935)
SZONTAGH, Tamás, (1851-1937)
SZTERÉNYI, Hugo, (1857-1909)

HUNGARY 1850-1899 Cont'd.

TRAXLER, László, (1868-1898)
TREITZ, Péter, (1866-1935)
VITÁLIS, István, (1871-1947)
VIZER, Vilmos, (1873-1941)
WÄHNER, Franz, (1856-1932)
WEINBERGER, Isidor, (1838-1915)
XÁNTUS, János, (1825-1894)
ZIMÁNYI, Károly, (1862-1941)
ZIPSER, Christian Andreas, (1783-1864)
ZSIGMONDY, Vilmos, (1821-1888)

1900 to present
ANDRÉANSZKY, Gábor, (1895-1967)
BALLENEGGER, Róbert, (1882-1969)
BALOGH, Ernö, (1882-1969)
BÁNYAI, János, (1886-1971)
BEM-BÖHM, Boleszláv, (1900-1964)
BEUDANT, François-Sulpice, (1787-1850)
BÖHM, Ferenc, (1881-1940)
BUKOWSKI, Gejza, (1858-1937)
CSAJÁGHY, Gábor, (1903-1972)
DARÁNYI, Ignác, (-1925)
DARNAY-DORNYAY, Béla, (1887-1965)
DU BOIS, Georges C., (1874-)
DUDICH, Endre, (1895-1971)
EMSZT, Kálmán, (1873-1957)
EÖTVÖS, Lorand, (1848-1919)
FEJÉRVÁRY, Géza Gyula Imre, Baron von Komlóskeresztes, (1894-1932)
FÖLDVÁRI, Aladár, (1906-1973)
FRANZENAU, Agoston, (1856-1919)
FUCHS, Theodor, (1842-1925)
GAÁL, István, (1877-1957)
GEDEON, Tihamér, (1898-1967)
GERBER, Frigyes, (1856-1897)
GROSSZ, Ádám, (1925-1973)
GRUNOW, Albert, (1826-1914)
GÜLL, Vilmos, (1876-1909)
HALAVÁTS, Gyula, (1853-1926)
HANUSZ, István, (1840-1909)
HERITSCH, Franz, (1882-1945)
HERRMANN, Margit, (-?1957)
HOLLENDONNER, Ferenc, (1882-1935)
HORUSITZKY, Ferenc, (1901-1971)
HORUSITZKY, Henrik, (1870-1945)
ILOSVAY, Lajos, (1851-1937)
INKEY BÉLA, Palini, (1847-1921)
KADIČ, Ottokár, (1876-1957)
KALECSINSZKY, Sandór, (1857-1911)
KERTAI, György, (1912-1968)
KOCH, Antal von Bodrog, (1843-1927)
KOCH, Sándor, (1906-)
KOECHLIN, Rudolf Ignatz, (1862-1939)
KOLOSVÁRY, Gábor, (1901-1968)
KORMOS, Tivadar, (1881-1948)

HUNGARY 1900 to present Cont'd.

KOZÁRY, Myron Theodore "Mirko", (1918-1966)
KRENNER, József Sándor, (1839-1920)
KULHAY, Gyula, (1910-1945)
KUTASSY, Endre, (1898-1938)
LACHMANN, Richard, (1885-1916)
LAMBRECHT, Kálmán, (1889-1936)
LIFFA, Aurél, (1872-1957)
LINGELSHEIM, Alexander von, (1874-1937)
LÓCZY, Lajos, (1849-1920)
LÖRENTHEY, Imre, (1867-1917)
LÖW, Márton, (1885-1943)
MAJZON, László, (1904-1973)
MAROS, Imre, (1882-1937)
MAURITZ, Béla, (1881-1971)
MELCZER, Gusztáv, (1869-1907)
MIHÁLTZ, István, (1897-1964)
NAGYSUR, Hugo Böckh von, (1874-1931)
NAGYSUR, János Böckh von, (1840-1909)
NÁRAY-SZABÓ, István, (1899-1972)
NOPCSA VON FELSÖ-SZILVÁS, Ferenc, baron, (1877-1933)
NORDEN, John Alexander Erdelyi-Fazekas, (né Erdélyi-Fazekas), (1912-1977)
NOSZKY, Jenö, Sr., (1880-1951)
NOSZKY, Jenö Jr., (1909-1970)
PANTÓ, Gábor, (1917-1972)
PAPP, Ferenc, (1901-1969)
PAPP, Károly, (1873-1963)
PAPP, Simon, (1886-1970)
PÁVAI, Vajna Ferenc, (1886-1964)
PERGENS, Eduard Willem Gerard, (1862-1917)
PETHÖ, Gyula, (1848-1902)
POSEWITZ, Tivadar, (1850-1917)
RAKUSZ, Gyula, (1896-1932)
REICHERT, Robert, (1901-1937)
ROZLOZSNIK, Pál, (1880-1940)
SAJÓHELYI, Frigyes, (1848-1940)
SCHAFARZIK, Ferenc, (1854-1927)
SCHEFFER, Viktor, (1900-1966)
SCHERF, Emil, (1889-1967)
SCHMIDT, Robert Eligius, (1902-1973)
SCHMIDT, Sándor, (1855-?1905)
SCHRÉTER, Zóltan, (1882-1970)
SEMSEY UR SEMSEI, Andor, (-1923)
SIGMOND, Elek, (1873-1939)
STAFF, Hans von, (1883-1915)
STAUB, Moric, (1842-1904)
SÜMEGHY, József, (1892-1955)
SZABÓ, Pál Zóltán, (1901-1965)
SZÁDECZKY, Gyula, Baron Szádeczky-Kardoss of Szádecsne and Kafdosfalva, (1860-1935)
SZONTAGH, Tamás, (1851-1937)
SZTERÉNYI, Hugo, (1857-1909)
TAEGER, Henrik, (1881-1939)
TASNÁDI-KUBACSKA, András, (1902-1977)

HUNGARY 1900 to present Cont'd.

TELEGDI ROTH, Károly, (1886-?1956)
TEMPLETON, James Clark, (1898-1950)
TOBORFFY, Zoltán, (1883-1927)
TOKODY, László, (1893-1964)
TÖRÖK, Zoltán, (1893-1963)
TREITZ, Péter, (1866-1935)
VADÁSZ, Elemér, (1885-1970)
VENDEL, Aladár, (1886-1971)
VENDL, Maria, (1890-1945)
VERTÉS, László, (1914-1968)
VIDACS, Aladár, (1909-?1965)
VÍGH, Gyula, (1889-1958)
VITÁLIS, István, (1871-1947)
VIZER, Vilmos, (1873-1941)
WÄHNER, Franz, (1856-1932)
WEINBERGER, Isidor, (1838-1915)
WIEDENMAYER, Carl, (1897-1951)
ZIMÁNYI, Károly, (1862-1941)

ICELAND

Up to 1800

BANKS, Sir Joseph, (1743-1820)
BARTHOLÍNUS, Erasmus, (1625-1698)
SOLANDER, Daniel Carl, (1733-1782)
TROIL, Uno von, (1746-1803)

1800-1849

BANKS, Sir Joseph, (1743-1820)
BARROW, John, Jr., (1808-1898)
BRYSON, Alexander, (1816-1866)
BUNSEN, Robert Wilhelm, (1811-1899)
GRESSLY, Amanz, (1814-1865)
HENDERSON, Ebenezer, (1784-1858)
HOOKER, William Jackson, Sir, (1785-1865)
JOHNSTRUP, Johannes Frederik, (1818-1894)
MENGE, Johann, (-)
STEENSTRUP, Johannes Japetus Smith, (1813-1897)
TORELL, Otto Martin, (1828-1900)
TROIL, Uno von, (1746-1803)

1850-1899

BALDAUF, Richard Julius, (1848-1931)
BARROW, John, Jr., (1808-1898)
BREWER, William Henry, (1828-1910)
BRYSON, Alexander, (1816-1866)
BUNSEN, Robert Wilhelm, (1811-1899)
CHANCOURTOIS, A. E. Béguyer de, (-1886?)
FLEURY, Ernest, (1878-1958)
FOX-STRANGWAYS, Charles Edward, (1844-1910)
GRESSLY, Amanz, (1814-1865)
HENDERSON, Ebenezer, (1784-1858)
HOOKER, William Jackson, Sir, (1785-1865)
JOHNSTRUP, Johannes Frederik, (1818-1894)
STEENSTRUP, Johannes Japetus Smith, (1813-1897)
THORODDSEN, Thorvaldur, (1855-1921)
TORELL, Otto Martin, (1828-1900)
ZIRKEL, Ferdinand, (1838-1912)

1900 to present

BALDAUF, Richard Julius, (1848-1931)
BÅRDHARSON, Gudhmundur G., (1880-1933)
BARTH, Thomas Frederik Weiby, (1899-1971)
BREWER, William Henry, (1828-1910)
EYTHORSSON, Jon, (1895-1968)
FLEURY, Ernest, (1878-1958)
FOX-STRANGWAYS, Charles Edward, (1844-1910)
GOETEL, Walery, (1889-1972)
HANNESSON, Pálmi, (1898-1956)
HENSON, Frederick Arthur, (1919-1957)
KOMOROWICZ, Maurycy, (1881-1923)
PÅLSSON, Sveinn, (-)
PEACOCK, Martin Alfred, (1898-1950)
RECK, Hans, (1886-1937)
RUTTEN, Martin Gerard, (1910-1970)
SCHWARZBACH, Martin, (-)

ICELAND 1900 to present Cont'd.

THORODDSEN, Thorvaldur, (1855-1921)
TODTMANN, Emmy Mercedes, (1888-1973)
TRYGGVASON, Tómas, (1907-1965)
TYRRELL, George Walter, (1883-1961)
WADELL, Hakon, (1895-1962)
ZIRKEL, Ferdinand, (1838-1912)

IFNI See MOROCCO.

INDIA

Up to 1800

DINWIDDIE, James, (-)
GREENOUGH, George Bellas, (1778-1855)
RENNELL, James, Major, (1742-1830)
TAVERNIER, Jean Baptiste, baron d'Aubonne, (1605-1689)

1800-1849

CARTER, Henry John, (1813-1895)
CAUTLEY, Proby T., Colonel Sir, (1802-1871)
CLARK, George Thomas, (1811-1898)
DINWIDDIE, James, (-)
DREW, Frederick, (1836-1891)
ETHERIDGE, Robert, (1819-1903)
FALCONER, Hugh, (1808-1865)
FERGUSSON, James, (1808-1885)
GREENOUGH, George Bellas, (1778-1855)
HENWOOD, William Jory, (1805-1875)
HISLOP, Stephen, Reverend, (1817-1863)
HOOKER, Joseph Dalton, Sir, (1817-1911)
HUNTER, Robert, Reverend, (1823-1897)
KAYE, Charles Turton, (1812-1846)
LAING, Samuel, (1810-1897)
LOFTUS, William Kennett, (1822-1859)
MEDLICOTT, Henry Benedict, (1829-1905)
OLDHAM, Thomas, (1816-1878)
RENNELL, James, Major, (1742-1830)
ROYLE, John Forbes, (-1857?)
SALTER, John William, (1820-1869)
SCHENK, Joseph August, (1815-1891)
SCHLAGINTWEIT, Adolf, (?1799-1857)
SCHLAGINTWEIT, Hermann, (-)
SCHLAGINTWEIT, Robert, (-)
STRACHEY, Richard, General Sir, (1817-1908)

1850-1899

ADAMS, Andrew Leith, (-1882)
ADYE, Ernest Howard, (1857-1918)
ANDERSON, William, (1860-1915)
BAIRD SMITH, R., Colonel, (-)
BALL, Valentine, (1843-1895)
BAUERMAN, Hilary, (1833-1909)
BEEBY-THOMPSON, Arthur, (1873-1968)
BLAKE, John Frederick, Reverend, (1839-1906)
BLANFORD, Henry Francis, (1834-1893)

INDIA 1850-1899 Cont'd.

BLANFORD, William Thomas, (1832-1905)
BOWEN, Charles Franklin, (1871-1953)
BOWMAN, Herbert Lister, (1874-1942)
BROWN, Barnum, (1873-1963)
CAMMIADE, Louis Aimé, (1872-?1951)
CARTER, Henry John, (1813-1895)
CAUTLEY, Proby T., Colonel Sir, (1802-1871)
CLARK, George Thomas, (1811-1898)
COTTEAU, Gustave Honoré, (1818-1894)
CRICK, George Charles, (1856-1917)
DE GEER, Gerard Jacob, Baron, (1858-1943)
DREW, Frederick, (1836-1891)
ETHERIDGE, Robert, (1819-1903)
EVANS, John William, (1857-1930)
FALCONER, Hugh, (1808-1865)
FEARNSIDES, William George, (1879-1968)
FEISTMANTEL, Otokar, (1848-1891)
FERGUSSON, James, (1808-1885)
FOOTE, R. Bruce, (-1912)
GARWOOD, Edmund Johnston, (1864-1949)
GREENOUGH, George Bellas, (1778-1855)
GREGORY, John Walter, (1864-1932)
GRIESBACH, Karl Ludolf, (1847-1907)
HAECKEL, Ernst Heinrich Phillipp August, (1834-1919)
HATCH, Frederick Henry, (?1864-1932)
HENWOOD, William Jory, (1805-1875)
HESLOP, Septimus, (?1860-1945)
HISLOP, Stephen, Reverend, (1817-1863)
HOLLAND, Thomas Henry, Sir, (1868-1947)
HOOKER, Joseph Dalton, Sir, (1817-1911)
HUGHES, Theodore W. H., (?1843-1907)
HUNTER, Robert, Reverend, (1823-1897)
JOWETT, Albert, (1875-?1958)
KAYE, Charles Turton, (1812-1846)
KRAFFT VON DELLMENSINGEN, Albrecht, (1871-1901)
LAING, Samuel, (1810-1897)
LAKE, Philip, (1865-1949)
LENOX-CONYNGHAM, Gerald Ponsonby, Colonel Sir, (1866-1956)
LÓCZY, Lajos, (1849-1920)
LOFTUS, William Kennett, (1822-1859)
LYDEKKER, Richard, (1849-1915)
MACLAREN, James Malcolm, (1873-1935)
MALLETT, Frederic Richard, (1841-1921)
MATLEY, Charles Alfred, (1866-1947)
McMAHON, Charles Alexander, Lieutenant-General, (1830-1904)
MEDLICOTT, Henry Benedict, (1829-1905)
MIDDLEMISS, Charles Stewart, (1859-1945)
MOJSISOVICS, Johann August Georg Edmund, Edler von Mojsvàr, (1839-1907)
NEWELL ARBER, Edward Alexander, (1870-1918)
OLDHAM, Charles Aemilius, (1831?-1869)
OLDHAM, Richard Dixon, (1858-1936)
OLDHAM, Thomas, (1816-1878)
OSMAN, Charles William, (1855-1932)

INDIA 1850-1899 Cont'd.

PICKERING, William Henry, (1858-1912)
PILGRIM, Henry Guy Ellcock, (1875-1943)
RAMA RAU, S. Sethu, Rao Bahadur, (1874-1929)
REED, Frederick Richard Cowper, (1869-1946)
ROCCATI, Alessandro, (1872-1928)
ROYLE, John Forbes, (-1857?)
SALTER, John William, (1820-1869)
SCHENK, Joseph August, (1815-1891)
SCHLAGINTWEIT, Hermann, (-)
SCHLAGINTWEIT, Robert, (-)
SEWARD, Albert Charles, Sir, (1863-1941)
SIMPSON, George Clarke, Sir, (1878-1965)
SLATER, Herbert Kelsall, (1875-1913)
STIFFE, Arthur William, Captain, (1849-1912)
STOLICZKA, Ferdinand, (1838-1874)
STRACHEY, Richard, General Sir, (1817-1908)
TRENCH, Richard, (1836-1861)
WALKER, Thomas Leonard, (1867-1942)
WOEIKOF, Alexander, (1842-1916)
WOOD-MASON, James, (1846-1893)
WYNNE, Arthur Beavor, (1835-1906)

1900 to present

ADYE, Ernest Howard, (1857-1918)
ANDERSON, William, (1860-1915)
ANDERTON, Russell, (-) [PROSPECTORS, DIVINERS AND MINING ENGINEERS]
BAJPAI, Mahesh Prasad, (1907-1938)
BARBER, Cecil Thomas, (-)
BAUERMAN, Hilary, (1833-1909)
BEEBY-THOMPSON, Arthur, (1873-1968)
BERNOUILLI, Walter, (1885-1946)
BION, Herbert Stanley, (1888-1915)
BLAKE, John Frederick, Reverend, (1839-1906)
BLANFORD, William Thomas, (1832-1905)
BORGES, Alexandre, (1898-1955)
BOWEN, Charles Franklin, (1871-1953)
BOWMAN, Herbert Lister, (1874-1942)
BROUGHTON-EDGE, Arthur Broughton, (1895-1953)
BROWN, Barnum, (1873-1963)
BROWN, John Coggin, (1884-1962)
BURTON, Richard Cooksey, (1890-1916)
CAMMIADE, Louis Aimé, (1872-?1951)
CHHIBBER, Harbans Lal, (1899-1955)
COLBERT, Edwin H., (1905-)
CONDIT, Daniel Dale, (1886-1955)
COTTER, Gerald de Purcell, (1881-1941)
COX, Leslie Reginald, (1897-1965)
CRICK, George Charles, (1856-1917)
CROOKSHANK, Henry, (1893-1972)
DAS-GUPTA, Hemchandra C., (1878-1933)
DE GEER, Gerard Jacob, Baron, (1858-1943)
DESIO, Ardito, (1897-)
DEWHURST, Thomas, (1881-1973)

INDIA 1900 to present Cont'd.

DVOŘÁK, Arnošt, (1904-)
ETHERIDGE, Robert, (1819-1903)
EVANS, John William, (1857-1930)
EVANS, Percy, (1892-1974)
FEARNSIDES, William George, (1879-1968)
FERMOR, Lewis Leigh, Sir, (1880-1954)
FINSTERWALDER, Richard, (1899-1963)
FOOTE, R. Bruce, (-1912)
FOX, Cyril Sankey, Sir, (1886-1951)
GARRETT, T. R. H., Major, (-1938)
GARWOOD, Edmund Johnston, (1864-1949)
GESTER, George Clark, (1884-1959)
GRANTHAM, Donald Robert, (1896-1969)
GREGORY, John Walter, (1864-1932)
GRIESBACH, Karl Ludolf, (1847-1907)
HAECKEL, Ernst Heinrich Philipp August, (1834-1919)
HATCH, Frederick Henry, (?1864-1932)
HAWLEY, James Edwin, (1897-1965)
HERON, Alexander Macmillan, (1884-1971)
HESLOP, Septimus, (?1860-1945)
HOFMANN, Walter, (1919-1975)
HOLLAND, Thomas Henry, Sir, (1868-1947)
HOOKER, Joseph Dalton, Sir, (1817-1911)
HUBMAN, Carl Wesley, (1905-1947)
HUGHES, Theodore W. H., (?1843-1907)
JOHNSTON, William Drumm, Jr., (1899-1972)
JONES, Hubert Cecil, (1881-1963)
JOWETT, Albert, (1875-?1958)
KRAFFT VON DELLMENSINGEN, Albrecht, (1871-1901)
KRISHNAN, Maharajapuram Sitaram, Padma Bhusan, (1898-1970)
LAKE, Philip, (1865-1949)
LEHNER, Ernst J., (1895-1968)
LENOX-CONYNGHAM, Gerald Ponsonby, Colonel Sir, (1866-1956)
LEPPER, Gordon Willis, (1889-1950)
LÓCZY, Lajos, (1849-1920)
LYDEKKER, Richard, (1849-1915)
MACLAREN, James Malcolm, (1873-1935)
MALLETT, Frederic Richard, (1841-1921)
MARR, Francis Alleyne, (1893-1942)
MATHUR, Krishna Kumar, (-1936)
MATLEY, Charles Alfred, (1866-1947)
McMAHON, Charles Alexander, Lieutenant-General, (1830-1904)
MEDLICOTT, Henry Benedict, (1829-1905)
METRE, Waman Bapu, (1906-1970)
MIDDLEMISS, Charles Stewart, (1859-1945)
MOJSISOVICS, Johann August Georg Edmund, Edler von Mojsvàr, (1839-1907)
MORRIS, Frederick Kuhne, (1885-1962)
MUIR-WOOD, Helen Marguerite, (1896-1968)
MUKERJI, Sushil Kumar, (1896-1934)
NAGAPPA, Yedatore, Shri, (1907-1960)
NAIDU, Purama Rangashayee Jagapathi, (1903-1970)
NARKE, Ganesh Govind, (-?1943)
NEWELL ARBER, Edward Alexander, (1870-1918)

INDIA 1900 to present Cont'd.

NIETHAMMER, Gottlob August, (1882-1915)
NIGHTINGALE, William Thomas, (1897-1964)
NOMLAND, Jorgen O., (1885-1943)
NOTESTEIN, Frank B., (1885-1973)
OLDHAM, Richard Dixon, (1858-1936)
OSMAN, Charles William, (1855-1932)
PALMER, Rupert William, (1890-1922)
PARSONS, Ernest, (?1887-1963)
PERCIVAL, Frederick George, (1887-1973)
PICKERING, William Henry, (1858-1912)
PILGRIM, Henry Guy Ellcock, (1875-1943)
RAJNATH, Shri, (1904-1972)
RAMAN, Chandrasekhara Venkata, Sir, (1888-1970)
RAMA RAO, Bellu, (1890-1970)
RAMA RAO, Lakshmeswar, (1896-1974)
RAMA RAU, S. Sethu, Rao Bahadur, (1874-1929)
REA, Henry Carter, (1900-1963)
REED, Frederick Richard Cowper, (1869-1946)
ROCCATI, Alessandro, (1872-1928)
SAHA, Baidya Nath, Babu, (-1929)
SAHNI, Birbal, (1891-1949)
SAMBASIVA IYER, Valiathur S., (-?1943)
SCHERMERHORN, W., (1894-)
SCOFIELD, Lloyd M., (1897-1969)
SEN GUPTA, Kiron Kumar, (1882-1943)
SEWARD, Albert Charles, Sir, (1863-1941)
SIMONS, Herbert Samuel, (1891-1956)
SIMPSON, George Clarke, Sir, (1878-1965)
SLATER, Herbert Kelsall, (1875-1913)
SPATH, Leonard Frank, (1882-1957)
SPENCER, Edmondson, (?1885-1970)
STIFFE, Arthur William, Captain, (1849-1912)
STRACHEY, Richard, General Sir, (1817-1908)
SUBRAMANIAM, Anantharama Parameshwara, (1921-1973)
SWAMY, Kolagotla Bhimeswara, Sri, (-1950)
TEILHARD DE CHARDIN, Pierre, Father, S.J., (1881-1955)
THOMPSON, Hugh R., (1926-1959)
TIPPER, George Howlett, (1881-1947)
TORRE DE ASSUNÇÃO, Carlos Fernando, (1901-)
TRASK, Parker Davies, (1899-1961)
TRÜMPY, Daniel, (1893-1971)
VERSLUYS, Jan, (1880-1935)
WADIA, Darashaw Nosherwan, (1883-1969)
WALKER, Thomas Leonard, (1867-1942)
WOEIKOF, Alexander, (1842-1916)
WYNNE, Arthur Beavor, (1835-1906)

INDOCHINA (unspecified)

Up to 1800

BARROW, John, Sir, (1764-1848)

1800-1849

BARROW, John, Sir, (1764-1848)

INDOCHINA (unspecified) Cont'd.

1850-1899
BRALY, Adrien, (1864-1927)
DOUVILLÉ, Joseph Henri Ferdinand, (1846-1937)
FUCHS, Philippe Jacques Edmond, (1837-1889)

1900 to present
BLONDEL, Fernand, (1894-1968)
BRALY, Adrien, (1864-1927)
CALAME, Jean Jacques, (1923-1973)
DOUVILLÉ, Joseph Henri Ferdinand, (1846-1937)
GUBLER, Jean, (1902-1940)
NOMLAND, Jorgen O., (1885-1943)
WARING, Gerald Ashley, (1883-1971)

INDONESIA
Up to 1800
TAVERNIER, Jean Baptiste, baron d'Aubonne, (1605-1689)

1800-1849
ARNOLD, Joseph, (1784-1818)
CAMPBELL, John Francis, (1821-1885)
ELLIS, H., Right Honourable Sir, (-1855)
JUKES, Joseph Beete, (1811-1869)
WALLACE, Alfred Russel, (1823-1913)

1850-1899
BAUMBERGER, Ernst, (1866-1935)
BLEEKRODE, S., (-)
BÖHM, Georg, (1854-1913)
BOWEN, Charles Franklin, (1871-1953)
BRADY, Henry Bowman, (1835-1891)
BULLEN, Robert Ashington, Reverend, (1850-1912)
BUTTGENBACH, Henri, (1874-1964)
CAMPBELL, John Francis, (1821-1885)
DENINGER, Karl, (1878-1917)
DOUVILLÉ, Joseph Henri Ferdinand, (1846-1937)
DUBOIS, Eugène, (1858-1940)
ELLIS, H., Right Honourable Sir, (-1855)
ERB, Josef Theodor, (1874-1934)
FENNEMA, Reinder, (1849-?1897)
GÜNTHER, Albert Charles Lewis Gotthilf, (1830-1914)
HAECKEL, Ernst Heinrich Philipp August, (1834-1919)
HORNEMAN, Hans Henrik, (1878-1945)
JENKINS, Henry Michael, (1841-1886)
JONGMANS, Wilhelmus Josephus, (1878-1957)
JUKES, Joseph Beete, (1811-1869)
JUNGHUHN, Franz Wilhelm, (1808-1864)
KISSLING, Ernst, (1865-1936)
KOKEN, Ernst Friedrich Rudolph Karl von, (1860-1912)
KORNITZER, Louis, (1873-) [PROSPECTORS, DIVINERS AND MINING ENGINEERS]
MARTIN, Johann Karl Ludwig, (1851-1942)
McGEE, William John, (1853-1912)
MILCH, Ludwig, (1867-1928)

INDONESIA 1850-1899 Cont'd.

MOLENGRAAFF, Gustaaf Adolf Frederik, (1860-1942)
MOON, Frederick William, (-1925)
PORRIO, Cesare, (1865-1940)
RICHTHOFEN, Paul Wilhelm Ferdinand von, Freiherr, (1833-1905)
RINNE, Friedrich Wilhelm Berthold, (1863-1933)
TOLWIŃSKI, Konstanty, (1876-1961)
TRUSCOTT, Samuel John, (?1870-1950)
VERBEEK, Rogier Diederick Marius, (1845-1926)
WALLACE, Alfred Russel, (1823-1913)
WANNER, Johannes, (1878-1957)
WICHMANN, Carl Ernst Arthur, (1851-1927)
WOEIKOF, Alexander, (1842-1916)
WOODWARD, Henry, (1832-1921)

1900 to present

ANDRAU, Evert Willem Karel, (1900-1951)
BARD, Richards J., (1900-1977)
BAREN, F. H. van, (-1975)
BASSINGER, William E., (1928-1977)
BAUMBERGER, Ernst, (1866-1935)
BELTZ, Edward W., (1891-1970)
BERNOUILLI, Walter, (1885-1946)
BÖHM, Georg, (1854-1913)
BONTZ, Conrad K., (1894-1927)
BOOBERG, Gunnar, (1892-1944)
BOWEN, Charles Franklin, (1871-1953)
BREMNER, Carl St. John, (1895-1944)
BROUWER, Hendrik Albertus, (1886-1973)
BULLEN, Robert Ashington, Reverend, (1850-1912)
BUTTERWORTH, Emerson McMillin, (1894-1961)
BUTTGENBACH, Henri, (1874-1964)
CLOOS, Hans, (1885-1951)
CONDIT, Daniel Dale, (1886-1955)
CUNNINGHAM, George Magee, (1895-1963)
DENINGER, Karl, (1878-1917)
DOORNINCK, Nico H. van, (1893-1956)
DOUVILLÉ, Joseph Henri Ferdinand, (1846-1937)
DUBOIS, Eugène, (1858-1940)
ERB, Josef Theodor, (1874-1934)
ESCHER, Berend George, (-1967)
FICHTER, Hans Jakob, (1908-1977)
FREI, Ernst, (1894-1974)
GARRETT, T. R. H., Major, (-1938)
GOETEL, Walery, (1889-1972)
GRUTTERINK, Jan Adolf, (1879-1949)
GÜNTHER, Albert Charles Lewis Gotthilf, (1830-1914)
HAECKEL, Ernst Heinrich Philipp August, (1834-1919)
HAZZARD, John Charles, (1903-1973)
HENNY, Gerhard, (1890-1944)
HILL, Robert G., (1914-1973)
HOLST PELLEKAAN, William van, (1880-1949)
HORNEMAN, Hans Henrik, (1878-1945)
HOTZ, Walter, (1884-1958)
JENNINGS, Philip H., (1899-1974)

INDONESIA 1900 to present Cont'd.

JONGMANS, Wilhelmus Josephus, (1878-1957)
JUNGHUHN, Franz Wilhelm, (1808-1864)
KEMMERLING, Georges Laure Louis, (1888-1932)
KINNEY, Harry Dill, (1886-1961)
KISSLING, Ernst, (1865-1936)
KLOMPE, Theodorus Hendricus Franciscus, (?1904-1963)
KNOWLES, Gerald Elgin, (1923-1976)
KOENIGSWALD, Gustav Heinrich Ralph von, (1902-)
KOKEN, Ernst Friedrich Rudolph Karl von, (1860-1912)
KOMOROWICZ, Maurycy, (1881-1923)
KORNITZER, Louis, (1873-) [PROSPECTORS, DIVINERS AND MINING ENGINEERS]
KUENEN, Phillip Henry, (1902-1976)
LE GROS CLARK, Wilfrid Edward, Sir, (1895-1971)
LEUTENEGGER, Willy Otto, (1903-1972)
LINDEN, Bernard Hugo "Ben" van der, (-1958)
MARTIN, Johann Karl Ludwig, (1851-1942)
McGEE, William John, (1853-1912)
MEYER, Edouard, (1899-1974)
MILCH, Ludwig, (1867-1928)
MOLENGRAAFF, Gerard Johan Hendrik, (1887-1961)
MOLENGRAAFF, Gustaaf Adolf Frederik, (1860-1942)
MOON, Frederick William, (-1925)
NELSON, Richard Newman, (1897-1964)
NIGHTINGALE, William Thomas, (1897-1964)
NOTESTEIN, Frank B., (1885-1973)
NUTTALL, Winfred Laurence Falkiner, (1897-1969)
NYGREN, Walter Eric, (1907-1971))
OSBERGER, Rudolf A. J., (1924-1972)
PET, Cees, (1900-1953)
PLESMAN, Albert, (1889-1953)
PORRIO, Cesare, (1865-1940)
REA, Henry Carter, (1900-1963)
RHEDEN, J. J. Pannekoek van, (-1945)
RICHTHOFEN, Paul Wilhelm Ferdinand von, Freiherr, (1833-1905)
RINNE, Friedrich Wilhelm Berthold, (1863-1933)
RUSSELL, Richard Joel, (1895-1971)
RUTTEN, Louis Martin Robert, (1884-1946)
RUTTEN, Martin Gerard, (1910-1970)
SCHAUB, Hans Peter, (1912-1965)
SCHROYER, Charles Ross, (1887-1932)
SMITH, G. Wendell, (1917-1974)
SMIT SIBINGA, Gerard Leonard, (1895-1963)
TAYLOR, Griffith, (1881-1963)
TEILHARD DE CHARDIN, Pierre, Father, S.J., (1881-1955)
THALMANN, Hans E., (1899-1975)
TOLWIŃSKI, Konstanty, (1876-1961)
TRUSCOTT, Samuel John, (?1870-1950)
UMBGROVE, Johannes Herman Frederik, (1899-1954)
VENING MEINESZ, Felix Andries, (1887-1966)
VERBEEK, Rogier Diederick Marius, (1845-1926)
VERSLUYS, Jan, (1880-1935)
VLERK, Isaak Martinus van der, (?1892-1974)
WALLACE, Alfred Russel, (1823-1913)

INDONESIA 1900 to present Cont'd.

WANNER, Johannes, (1878-1957)
WESTERVELD, J., (1905-1962)
WICHMANN, Carl Ernst Arthur, (1851-1927)
WILMOTH, Leslie James, (1881-1963)
WILSON, Eugene Jack, (1921-1976)
WIRZ, Albert Eduard, (1910-1973)
WOEIKOF, Alexander, (1842-1916)
WOODWARD, Henry, (1832-1921)
ZEIJLMANS VAN EMMICHOVEN, Cornelis Pieter Adriaan, (1897-1952)
ZIJLSTRA, Gerrit, (-1962)
ZIMERMAN, Sam, (1905-1972)
ZWIERZYCKI, Józef, (1888-1961)

INDONESIA *See also* TIMOR (listed separately since it is not wholly a part of Indonesia at present); and WEST IRIAN (early references listed under NEW GUINEA).

IRAN

Up to 1800

AVICENNA, (980-1037)

1800-1849

ELLIS, H., Right Honourable Sir, (-1855)
LOFTUS, William Kennett, (1822-1859)
NICOL, Germain Louis, (1922-1973)
VOSKOBOINIKOV, N. A., (-)

1850-1899

BLANFORD, William Thomas, (1832-1905)
DOUVILLÉ, Joseph Henri Ferdinand, (1846-1937)
ELLIS, H., Right Honourable Sir, (-1855)
GRIESBACH, Karl Ludolf, (1847-1907)
KITSON, Albert Ernest, Sir, (1868-1937)
LOFTUS, William Kennett, (1822-1859)
MACLAREN, James Malcolm, (1873-1935)
MORGAN, Jacques Jean Marie de, (1857-1924)
NAGYSUR, Hugo Böckh von, (1874-1931)
NICOL, Germain Louis, (1922-1973)
PORRIO, Cesare, (1865-1940)
TIETZE, Emil, (1845-1931)

1900 to present

ALDREDGE, Robert Frank, (1895-1961)
ARAMBOURG, Camille, (1885-1969)
BERNOUILLI, Walter, (1885-1946)
BLANFORD, William Thomas, (1832-1905)
BUSK, Henry Gould, (?1891-1956)
CHAPMAN, Sydney, (1888-1970)
CLAPP, Frederick Gardner, (1879-1944)
CURRIE, Ethel Dobbie, (-1963)
DOUGLAS, James Archibald, (1884-1978)
DOUVILLÉ, Joseph Henri Ferdinand, (1846-1937)
FARISH, Linn Markley, (1901-1944)
FURON, Raymond, (1898-)

IRAN 1900 to present Cont'd.

GRIESBACH, Karl Ludolf, (1847-1907)
HALSE, George Whitfield, (1885-1968)
HARRISON, John Vernon, (1892-1972)
HAZZARD, John Charles, (1903-1973)
HUBBARD, George Hulme, (1899-1969)
HUDSON, Robert George Spencer, (1895-1965)
HUXLEY, Julian Sorell, Sir, (1887-1975)
JONES, David Glynn, (1904-1974)
KITSON, Albert Ernest, Sir, (1868-1937)
LEES, George Martin, (1898-1955)
LEHNER, Ernst J., (1895-1968)
MACLAREN, James Malcolm, (1873-1935)
MAYO, Herbert Theodore, (1889-1965)
MEKEL, Johannes Antonius Alphonsus, (1891-1942)
MORGAN, Jacques Jean Marie de, (1857-1924)
MUIR-WOOD, Helen Marguerite, (1896-1968)
NAGYSUR, Hugo Böckh von, (1874-1931)
NEIGHBOURS, Glen Reed, (1917-1961)
PORRIO, Cesare, (1865-1940)
REA, Henry Carter, (1900-1963)
RICHARDSON, F. D. S. "Ricardo", (-1966)
SARJEANT, William Antony Swithin, (1935-)
SCHERMERHORN, W., (1894-)
SPATH, Leonard Frank, (1882-1957)
STONE, A. John F., (1917-1969)
TIETZE, Emil, (1845-1931)
TIPPER, George Howlett, (1881-1947)
TRÜMPY, Daniel, (1893-1971)
WASHINGTON-GRAY, Kenneth, (?1899-1963)
WIRZ, Albert Eduard, (1910-1973)

IRAQ

1800-1849

LOFTUS, William Kennett, (1822-1859)

1850-1899

BEEBY-THOMPSON, Arthur, (1873-1968)
BLANCKENHORN, Max Ludwig Paul, (1861-1947)
KISSLING, Ernst, (1865-1936)
LOFTUS, William Kennett, (1822-1859)
NAGYSUR, Hugo Böckh von, (1874-1931)
PORRIO, Cesare, (1865-1940)

1900 to present

AL-NAQIB, Khorshid Mohammed, (?1927-1972)
ANDRAU, Evert Willem Karel, (1900-1951)
BEEBY-THOMPSON, Arthur, (1873-1968)
BENTZ, Alfred, (1897-1964)
BIRAUD, François, (1899-1936)
BLANCKENHORN, Max Ludwig Paul, (1861-1947)
BOURCHIER, John Rickman, (1896-1927)
BRAMWYCH, Michael, (1911-1974)
CLOOS, Ernst, (1898-1974)
FALCON, Norman Leslie, (-)

IRAQ 1900 to present Cont'd.

HITCHEN, Charles Stansfield, (1904-1971)
HUBBARD, George Hulme, (1899-1969)
HUDSON, Robert George Spencer, (1895-1965)
HUXLEY, Julian Sorell, Sir, (1887-1975)
IRELAND, Hubert Andrew, (1904-1973)
JONES, David Glynn, (1904-1974)
KISSLING, Ernst, (1865-1936)
MUIR, John Malcolm, (1885-1938)
MUIR-WOOD, Helen Marguerite, (1896-1968)
NAGYSUR, Hugo Böckh von, (1874-1931)
PORRIO, Cesare, (1865-1940)
RICHARDSON, F. D. S. "Ricardo", (-1966)
SCHERMERHORN, W., (1894-)
SHAW, Eugene Wesley, (1881-1935)
SUGDEN, Walter "Bill", (1923-1977)
VAN COUVERING, Martin, (1888-1976)

IRELAND

Up to 1800

BOYLE, Robert, Honourable, (1627-1691)
GIESECKE, Charles Lewis, Sir, (né Metzle, Karl Ludwig), (1761-1833)
GREENOUGH, George Bellas, (1778-1855)
KIRWAN, Richard, (1733-1812)
McDONNELL, James, (?1763-1845)
PRIOR, Richard, (fl.1699)
RASPE, Rudolf Erich, (1737-1794)
USHER, James, Archbishop, (1581-1656)

1800-1849

AUSTIN, Thomas, Fort-Major, (1794-1881)
BAILY, William Hellier, (1819-1888)
BALD, William, (-?1857)
BALL, John, (1818-1889)
CLOSE, Maxwell Henry, Reverend, (1822-1903)
COLBY, Thomas F., Major-General, (?1783-1852)
DU NOYER, George Victor, (-1869)
FITTON, William Henry, (1780-1861)
GANLY, Patrick, (1809-1899)
GIESECKE, Charles Lewis, Sir, (né Metzle, Karl Ludwig), (1761-1833)
GREENOUGH, George Bellas, (1778-1855)
GRIFFITH, Richard John, Sir, (1784-1878)
HAMILTON, Sir William R., (1805-1865)
HAUGHTON, Samuel, Reverend, (1823-1897)
HUTTON, Robert, (1784-1870)
JUKES, Joseph Beete, (1811-1869)
KELVIN, William Thomson, Baron, (1824-1907)
KINAHAN, George Henry, (1829-1908)
KING, William, (1809-1886)
KIRWAN, Richard, (1733-1812)
MacADAM, James, (1801-1861)
McCOY, Frederick, Sir, (1823-1899)
McDONNELL, James, (?1763-1845)

IRELAND 1800-1849 Cont'd.

MEADE, William, (1789-1862)
NICOL, James, (1810-1879)
OGILBY, William, (1808-1873)
OLDHAM, Thomas, (1816-1878)
PENTLAND, Joseph Barclay, (1797-1873)
PHILLIPS, John, (1800-1874)
PORTLOCK, Joseph Ellison, Major-General, (1794-1864)
RÖMER, Carl Ferdinand von, (1818-1891)
SALMON, Henry Curwen, (1827-1873)
STOKES, George Gabriel, Sir, (1819-1903)
TYNDALL, John, (1820-1893)

1850-1899
ADAMS, Andrew Leith, (-1882)
AUSTIN, Thomas, Fort-Major, (1794-1881)
BAILY, William Hellier, (1819-1888)
BALL, John, (1818-1889)
BALL, Valentine, (1843-1895)
BARROIS, Charles Eugène, (1851-1939)
BELL, Robert, (1864-1934)
BRODRICK, Harold, (1872-1946)
CLELAND, Alexander M'Ivor, (1862-)
CLOSE, Maxwell Henry, Reverend, (1822-1903)
COLBY, Thomas F., Major-General, (?1783-1852)
COLE, Grenville Arthur James, (1859-1924)
CRICK, George Charles, (1856-1917)
DAVIS, James William, (1846-1893)
DU NOYER, George Victor, (-1869)
DWERRYHOUSE, Arthur Richard, (1867-1931)
EGAN, Frederick William, (1836-1901)
FEARNSIDES, William George, (1879-1968)
FITTON, William Henry, (1780-1861)
FOORD, Arthur Humphreys, (1845-1934)
FOOT, Frederick J., (-1867)
GANLY, Patrick, (1809-1899)
GARDINER, Charles Irving, (?1868-1940)
GREENOUGH, George Bellas, (1778-1855)
GRIFFITH, Richard John, Sir, (1784-1878)
HAMILTON, Sir William R., (1805-1865)
HARDMAN, Edward Townley, (1845-1887)
HARKNESS, Robert, (1816-1878)
HAUGHTON, Samuel, Reverend, (1823-1897)
HINCH, John de Witt, (?1875-1931)
HOBSON, Bernard, (1860-1933)
HULL, Edward, (1829-1917)
HUTTON, Robert, (1784-1870)
HYLAND, John Shearson, (1866-1898)
JOLY, John, (1857-1933)
JUKES, Joseph Beete, (1811-1869)
KELVIN, William Thomson, Baron, (1824-1907)
KILROE, James Robinson, (1848-1927)
KINAHAN, George Henry, (1829-1908)
KING, William, (1809-1886)
LACROIX, François Antoine Alfred, (1863-1948)

IRELAND 1850-1899 Cont'd.

LEONARD, Hugh, (1841-1909)
LEWIS, Henry Carvill, (1853-1888)
MacADAM, James, (1801-1861)
MacHENRY, Alexander, (1843-1919)
MALLETT, Frederic Richard, (1841-1921)
McCOY, Frederick, Sir, (1823-1899)
McHENRY, Alexander, (1843-1919)
MILLETT, Fortescue William, (1833-1915)
MOON, Frederick William, (-1925)
MOSS, Richard Jackson, (1847-1934)
NICOL, James, (1810-1879)
NOLAN, Joseph, (1841-1902)
OGILBY, William, (1808-1873)
O'KELLY, Joseph, (1832-1883)
OLDHAM, Thomas, (1816-1878)
O'REILLY, Joseph P., (1829-1905)
PENTLAND, Joseph Barclay, (1797-1873)
PHILLIPS, John, (1800-1874)
PORTLOCK, Joseph Ellison, Major-General, (1794-1864)
REED, Frederick Richard Cowper, (1869-1946)
REYNOLDS, James Emerson, (1844-1920)
REYNOLDS, Sidney Hugh, (1867-1949)
RÖMER, Carl Ferdinand von, (1818-1891)
SALMON, Henry Curwen, (1827-1873)
SCOTT, Robert Henry, (1833-1916)
SOLLAS, William Johnson, (1849-1936)
STOKES, George Gabriel, Sir, (1819-1903)
SYMES, Richard Glascott, (1840-1906)
TATE, Ralph, (1840-1901)
TYNDALL, John, (1820-1893)
VAUGHAN, Arthur, (1868-1915)
WATTS, William Whitehead, (1860-1947)
WHITLA, Francis, (fl.1830-1853)
WRIGHT, William Bourke, (1876-1939)
WYATT-EDGELL, Henry Adrian, (1847-1866)
WYNNE, Arthur Beavor, (1835-1906)
WYVILLE THOMSON, Charles, Sir, (1830-1882)

1900 to present

ASHBY, Donald Frederick, (?1914-1943)
BAILEY, Edward Battersby, Sir, (1881-1965)
BARROIS, Charles Eugène, (1851-1939)
BELL, Robert, (1864-1934)
BISHOPP, Douglas Wallace, (1900-1977)
BRODRICK, Harold, (1872-1946)
CHARLESWORTH, John Kaye, (1889-)
CLOSE, Maxwell Henry, Reverend, (1822-1903)
COLE, Grenville Arthur James, (1859-1924)
CRICK, George Charles, (1856-1917)
DIXON, Henry George Dacres, (1887-1947)
DOUGLAS, James Archibald, (1884-1978)
DWERRYHOUSE, Arthur Richard, (1867-1931)
EGAN, Frederick William, (1836-1901)
EYLES, Victor Ambrose, (1896-1978)

IRELAND 1900 to present Cont'd.

FEARNSIDES, William George, (1879-1968)
FOORD, Arthur Humphreys, (1845-1934)
GARDINER, Charles Irving, (?1868-1940)
HARPER, John Childs, (1913-1977)
HINCH, John de Witt, (?1875-1931)
HOBSON, Bernard, (1860-1933)
HUDSON, Robert George Spencer, (1895-1965)
HULL, Edward, (1829-1917)
JOLY, John, (1857-1933)
KELVIN, William Thomson, Baron, (1824-1907)
KENNEDY, Horas Tristram, Lieutenant, (1889-1917)
KILROE, James Robinson, (1848-1927)
KINAHAN, George Henry, (1829-1908)
LACROIX, François Antoine Alfred, (1863-1948)
LEONARD, Hugh, (1841-1909)
LEUTWEIN, Friedrich, (1911-1974)
MacHENRY, Alexander, (1843-1919)
MALLETT, Frederic Richard, (1841-1921)
MAUFE, Herbert Brantwood, (1879-1946)
McHENRY, Alexander, (1843-1919)
MILLETT, Fortescue William, (1833-1915)
MOON, Frederick William, (-1925)
MOSS, Richard Jackson, (1847-1934)
NOLAN, Joseph, (1841-1902)
O'REILLY, Joseph P., (1829-1905)
REED, Frederick Richard Cowper, (1869-1946)
REYNOLDS, James Emerson, (1844-1920)
REYNOLDS, Sidney Hugh, (1867-1949)
RICHEY, James Ernest, (1886-1968)
SCOTT, Robert Henry, (1833-1916)
SMYTH, Louis Bouvier, (1933-?1953)
SOLLAS, William Johnson, (1849-1936)
SPIKE, Wilfred Ivan, (1925-1969)
STOKES, George Gabriel, Sir, (1819-1903)
SYMES, Richard Glascott, (1840-1906)
TATE, Ralph, (1840-1901)
TOZER, Charles Fallows, (1918-1966)
VALENTINE, R. L., Lieutenant, (1890-1916)
VAUGHAN, Arthur, (1868-1915)
WATTS, William Whitehead, (1860-1947)
WRIGHT, William Bourke, (1876-1939)
WYNNE, Arthur Beavor, (1835-1906)

ISLE OF MAN (UNITED KINGDOM)

1800-1849

FORBES, Edward, (1815-1854)
STRICKLAND, Hugh Edwin, (1811-1853)

1850-1899

BIRDS, James Adey, (1831-1894)
BOLTON, Herbert, (1863-1936)
CRICK, George Charles, (1856-1917)
FORBES, Edward, (1815-1854)
STRICKLAND, Hugh Edwin, (1811-1853)
WATTS, William Whitehead, (1860-1947)

ISLE OF MAN (UNITED KINGDOM) Cont'd.

1900 to present
BOLTON, Herbert, (1863-1936)
CRICK, George Charles, (1856-1917)
WATTS, William Whitehead, (1860-1947)

ISRAEL

Up to 1800
VOLNEY, Constantin François Chasseboeuf, Comte de, (1757-1820)

1800-1849
EHRENBERG, Christian Gottfried, (1795-1876)
HOOKER, Joseph Dalton, Sir, (1817-1911)
HUDLESTON, Wilfrid Hudleston, (né Simpson), (1828-1909)
TAUNTON, John Hooke, (1820-1893)
VOLNEY, Constantin François Chasseboeuf, Comte de, (1757-1820)

1850-1899
BATE, Dorothea Minola Alice, (1879-1951)
BEEBY-THOMPSON, Arthur, (1873-1968)
BLANCKENHORN, Max Ludwig Paul, (1861-1947)
EHRENBERG, Christian Gottfried, (1795-1876)
HOOKER, Joseph Dalton, Sir, (1817-1911)
HUDLESTON, Wilfrid Hudleston, (né Simpson), (1828-1909)
HULL, Edward, (1829-1917)
LARTET, Louis, (1840-1899)
REED, Frederick Richard Cowper, (1869-1946)
TAUNTON, John Hooke, (1820-1893)
WADE, Arthur, (1878-1951)

1900 to present
BATE, Dorothea Minola Alice, (1879-1951)
BEEBY-THOMPSON, Arthur, (1873-1968)
BLAKE, George Stanfield, (-1940)
BLANCKENHORN, Max Ludwig Paul, (1861-1947)
CAMPBELL, Keir Arthur, (-1955)
CARROLL, Don Llewellyn, (1902-1954)
CLAPP, Frederick Gardner, (1879-1944)
FOHS, Ferdinand Julius, (1884-1965)
HOOKER, Joseph Dalton, Sir, (1817-1911)
HUDLESTON, Wilfrid Hudleston, (né Simpson), (1828-1909)
HULL, Edward, (1829-1917)
MORRIS, Frederick Kuhne, (1885-1962)
MUIR-WOOD, Helen Marguerite, (1896-1968)
PFANNENSTIEL, Max, (1902-1976)
REED, Frederick Richard Cowper, (1869-1946)
SOLOMONICA, Paul, (1908-1954)
VELIKOVSKY, Immanuel, (1895-)
WADE, Arthur, (1878-1951)
WING, Monta Eldo, (1897-1976)
ZIMERMAN, Sam, (1905-1972)

ITALY

Up to 1800

ALBERTI, Leon Battista, (1404-1472)
ALDROVANDI, Ulisse, (1522-1605)
ALLEN, William, (1770-1843)
ARDUINO, Giovanni, (1714-1795)
BASSI, Ferdinando, (1714-1774)
BECCARI, Jacopo Bartolommeo, (1682-1766)
BELGRADO, Jacopo, Padre, (fl.1738)
BIANCHI, Johannes-Janus Plancus, (1695-1775)
BIRINGUCCIO, Vannoccio, (1480-1538)
BOCCONE, Paolo, (1633-1704)
BORCH, Michael Jan, (1751-1810)
BREISLAK, Scipione, (1748-1826)
BROCCHI, Giovanni Battista (Giambattista), (1772-1826)
BRONGNIART, Alexandre, (1778-1847)
BRUNO, Filippo, called Giordano, (1548-1600)
BUCH, Christian Leopold von, Freiherr, (1774-1853)
CANALI, Luigi, (1759-1841)
CARDANO, Girolamo, (1501-1576)
CASSIUS, Dio Cocceianus, ([155-235] A.D.)
CIAMPINI, Giovanni Giustino, (1633-1698)
CIMARELLI, Vincenzo Maria, (-1660)
CLARKE, Edward Daniel, (1769-1822)
COLONNA, Fabio, (1567-1650)
DEGLI ALESSANDRI, Alessandro, (1461-1523)
DOLOMIEU, Dieudonné Sylvain Guy Tancrède, called Déodat, Gratet de, (1750-1801)
FAUJAS DE SAINT FOND, Barthélemy, (1741-1819)
FERBER, Johann Jacob, (1743-1790)
FRACASTORO, Girolamo Hieronymus, (1483-1553)
GALEAZZI, Domenico Mario Gusmano, (1686-ca.1775)
GREENOUGH, George Bellas, (1778-1855)
GRIMALDI, Francesco Antonio, (1740-1784)
GUALTIERI, Nicolaus, (1688-1744)
GUGLIELMINI, Domenico, (1655-1710)
HAMILTON, William, Sir, (1730-1803)
IMPERATO, Ferrante, (1550-1631)
LUCRETIUS CARUS, Titus, (98-55 B.C.)
MAILLET, Benoît de, (1656-1738)
MARASCHINI, Pietro, (1724-1825)
MONTI, Giuseppe, (1682-1760)
MONTICELLI, Teodoro, Abate, (1757-1845)
MORO, Anton Lazzaro, (1687-1740)
PINI, Ermenegildo, (1739-1825)
PLINY THE ELDER (CAIUS PLINIUS SECUNDUS), (?20 A.D.-79 A.D.)
SCILLA, Agostino, (1639-1700)
SENECA, Lucius Annaeus, (ca.4 B.C. - 65 A.D.)
SICULUS, Diodorus, (ca.85-21 B.C.)
SOLDANI, Ambrogio, Pater, (1736-1808)
SPALLANZANI, Lazzaro, Abate, (1729-1799)
STENO, Nicolaus, (1631-1687)
STRABO, (63B.C.-23 A.D.?)
TARGIONI TOZZETTI, Giovanni, (1712-1783)
TONDI, Matteo, (1762-1835)
VALLISNERI, Antonio, (1661-1730)

ITALY Up to 1800 Cont'd.

VANDELLI, Domenico, (1735-1816)
VINCI, Leonardo da, (1452-1519)
VIVENZIO, Giovanni, (-)

1800-1849

ALLEN, William, (1770-1843)
BABBAGE, Charles, (1792-1871)
BALSAMO-CRIVELLI, Giuseppe, Marchese, (1800-1874)
BARZANÒ, Gaetano, (-?1859)
BELLARDI, Luigi, (1818-1889)
BERTELLI, Petro Timoteo, (1826-1905)
BIANCONI, Gian Giuseppe, (1809-1878)
BORCH, Michael Jan, (1751-1810)
BREISLAK, Scipione, (1748-1826)
BROCCHI, Giovanni Battista (Giambattista), (1772-1826)
BRONGNIART, Alexandre, (1778-1847)
BRONN, Heinrich Georg, (1800-1862)
BUCH, Christian Leopold von, Freiherr, (1774-1853)
CALCARA, Pietro, (1819-1854)
CAMPANI, Giovanni, (1820-1891)
CANALI, Luigi, (1759-1841)
CARUS, Carl Gustav, (1789-1869)
CATULLO, Tommaso Antonio, (1782-1869)
CECCHI, Filippo, (1822-1887)
CHURCHILL, George Cheetham, (1822-1906)
CIOFALO, Saverio, (1825-1925)
CLARKE, Edward Daniel, (1769-1822)
COCCHI, Igino, (1827-1913)
CORNALIA, Emilio, (1824-1882)
COSSA, Alfonso, (-)
COSTA, Oronzio Gabriele, (1787-1867)
CRISTOFORI, Joseph de, (1803-1837)
CURIONI, Giulio, (1796-1878)
DA RICO, Nicolò, Conte, (-?1845)
DAUBENY, Charles Giles Brindle, (1795-1867)
DE VISIANI, Roberto, (1800-1878)
DE ZIGNO, Achille, Barone, (1813-1888)
DODERLEIN, Pietro, (1810-1895)
DOHRN, Felix Anton, (1840-1909)
DOLOMIEU, Dieudonné Sylvain Guy Tancrède, called Déodat, Gratet de, (1750-1801)
EHRENBERG, Christian Gottfried, (1795-1876)
ERLE, C., Reverend, (1790-1870?)
FAUJAS DE SAINT FOND, Barthélémy, (1741-1819)
FORESTI, Lodovico, (1829-1913)
GASTALDI, Bartolomeo, (1818-1879)
GAUDIN, Charles Théophile, (1822-1866)
GEMMELLARO, Carlo, (1787-1866)
GILBERT, Josiah, (1814-1892)
GIORDANO, Felice, (1825-1892)
GREENOUGH, George Bellas, (1778-1855)
GUIDONI, Gerolamo, (1794-1870)
GUISCARDI, Guglielmo, (1821-1885)
HAIDINGER, Wilhelm Karl von, Ritter, (1795-1871)

ITALY 1800-1849 Cont'd.

HAMILTON, William, Sir, (1730-1803)
HAMILTON, William John, (1805-1867)
HECKEL, Johann Jakob, (1790-1857)
HOFFMANN, Friedrich, (1797-1836)
KING, Samuel William, (1821-1868)
LAWLEY, Robert, (1818-1881)
MARASCHINI, Pietro, (1724-1825)
MARZARI PENCATI, Giuseppe Conte, (1779-1836)
MASSALONGO, Abramo Bartolomeo, (1824-1860)
MENEGHINI, Giuseppe Giovanni Antonio, (1811-1889)
MICHELOTTI, Giovanni, (1814-1898)
MILNE-EDWARDS, Henri, (1800-1885)
MONTICELLI, Teodoro, Abate, (1757-1845)
MURCHISON, Roderick Impey, Sir, (1792-1871)
OMBONI, Giovanni, (1829-1910)
PAGLIA, Enrico, (-1889)
PALMIERI, Luigi, (1807-1896)
PARETO, Lorenzo, (1800-1865)
PARKER, William Kitchen, (1823-1890)
PAROLINI, Alberto Cav., (1788-1867)
PASINI, Lodovico, (1804-1870)
PENTLAND, Joseph Barclay, (1797-1873)
PHILLIPS, John, (1800-1874)
PILLA, Leopoldo, (1805-1848)
PINI, Ermenegildo, (1739-1825)
PIRONA, Giulio Andrea, (1822-1895)
PONZI, Giuseppe di Tomaso, (1805-1885)
PRATT, Samuel Peace, (1789-1863)
RÖMER, Hermann, (1816-1894)
ROTH, Justus Ludwig Adolph, (1818-1892)
ROVASENDA, Luigi di, Conte, (1826-1917)
SARTORIUS VON WALTERSHAUSEN, Wolfgang, (1809-1876)
SAVI, Paolo, (1798-1871)
SCACCHI, Arcangelo, (1810-1893)
SCARABELLI GOMMI FLAMINI, Giuseppe, (1820-1905)
SCHAUROTH, Karl von, Freiherr, (1818-1893)
SCROPE, George Poulett, (né Thomson), (1797-1876)
SELLA, Quintino, (1827-1884)
SERPIERI, Alessandro, (1823-1885)
SISMONDA, Angelo, (1807-1878)
SISMONDA, Eugenio, (1815-1870)
SOLDANI, Ambrogio, Pater, (1736-1808)
STABILE, Giuseppe, Abate, (1827-1869)
STOPPANI, Antonio, (1824-1891)
STROBEL, Pellegrino, (1821-1895)
TACCHINI, Pietro, (1838-1905)
TENORE, Gaetano, (1826-1903)
TENORE, Michele, (-)
TONDI, Matteo, (1762-1835)
TREVELYAN, Walter Calverley, Sir, (1797-1879)
VANDELLI, Domenico, (1735-1816)

1850-1899

ARTINI, Ettore, (1866-1928)

ITALY 1850-1899 Cont'd.

BABBAGE, Charles, (1792-1871)
BALDACCI, Luigi, (1850-1927)
BALSAMO-CRIVELLI, Giuseppe, Marchese, (1800-1874)
BARAFFAEL, Angelo, (1866-1928)
BARATTA, Mario, (1868-1935)
BARETTI, Martino, (1841-1905)
BARGELLINI, Mariano, (1831-1906)
BARZANÒ, Gaetano, (-?1859)
BASSANI, Francesco, (1853-1916)
BATE, Dorothea Minola Alice, (1879-1951)
BELLARDI, Luigi, (1818-1889)
BELLINI, Raffaelo, (1874-1930)
BELLOTTI, Christoforo, (1823-1919)
BERGEAT, Alfred, Edmund, (1866-1924)
BERTELLI, Petro Timoteo, (1826-1905)
BERTOLIO, Leopoldo Sollman, (-?1922)
BIAGI, Giuseppi, (1859-1905)
BIANCONI, Gian Giuseppe, (1809-1878)
BIBOLINI, Aldo, (1876-1949)
BLAKE, John Frederick, Reverend, (1839-1906)
BOEGAN, Eugenio, (1875-1939)
BOERIS, Giovanni, (1867-1946)
BOMBICCI PORTA, Luigi, (1833-1903)
BONARELLI, Guido, (1871-1951)
BONETTI, Filippo, (1854-1911)
BONOMINI, Celestino, Don, (1866-?1946)
BORSARI, Ferdinando, (1858-1891)
BOTTI, Ulderigo, (1822-1905)
BRADY, Henry Bowman, (1835-1891)
BRANCA, Wilhelm von, (1844-1928)
BREZINA, Maria Aristides, (1848-1909)
BRONN, Heinrich Georg, (1800-1862)
BRUGNATELLI, Luigi, (1859-1928)
BRUNO, Carlo, (1831-1916)
BUCCA, Lorenzo, (1857-1930)
BUCH, Christian Leopold von, Freiherr, (1774-1853)
CACCIAMALI, Giambattista, (1856-1934)
CALCARA, Pietro, (1819-1854)
CAMPANI, Giovanni, (1820-1891)
CANAVARI, Mario, (1855-1928)
CAPACCI, Celso, (1854-1929)
CAPELLINI, Giovanni, (1833-1922)
CARUS, Carl Gustav, (1789-1869)
CATULLO, Tommaso Antonio, (1782-1869)
CECCHI, Filippo, (1822-1887)
CERMENATI, Mario, (1868-1924)
CERULLI-IRELLI, Serafino, (1873-1946)
CESÀRO, Guiseppe Raimondo Pio, (1849-1939)
CHECCHIA-RISPOLI, Giuseppe, (1877-1947)
CHURCHILL, George Cheetham, (1822-1906)
CIAMPI, Adolfo, (1876-1934)
CIOFALO, Saverio, (1825-1925)
CLERICI, Enrico, (1862-1938)
COCCHI, Igino, (1827-1913)

ITALY 1850-1899 Cont'd.

CONEDERA, Raimondo, (1859-1925)
CONTI, Piero Ginori, (1856-1939)
CORNALIA, Emilio, (1824-1882)
CORTESE, Emilio, (1856-1938)
CORTI, Benedetto, (1868-1907)
COSSA, Alfonso, (-)
COSTA, Oronzio Gabriele, (1787-1867)
COZZAGLIO, Arturo, (1862-)
CURIONI, Giulio, (1796-1878)
D'ACHIARDI, Antonio, (1839-1902)
D'ACHIARDI, Giovanni, (1872-1944)
DAL PIAZ, Giorgio, (1872-1962)
DAUBENY, Charles Giles Brindle, (1795-1867)
DE AGOSTINI, Giovanni, (-)
DE ALESSANDRI, Giulio, (-?1920)
DE GREGORIO, Antonio, Marchese, (1855-1930)
DE LORENZO, Giuseppe, (1871-1957)
DEMARCHI, Lamberto, (1841-1907)
DE NICOLIS, Enrico, (1841-1908)
DENINGER, Karl, (1878-1917)
DEO GRATIAS, Pietro Perrando, Don, (?1846-1889)
DE PRETTO, Olinto, (1857-1921)
DE REGNY, Paolo Vinassa, (1871-1957)
DE ROSSI, Michele Stefano, (1834-1898)
DE STEFANI, Carlo, (1851-1924)
DE VISIANI, Roberto, (1800-1878)
DE ZIGNO, Achille, Barone, (1813-1888)
DI MARCHI, Luigi, (1857-1936)
DI STEFANO, Giovanni, (1856-1918)
DODERLEIN, Pietro, (1810-1895)
DOHRN, Felix Anton, (1840-1909)
D'OSSAT, Gioacchino De Angelis, (1865-1957)
DU PASQUIER, Léon, (1864-1897)
EHRENBERG, Christian Gottfried, (1795-1876)
ERLE, C., Reverend, (1790-1870?)
FERRARIS, Erminio, (1851-1928)
FINO, Vincenzo, (1848-1914)
FOLCO, Carlo, (1866-1932)
FOOTE, Albert E., (1846-1895)
FORESTI, Lodovico, (1829-1913)
FORMA, Ernesto, (1869-1911)
FORNASINI, Carlo, (1854-1931)
FORSYTH MAJOR, Charles Immanuel, (1843-1923)
FOSTER, Clement Le Neve, Sir, (1841-1904)
FRANCO, Pasquale, (1852-1907)
FUCHS, Philippe Jacques Edmond, (1837-1889)
FUCINI, Alberto, (1864-1940)
GASTALDI, Bartolomeo, (1818-1879)
GAUDIN, Charles Théophile, (1822-1866)
GEMMELLARO, Carlo, (1787-1866)
GEMMELLARO, Gaetano Giorgio, (1832-1904)
GILBERT, Josiah, (1814-1892)
GIORDANO, Felice, (1825-1892)
GIULIETTI, G. M., (-?1928)

ITALY 1850-1899 Cont'd.

GRANDIDIER, Alfred, (1836-1921)
GREENOUGH, George Bellas, (1778-1855)
GUIDONI, Gerolamo, (1794-1870)
GUISCARDI, Guglielmo, (1821-1885)
HÄBERLE, Daniel, (1864-1934)
HAECKEL, Ernst Heinrich Philipp August, (1834-1919)
HAIDINGER, Wilhelm Karl von, Ritter, (1795-1871)
HAMILTON, William John, (1805-1867)
HARADA, Toyokitsi, (-1894)
HECKEL, Johann Jakob, (1790-1857)
ISSEL, Arturo, (1842-1922)
JAGGAR, Thomas Augustus, (1871-1953)
JERVIS, William Paget, (1831-1906)
JOHNSTON-LAVIS, Henry James, (1856-1914)
JULIAN, Henry Forbes, (1861-1912) [PROSPECTORS, DIVINERS AND MINING ENGINEERS]
KING, Samuel William, (1821-1868)
LACROIX, François Antoine Alfred, (1863-1948)
LAIS, Giuseppe, Padre, (1845-1921)
LANINO, Giuseppe, (1832-1907)
LASAULX, Arnold von, (1839-1886)
LAWLEY, Robert, (1818-1881)
LINCIO, Gabrielle, (1874-1938)
LIOY, Paolo, (1834-1911)
LOTTI, Bernardino, (1847-1933)
LOVISATO, Domenico, (1842-1916)
LYELL, Charles, Sir, (1797-1875)
MACLAREN, James Malcolm, (1873-1935)
MALLADRA, Alessandro, (1865-1945)
MANASSE, Ernesto, (1875-1922)
MARENGO, Paulo, (1863-1929)
MARGERIE, Emmanuel Marie Pierre Martin Jacquin de, (1862-1953)
MARIANI, Ernesto, (-?1939)
MARINELLI, Giovanni, (1846-1900)
MARINELLI, Olinto, (1874-1926)
MARINONI, Camillo, (1845-1883)
MARTELLI, Alessandro, (1877-1934)
MASSALONGO, Abramo Bartolomeo, (1824-1860)
MATTEUCCI, Raffaele Vittorio, (1862-1909)
MATTIROLO, Ettore, (1853-1923)
MELI, Romolo, (1852-1920)
MENEGHINI, Giuseppe Giovanni Antonio, (1811-1889)
MENEGUZZO, Giovanni, (-)
MENIN, Aurelio, Padre, (-)
MERCALLI, Giuseppe, (1850-1914)
MICHELOTTI, Giovanni, (1814-1898)
MILCH, Ludwig, (1867-1928)
MILLOSEVICH, Federico, (1875-1942)
MILNE-EDWARDS, Henri, (1800-1885)
MOJSISOVICS, Johann August Georg Edmund, Edler von Mojsvàr, (1839-1907)
MONETTI, Luigi, (1864-1926)
MONTEROSATO, Tommaso di Maria, Marchese di, (1841-1927)
MURCHISON, Roderick Impey, Sir, (1792-1871)

ITALY 1850-1899 Cont'd.

NEGRI, Arturo, (-1896)
NEGRI, Cristoforo, (-)
NICCOLI, Enrico, (1836-1915)
NIEVO, Ippolito, (1867-1916)
NOVARESE, Vittorio, (1861-1948)
OCHSENIUS, Carl Christian, (1830-)
OGILVIE GORDON, Maria Matilda, Dame, (1864-1939)
OMBONI, Giovanni, (1829-1910)
ORSINI, Antonio, (-1870)
PAGLIA, Enrico, (-1889)
PALAZZO, Luigi, (1861-1933)
PALMIERI, Luigi, (1807-1896)
PANEBIANCO, Ruggero, (1848-1930)
PANTANELLI, Dante, (1844-1913)
PARETO, Lorenzo, (1800-1865)
PARKER, William Kitchen, (1823-1890)
PARMA, Augusto, (1857-1926)
PAROLINI, Alberto Cav., (1788-1867)
PARONA, Carlo Fabrizio, (1855-1939)
PASINI, Lodovico, (1804-1870)
PAULUCCI, Marianna, Marchesa, (1835-1919)
PELLATI, Nicola, (1835-1907)
PELLOUX, Alberto, (-)
PENTLAND, Joseph Barclay, (1797-1873)
PHILLIPS, John, (1800-1874)
PICCINELLI, Piero, Conte, (1865-1937)
PIRONA, Giulio Andrea, (1822-1895)
PLATANIA, Gaetano, (1867-1938)
PONZI, Giuseppe di Tomaso, (1805-1885)
PORRIO, Cesare, (1865-1940)
PORTIS, Alessandro, (1853-1931)
POZZI, Giovanni, (-?1891)
PRATT, Samuel Peace, (1789-1863)
PRELLER, Charles du Riche, (1844-)
RAGAZZONI, Giuseppe, (-1898)
RATH, Gerhard vom, (1830-1888)
REPOSSI, Emilio, (1876-1931)
RISTORI, Giuseppe, (1856-1905)
RIVA, Carlo, (1872-1902)
RIVA-PALAZZI, Giovanni, (1838-1913)
ROCCATI, Alessandro, (1872-1928)
RÖMER, Hermann, (1816-1894)
ROSSI, Arturo, (1860-1892)
ROTH, Justus Ludwig Adolph, (1818-1892)
ROVASENDA, Luigi di, Conte, (1826-1917)
ROVERETO, Gaetano, (1870-1952)
SABATINI, Venturino, (1879-1922)
SACCO, Federico, (1864-1948)
SALMOJRAGHI, Francesco, (1837-?1910)
SARTORIUS VON WALTERSHAUSEN, Wolfgang, (1809-1876)
SAVI, Paolo, (1798-1871)
SCACCHI, Arcangelo, (1810-1893)
SCACCHI, Eugenio, (1854-1929)
SCALIA, Salvatore, (1874-1923)

ITALY 1850-1899 Cont'd.

SCARABELLI GOMMI FLAMINI, Giuseppe, (1820-1905)
SCHAUROTH, Karl von, Freiherr, (1818-1893)
SCROPE, George Poulett, (né Thomson), (1797-1876)
SECCO, Andrea, (-1889)
SEGRÉ, Claudio, (1853-1928)
SEGUENZA, Giuseppe, (1833-1889)
SEGUENZA, Luigi, (1873-1908)
SELLA, Quintino, (1827-1884)
SERPIERI, Alessandro, (1823-1885)
SILVESTRI, Orazio, (-1890)
SISMONDA, Angelo, (1807-1878)
SISMONDA, Eugenio, (1815-1870)
SOLMS-LAUBACH, Hermann zu, Graf, (1842-1915)
SOMIGLIANA, Carlo, (?1859-1955)
SORDELLI, Ferdinando, (?1837-1916)
SPEZIA, Giorgio, (1842-1912)
SPIREK, Vincenz, (1852-1907)
STABILE, Giuseppe, Abate, (1827-1869)
STATUTI, Augusto, (1829-1911)
STELLA, Augusto, (1863-1944)
STOPPANI, Antonio, (1824-1891)
STROBEL, Pellegrino, (1821-1895)
STRÜVER, Johannes Karl Theodor, called Giovanni, (1842-1915)
TACCHINI, Pietro, (1838-1905)
TARAMELLI, Torquato, (1845-1922)
TELLINI, Achille, (1866-1938)
TENORE, Gaetano, (1826-1903)
TERMIER, Pierre, (1859-1930)
TERRIGI, Guglielmo, (1831-1892)
TITTONI, Tommaso, (1855-1931)
TOMMASI, Annibale, (1858-1921)
TONINI, Lorenzo, ([1859]-1926)
TRABUCCO, Giacomo, (1845-1925)
TRAVERSO, Giovanni Battista, (1843-1914)
TRENTANOVE, Giorgio Morando, (1874-1914)
TREVELYAN, Walter Calverley, Sir, (1797-1879)
TUCCIMEI, Giuseppe, (1851-1915)
UGOLINI, Riccardo, (1874-1936)
UZIELLI, Gustavo, (1839-1911)
VACCHELLI, Nicola, Generale, (1870-1932)
VAGHI, Vittorio, (-1921)
VERRI, Antonio, (1839-1925)
VERRI, Pietro, (-)
VILLA, Antonio, (-1885)
VILLA, Giovanni Battista, (-1887)
VIOLA, Carlo Maria, (1855-1925)
WASHINGTON, Henry Stephens, (1867-1934)
WILLIAMS, John Francis, (1862-1891)
WYNDHAM, T. Heathcote, (?1842-1876)
ZÀMARA, Giuseppe, (1837-1917)

1900 to present

AIROLDI, Marco, (1900-1937)
ALOISI, Piero, (1881-1938)

ITALY 1900 to present Cont'd.

ANDREATTA, Benedettino Ciro, (1906-1960)
ANELLI, Mario, (1882-1953)
ARTINI, Ettore, (1866-1928)
AZARI, Mario, (-?1932)
BADEN-POWELL, Donald Ferlys Wilson, (1897-1973)
BAIRD, Chester Alvan, (1896-1959)
BALDACCI, Luigi, (1850-1927)
BARAFFAEL, Angelo, (1866-1928)
BARATTA, Mario, (1868-1935)
BARETTI, Martino, (1841-1905)
BARGELLINI, Mariano, (1831-1906)
BASSANI, Francesco, (1853-1916)
BATE, Dorothea Minola Alice, (1879-1951)
BATTAGLIA, Rafaello, (1896-1958)
BELLINI, Raffaelo, (1874-1930)
BELLOTTI, Christoforo, (1823-1919)
BERGEAT, Alfred, Edmund, (1866-1924)
BERTELLI, Petro Timoteo, (1826-1905)
BERTOLIO, Leopoldo Sollman, (-?1922)
BIAGI, Giuseppi, (1859-1905)
BIANCHI, Angelo, (1892-1970)
BIBOLINI, Aldo, (1876-1949)
BLAKE, George Stanfield, (-1940)
BLAKE, John Frederick, Reverend, (1839-1906)
BLANC, Alberto Carlo, (1906-1960)
BLANC, Gian Alberto, Barone, (1879-1966)
BOAGA, Giovanni, (1902-1961)
BOEGAN, Eugenio, (1875-1939)
BOERIS, Giovanni, (1867-1946)
BOMBICCI PORTA, Luigi, (1833-1903)
BONARELLI, Guido, (1871-1951)
BONETTI, Filippo, (1854-1911)
BONOMINI, Celestino, Don, (1866-?1946)
BOSWORTH, Thomas Owen, (1882-1928)
BOTTI, Ulderigo, (1822-1905)
BRANCA, Wilhelm von, (1844-1928)
BREZINA, Maria Aristides, (1848-1909)
BRUGNATELLI, Luigi, (1859-1928)
BRUNO, Carlo, (1831-1916)
BUCCA, Lorenzo, (1857-1930)
CACCIAMALI, Giambattista, (1856-1934)
CALCIATI, Cesare, (1885-1929)
CANAVARI, Mario, (1855-1928)
CAPACCI, Celso, (1854-1929)
CAPELLINI, Giovanni, (1833-1922)
CARPANESE, Tullio, (-?1940)
CASTIGLIONI, Bruno, (1898-1945)
CERMENATI, Mario, (1868-1924)
CERULLI-IRELLI, Serafino, (1873-1946)
CESÀRO, Guiseppe Raimondo Pio, (1849-1939)
CHECCHIA-RISPOLI, Giuseppe, (1877-1947)
CHURCHILL, George Cheetham, (1822-1906)
CIAMPI, Adolfo, (1876-1934)
CIOFALO, Saverio, (1825-1925)

ITALY 1900 to present Cont'd.

CLERICI, Enrico, (1862-1938)
CONEDERA, Raimondo, (1859-1925)
CONFORTO, Bruno, (1918-1963)
CONTI, Piero Ginori, (1856-1939)
CORTESE, Emilio, (1856-1938)
CORTI, Benedetto, (1868-1907)
COZZAGLIO, Arturo, (1862-)
CROSS, Rodman K., (1911-1970)
D'ACHIARDI, Antonio, (1839-1902)
D'ACHIARDI, Giovanni, (1872-1944)
DAL PIAZ, Giambattista, (1927-)
DAL PIAZ, Giorgio, (1872-1962)
DE ALESSANDRI, Giulio, (-?1920)
DE GREGORIO, Antonio, Marchese, (1855-1930)
DE LORENZO, Giuseppe, (1871-1957)
DEMARCHI, Lamberto, (1841-1907)
DE NICOLIS, Enrico, (1841-1908)
DENINGER, Karl, (1878-1917)
DE PRETTO, Olinto, (1857-1921)
D'ERASMO, Geremia, (1887-1962)
DE REGNY, Paolo Vinassa, (1871-1957)
DESIO, Ardito, (1897-)
DE STEFANI, Carlo, (1851-1924)
DI MARCHI, Luigi, (1857-1936)
DI STEFANO, Giovanni, (1856-1918)
DIXON, Henry George Dacres, (1887-1947)
D'OSSAT, Gioacchino De Angelis, (1865-1957)
DRAGO, Ciro, (1895-1960)
EDWARDS, Wilfred Norman, (1890-1956)
FABIANI, Ramiro, (1879-1954)
FALINI, Filippo, (1917-1963)
FAUSSONE, Luigi, (1880-1925)
FERASIN, Francesco, (1928-1958)
FERRARIS, Erminio, (1851-1928)
FINO, Vincenzo, (1848-1914)
FIOCCO, Giorgio, (1931-)
FOLCO, Carlo, (1866-1932)
FORESTI, Lodovico, (1829-1913)
FORMA, Ernesto, (1869-1911)
FORNASINI, Carlo, (1854-1931)
FORSYTH MAJOR, Charles Immanuel, (1843-1923)
FOSSA-MANCINI, Enrico, (1884-1950)
FOSTER, Clement Le Neve, Sir, (1841-1904)
FRANCHI, Secondo, (-?1939)
FRANCO, Pasquale, (1852-1907)
FUCINI, Alberto, (1864-1940)
GEMMELLARO, Gaetano Giorgio, (1832-1904)
GEMMELLARO, Mariano, (1879-1921)
GIANNOTTI, Carlo, (1897-1929)
GIULIETTI, G. M., (-?1928)
GORTANI, Michele, (1883-1966)
GRANDIDIER, Alfred, (1836-1921)
GRILL, Emanuele, (1884-1961)
HÄBERLE, Daniel, (1864-1934)

ITALY 1900 to present Cont'd.

HAECKEL, Ernst Heinrich Phillipp August, (1834-1919)
HARRISON, Thomas Samuel, (1881-1964)
HENDON, Bryan, (1896-1926)
HERITSCH, Franz, (1882-1945)
HLAUSCHEK, Hans, (1896-1971)
INEICHEN, Giovanni, (-1952)
ISSEL, Arturo, (1842-1922)
JAGGAR, Thomas Augustus, (1871-1953)
JERVIS, William Paget, (1831-1906)
JOHNSTON-LAVIS, Henry James, (1856-1914)
JULIAN, Henry Forbes, (1861-1912) [PROSPECTORS, DIVINERS AND MINING ENGINEERS]
KISLING, James William, Jr., (1898-1973)
KLOMPÉ, Theodorus Hendricus Franciscus, (?1904-1963)
LACROIX, François Antoine Alfred, (1863-1948)
LAIS, Giuseppe, Padre, (1845-1921)
LANINO, Giuseppe, (1832-1907)
LINCIO, Gabrielle, (1874-1938)
LIOY, Paolo, (1834-1911)
LORENZI, Arrigo, (-1953)
LOTTI, Bernardino, (1847-1933)
LOVISATO, Domenico, (1842-1916)
LYELL, Charles, Sir, (1797-1875)
MACKENZIE, Andrew Nicholls "Kelley", (1888-1951)
MACLAREN, James Malcolm, (1873-1935)
MAGISTRETTI, Luigi, (-?1959)
MALLADRA, Alessandro, (1865-1945)
MANASSE, Ernesto, (1875-1922)
MARCHESINI, Enrico, (1911-1968)
MARGERIE, Emmanuel Marie Pierre Martin Jacquin de, (1862-1953)
MARINELLI, Olinto, (1874-1926)
MARTELLI, Alessandro, (1877-1934)
MARTINELLI, Cesare, (-1929)
MARTINOTTI, Anna, (-1931)
MATTEUCCI, Raffaele Vittorio, (1862-1909)
MATTIROLO, Ettore, (1853-1923)
MELI, Romolo, (1852-1920)
MERCALLI, Giuseppe, (1850-1914)
MERCIAI, Giuseppe, (1877-1939)
MILCH, Ludwig, (1867-1928)
MILLOSEVICH, Federico, (1875-1942)
MOJSISOVICS, Johann August Georg Edmund, Edler von Mojsvàr, (1839-1907)
MONETTI, Luigi, (1864-1926)
MONTEROSATO, Tommaso di Maria, Marchese di, (1841-1927)
MORANDINI, Giuseppe, (1907-1969)
NICCOLI, Enrico, (1836-1915)
NICOLAS, Jean Paul, (1921-1967)
NIEVO, Ippolito, (1867-1916)
NOÉ, Emilio, (1881-1940)
NOVARESE, Vittorio, (1861-1948)
OGILVIE GORDON, Maria Matilda, Dame, (1864-1939)
OMBONI, Giovanni, (1829-1910)
ONORATO, Ettore, (-)

ITALY 1900 to present Cont'd.

PALAZZO, Luigi, (1861-1933)
PANEBIANCO, Ruggero, (1848-1930)
PANICHI, Ugo, (-?1967)
PANTANELLI, Dante, (1844-1913)
PARDÉ, Maurice, (1893-1973)
PARMA, Augusto, (1857-1926)
PARNISARI, Carlo, (-?1929)
PARONA, Carlo Fabrizio, (1855-1939)
PATRIZI, Saverio, (1902-1957)
PAULUCCI, Marianna, Marchesa, (1835-1919)
PELLATI, Nicola, (1835-1907)
PENTA, Francesco, (1899-1965)
PERRIER, Carlo, (1886-1948)
PICCINELLI, Piero, Conte, (1865-1937)
PIVETEAU, Jean, (1898-)
PLATANIA, Gaetano, (1867-1938)
PORRIO, Cesare, (1865-1940)
PORTIS, Alessandro, (1853-1931)
PRINCIPI, Paolo, (-)
QUERCIGH, Emanuele, (1885-1939)
RASMUSSEN, William Charles, (1917-1973)
REA, Henry Carter, (1900-1963)
REED, Lyman Coleman, (1900-1970)
REPOSSI, Emilio, (1876-1931)
RISTORI, Giuseppe, (1856-1905)
RIVA, Carlo, (1872-1902)
RIVA-PALAZZI, Giovanni, (1838-1913)
ROCCATI, Alessandro, (1872-1928)
ROCH, Edouard, (1901-)
RODIGHIERO, Andrea, (1892-1917)
ROVASENDA, Luigi di, Conte, (1826-1917)
ROVERETO, Gaetano, (1870-1952)
SABATINI, Venturino, (1879-1922)
SACCO, Federico, (1864-1948)
SALMOJRAGHI, Francesco, (1837-?1910)
SCACCHI, Eugenio, (1854-1929)
SCALIA, Salvatore, (1874-1923)
SCARABELLI GOMMI FLAMINI, Giuseppe, (1820-1905)
SCOTTI, Ariberto, (1885-1925)
SEGRÉ, Claudio, (1853-1928)
SEGUENZA, Luigi, (1873-1908)
SOLMS-LAUBACH, Hermann zu, Graf, (1842-1915)
SOMIGLIANA, Carlo, (?1859-1955)
SORDELLI, Ferdinando, (?1837-1916)
SPEZIA, Giorgio, (1842-1912)
SPIREK, Vincenz, (1852-1907)
STEFANINI, Giuseppe, (1882-1938)
STELLA, Augusto, (1863-1944)
STRÜVER, Johannes Karl Theodor, called Giovanni, (1842-1915)
TACCHINI, Pietro, (1838-1905)
TARAMELLI, Torquato, (1845-1922)
TAYLOR, Griffith, (1881-1963)
TAZIEFF, Haroun, (1914-)
TELLINI, Achille, (1866-1938)

ITALY 1900 to present Cont'd.

TENORE, Gaetano, (1826-1903)
TERMIER, Pierre, (1859-1930)
TITTONI, Tommaso, (1855-1931)
TOMMASI, Annibale, (1858-1921)
TONINI, Lorenzo, ([1859]-1926)
TONSO, Alessandro, (-)
TRABUCCO, Giacomo, (1845-1925)
TRAVERSO, Giovanni Battista, (1843-1914)
TRENER, Giovanni Battista, (-?1954)
TRENTANOVE, Giorgio Morando, (1874-1914)
TUCCIMEI, Giuseppe, (1851-1915)
UGOLINI, Riccardo, (1874-1936)
UZIELLI, Gustavo, (1839-1911)
VACCHELLI, Nicola, Generale, (1870-1932)
VAGHI, Vittorio, (-1921)
VARDABASSO, Silvio, (1891-1966)
VERCELLI, Francesco, (-?1952)
VERRI, Antonio, (1839-1925)
VERRI, Pietro, (-)
VIOLA, Carlo Maria, (1855-1925)
VISENTINI, Marco, (1885-1964)
WARING, Gerald Ashley, (1883-1971)
WASHINGTON, Henry Stephens, (1867-1934)
WIEDENMAYER, Carl, (1897-1951)
WILSON, Thomas Carroll, (1906-1969)
WRIGHT, Charles Will, (1879-)
ZACCAGNA, Domenico, (-?1940)
ZÀMARA, Giuseppe, (1837-1917)
ZAMBONINI, Ferruccio, (1880-1932)
ZENARI, Silvia, (1895-1956)
ZEZI, Pietro, (1884-1914)
ZUBER, Stanislaw, (1893-1947)
ZUFFARDI, Pietro, (1885-1915)

Time unknown
D'AREZZO, Restoro, (-)
D'ISENGARD, Luigi, Barone, (-)
LIPPI, Carmine Antonio, (-)

ITALY See also SARDINIA; SICILY.

IVORY COAST
1850-1899
BEEBY-THOMPSON, Arthur, (1873-1968)
BRALY, Adrien, (1864-1927)

1900 to present
BEEBY-THOMPSON, Arthur, (1873-1968)
BRALY, Adrien, (1864-1927)
MALAVOY, Jean, (1903-1945)
MILLOT, Georges, (1917-)

JAMAICA

1800-1849

DE LA BECHE, Henry Thomas, Sir, (1796-1855)
MOORE, John Carrick, (1804-1898)
SAWKINS, James Gay, (1806-1878)

1850-1899

BRADY, Henry Bowman, (1835-1891)
BROWN, Charles Barrington, (1839-1917)
COCKERELL, Theodore Dru Alison, (1866-1948)
DE LA BECHE, Henry Thomas, Sir, (1796-1855)
GUPPY, Robert John Lechmere, (1836-1916)
MATLEY, Charles Alfred, (1866-1947)
MOORE, John Carrick, (1804-1898)
SAWKINS, James Gay, (1806-1878)
THOMPSON, Beeby, (1848-1931)
WALL, George Parkes, (1832-1912)

1900 to present

BARRETT, Lucas, (1837-1862)
BOSWELL, Katherine Cumming, (1888-1952)
BROWN, Charles Barrington, (1839-1917)
BUTZ, Walter "Wally", (1905-1962)
CHUBB, Lawrence John, (?1888-1971)
COCKERELL, Theodore Dru Alison, (1866-1948)
CUSHMAN, Joseph Augustine, (1881-1949)
GOREAU, Thomas Fritz, (1924-1970)
GUPPY, Robert John Lechmere, (1836-1916)
HARRISON, John Vernon, (1892-1972)
HARRISON, Thomas Samuel, (1881-1964)
LAKE, Horace Alexander, (1887-1957)
MATLEY, Charles Alfred, (1866-1947)
ROCH, Edouard, (1901-)
SPATH, Leonard Frank, (1882-1957)
SULLINS, M. Joseph, (1930-1972)
THOMPSON, Beeby, (1848-1931)
TRECHMANN, Charles Taylor, (1885-1964)
WALL, George Parkes, (1832-1912)
ZANS, Verners Aleksandrs, (1904-1961)

JAN MAYEN ISLAND (ARCTIC OCEAN)

1850-1899

SCHARIZER, Rudolf, (1859-1935)

1900 to present

JOESTING, Henry Rochambeau, (1903-1965)
RENAUD, André, (1904-1964)
ROSSBY, Carl-Gustaf Arvid, (1898-1957)
SCHARIZER, Rudolf, (1859-1935)
TYRRELL, George Walter, (1883-1961)
WORDIE, James Mann, Sir, (?1890-1962)

JAPAN

Up to 1800

TAVERNIER, Jean Baptiste, baron d'Aubonne, (1605-1689)

JAPAN Cont'd.

1800-1849
BLAKE, William Phipps, (1826-1910)
CAMPBELL, John Francis, (1821-1885)
GEYLER, Hermann Theodor, (1834-1889)

1850-1899
BECHER, Harry Macdonald, (1855-1893)
BLAKE, William Phipps, (1826-1910)
BRADY, Henry Bowman, (1835-1891)
CAMPBELL, John Francis, (1821-1885)
DRUMMOND, Henry, (1851-1897)
FUKUCHI, Nobuyo, (1877-1934)
GEYLER, Hermann Theodor, (1834-1889)
GOTTSCHE, Carl Christian, (1855-1909)
GULICK, John Thomas, (1832-1923)
HANNAY, Harry Campbell, (1874-1961)
HARADA, Toyokitsi, (-1894)
HIKI, Tadasu, (1866-1927)
HOLLAND, William Jacob, (1848-1932)
JAGGAR, Thomas Augustus, (1871-1953)
JIMBO, Kotora, (1867-1924)
KIKUCHI, Dairoko, Baron, (1855-1917)
KINCH, Edward, (1848-1920)
KOTÔ, Bundjirô, (1856-1935)
KUSAKABE, Shiroka, (1872-1924)
LYMAN, Benjamin Smith, (1835-1920)
MENDENHALL, Thomas Corwin, (1841-1924)
MILNE, John, (1850-1913)
MOJSISOVICS, Johann August Georg Edmund, Edler von Mojsvàr, (1839-1907)
MOON, Frederick William, (-1925)
NAUMANN, Edmund, (1854-1927)
NORDENSKIÖLD, Nils Adolf Erik, (1832-1901)
OMORI, Fusakichi, (1868-1923)
OTANI, Ryokichi, (1875-1934)
PUMPELLY, Raphael, (1837-1923)
RICHTHOFEN, Paul Wilhelm Ferdinand von, Freiherr, (1833-1905)
SEIKEI, Sekiya, (1855-1896)
SHIDA, Toshi, (1876-1936)
STEPHENSON, Lloyd William, (1876-1962)
TERADA, Torahiko, (1878-1935)
TOKUNAGA, Shigeyasu, (1874-1940)
WADA, Tsunashirô, (1856-1920)
WOEIKOF, Alexander, (1842-1916)
YABE, Hisakatsu, (1878-1969)
YAMASAKI, Naomasa, (1870-1929)

1900 to present
ANDREWS, Roy Chapman, (1884-1960)
ASANO, Kiyoshi, (1910-)
BLAKE, William Phipps, (1826-1910)
BULLEN, Keith Edward, (1906-1976)
CHANEY, Ralph Works, (1890-1971)
FUKUCHI, Nobuyo, (1877-1934)

JAPAN 1900 to present Cont'd.

GOTTSCHE, Carl Christian, (1855-1909)
GULICK, John Thomas, (1832-1923)
HANNAY, Harry Campbell, (1874-1961)
HIKI, Tadasu, (1866-1927)
HOLLAND, William Jacob, (1848-1932)
ISHIMOTO, Mishio, (1893-1940)
JAGGAR, Thomas Augustus, (1871-1953)
JIMBŌ, Kotora, (1867-1924)
JOHNSON, Charles George, (1914-1969)
KIKUCHI, Dairoko, Baron, (1855-1917)
KINCH, Edward, (1848-1920)
KINNEY, Harry Dill, (1886-1961)
KOBAYASHI, Teiichi, (1901-)
KOTÔ, Bundjirô, (1856-1935)
KRYSHTOFOVICH, Afrikan Nikolaevich, (1885-1953)
KUNO, Hisashi, (1910-1969)
KUSAKABE, Shiroka, (1872-1924)
LEDINGHAM, Glen Wallace, (1911-1978)
LYMAN, Benjamin Smith, (1835-1920)
MASON, Arnold Caverly, (1906-1961)
MENDENHALL, Thomas Corwin, (1841-1924)
MILNE, John, (1850-1913)
MOJSISOVICS, Johann August Georg Edmund, Edler von Mojsvàr, (1839-1907)
MOON, Frederick William, (-1925)
NAGATA, Takesi, (1913-)
NAKAYA, Ukichiro, (1900-1962)
NAUMANN, Edmund, (1854-1927)
NININGER, Harvey Harlow, (-)
NORDENSKIÖLD, Nils Adolf Erik, (1832-1901)
ÒCURA, Hirobumi, (1921-1969)
OMORI, Fusakichi, (1868-1923)
OTANI, Ryokichi, (1875-1934)
POLLOCK, Charles Murray, (1898-1969)
PUMPELLY, Raphael, (1837-1923)
RICHTHOFEN, Paul Wilhelm Ferdinand von, Freiherr, (1833-1905)
SCHENCK, Hubert Gregory, (1897-1960)
SHIDA, Toshi, (1876-1936)
SHODA, Mikio, (1924-1974)
STEPHENSON, Lloyd William, (1876-1962)
STOPES, Marie Charlotte Carmichael, (1880-1958)
TAYLOR, Griffith, (1881-1963)
TAZIEFF, Haroun, (1914-)
TERADA, Torahiko, (1878-1935)
TOKUNAGA, Shigeyasu, (1874-1940)
WADA, Tsunashirô, (1856-1920)
WARING, Gerald Ashley, (1883-1971)
WING, Monta Eldo, (1897-1976)
WOEIKOF, Alexander, (1842-1916)
YABE, Hisakatsu, (1878-1969)
YAMASAKI, Naomasa, (1870-1929)

JAVA <u>See</u> INDONESIA.

JERSEY <u>See</u> CHANNEL ISLANDS (United Kingdom).

JORDAN

1800-1849

HUDLESTON, Wilfrid Hudleston, (né Simpson), (1828-1909)
TAUNTON, John Hooke, (1820-1893)

1850-1899

BEEBY-THOMPSON, Arthur, (1873-1968)
HUDLESTON, Wilfrid Hudleston, (né Simpson), (1828-1909)
HULL, Edward, (1829-1917)
LARTET, Louis, (1840-1899)
TAUNTON, John Hooke, (1820-1893)
WADE, Arthur, (1878-1951)

1900 to present

BEEBY-THOMPSON, Arthur, (1873-1968)
BLAKE, George Stanfield, (-1940)
CAMPBELL, Keir Arthur, (-1955)
CLAPP, Frederick Gardner, (1879-1944)
EDWARDS, Wilfred Norman, (1890-1956)
HUDLESTON, Wilfrid Hudleston, (né Simpson), (1828-1909)
HULL, Edward, (1829-1917)
HUXLEY, Julian Sorell, Sir, (1887-1975)
MORRIS, Frederick Kuhne, (1885-1962)
MUIR-WOOD, Helen Marguerite, (1896-1968)
WADE, Arthur, (1878-1951)
ZIMERMAN, Sam, (1905-1972)

JUAN FERNANDEZ ISLAND (CHILE)

1900 to present

QUENSEL, Percy Dudgeon, (1881-1966)

JUGOSLAVIA _See_ YUGOSLAVIA.

KEELING ISLANDS (PACIFIC OCEAN)

1800-1849

DARWIN, Charles Robert, (1809-1882)

1850-1899

DARWIN, Charles Robert, (1809-1882)

KENYA

1850-1899

BREUIL, Henri, Abbé, (1877-1961)
GREGORY, John Walter, (1864-1932)
HOBLEY, Charles William, (1867-1947)
KITSON, Albert Ernest, Sir, (1868-1937)
OSWALD, Felix, (1866-1958)
PRIOR, George Thurland, (1862-1936)
REED, Frederick Richard Cowper, (1869-1946)

1900 to present

BISHOP, Walter William, (1931-1977)
BOWEN, Norman Levi, (1887-1956)
BREUIL, Henri, Abbé, (1877-1961)
BUSK, Henry Gould, (?1891-1956)
CROWTHER, Arthur Frederick, (1929-1954)
DIXON, Henry George Dacres, (1887-1947)
FUCHS, Vivian Ernest, Sir, (1908-)
GREGORY, John Walter, (1864-1932)
HITCHEN, Charles Stansfield, (1904-1971)
HOBLEY, Charles William, (1867-1947)
HUGHES, William John, (?1899-1961)
HUXLEY, Julian Sorell, Sir, (1887-1975)
JÉRÉMINE, Elisabeth, (née Tschernaieff), (1879-1964)
KITSON, Albert Ernest, Sir, (1868-1937)
LEAKEY, Louis Seymour Bazett, (1903-1972)
MAUFE, Herbert Brantwood, (1879-1946)
OSWALD, Felix, (1866-1958)
PARSONS, Ernest, (?1887-1963)
PRIOR, George Thurland, (1862-1936)
REED, Frederick Richard Cowper, (1869-1946)
SIKES, Harold Lesley, (1881-1943)
SIMPSON, George Gaylord, (1902-)
TIDY, Anthony John Lintott, (1928-1969)
WALLACE, Maurice Hudiburg, (1917-1961)

KERGUELEN ISLAND (INDIAN OCEAN)

1900 to present

EDWARDS, Austin Burton, (1909-1960)
EDWARDS, Wilfred Norman, (1890-1956)
MAWSON, Douglas, Sir, (1882-1958)

KOMANDORSKI ISLANDS (ARCTIC U.S.S.R.)

1850-1899

MOROZEWICZ, Józef Marian, (1865-1941)

KOMANDORSKI ISLANDS (ARCTIC U.S.S.R.) Cont'd.

1900 to present
MOROZEWICZ, Józef Marian, (1865-1941)

KOREA

1850-1899
FUKUCHI, Nobuyo, (1877-1934)
GOTTSCHE, Carl Christian, (1855-1909)
JACK, Robert Lockhart, (1878-1964)
KOTÔ, Bundjirô, (1856-1935)
MACLAREN, James Malcolm, (1873-1935)
TOKUNAGA, Shigeyasu, (1874-1940)

1900 to present
ANDREWS, Roy Chapman, (1884-1960)
CARROLL, Don Llewellyn, (1902-1954)
FUKUCHI, Nobuyo, (1877-1934)
GALLAGHER, David, (1906-1968)
GOTTSCHE, Carl Christian, (1855-1909)
HALL, Thomas Clifford Fitzwilliam, (1886-1956)
JACK, Robert Lockhart, (1878-1964)
KOBAYASHI, Teiichi, (1901-)
KOTÔ, Bundjirô, (1856-1935)
MACLAREN, James Malcolm, (1873-1935)
MASON, Arnold Caverly, (1906-1961)
SCHULZ, Karl, (1885-1928)
TOKUNAGA, Shigeyasu, (1874-1940)
WILMOTH, Leslie James, (1881-1963)

KURILE ISLANDS (USSR)

1900 to present
GORSHKOV, Georgii Stepanovich, (1921-1975)

KUWAIT

1900 to present
BOOTS, Paul Heath, (1909-1947)
ISENBERGER, Nathan Perry, (1896-1956)
NEIGHBOURS, Glen Reed, (1917-1961)
RHOADES, Ralph Omer, (1895-1961)

LAOS
1850-1899
JACOB, Charles François Étienne, (1878-1962)

1900 to present
JACOB, Charles François Étienne, (1878-1962)

LAOS *See also* INDOCHINA (unspecified).

LATIN AMERICA (unspecified)
1900 to present
JABLONSKI, Eugene, (1892-1975)

LATVIAN SSR (USSR)
1800-1849
PANDER, Christian Heinrich, (1794-1865)

1850-1899
DOSS, Karl Bruno, (1861-1919)
MICKWITZ, Karl August von, (1849-1910)
PANDER, Christian Heinrich, (1794-1865)

1900 to present
DOSS, Karl Bruno, (1861-1919)
MELLIS, Otto, (1906-1970)
MICKWITZ, Karl August von, (1849-1910)
ZANS, Verners Aleksandrs, (1904-1961)

LEBANON
Up to 1800
BROCCHI, Giovanni Battista (Giambattista), (1772-1826)

1800-1849
BROCCHI, Giovanni Battista (Giambattista), (1772-1826)
EHRENBERG, Christian Gottfried, (1795-1876)

1850-1899
DOUVILLÉ, Joseph Henri Ferdinand, (1846-1937)
EHRENBERG, Christian Gottfried, (1795-1876)
WOODWARD, Henry, (1832-1921)

1900 to present
BOURCART, Jacques, (1891-1965)
DOUVILLÉ, Joseph Henri Ferdinand, (1846-1937)
HUXLEY, Julian Sorell, Sir, (1887-1975)
MILLOT, Georges, (1917-)
WILSON, Roy Arthur, (1891-1970)
WOODWARD, Henry, (1832-1921)

LEEWARD ISLANDS (WEST INDIES)
1850-1899
LENOX-CONYNGHAM, Gerald Ponsonby, Colonel Sir, (1866-1956)

1900 to present
EARLE, Kenneth Wilson, (1891-1944)
HUTTON, Colin Osborne, (1910-1971)

LEEWARD ISLANDS (WEST INDIES) 1900 to present Cont'd.

LENOX-CONYNGHAM, Gerald Ponsonby, Colonel Sir, (1866-1956)

LESOTHO

1850-1899

TOIT, Alexander Logie du, (1878-1948)

1900 to present

TOIT, Alexander Logie du, (1878-1948)

LIBERIA

1900 to present

NEWHOUSE, Walter Harry, (1897-1969)
SCOFIELD, Lloyd M., (1897-1969)
THIENHAUS, Rolf, (1913-1968)

LIBYA

1800-1849

EHRENBERG, Christian Gottfried, (1795-1876)
ZITTEL, Karl Alfred von, (1839-1904)

1850-1899

BALL, John, (-1941)
BEADNELL, Hugh John Llewellyn, Major, (1874-1944)
DE REGNY, Paolo Vinassa, (1871-1957)
EHRENBERG, Christian Gottfried, (1795-1876)
GREGORY, John Walter, (1864-1932)
PALAZZO, Luigi, (1861-1933)
PERVINQUIÈRE, Léon, (1873-1913)
SPENCER, Leonard James, (1870-1959)
ZITTEL, Karl Alfred von, (1839-1904)

1900 to present

ANDERSON, Robert Lee, (1912-1959)
BALL, John, (-1941)
BEADNELL, Hugh John Llewellyn, Major, (1874-1944)
BIANCHI, Angelo, (1892-1970)
BLOW, Walter Harry, (1924-1972)
BOAGA, Giovanni, (1902-1961)
CHLEBOWSKI, Tadeusz, (1908-1961)
CREWS, George Curtis, (1837-1977)
DE REGNY, Paolo Vinassa, (1871-1957)
DESIO, Ardito, (1897-)
DONOGHUE, David Patrick, (1891-1958)
GREGORY, John Walter, (1864-1932)
HERRING, Donald Grant, Jr., (1918-1962)
JOHNSON, Joseph Wright, (1917-1957)
KING, Robert Bruce, (1923-1977)
KOCH, Harold Kenneth, (1927-1973)
MARTIN, Lewis, (1920-1973)
McCLURE, Drexel Eugene, (1928-1977)
MORET, Léon, (1890-1972)
NELSON, Richard Newman, (1897-1964)
PALAZZO, Luigi, (1861-1933)
PERVINQUIÈRE, Léon, (1873-1913)

LIBYA 1900 to present Cont'd.

PETROCCHI, Carlo, (1898-1956)
SANDFORD, Kenneth Stuart, (1899-1971)
SMITH, G. Wendell, (1917-1974)
SPENCER, Leonard James, (1870-1959)
WALLACE, Maurice Hudiburg, (1917-1961)
WARING, Gerald Ashley, (1883-1971)
WILSON, Eugene Jack, (1921-1976)
ZITTEL, Karl Alfred von, (1839-1904)

LITHUANIAN SSR (USSR)

1800-1849

EICHWALD, Eduard Iwanowitsch von, (1795-1876)

1850-1899

EICHWALD, Eduard Iwanowitsch von, (1795-1876)

1900 to present

HALICKA, Antonina, (1908-1973)
KAVECKIS, Mykolas, (1889-)
PACHUCKI, Czeslaw, (1898-1965)

LUXEMBOURG

1800-1849

ENGELSPACH, Auguste, (LARIVIÈRE, pseud.), (1799-1831)
STEININGER, Johann, (1794-1874)

1850-1899

LUCIUS, Michel, (1876-1961)
STEININGER, Johann, (1794-1874)

1900 to present

KERCKHOFF, P. J. van, (-)
LUCIUS, Michel, (1876-1961)

MADAGASCAR *See* MALAGASY REPUBLIC.

MADEIRA (NORTH ATLANTIC OCEAN)

Up to 1800

HEBERDEN, Thomas, (1703-1769)

1800-1849

BUNBURY, Charles James Fox, Sir, (1809-1886)
SMITH, James, (1782-1867)

1850-1899

BERGT, Walther Adolf, (1864-1941)
BUNBURY, Charles James Fox, Sir, (1809-1886)
COCKERELL, Theodore Dru Alison, (1866-1948)
SMITH, James, (1782-1867)

1900 to present

BERGT, Walther Adolf, (1864-1941)
COCKERELL, Theodore Dru Alison, (1866-1948)
KOMOROWICZ, Maurycy, (1881-1923)
ROSSBY, Carl-Gustaf Arvid, (1898-1957)

MALAGASY REPUBLIC

1800-1849

CROSSE, Joseph Charles Hippolyte, (1826-1898)

1850-1899

BARON, Richard, Reverend, (1847-1907)
BERTRAND, Léon, (1869-1947)
BOULE, Marcellin, (1861-1942)
CROSSE, Joseph Charles Hippolyte, (1826-1898)
DOUVILLE, Joseph Henri Ferdinand, (1846-1937)
FORSYTH MAJOR, Charles Immanuel, (1843-1923)
GRANDIDIER, Alfred, (1836-1921)
LACROIX, Francois Antoine Alfred, (1863-1948)
LEMOINE, Paul, (1878-1940)
PISANI DE SERRES, Felix, (1831-1920)
THEVENIN, Armand, (1870-1918)
WADE, Arthur, (1878-1951)
ZEILLER, Charles René, (1847-1915)

1900 to present

ALLOITEAU, James Henri Ernest, (1890-1969)
ASTRE, Gaston, (1896-)
BARON, Richard, Reverend, (1847-1907)
BARRABÉ, Louis, (1895-1961)
BASSE DE MÉNORVAL, Éliane Marguerite, (1899-)
BELTZ, Edward W., (1891-1970)
BERTRAND, Léon, (1869-1947)
BIRAUD, Francois, (1899-1936)
BOULE, Marcellin, (1861-1942)

MALAGASY REPUBLIC 1900 to present Cont'd.

BUTTERWORTH, Emerson McMillin, (1894-1961)
CHUDOBA, Karl F., (1898-1976)
CURRIE, Philip John, (1949-)
DOUVILLE, Joseph Henri Ferdinand, (1846-1937)
ERHART, Henri, (-)
FORSYTH MAJOR, Charles Immanuel, (1843-1923)
GRANDIDIER, Alfred, (1836-1921)
HOURCQ, Victor, (1901-1961)
JÉRÉMINE, Elisabeth, (née Tschernaieff), (1879-1964)
JOLEAUD, Léonce, (1880-1938)
LACROIX, Francois Antoine Alfred, (1863-1948)
LEHNER, Ernst J., (1895-1968)
LEMOINE, Paul, (1878-1940)
McMATH, John Charles, (?1908-1954)
MILLOT, Georges, (1917-)
PISANI DE SERRES, Felix, (1831-1920)
PIVETEAU, Jean, (1898-)
ROCH, Edouard, (1901-)
ROUBAULT, Marcel Edouard, (1905-1974)
SPATH, Leonard Frank, (1882-1957)
THEVENIN, Armand, (1870-1918)
TRUMPY, Daniel, (1893-1971)
WADE, Arthur, (1878-1951)
WALLACE, Maurice Hudiburg, (1917-1961)
ZEILLER, Charles René, (1847-1915)

MALAWI

1850-1899

BAILEY, Tom Esmond Geoffrey, Captain, (-1919)
HYLAND, John Shearson, (1866-1898)
NEWELL ARBER, Edward Alexander, (1870-1918)

1900 to present

ANDREW, Arthur Robert, (1881-1941)
BAILEY, Tom Esmond Geoffrey, Captain, (-1919)
BISSET, Charles Barron, (-?1966)
CRAIG, Robert Meldrum, (?1882-1956)
GASKIN, Alan Robert James, (1927-1974)
HUGHES, William John, (?1899-1961)
NEWELL ARBER, Edward Alexander, (1870-1918)
SCHOLZ, Erich, (1884-1914)
TIDY, Anthony John Lintott, (1928-1969)
TREMLETT, Rex, (-) [PROSPECTORS, DIVINERS AND MINING ENGINEERS]

MALAYSIA

1800-1849

WALLACE, Alfred Russel, (1823-1913)

1850-1899

BALL, Francis Matthew, (1877-1956)
BECHER, Harry Macdonald, (1855-1893)

MALAYSIA 1850-1899 Cont'd.

HAECKEL, Ernst Heinrich Philipp August, (1834-1919)
HANNAY, Harry Campbell, (1874-1961)
JUNGHUHN, Franz Wilhelm, (1808-1864)
McCARTHY, Edward Thomas, (-) [PROSPECTORS, DIVINERS AND MINING ENGINEERS]
MILLETT, Fortescue William, (1833-1915)
MORGAN, Jacques Jean Marie de, (1857-1924)
RIDLEY, Henry Nicholas, (1855-1956)
SCHMIDT, Carl, (1862-1923)
STOLICZKA, Ferdinand, (1838-1874)
TRUSCOTT, Samuel John, (?1870-1950)
WALLACE, Alfred Russel, (1823-1913)

1900 to present
BALL, Francis Matthew, (1877-1956)
BROUGHTON-EDGE, Arthur Broughton, (1895-1953)
DOBBIE, Charles Herbert, (1905-1976)
HAECKEL, Ernst Heinrich Philipp August, (1834-1919)
HANNAY, Harry Campbell, (1874-1961)
HAZZARD, John Charles, (1903-1973)
HITCHEN, Charles Stansfield, (1904-1971)
HO, Robert, (1921-1972)
HUTCHISON, Arthur Gilbertson, (1902-1949)
HUTTON, Colin Osborne, (1910-1971)
JONES, William Richard, (1890-1970)
JUNGHUHN, Franz Wilhelm, (1808-1864)
KINNEY, Harry Dill, (1886-1961)
KLOMPÉ, Theodorus Hendricus Franciscus, (?1904-1963)
MACDONALD, Simon, (1927-1971)
McCARTHY, Edward Thomas, (-) [PROSPECTORS, DIVINERS AND MINING ENGINEERS]
MILLETT, Fortescue William, (1833-1915)
MORGAN, Jacques Jean Marie de, (1857-1924)
MUIR-WOOD, Helen Marguerite, (1896-1968)
RIDLEY, Henry Nicholas, (1855-1956)
ROE, Frank Walton "Peter", (?1914-1965)
SCHMIDT, Carl, (1862-1923)
SEN GUPTA, Kiron Kumar, (1882-1943)
SMIT SIBINGA, Gerard Leonard, (1895-1963)
TRÜMPY, Daniel, (1893-1971)
TRUSCOTT, Samuel John, (?1870-1950)
WALLACE, Alfred Russel, (1823-1913)

MALDIVE ISLANDS (INDIAN OCEAN)
1850-1899
GARDINER, John Stanley, (1872-1946)

1900 to present
GARDINER, John Stanley, (1872-1946)

MALI
1850-1899
CHUDEAU, René, (1864-1921)
DOUVILLÉ, Joseph Henri Ferdinand, (1846-1937)

MALI 1850-1899 Cont'd.

LACROIX, François Antoine Alfred, (1863-1948)
LEMOINE, Paul, (1878-1940)

1900 to present
CHUDEAU, René, (1864-1921)
DOUVILLÉ, Joseph Henri Ferdinand, (1846-1937)
ERHART, Henri, (-)
FURON, Raymond, (1898-)
JÉRÉMINE, Elisabeth, (née Tschernaieff), (1879-1964)
LACROIX, François Antoine Alfred, (1863-1948)
LEMOINE, Paul, (1878-1940)
MALAVOY, Jean, (1903-1945)
MILLOT, Georges, (1917-)
ROUBAULT, Marcel Edouard, (1905-1974)

MALTA
Up to 1800
BRYDONE, Patrick, (1736-1818)

1800-1849
BRYDONE, Patrick, (1736-1818)
PARKER, William Kitchen, (1823-1890)
WRIGHT, Thomas, (1809-1884)

1850-1899
ADAMS, Andrew Leith, (-1882)
BIRLEY, Caroline, (1851-1907)
PARKER, William Kitchen, (1823-1890)
WRIGHT, Thomas, (1809-1884)

1900 to present
BIRLEY, Caroline, (1851-1907)
BLOW, Walter Harry, (1924-1972)

MANCHURIA See CHINA.

MARIANA ISLANDS (PACIFIC OCEAN)
1900 to present
COOKE, Charles Wythe, (1887-1971)

MARTINIQUE (WEST INDIES)
1850-1899
ANDERSON, Tempest, (1846-1913)
GERLAND, Georg Karl Cornelius, (1833-1919)
HEILPRIN, Angelo, (1853-1907)
HOVEY, Edmund Otis, (1862-1924)
JAGGAR, Thomas Augustus, (1871-1953)
SAPPER, Karl Theodor, (1866-1945)

1900 to present
ANDERSON, Tempest, (1846-1913)
BARRABÉ, Louis, (1895-1961)
GERLAND, Georg Karl Cornelius, (1833-1919)
HEILPRIN, Angelo, (1853-1907)

MARTINIQUE (WEST INDIES) 1900 to present Cont'd.

HOVEY, Edmund Otis, (1862-1924)
JAGGAR, Thomas Augustus, (1871-1953)
SAPPER, Karl Theodor, (1866-1945)

MASCARENE ISLANDS (INDIAN OCEAN) See MAURITIUS; RÉUNION.

MAURITANIA

1850-1899

CHUDEAU, René, (1864-1921)
LEMOINE, Paul, (1878-1940)

1900 to present

CHUDEAU, René, (1864-1921)
JACQUET, Fernand, (1908-1937)
LEMOINE, Paul, (1878-1940)
MILLOT, Georges, (1917-)
PERCIVAL, Frederick George, (1887-1973)
ROUBAULT, Marcel Edouard, (1905-1974)
SCOFIELD, Lloyd M., (1897-1969)
SHAND, Samuel James, (1882-1957)
THIENHAUS, Rolf, (1913-1968)
VINCIENNE, Henri, (1898-1965)

MAURITIUS (INDIAN OCEAN)

1800-1849

DARWIN, Charles Robert, (1809-1882)
LLOYD, J. A., Colonel, (-1856?)

1850-1899

DARWIN, Charles Robert, (1809-1882)
GILL, David, Sir, (1843-1914)
LLOYD, J. A., Colonel, (-1856?)
MORGAN, Edward Delmar, (1840-1909)

1900 to present

GILL, David, Sir, (1843-1914)
MORGAN, Edward Delmar, (1840-1909)

MERGUI ARCHIPELAGO (BURMA)

1900 to present

RUDMOSE BROWN, Robert Neal, (1879-1957)

MESOPOTAMIA See IRAQ.

MEXICO

Up to 1800

BORDA, José de la, Don, (1699-1778) [PROSPECTORS, DIVINERS AND MINING ENGINEERS]
HUMBOLDT, Friedrich Wilhelm Heinrich Alexander von, Freiherr, (1769-1859)
MACLURE, William, (1763-1840)
RIO, Andrés Manuel del, (1764-1849)

1800-1849

BURKART, Hermann Josef, (1798-1874)
CASTILLO, Antonio del, Don, (1820-1895)
CHRISTY, Henry, (1810-1865)
FORBES, David, (1828-1876)
GALEOTTI, Henri Guillaume, (1814-1858)
HUMBOLDT, Friedrich Wilhelm Heinrich Alexander von, Freiherr, (1769-1859)
LINCECUM, Gideon, (1793-1874)
MACLURE, William, (1763-1840)
RIO, Andrés Manuel del, (1764-1849)
TAYLOR, John, (1780-1864)
VANUXEM, Lardner, (1792-1848)
XÁNTUS, János, (1825-1894)

1850-1899

ATTWOOD, George, (1845-1912)
BASTIN, Edson Sunderland, (1878-1953)
BAUERMAN, Hilary, (1833-1909)
BERGEAT, Alfred, Edmund, (1866-1924)
BERKEY, Charles Peter, (1867-1955)
BOND, Marshall, (1867-1941) [PROSPECTORS, DIVINERS AND MINING ENGINEERS]
BÖSE, Emil, (1868-1972)
BOUTWELL, John Mason, (1874-1968)
BOWEN, Charles Franklin, (1871-1953)
BROWN, Barnum, (1873-1963)
BURCKHARDT, Carl, (1869-1935)
BURKART, Hermann Josef, (1798-1874)
BUTTGENBACH, Henri, (1874-1964)
CASTILLO, Antonio del, Don, (1820-1895)
CHRISTY, Henry, (1810-1865)
CLELAND, Herdman Fitzgerald, (1869-1935)
COLLINS, Henry Francis, (?1864-1924)
COTTEAU, Gustave Honoré, (1818-1894)
CUMINGS, Willard Lawsen, (1874-1945)
DUMBLE, Edwin Theodore, (1852-1927)
EMMONS, Samuel Franklin, (1841-1911)
ERB, Josef Theodor, (1874-1934)
FENNER, Clarence Norman, (1870-1949)
FLORES REYES, Teodoro, (1873-1955)
FOOTE, Albert E., (1846-1895)
FORBES, David, (1828-1876)
GALEOTTI, Henri Guillaume, (1814-1858)
GETZENDANER, Frank Marshall, (1876-1957)
HAMMOND, John Hays, (1855-1936)
HAY, Oliver Perry, (1846-1930)

MEXICO 1850-1899 Cont'd.

HAYES, Charles Willard, (1858-1916)
HEILPRIN, Angelo, (1853-1907)
HENNEN, Ray Vernon, (1875-1958)
HESS, Frank L., (1871-1955)
HOFFMAN, Fr. Adolf, (1867-1898)
HOWE, Ernest, (1875-1932)
HUMBOLDT, Friedrich Wilhelm Heinrich Alexander von, Freiherr, (1769-1859)
JEBB, John Gladwyn, (1841-1893) [PROSPECTORS, DIVINERS AND MINING ENGINEERS]
JULIAN, Henry Forbes, (1861-1912) [PROSPECTORS, DIVINERS AND MINING ENGINEERS]
KHRUSHCHOV, Konstantin Dmitrievich, (? -1912)
LEDOUX, Albert Reid, (1852-1923) [PROSPECTORS, DIVINERS AND MINING ENGINEERS]
LENK, Hans, (1863-1938)
LINCECUM, Gideon, (1793-1874)
LINDGREN, Waldemar, (1860-1939)
MILLS, James Ellison, (1834-1901)
MORSE, William Clifford, (1874-1962)
ORDOÑEZ, Ezequiel, (1867-1950)
PIKE, John William, (1839-1874)
PILSBRY, Henry Augustus, (1862-1957)
RUTLEY, Frank, (1842-1904)
SCALIA, Salvatore, (1874-1923)
SPENCER, Joseph William Winthrop, (1851-1921)
STAPFF, Friedrich Moritz, (1836-1895)
STEPHENSON, Lloyd William, (1876-1962)
STEWART, Philip Charteris Anstruther, (1874-1923)
TAYLOR, John, (1780-1864)
TRELAWNEY-ANSELL, E. C., (-) [PROSPECTORS, DIVINERS AND MINING ENGINEERS]
WAITZ, Paul, (1876-1961)
WIELAND, George Reber, (1865-1953)
WINCHELL, Horace Vaughn, (1865-1923)
WOEIKOF, Alexander, (1842-1916)
WOODRUFF, Elmer Grant, (1872-1952)
XÁNTUS, János, (1825-1894)

1900 to present

ABERNATHY, George Elmer, (1888-1948)
AID, Harry, (1897-1927)
ASANO, Kiyoshi, (1910-)
ATHY, Lawrence Ferdinand, (1898-1955)
ATTWOOD, George, (1845-1912)
BAIRD, Chester Alvan, (1896-1959)
BANDY, Mark Chance, (1900-1963)
BASTIN, Edson Sunderland, (1878-1953)
BAUERMAN, Hilary, (1833-1909)
BECKER, Clyde M., (1882-1938)
BELL, Olin Grogan, (1893-1970)
BENTZ, Alfred, (1897-1964)
BERGEAT, Alfred, Edmund, (1866-1924)
BERKEY, Charles Peter, (1867-1955)

MEXICO 1900 to present Cont'd.

BEVIER, George Means, (1888-1972)
BISHOPP, Douglas Wallace, (1900-1977)
BLANCHARD, William Grant, Jr., (1894-1952)
BOND, Marshall, (1867-1941) [PROSPECTORS, DIVINERS AND MINING ENGINEERS]
BOOTS, Paul Heath, (1909-1947)
BÖSE, Emil, (1868-1972)
BOUTWELL, John Mason, (1874-1968)
BOWEN, Charles Franklin, (1871-1953)
BRACE, Orval L., (1893-1975)
BRAMLETTE, Milton Nunn, (1896-1977)
BROWN, Barnum, (1873-1963)
BURCKHARDT, Carl, (1869-1935)
BUTTGENBACH, Henri, (1874-1964)
CAMPBELL, Robert Burns, (1892-1955)
CARR, William L., (1919-1975)
CARSON, Carlton Mann "Kit", (1898-1972)
CASSEL, Chester A. "Chet", (1897-1971)
CASTER, Ermil Leslie, (1896-1955)
CLAPP, Frederick Gardner, (1879-1944)
CLELAND, Herdman Fitzgerald, (1869-1935)
COLLINS, Henry Francis, (?1864-1924)
CRUM, Harry Edwin, (1892-1955)
CUMINGS, Willard Lawsen, (1874-1945)
CUNNINGHAM, George Magee, (1895-1963)
CUSHMAN, Joseph Augustine, (1881-1949)
DAVILA, Salvador Ortiz, (1902-1931)
DAVIS, Ralph Emerson, (1884-1968)
DeGOLYER, Everette Lee, (1886-1956)
DERRY, Duncan Ramsay, (?1908-)
DRAKE, Cecil, (1899-1949)
DREHER, Otto, (1888-1945)
DUMBLE, Edwin Theodore, (1852-1927)
EBY, James Brian, (1896-)
EDWARDS, Everett C., (1897-1972)
ELY, Fred Burchard, (1883-1946)
EMMONS, Samuel Franklin, (1841-1911)
ERB, Josef Theodor, (1874-1934)
FARISH, Linn Markley, (1901-1944)
FENNER, Clarence Norman, (1870-1949)
FITZGERALD, Paul Eugene, (1903-1961)
FLORES REYES, Teodoro, (1873-1955)
FOLGER, James Anthony, (1896-1956)
FOSHAG, William Frederick, (1894-1956)
FOWLER, Charles Worth, Jr., (1899-1948)
FOXWORTH, Wyckliff Riley, (1936-1965)
FREI, Ernst, (1894-1974)
FREY, Alfred Paul, (1893-1963)
FRIES, Carl, Jr., (1910-1965)
FRYXELL, Roald Hilding, (1934-1974)
FULLER, Richard Eugene, (1897-1976)
GARFIAS, Valentin R., (-)
GESTER, George Clark, (1884-1959)
GETZENDANER, Frank Marshall, (1876-1957)

MEXICO 1900 to present Cont'd.

GIERHART, Guy B. "Joe", (1903-1968)
GOLDMAN, Marcus Isaac, (1881-1965)
GRIMM, Kenneth Edgar, (1912-1961)
HAKE, Benjamin Franklin, (1895-1973)
HALSE, George Whitfield, (1885-1968)
HAMILTON, Henry L., (1882-1925)
HAMMILL, Chester A., (1889-1941)
HAMMOND, John Hays, (1855-1936)
HAMMOND, Weldon Woolf, (1908-1964)
HANNA, G. Dallas, (1887-1970)
HARRISON, John Vernon, (1892-1972)
HARRISON, Thomas Samuel, (1881-1964)
HAY, Oliver Perry, (1846-1930)
HAYES, Charles Willard, (1858-1916)
HEILPRIN, Angelo, (1853-1907)
HENNEN, Ray Vernon, (1875-1958)
HERALD, Frank Edward, (1887-1965)
HESS, Frank L., (1871-1955)
HINDES, Earle P., (1888-1962)
HOLLISTER, Joseph Steffens, (1903-1953)
HOOD, Forrest Wilson, (1897-1963)
HOPKINS, Edwin Butcher, (1882-1940)
HOWE, Ernest, (1875-1932)
HOWELLS, Jenkyn Owen, (-1950)
HUNTLEY, Charles Stirling, (1895-1926)
HYDE, Clarence Edgar, (1894-1954)
ICKES, Eugene Law, (1885-1941)
ISENBERGER, Nathan Perry, (1896-1956)
JEFFREYS, Geoffrey, (1885-1953)
JENNINGS, Charles I. "Al", (1894-1968)
JOHNSON, Clifton Wood, (1905-1954)
JONES, William Francis, (1885-1941)
JUDSON, Wilber, (1880-1951)
JULIAN, Henry Forbes, (1861-1912) [PROSPECTORS, DIVINERS AND MINING ENGINEERS]
KAY, Fred Hall, (1885-1943)
KENNEDY, William, (-1926)
KHRUSHCHOV, Konstantin Dmitrievich, (? -1912)
KLEINPELL, William Darwin, (1898-1959)
KRIEGER, Philip, (1900-1940)
KRYNINE, Paul Dimitri, (1902-1964)
KUIPER, Willem Nicolaas, (1891-1952)
LEDOUX, Albert Reid, (1852-1923) [PROSPECTORS, DIVINERS AND MINING ENGINEERS]
LENK, Hans, (1863-1938)
LINDGREN, Waldemar, (1860-1939)
LÓPEZ VÁZQUEZ, Andrés, (1903-1954)
MALDONADO-KOERDELL, Manuel, (1908-1972)
MARTIN, Francis Ivan "Shorty", (1896-1952)
MARTYN, Phil F., (1903-1974)
MASON, Shirley Lowell, (1893-1975)
McKINSTRY, Hugh Exton, (1896-1961)
MEKEL, Johannes Antonius Alphonsus, (1891-1942)
MILLS, James Ellison, (1834-1901)

MEXICO 1900 to present Cont'd.

MOODY, Graham Blair, (1889-1975)
MORGAN, George Dillon, (1894-1950)
MORRIS, Frederick Kuhne, (1885-1962)
MORSE, William Clifford, (1874-1962)
MUIR, John Malcolm, (1885-1938)
MYERS, John Clark, (1894-1970)
NEIGHBOURS, Glen Reed, (1917-1961)
NELSON, Lloyd Alvino, (1895-1964)
NININGER, Harvey Harlow, (-)
NOBLE, Earl Bart, (1894-1967)
NOLAN, Philip Edward, (1887-1958)
NOMLAND, Jorgen O., (1885-1943)
NUTTALL, Winfred Laurence Falkiner, (1897-1969)
ORDOÑEZ, Ezequiel, (1867-1950)
PALMER, Robert Hastings, (1882-1948)
PETSCH, Arthur Henry, (1895-1969)
PILSBRY, Henry Augustus, (1862-1957)
POND, Walter Franklin, (1885-1974)
PORCH, Edwin Leight, Jr., (1889-1948)
PRETYMAN, Frank Remington, (1891?-1917)
RAMDOHR, Paul, (1890-)
REED, Eugene C., (1901-1976)
REED, Lyman Coleman, (1900-1970)
RENAUD, Charles Louis, (1898-1964)
RHOADES, Ralph Omer, (1895-1961)
RITER, Samuel W., (1896-1941)
ROBERTS, Hugh Marine, (1885-1965)
ROBERTS, Morgan Evan, (1894-1941)
ROW, Charles Herbert, (1895-1970)
RUSS, Leon Fair, (1886-1953)
RUTLEY, Frank, (1842-1904)
SANDS, John Melville, (1883-1962)
SANSOM, Charles Alan, (1902-1945)
SCALIA, Salvatore, (1874-1923)
SCHMITT, Harrison Ashley, (1896-1966)
SCHMITTOU, Maurice B., (1899-1929)
SEGALL, Julius, (1884-1928)
SHEARER, Harold Kurtz, (1889-1946)
SHERMAN, Roger Hiram, (1904-1954)
SHERRILL, Richard Ellis, (1899-1953)
SHORT, Maxwell Naylor, (1889-1952)
SINGEWALD, Joseph Theophilus, Jr., (1884-1963)
SNIDER, Luther Crocker, (1882-1947)
SOUTHER, John Barron, (1894-1964)
SPARKS, Dale Darrell Dolsun, (1894-1938)
SPENCER, Joseph William Winthrop, (1851-1921)
STAFF, Hans von, (1883-1915)
STEHR, Raymond Albert, (1903-1967)
STEIG, Maynard Herman, (1904-1965)
STEPHENSON, Lloyd William, (1876-1962)
STEWART, Philip Charteris Anstruther, (1874-1923)
SWARTZ, Joel Howard, (1893-1971)
TASHMAN, Djavad Eyoub, (1893-1956)
TEMPLETON, Eugene Crittenden, (1889-1932)

MEXICO 1900 to present Cont'd.

THALMANN, Hans E., (1899-1975)
THOMPSON, E. G., (1898-1973)
TRAGER, Earl A., (1893-1978)
TRELAWNEY-ANSELL, E. C., (-) [PROSPECTORS, DIVINERS AND MINING ENGINEERS]
TRÜMPY, Daniel, (1893-1971)
WADELL, Hakon, (1895-1962)
WAITZ, Paul, (1876-1961)
WANDKE, Alfred, (1887-1941)
WARING, Waldo W., (1903-1955)
WASHBURNE, Chester Wesley, (1883-1971)
WEAVER, Paul, (1888-1964)
WELLS, Francis Gerrit, (1898-1969)
WIELAND, George Reber, (1865-1953)
WILHELM, Oscar, (1897-1975)
WILSON, Bernard Wright, (1922-1976)
WILSON, Roy Arthur, (1891-1970)
WINCHELL, Horace Vaughn, (1865-1923)
WISSER, Edward Hollister, (1895-1970)
WOEIKOF, Alexander, (1842-1916)
WOLF, Albert G., (1885-1964)
WOODRUFF, Elmer Grant, (1872-1952)

MICRONESIA

1850-1899

GULICK, John Thomas, (1832-1923)

1900 to present

GULICK, John Thomas, (1832-1923)

MIDDLE EAST (unspecified)

Up to 1800

MAILLET, Benoît de, (1656-1738)

1850-1899

FLIEGEL, Gotthard, (1873-1947)
LIBBEY, William, (1855-1927)

1900 to present

COLBERT, Edwin H., (1905-)
DeGOLYER, Everette Lee, (1886-1956)
FLIEGEL, Gotthard, (1873-1947)
ISENBERGER, Nathan Perry, (1896-1956)
JABLONSKI, Eugene, (1892-1975)
LEES, George Martin, (1898-1955)
MARCHESINI, Enrico, (1911-1968)
SAPPINGTON, Chester, (1905-1968)

MOLUCCA ISLANDS See INDONESIA.

MONACO

1900 to present

BOURCART, Jacques, (1891-1965)

MONGOLIA

Up to 1800

TAVERNIER, Jean Baptiste, baron d'Aubonne, (1605-1689)

1900 to present

CHLEBOWSKI, Tadeusz, (1908-1961)
EFREMOV, Ivan Antonovich, (1907-1972)
FERSMAN, Aleksander Yevgenyevich, (1883-1945)
KIELAN-JAWOROWSKA, Zofia, (-)

MORAVIA See CZECHOSLOVAKIA.

MOROCCO

Up to 1800

MAILLET, Benoît de, (1656-1738)

1800-1849

BALL, John, (1818-1889)
HOOKER, Joseph Dalton, Sir, (1817-1911)

1850-1899

BALL, John, (1818-1889)
BEAUGÉ, Alfred, (1878-1935)
BRALY, Adrien, (1864-1927)
BUTTGENBACH, Henri, (1874-1964)
DELÉPINE, Gaston-Georges, Monseigneur, (1878-1963)
FLETCHER, Hugh, (1848-1909)
GENTIL, Louis, (1868-1925)
GÉRARD, Charles, Colonel, (1866-1953)
HOOKER, Joseph Dalton, Sir, (1817-1911)
LEMOINE, Paul, (1878-1940)
MACLAREN, James Malcolm, (1873-1935)
MAW, George, (1832-1912)
McCARTHY, Edward Thomas, (-) [PROSPECTORS, DIVINERS AND MINING ENGINEERS]
TERMIER, Pierre, (1859-1930)
VALLE DE LERSUNDI, Alfonso del, (1875-1944)

1900 to present

ABRARD, René, (1892-1973)
ARAMBOURG, Camille, (1885-1969)
BEAUGÉ, Alfred, (1878-1935)
BISAT, William Sawney, (1886-1973)
BOURCART, Jacques, (1891-1965)
BRALY, Adrien, (1864-1927)
BUTTGENBACH, Henri, (1874-1964)
DAGUIN, Fernand, (1889-1948)
DELÉPINE, Gaston-Georges, Monseigneur, (1878-1963)
DURAND DELGA, Michel, (1923-)

MOROCCO 1900 to present Cont'd.

FALLOT, Paul, (1889-1960)
FLETCHER, Hugh, (1848-1909)
FRIEDEL, Edmond, (1895-1972)
GENTIL, Louis, (1868-1925)
GÉRARD, Charles, Colonel, (1866-1953)
GROTH, Jean, (1899-1916)
GUBLER, Jean, (1902-1940)
HENDON, Bryan, (1896-1926)
HERNÁNDEZ-PACHECO, Francisco, (1899-1976)
HOOKER, Joseph Dalton, Sir, (1817-1911)
HUGHES, Richard, (1889-1969)
ICKES, Eugene Law, (1885-1941)
JOLEAUD, Léonce, (1880-1938)
LACOSTE, Jean Pierre Victor, (1901-1942)
LAPPARENT, Albert Félix de, (1905-1975)
LECKWIJCK, Willem Peter van, (1902-1975)
LECOINTRE, Georges, (1888-1972)
LEMOINE, Paul, (1878-1940)
LOTZE, Franz, (1903-1971)
LUTAUD, Léon, (1883-1964)
MACLAREN, James Malcolm, (1873-1935)
MAW, George, (1832-1912)
McCARTHY, Edward Thomas, (-) [PROSPECTORS, DIVINERS AND MINING ENGINEERS]
MILLOT, Georges, (1917-)
MORET, Léon, (1890-1972)
ROBERTS, Morgan Evan, (1894-1941)
ROCH, Edouard, (1901-)
SENN, Alfred, (1899-1949)
TERMIER, Pierre, (1859-1930)
TRÜMPY, Daniel, (1893-1971)
TRÜMPY, Rudolf D., (-)
VALLE DE LERSUNDI, Alfonso del, (1875-1944)
VINCIENNE, Henri, (1898-1965)
WALLACE, Maurice Hudiburg, (1917-1961)
WARING, Gerald Ashley, (1883-1971)
WESTERVELD, J., (1905-1962)

MOZAMBIQUE

1850-1899

CHOFFAT, Paul, (1849-1919)
GRIESBACH, Karl Ludolf, (1847-1907)
KAISER, Friedrich Wilhelm Erich, (1871-1934)
TEALE, Edmund Oswald, Sir, (1874-1971)
THORNTON, Richard, (1838-1863)

1900 to present

BANDY, Mark Chance, (1900-1963)
BORGES, Alexandre, (1898-1955)
CHOFFAT, Paul, (1849-1919)
CONDIT, Daniel Dale, (1886-1955)
FREIRE DE ANDRADE, Carlos Benot, (1893-1956)
GRIESBACH, Karl Ludolf, (1847-1907)
HOLMES, Arthur, (1890-1965)

MOZAMBIQUE 1900 to present Cont'd.

KAISER, Friedrich Wilhelm Erich, (1871-1934)
MENNELL, Frederick Philip, (1880-1966)
NOWACK, Ernst, (1891-1946)
PUTNAM, L. G., (1898-1936)
RAINIER, Peter W., (-) [PROSPECTORS, DIVINERS AND MINING ENGINEERS]
SCHÜLLER, Arno, (1908-1968)
SPATH, Leonard Frank, (1882-1957)
TEALE, Edmund Oswald, Sir, (1874-1971)
TREMLETT, Rex, (-) [PROSPECTORS, DIVINERS AND MINING ENGINEERS]
WILSON, Reginald Charles, (1888-1943)

NAMIBIA

1800-1849

GALTON, Francis, Sir, (1822-1911)

1850-1899

BREUIL, Henri, Abbé, (1877-1961)
GALTON, Francis, Sir, (1822-1911)
GURICH, Georg, (1859-1938)
KAISER, Friedrich Wilhelm Erich, (1871-1934)
PUFAHL, Otto Hermann August, (1855-1924)
SPENCER, Leonard James, (1870-1959)
STAPFF, Friedrich Moritz, (1836-1895)
TAYLOR, William P., (1859-) [PROSPECTORS, DIVINERS AND MINING ENGINEERS]
THOMPSON, Beeby, (1848-1931)

1900 to present

BARRON, William Ernest, (1880-1935)
BREUIL, Henri, Abbé, (1877-1961)
CLOOS, Hans, (1885-1951)
DERRY, Duncan Ramsay, (?1908-)
GALTON, Francis, Sir, (1822-1911)
GÜRICH, Georg, (1859-1938)
HODGE, Edwin Thomas, (1887-1970)
KAISER, Friedrich Wilhelm Erich, (1871-1934)
MAACK, Reinhard, (1892-1969)
MARTINUSSEN, Eugen, (1916-1942)
PARTRIDGE, Francis Chamberlain, (1903-1939)
PUFAHL, Otto Hermann August, (1855-1924)
RAINIER, Peter W., (-) [PROSPECTORS, DIVINERS AND MINING ENGINEERS]
RIMANN, Eberhard, (1882-1944)
SCHNEIDERHÖHN, Hans, (1887-1962)
SPENCER, Leonard James, (1870-1959)
STAFF, Hans von, (1883-1915)
TAYLOR, William P., (1859-) [PROSPECTORS, DIVINERS AND MINING ENGINEERS]
THOMPSON, Beeby, (1848-1931)
WAGNER, Percy Albert, (1885-1929)
WILSON, Eugene Jack, (1921-1976)

NEAR EAST (unspecified)

1900 to present

CATHCART, Stanley Holman, (1889-1953)
DE KOCH, William Gaspar, (1898-1954)

NEPAL

1800-1849

HOOKER, Joseph Dalton, Sir, (1817-1911)
JACQUEMONT, Victor, (1801-1832)

1850-1899

DIENER, Carl, (1862-1928)
HOOKER, Joseph Dalton, Sir, (1817-1911)
ROCCATI, Alessandro, (1872-1928)
ROMER, Eugeniusz, (1871-1954)

NEPAL Cont'd.

1900 to present

DIENER, Carl, (1862-1928)
HOOKER, Joseph Dalton, Sir, (1817-1911)
JURKOVIĆ, Ivan, (1917-)
LEHNER, Ernst J., (1895-1968)
ROCCATI, Alessandro, (1872-1928)
ROMER, Eugeniusz, (1871-1954)

NETHERLANDS

Up to 1800

CAMPER, Pieter (Petrus), (1722-1789)
DROUIN, Jean Baptiste, (1752-1792)
FAUJAS DE SAINT FOND, Barthélemy, (1741-1819)
FERBER, Johann Jacob, (1743-1790)
GODDING, Theodorus Joannes, Kanunnik, (1722-1797)
HOFFMANN, Jan Leonard, (1710-1782)
MARUM, Martin van, (1750-1838)
RASPE, Rudolf Erich, (1737-1794)
ROUX, Jean Marc, (?1736-1805)
SPILBERGEN, Joris van, (1568-1620)

1800-1849

BINKHOORST TOT DEN BINKHOORST, Johan Theodoor, (1810-1876)
BOGAERT, P. J. J., (1807-1894)
BREDA, Jacob Gisbertus Samuel van, (1788-1867)
DECHEN, Ernst Heinrich Carl von, (1800-1889)
ELST, E. van de, (1826-1897)
FAUJAS DE SAINT FOND, Barthélemy, (1741-1819)
HECKEL, Johann Jakob, (1790-1857)
HOEVEN, Jan van der, (1801-1868)
MARUM, Martin van, (1750-1838)
PANHUYS, George Emile Auguste van, (1796-1871)
RIEMSDIJK, A. W. G. van, (1803-1871)
ROUX, Jean Marc, (?1736-1805)
SCHLEGEL, Hermann, (1804-1884)
STARING, W. C. H., (1808-1874)
THIERENS, F. F., (1811-)
UBAGHS, Johann Casimir, (1829-1894)

1850-1899

BEMMELEN, Jacob Maarten van, (1830-1911)
BINKHOORST TOT DEN BINKHOORST, Johan Theodoor, (1810-1876)
BLEEKRODE, S., (-)
BOGAERT, P. J. J., (1807-1894)
BREDA, Jacob Gisbertus Samuel van, (1788-1867)
CALKER, Friedrich Julius Peter von, (1841-1913)
DECHEN, Ernst Heinrich Carl von, (1800-1889)
ELST, E. van de, (1826-1897)
ERB, Josef Theodor, (1874-1934)
HECKEL, Johann Jakob, (1790-1857)
HOEVEN, Jan van der, (1801-1868)
HONY, W. Edward, Archdeacon, (-)
JONGMANS, Wilhelmus Josephus, (1878-1957)
MARTIN, Johann Karl Ludwig, (1851-1942)

NETHERLANDS 1850-1899 Cont'd.

MOLENGRAAFF, Gustaaf Adolf Frederik, (1860-1942)
PANHUYS, George Emile Auguste van, (1796-1871)
RIEMSDIJK, A. W. G. van, (1803-1871)
SCHLEGEL, Hermann, (1804-1884)
SONNEVELD, Jan, (1877-195)
STARING, W. C. H., (1808-1874)
UBAGHS, Johann Casimir, (1829-1894)
VAN'T HOFF, Jacobus Hendricus, (1852-1911)
WATERSCHOOT VAN DER GRACHT, Willem Anton Josef Maria van, (1873-1943)
WICHMANN, Carl Ernst Arthur, (1851-1927)
WINKLER, Tiberius Cornelius, (1822-1897)

1900 to present

ANDREW, Arthur Robert, (1881-1941)
BARD, Richards J., (1900-1977)
BAREN, F. H. van, (-1975)
BEMMELEN, Jacob Maarten van, (1830-1911)
BERNSEN, J. J. A., Pater, (1888-1932)
BROUWER, Hendrik Albertus, (1886-1973)
BURGWAL, Rudolph Frans, (1917-1944)
CALKER, Friedrich Julius Peter von, (1841-1913)
CONKLING, Richard A., (1885-1952)
COSIJN, Adrianus Johannes, (1890-1945)
DREHER, Otto, (1888-1945)
EDELMANN, Cornelis Hendrik, (1901-1964)
ERB, Josef Theodor, (1874-1934)
FLORSCHÜTZ, Frans, (1887-1965)
GROND, Gerard Joseph Alfons, (1890-1970)
GROOTHOFF, Christiaan Theodor, (1887-1969)
GRUTTERINK, Jan Adolf, (1879-1949)
HEIMANS, E., (-?1937)
HENRY, Gerard, (1890-1943)
HLAUSCHEK, Hans, (1896-1971)
HOLST PELLEKAAN, William van, (1880-1949)
HUBRECHT, P. F., (-?1929)
HUTCHISON, Arthur Gilbertson, (1902-1949)
ITERSON, F. K. Th. van, (-1957)
JONGMANS, Wilhelmus Josephus, (1878-1957)
KLEIBER, Karl Ernst, (1912-1972)
KOENIGSWALD, Gustav Heinrich Ralph von, (1902-)
KOZÁRY, Myron Theodore "Mirko", (1918-1966)
KRUITZINGA, Pieter, (1885-1967)
KUENEN, Phillip Henry, (1902-1976)
KUIPER, Willem Nicolaas, (1891-1952)
LEDINGHAM, Glen Wallace, (1911-1978)
LEUTENEGGER, Willy Otto, (1903-1972)
LEYDS, Louis Willem, (1887-1963)
LINDEN, Bernard Hugo "Ben" van der, (-1958)
MARTIN, Johann Karl Ludwig, (1851-1942)
MEKEL, Johannes Antonius Alphonsus, (1891-1942)
MOLENGRAAFF, Gustaaf Adolf Frederik, (1860-1942)
MUNCK, Victor C. E. A. de, (1920-1956)
NELSON, Richard Newman, (1897-1964)

NETHERLANDS 1900 to present Cont'd.

NES, C. L. van, (?1878-1963)
NICOLAS, Jean Paul, (1921-1967)
PALM, Q. A., (1923-1960)
PET, Cees, (1900-1953)
PLUMMER, Frederick Byron, (1886-1947)
POLLOCK, Charles Murray, (1898-1969)
RAINWATER, Edward Harriman "Rainey", (1909-1972)
REA, Henry Carter, (1900-1963)
REED, Lyman Coleman, (1900-1970)
REINHOLD, Thomas, (1890-1955)
RUTTEN, Louis Martin Robert, (1884-1946)
RUTTEN, Martin Gerard, (1910-1970)
SATIJN, P. J. M., (?1907-1962)
SCHERMERHORN, W., (1894-)
SCHOUTEN, C., (1891-1963)
SCHRUEDER, Antje, (1887-1952)
SMIT SIBINGA, Gerard Leonard, (1895-1963)
SONNEVELD, Jan, (1877-195)
STEENHUIS, Jacob Frederik, (1883-1958)
STREIFF, Viktor, (1909-1973)
TER MEULEN, J. A. C., (1902-1960)
TESCH, P., (1879-1961)
THALMANN, Hans E., (1899-1975)
TRÜMPY, Daniel, (1893-1971)
VAN'T HOFF, Jacobus Hendricus, (1852-1911)
VENING MEINESZ, Felix Andries, (1887-1966)
VERSLUYS, Jan, (1880-1935)
VLERK, Isaak Martinus van der, (?1892-1974)
WATERSCHOOT VAN DER GRACHT, Willem Anton Josef Maria van, (1873-1943)
WICHMANN, Carl Ernst Arthur, (1851-1927)
WILHELM, Oscar, (1897-1975)
ZIJLSTRA, Gerrit, (-1962)

NEW BRITAIN

1800-1849

JUKES, Joseph Beete, (1811-1869)

1850-1899

JUKES, Joseph Beete, (1811-1869)

NEW CALEDONIA

1850-1899

BELL, James Macintosh, (1877-1934)
COCKERELL, Theodore Dru Alison, (1866-1948)
LEMOINE, Paul, (1878-1940)

1900 to present

BELL, James Macintosh, (1877-1934)
COCKERELL, Theodore Dru Alison, (1866-1948)
LEMOINE, Paul, (1878-1940)
TAZIEFF, Haroun, (1914-)
WHITEHEAD, Walter Lucius, (1891-1969)

NEW GUINEA
1800-1849
JUKES, Joseph Beete, (1811-1869)

1850-1899
BOWEN, Charles Franklin, (1871-1953)
ETHERIDGE, Robert Jr., (1847-1920)
JUKES, Joseph Beete, (1811-1869)
MACDONALD, Alexander, (-) [PROSPECTORS, DIVINERS AND MINING ENGINEERS]
MOON, Frederick William, (-1925)

1900 to present
ARGABRITE, William Greame, (1885-1963)
BARROW, Geoffrey, (1899-1963)
BOURCHIER, John Rickman, (1896-1927)
BOWEN, Charles Franklin, (1871-1953)
CARR, William L., (1919-1975)
EDWARDS, Austin Burton, (1909-1960)
ETHERIDGE, Robert Jr., (1847-1920)
FICHTER, Hans Jakob, (1908-1977)
LEAHY, Michael, (-) [PROSPECTORS, DIVINERS AND MINING ENGINEERS]
LEHNER, Ernst J., (1895-1968)
MOON, Frederick William, (-1925)
MUIR, John Malcolm, (1885-1938)
NUTTALL, Winfred Laurence Falkiner, (1897-1969)
PAPP, Simon, (1886-1970)
RAGGATT, Harold George, Sir, (1900-1968)
REED, Lyman Coleman, (1900-1970)

NEW GUINEA See also PAPUA; WEST IRIAN.

NEW HEBRIDES (PACIFIC OCEAN)
1900 to present
MAWSON, Douglas, Sir, (1882-1958)
TAZIEFF, Haroun, (1914-)

NEW ZEALAND
1800-1849
BENEDEN, Pierre Joseph van, (1809-1893)
BUCHANAN, John, (1819-1898)
DARWIN, Charles Robert, (1809-1882)
DIEFFENBACH, Ernst, (1811-1855)
ETTINGSHAUSEN, Constantin von, Freiherr, (1826-1897)
GREY, George, Sir, (1812-1898)
HAAST, John Francis Julius von, Sir, (1824-1887)
HOCHSTETTER, Ferdinand von, Ritter, (1829-1884)
HOOKER, Joseph Dalton, Sir, (1817-1911)
MANTELL, Walter Baldock Durant, Honourable, (1820-1895)
OWEN, Richard, Sir, (1804-1892)
PARKER, William Kitchen, (1823-1890)
THOMSON, John Turnbull, (1821-1884)

1850-1899
BELL, James Macintosh, (1877-1934)

NEW ZEALAND 1850-1899 Cont'd.

BENEDEN, Pierre Joseph van, (1809-1893)
BÖHM, Georg, (1854-1913)
BOWEN, Charles Franklin, (1871-1953)
BUCHANAN, John, (1819-1898)
CLIFFORD, William Hugh, Right Honourable, (1858-1943)
DARWIN, Charles Robert, (1809-1882)
DAVIS, William Morris, (1850-1934)
DIEFFENBACH, Ernst, (1811-1855)
ETTINGSHAUSEN, Constantin von, Freiherr, (1826-1897)
FINLAYSON, Alexander Moncreiff, (-1917)
GREY, George, Sir, (1812-1898)
GUPPY, Robert John Lechmere, (1836-1916)
HAAST, John Francis Julius von, Sir, (1824-1887)
HECTOR, James, Sir, (1834-1907)
HENDERSON, James Alexander Leo, (?1874-1940)
HOCHSTETTER, Ferdinand von, Ritter, (1829-1884)
HOGBEN, George, (1853-1920)
HOOKER, Joseph Dalton, Sir, (1817-1911)
HUTTON, Frederick Wollaston, (1836-1905)
JENNINGS, Albert Vaughan, (1864-1903)
JOHNSTON, Mary Sophia, (1875-1955)
LEEDS, Charles Edward, (1845-1912)
MANTELL, Walter Baldock Durant, Honourable, (1820-1895)
MARSHALL, Patrick, (1869-1950)
McCARTHY, Edward Thomas, (-) [PROSPECTORS, DIVINERS AND MINING ENGINEERS]
McKAY, Alexander, (1841-1917)
MORGAN, Peter (Percy) Gates, (1867-1927)
NEWELL ARBER, Edward Alexander, (1870-1918)
OWEN, Richard, Sir, (1804-1892)
PARK, James, (1857-1946)
PARKER, Thomas Jeffrey, (1850-1897)
PARKER, William Kitchen, (1823-1890)
RUSSELL, Israel Cook, (1852-1906)
SARGENT, Harry Lionel (1871-1955)
TEALE, Edmund Oswald, Sir, (1874-1971)
THOMSON, John Turnbull, (1821-1884)
ULRICH, George Henry Frederick, (1830-1900)
VEATCH, Arthur Clifford, (1878-1938)
WOODS, Henry, (1868-1952)

1900 to present

ALLAN, Robin Sutcliffe, (1900-1966)
ANDREW, Arthur Robert, (1881-1941)
BARTRUM, John Arthur, (1885-1949)
BELL, James Macintosh, (1877-1934)
BENSON, William Noel, (1885-1957)
BOHM, Georg, (1854-1913)
BOWEN, Charles Franklin, (1871-1953)
BREMNER, Carl St. John, (1895-1944)
BULLEN, Keith Edward, (1906-1976)
CARR, William L., (1919-1975)
CLAPP, Frederick Gardner, (1879-1944)
CLIFFORD, William Hugh, Right Honourable, (1858-1943)
COTTON, Charles Andrew, Sir, (1885-1970)

NEW ZEALAND 1900 to present Cont'd.

DAVIS, William Morris, (1850-1934)
EDWARDS, Wilfred Norman, (1890-1956)
FINLAY, Harold John, (1901-1951)
FINLAYSON, Alexander Moncreiff, (-1917)
GUPPY, Robert John Lechmere, (1836-1916)
HECTOR, James, Sir, (1834-1907)
HENDERSON, James Alexander Leo, (?1874-1940)
HOGBEN, George, (1853-1920)
HOOKER, Joseph Dalton, Sir, (1817-1911)
HUNT, Edwin Harold, (1897-1949)
HUTTON, Colin Osborne, (1910-1971)
HUTTON, Frederick Wollaston, (1836-1905)
JABLONSKI, Eugene, (1892-1975)
JENNINGS, Albert Vaughan, (1864-1903)
JOHNSTON, Mary Sophia, (1875-1955)
KINNEY, Harry Dill, (1886-1961)
LEEDS, Charles Edward, (1845-1912)
MACDONALD, Simon, (1927-1971)
MARSDEN, Ernest, (1889-1970)
MARSHALL, Patrick, (1869-1950)
MARWICK, John, (1891-1978)
MAY, Arthur Raymond, (1893-1948)
McCARTHY, Edward Thomas, (-) [PROSPECTORS, DIVINERS AND MINING ENGINEERS]
McKAY, Alexander, (1841-1917)
MORGAN, Peter (Percy) Gates, (1867-1927)
NEWELL ARBER, Edward Alexander, (1870-1918)
NUTTALL, Winfred Laurence Falkiner, (1897-1969)
PARK, James, (1857-1946)
RUSSELL, Israel Cook, (1852-1906)
SARGENT, Harry Lionel, (1871-1955)
SIMPSON, George Gaylord, (1902-)
TAYLOR, Griffith, (1881-1963)
TAZIEFF, Haroun, (1914-)
TEALE, Edmund Oswald, Sir, (1874-1971)
THOMSON, James Allan, (1881-1928)
TILLYARD, Robin John, (1881-1937)
TOEBES, Cornelius, (?-1975)
TRECHMANN, Charles Taylor, (1885-1964)
VEATCH, Arthur Clifford, (1878-1938)
WASHBURNE, Chester Wesley, (1883-1971)
WHETTEN, John T., (-)
WILLETT, Richard Wright, (1912-1974)
WOODS, Henry, (1868-1952)
WORLEY, W. F., (-1923)

NICARAGUA

1850-1899

BELT, Thomas, (1832-1878)
HAYES, Charles Willard, (1858-1916)

NICARAGUA 1850-1899 Cont'd.

McCARTHY, Edward Thomas, (-) [PROSPECTORS, DIVINERS AND MINING ENGINEERS]
SAPPER, Karl Theodor, (1866-1945)
TATE, Ralph, (1840-1901)

1900 to present
BASSINGER, William E., (1928-1977)
CRICKMAY, Geoffrey W., (1905-1971)
DAVILA, Salvador Ortiz, (1902-1931)
HALSE, George Whitfield, (1885-1968)
HAYES, Charles Willard, (1858-1916)
MACKENZIE, Andrew Nicholls "Kelley", (1888-1951)
McCARTHY, Edward Thomas, (-) [PROSPECTORS, DIVINERS AND MINING ENGINEERS]
PETERSON, Clarence J., (1885-1948)
SAPPER, Karl Theodor, (1866-1945)
TATE, Ralph, (1840-1901)
WILSON, Roy Arthur, (1891-1970)
WILSON, Thomas Carroll, (1906-1969)

NICOBAR ISLANDS (INDIAN OCEAN)
1850-1899
STOLICZKA, Ferdinand, (1838-1874)

NIGER
1850-1899
CHUDEAU, René, (1864-1921)

1900 to present
CHUDEAU, René, (1864-1921)
JÉRÉMINE, Elisabeth, (née Tschernaieff), (1879-1964)
JONES, Brynmor, (1902-1941)
LAPPARENT, Albert Félix de, (1905-1975)
MALAVOY, Jean, (1903-1945)
MILLOT, Georges, (1917-)
RENAULT, Philippe, (-)

NIGERIA
1850-1899
BEEBY-THOMPSON, Arthur, (1873-1968)
FALCONER, John Downie, (1876-1947)
HYLAND, John Shearson, (1866-1898)
KITSON, Albert Ernest, Sir, (1868-1937)
LONGBOTTOM, Arthur, (? -1910)
McCARTHY, Edward Thomas, (-) [PROSPECTORS, DIVINERS AND MINING ENGINEERS]
TEALE, Edmund Oswald, Sir, (1874-1971)
WOODS, Henry, (1868-1952)

1900 to present
ALEXANDER, Frances Elizabeth Somerville, (1908-1958)
BARRON, William Ernest, (1880-1935)
BEEBY-THOMPSON, Arthur, (1873-1968)
BELTZ, Edward W., (1891-1970)

NIGERIA 1900 to present Cont'd.

BENNETT, Walter Hellyer, (1892-1971)
BRAMWYCH, Michael, (1911-1974)
FALCONER, John Downie, (1876-1947)
HILL, Robert G., (1914-1973)
JENNINGS, Philip H., (1899-1974)
JONES, Brynmor, (1902-1941)
KITSON, Albert Ernest, Sir, (1868-1937)
LEDINGHAM, Glen Wallace, (1911-1978)
LONGBOTTOM, Arthur, (? -1910)
MACKAY, Robert Alexander, (?1904-1961)
McCARTHY, Edward Thomas, (-) [PROSPECTORS, DIVINERS AND MINING ENGINEERS]
PARGETER, Reginald Clive, (1919-1972)
RAEBURN, Colin, (1894-1970)
SMITH, G. Wendell, (1917-1974)
SPATH, Leonard Frank, (1882-1957)
STUDER, Joseph Paul, (1916-1951)
TEALE, Edmund Oswald, Sir, (1874-1971)
THOMAS, David, (1880-1934)
WHEELER, Charles Douglas, (?1922-1944)
WILSON, Reginald Charles, (1888-1943)
WOODS, Henry, (1868-1952)

NORTH AFRICA (unspecified)

1850-1899

HEILPRIN, Angelo, (1853-1907)

1900 to present

HARRISON, Thomas Samuel, (1881-1964)
HEILPRIN, Angelo, (1853-1907)
HIGGINS, Daniel Franklin, Jr., (1884-1930)
LECKWIJCK, Willem Peter van, (1902-1975)
UNGEMACH, Henri Léon, (1879-1936)
WAWRYK, Wlodzimierz, (1902-1963)

NORTH KOREA <u>See</u> KOREA.

NORTHERN IRELAND

Up to 1800

HAMILTON, William, Reverend, (1755-1797)

1800-1849

BRYCE, James, (1806-1877)
CONYBEARE, William Daniel, Reverend, (1787-1857)
McCOY, Frederick, Sir, (1823-1899)
YOUNG, Robert, Right Honourable, (1822-1917)

1850-1899

BRODRICK, Harold, (1872-1946)
BROWNRIDGE, Charles, Colonel, (?1862-1951)
BRYCE, James, (1806-1877)
CHRISTEN, Sydney Mary, (née THOMPSON), (-1923)
CONYBEARE, William Daniel, Reverend, (1787-1857)
DWERRYHOUSE, Arthur Richard, (1867-1931)

NORTHERN IRELAND 1850-1899 Cont'd.

KILROE, James Robinson, (1848-1927)
McCOY, Frederick, Sir, (1823-1899)
McHENRY, Alexander, (1843-1919)
NOLAN, Joseph, (1841-1902)
SYMES, Richard Glascott, (1840-1906)
TATE, Ralph, (1840-1901)
WYNNE, Arthur Beavor, (1835-1906)
WYVILLE THOMSON, Charles, Sir, (1830-1882)
YOUNG, Robert, Right Honourable, (1822-1917)

1900 to present

BRODRICK, Harold, (1872-1946)
BROWNRIDGE, Charles, Colonel, (?1862-1951)
CHRISTEN, Sydney Mary, (née THOMPSON), (-1923)
DWERRYHOUSE, Arthur Richard, (1867-1931)
HARTLEY, John Jerom, (1886-)
KILROE, James Robinson, (1848-1927)
McHENRY, Alexander, (1843-1919)
NOLAN, Joseph, (1841-1902)
ROBBIE, James Andrew, (1910-1977)
SYMES, Richard Glascott, (1840-1906)
TATE, Ralph, (1840-1901)
WYNNE, Arthur Beavor, (1835-1906)
YOUNG, Robert, Right Honourable, (1822-1917)

NORWAY

Up to 1800

BRONGNIART, Alexandre, (1778-1847)
BUCH, Christian Leopold von, Freiherr, (1774-1853)
ESMARCH, Jens, (1762-1839)
GROSSE, Karl Friedrich August, (1768-1847)
LINNE, Carl von, (1707-1778)

1800-1849

BRONGNIART, Alexandre, (1778-1847)
BUCH, Christian Leopold von, Freiherr, (1774-1853)
ESMARCH, Jens, (1762-1839)
FORBES, David, (1828-1876)
GRESSLY, Amanz, (1814-1865)
GROSSE, Karl Friedrich August, (1768-1847)
HAIDINGER, Wilhelm Karl von, Ritter, (1795-1871)
HAUSMANN, Johann Friedrich Ludwig, (1782-1859)
HOMAN, Christian Horrebow, (1826-1880)
KEILHAU, Baltazar Mathias, (1797-1858)
KJERULF, Theodor, (1825-1888)
MÜLLER, Carl Hermann, (1823-1907)
MURCHISON, Roderick Impey, Sir, (1792-1871)
NAUMANN, Carl Friedrich, (1797-1873)
PETTERSEN, Karl, (1826-1890)
SARS, Michael, (1805-1869)
ZITTEL, Karl Alfred von, (1839-1904)

1850-1899

BAUERMAN, Hilary, (1833-1909)

NORWAY 1850-1899 Cont'd.

BJERKNES, Vilhelm, (1862-1951)
BJØRLYKKE, Knut Olai, (1860-1946)
BLYTT, Axel, (1843-1898)
BRØGGER, Waldemar Christofer, (1851-1940)
BUCH, Christian Leopold von, Freiherr, (1774-1853)
CHANCOURTOIS, A. E. Béguyer de, (-1886?)
CODRINGTON, Thomas, (1829-1918)
COLLETT, Robert, (1842-1913)
COLLINS, Arthur L., (1868-1902)
COLLINS, Henry Francis, (?1864-1924)
DAMM, Carl Oluf Bernhard, (1869-1949)
DU PASQUIER, Léon, (1864-1897)
FEARNSIDES, William George, (1879-1968)
FEILDEN, Henry Wemyss, Colonel, (1838-1921)
FOOT, Frederick J., (-1867)
FORBES, David, (1828-1876)
FUCHS, Philippe Jacques Edmond, (1837-1889)
GRESSLY, Amanz, (1814-1865)
GRØNLIE, Ole Tobias, (1871-1954)
GURLDT, Adolf Friedrich, (1829-1902)
HAIDINGER, Wilhelm Karl von, Ritter, (1795-1871)
HALL, Marshall, Captain, (1831-1896)
HANSEN, Andreas M., (1857-1932)
HAUSMANN, Johann Friedrich Ludwig, (1782-1859)
HELLAND, Amund Theodor, (1846-1918)
HIORTDAHL, Thorstein Hallager, (1839-1925)
HOLMSEN, Andreas, (1869-1955)
HOMAN, Christian Horrebow, (1826-1880)
HORNEMAN, Hans Henrik, (1878-1945)
KEILHAU, Baltazar Mathias, (1797-1858)
KIAER, Johan Aschehong, (1869-1931)
KJERULF, Theodor, (1825-1888)
KOLDERUP, Carl Fredrik, (1869-1942)
MÜLLER, Carl Hermann, (1823-1907)
MÜNSTER, Thomas Georg, (1855-1938)
MURCHISON, Roderick Impey, Sir, (1792-1871)
NAUMANN, Carl Friedrich, (1797-1873)
ØYEN, Peter Arnaeus, (1863-1932)
PETTERSEN, Karl, (1826-1890)
REUSCH, Hans Henrik, (1852-1922)
SARS, Michael, (1805-1869)
SCHETELIG, Jakob Grubbe Cock, (1875-1935)
STELZNER, Alfred Wilhelm, (1840-1895)
VOGT, Johan Herman Lie, (1858-1932)
WILLIAMS, John Francis, (1862-1891)
WYVILLE THOMSON, Charles, Sir, (1830-1882)
ZITTEL, Karl Alfred von, (1839-1904)

1900 to present

ANDERSEN, Olaf, (1884-1941)
BARTH, Thomas Frederik Weiby, (1899-1971)
BATTLE, Walter Ravenhill Brown "Ben", (1919-1953)
BAUERMAN, Hilary, (1833-1909)
BJERKNES, Vilhelm, (1862-1951)

NORWAY 1900 to present Cont'd.

BJERRUM, Laurits, (1919-1973)
BJØRLYKKE, Knut Olai, (1860-1946)
BRØGGER, Waldemar Christofer, (1851-1940)
BROUGHTON-EDGE, Arthur Broughton, (1895-1953)
CODRINGTON, Thomas, (1829-1918)
COLLETT, Robert, (1842-1913)
COLLINS, Arthur L., (1868-1902)
COLLINS, Henry Francis, (?1864-1924)
DAMM, Carl Oluf Bernhard, (1869-1949)
ESKOLA, Pentti Eelis, (1883-1964)
FEARNSIDES, William George, (1879-1968)
FEILDEN, Henry Wemyss, Colonel, (1838-1921)
FOSLIE, Steinar, (1887-1951)
GOLDSCHMIDT, Victor Moritz, (1888-1947)
GRØNLIE, Ole Tobias, (1871-1954)
GURLDT, Adolf Friedrich, (1829-1902)
HANSEN, Andreas M., (1857-1932)
HEINTZ, Anatol, (1898-1975)
HELLAND, Amund Theodor, (1846-1918)
HIORTDAHL, Thorstein Hallager, (1839-1925)
HOLLINGWORTH, Sydney Ewart, (1899-1966)
HOLMES, George William, (1922-1970)
HOLMSEN, Andreas, (1869-1955)
HOLMSEN, Gunnar, (1880-1976)
HOLTEDAHL, Olaf, (1885-1975)
HORNEMAN, Hans Henrik, (1878-1945)
KIAER, Johan Aschehong, (1869-1931)
KOLDERUP, Carl Fredrik, (1869-1942)
KOLDERUP, Niels-Henrik, (1898-1971)
MACHATSCHKI, Felix Karl Ludwig, (1895-1970)
MacROBERT OF DOUNESIDE AND CROMAR, Rachel, Lady (née Workman), (-1954)
McCALL, John Gill, (?1923-1954)
MÜLLER, Carl Hermann, (1823-1907)
MÜNSTER, Thomas Georg, (1855-1938)
OXAAL, John Thomas Fearnley, (1887-1942)
ØYEN, Peter Arnaeus, (1863-1932)
REUSCH, Hans Henrik, (1852-1922)
ROSENDAHL, Halvor, (1890-1963)
SCHETELIG, Jakob Grubbe Cock, (1875-1935)
SMITH, Jeremy, (1929-1959)
STRØM, Kaare, (1902-1967)
THOMPSON, Hugh R., (1926-1959)
VOGT, Johan Herman Lie, (1858-1932)
VOGT, Thorolf, (1888-1958)
WERENSKIOLD, Werner, (-1962)
WESTERVELD, J., (1905-1962)
ZITTEL, Karl Alfred von, (1839-1904)

NOVAYA ZEMLYA (ARCTIC USSR)
1850-1899
FEILDEN, Henry Wemyss, Colonel, (1838-1921)
GRONLIE, Ole Tobias, (1871-1954)
MOROZEWICZ, Józef Marian, (1865-1941)

NOVAYA ZEMLYA (ARCTIC USSR) 1850-1899 Cont'd.

PAYER, Julius von, Ritter, (1841-1912)
PIWOWAR, Adam, (1874-1939)
WICHMANN, Carl Ernst Arthur, (1851-1927)

1900 to present
BODYLEVSKY, Vitaly Ivanovich, (1898-1968)
FEILDEN, Henry Wemyss, Colonel, (1838-1921)
GRØNLIE, Ole Tobias, (1871-1954)
HOLTEDAHL, Olaf, (1885-1975)
MOROZEWICZ, Józef Marian, (1865-1941)
PAYER, Julius von, Ritter, (1841-1912)
PIWOWAR, Adam, (1874-1939)
WICHMANN, Carl Ernst Arthur, (1851-1927)

NYASALAND *See* MALAWI.

OMAN

1850-1899

BEEBY-THOMPSON, Arthur, (1873-1968)
PILGRIM, Henry Guy Ellcock, (1875-1943)

1900 to present

BEEBY-THOMPSON, Arthur, (1873-1968)
FALCON, Norman Leslie, (-)
FOX, Cyril Sankey, Sir, (1886-1951)
HUDSON, Robert George Spencer, (1895-1965)
LEES, George Martin, (1898-1955)
LEHNER, Ernst J., (1895-1968)
PILGRIM, Henry Guy Ellcock, (1875-1943)
SUGDEN, Walter "Bill", (1923-1977)
WASHINGTON-GRAY, Kenneth, (?1899-1963)

PAKISTAN

Up to 1800

GREENOUGH, George Bellas, (1778-1855)

1800-1849

GREENOUGH, George Bellas, (1778-1855)
NICOL, Germain Louis, (1922-1973)

1850-1899

GREENOUGH, George Bellas, (1778-1855)
GROEBER, Paul Friedrich Carl, (1885-1968)
KOKEN, Ernst Friedrich Rudolph Karl von, (1860-1912)
MIDDLEMISS, Charles Stewart, (1859-1945)
NICOL, Germain Louis, (1922-1973)
OLDHAM, Richard Dixon, (1858-1936)
PILGRIM, Henry Guy Ellcock, (1875-1943)
REED, Frederick Richard Cowper, (1869-1946)
STOLICZKA, Ferdinand, (1838-1874)
WAAGEN, Wilhelm Heinrich, (1841-1900)
WYNNE, Arthur Beavor, (1835-1906)

1900 to present

CHHIBBER, Harbans Lal, (1899-1955)
CHLEBOWSKI, Tadeusz, (1908-1961)
COTTER, Gerald de Purcell, (1881-1941)
CROOKSHANK, Henry, (1893-1972)
DAVIES, Lewis Merson, Lieutenant-Colonel, (1882-1955)
DAVIS, Ralph Emerson, (1884-1968)
GROEBER, Paul Friedrich Carl, (1885-1968)
HAZZARD, John Charles, (1903-1973)
HERITSCH, Franz, (1882-1945)
KOKEN, Ernst Friedrich Rudolph Karl von, (1860-1912)
LAMSENS, O. Robert, (1922-1971)
LEHNER, Ernst J., (1895-1968)
MAYO, Herbert Theodore, (1889-1965)
MIDDLEMISS, Charles Stewart, (1859-1945)
MUHAMMAD, Wali, (1935-1963)
NUTTALL, Winfred Laurence Falkiner, (1897-1969)
OLDHAM, Richard Dixon, (1858-1936)
PILGRIM, Henry Guy Ellcock, (1875-1943)
PINFOLD, Ernest Sheppard, (1888-1969)
RASHID, Muhammad Abdur, (1938-1969)
REA, Henry Carter, (1900-1963)
REED, Frederick Richard Cowper, (1869-1946)
RHOADES, Roger F., (1905-1972)
SARJEANT, William Antony Swithin, (1935-)
SCHERMERHORN, W., (1894-)
SCHINDEWOLF, Otto Heinrich, (1896-1972)
SPATH, Leonard Frank, (1882-1957)
TEICHERT, Curt, (-)
TIPPER, George Howlett, (1881-1947)
WYNNE, Arthur Beavor, (1835-1906)

PAKISTAN <u>See</u> INDIA (for earlier references).

PALAU ISLANDS (PACIFIC OCEAN)
1850-1899
WICHMANN, Carl Ernst Arthur, (1851-1927)

1900 to present
ASANO, Kiyoshi, (1910-)
MASON, Arnold Caverly, (1906-1961)
WICHMANN, Carl Ernst Arthur, (1851-1927)

PALESTINE See ISRAEL; JORDAN.

PANAMA
1800-1849
HOPKINS, Evan, (?1810-1867)
LLOYD, J. A., Colonel, (-1856?)

1850-1899
BROWN, Amos Peaslee, (1864-1917)
CHAPER, Maurice Armand, (1834-1896)
DITMARS, Raymond Lee, (1876-1942)
DOUVILLÉ, Joseph Henri Ferdinand, (1846-1937)
GALE, Hoyt Stoddard, (1876-1952)
HOFFMAN, Fr. Adolf, (1867-1898)
HOPKINS, Evan, (?1810-1867)
HOWE, Ernest, (1875-1932)
LLOYD, J. A., Colonel, (-1856?)
WALTMAN, William Dewitt, (1875-1955)

1900 to present
BEVIER, George Means, (1888-1972)
BROWN, Amos Peaslee, (1864-1917)
COOKE, Charles Wythe, (1887-1971)
DITMARS, Raymond Lee, (1876-1942)
DOUVILLÉ, Joseph Henri Ferdinand, (1846-1937)
FOLGER, James Anthony, (1896-1956)
GALE, Hoyt Stoddard, (1876-1952)
HARRISON, Thomas Samuel, (1881-1964)
HIGGS, Morton Thomas, (1896-1956)
HOWE, Ernest, (1875-1932)
MACKENZIE, Andrew Nicholls "Kelley", (1888-1951)
MEAD, Warren Judson, (1883-1960)
MORRILL, Junius Hughes, (1913-1961)
NOMLAND, Jorgen O., (1885-1943)
POLLARD, Ernest Rex, (1922-1966)
SCHOLL, Louis August, Jr., (1887-1953)
SEARS, Julian Ducker, (1891-1970)
SINGEWALD, Joseph Theophilus, Jr., (1884-1963)
TRAGER, Earl A., (1893-1978)
WALLACE, Maurice Hudiburg, (1917-1961)
WALTMAN, William Dewitt, (1875-1955)
WEAVER, Charles Edwin, (1880-1958)

PAPUA
1800-1849
JUKES, Joseph Beete, (1811-1869)

PAPUA Cont'd.

1850-1899
BOWEN, Charles Franklin, (1871-1953)
JUKES, Joseph Beete, (1811-1869)
KORNITZER, Louis, (1873-) [PROSPECTORS, DIVINERS AND MINING ENGINEERS]
WADE, Arthur, (1878-1951)

1900 to present
BARROW, Geoffrey, (1899-1963)
BOWEN, Charles Franklin, (1871-1953)
CONDIT, Daniel Dale, (1886-1955)
EDWARDS, Austin Burton, (1909-1960)
KLEIBER, Karl Ernst, (1912-1972)
KORNITZER, Louis, (1873-) [PROSPECTORS, DIVINERS AND MINING ENGINEERS]
LEHNER, Ernst J., (1895-1968)
MAYO, Herbert Theodore, (1889-1965)
NUTTALL, Winfred Laurence Falkiner, (1897-1969)
PAPP, Simon, (1886-1970)
WADE, Arthur, (1878-1951)
WASHINGTON-GRAY, Kenneth, (?1899-1963)

PARAGUAY
1850-1899
KERR, John Graham, (1869-1957)
MILCH, Ludwig, (1867-1928)
TERNETZ, Carl, (1870-1928)

1900 to present
CARR, William L., (1919-1975)
CASSEL, Chester A. "Chet", (1897-1971)
HARRINGTON, Horacio Jaime, (1910-1973)
HAYES, Albert Orion, (1882-1965)
KERR, John Graham, (1869-1957)
MILCH, Ludwig, (1867-1928)
SCHMITTOU, Maurice B., (1899-1929)
TERNETZ, Carl, (1870-1928)

PELEW ISLANDS (PACIFIC OCEAN) <u>See</u> PALAU ISLANDS (Pacific Ocean).

PERIM (RED SEA)
1850-1899
MOORE, Henry Cecil, (1835-1908)

1900 to present
MOORE, Henry Cecil, (1835-1908)

PERSIA <u>See</u> IRAN.

PERU
Up to 1800
BOUGUER, Pierre, (1698-1758)
HELMS, Anton Zacharias, (1751-1803) [PROSPECTORS, DIVINERS AND MINING ENGINEERS]

PERU Up to 1800 Cont'd.

HUMBOLDT, Friedrich Wilhelm Heinrich Alexander von, Freiherr, (1769-1859)
MAWE, John, (1764-1829)
TREVITHICK, Richard, (1771-1833) [PROSPECTORS, DIVINERS AND MINING ENGINEERS]

1800-1849
DARWIN, Charles Robert, (1809-1882)
FORBES, David, (1828-1876)
HELMS, Anton Zacharias, (1751-1803) [PROSPECTORS, DIVINERS AND MINING ENGINEERS]
HUMBOLDT, Friedrich Wilhelm Heinrich Alexander von, Freiherr, (1769-1859)
LLOYD, J. A., Colonel, (-1856?)
MAWE, John, (1764-1829)
ORBIGNY, Alcide Charles Victor Dessalines d', (1802-1857)
TREVITHICK, Richard, (1771-1833) [PROSPECTORS, DIVINERS AND MINING ENGINEERS]
TSCHUDI, Johan Jacob von, (1818-1889)

1850-1899
ADAMS, George Irving, (1870-1932)
BALSTON, William Edward, (1848-1918)
BAUERMAN, Hilary, (1833-1909)
BEEBY-THOMPSON, Arthur, (1873-1968)
BERGT, Walther Adolf, (1864-1941)
BOUTWELL, John Mason, (1874-1968)
BOWMAN, Isaiah, (1882-1950)
BRAVO, José J., (1874-1927)
CHAPER, Maurice Armand, (1834-1896)
DARWIN, Charles Robert, (1809-1882)
DOUVILLÉ, Joseph Henri Ferdinand, (1846-1937)
FENNER, Clarence Norman, (1870-1949)
FORBES, David, (1828-1876)
GREGORY, Herbert E., (1869-1952)
HAUTHAL, Rudolf, (1854-1928)
HELMS, Anton Zacharias, (1751-1803) [PROSPECTORS, DIVINERS AND MINING ENGINEERS]
HENNEN, Ray Vernon, (1875-1958)
HUMBOLDT, Friedrich Wilhelm Heinrich Alexander von, Freiherr, (1769-1859)
LISSON, Carlos I., (1868?-1957)
LLOYD, J. A., Colonel, (-1856?)
MACLAREN, James Malcolm, (1873-1935)
NORDENSKIÖLD, Nils Otto Gustaf, (1869-1928)
OCHSENIUS, Carl Christian, (1830-)
ORBIGNY, Alcide Charles Victor Dessalines d', (1802-1857)
STEINMANN, Gustav, (1856-1929)
TSCHUDI, Johan Jacob von, (1818-1889)
WOODS, Henry, (1868-1952)

1900 to present
ADAMS, George Irving, (1870-1932)
ARMSTRONG, Harold Kerr, (1893-1962)

PERU 1900 to present Cont'd.

BALSTON, William Edward, (1848-1918)
BAUERMAN, Hilary, (1833-1909)
BEEBY-THOMPSON, Arthur, (1873-1968)
BEEKLEY, Albert Leon, (1883-1952)
BERGT, Walther Adolf, (1864-1941)
BOSWORTH, Thomas Owen, (1882-1928)
BOUTWELL, John Mason, (1874-1968)
BOWMAN, Isaiah, (1882-1950)
BRAVO, José J., (1874-1927)
BROGGI, Jorge Albert, (1888-1966)
BUSK, Henry Gould, (?1891-1956)
DERRY, Duncan Ramsay, (?1908-)
DOUGLAS, James Archibald, (1884-1978)
DOUVILLÉ, Joseph Henri Ferdinand, (1846-1937)
DRESSER, Myron Amos, (1894-1973)
DURHAM, Charles Albert, (1910-1954)
EDWARDS, Merwin Guy, (1889-1969)
FENNER, Clarence Norman, (1870-1949)
FRIZZELL, Donald Leslie, (1906-1972)
GESTER, Stephen Howell, (1889-1967)
GILLERMAN, Elliot "Bud", (1913-1974)
GORDON, Samuel George, (1897-1952)
GRATON, Louis Caryl, (1880-1970)
GREGORY, Herbert E., (1869-1952)
HARRISON, John Vernon, (1892-1972)
HAUTHAL, Rudolf, (1854-1928)
HEALD, Kenneth Conrad, (1888-1971)
HENNEN, Ray Vernon, (1875-1958)
HEWETT, Donnel Foster, (1881-1971)
HOLST PELLEKAAN, William van, (1880-1949)
HOTZ, Walter, (1884-1958)
HUBBARD, Bela, (1890-1959)
IDDINGS, Arthur, (1891-1957)
ISENBERGER, Nathan Perry, (1896-1956)
KREIDLER, William Lynn, (1911-1975)
LANDES, Robert William, (1910-1969)
LISSON, Carlos I., (1868?-1957)
MACKENZIE, Andrew Nicholls "Kelley", (1888-1951)
MACLAREN, James Malcolm, (1873-1935)
MAY, Arthur Raymond, (1893-1948)
McKINSTRY, Hugh Exton, (1896-1961)
MOORE, Gilbert Parvin, (1898-1970)
MORAN, Robert Breck, (1879-1961)
MYERS, William Marsh, (1892-1951)
NORDENSKIÖLD, Nils Otto Gustaf, (1869-1928)
NOTESTEIN, Frank B., (1885-1973)
PETERSEN, Georg, (1898-)
SCHILLING, Karl Howard, (1894-1945)
SCHOLL, Louis August, Jr., (1887-1953)
SCOFIELD, Lloyd M., (1897-1969)
SEVERY, Clarence Luther, (1885-1975)
SINCLAIR, Ewart Gladstone, (1884-1927)
SINGEWALD, Joseph Theophilus, Jr., (1884-1963)
SPROULE, John Campbell, (1905-1970)

PERU 1900 to present Cont'd.

STEINMANN, Gustav, (1856-1929)
SULLINS, M. Joseph, (1930-1972)
TEMPLETON, Eugene Crittenden, (1889-1932)
THOMAS, Ronald Hawksby, Major, (1885-?1940)
TRASK, Parker Davies, (1899-1961)
VEIGEL, Edward C., (1920-1949)
WEAVER, Charles Edwin, (1880-1958)
WHEELER, Orby Clinton, (1898-1970)
WILSON, Roy Arthur, (1891-1970)
WOODS, Henry, (1868-1952)
YOUNGQUIST, Walter, (-)

PHILIPPINE ISLANDS

1850-1899

ADAMS, George Irving, (1870-1932)
ANDERSON, Tempest, (1846-1913)
BECKER, George Ferdinand, (1847-1919)
BOWEN, Charles Franklin, (1871-1953)
RICHTHOFEN, Paul Wilhelm Ferdinand von, Freiherr, (1833-1905)
RINNE, Friedrich Wilhelm Berthold, (1863-1933)
WAY, Herbert W. L., (-) [PROSPECTORS, DIVINERS AND MINING ENGINEERS]

1900 to present

ABARQUEZ, Ramon Fabella, (1891-1959)
ADAMS, George Irving, (1870-1932)
ANDERSON, Tempest, (1846-1913)
ANDRES, Demetrio, (1889-1960)
BAIRD, Chester Alvan, (1896-1959)
BARD, Richards J., (1900-1977)
BECKER, George Ferdinand, (1847-1919)
BELTZ, Edward W., (1891-1970)
BOWEN, Charles Franklin, (1871-1953)
CARR, William L., (1919-1975)
CULVER, Harold Eugene, (1883-1970)
DOORNINCK, Nico H. van, (1893-1956)
DUGGLEBY, Arthur Francis, (1893-1945)
EATON, Arthur, (1886-1923)
FELICIANO, Jose Maria "Dako", (1887-1955)
HAZZARD, John Charles, (1903-1973)
HERRERA, Mariano B., (1918-1951)
HOLLISTER, Joseph Steffens, (1903-1953)
HUDSON, Frank Samuel, (1890-1967)
JOHNSON, Charles George, (1914-1969)
JOHNSTON, William Drumm, Jr., (1899-1972)
KIHLSTEDT, Folke H., (1901?-1956)
KRYSHTOFOVICH, Afrikan Nikolaevich, (1885-1953)
MARSMAN, J. H., (1892-1956)
MOODY, Graham Blair, (1889-1975)
PRATT, Wallace Everette, (1885-)
REED, Lyman Coleman, (1900-1970)
RICHTHOFEN, Paul Wilhelm Ferdinand von, Freiherr, (1833-1905)
RINNE, Friedrich Wilhelm Berthold, (1863-1933)
SCHENCK, Hubert Gregory, (1897-1960)
SCHERMERHORN, W., (1894-)

PHILIPPINE ISLANDS 1900 to present Cont'd.

SELGA, Miguel, Reverend Father, S. J., (1879-1956)
SMITH, Warren D., (1880-1950)
TAYLOR, Griffith, (1881-1963)
TEVES, Juan S., "Steve", (1902-1958)
THIENHAUS, Rolf, (1913-1968)
VENING MEINESZ, Felix Andries, (1887-1966)
WARING, Gerald Ashley, (1883-1971)
WISSER, Edward Hollister, (1895-1970)

POLAND

Up to 1800

BORCH, Michael Jan, (1751-1810)
CAROSI, Jan Filip, (1744-1799)
HACQUET, Belsazar, (1739-1815)
JAŚKIEWICZ, Jan D. P., (1749-1809)
JUNDZILL, Stanislaw Bonifacy, (1761-1847)
KLUK, Krzysztof, (1739-1796)
KOLLĄTAJ, Hugon, (1750-1812)
SAPIEHA, Aleksander, (1773-1812)
SENDIVOGIUS, Michael, (1556-1636 or 1646)
STASZIC, Stanislaw, (1755-1826)

1800-1849

ALTH, Alois von, Edler, (1819-1886)
BELZA, Józef, (1805-1888)
BESSER, Wilibald Swibert Joseph Gottlieb von, (1784-1842)
BORCH, Michael Jan, (1751-1810)
BRADETZKY, Samuel, (-)
CARNALL, Rudolf von, (1804-1874)
CHALUBIŃSKI, Tytus, (1820-1889)
CZYRNIAŃSKI, Emilian, (1824-)
DOMEYKO, Ignacy, (1802-1889)
ESTREICHER, Alojzy, (-?1852)
GÖPPERT, Heinrich Robert, (1801-1884)
HACQUET, Belsazar, (1739-1815)
HAUPT, Karl, (1829-1882)
JAŚKIEWICZ, Jan D. P., (1749-1809)
JUNDZILL, Stanislaw Bonifacy, (1761-1847)
KOLLĄTAJ, Hugon, (1750-1812)
LABĘCKI, Hieronim, (1809-1862)
LILL VON LILIENBACH, Karl, (1798-1831)
MENGE, Anton, (1808-1880)
MÜLLER, Carl Hermann, (1823-1907)
PUSCH, Georg Gottlieb, (1790-1846)
RÖMER, Carl Ferdinand von, (1818-1891)
SAPIEHA, Aleksander, (1773-1812)
SOLTYK, Michal, (-)
STASZIC, Stanislaw, (1755-1826)
TOROSIEWICZ, Teodor, (-)
ZBORZEWSKI, Wojciech, (-)

POLAND 1800-1849 Cont'd.

ZEJSZNER, Ludwik, (1805-1871)

1850-1899

AHLBURG, Johannes Heinrich Wilhelm, (1833-1919)
ALTH, Alois von, Edler, (1819-1886)
ARCTOWSKI, Henryk Bronislaw, (1871-1958)
BĄKOWSKI, Józef, (-1888)
BELZA, Józef, (1805-1888)
BERDAU, Feliks, (-?1888)
BEUTELL, Albert, (1859-1921)
BOHDANOWICZ, Karol, (1864-1947)
BOSNIASKI, Sigismund Grzymala, (1831-)
BUKOWSKI, Gejza, (1858-1937)
CARNALL, Rudolf von, (1804-1874)
CHALUBIŃSKI, Tytus, (1820-1889)
CHALUPCZYŃSKI, Adam, (-1887)
CHOROSZEWSKI, Wincenty Ignacy Jędrzej, (1845-1901)
CZARNOCKI, Stefan, (1878-1947)
DOMEYKO, Ignacy, (1802-1889)
DUBICKI, Józef, (-?1890)
DUNIN-BORKOWSKI, Stanislaw von, Graf, (1864-1934)
DYBOWSKI, Wladyslaw, (1838-1910)
ESTREICHER, Alojzy, (-?1852)
FALLAUX, Cornelius, (-1885)
FRIEDBERG, Wilhelm, (1873-1941)
GALLI, Edward Maryan de, (-?1893)
GIŽYCKI, Franciszek Ksawery, (-)
GLINKA, Konstantin Dmitrievich, (1867-1927)
GLINOJECKI, Jan, (-)
GÖPPERT, Heinrich Robert, (1801-1884)
GRZYBOWSKI, Józef, (1869-1922)
HASSE, Carl, (1841-1922)
HAUPT, Karl, (1829-1882)
HIRSZBERG, Franciszek, (1863-1933)
HORWITZ, Ludwik, (1875-1943)
HÜCKEL, Edward, (-?1896)
JAROSZ, Jan, (1877-1944)
JASTRZĘBOWSKI, Wojciech, (-)
JELSKI, Konstanty, (1837-1896)
KHRUSHCHOV, Konstantin Dmitrievich, (? -1912)
KONTKIEWICZ, Stanislaw, Sr., (1849-1924)
KREUTZ, Felix Franz Xaver, (1844-1910)
KREUTZ, Szczęsny, (-?1887)
KUŹNIAR, Wiktor, ([1871]-1935)
LABĘCKI, Hieronim, (1809-1862)
LEWIŃSKI, Jan Piotr, (1876-1939)
LIMANOWSKI, Mieczyslaw, (1876-1947)
LINGELSHEIM, Alexander von, (1874-1937)
LOMNICKI, Marian Alojzy R., (1845-1915)
LUBICZ-NIEZABITOWSKI, Edward, (1875-1946)
LUKASIEWICZ, Ignacy, (-)
LUKASZEWICZ, Józef, (1863-1928)
LUNIEWSKI, Adam, (1887-1945)
MENGE, Anton, (1808-1880)

POLAND 1850-1899 Cont'd.

MICHALSKI, Aleksander Octavianovich, (1855-1904)
MILCH, Ludwig, (1867-1928)
MOROZEWICZ, Józef Marian, (1865-1941)
MÜLLER, Carl Hermann, (1823-1907)
NIEWIADOMSKI, Wincenty, (-?1892)
OSSOWSKI, Godfryd, (-?1897)
RACIBORSKI, Maryan, (1863-1917)
REHMAN, Antoni, (-)
RÖMER, Carl Ferdinand von, (1818-1891)
ROMER, Eugeniusz, (1871-1954)
ROSLONSKI, Romuald, (1880-1956)
RUDZKI, Maurycy Pius, (-)
RUTKOWSKI, Feliks, (1878-1964)
SARJUSZ-MAKOWSKI, Arnold, (1876-1943)
SAWICKI, Ludomir, (-?1928)
ŚLOŚARSKI, Antoni, (-?1897)
SULIMIRSKI, Filip, (-?1885)
SZAJNOCHA, Wladyslaw, (1857-1928)
SZTOLCMAN, J., (-)
TEISSEYRE, Wawrzyniec, (1860-1939)
THOMPSON, Beeby, (1848-1931)
THUGUTT, Stanislaw Józef, (1862-1956)
TOLWIŃSKI, Konstanty, (1876-1961)
TOROSIEWICZ, Teodor, (-)
TREJDOSIEWICZ, Jan, (-)
UHLIG, Victor Karl, (1857-1911)
VULF, Georgii Viktorovich, (1863-1925)
WADE, Arthur, (1878-1951)
WAGA, Antoni Stanislaw Floryan, (-?1890)
WEYBERG, Zygmunt, (1872-1944)
WINDIAKIEWICZ, Edward, (1858-1942)
WIŚNIOWSKI, Sygurd, (-)
WIŚNIOWSKI, Tadeusz, (1865-1933)
YACHEVSKY, Leonard Antonovich, (1858-1916)
ZAPALOWICZ, Hugo, (-)
ZARĘCZNY, Stanislaw Wawrzyniec, (1848-1909)
ZAWISZA, Jan, (-)
ZDZITOWIECKI, Seweryn, (-?1881)
ŽEBROWSKI, Teofil, (-?1887)
ZEJSZNER, Ludwik, (1805-1871)
ZIELENIEWSKI, Michal, (-?1896)
ZUBER, Rudolf, (1858-1920)
ŽUKOWSKI, Wladyslaw, (1873-1952)

1900 to present

AHLBURG, Johannes Heinrich Wilhelm, (1833-1919)
ARCTOWSKI, Henryk Bronislaw, (1871-1958)
BADAK, Jerzy, (1932-1966)
BERES, Krzystztof, (-)
BEUTELL, Albert, (1859-1921)
BOCHEŃSKI, Tadeusz Antoni, (1901-1958)
BOHDANOWICZ, Karol, (1864-1947)
BUJALSKI, Boleslaw, (1888-1945)
BUKOWSKI, Gejza, (1858-1937)

POLAND 1900 to present Cont'd.

CHLEBOWSKI, Tadeusz, (1908-1961)
CHOROSZEWSKI, Wincenty Ignacy Jędrzej, (1845-1901)
CZARNOCKI, Jan, (1889-1951)
CZARNOCKI, Stefan, (1878-1947)
DANYSZ-FLESZAROWA, Regina, (1888-1969)
DE LA SAUCE, Wilhelm, (1882-1955)
DRATH, Adam, (1904-1942)
DUBYANSKY, Viktor Viktorovich, (1880-1925)
DUNIN-BORKOWSKI, Stanislaw von, Graf, (1864-1934)
DYBOWSKI, Wladyslaw, (1838-1910)
FORTUNAT, Wincenty Karol, (1889-1967)
FRIEDBERG, Wilhelm, (1873-1941)
FURMAŃSKI, Ignacy Jan, (1915-1972)
GĄSIOROWSKA, Zofia Halina, (1898-1957)
GLINKA, Konstantin Dmitrievich, (1867-1927)
GOETEL, Walery, (1889-1972)
GOLDSCHLAG, Maurycy, (1891-1917)
GORIZDRO-KULCZYCKA, Zinaida, (1884-1949)
GRZYBOWSKI, Józef, (1869-1922)
GUZIK, Kazimierz, (1911-1970)
HALICKA, Antonina, (1908-1973)
HASSE, Carl, (1841-1922)
HERALD, Frank Edward, (1887-1965)
HIRSZBERG, Franciszek, (1863-1933)
HLAUSCHEK, Hans, (1896-1971)
HORWITZ, Ludwik, (1875-1943)
HUXLEY, Julian Sorell, Sir, (1887-1975)
JABLONSKI, Eugene, (1892-1975)
JANCZEWSKI, Edward Walery, (1887-1959)
JARÓN, Bronislaw, (1905-1942)
JAROSZ, Jan, (1877-1944)
JOSSÉ, Kazimierz, (1892-1942)
KELUS, Aleksander, (1902-1946)
KHRUSHCHOV, Konstantin Dmitrievich, (? -1912)
KIELAN-JAWOROWSKA, Zofia, (-)
KITAJEWSKI, Adam Maksymilian, (-)
KOBYLECKI, Mieczyslaw, (1909-1944)
KOKOSZYŃSKA, Bronislawa, (1897-1959)
KONGIEL, Roman, (1904-1960)
KONTKIEWICZ, Stanislaw, Sr., (1849-1924)
KONTKIEWICZ, Stanislaw Jr., (1883-1951)
KOWALSKI, Ludwik, (1885-1943)
KOZLOWSKI, Roman, (1889-1977)
KRAJEWSKI, Stanislaw, (1890-1968)
KREUTZ, Felix Franz Xaver, (1844-1910)
KREUTZ, Stefan, (1883-1941)
KROPACZEK, Boleslaw, (1886-1914)
KRYZYŹKIEWICZ, Józef, (1890-1955)
KUŹNIAR, Czeslaw Faustyn, (1889-1955)
KUŹNIAR, Wiktor, ([1871]-1935)
KWOKEK, Stefan, (1923-1970)
LENCEWICZ, Stanislaw, (1899-1944)
LEWIŃSKI, Jan Piotr, (1876-1939)
LILPOP, Jerzy, (1888-1945)

POLAND 1900 to present Cont'd.

LIMANOWSKI, Mieczyslaw, (1876-1947)
LINGELSHEIM, Alexander von, (1874-1937)
LOMNICKI, Marian Alojzy R., (1845-1915)
LOZINSKI, Walery, (1880-1944)
LUBICZ-NIEZABITOWSKI, Edward, (1875-1946)
LUBICZ-SULIMIRSKI, Stefan, (1900-1953)
LUKASZEWICZ, Józef, (1863-1928)
LUNIEWSKI, Adam, (1887-1945)
MALKOWSKI, Stanislaw, (1889-1962)
MAZUREK, Alojzy, (1885-1944)
MICHALSKI, Aleksander Octavianovich, (1855-1904)
MILCH, Ludwig, (1867-1928)
MORAWIECKI, Antoni, (1903-1971)
MOROZEWICZ, Józef Marian, (1865-1941)
MOSSOCZY, Zbigniew, (1907-1963)
MÜLLER, Carl Hermann, (1823-1907)
NATURSKI, Jan, (-1936)
NECHAY, Wiktor, (1895-1940)
NIKLEWSKI, Jerzy, (1937-1971)
NOWAK, Jan, (1880-1940)
OBUCHOWICZ, Zbigniew Zygmunt, (1914-1965)
OPOLSKI, Zdzislaw, (1884-1941)
PACHUCKI, Czeslaw, (1898-1965)
PANOW, Eugeniusz, (1890-1958)
PARSONS, Ernest, (?1887-1963)
PAWLICA, Wladislaw, (1886-1919)
PAWLOWSKI, Stanislaw, (1882-1940)
PIECH, Kazimierz, (1893-1944)
PIOTROWSKI, Hugon Ludwik, (1897-1962)
POBORSKI, Czeslaw, (1916-1964)
POLANSKI, Jorge, (1892-1975)
PONIATOWSKI, Stanislaw Augusta, (1884-)
PREMIK, Józef, (1890-1963)
RABOWSKI, Ferdynand, (1884-1940)
RACIBORSKI, Maryan, (1863-1917)
REHMAN, Antoni, (-)
ROGALA, Wojciech, (1884-1947)
ROMANES, James, (1886-1942)
ROMER, Eugeniusz, (1871-1954)
ROSLONSKI, Romuald, (1880-1956)
ROZEN, Zygmunt, (-1936)
RUDZKI, Maurycy Pius, (-)
RUTKOWSKI, Feliks, (1878-1964)
RYCHLICKI, Jan, (-1942)
RYDZEWSKI, Bronislaw, (1884-1945)
RZACZYŃSKI, Gabriel, (-)
SAMSONOWICZ, Jan, (1888-1959)
SARJUSZ-MAKOWSKI, Arnold, (1876-1943)
SAWICKI, Ludomir, (-?1928)
SAWICKI, Ludwik, (1893-1972)
SKOCYLAS-CISZEWSKA, Kamila, (1902-1971)
ŚLIWA, Piotr, (1924-1959)
SMOLARSKA, Irena, (1928-1972)
SMOLEŃSKI, Jerzy, (1881-1940)

POLAND 1900 to present Cont'd.

SMOLUCHOWSKI, Marian, (-)
STENZ, Edward, (1897-1956)
SUJKOWSKI, Zbigniew Leliwa, (1898-1954)
ŚWIDERSKI, Bohdan, (1892-1943)
ŚWIDZINSKI, Henryk, (1904-1969)
SYNIEWSKA, Janina, (1893-1951)
SZAFER, Wladyslaw, (1886-1970)
SZAJNOCHA, Wladyslaw, (1857-1928)
TEISSEYRE, Henryk, (1903-1975)
TEISSEYRE, Wawrzyniec, (1860-1939)
THOMPSON, Beeby, (1848-1931)
THUGUTT, Stanislaw Józef, (1862-1956)
TOKARSKI, Julian, (1883-1961)
TOLWIŃSKI, Konstanty, (1876-1961)
TRZASKA-JURKIEWICZOWA, Irena, (1912-1966)
UHLIG, Victor Karl, (1857-1911)
VIGILEV, Boris, (1883-1924)
VULF, Georgii Viktorovich, (1863-1925)
WADE, Arthur, (1878-1951)
WALTON, John, (-1971)
WAWRYK, Wlodzimierz, (1902-1963)
WDOWIARZ, Jan, (1903-1956)
WEIGNER, Stanislaw Wiktor, (1886-1935)
WEYBERG, Zygmunt, (1872-1944)
WINDIAKIEWICZ, Edward, (1858-1942)
WIŚNIOWSKI, Tadeusz, (1865-1933)
WOJNO, Tadeusz Jerzy, (1884-1971)
YACHEVSKY, Leonard Antonovich, (1858-1916)
ZALESKI, Stanislaw, (-?1933)
ZAPALOWICZ, Hugo, (-)
ZAPART, Janusz Stanislaw, (1930-1961)
ZARĘCZNY, Stanislaw Wawrzyniec, (1848-1909)
ZUBER, Rudolf, (1858-1920)
ZUBER, Stanislaw, (1893-1947)
ŽUKOWSKI, Wladyslaw, (1873-1952)
ZWIERZYCKI, Józef, (1888-1961)
ŽYLA, Antoni, (1902-1966)

Time unknown

FONBERG, Ignacy, (-)
JURKIEWICZ, Karol, (-)
LASZKIEWICZ, Antoni, (-)
LUBOMĘSKI, Augusz, (-)
MALECKI, J., (-)
PISULEWSKI, Szymon, (-)
SAPALSKI, Jozef, (-)
SZAFARKIEWICZ, Józef, (-)

PORTUGAL

Up to 1800

DA COSTA, Emanuel Mendes, (1717-1791)
DOLOMIEU, Dieudonné Sylvain Guy Tancrède, called Déodat, Gratet de, (1750-1801)
ESCHWEGE, Wilhelm Ludwig, Baron, (1777-1855)

PORTUGAL Up to 1800 Cont'd.

HUNTER, John, (1728-1793)

1800-1849

BAPTISTA, Isidoro Emilio, (1815-1863)
COLLOMB, Edouard, (1801-1875)
COSTA, Pereira da, (1809-1889)
DOLOMIEU, Dieudonné Sylvain Guy Tancrède, called Déodat, Gratet de, (1750-1801)
ESCHWEGE, Wilhelm Ludwig, Baron, (1777-1855)
PEREIRA DA COSTA, Francisco Antonio, (1809-1889)
RIBEIRO, Carlos, General, (1813-1882)
SHARPE, Daniel, (1806-1856)
SMITH, James, (1782-1867)
VASCONCELLOS DE PEREIRA CABRAL, Frederico Augusto de, (-)

1850-1899

BAPTISTA, Isidoro Emilio, (1815-1863)
BEEBY-THOMPSON, Arthur, (1873-1968)
CARVALHO, Anselmo Ferraz de, (1878-1955)
CHOFFAT, Paul, (1849-1919)
COLLOMB, Edouard, (1801-1875)
COSTA, Pereira da, (1809-1889)
DAMASIO, José Victorio, (-)
ESCHWEGE, Wilhelm Ludwig, Baron, (1777-1855)
FLEURY, Ernest, (1878-1958)
GOMES, Jacinto Pedro, (1844-1910)
GONÇALVES GUIMARÃES, Antonio José, (1875-1919)
KOBY, Frédéric Louis, (1852-1930)
MACHADE E COSTA, Alfredo Augusto d'Olveira, (1870-?1952)
NERY DA ENCARNAÇAO DELGADO, Joaquim Filipe, (1835-1908)
PEREIRA DA COSTA, Francisco Antonio, (1809-1889)
PEREIRA DE LIMA, Wenceslau de Sousa, (1858-1920)
POMPECKJ, Josef Felix, (1867-1930)
RIBEIRO, Carlos, General, (1813-1882)
ROMAN, Albert Frédéric, (1871-1943)
SHARPE, Daniel, (1806-1856)
SMITH, James, (1782-1867)
SOUZA BRANDÃO, Vicente de, (1863-1916)
VASCONCELLOS DE PEREIRA CABRAL, Frederico Augusto de, (-)
WADE, Arthur, (1878-1951)

1900 to present

BEEBY-THOMPSON, Arthur, (1873-1968)
BOURCART, Jacques, (1891-1965)
BROUGHTON-EDGE, Arthur Broughton, (1895-1953)
CARVALHO, Anselmo Ferraz de, (1878-1955)
CHOFFAT, Paul, (1849-1919)
COX, Leslie Reginald, (1897-1965)
DAMASIO, José Victorio, (-)
DE SOUSA, Vasco Mendes, (1917-1962)
FLEURY, Ernest, (1878-1958)
FREIRE DE ANDRADE, Carlos Benot, (1893-1956)
GOMES, Jacinto Pedro, (1844-1910)
GONÇALVES GUIMARÃES, Antonio José, (1875-1919)

PORTUGAL 1900 to present Cont'd.

HALL, Thomas Clifford Fitzwilliam, (1886-1956)
JÉRÉMINE, Elisabeth, (née Tschernaieff), (1879-1964)
JESUS, Amílcar Mário de, (1895-1960)
KOBY, Frédéric Louis, (1852-1930)
LAPPARENT, Albert Félix de, (1905-1975)
LOTZE, Franz, (1903-1971)
MACHADE E COSTA, Alfredo Augusto d'Olveira, (1870-?1952)
NERY DA ENCARNAÇAO DELGADO, Joaquim Filipe, (1835-1908)
NICOLAS, Jean Paul, (1921-1967)
PEREIRA DE LIMA, Wenceslau de Sousa, (1858-1920)
PEREIRA DE SAMPAIO FORJAZ PIMENTEL, António Augusto Álvares, (1893-)
PILAR, Ludgero da Piedade, (1920-1967)
PIVETEAU, Jean, (1898-)
POMPECKJ, Josef Felix, (1867-1930)
ROMAN, Albert Frédéric, (1871-1943)
ROMANES, James, (1886-1942)
ROSSBY, Carl-Gustaf Arvid, (1898-1957)
SOUTO, Alberto, (1888-1961)
SOUZA BRANDÃO, Vicente de, (1863-1916)
TAYLOR, James Haward, (1909-1968)
VIANNA, António, (1887-1949)
WADE, Arthur, (1878-1951)
WATTISON, John Thomas, (1884-1974)
WESTERVELD, J., (1905-1962)

PORTUGUESE GUINEA
1900 to present
JACQUET, Fernand, (1908-1937)
TORRE DE ASSUNÇÃO, Carlos Fernando, (1901-)

PRIBILOF ISLANDS (NORTH PACIFIC OCEAN)
1900 to present
BARTH, Thomas Frederik Weiby, (1899-1971)

PUERTO RICO (WEST INDIES)
1900 to present
FETTKE, Charles Reinhard, (1888-1959)
GLASS, Jewell Jeanette, (1888-1966)
GUILLOU, Robert Barton, (1923-1965)
HOWELL, Benjamin Franklin, (1890-1976)
HUBBARD, Bela, (1890-1959)
LOBECK, Armin Kahl, (1886-1958)
McGUINNESS, Charles Lee, (1914-1971)
REYNOLDS, Thomas Emmett, (-1972)
TABER, Stephen, (1882-1963)
ZAPP, Alfred Dexter, (1916-1962)

QATAR

1900 to present

HUDSON, Robert George Spencer, (1895-1965)
SUGDEN, Walter "Bill", (1923-1977)

RÉUNION (INDIAN OCEAN)
1850-1899
LACROIX, François Antoine Alfred, (1863-1948)
MORGAN, Edward Delmar, (1840-1909)

1900 to present
LACROIX, François Antoine Alfred, (1863-1948)
MORGAN, Edward Delmar, (1840-1909)

RHODESIA
1850-1899
BALL, Francis Matthew, (1877-1956)
BREUIL, Henri, Abbé, (1877-1961)
GILL, David, Sir, (1843-1914)
HAMMOND, John Hays, (1855-1936)
HATCH, Frederick Henry, (?1864-1932)
KAISER, Friedrich Wilhelm Erich, (1871-1934)
MACLAREN, James Malcolm, (1873-1935)
MAUCH, Carl Gottlieb, (1837-1875)
MOLYNEUX, Arthur John Charles, (?1866-1921)
SARGENT, Harry Lionel, (1871-1955)

1900 to present
ANDERSON, Robert Beattie, (1904-1970)
BALL, Francis Matthew, (1877-1956)
BANDY, Mark Chance, (1900-1963)
BISHOPP, Douglas Wallace, (1900-1977)
BREUIL, Henri, Abbé, (1877-1961)
DERRY, Duncan Ramsay, (?1908-)
DOUGLAS, George Vibert, (1892?-1958)
GILL, David, Sir, (1843-1914)
HAMMOND, John Hays, (1855-1936)
HATCH, Frederick Henry, (?1864-1932)
HURLEY, Patrick Mason, (1912-)
JOHNSON, J. P., (1880-1918)
KAISER, Friedrich Wilhelm Erich, (1871-1934)
LIGHTFOOT, Ben, (1888-1966)
MACLAREN, James Malcolm, (1873-1935)
MAUFE, Herbert Brantwood, (1879-1946)
MENNELL, Frederick Philip, (1880-1966)
MOLYNEUX, Arthur John Charles, (?1866-1921)
RAINIER, Peter W., (-) [PROSPECTORS, DIVINERS AND MINING ENGINEERS]
SARGENT, Harry Lionel, (1871-1955)
TAYLOR, James Haward, (1909-1968)
TIDY, Anthony John Lintott, (1928-1969)
WAGNER, Percy Albert, (1885-1929)
WASSERSTEIN, Benno, (1906-1956)
ZEALLEY, Arthur Edmund Victor, (1886-1918)

Time unknown
MACGREGOR, A. M., (-)

RIO MUNI See EQUATORIAL GUINEA.

ROCKALL (NORTH ATLANTIC OCEAN)

1900 to present

TYRRELL, George Walter, (1883-1961)

ROUMANIA

Up to 1800

BORN, Ignaz von, Baron, (1742-1791)
HACQUET, Belsazar, (1739-1815)

1800-1849

BIELZ, Eduard Albert, (1827-1898)
BIELZ, Michael, (1787-1866)
COTTA, Carl Bernhard von, (1808-1879)
DAUBENY, Charles Giles Brindle, (1795-1867)
HACQUET, Belsazar, (1739-1815)
HAUER, Franz von, Ritter, (1822-1899)
LILL VON LILIENBACH, Karl, (1798-1831)
NEUGEBOREN, Johann Ludwig, (1806-1887)
STUR, Dénes, (1827-1893)
VERNEUIL, Philippe Edouard Poulletier de, (1805-1873)

1850-1899

ANDRUSSOV, Nikolai Ivanovich, (1861-1924)
BECKE, Friedrich Johann Karl, (1855-1931)
BEEBY-THOMPSON, Arthur, (1873-1968)
BIELZ, Eduard Albert, (1827-1898)
BIELZ, Michael, (1787-1866)
CĂDERE, Dimitrie M., (1876-1941)
CAPELLINI, Giovanni, (1833-1922)
COBÎLCESCU, Grigore, (1826-1892)
COTTA, Carl Bernhard von, (1808-1879)
DAUBENY, Charles Giles Brindle, (1795-1867)
DRĂGHICEANU, Mathei M., (1844-1939)
ERB, Josef Theodor, (1874-1934)
FONTANNES, Charles François, (1839-1886)
HAUER, Franz von, Ritter, (1822-1899)
ISTRATI, Constantin I., (1850-1918)
JOHN VON JOHNESBERG, Conrad Heinrich, Edler, (1852-1918)
KISSLING, Ernst, (1865-1936)
KOCH, Antal von Bodrog, (1843-1927)
MACLAREN, James Malcolm, (1873-1935)
MARTONNE, Emmanuel Louis Eugène de, (1873-1955)
MAUCH, Carl Gottlieb, (1837-1875)
MRAZEC, Lodovic, (1867-1944)
MUNTEANI-MURGOCI, Gheorghe, (1872-1925)
NEUGEBOREN, Johann Ludwig, (1806-1887)
PONI, Petru, (1841-1925)
POPOVICI-HAŢEG, Valeriu, (1868-1929)
PORRIO, Cesare, (1865-1940)
SCHARIZER, Rudolf, (1859-1935)
SONNEVELD, Jan, (1877-195)
STACHE, Guido, (1833-1921)
ŞTEFANESCU, Gregoriu, (1838-1911)
ŞTEFĂNESCU, Sabba, (1857-1931)
STUR, Dénes, (1827-1893)

ROUMANIA 1850-1899 Cont'd.

SZÁDECZKY, Gyula, Baron Szádeczky-Kardoss of Szádecsne and Kardosfalva, (1860-1935)
TEISSEYRE, Wawrzyniec, (1860-1939)
THOMPSON, Beeby, (1848-1931)
VERNEUIL, Philippe Edouard Poulletier de, (1805-1873)

1900 to present

ANDRAU, Evert Willem Karel, (1900-1951)
ANDRUSSOV, Nikolai Ivanovich, (1861-1924)
BECKE, Friedrich Johann Karl, (1855-1931)
BEEBY-THOMPSON, Arthur, (1873-1968)
BERTHELOT, Charles, (-) [PROSPECTORS, DIVINERS AND MINING ENGINEERS]
BÜRGL, Hans, (1907-1966)
CĂDERE, Dimitrie M., (1876-1941)
CAPELLINI, Giovanni, (1833-1922)
CERNESCU, Nicolae C., (1904-1967)
COBÎLCESCU, Grigore, (1826-1892)
CONSTANTINESCU, Valeriu, (1921-1970)
COSIJN, Adrianus Johannes, (1890-1945)
DE KOCH, William Gaspar, (1898-1954)
DRĂGHICEANU, Mathei M., (1844-1939)
DREHER, Otto, (1888-1945)
ERB, Josef Theodor, (1874-1934)
FREY, Alfred Paul, (1893-1963)
GARRETT, T. R. H., Major, (-1938)
GRIGORAŞ, Nicolae, (1913-1969)
HALSE, George Whitfield, (1885-1968)
HERALD, Frank Edward, (1887-1965)
HLAUSCHEK, Hans, (1896-1971)
ISTRATI, Constantin I., (1850-1918)
JABLONSKI, Eugene, (1892-1975)
JOHN VON JOHNESBERG, Conrad Heinrich, Edler, (1852-1918)
KISSLING, Ernst, (1865-1936)
KOCH, Antal von Bodrog, (1843-1927)
KUIPER, Willem Nicolaas, (1891-1952)
LEUTENEGGER, Willy Otto, (1903-1972)
LEYDS, Louis Willem, (1887-1963)
MACLAREN, James Malcolm, (1873-1935)
MACOVEI, Gheorghe, (1880-1969)
MARTONNE, Emmanuel Louis Eugène de, (1873-1955)
MAXIM, Ion Al., (1902-1968)
MEYER, Edouard, (1899-1974)
MIRĂUŢĂ, Orest, (1931-1967)
MRAZEC, Lodovic, (1867-1944)
MUNTEANI-MURGOCI, Gheorghe, (1872-1925)
NIETHAMMER, Gottlob August, (1882-1915)
PAPP, Simon, (1886-1970)
PENNY, Frederick Willoughby, (1887-1950)
POLLOCK, Charles Murray, (1898-1969)
PONI, Petru, (1841-1925)
POPOVICI, Viorica, (1918-1970)
POPOVICI-HAŢEG, Valeriu, (1868-1929)
PORRIO, Cesare, (1865-1940)

ROUMANIA 1900 to present Cont'd.

REED, Lyman Coleman, (1900-1970)
ROMAN, David, (1884-1927)
ROMANES, James, (1886-1942)
SCHARIZER, Rudolf, (1859-1935)
SEMAKA, Alexandru, (1924-1970)
SONNEVELD, Jan, (1877-195)
STACHE, Guido, (1833-1921)
ŞTEFANESCU, Gregoriu, (1838-1911)
ŞTEFĂNESCU, Sabba, (1857-1931)
SZÂDECZKY, Gyula, Baron Szádeczky-Kardoss of Szádecsne and Kardosfalva, (1860-1935)
TEISSEYRE, Wawrzyniec, (1860-1939)
THOMPSON, Beeby, (1848-1931)
VADÁSZ, Elemér, (1885-1970)
VÂLSAN, George, (1885-1935)
WIEDENMAYER, Carl, (1897-1951)

Time unknown
VOITEŞTI, Ion Popescu, (-)

RUMANIA See ROUMANIA.

RWANDA
1900 to present
DE DYCKER, Raymond, (1903-1947)
JAMOTTE, André, (1903-1951)

RYUKYU ISLANDS (PACIFIC OCEAN)
1850-1899
TOKUNAGA, Shigeyasu, (1874-1940)

1900 to present
JOHNSON, Charles George, (1914-1969)
TOKUNAGA, Shigeyasu, (1874-1940)

SAINT BARTHOLOMEW (WEST INDIES)
1850-1899
COTTEAU, Gustave Honoré, (1818-1894)
GUPPY, Robert John Lechmere, (1836-1916)

1900 to present
GUPPY, Robert John Lechmere, (1836-1916)

SAINT JOHNS ISLAND (RED SEA)
1850-1899
MOON, Frederick William, (-1925)

1900 to present
MOON, Frederick William, (-1925)

SAINT LUCIA (WEST INDIES)
1850-1899
HOVEY, Edmund Otis, (1862-1924)

1900 to present
HOVEY, Edmund Otis, (1862-1924)
LEHNER, Ernst J., (1895-1968)

SAINT VINCENT (WEST INDIES)
1850-1899
ANDERSON, Tempest, (1846-1913)
HOVEY, Edmund Otis, (1862-1924)
JAGGAR, Thomas Augustus, (1871-1953)
SAPPER, Karl Theodor, (1866-1945)

1900 to present
ANDERSON, Tempest, (1846-1913)
HOVEY, Edmund Otis, (1862-1924)
JAGGAR, Thomas Augustus, (1871-1953)
SAPPER, Karl Theodor, (1866-1945)

SAKHALIN (U.S.S.R.)
1900 to present
WEAVER, Paul, (1888-1964)

SALVADOR <u>See</u> EL SALVADOR.

SAMOA (PACIFIC OCEAN)
1900 to present
JOHNSON, Charles George, (1914-1969)

SANDWICH ISLANDS (NORTH PACIFIC OCEAN) <u>See</u> UNITED STATES OF AMERICA; HAWAII.

SANTO DOMINGO <u>See</u> DOMINCAN REPUBLIC; HAITI.

SARAWAK
1850-1899
ERB, Josef Theodor, (1874-1934)
HANNAY, Harry Campbell, (1874-1961)

SARAWAK Cont'd.

1900 to present
ANDREW, Arthur Robert, (1881-1941)
ERB, Josef Theodor, (1874-1934)
HANNAY, Harry Campbell, (1874-1961)
TRÜMPY, Daniel, (1893-1971)
WILHELM, Oscar, (1897-1975)

SARDINIA (ITALY)
1850-1899
MILNE-EDWARDS, Alphonse, (1835-1900)

1900 to present
DERRY, Duncan Ramsay, (?1908-)
FOSSA-MANCINI, Enrico, (1884-1950)
MILNE-EDWARDS, Alphonse, (1835-1900)

SAUDI ARABIA
1800-1849
EHRENBERG, Christian Gottfried, (1795-1876)

1850-1899
BAUERMAN, Hilary, (1833-1909)
BEEBY-THOMPSON, Arthur, (1873-1968)
EHRENBERG, Christian Gottfried, (1795-1876)
WADE, Arthur, (1878-1951)

1900 to present
BAUERMAN, Hilary, (1833-1909)
BEEBY-THOMPSON, Arthur, (1873-1968)
BERG, Ernest Lyle, (1910-1953)
BIRAUD, François, (1899-1936)
BRAMKAMP, Richard Allen, (1910-1958)
CUTTITTA, Frank, (1912-1974)
FOX, Cyril Sankey, Sir, (1886-1951)
FREY, M. Gordon, (1913-)
HIGGINS, Daniel Franklin, Jr., (1884-1930)
JONES, David Glynn, (1904-1974)
LAMARE, Pierre, (1894-1967)
LELUBRE, Maurice, (1916-)
McCLURE, Drexel Eugene, (1928-1977)
NOMLAND, Jorgen O., (1885-1943)
NOTESTEIN, Frank B., (1885-1973)
SAMPSON, David Noel, (1917-1971)
WADE, Arthur, (1878-1951)
WASHINGTON-GRAY, Kenneth, (?1899-1963)
ZIMERMAN, Sam, (1905-1972)

SCOTLAND
Up to 1800
ALLAN, Thomas, (1777-1833)
BALD, Robert, (1776-1861)
BROWN, Robert, (1773-1858)
CLARKE, Edward Daniel, (1769-1822)
CLERK, John, Jr., Lord Eldin, (1757-1812)

SCOTLAND Up to 1800 Cont'd.

DINWIDDIE, James, (-)
FAUJAS DE SAINT FOND, Barthélemy, (1741-1819)
FERBER, Johann Jacob, (1743-1790)
GREGOR, William, Reverend, (1761-1817)
GUTHRIE, Matthew, (1743-1807)
HALL, James, Sir, (1761-1832)
HUNTER, John, (1728-1793)
HUTTON, James, (1726-1797)
JAMESON, Robert, (1774-1854)
LETTSOM, John Coakley, (1744-1815)
MACCULLOCH, John, (1773-1835)
NICOL, William, (ca.1768-1851)
PLAYFAIR, John, (1748-1819)
RASPE, Rudolf Erich, (1737-1794)
SIBBALD, Robert, Sir, (1641-1772)
THOMSON, Thomas, (1773-1852)
TOWNSON, Robert, (1763-1827)
URE, Andrew, (1778-1857)
WALKER, John, Reverend, (1731-1803)

1800-1849

ALLAN, Thomas, (1777-1833)
ALLMAN, George James, (1812-1898)
ALLPORT, Samuel, (1816-1897)
ARGYLL, George Douglas, Duke of, (1823-1900)
BALD, Robert, (1776-1861)
BALD, William, (-?1857)
BELL, Dugald, (1827-1892)
BENNIE, James, (1821-1901)
BOUÉ, Ami, (1794-1881)
BROWN, Robert, (1773-1858)
BROWN, Thomas, Captain, (1785-1862)
BRYCE, James, (1806-1877)
BRYSON, Alexander, (1816-1866)
CAMPBELL, John Francis, (1821-1885)
CARUS, Carl Gustav, (1789-1869)
CHAMBERS, Robert, (1802-1871)
CLARKE, Edward Daniel, (1769-1822)
CLERK, John, Jr., Lord Eldin, (1757-1812)
CRAIG, Robert, (?1821-1901)
CROLL, James, (1821-1890)
DAVIDSON, Thomas, (1817-1885)
DICK, Robert, (1811-1866)
DINWIDDIE, James, (-)
FAUJAS DE SAINT FOND, Barthélemy, (1741-1819)
FERGUSSON, James, (1808-1885)
FLEMING, John, Reverend, (1785-1857)
FORBES, Edward, (1815-1854)
FORBES, James David, (1809-1868)
GLEN, David Corse, (1824-)
GORDON, George, Reverend, (1801-1893)
GRANT, Robert Edmond, (1793-1874)
GREGOR, William, Reverend, (1761-1817)
GUTHRIE, Matthew, (1743-1807)

SCOTLAND 1800-1849 Cont'd.

HALL, James, Sir, (1761-1832)
HEDDLE, Matthew Forster, (1828-1897)
HOME, David Milne, (1805-1890)
HOOKER, William Jackson, Sir, (1785-1865)
HUXLEY, Thomas Henry, (1825-1895)
JAMESON, Robert, (1774-1854)
JAMIESON, Thomas Francis, (1829-1913)
KELVIN, William Thomson, Baron, (1824-1907)
LAIDLAY, John Watson, (1808-1885)
LAUDER, Thomas Dick, Sir, (1784-1848)
LETTSOM, John Coakley, (1744-1815)
MACCULLOCH, John, (1773-1835)
MACGILLIVRAY, William, (1796-1852)
MACLAREN, Charles, (1782-1866)
MILLER, Hugh, (1802-1856)
MITCHELL, Hugh, Reverend, (1822-1894)
MOORE, John Carrick, (1804-1898)
MURCHISON, Roderick Impey, Sir, (1792-1871)
NECKER DE SAUSSURE, Louis Albert, (1786-1861)
NICOL, James, (1810-1879)
NICOL, William, (ca.1768-1851)
NORTHAMPTON, Joshua Alwyne Spencer, Marquis of, (1790-1851)
PAGE, David, (1814-1879)
PATTON, Andrew, (1821-1885)
PEACH, Charles William, (1800-1886)
PLAYFAIR, John, (1748-1819)
RAMSAY, Andrew Crombie, Sir, (1814-1891)
ROBERTSON, David, (1806-1896)
ROGERS, Henry Darwin, (1808-1866)
ROSE, Alexander, (1781-1860)
SELKIRK, Dunbar James Douglas, Earl of, (1809-1885)
SLIMON, Robert, (1803-1882)
SMITH, James, (1782-1867)
SMYTH, Warington Wilkinson, Sir, (1817-1890)
SOMERVILLE, Mary, (1780-1872)
STRICKLAND, Hugh Edwin, (1811-1853)
SYMONDS, William Samuel, Reverend, (1818-1887)
THOMSON, James, (1823-1900)
THOMSON, Thomas, (1773-1852)
TOWNSON, Robert, (1763-1827)
TRAILL, Thomas Stewart, (1781-1862)
TREVELYAN, Walter Calverley, Sir, (1797-1879)
URE, Andrew, (1778-1857)
WALKER, John, Reverend, (1731-1803)
WATSON, Robert Boog, Reverend, (?1824-1910)
YOUNG, John, (1823-1900)

1850-1899

ADAMS, Andrew Leith, (-1882)
ALLMAN, George James, (1812-1898)
ALLPORT, Samuel, (1816-1897)
ARGYLL, George Douglas, Duke of, (1823-1900)
ARMSTRONG, James, (1832-1892)
BALD, Robert, (1776-1861)

SCOTLAND 1850-1899 Cont'd.

BARROW, George, (1853-1932)
BELL, Dugald, (1827-1892)
BENNIE, James, (1821-1901)
BLAKE, John Frederick, Reverend, (1839-1906)
BLOXAM, Thomas, (1836-?1872)
BOUÉ, Ami, (1794-1881)
BREMNER, Alexander, (1863-1943)
BROWN, Robert, (1773-1858)
BROWN, Thomas, Captain, (1785-1862)
BRYCE, James, (1806-1877)
BRYSON, Alexander, (1816-1866)
BULLOWS, Wilfred Lowson, (?1869-1943)
CAMPBELL, John Francis, (1821-1885)
CARRUTHERS, William, (1830-1922)
CARUS, Carl Gustav, (1789-1869)
CHAMBERS, Robert, (1802-1871)
CHREE, Charles, (-)
CLOUGH, Charles Thomas, (1852-1916)
CONAN DOYLE, Arthur, Sir, (1859-1930)
COOMARASWAMY, Ananda K., (1877-1947)
CORSTORPHINE, George Steuart, (1865-1919)
CRAIG, Robert, (?1821-1901)
CROLL, James, (1821-1890)
CURRIE, James, (1863-1930)
DAKYNS, John Roche, (1836-1910)
DAVIDSON, Thomas, (1817-1885)
DAVIES, Thomas, (1837-1892)
DAY, Thomas Cuthbert, (1853-1935)
DICK, Allan Brugh, (1833-1926)
DICK, Robert, (1811-1866)
DRUMMOND, Henry, (1851-1897)
DUDGEON, Patrick, (-1895)
ELLES, Gertrude Lilian, (1872-1960)
ELLISON, John, (1859-1936)
ETHERIDGE, Robert Jr., (1847-1920)
FALCONER, John Downie, (1876-1947)
FERGUSSON, James, (1808-1885)
FLEMING, John, Reverend, (1785-1857)
FLETT, John Smith, Sir, (1869-1947)
FORBES, Edward, (1815-1854)
FORBES, James David, (1809-1868)
FRASER, James Pinkerton, (-)
GARDINER, Charles Irving, (?1868-1940)
GEIKIE, Archibald, Sir, (1835-1924)
GEIKIE, James, (1839-1915)
GIBB, Alfred William, (1864-1937)
GIBSON, Walcot, (1864-1941)
GILL, David, Sir, (1843-1914)
GLEN, David Corse, (1824-)
GOODCHILD, John George, (1844-1906)
GORDON, George, Reverend, (1801-1893)
GORDON, John More, Reverend, (1849-1922)
GRANT, Robert Edmond, (1793-1874)
GREEN, John Frederick Norman, (?1873-1949)

SCOTLAND 1850-1899 Cont'd.

GREENLY, Edward, (1861-1951)
GUNN, Robert Marcus, (?1850-1909)
GUNN, William, (1837-1902)
HARKER, Alfred, (1859-1939)
HARKNESS, Robert, (1816-1878)
HATCH, Frederick Henry, (?1864-1932)
HEDDLE, Matthew Forster, (1828-1897)
HILL, James Bastian, (1861-1927)
HOLLAND, Thomas Henry, Sir, (1868-1947)
HOME, David Milne, (1805-1890)
HOOKER, William Jackson, Sir, (1785-1865)
HORNE, John, (1848-1928)
HOWELL, Henry Hyatt, (1834-1915)
HULL, Edward, (1829-1917)
HUNTER-SELKIRK, John R. S., (1835-1898)
HUXLEY, Thomas Henry, (1825-1895)
JAMESON, Robert, (1774-1854)
JAMIESON, Thomas Francis, (1829-1913)
JEHU, Thomas John, (1871-?1943)
JENKIN, Henry Charles Fleeming, (1833-1885)
JOHNSTON, Robert Mackenzie, (-1918)
JUDD, John Wesley, (1840-1916)
JUKES-BROWNE, Alfred John, (1851-1914)
KELVIN, William Thomson, Baron, (1824-1907)
KERR, John Graham, (1869-1957)
KIDSTON, Robert, (1852-1924)
KIRKBY, James Walker, (1834-1901)
KYLE, John Joseph Jolly, called Juan José, (1838-1922)
KYNASTON, Herbert, (1868-1915)
LAIDLAY, John Watson, (1808-1885)
LANG, William Henry, (1874-1960)
LAPWORTH, Charles, (1842-1920)
LYONS, Henry George, (1864-?1943)
MACCONOCHIE, Arthur, (1850-1922)
MACGILLIVRAY, William, (1796-1852)
MACINTOSH, Charles, (1839-1922)
MACKIE, William, (1856-1932)
MACLAREN, Charles, (1782-1866)
MARGERIE, Emmanuel Marie Pierre Martin Jacquin de, (1862-1953)
MATLEY, Charles Alfred, (1866-1947)
McDONALD, William, (1876-1943)
McINTYRE, Peter, (1863-1940)
McLEAN, William, (?1877-1946)
McPHERSON, William, (-1928)
MILL, Hugh Robert, (1861-1950)
MILLER, Hugh, (1802-1856)
MILLER, Hugh Jr., (1850-1896)
MITCHELL, Hugh, Reverend, (1822-1894)
MOORE, John Carrick, (1804-1898)
MURCHISON, Roderick Impey, Sir, (1792-1871)
MURDOCH, James Barclay, (1831-1906)
MURRAY, John, Sir, (1841-1914)
MUSHET, David, (-?1859)
NECKER DE SAUSSURE, Louis Albert, (1786-1861)

SCOTLAND 1850-1899 Cont'd.

NICHOLSON, Henry Alleyne, (1844-1899)
NICOL, James, (1810-1879)
NICOL, William, (ca.1768-1851)
NORTHAMPTON, Joshua Alwyne Spencer, Marquis of, (1790-1851)
OGILVIE, Francis Grant, Sir, (1858-1930)
PAGE, David, (1814-1879)
PATTON, Andrew, (1821-1885)
PEACH, Benjamin Neeve, (1842-1926)
PEACH, Charles William, (1800-1886)
PRINGLE, John, (1877-1948)
RAMSAY, Andrew Crombie, Sir, (1814-1891)
REED, Frederick Richard Cowper, (1869-1946)
ROBERTSON, David, (1806-1896)
ROGERS, Henry Darwin, (1808-1866)
ROSE, Alexander, (1781-1860)
SCOTT, Robert Henry, (1833-1916)
SELKIRK, Dunbar James Douglas, Earl of, (1809-1885)
SLIMON, Robert, (1803-1882)
SMITH, James, (1782-1867)
SMITH, John, (1845-1930)
SMYTH, Warington Wilkinson, Sir, (1817-1890)
SOMERVILLE, Mary, (1780-1872)
STEWART, Thomas, (1857-1943)
STRICKLAND, Hugh Edwin, (1811-1853)
SYMES, Richard Glascott, (1840-1906)
SYMONDS, William Samuel, Reverend, (1818-1887)
TEALL, Jethro Justinian Harris, Sir, (1849-1924)
THOMAS, Herbert Henry, (1876-1935)
THOMSON, Alexander Gordon Milne, (1866-1919)
THOMSON, James, (1823-1900)
THOMSON, Thomas, (1773-1852)
TRAILL, Thomas Stewart, (1781-1862)
TRAQUAIR, Ramsay Heatley, (1840-1912)
TREVELYAN, Walter Calverley, Sir, (1797-1879)
URE, Andrew, (1778-1857)
WATSON, Robert Boog, Reverend, (?1824-1910)
WILSON, J. S. Grant, (1855-1908)
WOOD, Searles Valentine Jr., (1830-1884)
WRIGHT, James, (1878-1957)
WÜNSCH, Edward A., (1822-1895)
WYVILLE THOMSON, Charles, Sir, (1830-1882)
YOUNG, John, (1823-1900)
YOUNG, John, (1835-1902)

1900 to present

ANDERSON, Robert Beattie, (1904-1970)
BAILEY, Edward Battersby, Sir, (1881-1965)
BANDY, Mark Chance, (1900-1963)
BARBER, Cecil Thomas, (-)
BARROW, George, (1853-1932)
BENNIE, James, (1821-1901)
BISHOP, Walter William, (1931-1977)
BLAKE, John Frederick, Reverend, (1839-1906)
BOWEN, Norman Levi, (1887-1956)

SCOTLAND 1900 to present Cont'd.

BOYD, Alexander John, (1891-1969)
BOYLE, Robert, (1883-1917)
BREMNER, Alexander, (1863-1943)
BROUGHTON-EDGE, Arthur Broughton, (1895-1953)
BROWN, Peter Evans, (-)
BULLERWELL, William, (1916-1977)
BULLOWS, Wilfred Lowson, (?1869-1943)
BURNETT, George Alexander, (1892-1952)
CAMPBELL, Keir Arthur, (-1955)
CAMPBELL, Robert, (1881-1957)
CARRUTHERS, Robert George, (1880-1965)
CARRUTHERS, William, (1830-1922)
CHREE, Charles, (-)
CLOUGH, Charles Thomas, (1852-1916)
COCKBURN, Alexander Murray, (1901-1959)
CONAN DOYLE, Arthur, Sir, (1859-1930)
CONNOR, James, (1894-1977)
COOMARASWAMY, Ananda K., (1877-1947)
CORSTORPHINE, George Steuart, (1865-1919)
CRABTREE, John Edmund, (1940-1966)
CRAIG, Robert, (?1821-1901)
CRAIG, Robert Meldrum, (?1882-1956)
CROFT, William Noble, (1915-1953)
CURRIE, Ethel Dobbie, (-1963)
CURRIE, James, (1863-1930)
CURRIE, Leslie D., (1904-1925)
DAKYNS, John Roche, (1836-1910)
DAVIDSON, Charles Findlay, (1911-1967)
DAVIES, Thomas, (1837-1892)
DAY, Thomas Cuthbert, (1853-1935)
DICK, Allan Brugh, (1833-1926)
DINHAM, Charles Hawker, (1883-1955)
ELLES, Gertrude Lilian, (1872-1960)
ELLISON, John, (1859-1936)
ERDTMAN, Otto Gunnar Elias, (1897-1973)
ETHERIDGE, Robert Jr., (1847-1920)
EYLES, Victor Ambrose, (1896-1978)
FALCONER, John Downie, (1876-1947)
FLETT, John Smith, Sir, (1869-1947)
FORSTER-COOPER, Clive, (1880-1947)
GARDINER, Charles Irving, (?1868-1940)
GEIKIE, Archibald, Sir, (1835-1924)
GEIKIE, James, (1839-1915)
GIBB, Alfred William, (1864-1937)
GIBSON, Walcot, (1864-1941)
GILL, David, Sir, (1843-1914)
GOODCHILD, John George, (1844-1906)
GOODLET, George Anderson, (1910-1970)
GORDON, John More, Reverend, (1849-1922)
GORDON, William Thomas, (1884-1950)
GREEN, John Frederick Norman, (?1873-1949)
GREENLY, Edward, (1861-1951)
GUNN, Robert Marcus, (?1850-1909)
GUNN, William, (1837-1902)

HARKER, Alfred, (1859-1939)
HARRY, William Trevelyan, (?1922-1964)
HATCH, Frederick Henry, (?1864-1932)
HIBBERT-WARE, Samuel, (1782-1848)
HICKLING, Henry George Albert, (1883-1954)
HILL, Dorothy, (-)
HILL, James Bastian, (1861-1927)
HODGE, Matthew Burnop, (-1951)
HOLLAND, Thomas Henry, Sir, (1868-1947)
HOLLOWAY, Lionel E., (1898-1965)
HOLMES, Arthur, (1890-1965)
HORNE, John, (1848-1928)
HOWELL, Henry Hyatt, (1834-1915)
HULL, Edward, (1829-1917)
HUTCHISON, Arthur Gilbertson, (1902-1949)
JAMIESON, Thomas Francis, (1829-1913)
JEHU, Thomas John, (1871-?1943)
JOHNSTON, Robert Mackenzie, (-1918)
JUDD, John Wesley, (1840-1916)
JUKES-BROWNE, Alfred John, (1851-1914)
KELVIN, William Thomson, Baron, (1824-1907)
KERR, John Graham, (1869-1957)
KIDSTON, Robert, (1852-1924)
KIRKBY, James Walker, (1834-1901)
KYLE, John Joseph Jolly, called Juan José, (1838-1922)
KYNASTON, Herbert, (1868-1915)
LAMONT, Archie, (-)
LANG, William Henry, (1874-1960)
LAPWORTH, Charles, (1842-1920)
LARMAN, Mark Linington, (1899-1958)
LEITCH, Duncan, (1904-1956)
LEITCH, Patrick Arthur, (-1943)
LIGHTFOOT, Ben, (1888-1966)
LINTON, David Leslie, (1906-1971)
LOBJOIT, William Michael, (1932-1967)
LYONS, Henry George, (1864-?1943)
MACCONOCHIE, Arthur, (1850-1922)
MACGREGOR, Murray, (1884-1966)
MACINTOSH, Charles, (1839-1922)
MACKIE, William, (1856-1932)
MacROBERT OF DOUNESIDE AND CROMAR, Rachel, Lady (née Workman), (-1954)
MANSON, William, (1891-1973)
MARGERIE, Emmanuel Marie Pierre Martin Jacquin de, (1862-1953)
MARIES, Alan C., (1890-1971)
MARSHALL, Alan Edward, (1938-1970)
MATLEY, Charles Alfred, (1866-1947)
MAUFE, Herbert Brantwood, (1879-1946)
McDONALD, William, (1876-1943)
McINTYRE, Peter, (1863-1940)
McLEAN, William, (?1877-1946)
McMATH, John Charles, (?1908-1954)
McPHERSON, William, (-1928)
MILL, Hugh Robert, (1861-1950)

SCOTLAND 1900 to present Cont'd.

MITCHELL, George Hoole "Mick", (1902-1976)
MUIR-WOOD, Helen Marguerite, (1896-1968)
MURDOCH, James Barclay, (1831-1906)
MURRAY, John, Sir, (1841-1914)
OGILVIE, Francis Grant, Sir, (1858-1930)
PEACH, Benjamin Neeve, (1842-1926)
PIVETEAU, Jean, (1898-)
PRINGLE, John, (1877-1948)
READ, Herbert Harold, (1889-1970)
REED, Frederick Richard Cowper, (1869-1946)
RICHEY, James Ernest, (1886-1968)
ROBBIE, James Andrew, (1910-1977)
RUDMOSE BROWN, Robert Neal, (1879-1957)
RUTLEDGE, Harold, (1920-1954)
SCOTT, Alexander, (1890-1951)
SCOTT, James Frew, (1902-1956)
SCOTT, Robert Henry, (1833-1916)
SHAND, Samuel James, (1882-1957)
SMITH, John, (1845-1930)
SPATH, Leonard Frank, (1882-1957)
STENHOUSE, Andrew G., (-1950)
STEWART, Thomas, (1857-1943)
SUTTON, John, (-)
SYMES, Richard Glascott, (1840-1906)
TAYLOR, Griffith, (1881-1963)
TEALL, Jethro Justinian Harris, Sir, (1849-1924)
TEMPLETON, James Clark, (1898-1950)
THOMAS, Herbert Henry, (1876-1935)
THOMPSON, Hugh R., (1926-1959)
THOMSON, Alexander Gordon Milne, (1866-1919)
TRAQUAIR, Ramsay Heatley, (1840-1912)
TRUEMAN, Arthur Elijah, Sir, (1894-1956)
TYRRELL, George Walter, (1883-1961)
VOGT, Thorolf, (1888-1958)
WAGER, Lawrence Rickard, (1904-1965)
WALKER, Frederick, (1898-1968)
WALKER, George Forbes, (1916-1970)
WALTON, John, (-1971)
WATSON, Robert Boog, Reverend, (?1824-1910)
WHITEHEAD, Talbot Haes, (1890-1966)
WILSON, J. S. Grant, (1855-1908)
WRIGHT, James, (1878-1957)
WYATT, Michael, (-1956)
YOUNG, John, (1835-1902)

Time unknown
MACGREGOR, A. M., (-)

SENEGAL
1850-1899
CHUDEAU, René, (1864-1921)
DOUVILLÉ, Joseph Henri Ferdinand, (1846-1937)

SENEGAL Cont'd.

1900 to present
CHUDEAU, René, (1864-1921)
DOUVILLÉ, Joseph Henri Ferdinand, (1846-1937)
FURON, Raymond, (1898-)
JACQUET, Fernand, (1908-1937)
MALAVOY, Jean, (1903-1945)
MILLOT, Georges, (1917-)
NICOLAS, Jean Paul, (1921-1967)

SEYCHELLES ISLANDS (INDIAN OCEAN)
1900 to present
TAYLOR, James Haward, (1909-1968)

SHARJAH
1900 to present
BASSINGER, William E., (1928-1977)

SIAM <u>See</u> THAILAND.

SICILY (ITALY)
Up to 1800
BRYDONE, Patrick, (1736-1818)
DOLOMIEU, Dieudonné Sylvain Guy Tancrède, called Déodat, Gratet de, (1750-1801)

1800-1849
BRYDONE, Patrick, (1736-1818)
DODERLEIN, Pietro, (1810-1895)
DOLOMIEU, Dieudonné Sylvain Guy Tancrède, called Déodat, Gratet de, (1750-1801)
HOFFMANN, Friedrich, (1797-1836)

1850-1899
CARAPEZZA, E., (-1915)
DODERLEIN, Pietro, (1810-1895)
FOOTE, Albert E., (1846-1895)
SAUVAGE, Henri Émile, (1842-1917)
WILLIAMS, John Francis, (1862-1891)

1900 to present
BLOW, Walter Harry, (1924-1972)
CARAPEZZA, E., (-1915)
SAUVAGE, Henri Émile, (1842-1917)

SIERRA LEONE
1850-1899
HANNAY, Harry Campbell, (1874-1961)

1900 to present
BILLINGS, Katharine Stevens (neé Fowler-Lunn), (1902-)
BRAMWYCH, Michael, (1911-1974)
GASKIN, Alan Robert James, (1927-1974)
HANNAY, Harry Campbell, (1874-1961)
JUNNER, Norman Ross, (1891-1973)

SIERRA LEONE 1900 to present Cont'd.

MARMO, Vladi, (1914-1969)
SLAWSON, Chester Baker, (1898-1964)

SINGAPORE
1850-1899
NORDENSKIÖLD, Nils Adolf Erik, (1832-1901)
RIDLEY, Henry Nicholas, (1855-1956)

1900 to present
ALEXANDER, Frances Elizabeth Somèrville, (1908-1958)
HO, Robert, (1921-1972)
NORDENSKIÖLD, Nils Adolf Erik, (1832-1901)
RIDLEY, Henry Nicholas, (1855-1956)
SMITH, G. Wendell, (1917-1974)

SOCOTRA (INDIAN OCEAN)
1850-1899
SAUER, Gustav Adolf, (1825-1932)

1900 to present
SAUER, Gustav Adolf, (1825-1932)

SOLOMON ISLANDS (PACIFIC OCEAN)
1800-1849
JUKES, Joseph Beete, (1811-1869)

1850-1899
JUKES, Joseph Beete, (1811-1869)
WATTS, William Whitehead, (1860-1947)

1900 to present
FRASER, John A., (-) [PROSPECTORS, DIVINERS AND MINING ENGINEERS]
WATTS, William Whitehead, (1860-1947)

SOMALIA
1850-1899
BEEBY-THOMPSON, Arthur, (1873-1968)
CRICK, George Charles, (1856-1917)

1900 to present
BEEBY-THOMPSON, Arthur, (1873-1968)
CRAWFORD, David Johnston, (1899-1960)
CRICK, George Charles, (1856-1917)
LAMARE, Pierre, (1894-1967)
MUIR-WOOD, Helen Marguerite, (1896-1968)
NUTTALL, Winfred Laurence Falkiner, (1897-1969)
TEILHARD DE CHARDIN, Pierre, Father, S.J., (1881-1955)

SOMBRERO (WEST INDIES)
1850-1899
JULIEN, Alexis Anastay, (1840-1919)

SOMBRERO (WEST INDIES) Cont'd.

1900 to present
JULIEN, Alexis Anastay, (1840-1919)

SOUTH AFRICA
Up to 1800
BAIN, Andrew Geddes, (1797-1864)
BARROW, John, Sir, (1764-1848)

1800-1849
ATHERSTONE, William Guybon, Honourable, (-1898)
BAIN, Andrew Geddes, (1797-1864)
BARROW, John, Sir, (1764-1848)
BIGSBY, John Jeremiah, (1792-1881)
BOUSSINGAULT, Joseph Dieudonné Jean-Baptiste, (1807-1887)
BUNBURY, Charles James Fox, Sir, (1809-1886)
DARWIN, Charles Robert, (1809-1882)
GALTON, Francis, Sir, (1822-1911)
JONES, Thomas Rupert, (1819-1911)
MOLYNEUX, William, (1824-1882)
RUBIDGE, Richard Nathaniel, (1821?-1869)
SALTER, John William, (1820-1869)
STOW, George B. William, (1822-1882)
TENNANT, James, (1808-1881)

1850-1899
AMALITSKII, Vladimir Prochorovich, (1860-1917)
ANDERSON, William, (1860-1915)
ATHERSTONE, William Guybon, Honourable, (-1898)
BAIN, Thomas Charles John, (1829-1893)
BECK, Carl Richard, (1858-1919)
BIGSBY, John Jeremiah, (1792-1881)
BIRLEY, Caroline, (1851-1907)
BOND, Marshall, (1867-1941) [PROSPECTORS, DIVINERS AND MINING ENGINEERS]
BOUSSINGAULT, Joseph Dieudonné Jean-Baptiste, (1807-1887)
BREUIL, Henri, Abbé, (1877-1961)
BROOM, Robert, (1866-1951)
BROWN, Alfred "Gogga", (1834-1920)
BUNBURY, Charles James Fox, Sir, (1809-1886)
BUTTGENBACH, Henri, (1874-1964)
CHAPER, Maurice Armand, (1834-1896)
CHAPMAN, Frederick, (1864-1943)
COHEN, Emil Wilhelm, (1842-1905)
COLLINS, Henry Francis, (?1864-1924)
CORSTORPHINE, George Steuart, (1865-1919)
CRICK, George Charles, (1856-1917)
DARWIN, Charles Robert, (1809-1882)
DAVIS, William Morris, (1850-1934)
DRAPER, David, (1849-1929)
DUNN, Edward John, (1844-1937)
EXTON, Hugh, (1833-1903)
FOX-STRANGWAYS, Charles Edward, (1844-1910)
GALTON, Francis, Sir, (1822-1911)
GIBSON, Walcot, (1864-1941)

SOUTH AFRICA 1850-1899 Cont'd.

GILL, David, Sir, (1843-1914)
GREGORY, James Reynolds, (1832-1899)
GRIESBACH, Karl Ludolf, (1847-1907)
HALL, Arthur Lewis, (1872-1955)
HAMMOND, John Hays, (1855-1936)
HANNAY, Harry Campbell, (1874-1961)
HATCH, Frederick Henry, (?1864-1932)
HENDERSON, James Alexander Leo, (?1874-1940)
HOLUB, Emil, (1847-1902)
HUENE, Friedrich von, Freiherr, (1875-1969)
JACK, Robert Lockhart, (1878-1964)
JONES, Thomas Rupert, (1819-1911)
JULIAN, Henry Forbes, (1861-1912) [PROSPECTORS, DIVINERS AND MINING ENGINEERS]
KAISER, Friedrich Wilhelm Erich, (1871-1934)
KYNASTON, Herbert, (1868-1915)
LOMAS, Joseph, (1860-1908)
MACLAREN, James Malcolm, (1873-1935)
McCARTHY, Edward Thomas, (-) [PROSPECTORS, DIVINERS AND MINING ENGINEERS]
McKIM, John Walter, (1877-1927)
MERENSKY, Hans, (1871-1952)
MOLENGRAAFF, Gustaaf Adolf Frederik, (1860-1942)
MOLYNEUX, William, (1824-1882)
MORGAN, Conwy Lloyd, (?1852-1936)
NEWELL ARBER, Edward Alexander, (1870-1918)
PORRIO, Cesare, (1865-1940)
ROGERS, Arthur William, (1872-1946)
RUBIDGE, Richard Nathaniel, (1821?-1869)
RUTLEY, Frank, (1842-1904)
SALTER, John William, (1820-1869)
SARGENT, Harry Lionel, (1871-1955)
SCHWARZ, Ernest Hubert Lewis, (1873-1928)
SMITH, Edgar Albert, (1847-1916)
STELZNER, Alfred Wilhelm, (1840-1895)
STEWART, Thomas, (1857-1943)
STOW, George B. William, (1822-1882)
TAYLOR, William P., (1859-) [PROSPECTORS, DIVINERS AND MINING ENGINEERS]
TENNANT, James, (1808-1881)
TOIT, Alexander Logie du, (1878-1948)
TRELAWNEY-ANSELL, E. C., (-) [PROSPECTORS, DIVINERS AND MINING ENGINEERS]
TRUSCOTT, Samuel John, (?1870-1950)
WILLIAMS, Gardner Fred, (1842-1922)
WOODS, Henry, (1868-1952)

1900 to present

AHRENS, Louis Herman, (1918-)
AMALITSKII, Vladimir Prochorovich, (1860-1917)
ANDERSON, William, (1860-1915)
BALL, Harry Standish, (?1888-1941)
BARBOUR, George Brown, (1891-1977)
BARRON, William Ernest, (1880-1935)

SOUTH AFRICA 1900 to present Cont'd.

BECK, Carl Richard, (1858-1919)
BIRLEY, Caroline, (1851-1907)
BISHOPP, Douglas Wallace, (1900-1977)
BOARDMAN, Leslie Gray, (1913-1977)
BOND, Marshall, (1867-1941) [PROSPECTORS, DIVINERS AND MINING ENGINEERS]
BOWEN, Norman Levi, (1887-1956)
BREUIL, Henri, Abbé, (1877-1961)
BROCK, Byron Britton, (1904-1972)
BROOM, Robert, (1866-1951)
BROUWER, Hendrik Albertus, (1886-1973)
BROWN, Alfred "Gogga", (1834-1920)
BUTTGENBACH, Henri, (1874-1964)
CHAPMAN, Frederick, (1864-1943)
COHEN, Emil Wilhelm, (1842-1905)
COLLINS, Henry Francis, (?1864-1924)
CORSTORPHINE, George Steuart, (1865-1919)
CRICK, George Charles, (1856-1917)
DART, Raymond Arthur, (1893-)
DAVIS, William Morris, (1850-1934)
DRAPER, David, (1849-1929)
DUGGLEBY, Arthur Francis, (1893-1945)
DUNN, Edward John, (1844-1937)
EXTON, Hugh, (1833-1903)
FAIRBRIDGE, Kingsley, (1885-1924) [PROSPECTORS, DIVINERS AND MINING ENGINEERS]
FOX-STRANGWAYS, Charles Edward, (1844-1910)
GALTON, Francis, Sir, (1822-1911)
GIBSON, Walcot, (1864-1941)
GILL, David, Sir, (1843-1914)
GORTANI, Michele, (1883-1966)
GRATON, Louis Caryl, (1880-1970)
GRIESBACH, Karl Ludolf, (1847-1907)
HALL, Arthur Lewis, (1872-1955)
HAMMOND, John Hays, (1855-1936)
HANNAY, Harry Campbell, (1874-1961)
HARBORT, Erich, (1879-1929)
HATCH, Frederick Henry, (?1864-1932)
HENDERSON, James Alexander Leo, (?1874-1940)
HOLMES, Arthur, (1890-1965)
HOLUB, Emil, (1847-1902)
HOYT, John Harger, (1928-1970)
HUENE, Friedrich von, Freiherr, (1875-1969)
JACK, Robert Lockhart, (1878-1964)
JANENSCH, Werner, (1878-1969)
JOHNSON, J. P., (1880-1918)
JONES, Thomas Rupert, (1819-1911)
JULIAN, Henry Forbes, (1861-1912) [PROSPECTORS, DIVINERS AND MINING ENGINEERS]
KAISER, Friedrich Wilhelm Erich, (1871-1934)
KINNEY, Harry Dill, (1886-1961)
KITCHING, James, (-)
KOENIGSWALD, Gustav Heinrich Ralph von, (1902-)
KYNASTON, Herbert, (1868-1915)

SOUTH AFRICA 1900 to present Cont'd.

LAMSENS, O. Robert, (1922-1971)
LESLIE, Thomas Nicholas, (-1942)
LOMAS, Joseph, (1860-1908)
MACLAREN, James Malcolm, (1873-1935)
McCARTHY, Edward Thomas, (-) [PROSPECTORS, DIVINERS AND MINING ENGINEERS]
McKIM, John Walter, (1877-1927)
McKINSTRY, Hugh Exton, (1896-1961)
MERENSKY, Hans, (1871-1952)
MOLENGRAAFF, Gustaaf Adolf Frederik, (1860-1942)
MORGAN, Conwy Lloyd, (?1852-1936)
NEWELL ARBER, Edward Alexander, (1870-1918)
PARTRIDGE, Francis Chamberlain, (1903-1939)
PENNY, Frederick Willoughby, (1887-1950)
POLDERVAART, Arie, (1918-1964)
PORRIO, Cesare, (1865-1940)
ROGERS, Arthur William, (1872-1946)
ROMER, Alfred Sherwood, (1894-1973)
ROSE, Hermann, (1883-1976)
RUSSELL, William Low, (1897-1976)
RUTLEY, Frank, (1842-1904)
SARGENT, Harry Lionel, (1871-1955)
SCHNEIDERHÖHN, Hans, (1887-1962)
SCHOLTZ, Douglas Leinster, (1902-1977)
SCHWARZ, Ernest Hubert Lewis, (1873-1928)
SHAND, Samuel James, (1882-1957)
SLATER, George, (-1956)
SMITH, Edgar Albert, (1847-1916)
SMITH, James Leonard Brierley, (1897-)
STEWART, Thomas, (1857-1943)
TAYLOR, William P., (1859-) [PROSPECTORS, DIVINERS AND MINING ENGINEERS]
TEILHARD DE CHARDIN, Pierre, Father, S.J., (1881-1955)
THOMAS, Hugh Hamshaw, (1885-1962)
TOIT, Alexander Logie du, (1878-1948)
TRELAWNEY-ANSELL, E. C., (-) [PROSPECTORS, DIVINERS AND MINING ENGINEERS]
TREMLETT, Rex, (-) [PROSPECTORS, DIVINERS AND MINING ENGINEERS]
TRUSCOTT, Samuel John, (?1870-1950)
WADELL, Hakon, (1895-1962)
WAGNER, Percy Albert, (1885-1929)
WALKER, Frederick, (1898-1968)
WASSERSTEIN, Benno, (1906-1956)
WATSON, David Meredith Seares, (1886-1973)
WILLIAMS, Gardner Fred, (1842-1922)
WILLIAMSON, John Thorburn, (1907-1958)
WILMOTH, Leslie James, (1881-1963)
WOODS, Henry, (1868-1952)

SOUTH AMERICA (unspecified)

Up to 1800

HELMS, Anton Zacharias, (1751-1803) [PROSPECTORS, DIVINERS AND MINING ENGINEERS]
LA CONDAMINE, Charles-Marie de, (1701-1774)
NISSER, Pedro, (1799-1878) [PROSPECTORS, DIVINERS AND MINING ENGINEERS]

1800-1849

HELMS, Anton Zacharias, (1751-1803) [PROSPECTORS, DIVINERS AND MINING ENGINEERS]
HOPKINS, Evan, (?1810-1867)
LUND, Peter Wilhelm, (1801-1880)
NISSER, Pedro, (1799-1878) [PROSPECTORS, DIVINERS AND MINING ENGINEERS]

1850-1899

ATTWOOD, George, (1845-1912)
DICKERSON, Roy Ernest, (1877-1944)
HELMS, Anton Zacharias, (1751-1803) [PROSPECTORS, DIVINERS AND MINING ENGINEERS]
HOPKINS, Evan, (?1810-1867)
LUND, Peter Wilhelm (1801-1880)
ORDOÑEZ, Ezequiel, (1867-1950)
PILSBRY, Henry Augustus, (1862-1957)
RIGGS, Elmer Samuel, (1869-1963)
ROSENBUSCH, Karl Harry Ferdinand, (1836-1914)
TOIT, Alexander Logie du, (1878-1948)
WINCHELL, Horace Vaughn, (1865-1923)
WOODRUFF, Elmer Grant, (1872-1952)

1900 to present

ATTWOOD, George, (1845-1912)
BLANCHARD, William Grant, Jr., (1894-1952)
BRISCOE, Glenn Oliver, (1898-1974)
COLBERT, Edwin H., (1905-)
DICKERSON, Roy Ernest, (1877-1944)
DIETZ, Curt, (1882-1965)
GORDON, Samuel George, (1897-1952)
HARRISON, Thomas Samuel, (1881-1964)
HAWLEY, James Edwin, (1897-1965)
HOFMANN, Walter, (1919-1975)
ICKES, Eugene Law, (1885-1941)
KLEINPELL, William Darwin, (1898-1959)
ORDOÑEZ, Ezequiel, (1867-1950)
PILSBRY, Henry Augustus, (1862-1957)
ROSENBUSCH, Karl Harry Ferdinand, (1836-1914)
SMITH, Gerald Julian, (1903-1976)
STIRTON, Ruben Arthur, (1901-1966)
TOIT, Alexander Logie du, (1878-1948)
TYRRELL, George Walter, (1883-1961)
WEEK, Herbert John, (1897-1960)
WILSON, Walter Byron "Shorty", (1885-1951)

SOUTH AMERICA (unspecified) 1900 to present Cont'd.

WINCHELL, Horace Vaughn, (1865-1923)
WOODRUFF, Elmer Grant, (1872-1952)
WRIGHT, Charles Will, (1879-)

SOUTH ATLANTIC OCEAN
1900 to present
CAMPBELL, Robert, (1881-1957)

SOUTH KOREA *See* KOREA.

SOUTH SHETLAND ISLANDS (ANTARCTIC OCEAN)
1900 to present
HOLTEDAHL, Olaf, (1885-1975)

SOUTH YEMEN
1850-1899
MOORE, Henry Cecil, (1835-1908)
NORDENSKIÖLD, Nils Adolf Erik, (1832-1901)

1900 to present
MOORE, Henry Cecil, (1835-1908)
NORDENSKIÖLD, Nils Adolf Erik, (1832-1901)

SOUTHEAST ASIA (unspecified)
1850-1899
NAUMANN, Edmund, (1854-1927)

1900 to present
NAUMANN, Edmund, (1854-1927)

SOUTHWEST AFRICA *See* NAMIBIA.

SPAIN
Up to 1800
ELHUYAR Y DE ZUBICE, Fausto de, (1755-1833)
GIMBERNAT, Carlos de, Don, (1765-1839)
MACLURE, William, (1763-1840)
SICULUS, Diodorus, (ca.85-21 B.C.)

1800-1849
BAUZÂ, Felipe, (1802-1875)
CHARPENTIER, Jean de, (1786-1855)
COLLOMB, Edouard, (1801-1875)
DAUBENY, Charles Giles Brindle, (1795-1867)
DE PRADO, Casiano, Don, (1797-1866?)
ELHUYAR Y DE ZUBICE, Fausto de, (1755-1833)
GIMBERNAT, Carlos de, Don, (1765-1839)
MACLURE, William, (1763-1840)
MÜLLER, Carl Hermann, (1823-1907)
PARKER, William Kitchen, (1823-1890)

SPAIN 1800-1849 Cont'd.

PRATT, Samuel Peace, (1789-1863)
RÖMER, Carl Ferdinand von, (1818-1891)
TRAILL, Thomas Stewart, (1781-1862)
VERNEUIL, Philippe Edouard Poulletier de, (1805-1873)
VILANOVA Y PIERA, Juan, (1822-1893)

1850-1899

ALMERA, Jaume, Canon, (1854-1919)
AREITIO Y LARRINAGA, Alfonso de, (-1884)
BALL, John, (-1941)
BARROIS, Charles Eugène, (1851-1939)
BATE, Dorothea Minola Alice, (1879-1951)
BAUZÁ, Felipe, (1802-1875)
BEEBY-THOMPSON, Arthur, (1873-1968)
BERGERON, Jules, (1853-1919)
BERTRAND, Marcel Alexandre, (1847-1907)
BOFILL Y POCH, Arturo, (1846-1929)
BREUIL, Henri, Abbé, (1877-1961)
BULLEN, Robert Ashington, Reverend, (1850-1912)
CALDERON Y ARANA, Salvador, (1851-1911)
CAPELLINI, Giovanni, (1833-1922)
CASTRO Y SUERO, Manuel Fernandez de, (1825-1895)
CHAMPERKNOWNE, Arthur, (1839-1887)
CHARPENTIER, Jean de, (1786-1855)
COLLINS, Arthur L., (1868-1902)
COLLINS, Henry Francis, (?1864-1924)
COLLINS, Joseph Henry, (1841-1916)
COLLOMB, Edouard, (1801-1875)
COTTEAU, Gustave Honoré, (1818-1894)
DAUBENY, Charles Giles Brindle, (1795-1867)
DELÉPINE, Gaston-Georges, Monseigneur, (1878-1963)
DE PRADO, Casiano, Don, (1797-1866?)
DRUGMAN, Julien Léon Eugène, (1875-1950)
FINLAYSON, Alexander Moncreiff, (-1917)
FRIEDBERG, Wilhelm, (1873-1941)
GAURIER, Ludovic Charles, Abbé, (1874-1931)
GRANDIDIER, Alfred, (1836-1921)
JOHNSTON, Mary Sophia, (1875-1955)
KISSLING, Ernst, (1865-1936)
LLOYD, T. G. B., (1829-1876)
MACPHERSON, José, Don, (1839-1902)
MALLADA, Lucas, (1841-1921)
MARGERIE, Emmanuel Marie Pierre Martin Jacquin de, (1862-1953)
MERRICKS, Frank, (?1863-1936)
MÜLLER, Carl Hermann, (1823-1907)
O'REILLY, Joseph P., (1829-1905)
PARKER, William Kitchen, (1823-1890)
PRATT, Samuel Peace, (1789-1863)
ROMAN, Albert Frédéric, (1871-1943)
RÖMER, Carl Ferdinand von, (1818-1891)
SAUVAGE, Henri Émile, (1842-1917)
SCHMIDT, Carl, (1862-1923)
TRAILL, Thomas Stewart, (1781-1862)
VALLE DE LERSUNDI, Alfonso del, (1875-1944)

SPAIN 1850-1899 Cont'd.

VERNEUIL, Philippe Edouard Poulletier de, (1805-1873)
VIDAL I CARRERAS, Lluis Mariano, (1842-1922)
VILANOVA Y PIERA, Juan, (1822-1893)
WADE, Arthur, (1878-1951)

1900 to present

ALLOITEAU, James Henri Ernest, (1890-1969)
ALMERA, Jaume, Canon, (1854-1919)
ANDERSON, Robert Lee, (1912-1959)
ASTRE, Gaston, (1896-)
BALL, John, (-1941)
BARROIS, Charles Eugène, (1851-1939)
BATE, Dorothea Minola Alice, (1879-1951)
BEEBY-THOMPSON, Arthur, (1873-1968)
BERGERON, Jules, (1853-1919)
BERTRAND, Marcel Alexandre, (1847-1907)
BOFILL Y POCH, Arturo, (1846-1929)
BRAMWYCH, Michael, (1911-1974)
BREUIL, Henri, Abbé, (1877-1961)
BROUGHTON-EDGE, Arthur Broughton, (1895-1953)
BULLEN, Robert Ashington, Reverend, (1850-1912)
CALDERON Y ARANA, Salvador, (1851-1911)
CAPELLINI, Giovanni, (1833-1922)
CASTRO Y SUERO, Manuel Fernandez de, (1825-1895)
CHAUVE, Pierre, (1930-)
COLLINS, Arthur L., (1868-1902)
COLLINS, Henry Francis, (?1864-1924)
COLLINS, Joseph Henry, (1841-1916)
COSIJN, Adrianus Johannes, (1890-1945)
DELÉPINE, Gaston-Georges, Monseigneur, (1878-1963)
DIXON, Henry George Dacres, (1887-1947)
DOUGLAS, George Vibert, (1892?-1958)
DOUVILLÉ, Robert, (1881-1914)
DRUGMAN, Julien Léon Eugène, (1875-1950)
EDELMANN, Cornelis Hendrik, (1901-1964)
FALLOT, Paul, (1889-1960)
FATH, Arthur Earl, (1887-1976)
FINLAYSON, Alexander Moncreiff, (-1917)
FRIEDBERG, Wilhelm, (1873-1941)
FRYXELL, Roald Hilding, (1934-1974)
GAURIER, Ludovic Charles, Abbé, (1874-1931)
GRAMBAST, Louis, (-1976)
GRANDIDIER, Alfred, (1836-1921)
GROTH, Jean, (1899-1916)
HENDON, Bryan, (1896-1926)
HERNÁNDEZ-PACHECO, Eduardo, (-1965)
HERNÁNDEZ-PACHECO, Francisco, (1899-1976)
HLAUSCHEK, Hans, (1896-1971)
HOLLISTER, Joseph Steffens, (1903-1953)
HOLTEDAHL, Olaf, (1885-1975)
ICKES, Eugene Law, (1885-1941)
JEFFREYS, Geoffrey, (1885-1953)
JOHNSTON, Mary Sophia, (1875-1955)
KISSLING, Ernst, (1865-1936)

SPAIN 1900 to present Cont'd.

KOMOROWICZ, Maurycy, (1881-1923)
KUBIËNA, Walter L., (1897-1970)
LAPPARENT, Albert Félix de, (1905-1975)
LAPPARENT, Jacques de, (-?1949)
LOTZE, Franz, (1903-1971)
MACPHERSON, José, Don, (1839-1902)
MALLADA, Lucas, (1841-1921)
MARGERIE, Emmanuel Marie Pierre Martin Jacquin de, (1862-1953)
MERRICKS, Frank, (?1863-1936)
MÜLLER, Carl Hermann, (1823-1907)
NES, C. L. van, (?1878-1963)
NICOLAS, Jean Paul, (1921-1967)
O'REILLY, Joseph P., (1829-1905)
PARDILLO VAQUER, Francisco, Don, (1884-1955)
PARGETER, Reginald Clive, (1919-1972)
PIVETEAU, Jean, (1898-)
ROCH, Edouard, (1901-)
ROMAN, Albert Frédéric, (1871-1943)
ROYO Y GOMEZ, José, (1895-1958)
RUIZ DE GAONA, Máximo, Padre, (-1971)
SAUVAGE, Henri Émile, (1842-1917)
SCHMIDT, Carl, (1862-1923)
SCHRÖDER, Hans Adolf Karl Eckart, (1899-1968)
SCHÜLLER, Arno, (1908-1968)
SIMPSON, George Gaylord, (1902-)
TRÜMPY, Daniel, (1893-1971)
VALLE DE LERSUNDI, Alfonso del, (1875-1944)
VIDAL I CARRERAS, Lluis Mariano, (1842-1922)
WADE, Arthur, (1878-1951)
WILMOTH, Leslie James, (1881-1963)
ZEIJLMANS VAN EMMICHOVEN, Cornelis Pieter Adriaan, (1897-1952)

SPAIN <u>See also</u> BALEARIC ISLANDS.

SPANISH SAHARA (NOW JOINTLY CONTROLLED BY MOROCCO AND MAURITANIA)
1900 to present

HERNÁNDEZ-PACHECO, Francisco, (1899-1976)
JAMES, John William, (1921-1977)

SPITZBERGEN
1800-1849
SCHENK, Joseph August, (1815-1891)
TORELL, Otto Martin, (1828-1900)

1850-1899
BACKLUND, Helge Götrik, (1878-1958)
BÄCKSTROM, Helge Mattias, (1865-1932)
DE GEER, Gerard Jacob, Baron, (1858-1943)
FOX-STRANGWAYS, Charles Edward, (1844-1910)
GARWOOD, Edmund Johnston, (1864-1949)
GREGORY, John Walter, (1864-1932)

SPITZBERGEN 1850-1899 Cont'd.

HANNAY, Harry Campbell, (1874-1961)
KJELLSTRÖM, Carl Johan Otto, (1855-1913)
MARGERIE, Emmanuel Marie Pierre Martin Jacquin de, (1862-1953)
NATHORST, Alfred Gabriel, (1850-1921)
SCHENK, Joseph August, (1815-1891)
TCHERNYCHEV, Féodoss Nikolaievich, (1856-1914)
TORELL, Otto Martin, (1828-1900)
WIMAN, Carl Johan Josef Ernst, (1867-1944)

1900 to present
BACKLUND, Helge Götrik, (1878-1958)
BÄCKSTROM, Helge Mattias, (1865-1932)
BODYLEVSKY, Vitaly Ivanovich, (1898-1968)
BROWN, Bahngrell Walter "Bill", (1916-)
CAMPBELL, Robert, (1881-1957)
CORYELL, Horace Noble, (1888-1965)
CRAIG, Robert Meldrum, (?1882-1956)
CROFT, William Noble, (1915-1953)
DE GEER, Gerard Jacob, Baron, (1858-1943)
FOX-STRANGWAYS, Charles Edward, (1844-1910)
GARWOOD, Edmund Johnston, (1864-1949)
GREGORY, John Walter, (1864-1932)
HANNAY, Harry Campbell, (1874-1961)
HOLMSEN, Gunnar, (1880-1976)
HUTCHINS, Philip Frank, (1927-1955)
HUXLEY, Julian Sorell, Sir, (1887-1975)
KJELLSTRÖM, Carl Johan Otto, (1855-1913)
LAPPARENT, Albert Félix de, (1905-1975)
MAACK, Reinhard, (1892-1969)
MARGERIE, Emmanuel Marie Pierre Martin Jacquin de, (1862-1953)
NATHORST, Alfred Gabriel, (1850-1921)
NES, C. L. van, (?1878-1963)
NORBERG, Hans L., (-?1918)
ROMANES, James, (1886-1942)
RUDMOSE BROWN, Robert Neal, (1879-1957)
SÄVE-SÖDERBERGH, Gunnar, (1910-1948)
SCHYTT, Valter, (1919-)
SLATER, George, (-1956)
TCHERNYCHEV, Féodoss Nikolaievich, (1856-1914)
TODTMANN, Emmy Mercedes, (1888-1973)
TYRRELL, George Walter, (1883-1961)
VOGT, Thorolf, (1888-1958)
WIMAN, Carl Johan Josef Ernst, (1867-1944)
WORDIE, James Mann, Sir, (?1890-1962)

SRI LANKA
1800-1849
CAMPBELL, John Francis, (1821-1885)
KELAART, E. F., (1818-1860)

1850-1899
BROWN, Charles Barrington, (1839-1917)
CAMPBELL, John Francis, (1821-1885)
COOMARASWAMY, Ananda K., (1877-1947)

SRI LANKA 1850-1899 Cont'd.

GRANDIDIER, Alfred, (1836-1921)
GRÜNLING, Friedrich, (1857-1919)
HAECKEL, Ernst Heinrich Philipp August, (1834-1919)
KELAART, E. F., (1818-1860)
MELZI, Gilberto, Conte, (1868-1899)
NORDENSKIÖLD, Nils Adolf Erik, (1832-1901)
PARSONS, James, (1876-1908)
REED, Frederick Richard Cowper, (1869-1946)
ZIRKEL, Ferdinand, (1838-1912)

1900 to present
ANDERTON, Russell, (-) [PROSPECTORS, DIVINERS AND MINING ENGINEERS]
BROWN, Charles Barrington, (1839-1917)
COOMARASWAMY, Ananda K., (1877-1947)
GRANDIDIER, Alfred, (1836-1921)
GRÜNLING, Friedrich, (1857-1919)
HAECKEL, Ernst Heinrich Phillipp August, (1834-1919)
NORDENSKIÖLD, Nils Adolf Erik, (1832-1901)
PARSONS, James, (1876-1908)
REED, Frederick Richard Cowper, (1869-1946)
WADIA, Darashaw Nosherwan, (1883-1969)
ZIRKEL, Ferdinand, (1838-1912)

ST. BARTHOLOMEW <u>See</u> SAINT BARTHOLOMEW (West Indies).

ST. JOHN <u>See</u> SAINT JOHN (West Indies).

ST. LUCIA <u>See</u> SAINT LUCIA (West Indies).

ST. VINCENT <u>See</u> SAINT VINCENT (West Indies).

SUDAN

Up to 1800
BROCCHI, Giovanni Battista (Giambattista), (1772-1826)

1800-1849
BROCCHI, Giovanni Battista (Giambattista), (1772-1826)

1850-1899
BARRON, Thomas, (1867-1906)
BEEBY-THOMPSON, Arthur, (1873-1968)
CHUDEAU, René, (1864-1921)
GÜRICH, Georg, (1859-1938)

1900 to present
BARRON, Thomas, (1867-1906)
BEEBY-THOMPSON, Arthur, (1873-1968)
CHUDEAU, René, (1864-1921)
FOLEY, Edward J., (1908-1971)
FUCHS, Vivian Ernest, Sir, (1908-)
GÜRICH, Georg, (1859-1938)
MADIGAN, Cecil Thomas, (1889-1946)
SANDFORD, Kenneth Stuart, (1899-1971)

SUMATRA <u>See</u> INDONESIA.

SURINAM

1800-1849

WATERTON, Charles, (1782-1865)

1850-1899

BERGT, Walther Adolf, (1864-1941)
DU BOIS, Georges C., (1874-)
PEDRICK, Howard A. "Dad", (1863-?1930) [PROSPECTORS, DIVINERS AND MINING ENGINEERS]
WATERTON, Charles, (1782-1865)

1900 to present

BERGT, Walther Adolf, (1864-1941)
DERRY, Duncan Ramsay, (?1908-)
DU BOIS, Georges C., (1874-)
HALL, Thomas Clifford Fitzwilliam, (1886-1956)
LEHNER, Ernst J., (1895-1968)
NES, C. L. van, (?1878-1963)
PEDRICK, Howard A. "Dad", (1863-?1930) [PROSPECTORS, DIVINERS AND MINING ENGINEERS]
POLLARD, Ernest Rex, (1922-1966)
ROMANES, James, (1886-1942)
SCHOLS, Hendrikus, (1901-1957)
TER MEULEN, J. A. C., (1902-1960)
VERSLUYS, Jan, (1880-1935)

VALBARD See SPITZBERGEN.

WAZILAND

1900 to present

PERCIVAL, Frederick George, (1887-1973)

WEDEN

Up to 1800

ALLEN, William, (1770-1843)
BERGMAN, Torbern Olof, (1735-1784)
BRONGNIART, Alexandre, (1778-1847)
BUCH, Christian Leopold von, Freiherr, (1774-1853)
CEDERBAUM, Peter Christopher, (1733-1795)
CLARKE, Edward Daniel, (1769-1822)
COLLING, Olof, (1677-1754)
CRONSTEDT, Axel Fredrik, (1722-1765)
ESPLING, Olof, (1751-1821)
GROSSE, Karl Friedrich August, (1768-1847)
GYLLENBERG, Fredrik, (fl.1750)
GYLLENHAAL, Johan Gabriel, (1750-1788)
HERMELIN, Samuel Gustaf, (1744-1820)
HISINGER, Wilhelm, (1766-1852)
KALM, Pehr, (1716-1779)
LINNE, Carl von, (1707-1778)
SWEDENBORG, Emanuel, (1688-1772)
TIDSTRÖM, Andreas Philip, (1723-1779)
TILAS, Daniel, (1712-1772)

SWEDEN Up to 1800 Cont'd.

VALLERIUS, Göran, (1683-1744)
VON ENGESTRÖM, Gustaf, (1738-1813)
WALLERIUS, Johan Gottschalk, (1709-1785)

1800-1849

ALLEN, William, (1770-1843)
ANGELIN, Nils Peter, (1805-1876)
ARRHENIUS, Johan, (1811-1889)
BERZELIUS, Jöns Jakob, (1779-1848)
BLOMSTRAND, Christian Wilhelm, (1826-1897)
BRONGNIART, Alexandre, (1778-1847)
BUCH, Christian Leopold von, Freiherr, (1774-1853)
CLARKE, Edward Daniel, (1769-1822)
DICKSON, Oscar, (1823-1897)
EGGERTZ, Victor, (1817-1889)
ERDMANN, Axel Joachim, (-)
GROSSE, Karl Friedrich August, (1768-1847)
HAGLUND, Emil, (1877-1938)
HAIDINGER, Wilhelm Karl von, Ritter, (1795-1871)
HAUSMANN, Johann Friedrich Ludwig, (1782-1859)
HERMELIN, Samuel Gustaf, (1744-1820)
HISINGER, Wilhelm, (1766-1852)
IGELSTRÖM, Lars Johan, (1822-1897)
LINDSTRÖM, Jonas Gustaf Oscar, (1829-1901)
LOVÉN, Sven, (1809-1895)
MURCHISON, Roderick Impey, Sir, (1792-1871)
NILSSON, Sven, (1787-1883)
OEYNHAUSEN, Karl August Ludwig, (1795-1865)
OLBERS, Elam Willgott, (1816-1895)
SAXE, Sjur Aasmundsen, (1808-1887?)
SJÖGREN, Carl Anton Hjalmar, (1822-1893)
TORELL, Otto Martin, (1828-1900)
TROILIUS, Carl Oscar, (1813-[1899])
VON ENGESTRÖM, Gustaf, (1738-1813)
VON POST, Hampus Adolf, (1822-1911)
WALMSTEDT, Lars Edward, (1819-1892)
ZITTEL, Karl Alfred von, (1839-1904)

1850-1899

ALIN, Johan, (1879-1944)
AMINOFF, Gregori, (1883-1947)
ANDERSSON, Gunnar, (1865-1928)
ANDERSSON, Johann Gunnar, (1874-1960)
ANDERSSON, Oscar Frithiof, (1867-1911)
ANGELIN, Nils Peter, (1805-1876)
ARRHENIUS, Johan, (1811-1889)
ARRHENIUS, Svante August, (1859-1927)
ATTERBERG, Albert, (1846-1916)
BACKLUND, Helge Götrik, (1878-1958)
BÄCKSTROM, Helge Mattias, (1865-1932)
BATE, Dorothea Minola Alice, (1879-1951)
BAUERMAN, Hilary, (1833-1909)
BJERKNES, Vilhelm, (1862-1951)
BLOMBERG, Karl Albert, (1844-1922)

SWEDEN 1850-1899 Cont'd.

BLOMSTRAND, Christian Wilhelm, (1826-1897)
BORGSTRÖM, Leonard Henrik, (1876-?1954)
BRØGGER, Waldemar Christofer, (1851-1940)
BRÜCKNER, Eduard, (1862-1927)
BUCH, Christian Leopold von, Freiherr, (1774-1853)
CARLHEIM-GYLLENSKÖLD, Vilhelm, (1859-1934)
CARSTENS, Carl Wilhelm, (1877-?1950)
DE GEER, Gerard Jacob, Baron, (1858-1943)
DICKSON, Oscar, (1823-1897)
DUSÉN, Per, (1855-1926)
EGGERTZ, Victor, (1817-1889)
ERDMANN, Axel Joachim, (-)
ERDMANN, Edvard, (1840-1923)
FEARNSIDES, William George, (1879-1968)
FEGRAEUS, Torbern, (1853-1923)
FELLENIUS, Wolmar, (1876-?1957)
FLINK, Gustaf, (1849-1931)
FLODKVIST, Herman, (1875-)
FOOTE, Albert E., (1846-1895)
FUNKQUIST, Herman Per Anton, (1870-1952)
GAVELIN, Axel Olof, (1875-1947)
GELLERSTEDT, Göran, (1834-1913)
GERTZ, Otto Daniel, (1878-1948)
GRANSTRÖM, Gustaf Abraham, (1851-1941)
GRÖNWALL, Karl Anders Axel, (1869-1944)
GUMAELIUS, Otto Joël, (1835-1899)
HAGLUND, Emil, (1877-1938)
HAIDINGER, Wilhelm Karl von, Ritter, (1795-1871)
HAMBERG, Axel, (1863-1933)
HAUAN, Kund Møller, (1840-1898)
HAUSMANN, Johann Friedrich Ludwig, (1782-1859)
HEDSTROM, Herman Oskar, (1867-?1948)
HESSELMAN, Oscar August Henrik Wilhelm, (1874-1943)
HISINGER, Wilhelm, (1766-1852)
HÖGBOM, Arvid Gustaf, (1857-1940)
HOLLENDER, Artur, (-?1946)
HOLM, Edvard Johann Gerhard, (1853-1926)
HOLMQUIST, Per Johan, (1866-1946)
IGELSTRÖM, Lars Johan, (1822-1897)
JOHANSSON, Karl Fredrick, (1866-1933)
JÖNSSON, Jöns, (1848-1894)
KJELLMARK, Knut, (1866-?1944)
KJELLSTRÖM, Carl Johan Otto, (1855-1913)
KOLTHOFF, Gustaf, (1845-1914)
KURCK, Clas, (1849-1937)
LAGRELIUS, Axel, (1863-1944)
LINDSTRÖM, Axel Fredrik, (1839-1911)
LINDSTRÖM, Jonas Gustaf Oscar, (1829-1901)
LINNARSSON, Jonas Gustaf Oscar, (1841-1881)
LOVÉN, Sven, (1809-1895)
LUNDBÖHM, Hjalmar, (1855-1926)
LUNDGREN, Sven Anders Bernhard, (1843-1897)
MAUZELIUS, Robert, (1864-1921)
MOBERG, Johan Christian, (1854-1915)

SWEDEN 1850-1899 Cont'd.

MÖLLER, Hjalmar August, (1866-1941)
MUNTHE, Henrik Vilhelm, (1860-1958)
MURCHISON, Roderick Impey, Sir, (1792-1871)
NATHORST, Alfred Gabriel, (1850-1921)
NILSSON, Lars Albert, (1860-1906)
NILSSON, Sven, (1787-1883)
NORDENSKIÖLD, Nils Adolf Erik, (1832-1901)
NORDENSKIÖLD, Nils Otto Gustaf, (1869-1928)
NORELIUS, Carl Oskar, (1861-1941)
OEYNHAUSEN, Karl August Ludwig, (1795-1865)
OLBERS, Elam Willgott, (1816-1895)
PETERSSON, Walfrid, (1862-1933)
PHRAGMÉN, Gösta, (-?1944)
RETZIUS, Emilia, (-?1904)
RICHERT, Johann Gustaf, (1857-1954)
RICHTER, Paul Bogoslav, (1854-1911)
RINDELL, Artur, (1852-1936)
SAHLIN, Carl, (1861-1943)
SANTESSON, Gustaf Henrik, (1847-1915)
SANTESSON, Karl Olof Birger, (1845-1893)
SAXE, Sjur Aasmundsen, (1808-1887?)
SCHREIBER, Hans, (1859-?1936)
SERNANDER, Rutger, (1866-1944)
SJÖGREN, Carl Anton Hjalmar, (1822-1893)
SJÖGREN, Sten Anders Hjalmar, (1856-1922)
STAPFF, Friedrich Moritz, (1836-1895)
STOKES, Arthur Henry, (1842-1910)
SVEDMARK, Lennart Eugène, (1847-1922)
TIBERG, Björn Gustaf Georg, (1877-1948)
TIBERG, H. V., (1849-1913)
TORELL, Otto Martin, (1828-1900)
TÖRNEBOHM, Alfred Elis, (1838-1911)
TÖRNQUIST, Sven Leonhard, (1840-1920)
TROILIUS, Carl Oscar, (1813-[1899])
TULLBERG, Sven Axel Theodor, (1852-1886)
ULFFERS, Ernst, (1837-1919)
VON FEILITZEN, Carl Henrik Jobst, (1840?-1901)
VON POST, Hampus Adolf, (1822-1911)
WALLIN, Gustaf, (1871-1940)
WALMSTEDT, Lars Edward, (1819-1892)
WEIBULL, Kristian Oskar Mats, (1856-1923)
WESTLUND, Erik Wiktor, (-?1954)
WILLIAMS, John Francis, (1862-1891)
WIMAN, Carl Johan Josef Ernst, (1867-1944)
YNGSTRÖM, Lars, (1870-1944)
ZITTEL, Karl Alfred von, (1839-1904)

1900 to present

AHLMANN, Hans Wilhelmsson, (1889-1974)
ALIN, Johan, (1879-1944)
AMINOFF, Gregori, (1883-1947)
ANDERSSON, Gunnar, (1865-1928)
ANDERSSON, Johann Gunnar, (1874-1960)
ANDERSSON, Oscar Frithiof, (1867-1911)

SWEDEN 1900 to present Cont'd.

ARRHENIUS, Olof, (1895-)
ARRHENIUS, Svante August, (1859-1927)
ASSARSSON, Gunnar, (-1963)
ATTERBERG, Albert, (1846-1916)
BACKLUND, Erik, (1892-1938)
BACKLUND, Helge Götrik, (1878-1958)
BÄCKSTROM, Helge Mattias, (1865-1932)
BATE, Dorothea Minola Alice, (1879-1951)
BAUERMAN, Hilary, (1833-1909)
BENEDICKS, Carl, (-1958)
BERGSTRÖM, Gunnar, (-1954)
BJAREBY, Alfred Gunnar, (1899-1967)
BJERKNES, Vilhelm, (1862-1951)
BJÖRNSSON, Sven, (1905-1950)
BJORULF, Hjalmar, (-?1956)
BLOMBERG, Karl Albert, (1844-1922)
BODMAN, Gösta, (-1961)
BONTHRON, John A., (-?1948)
BOOBERG, Gunnar, (1892-1944)
BORGSTRÖM, Leonard Henrik, (1876-?1954)
BRØGGER, Waldemar Christofer, (1851-1940)
BROTZEN, Fritz, (1902-1968)
BROUGHTON-EDGE, Arthur Broughton, (1895-1953)
BROWN, Bahngrell Walter "Bill", (1916-)
BRÜCKNER, Eduard, (1862-1927)
BUBNOFF, Serge von, (1888-1957)
CALDENIUS, Carl, (-1961)
CARLHEIM-GYLLENSKÖLD, Vilhelm, (1859-1934)
CARSTENS, Carl Wilhelm, (1877-?1950)
CLAUSEN, Hans Carl Eisenreich, (1898-1971)
DANIELLSON, Johan Patrik, (-?1954)
DE GEER, Gerard Jacob, Baron, (1858-1943)
DE GEER, Sten, (1886-1933)
DUSÉN, Per, (1855-1926)
EKSTRÖM, Gunnar, (-1960)
ENEROTH, Olof, (-1945)
ERDMANN, Edvard, (1840-1923)
ERDTMAN, Otto Gunnar Elias, (1897-1973)
ERIKSSON, J. V., (1883-1930)
FEARNSIDES, William George, (1879-1968)
FEGRAEUS, Torbern, (1853-1923)
FELLENIUS, Wolmar, (1876-?1957)
FINNEMAN, Karl Johan, (1880-1953)
FLINK, Gustaf, (1849-1931)
FLODKVIST, Gunnar, (-?1946)
FLODKVIST, Herman, (1875-)
FLORIN, Sten, (-)
FRÖDIN, John, (-1961)
FUNKQUIST, Herman Per Anton, (1870-1952)
GAVELIN, Axel Olof, (1875-1947)
GEIJER, Per Adolf, (1886-1976)
GELLERSTEDT, Göran, (1834-1913)
GERTZ, Otto Daniel, (1878-1948)
GOLDKUTH, Algot, (-1957)

SWEDEN 1900 to present Cont'd.

GRANLUND, Erik, (1892-1938)
GRANÖ, Johannes Gabriel, (1882-1956)
GRANSTRÖM, Gustaf Abraham, (1851-1941)
GRÖNWALL, Karl Anders Axel, (1869-1944)
GUSTAFSSON, Johan Peter, (-?1943)
HADDING, Assar, (1886-1962)
HALLGREN, Gunnar, (1907-)
HAMBERG, Axel, (1863-1933)
HEDSTROM, Herman Oskar, (1867-?1948)
HELLAND-HANSEN, Björn, (-1958?)
HESSELMAN, Oscar August Henrik Wilhelm, (1874-1943)
HÖGBOM, Alvar, (1894-1939)
HÖGBOM, Arvid Gustaf, (1857-1940)
HÖGBOM, Bertil, (-1962)
HÖGBOM, Ivar, (1892-1962)
HOLLENDER, Artur, (-?1946)
HOLM, Edvard Johann Gerhard, (1853-1926)
HOLMBOE, Jens, (-?1943)
HOLMQUIST, Per Johan, (1866-1946)
HOLMSTRÖM, Leonard, (-?1922)
HORN AF RANTZIEN, Henning, (1922-1960)
HÖRNER, Nils Gustaf, (1896-1950)
HUMMEL-GUMAELIUS, Torsten, (-?1949)
JOHANSSON, Harald Elias, (1880-1931)
JOHANSSON, Karl Fredrick, (1866-1933)
JOHANSSON, Simon, (1881-1944)
JONSON, Per Abraham, (-?1957)
JUNGNER, Jonas Gustaf, (-?1942)
KAUDERN, Walter, (1881-1924)
KIHLSTEDT, Folke H., (1901?-1956)
KJELLMARK, Knut, (1866-?1944)
KJELLSTRÖM, Carl Johan Otto, (1855-1913)
KOLTHOFF, Gustaf, (1845-1914)
KROKSTRÖM, Torsten Hugo, (1905-1950)
KURCK, Clas, (1849-1937)
LAGRELIUS, Axel, (1863-1944)
LINDSTRÖM, Axel Fredrik, (1839-1911)
LINDSTRÖM, Jonas Gustaf Oscar, (1829-1901)
LJUNGDAHL, Gustaf Samuel, (1882-1940)
LJUNGNER, Erik, (1892-1954)
LUNDBERG, Sven, (1900-?1948)
LUNDBÖHM, Hjalmar, (1855-1926)
LUNDQVIST, Gösta, (1894-1967)
MAGNUSSON, Nils Harald, (1890-1976)
MARTINUSSEN, Eugen, (1916-1942)
MAUZELIUS, Robert, (1864-1921)
MELLIS, Otto, (1906-1970)
MOBERG, Johan Christian, (1854-1915)
MÖLLER, Hjalmar August, (1866-1941)
MOORE, Paul Brian, (1940-)
MÖRTSELL, Sture, (-?1952)
MOSSBERG, Karl Elis, (-?1941)
MUNCK, Victor C. E. A. de, (1920-1956)
MUNTHE, Henrik Vilhelm, (1860-1958)

SWEDEN 1900 to present Cont'd.

NATHORST, Alfred Gabriel, (1850-1921)
NAUMANN, Einar Christian Leonard, (1891-1934)
NILSSON, Lars Albert, (1860-1906)
NORBERG, Hans L., (-?1918)
NORDENSKIÖLD, Karl Erik, (-?1954)
NORDENSKIÖLD, Nils Adolf Erik, (1832-1901)
NORDENSKIÖLD, Nils Otto Gustaf, (1869-1928)
NORELIUS, Carl Oskar, (1861-1941)
NORIN, Erik, (-)
ORTON, Bror Natanael, (-?1942)
ÖSTER, Johannes, (1902-1975)
OSVALD, Hugo, (1892-)
PHRAGMÉN, Gösta, (-?1944)
QUENSEL, Percy Dudgeon, (1881-1966)
RETZIUS, Emilia, (-?1904)
RICHERT, Johann Gustaf, (1857-1954)
RICHTER, Paul Bogoslav, (1854-1911)
RINDELL, Artur, (1852-1936)
ROSSBY, Carl-Gustaf Arvid, (1898-1957)
RUTBERG, Karl, (-1961)
SAHLIN, Carl, (1861-1943)
SAMUELSSON, Frans Gustaf, (-?1948)
SAMUELSSON, Gunnar, (-?1945)
SANDEGREN, Ragnar, (1887-1966)
SANTESSON, Gustaf Henrik, (1847-1915)
SÄVE-SÖDERBERGH, Gunnar, (1910-1948)
SCHREIBER, Hans, (1859-?1936)
SCHYTT, Valter, (1919-)
SERNANDER, Rutger, (1866-1944)
SIDENBLADH, Elias, (1836-1914)
SIDENVALL, Karl Johan Ferdinand, (-?1957)
SIMONS, Herbert Samuel, (1891-1956)
SJÖGREN, Sten Anders Hjalmar, (1856-1922)
SPRINGORUM, Gerd, (1915-1941)
STOKES, Arthur Henry, (1842-1910)
STUTZER, Otto, (1881-1936)
SUNDBERG, Karl, (1891-1938)
SUNDELIN, Uno, (-1926)
SUNDHOLM, Olof Herman, (-?1946)
SVEDMARK, Lennart Eugène, (1847-1922)
TAMMEKANN, August, (1894-1959)
THOMASSON, Harald, (1893-1937)
THORNÉ, Sven Gustaf, (-?1950)
TIBERG, Björn Gustaf Georg, (1877-1948)
TIBERG, H. V., (1849-1913)
TÖRNEBOHM, Alfred Elis, (1838-1911)
TÖRNQUIST, Sven Leonhard, (1840-1920)
TROEDSSON, Gustaf T., (1891-1954)
TRYGGVASON, Tómas, (1907-1965)
ULFFERS, Ernst, (1837-1919)
VON ECKERMANN, Claes Walther Harry, (1886-1969)
VON POST, Ernst Jakob Lennart, (1884-1951)
VON POST, Hampus Adolf, (1822-1911)
WADELL, Hakon, (1895-1962)

SWEDEN 1900 to present Cont'd.

WALLÉN, Axel, (1877-1935)
WALLIN, Gustaf, (1871-1940)
WALLROTH, Gunnar, (-?1949)
WALTON, John, (-1971)
WEIBULL, Kristian Oskar Mats, (1856-1923)
WESSLAU, Eric, (-1958)
WESTERGÅRD, Anton Hilmer, (1880-1968)
WESTLUND, Erik Wiktor, (-?1954)
WIMAN, Carl Johan Josef Ernst, (1867-1944)
WINGE, Knut Ingel, (-1958)
WRÅK, Walter Olof, (-?1933)
YNGSTRÖM, Lars, (1870-1944)
ZANS, Verners Aleksandrs, (1904-1961)
ZENZÉN, Nils, (1883-1959)
ZITTEL, Karl Alfred von, (1839-1904)

Time unknown
ELZVIK, Erik, (-)
KAHLMETER, Henrik, (-)
TEGENGREN, Felix, (-)

SWITZERLAND
Up to 1800
BAUHIN, Jean, (1541-1613)
BERTRAND, Élie, (1712-1797)
BESSON, Henri, (fl.1786)
BRONGNIART, Alexandre, (1778-1847)
BUCH, Christian Leopold von, Freiherr, (1774-1853)
COULON, Paul Louis, (1777-1855)
CROUSAZ, Jean-Pierre de, (1663-1750)
ESCHER VON DER LINTH, Hans Conrad, (1767-1823)
GAGNEBIN DE LA FERRIÈRE, Abraham, (1707-1800)
GESSNER, Conrad, (1516-1565)
GIMBERNAT, Carlos de, Don, (1765-1839)
GRUNER, Gottlieb Sigmund, (1717-1778)
HOTTINGER, Johann Heinrich, (1680-1756)
KAPPELER, Moritz, (1685-1769)
KUHN, Bernhard Friedrich, (1762-1825)
LANG, Karl Nikolaus, (1670-1741)
LUC, Jean André de, (1727-1817)
MARTEL, Peter, (-)
MURALT, Johannes von, (1645-1733)
PARACELSUS, (pseud. of HOHENHEIM, Philippus Theophrastus Bombastus von), (1493-1541)
PERRAUDIN, Jean-Pierre, (1767-1858)
SAUSSURE, Horace Bénédict de, (1740-1799)
SAUSSURE, Nicolas Théodore de, (1765-1845)
SCHEUCHZER, Johann Gaspar, (1702-1729)
SCHEUCHZER, Johann Jakob, (1672-1733)
TREMBLEY, Abraham, (1710-1784)
VALLTRAVERS, Johann Rodolph von, (1723-)

1800-1849
AGASSIZ, Louis Jean Rodolphe, (1807-1873)

SWITZERLAND 1800-1849 Cont'd.

BALL, John, (1818-1889)
BIEDERMANN-IMHOOF, Adolf W. G., (1829-1900)
BOUÉ, Ami, (1794-1881)
BRONGNIART, Alexandre, (1778-1847)
BUCH, Christian Leopold von, Freiherr, (1774-1853)
CAMPICHE, Gustave, (1809-1870)
CARTIER, Robert, (1810-1886)
CARUS, Carl Gustav, (1789-1869)
COULON, Louis de, (1804-1894)
COULON, Paul Louis, (1777-1855)
CREDNER, Carl Friedrich Heinrich, (1809-1876)
DESOR, Pierre Jean Edouard, (1811-1882)
DUBOIS, Auguste, (1862-1923)
DUBOIS DE MONTPÉREUX, Frédéric, (1798-1850)
DUFOUR, Charles, (1827-1902)
EGERTON, William Henry, Reverend, (1811-1910)
ESCHER VON DER LINTH, Arnold, (1807-1872)
ESCHER VON DER LINTH, Hans Conrad, (1767-1823)
FAVRE, Jean Alphonse, (1815-1890)
FISCHER-OOSTER, Carl von, (1807-1875)
FONTAINE, Charles Aloyse, Chanoine, (1754-1834)
GAUDIN, Charles Théophile, (1822-1866)
GILLIÉRON, Victor, (1826-1890)
GIMBERNAT, Carlos de, Don, (1765-1839)
GREPPIN, Jean Baptiste, (-1881)
GRESSLY, Amanz, (1814-1865)
GUYOT, Arnold Henri, (1807-1884)
HEER, Oswald, (1809-1883)
HISELY, Charles, (1805-1873)
HITCHCOCK, Edward, (1793-1864)
HUGI, Franz Joseph, (1796-1855)
KENNGOTT, Gustav Adolf, (1818-1897)
KUHN, Bernhard Friedrich, (1762-1825)
LANG, Franz Vinzenz, (1821-1899)
LESQUEREUX, Charles Leo, (1806-1889)
LORIOL LE FORT, Charles Louis Perceval de, (1828-1908)
LUC, Jean André de, (1727-1817)
MARCOU, Jules, (1824-1898)
MAYER-EYMAR, Karl David Wilhelm, (1826-1907)
MERIAN, Peter, (1795-1883)
MOESCH, Casimir, (1827-1898)
MONTMOLLIN, Auguste de, (1808-1898)
MORLOT, Adolf von, (1820-1867)
MÜLLER, Albrecht, (1819-1890)
MURCHISON, Roderick Impey, Sir, (1792-1871)
NECKER DE SAUSSURE, Louis Albert, (1786-1861)
NICOL, Germain Louis, (1922-1973)
PARETO, Lorenzo, (1800-1865)
PERRAUDIN, Jean-Pierre, (1767-1858)
PICTET DE LA RIVE, François Jules, (1809-1872)
QUIQUEREZ, Auguste, (1801-1882)
RUSKIN, John, (1819-1900)
RÜTIMEYER, Karl Ludwig, (1825-1895)
SAUSSURE, Nicolas Théodore de, (1765-1845)

SWITZERLAND 1800-1849 Cont'd.

SHARPE, Daniel, (1806-1856)
STUDER, Bernhard Rudolf, (1794-1887)
TYNDALL, John, (1820-1893)
VALLTRAVERS, Johann Rodolph von, (1723-)
VENETZ, Ignaz, (1788-1859)
VOGT, Carl, (1817-1895)
VOLGER, Georg Heinrich Otto, (ca 1822-1897)
ZIEGLER, Jakob Melchior, (1801-1883)

1850-1899

AEBERHARDT, Berchtold, (1872-1912)
AGASSIZ, Louis Jean Rodolphe, (1807-1873)
ASHCROFT, Frederick Noel, (1878-1949)
BALL, John, (-1941)
BALL, John, (1818-1889)
BALLY-PRIOR, Peter Eduard (né Bally), (1847-1926)
BAUMBERGER, Ernst, (1866-1935)
BAUMHAUER, Heinrich Adolf, (1848-1926)
BEUST, Fritz von, (1856-1908)
BIEDERMANN-IMHOOF, Adolf W. G., (1829-1900)
BÖSE, Emil, (1868-1972)
BOUÉ, Ami, (1794-1881)
BREUIL, Henri, Abbé, (1877-1961)
BRÜCKNER, Eduard, (1862-1927)
BRUNNER VON WATTENWYL, Karl, (1823-1914)
BUCH, Christian Leopold von, Freiherr, (1774-1853)
BURCKHARDT, Carl, (1869-1935)
BURCKHARDT, Carl Rudolf, (1866-1908)
BUXTORF, August, (1877-)
CAMPICHE, Gustave, (1809-1870)
CARTIER, Robert, (1810-1886)
CARUS, Carl Gustav, (1789-1869)
CHOFFAT, Paul, (1849-1919)
COULON, Louis de, (1804-1894)
COULON, Paul Louis, (1777-1855)
CRAMER, Carl Edward, (1831-1901)
CREDNER, Carl Friedrich Heinrich, (1809-1876)
DAVIS, William Morris, (1850-1934)
DE LA HARPE, Philippe, (1830-1882)
DESOR, Pierre Jean Edouard, (1811-1882)
DUBOIS, Auguste, (1862-1923)
DU BOIS, Georges C., (1874-)
DUFOUR, Charles, (1827-1902)
DUPARC, Louis Claude, (1866-1932)
DU PASQUIER, Léon, (1864-1897)
EBEL, Johann Gottfried, (1764-1830)
EGERTON, William Henry, Reverend, (1811-1910)
ESCHER VON DER LINTH, Arnold, (1807-1872)
ÉTALLON, Claude Auguste, (1826-1862)
FAVRE, Ernest, (1845-1925)
FAVRE, Jean Alphonse, (1815-1890)
FISCHER-OOSTER, Carl von, (1807-1875)
FISCHLI, Heinrich, (1857-1932)
FONTAINE, Charles Aloyse, Chanoine, (1754-1834)

SWITZERLAND 1850-1899 Cont'd.

FOREL, François Alphonse, (1841-1912)
FORSYTH MAJOR, Charles Immanuel, (1843-1923)
FRIEDLAENDER, Immanuel, (-?1918)
FRÜH, Jakob, (1852-)
GAUDIN, Charles Théophile, (1822-1866)
GERBER, Eduard, (1876-1956)
GILLIÉRON, Victor, (1826-1890)
GOELDI, Emil August, (1859-1917)
GORDON, John More, Reverend, (1849-1922)
GREPPIN, Eduard, (1856-1927)
GREPPIN, Jean Baptiste, (-1881)
GRESSLY, Amanz, (1814-1865)
GRUBENMANN, Ulrich, (1850-1924)
GUTZWILLER, Andreas, (1845-1917)
GUYOT, Arnold Henri, (1807-1884)
HAGENBACH-BISCHOFF, Eduard, (1833-1910)
HALL, Marshall, Captain, (1831-1896)
HEER, Oswald, (1809-1883)
HEIM, Albert, (1849-1937)
HELBLING, Robert, (1874-1954)
HISELY, Charles, (1805-1873)
HITCHCOCK, Edward, (1793-1864)
HUENE, Friedrich von, Freiherr, (1875-1969)
HUGI, Emil, (1873-1937)
HUGI, Franz Joseph, (1796-1855)
HUMBERT, Alois, (1829-1887)
HUME, William Fraser, (1867-1949)
JACCARD, Samuel Auguste, (1833-1895)
JENNINGS, Albert Vaughan, (1864-1903)
JENNY, Fridolin, (1862-1933)
KAUFMANN, Franz Joseph, (-1892)
KENNGOTT, Gustav Adolf, (1818-1897)
KISSLING, Ernst, (1865-1936)
KOBY, Frédéric Louis, (1852-1930)
LACHAT, Hippolyte, (1829-1901)
LANG, Arnold, (1855-1914)
LANG, Franz Vinzenz, (1821-1899)
LAUE, Max Theodor Felix von, (1879-1960)
LESQUEREUX, Charles Leo, (1806-1889)
LEWIS, Henry Carvill, (1853-1888)
LOMAS, Joseph, (1860-1908)
LORENZ, Theodor, (1875-1909)
LORIOL LE FORT, Charles Louis Perceval de, (1828-1908)
LUBBOCK, John, Lord Avebury, (1834-1913)
LUGEON, Maurice, (1870-1953)
LYELL, Charles, Sir, (1797-1875)
MARCOU, Jules, (1824-1898)
MARGERIE, Emmanuel Marie Pierre Martin Jacquin de, (1862-1953)
MARTONNE, Emmanuel Louis Eugène de, (1873-1955)
MAYER-EYMAR, Karl David Wilhelm, (1826-1907)
McCONNEL, J. C., (1860-1890)
McINTYRE, Peter, (1863-1940)
MERCANTON, Paul Louis, (1876-1963)
MERIAN, Peter, (1795-1883)

SWITZERLAND 1850-1899 Cont'd.

MILNE-EDWARDS, Alphonse, (1835-1900)
MOESCH, Casimir, (1827-1898)
MONTMOLLIN, Auguste de, (1808-1898)
MORLOT, Adolf von, (1820-1867)
MÜHLBERG, Friedrich, (1840-1915)
MÜLLER, Albrecht, (1819-1890)
MURCHISON, Roderick Impey, Sir, (1792-1871)
NECKER DE SAUSSURE, Louis Albert, (1786-1861)
NICOL, Germain Louis, (1922-1973)
NÜESCH, Jakob, (1845-1915)
OPPLIGER, Fritz, (1861-1932)
PARETO, Lorenzo, (1800-1865)
PERRAUDIN, Jean-Pierre, (1767-1858)
PICTET DE LA RIVE, François Jules, (1809-1872)
PREISWERK, Heinrich, (1876-1940)
QUIQUEREZ, Auguste, (1801-1882)
RAISIN, Catherine Alice, (1855-1945)
RENEVIER, Eugène, (1831-1906)
ROBERTS, Thomas, (1856-1892)
ROLLIER, Louis, (1859-1931)
ROMER, Eugeniusz, (1871-1954)
RUSKIN, John, (1819-1900)
RÜTIMEYER, Karl Ludwig, (1825-1895)
RUTLEY, Frank, (1842-1904)
SARASIN, Paul, (1856-1929)
SCHARDT, W. Hans, (1858-1931)
SCHMIDT, Carl, (1862-1923)
SCHRÖTER, Carl, (1855-?1925)
SELIGMANN, Gustav, (1849-1920)
SHARPE, Daniel, (1806-1856)
SOLLY, Richard Harrison, (1851-1925)
SORET, Charles, (-?1903)
STEHLIN, Hans Georg, (1870-1942)
STRÜBIN, Karl, (1876-1916)
STUDER, Bernhard Rudolf, (1794-1887)
STUDER, Theophil, (1845-1922)
TAWNEY, Edward Bernard, (1840-1882)
TOBLER, August, (1872-1929)
TOLWIŃSKI, Konstanty, (1876-1961)
TRIBOLET, Georges de, (1830-1873)
TRIBOLET-HARDY, Frédéric Maurice de, (1852-1929)
TYNDALL, John, (1820-1893)
VENETZ, Ignaz, (1788-1859)
VOGT, Carl, (1817-1895)
VOLGER, Georg Heinrich Otto, (ca 1822-1897)
WEBER, Julius, (1864-1924)
WETTSTEIN, Alexander, (1861-1887)
WHYMPER, Edward, (1840-1911)
ZIEGLER, Jakob Melchior, (1801-1883)
ZYNDEL, Fortunat, (1882-1917)

1900 to present

AEBERHARDT, Berchtold, (1872-1912)
ANDRAU, Evert Willem Karel, (1900-1951)

SWITZERLAND 1900 to present Cont'd.

ARBENZ, Paul, (1880-1943)
ARGAND, Émile, (1879-1940)
ASHCROFT, Frederick Noel, (1878-1949)
BAILEY, Edward Battersby, Sir, (1881-1965)
BALL, John, (-1941)
BALLY-PRIOR, Peter Eduard (né Bally), (1847-1926)
BARBER, Cecil Thomas, (-)
BAUMBERGER, Ernst, (1866-1935)
BAUMHAUER, Heinrich Adolf, (1848-1926)
BECK, Paul, (1882-1958)
BEDER, Robert, (1885-1930)
BERNOUILLI, Walter, (1885-1946)
BEUST, Fritz von, (1856-1908)
BÖSE, Emil, (1868-1972)
BREUIL, Henri, Abbé, (1877-1961)
BRÜCKNER, Eduard, (1862-1927)
BRUNNER VON WATTENWYL, Karl, (1823-1914)
BUBNOFF, Serge von, (1888-1957)
BURCKHARDT, Carl, (1869-1935)
BURCKHARDT, Carl Rudolf, (1866-1908)
BUXTORF, August, (1877-)
CADISCH, Joos, (1895-)
CALAME, Jean Jacques, (1923-1973)
CHOFFAT, Paul, (1849-1919)
COLLET, Léon William, (1880-1957)
CRAMER, Carl Edward, (1831-1901)
DAVIS, William Morris, (1850-1934)
DU BOIS, Georges C., (1874-)
DUFOUR, Charles, (1827-1902)
DUPARC, Louis Claude, (1866-1932)
EBEL, Johann Gottfried, (1764-1830)
EDELMANN, Cornelis Hendrik, (1901-1964)
EGERTON, William Henry, Reverend, (1811-1910)
ERNI, Arthur, (1885-1945)
ESCHER, Berend George, (-1967)
FAVRE, Ernest, (1845-1925)
FISCHLI, Heinrich, (1857-1932)
FOREL, François Alphonse, (1841-1912)
FORSYTH MAJOR, Charles Immanuel, (1843-1923)
FREI, Ernst, (1894-1974)
FRIEDLAENDER, Immanuel, (-?1918)
FRÜH, Jakob, (1852-)
GAGNEBIN, Élie, (1891-1949)
GERBER, Eduard, (1876-1956)
GLASER, Theodor, (1894-1922)
GOELDI, Emil August, (1859-1917)
GORDON, John More, Reverend, (1849-1922)
GREPPIN, Eduard, (1856-1927)
GRUBENMANN, Ulrich, (1850-1924)
GUBLER, Jean, (1902-1940)
GÜNZLER-SEIFFERT, Hans, (1883-1957)
GUTZWILLER, Andreas, (1845-1917)
HAGENBACH-BISCHOFF, Eduard, (1833-1910)
HANDSCHIN, Eduard, (1894-1962)

SWITZERLAND 1900 to present Cont'd.

HEIM, Albert, (1849-1937)
HEIM, Arnold, (?1882-1965)
HELBING, Hermann, (1880-1938)
HELBLING, Robert, (1874-1954)
HEMPEL, Joachim, (1884-1955)
HERITSCH, Franz, (1882-1945)
HLAUSCHEK, Hans, (1896-1971)
HOLST PELLEKAAN, William van, (1880-1949)
HUENE, Friedrich von, Freiherr, (1875-1969)
HUGI, Emil, (1873-1937)
HUME, William Fraser, (1867-1949)
HUTTENLOCHER, Heinrich, (1890-1954)
JENNINGS, Albert Vaughan, (1864-1903)
JENNY, Fridolin, (1862-1933)
KISSLING, Ernst, (1865-1936)
KOBY, Fréderic Louis, (1852-1930)
KOCKEL, Carl Walter, (1898-1966)
KRAUS, Ernst, (1889-1970)
LACHAT, Hippolyte, (1829-1901)
LANG, Arnold, (1855-1914)
LAUE, Max Theodor Felix von, (1879-1960)
LEUTENEGGER, Willy Otto, (1903-1972)
LOMAS, Joseph, (1860-1908)
LORENZ, Theodor, (1875-1909)
LORIOL LE FORT, Charles Louis Perceval de, (1828-1908)
LUBBOCK, John, Lord Avebury, (1834-1913)
LUGEON, Maurice, (1870-1953)
LYELL, Charles, Sir, (1797-1875)
MACHATSCHEK, Fritz, (-1957)
MARGERIE, Emmanuel Marie Pierre Martin Jacquin de, (1862-1953)
MARTONNE, Emmanuel Louis Eugène de, (1873-1955)
MAYER-EYMAR, Karl David Wilhelm, (1826-1907)
McINTYRE, Peter, (1863-1940)
MERCANTON, Paul Louis, (1876-1963)
MEYER, Edouard, (1899-1974)
MILNE-EDWARDS, Alphonse, (1835-1900)
MÜHLBERG, Friedrich, (1840-1915)
NICOLAS, Jean Paul, (1921-1967)
NIETHAMMER, Gottlob August, (1882-1915)
NIGGLI, Paul, (1888-1953)
NÜESCH, Jakob, (1845-1915)
OERTEL, Walter, (1889-1924)
OPPLIGER, Fritz, (1861-1932)
PARKER, Robert Luling, (1893-1973)
PETERHANS, Emil, (1899-1931)
PICCARD, Auguste, (1884-1962)
PICCARD, Jacques Ernst, (1922-)
PREISWERK, Heinrich, (1876-1940)
RAISIN, Catherine Alice, (1855-1945)
REICHEL, Manfred, (1896-)
RENAUD, André, (1904-1964)
RENEVIER, Eugène, (1831-1906)
RHEDEN, J. J. Pannekoek van, (-1945)
ROBERTS, Thomas, (1856-1892)

SWITZERLAND 1900 to present Cont'd.

ROLLIER, Louis, (1859-1931)
ROMER, Eugeniusz, (1871-1954)
RUTLEY, Frank, (1842-1904)
RUTSCH, Rolf F., (1902-1975)
SARASIN, Paul, (1856-1929)
SCHARDT, W. Hans, (1858-1931)
SCHAUB, Hans Peter, (1912-1965)
SCHAUB, Samuel, (1882-1962)
SCHMID, Kurt, (1906-1957)
SCHMIDT, Carl, (1862-1923)
SCHRÖTER, Carl, (1855-?1925)
SELIGMANN, Gustav, (1849-1920)
SMITH, G. Wendell, (1917-1974)
SOLLY, Richard Harrison, (1851-1925)
STAUB, Rudolf, (1890-1961)
STEHLIN, Hans Georg, (1870-1942)
STEINER-BALZER, Alfred, (1879-1968)
STRÜBIN, Karl, (1876-1916)
STUDER, Theophil, (1845-1922)
ŚWIDERSKI, Bohdan, (1892-1943)
TOBLER, August, (1872-1929)
TOLWIŃSKI, Konstanty, (1876-1961)
TRIBOLET-HARDY, Frédéric Maurice de, (1852-1929)
TRÜMPY, Rudolf D., (-)
WEBER, Julius, (1864-1924)
WHYMPER, Edward, (1840-1911)
WIEDENMAYER, Carl, (1897-1951)
WILHELM, Oscar, (1897-1975)
WIRZ, Albert Eduard, (1910-1973)
WITZ, Albert Eduard, (1910-1973)
ZIJLSTRA, Gerrit, (-1962)
ZYNDEL, Fortunat, (1882-1917)

SYRIA

Up to 1800

CLARKE, Edward Daniel, (1769-1822)
VOLNEY, Constantin François Chasseboeuf, Comte de, (1757-1820)

1800-1849

CLARKE, Edward Daniel, (1769-1822)
EHRENBERG, Christian Gottfried, (1795-1876)
HOOKER, Joseph Dalton, Sir, (1817-1911)
STRICKLAND, Hugh Edwin, (1811-1853)
VOLNEY, Constantin François Chasseboeuf, Comte de, (1757-1820)

1850-1899

BLANCKENHORN, Max Ludwig Paul, (1861-1947)
BREUIL, Henri, Abbé, (1877-1961)
COTTEAU, Gustave Honoré, (1818-1894)
DOUVILLÉ, Joseph Henri Ferdinand, (1846-1937)
EHRENBERG, Christian Gottfried, (1795-1876)
FRAAS, Eberhard, (1862-1915)
FRECH, Fritz, (1861-1917)
HOOKER, Joseph Dalton, Sir, (1817-1911)

SYRIA 1850-1899 Cont'd.

STRICKLAND, Hugh Edwin, (1811-1853)

1900 to present

BERNOUILLI, Walter, (1885-1946)
BLANCKENHORN, Max Ludwig Paul, (1861-1947)
BREUIL, Henri, Abbé, (1877-1961)
DOUVILLÉ, Joseph Henri Ferdinand, (1846-1937)
EDWARDS, Wilfred Norman, (1890-1956)
FRAAS, Eberhard, (1862-1915)
FRECH, Fritz, (1861-1917)
HOOKER, Joseph Dalton, Sir, (1817-1911)
HUXLEY, Julian Sorell, Sir, (1887-1975)
JACQUET, Fernand, (1908-1937)
LEHNER, Ernst J., (1895-1968)
MORRIS, Frederick Kuhne, (1885-1962)
NIKLEWSKI, Jerzy, (1937-1971)
PFANNENSTIEL, Max, (1902-1976)
SCHERMERHORN, W., (1894-)

TAHITI (PACIFIC OCEAN)

1800-1849

DARWIN, Charles Robert, (1809-1882)

1850-1899

DARWIN, Charles Robert, (1809-1882)

1900 to present

JÉRÉMINE, Elisabeth, (née Tschernaieff), (1879-1964)

TAIWAN

1850-1899

OMORI, Fusakichi, (1868-1923)
TOKUNAGA, Shigeyasu, (1874-1940)

1900 to present

HAZZARD, John Charles, (1903-1973)
MASON, Arnold Caverly, (1906-1961)
OMORI, Fusakichi, (1868-1923)
SCHENCK, Hubert Gregory, (1897-1960)
TOKUNAGA, Shigeyasu, (1874-1940)

TANGANYIKA *See* TANZANIA.

TANGIER

Up to 1800

MAILLET, Benoît de, (1656-1738)

TANZANIA

1800-1849

KRAPF, Ludwig, (1810-1881)

1850-1899

BEEBY-THOMPSON, Arthur, (1873-1968)
BRANCA, Wilhelm von, (1844-1928)
DRUMMOND, Henry, (1851-1897)
FRAAS, Eberhard, (1862-1915)
GRANDIDIER, Alfred, (1836-1921)
GREGORY, John Walter, (1864-1932)
GÜRICH, Georg, (1859-1938)
HYLAND, John Shearson, (1866-1898)
KITCHIN, Finlay Lorimer, (1870-1934)
KRAPF, Ludwig, (1810-1881)
MEYER, Hans, (1858-1929)
MIGEOD, Frederick William Hugh, (1872-1952)
STAPFF, Friedrich Moritz, (1836-1895)
TEALE, Edmund Oswald, Sir, (1874-1971)
THORNTON, Richard, (1838-1863)
TORNAU, Friedrich Karl August, (1877-1914)
WALKER, Edward Eaton, (-1903)

1900 to present

BEEBY-THOMPSON, Arthur, (1873-1968)
BISHOPP, Douglas Wallace, (1900-1977)
BISSET, Charles Barron, (-?1966)
BRAMWYCH, Michael, (1911-1974)

TANZANIA 1900 to present Cont'd.

BRANCA, Wilhelm von, (1844-1928)
BUSK, Henry Gould, (?1891-1956)
CONNOR, James, (1894-1977)
COX, Leslie Reginald, (1897-1965)
FRAAS, Eberhard, (1862-1915)
GALLAGHER, David, (1906-1968)
GASKIN, Alan Robert James, (1927-1974)
GILLMAN, Clement, (?1883-1946)
GRANDIDIER, Alfred, (1836-1921)
GRANTHAM, Donald Robert, (1896-1969)
GREGORY, John Walter, (1864-1932)
GÜRICH, Georg, (1859-1938)
HENNIG, Edwin, (1882-)
HITCHEN, Charles Stansfield, (1904-1971)
HUGHES, William John, (?1899-1961)
HUTCHINS, Philip Frank, (1927-1955)
HUXLEY, Julian Sorell, Sir, (1887-1975)
JANENSCH, Werner, (1878-1969)
KENNERLEY, John Brian, (1937-1976)
KITCHIN, Finlay Lorimer, (1870-1934)
KOENIGSWALD, Gustav Heinrich Ralph von, (1902-)
LEAKEY, Louis Seymour Bazett, (1903-1972)
LEAKEY, Richard Erskine, (1944-)
MAACK, Reinhard, (1892-1969)
MACKAY, Robert Alexander, (?1904-1961)
McMATH, John Charles, (?1908-1954)
MENNELL, Frederick Philip, (1880-1966)
MEYER, Hans, (1858-1929)
MIGEOD, Frederick William Hugh, (1872-1952)
NOWACK, Ernst, (1891-1946)
OATES, Frank, (1890-1945)
PARSONS, Ernest, (?1887-1963)
PEET, Geoffrey Alan, (1922-1959)
RECK, Hans, (1886-1937)
SAMPSON, David Noel, (1917-1971)
SCHOLZ, Erich, (1884-1914)
SCHUMACHER, Friedrich, (1884-)
SLAWSON, Chester Baker, (1898-1964)
STAFF, Hans von, (1883-1915)
TAZIEFF, Haroun, (1914-)
TEALE, Edmund Oswald, Sir, (1874-1971)
TORNAU, Friedrich Karl August, (1877-1914)
TREMLETT, Rex, (-) [PROSPECTORS, DIVINERS AND MINING ENGINEERS]
WALKER, Edward Eaton, (-1903)
WILCOCKSON, William Howson, (1891-1976)

THAILAND
1850-1899
BECHER, Harry Macdonald, (1855-1893)
HAASE, Erich, (1857-1894)
KEITH, Arthur, Sir, (1866-1955)
KISSLING, Ernst, (1865-1936)
RICHTHOFEN, Paul Wilhelm Ferdinand von, Freiherr, (1833-1905)

THAILAND 1850-1899 Cont'd.

WAY, Herbert W. L., (-) [PROSPECTORS, DIVINERS AND MINING ENGINEERS]

1900 to present

HAZZARD, John Charles, (1903-1973)
HUTTON, Colin Osborne, (1910-1971)
JOHNSTON, William Drumm, Jr., (1899-1972)
KEITH, Arthur, Sir, (1866-1955)
KISSLING, Ernst, (1865-1936)
KLOMPÉ, Theodorus Hendricus Franciscus, (?1904-1963)
KOBAYASHI, Teiichi, (1901-)
RICHTHOFEN, Paul Wilhelm Ferdinand von, Freiherr, (1833-1905)
SCHERMERHORN, W., (1894-)
SUTTON, Frederick A., (1894-1950)
TRÜMPY, Daniel, (1893-1971)
VOLOSHIN, Vadim "Charlie", (1934-1970)

TIBET

1850-1899

GRANDIDIER, Alfred, (1836-1921)
GREGORY, John Walter, (1864-1932)

1900 to present

GRANDIDIER, Alfred, (1836-1921)
GREGORY, John Walter, (1864-1932)
WELLER, James Marvin, (1899-1976)

TIMOR

1850-1899

MOON, Frederick William, (-1925)

1900 to present

BONTZ, Conrad K., (1894-1927)
BROUWER, Hendrik Albertus, (1886-1973)
HERITSCH, Franz, (1882-1945)
MOON, Frederick William, (-1925)

TOBAGO (WEST INDIES)

1850-1899

GUPPY, Robert John Lechmere, (1836-1916)

1900 to present

GUPPY, Robert John Lechmere, (1836-1916)

TOGO

1900 to present

DEHÉE, René, (1898-1928)
MILLOT, Georges, (1917-)

TONGA ISLANDS (PACIFIC OCEAN)

1900 to present

TAYLOR, Griffith, (1881-1963)

TRINIDAD (WEST INDIES)

1850-1899

ARNOLD, Ralph, (1875-1961)
BEEBY-THOMPSON, Arthur, (1873-1968)
CROSBY, William Otis, (1850-1925)
DITMARS, Raymond Lee, (1876-1942)
GUPPY, Robert John Lechmere, (1836-1916)
HENDERSON, James Alexander Leo, (?1874-1940)
STEWART, Philip Charteris Anstruther, (1874-1923)
THOMPSON, Beeby, (1848-1931)
VEATCH, Arthur Clifford, (1878-1938)
WALL, George Parkes, (1832-1912)
ZYNDEL, Fortunat, (1882-1917)

1900 to present

ALDREDGE, Robert Frank, (1895-1961)
ARNOLD, Ralph, (1875-1961)
BALL, Harry Standish, (?1888-1941)
BARRINGTON, Thomas W., (-)
BEEBY-THOMPSON, Arthur, (1873-1968)
BELTZ, Edward W., (1891-1970)
BENNETT, Clyde M., (1881-1935)
BLOW, Walter Harry, (1924-1972)
BOSWORTH, Thomas Owen, (1882-1928)
BUSK, Henry Gould, (?1891-1956)
CARLSON, Charles Gordon, (1895-1977)
CARR, William L., (1919-1975)
COOKE, Charles Wythe, (1887-1971)
CROSBY, William Otis, (1850-1925)
DITMARS, Raymond Lee, (1876-1942)
FARISH, Linn Markley, (1901-1944)
GORMAN, Joseph M., (1910-1967)
GUPPY, Robert John Lechmere, (1836-1916)
HALSE, George Whitfield, (1885-1968)
HARRISON, John Vernon, (1892-1972)
HAWKINS, Glenn Dewayne, (1897-1970)
HENDERSON, James Alexander Leo, (?1874-1940)
HUTCHISON, Arthur Gilbertson, (1902-1949)
ILLING, Vincent Charles, (1890-1969)
LEHNER, Ernst, J., (1895-1968)
LIDDLE, Ralph Alexander, (1896-1963)
MACKENZIE, Andrew Nicholls "Kelley", (1888-1951)
MACREADY, George Alexander, (1885-1955)
MARTIN, Lewis, (1920-1973)
MORLEY, Harold Thompson "Midge", (1897-1976)
NOLAN, Edward Dana, (1890-1926)
NUTTALL, Winfred Laurence Falkiner, (1897-1969)
OLCUTT, David Perry, (1896-1958)
PENNY, Frederick Willougby, (1887-1973)
PERCIVAL, Frederick George, (1887-1973)
ROMANES, James, (1886-1942)
RUTSCH, Rolf F., (1902-1975)
SCHOLL, Louis August, Jr., (1887-1953)

TRINIDAD (WEST INDIES) 1900 to present Cont'd.

SPATH, Leonard Frank, (1882-1957)
STEWART, Philip Charteris Anstruther, (1874-1923)
THOMPSON, Beeby, (1848-1931)
VEATCH, Arthur Clifford, (1878-1938)
WALL, George Parkes, (1832-1912)
WARING, Gerald Ashley, (1883-1971)
WEAVER, Paul, (1888-1964)
WIRZ, Albert Eduard, (1910-1973)
ZYNDEL, Fortunat, (1882-1917)

TRUCIAL STATES (PERSIAN GULF)
1900 to present
HAZZARD, John Charles, (1903-1973)
JONES, David Glynn, (1904-1974)

TRUCIAL STATES (PERSIAN GULF) See also Oman; Qatar; Sharjah.

TUNISIA
1800-1849
MARÈS, Paul, (1826-1900)
POMEL, Nicolas Auguste, (1821-1898)
RÖMER, Hermann, (1816-1894)

1850-1899
AGAFONOFF, Valerian K., (1863-1955)
BUTTGENBACH, Henri, (1874-1964)
DOLLOT, Auguste, (1841-1924)
DOUVILLÉ, Joseph Henri Ferdinand, (1846-1937)
LE MESLE, George, (-1895)
MARÈS, Paul, (1826-1900)
MILNE-EDWARDS, Alphonse, (1835-1900)
PERVINQUIÈRE, Léon, (1873-1913)
POMEL, Nicolas Auguste, (1821-1898)
ROMAN, Albert Frédéric, (1871-1943)
RÖMER, Hermann, (1816-1894)
TERMIER, Pierre, (1859-1930)
THOMAS, Philippe, (1843-1910)

1900 to present
AGAFONOFF, Valerian K., (1863-1955)
ASTRE, Gaston, (1896-)
BASSE DE MÉNORVAL, Éliane Marguerite, (1899-)
BASSINGER, William E., (1928-1977)
BEEKLEY, Albert Leon, (1883-1952)
BELLAIR, Pierre, (1910-)
BIRAUD, François, (1899-1936)
BUTTGENBACH, Henri, (1874-1964)
DALLONI, Marius Gustave, (1880-1959)
DOLLOT, Auguste, (1841-1924)
DOUVILLÉ, Joseph Henri Ferdinand, (1846-1937)
DURAND DELGA, Michel, (1923-)
JOLEAUD, Léonce, (1880-1938)
JURKOVIČ, Ivan, (1917-)
KISLING, James William, Jr., (1898-1973)

TUNISIA 1900 to present Cont'd.

KLEIBER, Karl Ernst, (1912-1972)
LECKWIJCK, Willem Peter van, (1902-1975)
MARÈS, Paul, (1826-1900)
MILNE-EDWARDS, Alphonse, (1835-1900)
PERVINQUIÈRE, Léon, (1873-1913)
PIVETEAU, Jean, (1898-)
ROMAN, Albert Fréderic, (1871-1943)
SPATH, Leonard Frank, (1882-1957)
TERMIER, Pierre, (1859-1930)
THOMAS, Philippe, (1843-1910)
TSCHIKHACHEV, Peter A., (-)
WALLACE, Maurice Hudiburg, (1917-1961)
WARING, Gerald Ashley, (1883-1971)

TURKEY

Up to 1800

ALLEN, William, (1770-1843)
CASSIUS, Dio Cocceianus, ([155-235] A.D.)
CLARKE, Edward Daniel, (1769-1822)
TAVERNIER, Jean Baptiste, baron d'Aubonne, (1605-1689)

1800-1849

ALLEN, William, (1770-1843)
BOUÉ, Ami, (1794-1881)
CLARKE, Edward Daniel, (1769-1822)
DUBOIS DE MONTPÉREUX, Fréderic, (1798-1850)
GOBANZ, Alexander, (1826-1895)
HAMILTON, William John, (1805-1867)
LILL VON LILIENBACH, Karl, (1798-1831)
LLOYD, J. A., Colonel, (-1856?)
SMYTH, Warington Wilkinson, Sir, (1817-1890)
STRICKLAND, Hugh Edwin, (1811-1853)
VERNEUIL, Philippe Edouard Poulletier de, (1805-1873)
VIQUESNEL, Auguste, (1800-1867)

1850-1899

BOUÉ, Ami, (1794-1881)
BRALY, Adrien, (1864-1927)
DOUVILLÉ, Joseph Henri Ferdinand, (1846-1937)
FUCHS, Philippe Jacques Edmond, (1837-1889)
GOBANZ, Alexander, (1826-1895)
HAMILTON, William John, (1805-1867)
KISSLING, Ernst, (1865-1936)
LLOYD, J. A., Colonel, (-1856?)
OSWALD, Felix, (1866-1958)
PISANI DE SERRES, Felix, (1831-1920)
SALOMON-CALVI, Wilhelm, (1868-1941)
SMITH, Lawrence, (-1883)
SMYTH, Warington Wilkinson, Sir, (1817-1890)
STRICKLAND, Hugh Edwin, (1811-1853)
VERNEUIL, Philippe Edouard Poulletier de, (1805-1873)
VIQUESNEL, Auguste, (1800-1867)
WADE, Arthur, (1878-1951)
WASHINGTON, Henry Stephens, (1867-1934)

TURKEY Cont'd.

1900 to present

BARROW, Geoffrey, (1899-1963)
BISTRITSCHAN, Karl, (1916-1957)
BRALY, Adrien, (1864-1927)
CALAME, Jean Jacques, (1923-1973)
CHAPMAN, Sydney, (1888-1970)
CHAPUT, Jules Ernest, (1882-1943)
CRABTREE, John Edmund, (1940-1966)
CRAWFORD, David Johnston, (1899-1960)
DE LA SAUCE, Wilhelm, (1882-1955)
DOUVILLÉ, Joseph Henri Ferdinand, (1846-1937)
FOLEY, Edward J., (1908-1971)
FORTUNAT, Wincenty Karol, (1889-1967)
HUXLEY, Julian Sorell, Sir, (1887-1975)
JONES, David Glynn, (1904-1974)
KISLING, James William, Jr., (1898-1973)
KISSLING, Ernst, (1865-1936)
LEHNER, Ernst J., (1895-1968)
LOEWE, Fritz, (1895-1974)
MASON, Shirley Lowell, (1893-1975)
MAXSON, John Haviland, (1906-1966)
NOTESTEIN, Frank B., (1885-1973)
OSWALD, Felix, (1866-1958)
PAPP, Simon, (1886-1970)
PENCK, Walther, (1888-1923)
PFANNENSTIEL, Max, (1902-1976)
PISANI DE SERRES, Felix, (1831-1920)
RHOADES, Roger F., (1905-1972)
ROUBAULT, Marcel Edouard, (1905-1974)
SALOMON-CALVI, Wilhelm, (1868-1941)
SCHERMERHORN, W., (1894-)
SCHMID, Kurt, (1906-1957)
SCHNEIDERHÖHN, Hans, (1887-1962)
SCHWENNESEN, Alvin Theodore, (1883-1962)
SHAW, Eugene Wesley, (1881-1935)
SMITH, G. Wendell, (1917-1974)
SPARKS, Dale Darrell Dolsun, (1894-1938)
SPRAGUE, Harry Alvah, (1913-1950)
TASHMAN, Djavad Eyoub, (1893-1956)
TSCHIKHACHEV, Peter A., (-)
WADE, Arthur, (1878-1951)
WALLACE, Maurice Hudiburg, (1917-1961)
WASHINGTON, Henry Stephens, (1867-1934)

UGANDA

1850-1899

BEEBY-THOMPSON, Arthur, (1873-1968)
HOBLEY, Charles William, (1867-1947)
PRIOR, George Thurland, (1862-1936)
REED, Frederick Richard Cowper, (1869-1946)
ROCCATI, Alessandro, (1872-1928)

1900 to present

BEEBY-THOMPSON, Arthur, (1873-1968)
BISHOP, Walter William, (1931-1977)
BISSET, Charles Barron, (-?1966)
BROWN, James Macmillan, (1921-1970)
COMBE, A. D., (1893-1949)
FAIRBAIRN, Harold Williams, (1906-)
FUCHS, Vivian Ernest, Sir, (1908-)
GROVES, Arthur William, (1903-1956)
HOBLEY, Charles William, (1867-1947)
HOLMES, Arthur, (1890-1965)
HUXLEY, Julian Sorell, Sir, (1887-1975)
MOTTRAM, Brian Hunter, (1922-1976)
PARGETER, Reginald Clive, (1919-1972)
PARSONS, Ernest, (?1887-1963)
PENNY, Frederick Willoughby, (1887-1950)
PRIOR, George Thurland, (1862-1936)
REECE, Alan W., (1921-1960)
REED, Frederick Richard Cowper, (1869-1946)
ROCCATI, Alessandro, (1872-1928)
ROE, Frank Walton "Peter", (?1914-1965)
SIMMONS, William Charles, (1888-1951)
TIDY, Anthony John Lintott, (1928-1969)
TREMLETT, Rex, (-) [PROSPECTORS, DIVINERS AND MINING ENGINEERS]

ULSTER <u>See</u> NORTHERN IRELAND.

UNION OF SOVIET SOCIALIST REPUBLICS

Up to 1800

ALLEN, William, (1770-1843)
AVICENNA, (980-1037)
BIRUNI, (<u>pseud</u>. of Abu-al-Rayån Muhammad ibn-Ahmad al-Biruni), (ca.972-1048)
CLARKE, Edward Daniel, (1769-1822)
FERBER, Johann Jacob, (1743-1790)
FISCHER VON WALDHEIM, Johann Gotthelf Friedrich, (1771-1853)
FORSTER, Adolarius Jacob, (1739-1806)
FORSTER, Johann Georg Adam, (1754-1794)
HUMBOLDT, Friedrich Wilhelm Heinrich Alexander von, Freiherr, (1769-1859)
LAXMAN, Erik, (1737-1796)
LEHMANN, Johann Gottlob, (1719-1767)
LOMONOSOV, Mikhail Vasilyevich, (1711-1765)
NASIREDDIN, Mukhamed (called Tusi), (1201-1274)
PALLAS, Peter Simon, (1741-1811)
PARROT, Georg Friedrich, (1767-1852)
STELLER, Georg Wilhelm, (1709-1746)

UNION OF SOVIET SOCIALIST REPUBLICS Cont'd.

1800-1849
ABICH, Hermann Wilhelm, (1806-1886)
ALLEN, William, (1770-1843)
ANDRZEJOWSKI, Antoni, (1785-1868)
AUSTIN, Charles Edward, (1819-1893)
BAER, Karl Ernst von, (1792-1876)
BRANDT, Johann Friedrich von, (1802-1879)
CASPARY, Robert, (1818-1887)
CLARKE, Edward Daniel, (1769-1822)
COTTA, Carl Bernhard von, (1808-1879)
DARWIN, Francis Sacheverell, Sir, (1786-1859)
DUBOIS DE MONTPÉREUX, Frédéric, (1798-1850)
EICHWALD, Eduard Iwanowitsch von, (1795-1876)
ENGELHARDT, Otto Moritz Ludwig von, (1779-1842)
ERMAN, Georg Adolf, (1806-1877)
FISCHER VON WALDHEIM, Johann Gotthelf Friedrich, (1771-1853)
FORSTER, Adolarius Jacob, (1739-1806)
GADOLIN, Axel Wilhelm, General, (1828-1892)
GREWINGK, Konstantin, (1819-1887)
HAMILTON, William John, (1805-1867)
HELMERSEN, Gregor von, (1803-1885)
HIRIAKOFF, Modeste, (1812-1894)
HUMBOLDT, Friedrich Wilhelm Heinrich Alexander von, Freiherr, (1769-1859)
KEYSERLING, Alexander Andreevich, Count, (1815-1891)
KOKSHAROV, Nikola J. von, General, (1818-1893)
KUTORGA, Stepan Semenovich, (1805-1861)
LILL VON LILIENBACH, Karl, (1798-1831)
LLOYD, J. A., Colonel, (-1856?)
LUDWIG, Rudolph August Birminhold Sebastian, (1812-1880)
MÜLLER, Carl Hermann, (1823-1907)
MURCHISON, Roderick Impey, Sir, (1792-1871)
NORDMANN, Alexander von, (1803-1866)
PALLAS, Peter Simon, (1741-1811)
PANDER, Christian Heinrich, (1794-1865)
PARROT, Georg Friedrich, (1767-1852)
ROSE, Gustav, (1798-1873)
SCHLEIDEN, Matthias Jakob, (1804-1881)
SCHRENCK, Leopold von, (1826-1894)
TRAUTSCHOLD, Hermann Adolfovich, (1817-1902)
VERNEUIL, Philippe Edouard Poulletier de, (1805-1873)
VOLBORTH, Aleksandr Fedorovich, (1800-1876)
VOSKOBOINIKOV, N. A., (-)
ZADDACH, Ernst Gustav, (1817-1881)

1850-1899
ABICH, Hermann Wilhelm, (1806-1886)
ADAMS, Mikhail Ivanovich, (-)
AGAFONOFF, Valerian K., (1863-1955)
AMALITSKII, Vladimir Prochorovich, (1860-1917)
ANDRUSSOV, Nikolai Ivanovich, (1861-1924)
ANDRZEJOWSKI, Antoni, (1785-1868)
AUSTIN, Charles Edward, (1819-1893)
BACKLUND, Helge Götrik, (1878-1958)

UNION OF SOVIET SOCIALIST REPUBLICS 1850-1899 Cont'd.

BAER, Karl Ernst von, (1792-1876)
BALDAUF, Richard Julius, (1848-1931)
BARBOT DE MARNY, N. P., (1832-1877)
BATESON, William, (1861-1926)
BECHER, Harry Macdonald, (1855-1893)
BEEBY-THOMPSON, Arthur, (1873-1968)
BELL, James Macintosh, (1877-1934)
BELYANKIN, Dmitri Stepanovich, (1876-1953)
BOHDANOWICZ, Karol, (1864-1947)
BÖHM, Georg, (1854-1913)
BRADY, Henry Bowman, (1835-1891)
BRALY, Adrien, (1864-1927)
BRANDT, Alexander von, (1844-1932)
BRANDT, Johann Friedrich von, (1802-1879)
CASPARY, Robert, (1818-1887)
CHAPER, Maurice Armand, (1834-1896)
COTTA, Carl Bernhard von, (1808-1879)
CZARNOCKI, Stefan, (1878-1947)
DARWIN, Francis Sacheverell, Sir, (1786-1859)
DAVIS, William Morris, (1850-1934)
DELÉPINE, Gaston-Georges, Monseigneur, (1878-1963)
DUPARC, Louis Claude, (1866-1932)
DYBOWSKI, Benedykt, (1833-1930)
EICHWALD, Eduard Iwanowitsch von, (1795-1876)
ERB, Josef Theodor, (1874-1934)
ERMAN, Georg Adolf, (1806-1877)
FAVRE, Ernest, (1845-1925)
FEDEROV, Evgraf Stepanovich, (1853-1919)
FEILDEN, Henry Wemyss, Colonel, (1838-1921)
FISCHER VON WALDHEIM, Johann Gotthelf Friedrich, (1771-1853)
FOURNIER, Eugène, (1871-1941)
FUCHS, Philippe Jacques Edmond, (1837-1889)
GADOLIN, Axel Wilhelm, General, (1828-1892)
GLINKA, Konstantin Dmitrievich, (1867-1927)
GLINKA, Sergei Fedorovich, (1855-1933)
GOUBKIN, Ivan Mikhailovitch, (1871-1939)
GREWINGK, Konstantin, (1819-1887)
GRISHCHINSKY, Petr Iuustinovich, (-1920)
GROEBER, Paul Friedrich Carl, (1885-1968)
GUBKIN, Ivan Mikhailovich, (1874?-)
HAMILTON, William John, (1805-1867)
HAMMOND, John Hays, (1855-1936)
HELMERSEN, Gregor von, (1803-1885)
HELMHACKER, Rudolf Václav, (1840-1915)
HERZ, Otto, (1853-1905)
HIRIAKOFF, Modeste, (1812-1894)
HUENE, Friedrich von, Freiherr, (1875-1969)
HUMBOLDT, Friedrich Wilhelm Heinrich Alexander von, Freiherr, (1769-1859)
HUME, William Fraser, (1867-1949)
INOSTRANTZEV, Aleksandr Aleksandrovich, (1843-1920)
KAISER, Friedrich Wilhelm Erich, (1871-1934)
KARANDYEEV, Vissarion Vissarionovich, (-1916)
KARPINSKY, Alexander Petrovich, (1847-1936)

UNION OF SOVIET SOCIALIST REPUBLICS 1850-1899 Cont'd.

KEYSERLING, Alexander Andreevich, Count, (1815-1891)
KHARLICHKOV, Konstantin Vasil'evich, (-1921)
KHRUSHCHOV, Konstantin Dmitrievich, (? -1912)
KOKSHAROV, Nikola J. von, General, (1818-1893)
KONTKIEWICZ, Stanislaw, Sr., (1849-1924)
KOVALEVSKY, Alexander, (1840-1901)
KOVALEVSKY, Vladimir Onufrievich, (1843-1883)
KROTOV, Petr Ivanovich, (1852-1914)
KUPFFER, August Ernestovich, (1842-1918)
KURNAKOV, Nikolai Semenovich, (1860-1941)
KUTORGA, Stepan Semenovich, (1805-1861)
LIBBEY, William, (1855-1927)
LLOYD, J. A., Colonel, (-1856?)
LOEWINSSON-LESSING, Franc Julyevich, (1859-1939)
LUDWIG, Rudolph August Birminhold Sebastian, (1812-1880)
LUKASZEWICZ, Józef, (1863-1928)
MARGERIE, Emmanuel Marie Pierre Martin Jacquin de, (1862-1953)
MARINELLI, Olinto, (1874-1926)
MARKS, Edward Seaborn, (?1875-1950)
MICHALSKI, Aleksander Octavianovich, (1855-1904)
MILNE, John, (1850-1913)
MOROZEWICZ, Józef Marian, (1865-1941)
MÜLLER, Carl Hermann, (1823-1907)
MURCHISON, Roderick Impey, Sir, (1792-1871)
MUSHKETOV, Ivan Vassilievitch, (1850-1902)
NECHAEV, Aleksyei Vasil'evich, (1864-1915)
NIKITIN, Vasily Vasiliivich, (1867-1942)
NORDMANN, Alexander von, (1803-1866)
OBRUCHEV, Vladimir Afanas'evich, (1863-1956)
ORLOV, Nikolai Arsenevich, (-1919)
OSWALD, Felix, (1866-1958)
PAHLEN, Alexis von der Baron, (1850-1925)
PANDER, Christian Heinrich, (1794-1865)
PAPP, Károly, (1873-1963)
PARROT, Georg Friedrich, (1767-1852)
PERGENS, Eduard Willem Gerard, (1862-1917)
PFIZENMAYER, Eugen Wilhelm, (1869-)
PIWOWAR, Adam, (1874-1939)
POMPECKJ, Josef Felix, (1867-1930)
POPOFF, Boris Anfirovich, (1871-1950)
PORRIO, Cesare, (1865-1940)
RADDE, Gustav, (1831-1903)
RAMSAY, Wilhelm, (1865-1928)
RINNE, Friedrich Wilhelm Berthold, (1863-1933)
ROMER, Eugeniusz, (1871-1954)
ROSE, Gustav, (1798-1873)
SAMOILOV, Yakov Vladimirovich, (1870-1925)
SARGENT, Harry Lionel, (1871-1955)
SCHELLWIEN, Ernst, (1866-1906)
SCHLEIDEN, Matthias Jakob, (1804-1881)
SCHMALHAUSEN, Johannes Theodor, (1849-1894)
SCHMIDT, Friedrich Bogdanovich "Bohdan", (1832-1908)
SCHRENCK, Leopold von, (1826-1894)
SIEMIRADZKI, Józef, (1858-1933)

UNION OF SOVIET SOCIALIST REPUBLICS 1850-1899 Cont'd.

SJÖGREN, Sten Anders Hjalmar, (1856-1922)
STEINMANN, Gustav, (1856-1929)
SUSCHKIN, Peter Petrovich, (1868-1928)
TANNHÄUSER, Felix, (1874-1924)
TARASENKO, Vasily Efimovich, (1859-1926)
TCHERNYCHEV, Féodoss Nikolaievich, (1856-1914)
THOMPSON, Beeby, (1848-1931)
TOLL, Eduard von, Baron, (1858-1902)
TOLWÎNSKI, Konstanty, (1876-1961)
TRAUTSCHOLD, Hermann Adolfovich, (1817-1902)
TRELAWNEY-ANSELL, E. C., (-) [PROSPECTORS, DIVINERS AND MINING ENGINEERS]
TRUSCOTT, Samuel John, (?1870-1950)
TWELVETREES, William Harper, (1848-1919)
VANDERLIP, Washington B., (-) [PROSPECTORS, DIVINERS AND MINING ENGINEERS]
VERNADSKI, Vladimir Ivanovich, (1863-1945)
VERNEUIL, Philippe Edouard Poulletier de, (1805-1873)
VOLBORTH, Aleksandr Fedorovich, (1800-1876)
VULF, Georgii Viktorovich, (1863-1925)
WINCHELL, Horace Vaughn, (1865-1923)
WOEIKOF, Alexander, (1842-1916)
WYROUBOFF, Gr., (1843-1913)
YACHEVSKY, Leonard Antonovich, (1858-1916)
YAKOVLEV, N. N., (1870-1966)
ZADDACH, Ernst Gustav, (1817-1881)
ZEMYATCHENSKY, Petr Andreevich, (1856-1942)
ZHEMCHUZHNYI, Sergei Fedorovitch, (1873-1929)
ZUBER, Rudolf, (1858-1920)

1900 to present

ADAMS, Mikhail Ivanovich, (-)
AGAFONOFF, Valerian K., (1863-1955)
AMALITSKII, Vladimir Prochorovich, (1860-1917)
ANDRÉE, Karl, (1880-1959)
ANDRUSSOV, Nikolai Ivanovich, (1861-1924)
ARMASHEVSKY, Peter Jakovlevich, (-1919)
BACKLUND, Helge Götrik, (1878-1958)
BALDAUF, Richard Julius, (1848-1931)
BARSONOV, Georgyi Pavlovich, (1897-)
BATESON, William, (1861-1926)
BEEBY-THOMPSON, Arthur, (1873-1968)
BELL, James Macintosh, (1877-1934)
BELOUSSOV, Vladimir Vladimirovitch, (-)
BELYANKIN, Dmitri Stepanovich, (1876-1953)
BETECHTIN, Anatolii Georgievich, (1897-1962)
BODYLEVSKY, Vitaly Ivanovich, (1898-1968)
BOGDANOV, Alexei Alexeievitch, (1906-1971)
BOHDANOWICZ, Karol, (1864-1947)
BÖHM, Georg, (1854-1913)
BOLDYREV, Anatolii Kapitonovich, (1883-1946)
BRALY, Adrien, (1864-1927)
BRANDER, Gunnar, (1903-1940)
BRANDT, Alexander von, (1844-1932)

UNION OF SOVIET SOCIALIST REPUBLICS 1900 to present Cont'd.

BYSTROV, Alexei Petrovich, (1899-1959)
CHAPMAN, Sydney, (1888-1970)
CZARNOCKI, Stefan, (1878-1947)
DAVIS, William Morris, (1850-1934)
DE LA SAUCE, Wilhelm, (1882-1955)
DELÉPINE, Gaston-Georges, Monseigneur, (1878-1963)
DIGBY, Bassett, (-)
DOLIVO-DOBROVOLSKY, Vadim Vladimirovich, (1904-1936)
DUBYANSKY, Viktor Viktorovich, (1880-1925)
DUPARC, Louis Claude, (1866-1932)
DYBOWSKI, Benedykt, (1833-1930)
EFREMOV, Ivan Antonovich, (1907-1972)
ERB, Josef Theodor, (1874-1934)
FAVRE, Ernest, (1845-1925)
FEDEROV, Evgraf Stepanovich, (1853-1919)
FEILDEN, Henry Wemyss, Colonel, (1838-1921)
FERSMAN, Aleksander Yevgenyevich, (1883-1945)
FOURNIER, Eugène, (1871-1941)
GLINKA, Konstantin Dmitrievich, (1867-1927)
GLINKA, Sergei Fedorovich, (1855-1933)
GORIZDRO-KULCZYCKA, Zinaida, (1884-1949)
GORSKY, I. I., (1893-1975)
GOUBKIN, Ivan Mikhailovitch, (1871-1939)
GOUDKOFF, Paul Pavel, (1880-1955)
GRISHCHINSKY, Petr Iuustinovich, (-1920)
GROEBER, Paul Friedrich Carl, (1885-1968)
GROMOVA, Vera Isaacovna, (1891-1973)
GUBKIN, Ivan Mikhailovich, (1874?-)
HAMMOND, John Hays, (1855-1936)
HELMHACKER, Rudolf Václav, (1840-1915)
HEROLD, Stanley Carrollton, (1883-1958)
HERZ, Otto, (1853-1905)
HOTZ, Walter, (1884-1958)
HOWELL, Benjamin Franklin, (1890-1976)
HUENE, Friedrich von, Freiherr, (1875-1969)
HUME, William Fraser, (1867-1949)
HUXLEY, Julian Sorell, Sir, (1887-1975)
ICKES, Eugene Law, (1885-1941)
INOSTRANTZEV, Aleksandr Aleksandrovich, (1843-1920)
ISLAMOV, Oskar Islamovich, (1910-1978)
KAISER, Friedrich Wilhelm Erich, (1871-1934)
KALININ, Gennady Pavlovich, (1916-1975)
KARANDYEEV, Vissarion Vissarionovich, (-1916)
KARPINSKY, Alexander Petrovich, (1847-1936)
KHARLICHKOV, Konstantin Vasil'evich, (-1921)
KHRUSHCHOV, Konstantin Dmitrievich, (? -1912)
KINNEY, Harry Dill, (1886-1961)
KONTKIEWICZ, Stanislaw, Sr., (1849-1924)
KOVALEVSKY, Alexander, (1840-1901)
KRECHTER, Vladimir, (?1897-1967)
KROTOV, Petr Ivanovich, (1852-1914)
KRYSHTOFOVICH, Afrikan Nikolaevich, (1885-1953)
KULIK, Leonid Alekseevich, (1883-1942)
KUPFFER, August Ernestovich, (1842-1918)

UNION OF SOVIET SOCIALIST REPUBLICS 1900 to present Cont'd.

KURNAKOV, Nikolai Semenovich, (1860-1941)
KUŹNIAR, Czeslaw Faustyn, (1889-1955)
LEYDS, Louis Willem, (1887-1963)
LOEWINSSON-LESSING, Franc Julyevich, (1859-1939)
LUKASZEWICZ, Józef, (1863-1928)
MALAVOY, Jean, (1903-1945)
MARGERIE, Emmanuel Marie Pierre Martin Jacquin de, (1862-1953)
MARINELLI, Olinto, (1874-1926)
MARKHININ, Y., (-)
MARKS, Edward Seaborn, (?1875-1950)
MICHALSKI, Aleksander Octavianovich, (1855-1904)
MILNE, John, (1850-1913)
MOROZEWICZ, Józef Marian, (1865-1941)
MORRIS, Frederick Kuhne, (1885-1962)
MÜLLER, Carl Hermann, (1823-1907)
MUSHKETOV, Ivan Vassilievitch, (1850-1902)
NECHAEV, Aleksyei Vasil'evich, (1864-1915)
NEUBURG, Maria Feodorovna, (1894-1962)
NIETHAMMER, Gottlob August, (1882-1915)
NIKITIN, Vasily Vasiliivich, (1867-1942)
NOWAK, Jan, (1880-1940)
OBRUCHEV, Dmitri, (1900-1970)
OBRUCHEV, Vladimir Afanas'evich, (1863-1956)
ORLOV, Nikolai Arsenevich, (-1919)
ORLOV, Uri Aleksandrovich, (1893-1966)
OSWALD, Felix, (1866-1958)
PAHLEN, Alexis von der Baron, (1850-1925)
PAPANIN, Ivan Dmitrievich "Vanya", (1895-)
PAPP, Károly, (1873-1963)
PARDÉ, Maurice, (1893-1973)
PERGENS, Eduard Willem Gerard, (1862-1917)
PIWOWAR, Adam, (1874-1939)
POLANSKI, Jorge, (1892-1975)
POMPECKJ, Josef Felix, (1867-1930)
POPOFF, Boris Anfirovich, (1871-1950)
PORRIO, Cesare, (1865-1940)
RADDE, Gustav, (1831-1903)
RAMSAY, Wilhelm, (1865-1928)
RINNE, Friedrich Wilhelm Berthold, (1863-1933)
ROMER, Eugeniusz, (1871-1954)
SAMOILOV, Yakov Vladimirovich, (1870-1925)
SARGENT, Harry Lionel, (1871-1955)
SÄVE-SÖDERBERGH, Gunnar, (1910-1948)
SCHELLWIEN, Ernst, (1866-1906)
SCHMIDT, Friedrich Bogdanovich "Bohdan", (1832-1908)
SCHNEEGANS, Daniel, (1907-1949)
SCHOEP, Alfred Ferdinandus Maria Ghislenus, (1881-1966)
SIEMIRADZKI, Józef, (1858-1933)
SJÖGREN, Sten Anders Hjalmar, (1856-1922)
SMIRNOV, Sergei Sergei'evich, (1895-1947)
STEINMANN, Gustav, (1856-1929)
STRAKHOV, Nikolai Mikhailovich, (1900-)
SUSCHKIN, Peter Petrovich, (1868-1928)
SZATSKIY, Mikolai Sergiechewich, (1895-1960)

UNION OF SOVIET SOCIALIST REPUBLICS 1900 to present Cont'd.

TANNHÄUSER, Felix, (1874-1924)
TARASENKO, Vasily Efimovich, (1859-1926)
TCHERNYCHEV, Féodoss Nikolaievich, (1856-1914)
TEMPLETON, Eugene Crittenden, (1889-1932)
THOMAS, Hugh Hamshaw, (1885-1962)
THOMPSON, Beeby, (1848-1931)
TOLL, Eduard von, Baron, (1858-1902)
TOLWINSKI, Konstanty, (1876-1961)
TRAUTSCHOLD, Hermann Adolfovich, (1817-1902)
TRELAWNEY-ANSELL, E. C., (-) [PROSPECTORS, DIVINERS AND MINING ENGINEERS]
TRUSCOTT, Samuel John, (?1870-1950)
TWELVETREES, William Harper, (1848-1919)
TYRRELL, George Walter, (1883-1961)
VANDERLIP, Washington B., (-) [PROSPECTORS, DIVINERS AND MINING ENGINEERS]
VERNADSKI, Vladimir Ivanovich, (1863-1945)
VIGILEV, Boris, (1883-1924)
VINOGRADOV, A. P., (1895-1975)
VISSER, Philips Christiaan, (1882-1955)
VLASOV, Kuz'ma Alekseevich, (1905-1964)
VULF, Georgii Viktorovich, (1863-1925)
WASHBURNE, Chester Wesley, (1883-1971)
WEAVER, Paul, (1888-1964)
WINCHELL, Horace Vaughn, (1865-1923)
WOEIKOF, Alexander, (1842-1916)
WYROUBOFF, Gr., (1843-1913)
YACHEVSKY, Leonard Antonovich, (1858-1916)
YAKOVLEV, N. N., (1870-1966)
YEMTCHUKHNIKOV, G. A., (1885-1957)
ZEMYATCHENSKY, Petr Andreevich, (1856-1942)
ZHEMCHUZHNYI, Sergei Fedorovitch, (1873-1929)
ZUBER, Rudolf, (1858-1920)

Time unknown
GRIGORIEV, Dmitri Pavlovitch, (-)
SEVASTIANOV, D. P., (-)
YANISHEVSKY, M. E., (-)

UNION OF SOVIET SOCIALIST REPUBLICS <u>See also</u> ARMENIAN SSR; ESTONIAN SSR; FRANZ JOSEF LAND; KOMANDORSKI ISLANDS; LATVIAN SSR; LITHUANIAN SSR; NOVAYA ZEMLYA; PRIBILOF ISLANDS; SAKHALIN.

UNITED KINGDOM
Up to 1800
MARTIN, William, (1767-1810)

1800-1849
DECHEN, Ernst Heinrich Carl von, (1800-1889)
JUKES, Joseph Beete, (1811-1869)
MARTIN, William, (1767-1810)

1850-1899
DAVIS, William Morris, (1850-1934)

UNITED KINGDOM 1850-1899 Cont'd.

DECHEN, Ernst Heinrich Carl von, (1800-1889)
FLETT, John Smith, Sir, (1869-1947)
GEIKIE, Archibald, Sir, (1835-1924)
JUKES, Joseph Beete, (1811-1869)
VEATCH, Arthur Clifford, (1878-1938)

1900 to present
ANDREWS, Richard Wallis, (?1927-1967)
BRAMWYCH, Michael, (1911-1974)
DAVIS, William Morris, (1850-1934)
FLETT, John Smith, Sir, (1869-1947)
GEIKIE, Archibald, Sir, (1835-1924)
HITCHEN, Charles Stansfield, (1904-1971)
HUGHES, Harry, (1929-   )
LECKWIJCK, Willem Peter van, (1902-1975)
LEPPER, Gordon Willis, (1889-1950)
MOORE, James Augustus, (1910-1955)
MUIR, John Malcolm, (1885-1938)
ROMANES, James, (1886-1942)
TIDY, Anthony John Lintott, (1928-1969)
VEATCH, Arthur Clifford, (1878-1938)

UNITED KINGDOM See CHANNEL ISLANDS; ENGLAND; ISLE OF MAN; NORTHERN IRELAND; SCOTLAND; WALES.

UNITED STATES OF AMERICA
ALABAMA
1800-1849
LINCECUM, Gideon, (1793-1874)
TUOMEY, Michael, (1805-1857)
WINCHELL, Alexander, (1824-1891)

1850-1899
ADAMS, George Irving, (1870-1932)
ALDRICH, Truman Heminway, (1848-1932)
BURCHARD, Ernest Norman, (1875-1961)
BUTTS, Charles, (1863-1946)
LIEBER, Oscar M., (-1862)
LINCECUM, Gideon, (1793-1874)
McCALLEY, Henry, (1852-1904)
RIES, Heinrich, (1871-1951)
TUOMEY, Michael, (1805-1857)
WHITE, Charles David, (1862-1935)
WINCHELL, Alexander, (1824-1891)

1900 to present
ADAMS, George Irving, (1870-1932)
AIMER, James David, (1899-1973)
ALDRICH, Truman Heminway, (1848-1932)
BANDY, Orville Lee, (1917-1973)
BLAIR, Charles Scofield, (1883-1971)
BLANPIED, Bernerd William, (1897-1975)
BREMER, B. E. "Boots", (1904-1974)
BROWN, Olin Teter, (1900-1962)

UNITED STATES OF AMERICA (ALABAMA) 1900 to present Cont'd.

BURCHARD, Ernest Norman, (1875-1961)
BUTTS, Charles, (1863-1946)
COOKE, Charles Wythe, (1887-1971)
COPASS, Jack Malcolm, (1906-1963)
DALE, Nelson Clark, (1880-1962)
EARGLE, Dolan Hoye, (1905-1973)
EVANS, Vincent, (1907-1951)
FERGUSON, John Logan, (1893-1973)
FREED, Richard, (1908-1967)
GABRIEL, Alton, (1902-1954)
GAULT, Hugh Richard, (1915-1961)
GORDON, Dugald, (1886-1958)
GRIMES, Russell Winters, (1900-1969)
HAZZARD, Roy Thorpe, (1893-1976)
HOLMES, George William, (1922-1970)
JOHNSTON, William Drumm, Jr., (1899-1972)
KRALIS, Dan F., Jr., (1919-1962)
McCALLEY, Henry, (1852-1904)
McCLAIN, Alan Hartman, (1899-1960)
McDONALD, Stanley Matthews, (1915-1950)
McVAY, Thomas Newkirk, (1891-1971)
POOR, Russell Spurgeon, (1899-1972)
RIES, Heinrich, (1871-1951)
SANDIDGE, John Roy, (1897-1976)
SPOFFORD, Howard Nelson, (1881-1950)
THOMAS, George Dewey, (1898-1975)
WASHBURNE, Chester Wesley, (1883-1971)
WHITCOMB, Bruce, (1900-1950)
WHITE, Charles David, (1862-1935)

UNITED STATES OF AMERICA
ALASKA
Up to 1800
STELLER, Georg Wilhelm, (1709-1746)

1800-1849
BLAKE, William Phipps, (1826-1910)
FURUHJELM, Enoch Hjalmar, (1823-1886)

1850-1899
BANNISTER, Henry Martyn, (-)
BLAKE, William Phipps, (1826-1910)
BOND, Marshall, (1867-1941) [PROSPECTORS, DIVINERS AND MINING ENGINEERS]
BROOKS, Alfred Hulse, (1871-1924)
BURCHARD, Ernest Norman, (1875-1961)
EAKLE, Arthur Starr, (1862-1931)
ELDRIDGE, George Homans, (1854-1905)
EMERSON, Benjamin Kendall, (1843-1932)
FENNER, Clarence Norman, (1870-1949)
FURUHJELM, Enoch Hjalmar, (1823-1886)
GILMORE, Charles Whitney, (1874-1945)
GRANT, Ulysses Sherman, (1867-1932)
GRIGGS, Robert F., (-)

UNITED STATES OF AMERICA (ALASKA) 1850-1899 Cont'd.

HAYES, Charles Willard, (1858-1916)
HEILPRIN, Angelo, (1853-1907)
KENNICOTT, Robert, (1835-1866)
KINDLE, Edward Martin, (1869-1940)
LIBBEY, William, (1855-1927)
MARGERIE, Emmanuel Marie Pierre Martin Jacquin de, (1862-1953)
MARTIN, George C., (1875-1943)
MUIR, John, (1838-1914)
NORDENSKIÖLD, Nils Otto Gustaf, (1869-1928)
RICHARDSON, George Burr, (1872-1949)
ROMER, Eugeniusz, (1871-1954)
STONE, Ralph Walter, (1876-1964)
TARR, Ralph Stockman, (1864-1912)
TRELAWNEY-ANSELL, E. C., (-) [PROSPECTORS, DIVINERS AND MINING ENGINEERS]
WILLIAMS, George Huntington, (1856-1894)

1900 to present

ADKINS, John Nathaniel, (1911-)
BATEMAN, Alan Mara, (1889-1971)
BATES, Robert Glenn, (1923-1972)
BELTZ, Edward W., (1891-1970)
BESCHEL, Roland, (1928-1971)
BLAKE, William Phipps, (1826-1910)
BOND, Marshall, (1867-1941) [PROSPECTORS, DIVINERS AND MINING ENGINEERS]
BOSS, Richard F., (1923-1976)
BROOKS, Alfred Hulse, (1871-1924)
BROWN, Bahngrell Walter "Bill", (1916-)
BRYAN, Joseph Jefferson, (1905-1951)
BURCHARD, Ernest Norman, (1875-1961)
BUTTERWORTH, Emerson McMillin, (1894-1961)
CAMPBELL, Harry A., (1893-1951)
CHAPMAN, Sydney, (1888-1970)
CLARK, Leslie Madison, (1898-1976)
DORSEY, George Edwin, (1892-1953)
EAKLE, Arthur Starr, (1862-1931)
ELDRIDGE, George Homans, (1854-1905)
ELY, Fred Burchard, (1883-1946)
EMERSON, Benjamin Kendall, (1843-1932)
FENNER, Clarence Norman, (1870-1949)
FICHTER, Hans Jakob, (1908-1977)
FRITTS, Crawford Ellsworth, (1927-1972)
GAULT, Hugh Richard, (1915-1961)
GEIST, Frank Otto William, (-1963)
GILLULY, James, (-)
GILMORE, Charles Whitney, (1874-1945)
GOODING, Ansel Miller, (1924-1976)
GRANT, Ulysses Sherman, (1867-1932)
GRATON, Louis Caryl, (1880-1970)
GRIGGS, Robert F., (-)
HANNA, G. Dallas, (1887-1970)
HARRINGTON, George Leavitt, (1883-1972)
HARTWIG, Albert Ernest, Jr., (1922-1966)

UNITED STATES OF AMERICA (ALASKA) 1900 to present Cont'd.

HAYES, Charles Willard, (1858-1916)
HAZZARD, John Charles, (1903-1973)
HEILPRIN, Angelo, (1853-1907)
HENLEY, Arthur Sidney, (1879-1936)
HENNY, Gerhard, (1890-1944)
HEROLD, Stanley Carrollton, (1883-1958)
HERRING, Donald Grant, Jr., (1918-1962)
HEWITT, Robert Leigh, (1911-1956)
HIGGINS, Daniel Franklin, Jr., (1884-1930)
HOLMES, George William, (1922-1970)
HOWELLS, Jenkyn Owen, (-1950)
HUBLEY, Richard Carleton, (1926-1957)
HUSTON, Charles, (1908-) [PROSPECTORS, DIVINERS AND MINING ENGINEERS]
JABLONSKI, Eugene, (1892-1975)
JOESTING, Henry Rochambeau, (1903-1965)
JOHNSON, Bertrand Leroy, (1882-1962)
JOHNSON, Norman Lee, (1926-1970)
KAISEN, Ove C., (1902-1953)
KINDLE, Edward Martin, (1869-1940)
KNOPF, Adolph, (1882-1966)
MARGERIE, Emmanuel Marie Pierre Martin Jacquin de, (1862-1953)
MARTIN, George C., (1875-1943)
MASSON, Charles E., (-) [PROSPECTORS, DIVINERS AND MINING ENGINEERS]
MATHER, Kirtley Fletcher, (1888-1978)
McCALL, John Gill, (?1923-1954)
MILLER, Carroll C., (1899-1974)
MILLER, Don John, (1919-1961)
MILLER, John Preston, (1923-1961)
MORAN, Robert Breck, (1879-1961)
MORRILL, Junius Hughes, (1913-1961)
MUIR, John, (1838-1914)
NOLAN, Edward Dana, (1890-1926)
NORDENSKIÖLD, Nils Otto Gustaf, (1869-1928)
PECORA, William Thomas, (1913-1972)
RICH, John Lyon, (1884-1956)
RICHARDSON, George Burr, (1872-1949)
ROBINSON, G. D., (-)
ROMER, Eugeniusz, (1871-1954)
RUTZEBECK, Hjalmar, (-) [PROSPECTORS, DIVINERS AND MINING ENGINEERS]
SINCLAIR, Ewart Gladstone, (1884-1927)
STONE, Ralph Walter, (1876-1964)
SWARTZ, Joel Howard, (1893-1971)
TARR, Ralph Stockman, (1864-1912)
THORNBURG, Dwight Holbrook "Duke", (1895-1965)
TRELAWNEY-ANSELL, E. C., (-) [PROSPECTORS, DIVINERS AND MINING ENGINEERS]
WARING, Gerald Ashley, (1883-1971)
WASHBURNE, Chester Wesley, (1883-1971)
WEAVER, Charles Edwin, (1880-1958)
WHITTIER, William Harrison, (1889-1951)
WILKERSON, Albert Samuel, (1897-1963)

UNITED STATES OF AMERICA (ALASKA) 1900 to present Cont'd.

WILSON, Bernard Wright, (1922-1976)
YOUNGQUIST, Walter, (-)

UNITED STATES OF AMERICA
ARIZONA
1800-1849
BLAKE, William Phipps, (1826-1910)

1850-1899
ABEL, Othenio, (1875-1946)
BARRINGER, Daniel Moreau, Sr., (1860-1929)
BASTIN, Edson Sunderland, (1878-1953)
BLAKE, William Phipps, (1826-1910)
BOND, Marshall, (1867-1941) [PROSPECTORS, DIVINERS AND MINING ENGINEERS]
DARTON, Nelson Horatio, (1865-1948)
DOUGLAS, James, (1837-)
DUTTON, Clarence Edward, Major, (1841-1912)
GILBERT, Grove Karl, (1843-1918)
GILMORE, Charles Whitney, (1874-1945)
GREGORY, Herbert E., (1869-1952)
GUILD, Frank Nelson, (1870-1939)
MATTHES, François Èmile, (1874-1948)
POWELL, John Wesley, (1834-1902)
PUMPELLY, Raphael, (1837-1923)
RANSOME, Frederick Leslie, (1868-1935)
STONE, Ralph Walter, (1876-1964)
TOLMAN, Cyrus Fisher, (1873-1941)

1900 to present
ABEL, Othenio, (1875-1946)
ABERNATHY, George Elmer, (1888-1948)
ARMSTRONG, Harold Kerr, (1893-1962)
BARRINGER, Daniel Moreau, Sr., (1860-1929)
BASTIN, Edson Sunderland, (1878-1953)
BECKER, Clyde M., (1882-1938)
BLACKWELDER, Eliot, (1880-1969)
BLAKE, William Phipps, (1826-1910)
BOND, Marshall, (1867-1941) [PROSPECTORS, DIVINERS AND MINING ENGINEERS]
BRADY, Lionel Francis "Major", (1880-1963)
BRINKERHOFF, Ira A., (1904-1969)
BRODERICK, Thomas Monteith, (1889-1965)
BUTLER, Gurdon Montague, (1881-1961)
COX, William Edgerton, (1912-1957)
DARTON, Nelson Horatio, (1865-1948)

UNITED STATES OF AMERICA (ARIZONA) 1900 to present Cont'd.

DONAVAN, Jack Hilder, (1921-1953)
DUTTON, Clarence Edward, Major, (1841-1912)
GILBERT, Grove Karl, (1843-1918)
GILLSON, Joseph Lincoln, (1895-1964)
GILLULY, James, (-)
GILMORE, Charles Whitney, (1874-1945)
GORDON, Dugald, (1886-1958)
GREGORY, Herbert E., (1869-1952)
GUILD, Frank Nelson, (1870-1939)
HARRELL, Marshall Allen, (1900-1959)
JUDSON, Wilber, (1880-1951)
KITSON, Howard Walde, (1883-1931)
LANDWEHR, Walter Richard, (1893-1972)
LILLIGREN, Johnathan Martin, (1880-1956)
MATTHES, François Emile, (1874-1948)
McLAREN, Robert Lee, (1901-1953)
MOODY, Graham Blair, (1889-1975)
MYERS, William Marsh, (1892-1951)
NININGER, Harvey Harlow, (-)
NOBLE, Levi Fatzinger, (1882-1965)
NOWLAN, Harry Hackett, (1888-1958)
POWELL, John Wesley, (1834-1902)
PUMPELLY, Raphael, (1837-1923)
RANSOME, Frederick Leslie, (1868-1935)
REBER, Louis Ehrhart, Jr., (1889-1966)
SCHMITT, Harrison Ashley, (1896-1966)
SHERRILL, Richard Ellis, (1899-1953)
SHORT, Maxwell Naylor, (1889-1952)
SIMPSON, George Gaylord, (1902-)
SPOFFORD, Howard Nelson, (1881-1950)
STEUBING, William C., (1881-1943)
STONE, Ralph Walter, (1876-1964)
TAYLOR, James Carlton, (1924-1976)
TOLMAN, Cyrus Fisher, (1873-1941)
WHEALTON, Rowland Gardiner "Peg", (1896-1956)
WHITTIER, William Harrison, (1889-1951)
WILSON, Eldred Dewey, (1898-1965)
WILSON, Roy Arthur, (1891-1970)
WILSON, Thomas Carroll, (1906-1969)
WISSER, Edward Hollister, (1895-1970)

UNITED STATES OF AMERICA
ARKANSAS
1800-1849
LESQUEREUX, Charles Leo, (1806-1889)
NUTTALL, Thomas, (1786-1859)
OWEN, David Dale, (1807-1860)
SCHOOLCRAFT, Henry Rowe, (1793-1864)

1850-1899
BRANNER, John Casper, (1850-1922)
DRAKE, Noah Fields, (1864-1945)
GIRTY, George Herbert, (1869-1939)
KINDLE, Edward Martin, (1869-1940)

UNITED STATES OF AMERICA (ARKANSAS) 1850-1899 Cont'd.

LESQUEREUX, Charles Leo, (1806-1889)
NUTTALL, Thomas, (1786-1859)
OWEN, David Dale, (1807-1860)
PURDUE, Albert Homer, (1861-1917)
SCHOOLCRAFT, Henry Rowe, (1793-1864)
STEPHENSON, Lloyd William, (1876-1962)
VALERIUS, Claude N., (-)
VEATCH, Arthur Clifford, (1878-1938)
WILLIAMS, John Francis, (1862-1891)

1900 to present

AIMER, James David, (1899-1973)
BARTELL, Lawrence Denny, (1902-1955)
BLANPIED, Bernerd William, (1897-1975)
BLOOMER, Philip A., Jr., (1922-1967)
BRACE, Orval L., (1893-1975)
BRANNER, John Casper, (1850-1922)
BROWN, Bahngrell Walter "Bill", (1916-)
BURRESS, Walter March, (1893-1950)
BYRON, Clifford Winkler, (1901-1970)
CADY, Gilbert Haven, (1882-1970)
CAMPBELL, Robert Burns, (1892-1955)
CLARK, Howard, (1885-1972)
COCHRAN, Kenneth L., (1910-1973)
COTTINGHAM, Virgil Elisha, (1891-1958)
DANE, Carle Hamilton, (1900-1968)
DRAKE, Noah Fields, (1864-1945)
ELLEDGE, Emmett Russell, (1898-1960)
FAHLE, Paul Bartlett, (1910-1954)
FAIRCHILD, John Gifford, (1882-1965)
GALLAGHER, David, (1906-1968)
GALLIE, John Francis, III, (1913-1951)
GIRTY, George Herbert, (1869-1939)
GLASS, Jewell Jeanette, (1888-1966)
GOLDMAN, Marcus Isaac, (1881-1965)
HANSELL, James Myron, (1902-1970)
HOYLE, Charles Robert, Sr., (1891-1954)
KENNEDY, William, (-1926)
KINDLE, Edward Martin, (1869-1940)
KRIDLER, George Matthew, (1889-1953)
MEAD, Warren Judson, (1883-1960)
MOORE, James Augustus, (1910-1955)
PALMER, Leland L., (1906-1978)
PICKLESIMER, Harold Greeley, (1916-1952)
POWELL, Ralph Sterling, (1901-1951)
PURDUE, Albert Homer, (1861-1917)
RUBEY, William Walden, (1898-1974)
SCHNEIDER, George William, (1899-1963)
SCOTT, Horace Lafayette, (1901-1976)
SEVERY, Clarence Luther, (1885-1975)
SHEARER, Harold Kurtz, (1889-1946)
SPOFFORD, Howard Nelson, (1881-1950)
STEPHENSON, Lloyd William, (1876-1962)
THOMPSON, Robert Raymond, (1893-1954)

UNITED STATES OF AMERICA (ARKANSAS) 1900 to present Cont'd.

VEATCH, Arthur Clifford, (1878-1938)
WIER, Harvey James, Jr., (1924-1950)
WILBERT, Louis Joseph "Slick", Jr., (1919-1953)
WINES, Donald Bradford, (1902-1958)

UNITED STATES OF AMERICA
CALIFORNIA
1800-1849
ATTWOOD, Melville, (1812-1898)
HARGRAVES, Edward Hammond, (1816-1891)
LE CONTE, Joseph, (1823-1901)
NUTTALL, Thomas, (1786-1859)
PHILLIPS, John Arthur, (1822-1887)
SALMON, Henry Curwen, (1827-1873)
SILLIMAN, Benjamin Jr., (1816-1885)
WHITNEY, Josiah Dwight, (1819-1896)

1850-1899
ABEL, Othenio, (1875-1946)
ANDERSON, Frank Marion, (1863-1945)
ARNOLD, Ralph, (1875-1961)
ATTWOOD, George, (1845-1912)
ATTWOOD, Melville, (1812-1898)
BERKEY, Charles Peter, (1867-1955)
BOND, Marshall, (1867-1941) [PROSPECTORS, DIVINERS AND MINING ENGINEERS]
BOWMAN, Amos, (1839-1894)
BRANNER, John Casper, (1850-1922)
BREWER, William Henry, (1828-1910)
CHRISTMAN, Charles Gustave, (1873-1953)
CLAYPOLE, Edward Waller, (1835-1901)
COCKERELL, Theodore Dru Alison, (1866-1948)
DICKERSON, Roy Ernest, (1877-1944)
DUTTON, Clarence Edward, Major, (1841-1912)
EAKLE, Arthur Starr, (1862-1931)
ERB, Josef Theodor, (1874-1934)
FLORES REYES, Teodoro, (1873-1955)
GABB, William More, (1839-1878)
GALE, Hoyt Stoddard, (1876-1952)
GARDINER, Howard Calhoun, (1826-1917) [PROSPECTORS, DIVINERS AND MINING ENGINEERS]
HAMMOND, John Hays, (1855-1936)
HARGRAVES, Edward Hammond, (1816-1891)
HENDERSON, James Alexander Leo, (?1874-1940)
HESS, Frank L., (1871-1955)
HILGARD, Eugene Woldemar, (1833-1916)
HOLDEN, Edward Singleton, (1846-1914)
JORDAN, David Starr, (1851-1931)
JULIAN, Henry Forbes, (1861-1912) [PROSPECTORS, DIVINERS AND MINING ENGINEERS]
KHRUSHCHOV, Konstantin Dmitrievich, (? -1912)
LARSEN, Esper Signius Jr., (1879-1961)
LAWSON, Andrew Cowper, (1861-1952)
LE CONTE, Joseph, (1823-1901)

UNITED STATES OF AMERICA (CALIFORNIA) 1850-1899 Cont'd.

LEVERETT, Frank, (1859-1943)
LOUDERBACK, George Davis, (1874-1957)
MACKAY, John William, (1831-1902) [PROSPECTORS, DIVINERS AND MINING ENGINEERS]
MACLAREN, James Malcolm, (1873-1935)
MATTHEW, William Diller, (1871-1930)
MEAD, Roy Gibbons, (1881-1956)
MENDENHALL, Walter Curran, (1871-1957)
MERRILL, George Perkins, (1854-1929)
MILLER, Loye Holmes, (1874-1970)
MUIR, John, (1838-1914)
NUTTALL, Thomas, (1786-1859)
ORCUTT, William Warren, (1869-1942)
PETTEE, William Henry, (1836-1904)
PHILLIPS, John Arthur, (1822-1887)
PILSBRY, Henry Augustus, (1862-1957)
POLLOK, Rudolf R., (1876-1929)
POWELL, John Wesley, (1834-1902)
RANSOME, Frederick Leslie, (1868-1935)
RICHTHOFEN, Paul Wilhelm Ferdinand von, Freiherr, (1833-1905)
RICKARD, Thomas Arthur, (1864-1953)
SALMON, Henry Curwen, (1827-1873)
SILLIMAN, Benjamin Jr., (1816-1885)
SMITH, William Richard Tangier, (1869-1972)
STONE, Ralph Walter, (1876-1964)
TAFF, Joseph Alexander, (1862-1944)
TOLMAN, Cyrus Fisher, (1873-1941)
TRASK, John Boardman, (1824-1879)
VONSEN, Magnus, (1879-1954)
WHITNEY, Josiah Dwight, (1819-1896)
WILLARD, Daniel Everett, (1862-1947)
WILLIAMS, Gardner Fred, (1842-1922)
WILLIS, Bailey, (1857-1949)
WINCHELL, Horace Vaughn, (1865-1923)

1900 to present

ABEL, Othenio, (1875-1946)
ADKINS, John Nathaniel, (1911-)
AMBROSE, Arthur Warren, (1889-1952)
ANDERSON, Frank Marion, (1863-1945)
ANDERSON, George Harold, (1893-1956)
ANDRAU, Evert Willem Karel, (1900-1951)
ARMSTRONG, Harold Kerr, (1893-1962)
ARNOLD, Ralph, (1875-1961)
ATTWOOD, George, (1845-1912)
AUGUR, Irving V., (1887-1930)
BANDY, Orville Lee, (1917-1973)
BARNES, Roy Merrill, (1891-1960)
BASSINGER, William E., (1928-1977)
BERG, Ernest Lyle, (1910-1953)
BERKEY, Charles Peter, (1867-1955)
BOND, Lewis Alexander, (1896-1972)
BOND, Marshall, (1867-1941) [PROSPECTORS, DIVINERS AND MINING ENGINEERS]

UNITED STATES OF AMERICA (CALIFORNIA) 1900 to present Cont'd.

BOOTS, Paul Heath, (1909-1947)
BOSS, Richard F., (1923-1976)
BOUCOT, Arthur James, (1924-)
BOYD, Harold Edward, (1887-1952)
BRACE, Orval L., (1893-1975)
BRAMLETTE, Milton Nunn, (1896-1977)
BRANNER, John Casper, (1850-1922)
BRAVINDER, Kenneth Mason, (1907-1946)
BREMNER, Carl St. John, (1895-1944)
BREWER, William Henry, (1828-1910)
BRISCOE, Glenn Oliver, (1898-1974)
BRYAN, Joseph Jefferson, (1905-1951)
BURGER, Robert W. "Bush", (1906-1945)
BURNS, Russell Warren, (1912-1967)
BUTTERWORTH, Emerson McMillin, (1894-1961)
BUWALDA, John Peter, (1886-1954)
CAMPBELL, Harry A., (1893-1951)
CAMPBELL, Ian, (1900-1978)
CANFIELD, Charles Reiter, (1906-1964)
CARSON, Carlton Mann "Kit", (1898-1972)
CARTER, Frank B., (1898-1968)
CHANEY, Ralph Works, (1890-1971)
CHAPMAN, Sydney, (1888-1970)
CHRISTMAN, Charles Gustave, (1873-1953)
CLARK, Bruce Lawrence, (1880-1945)
CLARK, Leslie Madison, (1898-1976)
CLARK, Robert Watson, (1884-1948)
CLAYPOLE, Edward Waller, (1835-1901)
COCHRAN, Kenneth L., (1910-1973)
COCKERELL, Theodore Dru Alison, (1866-1948)
COLLOM, Roy Edward, (1881-1948)
CRANDALL, Richard Robbins, (1902-1949)
CROSS, Rodman K., (1911-1970)
CROWN, Walter James, (1901-1951)
CUNNINGHAM, George Magee, (1895-1963)
DALE, Nelson Clark, (1880-1962)
DARBY, Raymond Vern, (1896-1953)
DAVIS, Elmer Fred, (1887-1974)
DE KOCH, William Gaspar, (1898-1954)
DICKERSON, Roy Ernest, (1877-1944)
DIETZ, Curt, (1882-1965)
DRAKE, Cecil, (1899-1949)
DUTTON, Clarence Edward, Major, (1841-1912)
EAKLE, Arthur Starr, (1862-1931)
EATON, Joseph Edmund, (1895-1958)
EDWARDS, Everett C., (1897-1972)
ERB, Josef Theodor, (1874-1934)
FERGUSON, Roy Noble, (1883-1961)
FERRERO, Clarence de Labretonne, (1903-1954)
FINCH, Elmer Harrison, (1883-1953)
FITZGERALD, Thomas John, (1903-1973)
FLORES REYES, Teodoro, (1873-1955)
FOLGER, James Anthony, (1896-1956)
FORTINE, Dwight Hugh, (1904-1949)

UNITED STATES OF AMERICA (CALIFORNIA) 1900 to present Cont'd.

FRASER, Horace John "Dutch", (1905-1969)
GALE, Hoyt Stoddard, (1876-1952)
GALLAGHER, David, (1906-1968)
GARDINER, Howard Calhoun, (1826-1917) [PROSPECTORS, DIVINERS AND MINING ENGINEERS]
GARDINER, Thomas Moir, Jr., (1898-1930)
GARFIAS, Valentin R., (-)
GARIEPY, Glen Bernard, (1904-1962)
GEBELEIN, Conrad Dennis, (1945-1978)
GESTER, George Clark, (1884-1959)
GESTER, Stephen Howell, (1889-1967)
GILLES, Verner Arthur, (1886-1954)
GILLULY, James, (-)
GLASS, Jewell Jeanette, (1888-1966)
GOLLNICK, Robert Louis, (1918-1963)
GOUDKOFF, Paul Pavel, (1880-1955)
GRAHAM, David H., (1910-1944)
GRAHAM, Joseph John, (1909-1967)
GRIMM, Kenneth Edgar, (1912-1961)
GUTENBERG, Beno, (1889-1960)
HAKE, Benjamin Franklin, (1895-1973)
HALL, Elwin B., (1891-1949)
HAMMOND, John Hays, (1855-1936)
HANNA, G. Dallas, (1887-1970)
HARDISON, Harvey, (1895-1949)
HAWLEY, James Edwin, (1897-1965)
HAZZARD, John Charles, (1903-1973)
HEADLEE, Larry A., (1938-1970)
HENDERSON, James Alexander Leo, (?1874-1940)
HENLEY, Arthur Sidney, (1879-1936)
HEROLD, Stanley Carrollton, (1883-1958)
HESS, Frank L., (1871-1955)
HEWETT, Donnel Foster, (1881-1971)
HEWITT, Robert Leigh, (1911-1956)
HIDE, Raymond, (1929-)
HIGGS, Morton Thomas, (1896-1956)
HILGARD, Eugene Woldemar, (1833-1916)
HOFMANN, Walter, (1919-1975)
HOGAN, Dana, (1890-1945)
HOLCOMB, Leroy D., (1919-1974)
HOLDEN, Edward Singleton, (1846-1914)
HOLLISTER, Joseph Steffens, (1903-1953)
HOLST PELLEKAAN, William van, (1880-1949)
HORTON, Harold Merritt "Wog", (1897-1966)
HOWELLS, Jenkyn Owen, (-1950)
HUDSON, Frank Samuel, (1890-1967)
HUGHES, Donald Dudley, (1893-1946)
HUTTON, Colin Osborne, (1910-1971)
ICKES, Eugene Law, (1885-1941)
JACOB, Charles Edward, (1914-1970)
JENSEN, Christian "Cope", (1905-1961)
JOHNSON, Clifton Wood, (1905-1954)
JOHNSON, Earle Winston, (1902-1953)
JOHNSTON, William Drumm, Jr., (1899-1972)

UNITED STATES OF AMERICA (CALIFORNIA) 1900 to present Cont'd.

JORDAN, David Starr, (1851-1931)
JULIAN, Henry Forbes, (1861-1912) [PROSPECTORS, DIVINERS AND MINING ENGINEERS]
KEW, William Stephen Webster, (1890-1961)
KHRUSHCHOV, Konstantin Dmitrievich, (? -1912)
KITSON, Howard Walde, (1883-1931)
KLEINPELL, William Darwin, (1898-1959)
KNOPF, Adolph, (1882-1966)
KNOWLES, Gerald Elgin, (1923-1976)
KURTZ, Robert Gourdin, (1898-1958)
LAMMERS, Edward Chauncey Hinman, (1910-1971)
LARSEN, Esper Signius Jr., (1879-1961)
LAWSON, Andrew Cowper, (1861-1952)
LE CONTE, Joseph, (1823-1901)
LEDINGHAM, Glen Wallace, (1911-1978)
LEVERETT, Frank, (1859-1943)
LEVORSEN, Arville Irving, (1894-1965)
LILLIGREN, Johnathan Martin, (1880-1956)
LINDEN, Bernard Hugo "Ben" van der, (-1958)
LINES, Frederic George, (1921-1963)
LOSKAMP, Alvin Powell, (1900-1954)
LOUDERBACK, George Davis, (1874-1957)
LUPTON, Benjamin Charles, (1908-1964)
LUTH, William Clair, (1934-)
MACDONALD, Gordon James Fraser, (1929-)
MACLAREN, James Malcolm, (1873-1935)
MACREADY, George Alexander, (1885-1955)
MATHER, Kirtley Fletcher, (1888-1978)
MATTHEW, William Diller, (1871-1930)
MAXSON, John Haviland, (1906-1966)
MAY, Arthur Raymond, (1893-1948)
MEAD, Roy Gibbons, (1881-1956)
MENDENHALL, Walter Curran, (1871-1957)
MENKEN, Frederick Andrew, (1905-1965)
MERRILL, George Perkins, (1854-1929)
MILLER, Guy E., "Doc", (1893-1977)
MILLER, Loye Holmes, (1874-1970)
MILLER, William John, (1880-1965)
MIX, Charles Alexander, (1898-1954)
MONTGOMERY, Curtis Hall, (1897-1944)
MOODY, Graham Blair, (1889-1975)
MORAN, Robert Breck, (1879-1961)
MORGAN, Frank Albert, (1898-1976)
MUIR, John, (1838-1914)
MULLER, Siemon William, (1900-1970)
MURDOCK, Joseph, (1890-1973)
NARAMORE, Chester "Chet", (1879-1961)
NELSON, Richard Newman, (1897-1964)
NOBLE, Earl Bart, (1894-1967)
NOBLE, Levi Fatzinger, (1882-1965)
NOMLAND, Jorgen O., (1885-1943)
ORCUTT, William Warren, (1869-1942)
OWENS, Frith Cravens, (1898-1977)
PAINE, Paul, (1881-1957)

UNITED STATES OF AMERICA (CALIFORNIA) 1900 to present Cont'd.

PEACOCK, Martin Alfred, (1898-1950)
PEMBERTON, John Roy "Bill", (1884-1968)
PETERSON, Arthur Frederick, (1911-1967)
PETERSON, Clarence J., (1885-1948)
PETTEE, William Henry, (1836-1904)
PILSBRY, Henry Augustus, (1862-1957)
POLLOK, Rudolf R., (1876-1929)
POWELL, John Wesley, (1834-1902)
PRESS, Frank, (1924-)
PUTNAM, L. G., (1898-1936)
PUTNAM, William Clement, (1908-1963)
RANSOME, Frederick Leslie, (1868-1935)
REESE, Richard Griffith, (1902-1964)
RENICK, Brink Coleman, (1897-1971)
RHOADES, Roger F., (1905-1972)
RICHTHOFEN, Paul Wilhelm Ferdinand von, Freiherr, (1833-1905)
RICKARD, Thomas Arthur, (1864-1953)
ROBINSON, Ernest Guy, (1898-1973)
ROGERS, Austin Flint, (1887-1957)
RUBEY, William Walden, (1898-1974)
RUDOLPH, Joseph, (1897-1953)
SAUNDERS, Lowell Waller, (1901-1954)
SCHENCK, Hubert Gregory, (1897-1960)
SCHOLL, Louis August, Jr., (1887-1953)
SCOTT, Walter Winthrop, (1893-1939)
SEVERY, Clarence Luther, (1885-1975)
SHEPARD, Francis P., (-)
SHERBORNE, John Elliott, Sr., (1911-1974)
SICKLER, John Miller, Jr., (1891-1970)
SLICHTER, Louis Byrne, (1896-)
SMITH, William Richard Tangier, (1869-1972)
SNIVELEY, H. Norman, (1879-1959)
SPRAGUE, Harry Alvah, (1913-1950)
STALDER, Walter "Cap", (1881-1949)
STEHR, Raymond Albert, (1903-1967)
STEINY, Homer Julius, (1892-1974)
STEWART, Irvine E., (1893-1965)
STIRTON, Ruben Arthur, (1901-1966)
STOCK, Chester, (1892-1950)
STONE, Ralph Walter, (1876-1964)
TAFF, Joseph Alexander, (1862-1944)
TALLANT, Roy Leon, (1912-1961)
TAYLOR, James Carlton, (1924-1976)
TEMPLETON, Eugene Crittenden, (1889-1932)
THALMANN, Hans E., (1899-1975)
THOMAS, Harry George, (1929-1976)
THOMPSON, David Grosh, (1888-1943)
THORNBURG, Dwight Holbrook "Duke", (1895-1965)
TIEJE, Arthur Jerrold, (1891-1944)
TODD, Thomas W., (1927-1969)
TOLMAN, Cyrus Fisher, (1873-1941)
TRASK, Parker Davies, (1899-1961)
UREN, Lester Charles, (1888-1960)
VAN COUVERING, Martin, (1888-1976)

UNITED STATES OF AMERICA (CALIFORNIA) 1900 to present Cont'd.

VANDER HOOF, Vertress Lawrence, (1904-1964)
VANDER LECK, Lawrence, (1890-1954)
VOLOSHIN, Vadim "Charlie", (1934-1970)
VONSEN, Magnus, (1879-1954)
WAGNER, Carroll M., (1893-1965)
WANGSNESS, Orrin Jerold, (1917-1967)
WARING, Gerald Ashley, (1883-1971)
WASHBURNE, Chester Wesley, (1883-1971)
WASSON, Theron, (1887-1970)
WEAVER, Charles Edwin, (1880-1958)
WELLES, Samuel Paul, (-)
WHEALTON, Rowland Gardiner "Peg", (1896-1956)
WHEELER, Orby Clinton, (1898-1970)
WHITESIDE, Charles Edwin, (1897-1927)
WILLARD, Daniel Everett, (1862-1947)
WILLIAMS, Gardner Fred, (1842-1922)
WILLIS, Bailey, (1857-1949)
WILSON, Bernard Wright, (1922-1976)
WILSON, Eugene Jack, (1921-1976)
WILSON, James W., (?1925-1964)
WINCHELL, Horace Vaughn, (1865-1923)
WISSER, Edward Hollister, (1895-1970)
WOODS, Earl Hazen, (1902-1950)
WYATT, Michael, (-1956)
YULE, Frank Ingalls, (1926-1961)
ZAVOICO, Basil B., (1899-1975)

UNITED STATES OF AMERICA
COLORADO
1800-1849
CHAPMAN, Edward John, (1821-1904)
HAYDEN, Ferdinand Vandiveer, (1829-1887)
NEWBERRY, John Strong, (1822-1892)

1850-1899
ABEL, Othenio, (1875-1946)
AVERY, Charles Dwight, (1877-1947)
BASTIN, Edson Sunderland, (1878-1953)
BIRLEY, Caroline, (1851-1907)
CHAPMAN, Edward John, (1821-1904)
COCKERELL, Theodore Dru Alison, (1866-1948)
COLLINS, Arthur L., (1868-1902)
CRAWFORD, Ralph Dixon, (1873-1950)
DUTTON, Clarence Edward, Major, (1841-1912)
ELDRIDGE, George Homans, (1854-1905)
EMMONS, Samuel Franklin, (1841-1911)
FENNEMAN, Nevin Melancthan, (1865-1945)
FIGGINS, Jesse Dade, (1867-1944)
GREGORY, Herbert E., (1869-1952)
HAYDEN, Ferdinand Vandiveer, (1829-1887)
HENDERSON, Junius, (1865-1937)
HILLS, Richard Charles, (1848-1923)
HOWE, Ernest, (1875-1932)
IRVING, John Duer, (1874-1918)

UNITED STATES OF AMERICA (COLORADO) 1850-1899 Cont'd.

JEBB, John Gladwyn, (1841-1893) [PROSPECTORS, DIVINERS AND MINING ENGINEERS]
JOHANNSEN, Albert, (1871-1962)
KEYTE, Ivy Allen, (1878-1931)
KING, Clarence Rivers, (1842-1901)
KNOWLTON, Frank Hall, (1860-1926)
LARSEN, Esper Signius Jr., (1879-1961)
LEE, Willis Thomas, (1864-1926)
LEVERETT, Frank, (1859-1943)
LUPTON, Charles Thomas, (1878-1935)
MARSH, Othniel Charles, (1831-1899)
MARVINE, Archibald Robertson, (1848-1876)
McCARTHY, Edward Thomas, (-) [PROSPECTORS, DIVINERS AND MINING ENGINEERS]
NEWBERRY, John Strong, (1822-1892)
PEARCE, Richard, (1837-1927)
POWELL, John Wesley, (1834-1902)
RANSOME, Frederick Leslie, (1868-1935)
RICHARDSON, George Burr, (1872-1949)
RICKARD, Thomas Arthur, (1864-1953)
SCOTT, William Berryman, (1858-1947)
STANTON, Timothy William, (1860-1953)
STERNBERG, Charles Hazelius, (1850-1943)
ST. JOHN, Orestes Hawley, (1841-1921)
WALTMAN, William Dewitt, (1875-1955)
WAY, Herbert W. L., (-) [PROSPECTORS, DIVINERS AND MINING ENGINEERS]

1900 to present

ABEL, Othenio, (1875-1946)
ARMSTRONG, Harold Kerr, (1893-1962)
AVERY, Charles Dwight, (1877-1947)
BALLARD, William Norval, (1906-1976)
BASTIN, Edson Sunderland, (1878-1953)
BECKWITH, Radcliffe Harold, (1900-1964)
BEEKLEY, Albert Leon, (1883-1952)
BIRLEY, Caroline, (1851-1907)
BOSS, Richard F., (1923-1976)
BRADLEY, Wilmot Hyde, (1899-)
BROWN, John Albert, (1901-1954)
BROWN, Otto E., (1896-1973)
BUNTE, Arnold S., (1902-1968)
BURRESS, Walter March, (1893-1950)
CARLSON, Charles Gordon, (1895-1977)
CARR, William L., (1919-1975)
CHAPMAN, Edward John, (1821-1904)
CHAPMAN, Sydney, (1888-1970)
CLARK, Robert Watson, (1884-1948)
COCHRAN, Kenneth L., (1910-1973)
COCKERELL, Theodore Dru Alison, (1866-1948)
COFFIN, Reuben Clare, (1886-1972)
COLLINS, Arthur L., (1868-1902)
COLLINS, Melvin J., (1897-1936)
COOK, Harold James, (1887-?1962)

UNITED STATES OF AMERICA (COLORADO) 1900 to present Cont'd.

CRAWFORD, Ralph Dixon, (1873-1950)
CREWS, George Curtis, (1837-1977)
CRUM, Harry Edwin, (1892-1955)
DIETZ, Curt, (1882-1965)
DONOVAN, Percy Williams, (1879-1973)
DOUGLAS, James Mainwaring, (1887-1952)
DUTTON, Clarence Edward, Major, (1841-1912)
EARGLE, Dolan Hoye, (1905-1973)
EBY, James Brian, (1896-)
ELDRIDGE, George Homans, (1854-1905)
ELLEDGE, Emmett Russell, (1898-1960)
EMMONS, Samuel Franklin, (1841-1911)
ERDMANN, Charles Edgar, (1897-1971)
FENNEMAN, Nevin Melancthan, (1865-1945)
FERGUSON, John Logan, (1893-1973)
FICHTER, Hans Jakob, (1908-1977)
FIGGINS, Jesse Dade, (1867-1944)
FOLGER, James Anthony, (1896-1956)
FORD, Trevor David, (1932-)
FULTON, Loris John, (1906-1974)
GARRETT, Robert Ernest, (1888-1947)
GILL, James Rogers, (1922-1972)
GLASS, Jewell Jeanette, (1888-1966)
GOLDMAN, Marcus Isaac, (1881-1965)
GREGORY, Herbert E., (1869-1952)
HANSON, Ernest Albin, (1899-1973)
HARRISON, Thomas Samuel, (1881-1964)
HEILAND, Carl August, (1899-1956)
HENDERSON, Junius, (1865-1937)
HENDRICKSON, Victor James, (1892-1953)
HIGGINS, Daniel Franklin, Jr., (1884-1930)
HILLS, Richard Charles, (1848-1923)
HOLCOMB, Leroy D., (1919-1974)
HOOVER, James Barr, (1902-1953)
HOWE, Ernest, (1875-1932)
HOWELL, Benjamin Franklin, (1890-1976)
HUNT, Edwin Harold, (1897-1949)
IRVING, John Duer, (1874-1918)
JENNINGS, Charles I. "Al", (1894-1968)
JOHANNSEN, Albert, (1871-1962)
JOHNSON, Bertrand Leroy, (1882-1962)
JOHNSON, Charles George, (1914-1969)
JOHNSON, Earle Winston, (1902-1953)
JUDSON, Wilber, (1880-1951)
KELLY, Donald, (1898-1949)
KEYES, Wilson, (1901-1936)
KEYTE, Ivy Allen, (1878-1931)
KING, Clarence Rivers, (1842-1901)
KIVI, Wilho John, (1918-1950)
KNOPF, Adolph, (1882-1966)
KNOWLTON, Frank Hall, (1860-1926)
KOSCHMANN, Albert Herbert, (1896-1962)
KRALIS, Dan F., Jr., (1919-1962)
KURTZ, Robert Gourdin, (1898-1958)

UNITED STATES OF AMERICA (COLORADO) 1900 to present Cont'd.

LAMMERS, Edward Chauncey Hinman, (1910-1971)
LANDWEHR, Walter Richard, (1893-1972)
LARSEN, Esper Signius Jr., (1879-1961)
LEE, Willis Thomas, (1864-1926)
LEVERETT, Frank, (1859-1943)
LEWIS, Francis Leonard, Jr., (1921-1953)
LLOYD, Edwin Russell, (1882-1955)
LUPTON, Charles Thomas, (1878-1935)
MATHER, Kirtley Fletcher, (1888-1978)
MAUCINI, Joseph John, (1893-1955)
MAWDSLEY, James Buckland, (1894-1964)
MAXSON, John Haviland, (1906-1966)
McCARTHY, Edward Thomas, (-) [PROSPECTORS, DIVINERS AND MINING ENGINEERS]
McCOY, Alexander Watts, (1889-1944)
McLAREN, Robert Lee, (1901-1953)
McMINN, Paul Meloy, (1919-1974)
MILLER, Guy E., "Doc", (1893-1977)
MITCHELL, James Garland, (1908-1972)
MITCHELL, Kingsley Camden, (1903-1931)
MORRIS, Albert Ferd, (1898-1957)
MULLEN, John Timothy, Jr., (1906-1959)
MYERS, John Clark, (1894-1970)
NANCE, James W., (1908-1977)
NININGER, Harvey Harlow, (-)
OWEN, William Joseph, (1929-1959)
OWENS, Frith Cravens, (1898-1977)
PEARCE, Richard, (1837-1927)
PEPPER, James Franklin, (1898-1963)
POWELL, John Wesley, (1834-1902)
RANSOME, Frederick Leslie, (1868-1935)
REED, Lyman Coleman, (1900-1970)
RICHARDS, James Taylor "Dick", (1892-1953)
RICHARDSON, George Burr, (1872-1949)
RICKARD, Thomas Arthur, (1864-1953)
ROBERTS, Hugh Marine, (1885-1965)
RUSSOM, Vaughn Walter, (1891-1955)
SCOTT, William Berryman, (1858-1947)
SEARS, Julian Ducker, (1891-1970)
SERVISS, Frederick LeVerne, (1895-1954)
SHAW, Everett Simeon, (1885-1955)
SNIVELEY, H. Norman, (1879-1959)
STANTON, Timothy William, (1860-1953)
STERNBERG, Charles Hazelius, (1850-1943)
ST. JOHN, Orestes Hawley, (1841-1921)
STRANGWAY, David William, (1934-)
SUTTON, Arle Herbert, (1896-1972)
THOMAS, George Dewey, (1898-1975)
THOMPSON, Warren O., (1898-1968)
VAN TUYL, Francis M., (1887-1975)
WALTMAN, William Dewitt, (1875-1955)
WASHBURNE, Chester Wesley, (1883-1971)
WEEKS, Herbert John, (1897-1960)
WILLIAMS, George O. "Skipper", (1897-1967)

UNITED STATES OF AMERICA (COLORADO) 1900 to present Cont'd.

WINES, Donald Bradford, (1902-1958)
WORCESTER, Philip George, (1884-1970)

UNITED STATES OF AMERICA
CONNECTICUT
1800-1849
AGASSIZ, Louis Jean Rodolphe, (1807-1873)
BARRATT, Joseph, (1796-1882)
DANA, James Dwight, (1813-1895)
DEANE, James, (1801-1858)
FINCH, John, (fl.1823-1835)
HITCHCOCK, Edward, (1793-1864)
PERCIVAL, James Gates, (1795-1856)
POURTALÈS, Louis François de, (1823-1880)
SILLIMAN, Benjamin, Sr., (1779-1864)

1850-1899
ABEL, Othenio, (1875-1946)
AGASSIZ, Louis Jean Rodolphe, (1807-1873)
BARRATT, Joseph, (1796-1882)
BARRELL, Joseph, (1869-1919)
BAUR, Georg Hermann Carl Ludwig, (1859-1893)
BEECHER, Charles Emerson, (1856-1904)
BREWER, William Henry, (1828-1910)
BRUSH, George Jarvis, (1831-1912)
CHAPIN, James Henry, (1832-1892)
DANA, Edward Salisbury, (1849-1935)
DANA, James Dwight, (1813-1895)
DEANE, James, (1801-1858)
EAKLE, Arthur Starr, (1862-1931)
EMERSON, Benjamin Kendall, (1843-1932)
FOOTE, Harry Ward, (1875-1942)
FORD, William Ebenezer, (1878-1939)
GIBB, Hugh, (1860-1932)
HARGER, Oscar, (1843-1882)
HEILPRIN, Angelo, (1853-1907)
HITCHCOCK, Edward, (1793-1864)
IRVING, John Duer, (1874-1918)
LULL, Richard Swann, (1867-1957)
MARSH, Othniel Charles, (1831-1899)
PENFIELD, Samuel Lewis, (1856-1906)
PERCIVAL, James Gates, (1795-1856)
PIRSSON, Louis Valentine, (1860-1919)
POURTALÈS, Louis François de, (1823-1880)
RICE, William North, (1845-1928)
SCHUCHERT, Charles, (1858-1942)
SILLIMAN, Benjamin, Sr., (1779-1864)
VERRILL, Addison Emery, (1839-1926)
WARREN, Charles Hyde, (1876-1950)
WELLS, Horace Lemuel, (1855-1924)
WILLIAMS, Henry Shaler, (1847-1918)

1900 to present
ABEL, Othenio, (1875-1946)

UNITED STATES OF AMERICA (CONNECTICUT) 1900 to present Cont'd.

BANNERMAN, Harold MacColl, (1897-1976)
BARRELL, Joseph, (1869-1919)
BATEMAN, Alan Mara, (1889-1971)
BEECHER, Charles Emerson, (1856-1904)
BLACKWELDER, Eliot, (1880-1969)
BREWER, William Henry, (1828-1910)
BRUSH, George Jarvis, (1831-1912)
DANA, Edward Salisbury, (1849-1935)
EAKLE, Arthur Starr, (1862-1931)
EDINGER, Tilly, (1897-1967)
EMERSON, Benjamin Kendall, (1843-1932)
FLINT, Richard Foster, (1902-1976)
FOOTE, Harry Ward, (1875-1942)
FORD, William Ebenezer, (1878-1939)
FRITTS, Crawford Ellsworth, (1927-1972)
GIBB, Hugh, (1860-1932)
HEILPRIN, Angelo, (1853-1907)
IRVING, John Duer, (1874-1918)
KNOPF, Adolph, (1882-1966)
LOUGHLIN, Gerald Francis, (1880-1946)
LULL, Richard Swann, (1867-1957)
MATHER, Kirtley Fletcher, (1888-1978)
McKINSTRY, Hugh Exton, (1896-1961)
PEACOCK, Martin Alfred, (1898-1950)
PENFIELD, Samuel Lewis, (1856-1906)
PIRSSON, Louis Valentine, (1860-1919)
RICE, William North, (1845-1928)
SCHUCHERT, Charles, (1858-1942)
VERRILL, Addison Emery, (1839-1926)
WARING, Gerald Ashley, (1883-1971)
WARREN, Charles Hyde, (1876-1950)
WELLS, Horace Lemuel, (1855-1924)
WILLIAMS, Henry Shaler, (1847-1918)

UNITED STATES OF AMERICA
DELAWARE
Up to 1800
KALM, Pehr, (1716-1779)

1800-1849
EMMONS, Ebenezer, (1800-1863)

1850-1899
EMMONS, Ebenezer, (1800-1863)

1900 to present
RASMUSSEN, William Charles, (1917-1973)

UNITED STATES OF AMERICA
DISTRICT OF COLUMBIA
1850-1899
DAY, Arthur Louis, (1869-1960)
EAKLE, Arthur Starr, (1862-1931)
EGLESTON, Thomas, (1832-1900)

UNITED STATES OF AMERICA (DISTRICT OF COLUMBIA) 1850-1899 Cont'd.

LARSEN, Esper Signius Jr., (1879-1961)
RATHBURN, Richard, (1852-1918)
SCHUCHERT, Charles, (1858-1942)

1900 to present

BOWEN, Norman Levi, (1887-1956)
DAY, Arthur Louis, (1869-1960)
EAKLE, Arthur Starr, (1862-1931)
EGLESTON, Thomas, (1832-1900)
JONES, William Francis, (1885-1941)
LARSEN, Esper Signius Jr., (1879-1961)
MOREY, George Washington, (1888-1965)
RATHBURN, Richard, (1852-1918)
SCHAIRER, John Frank, (1904-1970)
SCHUCHERT, Charles, (1858-1942)
ZERFOSS, Samuel, (1912-1958)

UNITED STATES OF AMERICA

FLORIDA

Up to 1800

BARTRAM, William, (1739-1823)

1800-1849

BARTRAM, William, (1739-1823)
POURTALÈS, Louis François de, (1823-1880)

1850-1899

ABEL, Othenio, (1875-1946)
ALDRICH, Truman Heminway, (1848-1932)
BUTTGENBACH, Henri, (1874-1964)
DARTON, Nelson Horatio, (1865-1948)
GIDLEY, James Williams, (1866-1931)
MANSFIELD, Wendell Clay, (1874-1939)
POURTALÈS, Louis François de, (1823-1880)
SELLARDS, Elias Howard, (1875-1961)
SHALER, Nathaniel Southgate, (1841-1906)

1900 to present

ABEL, Othenio, (1875-1946)
ALDRICH, Truman Heminway, (1848-1932)
BELL, Olin Grogan, (1893-1970)
BREMER, B. E. "Boots", (1904-1974)
BURWELL, Edward Bouldin, Jr., (1894-1964)
BUTTGENBACH, Henri, (1874-1964)
CAMPBELL, Robert Burns, (1892-1955)
CARRITT, Dayton Ernest, (1915-)
COOKE, Charles Wythe, (1887-1971)
DARTON, Nelson Horatio, (1865-1948)
FARLEE, Richard Wilson, (1913-1961)
GEBELEIN, Conrad Dennis, (1945-1978)
GIDLEY, James Williams, (1866-1931)
GUNTER, Herman, (1885-1972)
GUT, H. James, (1904-1969)
HOWELL, Benjamin Franklin, (1890-1976)

UNITED STATES OF AMERICA (FLORIDA) 1900 to present Cont'd.

JEFFREYS, Geoffrey, (1885-1953)
JORDAN, Louise, (1908-1966)
LAW, John Melvin, (1914-1967)
MANSFIELD, Wendell Clay, (1874-1939)
McDONALD, Stanley Matthews, (1915-1950)
RAINWATER, Edward Harriman "Rainey", (1909-1972)
SELLARDS, Elias Howard, (1875-1961)
SHALER, Nathaniel Southgate, (1841-1906)
SHIDELER, William Henry, (1886-1958)
SMITH, Robert Hendee "Smokey", (1916-1946)
THOMAS, George Dewey, (1898-1975)
WHEELER, Orby Clinton, (1898-1970)
WIER, Harvey James, Jr., (1924-1950)
WILSON, Roy Arthur, (1891-1970)
WINCHESTER, John Widner, (1929-)
WINTER, Claud Victor, Jr., (1925-1968)

UNITED STATES OF AMERICA
GEORGIA
1800-1849
LE CONTE, Joseph, (1823-1901)

1850-1899
BEACHLER, Charles S., (1870-1894)
BUTTS, Charles, (1863-1946)
LE CONTE, Joseph, (1823-1901)
LOUGHRIDGE, Robert Hills, (1843-1917)
McCALLIE, Samuel Washington, (1856-)
SPENCER, Joseph William Winthrop, (1851-1921)
STEPHENSON, Lloyd William, (1876-1962)
WATSON, Thomas Leonard, (1871-1924)
YEATES, William Smith, (1856-1908)

1900 to present
BUTTS, Charles, (1863-1946)
CRICKMAY, Geoffrey W., (1905-1971)
EARGLE, Dolan Hoye, (1905-1973)
FURCRON, Aurelius Sydney, (1899-1971)
HOYT, John Harger, (1928-1970)
HULL, Joseph Poyer Deyo, (1889-1967)
JOHNSTON, William Drumm, Jr., (1899-1972)
LE CONTE, Joseph, (1823-1901)
LILLIGREN, Johnathan Martin, (1880-1956)
LOUGHRIDGE, Robert Hills, (1843-1917)
McCALLIE, Samuel Washington, (1856-)
PEYTON, Garland, (1892-1964)
PRETTYMAN, Thomas Mann, (1888-1940)
SHEARER, Harold Kurtz, (1889-1946)
SPENCER, Joseph William Winthrop, (1851-1921)
STEPHENSON, Lloyd William, (1876-1962)
WATSON, Thomas Leonard, (1871-1924)
WHITCOMB, Bruce, (1900-1950)
WILSON, Roy Arthur, (1891-1970)
YEATES, William Smith, (1856-1908)

UNITED STATES OF AMERICA
HAWAII
1800-1849
GREEN, William Lowthian, (1819-1890)
NUTTALL, Thomas, (1786-1859)
PARKER, Samuel, Reverend, (1779-1866)

1850-1899
ALEXANDER, Annie Montague, (1867-1950)
BARTON, George Hunt, (1852-1933)
DUTTON, Clarence Edward, Major, (1841-1912)
EAKLE, Arthur Starr, (1862-1931)
GREEN, William Lowthian, (1819-1890)
GREGORY, Herbert E., (1869-1952)
GULICK, John Thomas, (1832-1923)
HITCHCOCK, Charles Henry, (1836-1919)
JAGGAR, Thomas Augustus, (1871-1953)
LIBBEY, William, (1855-1927)
MEINZER, Oscar Edward, (1876-1948)
NUTTALL, Thomas, (1786-1859)
PARKER, Samuel, Reverend, (1779-1866)
PILSBRY, Henry Augustus, (1862-1957)

1900 to present
ALEXANDER, Annie Montague, (1867-1950)
BARTON, George Hunt, (1852-1933)
COCKBURN, Alexander Murray, (1901-1959)
DUTTON, Clarence Edward, Major, (1841-1912)
EAKLE, Arthur Starr, (1862-1931)
GREGORY, Herbert E., (1869-1952)
GULICK, John Thomas, (1832-1923)
HITCHCOCK, Charles Henry, (1836-1919)
JAGGAR, Thomas Augustus, (1871-1953)
JOHNSON, Charles George, (1914-1969)
MEINZER, Oscar Edward, (1876-1948)
PALMER, Harold Schjoeth, (1890-1959)
PILSBRY, Henry Augustus, (1862-1957)
SWARTZ, Joel Howard, (1893-1971)
WENTWORTH, Chester Keeler, (1891-1969)

UNITED STATES OF AMERICA
IDAHO
1850-1899
ELDRIDGE, George Homans, (1854-1905)
MARSH, Othniel Charles, (1831-1899)
RANSOME, Frederick Leslie, (1868-1935)
REINHART, Herman Francis, (1832-1889) [PROSPECTORS, DIVINERS AND MINING ENGINEERS]
ST. JOHN, Orestes Hawley, (1841-1921)
STONE, Ralph Walter, (1876-1964)

1900 to present
ANDERSON, Alfred Leonard, (1900-1964)
BECKWITH, Radcliffe Harold, (1900-1964)
CAMPBELL, Arthur Byron, (1924-1971)
CONDIT, Daniel Dale, (1886-1955)

UNITED STATES OF AMERICA (IDAHO) 1900 to present Cont'd.

ELDRIDGE, George Homans, (1854-1905)
ERDMANN, Charles Edgar, (1897-1971)
GILLES, Verner Arthur, (1886-1954)
GILLSON, Joseph Lincoln, (1895-1964)
GRIMM, Kenneth Edgar, (1912-1961)
HEWETT, Donnel Foster, (1881-1971)
JUDSON, Wilber, (1880-1951)
LILLIGREN, Johnathan Martin, (1880-1956)
McCONNEL, Roger H., (1908-1971)
MORAN, Robert Breck, (1879-1961)
MULLEN, John Timothy, Jr., (1906-1959)
RANSOME, Frederick Leslie, (1868-1935)
ST. JOHN, Orestes Hawley, (1841-1921)
STONE, Ralph Walter, (1876-1964)
WILSON, James W., (?1925-1964)
WILSON, Roy Arthur, (1891-1970)
YOUNGQUIST, Walter, (-)

UNITED STATES OF AMERICA
ILLINOIS
Up to 1800
BRADBURY, John, (1765-1823)

1800-1849
BRADBURY, John, (1765-1823)
BROADHEAD, Garland Carr, (1827-1912)
LESQUEREUX, Charles Leo, (1806-1889)
MARCY, Oliver, (1820-1899)
MEEK, Fielding Bradford, (1817-1876)
NORWOOD, Joseph Granville, (1807-1895)
OWEN, David Dale, (1807-1860)
WHITNEY, Josiah Dwight, (1819-1896)
WORTHEN, Amos Henry, (1812-1888)

1850-1899
BAKER, Frank Collins, (1867-1942)
BANNISTER, Henry Martyn, (-)
BASTIN, Edson Sunderland, (1878-1953)
BAUR, Georg Hermann Carl Ludwig, (1859-1893)
BROADHEAD, Garland Carr, (1827-1912)
CHAMBERLIN, Thomas Chrowder, (1843-1928)
CROOK, Alja Robinson, (1864-1930)
ENGELMANN, August Heinrich, (1831-1899)
FARRINGTON, Oliver Cummings, (1864-1933)
GORDON, Charles Henry, (1857-1934)
GURLEY, William Frank Eugene, (1854-1943)
HAY, Oliver Perry, (1846-1930)
HERRICK, Clarence Luther, (1858-1904)
HUBBARD, George David, (1871-1958)
IDDINGS, Joseph Paxson, (1857-1920)
JOHANNSEN, Albert, (1871-1962)
JONES, J. Claude, (1877-1932)
JORDAN, David Starr, (1851-1931)
KEYES, Charles Rollin, (1864-1942)

UNITED STATES OF AMERICA (ILLINOIS) 1850-1899 Cont'd.

LESQUEREUX, Charles Leo, (1806-1889)
LEVERETT, Frank, (1859-1943)
MARCY, Oliver, (1820-1899)
MEEK, Fielding Bradford, (1817-1876)
MILLER, Paul C., (1876-1945)
MILLER, Samuel A., (1837-1897)
MORSE, William Clifford, (1874-1962)
NOÉ VON ARCHENEGG, Adolph Carl, (1873-1939)
NORWOOD, Joseph Granville, (1807-1895)
OWEN, David Dale, (1807-1860)
RIGGS, Elmer Samuel, (1869-1963)
SALISBURY, Rollin D., (1858-1922)
UDDEN, Johan August, (1859-1932)
WELLER, Stuart, (1870-1927)
WHITNEY, Josiah Dwight, (1819-1896)
WILLISTON, Samuel Wendell, (1852-1918)
WORTHEN, Amos Henry, (1812-1888)

1900 to present

BAKER, Frank Collins, (1867-1942)
BALDWIN, William Felbert Jackson, Jr., (1927-1972)
BALL, John Rice, (1881-1953)
BALLARD, William Norval, (1906-1976)
BARTH, Thomas Frederik Weiby, (1899-1971)
BASTIN, Edson Sunderland, (1878-1953)
BELL, Alfred Hannam, (1895-1977)
BELL, John Arnold, (1939-1970)
BIEBER, Charles Leonard, (1901-1965)
BLACKWELDER, Eliot, (1880-1969)
BOWEN, Norman Levi, (1887-1956)
BRADLEY, William Frank, (1908-1973)
BRANSON, Carl Colton, (1906-1975)
BRANSON, Edwin Robert, (1908-1971)
BROWN, John Albert, (1901-1954)
BYRON, Clifford Winkler, (1901-1970)
CADY, Gilbert Haven, (1882-1970)
CARMAN, Joel Ernest, (1882-1966)
CARROLL, Don Llewellyn, (1902-1954)
CHAMBERLIN, Thomas Chrowder, (1843-1928)
CHARLES, Homer Herbert, (1900-1963)
CLARK, Leslie Madison, (1898-1976)
COCHRAN, Kenneth L., (1910-1973)
COOPER, Chalmer Lewis, (-)
COX, Guy Henry, (1881-1922)
CRONEIS, Carey, (1901-1972)
CROOK, Alja Robinson, (1864-1930)
CULVER, Harold Eugene, (1883-1970)
CURRIER, Louis Wade, (1890-1970)
DAWSON, Joseph Meadows, (1895-1946)
DAWSON, Thomas Albert, (1916-1974)
DeWOLF, Frank Walbridge, (1881-1957)
DRAKE, Cecil, (1899-1949)
EKBLAW, George Elbert, (1895-1972)
ELDER, Stanley Gordon, (1904-1957)

UNITED STATES OF AMERICA (ILLINOIS) 1900 to present Cont'd.

EMMONS, William Harvey, (1876-1948)
FARLEE, Richard Wilson, (1913-1961)
FARRINGTON, Oliver Cummings, (1864-1933)
GOODSON, Joseph Arthur, (1910-1946)
GORDON, Charles Henry, (1857-1934)
GRIMES, Russell Winters, (1900-1969)
GROUT, Frank Fitch, (1880-1958)
GURLEY, William Frank Eugene, (1854-1943)
HAY, Oliver Perry, (1846-1930)
HERALD, Frank Edward, (1887-1965)
HERRICK, Clarence Luther, (1858-1904)
HIDE, Raymond, (1929-)
HOOVER, James Barr, (1902-1953)
HORBERG, Carl Leland, (1910-1955)
HORTON, Harold Merritt "Wog", (1897-1966)
HOSKINS, John Hobart, (1896-1957)
HOWELL, Jesse V., (1891-1971)
HUBBARD, George David, (1871-1958)
IDDINGS, Joseph Paxson, (1857-1920)
JEWETT, John Mark, (1896-1970)
JOHANNSEN, Albert, (1871-1962)
JONES, J. Claude, (1877-1932)
JONES, Verner E., (1905-1977)
JORDAN, David Starr, (1851-1931)
JUST, Theodor Karl, (1904-1959)
KAY, Fred Hall, (1885-1943)
KEYES, Charles Rollin, (1864-1942)
KIMBALL, Kent K., (1892-1953)
KING, Robert Bruce, (1923-1977)
KLINE, Virginia Harriett, (1910-1959)
KOSCHMANN, Albert Herbert, (1896-1962)
KRIDLER, George Matthew, (1889-1953)
KROENLEIN, George Alfred, (1898-1940)
KURTZ, Robert Gourdin, (1898-1958)
LEIGHTON, Morris Morgan, (1887-1971)
LESTER, John Lawrence, (1914-1977)
LEVERETT, Frank, (1859-1943)
MASON, Arnold Caverly, (1906-1961)
McFARLAND, Leland R., (1904-1973)
McINTYRE, Paul Joseph, (1893-1963)
McNEELY, Robert Barnes, (1902-1966)
MEHL, Maurice Goldsmith, (1887-1966)
MILLER, Paul C., (1876-1945)
MITCHELL, James Garland, (1908-1972)
MOORE, Paul Brian, (1940-)
MORSE, William Clifford, (1874-1962)
MYLIUS, Louis Aubrey, (1890-1965)
NEWHOUSE, Walter Harry, (1897-1969)
NOË VON ARCHENEGG, Adolph Carl, (1873-1939)
PABST, Marie B., (1909-1963)
PAINE, William R., (-)
PLUMMER, Helen Jeanne (née Skewes), (1891-1951)
POOR, Russell Spurgeon, (1899-1972)
POTEET, James Harold, (1896-1965)

UNITED STATES OF AMERICA (ILLINOIS) 1900 to present Cont'd.

QUIRKE, Terence Thomas, (1886-1947)
RICH, John Lyon, (1884-1956)
RITTENHOUSE, Gordon, (1910-1974)
ROARK, Louis, (1890-1950)
ROBINSON, Ernest Guy, (1898-1973)
ROMER, Alfred Sherwood, (1894-1973)
ROSSBY, Carl-Gustaf Arvid, (1898-1957)
RUBEY, William Walden, (1898-1974)
RUNNER, Joseph James, (1885-1970)
SALISBURY, Rollin D., (1858-1922)
SCHOLL, Louis August, Jr., (1887-1953)
SCHYTT, Valter, (1919-)
SHEPARD, Francis P., (-)
SHERRILL, Richard Ellis, (1899-1953)
SIMMONS, Kenneth Albert, (1906-1971)
STEHR, Raymond Albert, (1903-1967)
STOCKDALE, Paris Buell, (1896-1962)
SUTTON, Arle Herbert, (1896-1972)
SWANN, David H., (1915-1968)
TETRICK, Paul Roderick, (1920-1956)
THOMAS, George Dewey, (1898-1975)
THOMAS, Harry George, (1929-1976)
UDDEN, Johan August, (1859-1932)
WADELL, Hakon, (1895-1962)
WANLESS, Harold Rollin, (1898-1970)
WASSON, Theron, (1887-1970)
WELLER, James Marvin, (1899-1976)
WELLER, Stuart, (1870-1927)
WHITAKER, Harvey, (1906-1964)
WHITE, George Willard, (1903-)
WILLIAMS, George O. "Skipper", (1897-1967)
WILLISTON, Samuel Wendell, (1852-1918)
WINGETT, Raymond Oscar, (1909-1953)

UNITED STATES OF AMERICA
INDIANA
Up to 1800
THOMAS, David, (1776-1859)

1800-1849
COX, Edward Travers, (1821-1907)
OWEN, David Dale, (1807-1860)
THOMAS, David, (1776-1859)

1850-1899
ASHLEY, George Hall, (1866-1951)
BEEDE, Joshua William, (1871-1940)
BRANNER, John Casper, (1850-1922)
COX, Edward Travers, (1821-1907)
JORDAN, David Starr, (1851-1931)
KINDLE, Edward Martin, (1869-1940)
MENDENHALL, Thomas Corwin, (1841-1924)
MILLER, Samuel A., (1837-1897)
OWEN, David Dale, (1807-1860)

UNITED STATES OF AMERICA (INDIANA) 1850-1899 Cont'd.

PURDUE, Albert Homer, (1861-1917)
THOMAS, David, (1776-1859)

1900 to present
ADDINGTON, Arch Rombough, (1894-1972)
ASHLEY, George Hall, (1866-1951)
BEEDE, Joshua William, (1871-1940)
BIEBER, Charles Leonard, (1901-1965)
BRANNER, John Casper, (1850-1922)
BROWN, John Albert, (1901-1954)
CORYELL, Horace Noble, (1888-1965)
COX, Edward Travers, (1821-1907)
DAWSON, Joseph Meadows, (1895-1946)
DAWSON, Thomas Albert, (1916-1974)
DEISS, Charles Frederick, (1903-1959)
ESAREY, Ralph Emerson, (1901-1975)
EVANS, Vincent, (1907-1951)
FREED, Richard, (1908-1967)
GALLOWAY, Jesse James, (1882-1962)
GOODING, Ansel Miller, (1924-1976)
HAKE, Benjamin Franklin, (1895-1973)
HERALD, Frank Edward, (1887-1965)
HONESS, Charles William, (1885-1949)
JORDAN, David Starr, (1851-1931)
KINDLE, Edward Martin, (1869-1940)
KURTZ, Robert Gourdin, (1898-1958)
MALOTT, Clyde A., (-1950)
McCLOY, Walter Lowrie, Jr., (1912-1954)
McGUINNESS, Charles Lee, (1914-1971)
MENDENHALL, Thomas Corwin, (1841-1924)
MORGAN, George Dillon, (1894-1950)
MYLIUS, Louis Aubrey, (1890-1965)
NOWLAN, Harry Hackett, (1888-1958)
O'BRIEN, Francis Eugene, (1916-1968)
PERRY, Thomas Gregory, (1919-1972)
PURDUE, Albert Homer, (1861-1917)
ROARK, Louis, (1890-1950)
SERVISS, Frederick LeVerne, (1895-1954)
SHROCK, Robert Rakes, (1904-)
SMITH, Ned Myron, (1923-1976)
STOCKDALE, Paris Buell, (1896-1962)
SWIGER, Rual Bower, (1897-1950)
THOMAS, George Dewey, (1898-1975)
WENTWORTH, Chester Keeler, (1891-1969)
WILLIAMS, George O. "Skipper", (1897-1967)
WINGETT, Raymond Oscar, (1909-1953)

UNITED STATES OF AMERICA
IOWA
1800-1849
MEEK, Fielding Bradford, (1817-1876)
NORWOOD, Joseph Granville, (1807-1895)
OWEN, David Dale, (1807-1860)
OWEN, Richard Dale, (1810-1890)

UNITED STATES OF AMERICA (IOWA) 1800-1849 Cont'd.

WACHSMUTH, Charles, (1829-1896)
WHITE, Charles Abiathar, (1826-1910)
WHITNEY, Josiah Dwight, (1819-1896)
WORTHEN, Amos Henry, (1812-1888)

1850-1899

BEYER, Samuel Walker, (1865-1931)
CALVIN, Samuel, (1840-1911)
GORDON, Charles Henry, (1857-1934)
HINRICHS, Gustavus Detlef, (-1922)
KEYES, Charles Rollin, (1864-1942)
LEONARD, Arthur Gray, (1865-1932)
LEVERETT, Frank, (1859-1943)
MACBRIDE, Thomas Huston, (1848-1934)
MEEK, Fielding Bradford, (1817-1876)
NORTON, William Harmon, (-)
NORWOOD, Joseph Granville, (1807-1895)
OWEN, David Dale, (1807-1860)
OWEN, Richard Dale, (1810-1890)
SHIMEK, Bohumil, (1861-1937)
THOMAS, Abram Owen, (1876-1931)
UDDEN, Johan August, (1859-1932)
WACHSMUTH, Charles, (1829-1896)
WHITE, Charles Abiathar, (1826-1910)
WHITNEY, Josiah Dwight, (1819-1896)
WILDER, Frank Alonzo, (1870-1930)
WILLIAMS, Ira Abraham, (1876-)
WORTHEN, Amos Henry, (1812-1888)

1900 to present

BEYER, Samuel Walker, (1865-1931)
CALVIN, Samuel, (1840-1911)
CARMAN, Joel Ernest, (1882-1966)
CHANEY, Ralph Works, (1890-1971)
CLINE, Lewis Manning, (1909-1971)
FOLGER, James Anthony, (1896-1956)
GALPIN, Sidney Longman, (1886-1962)
GEORGE, William Owsley, (1892-1966)
GORDON, Charles Henry, (1857-1934)
GWYNNE, Charles Sumner, (1885-1972)
HARRELL, Marshall Allen, (1900-1959)
HINRICHS, Gustavus Detlef, (-1922)
KEYES, Charles Rollin, (1864-1942)
LEONARD, Arthur Gray, (1865-1932)
LEVERETT, Frank, (1859-1943)
LONSDALE, John Tipton, (1895-1960)
MACBRIDE, Thomas Huston, (1848-1934)
NORTON, William Harmon, (-)
PENNEL, Harry David, (1907-1971)
RUNNER, Joseph James, (1885-1970)
SHIMEK, Bohumil, (1861-1937)
STAINBROOK, Merrill Addison, (1897-1956)
THOMAS, Abram Owen, (1876-1931)
THOMAS, Leo A., (1917-1965)

UNITED STATES OF AMERICA (IOWA) 1900 to present Cont'd.

UDDEN, Johan August, (1859-1932)
WHITE, Charles Abiathar, (1826-1910)
WILDER, Frank Alonzo, (1870-1930)
WILLIAMS, Ira Abraham, (1876-)
WILSON, Leonard Richard, (1906-)

UNITED STATES OF AMERICA
KANSAS
1800-1849
HAWN, Frederick, (1810-1898)
HAYDEN, Ferdinand Vandiveer, (1829-1887)
MUDGE, Benjamin Franklin, (1817-1879)
SWALLOW, George Clinton, (1817-1899)

1850-1899
BEEDE, Joshua William, (1871-1940)
CRIDER, Albert Foster, (1873-1945)
HAWN, Frederick, (1810-1898)
HAWORTH, Erasmus, (1855-1932)
HAY, Robert, (1835-1895)
HAYDEN, Ferdinand Vandiveer, (1829-1887)
LEVERETT, Frank, (1859-1943)
MARSH, Othniel Charles, (1831-1899)
MARTIN, Handel Tongue, (1862-1931)
MORSE, William Clifford, (1874-1962)
MUDGE, Benjamin Franklin, (1817-1879)
PROSSER, Charles Smith, (1860-1916)
SHALER, Nathaniel Southgate, (1841-1906)
SNOW, Francis Huntington, (1840-1908)
STERNBERG, Charles Hazelius, (1850-1943)
ST. JOHN, Orestes Hawley, (1841-1921)
SWALLOW, George Clinton, (1817-1899)
TODD, James Edward, (1846-1922)
TWENHOFEL, William Henry, (1875-1957)
VALERIUS, Claude N., (-)
WAY, Herbert W. L., (-) [PROSPECTORS, DIVINERS AND MINING ENGINEERS]
WILLISTON, Samuel Wendell, (1852-1918)

1900 to present
ALLEN, Donald Meredith, (1899-1971)
BEEDE, Joshua William, (1871-1940)
BEEKLEY, Albert Leon, (1883-1952)
BENSON, Dale L., (1899-1968)
BERG, Ernest Lyle, (1910-1953)
BLANPIED, Bernerd William, (1897-1975)
BOYD, Harold Edward, (1887-1952)
BRADLEY, Everett L., (1897-1969)
BRISCOE, Glenn Oliver, (1898-1974)
BROWN, Alton Roe, (1930-1975)
BROWN, John Albert, (1901-1954)
CARPENTER, Everett, (1884-1968)
CASTER, Ermil Leslie, (1896-1955)
CHARLES, Homer Herbert, (1900-1963)

UNITED STATES OF AMERICA (KANSAS) 1900 to present Cont'd.

CHENEY, George Monroe, (1893-1952)
CLARK, Leslie Madison, (1898-1976)
CLARK, Robert Watson, (1884-1948)
CLARKE, Carl W., (1896-1928)
CLIFTON, Roland Leroy, (1884-1962)
CLINKSCALES, Albert Sidney, (1894-1969)
COCHRAN, Kenneth L., (1910-1973)
COFFIN, Reuben Clare, (1886-1972)
COPASS, Jack Malcolm, (1906-1963)
CRIDER, Albert Foster, (1873-1945)
CRUM, Harry Edwin, (1892-1955)
DANIELS, James Ira, (1896-1967)
DRESSER, Myron Amos, (1894-1973)
DUTY, Richard Emerson, (1908-1966)
ELDER, Stanley Gordon, (1904-1957)
ELLEDGE, Emmett Russell, (1898-1960)
ELLISOR, Alva Christine, (1892-1964)
FATH, Arthur Earl, (1887-1976)
FITZGERALD, James, Jr., (1903-1951)
FOLGER, James Anthony, (1896-1956)
FOWLER, George Malcolm, (1885-1972)
FRALICH, Charles Eugene, (1902-1964)
GARRETT, Robert Ernest, (1888-1947)
GILL, James Rogers, (1922-1972)
GILLERMAN, Elliot "Bud", (1913-1974)
GOODELL, Horace N. "Hank", (1915-1971)
GORDON, Dugald, (1886-1958)
HANDLEY, Howard Walter, (1905-1940)
HAWKINS, Glenn Dewayne, (1897-1970)
HAWORTH, Erasmus, (1855-1932)
HERALD, Frank Edward, (1887-1965)
HESSE, Curtis Julian, (1905-1945)
HIBBARD, Claude W., (1905-1973)
HIGGS, Morton Thomas, (1896-1956)
HILL, Robert Luce, (1937-1964)
HONESS, Charles William, (1885-1949)
HOOVER, James Barr, (1902-1953)
HORNER, Seward Ellis, (1906-1954)
HOWELL, Jesse V., (1891-1971)
HOWELL, William Frederick, (1895-1965)
HUFFMAN, Edward Archibald, (1912-1954)
HUGHES, Richard, (1889-1969)
HUNT, Edwin Harold, (1897-1949)
HYDE, Clarence Edgar, (1894-1954)
IRELAND, Hubert Andrew, (1904-1973)
ISENBERGER, Nathan Perry, (1896-1956)
JEWETT, John Mark, (1896-1970)
JONES, Daniel J., (1889-1965)
JONES, Ogden Sherman, (1891-1952)
KEPPLER, Leo George, (1890-1971)
KIMBALL, Kent K., (1892-1953)
KINKEL, John Franklin, (1899-1936)
KNIGHT, Garold Louis "Jerry", (1894-1968)
KREPPS, William J., (1924-1976)

UNITED STATES OF AMERICA (KANSAS) 1900 to present Cont'd.

KRIDLER, George Matthew, (1889-1953)
KROENLEIN, George Alfred, (1898-1940)
LARSH, Walter W., (1893-1951)
LEE, Marvin, (1889-1953)
LEVERETT, Frank, (1859-1943)
LILIENBORG, Burton Armand, (1900-1949)
LILLIGREN, Johnathan Martin, (1880-1956)
MARTIN, Francis Ivan "Shorty", (1896-1952)
MARTIN, Handel Tongue, (1862-1931)
MASON, Shirley Lowell, (1893-1975)
MAUCINI, Joseph John, (1893-1955)
McCLOY, Walter Lowrie, Jr., (1912-1954)
McFERRON, George Irwin, (1895-1969)
McNEELY, Robert Barnes, (1902-1966)
McNEIL, Harold E., (1910-1965)
McNUTT, Vachel Harry, (1888-1936)
MILNER, Charles Albert, Jr., (1900-1971)
MOORE, James Augustus, (1910-1955)
MOORE, Raymond Cecil, (1892-1974)
MORSE, William Clifford, (1874-1962)
MYLIUS, Louis Aubrey, (1890-1965)
NEDOM, Henry Arthur, (1895-1940)
NININGER, Harvey Harlow, (-)
NISBET, John McHenry, (1894-1962)
NORVILLE, Glen Smith, (1899-1963)
OLTMAN, Roy E., (-1977)
OWENS, Frith Cravens, (1898-1977)
PEMBERTON, John Roy "Bill", (1884-1968)
PHILBRICK, Eugene Patrick, (1897-1951)
POOR, Russell Spurgeon, (1899-1972)
POTTER, Grover Cleveland, (1892-1975)
POWELL, Ralph Sterling, (1901-1951)
PRICE, Arthur Sebert, (1904-1948)
PROSSER, Charles Smith, (1860-1916)
RICHARDS, James Taylor "Dick", (1892-1953)
RIFE, Byron, (1892-1974)
RUBEY, William Walden, (1898-1974)
RUSSOM, Vaughn Walter, (1891-1955)
SCHOEWE, Walter Henry, (1891-1966)
SCHOLL, Guy Joseph, (1900-1968)
SCHOLL, Louis August, Jr., (1887-1953)
SCOTT, Horace Lafayette, (1901-1976)
SHALER, Nathaniel Southgate, (1841-1906)
SNIDER, Luther Crocker, (1882-1947)
SNOW, Francis Huntington, (1840-1908)
STERNBERG, Charles Hazelius, (1850-1943)
STERNBERG, Charles M., (1885-)
STERNBERG, George Fryer, (1883-1969)
STEWART, Irvine E., (1893-1965)
STIRTON, Ruben Arthur, (1901-1966)
ST. JOHN, Orestes Hawley, (1841-1921)
SYLVESTER-BRADLEY, Peter Colley, (1914?-1978)
TAYLOR, Garvin Lawrence, (1908-1974)
TAYLOR, W. Harlan, (1908-1972)

UNITED STATES OF AMERICA (KANSAS) 1900 to present Cont'd.

TODD, James Edward, (1846-1922)
TURNER, Joseph Harold, (1900-1950)
TWENHOFEL, William Henry, (1875-1957)
VERNON, Jesse, (1894-1972)
VER WIEBE, Walter August, (1887-1961)
WALKER, Lucian Hickman, (1894-1970)
WHEELER, Orby Clinton, (1898-1970)
WILLIAMS, George O. "Skipper", (1897-1967)
WILLISTON, Samuel Wendell, (1852-1918)
WILSON, James W., (?1925-1964)
WIMBISH, Forrest E., (1900-1972)
WINES, Donald Bradford, (1902-1958)
YEAGER, Lloyd Irvin, (1900-1938)
YEE, Bing Q., (1924-1973)

UNITED STATES OF AMERICA
KENTUCKY
Up to 1800
THOMAS, David, (1776-1859)

1800-1849
LESQUEREUX, Charles Leo, (1806-1889)
OWEN, David Dale, (1807-1860)
THOMAS, David, (1776-1859)

1850-1899
BUTTS, Charles, (1863-1946)
CALL, Richard Ellsworth, (1856-1917)
GLENN, Leonidas Chalmers, (1871-1951)
GOULD, Charles Newton, (1868-1949)
HUBBARD, George David, (1871-1958)
LESQUEREUX, Charles Leo, (1806-1889)
LOUGHRIDGE, Robert Hills, (1843-1917)
MORSE, William Clifford, (1874-1962)
NETTELROTH, Henry, (1835-1887)
NORWOOD, Charles Joseph, (1853-1927)
OWEN, David Dale, (1807-1860)
RIES, Heinrich, (1871-1951)
SHALER, Nathaniel Southgate, (1841-1906)
SMITH, William Richard Tangier, (1869-1972)
STONE, Ralph Walter, (1876-1964)
THOMAS, David, (1776-1859)
WELLER, Stuart, (1870-1927)
WHITE, Charles David, (1862-1935)

1900 to present
BOYD, Harold Edward, (1887-1952)
BRANSON, Carl Colton, (1906-1975)
BRANSON, Edwin Robert, (1908-1971)
BUTTS, Charles, (1863-1946)
CALL, Richard Ellsworth, (1856-1917)
CLARK, Leslie Madison, (1898-1976)
COCHRAN, Kenneth L., (1910-1973)
COOPER, Chalmer Lewis, (-)

UNITED STATES OF AMERICA (KENTUCKY) 1900 to present Cont'd.

CRUM, Harry Edwin, (1892-1955)
CURRIER, Louis Wade, (1890-1970)
DENNEN, William Henry, (1920-)
EVANS, Vincent, (1907-1951)
FOHS, Ferdinand Julius, (1884-1965)
GARDNER, James Henry, (1883-1964)
GLENN, Leonidas Chalmers, (1871-1951)
GOULD, Charles Newton, (1868-1949)
HAKE, Benjamin Franklin, (1895-1973)
HERALD, Frank Edward, (1887-1965)
HONESS, Charles William, (1885-1949)
HOWELL, Benjamin Franklin, (1890-1976)
HUBBARD, George David, (1871-1958)
JILLSON, Willard Rouse, (1890-)
JOHNSON, Charles George, (1914-1969)
KURTZ, Robert Gourdin, (1898-1958)
LILLIGREN, Johnathan Martin, (1880-1956)
LOUGHRIDGE, Robert Hills, (1843-1917)
MASON, Arnold Caverly, (1906-1961)
MAYFIELD, Samuel Martin, (1888?-1963)
McCLOY, Walter Lowrie, Jr., (1912-1954)
McINTYRE, Paul Joseph, (1893-1963)
MORSE, William Clifford, (1874-1962)
MYERS, Thurman H., (1890-1940)
MYLIUS, Louis Aubrey, (1890-1965)
NORWOOD, Charles Joseph, (1853-1927)
PEMBERTON, John Roy "Bill", (1884-1968)
PETERSON, Clarence J., (1885-1948)
POTEET, James Harold, (1896-1965)
RICHARDS, James Taylor "Dick", (1892-1953)
RIES, Heinrich, (1871-1951)
ROWLEY, Alden Bruce, (1888-1956)
RUSSELL, Philip G., (1891-1971)
SHALER, Nathaniel Southgate, (1841-1906)
SIMMONS, Kenneth Albert, (1906-1971)
SMITH, William Richard Tangier, (1869-1972)
STOCKDALE, Paris Buell, (1896-1962)
STONE, Ralph Walter, (1876-1964)
SUTTON, Arle Herbert, (1896-1972)
SWANN, David H., (1915-1968)
THOMAS, George Dewey, (1898-1975)
WELLER, James Marvin, (1899-1976)
WELLER, Stuart, (1870-1927)
WESLEY, George Rutherford, (1905-1965)
WHITE, Charles David, (1862-1935)
WHITEHEAD, Robert Brooks, (1889-1936)
WILDER, Newell Morris "Newt", (1908-1946)
WILLIAMS, George O. "Skipper", (1897-1967)
WOOD, Virgil O., (1892-1954)
WOODRUFF, John Grant, (1898-1971)

UNITED STATES OF AMERICA
LOUISIANA
1800-1849
DUMONT, André Hubert, (1809-1857)
RIDDELL, John Leonard, (1807-1865)

1850-1899
BURCHARD, Ernest Norman, (1875-1961)
CRIDER, Albert Foster, (1873-1945)
DUMONT, André Hubert, (1809-1857)
HAGER, Lee, (1874-1944)
HARRIS, Gilbert Dennison, (1864-1952)
HILGARD, Eugene Woldemar, (1833-1916)
LAWSON, Andrew Cowper, (1861-1952)
MORSE, William Clifford, (1874-1962)
RIDDELL, John Leonard, (1807-1865)
SNYDER, John Young, (1872-1939)
VALERIUS, Claude N., (-)
VEATCH, Arthur Clifford, (1878-1938)
WADE, Arthur, (1878-1951)

1900 to present
ADAMS, John Emery, (1899-1971)
AIMER, James David, (1899-1973)
ANDRAU, Evert Willem Karel, (1900-1951)
BALDWIN, William Felbert Jackson, Jr., (1927-1972)
BANKS, Thomas Russell, (1900-1948)
BARTELL, Lawrence Denny, (1902-1955)
BERNARD, Hugh Allen, (1915-1975)
BEVIER, George Means, (1888-1972)
BLANKENSHIP, Joseph Croxton, (1930-1964)
BLANPIED, Bernerd William, (1897-1975)
BLOOMER, Philip A., Jr., (1922-1967)
BOGARDUS, Robert Stephen "Bud", (1914-1953)
BRACE, Orval L., (1893-1975)
BREMER, B. E. "Boots", (1904-1974)
BRISCOE, Glenn Oliver, (1898-1974)
BROWN, Alton Roe, (1930-1975)
BURCHARD, Ernest Norman, (1875-1961)
BURRESS, Walter March, (1893-1950)
BYRON, Clifford Winkler, (1901-1970)
CAMPBELL, Robert Burns, (1892-1955)
CANADA, William Richards, (1909-1962)
CASTER, Ermil Leslie, (1896-1955)
CHENEY, George Monroe, (1893-1952)
CHISHOLM, William Francis, (1886-1944)
CLARK, Chester Charles, (1885-1953)
CONWAY, William Palen "Tony", Jr., (1914-1944)
CORTES, Henry Cornelius, (1892-1957)
COTTINGHAM, Virgil Elisha, (1891-1958)
CRIDER, Albert Foster, (1873-1945)

UNITED STATES OF AMERICA (LOUISIANA) 1900 to present Cont'd.

DACEY, Reginald Lynn, (1908-1951)
DAVIE, Thomas Early, (1906-1959)
DeRUDDER, Ronald D., (1932-1975)
DeWOLF, Frank Walbridge, (1881-1957)
DUCOTE, James Wardy, (1919-1964)
DURHAM, Charles Albert, (1910-1954)
DUTY, Richard Emerson, (1908-1966)
EARL, Eugene Leonard "Red", (1906-1956)
EBY, James Brian, (1896-)
FAHLE, Paul Bartlett, (1910-1954)
FERGUSON, John Logan, (1893-1973)
FERRERO, Clarence de Labretonne, (1903-1954)
FINCH, Elmer Harrison, (1883-1953)
FISK, Harold Norman, (1908-1964)
FRAZIER, David Edward, (1931-1976)
FREED, Richard, (1908-1967)
FREY, M. Gordon, (1913-)
FUELLHART, Donald Egbert, (1899-1945)
GOERNER, Frederic Stephen, (1915-1969)
GOLLNICK, Robert Louis, (1918-1963)
GORDON, Dugald, (1886-1958)
GOYER, Sidney Carroll, (1934-1965)
HAGER, Lee, (1874-1944)
HAMMILL, Chester A., (1889-1941)
HARRIS, Gilbert Dennison, (1864-1952)
HAZZARD, Roy Thorpe, (1893-1976)
HILGARD, Eugene Woldemar, (1833-1916)
HOOVER, James Barr, (1902-1953)
HOPPER, Walter Everett, (1886-1971)
HOSKINS, Baker, Jr., (1900-1941)
HOUSTON, Samuel Henry, (1905-1947)
HOYLE, Charles Robert, Sr., (1891-1954)
HUBMAN, Carl Wesley, (1905-1947)
HUTSON, Ezekiel Burney, (1897-1961)
HYDE, Clarence Edgar, (1894-1954)
IRWIN, Marvin Leroy "Mel", (1915-1964)
ISENBERGER, Nathan Perry, (1896-1956)
JENNINGS, Philip H., (1899-1974)
KENNEDY, William, (-1926)
KISLING, James William, Jr., (1898-1973)
KLINGER, Edgar Dewitt, (1899-1973)
KNOX, Thomas Knight, (1896-1949)
KREIDLER, William Lynn, (1911-1975)
KRIDLER, George Matthew, (1889-1953)
LAW, John Melvin, (1914-1967)
LAWSON, Andrew Cowper, (1861-1952)
LOPEZ, Abdon Joseph, (1924-1972)
MACDIARMID, Roy Angus, (1933-1966)
McCAULEY, Dan Wesley, (1916-1967)
McDONALD, Stanley Matthews, (1915-1950)
McDOUGALL, John Edward, (1914-1964)
McFARLAND, Leland R., (1904-1973)
McLAREN, Robert Lee, (1901-1953)
McMINN, Paul Meloy, (1919-1974)

UNITED STATES OF AMERICA (LOUISIANA) 1900 to present Cont'd.

MILLER, Willard L., (1896-1973)
MILLER, William F., (1929-1976)
MOORE, James Augustus, (1910-1955)
MORRIS, Henry Newton, Jr., (1920-1952)
MORSE, William Clifford, (1874-1962)
NEIGHBOURS, Glen Reed, (1917-1961)
NISBET, John McHenry, (1894-1962)
NOBLE, Earl Bart, (1894-1967)
OLCUTT, David Perry, (1896-1958)
OWENS, Frith Cravens, (1898-1977)
PAINE, William R., (-)
PENNEL, Harry David, (1907-1971)
PERRINE, Irving, (1884-1955)
PETTY, Olive Scott, (1895-)
PHILLIPS, Eldridge Douglas, (1894-1931)
PICKLESIMER, Harold Greeley, (1916-1952)
PORCH, Edwin Leight, Jr., (1889-1948)
POTTER, Grover Cleveland, (1892-1975)
RAINWATER, Edward Harriman "Rainey", (1909-1972)
RAMSEY, Thomas Yancey, (1910-1968)
REICHERT, Stanley Orville, (1912-1970)
RENICK, Brink Coleman, (1897-1971)
RICE, Elmer Michael, (1900-1971)
ROBINSON, Ernst Guy, (1898-1973)
RUKAS, Justin Matthew, (1910-1963)
RUSSELL, Richard Joel, (1895-1971)
RYAN, Russell Fossler, (1897-1935)
SAPPINGTON, Chester, (1905-1968)
SCHILLING, Karl Howard, (1894-1945)
SCHMIDT, Karl A. "Gus", (1900-1965)
SCHNEIDER, George William, (1899-1963)
SCHWABENLAND, James Richard, (1926-1968)
SEASHORE, Paul Theodore, (1898-1954)
SHEARER, Harold Kurtz, (1889-1946)
SMITH, Robert Hendee "Smokey", (1916-1946)
SNYDER, John Young, (1872-1939)
SPOFFORD, Howard Nelson, (1881-1950)
STEIG, Maynard Herman, (1904-1965)
SUMAN, John Robert, (1890-1972)
TAYLOR, W. Harlan, (1908-1972)
THIES, Roland K., (1905-1973)
THOMAS, George Dewey, (1898-1975)
THOMPSON, Robert Raymond, (1893-1954)
THORSEN, Carl Peter, (1927-1964)
VEATCH, Arthur Clifford, (1878-1938)
WADE, Arthur, (1878-1951)
WAPPLER, Ed Dickinson, (1908-1937)
WASSON, Theron, (1887-1970)
WHITTIER, William Harrison, (1889-1951)
WIER, Harvey James, Jr., (1924-1950)
WILBERT, Louis Joseph "Slick", Jr., (1919-1953)
WILSON, Malcolm Earl, Sr., (1885-1960)
WINGETT, Raymond Oscar, (1909-1953)
ZAVOICO, Basil B., (1899-1975)

UNITED STATES OF AMERICA
MAINE
Up to 1800
JOSSELYN, John, (ca.1608-1675)

1800-1849
JACKSON, Charles Thomas, (1805-1880)
WHITE, Charles Abiathar, (1826-1910)

1850-1899
BASTIN, Edson Sunderland, (1878-1953)
CHADWICK, George Halcott, (1876-1953)
COLONY, Roy Jed, (1870-1936)
HITCHCOCK, Charles Henry, (1836-1919)
JACKSON, Charles Thomas, (1805-1880)
WHITE, Charles Abiathar, (1826-1910)

1900 to present
BASTIN, Edson Sunderland, (1878-1953)
CHADWICK, George Halcott, (1876-1953)
COLONY, Roy Jed, (1870-1936)
DENNEN, William Henry, (1920-)
HITCHCOCK, Charles Henry, (1836-1919)
JOHNSON, Bertrand Leroy, (1882-1962)
WHITE, Charles Abiathar, (1826-1910)
WING, Monta Eldo, (1897-1976)

UNITED STATES OF AMERICA
MARYLAND
Up to 1800
HUTCHINS, Thomas, (1730-1789)

1800-1849
DUCATEL, Julius Timoleon, (1796-1849)

1850-1899
ABBE, Cleveland, Jr., (1872-1934)
CLARK, William Bullock, (1860-1917)
GIDLEY, James Williams, (1866-1931)
GLENN, Leonidas Chalmers, (1871-1951)
HAYES, Charles Willard, (1858-1916)
JOHANNSEN, Albert, (1871-1962)
KEYES, Charles Rollin, (1864-1942)
MATHEWS, Edward Bennett, (1869-1944)
MATTHEWS, Edward Bennett, (1869-1944)
MILLER, Benjamin Le Roy, (1874-1944)
PROSSER, Charles Smith, (1860-1916)
RIES, Heinrich, (1871-1951)
ROWE, Richard Burton, (1872-1902)
SHATTUCK, George Burbank, (1869-)
STEPHENSON, Lloyd William, (1876-1962)
TWITCHELL, Mayville William, (1866-1927)
WILLIAMS, George Huntington, (1856-1894)

1900 to present
ABBE, Cleveland, Jr., (1872-1934)

UNITED STATES OF AMERICA (MARYLAND) 1900 to present Cont'd.

CARRITT, Dayton Ernest, (1915-)
CLARK, William Bullock, (1860-1917)
CLOOS, Ernst, (1898-1974)
COOKE, Charles Wythe, (1887-1971)
DORSEY, George Edwin, (1892-1953)
GARDNER, Julia Anna, (1882-1960)
GIDLEY, James Williams, (1866-1931)
GLENN, Leonidas Chalmers, (1871-1951)
GOLDMAN, Marcus Isaac, (1881-1965)
HAYES, Charles Willard, (1858-1916)
HODGE, Matthew Burnop, (-1951)
HULL, Joseph Poyer Deyo, (1889-1967)
HUNT, Walter Frederick, (1882-1975)
JOHANNSEN, Albert, (1871-1962)
KEYES, Charles Rollin, (1864-1942)
MATHEWS, Edward Bennett, (1869-1944)
MATTHEWS, Edward Bennett, (1869-1944)
MILLER, Benjamin Le Roy, (1874-1944)
PROSSER, Charles Smith, (1860-1916)
RASMUSSEN, William Charles, (1917-1973)
RIES, Heinrich, (1871-1951)
ROWE, Richard Burton, (1872-1902)
SHATTUCK, George Burbank, (1869-)
SINGEWALD, Joseph Theophilus, Jr., (1884-1963)
STEPHENSON, Lloyd William, (1876-1962)
SWARTZ, Joel Howard, (1893-1971)
TRAINER, David Woolsey, Jr., (1899-1965)
TURNER, Joseph Harold, (1900-1950)
TWITCHELL, Mayville William, (1866-1927)

UNITED STATES OF AMERICA
MASSACHUSETTS
Up to 1800
JOSSELYN, John, (ca.1608-1675)
WARREN, John Collins, (1778-1856)
WATERHOUSE, Benjamin, (1753-1846)

1800-1849
AGASSIZ, Louis Jean Rodolphe, (1807-1873)
BOUVÉ, Thomas Tracy, (1815-1896)
DEANE, James, (1801-1858)
DEWEY, Chester, (1784-1867)
GUYOT, Arnold Henri, (1807-1884)
HAGEN, Hermann August, (1817-1893)
HITCHCOCK, Edward, (1793-1864)
ROGERS, William Barton, (1804-1882)
WARREN, John Collins, (1778-1856)
WATERHOUSE, Benjamin, (1753-1846)
WHITNEY, Josiah Dwight, (1819-1896)

1850-1899
AGASSIZ, Louis Jean Rodolphe, (1807-1873)
BARTON, George Hunt, (1852-1933)
BELL, James Macintosh, (1877-1934)

UNITED STATES OF AMERICA (MASSACHUSETTS) 1850-1899 Cont'd.

BERKEY, Charles Peter, (1867-1955)
BOUVÉ, Thomas Tracy, (1815-1896)
CHADWICK, George Halcott, (1876-1953)
CLELAND, Herdman Fitzgerald, (1869-1935)
CROSBY, William Otis, (1850-1925)
DALY, Reginald Aldworth, (1871-1957)
DEANE, James, (1801-1858)
DEWEY, Chester, (1784-1867)
EMERSON, Benjamin Kendall, (1843-1932)
GUYOT, Arnold Henri, (1807-1884)
HAGEN, Hermann August, (1817-1893)
HITCHCOCK, Edward, (1793-1864)
HYATT, Alpheus, (1838-1902)
JAGGAR, Thomas Augustus, (1871-1953)
JOHNSON, Douglas Wilson, (1878-1944)
LARSEN, Esper Signius Jr., (1879-1961)
LINDGREN, Waldemar, (1860-1939)
LOOMIS, Frederic Brewster, (1873-1937)
LULL, Richard Swann, (1867-1957)
NILES, William Harmon, (1838-1910)
RICKARD, Thomas Arthur, (1864-1953)
ROGERS, William Barton, (1804-1882)
SHALER, Nathaniel Southgate, (1841-1906)
SHEPARD, Edward Martin, (1854-1934)
SHIMER, Hervey Woodburn, (1872-1965)
UPHAM, Warren, (1850-1934)
VERRILL, Addison Emery, (1839-1926)
WADSWORTH, Marshman Edward, (1847-1921)
WARREN, Charles Hyde, (1876-1950)
WARREN, John Collins, (1778-1856)
WHITNEY, Josiah Dwight, (1819-1896)
WILLIAMS, John Francis, (1862-1891)
WOLFF, John Elict, (1857-1940)

1900 to present

AHRENS, Louis Herman, (1918-)
BARBOUR, Thomas, (1884-1946)
BARTH, Thomas Frederik Weiby, (1899-1971)
BARTON, George Hunt, (1852-1933)
BELL, James Macintosh, (1877-1934)
BERKEY, Charles Peter, (1867-1955)
BITTER, Francis, (1902-1967)
BJAREBY, Alfred Gunnar, (1899-1967)
BLACKWELDER, Eliot, (1880-1969)
BOUCOT, Arthur James, (1924-)
BOWEN, Norman Levi, (1887-1956)
BRACE, William Francis, (1926-)
BUERGER, Martin Julian, (-)
CANTWELL, Thomas, Jr., (1927-)
CARRITT, Dayton Ernest, (1915-)
CHADWICK, George Halcott, (1876-1953)
CLELAND, Herdman Fitzgerald, (1869-1935)
CRONEIS, Carey, (1901-1972)
CROSBY, William Otis, (1850-1925)

UNITED STATES OF AMERICA (MASSACHUSETTS) 1900 to present Cont'd.

CURRIER, Louis Wade, (1890-1970)
DALY, Reginald Aldworth, (1871-1957)
DEAN, Lee Wallace, III, (1932-)
DENNEN, William Henry, (1920-)
DOELL, Richard Rayman, (1923-)
DOUGLAS, George Vibert, (1892?-1958)
EMERSON, Benjamin Kendall, (1843-1932)
FAIRBAIRN, Harold Williams, (1906-)
FIELD, Richard Montgomery, (1885-1961)
FIOCCO, Giorgio, (1931-)
FRASER, Horace John "Dutch", (1905-1969)
GANGI, Anthony Frank, (1929-)
GILLSON, Joseph Lincoln, (1895-1964)
GOLDTHWAIT, James Walter, (1880-1947)
GRAHAM, John Warren, (1918-1971)
GRATON, Louis Caryl, (1880-1970)
HIDE, Raymond, (1929-)
HUGHES, Harry, (1929-)
HUTCHINS, Forbes M., (-)
HYATT, Alpheus, (1838-1902)
JAGGAR, Thomas Augustus, (1871-1953)
JOHNSON, Douglas Wilson, (1878-1944)
JONES, William Francis, (1885-1941)
LAHEE, Frederic Henry, (1884-1968)
LARSEN, Esper Signius Jr., (1879-1961)
LINDGREN, Waldemar, (1860-1939)
LOOMIS, Frederic Brewster, (1873-1937)
LULL, Richard Swann, (1867-1957)
LUTH, William Clair, (1934-)
MACDONALD, Gordon James Fraser, (1929-)
MADDEN, Theodore Richard, (1925-)
MENCHER, Ely, (1913-)
MILLER, John Preston, (1923-1961)
MILLER, William John, (1880-1965)
MORRIS, Frederick Kuhne, (1885-1962)
NEWHOUSE, Walter Harry, (1897-1969)
NILES, William Harmon, (1838-1910)
PALACHE, Charles, (1869-1954)
PARKS, Roland Dane, (1900-1972)
PEACOCK, Martin Alfred, (1898-1950)
PINSON, William Hamet, Jr., (1919-)
PRESS, Frank, (1924-)
RICKARD, Thomas Arthur, (1864-1953)
ROMER, Alfred Sherwood, (1894-1973)
ROSSBY, Carl-Gustaf Arvid, (1898-1957)
SHALER, Nathaniel Southgate, (1841-1906)
SHEPARD, Edward Martin, (1854-1934)
SHIMER, Hervey Woodburn, (1872-1965)
SHORT, Maxwell Naylor, (1889-1952)
SHROCK, Robert Rakes, (1904-)
SIMMONS, Marvin Gene, (1929-)
SIMPSON, Stephen Milton, Jr., (1929-)
SLICHTER, Louis Byrne, (1896-)
SPARKS, Dale Darrell Dolsun, (1894-1938)

UNITED STATES OF AMERICA (MASSACHUSETTS) 1900 to present Cont'd.

STRANGWAY, David William, (1934-)
TOKSÖZ, M. Nafi, (1934-)
UPHAM, Warren, (1850-1934)
VERRILL, Addison Emery, (1839-1926)
VON ARX, William Stelling, (1916-)
WADSWORTH, Marshman Edward, (1847-1921)
WARREN, Charles Hyde, (1876-1950)
WHITEHEAD, Walter Lucius, (1891-1969)
WILSON, Leonard Richard, (1906-)
WINCHESTER, John Widner, (1929-)
WOLFF, John Elict, (1857-1940)

Time unknown
WOLFE, Caleb Wroe, (-)

UNITED STATES OF AMERICA
MICHIGAN
Up to 1800
DARBY, William, (1775-1854)

1800-1849
BURT, William Austin, (1792-1858)
DARBY, William, (1775-1854)
HOUGHTON, Douglass, (1809-1845)
HUBBARD, Bela, (1814-1896)
NUTTALL, Thomas, (1786-1859)
ROMINGER, Carl Ludwig, (1820-1907)
WINCHELL, Alexander, (1824-1891)

1850-1899
BAUERMAN, Hilary, (1833-1909)
BOWLES, Oliver, (1877-1958)
BROOKS, Alfred Hulse, (1871-1924)
BROOKS, Thomas Benton, (1836-1900)
BURT, William Austin, (1792-1858)
CASE, Ermine Cowles, (1871-1953)
DARBY, William, (1775-1854)
GORDON, Charles Henry, (1857-1934)
HOBBS, William Herbert, (1864-1953)
HUBBARD, Bela, (1814-1896)
HUBBARD, Lucius Lee, (1849-1933)
KRAUS, Edward Henry, (1875-1973)
LANE, Alfred Church, (1863-1948)
NUTTALL, Thomas, (1786-1859)
PUMPELLY, Raphael, (1837-1923)
RIES, Heinrich, (1871-1951)
ROMINGER, Carl Ludwig, (1820-1907)
RUSSELL, Israel Cook, (1852-1906)
VAN HISE, Charles Richard, (1857-1918)
WADSWORTH, Marshman Edward, (1847-1921)
WINCHELL, Alexander, (1824-1891)
WRIGHT, Charles E., (1843-1888)
WRIGHT, Frederick Eugene, (1877-1953)

UNITED STATES OF AMERICA (MICHIGAN) Cont'd.

1900 to present

BALLARD, William Norval, (1906-1976)
BARRETT, Leslie Park, (1887-1972)
BAUERMAN, Hilary, (1833-1909)
BOWLES, Oliver, (1877-1958)
BOYD, Harold Edward, (1887-1952)
BRANSON, Edwin Robert, (1908-1971)
BRODERICK, Thomas Monteith, (1889-1965)
BROOKS, Alfred Hulse, (1871-1924)
BROWN, John Albert, (1901-1954)
BURTON, George Edgar, (1880-1953)
BUWALDA, John Peter, (1886-1954)
CARR, William L., (1919-1975)
CASE, Ermine Cowles, (1871-1953)
CHAPMAN, Sydney, (1888-1970)
CLARK, Leslie Madison, (1898-1976)
CLARK, Robert Watson, (1884-1948)
COOK, Charles Wilford, (1882-1933)
DAWSON, Joseph Meadows, (1895-1946)
DENNING, Reynolds McConnell, (1916-1967)
EARDLEY, Armand John, (1901-1972)
FRITTS, Crawford Ellsworth, (1927-1972)
GOODSON, Joseph Arthur, (1910-1946)
GORDON, Charles Henry, (1857-1934)
GRIMES, Russell Winters, (1900-1969)
HAKE, Benjamin Franklin, (1895-1973)
HIBBARD, Claude W., (1905-1973)
HOBBS, William Herbert, (1864-1953)
HUBBARD, Lucius Lee, (1849-1933)
HUNT, Walter Frederick, (1882-1975)
HUSSEY, Russell Claudius, (1888-1970)
KING, Robert Bruce, (1923-1977)
KRAUS, Edward Henry, (1875-1973)
KRIDLER, George Matthew, (1889-1953)
KURTZ, Robert Gourdin, (1898-1958)
LANE, Alfred Church, (1863-1948)
LESTER, John Lawrence, (1914-1977)
MAEBIUS, Jed B., (1907-1969)
MANDARINO, Joseph Anthony, (-)
McCLOY, Walter Lowrie, Jr., (1912-1954)
McGUINNESS, Charles Lee, (1914-1971)
McLAUGHIN, Dean Benjamin, (1901-1965)
MEAD, Warren Judson, (1883-1960)
MOSELEY, John Reed, (1900-1959)
NININGER, Harvey Harlow, (-)
PARKS, Roland Dane, (1900-1972)
PECK, Arthur Becker, (1892-1943)
PUMPELLY, Raphael, (1837-1923)
RAMSDELL, Lewis Stephen, (1895-1975)
RIES, Heinrich, (1871-1951)
ROMINGER, Carl Ludwig, (1820-1907)
RUNNER, Joseph James, (1885-1970)
RUSSELL, Israel Cook, (1852-1906)
SCHILLING, Karl Howard, (1894-1945)

UNITED STATES OF AMERICA (MICHIGAN) 1900 to present Cont'd.

SCOFIELD, Lloyd M., (1897-1969)
SLAWSON, Chester Baker, (1898-1964)
STEWART, Duncan, (1905-1969)
STUART, Duncan, VII, (1905-1969)
STUMM, Erwin Charles, (1908-1969)
TAGUE, Glenn Charles, (1905-1964)
VAN HISE, Charles Richard, (1857-1918)
VER WIEBE, Walter August, (1887-1961)
WADSWORTH, Marshman Edward, (1847-1921)
WASSON, Theron, (1887-1970)
WILKERSON, Albert Samuel, (1897-1963)
WINCHESTER, John Widner, (1929-)
WINGETT, Raymond Oscar, (1909-1953)
WRIGHT, Frederick Eugene, (1877-1953)
WRIGHT, "Put", (-1962)
ZAVOICO, Basil B., (1899-1975)

UNITED STATES OF AMERICA
MINNESOTA
1800-1849
MEEK, Fielding Bradford, (1817-1876)
NORWOOD, Joseph Granville, (1807-1895)
OWEN, David Dale, (1807-1860)
OWEN, Richard Dale, (1810-1890)

1850-1899
BERKEY, Charles Peter, (1867-1955)
BOWLES, Oliver, (1877-1958)
GRANT, Ulysses Sherman, (1867-1932)
HALL, Christopher Webber, (1845-1911)
HERRICK, Clarence Luther, (1858-1904)
LAWSON, Andrew Cowper, (1861-1952)
LONGYEAR, Edmund Joseph, (1864-1954) [PROSPECTORS, DIVINERS AND MINING ENGINEERS]
MEEK, Fielding Bradford, (1817-1876)
NORWOOD, Joseph Granville, (1807-1895)
OWEN, David Dale, (1807-1860)
OWEN, Richard Dale, (1810-1890)
PARSONS, Arthur Leonard, (1873-1957)
SARDESON, Frederick William, (1866-1959)
SCHUCHERT, Charles, (1858-1942)
UPHAM, Warren, (1850-1934)
WILLARD, Daniel Everett, (1862-1947)
WINCHELL, Horace Vaughn, (1865-1923)
WINCHELL, Newton Horace, (1829-1914)

1900 to present
BERKEY, Charles Peter, (1867-1955)
BOWLES, Oliver, (1877-1958)
BRODERICK, Thomas Monteith, (1889-1965)
EDWARDS, Merwin Guy, (1889-1969)
EMMONS, William Harvey, (1876-1948)
GILLSON, Joseph Lincoln, (1895-1964)
GORDON, Dugald, (1886-1958)

UNITED STATES OF AMERICA (MINNESOTA) 1900 to present Cont'd.

GRANT, Ulysses Sherman, (1867-1932)
GROUT, Frank Fitch, (1880-1958)
GWYNNE, Charles Sumner, (1885-1972)
HALL, Christopher Webber, (1845-1911)
HERRICK, Clarence Luther, (1858-1904)
HEWITT, Robert Leigh, (1911-1956)
HODGE, Edwin Thomas, (1887-1970)
HOPPER, Walter Everett, (1886-1971)
KIHLSTEDT, Folke H., (1901?-1956)
LAWSON, Andrew Cowper, (1861-1952)
LONGYEAR, Edmund Joseph, (1864-1954) [PROSPECTORS, DIVINERS AND MINING ENGINEERS]
PARSONS, Arthur Leonard, (1873-1957)
QUIRKE, Terence Thomas, (1886-1947)
ROBERTS, Hugh Marine, (1885-1965)
SARDESON, Frederick William, (1866-1959)
SCHUCHERT, Charles, (1858-1942)
STUART, Duncan, VII, (1905-1969)
THIEL, Edward, (1928-1962)
UPHAM, Warren, (1850-1934)
WILLARD, Daniel Everett, (1862-1947)
WINCHELL, Horace Vaughn, (1865-1923)
WINCHELL, Newton Horace, (1829-1914)

UNITED STATES OF AMERICA
MISSISSIPPI
1800-1849
BROADHEAD, Garland Carr, (1827-1912)
LINCECUM, Gideon, (1793-1874)
MILLINGTON, John, (1779-1886)
SCHOOLCRAFT, Henry Rowe, (1793-1864)
WAILES, Benjamin Leonard Couington, (1797-1862)

1850-1899
ALDRICH, Truman Heminway, (1848-1932)
BROADHEAD, Garland Carr, (1827-1912)
BURCHARD, Ernest Norman, (1875-1961)
CRIDER, Albert Foster, (1873-1945)
HARPER, Lewis, (-)
HILGARD, Eugene Woldemar, (1833-1916)
LIEBER, Oscar M., (-1862)
LINCECUM, Gideon, (1793-1874)
LITTLE, George, (1838-1924)
MILLINGTON, John, (1779-1886)
MORSE, William Clifford, (1874-1962)
SCHOOLCRAFT, Henry Rowe, (1793-1864)
SHIMEK, Bohumil, (1861-1937)
STEPHENSON, Lloyd William, (1876-1962)
WAILES, Benjamin Leonard Couington, (1797-1862)
WELLER, Stuart, (1870-1927)

1900 to present
AIMER, James David, (1899-1973)
ALDRICH, Truman Heminway, (1848-1932)

UNITED STATES OF AMERICA (MISSISSIPPI) 1900 to present Cont'd.

BLANPIED, Bernerd William, (1897-1975)
BREMER, B. E. "Boots", (1904-1974)
BRISCOE, Glenn Oliver, (1898-1974)
BROWN, Bahngrell Walter "Bill", (1916-)
BROWN, Olin Teter, (1900-1962)
BURCHARD, Ernest Norman, (1875-1961)
BURRESS, Walter March, (1893-1950)
BYRON, Clifford Winkler, (1901-1970)
CLARK, Chester Charles, (1885-1953)
COOKE, Charles Wythe, (1887-1971)
COPASS, Jack Malcolm, (1906-1963)
CRIDER, Albert Foster, (1873-1945)
DACEY, Reginald Lynn, (1908-1951)
DAWSON, Joseph Meadows, (1895-1946)
DOUGLAS, John Gray, (1900-1974)
EARGLE, Dolan Hoye, (1905-1973)
FARLEE, Richard Wilson, (1913-1961)
FERGUSON, John Logan, (1893-1973)
FERRERO, Clarence de Labretonne, (1903-1954)
GOLLNICK, Robert Louis, (1918-1963)
HAZZARD, Roy Thorpe, (1893-1976)
HILGARD, Eugene Woldemar, (1833-1916)
HOUSTON, Samuel Henry, (1905-1947)
HUGHES, Richard John, Jr., (1915-1968)
JEFFREYS, Geoffrey, (1885-1953)
KLINE, Virginia Harriett, (1910-1959)
KREIDLER, William Lynn, (1911-1975)
LITTLE, George, (1838-1924)
LOPEZ, Abdon Joseph, (1924-1972)
McARTHUR, Donald, (1895-1960)
McDONALD, Stanley Matthews, (1915-1950)
McFARLAND, Leland R., (1904-1973)
MORSE, Paul Franklin, (1897-1929)
MORSE, William Clifford, (1874-1962)
NEDOM, Henry Arthur, (1895-1940)
RAINWATER, Edward Harriman "Rainey", (1909-1972)
RICE, Elmer Michael, (1900-1971)
ROARK, Louis, (1890-1950)
RUSSELL, Richard Joel, (1895-1971)
SCRUTON, Philip Challacombe, (1922-1962)
SEASHORE, Paul Theodore, (1898-1954)
SEVERY, Clarence Luther, (1885-1975)
SHEARER, Harold Kurtz, (1889-1946)
SHIMEK, Bohumil, (1861-1937)
SMITH, Neal Johnstone, (1912-1971)
SMITH, Robert Hendee "Smokey", (1916-1946)
SPOFFORD, Howard Nelson, (1881-1950)
STEPHENSON, Lloyd William, (1876-1962)
THIES, Roland K., (1905-1973)
THOMAS, George Dewey, (1898-1975)
TOLER, Henry N., (1901-1970)
VESTAL, Franklin Earl, (1884-1969)
WARING, Gerald Ashley, (1883-1971)
WELLER, Stuart, (1870-1927)

UNITED STATES OF AMERICA (MISSISSIPPI) 1900 to present Cont'd.

WIER, Harvey James, Jr., (1924-1950)
WILLIAMS, George O. "Skipper", (1897-1967)
WINTER, Claud Victor, Jr., (1925-1968)

UNITED STATES OF AMERICA
MISSOURI
1800-1849
BROADHEAD, Garland Carr, (1827-1912)
MEEK, Fielding Bradford, (1817-1876)
NORWOOD, Joseph Granville, (1807-1895)
SCHOOLCRAFT, Henry Rowe, (1793-1864)
SHUMARD, Benjamin Franklin, (1820-1869)
SWALLOW, George Clinton, (1817-1899)

1850-1899
BRANSON, Edwin Bayer, (1877-1950)
BROADHEAD, Garland Carr, (1827-1912)
BUCKLEY, Ernest Robertson, (1872-1912)
BUEHLER, Henry Andrew, (1876-1944)
BURCHARD, Ernest Norman, (1875-1961)
CALL, Richard Ellsworth, (1856-1917)
GIRTY, George Herbert, (1869-1939)
HINRICHS, Gustavus Detlef, (-1922)
HUBBARD, George David, (1871-1958)
HUTCHINSON, Lon Lewis, (1877-1947)
KEYES, Charles Rollin, (1864-1942)
LEONARD, Arthur Gray, (1865-1932)
LEVERETT, Frank, (1859-1943)
MEEK, Fielding Bradford, (1817-1876)
MILLER, Samuel A., (1837-1897)
MORSE, William Clifford, (1874-1962)
NORWOOD, Charles Joseph, (1853-1927)
NORWOOD, Joseph Granville, (1807-1895)
PUMPELLY, Raphael, (1837-1923)
SCHOOLCRAFT, Henry Rowe, (1793-1864)
SHATTUCK, George Burbank, (1869-)
SHEPARD, Edward Martin, (1854-1934)
SHUMARD, Benjamin Franklin, (1820-1869)
SPENCER, Joseph William Winthrop, (1851-1921)
SWALLOW, George Clinton, (1817-1899)
WELLER, Stuart, (1870-1927)

1900 to present
BALL, John Rice, (1881-1953)
BORN, Kendall Eugene, (1908-1947)
BRANSON, Edwin Bayer, (1877-1950)
BRANSON, Edwin Robert, (1908-1971)
BUCKLEY, Ernest Robertson, (1872-1912)
BUEHLER, Henry Andrew, (1876-1944)
BURCHARD, Ernest Norman, (1875-1961)
BURFOOT, James Dabney "Dan", Jr., (1896-1966)
BURWELL, Edward Bouldin, Jr., (1894-1964)
CALL, Richard Ellsworth, (1856-1917)
CLARK, Chester Charles, (1885-1953)

UNITED STATES OF AMERICA (MISSOURI) 1900 to present Cont'd.

COCHRAN, Kenneth L., (1910-1973)
COX, Guy Henry, (1881-1922)
CURRIER, Louis Wade, (1890-1970)
DAKE, Charles Laurence, (1883-1934)
DUTY, Richard Emerson, (1908-1966)
FOLGER, James Anthony, (1896-1956)
FOWLER, George Malcolm, (1885-1972)
FRIZZELL, Donald Leslie, (1906-1972)
GIRTY, George Herbert, (1869-1939)
GLEASON, Charles D., (1908-1935)
GOODING, Ansel Miller, (1924-1976)
GRAWE, Oliver Rudolph, (1901-1965)
GREENE, Frank Cook, (1886-1975)
GRIMES, Russell Winters, (1900-1969)
HINRICHS, Gustavus Detlef, (-1922)
HUBBARD, George David, (1871-1958)
HUFFMAN, Edward Archibald, (1912-1954)
HUTCHINSON, Lon Lewis, (1877-1947)
KEYES, Charles Rollin, (1864-1942)
KITSON, Howard Walde, (1883-1931)
KNIGHT, Garold Louis "Jerry", (1894-1968)
KOSCHMANN, Albert Herbert, (1896-1962)
LEE, Marvin, (1889-1953)
LEONARD, Arthur Gray, (1865-1932)
LEVERETT, Frank, (1859-1943)
LILLIGREN, Johnathan Martin, (1880-1956)
MACELWANE, James Bernard, Father, S.J., (1883-1956)
MEHL, Maurice Goldsmith, (1887-1966)
MILLER, William D., (1930-1972)
MORSE, William Clifford, (1874-1962)
NININGER, Harvey Harlow, (-)
NORWOOD, Charles Joseph, (1853-1927)
POND, Walter Franklin, (1885-1974)
PUMPELLY, Raphael, (1837-1923)
RIFE, Byron, (1892-1974)
SHATTUCK, George Burbank, (1869-)
SHEPARD, Edward Martin, (1854-1934)
SIMMONS, Kenneth Albert, (1906-1971)
SIMPSON, George Gaylord, (1902-)
SPENCER, Joseph William Winthrop, (1851-1921)
SPOFFORD, Howard Nelson, (1881-1950)
TARR, William Arthur, (1881-1939)
WELLER, Stuart, (1870-1927)
WILKERSON, Albert Samuel, (1897-1963)
WILLIAMS, George O. "Skipper", (1897-1967)
WILSON, Malcolm Earl, Sr., (1885-1960)

UNITED STATES OF AMERICA
MONTANA
1800-1849
HAYDEN, Ferdinand Vandiveer, (1829-1887)

1850-1899
AVERY, Charles Dwight, (1877-1947)

UNITED STATES OF AMERICA (MONTANA) 1850-1899 Cont'd.

BOWEN, Charles Franklin, (1871-1953)
CAMPBELL, Marius Robinson, (1858-1940)
GOULD, Charles Newton, (1868-1949)
HAYDEN, Ferdinand Vandiveer, (1829-1887)
LARSEN, Esper Signius Jr., (1879-1961)
LUPTON, Charles Thomas, (1878-1935)
MANSFIELD, Wendell Clay, (1874-1939)
McKIM, John Walter, (1877-1927)
PIRSSON, Louis Valentine, (1860-1919)
REINHART, Herman Francis, (1832-1889) [PROSPECTORS, DIVINERS AND MINING ENGINEERS]
SCOTT, William Berryman, (1858-1947)
STONE, Ralph Walter, (1876-1964)
WARREN, Charles Hyde, (1876-1950)
WILLARD, Daniel Everett, (1862-1947)
WINCHELL, Alexander Newton, (1874-1958)
WINCHELL, Horace Vaughn, (1865-1923)

1900 to present
ABERNATHY, George Elmer, (1888-1948)
ALLEN, Donald Meredith, (1899-1971)
AVERY, Charles Dwight, (1877-1947)
BEEKLEY, Albert Leon, (1883-1952)
BILLINGSLEY, Paul, (1887-1962)
BOWEN, Charles Franklin, (1871-1953)
BROWN, Bahngrell Walter "Bill", (1916-)
BUTCHER, Walter William, (1916-1966)
CAMPBELL, Arthur Byron, (1924-1971)
CAMPBELL, Marius Robinson, (1858-1940)
CARR, William L., (1919-1975)
CASEY, John Hestwood, (1923-1967)
CASSEL, Chester A. "Chet", (1897-1971)
CLARK, Leslie Madison, (1898-1976)
CONDIT, Daniel Dale, (1886-1955)
CONWAY, William Palen "Tony", Jr., (1914-1944)
CRAWFORD, David Johnston, (1899-1960)
DAVIS, Ralph Emerson, (1884-1968)
DEISS, Charles Frederick, (1903-1959)
DOBBIN, Carroll Edward "Cy", (1892-1967)
DONAVAN, Jack Hilder, (1921-1953)
ELLIOTT, Harold Charles, (1910-1972)
ERDMANN, Charles Edgar, (1897-1971)
ESPACH, Ralph Homeward, (1900-1956)
EVANS, Vincent, (1907-1951)
FIELD, Richard Montgomery, (1885-1961)
FOLGER, James Anthony, (1896-1956)
FOWLER, George Malcolm, (1885-1972)
FULTON, Loris John, (1906-1974)
GILL, James Rogers, (1922-1972)
GOLDMAN, Marcus Isaac, (1881-1965)
GOULD, Charles Newton, (1868-1949)
GRIMES, Russell Winters, (1900-1969)
HALL, Elwin B., (1891-1949)
HANSON, Ernest Albin, (1899-1973)

UNITED STATES OF AMERICA (MONTANA) 1900 to present Cont'd.

HAYES, Albert Orion, (1882-1965)
HENDRICKSON, Victor James, (1892-1953)
HORNER, Jack, (-)
HOWELL, Benjamin Franklin, (1890-1976)
HOWELL, William Frederick, (1895-1965)
HOWER, John, Jr., (1927-)
HUNT, Walter Frederick, (1882-1975)
JONES, Verner E., (1905-1977)
JUDSON, Wilber, (1880-1951)
KNOPF, Adolph, (1882-1966)
LARSEN, Esper Signius Jr., (1879-1961)
LARSH, Walter W., (1893-1951)
LEE, Marvin, (1889-1953)
LEWIS, Francis Leonard, Jr., (1921-1953)
LILLIGREN, Johnathan Martin, (1880-1956)
LUPTON, Charles Thomas, (1878-1935)
MANSFIELD, Wendell Clay, (1874-1939)
MARKHAM, Edmund Olis, (1893-1960)
MAWDSLEY, James Buckland, (1894-1964)
MAYFIELD, Samuel Martin, (1888?-1963)
McKIM, John Walter, (1877-1927)
McMANNIS, William J., (1924-1973)
MITCHELL, James Garland, (1908-1972)
MIX, Charles Alexander, (1898-1954)
MULLEN, John Timothy, Jr., (1906-1959)
MUNCK, Victor C. E. A. de, (1920-1956)
MURRAY, George H., Jr., (1928-1971)
NOWLAN, Harry Hackett, (1888-1958)
PECORA, William Thomas, (1913-1972)
PIRSSON, Louis Valentine, (1860-1919)
RENICK, Brink Coleman, (1897-1971)
RICHARDS, James Taylor "Dick", (1892-1953)
ROWLEY, Alden Bruce, (1888-1956)
SCOTT, Walter Winthrop, (1893-1939)
SCOTT, William Berryman, (1858-1947)
SEASHORE, Paul Theodore, (1898-1954)
SMITH, Aylwin L., (1898-1965)
STEBINGER, Eugene, (1883-1951)
STEWART, Irvine E., (1893-1965)
STONE, Ralph Walter, (1876-1964)
SULLINS, M. Joseph, (1930-1972)
TAYLOR, W. Harlan, (1908-1972)
TEMPLETON, Eugene Crittenden, (1889-1932)
THOMAS, Leo A., (1917-1965)
WARREN, Charles Hyde, (1876-1950)
WASHBURNE, Chester Wesley, (1883-1971)
WILLARD, Daniel Everett, (1862-1947)
WILSEY, James Arthur, Jr., (1921-1949)
WILSON, Roy Arthur, (1891-1970)
WINCHELL, Alexander Newton, (1874-1958)
WINCHELL, Horace Vaughn, (1865-1923)

UNITED STATES OF AMERICA
NEBRASKA
1800-1849
EVANS, John, (1812-1861)
HAYDEN, Ferdinand Vandiveer, (1829-1887)
MARCOU, Jules, (1824-1898)
OWEN, David Dale, (1807-1860)

1850-1899
ABEL, Othenio, (1875-1946)
BARBOUR, Erwin Hinckley, (1856-1947)
EVANS, John, (1812-1861)
GIDLEY, James Williams, (1866-1931)
GORDON, Charles Henry, (1857-1934)
HAYDEN, Ferdinand Vandiveer, (1829-1887)
MARCOU, Jules, (1824-1898)
MARSH, Othniel Charles, (1831-1899)
OWEN, David Dale, (1807-1860)
SHIMEK, Bohumil, (1861-1937)

1900 to present
ABEL, Othenio, (1875-1946)
BARBOUR, Erwin Hinckley, (1856-1947)
CADY, Richard Carlysle, (1908-1943)
COOK, Harold James, (1887-?1962)
COOK, Margaret C., (1899-1968)
DRAKE, Cecil, (1899-1949)
ELLIOTT, Harold Charles, (1910-1972)
FOLGER, James Anthony, (1896-1956)
GIDLEY, James Williams, (1866-1931)
GORDON, Charles Henry, (1857-1934)
HUFFMAN, Edward Archibald, (1912-1954)
KREIDLER, William Lynn, (1911-1975)
LADNER, Albert L., (1905-1975)
MEFFERD, Ralph L., (1909-1964)
NININGER, Harvey Harlow, (-)
OLTMAN, Roy E., (-1977)
OWENS, Frith Cravens, (1898-1977)
PEPPERBERG, Leon J., (1883-1937)
REED, Eugene C., (1901-1976)
RUSSOM, Vaughn Walter, (1891-1955)
SCHRAMM, Eck Frank, (1883-1967)
SHIMEK, Bohumil, (1861-1937)
STIRTON, Ruben Arthur, (1901-1966)

UNITED STATES OF AMERICA
NEVADA
1850-1899
ATTWOOD, George, (1845-1912)
BASTIN, Edson Sunderland, (1878-1953)
BOND, Marshall, (1867-1941) [PROSPECTORS, DIVINERS AND MINING ENGINEERS]
GILBERT, Grove Karl, (1843-1918)
HAMMOND, John Hays, (1855-1936)
JONES, J. Claude, (1877-1932)
KING, Clarence Rivers, (1842-1901)

UNITED STATES OF AMERICA (NEVADA) 1850-1899 Cont'd.

LOUDERBACK, George Davis, (1874-1957)
MACLAREN, James Malcolm, (1873-1935)
OCHSENIUS, Carl Christian, (1830-)
POWELL, John Wesley, (1834-1902)
RANSOME, Frederick Leslie, (1868-1935)
RICHTHOFEN, Paul Wilhelm Ferdinand von, Freiherr, (1833-1905)
SMITH, William Richard Tangier, (1869-1972)
WILLIAMS, Gardner Fred, (1842-1922)
WINCHELL, Horace Vaughn, (1865-1923)

1900 to present

ALBERT, Herman W., (1882-) [PROSPECTORS, DIVINERS AND MINING ENGINEERS]
ATTWOOD, George, (1845-1912)
BASTIN, Edson Sunderland, (1878-1953)
BOND, Marshall, (1867-1941) [PROSPECTORS, DIVINERS AND MINING ENGINEERS]
BOSS, Richard F., (1923-1976)
CARLTON, Dave Pierce, (1890-1976)
CROWN, Walter James, (1901-1951)
FOWLER, George Malcolm, (1885-1972)
GILBERT, Grove Karl, (1843-1918)
GILLES, Verner Arthur, (1886-1954)
GILLSON, Joseph Lincoln, (1895-1964)
GILLULY, James, (-)
GOLDMAN, Marcus Isaac, (1881-1965)
GRAWE, Oliver Rudolph, (1901-1965)
HAMMOND, John Hays, (1855-1936)
HARRELL, Marshall Allen, (1900-1959)
HENLEY, Arthur Sidney, (1879-1936)
HEROLD, Stanley Carrollton, (1883-1958)
HEWETT, Donnel Foster, (1881-1971)
HEWITT, Robert Leigh, (1911-1956)
HOWELL, Benjamin Franklin, (1890-1976)
HUDSON, Frank Samuel, (1890-1967)
HUNT, Walter Frederick, (1882-1975)
JOHNSTON, William Drumm, Jr., (1899-1972)
JONES, J. Claude, (1877-1932)
JUDSON, Wilber, (1880-1951)
KENNERLEY, John Brian, (1937-1976)
KING, Clarence Rivers, (1842-1901)
KINNEY, Harry Dill, (1886-1961)
KNOPF, Adolph, (1882-1966)
KNOWLES, Gerald Elgin, (1923-1976)
LILLIGREN, Johnathan Martin, (1880-1956)
LOUDERBACK, George Davis, (1874-1957)
MACDIARMID, Roy Angus, (1933-1966)
MACLAREN, James Malcolm, (1873-1935)
MAXEY, George Burke, (1917-1977)
MULLER, Siemon William, (1900-1970)
NOBLE, Levi Fatzinger, (1882-1965)
PECORA, William Thomas, (1913-1972)
PETERSON, Clarence J., (1885-1948)
POWELL, John Wesley, (1834-1902)

UNITED STATES OF AMERICA (NEVADA) 1900 to present Cont'd.

RANSOME, Frederick Leslie, (1868-1935)
RICHTHOFEN, Paul Wilhelm Ferdinand von, Freiherr, (1833-1905)
SMITH, William Richard Tangier, (1869-1972)
STALDER, Walter "Cap", (1881-1949)
THOMAS, George Dewey, (1898-1975)
WARING, Gerald Ashley, (1883-1971)
WHEALTON, Rowland Gardiner "Peg", (1896-1956)
WILLIAMS, Gardner Fred, (1842-1922)
WINCHELL, Horace Vaughn, (1865-1923)
WOLF, Albert G., (1885-1964)

UNITED STATES OF AMERICA
NEW ENGLAND
1900 to present
HOLMES, George William, (1922-1970)

UNITED STATES OF AMERICA
NEW HAMPSHIRE
1800-1849
HUBBARD, Oliver Payson, (1809-1900)
JACKSON, Charles Thomas, (1805-1880)
WHITNEY, Josiah Dwight, (1819-1896)

1850-1899
HITCHCOCK, Charles Henry, (1836-1919)
HUBBARD, Oliver Payson, (1809-1900)
JACKSON, Charles Thomas, (1805-1880)
PATTEN, William, (1861-1932)
PIRSSON, Louis Valentine, (1860-1919)
UPHAM, Warren, (1850-1934)
WADSWORTH, Marshman Edward, (1847-1921)
WHITNEY, Josiah Dwight, (1819-1896)

1900 to present
BANNERMAN, Harold MacColl, (1897-1976)
GOLDTHWAIT, James Walter, (1880-1947)
HITCHCOCK, Charles Henry, (1836-1919)
MOORE, Paul Brian, (1940-)
PATTEN, William, (1861-1932)
PIRSSON, Louis Valentine, (1860-1919)
UPHAM, Warren, (1850-1934)
WADSWORTH, Marshman Edward, (1847-1921)
WHITE, George Willard, (1903-)

UNITED STATES OF AMERICA
NEW JERSEY
Up to 1800
KALM, Pehr, (1716-1779)

1800-1849
COOK, George Hammell, (1818-1889)
FINCH, John, (fl.1823-1835)
GREEN, Jacob, (1790-1841)
ROGERS, Henry Darwin, (1808-1866)

UNITED STATES OF AMERICA (NEW JERSEY) Cont'd.

1850-1899

BAYLEY, William Shirley, (1861-1943)
BROOKS, Thomas Benton, (1836-1900)
BUTLER, Bertram Theodore, (1874-1958)
CHADWICK, George Halcott, (1876-1953)
COOK, George Hammell, (1818-1889)
DARTON, Nelson Horatio, (1865-1948)
JULIAN, Henry Forbes, (1861-1912) [PROSPECTORS, DIVINERS AND MINING ENGINEERS]
KUMMEL, Henry Barnard, (1867-1945)
LANDES, Henry, (1867-1936)
LEVERETT, Frank, (1859-1943)
MARSH, Othniel Charles, (1831-1899)
MORTON, James F., (1870-1941)
OSBORN, Henry Fairfield, (1857-1935)
PHILLIPS, Alexander Hamilton, (1866-1937)
RIES, Heinrich, (1871-1951)
ROGERS, Henry Darwin, (1808-1866)
SALISBURY, Rollin D., (1858-1922)
SCOTT, William Berryman, (1858-1947)
SHATTUCK, George Burbank, (1869-)
SMOCK, John Conover, (1842-1925)
TWITCHELL, Mayville William, (1866-1927)

1900 to present

BAUER, Lawson H., (1889-1954)
BAYLEY, William Shirley, (1861-1943)
BUTLER, Bertram Theodore, (1874-1958)
CHADWICK, George Halcott, (1876-1953)
CONDIT, Daniel Dale, (1886-1955)
COOKE, Charles Wythe, (1887-1971)
DARTON, Nelson Horatio, (1865-1948)
FIELD, Richard Montgomery, (1885-1961)
GABRIEL, Alton, (1902-1954)
HAWKINS, Alfred Cary, (1887-1954)
HAYES, Albert Orion, (1882-1965)
HERPERS, Henry F., Jr., (1915-1952)
HESS, Harry Hammond, (1906-1969)
HORNER, Jack, (-)
HOWELL, Benjamin Franklin, (1890-1976)
JULIAN, Henry Forbes, (1861-1912) [PROSPECTORS, DIVINERS AND MINING ENGINEERS]
KUMMEL, Henry Barnard, (1867-1945)
LANDES, Henry, (1867-1936)
LEE, Olan Ivan, (1888-1952)
LEVERETT, Frank, (1859-1943)
LILLIGREN, Johnathan Martin, (1880-1956)
McLAUGHIN, Dean Benjamin, (1901-1965)
MORTON, James F., (1870-1941)
OSBORN, Henry Fairfield, (1857-1935)
PHILLIPS, Alexander Hamilton, (1866-1937)
RIES, Heinrich, (1871-1951)
SALISBURY, Rollin D., (1858-1922)
SANDIDGE, John Roy, (1897-1976)

UNITED STATES OF AMERICA (NEW JERSEY) 1900 to present Cont'd.

SCOTT, William Berryman, (1858-1947)
SHATTUCK, George Burbank, (1869-)
SMOCK, John Conover, (1842-1925)
TWITCHELL, Mayville William, (1866-1927)
WALDBAUM, David Robert, (1937-1974)
WALKER, Frederick, (1898-1968)
WASSON, Theron, (1887-1970)
WILKERSON, Albert Samuel, (1897-1963)

UNITED STATES OF AMERICA
NEW MEXICO
1800-1849
ABERT, James W., Sr., (1820-1897)
HAYDEN, Ferdinand Vandiveer, (1829-1887)
MARCY, Randolph Barnes, (1812-1887)
SCHIEL, Jacob Heinrich, (1813-)

1850-1899
ABERT, James W., Sr., (1820-1897)
BOND, Marshall, (1867-1941) [PROSPECTORS, DIVINERS AND MINING ENGINEERS]
COCKERELL, Theodore Dru Alison, (1866-1948)
DARTON, Nelson Horatio, (1865-1948)
GORDON, Charles Henry, (1857-1934)
GOULD, Charles Newton, (1868-1949)
GREGORY, Herbert E., (1869-1952)
HAYDEN, Ferdinand Vandiveer, (1829-1887)
HERRICK, Clarence Luther, (1858-1904)
HESS, Frank L., (1871-1955)
KEYES, Charles Rollin, (1864-1942)
LEE, Willis Thomas, (1864-1926)
MARCY, Randolph Barnes, (1812-1887)
POWELL, John Wesley, (1834-1902)
RICHARDSON, George Burr, (1872-1949)
SPRINGER, Frank, (1848-1924)
ST. JOHN, Orestes Hawley, (1841-1921)
VALERIUS, Claude N., (-)

1900 to present
ADAMS, John Emery, (1899-1971)
ALCORN, Avary Hunt, (1902-1967)
BLANCHARD, William Grant, Jr., (1894-1952)
BOGARDUS, Robert Stephen "Bud", (1914-1953)
BOND, Marshall, (1867-1941) [PROSPECTORS, DIVINERS AND MINING ENGINEERS]
BROWN, John Albert, (1901-1954)
BUNTE, Arnold S., (1902-1968)
BURRESS, Walter March, (1893-1950)
BUTCHER, Cary Preston, (1898-1950)
BYRON, Clifford Winkler, (1901-1970)
CARTER, Thomas Ray, (1927-1959)
CHENEY, George Monroe, (1893-1952)
CLARK, Robert Watson, (1884-1948)
CLIFTON, Roland Leroy, (1884-1962)

UNITED STATES OF AMERICA (NEW MEXICO) 1900 to present Cont'd.

COCKERELL, Theodore Dru Alison, (1866-1948)
COFFIN, Reuben Clare, (1886-1972)
CONKLING, Russell C., (1897-1955)
COTTINGHAM, Virgil Elisha, (1891-1958)
COX, William Edgerton, (1912-1957)
DANE, Carle Hamilton, (1900-1968)
DARTON, Nelson Horatio, (1865-1948)
DAVIE, Thomas Early, (1906-1959)
EYSSELL, Alfred Raab, (1898-1944)
FAIRCHILD, John Gifford, (1882-1965)
FERGUSON, John Logan, (1893-1973)
FERRERO, Clarence de Labretonne, (1903-1954)
FITZGERALD, James, Jr., (1903-1951)
FLESCH, Gary Allen, (1949-1976)
FOLEY, Edward J., (1908-1971)
GARDNER, James Henry, (1883-1964)
GARRETT, Robert Ernest, (1888-1947)
GILLERMAN, Elliot "Bud", (1913-1974)
GILLES, Verner Arthur, (1886-1954)
GLASS, Jewell Jeanette, (1888-1966)
GOLLNICK, Robert Louis, (1918-1963)
GORDON, Charles Henry, (1857-1934)
GORDON, Dugald, (1886-1958)
GOULD, Charles Newton, (1868-1949)
GREGORY, Herbert E., (1869-1952)
GRIMES, Russell Winters, (1900-1969)
GUINN, Delmar Richard, (1902-1955)
HANDLEY, Howard Walter, (1905-1940)
HANSON, Ernest Albin, (1899-1973)
HANSON, Wallace W., (1921-1971)
HEDRICK, Oley Foster, (1891-1950)
HERRICK, Clarence Luther, (1858-1904)
HESS, Frank L., (1871-1955)
HIGGINS, Daniel Franklin, Jr., (1884-1930)
HOLCOMB, Leroy D., (1919-1974)
HOWARD, Edgar Billings, (1887-1943)
HUBMAN, Carl Wesley, (1905-1947)
HYDE, Clarence Edgar, (1894-1954)
JACOB, Charles Edward, (1914-1970)
JOESTING, Henry Rochambeau, (1903-1965)
KEYES, Charles Rollin, (1864-1942)
KISLING, James William, Jr., (1898-1973)
KITSON, Howard Walde, (1883-1931)
KOCH, Harold Kenneth, (1927-1973)
KOSCHMANN, Albert Herbert, (1896-1962)
KROENLEIN, George Alfred, (1898-1940)
KURTZ, Robert Gourdin, (1898-1958)
LARSH, Napoleon Bonaparte "Po", (1897-1967)
LEE, Willis Thomas, (1864-1926)
LILLIGREN, Johnathan Martin, (1880-1956)
LLOYD, Edwin Russell, (1882-1955)
LOSKAMP, Alvin Powell, (1900-1954)
MATHER, Kirtley Fletcher, (1888-1978)
McLAREN, Robert Lee, (1901-1953)

UNITED STATES OF AMERICA (NEW MEXICO) 1900 to present Cont'd.

McMINN, Paul Meloy, (1919-1974)
McNEELY, Robert Barnes, (1902-1966)
McNUTT, Vachel Harry, (1888-1936)
MILLER, John Preston, (1923-1961)
MITCHELL, Kingsley Camden, (1903-1931)
MIX, Charles Alexander, (1898-1954)
MOODY, Graham Blair, (1889-1975)
MOORE, James Augustus, (1910-1955)
MYERS, John Clark, (1894-1970)
NEEDHAM, Claude Ervin, (1894-1950)
NININGER, Harvey Harlow, (-)
PENNEL, Harry David, (1907-1971)
PETSCH, Arthur Henry, (1895-1969)
PETTY, Thomas Franklin, (1900-1962)
PILCHER, Benjamin Luther, Jr., (1906-1946)
PORCH, Edwin Leight, Jr., (1889-1948)
POTTER, Grover Cleveland, (1892-1975)
POWELL, John Wesley, (1834-1902)
POWELL, Ralph Sterling, (1901-1951)
PROUT, Franklin S. "Shorty", (1889-1940)
PUTNAM, L. G., (1898-1936)
RENICK, Brink Coleman, (1897-1971)
REYNOLDS, Thomas Emmett, (-1972)
RICHARDSON, George Burr, (1872-1949)
RIGGS, Robert Jennings, (1888-1965)
ROMER, Alfred Sherwood, (1894-1973)
RUDOLPH, Joseph, (1897-1953)
SCHALLER, Waldemar Theodore, (1882-1967)
SCHMITT, Harrison Ashley, (1896-1966)
SEASHORE, Paul Theodore, (1898-1954)
SEVERY, Clarence Luther, (1885-1975)
SPOFFORD, Howard Nelson, (1881-1950)
SPRINGER, Frank, (1848-1924)
ST. JOHN, Orestes Hawley, (1841-1921)
SULLINS, M. Joseph, (1930-1972)
TANNER, Hugh Alexander, (1904-1954)
TAYLOR, James Carlton, (1924-1976)
VANDER LECK, Lawrence, (1890-1954)
WHEELER, Orby Clinton, (1898-1970)
WHITAKER, George O., (-)
WILLIAMS, George O. "Skipper", (1897-1967)

UNITED STATES OF AMERICA
NEW YORK
Up to 1800
DARBY, William, (1775-1854)
EATON, Amos, (1776-1842)
KALM, Pehr, (1716-1779)
THOMAS, David, (1776-1859)

1800-1849
BARNES, Daniel Henry, (1785-1828)
BECK, Theodoric Romeyn, (1791-1855)
CONRAD, Timothy Abbott, (1803-1877)

UNITED STATES OF AMERICA (NEW YORK) 1800-1849 Cont'd.

COZZENS, Issachar, Jr., (1780-1865)
DARBY, William, (1775-1854)
DEWEY, Chester, (1784-1867)
EATON, Amos, (1776-1842)
EMMONS, Ebenezer, (1800-1863)
HALL, James, (1811-1898)
HENWOOD, William Jory, (1805-1875)
MEADE, William, (1789-1862)
MEEK, Fielding Bradford, (1817-1876)
MUDGE, Benjamin Franklin, (1817-1879)
NEWBERRY, John Strong, (1822-1892)
THOMAS, David, (1776-1859)
WHITFIELD, Robert Parr, (1828-1910)
WINCHELL, Alexander, (1824-1891)

1850-1899

BECK, Theodoric Romeyn, (1791-1855)
BERKEY, Charles Peter, (1867-1955)
BURROUGHS, John, (1837-1921)
BUTLER, Bertram Theodore, (1874-1958)
CALL, Richard Ellsworth, (1856-1917)
CASWELL, John Henry, (1846-1909)
CHADWICK, George Halcott, (1876-1953)
CHANDLER, Charles Frederick, (1836-1925)
CHRISTMAN, Charles Gustave, (1873-1953)
CLARKE, John Mason, (1857-1925)
CLELAND, Herdman Fitzgerald, (1869-1935)
COLONY, Roy Jed, (1870-1936)
CONRAD, Timothy Abbott, (1803-1877)
COZZENS, Issachar, Jr., (1780-1865)
CUSHING, Henry Platt, (1860-1921)
DARBY, William, (1775-1854)
DARTON, Nelson Horatio, (1865-1948)
DEWEY, Chester, (1784-1867)
DWIGHT, William Buck, (1833-1906)
EAKLE, Arthur Starr, (1862-1931)
EGLESTON, Thomas, (1832-1900)
EMMONS, Ebenezer, (1800-1863)
GALE, Hoyt Stoddard, (1876-1952)
GILBERT, Grove Karl, (1843-1918)
GLENN, Leonidas Chalmers, (1871-1951)
GREGORY, William King, (1876-1970)
HALL, James, (1811-1898)
HARTNAGEL, Chris Andrew, (1874-1962)
HARTT, Charles Frederic, (1840-1878)
HENWOOD, William Jory, (1805-1875)
HOLLICK, Arthur, (1857-1933)
HUDSON, George Henry, (1855-1934)
JOHNSON, Douglas Wilson, (1878-1944)
JULIAN, Henry Forbes, (1861-1912) [PROSPECTORS, DIVINERS AND MINING ENGINEERS]
JULIEN, Alexis Anastay, (1840-1919)
KEMP, James Furman, (1859-1926)
KHRUSHCHOV, Konstantin Dmitrievich, (?   -1912)

UNITED STATES OF AMERICA (NEW YORK) 1850-1899 Cont'd.

KRAUS, Edward Henry, (1875-1973)
LEDOUX, Albert Reid, (1852-1923) [PROSPECTORS, DIVINERS AND MINING ENGINEERS]
LLOYD, T. G. B., (1829-1876)
LUQUER, Lea McIlvaine, (1864-1930)
LUTHER, Daniel Dana, (1840-1923)
MEEK, Fielding Bradford, (1817-1876)
MOSES, Alfred Joseph, (1859-1920)
MUDGE, Benjamin Franklin, (1817-1879)
NEWBERRY, John Strong, (1822-1892)
NEWLAND, David Hale, (1872-1943)
OGILVIE, Ida Helen, (1874-1963)
OSBORN, Henry Fairfield, (1857-1935)
PROSSER, Charles Smith, (1860-1916)
RICHARDSON, Charles Henry, (1862-1935)
RICHARDSON, George Burr, (1872-1949)
RIES, Heinrich, (1871-1951)
RUEDEMANN, Rudolf, (1864-1956)
RUSSELL, Israel Cook, (1852-1906)
SCHUCHERT, Charles, (1858-1942)
SHATTUCK, George Burbank, (1869-)
SMOCK, John Conover, (1842-1925)
STEVENSON, John James, (1841-1924)
STONE, Ralph Walter, (1876-1964)
THOMAS, David, (1776-1859)
VEATCH, Arthur Clifford, (1878-1938)
WARD, Henry Augustus, (1834-1906)
WHITE, Theodore Greely, (1872-1901)
WHITFIELD, Robert Parr, (1828-1910)
WHITLOCK, Herbert Percy, (1868-1948)
WILLIAMS, George Huntington, (1856-1894)
WILLIAMS, Henry Shaler, (1847-1918)
WILLIAMS, John Francis, (1862-1891)
WINCHELL, Alexander, (1824-1891)
WINCHELL, Alexander Newton, (1874-1958)
WOODWARD, Robert Simpson, (1849-1924)
WOODWORTH, Jay Backus, (1865-1925)

1900 to present

ABERNATHY, George Elmer, (1888-1948)
ALLING, Harold Lattimore, (1888-1960)
ANDERSON, Alfred Leonard, (1900-1964)
BARTH, Thomas Frederik Weiby, (1899-1971)
BEACH, Hugh Hamilton, (1909-1973)
BERKEY, Charles Peter, (1867-1955)
BREWER, Charles, Jr., (1900-1944)
BRIGHAM, Albert Perry, (1855-1932)
BROWN, Noel K., Jr., (1925-1968)
BUCHER, Walter Herman, (1888-1965)
BURFOOT, James Dabney "Dan", Jr., (1896-1966)
BURROUGHS, John, (1837-1921)
BURWELL, Edward Bouldin, Jr., (1894-1964)
BUTLER, Bertram Theodore, (1874-1958)
CALL, Richard Ellsworth, (1856-1917)

CANTWELL, Thomas, Jr., (1927-)
CASWELL, John Henry, (1846-1909)
CHADWICK, George Halcott, (1876-1953)
CHANDLER, Charles Frederick, (1836-1925)
CHRISTMAN, Charles Gustave, (1873-1953)
CLARKE, John Mason, (1857-1925)
CLELAND, Herdman Fitzgerald, (1869-1935)
COLBERT, Edwin H., (1905-)
COLONY, Roy Jed, (1870-1936)
CONDIT, Daniel Dale, (1886-1955)
COPASS, Jack Malcolm, (1906-1963)
CORYELL, Horace Noble, (1888-1965)
CURRIER, Louis Wade, (1890-1970)
CUSHING, Henry Platt, (1860-1921)
DALE, Nelson Clark, (1880-1962)
DARTON, Nelson Horatio, (1865-1948)
DORSEY, George Edwin, (1892-1953)
DRESSER, Myron Amos, (1894-1973)
EAKLE, Arthur Starr, (1862-1931)
EGLESTON, Thomas, (1832-1900)
EHRMANN, Martin L., (1903-1972)
ELLIS, Brooks Fleming, (1897-1976)
FETTKE, Charles Reinhard, (1888-1959)
FIOCCO, Giorgio, (1931-)
FRALICH, Charles Eugene, (1902-1964)
GALE, Hoyt Stoddard, (1876-1952)
GALLOWAY, Jesse James, (1882-1962)
GARRETSON, Mary Welleck, (1896-1971)
GEBELEIN, Conrad Dennis, (1945-1978)
GILBERT, Grove Karl, (1843-1918)
GLENN, Leonidas Chalmers, (1871-1951)
GOLDRING, Winifred, (1888-1971)
GREGORY, William King, (1876-1970)
HAHN, Friedrich Felix (Fritz), (1885-1914)
HAKE, Benjamin Franklin, (1895-1973)
HAMILTON, Peggy-Kay, (1922-1959)
HARTNAGEL, Chris Andrew, (1874-1962)
HAWKINS, Alfred Cary, (1887-1954)
HEEZEN, Bruce Charles, (1924-1977)
HILLS, Thomas McDougall, (1881-1970)
HOLLICK, Arthur, (1857-1933)
HOWELL, Benjamin Franklin, (1890-1976)
HUDSON, George Henry, (1855-1934)
JAMES, John William, (1921-1977)
JEFFREYS, Geoffrey, (1885-1953)
JOHNSON, Bertrand Leroy, (1882-1962)
JOHNSON, Douglas Wilson, (1878-1944)
JONES, Verner E., (1905-1977)
JULIAN, Henry Forbes, (1861-1912) [PROSPECTORS, DIVINERS AND MINING ENGINEERS]
JULIEN, Alexis Anastay, (1840-1919)
KEMP, James Furman, (1859-1926)
KHRUSHCHOV, Konstantin Dmitrievich, (? -1912)
KIHLSTEDT, Folke H., (1901?-1956)

UNITED STATES OF AMERICA (NEW YORK) 1900 to present Cont'd.

KNEBEL, G. Moses, (1899-1974)
KOZÂRY, Myron Theodore "Mirko", (1918-1966)
KRAUS, Edward Henry, (1875-1973)
KREIDLER, William Lynn, (1911-1975)
KRIEGER, Philip, (1900-1940)
KURTZ, Robert Gourdin, (1898-1958)
LEDOUX, Albert Reid, (1852-1923) [PROSPECTORS, DIVINERS AND MINING ENGINEERS]
LEIGHTON, Henry, (1884-1963)
LILLIGREN, Johnathan Martin, (1880-1956)
LOBECK, Armin Kahl, (1886-1958)
LUQUER, Lea McIlvaine, (1864-1930)
LUTHER, Daniel Dana, (1840-1923)
MACDIARMID, Roy Angus, (1933-1966)
MASON, Shirley Lowell, (1893-1975)
MENCHER, Ely, (1913-)
MILLER, William John, (1880-1965)
MOORE, Carl Allphin, (1911-1973)
MOSES, Alfred Joseph, (1859-1920)
NEWLAND, David Hale, (1872-1943)
O'CONNELL, Daniel Trugott, (1901-1971)
OGILVIE, Ida Helen, (1874-1963)
OSBORN, Henry Fairfield, (1857-1935)
PEPPER, James Franklin, (1898-1963)
POLDERVAART, Arie, (1918-1964)
POSTEL, Albert Williams, (1909-1966)
PROSSER, Charles Smith, (1860-1916)
RICHARDS, James Taylor "Dick", (1892-1953)
RICHARDSON, Charles Henry, (1862-1935)
RICHARDSON, George Burr, (1872-1949)
RIES, Heinrich, (1871-1951)
ROMER, Alfred Sherwood, (1894-1973)
RUEDEMANN, Rudolf, (1864-1956)
RUSSELL, Israel Cook, (1852-1906)
RUSSELL, William Low, (1897-1976)
SCHLAIKJER, Erich Maren, (1905-1972)
SCHUCHERT, Charles, (1858-1942)
SHAND, Samuel James, (1882-1957)
SHATTUCK, George Burbank, (1869-)
SHERMAN, Roger Hiram, (1904-1954)
SMOCK, John Conover, (1842-1925)
STEBINGER, Eugene, (1883-1951)
STEVENSON, John James, (1841-1924)
STONE, Ralph Walter, (1876-1964)
STOW, Marcellus Henry, (1902-1957)
TAGUE, Glenn Charles, (1905-1964)
TEILHARD DE CHARDIN, Pierre, Father, S.J., (1881-1955)
TRAINER, David Woolsey, Jr., (1899-1965)
TURNER, Joseph Harold, (1900-1950)
VEATCH, Arthur Clifford, (1878-1938)
VON ENGELN, Oskar Dietrich, (1880-1965)
WADELL, Hakon, (1895-1962)
WALDBAUM, David Robert, (1937-1974)
WARD, Henry Augustus, (1834-1906)

UNITED STATES OF AMERICA (NEW YORK) 1900 to present Cont'd.

WASSON, Theron, (1887-1970)
WEDEL, Arthur Albert, (1898-1941)
WHITE, Theodore Greely, (1872-1901)
WHITFIELD, Robert Parr, (1828-1910)
WHITLOCK, Herbert Percy, (1868-1948)
WILLIAMS, Henry Shaler, (1847-1918)
WINCHELL, Alexander Newton, (1874-1958)
WOODRUFF, John Grant, (1898-1971)
WOODWARD, Robert Simpson, (1849-1924)
WOODWORTH, Jay Backus, (1865-1925)
ZAVOICO, Basil B., (1899-1975)

UNITED STATES OF AMERICA
NORTH CAROLINA
Up to 1800
BARTRAM, William, (1739-1823)
HUTCHINS, Thomas, (1730-1789)

1800-1849
BARTRAM, William, (1739-1823)
EMMONS, Ebenezer, (1800-1863)
MITCHELL, Elisha, Reverend, (1793-1857)
OLMSTED, Denison, (1791-1859)

1850-1899
BROWN, Charles Barrington, (1839-1917)
EMMONS, Ebenezer, (1800-1863)
GLENN, Leonidas Chalmers, (1871-1951)
HESS, Frank L., (1871-1955)
HOLMES, Joseph Austin, (1859-1915)
JOHNSON, Douglas Wilson, (1878-1944)
LEDOUX, Albert Reid, (1852-1923) [PROSPECTORS, DIVINERS AND MINING ENGINEERS]
LEWIS, Joseph Volney, (1869-1969)
MARSH, Othniel Charles, (1831-1899)
MITCHELL, Elisha, Reverend, (1793-1857)
OLMSTED, Denison, (1791-1859)
PRATT, Joseph Hyde, (1870-1942)
RIES, Heinrich, (1871-1951)
STEPHENSON, Lloyd William, (1876-1962)
STONE, Ralph Walter, (1876-1964)

1900 to present
BERRY, Edward Willard, (1900-1968)
BRANNOCK, Kent Combs "Casey", (1923-1973)
BROWN, Charles Barrington, (1839-1917)
DOUGLAS, John Gray, (1900-1974)
FRITTS, Crawford Ellsworth, (1927-1972)
GLENN, Leonidas Chalmers, (1871-1951)
HESS, Frank L., (1871-1955)
HOLMES, Joseph Austin, (1859-1915)
HUDDLE, John Warfield, (1907-1975)
JOHNSON, Bertrand Leroy, (1882-1962)
JOHNSON, Douglas Wilson, (1878-1944)

UNITED STATES OF AMERICA (NORTH CAROLINA) 1900 to present Cont'd.

JOHNSON, W. Ray, (-1952)
KIMBALL, Kent K., (1892-1953)
LEDOUX, Albert Reid, (1852-1923) [PROSPECTORS, DIVINERS AND MINING ENGINEERS]
LEWIS, Joseph Volney, (1869-1969)
PRATT, Joseph Hyde, (1870-1942)
RIES, Heinrich, (1871-1951)
SCHOLL, Louis August, Jr., (1887-1953)
STEPHENSON, Lloyd William, (1876-1962)
STONE, Ralph Walter, (1876-1964)
SWARTZ, Joel Howard, (1893-1971)

UNITED STATES OF AMERICA
NORTH DAKOTA
1800-1849
HAYDEN, Ferdinand Vandiveer, (1829-1887)
MEEK, Fielding Bradford, (1817-1876)
NICOLLET, Joseph Nicolas, (1786-1843)

1850-1899
BAYLEY, William Shirley, (1861-1943)
HALL, Charles Monroe, (1870-1903)
HAYDEN, Ferdinand Vandiveer, (1829-1887)
HESNARD, Emile E., (1867-1929)
HUBBARD, George David, (1871-1958)
LEONARD, Arthur Gray, (1865-1932)
MEEK, Fielding Bradford, (1817-1876)
RICHARDSON, George Burr, (1872-1949)
UPHAM, Warren, (1850-1934)
WILDER, Frank Alonzo, (1870-1930)
WILLARD, Daniel Everett, (1862-1947)

1900 to present
ARMSTRONG, Harold Kerr, (1893-1962)
ASHMORE, Herman T., (1916-1976)
BALL, John Rice, (1881-1953)
BALLARD, William Norval, (1906-1976)
BAYLEY, William Shirley, (1861-1943)
BECKWITH, Radcliffe Harold, (1900-1964)
BEEKLEY, Albert Leon, (1883-1952)
CARR, William L., (1919-1975)
ERDMANN, Charles Edgar, (1897-1971)
EVANS, Vincent, (1907-1951)
FERGUSON, John Logan, (1893-1973)
FOLGER, James Anthony, (1896-1956)
FULTON, Loris John, (1906-1974)
GILL, James Rogers, (1922-1972)
GOLLNICK, Robert Louis, (1918-1963)
HALL, Charles Monroe, (1870-1903)
HARES, Charles Joseph, (1881-1970)
HESNARD, Emile E., (1867-1929)
HUBBARD, George David, (1871-1958)
LEACH, Thomas Witt, (1896-1966)
LEONARD, Arthur Gray, (1865-1932)

UNITED STATES OF AMERICA (NORTH DAKOTA) 1900 to present Cont'd.

MULLEN, John Timothy, Jr., (1906-1959)
OWEN, William Joseph, (1929-1959)
RASMUSSEN, William Charles, (1917-1973)
RICHARDSON, George Burr, (1872-1949)
SEASHORE, Paul Theodore, (1898-1954)
TAYLOR, W. Harlan, (1908-1972)
TETRICK, Paul Roderick, (1920-1956)
TOWSE, Donald Frederick, (1924-)
TYLER, Stanley Allen, (1906-1963)
UPHAM, Warren, (1850-1934)
VANDER LECK, Lawrence, (1890-1954)
WILDER, Frank Alonzo, (1870-1930)
WILLARD, Daniel Everett, (1862-1947)

UNITED STATES OF AMERICA
OHIO
Up to 1800
THOMAS, David, (1776-1859)

1800-1849
HARRIS, Israel Hopkins, (1832-1897)
HILDRETH, Samuel Prescott, (1783-1863)
JAMES, Uriah Pierson, (1811-1889)
LESQUEREUX, Charles Leo, (1806-1889)
LOCKE, John, (1792-1856)
MATHER, William Williams, (1804-1859)
NEFF, Peter, (1827-1903)
NEWBERRY, John Strong, (1822-1892)
RIDDELL, John Leonard, (1807-1865)
ROMINGER, Carl Ludwig, (1820-1907)
THOMAS, David, (1776-1859)
WHITTLESEY, Charles, Colonel, (1808-1886)
WINCHELL, Alexander, (1824-1891)

1850-1899
CLAYPOLE, Edward Waller, (1835-1901)
FENNEMAN, Nevin Melancthan, (1865-1945)
FOERSTE, August Frederick, (1862-1936)
GILBERT, Grove Karl, (1843-1918)
HARRIS, Israel Hopkins, (1832-1897)
HERRICK, Clarence Luther, (1858-1904)
HILDRETH, Samuel Prescott, (1783-1863)
HUBBARD, George David, (1871-1958)
JAMES, Joseph Francis, (1857-1897)
JAMES, Uriah Pierson, (1811-1889)
LESQUEREUX, Charles Leo, (1806-1889)
LEVERETT, Frank, (1859-1943)
MATHER, William Williams, (1804-1859)
McCAUGHEY, William John, (1882-1962)
MENDENHALL, Thomas Corwin, (1841-1924)
MILLER, Samuel A., (1837-1897)
MORSE, William Clifford, (1874-1962)
NEFF, Peter, (1827-1903)
NEWBERRY, John Strong, (1822-1892)

UNITED STATES OF AMERICA (OHIO) 1850-1899 Cont'd.

ODENBACH, Frederick L., Father, (1857-1933)
ORTON, Edward Francis Baxter, (1829-1899)
PROSSER, Charles Smith, (1860-1916)
RIDDELL, John Leonard, (1807-1865)
ROMINGER, Carl Ludwig, (1820-1907)
STEVENSON, John James, (1841-1924)
STONE, Ralph Walter, (1876-1964)
THOMAS, David, (1776-1859)
TIGHT, William George, (1865-1910)
UPHAM, Warren, (1850-1934)
VAN HORN, Frank Robertson, (1872-1933)
WESTGATE, Lewis Gardner, (1868-1948)
WHITTLESEY, Charles, Colonel, (1808-1886)
WINCHELL, Alexander, (1824-1891)

1900 to present
BAIRD, Chester Alvan, (1896-1959)
BARBOUR, George Brown, (1891-1977)
BECKWITH, Radcliffe Harold, (1900-1964)
BERRY, Edward Willard, (1900-1968)
BUCHER, Walter Herman, (1888-1965)
BURWELL, Edward Bouldin, Jr., (1894-1964)
CARMAN, Joel Ernest, (1882-1966)
CLARK, Leslie Madison, (1898-1976)
CLAYPOLE, Edward Waller, (1835-1901)
CONDIT, Daniel Dale, (1886-1955)
CRUM, Harry Edwin, (1892-1955)
DUTY, Richard Emerson, (1908-1966)
FENNEMAN, Nevin Melancthan, (1865-1945)
FOERSTE, August Frederick, (1862-1936)
FREY, M. Gordon, (1913-)
FURCRON, Aurelius Sydney, (1899-1971)
GILBERT, Grove Karl, (1843-1918)
GOODING, Ansel Miller, (1924-1976)
GRAHAM, William Armstrong Patterson, (1899-1934)
HAKE, Benjamin Franklin, (1895-1973)
HERALD, Frank Edward, (1887-1965)
HERRICK, Clarence Luther, (1858-1904)
HILLS, Thomas McDougall, (1881-1970)
HOWER, John, Jr., (1927-)
HUBBARD, George David, (1871-1958)
HUSSAKOF, Louis, (1881-1965)
KURTZ, Robert Gourdin, (1898-1958)
LEIGHTON, Morris Morgan, (1887-1971)
LEVERETT, Frank, (1859-1943)
LILLIGREN, Johnathan Martin, (1880-1956)
MARPLE, Mildred Fisher, (1898-1966)
MASON, Shirley Lowell, (1893-1975)
MATHER, Kirtley Fletcher, (1888-1978)
MAYFIELD, Samuel Martin, (1888?-1963)
McCAUGHEY, William John, (1882-1962)
McCLOY, Walter Lowrie, Jr., (1912-1954)
MENDENHALL, Thomas Corwin, (1841-1924)
MILLER, Wendell Zerbe, (1892-1958)

UNITED STATES OF AMERICA (OHIO) 1900 to present Cont'd.

MORSE, William Clifford, (1874-1962)
NEFF, Peter, (1827-1903)
O'BRIEN, Francis Eugene, (1916-1968)
ODENBACH, Frederick L., Father, (1857-1933)
ORTON, Edward Francis Baxter, (1829-1899)
PAINE, William R., (-)
PANYITY, Louis Samuel, (1890-1943)
PEPPER, James Franklin, (1898-1963)
PROSSER, Charles Smith, (1860-1916)
ROMINGER, Carl Ludwig, (1820-1907)
SCHLICHTEN, Otto Charles von, (1886-1950)
SHIDELER, William Henry, (1886-1958)
STEVENSON, John James, (1841-1924)
STEWART, Grace Anne, (1893-1970)
STOCKDALE, Paris Buell, (1896-1962)
STONE, Ralph Walter, (1876-1964)
STUMM, Erwin Charles, (1908-1969)
SULLINS, M. Joseph, (1930-1972)
SZMUC, Eugene Joseph, (1927-1976)
TIGHT, William George, (1865-1910)
UPHAM, Warren, (1850-1934)
VAN HORN, Frank Robertson, (1872-1933)
WEBB, David Knowlton, Jr., (1928-1973)
WESTGATE, Lewis Gardner, (1868-1948)
WHITE, George Willard, (1903-)
WHITE, Stanley Battershill, (1899-1966)
WILLIAMS, George O. "Skipper", (1897-1967)
WING, Monta Eldo, (1897-1976)
WRIGHT, Frank James, (1888-1954)

UNITED STATES OF AMERICA
OKLAHOMA
1800-1849
BRUMBY, Richard T., (1804-1856)
MARCY, Randolph Barnes, (1812-1887)

1850-1899
BEEBY-THOMPSON, Arthur, (1873-1968)
BEEDE, Joshua William, (1871-1940)
BRUMBY, Richard T., (1804-1856)
DECKER, Charles Elijah, (1868-1958)
DRAKE, Noah Fields, (1864-1945)
ERB, Josef Theodor, (1874-1934)
GIDLEY, James Williams, (1866-1931)
GIRTY, George Herbert, (1869-1939)
GOULD, Charles Newton, (1868-1949)
HENDERSON, James Alexander Leo, (?1874-1940)
HENNEN, Ray Vernon, (1875-1958)
HUTCHINSON, Lon Lewis, (1877-1947)
MANKIN, Charles John, (1932-)
MARCY, Randolph Barnes, (1812-1887)
MORSE, William Clifford, (1874-1962)
MUNN, Malcolm John, (1874-1949)
OHERN, Daniel Webster, (1870-1953)

UNITED STATES OF AMERICA (OKLAHOMA) 1850-1899 Cont'd.

RICHARDSON, George Burr, (1872-1949)
SMITH, Eugene Allen, (1841-1927)
TAYLOR, Charles Henry, (1877-1964)
VALERIUS, Claude N., (-)
VALERIUS, Michael Marcus, (1872-1947)
WADE, Arthur, (1878-1951)
WEIDMAN, Samuel, (1870-1945)
WILLIAMS, Arthur James, (1877-1954)
WOODRUFF, Elmer Grant, (1872-1952)

1900 to present

ADAMS, William Clyde, (1892-1957)
ALLEN, Donald Meredith, (1899-1971)
ALLEN, Walter J., (1890-1961)
ANDERSON, Gustavus Edwin, (1879-1940)
ARGABRITE, William Greame, (1885-1963)
ARMSTRONG, William Burrus, (1917-1954)
ATHY, Lawrence Ferdinand, (1898-1955)
ATKINSON, William H., (1898-1972)
AURIN, Fritz Love, (1891-1969)
BALDWIN, William Felbert Jackson, Jr., (1927-1972)
BALE, Hubert Eugene, (1898-1969)
BALLARD, William Norval, (1906-1976)
BARTRAM, John Greer, (1893-1955)
BEACH, John Osa, (1895-1964)
BECKER, Clyde M., (1882-1938)
BECKER, Robert M., (1919-1975)
BECKMANN, Lawrence John, (1912-1976)
BEEBY-THOMPSON, Arthur, (1873-1968)
BEEDE, Joshua William, (1871-1940)
BEEKLEY, Albert Leon, (1883-1952)
BISAT, William Sawney, (1886-1973)
BLANCHARD, William Grant, Jr., (1894-1952)
BLANKENSHIP, Joseph Croxton, (1930-1964)
BLANPIED, Bernerd William, (1897-1975)
BOWEN, James P., (1889-1960)
BOYD, Harold Edward, (1887-1952)
BOYLE, Walter J., (1897-1973)
BRANSON, Carl Colton, (1906-1975)
BRANSON, Edwin Robert, (1908-1971)
BRANT, Ralph Allen, (1899-1958)
BRAUCHLI, R. W., (1895-1975)
BRISCOE, Glenn Oliver, (1898-1974)
BROWN, Alton Roe, (1930-1975)
BROWN, John Albert, (1901-1954)
BROWN, Otto E., (1896-1973)
BURRESS, Walter March, (1893-1950)
BURTON, George Edgar, (1880-1953)
BUTCHER, Walter William, (1916-1966)
BUTTRAM, George Franklin, (1886-1966)
BYRON, Clifford Winkler, (1901-1970)
CANFIELD, Charles Reiter, (1906-1964)
CARLSON, Charles Gordon, (1895-1977)
CARPENTER, Everett, (1884-1968)

UNITED STATES OF AMERICA (OKLAHOMA) 1900 to present Cont'd.

CHARLES, Homer Herbert, (1900-1963)
CLARK, Howard, (1885-1972)
CLARK, Robert Watson, (1884-1948)
CLARKE, Carl W., (1896-1928)
CLIFTON, Roland Leroy, (1884-1962)
CLINE, Lewis Manning, (1909-1971)
CLINKSCALES, Albert Sidney, (1894-1969)
COCHRAN, Kenneth L., (1910-1973)
COFFIN, Reuben Clare, (1886-1972)
CONKLING, Richard A., (1885-1952)
CONKLING, Russell C., (1897-1955)
COOPER, Chalmer Lewis, (-)
COOPER, Herschel H., (1894-1954)
COPASS, Jack Malcolm, (1906-1963)
COTTINGHAM, Virgil Elisha, (1891-1958)
CRAWFORD, David Johnston, (1899-1960)
CRUM, Harry Edwin, (1892-1955)
DAMON, Fred T., (1912-1977)
DANIELS, James Ira, (1896-1967)
DARBY, Raymond Vern, (1896-1953)
DECKER, Charles Elijah, (1868-1958)
DECKER, Charles LaVerne, (1898-1955)
DeGOLYER, Everette Lee, (1886-1956)
DILLÉ, Glenn Scott, (1896-1963)
DORSEY, George Edwin, (1892-1953)
DRAKE, Cecil, (1899-1949)
DRAKE, Noah Fields, (1864-1945)
DRESSER, Myron Amos, (1894-1973)
EBY, James Brian, (1896-)
EDSON, Fanny Carter, (1887-1952)
ELDER, Stanley Gordon, (1904-1957)
ELLEDGE, Emmett Russell, (1898-1960)
ELLISON, Charles Walker, (1897-1953)
ELLISON, Kenneth Alfred, (1902-1963)
ELSON, William J., (1895-1965)
ERB, Josef Theodor, (1874-1934)
ESPACH, Ralph Homeward, (1900-1956)
EVANS, Vincent, (1907-1951)
FATH, Arthur Earl, (1887-1976)
FERGUSON, John Logan, (1893-1973)
FICHTER, Hans Jakob, (1908-1977)
FITTS, John, (1879-1942)
FITZGERALD, James, Jr., (1903-1951)
FLESCH, Gary Allen, (1949-1976)
FOHS, Ferdinand Julius, (1884-1965)
FOLGER, James Anthony, (1896-1956)
FOSSHAGE, Ernest Willard, (1908-1955)
FOWLER, George Malcolm, (1885-1972)
FULTON, Loris John, (1906-1974)
GARDNER, James Henry, (1883-1964)
GARRETT, Robert Ernest, (1888-1947)
GIDLEY, James Williams, (1866-1931)
GIRTY, George Herbert, (1869-1939)
GORDON, Dugald, (1886-1958)

UNITED STATES OF AMERICA (OKLAHOMA) 1900 to present Cont'd.

GOULD, Charles Newton, (1868-1949)
GREENE, Frank Cook, (1886-1975)
GRIFFITH, Glenn Rodgers Vickery, (1891-1964)
GRIMES, Russell Winters, (1900-1969)
GROSS, Arthur Bunert "Abie", (1904-1965)
GROUT, Frank Fitch, (1880-1958)
GUINN, Delmar Richard, (1902-1955)
HAM, William Eugene, (1916-1970)
HANDLEY, Howard Walter, (1905-1940)
HANSON, Ernest Albin, (1899-1973)
HARRIS, Reginald W., (-)
HAWKINS, Glenn Dewayne, (1897-1970)
HENDERSON, James Alexander Leo, (?1874-1940)
HENLEY, Arthur Sidney, (1879-1936)
HENNEN, Ray Vernon, (1875-1958)
HERALD, Frank Edward, (1887-1965)
HEROLD, Stanley Carrollton, (1883-1958)
HONESS, Charles William, (1885-1949)
HOOVER, James Barr, (1902-1953)
HOPPER, Walter Everett, (1886-1971)
HOWELL, Jesse V., (1891-1971)
HOWELL, William Frederick, (1895-1965)
HOYLE, Charles Robert, Sr., (1891-1954)
HUGHES, Richard, (1889-1969)
HUNT, Edwin Harold, (1897-1949)
HUTCHINSON, Lon Lewis, (1877-1947)
HYDE, Clarence Edgar, (1894-1954)
ISENBERGER, Nathan Perry, (1896-1956)
JOHNSON, Norman Lee, (1926-1970)
JONES, William Francis, (1885-1941)
JORDAN, Louise, (1908-1966)
KELLY, Donald, (1898-1949)
KENNEDY, Luther Eugene, (1891-1966)
KENNEDY, William, (-1926)
KEPPLER, Leo George, (1890-1971)
KIMBALL, Kent K., (1892-1953)
KISLING, James William, Jr., (1898-1973)
KNIGHT, Garold Louis "Jerry", (1894-1968)
KOSCHMANN, Albert Herbert, (1896-1962)
KREIDLER, William Lynn, (1911-1975)
KREPPS, William J., (1924-1976)
KRIDLER, George Matthew, (1889-1953)
LANCASTER, L. Kenneth, (1906-1971)
LARSH, Napoleon Bonaparte "Po", (1897-1967)
LATIMER, Frank Parkes, (1903-1928)
LATTA, Albert Lloyd, (1907-1960)
LEACH, Thomas Witt, (1896-1966)
LEATHEROCK, Otto, (1901-1941)
LEE, Marvin, (1889-1953)
LEVORSEN, Arville Irving, (1894-1965)
LILIENBORG, Burton Armand, (1900-1949)
LILLIGREN, Johnathan Martin, (1880-1956)
LONSDALE, John Tipton, (1895-1960)
LÓPEZ VÁZQUEZ, Andrés, (1903-1954)

UNITED STATES OF AMERICA (OKLAHOMA) 1900 to present Cont'd.

LOWMAN, Shepard Wetmore, (1899-1968)
LUCAS, Elmer Lawrence, (1898-1966)
MACKENZIE, Andrew Nicholls "Kelley", (1888-1951)
MAEBIUS, Jed B., (1907-1969)
MARKHAM, Edmund Olis, (1893-1960)
MASON, Shirley Lowell, (1893-1975)
MATHER, Kirtley Fletcher, (1888-1978)
McCLOY, Walter Lowrie, Jr., (1912-1954)
McCOY, Alexander Watts, (1889-1944)
McFARLAND, Leland R., (1904-1973)
McFARLAND, Russell Scott, (1893-1968)
McFERRON, George Irwin, (1895-1969)
McGEHEE, Julius Rex, (1905-1963)
McINTYRE, Paul Joseph, (1893-1963)
McNEELY, Robert Barnes, (1902-1966)
MEHL, Maurice Goldsmith, (1887-1966)
MELTON, Frank A., (-)
MERRITT, Clifford A., (-)
MERRITT, John Wesley, (1885-1958)
MILLER, Carroll C., (1899-1974)
MILLER, John Cummins, (1895-1951)
MILLER, Willard L., (1896-1973)
MILNER, Charles Albert, Jr., (1900-1971)
MISER, Hugh Dinsmore, (1884-1969)
MITCHELL, James Garland, (1908-1972)
MONNETT, Victor Elvert, (1889-1972)
MOORE, Carl Allphin, (1911-1973)
MOORE, James Augustus, (1910-1955)
MORGAN, George Dillon, (1894-1950)
MORRIS, Albert Ferd, (1898-1957)
MORSE, William Clifford, (1874-1962)
MUNN, Malcolm John, (1874-1949)
MURPHY, James Kenneth, (1900-1955)
MURRAY, Albert Nelson, (1894-1961)
MYERS, John Clark, (1894-1970)
MYLIUS, Louis Aubrey, (1890-1965)
NEARHOOD, Clinton Elsworth, (1915-1957)
NEDOM, Henry Arthur, (1895-1940)
NETTERSTROM, Paul Wagner, (1914-1960)
NORDEN, John Alexander Erdelyi-Fazekas, (né Erdélyi-Fazekas), (1912-1977)
NORVILLE, Glen Smith, (1899-1963)
NOWLAN, Harry Hackett, (1888-1958)
OAKES, Malcolm Christie, (1890-1977)
O'BRIEN, Francis Eugene, (1916-1968)
OHERN, Daniel Webster, (1870-1953)
OWENS, Frith Cravens, (1898-1977)
PALMER, Leland L., (1906-1978)
PEATTIE, Roderick, (1891-)
PEMBERTON, John Roy "Bill", (1884-1968)
PENNY, Frederick Willoughby, (1887-1950)
PERRINE, Irving, (1884-1955)
PHELPS, Frederick George, (1913-1964)
PHILBRICK, Eugene Patrick, (1897-1951)

UNITED STATES OF AMERICA (OKLAHOMA) 1900 to present Cont'd.

PLUMMER, Helen Jeanne (née Skewes), (1891-1951)
POTEET, James Harold, (1896-1965)
POTTER, Grover Cleveland, (1892-1975)
POTTER, Nelson Bryan, (1897-1965)
POWELL, Ralph Sterling, (1901-1951)
PRICE, Arthur Sebert, (1904-1948)
RASHID, Muhammad Abdur, (1938-1969)
REIFF, Allan Eugene, (1908-1933)
RENICK, Brink Coleman, (1897-1971)
REYNOLDS, Karl William, (1894-1970)
RICHARDS, James Taylor "Dick", (1892-1953)
RICHARDSON, George Burr, (1872-1949)
RIGGS, Robert Jennings, (1888-1965)
RITTENHOUSE, Gordon, (1910-1974)
ROARK, Louis, (1890-1950)
ROBINSON, Bryan Franklin, (1897-1949)
ROSEN, George D., (1914-1978)
ROWLEY, Alden Bruce, (1888-1956)
RUBEY, William Walden, (1898-1974)
RUDOLPH, Joseph, (1897-1953)
RUSSELL, Philip G., (1891-1971)
RUSSOM, Vaughn Walter, (1891-1955)
SANDS, John Melville, (1883-1962)
SARJEANT, William Antony Swithin, (1935-)
SAWYER, Roger W., (1895-1941)
SCHILLING, Karl Howard, (1894-1945)
SCHMIDT, Karl A. "Gus", (1900-1965)
SCHOLL, Louis August, Jr., (1887-1953)
SCHROYER, Charles Ross, (1887-1932)
SCOTT, Horace Lafayette, (1901-1976)
SCOTT, Walter Winthrop, (1893-1939)
SEVERY, Clarence Luther, (1885-1975)
SHERWOOD, Theodore C., Jr., (1896-1928)
SICKLER, John Miller, Jr., (1891-1970)
SIMMONS, Kenneth Albert, (1906-1971)
SMITH, Eugene Allen, (1841-1927)
SMITH, G. Wendell, (1917-1974)
SMITH, Gerald Julian, (1903-1976)
SMITH, Homer James, (1901-1954)
SNIDER, Luther Crocker, (1882-1947)
SPOFFORD, Howard Nelson, (1881-1950)
STEELE, Horace C., (1930-1977)
STEHR, Raymond Albert, (1903-1967)
STEIG, Maynard Herman, (1904-1965)
STEWART, Irvine E., (1893-1965)
SUFFIELD, Harry F., (1916-1966)
TANDY, James Hiram "Spot", (1892-1962)
TAYLOR, Charles Henry, (1877-1964)
TAYLOR, Garvin Lawrence, (1908-1974)
TAYLOR, W. Harlan, (1908-1972)
THOMPSON, Jean Cleveland, (1887-1929)
THOMPSON, Robert Raymond, (1893-1954)
THORNBURG, Dwight Holbrook "Duke", (1895-1965)
TILLOTSON, Allen Walter, (1905-1964)

UNITED STATES OF AMERICA (OKLAHOMA) 1900 to present Cont'd.

TOMLINSON, Charles Weldon, (1892-1960)
TRAGER, Earl A., (1893-1978)
TROXELL, John N., (1895-1970)
TURNER, Joseph Harold, (1900-1950)
VALERIUS, Michael Marcus, (1872-1947)
VERNON, Jesse, (1894-1972)
WADE, Arthur, (1878-1951)
WADELL, Hakon, (1895-1962)
WAGENER, Charles Henry "Chick", (1897-1950)
WAGNER, Cary Richard, Jr., (1917-1962)
WALKER, Lucian Hickman, (1894-1970)
WALLACE, Maurice Hudiburg, (1917-1961)
WARE, John McKee, (1905-1973)
WARING, Gerald Ashley, (1883-1971)
WEICHBRODT, Harold Tague, (1921-1966)
WEIDMAN, Samuel, (1870-1945)
WHEELER, Orby Clinton, (1898-1970)
WHITE, Stanley Battershill, (1899-1966)
WILLIAMS, Arthur James, (1877-1954)
WILLIAMS, George O. "Skipper", (1897-1967)
WILSON, Joseph Garland, (1902-1955)
WILSON, Leonard Richard, (1906-)
WILSON, Roy Arthur, (1891-1970)
WILSON, Walter Byron "Shorty", (1885-1951)
WINES, Donald Bradford, (1902-1958)
WINGETT, Raymond Oscar, (1909-1953)
WOLF, Key, (1886-1960)
WOOD, Robert Harrison, (1887-1956)
WOOD, Virgil O., (1892-1954)
WOODRUFF, Elmer Grant, (1872-1952)
WOODS, Earl Hazen, (1902-1950)
WRIGHT, Harry Favill, (1888-1944)
YEAGER, Lloyd Irvin, (1900-1938)
YEE, Bing Q., (1924-1973)
ZAVOICO, Basil B., (1899-1975)
ZOLLER, Lawrence J., (1895-1973)

UNITED STATES OF AMERICA
OREGON
1800-1849
CONDON, Thomas, (1822-1907)
EVANS, John, (1812-1861)
NUTTALL, Thomas, (1786-1859)
PARKER, Samuel, Reverend, (1779-1866)
SALTER, John William, (1820-1869)
SHUMARD, Benjamin Franklin, (1820-1869)

1850-1899
ANDERSON, Frank Marion, (1863-1945)
CONDON, Thomas, (1822-1907)
DUTTON, Clarence Edward, Major, (1841-1912)
EVANS, John, (1812-1861)
GRANT, Ulysses Sherman, (1867-1932)
NUTTALL, Thomas, (1786-1859)

UNITED STATES OF AMERICA (OREGON) 1850-1899 Cont'd.

PARKER, Samuel, Reverend, (1779-1866)
REINHART, Herman Francis, (1832-1889) [PROSPECTORS, DIVINERS AND MINING ENGINEERS]
RUSSELL, Israel Cook, (1852-1906)
SALTER, John William, (1820-1869)
SCOTT, William Berryman, (1858-1947)
SHUMARD, Benjamin Franklin, (1820-1869)
STERNBERG, Charles Hazelius, (1850-1943)
WILLIAMS, Ira Abraham, (1876-)

1900 to present
AHO, Aaro E., (-1977)
ANDERSON, Frank Marion, (1863-1945)
BELTZ, Edward W., (1891-1970)
BOSS, Richard F., (1923-1976)
BOUCOT, Arthur James, (1924-)
BURGER, Robert W. "Bush", (1906-1945)
CONDON, Thomas, (1822-1907)
CONKLING, Russell C., (1897-1955)
CROSS, Rodman K., (1911-1970)
DUTTON, Clarence Edward, Major, (1841-1912)
FISK, Harold Norman, (1908-1964)
FRY, Willis E., (1926-1973)
GILLULY, James, (-)
GOODSPEED, George Edward, (1887-1974)
GRANT, Ulysses Sherman, (1867-1932)
HEWETT, Donnel Foster, (1881-1971)
HODGE, Edwin Thomas, (1887-1970)
LINDSLEY, Thayer, (1882-) [PROSPECTORS, DIVINERS AND MINING ENGINEERS]
MAYFIELD, Samuel Martin, (1888?-1963)
PECORA, William Thomas, (1913-1972)
RUSSELL, Israel Cook, (1852-1906)
SCOTT, William Berryman, (1858-1947)
SMITH, Warren D., (1880-1950)
STERNBERG, Charles Hazelius, (1850-1943)
WASHBURNE, Chester Wesley, (1883-1971)
WEAVER, Charles Edwin, (1880-1958)
WHITTIER, William Harrison, (1889-1951)
WILKINSON, William Donald, (1901-1969)
WILLIAMS, Ira Abraham, (1876-)
YOUNGQUIST, Walter, (-)

UNITED STATES OF AMERICA
PENNSYLVANIA
Up to 1800
COOPER, Thomas, (1759-1840)
HUTCHINS, Thomas, (1730-1789)
KALM, Pehr, (1716-1779)
THOMAS, David, (1776-1859)

1800-1849
COOPER, Thomas, (1759-1840)
FINCH, John, (fl.1823-1835)

UNITED STATES OF AMERICA (PENNSYLVANIA) 1800-1849 Cont'd.

GENTH, Friedrick August Ludwig Karl Wilhelm, later called Frederick Augustus, (1820-1893)
GREEN, Jacob, (1790-1841)
HARE, Robert, (1781-1858)
HOLL, Harvey Buchanan, (1820-1886)
HUNT, Thomas Sterry, (1826-1892)
LESLEY, J. Peter, (1819-1903)
LESQUEREUX, Charles Leo, (1806-1889)
LOGAN, William Edmond, Sir, (1798-1875)
ROGERS, Henry Darwin, (1808-1866)
SILLIMAN, Benjamin Jr., (1816-1885)
THOMAS, David, (1776-1859)
WHITTLESEY, Charles, Colonel, (1808-1886)

1850-1899

ASHBURNER, Charles Albert, (1854-1889)
ASHLEY, George Hall, (1866-1951)
BRANNER, John Casper, (1850-1922)
BREWER, William Henry, (1828-1910)
BROWN, Amos Peaslee, (1864-1917)
BUTTS, Charles, (1863-1946)
CAMPBELL, Marius Robinson, (1858-1940)
CHADWICK, George Halcott, (1876-1953)
CLAYPOLE, Edward Waller, (1835-1901)
DECKER, Charles Elijah, (1868-1958)
DOUGLAS, James, (1837-)
DOUGLASS, Earl, (1862-1931)
FRAZER, Persifor, (1844-1909)
FRAZIER, Benjamin West, (1841-1905)
GENTH, Friedrick August Ludwig Karl Wilhelm, later called Frederick Augustus, (1820-1893)
GILMORE, Charles Whitney, (1874-1945)
HARE, Robert, (1781-1858)
HEILPRIN, Angelo, (1853-1907)
HENNEN, Ray Vernon, (1875-1958)
HICE, Richard Robert, (1865-1925)
HILL, Frank A., (1858-1915)
HOLL, Harvey Buchanan, (1820-1886)
HOLLAND, William Jacob, (1848-1932)
HUNT, Thomas Sterry, (1826-1892)
INVILLIERS, Edward Vincent d', (1857-1928)
LESLEY, J. Peter, (1819-1903)
LESQUEREUX, Charles Leo, (1806-1889)
LEWIS, Henry Carvill, (1853-1888)
LOGAN, William Edmond, Sir, (1798-1875)
MILLER, Benjamin Le Roy, (1874-1944)
MOORE, Elwood S., (1879-1966)
PECK, Frederick Burritt, (1860-1925)
PETERSON, Olof August, (1865-1933)
PILSBRY, Henry Augustus, (1862-1957)
PLATT, Franklin, (1844-1900)
RICHARDSON, George Burr, (1872-1949)
ROGERS, Henry Darwin, (1808-1866)
SILLIMAN, Benjamin Jr., (1816-1885)

UNITED STATES OF AMERICA (PENNSYLVANIA) 1850-1899 Cont'd.

SMITH, William Richard Tangier, (1869-1972)
STEVENSON, John James, (1841-1924)
STONE, Ralph Walter, (1876-1964)
THOMAS, David, (1776-1859)
TOOTHAKER, Charles Robinson, (1873-1952)
WADSWORTH, Marshman Edward, (1847-1921)
WHITE, Charles David, (1862-1935)
WHITTLESEY, Charles, Colonel, (1808-1886)
WILLIAMS, George Huntington, (1856-1894)

1900 to present

ANDERSON, Gustavus Edwin, (1879-1940)
ASHLEY, George Hall, (1866-1951)
BELL, Olin Grogan, (1893-1970)
BEVIER, George Means, (1888-1972)
BOUCOT, Arthur James, (1924-)
BRANNER, John Casper, (1850-1922)
BREWER, William Henry, (1828-1910)
BROWN, Amos Peaslee, (1864-1917)
BUTTS, Charles, (1863-1946)
CAMPBELL, Marius Robinson, (1858-1940)
CATHCART, Stanley Holman, (1889-1953)
CHADWICK, George Halcott, (1876-1953)
CLAYPOLE, Edward Waller, (1835-1901)
CONDIT, Daniel Dale, (1886-1955)
DECKER, Charles Elijah, (1868-1958)
DOUGLASS, Earl, (1862-1931)
FAIRCHILD, John Gifford, (1882-1965)
FETTKE, Charles Reinhard, (1888-1959)
FRALICH, Charles Eugene, (1902-1964)
FRAZER, Persifor, (1844-1909)
FRAZIER, Benjamin West, (1841-1905)
GARDNER, James Henry, (1883-1964)
GAULT, Hugh Richard, (1915-1961)
GILLES, Verner Arthur, (1886-1954)
GILMORE, Charles Whitney, (1874-1945)
GLASS, Jewell Jeanette, (1888-1966)
GOLDMAN, Marcus Isaac, (1881-1965)
GORDON, Samuel George, (1897-1952)
GORMAN, Joseph M., (1910-1967)
HAKE, Benjamin Franklin, (1895-1973)
HANSON, Wallace W., (1921-1971)
HEALD, Kenneth Conrad, (1888-1971)
HEILPRIN, Angelo, (1853-1907)
HENNEN, Ray Vernon, (1875-1958)
HICE, Richard Robert, (1865-1925)
HILL, Frank A., (1858-1915)
HODGE, Edwin Thomas, (1887-1970)
HOLLAND, William Jacob, (1848-1932)
HONESS, Arthur Pharaoh, (1887-1942)
HOWARD, Edgar Billings, (1887-1943)
HOWELL, Benjamin Franklin, (1890-1976)
INVILLIERS, Edward Vincent d', (1857-1928)
KAY, James LeFoy, (1892-1971)

UNITED STATES OF AMERICA (PENNSYLVANIA) 1900 to present Cont'd.

KIMBALL, Kent K., (1892-1953)
KRYNINE, Paul Dimitri, (1902-1964)
KURTZ, Robert Gourdin, (1898-1958)
LAPHAM, Davis M., (1931-1974)
LEIGHTON, Henry, (1884-1963)
LESLEY, J. Peter, (1819-1903)
LILLIGREN, Johnathan Martin, (1880-1956)
MARPLE, Mildred Fisher, (1898-1966)
MASON, Shirley Lowell, (1893-1975)
McLAUGHIN, Dean Benjamin, (1901-1965)
MILLER, Benjamin Le Roy, (1874-1944)
MILLER, John Preston, (1923-1961)
MOORE, Elwood S., (1879-1966)
MOSELEY, John Reed, (1900-1959)
MYERS, Thurman H., (1890-1940)
MYERS, William Marsh, (1892-1951)
NEWHOUSE, Walter Harry, (1897-1969)
NICOLAS, Jean Paul, (1921-1967)
PATTERSON, Arthur Lindo, (1902-1966)
PECK, Frederick Burritt, (1860-1925)
PETERSON, Olof August, (1865-1933)
PILSBRY, Henry Augustus, (1862-1957)
PLATT, Franklin, (1844-1900)
POSTEL, Albert Williams, (1909-1966)
RICHARDS, James Taylor "Dick", (1892-1953)
RICHARDSON, George Burr, (1872-1949)
RONES, Morris, (1922-1965)
SHEPPS, Vincent Chester, (1928-1967)
SHERRILL, Richard Ellis, (1899-1953)
SMITH, William Richard Tangier, (1869-1972)
STEVENSON, John James, (1841-1924)
STONE, Ralph Walter, (1876-1964)
SWAIN, James Fulton, (1903-1948)
TAYLOR, W. Harlan, (1908-1972)
TOOTHAKER, Charles Robinson, (1873-1952)
TURNER, Joseph Harold, (1900-1950)
WADSWORTH, Marshman Edward, (1847-1921)
WAGNER, Walter Richard, (1928-1974)
WALTERS, Mason G., (1903-1967)
WANLESS, Harold Rollin, (1898-1970)
WARD, Freeman, (1879-1943)
WASSON, Theron, (1887-1970)
WHITE, Charles David, (1862-1935)
WINGETT, Raymond Oscar, (1909-1953)
WRIGHT, Harold Douglas, (1921-1969)

UNITED STATES OF AMERICA
RHODE ISLAND
1800-1849
JACKSON, Charles Thomas, (1805-1880)

1850-1899
EMERSON, Benjamin Kendall, (1843-1932)
JACKSON, Charles Thomas, (1805-1880)

UNITED STATES OF AMERICA (RHODE ISLAND) 1850-1899 Cont'd.

WARD, Lester Frank, (1841-1913)
WARREN, Charles Hyde, (1876-1950)

1900 to present
BRANSON, Carl Colton, (1906-1975)
DALE, Nelson Clark, (1880-1962)
EMERSON, Benjamin Kendall, (1843-1932)
FIELD, Richard Montgomery, (1885-1961)
HAWKINS, Alfred Cary, (1887-1954)
WARD, Lester Frank, (1841-1913)
WARREN, Charles Hyde, (1876-1950)

UNITED STATES OF AMERICA
SOUTH CAROLINA
Up to 1800
BARTRAM, William, (1739-1823)
COOPER, Thomas, (1759-1840)

1800-1849
BARTRAM, William, (1739-1823)
BRUMBY, Richard T., (1804-1856)
COOPER, Thomas, (1759-1840)
GIBBES, Robert W., (1809-)
LE CONTE, Joseph, (1823-1901)
TUOMEY, Michael, (1805-1857)
VANUXEM, Lardner, (1792-1848)
WOODROW, James, (1828-1907)

1850-1899
ABBE, Cleveland, Jr., (1872-1934)
ALDRICH, Truman Heminway, (1848-1932)
BRUMBY, Richard T., (1804-1856)
HOLMES, Francis S., (1815-1882)
JOHNSON, Douglas Wilson, (1878-1944)
LEATHERS, Walter S., (1874-)
LE CONTE, Joseph, (1823-1901)
LIEBER, Oscar M., (-1862)
MARTIN, Daniel Strobel, (1842-1925)
MOORE, Andrew Charles, (1866-1928)
SLOAN, Earle, (1858-1926)
TUOMEY, Michael, (1805-1857)
TWITCHELL, Mayville William, (1866-1927)
WOODROW, James, (1828-1907)

1900 to present
ABBE, Cleveland, Jr., (1872-1934)
ALDRICH, Truman Heminway, (1848-1932)
COOKE, Charles Wythe, (1887-1971)
GARDNER, Julia Anna, (1882-1960)
GLASS, Jewell Jeanette, (1888-1966)
HOLMES, Francis S., (1815-1882)
JOHNSON, Douglas Wilson, (1878-1944)
LEATHERS, Walter S., (1874-)
LE CONTE, Joseph, (1823-1901)

UNITED STATES OF AMERICA (SOUTH CAROLINA) 1900 to present Cont'd.

MELTON, Frank A., (-)
MOORE, Andrew Charles, (1866-1928)
SEASHORE, Paul Theodore, (1898-1954)
SLOAN, Earle, (1858-1926)
TABER, Stephen, (1882-1963)
TWITCHELL, Mayville William, (1866-1927)
WOODROW, James, (1828-1907)

UNITED STATES OF AMERICA
SOUTH DAKOTA
1800-1849
EVANS, John, (1812-1861)
HAYDEN, Ferdinand Vandiveer, (1829-1887)
MEEK, Fielding Bradford, (1817-1876)
NICOLLET, Joseph Nicolas, (1786-1843)

1850-1899
ABEL, Othenio, (1875-1946)
CARPENTER, Franklin Reuben, (1848-1910)
CASE, Ermine Cowles, (1871-1953)
CASWELL, John Henry, (1846-1909)
EVANS, John, (1812-1861)
GIDLEY, James Williams, (1866-1931)
HAYDEN, Ferdinand Vandiveer, (1829-1887)
HESNARD, Emile E., (1867-1929)
JOHANNSEN, Albert, (1871-1962)
LEE, Henry E., (1875-1947)
MEEK, Fielding Bradford, (1817-1876)
RICHARDSON, George Burr, (1872-1949)
SCOTT, William Berryman, (1858-1947)
SMITH, William Richard Tangier, (1869-1972)
TODD, James Edward, (1846-1922)
UPHAM, Warren, (1850-1934)
WIELAND, George Reber, (1865-1953)

1900 to present
ABEL, Othenio, (1875-1946)
BALLARD, William Norval, (1906-1976)
BEEKLEY, Albert Leon, (1883-1952)
BUMP, James Dye, (1903-1959)
CARMAN, Joel Ernest, (1882-1966)
CARPENTER, Franklin Reuben, (1848-1910)
CARR, William L., (1919-1975)
CASE, Ermine Cowles, (1871-1953)
CASWELL, John Henry, (1846-1909)
CONNOLLY, Joseph Peter, (1890-1947)
FOLGER, James Anthony, (1896-1956)
GIDLEY, James Williams, (1866-1931)
GWYNNE, Charles Sumner, (1885-1972)
HARES, Charles Joseph, (1881-1970)
HESNARD, Emile E., (1867-1929)
HIGAZY, Riad Abdel-Megid, (1919-1967)
JOHANNSEN, Albert, (1871-1962)
KIRBY, Maurice E. "Bud", (1909-1963)

UNITED STATES OF AMERICA (SOUTH DAKOTA) 1900 to present Cont'd.

LEE, Henry E., (1875-1947)
MEFFERD, Ralph L., (1909-1964)
MOORE, Paul Brian, (1940-)
REYNOLDS, Thomas Emmett, (-1972)
RICHARDSON, George Burr, (1872-1949)
RUBEY, William Walden, (1898-1974)
RUNNER, Joseph James, (1885-1970)
RUSSELL, William Low, (1897-1976)
SCOTT, William Berryman, (1858-1947)
SMITH, William Richard Tangier, (1869-1972)
STIRTON, Ruben Arthur, (1901-1966)
TETRICK, Paul Roderick, (1920-1956)
TODD, James Edward, (1846-1922)
TURNER, Joseph Harold, (1900-1950)
UPHAM, Warren, (1850-1934)
WARD, Freeman, (1879-1943)
WIELAND, George Reber, (1865-1953)
WILSON, Roy Arthur, (1891-1970)

UNITED STATES OF AMERICA
TENNESSEE
Up to 1800
KAIN, John H., (1759-1831)
TROOST, Gerard, (1776-1850)

1800-1849
BOWEN, George Thomas, (1803-1828)
CORNELIUS, Elias C., (1794-1832)
ERNI, Henri, (1822-1885)
KAIN, John H., (1759-1831)
LONG, Stephen Harriman, (1784-1864)
OWEN, David Dale, (1807-1860)
RHEA, Matthew, (1795-1879)
RÖMER, Carl Ferdinand von, (1818-1891)
TROOST, Gerard, (1776-1850)

1850-1899
ASHLEY, George Hall, (1866-1951)
BRADLEY, Frank Howe, (1838-1879)
ERNI, Henri, (1822-1885)
GLENN, Leonidas Chalmers, (1871-1951)
GORDON, Charles Henry, (1857-1934)
INVILLIERS, Edward Vincent d', (1857-1928)
LONG, Stephen Harriman, (1784-1864)
McCALLIE, Samuel Washington, (1856-)
OWEN, David Dale, (1807-1860)
PURDUE, Albert Homer, (1861-1917)
RHEA, Matthew, (1795-1879)
RÖMER, Carl Ferdinand von, (1818-1891)
SAFFORD, James Merrill, (1822-1907)
WHITE, Charles David, (1862-1935)
WILSON, Charles Henry, (1851-1887)

UNITED STATES OF AMERICA (TENNESSEE) Cont'd.

1900 to present
ASHLEY, George Hall, (1866-1951)
BORN, Kendall Eugene, (1908-1947)
BRANNOCK, Kent Combs "Casey", (1923-1973)
BURFOOT, James Dabney "Dan", Jr., (1896-1966)
CLARK, Leslie Madison, (1898-1976)
FRALICH, Charles Eugene, (1902-1964)
FRITTS, Crawford Ellsworth, (1927-1972)
GLENN, Leonidas Chalmers, (1871-1951)
GORDON, Charles Henry, (1857-1934)
HALL, George Martin, (1891-1941)
HIGGINS, Daniel Franklin, Jr., (1884-1930)
INVILLIERS, Edward Vincent d', (1857-1928)
JEWELL, Willard B., (-)
KIMBALL, Kent K., (1892-1953)
LILLIGREN, Johnathan Martin, (1880-1956)
MATHER, Kirtley Fletcher, (1888-1978)
McCALLIE, Samuel Washington, (1856-)
McCLAIN, Alan Hartman, (1899-1960)
McNEELY, Robert Barnes, (1902-1966)
MISER, Hugh Dinsmore, (1884-1969)
NELSON, Wilbur Armistead, (1890-1969)
PERRY, Thomas Gregory, (1919-1972)
POND, Walter Franklin, (1885-1974)
PURDUE, Albert Homer, (1861-1917)
RHOADES, Roger F., (1905-1972)
RICHARDS, James Taylor "Dick", (1892-1953)
SAFFORD, James Merrill, (1822-1907)
STOCKDALE, Paris Buell, (1896-1962)
SWARTZ, Joel Howard, (1893-1971)
WALLS, James G., (1907-)
WELLS, Francis Gerrit, (1898-1969)
WHITE, Charles David, (1862-1935)
WILLIAMS, George O. "Skipper", (1897-1967)

UNITED STATES OF AMERICA
TEXAS
1800-1849
BOLL, Jacob, (1828-1880)
BUCKLEY, Samuel Botsford, (1809-1884)
LINCECUM, Gideon, (1793-1874)
MOORE, Francis, Jr., (1808-1864)
RÖMER, Carl Ferdinand von, (1818-1891)
SHUMARD, Benjamin Franklin, (1820-1869)
SHUMARD, George Getz, (1826-1867)

1850-1899
ALDRICH, Truman Heminway, (1848-1932)
BEEDE, Joshua William, (1871-1940)
BOLL, Jacob, (1828-1880)
BÖSE, Emil, (1868-1972)
BROILI, Ferdinand, (1874-1946)
BUCKLEY, Samuel Botsford, (1809-1884)
BURCHARD, Ernest Norman, (1875-1961)

UNITED STATES OF AMERICA (TEXAS) 1850-1899 Cont'd.

CARNEY, Frank, (1868-1934)
CHADWICK, George Halcott, (1876-1953)
DARTON, Nelson Horatio, (1865-1948)
DUMBLE, Edwin Theodore, (1852-1927)
GETZENDANER, Frank Marshall, (1876-1957)
GIDLEY, James Williams, (1866-1931)
GIRTY, George Herbert, (1869-1939)
GLENN, John, (-)
GORDON, Charles Henry, (1857-1934)
GOULD, Charles Newton, (1868-1949)
HAGER, Lee, (1874-1944)
HARRIS, Gilbert Dennison, (1864-1952)
HENDERSON, James Alexander Leo, (?1874-1940)
HENNEN, Ray Vernon, (1875-1958)
HESS, Frank L., (1871-1955)
HILL, Robert Thomas, (1858-1941)
KOCH, Walter Edward, (1848-1916)
LINCECUM, Gideon, (1793-1874)
MOORE, Francis, Jr., (1808-1864)
MORSE, William Clifford, (1874-1962)
PHILLIPS, William Battle, (1857-1918)
RICHARDSON, George Burr, (1872-1949)
RIES, Heinrich, (1871-1951)
RÖMER, Carl Ferdinand von, (1818-1891)
SELLARDS, Elias Howard, (1875-1961)
SHUMARD, Benjamin Franklin, (1820-1869)
SHUMARD, George Getz, (1826-1867)
STANTON, Timothy William, (1860-1953)
STEPHENSON, Lloyd William, (1876-1962)
STERNBERG, Charles Hazelius, (1850-1943)
TAFF, Joseph Alexander, (1862-1944)
UDDEN, Johan August, (1859-1932)
VALERIUS, Claude N., (-)

1900 to present

ADAMS, John Emery, (1899-1971)
ADKINS, Walter Scott, (1890-1956)
AID, Harry, (1897-1927)
AIMER, James David, (1899-1973)
ALDRICH, Truman Heminway, (1848-1932)
ALLEN, Walter J., (1890-1961)
AMBROSE, Arthur Warren, (1889-1952)
ANDERSON, George Harold, (1893-1956)
ARGABRITE, William Greame, (1885-1963)
ASHMORE, Herman T., (1916-1976)
ATHY, Lawrence Ferdinand, (1898-1955)
ATKINSON, William H., (1898-1972)
BALLARD, James Logan, (1902-1974)
BANKS, Thomas Russell, (1900-1948)
BARLOW, Maerlant M. "Pete", (1906-1950)
BARROW, Leonidas Theodore "Slim", (1895-1978)
BARTELL, Lawrence Denny, (1902-1955)
BASSINGER, William E., (1928-1977)
BEACH, Hugh Hamilton, (1909-1973)

UNITED STATES OF AMERICA (TEXAS) 1900 to present Cont'd.

BECKER, Ted G., (1921-1975)
BECKWITH, Radcliffe Harold, (1900-1964)
BEEDE, Joshua William, (1871-1940)
BEEKLEY, Albert Leon, (1883-1952)
BEERY, Roy F., Jr., (1917-1964)
BELL, Olin Grogan, (1893-1970)
BENSON, Dale L., (1899-1968)
BERG, Ernest Lyle, (1910-1953)
BERNARD, Hugh Allen, (1915-1975)
BEVIER, George Means, (1888-1972)
BLANCHARD, William Grant, Jr., (1894-1952)
BLANPIED, Bernerd William, (1897-1975)
BLOODWORTH, Bill Lloyd, (1916-1971)
BLOOMER, Philip A., Jr., (1922-1967)
BLUMENTHAL, Milton Diez, (1922-1964)
BOCKELMAN, Frederick H., Jr., (1923-1968)
BOGARDUS, Robert Stephen "Bud", (1914-1953)
BÖSE, Emil, (1868-1972)
BOSS, Richard F., (1923-1976)
BOWEN, James P., (1889-1960)
BOYLE, Walter J., (1897-1973)
BRACE, Orval L., (1893-1975)
BRADLEY, William Frank, (1908-1973)
BREMER, B. E. "Boots", (1904-1974)
BRILL, Virgil August, (1905-1972)
BRINKERHOFF, Ira A., (1904-1969)
BRISCOE, Glenn Oliver, (1898-1974)
BROILI, Ferdinand, (1874-1946)
BRYAN, Roger Wesley, (1927-1971)
BURCHARD, Ernest Norman, (1875-1961)
BURNS, Russell Warren, (1912-1967)
BURRESS, Walter March, (1893-1950)
BURTON, George Edgar, (1880-1953)
BUTCHER, Cary Preston, (1898-1950)
BYBEE, Halbert Pleasant, (1888-1957)
BYRON, Clifford Winkler, (1901-1970)
CAMP, John F. Jr., (1915-1976)
CAMPBELL, Harry A., (1893-1951)
CAMPBELL, Robert Burns, (1892-1955)
CANADA, William Richards, (1909-1962)
CANFIELD, Charles Reiter, (1906-1964)
CANNON, Robert Lee, (1899-1961)
CARLTON, Dave Pierce, (1890-1976)
CARNEY, Frank, (1868-1934)
CARR, George Watson, (1899-1950)
CASTEEL, Sanford Henry, (1898-1952)
CASTER, Ermil Leslie, (1896-1955)
CHADWICK, George Halcott, (1876-1953)
CHENEY, George Monroe, (1893-1952)
CLARK, Frank Rinker, (1881-1974)
CLARK, Paul Henry, (1909-1951)
CLIFTON, Roland Leroy, (1884-1962)
CLINE, Lewis Manning, (1909-1971)
COFFIN, Reuben Clare, (1886-1972)

UNITED STATES OF AMERICA (TEXAS) 1900 to present Cont'd.

COLLINS, Melvin J., (1897-1936)
CONKLING, Richard A., (1885-1952)
CONKLING, Russell C., (1897-1955)
CONWAY, William Palen "Tony", Jr., (1914-1944)
COOPER, Herschel H., (1894-1954)
COPASS, Jack Malcolm, (1906-1963)
CORTES, Henry Cornelius, (1892-1957)
COTTINGHAM, Virgil Elisha, (1891-1958)
COX, William Edgerton, (1912-1957)
CRAIG, William Henry, (1897-1951)
CRONEIS, Carey, (1901-1972)
CRUM, Harry Edwin, (1892-1955)
CUYLER, Robert Hamilton, (1908-1944)
DACEY, Reginald Lynn, (1908-1951)
DAMON, Henry Gordon, (1883-1960)
DARTON, Nelson Horatio, (1865-1948)
DAVIE, Thomas Early, (1906-1959)
DAVIS, Thornton, (1892-1969)
DAWSON, Joseph Meadows, (1895-1946)
DECKER, Charles LaVerne, (1898-1955)
DeGOLYER, Everette Lee, (1886-1956)
DENISON, Albert Rodger, (1897-1962)
DeRUDDER, Ronald D., (1932-1975)
DEUSSEN, Alexander, (1882-1959)
DeWOLF, Frank Walbridge, (1881-1957)
DICKSON, Fielding Breeden, (1912-1972)
DODSON, Floyd Cummings, (1895-1949)
DONOGHUE, David Patrick, (1891-1958)
DORSEY, George Edwin, (1892-1953)
DOYLE, Robert, (1913-1963)
DUMBLE, Edwin Theodore, (1852-1927)
DURHAM, Charles Albert, (1910-1954)
EARGLE, Dolan Hoye, (1905-1973)
EARL, Eugene Leonard "Red", (1906-1956)
EBY, James Brian, (1896-)
EDSON, Dwight J., (1896-1976)
EDWARDS, Everett C., (1897-1972)
ELLEDGE, Emmett Russell, (1898-1960)
ERDMANN, Charles Edgar, (1897-1971)
ESAREY, Ralph Emerson, (1901-1975)
EVANS, Vincent, (1907-1951)
EYSSELL, Alfred Raab, (1898-1944)
FAHLE, Paul Bartlett, (1910-1954)
FATH, Arthur Earl, (1887-1976)
FERGUSON, John Logan, (1893-1973)
FERRERO, Clarence de Labretonne, (1903-1954)
FICHTER, Hans Jakob, (1908-1977)
FINCH, Elmer Harrison, (1883-1953)
FITZGERALD, James, Jr., (1903-1951)
FLY, Paul J., (1907-1967)
FOHS, Ferdinand Julius, (1884-1965)
FOLGER, James Anthony, (1896-1956)
FOSSHAGE, Ernest Willard, (1908-1955)
FREED, Richard, (1908-1967)

UNITED STATES OF AMERICA (TEXAS) 1900 to present Cont'd.

FRIZZELL, Donald Leslie, (1906-1972)
FROST, Jack, (1900-1970)
GAITHER, Benjamin Mason, (1919-1960)
GALLIE, John Francis, III, (1913-1951)
GANGI, Anthony Frank, (1929-)
GARDNER, Frank J., (1915-1975)
GARRETT, Melvin Miller "Dutch", (1894-1969)
GARRETT, Robert Ernest, (1888-1947)
GEORGE, William Owsley, (1892-1966)
GETZENDANER, Avalette Edmond, (1902-1962)
GETZENDANER, Frank Marshall, (1876-1957)
GIDLEY, James Williams, (1866-1931)
GIERHART, Guy B. "Joe", (1903-1968)
GIRTY, George Herbert, (1869-1939)
GOERNER, Frederic Stephen, (1915-1969)
GOLDMAN, Marcus Isaac, (1881-1965)
GOLLNICK, Robert Louis, (1918-1963)
GOODELL, Horace N. "Hank", (1915-1971)
GORDON, Charles Henry, (1857-1934)
GORDON, Dugald, (1886-1958)
GOULD, Charles Newton, (1868-1949)
GRAHAM, Daniel Willard, (1921-1954)
GRAHAM, John Warren, (1918-1971)
GRAHAM, Joseph John, (1909-1967)
GREEN, Guy Emmett "Squire", (1896-1965)
GRIMES, Russell Winters, (1900-1969)
GROSS, Arthur Bunert "Abie", (1904-1965)
GUINN, Delmar Richard, (1902-1955)
HAGER, Lee, (1874-1944)
HALEPESKA, Dennis Leonard, (1920-1958)
HALL, Thomas O., (1906-1972)
HAMMOND, Weldon Woolf, (1908-1964)
HAMMONDS, G. Scott, (1910-1965)
HANDLEY, Howard Walter, (1905-1940)
HANSON, Ernest Albin, (1899-1973)
HANSON, Wallace W., (1921-1971)
HARDISON, Harvey, (1895-1949)
HARPER, O. C., (1897-1973)
HARRELL, Marshall Allen, (1900-1959)
HARRINGTON, Horacio Jaime, (1910-1973)
HARRIS, Gilbert Dennison, (1864-1952)
HARRISON, George Haymond, (1903-1949)
HARTWIG, Albert Ernest, Jr., (1922-1966)
HAYWARD, Raymond Edward, (1932-1964)
HAZZARD, Roy Thorpe, (1893-1976)
HEALD, Kenneth Conrad, (1888-1971)
HEDRICK, Oley Foster, (1891-1950)
HENDERSON, Harold H. "Heinie", (1905-1949)
HENDERSON, James Alexander Leo, (?1874-1940)
HENLEY, Arthur Sidney, (1879-1936)
HENNEN, Ray Vernon, (1875-1958)
HERALD, Frank Edward, (1887-1965)
HEROY, William Bayard, Sr., (1883-1971)
HESS, Frank L., (1871-1955)

UNITED STATES OF AMERICA (TEXAS) 1900 to present Cont'd.

HESSE, Curtis Julian, (1905-1945)
HIGGS, Morton Thomas, (1896-1956)
HILL, Robert Thomas, (1858-1941)
HLAUSCHEK, Hans, (1896-1971)
HOLCOMB, Leroy D., (1919-1974)
HOOKWAY, Lozell Charles, (1905-1938)
HOOVER, James Barr, (1902-1953)
HORTON, Harold Merritt "Wog", (1897-1966)
HOUSTON, Samuel Henry, (1905-1947)
HOWELL, Benjamin Franklin, (1890-1976)
HOYLE, Charles Robert, Sr., (1891-1954)
HUBMAN, Carl Wesley, (1905-1947)
HUGHES, Richard, (1889-1969)
HUNT, Edwin Harold, (1897-1949)
HUTCHINSON, Frederick McAlpine, (1884-1952)
IDDINGS, Arthur, (1891-1957)
IRWIN, Marvin Leroy "Mel", (1915-1964)
ISENBERGER, Nathan Perry, (1896-1956)
JAMISON, Claude E., (1879-1939)
JANES, Charles Arthur, (1928-1966)
JANOSKY, Robert A., (1923-1955)
JARVIS, Daniel, (1920-1973)
JEFFREYS, Geoffrey, (1885-1953)
JENNINGS, Charles I. "Al", (1894-1968)
JENSEN, Christian "Cope", (1905-1961)
JOHNSON, Earle Winston, (1902-1953)
JONES, Daniel J., (1889-1965)
JUDSON, Sidney Arthur, (1887-1971)
JUDSON, Wilber, (1880-1951)
KELLY, Donald, (1898-1949)
KENNEDY, William, (-1926)
KEPPLER, Leo George, (1890-1971)
KEYES, Wilson, (1901-1936)
KIDD, Major Gentry, (1904-1974)
KING, Jack Joseph, (1902-1929)
KIRBY, Grady, (1889-1969)
KISLING, James William, Jr., (1898-1973)
KITSON, Howard Walde, (1883-1931)
KLINGER, Edgar Dewitt, (1899-1973)
KNEBEL, G. Moses, (1899-1974)
KNOX, Thomas Knight, (1896-1949)
KOCH, Harold Kenneth, (1927-1973)
KOCH, Walter Edward, (1848-1916)
KRALIS, Dan F., Jr., (1919-1962)
KREIDLER, William Lynn, (1911-1975)
KRIDLER, George Matthew, (1889-1953)
KURTZ, Robert Gourdin, (1898-1958)
LADNER, Albert L., (1905-1975)
LANCASTER, L. Kenneth, (1906-1971)
LARSH, Napoleon Bonaparte "Po", (1897-1967)
LAW, John Melvin, (1914-1967)
LEDBETTER, William Robert, (1921-1971)
LIDDLE, Ralph Alexander, (1896-1963)
LILLIGREN, Johnathan Martin, (1880-1956)

UNITED STATES OF AMERICA (TEXAS) 1900 to present Cont'd.

LIND, Charles McClelland, (1911-1957)
LINEHAN, Charles M., (1913-1973)
LLOYD, Edwin Russell, (1882-1955)
LLOYD, Edwin Russell "Rusty", Jr., (1924-1951)
LONSDALE, John Tipton, (1895-1960)
LOSKAMP, Alvin Powell, (1900-1954)
MACKENZIE, Andrew Nicholls "Kelley", (1888-1951)
MacNAUGHTON, Lewis Winslow, (1902-1969)
MANN, Robert Scott, (1915-1956)
MARSHALL, Alan Edward, (1938-1970)
MARTYN, Phil F., (1903-1974)
MASON, Shirley Lowell, (1893-1975)
MAYFIELD, William I., (1912-1972)
McCAULEY, Dan Wesley, (1916-1967)
McCLOY, Walter Lowrie, Jr., (1912-1954)
McCOLLOUGH, Edward Lamb, (1931-1955)
McFARLAND, Russell Scott, (1893-1968)
McFERRON, George Irwin, (1895-1969)
McGLOTHLIN, William Claude "Big Mack", (1895-1942)
McLAREN, Robert Lee, (1901-1953)
McNUTT, Vachel Harry, (1888-1936)
METCALF, Roy J., (1889-1941)
MILLER, Buford, (1906-1962)
MILLER, Carroll C., (1899-1974)
MILLER, John Cummins, (1895-1951)
MILLER, William D., (1930-1972)
MILLER, William F., (1929-1976)
MILNER, Charles Albert, Jr., (1900-1971)
MIX, Charles Alexander, (1898-1954)
MOODIE, Roy Lee, (1880-1934)
MOODY, Graham Blair, (1889-1975)
MOORE, Gilbert Parvin, (1898-1970)
MOORE, James Augustus, (1910-1955)
MOORE, Prentiss D., (1895-1955)
MORGAN, George Dillon, (1894-1950)
MORRILL, Junius Hughes, (1913-1961)
MORRIS, Albert Ferd, (1898-1957)
MORSE, William Clifford, (1874-1962)
MURCHISON, Eugene Alexander, III, (1931-1953)
MURPHY, James Kenneth, (1900-1955)
MYERS, Julian Q., (1889-1950)
MYERS, Thurman H., (1890-1940)
MYLIUS, Louis Aubrey, (1890-1965)
NEIGHBOURS, Glen Reed, (1917-1961)
NELSON, Lloyd Alvino, (1895-1964)
NEUMAN, Leonard Joseph, (1900-1953)
NININGER, Harvey Harlow, (-)
NISBET, John McHenry, (1894-1962)
NOBLE, Earl Bart, (1894-1967)
NYGREN, Walter Eric, (1907-1971))
OLCUTT, David Perry, (1896-1958)
OWENS, Frith Cravens, (1898-1977)
PALMER, Leland L., (1906-1978)
PATTON, Leroy Thompson, (1880-1957)

UNITED STATES OF AMERICA (TEXAS) 1900 to present Cont'd.

PEMBERTON, John Roy "Bill", (1884-1968)
PENNEL, Harry David, (1907-1971)
PENNY, Frederick Willoughby, (1887-1950)
PENROSE, Evelyn, (-) [PROSPECTORS, DIVINERS AND MINING ENGINEERS]
PEPPER, James Franklin, (1898-1963)
PERINI, Vincent Charles "Bud", Jr., (1895-1965)
PETSCH, Arthur Henry, (1895-1969)
PETTY, Dabney E., (1891-1963)
PETTY, Olive Scott, (1895-)
PETTY, Thomas Franklin, (1900-1962)
PHILBRICK, Eugene Patrick, (1897-1951)
PHILLIPS, William Battle, (1857-1918)
PICKLESIMER, Harold Greeley, (1916-1952)
PILCHER, Benjamin Luther, Jr., (1906-1946)
PLUMMER, Frederick Byron, (1886-1947)
PORCH, Edwin Leight, Jr., (1889-1948)
POTEET, James Harold, (1896-1965)
POTTER, Grover Cleveland, (1892-1975)
POWELL, Ralph Sterling, (1901-1951)
PRETTYMAN, Thomas Mann, (1888-1940)
PROUT, Franklin S. "Shorty", (1889-1940)
RAINWATER, Edward Harriman "Rainey", (1909-1972)
RASMUSSEN, William Charles, (1917-1973)
REIFF, Allan Eugene, (1908-1933)
RENAUD, Charles Louis, (1898-1964)
RENICK, Brink Coleman, (1897-1971)
REYNOLDS, Roy A., (1889-1943)
RICE, Elmer Michael, (1900-1971)
RICHARDSON, George Burr, (1872-1949)
RIES, Heinrich, (1871-1951)
RIFE, Byron, (1892-1974)
RITTENHOUSE, Gordon, (1910-1974)
ROBERTS, Morgan Evan, (1894-1941)
ROBINSON, Bryan Franklin, (1897-1949)
ROBINSON, Ernest Guy, (1898-1973)
ROGATZ, Henry, (1902-1974)
ROMER, Alfred Sherwood, (1894-1973)
ROW, Charles Herbert, (1895-1970)
RUDOLPH, Joseph, (1897-1953)
RUSS, Leon Fair, (1886-1953)
RUSSELL, Philip G., (1891-1971)
RUSSELL, Richard Joel, (1895-1971)
RUSSELL, William Low, (1897-1976)
RYAN, Russell Fossler, (1897-1935)
SANDIDGE, John Roy, (1897-1976)
SANFORD, Wendell Glenn, (1914-1958)
SAPPINGTON, Chester, (1905-1968)
SARJEANT, William Antony Swithin, (1935-)
SAVILLE, Wilson Gordon, (1897-1954)
SCHILLING, Karl Howard, (1894-1945)
SCHMIDT, Karl A. "Gus", (1900-1965)
SCHOLL, Guy Joseph, (1900-1968)
SCHOLL, Louis August, Jr., (1887-1953)

UNITED STATES OF AMERICA (TEXAS) 1900 to present Cont'd.

SCHROYER, Charles Ross, (1887-1932)
SCHWABENLAND, James Richard, (1926-1968)
SCHWARZ, Melbert Edgar, (1898-1957)
SCOTT, Gayle, (1894-1948)
SCOTT, Horace Lafayette, (1901-1976)
SCOTT, Walter Winthrop, (1893-1939)
SEASHORE, Paul Theodore, (1898-1954)
SELLARDS, Elias Howard, (1875-1961)
SEVERY, Clarence Luther, (1885-1975)
SHEARER, Harold Kurtz, (1889-1946)
SHERMAN, Roger Hiram, (1904-1954)
SHULER, Ellis William, (1881-1954)
SIMMONS, Kenneth Albert, (1906-1971)
SMITH, G. Wendell, (1917-1974)
SMITH, Gerald Julian, (1903-1976)
SNIDER, Luther Crocker, (1882-1947)
SOUTHER, John Barron, (1894-1964)
SPALDING, George M., (1930-1977)
SPOFFORD, Howard Nelson, (1881-1950)
STAINBROOK, Merrill Addison, (1897-1956)
STANTON, Timothy William, (1860-1953)
STEELE, Horace C., (1930-1977)
STEHR, Raymond Albert, (1903-1967)
STEIG, Maynard Herman, (1904-1965)
STEPHENSON, Lloyd William, (1876-1962)
STERNBERG, Charles Hazelius, (1850-1943)
STEUBING, William C., (1881-1943)
STEVENS, George D., (1898-1965)
STEWART, Irvine E., (1893-1965)
STONE, A. John F., (1917-1969)
SULLINS, M. Joseph, (1930-1972)
SUMAN, John Robert, (1890-1972)
SWEENEY, Henry Nunn, (1914-1972)
SWIGER, Rual Bower, (1897-1950)
TAFF, Joseph Alexander, (1862-1944)
TANNER, Hugh Alexander, (1904-1954)
TAYLOR, Garvin Lawrence, (1908-1974)
TAYLOR, W. Harlan, (1908-1972)
THAEMAR, James T., Sr., (1911-1972)
THIES, Roland K., (1905-1973)
THOMAS, Harry George, (1929-1976)
THOMAS, John Elmer, (1892-1949)
THOMAS, Norman Louis, (1897-1970)
THOMPSON, Robert Raymond, (1893-1954)
THORNBURG, Dwight Holbrook "Duke", (1895-1965)
THORNTON, Owen Fuller, (1916-1953)
TOLER, Henry N., (1901-1970)
TOMPKINS, Joseph D., (1905-1974)
TROXELL, John N., (1895-1970)
TURNER, Joseph Harold, (1900-1950)
UDDEN, Johan August, (1859-1932)
VANDER LECK, Lawrence, (1890-1954)
WAGENER, Charles Henry "Chick", (1897-1950)
WAGNER, Cary Richard, Jr., (1917-1962)

UNITED STATES OF AMERICA (TEXAS) 1900 to present Cont'd.

WALKER, Lucian Hickman, (1894-1970)
WARE, Thomas Preston, (1909-1972)
WARING, Gerald Ashley, (1883-1971)
WASHBURNE, Chester Wesley, (1883-1971)
WASSON, Theron, (1887-1970)
WEEKS, Herbert John, (1897-1960)
WEIGELT, Johannes, (1890-1948)
WEISS, Eriol Joseph, (1918-1967)
WHEELER, Orby Clinton, (1898-1970)
WHITAKER, Harvey, (1906-1964)
WHITCOMB, Bruce, (1900-1950)
WHITEHEAD, Robert Brooks, (1889-1936)
WHITTIER, William Harrison, (1889-1951)
WIER, Harvey James, Jr., (1924-1950)
WILHELM, Oscar, (1897-1975)
WILKERSON, Albert Samuel, (1897-1963)
WILLIAMS, George O. "Skipper", (1897-1967)
WILSON, Bernard Wright, (1922-1976)
WILSON, Joseph Garland, (1902-1955)
WILSON, Malcolm Earl, Sr., (1885-1960)
WILSON, William Jay, (1922-1971)
WIMBISH, Forrest E., (1900-1972)
WINGETT, Raymond Oscar, (1909-1953)
WINTON, Will McClain, (1885-1960)
WOLF, Albert G., (1885-1964)
WOODS, Earl Hazen, (1902-1950)
WRATHER, William Embry, (1883-1963)
WRIGHT, Andrew Clemmons, (1893-1968)
ZAVOICO, Basil B., (1899-1975)
ZIMMERMAN, Charles Collingwood, (1905-1961)

UNITED STATES OF AMERICA
UTAH
1800-1849
HAYDEN, Ferdinand Vandiveer, (1829-1887)
SCHIEL, Jacob Heinrich, (1813-)

1850-1899
AVERY, Charles Dwight, (1877-1947)
DOUGLASS, Earl, (1862-1931)
DUTTON, Clarence Edward, Major, (1841-1912)
EMMONS, Samuel Franklin, (1841-1911)
GILBERT, Grove Karl, (1843-1918)
GREGORY, Herbert E., (1869-1952)
GURLEY, William Frank Eugene, (1854-1943)
HAYDEN, Ferdinand Vandiveer, (1829-1887)
HESS, Frank L., (1871-1955)
KING, Clarence Rivers, (1842-1901)
OCHSENIUS, Carl Christian, (1830-)
POWELL, John Wesley, (1834-1902)
REINHART, Herman Francis, (1832-1889) [PROSPECTORS, DIVINERS AND MINING ENGINEERS]
STERNBERG, Charles Hazelius, (1850-1943)
STONE, Ralph Walter, (1876-1964)

UNITED STATES OF AMERICA (UTAH) 1850-1899 Cont'd.

WILLIAMS, Gardner Fred, (1842-1922)
WOODRUFF, Elmer Grant, (1872-1952)

1900 to present

ABERNATHY, George Elmer, (1888-1948)
AVERY, Charles Dwight, (1877-1947)
BANDY, Mark Chance, (1900-1963)
BECKWITH, Radcliffe Harold, (1900-1964)
BILLINGSLEY, Paul, (1887-1962)
BOSS, Richard F., (1923-1976)
BRADLEY, Wilmot Hyde, (1899-)
COCHRAN, Kenneth L., (1910-1973)
CONDIT, Daniel Dale, (1886-1955)
CRUM, Harry Edwin, (1892-1955)
DONAVAN, Jack Hilder, (1921-1953)
DOUGLASS, Earl, (1862-1931)
DUTTON, Clarence Edward, Major, (1841-1912)
EARDLEY, Armand John, (1901-1972)
EBY, James Brian, (1896-)
EMMONS, Samuel Franklin, (1841-1911)
ERDMANN, Charles Edgar, (1897-1971)
EVANS, Vincent, (1907-1951)
FINCH, Elmer Harrison, (1883-1953)
FOWLER, George Malcolm, (1885-1972)
GILBERT, Grove Karl, (1843-1918)
GILL, James Rogers, (1922-1972)
GILLES, Verner Arthur, (1886-1954)
GILLULY, James, (-)
GOLDTHWAIT, James Walter, (1880-1947)
GREGORY, Herbert E., (1869-1952)
GURLEY, William Frank Eugene, (1854-1943)
HANSON, Ernest Albin, (1899-1973)
HAZZARD, John Charles, (1903-1973)
HESS, Frank L., (1871-1955)
HUNT, Walter Frederick, (1882-1975)
JACOB, Charles Edward, (1914-1970)
JONES, Verner E., (1905-1977)
KAY, James LeFoy, (1892-1971)
KIMBALL, Kent K., (1892-1953)
KING, Clarence Rivers, (1842-1901)
KIVI, Wilho John, (1918-1950)
LAMMERS, Edward Chauncey Hinman, (1910-1971)
MARSELL, Ray Everett, (1893-1971)
McMINN, Paul Meloy, (1919-1974)
MOODY, Graham Blair, (1889-1975)
NOBLE, Levi Fatzinger, (1882-1965)
OWEN, William Joseph, (1929-1959)
PICARD, Meredith Dane, (1927-)
POWELL, John Wesley, (1834-1902)
PRATLEY, Henry Hart, (1893-1941)
RUDOLPH, Joseph, (1897-1953)
SCHILLING, Karl Howard, (1894-1945)
SPOFFORD, Howard Nelson, (1881-1950)
STERNBERG, Charles Hazelius, (1850-1943)

UNITED STATES OF AMERICA (UTAH) 1900 to present Cont'd.

STONE, Ralph Walter, (1876-1964)
STRINGHAM, Bronson F., (1907-1968)
SUTTON, Frederick A., (1894-1950)
THIEL, Edward, (1928-1962)
UNTERMANN, Billie R., (1906-1973)
UNTERMANN, Gerhard Ernest, Jr., (1898-1975)
WHITE, Stanley Battershill, (1899-1966)
WHITTIER, William Harrison, (1889-1951)
WILLIAMS, Gardner Fred, (1842-1922)
WILLIAMS, George O. "Skipper", (1897-1967)
WOODRUFF, Elmer Grant, (1872-1952)

UNITED STATES OF AMERICA
VERMONT
1800-1849
SEELY, Henry Martyn, (1828-1917)
THOMPSON, Zadock, (1796-1856)
WING, Augustus, Reverend, (1808-1876)

1850-1899
HITCHCOCK, Charles Henry, (1836-1919)
HUDSON, George Henry, (1855-1934)
JACOBS, Elbridge Churchill, (1873-1957)
PERKINS, George Henry, (1844-1933)
RICHARDSON, Charles Henry, (1862-1935)
SEELY, Henry Martyn, (1828-1917)
THOMPSON, Zadock, (1796-1856)
UPHAM, Warren, (1850-1934)
WING, Augustus, Reverend, (1808-1876)

1900 to present
BROWN, Thomas Clachar, (1882-1934)
CURRIER, Louis Wade, (1890-1970)
HITCHCOCK, Charles Henry, (1836-1919)
HOWELL, Benjamin Franklin, (1890-1976)
HUDSON, George Henry, (1855-1934)
JACOBS, Elbridge Churchill, (1873-1957)
McKINSTRY, Hugh Exton, (1896-1961)
PERKINS, George Henry, (1844-1933)
RICHARDSON, Charles Henry, (1862-1935)
SEELY, Henry Martyn, (1828-1917)
UPHAM, Warren, (1850-1934)

UNITED STATES OF AMERICA
VIRGINIA
Up to 1800
HARIOT, Thomas, (1560-1621)
HUTCHINS, Thomas, (1730-1789)
JEFFERSON, Thomas, (1743-1826)
SMITH, John, Captain, (1580-1631)

1800-1849
CORNELIUS, Elias C., (1794-1832)
HARE, Robert, (1781-1858)

UNITED STATES OF AMERICA (VIRGINIA) 1800-1849 Cont'd.

JEFFERSON, Thomas, (1743-1826)
ROGERS, William Barton, (1804-1882)

1850-1899

BUTTS, Charles, (1863-1946)
CAMPBELL, Henry Donald, (1862-1934)
CHADWICK, George Halcott, (1876-1953)
FONTAINE, William Morris, (1835-1913)
GILMORE, Charles Whitney, (1874-1945)
HARE, Robert, (1781-1858)
HESS, Frank L., (1871-1955)
HOLDEN, Roy Jay, (1870-1945)
INVILLIERS, Edward Vincent d', (1857-1928)
JOHNSON, Douglas Wilson, (1878-1944)
JULIAN, Henry Forbes, (1861-1912) [PROSPECTORS, DIVINERS AND MINING ENGINEERS]
MANSFIELD, Wendell Clay, (1874-1939)
RIES, Heinrich, (1871-1951)
ROGERS, William Barton, (1804-1882)
SHALER, Nathaniel Southgate, (1841-1906)
WATSON, Thomas Leonard, (1871-1924)
WINCHELL, Alexander Newton, (1874-1958)

1900 to present

BURFOOT, James Dabney "Dan", Jr., (1896-1966)
BURWELL, Edward Bouldin, Jr., (1894-1964)
BUTTS, Charles, (1863-1946)
CAMPBELL, Henry Donald, (1862-1934)
CANTWELL, Thomas, Jr., (1927-)
CARR, William L., (1919-1975)
CHADWICK, George Halcott, (1876-1953)
COOPER, Byron Nelson, (1912-1971)
COPASS, Jack Malcolm, (1906-1963)
CURRIER, Louis Wade, (1890-1970)
EBY, James Brian, (1896-)
FONTAINE, William Morris, (1835-1913)
FURCRON, Aurelius Sydney, (1899-1971)
GALLAGHER, David, (1906-1968)
GALPIN, Sidney Longman, (1886-1962)
GARDNER, Julia Anna, (1882-1960)
GILMORE, Charles Whitney, (1874-1945)
GLASS, Jewell Jeanette, (1888-1966)
HESS, Frank L., (1871-1955)
HOLDEN, Roy Jay, (1870-1945)
INVILLIERS, Edward Vincent d', (1857-1928)
JOHNSON, Douglas Wilson, (1878-1944)
JULIAN, Henry Forbes, (1861-1912) [PROSPECTORS, DIVINERS AND MINING ENGINEERS]
KIMBALL, Kent K., (1892-1953)
KREIDLER, William Lynn, (1911-1975)
LILLIGREN, Johnathan Martin, (1880-1956)
LONSDALE, John Tipton, (1895-1960)
MANSFIELD, Wendell Clay, (1874-1939)
McCLAIN, Alan Hartman, (1899-1960)

UNITED STATES OF AMERICA (VIRGINIA) 1900 to present Cont'd.

NELSON, Wilbur Armistead, (1890-1969)
PEGAU, Albert August, (1894-1963)
RIES, Heinrich, (1871-1951)
SHALER, Nathaniel Southgate, (1841-1906)
SHERRILL, Richard Ellis, (1899-1953)
STOW, Marcellus Henry, (1902-1957)
WATSON, Thomas Leonard, (1871-1924)
WINCHELL, Alexander Newton, (1874-1958)
WRIGHT, Frank James, (1888-1954)

UNITED STATES OF AMERICA
WASHINGTON
1800-1849
EVANS, John, (1812-1861)
SHUMARD, Benjamin Franklin, (1820-1869)

1850-1899
BASSLER, Raymond Smith, (1878-1961)
ENDLICH, Frederic Miller, (1851-1899)
EVANS, John, (1812-1861)
GORDON, Charles Henry, (1857-1934)
HAY, Oliver Perry, (1846-1930)
HILLEBRAND, William Francis, (1853-1925)
LANDES, Henry, (1867-1936)
LEVERETT, Frank, (1859-1943)
MARTIN, Daniel Strobel, (1842-1925)
MORSE, William Clifford, (1874-1962)
REINHART, Herman Francis, (1832-1889) [PROSPECTORS, DIVINERS AND MINING ENGINEERS]
SHUMARD, Benjamin Franklin, (1820-1869)
STERNBERG, Charles Hazelius, (1850-1943)
WASHINGTON, Henry Stephens, (1867-1934)
WILLIAMS, Ira Abraham, (1876-)
WILLIS, Bailey, (1857-1949)

1900 to present
BARTH, Thomas Frederik Weiby, (1899-1971)
BASSLER, Raymond Smith, (1878-1961)
BELTZ, Edward W., (1891-1970)
BILLINGSLEY, Paul, (1887-1962)
BRANSON, Carl Colton, (1906-1975)
BURGER, Robert W. "Bush", (1906-1945)
CAMPBELL, Arthur Byron, (1924-1971)
CASSEL, Chester A. "Chet", (1897-1971)
COOMBS, Howard A., (-)
CROSS, Rodman K., (1911-1970)
CULVER, Harold Eugene, (1883-1970)
ERDMANN, Charles Edgar, (1897-1971)
FICHTER, Hans Jakob, (1908-1977)
FRY, Willis E., (1926-1973)
FRYXELL, Roald Hilding, (1934-1974)
FULLER, Richard Eugene, (1897-1976)
GOODSPEED, George Edward, (1887-1974)
GORDON, Charles Henry, (1857-1934)

UNITED STATES OF AMERICA (WASHINGTON) 1900 to present Cont'd.

HAY, Oliver Perry, (1846-1930)
HILLEBRAND, William Francis, (1853-1925)
HOLCOMB, Leroy D., (1919-1974)
HOWARD, Edgar Billings, (1887-1943)
HOWELL, Benjamin Franklin, (1890-1976)
LANDES, Henry, (1867-1936)
LEIGHTON, Morris Morgan, (1887-1971)
LEVERETT, Frank, (1859-1943)
LILLIGREN, Johnathan Martin, (1880-1956)
McCONNEL, Roger H., (1908-1971)
MORAN, Robert Breck, (1879-1961)
MORSE, William Clifford, (1874-1962)
MULLEN, John Timothy, Jr., (1906-1959)
PABST, Marie B., (1909-1963)
PALMER, Robert Hastings, (1882-1948)
SCHOLL, Louis August, Jr., (1887-1953)
SERVISS, Frederick LeVerne, (1895-1954)
SHERRILL, Richard Ellis, (1899-1953)
STERNBERG, Charles Hazelius, (1850-1943)
WASHINGTON, Henry Stephens, (1867-1934)
WEAVER, Charles Edwin, (1880-1958)
WHETTEN, John T., (-)
WILLIAMS, Ira Abraham, (1876-)
WILLIS, Bailey, (1857-1949)
WILSON, James W., (?1925-1964)

UNITED STATES OF AMERICA
WEST VIRGINIA
1850-1899
HENNEN, Ray Vernon, (1875-1958)
TILTON, John Littlefield, (1863-1930)
WHITE, Israel Charles, (1848-1927)
WILLIAMS, George Huntington, (1856-1894)

1900 to present
CRUM, Harry Edwin, (1892-1955)
GARDNER, James Henry, (1883-1964)
GROUT, Frank Fitch, (1880-1958)
HAKE, Benjamin Franklin, (1895-1973)
HENNEN, Ray Vernon, (1875-1958)
KURTZ, Robert Gourdin, (1898-1958)
LILLIGREN, Johnathan Martin, (1880-1956)
MASON, Shirley Lowell, (1893-1975)
McCLAIN, Alan Hartman, (1899-1960)
MYERS, Thurman H., (1890-1940)
PEPPER, James Franklin, (1898-1963)
PRICE, Paul Holland, (1898-)
RICHARDS, James Taylor "Dick", (1892-1953)
ROMER, Alfred Sherwood, (1894-1973)
SCHROYER, Charles Ross, (1887-1932)
SWIGER, Rual Bower, (1897-1950)
TAYLOR, W. Harlan, (1908-1972)
TILTON, John Littlefield, (1863-1930)
TURNER, Joseph Harold, (1900-1950)

UNITED STATES OF AMERICA (WEST VIRGINIA) 1900 to present Cont'd.

WHITE, Israel Charles, (1848-1927)
WOODWARD, Herbert Preston, (1899-1968)

UNITED STATES OF AMERICA
WISCONSIN
1800-1849
LAPHAM, Increase Allen, (1811-1875)
MEEK, Fielding Bradford, (1817-1876)
NICOLLET, Joseph Nicolas, (1786-1843)
NORWOOD, Joseph Granville, (1807-1895)
NUTTALL, Thomas, (1786-1859)
OWEN, David Dale, (1807-1860)
OWEN, Richard Dale, (1810-1890)
PERCIVAL, James Gates, (1795-1856)

1850-1899
BUCKLEY, Ernest Robertson, (1872-1912)
BURCHARD, Ernest Norman, (1875-1961)
CASE, Ermine Cowles, (1871-1953)
CHAMBERLIN, Thomas Chrowder, (1843-1928)
CLELAND, Herdman Fitzgerald, (1869-1935)
FENNEMAN, Nevin Melancthan, (1865-1945)
GRANT, Ulysses Sherman, (1867-1932)
HOBBS, William Herbert, (1864-1953)
IRVING, Roland Duer, (1847-1888)
JORDAN, David Starr, (1851-1931)
LAPHAM, Increase Allen, (1811-1875)
MEEK, Fielding Bradford, (1817-1876)
MILLER, Samuel A., (1837-1897)
MUIR, John, (1838-1914)
NORWOOD, Joseph Granville, (1807-1895)
NUTTALL, Thomas, (1786-1859)
OWEN, David Dale, (1807-1860)
OWEN, Richard Dale, (1810-1890)
PERCIVAL, James Gates, (1795-1856)
RIES, Heinrich, (1871-1951)
SALISBURY, Rollin D., (1858-1922)
TWENHOFEL, William Henry, (1875-1957)
UPHAM, Warren, (1850-1934)
VAN HISE, Charles Richard, (1857-1918)
VERRILL, Addison Emery, (1839-1926)
WEIDMAN, Samuel, (1870-1945)
WINCHELL, Alexander Newton, (1874-1958)

1900 to present
BIEBER, Charles Leonard, (1901-1965)
BILLINGS, Katharine Stevens (neé Fowler-Lunn), (1902-)
BLACKWELDER, Eliot, (1880-1969)
BRADY, Lionel Francis "Major", (1880-1963)
BRODERICK, Thomas Monteith, (1889-1965)
BUCKLEY, Ernest Robertson, (1872-1912)
BURCHARD, Ernest Norman, (1875-1961)
CADY, Gilbert Haven, (1882-1970)
CASE, Ermine Cowles, (1871-1953)

UNITED STATES OF AMERICA (WISCONSIN) 1900 to present Cont'd.

CHAMBERLIN, Thomas Chrowder, (1843-1928)
CLARK, Robert Watson, (1884-1948)
CLELAND, Herdman Fitzgerald, (1869-1935)
CLINE, Lewis Manning, (1909-1971)
COX, Guy Henry, (1881-1922)
CRONEIS, Carey, (1901-1972)
DAVIS, Ralph Emerson, (1884-1968)
FENNEMAN, Nevin Melancthan, (1865-1945)
GRANT, Ulysses Sherman, (1867-1932)
HANSELL, James Myron, (1902-1970)
HAWLEY, James Edwin, (1897-1965)
HOBBS, William Herbert, (1864-1953)
HODGE, Edwin Thomas, (1887-1970)
HONESS, Charles William, (1885-1949)
HOWELL, Benjamin Franklin, (1890-1976)
JORDAN, David Starr, (1851-1931)
KNIGHT, Garold Louis "Jerry", (1894-1968)
KOSCHMANN, Albert Herbert, (1896-1962)
LANCASTER, L. Kenneth, (1906-1971)
LEITH, Charles Kenneth, (1875-1956)
LOBECK, Armin Kahl, (1886-1958)
MARTIN, Francis Ivan "Shorty", (1896-1952)
McKINSTRY, Hugh Exton, (1896-1961)
MEAD, Warren Judson, (1883-1960)
MEHL, Maurice Goldsmith, (1887-1966)
MUIR, John, (1838-1914)
PLUMMER, Frederick Byron, (1886-1947)
PRETYMAN, Frank Remington, (1891?-1917)
RIES, Heinrich, (1871-1951)
ROARK, Louis, (1890-1950)
SALISBURY, Rollin D., (1858-1922)
SHROCK, Robert Rakes, (1904-)
SLICHTER, Louis Byrne, (1896-)
STEHR, Raymond Albert, (1903-1967)
THOMAS, Harry George, (1929-1976)
TRAINER, David Woolsey, Jr., (1899-1965)
TWENHOFEL, William Henry, (1875-1957)
TYLER, Stanley Allen, (1906-1963)
UPHAM, Warren, (1850-1934)
VAN HISE, Charles Richard, (1857-1918)
VERRILL, Addison Emery, (1839-1926)
WEIDMAN, Samuel, (1870-1945)
WILSON, Leonard Richard, (1906-)
WINCHELL, Alexander Newton, (1874-1958)
WING, Monta Eldo, (1897-1976)

UNITED STATES OF AMERICA
WYOMING
1800-1849
HAYDEN, Ferdinand Vandiveer, (1829-1887)

1850-1899
ABEL, Othenio, (1875-1946)
AVERY, Charles Dwight, (1877-1947)

UNITED STATES OF AMERICA (WYOMING) 1850-1899 Cont'd.

COOK, James, (fl. 1870)
DARTON, Nelson Horatio, (1865-1948)
ELDRIDGE, George Homans, (1854-1905)
FENNER, Clarence Norman, (1870-1949)
GRANT, Ulysses Sherman, (1867-1932)
HAGUE, Arnold, (1840-1917)
HAYDEN, Ferdinand Vandiveer, (1829-1887)
IRVING, John Duer, (1874-1918)
KNIGHT, Wilbur Clinton, (1858-1903)
LULL, Richard Swann, (1867-1957)
MARSH, Othniel Charles, (1831-1899)
MATTHES, François Emile, (1874-1948)
McKIM, John Walter, (1877-1927)
MORGAN, Glenn Beckley, (1876-1925)
PORRIO, Cesare, (1865-1940)
REINHART, Herman Francis, (1832-1889) [PROSPECTORS, DIVINERS AND MINING ENGINEERS]
RICHARDSON, George Burr, (1872-1949)
SCOTT, William Berryman, (1858-1947)
SMITH, William Richard Tangier, (1869-1972)
STANTON, Timothy William, (1860-1953)
STONE, Ralph Walter, (1876-1964)
WALTMAN, William Dewitt, (1875-1955)
WIELAND, George Reber, (1865-1953)
WOODRUFF, Elmer Grant, (1872-1952)

1900 to present

ABEL, Othenio, (1875-1946)
AGOGINO, George, (-)
AVERY, Charles Dwight, (1877-1947)
BALLARD, William Norval, (1906-1976)
BARTRAM, John Greer, (1893-1955)
BECKMANN, Lawrence John, (1912-1976)
BECKWITH, Radcliffe Harold, (1900-1964)
BRACE, Orval L., (1893-1975)
BRADLEY, Wilmot Hyde, (1899-)
BROWN, Bahngrell Walter "Bill", (1916-)
BROWN, John Albert, (1901-1954)
CARLSON, Charles Gordon, (1895-1977)
CARMAN, Joel Ernest, (1882-1966)
CARR, William L., (1919-1975)
CASSEL, Chester A. "Chet", (1897-1971)
COCHRAN, Kenneth L., (1910-1973)
COMSTOCK, Theodore Bryant, (1849-1915)
CONDIT, Daniel Dale, (1886-1955)
COOK, Harold James, (1887-?1962)
CRAWFORD, David Johnston, (1899-1960)
CREWS, George Curtis, (1837-1977)
DAKE, Charles Laurence, (1883-1934)
DARTON, Nelson Horatio, (1865-1948)
DOBBIN, Carroll Edward "Cy", (1892-1967)
DONAVAN, Jack Hilder, (1921-1953)
DOUGLAS, James Mainwaring, (1887-1952)
DOYLE, Robert, (1913-1963)

UNITED STATES OF AMERICA (WYOMING) 1900 to present Cont'd.

EARGLE, Dolan Hoye, (1905-1973)
ELDRIDGE, George Homans, (1854-1905)
ESPACH, Ralph Homeward, (1900-1956)
FENNER, Clarence Norman, (1870-1949)
FINCH, Elmer Harrison, (1883-1953)
FOLGER, James Anthony, (1896-1956)
FREY, M. Gordon, (1913-)
FRITTS, Crawford Ellsworth, (1927-1972)
GILL, James Rogers, (1922-1972)
GOLDMAN, Marcus Isaac, (1881-1965)
GRANT, Ulysses Sherman, (1867-1932)
HAGUE, Arnold, (1840-1917)
HALL, Elwin B., (1891-1949)
HANSON, Ernest Albin, (1899-1973)
HARES, Charles Joseph, (1881-1970)
HARRISON, John Wiley, (1916-1952)
HARRISON, Thomas Samuel, (1881-1964)
HEROLD, Stanley Carrollton, (1883-1958)
HEWETT, Donnel Foster, (1881-1971)
HIGGINS, Daniel Franklin, Jr., (1884-1930)
HORBERG, Carl Leland, (1910-1955)
HOWELL, Jesse V., (1891-1971)
IRVING, John Duer, (1874-1918)
JAMISON, Claude E., (1879-1939)
JOHNSON, Bertrand Leroy, (1882-1962)
KIVI, Wilho John, (1918-1950)
KNIGHT, Samuel Howell, (1892-1975)
KNIGHT, Wilbur Clinton, (1858-1903)
KURTZ, Robert Gourdin, (1898-1958)
LEWIS, Francis Leonard, Jr., (1921-1953)
LLOYD, Edwin Russell, (1882-1955)
LULL, Richard Swann, (1867-1957)
MACKENZIE, Andrew Nicholls "Kelley", (1888-1951)
MAEBIUS, Jed B., (1907-1969)
MATTHES, François Emile, (1874-1948)
MAUCINI, Joseph John, (1893-1955)
McKIM, John Walter, (1877-1927)
McMINN, Paul Meloy, (1919-1974)
MILLER, John Preston, (1923-1961)
MITCHELL, Kingsley Camden, (1903-1931)
MIX, Charles Alexander, (1898-1954)
MOORE, James Augustus, (1910-1955)
MORAN, Robert Breck, (1879-1961)
MORGAN, George Dillon, (1894-1950)
MORGAN, Glenn Beckley, (1876-1925)
MULLEN, John Timothy, Jr., (1906-1959)
MURRAY, George H., Jr., (1928-1971)
NANCE, James W., (1908-1977)
NETTERSTROM, Paul Wagner, (1914-1960)
NININGER, Harvey Harlow, (-)
NOWLAN, Harry Hackett, (1888-1958)
OWEN, William Joseph, (1929-1959)
PORRIO, Cesare, (1865-1940)
RATH, Charles Martin, (1881-1959)

UNITED STATES OF AMERICA (WYOMING) 1900 to present Cont'd.

RICHARDS, James Taylor "Dick", (1892-1953)
RICHARDSON, George Burr, (1872-1949)
RUBEY, William Walden, (1898-1974)
SCHILLING, Karl Howard, (1894-1945)
SCHRAMM, Eck Frank, (1883-1967)
SCOTT, Walter Winthrop, (1893-1939)
SCOTT, William Berryman, (1858-1947)
SMITH, Aylwin L., (1898-1965)
SMITH, William Richard Tangier, (1869-1972)
SNIVELEY, H. Norman, (1879-1959)
STANTON, Timothy William, (1860-1953)
STERNBERG, Charles M., (1885-)
STERNBERG, George Fryer, (1883-1969)
STEWART, Irvine E., (1893-1965)
STONE, Ralph Walter, (1876-1964)
TAYLOR, W. Harlan, (1908-1972)
TETRICK, Paul Roderick, (1920-1956)
THOMAS, Harry George, (1929-1976)
THOMAS, Horace David "Bill", (1905-1967)
TILLOTSON, Allen Walter, (1905-1964)
TURNER, Joseph Harold, (1900-1950)
WALTMAN, William Dewitt, (1875-1955)
WASHBURNE, Chester Wesley, (1883-1971)
WHITE, Stanley Battershill, (1899-1966)
WIELAND, George Reber, (1865-1953)
WILLIAMS, George O. "Skipper", (1897-1967)
WILSEY, James Arthur, Jr., (1921-1949)
WOOD, Virgil O., (1892-1954)
WOODRUFF, Elmer Grant, (1872-1952)

UNITED STATES OF AMERICA
STATE unspecified
Up to 1800
ANNAN, Robert, Reverend, (fl.1782)
BARTRAM, John, (1699-1777)
CLARK, William, (1770-1838)
CLINTON, DeWitt, (1769-1828)
COLLOT, Georges Henri Victor, (1750-1805)
COOPER, Thomas, (1759-1840)
DRAKE, Daniel, (1785-1852)
EVANS, Lewis, (1700-1756)
GIBBS, George, Colonel, (1776-1833)
LESUEUR, (1778-1846)
MACKENZIE, Alexander, Sir, (ca.1767-1830)
MACLURE, William, (1763-1840)
MICHAELIS, Christian Friedrich, (fl.1766-1780)
MITCHILL, Samuel Latham, (1764-1831)
MORGAN, George, (fl.1766)
MORGAN, John, (fl.1765)
PEALE, Charles Willson, (1741-1827)
PERKINS, Benjamin Douglas, (1774-1810)
PETIVER, James, (1658-1718)
SCHOEPF, Johann David, (1752-1800)
VOLNEY, Constantin Francois Chasseboeuf, Comte de, (1757-1820)

UNITED STATES OF AMERICA (STATE unspecified) Up to 1800 Cont'd.

WELD, Isaac, Jr., (1774-1856)
WINTHROP, John, (1714-1779)
WISTAR, Caspar, (1761-1818)

1800-1849

ADAMS, Charles Baker, (1814-1853)
BACHE, Alexander Dallas, (1806-1867)
BAILEY, Jacob Whitman, (1811-1857)
BAIRD, Spencer Fullerton, (1823-1887)
BARRIS, Willis Hervey, (1820-1901)
BARTRAM, William, (1739-1823)
BLAKE, William Phipps, (1826-1910)
BOUVÉ, Thomas Tracy, (1815-1896)
CATLIN, George, (1796-1872)
CHRISTY, Henry, (1810-1865)
CLARK, William, (1770-1838)
CLINTON, DeWitt, (1769-1828)
COLLOT, Georges Henri Victor, (1750-1805)
CONRAD, Timothy Abbott, (1803-1877)
COOK, George Hammell, (1818-1889)
COOPER, Thomas, (1759-1840)
COUTHONY, Joseph Pitty, (1808-1864)
CROGHAN, George, (fl. 1765)
DANA, James Freeman, (1793-1827)
DAVIDSON, George, (1825-1911)
DEAN, Bashford, (1867-1928)
DOROSZHIN, Peter P., (-)
DRAKE, Daniel, (1785-1852)
EATON, T. D., (-?1829)
EMORY, William Hemsley, (1811-1887)
ENGELMANN, George, (1809-1884)
ERNI, Henri, (1822-1885)
FANNING, Tolbert, (1810-1874)
FERREL, William, (1817-1891)
FRÉMONT, John Charles, (1813-1890)
GAY, Martin, (1803-1850)
GIBBS, George, Colonel, (1776-1833)
GODMAN, John Davidson, (1794-1830)
GRAY, Asa, (1810-1888)
HALL, James, (1811-1898)
HAYS, Isaac, (1796-1879)
HOOKER, Joseph Dalton, Sir, (1817-1911)
JAMES, Edwin, (1797-1861)
JOHNSTON, James Finlay Weir, (1796-1855)
KEATING, William Hypolitus, (1799-1840)
KERR, Washington Caruthers, (1827-1885)
LACOE, Ralph Dupuy, (1824-1901)
LEA, Isaac, (1792-1886)
LEIDY, Joseph, (1823-1891)
LESQUEREUX, Charles Leo, (1806-1889)
LESUEUR, (1778-1846)
LOOMIS, Elias, (1811-1890)
MACFARLANE, James, (1819-1885)
MACKENZIE, Alexander, Sir, (ca.1767-1830)

UNITED STATES OF AMERICA (STATE unspecified) 1800-1849 Cont'd.

MACLURE, William, (1763-1840)
MARCOU, Jules, (1824-1898)
MAURY, Matthew Fontaine, (1806-1873)
MITCHILL, Samuel Latham, (1764-1831)
MOORE, Nathaniel Fish, (1782-1872)
MORTON, Samuel George, (1799-1851)
O'SULLIVAN, T. H., (-1882)
PEALE, Charles Willson, (1741-1827)
PEALE, Titian Ramsay, (1799-1885)
PERKINS, Benjamin Douglas, (1774-1810)
RAFINESQUE-SCHMALTZ, Constantine Samuel, (1783-1840)
REDFIELD, John Howard, (1815-1895)
REDFIELD, William C., (1789-1857)
RENSSELAER, Jeremiah van, (1793-1871)
SAY, Thomas, (1787-1834)
SCHOEPF, Johann David, (1752-1800)
SILLIMAN, Benjamin, Sr., (1779-1864)
ST. JOHN, Samuel, (1813-1876)
SYMMES, John Cleves, (1780-1829)
TAYLOR, Richard Cowling, (1789-1851)
TORREY, John, (1796-1873)
VERNEUIL, Philippe Edouard Poulletier de, (1805-1873)
VOLNEY, Constantin François Chasseboeuf, Comte de, (1757-1820)
WEBSTER, John White, (1793-1850)
WELD, Isaac, Jr., (1774-1856)
WHITE, Charles Abiathar, (1826-1910)
WISTAR, Caspar, (1761-1818)
WYMAN, Jeffries, (1814-1874)
XÁNTUS, János, (1825-1894)

1850-1899

ABBOT, Henry Larcom, (1831-1927)
ADAMS, Charles Baker, (1814-1853)
ADAMS, George Whitney, (1872-1930)
AGASSIZ, Alexander, (1835-1910)
ALLEN, Joel Asaph, (1838-1921)
ARCTOWSKI, Henryk Bronislaw, (1871-1958)
ARNOLD, Ralph, (1875-1961)
ASHBURNER, Charles Albert, (1854-1889)
ASHLEY, George Hall, (1866-1951)
AVERY, Charles Dwight, (1877-1947)
BACHE, Alexander Dallas, (1806-1867)
BAGG, Rufus Mather, (1869-1946)
BAILEY, Jacob Whitman, (1811-1857)
BAILEY, Loring Woart, (1839-1925)
BAIRD, Spencer Fullerton, (1823-1887)
BALDWIN, David, (fl.1874-1881)
BALLOU, William Hosea, (1857-1937)
BARBER, Charles M, (1877-1954)
BARRINGER, Daniel Moreau, Sr., (1860-1929)
BARRIS, Willis Hervey, (1820-1901)
BARROIS, Charles Eugène, (1851-1939)
BASCOM, Florence, (1862-1945)
BATES, Mowry, (1875-1926)

UNITED STATES OF AMERICA (STATE unspecified) 1850-1899 Cont'd.

BAYLEY, William Shirley, (1861-1943)
BECKER, George Ferdinand, (1847-1919)
BEMENT, Clarence S., (1843-1923)
BICKMORE, Albert Smith, (1839-1914)
BLAKE, John Marcus, (1838-1920)
BLAKE, William Phipps, (1826-1910)
BOLTON, Sarah Elizabeth Mary (née Knowles), (1841-1916)
BOUTWELL, John Mason, (1874-1968)
BOUVÉ, Thomas Tracy, (1815-1896)
BOWEN, Charles Franklin, (1871-1953)
BOWMAN, Isaiah, (1882-1950)
BRADY, Henry Bowman, (1835-1891)
BRANSON, Edwin Bayer, (1877-1950)
BROOKS, Alfred Hulse, (1871-1924)
BROWN, Amos Peaslee, (1864-1917)
BROWN, Barnum, (1873-1963)
BRYAN, William Jennings, (1860-1925)
BUEHLER, Henry Andrew, (1876-1944)
CAHN, Lazard, (1865-1940)
CAMPBELL, Douglas Houghton, (1859-1953)
CAMPBELL, Guy, (1877-1970)
CAMPBELL, Marius Robinson, (1858-1940)
CANFIELD, Frederick Alexander, (1849-1926)
CAPELLINI, Giovanni, (1833-1922)
CARNEY, Frank, (1868-1934)
CARPENTER, Franklin Reuben, (1848-1910)
CATLIN, George, (1796-1872)
CHAMBERLAIN, Arthur, (-)
CHAMBERLIN, Thomas Chrowder, (1843-1928)
CHEVALIER, Jerome Archibald, (1877-1940)
CHRISTY, Henry, (1810-1865)
CHURCH, James Edward, (1869-1959)
CLARK, William Bullock, (1860-1917)
CLARKE, Frank Wigglesworth, (1847-1931)
CLARKE, John Mason, (1857-1925)
COLBURN, Burnham Standish, (1872-1959)
COLONY, Roy Jed, (1870-1936)
CONRAD, Timothy Abbott, (1803-1877)
COOK, George Hammell, (1818-1889)
COOPER, James Graham, (1830-1902)
COPE, Edward Drinker, (1840-1897)
COUPER, James Graham, (1830-1902)
COUTHONY, Joseph Pitty, (1808-1864)
CREDNER, Carl Hermann Georg, (1841-1913)
CRIDER, Albert Foster, (1873-1945)
CROSBY, William Otis, (1850-1925)
CUMINGS, Willard Lawsen, (1874-1945)
DALL, William Healey, (1845-1927)
DARTON, Nelson Horatio, (1865-1948)
DAVIDSON, George, (1825-1911)
DAVIS, Charles Albert, (1861-1916)
DAVIS, Leander S., (fl. 1871)
DAVIS, William Morris, (1850-1934)
DEAN, Bashford, (1867-1928)

UNITED STATES OF AMERICA (STATE unspecified) 1850-1899 Cont'd.

DELÉPINE, Gaston-Georges, Monseigneur, (1878-1963)
DERBY, Orville Adelbert, (1851-1915)
DITMARS, Raymond Lee, (1876-1942)
DONNELLY, Ignatius, (1831-1901)
DORSEY, Noah Ernest, (1873-1959)
DRAKE, Daniel, (1785-1852)
DRUMMOND, Henry, (1851-1897)
EASTMAN, Charles Rochester, (1868-1918)
EMMONS, Samuel Franklin, (1841-1911)
EMORY, William Hemsley, (1811-1887)
ENDLICH, Frederic Miller, (1851-1899)
ENGELMANN, George, (1809-1884)
ENGLISH, George Letchworth, (1864-1944)
ERNI, Henri, (1822-1885)
FANNING, Tolbert, (1810-1874)
FERREL, William, (1817-1891)
FISCHER, Otto, (1877-1943)
FISHER, Cassius Asa, (1872-1930)
FISKE, John, (1842-1901)
FLEMING, John Adam, (1877-1956)
FONTAINE, William Morris, (1835-1913)
FOOTE, Albert E., (1846-1895)
FORT, Charles, (1874-1932)
FOX-STRANGWAYS, Charles Edward, (1844-1910)
FRAAS, Eberhard, (1862-1915)
FRAZER, Persifor, (1844-1909)
FRÉMONT, John Charles, (1813-1890)
FULLER, Homer Taylor, (1838-1910)
GALE, Hoyt Stoddard, (1876-1952)
GLENN, Leonidas Chalmers, (1871-1951)
GOODE, George Brown, (1851-1896)
GRABAU, Amadeus William, (1870-1946)
GRANGER, Walter, (1872-1941)
GRANT, Ulysses Sherman, (1867-1932)
GRATACAP, Louis Pope, (1850-1917)
GRAY, Asa, (1810-1888)
GRINNELL, George Bird, (1849-1938)
HAGUE, Arnold, (1840-1917)
HAGUE, James Duncan, (1836-1908)
HALL, James, (1811-1898)
HARRIS, Gilbert Dennison, (1864-1952)
HASEMAN, William Peter, (1878-1932)
HATCH, Frederick Henry, (?1864-1932)
HATCHER, John Bell, (1861-1904)
HAY, Oliver Perry, (1846-1930)
HAY, Robert, (1835-1895)
HAYES, Charles Willard, (1858-1916)
HAYS, Isaac, (1796-1879)
HEATH, Daisy Winifred, (1875-1954)
HELLMAN, Milo, (1872-1947)
HENDERSON, James Alexander Leo, (?1874-1940)
HENDERSON, Junius, (1865-1937)
HENNEN, Ray Vernon, (1875-1958)
HERMANN, Adam, (-)

UNITED STATES OF AMERICA (STATE unspecified) 1850-1899 Cont'd.

HIDDEN, William Earl, (1853-1918)
HILLIERS, John K., (-)
HILLS, Richard Charles, (1848-1923)
HOLBROOK, Levi, (1836-1922)
HOLDEN, Roy Jay, (1870-1945)
HOLMES, Joseph Austin, (1859-1915)
HOOKER, Joseph Dalton, Sir, (1817-1911)
HOVEY, Horace Carter, Reverend, (1833-1914)
HOWE, Marshall Avery, (1867-1936)
HOWELL, Edwin Eugene, (1845-1911)
HYLAND, John Shearson, (1866-1898)
JACKSON, William Henry, (1843-1942)
JAMES, Edwin, (1797-1861)
JOHNSTON, James Finlay Weir, (1796-1855)
JULIEN, Alexis Anastay, (1840-1919)
KEITH, Arthur, (1864-1944)
KERR, Washington Caruthers, (1827-1885)
KEYES, Charles Rollin, (1864-1942)
KNIGHT, Charles Robert, (1875-1953)
KUNZ, George Frederick, (1856-1932)
LACOE, Ralph Dupuy, (1824-1901)
LANGTON, David W., Jr, (1864-?1908)
LEA, Isaac, (1792-1886)
LEIDY, Joseph, (1823-1891)
LESQUEREUX, Charles Leo, (1806-1889)
LEVISON, Wallace Goold, (1846-1924)
LEWIS, Joseph Volney, (1869-1969)
LOOMIS, Elias, (1811-1890)
LOWELL, Percival, (1855-1916)
LUCAS, Frederick Augustus, (1852-1929)
LYELL, Charles, Sir, (1797-1875)
LYMAN, Benjamin Smith, (1835-1920)
MacCURDY, George Grant, (1863-1947)
MACFARLANE, James, (1819-1885)
MANCHESTER, James Greenfield, (1871-1948)
MARBUT, Curtis Fletcher, (1863-1935)
MARCOU, Jules, (1824-1898)
MARGERIE, Emmanuel Marie Pierre Martin Jacquin de, (1862-1953)
MARKS, Edward Seaborn, (?1875-1950)
MARTIN, George C., (1875-1943)
MARVINE, Archibald Robertson, (1848-1876)
MATSON, George Charlton, (1873-1940)
MATTHES, François Emile, (1874-1948)
MAURY, Matthew Fontaine, (1806-1873)
McCLUNG, Clarence Erwin, (1870-1946)
McGEE, William John, (1853-1912)
MELL, Patrick Hues, (1850-1918)
MERRIAM, John Campbell, (1869-1945)
MERRILL, George Perkins, (1854-1929)
MERRILL, Loren B., (?1853-1930)
MILLER, Benjamin Le Roy, (1874-1944)
MILLS, Enos Abijah, (1870-1922)
MILLS, James Ellison, (1834-1901)
MOORE, Nathaniel Fish, (1782-1872)

UNITED STATES OF AMERICA (STATE unspecified) 1850-1899 Cont'd.

MORAN, Thomas, (1837-1926)
MORTON, Samuel George, (1799-1851)
NASON, Henry Bradford, (1831-1895)
ORTMANN, Arnold Edward, (1863-1927)
O'SULLIVAN, T. H., (-1882)
PACKARD, Alpheus Spring, Jr., (1839-1905)
PEALE, Titian Ramsay, (1799-1885)
PEDRICK, Howard A. "Dad", (1863-?1930) [PROSPECTORS, DIVINERS AND MINING ENGINEERS]
PENROSE, Richard Alexander Fullerton, Jr., (1863-1931)
PROSSER, Charles Smith, (1860-1916)
RAYMOND, Rossiter Worthington, (1840-1918)
REDFIELD, John Howard, (1815-1895)
REDFIELD, William C., (1789-1857)
RENSSELAER, Jeremiah van, (1793-1871)
ROCKWOOD, Charles Greene, (1843-1913)
ROEBLING, Washington Augustus, Colonel, (1837-1926)
ROMEU, Albert de, (1875-1915)
RUEDEMANN, Rudolf, (1864-1956)
SALISBURY, Rollin D., (1858-1922)
SARGENT, Harry Lionel, (1871-1955)
SCHAEFFER, Charles Frederic August, (1860-1903)
SCHRADER, Frank Charles, (1860-1944)
SCUDDER, Samuel Hubbard, (1837-1911)
SEE, Thomas Jefferson Jackson, (1866-)
SHALER, Nathaniel Southgate, (1841-1906)
SHEPARD, Edward Martin, (1854-1934)
SHUFELDT, Robert Wilson, (1850-1934)
SILLIMAN, Benjamin, Sr., (1779-1864)
SLOAN, Earle, (1858-1926)
SMITH, George Otis, (1871-1944)
SMITH, James Perrin, (1864-1931)
SMITH, Lawrence, (-1883)
STAINER, Xavier, (1865-)
STAPFF, Friedrich Moritz, (1836-1895)
STEIGER, George, (1869-1944)
STEINMANN, Gustav, (1856-1929)
STEWART, Philip Charteris Anstruther, (1874-1923)
ST. JOHN, Samuel, (1813-1876)
SWEET, Sidney R., (1878-1946)
THIESSEN, Reinhardt, (1867-1938)
TORREY, John, (1796-1873)
ULRICH, Edward Oscar, (1857-1944)
VAN INGEN, Gilbert, (1869-1928)
VAUGHAN, Thomas Wayland, (1870-1952)
VAUX, George, Jr., (1863-1927)
VAUX, William S., Jr., (1872-1908)
VERNEUIL, Philippe Edouard Poulletier de, (1805-1873)
VOGDES, Anthony Wayne, General, (1843-1923)
WADE, Arthur, (1878-1951)
WALCOTT, Charles Doolittle, (1850-1927)
WALL, George Parkes, (1832-1912)
WALTMAN, William Dewitt, (1875-1955)
WARD, Lester Frank, (1841-1913)

UNITED STATES OF AMERICA (STATE unspecified) 1850-1899 Cont'd.

WATERSCHOOT VAN DER GRACHT, Willem Anton Josef Maria van, (1873-1943)
WATSON, Thomas Leonard, (1871-1924)
WEBSTER, John White, (1793-1850)
WELD, Isaac, Jr., (1774-1856)
WELLS, Roger Clark, (1877-1944)
WESTGATE, Lewis Gardner, (1868-1948)
WHITE, Charles Abiathar, (1826-1910)
WHITE, Charles David, (1862-1935)
WHITE, Israel Charles, (1848-1927)
WOEIKOF, Alexander, (1842-1916)
WOODRUFF, Elmer Grant, (1872-1952)
WOODWARD, Robert Simpson, (1849-1924)
WOOLNOUGH, Walter George, (1876-1958)
WORTMAN, Jacob Lawson, (1856-1926)
WRIGHT, Albert Allen, (1846-1905)
WRIGHT, Frederick Eugene, (1877-1953)
WRIGHT, George Frederick, (1838-1921)
WYMAN, Jeffries, (1814-1874)
XÁNTUS, János, (1825-1894)

1900 to present
ADAMS, George Whitney, (1872-1930)
ADAMS, Noah C., (1891-1927)
ADKINS, John Nathaniel, (1911-)
AGASSIZ, Alexander, (1835-1910)
AINSWORTH, David, (1898-1967)
ALDREDGE, Robert Frank, (1895-1961)
ALLEN, Joel Asaph, (1838-1921)
ANDERSEN, Olaf, (1884-1941)
ANDERSON, Alfred Leonard, (1900-1964)
ANDERSON, Robert Lee, (1912-1959)
ANDERTON, Russell, (-) [PROSPECTORS, DIVINERS AND MINING ENGINEERS]
ANDREEN, Harry Mayo, (1891-1933)
ANDREWS, Richard Wallis, (?1927-1967)
ANSBERRY, Leroy Eugene, (1923-1961)
APPLIN, Esther Richards, (1895-1972)
ARCTOWSKI, Henryk Bronislaw, (1871-1958)
ARGABRITE, William Greame, (1885-1963)
ARNOLD, Ralph, (1875-1961)
ASHLEY, George Hall, (1866-1951)
ASHMORE, Leland Wayne, (1916-1964)
AURIN, Fritz Love, (1891-1969)
AVERY, Charles Dwight, (1877-1947)
BAGG, Rufus Mather, (1869-1946)
BAILEY, Loring Woart, (1839-1925)
BAIN, Samuel McCutcheon, (-)
BAIRD, Lucille Bailey, (1926-1963)
BALK, Robert, (1899-1955)
BALL, Max Waite, (1885-1954)
BALLARD, James Logan, (1902-1974)
BALLARD, William Norval, (1906-1976)
BALLOU, Albert Lorenzo, Jr., (1917-1962)

UNITED STATES OF AMERICA (STATE unspecified) 1900 to present Cont'd.

BALLOU, William Hosea, (1857-1937)
BANDY, Mark Chance, (1900-1963)
BANKS, Thomas Russell, (1900-1948)
BARBER, Charles M, (1877-1954)
BARD, Richards J., (1900-1977)
BARLOW, Maerlant M. "Pete", (1906-1950)
BARRINGER, Daniel Moreau, Sr., (1860-1929)
BARRINGTON, Thomas W., (-)
BARRIS, Willis Hervey, (1820-1901)
BARROIS, Charles Eugène, (1851-1939)
BARROW, Geoffrey, (1899-1963)
BARTLETT, Helen Blair, (1901-1969)
BARTON, Donald Clinton, (1889-1939)
BARTRAM, John Greer, (1893-1955)
BASCOM, Florence, (1862-1945)
BASCOM, Willard, (1917-)
BATES, Mowry, (1875-1926)
BATES, Robert Glenn, (1923-1972)
BAYLEY, William Shirley, (1861-1943)
BEACH, Floyd Kellogg, (1884-1964)
BECKER, George Ferdinand, (1847-1919)
BECKWITH, Frank Asahel, (-1951)
BEEKLEY, Albert Leon, (1883-1952)
BEERY, Roy F., Jr., (1917-1964)
BELL, Olin Grogan, (1893-1970)
BEMENT, Clarence S., (1843-1923)
BENNETT, Clyde M., (1881-1935)
BENNETT, Russell Hoadley, (1896-)
BENSON, Dale L., (1899-1968)
BENTZ, Alfred, (1897-1964)
BENTZ, Ivan Vincent, (1896-1938)
BERMAN, Harry, (1902-1944)
BERNOUILLI, Walter, (1885-1946)
BICKMORE, Albert Smith, (1839-1914)
BIGGS, Charles L. "Tib", (1919-1971)
BINNEY, Edwin, Jr., (1899-1928)
BIRD, Roland T., (-1978)
BLAIR, Charles Scofield, (1883-1971)
BLAKE, George Stanfield, (-1940)
BLAKE, John Marcus, (1838-1920)
BLAKE, William Phipps, (1826-1910)
BLEDSOE, Lucybelle, (1923-1966)
BLOODWORTH, Bill Lloyd, (1916-1971)
BLUMENTHAL, Milton Diez, (1922-1964)
BOCKELMAN, Frederick H., Jr., (1923-1968)
BOLTON, Sarah Elizabeth Mary (née Knowles), (1841-1916)
BOLYARD, Garrett Lincoln, (1897-1963)
BONTZ, Conrad K., (1894-1927)
BOSWORTH, Thomas Owen, (1882-1928)
BOUCOT, Arthur James, (1924-)
BOUTWELL, John Mason, (1874-1968)
BOWEN, Charles Franklin, (1871-1953)
BOWLING, Leslie, (1908-1977)
BOWMAN, Isaiah, (1882-1950)

UNITED STATES OF AMERICA (STATE unspecified) 1900 to present Cont'd.

BRANSON, Edwin Bayer, (1877-1950)
BRANT, Ralph Allen, (1899-1958)
BRAUCHLI, R. W., (1895-1975)
BREWER, Charles, Jr., (1900-1944)
BRILL, Virgil August, (1905-1972)
BRINKERHOFF, Ira A., (1904-1969)
BROOKS, Alfred Hulse, (1871-1924)
BROUWER, Hendrik Albertus, (1886-1973)
BROWN, Amos Peaslee, (1864-1917)
BROWN, Barnum, (1873-1963)
BROWN, Noel K., Jr., (1925-1968)
BRYAN, Roger Wesley, (1927-1971)
BRYAN, William Jennings, (1860-1925)
BUEHLER, Henry Andrew, (1876-1944)
BURDICK, Clifford, (-)
BURNS, Russell Warren, (1912-1967)
BURTON, George Edgar, (1880-1953)
BUTCHER, Cary Preston, (1898-1950)
BUTCHER, Walter William, (1916-1966)
BUTTERWORTH, Emerson McMillin, (1894-1961)
BUTZ, Walter "Wally", (1905-1962)
BUWALDA, John Peter, (1886-1954)
BYBEE, Halbert Pleasant, (1888-1957)
CADY, Richard Carlysle, (1908-1943)
CAHN, Lazard, (1865-1940)
CAIN, James Walter, Jr., (1917-1941)
CALAME, Jean Jacques, (1923-1973)
CAMPBELL, Douglas Houghton, (1859-1953)
CAMPBELL, Guy, (1877-1970)
CAMPBELL, Marius Robinson, (1858-1940)
CANFIELD, Frederick Alexander, (1849-1926)
CANNON, Robert Lee, (1899-1961)
CAPELLINI, Giovanni, (1833-1922)
CARNEY, Frank, (1868-1934)
CARPENTER, Everett, (1884-1968)
CARPENTER, Franklin Reuben, (1848-1910)
CARROLL, Don Llewellyn, (1902-1954)
CARTER, Frank B., (1898-1968)
CASEBEER, Charles Theodore, (1905-1941)
CASEY, John Hestwood, (1923-1967)
CASTEEL, Sanford Henry, (1898-1952)
CATHCART, Stanley Holman, (1889-1953)
CHAMBERLAIN, Arthur, (-)
CHAMBERLIN, Rollin Thomas, (1881-1948)
CHAMBERLIN, Thomas Chrowder, (1843-1928)
CHATMAN, Cecil La Mar, (1902-1938)
CHENEY, Charles Albert, (1885-1937)
CHENEY, George Monroe, (1893-1952)
CHEVALIER, Jerome Archibald, (1877-1940)
CHISHOLM, William Francis, (1886-1944)
CHURCH, James Edward, (1869-1959)
CHURCHILL, Frederick Holmes, (1901-1959)
CLAPP, Frederick Gardner, (1879-1944)
CLARK, Frank Rinker, (1881-1974)

UNITED STATES OF AMERICA (STATE unspecified) 1900 to present Cont'd.

CLARK, Howard, (1885-1972)
CLARK, William Bullock, (1860-1917)
CLARKE, Frank Wigglesworth, (1847-1931)
CLARKE, John Mason, (1857-1925)
CLIFTON, Roland Leroy, (1884-1962)
CLOOS, Ernst, (1898-1974)
CLOOS, Hans, (1885-1951)
CLOUD, Preston Ercelle, Jr., (1912-)
CLOUGH, Kenneth H., (1903-1966)
COFFIN, Reuben Clare, (1886-1972)
COIL, Clarence, (-)
COLBURN, Burnham Standish, (1872-1959)
COLLOM, Roy Edward, (1881-1948)
COLONY, Roy Jed, (1870-1936)
COMSTOCK, Theodore Bryant, (1849-1915)
CONDIT, Daniel Dale, (1886-1955)
CONKLING, Richard A., (1885-1952)
CONNOLLY, Joseph Peter, (1890-1947)
COOKE, Charles Wythe, (1887-1971)
COOPER, James Graham, (1830-1902)
CORTES, Henry Cornelius, (1892-1957)
COUPER, James Graham, (1830-1902)
CREDNER, Carl Hermann Georg, (1841-1913)
CRICKMAY, Geoffrey W., (1905-1971)
CRIDER, Albert Foster, (1873-1945)
CROSBY, Irving Ballard, (1891-1959)
CROSBY, William Otis, (1850-1925)
CULVER, Harold Eugene, (1883-1970)
CUMINGS, Willard Lawsen, (1874-1945)
CUSHMAN, Joseph Augustine, (1881-1949)
CUTTITTA, Frank, (1912-1974)
DAKE, Charles Laurence, (1883-1934)
DALL, William Healey, (1845-1927)
DANE, Carle Hamilton, (1900-1968)
DARTON, Nelson Horatio, (1865-1948)
DAVIDSON, George, (1825-1911)
DAVIS, Charles Albert, (1861-1916)
DAVIS, William Morris, (1850-1934)
DeGOLYER, Everette Lee, (1886-1956)
DE KOCH, William Gaspar, (1898-1954)
DELÉPINE, Gaston-Georges, Monseigneur, (1878-1963)
DENISON, Albert Rodger, (1897-1962)
DERBY, Orville Adelbert, (1851-1915)
DERDEN, Jesse Homer, (1890-1944)
DeRUDDER, Ronald D., (1932-1975)
DEUSSEN, Alexander, (1882-1959)
DeWOLF, Frank Walbridge, (1881-1957)
DIETZ, Robert Sinclair, (1914-)
DILLÉ, Glenn Scott, (1896-1963)
DITMARS, Raymond Lee, (1876-1942)
DOBBIN, Carroll Edward "Cy", (1892-1967)
DOBSON, Peter, (-)
DOELL, Richard Rayman, (1923-)
DONNELLY, Ignatius, (1831-1901)

UNITED STATES OF AMERICA (STATE unspecified) 1900 to present Cont'd.

DONOGHUE, David Patrick, (1891-1958)
DONOVAN, Percy Williams, (1879-1973)
DORSEY, Noah Ernest, (1873-1959)
DOUGLAS, James Mainwaring, (1887-1952)
DREHER, Otto, (1888-1945)
DUCOTE, James Wardy, (1919-1964)
DUGGLEBY, Arthur Francis, (1893-1945)
DUSTON, Arthur W., (1886-1938)
DUTY, Richard Emerson, (1908-1966)
EARDLEY, Armand John, (1901-1972)
EARGLE, Dolan Hoye, (1905-1973)
EASTMAN, Charles Rochester, (1868-1918)
EATON, Arthur, (1886-1923)
EDSON, Fanny Carter, (1887-1952)
EDWARDS, Merwin Guy, (1889-1969)
ELEY, Hugh Moore, (1902-1942)
ELLIOTT, Harold Charles, (1910-1972)
ELLIS, Charles W., Jr., (1934-1970)
ELLISOR, Alva Christine, (1892-1964)
ELY, Fred Burchard, (1883-1946)
EMMONS, Samuel Franklin, (1841-1911)
EMMONS, William Harvey, (1876-1948)
ENGLISH, George Letchworth, (1864-1944)
ESAREY, Ralph Emerson, (1901-1975)
ESKOLA, Pentti Eelis, (1883-1964)
EVANS, Vincent, (1907-1951)
EWING, William Maurice, (1906-1974)
FAIRBAIRN, Harold Williams, (1906-)
FAIRCHILD, John Gifford, (1882-1965)
FANGUAY, Nolan Anthony, (1916-1942)
FARISH, Linn Markley, (1901-1944)
FERGUSON, Henry Gardiner, (1882-1966)
FERGUSON, Roy Noble, (1883-1961)
FIEDLER, William M., (1914-1970)
FINCH, Elmer Harrison, (1883-1953)
FISCHER, Otto, (1877-1943)
FISHER, Bernard, (1918-1946)
FISHER, Cassius Asa, (1872-1930)
FISKE, John, (1842-1901)
FITZGERALD, Paul Eugene, (1903-1961)
FLEMING, John Adam, (1877-1956)
FLETCHER, Corbin Drummond, (1900-1931)
FLINT, Richard Foster, (1902-1976)
FOLEY, Edward J., (1908-1971)
FONTAINE, William Morris, (1835-1913)
FOOTE, Warren, (-1936)
FORT, Charles, (1874-1932)
FOSHAG, William Frederick, (1894-1956)
FOSTER, Margaret Dorothy, (1895-1970)
FOX-STRANGWAYS, Charles Edward, (1844-1910)
FRAAS, Eberhard, (1862-1915)
FRAZER, Persifor, (1844-1909)
FRAZIER, David Edward, (1931-1976)
FREY, Alfred Paul, (1893-1963)

UNITED STATES OF AMERICA (STATE unspecified) 1900 to present Cont'd.

FRIES, Carl, Jr., (1910-1965)
FRITTS, Crawford Ellsworth, (1927-1972)
FRIZZELL, Donald Leslie, (1906-1972)
FULLER, Homer Taylor, (1838-1910)
GALE, Hoyt Stoddard, (1876-1952)
GALLAGHER, William G., Jr., (1898-1932)
GARDNER, James Henry, (1883-1964)
GARDNER, Julia Anna, (1882-1960)
GARIEPY, Glen Bernard, (1904-1962)
GARRETSON, Mary Welleck, (1896-1971)
GARRETT, Lovic Pierce, (1880-1943)
GEIS, Wilfrid Hoy "Bill", (1892-1961)
GEORGE, William Owsley, (1892-1966)
GERMANN, John Christian, (1903-1954)
GESTER, Stephen Howell, (1889-1967)
GILL, James Rogers, (1922-1972)
GILLES, Verner Arthur, (1886-1954)
GILLULY, James, (-)
GINTER, Roy Lamont, (1892-1958)
GLENN, Leonidas Chalmers, (1871-1951)
GOLDEN, John Muro, (1900-1945)
GOODELL, Horace N. "Hank", (1915-1971)
GOODSON, Joseph Arthur, (1910-1946)
GRABAU, Amadeus William, (1870-1946)
GRANGER, Walter, (1872-1941)
GRANQUIST, William T., (1923-1974)
GRANT, Ulysses Sherman, (1867-1932)
GRATACAP, Louis Pope, (1850-1917)
GRATON, Louis Caryl, (1880-1970)
GRIFFITH, Glenn Rodgers Vickery, (1891-1964)
GRIMES, Russell Winters, (1900-1969)
GRINNELL, George Bird, (1849-1938)
GROSS, Henry, (-) [PROSPECTORS, DIVINERS AND MINING ENGINEERS]
GUILLOU, Robert Barton, (1923-1965)
GÜMBEL, Emil J., (-1966)
HAGUE, Arnold, (1840-1917)
HAGUE, James Duncan, (1836-1908)
HALL, George Martin, (1891-1941)
HAMILTON, Henry L., (1882-1925)
HAMMILL, Chester A., (1889-1941)
HANSELL, James Myron, (1902-1970)
HANSON, Ernest Albin, (1899-1973)
HARES, Charles Joseph, (1881-1970)
HARRINGTON, George Leavitt, (1883-1972)
HARRIS, Gilbert Dennison, (1864-1952)
HARRISON, George Haymond, (1903-1949)
HARRISON, John Wiley, (1916-1952)
HASEMAN, William Peter, (1878-1932)
HATCH, Frederick Henry, (?1864-1932)
HATCHER, John Bell, (1861-1904)
HAUPTMAN, Charles R., (1891-1971)
HAVARD, Charles Gentry, (1921-1950)
HAY, Oliver Perry, (1846-1930)

UNITED STATES OF AMERICA (STATE unspecified) 1900 to present Cont'd.

HAYES, Charles Willard, (1858-1916)
HAYES, George Gibbons, (1889-)
HAZELTINE, Roy S., (1890-1972)
HEALD, Kenneth Conrad, (1888-1971)
HEATH, Daisy Winifred, (1875-1954)
HECK, Nicholas Hunter, (1882-1953)
HEILAND, Carl August, (1899-1956)
HELLMAN, Milo, (1872-1947)
HENDERSON, James Alexander Leo, (?1874-1940)
HENDERSON, Junius, (1865-1937)
HENDRICKSON, Victor James, (1892-1953)
HENNEN, Ray Vernon, (1875-1958)
HERALD, Frank Edward, (1887-1965)
HERMANN, Adam, (-)
HEROY, William Bayard, Sr., (1883-1971)
HERRING, Donald Grant, Jr., (1918-1962)
HESSE, Curtis Julian, (1905-1945)
HEWETT, Donnel Foster, (1881-1971)
HIDDEN, William Earl, (1853-1918)
HILL, James Wilcott, (1920-1954)
HILLS, Richard Charles, (1848-1923)
HINDES, Earle P., (1888-1962)
HODGE, Edwin Thomas, (1887-1970)
HOFFMEISTER, William S., (-)
HOGAN, Dana, (1890-1945)
HOLBROOK, Levi, (1836-1922)
HOLDEN, Edward Fuller, (1901-1925)
HOLDEN, Roy Jay, (1870-1945)
HOLMAN, Eugene, (1895-1962)
HOLMES, Joseph Austin, (1859-1915)
HOOD, Forrest Wilson, (1897-1963)
HOOKER, Joseph Dalton, Sir, (1817-1911)
HOOKER, Marjorie, (1908-1976)
HOPKINS, Edwin Butcher, (1882-1940)
HORTON, Harold Merritt "Wog", (1897-1966)
HOUSER, Gertrude, (-)
HOVEY, Horace Carter, Reverend, (1833-1914)
HOWE, Marshall Avery, (1867-1936)
HOWELL, Edwin Eugene, (1845-1911)
HOYT, John Harger, (1928-1970)
HUBBARD, Bela, (1890-1959)
HUDDLE, John Warfield, (1907-1975)
HUGHES, Richard John, Jr., (1915-1968)
HULL, Joseph Poyer Deyo, (1889-1967)
HUMMEL, Carl Bernard, (1896-1923)
HUNTLEY, Charles Stirling, (1895-1926)
HURLEY, Patrick Mason, (1912-)
HURNDALL, John P., (1902-1966)
HUTSON, Ezekiel Burney, (1897-1961)
HUXLEY, Julian Sorell, Sir, (1887-1975)
JACKSON, William Henry, (1843-1942)
JEFFREYS, Geoffrey, (1885-1953)
JEFFRIES, Charles Davis, (1896-1965)
JENNI, Clarence M., Colonel, (1896-)

UNITED STATES OF AMERICA (STATE unspecified) 1900 to present Cont'd.

JOESTING, Henry Rochambeau, (1903-1965)
JOHNS, Walter Wendell Schell, (1911-1963)
JOHNSON, Norman Lee, (1926-1970)
JOHNSON, Ralph Gordon, (1927-1976)
JULIEN, Alexis Anastay, (1840-1919)
KAY, George Marshall, (1904-1975)
KEELER, William Weaver, (1897-1943)
KEITH, Arthur, (1864-1944)
KELLOGG, Arthur Remington, (1892-1969)
KELLY, Donald, (1898-1949)
KENNEDY, William, (-1926)
KENNERLEY, John Brian, (1937-1976)
KEYES, Charles Rollin, (1864-1942)
KIMBALL, Kent K., (1892-1953)
KINNEY, Harry Dill, (1886-1961)
KIRBY, Grady, (1889-1969)
KLEIN, A. Albert, (1889-1953)
KNIGHT, Charles Robert, (1875-1953)
KRAMER, William Baltser, III, (1896-1963)
KRIEGER, Philip, (1900-1940)
KRYNINE, Paul Dimitri, (1902-1964)
KUNZ, George Frederick, (1856-1932)
LACOE, Ralph Dupuy, (1824-1901)
LAKE, Merrill Evans, (1902-1938)
LAMSENS, O. Robert, (1922-1971)
LANDWEHR, Walter Richard, (1893-1972)
LARSEN, Esper Signius, III, (1912-1961)
LEVISON, Wallace Goold, (1846-1924)
LEWIS, G. Edward, (-)
LEWIS, Joseph Volney, (1869-1969)
LEYDS, Louis Willem, (1887-1963)
LIBBY, Willard Frank, (1908-)
LIDZ, Louis, (1938-1967)
LIND, Charles McClelland, (1911-1957)
LOUGHLIN, Gerald Francis, (1880-1946)
LOW, Julian W., (1904-1968)
LOWELL, Percival, (1855-1916)
LUCAS, Frederick Augustus, (1852-1929)
LUCKY, Maurice Cecil, (1895-1959)
LYELL, Charles, Sir, (1797-1875)
LYMAN, Benjamin Smith, (1835-1920)
MacCLINTOCK, Paul, (1891-1970)
MacCURDY, George Grant, (1863-1947)
MACREADY, George Alexander, (1885-1955)
MALOTT, Clyde A., (-1950)
MANCHESTER, James Greenfield, (1871-1948)
MARBLE, John Putnam, (1897-1955)
MARBUT, Curtis Fletcher, (1863-1935)
MARGERIE, Emmanuel Marie Pierre Martin Jacquin de, (1862-1953)
MARKHAM, Edmund Olis, (1893-1960)
MARKS, Edward Seaborn, (?1875-1950)
MARTI, Ernest John, (1913-1964)
MARTIN, George C., (1875-1943)
MARTIN, Lewis, (1920-1973)

UNITED STATES OF AMERICA (STATE unspecified) 1900 to present Cont'd.

MASON, Arnold Caverly, (1906-1961)
MATSON, George Charlton, (1873-1940)
MATTHES, François Emile, (1874-1948)
MAYFIELD, William I., (1912-1972)
McCLUNG, Clarence Erwin, (1870-1946)
McGEE, William John, (1853-1912)
McLAUGHLIN, R. P., (-)
McMANNIS, William J., (1924-1973)
MELL, Patrick Hues, (1850-1918)
MERRIAM, John Campbell, (1869-1945)
MERRICK, F. Kermit, (1915-1963)
MERRILL, George Perkins, (1854-1929)
MERRILL, Loren B., (?1853-1930)
MESSINA, Angelina Rose, (1910-1968)
MILLER, Benjamin Le Roy, (1874-1944)
MILLER, Carroll C., (1899-1974)
MILLER, Wendell Zerbe, (1892-1958)
MILLS, Enos Abijah, (1870-1922)
MILLS, James Ellison, (1834-1901)
MONGIN, Denise, (-)
MOODY, Clarence Lemuel, (1888-1963)
MORAN, Thomas, (1837-1926)
MORLEY, Harold Thompson "Midge", (1897-1976)
MUIR, John Malcolm, (1885-1938)
MULFORD, John W., (1902-1964)
MYERS, John Clark, (1894-1970)
NEWBY, Warner Wilson, (1898-1926)
NIGHTINGALE, William Thomas, (1897-1964)
NININGER, Robert D., (-)
NOTESTEIN, Frank B., (1885-1973)
NUTTALL, Winfred Laurence Falkiner, (1897-1969)
NYGREN, Walter Eric, (1907-1971))
OFFICER, Herbert George, (1889-1937)
OLDS, Thomas Hartman, (1883-1932)
ORTMANN, Arnold Edward, (1863-1927)
OWENS, John Edward, (1912-1963)
PACKARD, Alpheus Spring, Jr., (1839-1905)
PACKARD, Henry Jessup, (1883-1930)
PARDÉ, Maurice, (1893-1973)
PARKER, Ben Hutchinson, (1902-1969)
PEABODY, Frank Elmer, (1914-1958)
PEABODY, Harlan William, (1881-1960)
PECORA, William Thomas, (1913-1972)
PEDRICK, Howard A. "Dad", (1863-?1930) [PROSPECTORS, DIVINERS AND MINING ENGINEERS]
PENCK, Walther, (1888-1923)
PENROSE, Richard Alexander Fullerton, Jr., (1863-1931)
PEPPERBERG, Leon J., (1883-1937)
PETERSON, Clarence J., (1885-1948)
PHELPS, Frederick George, (1913-1964)
PHILBRICK, Eugene Patrick, (1897-1951)
POLLOCK, Charles Murray, (1898-1969)
POTEET, James Harold, (1896-1965)
POTTER, Nelson Bryan, (1897-1965)

UNITED STATES OF AMERICA (STATE unspecified) 1900 to present Cont'd.

POWERS, Sidney, (1890-1932)
PRATT, Wallace Everette, (1885-)
PREECE, Rae, (1897-1961)
PROSSER, Charles Smith, (1860-1916)
RABBITT, John Charles, (1907-1957)
RAINIER, Peter W., (-) [PROSPECTORS, DIVINERS AND MINING ENGINEERS]
RAINWATER, Edward Harriman "Rainey", (1909-1972)
RAMDOHR, Paul, (1890-)
RAVEN, Henry Cushier, (1889-1944)
RAY, Robert Hillyer, (1902-1968)
RAYMOND, Rossiter Worthington, (1840-1918)
REA, Henry Carter, (1900-1963)
REED, Lyman Coleman, (1900-1970)
REED, Ralph Daniel, (1889-1940)
REESIDE, John Bernard, Jr., (1889-1958)
REID, William McCormick, (1911-1943)
REISS, John W., (1920-1943)
RENICK, Brink Coleman, (1897-1971)
RHOADES, Ralph Omer, (1895-1961)
RHOADES, Roger F., (1905-1972)
RICH, Charles Irvin, (1918-1975)
RICHARDS, George L., Jr., (1908-1944)
RICHARDS, James Taylor "Dick", (1892-1953)
RIDGEWAY, Bertrand S., (1899-1942)
RIEBER, Frank, (1891-1948)
RIGGS, Robert Jennings, (1888-1965)
RITER, Samuel W., (1896-1941)
RITTENHOUSE, Gordon, (1910-1974)
ROARK, Louis, (1890-1950)
ROBERTS, Marion Stanton, (1909-1963)
ROBERTS, Morgan Evan, (1894-1941)
ROCKWOOD, Charles Greene, (1843-1913)
ROEBLING, Washington Augustus, Colonel, (1837-1926)
ROGERS, Gaillard Sherburne, (1889-1919)
ROGERS, Jesse Armstead, (1913-1962)
ROMEU, Albert de, (1875-1915)
RONES, Morris, (1922-1965)
ROSENHOLTZ, Joseph Leon, (1899-1963)
ROSS, Clarence Samuel, (1880-1975)
ROUBAULT, Marcel Edouard, (1905-1974)
ROUNDY, Paul Vere, (1884-1937)
RUEDEMANN, Rudolf, (1864-1956)
SALISBURY, Rollin D., (1858-1922)
SANFORD, Wendell Glenn, (1914-1958)
SAPPINGTON, Chester, (1905-1968)
SARGENT, Harry Lionel, (1871-1955)
SAVILLE, Wilson Gordon, (1897-1954)
SAWYER, Roger W., (1895-1941)
SCHAEFFER, Charles Frederic August, (1860-1903)
SCHALLER, Waldemar Theodore, (1882-1967)
SCHLAIKJER, Erich Maren, (1905-1972)
SCHMITT, Harrison Ashley, (1896-1966)
SCHNEIDERHÖHN, Hans, (1887-1962)

UNITED STATES OF AMERICA (STATE unspecified) 1900 to present Cont'd.

SCHOLTZ, Douglas Leinster, (1902-1977)
SCHRADER, Frank Charles, (1860-1944)
SCHUMACHER, Friedrich, (1884-)
SCHWENNESEN, Alvin Theodore, (1883-1962)
SCOFIELD, Lloyd M., (1897-1969)
SCOPES, John Thomas, (1900-)
SCRUTON, Philip Challacombe, (1922-1962)
SCUDDER, Samuel Hubbard, (1837-1911)
SEARS, Julian Ducker, (1891-1970)
SEE, Thomas Jefferson Jackson, (1866-)
SEGALL, Julius, (1884-1928)
SHALER, Nathaniel Southgate, (1841-1906)
SHAW, Eugene Wesley, (1881-1935)
SHEPARD, Edward Martin, (1854-1934)
SHERWOOD, Theodore C., Jr., (1896-1928)
SHIPEK, Carl Joseph, (1917-1969)
SHUFELDT, Robert Wilson, (1850-1934)
SIMMONS, Jesse Elmore, (1907-1941)
SINCLAIR, Ewart Gladstone, (1884-1927)
SINGEWALD, Joseph Theophilus, Jr., (1884-1963)
SISLER, James Donaldson, (1896-1935)
SLOAN, Earle, (1858-1926)
SMITH, Aylwin L., (1898-1965)
SMITH, George Otis, (1871-1944)
SMITH, James Perrin, (1864-1931)
SMITH, Neal Johnstone, (1912-1971)
SORENSEN, Carl Christian, (1893-1965)
SOVINSKY, Viacheslav Nicholas, (1894-1962)
SPARKS, Dale Darrell Dolsun, (1894-1938)
SPEED, Carleton D., Jr., (1903-1970)
STAFF, Hans von, (1883-1915)
STAINER, Xavier, (1865-)
STANLEY, M. Wood, (1893-1929)
STEBINGER, Eugene, (1883-1951)
STEHR, Raymond Albert, (1903-1967)
STEIGER, George, (1869-1944)
STEINMANN, Gustav, (1856-1929)
STELFOX, Reichie S., (1904-1932)
STETSON, Henry Crosby, (1900-1955)
STEWART, Philip Charteris Anstruther, (1874-1923)
STONE, Robert L., (1912-1969)
STONELEY, Robert, (1894-1976)
STONER, Oscar E., Sr., (1892-1965)
STOPES, Marie Charlotte Carmichael, (1880-1958)
SULLINS, M. Joseph, (1930-1972)
SWARTZ, Joel Howard, (1893-1971)
SWEET, Sidney R., (1878-1946)
TANDY, James Hiram "Spot", (1892-1962)
TASHMAN, Djavad Eyoub, (1893-1956)
TAYLOR, Griffith, (1881-1963)
TAYLOR, James Carlton, (1924-1976)
TEAS, Livingstone Pierson, (1894-1970)
TEICHERT, Curt, (-)
THIESSEN, Reinhardt, (1867-1938)

UNITED STATES OF AMERICA (STATE unspecified) 1900 to present Cont'd.

THOMPSON, David Grosh, (1888-1943)
THOMPSON, E. G., (1898-1973)
THORNBURG, Dwight Holbrook "Duke", (1895-1965)
THORPE, Malcolm Rutherford, (1891-1958)
TOOMEY, Charles Calvin, (1892-1957)
TRICE, Edwin Leslie, Jr., (1929-1967)
TROXELL, Edward Leffingwell, (1884-1972)
TURNER, Joseph Harold, (1900-1950)
ULRICH, Edward Oscar, (1857-1944)
UNGEMACH, Henri Léon, (1879-1936)
UREY, Harold Clayton, (1893-)
VAN AMRINGE, Edwin Verne, (1899-1956)
VAN INGEN, Gilbert, (1869-1928)
VAUGHAN, Thomas Wayland, (1870-1952)
VAUX, George, Jr., (1863-1927)
VAUX, William S., Jr., (1872-1908)
VOGDES, Anthony Wayne, General, (1843-1923)
WADE, Arthur, (1878-1951)
WAGNER, Carroll M., (1893-1965)
WALCOTT, Charles Doolittle, (1850-1927)
WALCOTT, Mary Vaux, née Vaux, (-1940)
WALL, George Parkes, (1832-1912)
WALTERS, Mason G., (1903-1967)
WALTMAN, William Dewitt, (1875-1955)
WARD, Lester Frank, (1841-1913)
WARING, Gerald Ashley, (1883-1971)
WARREN, Percival Sidney, (1890-1970)
WASSERSTEIN, Benno, (1906-1956)
WASSON, Theron, (1887-1970)
WATERMAN, Willis D., (1925-1968)
WATERSCHOOT VAN DER GRACHT, Willem Anton Josef Maria van, (1873-1943)
WATSON, Thomas Leonard, (1871-1924)
WEAVER, Paul, (1888-1964)
WEBB, David Knowlton, Jr., (1928-1973)
WEDEL, Arthur Albert, (1898-1941)
WEEKS, Herbert John, (1897-1960)
WEINZIERL, Laura Lee, (née Lane), (1900-1928)
WELLS, Roger Clark, (1877-1944)
WESTGATE, Lewis Gardner, (1868-1948)
WETMORE, Alexander, (1886-)
WEXLER, Harry, (1911-1962)
WHEELER, Orby Clinton, (1898-1970)
WHITE, Charles Abiathar, (1826-1910)
WHITE, Charles David, (1862-1935)
WHITE, George Willard, (1903-)
WHITE, Israel Charles, (1848-1927)
WHITE, Maynard Pressley, (1896-1961)
WHITEHEAD, Walter Lucius, (1891-1969)
WHITESIDE, Robert Massie, (1895-1936)
WHITTIER, William Harrison, (1889-1951)
WHITWELL, Earl Van Dorn, (1892-1963)
WIGGLESWORTH, Edward, (1885-1945)
WILLIAMSON, Erskine Douglas, (1886-1923)

UNITED STATES OF AMERICA (STATE unspecified) 1900 to present Cont'd.

WILSON, Harry H., (1902-1929)
WILSON, Thomas Carroll, (1906-1969)
WINCHESTER, Dean Eddy, (1883-1936)
WINTERER, Edward Virgil "Tubby", (1897-1944)
WOEIKOF, Alexander, (1842-1916)
WOODRUFF, Elmer Grant, (1872-1952)
WOODWARD, Robert Simpson, (1849-1924)
WOOLNOUGH, Walter George, (1876-1958)
WORTMAN, Jacob Lawson, (1856-1926)
WRIGHT, Albert Allen, (1846-1905)
WRIGHT, Charles Will, (1879-)
WRIGHT, Frederick Eugene, (1877-1953)
WRIGHT, George Frederick, (1838-1921)
WRIGHT, Jean Davies, (-)
YAGODA, Herman, (1908-1964)
YOUNG, Karl Etienne "Happy", (1903-1936)
ZAPP, Alfred Dexter, (1916-1962)
ZODAC, Peter, (1894-1967)

Time unknown
ANGELL, Cyrus D., (-)
JENSEN, David E., (-)
PERHAM, Frank, (-)
SCHORTMANN, Alvin, (-)
SCHORTMANN, Raymond, (-)
THOMPSON, Frederick Oliver, (_)
YEDLIN, Neal, (_)

UPPER VOLTA
1900 to present
MILLOT, Georges, (1917-)

URUGUAY
1800-1849
DARWIN, Charles Robert, (1809-1882)

1850-1899
DARWIN, Charles Robert, (1809-1882)
FALCONER, John Downie, (1876-1947)
GOULD, Charles, (-1895)
McCARTHY, Edward Thomas, (-) [PROSPECTORS, DIVINERS AND MINING ENGINEERS]

1900 to present
CASTELLANOS, Alberto, (1896-1968)
FALCONER, John Downie, (1876-1947)
FERUGLIO, Egidio, (1897-1954)
HARRINGTON, Horacio Jaime, (1910-1973)
HAYES, Albert Orion, (1882-1965)
KRAGLIEVICH, Lucas, (1886-1932)
McCARTHY, Edward Thomas, (-) [PROSPECTORS, DIVINERS AND MINING ENGINEERS]

VENEZUELA

Up to 1800

HUMBOLDT, Friedrich Wilhelm Heinrich Alexander von, Freiherr, (1769-1859)

1800-1849

HUMBOLDT, Friedrich Wilhelm Heinrich Alexander von, Freiherr, (1769-1859)

1850-1899

ARNOLD, Ralph, (1875-1961)
BALL, Francis Matthew, (1877-1956)
BOWEN, Charles Franklin, (1871-1953)
BRALY, Adrien, (1864-1927)
BURCHARD, Ernest Norman, (1875-1961)
CHAPER, Maurice Armand, (1834-1896)
CUMINGS, Willard Lawsen, (1874-1945)
DARTON, Nelson Horatio, (1865-1948)
DICKERSON, Roy Ernest, (1877-1944)
ERB, Josef Theodor, (1874-1934)
FOSTER, Clement Le Neve, Sir, (1841-1904)
GUPPY, Robert John Lechmere, (1836-1916)
HENNEN, Ray Vernon, (1875-1958)
HUMBOLDT, Friedrich Wilhelm Heinrich Alexander von, Freiherr, (1769-1859)
MACLAREN, James Malcolm, (1873-1935)
STEPHENSON, Lloyd William, (1876-1962)
TATE, Ralph, (1840-1901)
VEATCH, Arthur Clifford, (1878-1938)

1900 to present

ALDREDGE, Robert Frank, (1895-1961)
ARGABRITE, William Greame, (1885-1963)
ARNOLD, Ralph, (1875-1961)
BAIRD, Chester Alvan, (1896-1959)
BALDWIN, William Felbert Jackson, Jr., (1927-1972)
BALL, Francis Matthew, (1877-1956)
BALL, Harry Standish, (?1888-1941)
BANDY, Mark Chance, (1900-1963)
BARRINGTON, Thomas W., (-)
BARROW, Geoffrey, (1899-1963)
BECKMANN, Lawrence John, (1912-1976)
BELL, Olin Grogan, (1893-1970)
BERNOUILLI, Walter, (1885-1946)
BLOW, Walter Harry, (1924-1972)
BONTZ, Conrad K., (1894-1927)
BOOTS, Paul Heath, (1909-1947)
BOSWORTH, Thomas Owen, (1882-1928)
BOWEN, Charles Franklin, (1871-1953)
BOYD, Harold Edward, (1887-1952)
BRALY, Adrien, (1864-1927)
BRAMLETTE, Milton Nunn, (1896-1977)
BREMNER, Carl St. John, (1895-1944)
BURCHARD, Ernest Norman, (1875-1961)
BUSK, Henry Gould, (?1891-1956)
BUTCHER, Walter William, (1916-1966)

VENEZUELA 1900 to present Cont'd.

BUTTERWORTH, Emerson McMillin, (1894-1961)
CARLSON, Charles Gordon, (1895-1977)
CARR, William L., (1919-1975)
CARSON, Carlton Mann "Kit", (1898-1972)
CASSEL, Chester A. "Chet", (1897-1971)
CASTON, John E. F., (1925-1972)
CONKLING, Richard A., (1885-1952)
COOKE, Charles Wythe, (1887-1971)
CRICKMAY, Geoffrey W., (1905-1971)
CUMINGS, Willard Lawsen, (1874-1945)
DARTON, Nelson Horatio, (1865-1948)
DAWSON, Joseph Meadows, (1895-1946)
DAWSON, Thomas Albert, (1916-1974)
DICKERSON, Roy Ernest, (1877-1944)
DIXON, Henry George Dacres, (1887-1947)
DOUGLAS, James Mainwaring, (1887-1952)
DOUGLAS, John Gray, (1900-1974)
EDWARDS, Everett C., (1897-1972)
ELY, Fred Burchard, (1883-1946)
ERB, Josef Theodor, (1874-1934)
FARISH, Linn Markley, (1901-1944)
FATH, Arthur Earl, (1887-1976)
FICHTER, Hans Jakob, (1908-1977)
FINCH, Elmer Harrison, (1883-1953)
FOLEY, Edward J., (1908-1971)
FORTINE, Dwight Hugh, (1904-1949)
FOSTER, Clement Le Neve, Sir, (1841-1904)
FOWLER, Charles Worth, Jr., (1899-1948)
FREY, Alfred Paul, (1893-1963)
GESTER, George Clark, (1884-1959)
GETZENDANER, Avalette Edmond, (1902-1962)
GORMAN, Joseph M., (1910-1967)
GUPPY, Robert John Lechmere, (1836-1916)
HALSE, George Whitfield, (1885-1968)
HAWKINS, Glenn Dewayne, (1897-1970)
HENDON, Bryan, (1896-1926)
HENLEY, Arthur Sidney, (1879-1936)
HENNEN, Ray Vernon, (1875-1958)
HEROLD, Stanley Carrollton, (1883-1958)
HILL, Robert G., (1914-1973)
HOLMAN, Eugene, (1895-1962)
HOPKINS, Edwin Butcher, (1882-1940)
HORTON, Harold Merritt "Wog", (1897-1966)
HUBBARD, Bela, (1890-1959)
HUBMAN, Carl Wesley, (1905-1947)
HUNTLEY, Charles Stirling, (1895-1926)
HUTCHISON, Arthur Gilbertson, (1902-1949)
HYDE, Clarence Edgar, (1894-1954)
IDDINGS, Arthur, (1891-1957)
ILLING, Vincent Charles, (1890-1969)
IRWIN, Marvin Leroy "Mel", (1915-1964)
ISENBERGER, Nathan Perry, (1896-1956)
JEFFREYS, Geoffrey, (1885-1953)
JOESTING, Henry Rochambeau, (1903-1965)

VENEZUELA 1900 to present Cont'd.

JOHNSON, Joseph Wright, (1917-1957)
JOHNSTON, William Drumm, Jr., (1899-1972)
JONES, Verner E., (1905-1977)
JONES, William Francis, (1885-1941)
JURKOVIĆ, Ivan, (1917-)
KAY, Fred Hall, (1885-1943)
KINKEL, John Franklin, (1899-1936)
KIRBY, Grady, (1889-1969)
KNEBEL, G. Moses, (1899-1974)
LAMARE, Pierre, (1894-1967)
LEHNER, Ernst J., (1895-1968)
LEPPER, Gordon Willis, (1889-1950)
LIDDLE, Ralph Alexander, (1896-1963)
LUCKY, Maurice Cecil, (1895-1959)
MACKENZIE, Andrew Nicholls "Kelley", (1888-1951)
MACLAREN, James Malcolm, (1873-1935)
MacNAUGHTON, Lewis Winslow, (1902-1969)
MACREADY, George Alexander, (1885-1955)
MARTIN, Francis Ivan "Shorty", (1896-1952)
MARTYN, Phil F., (1903-1974)
MASON, Shirley Lowell, (1893-1975)
McARTHUR, Donald, (1895-1960)
McCLURE, Drexel Eugene, (1928-1977)
MENCHER, Ely, (1913-)
MEYER, Edouard, (1899-1974)
MILLER, John Cummins, (1895-1951)
MILLER, Wendell Zerbe, (1892-1958)
MOORE, Gilbert Parvin, (1898-1970)
NANCE, James W., (1908-1977)
NEDOM, Henry Arthur, (1895-1940)
NEWHOUSE, Walter Harry, (1897-1969)
NOLAN, Philip Edward, (1887-1958)
NOMLAND, Jorgen O., (1885-1943)
NOTESTEIN, Frank B., (1885-1973)
NUTTALL, Winfred Laurence Falkiner, (1897-1969)
NYGREN, Walter Eric, (1907-1971))
OLCUTT, David Perry, (1896-1958)
PACKARD, Henry Jessup, (1883-1930)
PARSONS, Ernest, (?1887-1963)
PECORA, William Thomas, (1913-1972)
PERCIVAL, Frederick George, (1887-1973)
PETERSON, Clarence J., (1885-1948)
POLLARD, Ernest Rex, (1922-1966)
PROUT, Franklin S. "Shorty", (1889-1940)
RAINWATER, Edward Harriman "Rainey", (1909-1972)
REED, Eugene C., (1901-1976)
REED, Lyman Coleman, (1900-1970)
RITER, Samuel W., (1896-1941)
ROYO Y GOMEZ, José, (1895-1958)
RUSSELL, William Low, (1897-1976)
RUTSCH, Rolf F., (1902-1975)
SCHAUB, Hans Peter, (1912-1965)
SCHILLING, Karl Howard, (1894-1945)
SCHNEEGANS, Daniel, (1907-1949)

VENEZUELA 1900 to present Cont'd.

SCHOLL, Louis August, Jr., (1887-1953)
SENN, Alfred, (1899-1949)
SHERMAN, Roger Hiram, (1904-1954)
SIMPSON, George Gaylord, (1902-)
SINCLAIR, Ewart Gladstone, (1884-1927)
SPROULE, John Campbell, (1905-1970)
STEHR, Raymond Albert, (1903-1967)
STEPHENSON, Lloyd William, (1876-1962)
STONE, A. John F., (1917-1969)
STREIFF, Viktor, (1909-1973)
SUGDEN, Walter "Bill", (1923-1977)
SUTTON, Frederick A., (1894-1950)
TATE, Ralph, (1840-1901)
TEMPLETON, Eugene Crittenden, (1889-1932)
THALMANN, Hans E., (1899-1975)
VEATCH, Arthur Clifford, (1878-1938)
WARING, Gerald Ashley, (1883-1971)
WASHBURNE, Chester Wesley, (1883-1971)
WEAVER, Paul, (1888-1964)
WEEKS, Herbert John, (1897-1960)
WHEELER, Orby Clinton, (1898-1970)
WHITEHEAD, Walter Lucius, (1891-1969)
WIEDENMAYER, Carl, (1897-1951)
WILSON, Joseph Garland, (1902-1955)
WILSON, Roy Arthur, (1891-1970)
WILSON, Thomas Carroll, (1906-1969)
WIRZ, Albert Eduard, (1910-1973)

VIETNAM

1850-1899

JACOB, Charles François Étienne, (1878-1962)

1900 to present

JACOB, Charles François Étienne, (1878-1962)
NININGER, Harvey Harlow, (-)
RASMUSSEN, William Charles, (1917-1973)
ROCH, Edouard, (1901-)

VIETNAM See also INDOCHINA (unspecified).

VIRGIN ISLANDS (WEST INDIES)

Up to 1800

LETTSOM, John Coakley, (1744-1815)

1800-1849

LETTSOM, John Coakley, (1744-1815)

1900 to present

HUBBARD, Bela, (1890-1959)
McGUINNESS, Charles Lee, (1914-1971)
WHETTEN, John T., (-)

WALES

Up to 1800

AIKIN, Arthur, (1773-1854)
HUTTON, James, (1726-1797)
LHUYD, Edward, (1660-1709)
OWEN, George, (1552-1613)
RASPE, Rudolf Erich, (1737-1794)
SMITH, William, (1769-1839)

1800-1849

AIKIN, Arthur, (1773-1854)
AVELINE, William Talbot, (1822-1903)
CLARK, George Thomas, (1811-1898)
HOMFRAY, David, (1822-1893)
JEFFREYS, John Gwyn, (1809-1885)
LOGAN, William Edmond, Sir, (1798-1875)
LUXMOORE, Edward Bouverie, (1829-1893)
MACKINTOSH, Daniel, (1815-1891)
MURCHISON, Roderick Impey, Sir, (1792-1871)
PHILLIPS, William, (1800-1834)
PLANT, Major John, (1819-?1892)
RAMSAY, Andrew Crombie, Sir, (1814-1891)
SALTER, John William, (1820-1869)
SEDGWICK, Adam, Reverend, (1785-1873)
SELWYN, Alfred Richard Cecil, (1824-1902)
SHARPE, Daniel, (1806-1856)
SHRUBSOLE, George William, (1827-1893)
SMITH, Gilbert Nicholas, Reverend, (1796-1877)
SMITH, William, (1769-1839)
SMYTH, Warington Wilkinson, Sir, (1817-1890)
WILLIAMS, Thomas, (1737-1802) [PROSPECTORS, DIVINERS AND MINING ENGINEERS]

1850-1899

AIKIN, Arthur, (1773-1854)
AVELINE, William Talbot, (1822-1903)
BLAKE, John Frederick, Reverend, (1839-1906)
BOULTON, William Savage, (1867-ca.1954)
CALLAWAY, Charles, (1838-1915)
CLARK, George Thomas, (1811-1898)
CODRINGTON, Thomas, (1829-1918)
COLLINS, Henry Francis, (?1864-1924)
CROSFIELD, Margaret Chorley, (1859-1952)
DAKYNS, John Roche, (1836-1910)
DAVIES, Arthur Morley, (1869-1959)
DAVIES, David, (1870-1931)
DAVIES, David Christopher, (1827-1885)
DAVIES, David Farr, (1871-1935)
DAVIES, Thomas, (1837-1892)
ELLES, Gertrude Lilian, (1872-1960)
ELLISON, John, (1859-1936)
GARDINER, Charles Irving, (?1868-1940)
GLASS, Norman, Reverend, (1832-1895)
GOODYEAR, Edith, (1879-1959)
GREENLY, Edward, (1861-1951)
GROOM, Theodore Thomas, (1863-)

WALES 1850-1899 Cont'd.

HICKS, Henry, (1837-1899)
HOMFRAY, David, (1822-1893)
HULL, Edward, (1829-1917)
JEFFREYS, John Gwyn, (1809-1885)
JOHNS, Cosmo, (?1866-1951)
JONES, Ernest Lloyd, (?1863-1942)
JONES, Owen Thomas, (1878-1967)
KEEPING, Walter, (1854-1888)
LAPWORTH, Charles, (1842-1920)
LEACH, Arthur Leonard, (1869-1957)
LEWIS, Henry Carvill, (1853-1888)
LOGAN, William Edmond, Sir, (1798-1875)
LUXMOORE, Edward Bouverie, (1829-1893)
MACKINTOSH, Daniel, (1815-1891)
MATLEY, Charles Alfred, (1866-1947)
MAW, George, (1832-1912)
MONTAG, Emil, (1875-1960)
MORGAN, Conwy Lloyd, (?1852-1936)
MORGAN, Joseph Bickerton, (1859-1894)
MURCHISON, Roderick Impey, Sir, (1792-1871)
PHILLIPS, William, (1800-1834)
PLANT, Major John, (1819-?1892)
PRINGLE, John, (1877-1948)
RAISIN, Catherine Alice, (1855-1945)
RAMSAY, Andrew Crombie, Sir, (1814-1891)
REED, Frederick Richard Cowper, (1869-1946)
ROBERTS, Robert Davis, (1851-1911)
RUDLER, Frederick William, (1840-1915)
RUTLEY, Frank, (1842-1904)
SALTER, John William, (1820-1869)
SEDGWICK, Adam, Reverend, (1785-1873)
SELWYN, Alfred Richard Cecil, (1824-1902)
SHARPE, Daniel, (1806-1856)
SHRUBSOLE, George William, (1827-1893)
SMALL, Evan William, (1859-1933)
SMITH, Gilbert Nicholas, Reverend, (1796-1877)
SMYTH, Warington Wilkinson, Sir, (1817-1890)
SOLLAS, William Johnson, (1849-1936)
STIRRUP, Mark, (1831-1907)
STORRIE, John, (1844-1901)
STRAHAN, Aubrey, Sir, (1852-1928)
TAWNEY, Edward Bernard, (1840-1882)
THOMAS, Herbert Henry, (1876-1935)
TIDDEMAN, Richard Hill, (1842-1917)
VAUGHAN, Arthur, (1868-1915)
WOODWARD, Horace Bolingbroke, (1848-1914)

1900 to present

AVELINE, William Talbot, (1822-1903)
BLAKE, John Frederick, Reverend, (1839-1906)
BOULTON, William Savage, (1867-ca.1954)
BROUGHTON-EDGE, Arthur Broughton, (1895-1953)
CALLAWAY, Charles, (1838-1915)
CODRINGTON, Thomas, (1829-1918)

WALES 1900 to present Cont'd.

COLLINS, Henry Francis, (?1864-1924)
CONNOR, James, (1894-1977)
CROFT, William Noble, (1915-1953)
CROSFIELD, Margaret Chorley, (1859-1952)
CUNNINGTON, C. H., Lieutenant, (1889-1918)
DAKYNS, John Roche, (1836-1910)
DAVIES, Arthur Morley, (1869-1959)
DAVIES, David, (1870-1931)
DAVIES, David Farr, (1871-1935)
DAVIES, John Henry, (1887-1971)
DAVIES, Thomas, (1837-1892)
DIX, Emily, (1904-1973)
DODD, C., (-1928)
EDMONDS, Frederic Harrison, (1898-1965)
ELLES, Gertrude Lilian, (1872-1960)
ELLISON, John, (1859-1936)
GARDINER, Charles Irving, (?1868-1940)
GOODE, R. H., Reverend, (1888-1969)
GOODYEAR, Edith, (1879-1959)
GREENLY, Edward, (1861-1951)
GROOM, Theodore Thomas, (1863-)
GROVES, Arthur William, (1903-1956)
HARPER, John Childs, (1913-1977)
HARRISON, Janet Mitchell Marr, (1893-1971)
HEARD, Albert, (1894-1944)
HIBBERT, Frank, Colonel, (?1889-1951)
HOWARTH, W. E., (-1956)
HOWELLS, Jenkyn Owen, (-1950)
HUGHES, William John, (?1899-1961)
HULL, Edward, (1829-1917)
JACKSON, James Frederick, (1896-1966)
JOHNS, Cosmo, (?1866-1951)
JONES, Brynmor, (1902-1941)
JONES, Ernest Lloyd, (?1863-1942)
JONES, Owen Thomas, (1878-1967)
KELLOW, Moses, (-1943)
LAPWORTH, Charles, (1842-1920)
LEACH, Arthur Leonard, (1869-1957)
LEITCH, Duncan, (1904-1956)
LeLACHEUR, W. J., (-1923)
LEWIS, Herbert Price, (?1895-1947)
LEWIS, William Vaughan, (1907-1961)
MATLEY, Charles Alfred, (1866-1947)
MAW, George, (1832-1912)
MONTAG, Emil, (1875-1960)
MORGAN, Conwy Lloyd, (?1852-1936)
MUIR-WOOD, Helen Marguerite, (1896-1968)
OWENS, Eifion Wyn, (1951-1975)
PHILLIPS, David Watkins, (1899-1962)
POCOCK, Roy Woodhouse, (1887-1954)
PRINGLE, John, (1877-1948)
PUGH, William John, Sir, (1892-1974)
RAISIN, Catherine Alice, (1855-1945)
REED, Frederick Richard Cowper, (1869-1946)

WALES 1900 to present Cont'd.

ROBERTS, Robert Davis, (1851-1911)
RUDLER, Frederick William, (1840-1915)
RUTLEY, Frank, (1842-1904)
SELWYN, Alfred Richard Cecil, (1824-1902)
SIBLY, Thomas Franklin, Sir, (1883-1948)
SMALL, Evan William, (1859-1933)
SMITH, Bernard, (1881-1936)
SMITH, Herbert Gladstone, (-1950)
SMITH, Stanley, (1883-1955)
SOLLAS, William Johnson, (1849-1936)
STAMP, Laurence Dudley, Sir, (1898-1966)
STIRRUP, Mark, (1831-1907)
STORRIE, John, (1844-1901)
STRAHAN, Aubrey, Sir, (1852-1928)
STRAW, Stephen Henry, (?1892-1963)
THOMAS, Herbert Henry, (1876-1935)
THOMAS, Trevor Morgan, (?1919-1974)
TIDDEMAN, Richard Hill, (1842-1917)
VAUGHAN, Arthur, (1868-1915)
WILLIAMS, Mervyn, (1934-1970)
WOODS, Ethel Gertrude, (-1939)
WOODWARD, Horace Bolingbroke, (1848-1914)
WOOLDRIDGE, Sidney William, (1900-1963)

WEST AFRICA (unspecified)

1850-1899

TRUSCOTT, Samuel John, (?1870-1950)

1900 to present

FAULKNER, Joseph Stanislaus "Ginger", (1928-1963)
JOYCE, John Raymond Foggan, (?1907-1963)
TRUSCOTT, Samuel John, (?1870-1950)

WEST INDIES (unspecified)

Up to 1800

LETTSOM, John Coakley, (1744-1815)

1800-1849

LETTSOM, John Coakley, (1744-1815)
SAINTE-CLAIRE DEVILLE, Charles, (1814-1876)

1850-1899

ABEL, Othenio, (1875-1946)
BROWN, Amos Peaslee, (1864-1917)
DAVIS, William Morris, (1850-1934)
FLETT, John Smith, Sir, (1869-1947)
HARRISON, John Burchmore, Sir, (1856-1926)
RINNE, Friedrich Wilhelm Berthold, (1863-1933)
SAINTE-CLAIRE DEVILLE, Charles, (1814-1876)

1900 to present

ABEL, Othenio, (1875-1946)
BROWN, Amos Peaslee, (1864-1917)
CLAPP, Frederick Gardner, (1879-1944)

WEST INDIES (unspecified) 1900 to present Cont'd.

COX, Leslie Reginald, (1897-1965)
DAVIS, William Morris, (1850-1934)
EARLE, Kenneth Wilson, (1891-1944)
FLETT, John Smith, Sir, (1869-1947)
HALSE, George Whitfield, (1885-1968)
HAMMOND, Weldon Woolf, (1908-1964)
HARRISON, John Burchmore, Sir, (1856-1926)
JONES, William Francis, (1885-1941)
RICH, John Lyon, (1884-1956)
RINNE, Friedrich Wilhelm Berthold, (1863-1933)
TRECHMANN, Charles Taylor, (1885-1964)
TRIEBEL, Erich, (1894-1971)
WINCHESTER, Dean Eddy, (1883-1936)

WEST INDIES (unspecified) See also ANGUILLA; ANTIGUA; BAHAMA ISLANDS; BARBADOS; CURAÇAO; DUTCH WEST INDIES; GUADELOUPE; JAMAICA;

WEST INDIES (unspecified) See also MARTINIQUE; LEEWARD ISLANDS; PUERTO RICO; SAINT BARTHOLOMEW; SAINT LUCIA; SAINT VINCENT; SOMBRERO; TRINIDAD.

WEST IRIAN (INDONESIA)

1850-1899

DAVID, Tannatt William Edgeworth, Sir, (1858-1934)

1900 to present

DAVID, Tannatt William Edgeworth, Sir, (1858-1934)
KLEIBER, Karl Ernst, (1912-1972)
KLOMPÉ, Theodorus Hendricus Franciscus, (?1904-1963)
NES, C. L. van, (?1878-1963)
RUTTEN, Martin Gerard, (1910-1970)
SCHERMERHORN, W., (1894-)
TER MEULEN, J. A. C., (1902-1960)

WINDWARD ISLANDS (WEST INDIES)

1850-1899

FRANKS, Godfrey Firth, (1853-1917)

1900 to present

EARLE, Kenneth Wilson, (1891-1944)
FRANKS, Godfrey Firth, (1853-1917)

YEMEN

1900 to present

ALLOITEAU, James Henri Ernest, (1890-1969)
LAMARE, Pierre, (1894-1967)
ROMAN, David, (1884-1927)

YUGOSLAVIA

Up to 1800

FORTIS, Giovanni Battista Alberto, Abate, (1741-1803)
SAPIEHA, Aleksander, (1773-1812)

1800-1849

BUCCHICH, Gregorio, (1828-1911)
FORTIS, Giovanni Battista Alberto, Abate, (1741-1803)
FREYER, Heinrich, (1802-1865)
FUCHS, Wilhelm, (1802-1853)
LIPOLD, Markus Vincenc, (1816-1883)
PANČIČ, Joseph, (-1888)
PŎSEPNÝ, František, (1836-1895)
SAPIEHA, Aleksander, (1773-1812)
SEELAND, Ferdinand, (1822-1901)
STUR, Dénes, (1827-1893)
VIQUESNEL, Auguste, (1800-1867)

1850-1899

ANDRIAN-WERBURG, Ferdinand von, Freiherr, (1835-1914)
BRUSINA, Spiridion, (1845-1908)
BUCCHICH, Gregorio, (1828-1911)
BUKOWSKI, Gejza, (1858-1937)
CVICIĆ, Jovan, (1865-1927)
DAVIS, William Morris, (1850-1934)
DE REGNY, Paolo Vinassa, (1871-1957)
FREYER, Heinrich, (1802-1865)
FUCHS, Wilhelm, (1802-1853)
GORJANOVIĆ-KRAMBERGER, Dragutin, (1856-1937)
HINTERLECHNER, Karl, (1874-1932)
KATZER, Bedřich, (1861-1925)
KIŠPATIĆ, Mišo, (1851-1926)
KITTL, Ernst Anton Leopold, (1854-1913)
KOCH, Ferdo, (1874-?1936)
KOSSMAT, Franz, (1871-1938)
LIEBUS, Adalbert, (1876-1945)
LIPOLD, Markus Vincenc, (1816-1883)
LÓCZY, Lajos, (1849-1920)
MACLAREN, James Malcolm, (1873-1935)
MARTELLI, Alessandro, (1877-1934)
MOHOROVIČIĆ, Andrya, (1857-1936)
MOJSISOVICS, Johann August Georg Edmund, Edler von Mojsvàr, (1839-1907)
NIKITIN, Vasily Vasiliivich, (1867-1942)
OGILVIE GORDON, Maria Matilda, Dame, (1864-1939)
PANČIČ, Joseph, (-1888)
PERGENS, Eduard Willem Gerard, (1862-1917)
PETKOVIĆ, Vladamir K., (1873-1934)
PILAR, Georg, (1847-1893)
PŎSEPNÝ, František, (1836-1895)

YUGOSLAVIA 1850-1899 Cont'd.

RADIMSKY, Wenzel, (-1895)
RADOVANOVIĆ, Svetolik A., (1863-1928)
SCHLOENBACH, Georg Justin Carl Urban, (1841-1870)
SEELAND, Ferdinand, (1822-1901)
STUR, Dénes, (1827-1893)
TUČAN, Fran, (1878-1954)
VIQUESNEL, Auguste, (1800-1867)
VIZER, Vilmos, (1873-1941)
ŽUKOVIĆ, Jovan M., (1856-1937)

1900 to present
ANDREW, Arthur Robert, (1881-1941)
ANDRIAN-WERBURG, Ferdinand von, Freiherr, (1835-1914)
ANIĆ, Dragutin, (1898-1961)
ATHANASIU, Sava, (-?1946)
AUBOUIN, Jean, (1928-)
BABIĆ, Bogdan, (1891-1961)
BASSINGER, William E., (1928-1977)
BROCK, Byron Britton, (1904-1972)
BRUSINA, Spiridion, (1845-1908)
BUCCHICH, Gregorio, (1828-1911)
BUKOWSKI, Gejza, (1858-1937)
ČELEBRIĆ, Dorde, (1929-1970)
CVICIĆ, Jovan, (1865-1927)
DAVIS, William Morris, (1850-1934)
DE REGNY, Paolo Vinassa, (1871-1957)
GORJANOVIĆ-KRAMBERGER, Dragutin, (1856-1937)
GRČEV, Kočo, (1929-1963)
HALL, Thomas Clifford Fitzwilliam, (1886-1956)
HENDON, Bryan, (1896-1926)
HERAK, Milan, (1917-)
HERITSCH, Franz, (1882-1945)
HINTERLECHNER, Karl, (1874-1932)
HORN, Hans Lorenz, (1918-1941)
JURKOVIĆ, Ivan, (1917-)
KATZER, Bedřich, (1861-1925)
KIŠPATIĆ, Mišo, (1851-1926)
KITTL, Ernst Anton Leopold, (1854-1913)
KOCH, Ferdo, (1874-?1936)
KOCHANSKY-DEVIDÉ, Vanda, (1915-)
KOSSMAT, Franz, (1871-1938)
LAPPARENT, Jacques de, (-?1949)
LIEBUS, Adalbert, (1876-1945)
LÓCZY, Lajos, (1849-1920)
LUKOVIĆ, Milan, (1889-)
MACKAY, Robert Alexander, (?1904-1961)
MACLAREN, James Malcolm, (1873-1935)
MARGETIČ, Marko, (1891-1955)
MARIĆ, Luka, (1899-)
MARTELLI, Alessandro, (1877-1934)
MIHOLIĆ, Stanko, (1891-1960)
MOHOROVIČIĆ, Andrya, (1857-1936)
MOJSISOVICS, Johann August Georg Edmund, Edler von Mojsvàr, (1839-1907)

YUGOSLAVIA 1900 to present Cont'd.

NIKITIN, Vasily Vasiliivich, (1867-1942)
OGILVIE GORDON, Maria Matilda, Dame, (1864-1939)
PAPP, Simon, (1886-1970)
PAVLOVIĆ, Stojan, (1903-)
PERGENS, Eduard Willem Gerard, (1862-1917)
PETKOVIĆ, Kosta, (1903-)
PETKOVIĆ, Vladamir K., (1873-1934)
PEVALEK, Ivo, (1893-)
POLJAK, Josip, (1882-1962)
RADOVANOVIĆ, Svetolik A., (1863-1928)
RAKOVEC, Ivan, (1899-)
SALOPEK, Marijan, (1883-1967)
SEELAND, Ferdinand, (1822-1901)
ŠIMEK, Antonín, (1887-1942)
SIMONOVIĆ, Radivoj, (-1950)
TAJDER, Miroslav, (1909-)
TRIBUSSON, Boris, (1924-1957)
TRÜMPY, Daniel, (1893-1971)
TUČAN, Fran, (1878-1954)
VIZER, Vilmos, (1873-1941)
WASHINGTON-GRAY, Kenneth, (?1899-1963)
ZUBER, Stanislaw, (1893-1947)
ŽUKOVIĆ, Jovan M., (1856-1937)

ZAIRE
1850-1899
BOND, Marshall, (1867-1941) [PROSPECTORS, DIVINERS AND MINING ENGINEERS]
BUTTGENBACH, Henri, (1874-1964)
CORNET, René Jules, (1865-1929)
DUPONT, Édouard François, (1841-1911)
FOURMARIER, Paul Frédéric Joseph Lucien Lauren, (1877-1970)
HORNEMAN, Hans Henrik, (1878-1945)
ROMEU, Albert de, (1875-1915)
VEATCH, Arthur Clifford, (1878-1938)

1900 to present
BILLIET, Valère Louis, (1903-1945)
BOND, Marshall, (1867-1941) [PROSPECTORS, DIVINERS AND MINING ENGINEERS]
BOWEN, Norman Levi, (1887-1956)
BUTTGENBACH, Henri, (1874-1964)
CORNET, René Jules, (1865-1929)
DE DYCKER, Raymond, (1903-1947)
DOORNINCK, Nico H. van, (1893-1956)
DUPONT, Édouard François, (1841-1911)
FOURMARIER, Paul Frédéric Joseph Lucien Lauren, (1877-1970)
HATFIELD, Willis Charles, "Luke", (1898-1970)
HERNÁNDEZ-PACHECO, Francisco, (1899-1976)
HORNEMAN, Hans Henrik, (1878-1945)
JAMOTTE, André, (1903-1951)
ROMEU, Albert de, (1875-1915)
SALÉE, Achille, Chanoine, (1883-1932)
SCHNEEGANS, Daniel, (1907-1949)
SCHOEP, Alfred Ferdinandus Maria Ghislenus, (1881-1966)
SHALER, Millard King, (1880-1942)
SLAWSON, Chester Baker, (1898-1964)
STRAELEN, Victor Emile van, (1889-1964)
STUTZER, Otto, (1881-1936)
TAZIEFF, Haroun, (1914-)
VANDENDRIESSCHE, Adrien, (1914-1940)
VEATCH, Arthur Clifford, (1878-1938)
VERHOOGEN, Jean, (-)
WALLACE, Maurice Hudiburg, (1917-1961)

ZAMBIA
1850-1899
CODRINGTON, Thomas, (1829-1918)
FEARNSIDES, William George, (1879-1968)
HYLAND, John Shearson, (1866-1898)
MOLYNEUX, Arthur John Charles, (?1866-1921)
THORNTON, Richard, (1838-1863)

1900 to present
ANDERSON, Robert Beattie, (1904-1970)
BARRON, William Ernest, (1880-1935)
BATEMAN, Alan Mara, (1889-1971)
BROCK, Byron Britton, (1904-1972)
CODRINGTON, Thomas, (1829-1918)
DONOVAN, Percy Williams, (1879-1973)

ZAMBIA 1900 to present Cont'd.

FEARNSIDES, William George, (1879-1968)
FUCHS, Vivian Ernest, Sir, (1908-)
GALLAGHER, David, (1906-1968)
GILLANDERS, Earl B., (-)
HATFIELD, Willis Charles, "Luke", (1898-1970)
HITCHEN, Charles Stansfield, (1904-1971)
MENNELL, Frederick Philip, (1880-1966)
MOLYNEUX, Arthur John Charles, (?1866-1921)
WASSERSTEIN, Benno, (1906-1956)
WILLIAMSON, John Thorburn, (1907-1958)

ZANZIBAR <u>See</u> TANZANIA.

ZIMBABWE <u>See</u> RHODESIA.

COUNTRY unspecified

Up to 1800

ASCOLI, Cacco d', (1257-1327)
ASHMOLE, Elias, (1617-1692)
BLACK, Joseph, (1728-1799)
BRAHE, Tyge (Tycho), (1546-1601)
FORSIUS (HELSINGFORSIUS), Sigfrid Aronius, (ca.1550-1624)
FOURIER, Jean Baptiste Joseph, (1768-1830)
HELMONT, Johann Baptista van, (1578-1644)
HOBBS, William, (fl. 1700-1715)
LEEUWENHOEK, Antoni van, (1632-1723)
LESSER, Friedrich Christian, (1692-1754)
MILLER, Samuel, (1769-1850)
MONBODDO, James Burnett, Lord, (1714-1799)
PALEY, William, Archdeacon, (1743-1805)
PHILLIPS, Richard, (1778?-1851)
TOULMIN, George Hoggart, (ca.1750-1817)
WHISTON, William, (1666-1753)

1800-1849

ADAMS, John Couch, (1819-1892)
BAKER, William, (1787-1853)
FOURIER, Jean Baptiste Joseph, (1768-1830)
LEWIS, Tayler, (1802-)
LOWRY, Joseph Wilson, (1803-1879)
MILLER, Samuel, (1769-1850)
MIVART, St. George, (1827-1900)
OKEN, Lorenz, (1779-1851)
PALEY, William, Archdeacon, (1743-1805)
PHILLIPS, Richard, (1778?-1851)
SEUBERT, Moritz August, (1818-1878)
SPITZWEG, Carl, (1808-1885)
TOULMIN, George Hoggart, (ca.1750-1817)
WAGNER, William, (1796-1885)

1850-1899

ABBE, Ernst, (1840-1905)
ADAMS, John Couch, (1819-1892)
ANDERSSON, Axel R., (1858-1952)
ANDREWS, W. S., (1847-1929)
BAKER, William, (1787-1853)
BALL, Robert Stawell, Sir, (1840-1913)
BUTLER, Samuel, (1835-1902)
CLARK, John Willis, (1833-1910)
DAMPIER, William Cecil, Sir, (1867-1952)
DARROW, Clarence, (1857-1938)
HAITINGER, Max, (1868-1946)
HOLMES, William Henry, (1846-1933)
HOWORTH, Henry Hoyle, Sir, (1842-1923)
LEWIS, Tayler, (1802-)
LILJEVALL, Georg Gideon, (1848-1928)
LOWRY, Joseph Wilson, (1803-1879)
MACOUN, John, (1831-1920)
MICHALSKI, Stanislaw, (1865-1949)
MILLER, Samuel, (1769-1850)
MIVART, St. George, (1827-1900)

COUNTRY unspecified 1850-1899 Cont'd.

MOY-THOMAS, James Alan, (-1944)
OKEN, Lorenz, (1779-1851)
ROMANES, George John, (1848-1894)
SEAMAN, Elizabeth Cochrane, (1865-1922)
SEUBERT, Moritz August, (1818-1878)
SPITZWEG, Carl, (1808-1885)
WAGNER, William, (1796-1885)
WEIDENREICH, Franz, (1873-1947)

1900 to present

ANDERSSON, Axel R., (1858-1952)
ANDREWS, W.S., (1847-1929)
BALL, Robert Stawell, Sir, (1840-1913)
BUTLER, Samuel, (1835-1902)
CAMP, Russell R., (1904-1967)
CANNON, Herbert Graham, (1897-1963)
CLARK, John Willis, (1833-1910)
DAMPIER, William Cecil, Sir, (1867-1952)
DARROW, Clarence, (1857-1938)
DRURY, James, (1879-   )
EDDINGTON, Arthur Stanley, Sir, (1882-1944)
FISHER, Ronald Aylmer, (1890-1962)
HAHN, Emily, (1904-   )
HAITINGER, Max, (1868-1946)
HOLMES, William Henry, (1846-1933)
HOWORTH, Henry Hoyle, Sir, (1841-1923)
LILJEVALL, Georg Gideon, (1848-1928)
MACOUN, John, (1831-1920)
MICHALSKI, Stanislaw, (1865-1949)
MOY-THOMAS, James Alan, (-1944)
NAGELE, Julius, (1906-1973)
NASH, Carlton S., (1914-   )
PRICE, George McCready, (1870-   )
SEAMAN, Elizabeth Cochrane, (1865-1922)
TOLANSKY, Samuel, (?1908-1973)
WEIDENREICH, Franz, (1873-1947)

Time unknown

CAPORIACCO, Ludovico di, (-   )

Section IX

INDEX OF GEOLOGISTS BY SPECIALITY

Introduction

In the brief biographies in Section VI, the special fields in which geologists undertook research have, wherever possible, been particularised (see also discussion, Introduction to Section VI, Vol. 2). These specialities are indexed here.

In deciding what subdivisions to recognize, I have, to some extent followed my personal prejudices. As noted earlier (Vol. 2, p. 380), I prefer "geomorphology" to "physical geology", since the latter overlaps so markedly in coverage with structural geology. "Dynamic geology" is included under the heading of structural geology; and "petrography" under petrology, of which it is a sub-discipline. For convenience, sedimentary petrologists are not distinguished from sedimentologists, the latter name being consistently applied. "Regional geology" is not distinguished as a specialism, being considered under the headings of petrology, structural geology or stratigraphy as appropriate. "Historical geology" is considered synonymous with stratigraphy. Geological mapping is regarded as an aspect of work in stratigraphy, except where the terrain mapped consisted exclusively of igneous or metamorphic rocks; it is not regarded as a separate discipline within geology. Much geological work involves work with photographs; geologists are listed under "photogeology" only when this was one their principal preoccupations. Similarly, the only palaeoclimatologists listed are those for whom this was an especial, rather than an incidental, concern.

Because I am a palaeontologist, I have found it easy to be precise with regard to the concerns of other palaeontologists; thus I have discriminated between invertebrate and vertebrate palaeontologists, between micropalaeontologist and palynologists, and I have distinguished palaeobotanists, palaeoichnologists and palaeoecologists. I am not competent, however, to make similarly precise identifications among petrologists (though I have endeavoured, at least, to distinguish igneous and metamorphic petrologists , vulcanologists and sedimentologists). I have distinguished seismologists but have not otherwise attempted to particularise the different branches of geophysics.

A variety of concerns are embraced under the umbrella term "Quaternary geologist"; this implies an interest in a number of different aspects of the study of post-Pliocene environments, sediments, erosional processes and land forms.

Gemmology can only marginally be regarded, if at all, as a branch of geology. I have not listed in this bibliography those individuals for whom gemmology was the sole approach to geology; the gemmologists listed here are those who included this subject among other geological concerns.

As in the last section, the listings are subdivided chronologically for the sake of convenience, the same four time-periods being distinguished. As with the listing by countries worked on and with the same reservations (see Introduction to Section VIII), the name of the same geologist may appear in the listings for two or, rarely, three successive time periods.

AGRICULTURAL GEOLOGY

1800-1849

JOHNSTON, James Finlay Weir, (1796-1855)
MILLINGTON, John, (1779-1886)
TRIMMER, Joshua, (1797-1857?)
VANUXEM, Lardner, (1792-1848)
VON POST, Hampus Adolf, (1822-1911)
WAILES, Benjamin Leonard Couington, (1797-1862)
WHITTLESEY, Charles, Colonel, (1808-1886)

1850-1899

ATTERBERG, Albert, (1846-1916)
BJØRLYKKE, Knut Olai, (1860-1946)
GLINKA, Konstantin Dmitrievich, (1867-1927)
HARRISON, John Burchmore, Sir, (1856-1926)
HORUSITZKY, Henrik, (1870-1945)
INKEY BÉLA, Palini, (1847-1921)
JOHNSTON, James Finlay Weir, (1796-1855)
JÖNSSON, Jöns, (1848-1894)
LOUGHRIDGE, Robert Hills, (1843-1917)
MARTELLI, Alessandro, (1877-1934)
MILLINGTON, John, (1779-1886)
MUNTEANI-MURGOCI, Gheorghe, (1872-1925)
ST. JOHN, Orestes Hawley, (1841-1921)
TRABUCCO, Giacomo, (1845-1925)
TREITZ, Péter, (1866-1935)
TRIMMER, Joshua, (1797-1857?)
VON POST, Hampus Adolf, (1822-1911)
WADSWORTH, Marshman Edward, (1847-1921)
WAHNSCHAFFE, Gustav Albert Bruno Felix, (1851-1914)
WAILES, Benjamin Leonard Couington, (1797-1862)
WHITTLESEY, Charles, Colonel, (1808-1886)
ŽUKOVIĆ, Jovan M., (1856-1937)

1900 to present

ATTERBERG, Albert, (1846-1916)
BALLENEGGER, Róbert, (1882-1969)
BAREN, F. H. van, (-1975)
BJØRLYKKE, Knut Olai, (1860-1946)
CERNESCU, Nicolae C., (1904-1967)
EDELMANN, Cornelis Hendrik, (1901-1964)
FELICIANO, Jose Maria "Dako", (1887-1955)
GLINKA, Konstantin Dmitrievich, (1867-1927)
HARRISON, John Burchmore, Sir, (1856-1926)
HORUSITZKY, Henrik, (1870-1945)
INKEY BÉLA, Palini, (1847-1921)
LONSDALE, John Tipton, (1895-1960)
LOUGHRIDGE, Robert Hills, (1843-1917)
MAROS, Imre, (1882-1937)
MARTELLI, Alessandro, (1877-1934)
MUNTEANI-MURGOCI, Gheorghe, (1872-1925)
ROARK, Louis, (1890-1950)
ST. JOHN, Orestes Hawley, (1841-1921)
TRABUCCO, Giacomo, (1845-1925)
TREITZ, Péter, (1866-1935)
VON POST, Hampus Adolf, (1822-1911)

AGRICULTURAL GEOLOGY 1900 to present Cont'd.

WADSWORTH, Marshman Edward, (1847-1921)
WAHNSCHAFFE, Gustav Albert Bruno Felix, (1851-1914)
ŽUKOVIĆ, Jovan M., (1856-1937)

AGRICULTURAL GEOLOGY See also PEDOLOGY

CRYSTALLOGRAPHY

Up to 1800

BARTHOLINUS, Erasmus, (1625-1698)
BIRINGUCCIO, Vannoccio, (1480-1538)
BOURGUET, Louis, (1678-1742)
BOURNON, Jacques Louis de, Comte, (1751-1825)
BROOKE, Henry James, (1771-1857)
DU FAY, Charles François, (1698-1739)
GUGLIELMINI, Domenico, (1655-1710)
HAÜY, René-Just, Abbé, (1743-1822)
HOTTINGER, Johann Heinrich, (1680-1756)
JAŚKIEWICZ, Jan D. P., (1749-1809)
LINNE, Carl von, (1707-1778)
PLINY THE ELDER (CAIUS PLINIUS SECUNDUS), (?20 A.D.-79 A.D.)
STENO, Nicolaus, (1631-1687)
WOLLASTON, William Hyde, (1766-1828)

1800-1849

BLUM, Johann Reinhard, (1802-1883)
BOURNON, Jacques Louis de, Comte, (1751-1825)
BRAVAIS, Auguste, (1811-1863)
BREITHAUPT, Johann Friedrich August, (1791-1873)
BROOKE, Henry James, (1771-1857)
CASTELLI, Albin Benedict, (1822-1892)
DANA, James Dwight, (1813-1895)
DELAFOSSE, Gabriel, (1796-1878)
DES CLOIZEAUX, Alfred Louis Olivier Legrand, (1817-1897)
GERMAR, Ernst Friedrich, (1786-1853)
HAIDINGER, Wilhelm Karl von, Ritter, (1795-1871)
HAÜY, René-Just, Abbé, (1743-1822)
HERSCHEL, John Frederick William, Sir, (1792-1871)
JAŚKIEWICZ, Jan D. P., (1749-1809)
KARSTEN, Hermann, (1809-1877)
KENNGOTT, Gustav Adolf, (1818-1897)
KOBELL, Franz von, (1803-1882)
KOKSHAROV, Nikola J. von, General, (1818-1893)
LARKIN, Nathaniel John, (1781-1855)
LÉVY, Armand, (1795-1861)
MILLER, William Hallowes, (1801-1880)
NAUMANN, Carl Friedrich, (1797-1873)
PHILLIPS, William, (1800-1834)
PRESL, Jan Svatopluk, (1791-1849)
RAMMELSBERG, Karl Friedrich, (1813-1899)
ROSE, Gustav, (1798-1873)
SCHRAUF, Albrecht, (1837-1897)
SELLA, Quintino, (1827-1884)
SZABÓ, József, (1822-1894)
TYNDALL, John, (1820-1893)
WEBSKY, Christian Friedrich Martin, (1824-1886)
WEISS, Christian Samuel, (1780-1856)
WHEWELL, William, Reverend, (1794-1866)
WOLLASTON, William Hyde, (1766-1828)
ZIPPE, František Xaver Maximilian, (1791-1863)

1850-1899

AF SCHULTÉN, August Benjamin, (1856-1913)

CRYSTALLOGRAPHY 1850-1899 Cont'd.

AMINOFF, Gregori, (1883-1947)
ANDERSON, Charles, (1876-1944)
BÄCKSTROM, Helge Mattias, (1865-1932)
BARLOW, William, (1845-1934)
BAUER, Max Hermann, (1844-1917)
BAUMHAUER, Heinrich Adolf, (1848-1926)
BEALE, William Phipson, Sir, (1839-1922)
BECKE, Friedrich Johann Karl, (1855-1931)
BECKENKAMP, Jakob, (1855-1931)
BECKER, Ewald, (-1873)
BERTRAND, Émile, (1844-1909)
BLAKE, John Marcus, (1838-1920)
BLUM, Johann Reinhard, (1802-1883)
BOERIS, Giovanni, (1867-1946)
BØGGILD, Ove Balthazar, (1872-1956)
BOWMAN, Herbert Lister, (1874-1942)
BRAGG, William Henry, Sir, (1862-1942)
BRAVAIS, Auguste, (1811-1863)
BREITHAUPT, Johann Friedrich August, (1791-1873)
BREZINA, Maria Aristides, (1848-1909)
BROGLIE, Maurice de, (1875-1960)
BROWN, Amos Peaslee, (1864-1917)
BRUGNATELLI, Luigi, (1859-1928)
BRUHNS, Franz Siegfried Willy, (1864-1929)
BUCCA, Lorenzo, (1857-1930)
CAHN, Lazard, (1865-1940)
CALKER, Friedrich Julius Peter von, (1841-1913)
CARTER, William Lower, (1855-1918)
CASTELLI, Albin Benedict, (1822-1892)
CESÀRO, Guiseppe Raimondo Pio, (1849-1939)
CHRISTA, Emanuel, (1874-1948)
COPAUX, Hippolyte, (1872-1934)
DANA, James Dwight, (1813-1895)
DELAFOSSE, Gabriel, (1796-1878)
DES CLOIZEAUX, Alfred Louis Olivier Legrand, (1817-1897)
DRUGMAN, Julien Léon Eugène, (1875-1950)
DUFFOUR, Alexis, (1877-1948)
DUPARC, Louis Claude, (1866-1932)
EVANS, John William, (1857-1930)
FEDEROV, Evgraf Stepanovich, (1853-1919)
FLETCHER, Lazarus, Sir, (1854-1921)
FLETCHER, Mark, Reverend, (1870-1956)
FORD, William Ebenezer, (1878-1939)
FRANZENAU, Agoston, (1856-1919)
FRIEDEL, Charles, (1832-1899)
FRIEDEL, Georges, (1865-1933)
GAUBERT, Marie Benoit Paul, (1865-)
GAUDEFROY, Christophe Léon, Chanoine, (1878-1971)
GERMAR, Ernst Friedrich, (1786-1853)
GLINKA, Sergei Fedorovich, (1855-1933)
GOLDSCHMIDT, Victor, (1853-1933)
GONÇALVES GUIMARÃES, Antonio José, (1875-1919)
GOSSNER, Balthasar, (1877-1937)
GROSSPIETSCH, Oskar, (-1920)

CRYSTALLOGRAPHY 1850-1899 Cont'd.

GROTH, Paul Heinrich von, Ritter, (1843-1927)
GRÜNLING, Friedrich, (1857-1919)
GURNEY, Henry Palin, Reverend, (1847-1904)
HAIDINGER, Wilhelm Karl von, Ritter, (1795-1871)
HAUTEFEUILLE, Paul Gabriel, (1836-1902)
HERSCHEL, John Frederick William, Sir, (1792-1871)
HIKI, Tadasu, (1866-1927)
HINRICHS, Gustavus Detlef, (-1922)
HINTZE, Carl Adolf Ferdinand, (1851-1916)
HUTCHINSON, Arthur, (1866-1937)
JANNETTAZ, Edouard, (-1900)
JOHNSEN, Arrien, (1877-1934)
JOLY, John, (1857-1933)
JORDAN, James Brown, (1838-1915)
KAISER, Friedrich Wilhelm Erich, (1871-1934)
KARANDYEEV, Vissarion Vissarionovich, (-1916)
KARSTEN, Hermann, (1809-1877)
KENNGOTT, Gustav Adolf, (1818-1897)
KLEIN, Johann Friedrich Carl, (1842-1907)
KLOCKE, Friedrich, (1847-1885)
KOBELL, Franz von, (1803-1882)
KOENIGSBERGER, Johann Georg, (1874-1946)
KOKSHAROV, Nikola J. von, General, (1818-1893)
KRANTZ, Friedrich Ludwig Robert, (1859-1926)
KRAUS, Edward Henry, (1875-1973)
KRENNER, József Sándor, (1839-1920)
LANG, Victor von, Edler, (1838-1921)
LARKIN, Nathaniel John, (1781-1855)
LAUE, Max Theodor Felix von, (1879-1960)
LEHMANN, Otto, (1855-1922)
LÉVY, Armand, (1795-1861)
LIEBISCH, Theodor, (1852-1922)
LIFFA, Aurél, (1872-1957)
LINCIO, Gabrielle, (1874-1938)
LIVEING, George Downing, (1827-1924)
LOEWINSSON-LESSING, Franc Julyevich, (1859-1939)
MALLARD, François Ernest, (1833-1894)
MAUGIN, Charles, (1878-1958)
MELCZER, Gusztáv, (1869-1907)
MICHEL, Léopold Jean, (1846-1919)
MIERS, Henry Alexander, Sir, (1858-1942)
MILLER, William Hallowes, (1801-1880)
MILLOSEVICH, Federico, (1875-1942)
MOSES, Alfred Joseph, (1859-1920)
MÜGGE, Johannes Otto Konrad, (1858-1932)
NAUMANN, Carl Friedrich, (1797-1873)
NECHAEV, Aleksyei Vasil'evich, (1864-1915)
NIKITIN, Vasily Vasiliivich, (1867-1942)
PANEBIANCO, Ruggero, (1848-1930)
PECK, Frederick Burritt, (1860-1925)
PENFIELD, Samuel Lewis, (1856-1906)
PHILLIPS, Alexander Hamilton, (1866-1937)
PHILLIPS, William, (1800-1834)
PIRSSON, Louis Valentine, (1860-1919)

CRYSTALLOGRAPHY 1850-1899 Cont'd.

RAMMELSBERG, Karl Friedrich, (1813-1899)
RATH, Gerhard vom, (1830-1888)
REPOSSI, Emilio, (1876-1931)
RINNE, Friedrich Wilhelm Berthold, (1863-1933)
ROSE, Gustav, (1798-1873)
ROSENBUSCH, Karl Harry Ferdinand, (1836-1914)
SAINTE-CLAIRE DEVILLE, Henri, (-1881)
SAMOILOV, Yakov Vladimirovich, (1870-1925)
SCACCHI, Eugenio, (1854-1929)
SCHARIZER, Rudolf, (1859-1935)
SCHOENFLIES, Arthur Moritz, (1853-1928)
SCHRAUF, Albrecht, (1837-1897)
SELIGMANN, Gustav, (1849-1920)
SELLA, Quintino, (1827-1884)
SJÖGREN, Sten Anders Hjalmar, (1856-1922)
SMOLAŘ, Gotthard, (1863-1926)
SOLLAS, William Johnson, (1849-1936)
SOLLY, Richard Harrison, (1851-1925)
SOUZA BRANDÃO, Vicente de, (1863-1916)
SPENCER, Leonard James, (1870-1959)
SZABÓ, József, (1822-1894)
TAMMANN, Gustav Heinrich Johann Apoll, (1861-1938)
TOPSØE, Haldor Frederik Axel, (1842-1935)
TUTTON, Alfred Edwin Howard, (1864-1938)
TYNDALL, John, (1820-1893)
VERNADSKI, Vladimir Ivanovich, (1863-1945)
VIOLA, Carlo Maria, (1855-1925)
VOGELSANG, Hermann, (1838-1874)
VOIGT, Waldemar, (1850-1919)
VULF, Georgii Viktorovich, (1863-1925)
WALLERANT, Frédéric Félix Auguste, (1858-1936)
WEBSKY, Christian Friedrich Martin, (1824-1886)
WEIBULL, Kristian Oskar Mats, (1856-1923)
WEISBACH, Albin, (1833-1901)
WEISS, Christian Ernst, (1833-1890)
WEISS, Christian Samuel, (1780-1856)
WEYBERG, Zygmunt, (1872-1944)
WINCHELL, Alexander Newton, (1874-1958)
WRIGHT, Frederick Eugene, (1877-1953)
WÜLFING, Ernst Anton, (1860-1930)
WYROUBOFF, Gr., (1843-1913)
ZEMYATCHENSKY, Petr Andreevich, (1856-1942)
ZEPHAROVICH, Victor Leopold von, Ritter, (1830-1890)
ZIPPE, František Xaver Maximilian, (1791-1863)
ZYNDEL, Fortunat, (1882-1917)

1900 to present

AF SCHULTÉN, August Benjamin, (1856-1913)
AMINOFF, Gregori, (1883-1947)
ANDERSON, Charles, (1876-1944)
ARMASHEVSKY, Peter Jakovlevich, (-1919)
BÄCKSTROM, Helge Mattias, (1865-1932)
BALOGH, Ernö, (1882-1969)
BARKER, Thomas Vipond, (1881-1931)

CRYSTALLOGRAPHY 1900 to present Cont'd.

BARLOW, William, (1845-1934)
BARTH, Thomas Frederik Weiby, (1899-1971)
BAUER, Max Hermann, (1844-1917)
BAUHANS, Hans, (1888-1918)
BAUMHAUER, Heinrich Adolf, (1848-1926)
BEALE, William Phipson, Sir, (1839-1922)
BECKE, Friedrich Johann Karl, (1855-1931)
BECKENKAMP, Jakob, (1855-1931)
BERMAN, Harry, (1902-1944)
BERTRAND, Émile, (1844-1909)
BEUDANT, François-Sulpice, (1787-1850)
BILLIET, Valère Louis, (1903-1945)
BLAKE, John Marcus, (1838-1920)
BOERIS, Giovanni, (1867-1946)
BØGGILD, Ove Balthazar, (1872-1956)
BOLDYREV, Anatolii Kapitonovich, (1883-1946)
BOWMAN, Herbert Lister, (1874-1942)
BRADLEY, William Frank, (1908-1973)
BRAGG, William Henry, Sir, (1862-1942)
BRAGG, William Lawrence, Sir, (1890-1971)
BREZINA, Maria Aristides, (1848-1909)
BROGLIE, Maurice de, (1875-1960)
BROWN, Amos Peaslee, (1864-1917)
BRUGNATELLI, Luigi, (1859-1928)
BRUHNS, Franz Siegfried Willy, (1864-1929)
BUCCA, Lorenzo, (1857-1930)
BUCKLEY, Harold Eugene, (1897-1959)
BUERGER, Martin Julian, (-)
CAHN, Lazard, (1865-1940)
CALKER, Friedrich Julius Peter von, (1841-1913)
CARTER, William Lower, (1855-1918)
CESÀRO, Guiseppe Raimondo Pio, (1849-1939)
CHERVET, Jean, (1904-1962)
CHEVENARD, Pierre, (1888-1960)
CHRISTA, Emanuel, (1874-1948)
CHUDOBA, Karl F., (1898-1976)
CLAUSEN, Hans Carl Eisenreich, (1898-1971)
COPAUX, Hippolyte, (1872-1934)
DENNING, Reynolds McConnell, (1916-1967)
DOLIVO-DOBROVOLSKY, Vadim Vladimirovich, (1904-1936)
DRUGMAN, Julien Léon Eugène, (1875-1950)
DUFFOUR, Alexis, (1877-1948)
DUPARC, Louis Claude, (1866-1932)
EVANS, John William, (1857-1930)
FANKUCHEN, Isidor, (1905-1964)
FEDEROV, Evgraf Stepanovich, (1853-1919)
FERSMAN, Aleksander Yevgenyevich, (1883-1945)
FLETCHER, Lazarus, Sir, (1854-1921)
FLETCHER, Mark, Reverend, (1870-1956)
FÖLDVÁRI, Aladár, (1906-1973)
FORD, William Ebenezer, (1878-1939)
FRANZENAU, Agoston, (1856-1919)
FRIEDEL, Edmond, (1895-1972)
FRIEDEL, Georges, (1865-1933)

CRYSTALLOGRAPHY 1900 to present Cont'd.

GAUDEFROY, Christophe Léon, Chanoine, (1878-1971)
GLINKA, Sergei Fedorovich, (1855-1933)
GOLDSCHMIDT, Victor, (1853-1933)
GONÇALVES GUIMARÃES, Antonio José, (1875-1919)
GORDON, Samuel George, (1897-1952)
GOSSNER, Balthasar, (1877-1937)
GRAHAM, Richard Percival Devereux, (1880-1965)
GROSSPIETSCH, Oskar, (-1920)
GROTH, Paul Heinrich von, Ritter, (1843-1927)
GRÜNLING, Friedrich, (1857-1919)
GURNEY, Henry Palin, Reverend, (1847-1904)
HAUTEFEUILLE, Paul Gabriel, (1836-1902)
HAWKINS, Alfred Cary, (1887-1954)
HIKI, Tadasu, (1866-1927)
HINRICHS, Gustavus Detlef, (-1922)
HINTZE, Carl Adolf Ferdinand, (1851-1916)
HUNT, Walter Frederick, (1882-1975)
HUTCHINSON, Arthur, (1866-1937)
JANNETTAZ, Edouard, (-1900)
JOHNSEN, Arrien, (1877-1934)
JOLY, John, (1857-1933)
JORDAN, James Brown, (1838-1915)
KAISER, Friedrich Wilhelm Erich, (1871-1934)
KARANDYEEV, Vissarion Vissarionovich, (-1916)
KAŠPAR, Jan, (1898-)
KLEBER, Will, (-1970)
KLEIN, Johann Friedrich Carl, (1842-1907)
KOENIGSBERGER, Johann Georg, (1874-1946)
KÖHLER, Alexander, (1893-1955)
KRANTZ, Friedrich Ludwig Robert, (1859-1926)
KRAUS, Edward Henry, (1875-1973)
KRENNER, József Sándor, (1839-1920)
KREUTZ, Stefan, (1883-1941)
LANG, Victor von, Edler, (1838-1921)
LAUE, Max Theodor Felix von, (1879-1960)
LEDOUX, Auguste, (1887-1918)
LEHMANN, Otto, (1855-1922)
LEUTWEIN, Friedrich, (1911-1974)
LIEBISCH, Theodor, (1852-1922)
LIFFA, Aurél, (1872-1957)
LINCIO, Gabrielle, (1874-1938)
LIVEING, George Downing, (1827-1924)
LOEWINSSON-LESSING, Franc Julyevich, (1859-1939)
LONSDALE, Kathleen, Dame, (née Yardley), (1903-1971)
LÖW, Márton, (1885-1943)
MACHATSCHKI, Felix Karl Ludwig, (1895-1970)
MAUGIN, Charles, (1878-1958)
MELCZER, Gusztáv, (1869-1907)
MELLIS, Otto, (1906-1970)
MICHEL, Léopold Jean, (1846-1919)
MIERS, Henry Alexander, Sir, (1858-1942)
MILLOSEVICH, Federico, (1875-1942)
MITSCHERLICH, Eilhard, (1794-1863)
MOSES, Alfred Joseph, (1859-1920)

CRYSTALLOGRAPHY 1900 to present Cont'd.

MÜGGE, Johannes Otto Konrad, (1858-1932)
MURDOCK, Joseph, (1890-1973)
NÁRAY-SZABÓ, István, (1899-1972)
NECHAEV, Aleksyei Vasil'evich, (1864-1915)
NEUHAUS, Alfred, (1903-1975)
NIGGLI, Paul, (1888-1953)
NIKITIN, Vasily Vasiliivich, (1867-1942)
NOVÁK, Jiří, (?1902-1971)
PALACHE, Charles, (1869-1954)
PANEBIANCO, Ruggero, (1848-1930)
PARDILLO VAQUER, Francisco, Don, (1884-1955)
PARKER, Robert Luling, (1893-1973)
PATTERSON, Arthur Lindo, (1902-1966)
PEACOCK, Martin Alfred, (1898-1950)
PECK, Frederick Burritt, (1860-1925)
PENFIELD, Samuel Lewis, (1856-1906)
PHILLIPS, Alexander Hamilton, (1866-1937)
PIOTROWSKI, Hugon Ludwik, (1897-1962)
PIRSSON, Louis Valentine, (1860-1919)
RAMAN, Chandrasekhara Venkata, Sir, (1888-1970)
RAMSDELL, Lewis Stephen, (1895-1975)
REPOSSI, Emilio, (1876-1931)
RINNE, Friedrich Wilhelm Berthold, (1863-1933)
ROGERS, Austin Flint, (1887-1957)
ROMAN, David, (1884-1927)
ROSE, Hermann, (1883-1976)
ROSENBUSCH, Karl Harry Ferdinand, (1836-1914)
ROSICKÝ, Vojtěch, (1880-1942)
ROST, Rudolf, (1902-)
SAMOILOV, Yakov Vladimirovich, (1870-1925)
SCACCHI, Eugenio, (1854-1929)
SCHALLER, Waldemar Theodore, (1882-1967)
SCHARIZER, Rudolf, (1859-1935)
SCHOENFLIES, Arthur Moritz, (1853-1928)
SEKANINA, Josef, (1900-)
SELIGMANN, Gustav, (1849-1920)
ŠIMEK, Antonín, (1887-1942)
SIMONS, Herbert Samuel, (1891-1956)
SJÖGREN, Sten Anders Hjalmar, (1856-1922)
SLAVÍKOVÁ, Ludmila, (1890-1943)
SLAWSON, Chester Baker, (1898-1964)
SMITH, Thomas Hodge, (1894-1945)
SMOLAŘ, Gotthard, (1863-1926)
SOLLAS, William Johnson, (1849-1936)
SOLLY, Richard Harrison, (1851-1925)
SOUZA BRANDÃO, Vicente de, (1863-1916)
SPANGENBERG, Kurt, (1889-1957)
SPENCER, Leonard James, (1870-1959)
STEINMETZ, Hermann, (1879-1964)
TAMMANN, Gustav Heinrich Johann Apoll, (1861-1938)
TERTSCH, Hermann, (1880-1963)
TOKODY, László, (1893-1964)
TOPSØE, Haldor Frederik Axel, (1842-1935)
TUTTON, Alfred Edwin Howard, (1864-1938)

CRYSTALLOGRAPHY 1900 to present Cont'd.

ULRICH, František, (1899-1941)
UNGEMACH, Henri Léon, (1879-1936)
VENDL, Maria, (1890-1945)
VERNADSKI, Vladimir Ivanovich, (1863-1945)
VIOLA, Carlo Maria, (1855-1925)
VOIGT, Waldemar, (1850-1919)
VULF, Georgii Viktorovich, (1863-1925)
WALLERANT, Frédéric Félix Auguste, (1858-1936)
WEIBULL, Kristian Oskar Mats, (1856-1923)
WEISBACH, Albin, (1833-1901)
WEYBERG, Zygmunt, (1872-1944)
WINCHELL, Alexander Newton, (1874-1958)
WOJNO, Tadeusz Jerzy, (1884-1971)
WRIGHT, Frederick Eugene, (1877-1953)
WÜLFING, Ernst Anton, (1860-1930)
WYROUBOFF, Gr., (1843-1913)
ZEMYATCHENSKY, Petr Andreevich, (1856-1942)
ZYNDEL, Fortunat, (1882-1917)

Time unknown
WOLFE, Caleb Wroe, (-)

DIVINING

1850-1899

MULLINS, John, (1838-) [PROSPECTORS, DIVINERS AND MINING ENGINEERS]

1900 to present

GROSS, Henry, (-) [PROSPECTORS, DIVINERS AND MINING ENGINEERS]

PENROSE, Evelyn, (-) [PROSPECTORS, DIVINERS AND MINING ENGINEERS]

ECONOMIC GEOLOGY

Up to 1800

BOYLE, Robert, Honourable, (1627-1691)
BROCHANT DE VILLIERS, André Jean Marie, (1772-1840)
CHARPENTIER, Johann Friedrich Wilhelm Toussaint von, (1728-1805)
CORDIER, Pierre Louis Antoine, (1777-1861)
CRONSTEDT, Axel Fredrik, (1722-1765)
FREIESLEBEN, Johann Carl, (1774-1846)
HARDENBERG, Friedrich Leopold von, Freiherr, (1772-1801)
HERDER, Siegmund August Wolfgang von, Freiherr, (1776-1838)
JAMESON, Robert, (1774-1854)
JAŚKIEWICZ, Jan D. P., (1749-1809)
LOMONOSOV, Mikhail Vasilyevich, (1711-1765)
SWEDENBORG, Emanuel, (1688-1772)
TELEKI, Domokos, Graf, (-1803)
TREBRA, Friedrich Wilhelm Heinrich von, (1740-1819)
WERNER, Abraham Gottlob, (1749-1817)
WHITEHURST, John, (1713-1788)

1800-1849

ANSTED, David Thomas, (1814-1880)
APLIN, Christopher d'Oyley Hale, (1819-1875)
BEUST, Friedrich Constantin von, Freiherr, (1806-)
BLAKE, William Phipps, (1826-1910)
BOGAERT, P. J. J., (1807-1894)
BROADHEAD, Garland Carr, (1827-1912)
BROCHANT DE VILLIERS, André Jean Marie, (1772-1840)
CASTILLO, Antonio del, Don, (1820-1895)
CHARPENTIER, Johann Friedrich Wilhelm Toussaint von, (1728-1805)
CONDON, Thomas, (1822-1907)
COOK, George Hammell, (1818-1889)
CORDIER, Pierre Louis Antoine, (1777-1861)
COTTA, Carl Bernhard von, (1808-1879)
DAUBRÉE, Gabriel Auguste, (1814-1896)
DECHEN, Ernst Heinrich Carl von, (1800-1889)
DREW, Frederick, (1836-1891)
DUNSTAN, Benjamin, (1864-1933)
EGGERTZ, Victor, (1817-1889)
ENGELHARDT, Otto Moritz Ludwig von, (1779-1842)
ERDMANN, Axel Joachim, (-)
FORBES, David, (1828-1876)
FREIESLEBEN, Johann Carl, (1774-1846)
FUCHS, Wilhelm, (1802-1853)
FURUHJELM, Enoch Hjalmar, (1823-1886)
GIORDANO, Felice, (1825-1892)
GRIFFITH, Richard John, Sir, (1784-1878)
GÜMBEL, Carl Wilhelm von, Ritter, (1823-1898)
HAGLUND, Emil, (1877-1938)
HARDENBERG, Friedrich Leopold von, Freiherr, (1772-1801)
HAUSMANN, Johann Friedrich Ludwig, (1782-1859)
HAYDEN, Ferdinand Vandiveer, (1829-1887)
HERBICH, Franz, (1821-1887)
HERDER, Siegmund August Wolfgang von, Freiherr, (1776-1838)
HILDRETH, Samuel Prescott, (1783-1863)
HOME, David Milne, (1805-1890)
HOUGHTON, Douglass, (1809-1845)

ECONOMIC GEOLOGY 1800-1849 Cont'd.

HOW, Henry, (1828-1879)
HUBBARD, Bela, (1814-1896)
IGELSTRÖM, Lars Johan, (1822-1897)
JAMESON, Robert, (1774-1854)
JAŚKIEWICZ, Jan D. P., (1749-1809)
JOHNSTRUP, Johannes Frederik, (1818-1894)
JONES, Thomas Rupert, (1819-1911)
KEENE, William, (1798-1872)
KELAART, E. F., (1818-1860)
KINAHAN, George Henry, (1829-1908)
LESLEY, J. Peter, (1819-1903)
MEDLICOTT, Henry Benedict, (1829-1905)
MOORE, Francis, Jr., (1808-1864)
MÜLLER, Carl Hermann, (1823-1907)
MURRAY, Alexander, (1810-1884)
OLDHAM, Thomas, (1816-1878)
OMALIUS D'HALLOY, Jean-Baptist Julien d', (1783-1875)
OWEN, David Dale, (1807-1860)
PISSIS, Aimé, (1812-1889)
POSEPNÝ, František, (1836-1895)
PUSCH, Georg Gottlieb, (1790-1846)
RAMSAY, Andrew Crombie, Sir, (1814-1891)
RANDALL, John, (1810-1910)
ROGERS, Henry Darwin, (1808-1866)
RÖMER, Hermann, (1816-1894)
ROSALES, Henry, (?1822-1916)
SAWKINS, James Gay, (1806-1878)
SELWYN, Alfred Richard Cecil, (1824-1902)
STARING, W. C. H., (1808-1874)
STUTCHBURY, Samuel, (1798-1859)
TELEKI, Domokos, Graf, (-1803)
TENORE, Gaetano, (1826-1903)
TOPLEY, William, (1841-1894)
TRAILL, Thomas Stewart, (1781-1862)
TREBRA, Friedrich Wilhelm Heinrich von, (1740-1819)
WERNER, Abraham Gottlob, (1749-1817)
WHITNEY, Josiah Dwight, (1819-1896)
WHITTLESEY, Charles, Colonel, (1808-1886)
WINCHELL, Alexander, (1824-1891)
WOLF, Heinrich, (1825-1882)
WORTHEN, Amos Henry, (1812-1888)

1850-1899

ADYE, Ernest Howard, (1857-1918)
ALIN, Johan, (1879-1944)
ANDERSON, Frank Marion, (1863-1945)
ANDERSON, William, (1860-1915)
ANDERSSON, Oscar Frithiof, (1867-1911)
ANDREWS, Ernest Clayton, (1870-1948)
ANSTED, David Thomas, (1814-1880)
APLIN, Christopher d'Oyley Hale, (1819-1875)
ASHLEY, George Hall, (1866-1951)
BABÁNEK, František, (1836-1910)
BAGG, Rufus Mather, (1869-1946)

ECONOMIC GEOLOGY 1850-1899 Cont'd.

BAILEY, Loring Woart, (1839-1925)
BAIN, Thomas Charles John, (1829-1893)
BAKER, Benjamin Alfred, (1859-1939)
BALDAUF, Richard Julius, (1848-1931)
BALL, John, (-1941)
BALL, Valentine, (1843-1895)
BARLOW, Alfred Ernest, (1861-1914)
BARROIS, Charles Eugène, (1851-1939)
BASTIN, Edson Sunderland, (1878-1953)
BAUERMAN, Hilary, (1833-1909)
BAUR, Carl Theodor von, (1836-1911)
BAYLEY, William Shirley, (1861-1943)
BEAUGÉ, Alfred, (1878-1935)
BECK, Carl Richard, (1858-1919)
BEEDE, Joshua William, (1871-1940)
BEHR, Johannes, (1875-1973)
BELL, James Macintosh, (1877-1934)
BELL, Robert, (1841-1917)
BERGEAT, Alfred, Edmund, (1866-1924)
BERTRAND, C. Eg., (1851-1920)
BEUST, Friedrich Constantin von, Freiherr, (1806-)
BEUTELL, Albert, (1859-1921)
BEYER, Samuel Walker, (1865-1931)
BIBOLINI, Aldo, (1876-1949)
BLAKE, William Phipps, (1826-1910)
BOGAERT, P. J. J., (1807-1894)
BOULTON, William Savage, (1867-ca.1954)
BOUTWELL, John Mason, (1874-1968)
BOWEN, Charles Franklin, (1871-1953)
BOWLES, Oliver, (1877-1958)
BRANNER, John Casper, (1850-1922)
BROADHEAD, Garland Carr, (1827-1912)
BROCK, Reginald Walter, (1874-1935)
BROOKS, Thomas Benton, (1836-1900)
BROWN, Henry Yorke Lyell "Stoney", (1844-1928)
BÜCKING, Ferdinand Carl Bertram Hugo, (1851-1932)
BUEHLER, Henry Andrew, (1876-1944)
BURCHARD, Ernest Norman, (1875-1961)
BUTTGENBACH, Henri, (1874-1964)
CAMERON, A. E., (-)
CAMPBELL, Henry Donald, (1862-1934)
CAMPBELL, Marius Robinson, (1858-1940)
CAPACCI, Celso, (1854-1929)
CASTILLO, Antonio del, Don, (1820-1895)
CHAMBERLIN, Thomas Chrowder, (1843-1928)
CHANCOURTOIS, A. E. Béguyer de, (-1886?)
CHAPER, Maurice Armand, (1834-1896)
CIAMPI, Adolfo, (1876-1934)
CLOUGH, Charles Thomas, (1852-1916)
COLONY, Roy Jed, (1870-1936)
CONDON, Thomas, (1822-1907)
CONEDERA, Raimondo, (1859-1925)
COOK, George Hammell, (1818-1889)
CORDIER, Pierre Louis Antoine, (1777-1861)

ECONOMIC GEOLOGY 1850-1899 Cont'd.

CORSTORPHINE, George Steuart, (1865-1919)
COTTA, Carl Bernhard von, (1808-1879)
CRAWFORD, Ralph Dixon, (1873-1950)
CRIDER, Albert Foster, (1873-1945)
CROOK, Alja Robinson, (1864-1930)
CROOK, Thomas, (1876-1937)
CROSBY, William Otis, (1850-1925)
DAINTREE, Richard, (1831-1878)
DAKYNS, John Roche, (1836-1910)
DAUBRÉE, Gabriel Auguste, (1814-1896)
DAVISON, Ernest Henry, (1877-1944)
DAWKINS, William Boyd, Sir, (1837-1929)
DAWSON, George Mercer, (1849-1901)
DECHEN, Ernst Heinrich Carl von, (1800-1889)
DERBY, Orville Adelbert, (1851-1915)
DIXON, Ernest Edward Leslie, (1876-1963)
DOWLING, Donaldson Bogart, (1858-1925)
DRAKE, Noah Fields, (1864-1945)
DREW, Frederick, (1836-1891)
DUMBLE, Edwin Theodore, (1852-1927)
DUNLOP, Andrew, (1845-1915)
DUNN, Edward John, (1844-1937)
DUNSTAN, Benjamin, (1864-1933)
DUPARC, Louis Claude, (1866-1932)
EGGERTZ, Victor, (1817-1889)
ELLS, Robert Wheelock, (1845-1911)
ELSDEN, James Vincent, (?1856-1930)
EMMONS, Samuel Franklin, (1841-1911)
ENGELMANN, August Heinrich, (1831-1899)
ERDMANN, Axel Joachim, (-)
ERDMANN, Edvard, (1840-1923)
ETHERIDGE, Robert Jr., (1847-1920)
FAYOL, Henri, (1841-1925)
FEARNSIDES, William George, (1879-1968)
FEGRAEUS, Torbern, (1853-1923)
FENNER, Clarence Norman, (1870-1949)
FERRARIS, Erminio, (1851-1928)
FIRKET, Adolphe, (1837-1905)
FISCHER, Otto, (1877-1943)
FISHER, Cassius Asa, (1872-1930)
FLETCHER, Hugh, (1848-1909)
FLIEGEL, Gotthard, (1873-1947)
FOLCO, Carlo, (1866-1932)
FORBES, David, (1828-1876)
FOSTER, Clement Le Neve, Sir, (1841-1904)
FROSTERUS, Gustaf Benjamin, (1866-1931)
FUCHS, Alexander, (1874-1935)
FUCHS, Philippe Jacques Edmond, (1837-1889)
FUCHS, Wilhelm, (1802-1853)
FURUHJELM, Enoch Hjalmar, (1823-1886)
GALE, Hoyt Stoddard, (1876-1952)
GATTNAR, Josef, (1854-1928)
GIBSON, Walcot, (1864-1941)
GIORDANO, Felice, (1825-1892)

ECONOMIC GEOLOGY 1850-1899 Cont'd.

GLENN, Leonidas Chalmers, (1871-1951)
GLINKA, Sergei Fedorovich, (1855-1933)
GORDON, Charles Henry, (1857-1934)
GOULD, Charles, (-1895)
GOULD, Charles Newton, (1868-1949)
GRAND'EURY, François Cyrille, (1839-1917)
GRANT, Ulysses Sherman, (1867-1932)
GREEN, Alexander Henry, (1832-1896)
GREEN, Upfield, (1834-1917)
GREGORY, John Walter, (1864-1932)
GRIESBACH, Karl Ludolf, (1847-1907)
GRIFFITH, Richard John, Sir, (1784-1878)
GRODDECK, Albrecht Ludwig von, (1837-1887)
GROEBER, Paul Friedrich Carl, (1885-1968)
GÜMBEL, Carl Wilhelm von, Ritter, (1823-1898)
GÜRICH, Georg, (1859-1938)
GURLDT, Adolf Friedrich, (1829-1902)
HABETS, Alfred, (1839-1908)
HAGLUND, Emil, (1877-1938)
HAGUE, James Duncan, (1836-1908)
HALL, Arthur Lewis, (1872-1955)
HAMMER, Wilhelm, (1875-1942)
HARADA, Toyokitsi, (-1894)
HARDMAN, Edward Townley, (1845-1887)
HARPER, Lewis, (-)
HARRIS, George Frederick, (1862-1906)
HARTNAGEL, Chris Andrew, (1874-1962)
HAUSMANN, Johann Friedrich Ludwig, (1782-1859)
HAYDEN, Ferdinand Vandiveer, (1829-1887)
HAYES, Charles Willard, (1858-1916)
HECTOR, James, Sir, (1834-1907)
HELLAND, Amund Theodor, (1846-1918)
HERBICH, Franz, (1821-1887)
HIKI, Tadasu, (1866-1927)
HILDRETH, Samuel Prescott, (1783-1863)
HILGARD, Eugene Woldemar, (1833-1916)
HILL, Frank A., (1858-1915)
HILL, James Bastian, (1861-1927)
HILL, Robert Thomas, (1858-1941)
HILLS, Richard Charles, (1848-1923)
HOBLEY, Charles William, (1867-1947)
HÖFER-HEIMHALT, Hans, (1843-1924)
HOLDEN, Roy Jay, (1870-1945)
HOLLAND, Thomas Henry, Sir, (1868-1947)
HOLMES, Joseph Austin, (1859-1915)
HOLMQUIST, Per Johan, (1866-1946)
HOME, David Milne, (1805-1890)
HORNE, John, (1848-1928)
HOW, Henry, (1828-1879)
HOWE, Ernest, (1875-1932)
HOWELL, Henry Hyatt, (1834-1915)
HUBBARD, Bela, (1814-1896)
HUBBARD, Lucius Lee, (1849-1933)
HUGHES, Theodore W. H., (?1843-1907)

ECONOMIC GEOLOGY 1850-1899 Cont'd.

HUGHES, Thomas McKenny, (1832-1917)
HULL, Edward, (1829-1917)
HUME, William Fraser, (1867-1949)
HUTTON, Frederick Wollaston, (1836-1905)
IGELSTRÖM, Lars Johan, (1822-1897)
INVILLIERS, Edward Vincent d', (1857-1928)
IRVING, John Duer, (1874-1918)
JACK, Robert Lockhart, (1878-1964)
JACOBS, Elbridge Churchill, (1873-1957)
JAMESON, Robert, (1774-1854)
JOHNSTRUP, Johannes Frederik, (1818-1894)
JONES, Thomas Rupert, (1819-1911)
JOWETT, Albert, (1875-?1958)
JULIEN, Alexis Anastay, (1840-1919)
JULIN, Albert Lindsay von, (1876-1918)
KEENE, William, (1798-1872)
KELAART, E. F., (1818-1860)
KEMP, James Furman, (1859-1926)
KENDALL, Percy Fry, (1856-1936)
KINAHAN, George Henry, (1829-1908)
KING, Clarence Rivers, (1842-1901)
KITSON, Albert Ernest, Sir, (1868-1937)
KLEMM, Gustav, (1858-1938)
KNETT, Josef, (1869-1946)
KNIGHT, Wilbur Clinton, (1858-1903)
KONTKIEWICZ, Stanislaw, Sr., (1849-1924)
KOSSMAT, Franz, (1871-1938)
KRUSCH, Paul, (1869-1939)
KUMMEL, Henry Barnard, (1867-1945)
KYNASTON, Herbert, (1868-1915)
LAFLAMME, Joseph Clovis Kemner, Monseigneur, (1849-1910)
LANDES, Henry, (1867-1936)
LANE, Alfred Church, (1863-1948)
LAUNAY, Louis Alphonse Auguste de, (1860-1938)
LAWSON, Andrew Cowper, (1861-1952)
LEONARD, Arthur Gray, (1865-1932)
LESLEY, J. Peter, (1819-1903)
LEWIS, Henry Carvill, (1853-1888)
LEWIS, Joseph Volney, (1869-1969)
LIBERT, Joseph, (1853-?1920)
LIEBER, Oscar M., (-1862)
LINDGREN, Waldemar, (1860-1939)
LOHEST, Max, (1857-1926)
LONGDEN, Geoffrey Appleby, (1876-1964)
LUCIUS, Michel, (1876-1961)
LUNDBÖHM, Hjalmar, (1855-1926)
LUPTON, Charles Thomas, (1878-1935)
MACLAREN, James Malcolm, (1873-1935)
MALHERBE, Renier, (-1891)
MANKIN, Charles John, (1932-)
MARENGO, Paulo, (1863-1929)
MARKS, Edward Seaborn, (?1875-1950)
MARTELLI, Alessandro, (1877-1934)
MATHEWS, Edward Bennett, (1869-1944)

ECONOMIC GEOLOGY 1850-1899 Cont'd.

MATSON, George Charlton, (1873-1940)
McCALLEY, Henry, (1852-1904)
McCALLIE, Samuel Washington, (1856-)
McCONNELL, Richard George, (1857-1942)
McHENRY, Alexander, (1843-1919)
MEDLICOTT, Henry Benedict, (1829-1905)
MENDENHALL, Walter Curran, (1871-1957)
MIDDLEMISS, Charles Stewart, (1859-1945)
MILLER, Benjamin Le Roy, (1874-1944)
MONETTI, Luigi, (1864-1926)
MOORE, Elwood S., (1879-1966)
MOORE, Francis, Jr., (1808-1864)
MORGAN, Glenn Beckley, (1876-1925)
MORGAN, Jacques Jean Marie de, (1857-1924)
MORSE, William Clifford, (1874-1962)
MOURLON, Michel Félix, (1845-1915)
MÜLLER, Carl Hermann, (1823-1907)
MUNTHE, Henrik Vilhelm, (1860-1958)
MURRAY, Alexander, (1810-1884)
NAUMANN, Edmund, (1854-1927)
NEWLAND, David Hale, (1872-1943)
NORWOOD, Charles Joseph, (1853-1927)
NOVARESE, Vittorio, (1861-1948)
OCHSENIUS, Carl Christian, (1830-)
OCKENDEN, Maurice Albion, (1874-1954)
OHERN, Daniel Webster, (1870-1953)
OLDHAM, Charles Aemilius, (1831?-1869)
OLDHAM, Richard Dixon, (1858-1936)
OLDHAM, Thomas, (1816-1878)
OLRY, Albert, (1847-1913)
OMALIUS D'HALLOY, Jean-Baptist Julien d', (1783-1875)
ORDOÑEZ, Ezequiel, (1867-1950)
ORTON, Edward Francis Baxter, (1829-1899)
OWEN, David Dale, (1807-1860)
PAPP, Károly, (1873-1963)
PARMA, Augusto, (1857-1926)
PARSONS, James, (1876-1908)
PAULCKE, Wilhelm, (1873-1949)
PECK, Frederick Burritt, (1860-1925)
PERKINS, George Henry, (1844-1933)
PETRASCHEK, Wilhelm, (1876-1967)
PISSIS, Aimé, (1812-1889)
PLATT, Franklin, (1844-1900)
POŠEPNÝ, František, (1836-1895)
POSTLETHWAITE, John, (1840-1925)
POTONIÉ, Henry Edmond, (1857-1913)
PRATT, Joseph Hyde, (1870-1942)
PURDUE, Albert Homer, (1861-1917)
RAMSAY, Andrew Crombie, Sir, (1814-1891)
RANDALL, John, (1810-1910)
RANSOME, Frederick Leslie, (1868-1935)
REDLICH, Karl A., (1869-1942)
RENIER, Armand Marie Vincent Joseph, (1876-1951)
RICHARDSON, George Burr, (1872-1949)

ECONOMIC GEOLOGY 1850-1899 Cont'd.

RIES, Heinrich, (1871-1951)
ROGERS, Henry Darwin, (1808-1866)
RÖMER, Hermann, (1816-1894)
ROMEU, Albert de, (1875-1915)
ROSALES, Henry, (?1822-1916)
ROTKY, Otto, (1870-1951)
RUDLER, Frederick William, (1840-1915)
SAARINEN, Juho Hugo, (-?1908)
SAFFORD, James Merrill, (1822-1907)
SAHLIN, Carl, (1861-1943)
SALOMON-CALVI, Wilhelm, (1868-1941)
SAMOILOV, Yakov Vladimirovich, (1870-1925)
SANTESSON, Karl Olof Birger, (1845-1893)
SARJUSZ-MAKOWSKI, Arnold, (1876-1943)
SAWKINS, James Gay, (1806-1878)
SCHALCH, Ferdinand, (1848-1918)
SCHOTTLER, Wilhelm, (1869-1932)
SCHRADER, Frank Charles, (1860-1944)
SEGER, Hermann, (-1893)
SELLARDS, Elias Howard, (1875-1961)
SELWYN, Alfred Richard Cecil, (1824-1902)
SHALER, Nathaniel Southgate, (1841-1906)
SHEPARD, Edward Martin, (1854-1934)
SHERLOCK, R. L., (1875-1948)
SLOAN, Earle, (1858-1926)
SMEYSTERS, Joseph, (1837-?1908)
SMITH, Eugene Allen, (1841-1927)
SMITH, George Otis, (1871-1944)
SPIREK, Vincenz, (1852-1907)
STAPFF, Friedrich Moritz, (1836-1895)
STARING, W. C. H., (1808-1874)
STELZNER, Alfred Wilhelm, (1840-1895)
STEVENSON, John James, (1841-1924)
STOBBS, John Thomas, (?1866-1942)
STOKES, Arthur Henry, (1842-1910)
STONE, Ralph Walter, (1876-1964)
STRAHAN, Aubrey, Sir, (1852-1928)
STUTCHBURY, Samuel, (1798-1859)
SYMES, Richard Glascott, (1840-1906)
TEALE, Edmund Oswald, Sir, (1874-1971)
TENORE, Gaetano, (1826-1903)
THIESSEN, Reinhardt, (1867-1938)
THOMAS, Herbert Henry, (1876-1935)
TIBERG, Björn Gustaf Georg, (1877-1948)
TIBERG, H. V., (1849-1913)
TIETZE, Emil, (1845-1931)
TODD, James Edward, (1846-1922)
TOLMAN, Cyrus Fisher, (1873-1941)
TOPLEY, William, (1841-1894)
TORNAU, Friedrich Karl August, (1877-1914)
TRABUCCO, Giacomo, (1845-1925)
TRAILL, Thomas Stewart, (1781-1862)
TRASK, John Boardman, (1824-1879)
TRÜSTEDT, Otto, (1866-1929)

ECONOMIC GEOLOGY 1850-1899 Cont'd.

TYRRELL, Joseph Burr, (1858-1957)
UDDEN, Johan August, (1859-1932)
ULRICH, George Henry Frederick, (1830-1900)
USSHER, William Augustus Edmond, (1849-1920)
VALLE DE LERSUNDI, Alfonso del, (1875-1944)
VITÁLIS, István, (1871-1947)
VIZER, Vilmos, (1873-1941)
WALL, George Parkes, (1832-1912)
WALTMAN, William Dewitt, (1875-1955)
WARD, Henry Augustus, (1834-1906)
WARD, Leonard Keith, (1879-1964)
WARNES, Arthur Robert, (?1877-1942)
WARREN, Charles Hyde, (1876-1950)
WATSON, John, (1842-1918)
WATSON, Thomas Leonard, (1871-1924)
WEIDMAN, Samuel, (1870-1945)
WESTGATE, Lewis Gardner, (1868-1948)
WHITE, Charles David, (1862-1935)
WHITE, Israel Charles, (1848-1927)
WHITNEY, Josiah Dwight, (1819-1896)
WHITTLESEY, Charles, Colonel, (1808-1886)
WILDER, Frank Alonzo, (1870-1930)
WILLARD, Daniel Everett, (1862-1947)
WILLIAMS, Ira Abraham, (1876-)
WILLIS, Bailey, (1857-1949)
WILLMOTT, Arthur Brown, (1866-1914)
WINCHELL, Alexander, (1824-1891)
WINCHELL, Newton Horace, (1829-1914)
WOLF, Heinrich, (1825-1882)
WOODRUFF, Elmer Grant, (1872-1952)
WOODWARD, Harry Page, (1858-1917)
WOOLNOUGH, Walter George, (1876-1958)
WORTHEN, Amos Henry, (1812-1888)
YACHEVSKY, Leonard Antonovich, (1858-1916)
YOUNG, George Albert, (1898-1947)
ZEILLER, Charles René, (1847-1915)

1900 to present

ADYE, Ernest Howard, (1857-1918)
ALIN, Johan, (1879-1944)
ALLAN, John Andrew, (1884-1955)
AMBROSE, John Willis, (-1974)
ANDERSON, Alfred Leonard, (1900-1964)
ANDERSON, Frank Marion, (1863-1945)
ANDERSON, George Harold, (1893-1956)
ANDERSON, Gustavus Edwin, (1879-1940)
ANDERSON, William, (1860-1915)
ANDERSSON, Oscar Frithiof, (1867-1911)
ANDREEN, Harry Mayo, (1891-1933)
ANDRES, Demetrio, (1889-1960)
ANDREWS, Ernest Clayton, (1870-1948)
ANDREWS, Richard Wallis, (?1927-1967)
ASHLEY, George Hall, (1866-1951)
BABÁNEK, František, (1836-1910)

ECONOMIC GEOLOGY 1900 to present Cont'd.

BADAK, Jerzy, (1932-1966)
BAGG, Rufus Mather, (1869-1946)
BAILEY, Loring Woart, (1839-1925)
BAJPAI, Mahesh Prasad, (1907-1938)
BAKER, Benjamin Alfred, (1859-1939)
BALDAUF, Richard Julius, (1848-1931)
BALK, Robert, (1899-1955)
BALL, John, (-1941)
BALL, Max Waite, (1885-1954)
BANNERMAN, Harold MacColl, (1897-1976)
BARLOW, Alfred Ernest, (1861-1914)
BARROIS, Charles Eugène, (1851-1939)
BASCOM, Willard, (1917-)
BASTIN, Edson Sunderland, (1878-1953)
BATEMAN, Alan Mara, (1889-1971)
BATEMAN, John Danvers, (1910-1965)
BATES, Robert Glenn, (1923-1972)
BAUERMAN, Hilary, (1833-1909)
BAUR, Carl Theodor von, (1836-1911)
BAYLEY, William Shirley, (1861-1943)
BEACH, John Osa, (1895-1964)
BEAUGÉ, Alfred, (1878-1935)
BECK, Carl Richard, (1858-1919)
BECKWITH, Radcliffe Harold, (1900-1964)
BEEDE, Joshua William, (1871-1940)
BEEKLEY, Albert Leon, (1883-1952)
BEHR, Johannes, (1875-1973)
BELL, James Macintosh, (1877-1934)
BELL, John Arnold, (1939-1970)
BELL, Robert, (1841-1917)
BELOUSSOV, Vladimir Vladimirovitch, (-)
BENNETT, Russell Hoadley, (1896-)
BERGEAT, Alfred, Edmund, (1866-1924)
BERGSTRÖM, Gunnar, (-1954)
BETECHTIN, Anatolii Georgievich, (1897-1962)
BEUTELL, Albert, (1859-1921)
BEYER, Samuel Walker, (1865-1931)
BIBOLINI, Aldo, (1876-1949)
BILLINGS, Katharine Stevens (neé Fowler-Lunn), (1902-)
BIRD, Roland T., (-1978)
BISHOPP, Douglas Wallace, (1900-1977)
BISTRITSCHAN, Karl, (1916-1957)
BLAIR, Charles Scofield, (1883-1971)
BLAKE, George Stanfield, (-1940)
BLAKE, William Phipps, (1826-1910)
BLONDEL, Fernand, (1894-1968)
BOARDMAN, Leslie Gray, (1913-1977)
BOCHEŃSKI, Tadeusz Antoni, (1901-1958)
BØGVAD, Richard, (1897-1952)
BONTHRON, John A., (-?1948)
BORGES, Alexandre, (1898-1955)
BORN, Kendall Eugene, (1908-1947)
BORRELLO, Angel V., (1918-1971)
BOSWORTH, Thomas Owen, (1882-1928)

ECONOMIC GEOLOGY 1900 to present Cont'd.

BOULTON, William Savage, (1867-ca.1954)
BOUTWELL, John Mason, (1874-1968)
BOWEN, Charles Franklin, (1871-1953)
BOWLES, Oliver, (1877-1958)
BRACE, Orval L., (1893-1975)
BRADLEY, Wilmot Hyde, (1899-)
BRAMLETTE, Milton Nunn, (1896-1977)
BRAMWYCH, Michael, (1911-1974)
BRANNER, John Casper, (1850-1922)
BRANSON, Edwin Robert, (1908-1971)
BROCK, Reginald Walter, (1874-1935)
BRODERICK, Thomas Monteith, (1889-1965)
BROMEHEAD, Cyril Edward Nowill, (1885-1952)
BROTZEN, Fritz, (1902-1968)
BROUGHTON-EDGE, Arthur Broughton, (1895-1953)
BROWN, Bahngrell Walter "Bill", (1916-)
BROWN, Henry Yorke Lyell "Stoney", (1844-1928)
BROWN, John Coggin, (1884-1962)
BRUCE, Everend Lester "Louie", (1884-1949)
BRYAN, Joseph Jefferson, (1905-1951)
BRYAN, Walter Heywood, (1891-1966)
BÜCKING, Ferdinand Carl Bertram Hugo, (1851-1932)
BUEHLER, Henry Andrew, (1876-1944)
BURCHARD, Ernest Norman, (1875-1961)
BURTON, George Edgar, (1880-1953)
BUTCHER, Walter William, (1916-1966)
BUTTGENBACH, Henri, (1874-1964)
BUTZ, Walter "Wally", (1905-1962)
CADY, Gilbert Haven, (1882-1970)
CAIRNES, Delorme Donaldson, (1879-1917)
CAMERON, A. E., (-)
CAMPBELL, Henry Donald, (1862-1934)
CAMPBELL, Ian, (1900-1978)
CAMPBELL, Marius Robinson, (1858-1940)
CAPACCI, Celso, (1854-1929)
CARLSON, Charles Gordon, (1895-1977)
CARLTON, Dave Pierce, (1890-1976)
CARRUTHERS, Robert George, (1880-1965)
CATHCART, Stanley Holman, (1889-1953)
ČECH, Vladimír, (1901-)
ČELEBRIĆ, Dorde, (1929-1970)
CHAMBERLIN, Thomas Chrowder, (1843-1928)
CHIPMAN, Kenneth G., (-)
CIAMPI, Adolfo, (1876-1934)
CLARK, Howard, (1885-1972)
CLARK, Robert Watson, (1884-1948)
CLOUGH, Charles Thomas, (1852-1916)
COLLINS, William Ion, (1897-1964)
COLLOM, Roy Edward, (1881-1948)
COLONY, Roy Jed, (1870-1936)
COMBE, A. D., (1893-1949)
COMSTOCK, Theodore Bryant, (1849-1915)
CONDON, Thomas, (1822-1907)
CONEDERA, Raimondo, (1859-1925)

ECONOMIC GEOLOGY 1900 to present Cont'd.

CONNOLLY, Joseph Peter, (1890-1947)
COOK, Charles Wilford, (1882-1933)
COOK, Harold James, (1887-?1962)
COOKE, Harold Caswell, (1884-1956)
COOPER, Byron Nelson, (1912-1971)
COOPER, Chalmer Lewis, (-)
CORSTORPHINE, George Steuart, (1865-1919)
COTTER, Gerald de Purcell, (1881-1941)
CRABTREE, John Edmund, (1940-1966)
CRAIG, Robert Meldrum, (?1882-1956)
CRAWFORD, Ralph Dixon, (1873-1950)
CRICKMAY, Geoffrey W., (1905-1971)
CRIDER, Albert Foster, (1873-1945)
CROOK, Alja Robinson, (1864-1930)
CROOK, Thomas, (1876-1937)
CROSBY, William Otis, (1850-1925)
CROWN, Walter James, (1901-1951)
CRUM, Harry Edwin, (1892-1955)
CULVER, Harold Eugene, (1883-1970)
CUNNINGTON, C. H., Lieutenant, (1889-1918)
CURRIER, Louis Wade, (1890-1970)
DAKYNS, John Roche, (1836-1910)
DALE, Nelson Clark, (1880-1962)
DAMON, Henry Gordon, (1883-1960)
DANE, Carle Hamilton, (1900-1968)
DAVIDSON, Charles Findlay, (1911-1967)
DAVIS, Elmer Fred, (1887-1974)
DAVISON, Ernest Henry, (1877-1944)
DAWKINS, William Boyd, Sir, (1837-1929)
DAWSON, George Mercer, (1849-1901)
DE DYCKER, Raymond, (1903-1947)
DEISS, Charles Frederick, (1903-1959)
DE LA SAUCE, Wilhelm, (1882-1955)
DERBY, Orville Adelbert, (1851-1915)
DE SOUSA, Vasco Mendes, (1917-1962)
DeWOLF, Frank Walbridge, (1881-1957)
DINES, Henry George, (1891-?1964)
DINHAM, Charles Hawker, (1883-1955)
DIXON, Ernest Edward Leslie, (1876-1963)
DOBBIN, Carroll Edward "Cy", (1892-1967)
DORLODOT, Henry de, Chanoîne, (1855-1929)
DOUGLAS, George Vibert, (1892?-1958)
DOWLING, Donaldson Bogart, (1858-1925)
DRAKE, Noah Fields, (1864-1945)
DRATH, Adam, (1904-1942)
DUMBLE, Edwin Theodore, (1852-1927)
DUNLOP, Andrew, (1845-1915)
DUNN, Edward John, (1844-1937)
DUPARC, Louis Claude, (1866-1932)
DUPARQUE, André, (1892-1960)
DUTY, Richard Emerson, (1908-1966)
DYER, William Spafford, (1894-1941)
EARGLE, Dolan Hoye, (1905-1973)
EARLE, Kenneth Wilson, (1891-1944)

ECONOMIC GEOLOGY 1900 to present Cont'd.

EASTWOOD, Tom, (1889?-1970)
EBY, James Brian, (1896-)
ELLS, Robert Wheelock, (1845-1911)
ELSDEN, James Vincent, (?1856-1930)
ELY, Fred Burchard, (1883-1946)
EMMONS, Samuel Franklin, (1841-1911)
EMMONS, William Harvey, (1876-1948)
ERDMANN, Edvard, (1840-1923)
ETHERIDGE, Robert Jr., (1847-1920)
EYLES, Victor Ambrose, (1896-1978)
FABIANI, Ramiro, (1879-1954)
FAYOL, Henri, (1841-1925)
FEARNSIDES, William George, (1879-1968)
FEGRAEUS, Torbern, (1853-1923)
FELICIANO, Jose Maria "Dako", (1887-1955)
FENNER, Clarence Norman, (1870-1949)
FERGUSON, Henry Gardiner, (1882-1966)
FERMOR, Lewis Leigh, Sir, (1880-1954)
FERRARIS, Erminio, (1851-1928)
FERSMAN, Aleksander Yevgenyevich, (1883-1945)
FETTKE, Charles Reinhard, (1888-1959)
FIEDLER, William M., (1914-1970)
FIRKET, Adolphe, (1837-1905)
FISCHER, Otto, (1877-1943)
FISHER, Cassius Asa, (1872-1930)
FLAHERTY, Robert J., (1884-1951)
FLETCHER, Hugh, (1848-1909)
FLIEGEL, Gotthard, (1873-1947)
FOLCO, Carlo, (1866-1932)
FÖLDVÁRI, Aladár, (1906-1973)
FOLEY, Edward J., (1908-1971)
FORD, Trevor David, (1932-)
FOSTER, Clement Le Neve, Sir, (1841-1904)
FOX, Cyril Sankey, Sir, (1886-1951)
FRASER, Horace John "Dutch", (1905-1969)
FRIEDEL, Edmond, (1895-1972)
FROSTERUS, Gustaf Benjamin, (1866-1931)
FUCHS, Alexander, (1874-1935)
FURCRON, Aurelius Sydney, (1899-1971)
GÄBERT, Carl, (-1931)
GALE, Hoyt Stoddard, (1876-1952)
GALPIN, Sidney Longman, (1886-1962)
GARDNER, James Henry, (1883-1964)
GARNETT, Cecil Stevenson, (1896-1951)
GATTNAR, Josef, (1854-1928)
GAULT, Hugh Richard, (1915-1961)
GEDEON, Tihamér, (1898-1967)
GEIJER, Per Adolf, (1886-1976)
GIBSON, Walcot, (1864-1941)
GILL, James Rogers, (1922-1972)
GILLERMAN, Elliot "Bud", (1913-1974)
GILLES, Verner Arthur, (1886-1954)
GILLSON, Joseph Lincoln, (1895-1964)
GLENN, Leonidas Chalmers, (1871-1951)

ECONOMIC GEOLOGY 1900 to present Cont'd.

GLINKA, Sergei Fedorovich, (1855-1933)
GOETEL, Walery, (1889-1972)
GOODELL, Horace N. "Hank", (1915-1971)
GOODLET, George Anderson, (1910-1970)
GORDON, Charles Henry, (1857-1934)
GORMAN, Joseph M., (1910-1967)
GORSKY, I. I., (1893-1975)
GOUDKOFF, Paul Pavel, (1880-1955)
GOULD, Charles Newton, (1868-1949)
GOWER, John A., (1921-1972)
GRABHAM, George Walter, (1882-1955)
GRAHAM, Richard Percival Devereux, (1880-1965)
GRAND'EURY, François Cyrille, (1839-1917)
GRANIGG, Bartel, (1883-1951)
GRANT, Ulysses Sherman, (1867-1932)
GRANTHAM, Donald Robert, (1896-1969)
GRATON, Louis Caryl, (1880-1970)
GRAWE, Oliver Rudolph, (1901-1965)
GREEN, Upfield, (1834-1917)
GREENE, Frank Cook, (1886-1975)
GREGORY, John Walter, (1864-1932)
GRIESBACH, Karl Ludolf, (1847-1907)
GRIFFITH, Glenn Rodgers Vickery, (1891-1964)
GRIMM, Kenneth Edgar, (1912-1961)
GROEBER, Paul Friedrich Carl, (1885-1968)
GROOTHOFF, Christiaan Theodor, (1887-1969)
GROVES, Arthur William, (1903-1956)
GUILLOU, Robert Barton, (1923-1965)
GUNTER, Herman, (1885-1972)
GÜRICH, Georg, (1859-1938)
GURLDT, Adolf Friedrich, (1829-1902)
GUZIK, Kazimierz, (1911-1970)
GWYNNE, Charles Sumner, (1885-1972)
HABETS, Alfred, (1839-1908)
HAGUE, James Duncan, (1836-1908)
HAIG, David Salusbury, (?1913-1955)
HALICKA, Antonina, (1908-1973)
HALL, Arthur Lewis, (1872-1955)
HAM, William Eugene, (1916-1970)
HAMMER, Wilhelm, (1875-1942)
HANIEL, Klaus, (1914-1942)
HARBORT, Erich, (1879-1929)
HARRINGTON, George Leavitt, (1883-1972)
HARRIS, George Frederick, (1862-1906)
HARTNAGEL, Chris Andrew, (1874-1962)
HATTEY, Hans, (1889-1955)
HAUPTMAN, Charles R., (1891-1971)
HAYCOCK, Maurice, (1900-)
HAYES, Albert Orion, (1882-1965)
HAYES, Charles Willard, (1858-1916)
HEARD, Albert, (1894-1944)
HECTOR, James, Sir, (1834-1907)
HEIM, Arnold, (?1882-1965)
HELLAND, Amund Theodor, (1846-1918)

ECONOMIC GEOLOGY 1900 to present Cont'd.

HEMPEL, Joachim, (1884-1955)
HENRY, Gerard, (1890-1943)
HERALD, Frank Edward, (1887-1965)
HERNÁNDEZ-PACHECO, Francisco, (1899-1976)
HERZENBERG, Robert, (1885-1955)
HEWETT, Donnel Foster, (1881-1971)
HEWITT, Robert Leigh, (1911-1956)
HIBBERT-WARE, Samuel, (1782-1848)
HICKLING, Henry George Albert, (1883-1954)
HIGGINS, Daniel Franklin, Jr., (1884-1930)
HIKI, Tadasu, (1866-1927)
HILGARD, Eugene Woldemar, (1833-1916)
HILL, Frank A., (1858-1915)
HILL, James Bastian, (1861-1927)
HILL, James Wilcott, (1920-1954)
HILL, Robert Thomas, (1858-1941)
HILLS, Richard Charles, (1848-1923)
HITCHEN, Charles Stansfield, (1904-1971)
HOBLEY, Charles William, (1867-1947)
HODGE, Edwin Thomas, (1887-1970)
HÖFER-HEIMHALT, Hans, (1843-1924)
HOLDEN, Roy Jay, (1870-1945)
HOLLAND, Thomas Henry, Sir, (1868-1947)
HOLMES, Joseph Austin, (1859-1915)
HOLMQUIST, Per Johan, (1866-1946)
HONESS, Arthur Pharaoh, (1887-1942)
HOPPER, Walter Everett, (1886-1971)
HORNE, John, (1848-1928)
HOTZ, Walter, (1884-1958)
HOWE, Ernest, (1875-1932)
HOWELL, Henry Hyatt, (1834-1915)
HUBBARD, Lucius Lee, (1849-1933)
HUDDLE, John Warfield, (1907-1975)
HUGHES, Theodore W. H., (?1843-1907)
HUGHES, Thomas McKenny, (1832-1917)
HULL, Edward, (1829-1917)
HULL, Joseph Poyer Deyo, (1889-1967)
HUME, William Fraser, (1867-1949)
HUTTON, Colin Osborne, (1910-1971)
HUTTON, Frederick Wollaston, (1836-1905)
ILLING, Vincent Charles, (1890-1969)
INVILLIERS, Edward Vincent d', (1857-1928)
IRELAND, Hubert Andrew, (1904-1973)
IRVING, John Duer, (1874-1918)
JACK, Robert Lockhart, (1878-1964)
JACOBS, Elbridge Churchill, (1873-1957)
JEFFREYS, Geoffrey, (1885-1953)
JOHNSON, Bertrand Leroy, (1882-1962)
JOHNSTON, William Drumm, Jr., (1899-1972)
JONES, Daniel J., (1889-1965)
JONES, Hubert Cecil, (1881-1963)
JONES, Thomas Rupert, (1819-1911)
JONES, Verner E., (1905-1977)
JONES, William Francis, (1885-1941)

ECONOMIC GEOLOGY 1900 to present Cont'd.

JOWETT, Albert, (1875-?1958)
JUDSON, Wilber, (1880-1951)
JULIEN, Alexis Anastay, (1840-1919)
JULIN, Albert Lindsay von, (1876-1918)
JUNNER, Norman Ross, (1891-1973)
JURASKY, Karl Alfons, (1903-1945)
JURKOVIĆ, Ivan, (1917-)
KAVECKIS, Mykolas, (1889-)
KEMP, James Furman, (1859-1926)
KENDALL, Percy Fry, (1856-1936)
KENNERLEY, John Brian, (1937-1976)
KETTNER, Radim, (1891-1967)
KIMBALL, Kent K., (1892-1953)
KINAHAN, George Henry, (1829-1908)
KING, Clarence Rivers, (1842-1901)
KITSON, Albert Ernest, Sir, (1868-1937)
KITSON, Howard Walde, (1883-1931)
KLEMM, Gustav, (1858-1938)
KLOMPÉ, Theodorus Hendricus Franciscus, (?1904-1963)
KNETT, Josef, (1869-1946)
KNIGHT, Wilbur Clinton, (1858-1903)
KNOPF, Adolph, (1882-1966)
KONTKIEWICZ, Stanislaw, Sr., (1849-1924)
KONTKIEWICZ, Stanislaw Jr., (1883-1951)
KOSCHMANN, Albert Herbert, (1895-1962)
KOSSMAT, Franz, (1871-1938)
KOUTEK, Jaromír, (1902-)
KOZLOWSKI, Roman, (1889-1977)
KRIEGER, Philip, (1900-1940)
KRISHNAN, Maharajapuram Sitaram, Padma Bhusan, (1898-1970)
KRUSCH, Paul, (1869-1939)
KRYZYŻKIEWICZ, Józef, (1890-1955)
KUMMEL, Henry Barnard, (1867-1945)
KUŹNIAR, Czeslaw Faustyn, (1889-1955)
KYNASTON, Herbert, (1868-1915)
LAFLAMME, Joseph Clovis Kemner, Monseigneur, (1849-1910)
LANCASTER, L. Kenneth, (1906-1971)
LANDES, Henry, (1867-1936)
LANE, Alfred Church, (1863-1948)
LAPPARENT, Jacques de, (-?1949)
LAUNAY, Louis Alphonse Auguste de, (1860-1938)
LAUTERBACH, Robert, (1915-)
LAWSON, Andrew Cowper, (1861-1952)
LEIGHTON, Henry, (1884-1963)
LEITH, Charles Kenneth, (1875-1956)
LEONARD, Arthur Gray, (1865-1932)
LESLEY, J. Peter, (1819-1903)
LEUTWEIN, Friedrich, (1911-1974)
LEWIS, G. Edward, (-)
LEWIS, Joseph Volney, (1869-1969)
LIGHTFOOT, Ben, (1888-1966)
LILLIGREN, Johnathan Martin, (1880-1956)
LILPOP, Jerzy, (1888-1945)
LINDGREN, Waldemar, (1860-1939)

ECONOMIC GEOLOGY 1900 to present Cont'd.

LOHEST, Max, (1857-1926)
LONGDEN, Geoffrey Appleby, (1876-1964)
LONSDALE, John Tipton, (1895-1960)
LOTZE, Franz, (1903-1971)
LOUGHLIN, Gerald Francis, (1880-1946)
LUCAS, Elmer Lawrence, (1898-1966)
LUCIUS, Michel, (1876-1961)
LUNDBÖHM, Hjalmar, (1855-1926)
LUPTON, Charles Thomas, (1878-1935)
MACDIARMID, Roy Angus, (1933-1966)
MACDONALD, Simon, (1927-1971)
MACGREGOR, Murray, (1884-1966)
MACLAREN, James Malcolm, (1873-1935)
MADIGAN, Cecil Thomas, (1889-1946)
MAGNUSSON, Nils Harald, (1890-1976)
MARIES, Alan C., (1890-1971)
MARKS, Edward Seaborn, (?1875-1950)
MARMO, Vladi, (1914-1969)
MARTELLI, Alessandro, (1877-1934)
MARTINELLI, Cesare, (-1929)
MARTINI, Hans Joachim, (1908-1969)
MARTINUSSEN, Eugen, (1916-1942)
MATHER, Kirtley Fletcher, (1888-1978)
MATHEWS, Edward Bennett, (1869-1944)
MATSON, George Charlton, (1873-1940)
MAUFE, Herbert Brantwood, (1879-1946)
MAWDSLEY, James Buckland, (1894-1964)
MAYFIELD, Samuel Martin, (1888?-1963)
McCALLEY, Henry, (1852-1904)
McCALLIE, Samuel Washington, (1856-)
McCONNELL, Richard George, (1857-1942)
McHENRY, Alexander, (1843-1919)
McMATH, John Charles, (?1908-1954)
MEAD, Warren Judson, (1883-1960)
MEDLICOTT, Henry Benedict, (1829-1905)
MENDENHALL, Walter Curran, (1871-1957)
MENNELL, Frederick Philip, (1880-1966)
MERRITT, Clifford A., (-)
METRE, Waman Bapu, (1906-1970)
MICHELS, Franz-Xaver, (1900-1973)
MIDDLEMISS, Charles Stewart, (1859-1945)
MILES, Keith Rodney, (1915-1972)
MILLER, Benjamin Le Roy, (1874-1944)
MILLER, William John, (1880-1965)
MISER, Hugh Dinsmore, (1884-1969)
MITCHELL, George Hoole "Mick", (1902-1976)
MONETTI, Luigi, (1864-1926)
MOORE, Elwood S., (1879-1966)
MORAWIECKI, Antoni, (1903-1971)
MORGAN, Eilir Evans, (?1911-1969)
MORGAN, Glenn Beckley, (1876-1925)
MORGAN, Jacques Jean Marie de, (1857-1924)
MORSE, William Clifford, (1874-1962)
MÖRTSELL, Sture, (-?1952)

ECONOMIC GEOLOGY 1900 to present Cont'd.

MOSSBERG, Karl Elis, (-?1941)
MOURLON, Michel Félix, (1845-1915)
MULLEN, John Timothy, Jr., (1906-1959)
MÜLLER, Carl Hermann, (1823-1907)
MUNTHE, Henrik Vilhelm, (1860-1958)
MURRAY, Albert Nelson, (1894-1961)
NAUMANN, Edmund, (1854-1927)
NEEDHAM, Claude Ervin, (1894-1950)
NELSON, Wilbur Armistead, (1890-1969)
NEWHOUSE, Walter Harry, (1897-1969)
NEWLAND, David Hale, (1872-1943)
NICOLAS, Jean, (1925-)
NININGER, Robert D., (-)
NOBLE, Levi Fatzinger, (1882-1965)
NØRREGAARD, Erik Martin, (1880-1959)
NORWOOD, Charles Joseph, (1853-1927)
NOSZKY, Jenö Jr., (1909-1970)
NOVARESE, Vittorio, (1861-1948)
NOWLAN, Harry Hackett, (1888-1958)
OAKES, Malcolm Christie, (1890-1977)
OATES, Frank, (1890-1945)
OBUCHOWICZ, Zbigniew Zygmunt, (1914-1965)
OCKENDEN, Maurice Albion, (1874-1954)
OHERN, Daniel Webster, (1870-1953)
OLDHAM, Richard Dixon, (1858-1936)
OLIVEIRA, Euzebio Paulo de, (1882-1939)
OLRY, Albert, (1847-1913)
ORDOÑEZ, Ezequiel, (1867-1950)
ORTON, Bror Natanael, (-?1942)
ORTON, Edward Francis Baxter, (1829-1899)
OSBERGER, Rudolf A. J., (1924-1972)
OXAAL, John Thomas Fearnley, (1887-1942)
PALMER, Leland L., (1906-1978)
PAPP, Károly, (1873-1963)
PARMA, Augusto, (1857-1926)
PARNISARI, Carlo, (-?1929)
PARSONS, Ernest, (?1887-1963)
PARSONS, James, (1876-1908)
PAULCKE, Wilhelm, (1873-1949)
PÁVAI, Vajna Ferenc, (1886-1964)
PAVLOVIĆ, Stojan, (1903-)
PECK, Frederick Burritt, (1860-1925)
PECORA, William Thomas, (1913-1972)
PENTA, Francesco, (1899-1965)
PERCIVAL, Frederick George, (1887-1973)
PERKINS, George Henry, (1844-1933)
PET, Cees, (1900-1953)
PETERSON, Clarence J., (1885-1948)
PETRASCHEK, Wilhelm, (1876-1967)
PETTY, Dabney E., (1891-1963)
PEYTON, Garland, (1892-1964)
PLATT, Franklin, (1844-1900)
PLUMMER, Frederick Byron, (1886-1947)
POCOCK, Roy Woodhouse, (1887-1954)

ECONOMIC GEOLOGY 1900 to present Cont'd.

POND, Walter Franklin, (1885-1974)
POOR, Russell Spurgeon, (1899-1972)
POPOVICI, Viorica, (1918-1970)
POSTLETHWAITE, John, (1840-1925)
POTONIÉ, Henry Edmond, (1857-1913)
POWERS, Sidney, (1890-1932)
PRATT, Joseph Hyde, (1870-1942)
PRATT, Wallace Everette, (1885-)
PRETTYMAN, Thomas Mann, (1888-1940)
PRETYMAN, Frank Remington, (1891?-1917)
PRICE, Paul Holland, (1898-)
PRUVOST, Pierre Eugéne Marie Joseph, (1890-1967)
PURDUE, Albert Homer, (1861-1917)
QUIRKE, Terence Thomas, (1886-1947)
RAEBURN, Colin, (1894-1970)
RAGGATT, Harold George, Sir, (1900-1968)
RAJNATH, Shri, (1904-1972)
RANDALL, John, (1810-1910)
RANSOME, Frederick Leslie, (1868-1935)
REDLICH, Karl A., (1869-1942)
REECE, Alan W., (1921-1960)
REICHERT, Stanley Orville, (1912-1970)
REINHOLD, Thomas, (1890-1955)
RENIER, Armand Marie Vincent Joseph, (1876-1951)
RICHARDSON, George Burr, (1872-1949)
RICHEY, James Ernest, (1886-1968)
RIES, Heinrich, (1871-1951)
RITTENHOUSE, Gordon, (1910-1974)
ROARK, Louis, (1890-1950)
ROBBIE, James Andrew, (1910-1977)
ROE, Frank Walton "Peter", (?1914-1965)
ROMEU, Albert de, (1875-1915)
RONES, Morris, (1922-1965)
ROSALES, Henry, (?1822-1916)
ROTKY, Otto, (1870-1951)
ROUBAULT, Marcel Edouard, (1905-1974)
ROUNDY, Paul Vere, (1884-1937)
ROYO Y GOMEZ, José, (1895-1958)
RUDLER, Frederick William, (1840-1915)
RUSSELL, Frank Scott, (-1943)
RUSSELL, Philip G., (1891-1971)
SAARINEN, Juho Hugo, (-?1908)
SAFFORD, James Merrill, (1822-1907)
SAHLIN, Carl, (1861-1943)
SALOMON-CALVI, Wilhelm, (1868-1941)
SAMBASIVA IYER, Valiathur S., (-?1943)
SAMOILOV, Yakov Vladimirovich, (1870-1925)
SAMUELSSON, Frans Gustaf, (-?1948)
SAPPINGTON, Chester, (1905-1968)
SARJUSZ-MAKOWSKI, Arnold, (1876-1943)
SCHALCH, Ferdinand, (1848-1918)
SCHENCK, Hubert Gregory, (1897-1960)
SCHIENER, Alfred, (1906-1962)
SCHNEIDERHÖHN, Hans, (1887-1962)

ECONOMIC GEOLOGY 1900 to present Cont'd.

SCHOLTZ, Douglas Leinster, (1902-1977)
SCHOLZ, Erich, (1884-1914)
SCHOTTLER, Wilhelm, (1869-1932)
SCHRADER, Frank Charles, (1860-1944)
SCHRAMM, Eck Frank, (1883-1967)
SCOTT, Alexander, (1890-1951)
SEARS, Julian Ducker, (1891-1970)
SEGALL, Julius, (1884-1928)
SELLARDS, Elias Howard, (1875-1961)
SELWYN, Alfred Richard Cecil, (1824-1902)
SEN GUPTA, Kiron Kumar, (1882-1943)
SERVISS, Frederick LeVerne, (1895-1954)
SHALER, Millard King, (1880-1942)
SHALER, Nathaniel Southgate, (1841-1906)
SHAW, Eugene Wesley, (1881-1935)
SHEPARD, Edward Martin, (1854-1934)
SHERLOCK, R. L., (1875-1948)
SIBLY, Thomas Franklin, Sir, (1883-1948)
SIDENVALL, Karl Johan Ferdinand, (-?1957)
SIMMONS, William Charles, (1888-1951)
SINCLAIR, Ewart Gladstone, (1884-1927)
SINGEWALD, Joseph Theophilus, Jr., (1884-1963)
SISLER, James Donaldson, (1896-1935)
SLAVÍKOVÁ, Ludmila, (1890-1943)
SLOAN, Earle, (1858-1926)
SMEYSTERS, Joseph, (1837-?1908)
SMIRNOV, Sergei Sergei'evich, (1895-1947)
SMITH, Eugene Allen, (1841-1927)
SMITH, George Otis, (1871-1944)
SMOLARSKA, Irena, (1928-1972)
SOBRAL, José María, (-1961)
SORGENFREI, Theodor, (1915-1972)
SPENCER, Edmondson, (?1885-1970)
SPIREK, Vincenz, (1852-1907)
SPRINGORUM, Gerd, (1915-1941)
STEBINGER, Eugene, (1883-1951)
STEJSKAL, Jan, (1902-1972)
STEVENSON, John James, (1841-1924)
STILLWELL, Frank Leslie, (1888-1963)
STOBBS, John Thomas, (?1866-1942)
STOKES, Arthur Henry, (1842-1910)
STONE, Ralph Walter, (1876-1964)
STRAHAN, Aubrey, Sir, (1852-1928)
STRINGHAM, Bronson F., (1907-1968)
STUTZER, Otto, (1881-1936)
SUBRAMANIAM, Anantharama Parameshwara, (1921-1973)
SUNDHOLM, Olof Herman, (-?1946)
SUTTON, Arle Herbert, (1896-1972)
SVOBODA, J. V., (1921-1972)
ŚWIDZINSKI, Henryk, (1904-1969)
SYMES, Richard Glascott, (1840-1906)
SZATSKIY, Mikolai Sergiechewich, (1895-1960)
TANTON, Thomas Leslie, (1890-1971)
TARR, William Arthur, (1881-1939)

ECONOMIC GEOLOGY 1900 to present Cont'd.

TAYLOR, James Haward, (1909-1968)
TEALE, Edmund Oswald, Sir, (1874-1971)
TEAS, Livingstone Pierson, (1894-1970)
TELEGDI ROTH, Károly, (1886-?1956)
TENORE, Gaetano, (1826-1903)
TER MEULEN, J. A. C., (1902-1960)
TEZON, Roberto Vitelmo, (1921-1973)
THIENHAUS, Rolf, (1913-1968)
THIESSEN, Reinhardt, (1867-1938)
THOMAS, Herbert Henry, (1876-1935)
THOMAS, Trevor Morgan, (?1919-1974)
THOMPSON, Robert Mitchell, (1918-1967)
THOMSON, Joseph Ellis, (1882-1944)
THOMSON, Paul William, (1891-1957)
THOMSON, Robert, (1902-1970)
THORNÉ, Sven Gustaf, (-?1950)
TIBERG, Björn Gustaf Georg, (1877-1948)
TIBERG, H. V., (1849-1913)
TIDY, Anthony John Lintott, (1928-1969)
TIETZE, Emil, (1845-1931)
TING, Ven Kiang, (?1887-1936)
TODD, James Edward, (1846-1922)
TOLMAN, Cyrus Fisher, (1873-1941)
TORNAU, Friedrich Karl August, (1877-1914)
TRABUCCO, Giacomo, (1845-1925)
TRAGER, Earl A., (1893-1978)
TRASK, Parker Davies, (1899-1961)
TRÖGER, Walter Ehrenreich, (1901-1963)
TRÜMPY, Daniel, (1893-1971)
TRÜSTEDT, Otto, (1866-1929)
TYLER, Stanley Allen, (1906-1963)
TYRRELL, Joseph Burr, (1858-1957)
UDDEN, Johan August, (1859-1932)
UGLOW, William Lawrence, (1884-1926)
ULRICH, František, (1899-1941)
USSHER, William Augustus Edmond, (1849-1920)
VADÁSZ, Elemér, (1885-1970)
VALLE DE LERSUNDI, Alfonso del, (1875-1944)
VAN TUYL, Francis M., (1887-1975)
VENDEL, Aladár, (1886-1971)
VERSLUYS, Jan, (1880-1935)
VESTAL, Franklin Earl, (1884-1969)
VIANNA, António, (1887-1949)
VIDACS, Aladár, (1909-?1965)
VITÁLIS, István, (1871-1947)
VIZER, Vilmos, (1873-1941)
WADELL, Hakon, (1895-1962)
WADIA, Darashaw Nosherwan, (1883-1969)
WAGNER, Georg, (1885-1972)
WAGNER, Walter Richard, (1928-1974)
WALL, George Parkes, (1832-1912)
WALLACE, Robert Charles, (1881-1955)
WALLS, James G., (1907-)
WALTMAN, William Dewitt, (1875-1955)

ECONOMIC GEOLOGY 1900 to present Cont'd.

WARBURG, Robert Oscar, (1915-1965)
WARD, Henry Augustus, (1834-1906)
WARD, Leonard Keith, (1879-1964)
WARNES, Arthur Robert, (?1877-1942)
WARREN, Charles Hyde, (1876-1950)
WASHBURNE, Chester Wesley, (1883-1971)
WATSON, John, (1842-1918)
WATSON, Thomas Leonard, (1871-1924)
WAWRYK, Wlodzimierz, (1902-1963)
WEBB, John Benwell, (1901-1975)
WEEKS, Ludlow Jackson, (1899-1964)
WEIDMAN, Samuel, (1870-1945)
WELLER, James Marvin, (1899-1976)
WELLS, Francis Gerrit, (1898-1969)
WESLEY, George Rutherford, (1905-1965)
WESSLAU, Eric, (-1958)
WESTERVELD, J., (1905-1962)
WESTGATE, Lewis Gardner, (1868-1948)
WHEALTON, Rowland Gardiner "Peg", (1896-1956)
WHEELER, Orby Clinton, (1898-1970)
WHEELER, Richard Vernon, (1883-1939)
WHITE, Charles David, (1862-1935)
WHITE, Israel Charles, (1848-1927)
WHITEHEAD, Talbot Haes, (1890-1966)
WILCOCKSON, William Howson, (1891-1976)
WILDER, Frank Alonzo, (1870-1930)
WILKERSON, Albert Samuel, (1897-1963)
WILLARD, Daniel Everett, (1862-1947)
WILLETT, Richard Wright, (1912-1974)
WILLIAMS, Ira Abraham, (1876-)
WILLIAMSON, John Thorburn, (1907-1958)
WILLIS, Bailey, (1857-1949)
WILLMOTT, Arthur Brown, (1866-1914)
WILSON, Eldred Dewey, (1898-1965)
WILSON, Reginald Charles, (1888-1943)
WILSON, Roy Arthur, (1891-1970)
WILSON, Vernon, (1907-1971)
WINCHELL, Newton Horace, (1829-1914)
WINCHESTER, Dean Eddy, (1883-1936)
WING, Monta Eldo, (1897-1976)
WISSER, Edward Hollister, (1895-1970)
WOODRUFF, Elmer Grant, (1872-1952)
WOODWARD, Harry Page, (1858-1917)
WOODWARD, Herbert Preston, (1899-1968)
WOOLNOUGH, Walter George, (1876-1958)
WRIGHT, Harold Douglas, (1921-1969)
YACHEVSKY, Leonard Antonovich, (1858-1916)
YEMTCHUKHNIKOV, G. A., (1885-1957)
YOUNG, George Albert, (1898-1947)
ZANS, Verners Aleksandrs, (1904-1961)
ZAPP, Alfred Dexter, (1916-1962)
ZEALLEY, Arthur Edmund Victor, (1886-1918)
ZEILLER, Charles René, (1847-1915)
ZIJLSTRA, Gerrit, (-1962)

ECONOMIC GEOLOGY 1900 to present Cont'd.

ZÖLLICH, Martin Sylvester, (1913-1940)
ZWIERZYCKI, Józef, (1888-1961)

Time unknown
MACGREGOR, A. M., (-)
TEGENGREN, Felix, (-)

ECONOMIC GEOLOGY <u>See also</u> MINING GEOLOGY, PETROLEUM GEOLOGY

ENGINEERING GEOLOGY
1800-1849
BURAT, Amédée, (1809-1883)
SOPWITH, Thomas, (1803-1879)
TOMLINSON, Charles, (1806-1897)
TOPLEY, William, (1841-1894)

1850-1899
ASCHER, Hans, (1848-1953)
BECHER, Harry Macdonald, (1855-1893)
BECKER, George Ferdinand, (1847-1919)
BENNETT, Francis James, (1845-1920)
BERKEY, Charles Peter, (1867-1955)
BURAT, Amédée, (1809-1883)
COLONY, Roy Jed, (1870-1936)
CROSBY, William Otis, (1850-1925)
GLENN, Leonidas Chalmers, (1871-1951)
HIRSCHWALD, Julius, (1845-1928)
LAPWORTH, Herbert, (1875-1933)
LEBOUR, George Alexander Louis, (1847-1918)
PENNING, William Henry, (1838-1902)
REDLICH, Karl A., (1869-1942)
SCHMIDT, Carl, (1862-1923)
SEGRÉ, Claudio, (1853-1928)
SINGER, Max, (1874-1952)
SMALL, Evan William, (1859-1933)
SOPWITH, Thomas, (1803-1879)
SPACEK, Karel, (1866-1937)
TOMLINSON, Charles, (1806-1897)
TOPLEY, William, (1841-1894)
VIOLA, Carlo Maria, (1855-1925)
WATSON, Thomas Leonard, (1871-1924)
WATTS, William, (-1920)
WILLIAMS, Ira Abraham, (1876-)
WINDIAKIEWICZ, Edward, (1858-1942)
ŽUKOVIĆ, Jovan M., (1856-1937)

1900 to present
ASCHER, Hans, (1848-1953)
BALDWIN, William Felbert Jackson, Jr., (1927-1972)
BARTON, Donald Clinton, (1889-1939)
BECKER, George Ferdinand, (1847-1919)
BENNETT, Francis James, (1845-1920)
BENSON, William Noel, (1885-1957)
BERES, Krzystztof, (-)

ENGINEERING GEOLOGY 1900 to present Cont'd.

BERKEY, Charles Peter, (1867-1955)
BISSET, Charles Barron, (-?1966)
BJERRUM, Laurits, (1919-1973)
BRACE, William Francis, (1926-)
BURWELL, Edward Bouldin, Jr., (1894-1964)
CAMPBELL, Arthur Byron, (1924-1971)
ČEPEK, Ladislav, (1899-1974)
COLONY, Roy Jed, (1870-1936)
CROSBY, Irving Ballard, (1891-1959)
CROSBY, William Otis, (1850-1925)
DIPROSE, Kenward Thomas, (1949-1977)
DOBBIE, Charles Herbert, (1905-1976)
DÜCKER, Alfred, (1908-1969)
DVOŘÁK, Arnošt, (1904-)
EDMUNDS, Francis Hereward, (1893-1960)
EKBLAW, George Elbert, (1895-1972)
ELLIOTT, Harold Charles, (1910-1972)
FOLEY, Edward J., (1908-1971)
FORTUNAT, Wincenty Karol, (1889-1967)
FOX, Cyril Sankey, Sir, (1886-1951)
FURMAŃSKI, Ignacy Jan, (1915-1972)
GARDNER, Frank J., (1915-1975)
GARTNER, Otokar, (1889-1961)
GASKIN, Alan Robert James, (1927-1974)
GILLES, Verner Arthur, (1886-1954)
GLENN, Leonidas Chalmers, (1871-1951)
GOLDTHWAIT, James Walter, (1880-1947)
GOODELL, Horace N. "Hank", (1915-1971)
GRABHAM, George Walter, (1882-1955)
GRENGG, Roman, (1884-1972)
HAUSER, Alois, (1899-1955)
HAWKINS, Glenn Dewayne, (1897-1970)
HIRSCHWALD, Julius, (1845-1928)
HITCHEN, Charles Stansfield, (1904-1971)
HORNER, Seward Ellis, (1906-1954)
HROMADA, Karel, (1919-1974)
ISLAMOV, Oskar Islamovich, (1910-1978)
JEWETT, John Mark, (1896-1970)
JOHNSON, Charles George, (1914-1969)
JONES, Hubert Cecil, (1881-1963)
KAVECKIS, Mykolas, (1889-)
LAPWORTH, Herbert, (1875-1933)
LEBOUR, George Alexander Louis, (1847-1918)
LUBICZ-SULIMIRSKI, Stefan, (1900-1953)
LUCAS, Elmer Lawrence, (1898-1966)
LUCEK, Josef, (1930-1969)
MARSH, Thomas William, (-1957)
MASON, Arnold Caverly, (1906-1961)
McCALLUM, Robert Rowson, Lieutenant-Colonel, (1882-1934)
MEAD, Warren Judson, (1883-1960)
MOHR, Hannes, (1882-1967)
NARKE, Ganesh Govind, (-?1943)
PENNING, William Henry, (1838-1902)
QUIRKE, Terence Thomas, (1886-1947)

ENGINEERING GEOLOGY 1900 to present Cont'd.

REDLICH, Karl A., (1869-1942)
RHOADES, Roger F., (1905-1972)
SAMBASIVA IYER, Valiathur S., (-?1943)
SCHLICHTEN, Otto Charles von, (1886-1950)
SCHMIDT, Carl, (1862-1923)
SCHMIDT, Robert Eligius, (1902-1973)
SEGRÉ, Claudio, (1853-1928)
SEN GUPTA, Kiron Kumar, (1882-1943)
SINGER, Max, (1874-1952)
ŚLIWA, Piotr, (1924-1959)
SMALL, Evan William, (1859-1933)
SMITH, Ned Myron, (1923-1976)
SORGENFREI, Theodor, (1915-1972)
SPACEK, Karel, (1866-1937)
STEJSKAL, Jan, (1902-1972)
STINY, Josef, (1880-1958)
STROBEL, Ernst Gustav, (1903-1974)
TEMPLEMAN, Andrew, (1887-1945)
TRASK, Parker Davies, (1899-1961)
TRYGGVASON, Tómas, (1907-1965)
VAN TUYL, Francis M., (1887-1975)
VERSLUYS, Jan, (1880-1935)
VIOLA, Carlo Maria, (1855-1925)
VOLOSHIN, Vadim "Charlie", (1934-1970)
VOSYKA, Stanislav, (1902-)
WADELL, Hakon, (1895-1962)
WATSON, Thomas Leonard, (1871-1924)
WATTS, William, (-1920)
WILLIAMS, Ira Abraham, (1876-)
WILSON, William Jay, (1922-1971)
WINDIAKIEWICZ, Edward, (1858-1942)
ZÁRUBA, Quido, (1899-)
ŽUKOVIČ, Jovan M., (1856-1937)

FOLKLORE OF GEOLOGY
Up to 1800
VALENTINI, Michael Bernhard, (1657-1729)

1800-1849
VERNE, Jules, (1828-1905)

1850-1899
ABEL, Othenio, (1875-1946)
CONAN DOYLE, Arthur, Sir, (1859-1930)
VERNE, Jules, (1828-1905)
WELLS, Herbert George, (1866-1946)

1900 to present
ABEL, Othenio, (1875-1946)
CARRINGTON, Richard, (-)
CONAN DOYLE, Arthur, Sir, (1859-1930)
OAKLEY, Kenneth Page, (1911-)
VERNE, Jules, (1828-1905)
WELLS, Herbert George, (1866-1946)

FOSSIL COLLECTING
Up to 1800
ANNAN, Robert, Reverend, (fl. 1782)
COULON, Paul Louis, (1777-1855)
DROUIN, Jean Baptiste, (1752-1792)
DUFRESNE, Louis, (1752-1832)
GODDING, Theodorus Joannes, Kanunnik, (1722-1797)
HARTMANN, Ernst Gustav Friedrich von, (1767-1851)
HOFFMANN, Jan Leonard, (1710-1782)
MORGAN, George, (fl.1766)
MORGAN, John, (fl.1765)
PENNANT, Thomas, (1726-1798)
PETIVER, James, (1658-1718)
PHILPOT, Mary, (1777-1838)
ROUX, Jean Marc, (?1736-1805)
SCHAFFER, Jacob Christian, (1718-1790)
SCILLA, Agostino, (1639-1700)
SIBBALD, Robert, Sir, (1641-1772)
TRADESCANT, John, The Elder, (1587-1638)
TU WAN, (fl.1115-1133)
WEISMANN, Gottlieb, (1798-1859)
WORM, Olle, (1588-1654)

1800-1849
ANNING, Mary, (1799-1847)
ARNOLD, Joseph, (1784-1818)
BENNIE, James, (1821-1901)
BENSTED, William Harding, (1802-1873)
BOXBERG, Ida von, (1806-1893)
BRACKENRIDGE, George Weare, (1775-1856)
BRADY, Antonio, Sir, (1811-1881)
BREDA, Jacob Gisbertus Samuel van, (1788-1867)
BROWN, John, (1780-1859)
BUCCHICH, Gregorio, (1828-1911)
BUHLER, Georg Wilhelm Christian von, (1797-1860)
CARTIER, Robert, (1810-1886)

FOSSIL COLLECTING 1800-1849 Cont'd.

COLCHESTER, William, (1813-1898)
COULON, Paul Louis, (1777-1855)
CRISTOFORI, Joseph de, (1803-1837)
CROGHAN, George, (fl. 1765)
CROIZET, Abbé, (-1863)
DAMON, Robert, (1814-1889)
DIXON, Frederick, (1799-1849)
DUBOIS DE MONTPÉREUX, Fréderic, (1798-1850)
DUFRESNE, Louis, (1752-1832)
EISEL, Robert, (1826-1917)
ELLIS, H., Right Honourable Sir, (-1855)
ENNISKILLEN, William Willoughby Cole, 3rd Earl of, (1807-1886)
FACK, M. W., (1823-)
FISCHER, Carl Ernst, (1818-1886)
FLETCHER, Thomas William, (1808-1893)
FONTAINE, Charles Aloyse, Chanoine, (1754-1834)
GEMMING, Carl Emil von, (1794-1880)
GLEN, David Corse, (1824-)
GREY, George, Sir, (1812-1898)
GUIRAND, Edmond, (1812-1888)
HÄBERLEIN, Ernst, (1819-1895)
HÄBERLEIN, Friedrich Karl, (1787-1871)
HARRIS, Israel Hopkins, (1832-1897)
HARRIS, William, (1797-1877)
HARRISON, James, (1819-1868)
HARTMANN, Ernst Gustav Friedrich von, (1767-1851)
HISELY, Charles, (1805-1873)
HOLLER, Anton, (1826-1909)
HOMFRAY, David, (1822-1893)
HUTTON, Robert, (1784-1870)
KAPFF, Sixt Friedrich Jakob, (1809-1887)
KEEPING, Henry, (1827-1924)
KRAPF, Ludwig, (1810-1881)
LACOE, Ralph Dupuy, (1824-1901)
LANG, Franz Vinzenz, (1821-1899)
LAWLEY, Robert, (1818-1881)
LECKENBY, John, (1814-1877)
LINCECUM, Gideon, (1793-1874)
MacADAM, James, (1801-1861)
MANDELSLOH, Friedrich von, Graf, (1795-1870)
MÜLLER, Moritz, (1791-1858)
PASINI, Lodovico, (1804-1870)
PATTON, Andrew, (1821-1885)
PHILPOT, Elizabeth, (1780-1857)
PHILPOT, Margaret, (? -1845)
PHILPOT, Mary, (1777-1838)
POPPELACK, Joseph, (?1780-1859)
RANDALL, John, (1810-1910)
REED, William, (1810-1892)
RICHARDSON, Benjamin, Reverend, (-1832)
RIEMSDIJK, A. W. G. van, (1803-1871)
ROCHE, Auguste, (1827-1905)
ROUX, Jean Marc, (?1736-1805)
ROVASENDA, Luigi di, Conte, (1826-1917)

FOSSIL COLLECTING 1800-1849 Cont'd.

SAEMANN, Louis, (1821-1866)
SLIMON, Robert, (1803-1882)
THIERENS, F. F., (1811-)
VALET, August Friedrich, (1811-1889)
VOELKEL, Josef, (1828-1906)
WEISMANN, Gottlieb, (1798-1859)

1850-1899
ALEXANDER, Annie Montague, (1867-1950)
ARCHER, Francis, (1839-1892)
ARMSTRONG, James, (1832-1892)
BALDWIN, David, (fl.1874-1881)
BALSTON, William Edward, (1848-1918)
BARBER, Charles M, (1877-1954)
BARCLAY, Francis Hubert, (1869-1935)
BARGELLINI, Mariano, (1831-1906)
BELL, Robert, (1864-1934)
BENNIE, James, (1821-1901)
BENSTED, William Harding, (1802-1873)
BOXBERG, Ida von, (1806-1893)
BRACKENRIDGE, George Weare, (1775-1856)
BRADY, Antonio, Sir, (1811-1881)
BREDA, Jacob Gisbertus Samuel van, (1788-1867)
BROWN, Alfred "Gogga", (1834-1920)
BROWN, John, (1780-1859)
BUCAILLE, Ernest-Lucien, (1835-1891)
BUCCHICH, Gregorio, (1828-1911)
BÜHLER, Georg Wilhelm Christian von, (1797-1860)
BULLOCK, William, (1795-1840)
CAIRNS, Robert, (1854-1911)
CARTIER, Robert, (1810-1886)
CHADWICK, Samuel, (1845-1903)
COLCHESTER, William, (1813-1898)
COOK, James, (fl. 1870)
COULON, Paul Louis, (1777-1855)
CROIZET, Abbé, (-1863)
DAMON, Robert, (1814-1889)
DAVIS, Leander S., (fl. 1871)
DAWSON, Charles, (1864-1916)
DEO GRATIAS, Pietro Perrando, Don, (?1846-1889)
DIETLEN, Karl Rudolf, (1857-1931)
DRUGČEVIČ, Ferdinand, (1856-1928)
EISEL, Robert, (1826-1917)
ELLIS, H., Right Honourable Sir, (-1855)
ENNISKILLEN, William Willoughby Cole, 3rd Earl of, (1807-1886)
FACK, M. W., (1823-)
FISCHER, Carl Ernst, (1818-1886)
FLETCHER, Thomas William, (1808-1893)
FONTAINE, Charles Aloyse, Chanoine, (1754-1834)
FRANCIS, James, (1843-1925)
GEMMING, Carl Emil von, (1794-1880)
GLEN, David Corse, (1824-)
GORE, Charles Herbert Henry, (1866-1951)
GREY, George, Sir, (1812-1898)

FOSSIL COLLECTING 1850-1899 Cont'd.

GROSSOUVRE, Albert de, (1849-1932)
GUIRAND, Edmond, (1812-1888)
GUNN, Robert Marcus, (?1850-1909)
HÄBERLEIN, Ernst, (1819-1895)
HÄBERLEIN, Friedrich Karl, (1787-1871)
HARRIS, Israel Hopkins, (1832-1897)
HARRIS, William, (1797-1877)
HARRISON, Benjamin, (1837-1921)
HARRISON, James, (1819-1868)
HARTMANN, Ernst Gustav Friedrich von, (1767-1851)
HISELY, Charles, (1805-1873)
HOLLER, Anton, (1826-1909)
HOMFRAY, David, (1822-1893)
HORNE, William, (?1836-1928)
HUNTER-SELKIRK, John R. S., (1835-1898)
HUTTON, Robert, (1784-1870)
JESSON, Thomas, (?1850-1928)
KAPFF, Sixt Friedrich Jakob, (1809-1887)
KEEPING, Henry, (1827-1924)
KIESOW, Johann, (1846-1901)
KOCH, Eduard, (1838-1897)
KRAPF, Ludwig, (1810-1881)
LACOE, Ralph Dupuy, (1824-1901)
LANG, Franz Vinzenz, (1821-1899)
LAWLEY, Robert, (1818-1881)
LECKENBY, John, (1814-1877)
LEE, Henry E., (1875-1947)
LEEDS, Alfred Nicholson, (1847-1917)
LEEDS, Charles Edward, (1845-1912)
LINCECUM, Gideon, (1793-1874)
MacADAM, James, (1801-1861)
MACCONOCHIE, Arthur, (1850-1922)
MANDELSLOH, Friedrich von, Graf, (1795-1870)
MARKGRAF, Richard, (1856-1916)
MARSHAM-TOWNSHEND, Robert Honourable, (1834-1915)
MAWSON, Joseph, (1830-1927)
MITCHINSON, John, Right Reverend Bishop, (1833-1918)
MOYSEY, Lewis, (1869-1918)
MÜLLER, Moritz, (1791-1858)
NETTELROTH, Henry, (1835-1887)
NORDENSKIÖLD, Nils Otto Gustaf, (1869-1928)
NORWOOD, Thomas Wilkinson, Reverend, (1829-1908)
ORSINI, Antonio, (-1870)
ORTH, Albert, (1835-1915)
PARKER, James, (1833-1912)
PASINI, Lodovico, (1804-1870)
PATTON, Andrew, (1821-1885)
PERCEVAL, Spencer George, (1838-1922)
PHILPOT, Elizabeth, (1780-1857)
PLANTÉ, Gaston, (1834-1889)
POPPELACK, Joseph, (?1780-1859)
RACHOY, Josef, (-1902)
RADDE, Gustav, (1831-1903)
RANDALL, John, (1810-1910)

FOSSIL COLLECTING 1850-1899 Cont'd.

REED, William, (1810-1892)
RIAZ, Auguste de, (-1920)
RICHTER, Ernst Julius, (-1868)
RIEMSDIJK, A. W. G. van, (1803-1871)
RIGGS, Elmer Samuel, (1869-1963)
ROCHE, Auguste, (1827-1905)
ROTH, Santiago, (1850-1924)
ROVASENDA, Luigi di, Conte, (1826-1917)
RUFFORD, Philip James, (1852-1902)
SAEMANN, Louis, (1821-1866)
SCHALCH, Ferdinand, (1848-1918)
SCHWEINFURTH, Georg, (1836-1925)
SLATTER, Thomas James, (1834-1895)
SLIMON, Robert, (1803-1882)
SOHM, Bonifaz, Pater, (1847-1923)
STERNBERG, Charles Hazelius, (1850-1943)
SWEET, Sidney R., (1878-1946)
THACKERAY, Francis St. John, Reverend, (1832-1919)
THOMPSON, Beeby, (1848-1931)
TYLER, Charles, Captain, (1826-1895)
VALET, August Friedrich, (1811-1889)
VASSALL, Henry, (1860-1926)
VILLA, Antonio, (-1885)
VILLA, Giovanni Battista, (-1887)
VOELKEL, Josef, (1828-1906)
WEISMANN, Gottlieb, (1798-1859)
WIELAND, George Reber, (1865-1953)
WINDOES, James, (1839-1897)

1900 to present

ALEXANDER, Annie Montague, (1867-1950)
BAIRD, Lucille Bailey, (1926-1963)
BALSTON, William Edward, (1848-1918)
BARBER, Charles M, (1877-1954)
BARCLAY, Francis Hubert, (1869-1935)
BARGELLINI, Mariano, (1831-1906)
BECKWITH, Frank Asahel, (-1951)
BELL, Robert, (1864-1934)
BENNIE, James, (1821-1901)
BRÖSAMLEN, Otto, (1885-1966)
BROWN, Alfred "Gogga", (1834-1920)
BUCCHICH, Gregorio, (1828-1911)
CAIRNS, Robert, (1854-1911)
CHADWICK, Samuel, (1845-1903)
CHURCHILL, Frederick Holmes, (1901-1959)
DAWSON, Charles, (1864-1916)
DIETLEN, Karl Rudolf, (1857-1931)
DRUGČEVIČ, Ferdinand, (1856-1928)
EISEL, Robert, (1826-1917)
FINCKH, Alfred, (-1914)
FRANCIS, James, (1843-1925)
GEIST, Frank Otto William, (-1963)
GORE, Charles Herbert Henry, (1866-1951)
GROSSOUVRE, Albert de, (1849-1932)

FOSSIL COLLECTING 1900 to present Cont'd.

GUNN, Robert Marcus, (?1850-1909)
HAMLING, Joseph Green, (-?1935)
HARRISON, Benjamin, (1837-1921)
HOLLER, Anton, (1826-1909)
HORNE, William, (?1836-1928)
HOUSER, Gertrude, (-)
JENNI, Clarence M., Colonel, (1896-)
JESSON, Thomas, (?1850-1928)
KEEPING, Henry, (1827-1924)
KIESOW, Johann, (1846-1901)
LACOE, Ralph Dupuy, (1824-1901)
LEE, Henry E., (1875-1947)
LEEDS, Alfred Nicholson, (1847-1917)
LEEDS, Charles Edward, (1845-1912)
LISSAJOUS, Marcel, (-1921)
MACCONOCHIE, Arthur, (1850-1922)
MACMILLAN, William Edward Frank, (-1954)
MARKGRAF, Richard, (1856-1916)
MARSHAM-TOWNSHEND, Robert Honourable, (1834-1915)
MAWSON, Joseph, (1830-1927)
MITCHINSON, John, Right Reverend Bishop, (1833-1918)
MOYSEY, Lewis, (1869-1918)
NORDENSKIÖLD, Nils Otto Gustaf, (1869-1928)
NORWOOD, Thomas Wilkinson, Reverend, (1829-1908)
NUTTALL, Ernest, (-?1953)
ORTH, Albert, (1835-1915)
PARKER, James, (1833-1912)
PERCEVAL, Spencer George, (1838-1922)
POZZI, Santiago, (-1929)
RACHOY, Josef, (-1902)
RADDE, Gustav, (1831-1903)
RANDALL, John, (1810-1910)
RAVEN, Henry Cushier, (1889-1944)
RIAZ, Auguste de, (-1920)
ROCHE, Auguste, (1827-1905)
ROSS, Frank Stenhouse, (1890-1975)
ROTH, Santiago, (1850-1924)
ROVASENDA, Luigi di, Conte, (1826-1917)
RUFFORD, Philip James, (1852-1902)
SCHALCH, Ferdinand, (1848-1918)
SCHWEINFURTH, Georg, (1836-1925)
SOHM, Bonifaz, Pater, (1847-1923)
SOUTHWORTH, Charles, (1880-1968)
STERNBERG, Charles Hazelius, (1850-1943)
STERNBERG, George Fryer, (1883-1969)
SWEET, Sidney R., (1878-1946)
THACKERAY, Francis St. John, Reverend, (1832-1919)
THOMPSON, Beeby, (1848-1931)
TOURNOUËR, André, (-?1930)
VASSALL, Henry, (1860-1926)
VOELKEL, Josef, (1828-1906)
WATTISON, John Thomas, (1884-1974)
WHITAKER, George O., (-)
WIELAND, George Reber, (1865-1953)

FOSSIL PREPARATION

1800-1849

COULON, Louis de, (1804-1894)
KAPFF, Sixt Friedrich Jakob, (1809-1887)
LAURILLARD, Charles Léopold, (1783-1853)
STAHL, Jean Benjamin, (1816-1893)

1850-1899

BARLOW, Caleb, (1840-1908)
BORCHERT, Gustav, (1863-1934)
CHRISTMAN, Charles Gustave, (1873-1953)
COULON, Louis de, (1804-1894)
DRUGČEVIČ, Ferdinand, (1856-1928)
GIBB, Hugh, (1860-1932)
HERMANN, Adam, (-)
KAPFF, Sixt Friedrich Jakob, (1809-1887)
LAURILLARD, Charles Léopold, (1783-1853)
LUCAS, Frederick Augustus, (1852-1929)
STAHL, Jean Benjamin, (1816-1893)

1900 to present

BARLOW, Caleb, (1840-1908)
BORCHERT, Gustav, (1863-1934)
CAMP, Russell R., (1904-1967)
CHRISTMAN, Charles Gustave, (1873-1953)
DRUGČEVIČ, Ferdinand, (1856-1928)
GIBB, Hugh, (1860-1932)
HERMANN, Adam, (-)
LUCAS, Frederick Augustus, (1852-1929)
SHEARMAN, Harold Leigh, (1921-1974)
SKOU, Kristian, (1912-1967)

GEMMOLOGY

Up to 1800

BIRUNI, (pseud. of Abu-al-Rayán Muhammad ibn-Ahmad al-Biruni), (ca.972-1048)
BOODTS, Anselmus Boëtius de, (1550-1634)
GUTHRIE, Matthew, (1743-1807)
NICOLS, Thomas, (fl. 1652)

1800-1849

BOUVÉ, Thomas Tracy, (1815-1896)
GUTHRIE, Matthew, (1743-1807)

1850-1899

ABBOTT, William James Lewis, (1853-1933)
BOUVÉ, Thomas Tracy, (1815-1896)
BRAUNS, Reinhard Anton, (1861-1937)
FARRINGTON, Oliver Cummings, (1864-1933)
GREGORY, James Reynolds, (1832-1899)
JOHNSEN, Arrien, (1877-1934)
KAISER, Friedrich Wilhelm Erich, (1871-1934)
KITSON, Albert Ernest, Sir, (1868-1937)
KORNITZER, Louis, (1873-) [PROSPECTORS, DIVINERS AND MINING ENGINEERS]
KRAUS, Edward Henry, (1875-1973)
KUNZ, George Frederick, (1856-1932)
LOEHR, August von, Ritter, (1847-1917)
MERENSKY, Hans, (1871-1952)
MERRILL, Loren B., (?1853-1930)
MOSES, Alfred Joseph, (1859-1920)
STELZNER, Alfred Wilhelm, (1840-1895)
SUTTON, John Richard, (1864-1937)
VÉSIGNIÉ, Louis, Colonel, (1870-1954)

1900 to present

ABBOTT, William James Lewis, (1853-1933)
BRAUNS, Reinhard Anton, (1861-1937)
EHRMANN, Martin L., (1903-1972)
FARRINGTON, Oliver Cummings, (1864-1933)
FOSHAG, William Frederick, (1894-1956)
GORDON, William Thomas, (1884-1950)
JOHNSEN, Arrien, (1877-1934)
JUNNER, Norman Ross, (1891-1973)
KAISER, Friedrich Wilhelm Erich, (1871-1934)
KITSON, Albert Ernest, Sir, (1868-1937)
KORNITZER, Louis, (1873-) [PROSPECTORS, DIVINERS AND MINING ENGINEERS]
KRAUS, Edward Henry, (1875-1973)
KUNZ, George Frederick, (1856-1932)
LOEHR, August von, Ritter, (1847-1917)
MEEN, Victor Ben, (1910-1971)
MERENSKY, Hans, (1871-1952)
MERRILL, Loren B., (?1853-1930)
MOSES, Alfred Joseph, (1859-1920)
POLLARD, Ernest Rex, (1922-1966)
SLAWSON, Chester Baker, (1898-1964)
SMITH, Denis Raymond James, (1937-1964)

GEMMOLOGY 1900 to present Cont'd.

SUTTON, John Richard, (1864-1937)
VÉSIGNIÉ, Louis, Colonel, (1870-1954)
VLASOV, Kuz'ma Alekseevich, (1905-1964)
WIGGLESWORTH, Edward, (1885-1945)

GEOCHEMISTRY

Up to 1800

GADOLIN, Johan, (1760-1852)

1800-1849

BISCHOFF, Karl Gustav, (1792-1870)
BLOMSTRAND, Christian Wilhelm, (1826-1897)
BUNSEN, Robert Wilhelm, (1811-1899)
FORBES, David, (1828-1876)
GADOLIN, Johan, (1760-1852)
GESNER, Abraham, (1797-1864)
HAUGHTON, Samuel, Reverend, (1823-1897)
JACKSON, Charles Thomas, (1805-1880)
KELVIN, William Thomson, Baron, (1824-1907)
NESBIT, John Collis, (1819-1862)
PUSCH, Georg Gottlieb, (1790-1846)
ROTH, Justus Ludwig Adolph, (1818-1892)

1850-1899

ADAMS, Frank Dawson, (1859-1942)
BISCHOFF, Karl Gustav, (1792-1870)
BLOMSTRAND, Christian Wilhelm, (1826-1897)
BLOXAM, Thomas, (1836-?1872)
BOŘICKÝ, Emanuel, (1840-1881)
BUNSEN, Robert Wilhelm, (1811-1899)
CARNOT, Marie Adolphe, (1839-1920)
CLARKE, Frank Wigglesworth, (1847-1931)
DOELTER y CISTERLICH, Cornelio August, (1850-1930)
ELSDEN, James Vincent, (?1856-1930)
EMSZT, Kálmán, (1873-1957)
FORBES, David, (1828-1876)
GADOLIN, Johan, (1760-1852)
GESNER, Abraham, (1797-1864)
HAUGHTON, Samuel, Reverend, (1823-1897)
HENRICH, Ferdinand, (1837-1915)
HILLEBRAND, William Francis, (1853-1925)
HÖGBOM, Arvid Gustaf, (1857-1940)
HUBBARD, Lucius Lee, (1849-1933)
ILOSVAY, Lajos, (1851-1937)
ISTRATI, Constantin I., (1850-1918)
JACKSON, Charles Thomas, (1805-1880)
JOLY, John, (1857-1933)
KELVIN, William Thomson, Baron, (1824-1907)
KHRUSHCHOV, Konstantin Dmitrievich, (? -1912)
KLVAŇA, Josef, (1857-1919)
KUPFFER, August Ernestovich, (1842-1918)
KYLE, John Joseph Jolly, called Juan José, (1838-1922)
LEMBERG, Johannes, (1842-1902)
LOEWINSSON-LESSING, Franc Julyevich, (1859-1939)

GEOCHEMISTRY 1850-1899 Cont'd.

MAUZELIUS, Robert, (1864-1921)
McCAUGHEY, William John, (1882-1962)
MICHEL-LÉVY, Albert, (1877-1955)
NESBIT, John Collis, (1819-1862)
OSANN, Carl Alfred, (1859-1923)
PLAGEMANN, A., (1858-1908)
PLAYER, Jacob Hort, (1834-1908)
POLLARD, William, (1870-1946)
ROTH, Justus Ludwig Adolph, (1818-1892)
SANTESSON, Gustaf Henrik, (1847-1915)
SCHARIZER, Rudolf, (1859-1935)
SKEATS, Ernest Willington, (1875-1953)
SLAVÍK, František, (1876-1957)
SOLLAS, William Johnson, (1849-1936)
SPEZIA, Giorgio, (1842-1912)
TATE, Alexander Norman, (1837-1892)
THUGUTT, Stanislaw Józef, (1862-1956)
VAN'T HOFF, Jacobus Hendricus, (1852-1911)
VERNADSKI, Vladimir Ivanovich, (1863-1945)
VON FEILITZEN, Carl Henrik Jobst, (1840?-1901)
WAITZ, Paul, (1876-1961)
WYNDHAM, T. Heathcote, (?1842-1876)
ZIRKEL, Ferdinand, (1838-1912)

1900 to present

ADAMS, Frank Dawson, (1859-1942)
AHRENS, Louis Herman, (1918-)
ALLING, Harold Lattimore, (1888-1960)
BARSONOV, Georgyi Pavlovich, (1897-)
BARTH, Thomas Frederik Weiby, (1899-1971)
BAUHANS, Hans, (1888-1918)
BLAKE, George Stanfield, (-1940)
BLANC, Gian Alberto, Barone, (1879-1966)
BOWEN, Norman Levi, (1887-1956)
BROWN, Bahngrell Walter "Bill", (1916-)
CARNOT, Marie Adolphe, (1839-1920)
CARRITT, Dayton Ernest, (1915-)
CARROLL, Dorothy, (1907-1970)
CLARKE, Frank Wigglesworth, (1847-1931)
CSAJÁGHY, Gábor, (1903-1972)
CUTTITTA, Frank, (1912-1974)
DENNEN, William Henry, (1920-)
DITTLER, Emil, (1882-1945)
DOELTER y CISTERLICH, Cornelio August, (1850-1930)
DUPARQUE, André, (1892-1960)
ELLSWORTH, Hardy Vincent, (1889-1952)
ELSDEN, James Vincent, (?1856-1930)
EMSZT, Kálmán, (1873-1957)
ESKOLA, Pentti Eelis, (1883-1964)
FAIRCHILD, John Gifford, (1882-1965)
FERSMAN, Aleksander Yevgenyevich, (1883-1945)
FOSTER, Margaret Dorothy, (1895-1970)
FRASER, Horace John "Dutch", (1905-1969)
GOLDSCHMIDT, Victor Moritz, (1888-1947)

GEOCHEMISTRY 1900 to present Cont'd.

GROSSZ, Ádám, (1925-1973)
HEIDE, Friedrich, (1891-)
HENRICH, Ferdinand, (1837-1915)
HILLEBRAND, William Francis, (1853-1925)
HÖGBOM, Arvid Gustaf, (1857-1940)
HOWER, John, Jr., (1927-)
HUBBARD, Lucius Lee, (1849-1933)
HURLEY, Patrick Mason, (1912-)
HUTTON, Colin Osborne, (1910-1971)
ILOSVAY, Lajos, (1851-1937)
ISTRATI, Constantin I., (1850-1918)
JOLY, John, (1857-1933)
KAŠPAR, Jan, (1898-)
KELVIN, William Thomson, Baron, (1824-1907)
KHRUSHCHOV, Konstantin Dmitrievich, (? -1912)
KLVAŇA, Josef, (1857-1919)
KOKTA, Jaroslav, (1894-)
KOLDERUP, Niels-Henrik, (1898-1971)
KOWALSKI, Ludwik, (1885-1943)
KUNO, Hisashi, (1910-1969)
KUPFFER, August Ernestovich, (1842-1918)
KUŹNIAR, Czeslaw Faustyn, (1889-1955)
KYLE, John Joseph Jolly, called Juan José, (1838-1922)
LARSEN, Esper Signius, III, (1912-1961)
LEMBERG, Johannes, (1842-1902)
LEUTWEIN, Friedrich, (1911-1974)
LOEWINSSON-LESSING, Franc Julyevich, (1859-1939)
LUBICZ-SULIMIRSKI, Stefan, (1900-1953)
LUTH, William Clair, (1934-)
MACDIARMID, Roy Angus, (1933-1966)
MAUZELIUS, Robert, (1864-1921)
McCAUGHEY, William John, (1882-1962)
MICHEL-LÉVY, Albert, (1877-1955)
MIHOLIĆ, Stanko, (1891-1960)
MILLOT, Georges, (1917-)
MOREY, George Washington, (1888-1965)
NIGGLI, Paul, (1888-1953)
NOVÁČK, Radim, (1905-1942)
NOVÁK, Jiří, (?1902-1971)
NOVOTNÝ, Miroslav, (1921-1971)
OSANN, Carl Alfred, (1859-1923)
PECORA, William Thomas, (1913-1972)
PINSON, William Hamet, Jr., (1919-)
PLAGEMANN, A., (1858-1908)
PLAYER, Jacob Hort, (1834-1908)
POLDERVAART, Arie, (1918-1964)
POLLARD, William, (1870-1946)
RABBITT, John Charles, (1907-1957)
RADLEY, Ernest George, (1880-1945)
ROSENDAHL, Halvor, (1890-1963)
ROSENHOLTZ, Joseph Leon, (1899-1963)
ROSS, Clarence Samuel, (1880-1975)
ROSSBY, Carl-Gustaf Arvid, (1898-1957)
ROST, Rudolf, (1902-)

GEOCHEMISTRY 1900 to present Cont'd.

ROUBAULT, Marcel Edouard, (1905-1974)
SALÁT, Ján, (1919-1973)
SANTESSON, Gustaf Henrik, (1847-1915)
SCHAIRER, John Frank, (1904-1970)
SCHALLER, Waldemar Theodore, (1882-1967)
SCHARIZER, Rudolf, (1859-1935)
SHAND, Samuel James, (1882-1957)
SKEATS, Ernest Willington, (1875-1953)
SLAVÍK, František, (1876-1957)
SOLLAS, William Johnson, (1849-1936)
SPEZIA, Giorgio, (1842-1912)
STEINOCHER, Vladimír, (1902-1972)
THUGUTT, Stanislaw Józef, (1862-1956)
TIPPER, George Howlett, (1881-1947)
TRÖGER, Walter Ehrenreich, (1901-1963)
UHLIG, Johannes, (1883-1918)
URBAIN, Pierre, (1895-1968)
VANDENDRIESSCHE, Adrien, (1914-1940)
VAN'T HOFF, Jacobus Hendricus, (1852-1911)
VERNADSKI, Vladimir Ivanovich, (1863-1945)
VESELÝ, Václav, (1896-1956)
VINOGRADOV, A. P., (1895-1975)
VLASOV, Kuz'ma Alekseevich, (1905-1964)
VRABKA, František, (1924-1957)
WAITZ, Paul, (1876-1961)
WALDBAUM, David Robert, (1937-1974)
WILLIAMS, Mervyn, (1934-1970)
WINCHESTER, John Widner, (1929-)
WRIGHT, Harold Douglas, (1921-1969)
ZERFOSS, Samuel, (1912-1958)
ZEUNER, Frederick Everard, (1905-1963)
ZIRKEL, Ferdinand, (1838-1912)

GEOCHRONOLOGY

1850-1899

BULL, Alfred Joseph, (1875-1950)
DE GEER, Gerard Jacob, Baron, (1858-1943)
LANE, Alfred Church, (1863-1948)
LARSEN, Esper Signius Jr., (1879-1961)

1900 to present

AHRENS, Louis Herman, (1918-)
BULL, Alfred Joseph, (1875-1950)
CUTTITTA, Frank, (1912-1974)
DE GEER, Gerard Jacob, Baron, (1858-1943)
FAIRBAIRN, Harold Williams, (1906-)
FRIES, Carl, Jr., (1910-1965)
HIGAZY, Riad Abdel-Megid, (1919-1967)
HOLMES, Arthur, (1890-1965)
HOWER, John, Jr., (1927-)
HURLEY, Patrick Mason, (1912-)
KNOPF, Adolph, (1882-1966)
LANE, Alfred Church, (1863-1948)
LARSEN, Esper Signius Jr., (1879-1961)

GEOCHRONOLOGY 1900 to present Cont'd.

LEUTWEIN, Friedrich, (1911-1974)
LIBBY, Willard Frank, (1908-)
MARBLE, John Putnam, (1897-1955)
OAKLEY, Kenneth Page, (1911-)
PINSON, William Hamet, Jr., (1919-)
VINOGRADOV, A. P., (1895-1975)

GEOMORPHOLOGY
Up to 1800
ALBERTI, Leon Battista, (1404-1472)
ARISTOTLE, (384-322 B.C.)
BAKEWELL, Robert, (1768-1843)
BARTRAM, John, (1699-1777)
BARTRAM, William, (1739-1823)
BRUNO, Filippo, called Giordano, (1548-1600)
CLARK, William, (1770-1838)
COLLOT, Georges Henri Victor, (1750-1805)
DAMPIER, William, (1652-1715)
DESMAREST, Nicolas, (1725-1815)
EATON, Amos, (1776-1842)
EVANS, Lewis, (1700-1756)
FERBER, Johann Jacob, (1743-1790)
FORSTER, Johann Georg Adam, (1754-1794)
FORSTER, Johann Reinhold, (1729-1798)
GREENOUGH, George Bellas, (1778-1855)
GUETTARD, Jean Étienne, (1715-1786)
HUNTER, John, (1728-1793)
HUTTON, James, (1726-1797)
IMPERATO, Ferrante, (1550-1631)
JEFFERSON, Thomas, (1743-1826)
KALM, Pehr, (1716-1779)
KANT, Immanuel, (1724-1804)
KIRCHER, Athanasius, (1601-1680)
LAMARCK, Jean-Baptiste Pierre Antoine de Monet, Chevalier de, (1744-1829)
LINNE, Carl von, (1707-1778)
LUC, Jean André de, (1727-1817)
MACCULLOCH, John, (1773-1835)
MACKENZIE, Alexander, Sir, (ca.1767-1830)
MAILLET, Benoît de, (1656-1738)
MORITZ, Karl Philipp, (1757-1793)
PALISSY, Bernard, (1510-1589)
PAUSANIAS, (ca.130-180 A.D.)
PINGEL, Christian, (1793-1852)
PLAYFAIR, John, (1748-1819)
PRIOR, Richard, (fl.1699)
RAY, John, (1627-1705)
RENNELL, James, Major, (1742-1830)
SAUSSURE, Horace Bénédict de, (1740-1799)
SCHEUCHZER, Johann Jakob, (1672-1733)
STRABO, (63B.C.-23 A.D.?)
TARGIONI TOZZETTI, Giovanni, (1712-1783)
TROIL, Uno von, (1746-1803)
VINCI, Leonardo da, (1452-1519)

GEOMORPHOLOGY Cont'd.

1800-1849

ABERT, James W., Sr., (1820-1897)
AGASSIZ, Louis Jean Rodolphe, (1807-1873)
ANSTED, David Thomas, (1814-1880)
ARGYLL, George Douglas, Duke of, (1823-1900)
AUSTIN, Thomas, Fort-Major, (1794-1881)
BAKEWELL, Robert, (1768-1843)
BALL, John, (1818-1889)
BARTRAM, William, (1739-1823)
BLACK, Samuel, (1785-1841)
BOLTON, John, (1788-1873)
BRAVAIS, Auguste, (1811-1863)
CHAMBERS, Robert, (1802-1871)
CHURCHILL, George Cheetham, (1822-1906)
CLARK, William, (1770-1838)
CLOSE, Maxwell Henry, Reverend, (1822-1903)
COLLOT, Georges Henri Victor, (1750-1805)
CONYBEARE, William Daniel, Reverend, (1787-1857)
DANA, James Dwight, (1813-1895)
DARWIN, Charles Robert, (1809-1882)
DAUBENY, Charles Giles Brindle, (1795-1867)
DAVIDSON, George, (1825-1911)
DESMAREST, Nicolas, (1725-1815)
DIEFFENBACH, Ernst, (1811-1855)
DUFOUR, Charles, (1827-1902)
EATON, Amos, (1776-1842)
EGERTON, William Henry, Reverend, (1811-1910)
FAVRE, Jean Alphonse, (1815-1890)
FERGUSSON, James, (1808-1885)
FISHER, Osmond, Reverend, (1817-1914)
FORBES, James David, (1809-1868)
GILBERT, Josiah, (1814-1892)
GODWIN-AUSTEN, Robert Alfred Cloyne, (1808-1884)
GREENOUGH, George Bellas, (1778-1855)
GREENWOOD, George, Colonel, (1799-1875)
HAAST, John Francis Julius von, Sir, (1824-1887)
HELMERSEN, Gregor von, (1803-1885)
HOFF, Karl Ernst Adolf von, (1771-1837)
HOME, David Milne, (1805-1890)
HUBBARD, Oliver Payson, (1809-1900)
JAMIESON, Thomas Francis, (1829-1913)
JEFFERSON, Thomas, (1743-1826)
JUKES, Joseph Beete, (1811-1869)
KANE, Elisha Kent, (1820-1857)
KANT, Immanuel, (1724-1804)
KING, Samuel William, (1821-1868)
LAIDLAY, John Watson, (1808-1885)
LAMARCK, Jean-Baptiste Pierre Antoine de Monet, Chevalier de, (1744-1829)
LAMOUROUX, Jean Vincent Félix, (1779-1825)
LAUDER, Thomas Dick, Sir, (1784-1848)
LEICHHARDT, Friedrich Wilhelm Ludwig, (1813-1848)
LUC, Jean André de, (1727-1817)
MACCULLOCH, John, (1773-1835)

GEOMORPHOLOGY 1800-1849 Cont'd.

MACFARLANE, James, (1819-1885)
MACKENZIE, Alexander, Sir, (ca.1767-1830)
MACKINTOSH, Daniel, (1815-1891)
MARTEN, Henry John, (1827-1892)
MARTIN, Peter John, (1786?-1860)
MERIAN, Peter, (1795-1883)
MOLYNEUX, William, (1824-1882)
PENGELLY, William, (1812-1894)
PHILLIPS, John, (1800-1874)
PINGEL, Christian, (1793-1852)
PLAYFAIR, John, (1748-1819)
RAMSAY, Andrew Crombie, Sir, (1814-1891)
RAULIN, Félix Victor, (1815-1905)
REDMAN, John Baldry, (1816-1899)
RENNELL, James, Major, (1742-1830)
RICHARDSON, John, Sir, (1787-1865)
ROY, Thomas, (-1842)
RUSKIN, John, (1819-1900)
RÜTIMEYER, Karl Ludwig, (1825-1895)
SAXE, Sjur Aasmundsen, (1808-1887?)
SCHIEL, Jacob Heinrich, (1813-)
SCHIMPER, Karl Friedrich, (1803-1867)
SCHOOLCRAFT, Henry Rowe, (1793-1864)
SCROPE, George Poulett, (né Thomson), (1797-1876)
SIMONY, Friedrich, (1813-1896)
SOMERVILLE, Mary, (1780-1872)
STRACHEY, Richard, General Sir, (1817-1908)
SÜSSMILCH, Carl Adolf, (?1874-1946)
THOMPSON, Zadock, (1796-1856)
TOPLEY, William, (1841-1894)
TORELL, Otto Martin, (1828-1900)
TROIL, Uno von, (1746-1803)
TSCHUDI, Johan Jacob von, (1818-1889)
VALPY, Robert Harris, (1819-1904)
VIQUESNEL, Auguste, (1800-1867)
WITCHELL, Edwin, (1823-1887)
YOUNG, Robert, Right Honourable, (1822-1917)
ZITTEL, Karl Alfred von, (1839-1904)

1850-1899

ABBE, Cleveland, Jr., (1872-1934)
ABBOTT, George, (1844-1925)
ABERT, James W., Sr., (1820-1897)
ABSOLON, Karel, (1877-1960)
AGASSIZ, Louis Jean Rodolphe, (1807-1873)
AILIO, Julius, (1872-1933)
ANSTED, David Thomas, (1814-1880)
ARGYLL, George Douglas, Duke of, (1823-1900)
AUSTIN, Thomas, Fort-Major, (1794-1881)
BALL, John, (-1941)
BALL, John, (1818-1889)
BALTZER, Richard Armin, (1842-1913)
BARRELL, Joseph, (1869-1919)
BARTON, George Hunt, (1852-1933)

GEOMORPHOLOGY 1850-1899 Cont'd.

BEADNELL, Hugh John Llewellyn, Major, (1874-1944)
BEEDE, Joshua William, (1871-1940)
BELT, Thomas, (1832-1878)
BENNETT, Francis James, (1845-1920)
BIGOT, Alexandre, (1863-1953)
BLAYAC, Joseph, (1865-1936)
BÖHMERSHEIM, August Böhm von, Edler, (1858-1930)
BOLTON, John, (1788-1873)
BONNEY, Thomas George, Reverend, (1833-1923)
BOWMAN, Isaiah, (1882-1950)
BOYD, Walter Halcro, (1878-1960)
BRAVAIS, Auguste, (1811-1863)
BREMNER, Alexander, (1863-1943)
BRÜCKNER, Eduard, (1862-1927)
BULL, Alfred Joseph, (1875-1950)
BULLOWS, Wilfred Lowson, (?1869-1943)
BURY, Henry, (?1862-1958)
BUTLER, Bertram Theodore, (1874-1958)
CALL, Richard Ellsworth, (1856-1917)
CAMPBELL, Marius Robinson, (1858-1940)
CARNEY, Frank, (1868-1934)
CARVALHO, Anselmo Ferraz de, (1878-1955)
CHAMBERS, Robert, (1802-1871)
CHOFFAT, Paul, (1849-1919)
CHUDEAU, René, (1864-1921)
CHURCHILL, George Cheetham, (1822-1906)
CLELAND, Herdman Fitzgerald, (1869-1935)
CLOSE, Maxwell Henry, Reverend, (1822-1903)
CODRINGTON, Thomas, (1829-1918)
CONYBEARE, William Daniel, Reverend, (1787-1857)
CVIJIĆ, Jovan, (1865-1927)
DALY, Reginald Aldworth, (1871-1957)
DANA, James Dwight, (1813-1895)
DARWIN, Charles Robert, (1809-1882)
DARWIN, Leonard, (1850-1943)
DAUBENY, Charles Giles Brindle, (1795-1867)
DAVIDSON, George, (1825-1911)
DAVIS, William Morris, (1850-1934)
DĚDINA, Václav, (1870-1956)
DEECKE, Johannes Ernst Wilhelm, (1862-1934)
DE LORENZO, Giuseppe, (1871-1957)
DE PRETTO, Olinto, (1857-1921)
DE RANCE, Charles Eugène, (1847-1906)
DIEFFENBACH, Ernst, (1811-1855)
DOUXAMI, Henri, (1871-1913)
DUFOUR, Charles, (1827-1902)
DUNLOP, Andrew, (1845-1915)
DU PASQUIER, Léon, (1864-1897)
DWERRYHOUSE, Arthur Richard, (1867-1931)
EGERTON, William Henry, Reverend, (1811-1910)
FAVRE, Jean Alphonse, (1815-1890)
FEILDEN, Henry Wemyss, Colonel, (1838-1921)
FENNEMAN, Nevin Melancthan, (1865-1945)
FERGUSSON, James, (1808-1885)

GEOMORPHOLOGY 1850-1899 Cont'd.

FINSTERWALDER, Sebastian, (1862-?1931)
FISHER, Osmond, Reverend, (1817-1914)
FLEURY, Ernest, (1878-1958)
FORBES, James David, (1809-1868)
FOREL, François Alphonse, (1841-1912)
FOWLER, Joseph, (1872-1958)
GARDINER, John Stanley, (1872-1946)
GARWOOD, Edmund Johnston, (1864-1949)
GEIKIE, Archibald, Sir, (1835-1924)
GEIKIE, James, (1839-1915)
GERLAND, Georg Karl Cornelius, (1833-1919)
GILBERT, Grove Karl, (1843-1918)
GILBERT, Josiah, (1814-1892)
GIŽYCKI, Franciszek Ksawery, (-)
GLENN, Leonidas Chalmers, (1871-1951)
GODWIN-AUSTEN, Robert Alfred Cloyne, (1808-1884)
GORDON, Charles Henry, (1857-1934)
GOSSLING, Frank Newbery, (1888-1946)
GREEN, Alexander Henry, (1832-1896)
GREEN, John Frederick Norman, (?1873-1949)
GREENOUGH, George Bellas, (1778-1855)
GREENWOOD, George, Colonel, (1799-1875)
GREGORY, Herbert E., (1869-1952)
GREGORY, John Walter, (1864-1932)
HAAST, John Francis Julius von, Sir, (1824-1887)
HALL, Marshall, Captain, (1831-1896)
HALL, Townshend Monckton, (1845-1899)
HANSEN, Andreas M., (1857-1932)
HARDMAN, Edward Townley, (1845-1887)
HARKER, Alfred, (1859-1939)
HARTT, Charles Frederic, (1840-1878)
HEIM, Albert, (1849-1937)
HELMERSEN, Gregor von, (1803-1885)
HENDERSON, Junius, (1865-1937)
HIND, Wheelton, (1859-1920)
HINDE, George Jennings, (1839-1918)
HOBBS, William Herbert, (1864-1953)
HOFF, Karl Ernst Adolf von, (1771-1837)
HOME, David Milne, (1805-1890)
HOOLEY, Reginald Walter, (1866-1923)
HORUSITZKY, Henrik, (1870-1945)
HOVEY, Edmund Otis, (1862-1924)
HUBBARD, George David, (1871-1958)
HUBBARD, Oliver Payson, (1809-1900)
JACOBS, Elbridge Churchill, (1873-1957)
JAMIESON, Thomas Francis, (1829-1913)
JOHNSON, Douglas Wilson, (1878-1944)
JOHNSON, Walter, (1867-1950)
JOHNSTON, Mary Sophia, (1875-1955)
JONES, Owen Thomas, (1878-1967)
JUKES, Joseph Beete, (1811-1869)
JULIEN, Alexis Anastay, (1840-1919)
KANE, Elisha Kent, (1820-1857)
KATZER, Bedřich, (1861-1925)

GEOMORPHOLOGY 1850-1899 Cont'd.

KENDALL, Percy Fry, (1856-1936)
KING, Clarence Rivers, (1842-1901)
KING, Samuel William, (1821-1868)
KŘIŽ, Martin, (1841-1916)
LAFLAMME, Joseph Clovis Kemner, Monseigneur, (1849-1910)
LAIDLAY, John Watson, (1808-1885)
LAKE, Philip, (1865-1949)
LANINO, Giuseppe, (1832-1907)
LAPPARENT, Albert Auguste de, (1839-1908)
LAWSON, Andrew Cowper, (1861-1952)
LEACH, Arthur Leonard, (1869-1957)
LEMOINE, Paul, (1878-1940)
LEVERETT, Frank, (1859-1943)
LORY, Pierre, (1866-1956)
LUBBOCK, John, Lord Avebury, (1834-1913)
LUGEON, Maurice, (1870-1953)
LYELL, Charles, Sir, (1797-1875)
LYONS, Henry George, (1864-?1943)
MACFARLANE, James, (1819-1885)
MACKINTOSH, Daniel, (1815-1891)
MALLADRA, Alessandro, (1865-1945)
MARGERIE, Emmanuel Marie Pierre Martin Jacquin de, (1862-1953)
MARINELLI, Olinto, (1874-1926)
MARKHAM, Clements Robert, Sir, (1830-1916)
MARR, John Edward, (1857-1933)
MARTEN, Henry John, (1827-1892)
MARTIN, Edward Alfred, (1864-1943)
MARTIN, Peter John, (1786?-1860)
MARTONNE, Emmanuel Louis Eugène de, (1873-1955)
MATTHES, François Emile, (1874-1948)
MELLO, John Magens, Reverend, (1836-1914)
MERIAN, Peter, (1795-1883)
MERRILL, George Perkins, (1854-1929)
MEYER, Hans, (1858-1929)
MILL, Hugh Robert, (1861-1950)
MILLER, Hugh Jr., (1850-1896)
MILLS, Enos Abijah, (1870-1922)
MILNE, John, (1850-1913)
MOLYNEUX, Arthur John Charles, (?1866-1921)
MOLYNEUX, William, (1824-1882)
MORSE, William Clifford, (1874-1962)
MUIR, John, (1838-1914)
NEWELL ARBER, Edward Alexander, (1870-1918)
NILES, William Harmon, (1838-1910)
OBRUCHEV, Vladimir Afanas'evich, (1863-1956)
OGILVIE, Francis Grant, Sir, (1858-1930)
OSMAN, Charles William, (1855-1932)
PAYER, Julius von, Ritter, (1841-1912)
PENCK, Friedrich Carl Albrecht, (1858-1945)
PENGELLY, William, (1812-1894)
PENNING, William Henry, (1838-1902)
PHILLIPS, John, (1800-1874)
PINGEL, Christian, (1793-1852)
PIRSSON, Louis Valentine, (1860-1919)

GEOMORPHOLOGY 1850-1899 Cont'd.

PITTIER, Henry François, (1857-1950)
POWELL, John Wesley, (1834-1902)
PURDUE, Albert Homer, (1861-1917)
PURKYNĚ, Cyril, (1862-1937)
RAMSAY, Andrew Crombie, Sir, (1814-1891)
RAMSAY, Wilhelm, (1865-1928)
RAULIN, Félix Victor, (1815-1905)
READE, Thomas Mellard, (1832-1909)
REDMAN, John Baldry, (1816-1899)
REED, Frederick Richard Cowper, (1869-1946)
REICHERT, Friedrich, (1878-)
RICHARDSON, John, Sir, (1787-1865)
RICHTHOFEN, Paul Wilhelm Ferdinand von, Freiherr, (1833-1905)
ROMER, Eugeniusz, (1871-1954)
RUSKIN, John, (1819-1900)
RÜTIMEYER, Karl Ludwig, (1825-1895)
RUTKOWSKI, Feliks, (1878-1964)
SALISBURY, Rollin D., (1858-1922)
SAXE, Sjur Aasmundsen, (1808-1887?)
SCHIMPER, Karl Friedrich, (1803-1867)
SCHOOLCRAFT, Henry Rowe, (1793-1864)
SCHRADER, François Jean Daniel "Franz", (1844-1924)
SCHWARZ, Ernest Hubert Lewis, (1873-1928)
SCOTT, William Berryman, (1858-1947)
SCROPE, George Poulett, (né Thomson), (1797-1876)
SEDERHOLM, Jakob Johannes, (1863-1934)
SELLARDS, Elias Howard, (1875-1961)
SHATTUCK, George Burbank, (1869-)
SHORE, Thomas William, (1840-1905)
SMITH, George Otis, (1871-1944)
SMITH, Reginald Allender, (?1847-1940)
SMITH, William Richard Tangier, (1869-1972)
SOLLAS, William Johnson, (1849-1936)
SOMERVILLE, Mary, (1780-1872)
STIFFE, Arthur William, Captain, (1849-1912)
STIRRUP, Mark, (1831-1907)
STRACHEY, Richard, General Sir, (1817-1908)
STRZELECKI, Pawel Edmund, Hrabia, (1797-1873)
SUESS, Eduard, (1831-1914)
SÜSSMILCH, Carl Adolf, (?1874-1946)
SWINNERTON, Henry Hurd, (1875-1967)
TARR, Ralph Stockman, (1864-1912)
THOMPSON, Zadock, (1796-1856)
THORNTON, Richard, (1838-1863)
TIGHT, William George, (1865-1910)
TOPLEY, William, (1841-1894)
TORELL, Otto Martin, (1828-1900)
TSCHUDI, Johan Jacob von, (1818-1889)
UPHAM, Warren, (1850-1934)
VALPY, Robert Harris, (1819-1904)
VIQUESNEL, Auguste, (1800-1867)
WAITZ, Paul, (1876-1961)
WALTHER, Johannes, (1860-1937)
WHYMPER, Edward, (1840-1911)

GEOMORPHOLOGY 1850-1899 Cont'd.

WILKS, Mark, (?1862-1945)
WILLARD, Daniel Everett, (1862-1947)
WILLIAMS, Ira Abraham, (1876-)
WITCHELL, Edwin, (1823-1887)
WOEIKOF, Alexander, (1842-1916)
WOOD, Searles Valentine Jr., (1830-1884)
WOODWORTH, Jay Backus, (1865-1925)
WOOLACOTT, David, (1872-1924)
WORSFORD, Frederick H., (1870-1954)
WÜNSCH, Edward A., (1822-1895)
WYVILLE THOMSON, Charles, Sir, (1830-1882)
YABE, Hisakatsu, (1878-1969)
YAMASAKI, Naomasa, (1870-1929)
YOUNG, Robert, Right Honourable, (1822-1917)
ZIMÂNYI, Károly, (1862-1941)
ZITTEL, Karl Alfred von, (1839-1904)

1900 to present

ABBE, Cleveland, Jr., (1872-1934)
ABBOTT, George, (1844-1925)
ABSOLON, Karel, (1877-1960)
ADDINGTON, Arch Rombough, (1894-1972)
AHLMANN, Hans Wilhelmsson, (1889-1974)
AILIO, Julius, (1872-1933)
ALEXANDER, Frances Elizabeth Somerville, (1908-1958)
ANTHONY, Herbert Douglas, (1892-1968)
ARKELL, William Joscelyn, (1904-1958)
BALL, John, (-1941)
BALTZER, Richard Armin, (1842-1913)
BARRELL, Joseph, (1869-1919)
BARTON, Donald Clinton, (1889-1939)
BARTON, George Hunt, (1852-1933)
BARTRUM, John Arthur, (1885-1949)
BEADNELL, Hugh John Llewellyn, Major, (1874-1944)
BECK, Paul, (1882-1958)
BEEDE, Joshua William, (1871-1940)
BENNETT, Francis James, (1845-1920)
BENSON, William Noel, (1885-1957)
BIGOT, Alexandre, (1863-1953)
BISHOPP, Douglas Wallace, (1900-1977)
BLACKWELDER, Eliot, (1880-1969)
BLAYAC, Joseph, (1865-1936)
BŎCEK, Antonîn, (1880-1956)
BÖHMERSHEIM, August Böhm von, Edler, (1858-1930)
BONNEY, Thomas George, Reverend, (1833-1923)
BOSWELL, Katherine Cumming, (1888-1952)
BOURCART, Jacques, (1891-1965)
BOWMAN, Isaiah, (1882-1950)
BOYD, Alexander John, (1891-1969)
BOYD, Walter Halcro, (1878-1960)
BRADLEY, Wilmot Hyde, (1899-)
BREMNER, Alexander, (1863-1943)
BRIGHAM, Albert Perry, (1855-1932)
BROGGI, Jorge Albert, (1888-1966)

GEOMORPHOLOGY 1900 to present Cont'd.

BROMEHEAD, Cyril Edward Nowill, (1885-1952)
BROWN, Olin Teter, (1900-1962)
BRÜCKNER, Eduard, (1862-1927)
BUCHER, Walter Herman, (1888-1965)
BULL, Alfred Joseph, (1875-1950)
BULLOWS, Wilfred Lowson, (?1869-1943)
BURY, Henry, (?1862-1958)
BUTLER, Bertram Theodore, (1874-1958)
CADY, Richard Carlysle, (1908-1943)
CALCIATI, Cesare, (1885-1929)
CALL, Richard Ellsworth, (1856-1917)
CAMPBELL, Marius Robinson, (1858-1940)
CARNEY, Frank, (1868-1934)
CARVALHO, Anselmo Ferraz de, (1878-1955)
CASTERET, Norbert, (1897-)
CASTIGLIONI, Bruno, (1898-1945)
CAYEUX DE SENARPONT, André, called CAILLEUX, André, (1907-)
ČERMÁK, Jiři, (1884-1955)
CHAPUT, Jules Ernest, (1882-1943)
CHARLESWORTH, John Kaye, (1889-)
CHOFFAT, Paul, (1849-1919)
CHUDEAU, René, (1864-1921)
CHURCHILL, George Cheetham, (1822-1906)
CLELAND, Herdman Fitzgerald, (1869-1935)
CLOOS, Hans, (1885-1951)
CLOSE, Maxwell Henry, Reverend, (1822-1903)
CODRINGTON, Thomas, (1829-1918)
COLLET, Léon William, (1880-1957)
COLLINS, William Ion, (1897-1964)
COMBAZ, Paul, Chanoine, (-1930)
COOKE, Charles Wythe, (1887-1971)
COTTON, Charles Andrew, Sir, (1885-1970)
CVIJIĆ, Jovan, (1865-1927)
DAKE, Charles Laurence, (1883-1934)
DALY, Reginald Aldworth, (1871-1957)
DARWIN, Leonard, (1850-1943)
DAVIDSON, George, (1825-1911)
DAVIS, William Morris, (1850-1934)
DĚDINA, Václav, (1870-1956)
DEECKE, Johannes Ernst Wilhelm, (1862-1934)
DE LORENZO, Giuseppe, (1871-1957)
DE PRETTO, Olinto, (1857-1921)
DE RANCE, Charles Eugène, (1847-1906)
DOBBIE, Charles Herbert, (1905-1976)
DOUXAMI, Henri, (1871-1913)
DUFOUR, Charles, (1827-1902)
DUNLOP, Andrew, (1845-1915)
DWERRYHOUSE, Arthur Richard, (1867-1931)
EDWARDS, Austin Burton, (1909-1960)
EGERTON, William Henry, Reverend, (1811-1910)
EKBLAW, George Elbert, (1895-1972)
EVANS, Emile, (1883-1966)
EWING, William Maurice, (1906-1974)
FAGG, Christopher Charles, (1883-1965)

GEOMORPHOLOGY 1900 to present Cont'd.

FEILDEN, Henry Wemyss, Colonel, (1838-1921)
FENNEMAN, Nevin Melancthan, (1865-1945)
FINSTERWALDER, Richard, (1899-1963)
FINSTERWALDER, Sebastian, (1862-?1931)
FISHER, Osmond, Reverend, (1817-1914)
FLEURY, Ernest, (1878-1958)
FORD, Trevor David, (1932-)
FOREL, François Alphonse, (1841-1912)
FOWLER, Gordon, Major, (-1953)
FOWLER, Joseph, (1872-1958)
FRYXELL, Roald Hilding, (1934-1974)
GAGNEBIN, Élie, (1891-1949)
GARDINER, John Stanley, (1872-1946)
GARNETT, Alice, (-)
GARWOOD, Edmund Johnston, (1864-1949)
GEIKIE, Archibald, Sir, (1835-1924)
GEIKIE, James, (1839-1915)
GERLAND, Georg Karl Cornelius, (1833-1919)
GIGNOUX, Maurice Irénée Marie, (1881-1955)
GILBERT, Grove Karl, (1843-1918)
GILLMAN, Clement, (?1883-1946)
GLENN, Leonidas Chalmers, (1871-1951)
GOLDTHWAIT, James Walter, (1880-1947)
GORDON, Charles Henry, (1857-1934)
GOREAU, Thomas Fritz, (1924-1970)
GOSSLING, Frank Newbery, (1888-1946)
GRANLUND, Erik, (1892-1938)
GREEN, John Frederick Norman, (?1873-1949)
GREGORY, Herbert E., (1869-1952)
GREGORY, John Walter, (1864-1932)
GRESSWELL, Ronald Kay, (1905-1969)
GRIPP, Karl, (1891-)
GWYNNE, Charles Sumner, (1885-1972)
HANNESSON, Pálmi, (1898-1956)
HANSEN, Andreas M., (1857-1932)
HARKER, Alfred, (1859-1939)
HASSINGER, Hugo, (-1952)
HAYWARD, John, (-1958)
HEIM, Albert, (1849-1937)
HENDERSON, Junius, (1865-1937)
HENNIG, Edwin, (1882-)
HERNÁNDEZ-PACHECO, Francisco, (1899-1976)
HILLS, Thomas McDougall, (1881-1970)
HIND, Wheelton, (1859-1920)
HINDE, George Jennings, (1839-1918)
HINTON, Martin Alister Campbell, (1883-1961)
HO, Robert, (1921-1972)
HOBBS, William Herbert, (1864-1953)
HODGE, Edwin Thomas, (1887-1970)
HOLLINGWORTH, Sydney Ewart, (1899-1966)
HOLMES, Arthur, (1890-1965)
HOLMES, George William, (1922-1970)
HOLMSEN, Gunnar, (1880-1976)
HOOLEY, Reginald Walter, (1866-1923)

GEOMORPHOLOGY 1900 to present Cont'd.

HORBERG, Carl Leland, (1910-1955)
HORUSITZKY, Henrik, (1870-1945)
HOVEY, Edmund Otis, (1862-1924)
HOYT, John Harger, (1928-1970)
HUBBARD, George David, (1871-1958)
JACOBS, Elbridge Churchill, (1873-1957)
JAMIESON, Thomas Francis, (1829-1913)
JOHNSON, Douglas Wilson, (1878-1944)
JOHNSON, Walter, (1867-1950)
JOHNSTON, Mary Sophia, (1875-1955)
JONES, Owen Thomas, (1878-1967)
JONES, William Francis, (1885-1941)
JULIEN, Alexis Anastay, (1840-1919)
KATZER, Bedřich, (1861-1925)
KENDALL, Percy Fry, (1856-1936)
KING, Clarence Rivers, (1842-1901)
KING, William Bernard Robinson, (1899-1963)
KLEBELSBERG, Raimund von, (1886-1957)
KOMOROWICZ, Maurycy, (1881-1923)
KŘÍŽ, Martin, (1841-1916)
KUENEN, Phillip Henry, (1902-1976)
KULIK, Leonid Alekseevich, (1883-1942)
KUNSKÝ, Josef, (1903-)
LAFLAMME, Joseph Clovis Kemner, Monseigneur, (1849-1910)
LAKE, Philip, (1865-1949)
LANINO, Giuseppe, (1832-1907)
LAPPARENT, Albert Auguste de, (1839-1908)
LAWSON, Andrew Cowper, (1861-1952)
LEACH, Arthur Leonard, (1869-1957)
LeLACHEUR, W. J., (-1923)
LEMOINE, Paul, (1878-1940)
LENCEWICZ, Stanislaw, (1899-1944)
LEVERETT, Frank, (1859-1943)
LEWIS, William Vaughan, (1907-1961)
LICHTENECKER, Norbert, (1897-1938)
LINTON, David Leslie, (1906-1971)
LJUNGNER, Erik, (1892-1954)
LOBECK, Armin Kahl, (1886-1958)
LORY, Pierre, (1866-1956)
LOZINSKI, Walery, (1880-1944)
LUBBOCK, John, Lord Avebury, (1834-1913)
LUGEON, Maurice, (1870-1953)
LYELL, Charles, Sir, (1797-1875)
LYONS, Henry George, (1864-?1943)
MacCLINTOCK, Paul, (1891-1970)
MACHATSCHEK, Fritz, (-1957)
MALLADRA, Alessandro, (1865-1945)
MALOTT, Clyde A., (-1950)
MARGERIE, Emmanuel Marie Pierre Martin Jacquin de, (1862-1953)
MARINELLI, Olinto, (1874-1926)
MARKHAM, Clements Robert, Sir, (1830-1916)
MARR, John Edward, (1857-1933)
MARSELL, Ray Everett, (1893-1971)
MARSH, Thomas William, (-1957)

GEOMORPHOLOGY 1900 to present Cont'd.

MARTIN, Edgar Charles, (1884-1955)
MARTIN, Edward Alfred, (1864-1943)
MARTONNE, Emmanuel Louis Eugène de, (1873-1955)
MARWICK, John, (1891-1978)
MASON, Arnold Caverly, (1906-1961)
MATHER, Kirtley Fletcher, (1888-1978)
MATTHES, François Emile, (1874-1948)
MELLO, John Magens, Reverend, (1836-1914)
MELTON, Frank A., (-)
MERCIAI, Giuseppe, (1877-1939)
MERRILL, George Perkins, (1854-1929)
MEYER, Hans, (1858-1929)
MILL, Hugh Robert, (1861-1950)
MILLER, Arthur Austin, (1900-1968)
MILLER, Don John, (1919-1961)
MILLER, John Preston, (1923-1961)
MILLER, William John, (1880-1965)
MILLS, Enos Abijah, (1870-1922)
MILNE, John, (1850-1913)
MILTHERS, Keld, (1907-1960)
MOLYNEUX, Arthur John Charles, (?1866-1921)
MORRIS, Frederick Kuhne, (1885-1962)
MORSE, William Clifford, (1874-1962)
MOSCHELESOVÁ, Julie, (1892-1956)
MOTTRAM, Brian Hunter, (1922-1976)
MUIR, John, (1838-1914)
NAUMANN, Einar Christian Leonard, (1891-1934)
NECHAY, Wiktor, (1895-1940)
NEWELL ARBER, Edward Alexander, (1870-1918)
NILES, William Harmon, (1838-1910)
NØRVANG, Aksel, (1914-1968)
NOWACK, Ernst, (1891-1946)
OBRUCHEV, Vladimir Afanas'evich, (1863-1956)
OGILVIE, Francis Grant, Sir, (1858-1930)
OSMAN, Charles William, (1855-1932)
ÖSTER, Johannes, (1902-1975)
OVERY, Charles, Reverend, (1882-)
OWENS, Eifion Wyn, (1951-1975)
PALMER, L. S., (1890-1962)
PAPANIN, Ivan Dmitrievich "Vanya", (1895-)
PARDÉ, Maurice, (1893-1973)
PATTON, Leroy Thompson, (1880-1957)
PAWLOWSKI, Stanislaw, (1882-1940)
PAYER, Julius von, Ritter, (1841-1912)
PEATTIE, Roderick, (1891-)
PENCK, Friedrich Carl Albrecht, (1858-1945)
PENCK, Walther, (1888-1923)
PENNING, William Henry, (1838-1902)
PFANNENSTIEL, Max, (1902-1976)
PIRSSON, Louis Valentine, (1860-1919)
PITTIER, Henry François, (1857-1950)
POLANSKI, Jorge, (1892-1975)
POLJAK, Josip, (1882-1962)
POWELL, John Wesley, (1834-1902)

GEOMORPHOLOGY 1900 to present Cont'd.

PRIESTLEY, Raymond Edward, Sir, (1886-1974)
PURDUE, Albert Homer, (1861-1917)
PURKYNĚ, Cyril, (1862-1937)
PUTNAM, William Clomont, (1908-1963)
RAKOVEC, Ivan, (1899-)
RAMSAY, Wilhelm, (1865-1928)
RAULIN, Félix Victor, (1815-1905)
READE, Thomas Mellard, (1832-1909)
REED, Frederick Richard Cowper, (1869-1946)
REICHERT, Friedrich, (1878-)
RENAULT, Philippe, (-)
RHEDEN, J. J. Pannekoek van, (-1945)
RICHTHOFEN, Paul Wilhelm Ferdinand von, Freiherr, (1833-1905)
ROMER, Eugeniusz, (1871-1954)
RUBEY, William Walden, (1898-1974)
RUDMOSE BROWN, Robert Neal, (1879-1957)
RUSSELL, Richard Joel, (1895-1971)
RUTKOWSKI, Feliks, (1878-1964)
RUTTEN, Martin Gerard, (1910-1970)
SALISBURY, Rollin D., (1858-1922)
SANDFORD, Kenneth Stuart, (1899-1971)
SCHOEWE, Walter Henry, (1891-1966)
SCHOFIELD, Eric, (1911-1944)
SCHRADER, François Jean Daniel "Franz", (1844-1924)
SCHWARZ, Ernest Hubert Lewis, (1873-1928)
SCOTT, James Frew, (1902-1956)
SCOTT, William Berryman, (1858-1947)
SEARS, Julian Ducker, (1891-1970)
SEDERHOLM, Jakob Johannes, (1863-1934)
SELLARDS, Elias Howard, (1875-1961)
SENN, Alfred, (1899-1949)
SHATTUCK, George Burbank, (1869-)
SHAW, Eugene Wesley, (1881-1935)
SHEPPS, Vincent Chester, (1928-1967)
SHORE, Thomas William, (1840-1905)
SHULER, Ellis William, (1881-1954)
SLATER, George, (-1956)
SMITH, Bernard, (1881-1936)
SMITH, George Otis, (1871-1944)
SMITH, Reginald Allender, (?1847-1940)
SMITH, William Richard Tangier, (1869-1972)
SMIT SIBINGA, Gerard Leonard, (1895-1963)
SMOLEŃSKI, Jerzy, (1881-1940)
SOLLAS, William Johnson, (1849-1936)
SPREITZER, Hans, (1897-1973)
STIFFE, Arthur William, Captain, (1849-1912)
STINY, Josef, (1880-1958)
STIRRUP, Mark, (1831-1907)
STONELEY, Robert, (1894-1976)
STRACHEY, Richard, General Sir, (1817-1908)
STRØM, Kaare, (1902-1967)
SUESS, Eduard, (1831-1914)
SWINNERTON, Henry Hurd, (1875-1967)
SZABÓ, Pál Zóltán, (1901-1965)

GEOMORPHOLOGY 1900 to present Cont'd.

TARR, Ralph Stockman, (1864-1912)
TEISSEYRE, Henryk, (1903-1975)
TESCH, P., (1879-1961)
THOMAS, Trevor Morgan, (?1919-1974)
THOMPSON, Hugh R., (1926-1959)
THOMPSON, Warren O., (1898-1968)
TIGHT, William George, (1865-1910)
TRAINER, David Woolsey, Jr., (1899-1965)
TROLL, Carl, (1899-)
TROTTER, Frederick Murray, (1897-1968)
UMBGROVE, Johannes Herman Frederik, (1899-1954)
UPHAM, Warren, (1850-1934)
VACHTL, Josef, (1907-)
VALPY, Robert Harris, (1819-1904)
VÂLSAN, George, (1885-1935)
VISSER, Philips Christiaan, (1882-1955)
VON ENGELN, Oskar Dietrich, (1880-1965)
WAITZ, Paul, (1876-1961)
WALTHER, Johannes, (1860-1937)
WENTWORTH, Chester Keeler, (1891-1969)
WHITE, George Willard, (1903-)
WHYMPER, Edward, (1840-1911)
WICHE, Konrad, (1913-1969)
WILKS, Mark, (?1862-1945)
WILLARD, Daniel Everett, (1862-1947)
WILLIAMS, Ira Abraham, (1876-)
WOEIKOF, Alexander, (1842-1916)
WOODWORTH, Jay Backus, (1865-1925)
WOOLACOTT, David, (1872-1924)
WOOLDRIDGE, Sidney William, (1900-1963)
WORCESTER, Philip George, (1884-1970)
WORSFORD, Frederick H., (1870-1954)
WRIGHT, Frank James, (1888-1954)
YABE, Hisakatsu, (1878-1969)
YAMASAKI, Naomasa, (1870-1929)
YOUNG, Robert, Right Honourable, (1822-1917)
ZANS, Verners Aleksandrs, (1904-1961)
ZIMÁNYI, Károly, (1862-1941)
ZITTEL, Karl Alfred von, (1839-1904)

GEOMORPHOLOGY See also GLACIOLOGY, SPELAEOLOGY, QUATERNARY GEOLOGY, VULCANOLOGY

GEOPHYSICS

Up to 1800

BOUGUER, Pierre, (1698-1758)
CAVENDISH, Henry, (1731-1810)
GAUSS, Carl Friedrich, (1777-1855)
GILBERT, William, (1544-1603)
HOOKE, Robert, (1635-1703)
KEPLER, Johannes, (1571-1630)
KIELMEYER, Carl Heinrich, (1765-1844)
LA CONDAMINE, Charles-Marie de, (1701-1774)
MALTHUS, Thomas Robert, (1766-1834)

GEOPHYSICS Up to 1800 Cont'd.

WEBER, Wilhelm, (1804-1891)

1800-1849

BACHE, Alexander Dallas, (1806-1867)
CAVENDISH, Henry, (1731-1810)
FERREL, William, (1817-1891)
FOUCAULT, Jean Bernard Léon, (1819-1868)
GAUSS, Carl Friedrich, (1777-1855)
HAMILTON, Sir William R., (1805-1865)
HERSCHEL, John Frederick William, Sir, (1792-1871)
KELVIN, William Thomson, Baron, (1824-1907)
KIELMEYER, Carl Heinrich, (1765-1844)
LOCKE, John, (1792-1856)
LOOMIS, Elias, (1811-1890)
MALTHUS, Thomas Robert, (1766-1834)
NICOLLET, Joseph Nicolas, (1786-1843)
SAINTE-CLAIRE DEVILLE, Charles, (1814-1876)
WEBER, Wilhelm, (1804-1891)

1850-1899

ARCTOWSKI, Henryk Bronislaw, (1871-1958)
ARRHENIUS, Svante August, (1859-1927)
BACHE, Alexander Dallas, (1806-1867)
BECKER, George Ferdinand, (1847-1919)
BJERKNES, Vilhelm, (1862-1951)
BONETTI, Filippo, (1854-1911)
CARLHEIM-GYLLENSKÖLD, Vilhelm, (1859-1934)
CARVALHO, Anselmo Ferraz de, (1878-1955)
CHREE, Charles, (-)
DARWIN, George Howard, Sir, (1845-1912)
DAY, Arthur Louis, (1869-1960)
EÖTVÖS, Lorand, (1848-1919)
EVANS, John William, (1857-1930)
FENNER, Clarence Norman, (1870-1949)
FERREL, William, (1817-1891)
FLEMING, John Adam, (1877-1956)
FOUCAULT, Jean Bernard Léon, (1819-1868)
GAUSS, Carl Friedrich, (1777-1855)
HAMILTON, Sir William R., (1805-1865)
HASEMAN, William Peter, (1878-1932)
HERSCHEL, John Frederick William, Sir, (1792-1871)
JOLY, John, (1857-1933)
KELVIN, William Thomson, Baron, (1824-1907)
KOENIGSBERGER, Johann Georg, (1874-1946)
KUSAKABE, Shiroka, (1872-1924)
LENOX-CONYNGHAM, Gerald Ponsonby, Colonel Sir, (1866-1956)
LOOMIS, Elias, (1811-1890)
LOWELL, Percival, (1855-1916)
MENDENHALL, Thomas Corwin, (1841-1924)
MERCANTON, Paul Louis, (1876-1963)
OTANI, Ryokichi, (1875-1934)
PLATANIA, Gaetano, (1867-1938)
POCKELS, Friedrich Carl Alwin, (1865-1913)
REBEUR-PASCHWITZ, Ernst von, (1861-1895)

GEOPHYSICS 1850-1899 Cont'd.

REYER, Eduard, (1849-1914)
RODOLPH, Emil, (1853-1915)
RUDOLPH, Emil, (1853-1915)
RUDZKI, Maurycy Pius, (-)
SAINTE-CLAIRE DÉVILLE, Charles, (1814-1876)
SCHERING, Ernst, (1833-1897)
SCHLUMBERGER, Conrad, (1878-1936)
SCHWINNER, Robert Gangolf, (1878-1953)
SHIDA, Toshi, (1876-1936)
SPRING, Walthère, (-)
TERADA, Torahiko, (1878-1935)
WHITE, Charles David, (1862-1935)
WOODWARD, Robert Simpson, (1849-1924)

1900 to present

ADKINS, John Nathaniel, (1911-)
ALDREDGE, Robert Frank, (1895-1961)
ANDRAU, Evert Willem Karel, (1900-1951)
ARCTOWSKI, Henryk Bronislaw, (1871-1958)
ARRHENIUS, Svante August, (1859-1927)
BALLOU, Albert Lorenzo, Jr., (1917-1962)
BARTH, Thomas Frederik Weiby, (1899-1971)
BASSINGER, William E., (1928-1977)
BATES, Robert Glenn, (1923-1972)
BECKER, George Ferdinand, (1847-1919)
BĚHOUNEK, Rudolf, (1902-1974)
BITTER, Francis, (1902-1967)
BJERKNES, Vilhelm, (1862-1951)
BLACKETT, Patrick Maynard Stuart, Baron of Chelsea, (1897-1974)
BOGARDUS, Robert Stephen "Bud", (1914-1953)
BOND, Lewis Alexander, (1896-1972)
BONETTI, Filippo, (1854-1911)
BOOTS, Paul Heath, (1909-1947)
BRAMWYCH, Michael, (1911-1974)
BREMER, B. E. "Boots", (1904-1974)
BROGGI, Jorge Albert, (1888-1966)
BROUGHTON-EDGE, Arthur Broughton, (1895-1953)
BROWN, James Macmillan, (1921-1970)
BULLERWELL, William, (1916-1977)
BURGWAL, Rudolph Frans, (1917-1944)
CAIN, James Walter, Jr., (1917-1941)
CANADA, William Richards, (1909-1962)
CANTWELL, Thomas, Jr., (1927-)
CARLHEIM-GYLLENSKÖLD, Vilhelm, (1859-1934)
CARLTON, Dave Pierce, (1890-1976)
CARR, George Watson, (1899-1950)
CARVALHO, Anselmo Ferraz de, (1878-1955)
CHAPMAN, Sydney, (1888-1970)
CHREE, Charles, (-)
COFFIN, Reuben Clare, (1886-1972)
CORTES, Henry Cornelius, (1892-1957)
COTTON, Leo Arthur, (1883-1963)
CUNNINGHAM, George Magee, (1895-1963)
DARWIN, George Howard, Sir, (1845-1912)

GEOPHYSICS 1900 to present Cont'd.

DAVIE, Thomas Early, (1906-1959)
DAY, Arthur Louis, (1869-1960)
DEAN, Lee Wallace, III, (1932-)
DeGOLYER, Everette Lee, (1886-1956)
DOELL, Richard Rayman, (1923-)
DRATH, Adam, (1904-1942)
DURHAM, Charles Albert, (1910-1954)
DVOŘÁK, Arnošt, (1904-)
EDWARDS, Merwin Guy, (1889-1969)
EÖTVÖS, Lorand, (1848-1919)
ERTEL, Hans, (1904-1971)
EVANS, John William, (1857-1930)
EVANS, Percy, (1892-1974)
EWING, William Maurice, (1906-1974)
FAULKNER, Joseph Stanislaus "Ginger", (1928-1963)
FENNER, Clarence Norman, (1870-1949)
FERGUSON, John Logan, (1893-1973)
FIELD, Richard Montgomery, (1885-1961)
FIOCCO, Giorgio, (1931-)
FITZGERALD, Thomas John, (1903-1973)
FLEMING, John Adam, (1877-1956)
FOSSHAGE, Ernest Willard, (1908-1955)
FOSTER, Margaret Dorothy, (1895-1970)
FREY, M. Gordon, (1913-)
GALLIE, John Francis, III, (1913-1951)
GANGI, Anthony Frank, (1929-)
GARRETT, Lovic Pierce, (1880-1943)
GENTNER, Wolfgang, (1906-)
GOGUEL, Jean, (1908-)
GOLDEN, John Muro, (1900-1945)
GRAHAM, John Warren, (1918-1971)
GUILLOU, Robert Barton, (1923-1965)
GUTENBERG, Beno, (1889-1960)
HALL, Thomas O., (1906-1972)
HALLEY, Edmond, (1656-1742)
HALLIMOND, Arthur Francis, (1890-1968)
HANDLEY, Howard Walter, (1905-1940)
HASEMAN, William Peter, (1878-1932)
HECK, Nicholas Hunter, (1882-1953)
HEILAND, Carl August, (1899-1956)
HENSON, Frederick Arthur, (1919-1957)
HESS, Harry Hammond, (1906-1969)
HIDE, Raymond, (1929-)
HILL, Maurice Neville, (1919-1966)
HILLER, Wilhelm, (1899-)
HLAUSCHEK, Hans, (1896-1971)
HOCHSTETTER, Harald Egbert von, Ritter, (1902-1972)
HOINKES, Herfried, (1916-1973)
HOLMES, Arthur, (1890-1965)
HUBBARD, Bela, (1890-1959)
HUGHES, Harry, (1929-)
HURLEY, Patrick Mason, (1912-)
JANCZEWSKI, Edward Walery, (1887-1959)
JOESTING, Henry Rochambeau, (1903-1965)

GEOPHYSICS 1900 to present Cont'd.

JOHNSON, Earle Winston, (1902-1953)
JOHNSON, W. Ray, (-1952)
JOLY, John, (1857-1933)
JUNG, Karl, (-?1972)
KAUENHOWEN, Walter, (1900-1942)
KELVIN, William Thomson, Baron, (1824-1907)
KIHLSTEDT, Folke H., (1901?-1956)
KISLING, James William, Jr., (1898-1973)
KOENIGSBERGER, Johann Georg, (1874-1946)
KRISHNAN, Maharajapuram Sitaram, Padma Bhusan, (1898-1970)
KROENLEIN, George Alfred, (1898-1940)
KUSAKABE, Shiroka, (1872-1924)
LADNER, Albert L., (1905-1975)
LAUTERBACH, Robert, (1915-)
LEDERSTEGER, Karl, (-1973)
LENOX-CONYNGHAM, Gerald Ponsonby, Colonel Sir, (1866-1956)
LJUNGDAHL, Gustaf Samuel, (1882-1940)
LOWELL, Percival, (1855-1916)
MACDONALD, Gordon James Fraser, (1929-)
MACELWANE, James Bernard, Father, S.J., (1883-1956)
MADDEN, Theodore Richard, (1925-)
MARSDEN, Ernest, (1889-1970)
MARTI, Ernest John, (1913-1964)
MAYFIELD, William I., (1912-1972)
McDOUGALL, John Edward, (1914-1964)
McLAREN, Robert Lee, (1901-1953)
MEKEL, Johannes Antonius Alphonsus, (1891-1942)
MENDENHALL, Thomas Corwin, (1841-1924)
MENKEN, Frederick Andrew, (1905-1965)
MERCANTON, Paul Louis, (1876-1963)
MERRITT, John Wesley, (1885-1958)
MONNETT, Victor Elvert, (1889-1972)
MOORE, James Augustus, (1910-1955)
MURCHISON, Eugene Alexander, III, (1931-1953)
MURRAY, George H., Jr., (1928-1971)
NAGATA, Takesi, (1913-)
NEUMAN, Leonard Joseph, (1900-1953)
NORDEN, John Alexander Erdelyi-Fazekas, (né Erdélyi-Fazekas), (1912-1977)
O'BRIEN, Francis Eugene, (1916-1968)
OBUCHOWICZ, Zbigniew Zygmunt, (1914-1965)
OTANI, Ryokichi, (1875-1934)
PALMER, Leland L., (1906-1978)
PETERSON, Arthur Frederick, (1911-1967)
PETTY, Dabney E., (1891-1963)
PETTY, Olive Scott, (1895-)
PLATANIA, Gaetano, (1867-1938)
POCKELS, Friedrich Carl Alwin, (1865-1913)
PRATLEY, Henry Hart, (1893-1941)
PRESS, Frank, (1924-)
RAY, Robert Hillyer, (1902-1968)
REYER, Eduard, (1849-1914)
RIEBER, Frank, (1891-1948)
RODOLPH, Emil, (1853-1915)

GEOPHYSICS 1900 to present Cont'd.

RUDOLPH, Emil, (1853-1915)
RUDZKI, Maurycy Pius, (-)
RUTTEN, Martin Gerard, (1910-1970)
SAPPINGTON, Chester, (1905-1968)
SAVILLE, Wilson Gordon, (1897-1954)
SCHAIRER, John Frank, (1904-1970)
SCHEFFER, Viktor, (1900-1966)
SCHLUMBERGER, Conrad, (1878-1936)
SCHLUMBERGER, Marcel, (1884-1953)
SCHWINNER, Robert Gangolf, (1878-1953)
SELGA, Miguel, Reverend Father, S. J., (1879-1956)
SHIDA, Toshi, (1876-1936)
SHROCK, Robert Rakes, (1904-)
SIMMONS, Marvin Gene, (1929-)
SIMPSON, Stephen Milton, Jr., (1929-)
SLICHTER, Louis Byrne, (1896-)
SMITH, Aylwin L., (1898-1965)
SMITH, Neal Johnstone, (1912-1971)
SMOLUCHOWSKI, Marian, (-)
STENZ, Edward, (1897-1956)
STRANGWAY, David William, (1934-)
SUNDBERG, Karl, (1891-1938)
SWARTZ, Joel Howard, (1893-1971)
TAYLOR, Garvin Lawrence, (1908-1974)
TAYLOR, W. Harlan, (1908-1972)
TEMPLETON, James Clark, (1898-1950)
TERADA, Torahiko, (1878-1935)
THIEL, Edward, (1928-1962)
THIES, Roland K., (1905-1973)
THORSEN, Carl Peter, (1927-1964)
TOKSÖZ, M. Nafi, (1934-)
VENING MEINESZ, Felix Andries, (1887-1966)
VERCELLI, Francesco, (-?1952)
VON ARX, William Stelling, (1916-)
WARING, Waldo W., (1903-1955)
WEAVER, Paul, (1888-1964)
WEGENER, Alfred Lothar, (1880-1930)
WEIGELT, Johannes, (1890-1948)
WHITCOMB, Bruce, (1900-1950)
WHITE, Charles David, (1862-1935)
WIER, Harvey James, Jr., (1924-1950)
WILLIAMSON, Erskine Douglas, (1886-1923)
WILSON, John Tuzo, (1908-)
WINGETT, Raymond Oscar, (1909-1953)
WOODWARD, Robert Simpson, (1849-1924)
ZIJLSTRA, Gerrit, (-1962)
ZIMERMAN, Sam, (1905-1972)
ZIMMERMAN, Charles Collingwood, (1905-1961)

GEOPHYSICS <u>See also</u> SEISMOLOGY

GLACIOLOGY
Up to 1800
BESSON, Henri, (fl.1786)

GLACIOLOGY Up to 1800 Cont'd.

GRUNER, Gottlieb Sigmund, (1717-1778)
KUHN, Bernhard Friedrich, (1762-1825)
MARTEL, Peter, (-)
PERRAUDIN, Jean-Pierre, (1767-1858)
SAUSSURE, Horace Bénédict de, (1740-1799)
THOMAS, David, (1776-1859)

1800-1849
AGASSIZ, Louis Jean Rodolphe, (1807-1873)
ARGYLL, George Douglas, Duke of, (1823-1900)
BELL, Dugald, (1827-1892)
CAMPBELL, John Francis, (1821-1885)
CROLL, James, (1821-1890)
CUMMING, Joseph G., Reverend, (1812-1868)
DUFOUR, Charles, (1827-1902)
FAVRE, Jean Alphonse, (1815-1890)
FORBES, James David, (1809-1868)
HITCHCOCK, Edward, (1793-1864)
HONEYMAN, David, Reverend, (1814-1889)
KANE, Elisha Kent, (1820-1857)
KUHN, Bernhard Friedrich, (1762-1825)
MORTON, George Highfield, (1826-1900)
MURRAY, Alexander, (1810-1884)
PERRAUDIN, Jean-Pierre, (1767-1858)
RÜTIMEYER, Karl Ludwig, (1825-1895)
SCHLAGINTWEIT, Adolf, (?1799-1857)
SCHLAGINTWEIT, Hermann, (-)
SCHLAGINTWEIT, Robert, (-)
SIMONY, Friedrich, (1813-1896)
STRACHEY, Richard, General Sir, (1817-1908)
THOMAS, David, (1776-1859)
THOMSON, John Turnbull, (1821-1884)
TREVELYAN, Walter Calverley, Sir, (1797-1879)
TYNDALL, John, (1820-1893)
VENETZ, Ignaz, (1788-1859)
ZITTEL, Karl Alfred von, (1839-1904)

1850-1899
AGASSIZ, Louis Jean Rodolphe, (1807-1873)
ARGYLL, George Douglas, Duke of, (1823-1900)
BARTON, George Hunt, (1852-1933)
BEASLEY, Henry Charles, (1836-1919)
BELL, Dugald, (1827-1892)
BEMROSE, Henry Howe, (1857-1939)
BLAKE, John Frederick, Reverend, (1839-1906)
BLÜMCKE, Adolf, (1857-1914)
BREMNER, Alexander, (1863-1943)
BRÜCKNER, Eduard, (1862-1927)
BURROUGHS, John, (1837-1921)
CAMPBELL, John Francis, (1821-1885)
CHAMBERLIN, Thomas Chrowder, (1843-1928)
CHRISTEN, Sydney Mary, (née THOMPSON), (-1923)
CHURCH, James Edward, (1869-1959)
CODRINGTON, Thomas, (1829-1918)

GLACIOLOGY 1850-1899 Cont'd.

COTTEAU, Gustave Honoré, (1818-1894)
CROLL, James, (1821-1890)
CUMMING, Joseph G., Reverend, (1812-1868)
DALY, Reginald Aldworth, (1871-1957)
DAVIS, William Morris, (1850-1934)
DEECKE, Johannes Ernst Wilhelm, (1862-1934)
DORSEY, Noah Ernest, (1873-1959)
DUFOUR, Charles, (1827-1902)
DU PASQUIER, Léon, (1864-1897)
DWERRYHOUSE, Arthur Richard, (1867-1931)
FAVRE, Jean Alphonse, (1815-1890)
FINSTERWALDER, Sebastian, (1862-?1931)
FORBES, James David, (1809-1868)
FOREL, François Alphonse, (1841-1912)
GAURIER, Ludovic Charles, Abbé, (1874-1931)
GILBERT, Grove Karl, (1843-1918)
GOODCHILD, John George, (1844-1906)
GRANT, Ulysses Sherman, (1867-1932)
GREGORY, John Walter, (1864-1932)
HAGENBACH-BISCHOFF, Eduard, (1833-1910)
HALL, Marshall, Captain, (1831-1896)
HARKER, Alfred, (1859-1939)
HARMER, Frederic William, (1835-1923)
HEIM, Albert, (1849-1937)
HESS, Hans, (1864-1940)
HINDE, George Jennings, (1839-1918)
HITCHCOCK, Edward, (1793-1864)
HOBBS, William Herbert, (1864-1953)
HONEYMAN, David, Reverend, (1814-1889)
JACOB, Charles François Étienne, (1878-1962)
JULIEN, Alexis Anastay, (1840-1919)
KANE, Elisha Kent, (1820-1857)
LEVERETT, Frank, (1859-1943)
LEWIS, Henry Carvill, (1853-1888)
LOMAS, Joseph, (1860-1908)
LORY, Pierre, (1866-1956)
MACBRIDE, Thomas Huston, (1848-1934)
MARINELLI, Olinto, (1874-1926)
MARTONNE, Emmanuel Louis Eugène de, (1873-1955)
MATTHES, François Emile, (1874-1948)
MATTHEW, George Frederic, (1837-1923)
McCONNEL, J. C., (1860-1890)
MERCANTON, Paul Louis, (1876-1963)
MEYER, Hans, (1858-1929)
MONTAG, Emil, (1875-1960)
MORTON, George Highfield, (1826-1900)
MUIR, John, (1838-1914)
MURRAY, Alexander, (1810-1884)
NILES, William Harmon, (1838-1910)
OBRUCHEV, Vladimir Afanas'evich, (1863-1956)
OGILVIE, Ida Helen, (1874-1963)
PAYER, Julius von, Ritter, (1841-1912)
PENCK, Friedrich Carl Albrecht, (1858-1945)
PERRAUDIN, Jean-Pierre, (1767-1858)

GLACIOLOGY 1850-1899 Cont'd.

RÜTIMEYER, Karl Ludwig, (1825-1895)
SACCO, Federico, (1864-1948)
SCHLAGINTWEIT, Hermann, (-)
SCHLAGINTWEIT, Robert, (-)
SCHRADER, François Jean Daniel "Franz", (1844-1924)
SOMIGLIANA, Carlo, (?1859-1955)
STRACHEY, Richard, General Sir, (1817-1908)
STREIFF-BECKER, Rudolf, (1873-1959)
TARR, Ralph Stockman, (1864-1912)
TAYLER, Joseph Walter, (-?1896)
TCHERNYCHEV, Féodoss Nikolaievich, (1856-1914)
THOMAS, David, (1776-1859)
THOMSON, John Turnbull, (1821-1884)
TREVELYAN, Walter Calverley, Sir, (1797-1879)
TYNDALL, John, (1820-1893)
VAUX, George, Jr., (1863-1927)
VAUX, William S., Jr., (1872-1908)
VENETZ, Ignaz, (1788-1859)
VOGT, Johan Herman Lie, (1858-1932)
WOEIKOF, Alexander, (1842-1916)
WRIGHT, George Frederick, (1838-1921)
YAMASAKI, Naomasa, (1870-1929)
ZITTEL, Karl Alfred von, (1839-1904)

1900 to present

BARTON, George Hunt, (1852-1933)
BATTLE, Walter Ravenhill Brown "Ben", (1919-1953)
BEASLEY, Henry Charles, (1836-1919)
BEMROSE, Henry Howe, (1857-1939)
BENNETT, Walter Hellyer, (1892-1971)
BLAKE, John Frederick, Reverend, (1839-1906)
BLÜMCKE, Adolf, (1857-1914)
BREMNER, Alexander, (1863-1943)
BRÜCKNER, Eduard, (1862-1927)
BURROUGHS, John, (1837-1921)
CARRUTHERS, Robert George, (1880-1965)
CAYEUX DE SENARPONT, André, called CAILLEUX, André, (1907-)
CHAMBERLIN, Thomas Chrowder, (1843-1928)
CHARLESWORTH, John Kaye, (1889-)
CHRISTEN, Sydney Mary, (née THOMPSON), (-1923)
CHURCH, James Edward, (1869-1959)
CODRINGTON, Thomas, (1829-1918)
COMBAZ, Paul, Chanoine, (-1930)
DALY, Reginald Aldworth, (1871-1957)
DAVIS, William Morris, (1850-1934)
DEECKE, Johannes Ernst Wilhelm, (1862-1934)
DORSEY, Noah Ernest, (1873-1959)
DUFOUR, Charles, (1827-1902)
DWERRYHOUSE, Arthur Richard, (1867-1931)
EDMONDS, Frederic Harrison, (1898-1965)
EKBLAW, George Elbert, (1895-1972)
EYTHORSSON, Jon, (1895-1968)
FINSTERWALDER, Richard, (1899-1963)
FINSTERWALDER, Sebastian, (1862-?1931)

GLACIOLOGY 1900 to present Cont'd.

FOREL, François Alphonse, (1841-1912)
FUCHS, Vivian Ernest, Sir, (1908-)
GAURIER, Ludovic Charles, Abbé, (1874-1931)
GILBERT, Grove Karl, (1843-1918)
GOODCHILD, John George, (1844-1906)
GRANT, Ulysses Sherman, (1867-1932)
GREGORY, John Walter, (1864-1932)
GRESSWELL, Ronald Kay, (1905-1969)
GRIPP, Karl, (1891-)
HAGENBACH-BISCHOFF, Eduard, (1833-1910)
HARKER, Alfred, (1859-1939)
HARMER, Frederic William, (1835-1923)
HEIM, Albert, (1849-1937)
HEINSHEIMER, Georg J., (1887-1970)
HESS, Hans, (1864-1940)
HINDE, George Jennings, (1839-1918)
HOBBS, William Herbert, (1864-1953)
HOINKES, Herfried, (1916-1973)
HOLMES, George William, (1922-1970)
HUBLEY, Richard Carleton, (1926-1957)
JACOB, Charles François Étienne, (1878-1962)
JULIEN, Alexis Anastay, (1840-1919)
KLEBELSBERG, Raimund von, (1886-1957)
KOCH, Lauge, (1892-1964)
LEIGHTON, Morris Morgan, (1887-1971)
LEVERETT, Frank, (1859-1943)
LEWIS, William Vaughan, (1907-1961)
LOEWE, Fritz, (1895-1974)
LOMAS, Joseph, (1860-1908)
LORY, Pierre, (1866-1956)
MACBRIDE, Thomas Huston, (1848-1934)
MACHATSCHEK, Fritz, (-1957)
MARINELLI, Olinto, (1874-1926)
MARTONNE, Emmanuel Louis Eugène de, (1873-1955)
MATTHES, François Emile, (1874-1948)
MATTHEW, George Frederic, (1837-1923)
McCALL, John Gill, (?1923-1954)
MERCANTON, Paul Louis, (1876-1963)
MEYER, Hans, (1858-1929)
MILLER, Arthur Austin, (1900-1968)
MILLER, Don John, (1919-1961)
MONTAG, Emil, (1875-1960)
MORTON, George Highfield, (1826-1900)
MUIR, John, (1838-1914)
NAKAYA, Ukichiro, (1900-1962)
NILES, William Harmon, (1838-1910)
OBRUCHEV, Vladimir Afanas'evich, (1863-1956)
ŌCURA, Hirobumi, (1921-1969)
OGILVIE, Ida Helen, (1874-1963)
ÖSTER, Johannes, (1902-1975)
PAPANIN, Ivan Dmitrievich "Vanya", (1895-)
PAWLOWSKI, Stanislaw, (1882-1940)
PAYER, Julius von, Ritter, (1841-1912)
PENCK, Friedrich Carl Albrecht, (1858-1945)

GLACIOLOGY 1900 to present Cont'd.

PREMIK, Józef, (1890-1963)
PUTNAM, William Clement, (1908-1963)
RATHSBURG, Alfred, (-)
REECE, Alan W., (1921-1960)
RENAUD, André, (1904-1964)
SACCO, Federico, (1864-1948)
SCHRADER, François Jean Daniel "Franz", (1844-1924)
SCHYTT, Valter, (1919-)
SELIGMAN, Gerald, (1886-1973)
SHODA, Mikio, (1924-1974)
SLATER, George, (-1956)
SMITH, Bernard, (1881-1936)
SMITH, Jeremy, (1929-1959)
SMIT SIBINGA, Gerard Leonard, (1895-1963)
SOMIGLIANA, Carlo, (?1859-1955)
STRACHEY, Richard, General Sir, (1817-1908)
STREIFF-BECKER, Rudolf, (1873-1959)
TARR, Ralph Stockman, (1864-1912)
TCHERNYCHEV, Féodoss Nikolaievich, (1856-1914)
THOMPSON, Hugh R., (1926-1959)
TODTMANN, Emmy Mercedes, (1888-1973)
TRENER, Giovanni Battista, (-?1954)
TYRRELL, George Walter, (1883-1961)
VAUX, George, Jr., (1863-1927)
VAUX, William S., Jr., (1872-1908)
VISSER, Philips Christiaan, (1882-1955)
VOGT, Johan Herman Lie, (1858-1932)
WALCOTT, Mary Vaux, née Vaux, (-1940)
WALKER, Paul, (1934-1959)
WEINBERGER, Ludwig, (1911-1966)
WEXLER, Harry, (1911-1962)
WICHE, Konrad, (1913-1969)
WIGGLESWORTH, Edward, (1885-1945)
WOEIKOF, Alexander, (1842-1916)
WRIGHT, George Frederick, (1838-1921)
YAMASAKI, Naomasa, (1870-1929)
ZITTEL, Karl Alfred von, (1839-1904)

HISTORY OF GEOLOGY

Up to 1800

CLINTON, DeWitt, (1769-1828)
CUVIER, Jean Léopold Nicolas Frédéric, (1769-1832)
LUCRETIUS CARUS, Titus, (98-55 B.C.)
MILLER, Samuel, (1769-1850)

1800-1849

ARCHIAC, Étienne Jules Adolphe Dexmier de Simon, Vicomte d', (1802-1868)
BUCKLAND, Francis Trevelyan, (1826-1880)
CAMPANI, Giovanni, (1820-1891)
CAUCHY, F. P., (-?1842)
CLINTON, DeWitt, (1769-1828)
CUVIER, Jean Léopold Nicolas Frédéric, (1769-1832)
DEWALQUE, Gilles Joseph Gustave, (1828-1905)
DRAPER, John William, (1811-1882)
GAUDRY, Jean Albert, (1827-1908)
MACGILLIVRAY, William, (1796-1852)
MILLER, Samuel, (1769-1850)
RENSSELAER, Jeremiah van, (1793-1871)
SCHWIPPEL, Karl, (1821-1911)
ZITTEL, Karl Alfred von, (1839-1904)

1850-1899

ADAMS, Frank Dawson, (1859-1942)
ARCHIAC, Étienne Jules Adolphe Dexmier de Simon, Vicomte d', (1802-1868)
BERKEY, Charles Peter, (1867-1955)
BOLTON, Sarah Elizabeth Mary (née Knowles), (1841-1916)
BUCKLAND, Francis Trevelyan, (1826-1880)
CAMPANI, Giovanni, (1820-1891)
CERMENATI, Mario, (1868-1924)
CLARKE, John Mason, (1857-1925)
CLÉMENT, Charles Adolphe Antoine, (1860-1933)
CLODD, Edward, (1840-1930)
CORSTORPHINE, George Steuart, (1865-1919)
CROOK, Thomas, (1876-1937)
DARWIN, Francis, (1848-1925)
DEWALQUE, Gilles Joseph Gustave, (1828-1905)
DRAPER, John William, (1811-1882)
FLETT, John Smith, Sir, (1869-1947)
GAUDRY, Jean Albert, (1827-1908)
GEIKIE, Archibald, Sir, (1835-1924)
GOODE, George Brown, (1851-1896)
HARTLEY, Harold Brewer, Sir, (1878-1972)
HUXLEY, Leonard, (1860-1933)
JUKES-BROWNE, Alfred John, (1851-1914)
JULIAN, Hester Forbes, Mrs., (1860-1934)
KEYES, Charles Rollin, (1864-1942)
LANG, William Dickson, (1878-1966)
MACGILLIVRAY, William, (1796-1852)
MARGERIE, Emmanuel Marie Pierre Martin Jacquin de, (1862-1953)
MAY, Walther, (1868-)
MERRILL, George Perkins, (1854-1929)
MIELEITNER, Karl, (1890-1923)

HISTORY OF GEOLOGY 1850-1899 Cont'd.

MILLER, Samuel, (1769-1850)
MITCHELL, Peter Chalmers, Sir, (1864-1945)
OSBORN, Henry Fairfield, (1857-1935)
RENSSELAER, Jeremiah van, (1793-1871)
ROGERS, Arthur William, (1872-1946)
SCHUCHERT, Charles, (1858-1942)
SCHWIPPEL, Karl, (1821-1911)
SPENCER, Leonard James, (1870-1959)
TRIBOLET-HARDY, Frédéric Maurice de, (1852-1929)
WATTS, William Whitehead, (1860-1947)
WICHMANN, Carl Ernst Arthur, (1851-1927)
WILLS, Robert Glover, (1878-1950)
WIŚNIOWSKI, Tadeusz, (1865-1933)
WOODWARD, Horace Bolingbroke, (1848-1914)
ZITTEL, Karl Alfred von, (1839-1904)

1900 to present

ADAMS, Frank Dawson, (1859-1942)
ALCOCK, Frederick James, (1888-1972)
ANDRADE, Edward Neville da Costa, (1887-1971)
BAILEY, Edward Battersby, Sir, (1881-1965)
BARSONOV, Georgyi Pavlovich, (1897-)
BERKEY, Charles Peter, (1867-1955)
BOLTON, Sarah Elizabeth Mary (née Knowles), (1841-1916)
BROMEHEAD, Cyril Edward Nowill, (1885-1952)
BROUGHTON, Derek George, (-1971)
BROWN, Bahngrell Walter "Bill", (1916-)
CERMENATI, Mario, (1868-1924)
CHAPMAN, Sydney, (1888-1970)
CHUBB, Lawrence John, (?1888-1971)
CLARKE, John Mason, (1857-1925)
CLÉMENT, Charles Adolphe Antoine, (1860-1933)
CLODD, Edward, (1840-1930)
CORSTORPHINE, George Steuart, (1865-1919)
CROOK, Thomas, (1876-1937)
DANYSZ-FLESZAROWA, Regina, (1888-1969)
DARWIN, Francis, (1848-1925)
DEWALQUE, Gilles Joseph Gustave, (1828-1905)
EDWARDS, Wilfred Norman, (1890-1956)
EYLES, Victor Ambrose, (1896-1978)
FLETT, John Smith, Sir, (1869-1947)
FORD, Trevor David, (1932-)
GĄSIOROWSKA, Zofia Halina, (1898-1957)
GAUDRY, Jean Albert, (1827-1908)
GEIKIE, Archibald, Sir, (1835-1924)
HARTLEY, Harold Brewer, Sir, (1878-1972)
HENNIG, Edwin, (1882-)
HIBBARD, Claude W., (1905-1973)
HOWELL, Jesse V., (1891-1971)
HUXLEY, Leonard, (1860-1933)
JUKES-BROWNE, Alfred John, (1851-1914)
JULIAN, Hester Forbes, Mrs., (1860-1934)
KETTNER, Radim, (1891-1967)
KEYES, Charles Rollin, (1864-1942)

HISTORY OF GEOLOGY 1900 to present Cont'd.

LAMBRECHT, Kálmán, (1889-1936)
LANG, William Dickson, (1878-1966)
MARGERIE, Emmanuel Marie Pierre Martin Jacquin de, (1862-1953)
MATHER, Kirtley Fletcher, (1888-1978)
MAXIM, Ion Al., (1902-1968)
MERRILL, George Perkins, (1854-1929)
MILLER, William John, (1880-1965)
MITCHELL, Peter Chalmers, Sir, (1864-1945)
MUIR-WOOD, Helen Marguerite, (1896-1968)
NANCE, James W., (1908-1977)
NICOLAS, Jean Paul, (1921-1967)
ORCEL, Jean, (1896-)
OSBORN, Henry Fairfield, (1857-1935)
PFANNENSTIEL, Max, (1902-1976)
RAVEN, Charles, (1885-1964)
ROGERS, Arthur William, (1872-1946)
RUTTEN, Louis Martin Robert, (1884-1946)
SARJEANT, William Antony Swithin, (1935-)
SCHUCHERT, Charles, (1858-1942)
SCHWIPPEL, Karl, (1821-1911)
SHROCK, Robert Rakes, (1904-)
SPENCER, Leonard James, (1870-1959)
STEENHUIS, Jacob Frederik, (1883-1958)
SWEETING, George Scotland, (1889-1977)
TASNÁDI-KUBACSKA, András, (1902-1977)
TOMKEIEFF, Sergei Ivanovich, (1892-1968)
TRIBOLET-HARDY, Frédéric Maurice de, (1852-1929)
VANDER HOOF, Vertress Lawrence, (1904-1964)
WATTS, William Whitehead, (1860-1947)
WHITE, George Willard, (1903-)
WICHMANN, Carl Ernst Arthur, (1851-1927)
WILLS, Robert Glover, (1878-1950)
WIŚNIOWSKI, Tadeusz, (1865-1933)
WOODWARD, Horace Bolingbroke, (1848-1914)
WRIGHT, Jean Davies, (-)
ZENZÉN, Nils, (1883-1959)
ZITTEL, Karl Alfred von, (1839-1904)

Time unknown
BEER, Gavin R. de, Sir, (-)

HYDROGEOLOGY
Up to 1800
KEILL, John, (1671-1721)
LAMARCK, Jean-Baptiste Pierre Antoine de Monet, Chevalier de, (1744-1829)
MILLER, Samuel, (1769-1850)
PERRAULT, Pierre, (1611-1680)
SENECA, Lucius Annaeus, (ca.4 B.C. - 65 A.D.)
VALLISNERI, Antonio, (1661-1730)

1800-1849
BEARDMORE, Nathaniel, (1816-1872)
BELGRAND, Eugène, (1810-1878)

HYDROGEOLOGY 1800-1849 Cont'd.

BISCHOFF, Karl Gustav, (1792-1870)
DARCY, Henri Philibert Gaspard, (1803-1858)
DOWKER, George, (1828-1899)
ETHERIDGE, Robert, (1819-1903)
HARRISON, John Thornhill, (1815-1891)
HAWKSHAW, John, Sir, (1811-1891)
LAMARCK, Jean-Baptiste Pierre Antoine de Monet, Chevalier de, (1744-1829)
MILLER, Samuel, (1769-1850)
MYLNE, Robert William, (1816-1890)
PRESTWICH, Joseph, Sir, (1812-1896)
RIDDELL, John Leonard, (1807-1865)
SANDERS, William, (1799-1875)
TAUNTON, John Hooke, (1820-1893)
TOPLEY, William, (1841-1894)

1850-1899

ADAMS, George Irving, (1870-1932)
ASCHER, Hans, (1848-1953)
BAIN, Thomas Charles John, (1829-1893)
BEADNELL, Hugh John Llewellyn, Major, (1874-1944)
BEARDMORE, Nathaniel, (1816-1872)
BELGRAND, Eugène, (1810-1878)
BENNETT, Arthur, (1873-1941)
BIGOT, Alexandre, (1863-1953)
BINET, Auguste, (-1900)
BISCHOFF, Karl Gustav, (1792-1870)
BLAKE, John Hopwood, (1843-1901)
BOULTON, William Savage, (1867-ca.1954)
BRÜCKNER, Eduard, (1862-1927)
CHOFFAT, Paul, (1849-1919)
DARCY, Henri Philibert Gaspard, (1803-1858)
DARTON, Nelson Horatio, (1865-1948)
DE RANCE, Charles Eugène, (1847-1906)
DOLLÉ, Louis, (1878-1965)
DORSEY, Noah Ernest, (1873-1959)
DOUXAMI, Henri, (1871-1913)
DOWKER, George, (1828-1899)
DOWLING, Donaldson Bogart, (1858-1925)
DRĂGHICEANU, Mathei M., (1844-1939)
DUNCAN, Cecil Cooke, (1868-1948)
DWERRYHOUSE, Arthur Richard, (1867-1931)
ETHERIDGE, Robert, (1819-1903)
FOURNIER, Eugène, (1871-1941)
FOX-STRANGWAYS, Charles Edward, (1844-1910)
GRAHAM, William Vaux, (1859-1940)
GREGORY, Herbert E., (1869-1952)
HALL, Charles Monroe, (1870-1903)
HARRIS, George Frederick, (1862-1906)
HARRISON, John Thornhill, (1815-1891)
HAWKSHAW, John, Sir, (1811-1891)
HAY, Robert, (1835-1895)
HILL, James Bastian, (1861-1927)
HORUSITZKY, Henrik, (1870-1945)

HYDROGEOLOGY 1850-1899 Cont'd.

HUME, William Fraser, (1867-1949)
JACK, Robert Lockhart, (1878-1964)
KAISER, Friedrich Wilhelm Erich, (1871-1934)
KATZER, Bedřich, (1861-1925)
KEILHACK, Konrad, (1858-1944)
LALOY, Roger, (1851-1909)
LANE, Alfred Church, (1863-1948)
LAPWORTH, Herbert, (1875-1933)
LIFFA, Aurél, (1872-1957)
MATLEY, Charles Alfred, (1866-1947)
MATSON, George Charlton, (1873-1940)
McCALLIE, Samuel Washington, (1856-)
McGEE, William John, (1853-1912)
MEINZER, Oscar Edward, (1876-1948)
MENDENHALL, Walter Curran, (1871-1957)
MILLER, Samuel, (1769-1850)
MORSE, William Clifford, (1874-1962)
MYLNE, Robert William, (1816-1890)
NAGYSUR, János Böckh von, (1840-1909)
NICCOLI, Enrico, (1836-1915)
NORTON, William Harmon, (-)
OCKENDEN, Maurice Albion, (1874-1954)
POWELL, John Wesley, (1834-1902)
PRESTON, Henry, (1852-1940)
PRESTWICH, Joseph, Sir, (1812-1896)
PURKYNĚ, Cyril, (1862-1937)
RANSOME, Frederick Leslie, (1868-1935)
RIDDELL, John Leonard, (1807-1865)
ROSLONSKI, Romuald, (1880-1956)
RUTKOWSKI, Feliks, (1878-1964)
SALMOJRAGHI, Francesco, (1837-?1910)
SANDERS, William, (1799-1875)
SCHARDT, W. Hans, (1858-1931)
SELLARDS, Elias Howard, (1875-1961)
SINGER, Max, (1874-1952)
SMITH, William Richard Tangier, (1869-1972)
STEPHENSON, Lloyd William, (1876-1962)
STEWART, Thomas, (1857-1943)
STOW, George, (1876-?1955)
STRAHAN, Aubrey, Sir, (1852-1928)
SZONTAGH, Tamás, (1851-1937)
TAUNTON, John Hooke, (1820-1893)
TODD, James Edward, (1846-1922)
TOLMAN, Cyrus Fisher, (1873-1941)
TOPLEY, William, (1841-1894)
VALLE DE LERSUNDI, Alfonso del, (1875-1944)
VEATCH, Arthur Clifford, (1878-1938)
WAAGEN, Lukas, (1877-1959)
WAITZ, Paul, (1876-1961)
WHITAKER, William, (1836-1925)
WOODWARD, Horace Bolingbroke, (1848-1914)

1900 to present
ADAMS, George Irving, (1870-1932)

HYDROGEOLOGY 1900 to present Cont'd.

ALEXANDER, Geoffrey Newman, (1908-1975)
ASCHER, Hans, (1848-1953)
BADAK, Jerzy, (1932-1966)
BEADNELL, Hugh John Llewellyn, Major, (1874-1944)
BENNETT, Arthur, (1873-1941)
BIGOT, Alexandre, (1863-1953)
BISSET, Charles Barron, (-?1966)
BLAKE, George Stanfield, (-1940)
BLAKE, John Hopwood, (1843-1901)
BOSS, Richard F., (1923-1976)
BOULTON, William Savage, (1867-ca.1954)
BRAMWYCH, Michael, (1911-1974)
BRÜCKNER, Eduard, (1862-1927)
BURROWS, William Alfred, (1903-1970)
CADY, Richard Carlysle, (1908-1943)
CALEY, John Fletcher, (1903-1971)
CHOFFAT, Paul, (1849-1919)
CLAPP, Frederick Gardner, (1879-1944)
CONFORTO, Bruno, (1918-1963)
CONSTANTINESCU, Valeriu, (1921-1970)
DANGEARD, Louis, (-)
DARTON, Nelson Horatio, (1865-1948)
DE RANCE, Charles Eugène, (1847-1906)
DEUSSEN, Alexander, (1882-1959)
DINHAM, Charles Hawker, (1883-1955)
DOBBIN, Carroll Edward "Cy", (1892-1967)
DOCWRA, Colin, (-1942)
DOLLÉ, Louis, (1878-1965)
DORSEY, Noah Ernest, (1873-1959)
DOUXAMI, Henri, (1871-1913)
DOWLING, Donaldson Bogart, (1858-1925)
DRĂGHICEANU, Mathei M., (1844-1939)
DUNCAN, Cecil Cooke, (1868-1948)
DVOŘÁK, Josef, (1904-)
DWERRYHOUSE, Arthur Richard, (1867-1931)
EDMUNDS, Francis Hereward, (1893-1960)
ERTEL, Hans, (1904-1971)
ESCHER, Berend George, (-1967)
ETHERIDGE, Robert, (1819-1903)
FÖLDVÁRI, Aladár, (1906-1973)
FOSTER, Margaret Dorothy, (1895-1970)
FOURNIER, Eugène, (1871-1941)
FOX-STRANGWAYS, Charles Edward, (1844-1910)
FURMAŃSKI, Ignacy Jan, (1915-1972)
GĂBERT, Carl, (-1931)
GALPIN, Sidney Longman, (1886-1962)
GASKIN, Alan Robert James, (1927-1974)
GEORGE, William Owsley, (1892-1966)
GILLMAN, Clement, (?1883-1946)
GLEASON, Charles D., (1908-1935)
GORIZDRO-KULCZYCKA, Zinaida, (1884-1949)
GOURLEY, Harold John Frederick, (1886-1956)
GRABHAM, George Walter, (1882-1955)
GRAHAM, William Vaux, (1859-1940)

HYDROGEOLOGY 1900 to present Cont'd.

GREENE, Frank Cook, (1886-1975)
GREGORY, Herbert E., (1869-1952)
GÜMBEL, Emil J., (-1966)
GUNTER, Herman, (1885-1972)
HADJISTAVRINOU, Yangos, (1934-1978)
HALL, Charles Monroe, (1870-1903)
HALL, George Martin, (1891-1941)
HARRELL, Marshall Allen, (1900-1959)
HARRIS, George Frederick, (1862-1906)
HARWOOD, Harold Benjamin, (-1956)
HEINSHEIMER, Georg J., (1887-1970)
HIBBERT, Frank, Colonel, (?1889-1951)
HILL, James Bastian, (1861-1927)
HOFMANN, Walter, (1919-1975)
HORUSITZKY, Ferenc, (1901-1971)
HORUSITZKY, Henrik, (1870-1945)
HUFFMAN, Edward Archibald, (1912-1954)
HUME, William Fraser, (1867-1949)
HYNIE, Ota, (1899-1968)
INESON, Jack, (1917-1970)
JACK, Robert Lockhart, (1878-1964)
JACOB, Charles Edward, (1914-1970)
JELÍNEK, František, (1911-1966)
JOHNSON, Bertrand Leroy, (1882-1962)
JOHNSON, Charles George, (1914-1969)
JOHNSTON, William Drumm, Jr., (1899-1972)
JONES, Ogden Sherman, (1891-1952)
KAISER, Friedrich Wilhelm Erich, (1871-1934)
KALININ, Gennady Pavlovich, (1916-1975)
KATZER, Bedřich, (1861-1925)
KAVECKIS, Mykolas, (1889-)
KEILHACK, Konrad, (1858-1944)
KERCKHOFF, P. J. van, (-)
KING, William Bernard Robinson, (1899-1963)
LALOY, Roger, (1851-1909)
LANE, Alfred Church, (1863-1948)
LAPWORTH, Herbert, (1875-1933)
LEWIS, William Vaughan, (1907-1961)
LIFFA, Aurél, (1872-1957)
LITTLE, Otway Henry, (-1956)
LONSDALE, John Tipton, (1895-1960)
LUCEK, Josef, (1930-1969)
LUKOVIĆ, Milan, (1889-)
LYSHEDE, Jens Martinus, (1911-1961)
MAROS, Imre, (1882-1937)
MARSELL, Ray Everett, (1893-1971)
MASON, Arnold Caverly, (1906-1961)
MATLEY, Charles Alfred, (1866-1947)
MATSON, George Charlton, (1873-1940)
MAUFE, Herbert Brantwood, (1879-1946)
MAXEY, George Burke, (1917-1977)
MAXIM, Ion Al., (1902-1968)
McCALLIE, Samuel Washington, (1856-)
McGEE, William John, (1853-1912)

HYDROGEOLOGY 1900 to present Cont'd.

McGUINNESS, Charles Lee, (1914-1971)
MEFFERD, Ralph L., (1909-1964)
MEINZER, Oscar Edward, (1876-1948)
MEISTER, Ernst, (1887-1939)
MENDENHALL, Walter Curran, (1871-1957)
MILES, Keith Rodney, (1915-1972)
MILLER, William D., (1930-1972)
MOLENGRAAFF, Gerard Johan Hendrik, (1887-1961)
MORGAN, Eilir Evans, (?1911-1969)
MORRIS, Frederick Kuhne, (1885-1962)
MORSE, William Clifford, (1874-1962)
NAGYSUR, János Böckh von, (1840-1909)
NICCOLI, Enrico, (1836-1915)
NORTON, William Harmon, (-)
NOSZKY, Jenö, Sr., (1880-1951)
OCKENDEN, Maurice Albion, (1874-1954)
OLTMAN, Roy E., (-1977)
PALMER, Harold Schjoeth, (1890-1959)
PAPP, Ferenc, (1901-1969)
PARDÉ, Maurice, (1893-1973)
PENNEL, Harry David, (1907-1971)
PETERSON, Arthur Frederick, (1911-1967)
PEYTON, Garland, (1892-1964)
POND, Walter Franklin, (1885-1974)
POTEET, James Harold, (1896-1965)
POWELL, John Wesley, (1834-1902)
PRESTON, Henry, (1852-1940)
PRICE, Paul Holland, (1898-)
PURKYNĚ, Cyril, (1862-1937)
RAEBURN, Colin, (1894-1970)
RAJNATH, Shri, (1904-1972)
RANSOME, Frederick Leslie, (1868-1935)
RASMUSSEN, William Charles, (1917-1973)
REED, Eugene C., (1901-1976)
RENAULT, Philippe, (-)
RENICK, Brink Coleman, (1897-1971)
RICHARDSON, Linsdall, (1881-1967)
ROMANES, James, (1886-1942)
ROSLONSKI, Romuald, (1880-1956)
ROSSBY, Carl-Gustaf Arvid, (1898-1957)
RUSSELL, William Low, (1897-1976)
RUTKOWSKI, Feliks, (1878-1964)
RUTSCH, Rolf F., (1902-1975)
SALMOJRAGHI, Francesco, (1837-?1910)
SCHARDT, W. Hans, (1858-1931)
SCHMIDT, Robert Eligius, (1902-1973)
SCHWENNESEN, Alvin Theodore, (1883-1962)
SELLARDS, Elias Howard, (1875-1961)
SENN, Alfred, (1899-1949)
SIKES, Harold Lesley, (1881-1943)
SINGER, Max, (1874-1952)
SMITH, Bernard, (1881-1936)
SMITH, William Richard Tangier, (1869-1972)
STEENHUIS, Jacob Frederik, (1883-1958)

HYDROGEOLOGY 1900 to present Cont'd.

STEPHENSON, Lloyd William, (1876-1962)
STEWART, Thomas, (1857-1943)
STINY, Josef, (1880-1958)
STOW, George, (1876-?1955)
STRAHAN, Aubrey, Sir, (1852-1928)
SÜMEGHY, József, (1892-1955)
SZONTAGH, Tamás, (1851-1937)
TEZON, Roberto Vitelmo, (1921-1973)
THOMPSON, David Grosh, (1888-1943)
THOROLD, William George Preston, (?1892-1943)
TODD, James Edward, (1846-1922)
TOEBES, Cornelius, (?-1975)
TOLMAN, Cyrus Fisher, (1873-1941)
URBAIN, Pierre, (1895-1968)
VALLE DE LERSUNDI, Alfonso del, (1875-1944)
VEATCH, Arthur Clifford, (1878-1938)
VENDEL, Aladár, (1886-1971)
VERSLUYS, Jan, (1880-1935)
VISENTINI, Marco, (1885-1964)
WAAGEN, Lukas, (1877-1959)
WAITZ, Paul, (1876-1961)
WARING, Gerald Ashley, (1883-1971)
WATERMAN, Willis D., (1925-1968)
WELLS, Francis Gerrit, (1898-1969)
WENTWORTH, Chester Keeler, (1891-1969)
WHITE, George Willard, (1903-)
WILLIAMS, Mervyn, (1934-1970)
WINTON, Will McClain, (1885-1960)
WOODWARD, Horace Bolingbroke, (1848-1914)

IGNEOUS PETROLOGY

Up to 1800

ARDUINO, Giovanni, (1714-1795)
BEDDOES, Thomas, (1760-1808)
BUCH, Christian Leopold von, Freiherr, (1774-1853)
DESMAREST, Nicolas, (1725-1815)
HALL, James, Sir, (1761-1832)
HAMILTON, William, Reverend, (1755-1797)
JAŚKIEWICZ, Jan D. P., (1749-1809)
KEIR, James, (1735-1820)
McDONNELL, James, (?1763-1845)
OTLEY, Jonathan, (1766-1856)
PETTKO, Ján, (1812-1890)
PINGEL, Christian, (1793-1852)
SAUSSURE, Horace Bénédict de, (1740-1799)
SPALLANZANI, Lazzaro, Abate, (1729-1799)
WHITEHURST, John, (1713-1788)

1800-1849

ALLPORT, Samuel, (1816-1897)
BEAUMONT, Jean Baptiste Armand Louis Léonce Élie de, (1798-1874)
BEDDOES, Thomas, (1760-1808)
BOUÉ, Ami, (1794-1881)
BUCH, Christian Leopold von, Freiherr, (1774-1853)
CLARK, George Thomas, (1811-1898)
DAUBENY, Charles Giles Brindle, (1795-1867)
DELESSE, Achille Ernest Oscar Joseph, (1817-1881)
DESMAREST, Nicolas, (1725-1815)
FORBES, David, (1828-1876)
HALL, James, Sir, (1761-1832)
HAUGHTON, Samuel, Reverend, (1823-1897)
HUNT, Thomas Sterry, (1826-1892)
JAŚKIEWICZ, Jan D. P., (1749-1809)
JOKÉLY, Johann, (1826-1862)
KEIR, James, (1735-1820)
McDONNELL, James, (?1763-1845)
OTLEY, Jonathan, (1766-1856)
PETTKO, Ján, (1812-1890)
PHILLIPS, John Arthur, (1822-1887)
PINGEL, Christian, (1793-1852)
ROSE, Gustav, (1798-1873)
SCHMID, Ernst Erhard, (1815-1885)
SCROPE, George Poulett, (né Thomson), (1797-1876)
SLADE, Jeremiah, (1828-1905)
SZABÓ, József, (1822-1894)

1850-1899

ALLPORT, Samuel, (1816-1897)
ANDERSON, William, (1860-1915)
BAIRD SMITH, R., Colonel, (-)
BARON, Richard, Reverend, (1847-1907)
BARVÍŘ, Jindřich Ladislav, (1863-1952)
BASTIN, Edson Sunderland, (1878-1953)
BEAUMONT, Jean Baptiste Armand Louis Léonce Élie de, (1798-1874)
BECKE, Friedrich Johann Karl, (1855-1931)
BELYANKIN, Dmitri Stepanovich, (1876-1953)

IGNEOUS PETROLOGY 1850-1899 Cont'd.

BEMROSE, Henry Howe, (1857-1939)
BERGEAT, Alfred, Edmund, (1866-1924)
BERGT, Walther Adolf, (1864-1941)
BEUTELL, Albert, (1859-1921)
BOŘICKÝ, Emanuel, (1840-1881)
BOUÉ, Ami, (1794-1881)
BRUHNS, Franz Siegfried Willy, (1864-1929)
BUCH, Christian Leopold von, Freiherr, (1774-1853)
CĂDERE, Dimitrie M., (1876-1941)
CAMPBELL, Henry Donald, (1862-1934)
CASWELL, John Henry, (1846-1909)
CLARK, George Thomas, (1811-1898)
CORSTORPHINE, George Steuart, (1865-1919)
CRAWFORD, Ralph Dixon, (1873-1950)
DALY, Reginald Aldworth, (1871-1957)
DANA, Edward Salisbury, (1849-1935)
DAUBENY, Charles Giles Brindle, (1795-1867)
DAVID, Tannatt William Edgeworth, Sir, (1858-1934)
DAVIES, Thomas, (1837-1892)
DAY, Thomas Cuthbert, (1853-1935)
DELESSE, Achille Ernest Oscar Joseph, (1817-1881)
DRESSER, John Alexander, (1866-1954)
EMERSON, Benjamin Kendall, (1843-1932)
ERDMANNSDÖFFER, Otto Heinrich, (1876-1955)
FORBES, David, (1828-1876)
FRANCO, Pasquale, (1852-1907)
GARDINER, Charles Irving, (?1868-1940)
GEIKIE, Archibald, Sir, (1835-1924)
GIBB, Alfred William, (1864-1937)
GORDON, Charles Henry, (1857-1934)
GREENLY, Edward, (1861-1951)
GRUBENMANN, Ulrich, (1850-1924)
HACKMAN, Victor Axel, (1866-1941)
HARKER, Alfred, (1859-1939)
HATCH, Frederick Henry, (?1864-1932)
HAUGHTON, Samuel, Reverend, (1823-1897)
HENDERSON, James Alexander Leo, (?1874-1940)
HIBSCH, Josef Emanuel, (1852-1940)
HOBBS, William Herbert, (1864-1953)
HOFFMAN, Fr. Adolf, (1867-1898)
HÖGBOM, Arvid Gustaf, (1857-1940)
HUNT, Thomas Sterry, (1826-1892)
HYLAND, John Shearson, (1866-1898)
IDDINGS, Joseph Paxson, (1857-1920)
IRVING, Roland Duer, (1847-1888)
JAGGAR, Thomas Augustus, (1871-1953)
JOHANNSEN, Albert, (1871-1962)
JOHN VON JOHNESBERG, Conrad Heinrich, Edler, (1852-1918)
JOKÉLY, Johann, (1826-1862)
JUDD, John Wesley, (1840-1916)
KARPINSKY, Alexander Petrovich, (1847-1936)
KEMP, James Furman, (1859-1926)
KIŠPATIĆ, Mišo, (1851-1926)
KOECHLIN, Rudolf Ignatz, (1862-1939)

IGNEOUS PETROLOGY 1850-1899 Cont'd.

KOLDERUP, Carl Fredrik, (1869-1942)
KUPFFER, August Ernestovich, (1842-1918)
LARSEN, Esper Signius Jr., (1879-1961)
LOEWINSSON-LESSING, Franc Julyevich, (1859-1939)
LOUDERBACK, George Davis, (1874-1957)
MACKIE, William, (1856-1932)
MARSHALL, James Garth, (-1873)
MARSHALL, Patrick, (1869-1950)
MARTELLI, Alessandro, (1877-1934)
MIDDLEMISS, Charles Stewart, (1859-1945)
MILCH, Ludwig, (1867-1928)
MOON, Frederick William, (-1925)
MOORE, Elwood S., (1879-1966)
MOROZEWICZ, Józef Marian, (1865-1941)
O'KELLY, Joseph, (1832-1883)
OSANN, Carl Alfred, (1859-1923)
OSMAN, Charles William, (1855-1932)
OTLEY, Jonathan, (1766-1856)
PHILLIPS, John Arthur, (1822-1887)
PINGEL, Christian, (1793-1852)
PIRSSON, Louis Valentine, (1860-1919)
RAISIN, Catherine Alice, (1855-1945)
RENARD, Alphonse François, (1842-1903)
RIVA, Carlo, (1872-1902)
ROCCATI, Alessandro, (1872-1928)
ROSE, Gustav, (1798-1873)
ROSENBUSCH, Karl Harry Ferdinand, (1836-1914)
RUTLEY, Frank, (1842-1904)
SABATINI, Venturino, (1879-1922)
SAUER, Gustav Adolf, (1825-1932)
SCHMID, Ernst Erhard, (1815-1885)
SCROPE, George Poulett, (né Thomson), (1797-1876)
SILVESTRI, Orazio, (-1890)
SJÖGREN, Sten Anders Hjalmar, (1856-1922)
SKEATS, Ernest Willington, (1875-1953)
SLADE, Jeremiah, (1828-1905)
SOLLAS, William Johnson, (1849-1936)
SOUZA BRANDÃO, Vicente de, (1863-1916)
SUESS, Franz Eduard, (1867-1941)
SZABÓ, József, (1822-1894)
TANNHÄUSER, Felix, (1874-1924)
TARASENKO, Vasily Efimovich, (1859-1926)
TEALL, Jethro Justinian Harris, Sir, (1849-1924)
THOMAS, Herbert Henry, (1876-1935)
TRECHMANN, Charles Otto, (1851?-1917)
TRÜSTEDT, Otto, (1866-1929)
TSCHERMAK VON SEYSENEGG, Gustav, (1836-1927)
VAN HISE, Charles Richard, (1857-1918)
WADSWORTH, Marshman Edward, (1847-1921)
WAITZ, Paul, (1876-1961)
WARREN, Charles Hyde, (1876-1950)
WASHINGTON, Henry Stephens, (1867-1934)
WATTS, William Whitehead, (1860-1947)
WEINSCHENK, Ernst Heinrich Oskar Kasimir, (1865-1921)

IGNEOUS PETROLOGY 1850-1899 Cont'd.

WEYBERG, Zygmunt, (1872-1944)
WICHMANN, Carl Ernst Arthur, (1851-1927)
WILKMAN, Wanold Wrydon, (1872-?1938)
WILLIAMS, George Huntington, (1856-1894)
WILLIAMS, John Francis, (1862-1891)
WILSON, J. S. Grant, (1855-1908)
WORTH, Richard Nicholas, (-1896)
WYNDHAM, T. Heathcote, (?1842-1876)
ZIRKEL, Ferdinand, (1838-1912)

1900 to present
ALLING, Harold Lattimore, (1888-1960)
ANDERSON, William, (1860-1915)
ANDREATTA, Benedettino Ciro, (1906-1960)
ASHBY, Donald Frederick, (?1914-1943)
BAILEY, Edward Battersby, Sir, (1881-1965)
BAKER, George, (1908-1975)
BARBER, Cecil Thomas, (-)
BARON, Richard, Reverend, (1847-1907)
BARTH, Thomas Frederik Weiby, (1899-1971)
BARVÍŘ, Jindřich Ladislav, (1863-1952)
BASTIN, Edson Sunderland, (1878-1953)
BECKE, Friedrich Johann Karl, (1855-1931)
BELLAIR, Pierre, (1910-)
BELYANKIN, Dmitri Stepanovich, (1876-1953)
BEMROSE, Henry Howe, (1857-1939)
BENSON, William Noel, (1885-1957)
BERGEAT, Alfred, Edmund, (1866-1924)
BERGT, Walther Adolf, (1864-1941)
BEUTELL, Albert, (1859-1921)
BOWEN, Norman Levi, (1887-1956)
BOYLE, Robert, (1883-1917)
BRAMMALL, Alfred, (?1879-1954)
BROWN, Peter Evans, (-)
BRUHNS, Franz Siegfried Willy, (1864-1929)
BURTON, Richard Cooksey, (1890-1916)
CĂDERE, Dimitrie M., (1876-1941)
CAMPBELL, Henry Donald, (1862-1934)
CAMPBELL, Robert, (1881-1957)
CASWELL, John Henry, (1846-1909)
CHHIBBER, Harbans Lal, (1899-1955)
COCKBURN, Alexander Murray, (1901-1959)
COOKE, Harold Caswell, (1884-1956)
COOMBS, Howard A., (-)
CORNELIUS, Hans Peter, (1888-)
CORSTORPHINE, George Steuart, (1865-1919)
CRAIG, Robert Meldrum, (?1882-1956)
CRAWFORD, Ralph Dixon, (1873-1950)
DALY, Reginald Aldworth, (1871-1957)
DANA, Edward Salisbury, (1849-1935)
DAVID, Tannatt William Edgeworth, Sir, (1858-1934)
DAVIES, Thomas, (1837-1892)
DAY, Thomas Cuthbert, (1853-1935)
DE DYCKER, Raymond, (1903-1947)

IGNEOUS PETROLOGY 1900 to present Cont'd.

DOUGLAS, James Archibald, (1884-1978)
DRESSER, John Alexander, (1866-1954)
DUBYANSKY, Viktor Viktorovich, (1880-1925)
EDWARDS, Austin Burton, (1909-1960)
EMERSON, Benjamin Kendall, (1843-1932)
ERDMANNSDÖFFER, Otto Heinrich, (1876-1955)
ESCHER, Berend George, (-1967)
EYLES, Victor Ambrose, (1896-1978)
FAIRBAIRN, Harold Williams, (1906-)
FIALA, František, (1903-)
FISHER, Bernard, (1918-1946)
FISK, Harold Norman, (1908-1964)
FOLEY, Mary Cecilia, (-1925)
FOSHAG, William Frederick, (1894-1956)
FRANCO, Pasquale, (1852-1907)
FULLER, Richard Eugene, (1897-1976)
GARDINER, Charles Irving, (?1868-1940)
GAULT, Hugh Richard, (1915-1961)
GEIKIE, Archibald, Sir, (1835-1924)
GIBB, Alfred William, (1864-1937)
GILLSON, Joseph Lincoln, (1895-1964)
GILLULY, James, (-)
GLANGEAUD, Louis, (1903-)
GOLDSCHLAG, Maurycy, (1891-1917)
GOODSPEED, George Edward, (1887-1974)
GORDON, Charles Henry, (1857-1934)
GREENLY, Edward, (1861-1951)
GROUT, Frank Fitch, (1880-1958)
GROVES, Arthur William, (1903-1956)
GRUBENMANN, Ulrich, (1850-1924)
HACKMAN, Victor Axel, (1866-1941)
HARKER, Alfred, (1859-1939)
HARRY, William Trevelyan, (?1922-1964)
HATCH, Frederick Henry, (?1864-1932)
HEJTMAN, Bohuslav, (1901-)
HENDERSON, James Alexander Leo, (?1874-1940)
HENSON, Frederick Arthur, (1919-1957)
HESLOP, Mary Kingdon, (?1884-?1954)
HIBSCH, Josef Emanuel, (1852-1940)
HIGAZY, Riad Abdel-Megid, (1919-1967)
HOBBS, William Herbert, (1864-1953)
HODGE, Matthew Burnop, (-1951)
HOERNES, Dieter, (1912-1955)
HÖGBOM, Arvid Gustaf, (1857-1940)
HOLLINGWORTH, Sydney Ewart, (1899-1966)
HOLMES, Arthur, (1890-1965)
HOOKER, Marjorie, (1908-1976)
HUNT, Walter Frederick, (1882-1975)
HUTCHINSON, William W., (-)
IDDINGS, Joseph Paxson, (1857-1920)
JAGGAR, Thomas Augustus, (1871-1953)
JOHANNSEN, Albert, (1871-1962)
JOHN VON JOHNESBERG, Conrad Heinrich, Edler, (1852-1918)
JUDD, John Wesley, (1840-1916)

IGNEOUS PETROLOGY 1900 to present Cont'd.

KAMENICKÝ, Jakub, (1917-)
KARPINSKY, Alexander Petrovich, (1847-1936)
KEMP, James Furman, (1859-1926)
KIŠPATIĆ, Mišo, (1851-1926)
KNOPF, Adolph, (1882-1966)
KOECHLIN, Rudolf Ignatz, (1862-1939)
KOLDERUP, Carl Fredrik, (1869-1942)
KUNO, Hisashi, (1910-1969)
KUPFFER, August Ernestovich, (1842-1918)
LAPHAM, Davis M., (1931-1974)
LARMAN, Mark Linington, (1899-1958)
LARSEN, Esper Signius, III, (1912-1961)
LARSEN, Esper Signius Jr., (1879-1961)
LEITH, Charles Kenneth, (1875-1956)
LOBJOIT, William Michael, (1932-1967)
LOEWINSSON-LESSING, Franc Julyevich, (1859-1939)
LOUDERBACK, George Davis, (1874-1957)
LUTH, William Clair, (1934-)
MACKIE, William, (1856-1932)
MacROBERT OF DOUNESIDE AND CROMAR, Rachel, Lady (née Workman), (-1954)
MALKOWSKI, Stanislaw, (1889-1962)
MARIĆ, Luka, (1899-)
MARSHALL, Patrick, (1869-1950)
MARTELLI, Alessandro, (1877-1934)
MATHUR, Krishna Kumar, (-1936)
McMATH, John Charles, (?1908-1954)
MIDDLEMISS, Charles Stewart, (1859-1945)
MILCH, Ludwig, (1867-1928)
MILLER, William John, (1880-1965)
MILLOT, Georges, (1917-)
MITCHELL, George Hoole "Mick", (1902-1976)
MOON, Frederick William, (-1925)
MOORE, Elwood S., (1879-1966)
MOORHOUSE, Walter Wilson, (1913-1969)
MOROZEWICZ, Józef Marian, (1865-1941)
MORRIS, Frederick Kuhne, (1885-1962)
MORTOLA, Edelmira Ines, (-1973)
NEALE, Ernest Richard Ward, (1923-)
NĚMEC, František, (1905-1972)
NOVÁK, Jiří, (?1902-1971)
NOVOTNÝ, Miroslav, (1921-1971)
ORCEL, Jean, (1896-)
OSANN, Carl Alfred, (1859-1923)
OSMAN, Charles William, (1855-1932)
PALM, Q. A., (1923-1960)
PALM-LAZARD, C. B. B., (1933-1960)
PASTORE, Franco, (1885-1958)
PAUK, František, (1905-)
PAVLOVIĆ, Stojan, (1903-)
PAWLICA, Wladislaw, (1886-1919)
PEACOCK, Martin Alfred, (1898-1950)
PEGAU, Albert August, (1894-1963)
PERRIN, René, (1893-1966)

IGNEOUS PETROLOGY 1900 to present Cont'd.

PIETZSCH, Kurt, (1884-1964)
PINSON, William Hamet, Jr., (1919-)
PIRSSON, Louis Valentine, (1860-1919)
POLDERVAART, Arie, (1918-1964)
POSTEL, Albert Williams, (1909-1966)
PROKOP, František, (1898-)
QUENSEL, Percy Dudgeon, (1881-1966)
RAISIN, Catherine Alice, (1855-1945)
RAMDOHR, Paul, (1890-)
RAVIER, Jean, (1922-)
READ, Herbert Harold, (1889-1970)
RENARD, Alphonse François, (1842-1903)
RICHEY, James Ernest, (1886-1968)
RIMANN, Eberhard, (1882-1944)
RIVA, Carlo, (1872-1902)
ROBINSON, G. D., (-)
ROCCATI, Alessandro, (1872-1928)
ROMAN, David, (1884-1927)
ROSENBUSCH, Karl Harry Ferdinand, (1836-1914)
ROSICKÝ, Vojtěch, (1880-1942)
ROSS, Clarence Samuel, (1880-1975)
ROUBAULT, Marcel Edouard, (1905-1974)
RUNNER, Joseph James, (1885-1970)
RUTLEDGE, Harold, (1920-1954)
RUTLEY, Frank, (1842-1904)
SABATINI, Venturino, (1879-1922)
SALÁT, Ján, (1919-1973)
SAUER, Gustav Adolf, (1825-1932)
SCHAIRER, John Frank, (1904-1970)
SCHOLTZ, Douglas Leinster, (1902-1977)
SCHÜLLER, Arno, (1908-1968)
SCHULZ, Karl, (1885-1928)
SCHWARZBACH, Martin, (-)
SCOTT, Alexander, (1890-1951)
SCOTT, James Frew, (1902-1956)
SEKANINA, Josef, (1900-)
SHAND, Samuel James, (1882-1957)
SIMMONS, William Charles, (1888-1951)
SJÖGREN, Sten Anders Hjalmar, (1856-1922)
SKEATS, Ernest Willington, (1875-1953)
SLADE, Jeremiah, (1828-1905)
SMITH, Herbert Gladstone, (-1950)
SOLLAS, William Johnson, (1849-1936)
SOUZA BRANDÃO, Vicente de, (1863-1916)
STEINOCHER, Vladimír, (1902-1972)
STENHOUSE, Andrew G., (-1950)
STILLWELL, Frank Leslie, (1888-1963)
STUART, Duncan, VII, (1905-1969)
SUESS, Franz Eduard, (1867-1941)
SUTTON, John, (-)
SVOBODA, Josef, (1898-)
TAJDER, Miroslav, (1909-)
TANNHÄUSER, Felix, (1874-1924)
TARASENKO, Vasily Efimovich, (1859-1926)

IGNEOUS PETROLOGY 1900 to present Cont'd.

TEALL, Jethro Justinian Harris, Sir, (1849-1924)
THOMAS, Herbert Henry, (1876-1935)
THOMAS, Trevor Morgan, (?1919-1974)
THOMSON, James Allan, (1881-1928)
TILLEY, Cecil Edgar, (1894-1973)
TOKARSKI, Julian, (1883-1961)
TOMKEIEFF, Sergei Ivanovich, (1892-1968)
TOZER, Charles Fallows, (1918-1966)
TRECHMANN, Charles Otto, (1851?-1917)
TRÜSTEDT, Otto, (1866-1929)
TSCHERMAK VON SEYSENEGG, Gustav, (1836-1927)
TYRRELL, George Walter, (1883-1961)
URBAN, Karel, (1905-)
VACHTL, Josef, (1907-)
VAN HISE, Charles Richard, (1857-1918)
VARDABASSO, Silvio, (1891-1966)
VINOGRADOV, A. P., (1895-1975)
VOGT, Thorolf, (1888-1958)
VON ECKERMANN, Claes Walther Harry, (1886-1969)
WADSWORTH, Marshman Edward, (1847-1921)
WAGER, Lawrence Rickard, (1904-1965)
WAITZ, Paul, (1876-1961)
WARREN, Charles Hyde, (1876-1950)
WASHINGTON, Henry Stephens, (1867-1934)
WATTS, William Whitehead, (1860-1947)
WEINSCHENK, Ernst Heinrich Oskar Kasimir, (1865-1921)
WESTERVELD, J., (1905-1962)
WEYBERG, Zygmunt, (1872-1944)
WHETTEN, John T., (-)
WICHMANN, Carl Ernst Arthur, (1851-1927)
WILKERSON, Albert Samuel, (1897-1963)
WILKMAN, Wanold Wrydon, (1872-?1938)
WILSON, J. S. Grant, (1855-1908)
WOOLDRIDGE, Sidney William, (1900-1963)
WYATT, Michael, (-1956)
ZAPLETAL, Karel, (1903-1972)
ZIRKEL, Ferdinand, (1838-1912)
ZOUBEK, Vladimír, (1903-)

INVERTEBRATE PALAEONTOLOGY

Up to 1800

ASTRUC, Jean, (1684-1766)
AYCKE, Johann Christian, (1766-1854)
BAIN, Andrew Geddes, (1797-1864)
BERTRAND, Élie, (1712-1797)
BLAINVILLE, Henri Marie Ducrotay de, (1777-1850)
BLUMENBACH, Johann Friedrich, (1752-1840)
BORN, Ignaz von, Baron, (1742-1791)
BROCCHI, Giovanni Battista (Giambattista), (1772-1826)
BRONGNIART, Alexandre, (1778-1847)
BRUGUIÈRES, Jean Guillaume, (1750-1799)
BRÜNNICH, Morten Thrane, (1737-1827)
BUCH, Christian Leopold von, Freiherr, (1774-1853)
COLLINI, Cosimo Alessandro, (1727-1806)

INVERTEBRATE PALAEONTOLOGY Up to 1800 Cont'd.

COLONNA, Fabio, (1567-1650)
CUMBERLAND, George, (1754-1848)
DALE, Samuel, (1659-1739)
DEFRANCE, Jacques Louis Marie, (1758-1850)
DILLWYN, Lewis Weston, (1778-1855)
EHRHART, Balthazar, (1700-1756)
ESCHWEGE, Wilhelm Ludwig, Baron, (1777-1855)
FAUJAS DE SAINT FOND, Barthélémy, (1741-1819)
FICHTEL, Leopold von, (fl.1798)
FONTENELLE, Bernard Le Bovier de, (1657-1757)
GAGNEBIN DE LA FERRIÈRE, Abraham, (1707-1800)
GUETTARD, Jean Étienne, (1715-1786)
HEHL, Johann Carl Ludwig, (1774-1853)
HENNAH, Richard, Reverend, (?1765-1846)
HILL, John, Sir, (1706-1775)
HISINGER, Wilhelm, (1766-1852)
HOENINGHAUS, Friedrich Wilhelm, (1770-1854)
KIRCHER, Athanasius, (1601-1680)
LAMARCK, Jean-Baptiste Pierre Antoine de Monet, Chevalier de, (1744-1829)
LANG, Karl Nikolaus, (1670-1741)
LESUEUR, (1778-1846)
LHUYD, Edward, (1660-1709)
LINNE, Carl von, (1707-1778)
LISTER, Martin, (1638-1712)
MARTIN, William, (1767-1810)
MÉTHERIE, Jean Claude de la, (1743-1817)
MONTI, Giuseppe, (1682-1760)
PARKINSON, James, (1755-1824)
REINECKE, Johann Christoph Matthias, (1769-1818)
SCHEUCHZER, Johann Gaspar, (1702-1729)
SCHLOTHEIM, Ernst Friedrich von, Freiherr, (1764-1832)
SERRES, Marcel de, (1783-1862)
SMITH, William, (1769-1839)
SOLANDER, Daniel Carl, (1733-1782)
SOLDANI, Ambrogio, Pater, (1736-1808)
SWEDENBORG, Emanuel, (1688-1772)
TROOST, Gerard, (1776-1850)
WALLERIUS, Johan Gottschalk, (1709-1785)
WEBSTER, Thomas, (1773-1844)

1800-1849

ADAMS, Charles Baker, (1814-1853)
AGASSIZ, Louis Jean Rodolphe, (1807-1873)
ALBERTI, Friedrich August von, (1795-1878)
ALLMAN, George James, (1812-1898)
ANDRZEJOWSKI, Antoni, (1785-1868)
ANGELIN, Nils Peter, (1805-1876)
ARCHIAC, Étienne Jules Adolphe Dexmier de Simon, Vicomte d', (1802-1868)
AUDOUIN, Jean Victor, (1797-1841)
AUINGER, Mathias, (?1810-1890)
AUSTIN, Charles Edward, (1819-1893)
AUSTIN, Thomas, Fort-Major, (1794-1881)

INVERTEBRATE PALAEONTOLOGY 1800-1849 Cont'd.

AYCKE, Johann Christian, (1766-1854)
BAILY, William Hellier, (1819-1888)
BAIN, Andrew Geddes, (1797-1864)
BARKAS, Thomas Pallister, (1819-1891)
BARRANDE, Joachim, (1799-1883)
BARRIS, Willis Hervey, (1820-1901)
BAYFIELD, Thomas Gabriel, (1817-1893)
BAYLE, Émile, (1819-1895)
BEESLEY, Thomas, (1818-1896)
BEISSEL, Ignaz, (1820-1887)
BELL, Dugald, (1827-1892)
BELLARDI, Luigi, (1818-1889)
BEYRICH, August Heinrich Ernst, (1815-1896)
BIANCONI, Gian Giuseppe, (1809-1878)
BIELZ, Michael, (1787-1866)
BIGSBY, John Jeremiah, (1792-1881)
BILLINGS, Elkanah, (1820-1876)
BINKHOORST TOT DEN BINKHOORST, Johan Theodoor, (1810-1876)
BLAINVILLE, Henri Marie Ducrotay de, (1777-1850)
BLUMENBACH, Johann Friedrich, (1752-1840)
BOISDUVAL, Jean-Baptiste Alphonse, (1799-1879)
BOSQUET, Joseph Augustin Hubert, (1814-1880)
BOURDOT, Jules Damase, (1837-1906)
BOURGUIGNAT, Jules René, (1829-?1919)
BOWERBANK, James Scott, (1797-1877)
BRARD, Cyprien Prosper, (1788-1838)
BROCCHI, Giovanni Battista (Giambattista), (1772-1826)
BRONGNIART, Alexandre, (1778-1847)
BRONN, Heinrich Georg, (1800-1862)
BROWN, Thomas, Captain, (1785-1862)
BRÜNNICH, Morten Thrane, (1737-1827)
BUCH, Christian Leopold von, Freiherr, (1774-1853)
BUCKMAN, James, (1814-1884)
BURMEISTER, Karl Hermann Konrad, (1807-1892)
BUSK, George, (1807-1886)
CARPENTER, William Benjamin, (1813-1885)
CARTER, Henry John, (1813-1895)
CARTER, James, (1813-1895)
CATULLO, Tommaso Antonio, (1782-1869)
CHARPENTIER, Jean de, (1786-1855)
CHENU, Jean Charles, (1808-1879)
COLLENOT, Jean Jacques, (1814-1893)
COLLINI, Cosimo Alessandro, (1727-1806)
CONRAD, Timothy Abbott, (1803-1877)
CONYBEARE, William Daniel, Reverend, (1787-1857)
COQUAND, Henri, (1813-1881)
CORDA, August Josef, (1811-1849)
CORNALIA, Emilio, (1824-1882)
COUTHONY, Joseph Pitty, (1808-1864)
CRAHAY, Jacques Guillaume, (1789-1855)
CRANE, Edward, (1822-1901)
CROSSE, Joseph Charles Hippolyte, (1826-1898)
CROSSKEY, H. W., Reverend, (1826-1893)
CUMBERLAND, George, (1754-1848)

INVERTEBRATE PALAEONTOLOGY 1800-1849 Cont'd.

CUNNINGTON, William, (1813-1906)
CURTIS, John, (1791-1862)
DANA, James Dwight, (1813-1895)
DAVIDSON, Thomas, (1817-1885)
DAWSON, John William, Sir, (1820-1899)
DEFRANCE, Jacques Louis Marie, (1758-1850)
DESHAYES, Gérard Paul, (1796-1875)
DES MOULINS, Charles, (1798-1875)
DESOR, Pierre Jean Edouard, (1811-1882)
DEWALQUE, Gilles Joseph Gustave, (1828-1905)
DILLWYN, Lewis Weston, (1778-1855)
DODERLEIN, Pietro, (1810-1895)
DOHRN, Felix Anton, (1840-1909)
DOVER, William Kinsey, (?1816-1891)
DUNCAN, Peter Martin, (1824-1891)
DUVAL-JOUVE, Joseph, (1810-1883)
EDWARDS, Frederic Erasmus, (1799-1875)
EICHWALD, Eduard Iwanowitsch von, (1795-1876)
EISEL, Robert, (1826-1917)
EMMONS, Ebenezer, (1800-1863)
EMMRICH, Hermann Friedrich, (1815-1879)
ESCHER VON DER LINTH, Arnold, (1807-1872)
ESCHWEGE, Wilhelm Ludwig, Baron, (1777-1855)
ESER, Anton Friedrich, (1798-1873)
ETHERIDGE, Robert, (1819-1903)
EUDES-DESLONGCHAMPS, Jacques Amand, (1794-1867)
EWALD, Julius Wilhelm, (1811-1891)
FAUJAS DE SAINT FOND, Barthélémy, (1741-1819)
FEISTMANTEL, Karel, (1819-1885)
FÉRUSSAC, André Étienne Just Pascal Joseph François d'Audebard, Baron de, (1786-1836)
FICHTEL, Leopold von, (fl.1798)
FISHER, Osmond, Reverend, (1817-1914)
FORBES, Edward, (1815-1854)
FORESTI, Lodovico, (1829-1913)
GASTALDI, Bartolomeo, (1818-1879)
GEINITZ, Hanns Bruno, (1814-1900)
GEMMELLARO, Carlo, (1787-1866)
GIBBES, Robert W., (1809-)
GILLIÉRON, Victor, (1826-1890)
GIRARD, Carl Adolph, (1814-1878)
GOLDENBERG, Friedrich, (1799-1881)
GOLDFUSS, Georg August, (1782-1848)
GRAS, Albin, (-1854)
GRATELOUP, Jean Pierre Sylvestre de, (1782-1861)
GREEN, Jacob, (1790-1841)
GRESSLY, Amanz, (1814-1865)
GREWINGK, Konstantin, (1819-1887)
GRIEPENKERL, Otto, (1820-1888)
GUIDONI, Gerolamo, (1794-1870)
GÜMBEL, Carl Wilhelm von, Ritter, (1823-1898)
HAGEN, Hermann August, (1817-1893)
HAGENOW, Friedrich von, (1797-1865)
HAIME, Jules, (1824-1856)

INVERTEBRATE PALAEONTOLOGY 1800-1849 Cont'd.

HANCOCK, Albany, (1806-1873)
HASSENCAMP, Ernst, (1824-1881)
HAUER, Franz von, Ritter, (1822-1899)
HAUPT, Karl, (1829-1882)
HAZSLINSZKY, Hazslini Frigyes Ágost, (1818-1896)
HÉBERT, Edmond, (1812-1890)
HEER, Oswald, (1809-1883)
HEHL, Johann Carl Ludwig, (1774-1853)
HENNAH, Richard, Reverend, (?1765-1846)
HERBICH, Franz, (1821-1887)
HISINGER, Wilhelm, (1766-1852)
HOENINGHAUS, Friedrich Wilhelm, (1770-1854)
HOERNES, Moritz, (1815-1868)
HOEVEN, Jan van der, (1801-1868)
HOLL, Harvey Buchanan, (1820-1886)
HOPE, Frederick William, Reverend, (1797-1862)
HOSIUS, August, (1825-1896)
HUDLESTON, Wilfrid Hudleston, (né Simpson), (1828-1909)
HUOT, Jean Jacques Nicolas, (1790-1845)
HUXLEY, Thomas Henry, (1825-1895)
JAMES, Uriah Pierson, (1811-1889)
JEFFREYS, John Gwyn, (1809-1885)
JONES, Thomas Rupert, (1819-1911)
KAUP, Johann Jakob, (1803-1873)
KEYSERLING, Alexander Andreevich, Count, (1815-1891)
KING, William, (1809-1886)
KJERULF, Theodor, (1825-1888)
KLIPSTEIN, August Wilhelm, (1801-1894)
KOCH, Carl, (1827-1882)
KONINCK, Laurent Guillaume de, (1809-1887)
KUDERNATSCH, Johann, (1819-1856)
KUTORGA, Stepan Semenovich, (1805-1861)
LAMARCK, Jean-Baptiste Pierre Antoine de Monet, Chevalier de, (1744-1829)
LEA, Isaac, (1792-1886)
LECKENBY, John, (1814-1877)
LEFEBVRE, Alexandre, (1798-1868)
LE HON, Henri Sebastian, (1809-1872)
LESUEUR, (1778-1846)
LEYMERIE, Alexander, (1801-1878)
LINDSTRÖM, Jonas Gustaf Oscar, (1829-1901)
LOCKE, John, (1792-1856)
LONSDALE, William, (1794-1871)
LORIOL LE FORT, Charles Louis Perceval de, (1828-1908)
LOVÉN, Sven, (1809-1895)
LUDWIG, Rudolph August Birminhold Sebastian, (1812-1880)
LÜTKEN, Christian Fredrik, (1827-1901)
MANTELL, Gideon Algernon, (1790-1852)
MARCOU, Jules, (1824-1898)
MARCY, Oliver, (1820-1899)
MARÈS, Paul, (1826-1900)
MARSSON, Theodor, (1816-1892)
MARTIN, William, (1767-1810)
MATHERON, Pierre Émile Philippe, (1807-1899)

INVERTEBRATE PALAEONTOLOGY 1800-1849 Cont'd.

MAYER-EYMAR, Karl David Wilhelm, (1826-1907)
McCOY, Frederick, Sir, (1823-1899)
MEEK, Fielding Bradford, (1817-1876)
MENEGHINI, Giuseppe Giovanni Antonio, (1811-1889)
MENGE, Anton, (1808-1880)
MÉTHERIE, Jean Claude de la, (1743-1817)
MEYER, Christian Erich Hermann von, (1801-1869)
MICHELIN, Jean Louis Hardouin, (1786-1867)
MICHELOTTI, Giovanni, (1814-1898)
MILNE-EDWARDS, Henri, (1800-1885)
MOORE, John Carrick, (1804-1898)
MORIÈRE, Jules Pierre Gilles, (1817-1888)
MORREN, Charles, (1807-1858)
MORTON, Samuel George, (1799-1851)
MURCHISON, Roderick Impey, Sir, (1792-1871)
NEUGEBOREN, Johann Ludwig, (1806-1887)
NORTON, Henry, (?1812-1892)
NORWOOD, Joseph Granville, (1807-1895)
NOULET, Jean Baptiste, (1802-1890)
NYST, Pierre Henri, (1813-1880)
OMBONI, Giovanni, (1829-1910)
PAGENSTECHER, Heinrich Alexander, (1825-1889)
PANDER, Christian Heinrich, (1794-1865)
PARKINSON, James, (1755-1824)
PARTSCH, Paul, (1791-1856)
PEACH, Charles William, (1800-1886)
PEARCE, Chaning, (1811-1847)
PENGELLY, William, (1812-1894)
PEREIRA DA COSTA, Francisco Antonio, (1809-1889)
PETERS, Karl, (1825-1881)
PHILIPPI, Rudolph Amandus, (1808-1904)
PHILLIPS, John, (1800-1874)
PICTET DE LA RIVE, François Jules, (1809-1872)
PIPER, George Harry, (1819-1897)
PIRONA, Giulio Andrea, (1822-1895)
POMEL, Nicolas Auguste, (1821-1898)
PONZI, Giuseppe di Tomaso, (1805-1885)
PRESL, Jan Svatopluk, (1791-1849)
REEVE, James, (1833-1920)
REEVE, Lovell, (1814-1865)
REINECKE, Johann Christoph Matthias, (1769-1818)
REUSS, August Emanuel von, Ritter, (1811-1873)
RICHTER, Reinhard, (1813-1884)
ROBERTSON, David, (1806-1896)
ROBINEAU-DESVOIDY, André Jean Baptiste, (1799-1857)
ROFE, John, (1801-1878)
ROLLE, Friedrich, (1827-1887)
RÖMER, Carl Ferdinand von, (1818-1891)
ROMINGER, Carl Ludwig, (1820-1907)
RÜPPELL, Eduard, (1794-1884)
SALTER, John William, (1820-1869)
SALWEY, Humphry, (1803-1877)
SANDBERGER, Fridolin Carl Ludwig von, (1826-1898)
SAY, Thomas, (1787-1834)

INVERTEBRATE PALAEONTOLOGY 1800-1849 Cont'd.

SCACCHI, Arcangelo, (1810-1893)
SCHIMPER, Karl Friedrich, (1803-1867)
SCHLOTHEIM, Ernst Friedrich von, Freiherr, (1764-1832)
SCHMID, Ernst Erhard, (1815-1885)
SCOULER, John, (-1874)
SEDGWICK, Adam, Reverend, (1785-1873)
SELYS-LONGCHAMPS, Michel Edmond de, Baron, (1813-1900)
SERRES, Marcel de, (1783-1862)
SHRUBSOLE, George William, (1827-1893)
SHUMARD, Benjamin Franklin, (1820-1869)
SIMPSON, Martin, (1798-1892)
SISMONDA, Eugenio, (1815-1870)
SMITH, William, (1769-1839)
SMITHE, Frederick, Reverend, (1822-1900)
SOLDANI, Ambrogio, Pater, (1736-1808)
SOWERBY, George Brettingham, I, (1788-1854)
SPEYER, Oscar, (1827-1882)
STABILE, Giuseppe, Abate, (1827-1869)
STEIN, Johannes Philip Emil Friedrich, (1814-1882)
STEININGER, Johann, (1794-1874)
STOKES, Charles, (1783-1853)
STOPPANI, Antonio, (1824-1891)
STRICKLAND, Hugh Edwin, (1811-1853)
SZABŐ, József, (1822-1894)
TERQUEM, Olry, (1797-1887)
THOMSON, James, (1823-1900)
TOURNOÜER, Jacques Raoul, (1822-1882)
TRAUTSCHOLD, Hermann Adolfovich, (1817-1902)
TROOST, Gerard, (1776-1850)
TROSCHEL, Franz Hermann, (1810-1882)
UBAGHS, Johann Casimir, (1829-1894)
UNGER, Franz Joseph Andreas Nicolas, (1800-1870)
VALPY, Robert Harris, (1819-1904)
VERNEUIL, Philippe Edouard Poulletier de, (1805-1873)
VINCENT, Gérard, (?1824-1899)
VOGT, Carl, (1817-1895)
VOLBORTH, Aleksandr Fedorovich, (1800-1876)
WACHSMUTH, Charles, (1829-1896)
WAGNER, William, (1796-1885)
WATELET, Adolphe, (1811-1880)
WEBSTER, Thomas, (1773-1844)
WEINKAUFF, H. C., (1817-1886)
WESTWOOD, John Obadiah, (1805-1893)
WETHERELL, Nathaniel Thomas, (1800-1875)
WHITE, Charles Abiathar, (1826-1910)
WHITFIELD, Robert Parr, (1828-1910)
WIECHMANN, Carl Michael, (1828-1883)
WINCHELL, Alexander, (1824-1891)
WOOD, Searles Valentine, (1798-1880)
WOODWARD, Samuel, (1790-1838)
WOODWARD, Samuel P., (1821-1865)
WORTHEN, Amos Henry, (1812-1888)
WRIGHT, Thomas, (1809-1884)
YOUNG, John, (1823-1900)

INVERTEBRATE PALAEONTOLOGY 1800-1849 Cont'd.

ZITTEL, Karl Alfred von, (1839-1904)

1850-1899

ABBOTT, William James Lewis, (1853-1933)
ABEL, Othenio, (1875-1946)
ADAMS, Andrew Leith, (-1882)
ADAMS, Charles Baker, (1814-1853)
AGASSIZ, Alexander, (1835-1910)
AGASSIZ, Louis Jean Rodolphe, (1807-1873)
AIM, William Lennie, (?1872-1938)
ALBERTI, Friedrich August von, (1795-1878)
ALDRICH, Truman Heminway, (1848-1932)
ALLEN, Henry Atwool, (?1854-1934)
ALLMAN, George James, (1812-1898)
AMALITSKII, Vladimir Prochorovich, (1860-1917)
AMI, Henry Mark, (1858-1931)
AMMON, Ludwig von, (1850-1922)
ANDERSSON, Oscar Frithiof, (1867-1911)
ANDRUSSOV, Nikolai Ivanovich, (1861-1924)
ANDRZEJOWSKI, Antoni, (1785-1868)
ANGELIN, Nils Peter, (1805-1876)
ARCHIAC, Étienne Jules Adolphe Dexmier de Simon, Vicomte d', (1802-1868)
ARNOLD, Ralph, (1875-1961)
AUINGER, Mathias, (?1810-1890)
AUSTIN, Charles Edward, (1819-1893)
AUSTIN, Thomas, Fort-Major, (1794-1881)
AYCKE, Johann Christian, (1766-1854)
BAILY, William Hellier, (1819-1888)
BAIN, Francis, (1842-1894)
BAKER, Frank Collins, (1867-1942)
BALDWIN, Walter, (?1878-1956)
BARBOT DE MARNY, N. P., (1832-1877)
BARKAS, Thomas Pallister, (1819-1891)
BARON, Richard, Reverend, (1847-1907)
BARRANDE, Joachim, (1799-1883)
BARRIS, Willis Hervey, (1820-1901)
BARROIS, Charles Eugène, (1851-1939)
BASSLER, Raymond Smith, (1878-1961)
BATHER, Francis Arthur, (1863-1934)
BAUMBERGER, Ernst, (1866-1935)
BAUR, Carl Theodor von, (1836-1911)
BAYAN, Joseph Félix Ferdinand, (1845-1874)
BAYFIELD, Thomas Gabriel, (1817-1893)
BAYLE, Émile, (1819-1895)
BEACHLER, Charles S., (1870-1894)
BEECHER, Charles Emerson, (1856-1904)
BEESLEY, Thomas, (1818-1896)
BEISSEL, Ignaz, (1820-1887)
BELL, Dugald, (1827-1892)
BELL, Robert George, (1833-1888)
BELLARDI, Luigi, (1818-1889)
BELLINI, Raffaelo, (1874-1930)
BENEDEN, Edouard van, (1846-1910)

INVERTEBRATE PALAEONTOLOGY 1850-1899 Cont'd.

BERENDT, Gottlieb Michael, (1836-1920)
BERGERON, Jules, (1853-1919)
BERNARD, Félix, (1863-1898)
BERTHELIN, Jean Georges, (1840-1897)
BERTKAU, Philipp, (1849-1895)
BEUSHAUSEN, Hermann Emil Louis, (1863-1904)
BEYRICH, August Heinrich Ernst, (1815-1896)
BIANCONI, Gian Giuseppe, (1809-1878)
BIELZ, Michael, (1787-1866)
BIGOT, Alexandre, (1863-1953)
BIGSBY, John Jeremiah, (1792-1881)
BILLINGS, Elkanah, (1820-1876)
BILLINGS, Walter R., (?1849-1920)
BINKHOORST TOT DEN BINKHOORST, Johan Theodoor, (1810-1876)
BIRKMAIER, Anton, (-1926)
BIRLEY, Caroline, (1851-1907)
BLACKMORE, Humphrey Purnell, (-1928)
BLAKE, John Frederick, Reverend, (1839-1906)
BLANCKENHORN, Max Ludwig Paul, (1861-1947)
BLANFORD, Henry Francis, (1834-1893)
BOFILL Y POCH, Arturo, (1846-1929)
BÖHMERSHEIM, August Böhm von, Edler, (1858-1930)
BOISDUVAL, Jean-Baptiste Alphonse, (1799-1879)
BOLTON, Herbert, (1863-1936)
BÖSE, Emil, (1868-1972)
BOSQUET, Joseph Augustin Hubert, (1814-1880)
BOURDOT, Jules Damase, (1837-1906)
BOURGUIGNAT, Jules René, (1829-?1919)
BOWERBANK, James Scott, (1797-1877)
BRADLEY, Frank Howe, (1838-1879)
BRANDES, Theodor, (-1916)
BRANDT, Alexander von, (1844-1932)
BRAUER, Friedrich Moritz, (1832-1904)
BRØGGER, Waldemar Christofer, (1851-1940)
BROILI, Ferdinand, (1874-1946)
BRONGNIART, Charles Jules Edme, (1859-1899)
BRONN, Heinrich Georg, (1800-1862)
BROWN, Thomas, Captain, (1785-1862)
BRUSINA, Spiridion, (1845-1908)
BUCH, Christian Leopold von, Freiherr, (1774-1853)
BUCKMAN, James, (1814-1884)
BUCKMAN, Sidney Savory, (1860-1929)
BULLEN, Robert Ashington, Reverend, (1850-1912)
BURCKHARDT, Carl Rudolf, (1866-1908)
BURMEISTER, Karl Hermann Konrad, (1807-1892)
BUSK, George, (1807-1886)
CALL, Richard Ellsworth, (1856-1917)
CALVIN, Samuel, (1840-1911)
CAMPBELL, Henry Donald, (1862-1934)
CAPELLINI, Giovanni, (1833-1922)
CARPENTER, Philip Herbert, (1852-1891)
CARPENTER, William Benjamin, (1813-1885)
CARRUTHERS, William, (1830-1922)
CARTER, Henry John, (1813-1895)

INVERTEBRATE PALAEONTOLOGY 1850-1899 Cont'd.

CARTER, James, (1813-1895)
CATULLO, Tommaso Antonio, (1782-1869)
CAZIOT, Eugène, (1844-1931)
CHADWICK, George Halcott, (1876-1953)
CHAMPERKNOWNE, Arthur, (1839-1887)
CHAPER, Maurice Armand, (1834-1896)
CHAPMAN, Frederick, (1864-1943)
CHARLESWORTH, Edward, (1813-1893)
CHARPENTIER, Jean de, (1786-1855)
CHELOT, Émile, (1862-1930)
CHENU, Jean Charles, (1808-1879)
CHOFFAT, Paul, (1849-1919)
CLARK, William Bullock, (1860-1917)
CLARKE, John Mason, (1857-1925)
CLELAND, Herdman Fitzgerald, (1869-1935)
COCKERELL, Theodore Dru Alison, (1866-1948)
COLENUTT, George William, (?1862-1944)
COLLENOT, Jean Jacques, (1814-1893)
COLLOT, Louis Marie François, (1846-1915)
CONRAD, Timothy Abbott, (1803-1877)
CONYBEARE, William Daniel, Reverend, (1787-1857)
COOPER, James Graham, (1830-1902)
COQUAND, Henri, (1813-1881)
CORNALIA, Emilio, (1824-1882)
CORNET, François Léopold, (1834-1887)
CORNWALL, Ira Edmund, (1875-1962)
CORTI, Benedetto, (1868-1907)
COSSMANN, Maurice, (1850-1924)
COTTEAU, Gustave Honoré, (1818-1894)
COTTREAU, Jean, (1877-1945)
COUPER, James Graham, (1830-1902)
COUTHONY, Joseph Pitty, (1808-1864)
CRAHAY, Jacques Guillaume, (1789-1855)
CRANE, Edward, (1822-1901)
CRICK, George Charles, (1856-1917)
CROSSE, Joseph Charles Hippolyte, (1826-1898)
CROSSKEY, H. W., Reverend, (1826-1893)
CUNNINGTON, William, (1813-1906)
CURTIS, John, (1791-1862)
D'ACHIARDI, Antonio, (1839-1902)
DALL, William Healey, (1845-1927)
DAMES, Wilhelm Barnim, (1843-1898)
DANA, James Dwight, (1813-1895)
DARELL, Robert Darell Smythe, (1851-1936)
DAUTZENBERG, Philippe, (1849-1935)
DAVIDSON, Thomas, (1817-1885)
DAVIES, Arthur Morley, (1869-1959)
DAWSON, John William, Sir, (1820-1899)
DECKER, Charles Elijah, (1868-1958)
DEECKE, Johannes Ernst Wilhelm, (1862-1934)
DE GREGORIO, Antonio, Marchese, (1855-1930)
DEICHMÜLLER, Johannes, (1854-)
DENCKMANN, August, (1860-1925)
DE NICOLIS, Enrico, (1841-1908)

INVERTEBRATE PALAEONTOLOGY 1850-1899 Cont'd.

DERBY, Orville Adelbert, (1851-1915)
DESHAYES, Gérard Paul, (1796-1875)
DES MOULINS, Charles, (1798-1875)
DESOR, Pierre Jean Edouard, (1811-1882)
DEWALQUE, Gilles Joseph Gustave, (1828-1905)
DIBLEY, George Edward, (1857-1943)
DIENER, Carl, (1862-1928)
DILLWYN, Lewis Weston, (1778-1855)
DI STEFANO, Giovanni, (1856-1918)
DODERLEIN, Pietro, (1810-1895)
DOHRN, Felix Anton, (1840-1909)
DOLLÉ, Louis, (1878-1965)
DOLLFUS, Auguste, (1840-1869)
DOLLFUS, Gustave Frédéric, (1850-1931)
DONCIEUX, Louis, (1874-1960)
DONNEZAN, Albert, (1846-1914)
DORMAL, Victor, (-1900)
DOUVILLÉ, Joseph Henri Ferdinand, (1846-1937)
DOVER, William Kinsey, (?1816-1891)
DREVERMANN, Fritz, (1875-1932)
DUNCAN, Peter Martin, (1824-1891)
DUPONT, Édouard François, (1841-1911)
DUVAL-JOUVE, Joseph, (1810-1883)
DWERRYHOUSE, Arthur Richard, (1867-1931)
EBERT, Theodor, (1857-1899)
EDWARDS, Frederic Erasmus, (1799-1875)
EICHWALD, Eduard Iwanowitsch von, (1795-1876)
EISEL, Robert, (1826-1917)
ELLES, Gertrude Lilian, (1872-1960)
EMMONS, Ebenezer, (1800-1863)
EMMRICH, Hermann Friedrich, (1815-1879)
ESCHER VON DER LINTH, Arnold, (1807-1872)
ESCHWEGE, Wilhelm Ludwig, Baron, (1777-1855)
ESER, Anton Friedrich, (1798-1873)
ÉTALLON, Claude Auguste, (1826-1862)
ETHERIDGE, Robert, (1819-1903)
ETHERIDGE, Robert Jr., (1847-1920)
EUDES-DESLONGCHAMPS, Eugène, (1830-1889)
EUDES-DESLONGCHAMPS, Jacques Amand, (1794-1867)
EWALD, Julius Wilhelm, (1811-1891)
FAVRE, Ernest, (1845-1925)
FEISTMANTEL, Karel, (1819-1885)
FELIX, Johannes Paul, (1859-1941)
FENDRYCH, Otakar, (1860-1945)
FIEBELKORN, Max, (1869-1912)
FISCHER, Paul Henri, (1835-1893)
FISHER, Osmond, Reverend, (1817-1914)
FLEURY, Ernest, (1878-1958)
FOERSTE, August Frederick, (1862-1936)
FONTANNES, Charles François, (1839-1886)
FOORD, Arthur Humphreys, (1845-1934)
FORBES, Edward, (1815-1854)
FORESTI, Lodovico, (1829-1913)
FORMA, Ernesto, (1869-1911)

INVERTEBRATE PALAEONTOLOGY 1850-1899 Cont'd.

FORNASINI, Carlo, (1854-1931)
FORTAU, René, (1867-1920)
FRAIPONT, Julien, (1857-1910)
FRAUSCHER, Karl, (1852-1914)
FRECH, Fritz, (1861-1917)
FRIEDBERG, Wilhelm, (1873-1941)
FRITSCH, Anton, (1832-1913)
FRITSCH, Karl von, Freiherr, (1838-1906)
FUCHS, Theodor, (1842-1925)
FUGGER, Friedrich Eberhard, (1842-1919)
GABB, William More, (1839-1878)
GAGEL, Kurt, (1865-1927)
GARWOOD, Edmund Johnston, (1864-1949)
GASTALDI, Bartolomeo, (1818-1879)
GASTER, Christopher Theodore Augustine, (1878-1963)
GAUTHIER, Victor Auguste, (1837-1911)
GEINITZ, Hanns Bruno, (1814-1900)
GEMMELLARO, Carlo, (1787-1866)
GÉRARD, Charles, Colonel, (1866-1953)
GERBER, Eduard, (1876-1956)
GEYER, David, (1855-1932)
GEYER, Georg, (1857-1938)
GIARD, Alfred, (1846-1908)
GILLIÉRON, Victor, (1826-1890)
GIRARD, Carl Adolph, (1814-1878)
GIRTY, George Herbert, (1869-1939)
GLASS, Norman, Reverend, (1832-1895)
GOBANZ, Josef, (-1899)
GOLDENBERG, Friedrich, (1799-1881)
GOMES, Jacinto Pedro, (1844-1910)
GORDON, Charles Henry, (1857-1934)
GORJANOVIĆ-KRAMBERGER, Dragutin, (1856-1937)
GOSSELET, Jules Auguste, (1832-1916)
GOTTSCHE, Carl Christian, (1855-1909)
GOTTSCHICK, Franz, (1865-1927)
GOULD, Charles, (-1895)
GRABAU, Amadeus William, (1870-1946)
GRAS, Albin, (-1854)
GRATELOUP, Jean Pierre Sylvestre de, (1782-1861)
GREGORY, John Walter, (1864-1932)
GREPPIN, Eduard, (1856-1927)
GRESSLY, Amanz, (1814-1865)
GREWINGK, Konstantin, (1819-1887)
GRIEPENKERL, Otto, (1820-1888)
GROOM, Theodore Thomas, (1863-)
GRZYBOWSKI, Józef, (1869-1922)
GUIDONI, Gerolamo, (1794-1870)
GULICK, John Thomas, (1832-1923)
GÜMBEL, Carl Wilhelm von, Ritter, (1823-1898)
GUPPY, Robert John Lechmere, (1836-1916)
GURLEY, William Frank Eugene, (1854-1943)
HAASE, Erich, (1857-1894)
HÄBERLE, Daniel, (1864-1934)
HAECKEL, Ernst Heinrich Philipp August, (1834-1919)

INVERTEBRATE PALAEONTOLOGY 1850-1899 Cont'd.

HAGEN, Hermann August, (1817-1893)
HAGENOW, Friedrich von, (1797-1865)
HAIME, Jules, (1824-1856)
HALAVÁTS, Gyula, (1853-1926)
HALFAR, Anton, (1836-1893)
HALL, Thomas Sergeant, (1858-1915)
HALL, Townshend Monckton, (1845-1899)
HANCOCK, Albany, (1806-1873)
HANDLIRSCH, Anton, (1865-1935)
HARGER, Oscar, (1843-1882)
HARMER, Frederic William, (1835-1923)
HARRIS, George Frederick, (1862-1906)
HARRIS, Gilbert Dennison, (1864-1952)
HARTT, Charles Frederic, (1840-1878)
HASSE, Carl, (1841-1922)
HASSENCAMP, Ernst, (1824-1881)
HAUER, Franz von, Ritter, (1822-1899)
HAUPT, Karl, (1829-1882)
HAZSLINSZKY, Hazslini Frigyes Ágost, (1818-1896)
HÉBERT, Edmond, (1812-1890)
HEER, Oswald, (1809-1883)
HERBICH, Franz, (1821-1887)
HERON-ALLEN, Edward, (1861-1943)
HERRICK, Clarence Luther, (1858-1904)
HERRIES, Robert Stansfield, (?1860-1941)
HEYDEN, Lucas von, (1838-1915)
HILBER, Vincenz, (1853-1931)
HILGENDORF, Franz, (1839-1904)
HIND, Wheelton, (1859-1920)
HINDE, George Jennings, (1839-1918)
HISINGER, Wilhelm, (1766-1852)
HOERNES, Moritz, (1815-1868)
HOEVEN, Jan van der, (1801-1868)
HOLL, Harvey Buchanan, (1820-1886)
HOLM, Edvard Johann Gerhard, (1853-1926)
HOLMES, Francis S., (1815-1882)
HOLUB, Emil, (1847-1902)
HOLZAPFEL, Eduard, (1853-1913)
HOPE, Frederick William, Reverend, (1797-1862)
HOPKINSON, John, (1844-1919)
HORUSITZKY, Henrik, (1870-1945)
HOSIUS, August, (1825-1896)
HUDLESTON, Wilfrid Hudleston, (né Simpson), (1828-1909)
HUDSON, George Henry, (1855-1934)
HUENE, Friedrich von, Freiherr, (1875-1969)
HUMBERT, Alois, (1829-1887)
HUNTER-SELKIRK, John R. S., (1835-1898)
HUOT, Jean Jacques Nicolas, (1790-1845)
HUTTON, Frederick Wollaston, (1836-1905)
HUTTON, Harriet Mary, (1873-1937)
HUXLEY, Thomas Henry, (1825-1895)
ISSEL, Arturo, (1842-1922)
JACCARD, Samuel Auguste, (1833-1895)
JAECKEL, Otto, (1863-1929)

INVERTEBRATE PALAEONTOLOGY 1850-1899 Cont'd.

JAMES, Uriah Pierson, (1811-1889)
JAROSZ, Jan, (1877-1944)
JEFFREYS, John Gwyn, (1809-1885)
JENKINS, Henry Michael, (1841-1886)
JOHNSTON, Mary Sophia, (1875-1955)
JOHNSTON, Robert Mackenzie, (-1918)
JONES, Ernest Lloyd, (?1863-1942)
JONES, Thomas Rupert, (1819-1911)
JUKES-BROWNE, Alfred John, (1851-1914)
JULIEN, Pierre Alphonse, (1838-1905)
KAFKA, Josef, (1858-1929)
KARPINSKY, Alexander Petrovich, (1847-1936)
KARSCH, Ferdinand, (1853-1936)
KASTNER, Karl, (1847-1907)
KATZER, Bedřich, (1861-1925)
KAUP, Johann Jakob, (1803-1873)
KAYSER, Friedrich Heinrich Emanuel, (1845-1927)
KEEPING, Walter, (1854-1888)
KENNARD, Alfred Santer, (1870-1948)
KESSLER, Paul, (?1882-1927)
KEYES, Charles Rollin, (1864-1942)
KEYSERLING, Alexander Andreevich, Count, (1815-1891)
KEYTE, Ivy Allen, (1878-1931)
KIAER, Johan Aschehong, (1869-1931)
KINDLE, Edward Martin, (1869-1940)
KING, William, (1809-1886)
KIRKBY, James Walker, (1834-1901)
KITCHIN, Finlay Lorimer, (1870-1934)
KJERULF, Theodor, (1825-1888)
KLIPSTEIN, August Wilhelm, (1801-1894)
KLOUČEK, Celda, (1855-1935)
KOBY, Fréderic Louis, (1852-1930)
KOCH, Carl, (1827-1882)
KOCH, Gottlieb Karl David von, (1849-1914)
KOERT, Willi, (1875-1927)
KOKEN, Ernst Friedrich Rudolph Karl von, (1860-1912)
KOLBE, Hermann Julius, (1855-)
KONINCK, Laurent Guillaume de, (1809-1887)
KUDERNATSCH, Johann, (1819-1856)
KUNTH, Karl Ernst Albrecht, (1842-1871)
KUSTA, Johann, (1845-1900)
KUTORGA, Stepan Semenovich, (1805-1861)
LAKE, Philip, (1865-1949)
LAMBERT, Jules Mathieu, (1848-1940)
LANDOIS, Hermann, (1835-1905)
LANG, William Dickson, (1878-1966)
LAPPARENT, Albert Auguste de, (1839-1908)
LAPWORTH, Charles, (1842-1920)
LARTET, Louis, (1840-1899)
LAUBE, Gustav C., (1839-1923)
LAW, Robert, (1840-1907)
LEA, Isaac, (1792-1886)
LECKENBY, John, (1814-1877)
LEFEBVRE, Alexandre, (1798-1868)

INVERTEBRATE PALAEONTOLOGY 1850-1899 Cont'd.

LE HON, Henri Sebastian, (1809-1872)
LEONHARD, Richard, (1870-1916)
LERICHE, Maurice, (1875-1948)
LEYMERIE, Alexander, (1801-1878)
LIENENKLAUS, Ernst, (1849-1905)
LINDSTRÖM, Jonas Gustaf Oscar, (1829-1901)
LINSTOW, Otto von, (1872-1929)
LISSON, Carlos I., (1868?-1957)
LOCARD, Arnould, (1841-1904)
LONGSTAFF, Mary Jane, (née Donald), (1856-1935)
LONSDALE, William, (1794-1871)
LOOMIS, Theodor, (1875-1909)
LÖRENTHEY, Imre, (1867-1917)
LORENZ, Theodor, (1875-1909)
LORIOL LE FORT, Charles Louis Perceval de, (1828-1908)
LOVÉN, Sven, (1809-1895)
LOVISATO, Domenico, (1842-1916)
LUDWIG, Rudolph August Birminhold Sebastian, (1812-1880)
LUNIEWSKI, Adam, (1887-1945)
LÜTKEN, Christian Fredrik, (1827-1901)
LYELL, Charles, Sir, (1797-1875)
MAAS, Otto, (1867-1916)
MACPHERSON, José, Don, (1839-1902)
MAILLARD, Gustave, (1860-1891)
MAIRE, Julien Victor, (1856-1941)
MAKOWSKY, Alexander, (1833-1908)
MALAISE, Constantin H. L., (1834-1916)
MANSFIELD, Wendell Clay, (1874-1939)
MANTELL, Gideon Algernon, (1790-1852)
MARCOU, Jules, (1824-1898)
MARCY, Oliver, (1820-1899)
MARÈS, Paul, (1826-1900)
MARSHALL, Patrick, (1869-1950)
MARSSON, Theodor, (1816-1892)
MARTELLI, Alessandro, (1877-1934)
MARTIN, Handel Tongue, (1862-1931)
MARTIN, Johann Karl Ludwig, (1851-1942)
MATHERON, Pierre Émile Philippe, (1807-1899)
MAYER-EYMAR, Karl David Wilhelm, (1826-1907)
McCOY, Frederick, Sir, (1823-1899)
MEEK, Fielding Bradford, (1817-1876)
MELI, Romolo, (1852-1920)
MENEGHINI, Giuseppe Giovanni Antonio, (1811-1889)
MENGE, Anton, (1808-1880)
MENZEL, Karl Gustav Johannes, (1875-1914)
MEUNIER, Fernand, (1868-1926)
MEŸER, Charles John Adrian, (1832-1900)
MEYER, Christian Erich Hermann von, (1801-1869)
MICHAEL, Richard, (1869-1925)
MICHELIN, Jean Louis Hardouin, (1786-1867)
MICHELOTTI, Giovanni, (1814-1898)
MICKWITZ, Karl August von, (1849-1910)
MILNE-EDWARDS, Alphonse, (1835-1900)
MILNE-EDWARDS, Henri, (1800-1885)

INVERTEBRATE PALAEONTOLOGY 1850-1899 Cont'd.

MILTON, John Herbert, (?1862-1925)
MOBERG, Johan Christian, (1854-1915)
MOJSISOVICS, Johann August Georg Edmund, Edler von Mojsvàr, (1839-1907)
MONTEROSATO, Tommaso di Maria, Marchese di, (1841-1927)
MOORE, John Carrick, (1804-1898)
MORGAN, Joseph Bickerton, (1859-1894)
MÖRICKE, Wilhelm, (1861-1897)
MORIÈRE, Jules Pierre Gilles, (1817-1888)
MORREN, Charles, (1807-1858)
MORTON, Samuel George, (1799-1851)
MOSELEY, Henry Nottidge, (1845-1891)
MÜLLER, Karl Christian Gottfried Ludwig, (1862-1906)
MUNIER-CHALMAS, Ernest Philippe Auguste, (1843-1903)
MURCHISON, Roderick Impey, Sir, (1792-1871)
NEUGEBOREN, Johann Ludwig, (1806-1887)
NEUMAYR, Melchior, (1845-1890)
NEWTON, Richard Bullen, (1850-1926)
NICHOLSON, Henry Alleyne, (1844-1899)
NORTON, Henry, (?1812-1892)
NORWOOD, Joseph Granville, (1807-1895)
NOULET, Jean Baptiste, (1802-1890)
NOVÁK, Otomar Pravoslav, (1851-1892)
NYST, Pierre Henri, (1813-1880)
OEHLERT, Daniel Victor, (1849-1920)
OMBONI, Giovanni, (1829-1910)
OPPEL, Albert, (1831-1865)
OPPENHEIM, Leo Paul, (1863-1934)
OPPLIGER, Fritz, (1861-1932)
ORTMANN, Arnold Edward, (1863-1927)
OUSTALET, Émile, (1844-1905)
PACKARD, Alpheus Spring, Jr., (1839-1905)
PAGENSTECHER, Heinrich Alexander, (1825-1889)
PANDER, Christian Heinrich, (1794-1865)
PANTANELLI, Dante, (1844-1913)
PAQUIER, Victor Lucien, (1870-1911)
PARKER, William Albert, (1855-1918)
PARKS, William Arthur, (1868-1936)
PARONA, Carlo Fabrizio, (1855-1939)
PARTSCH, Paul, (1791-1856)
PATRUNKY, H., (1854-1930)
PAULCKE, Wilhelm, (1873-1949)
PAULUCCI, Marianna, Marchesa, (1835-1919)
PEACH, Charles William, (1800-1886)
PENGELLY, William, (1812-1894)
PEREIRA DA COSTA, Francisco Antonio, (1809-1889)
PEREIRA DE LIMA, Wenceslau de Sousa, (1858-1920)
PERGENS, Eduard Willem Gerard, (1862-1917)
PERNER, Jaroslav, (1869-1947)
PERON, Alphonse, (1834-1908)
PERVINQUIÈRE, Léon, (1873-1913)
PETERS, Karl, (1825-1881)
PETKOVIĆ, Vladamir K., (1873-1934)
PHILIPPI, Rudolph Amandus, (1808-1904)

INVERTEBRATE PALAEONTOLOGY 1850-1899 Cont'd.

PHILLIPS, John, (1800-1874)
PICTET DE LA RIVE, François Jules, (1809-1872)
PILSBRY, Henry Augustus, (1862-1957)
PIPER, George Harry, (1819-1897)
PIRONA, Giulio Andrea, (1822-1895)
POČTA, Filip, (1859-1924)
POMEL, Nicolas Auguste, (1821-1898)
POMPECKJ, Josef Felix, (1867-1930)
PONZI, Giuseppe di Tomaso, (1805-1885)
POPOVICI-HAŢEG, Valeriu, (1868-1929)
PRINGLE, John, (1877-1948)
RADOVANOVIĆ, Svetolik A., (1863-1928)
RATHBURN, Richard, (1852-1918)
RAVN, Jesper Peter Johansen, (1866-1951)
RAW, Frank, (?1875-1961)
REBEL, Hans, (1861-)
REED, Frederick Richard Cowper, (1869-1946)
REEVE, James, (1833-1920)
REEVE, Lovell, (1814-1865)
REIS, Otto Maria, (1862-1934)
REMEŠ, Mořic, (1867-1959)
RENEVIER, Eugène, (1831-1906)
REUSS, August Emanuel von, Ritter, (1811-1873)
REYNOLDS, Sidney Hugh, (1867-1949)
RIAZ, Auguste de, (-1920)
RICHTER, Reinhard, (1813-1884)
RIDLEY, Henry Nicholas, (1855-1956)
RIGAUX, Edmond, (1839-1915)
RISTORI, Giuseppe, (1856-1905)
ROBERTS, George Edward, (1831-1865)
ROBERTS, Thomas, (1856-1892)
ROBERTSON, David, (1806-1896)
ROBINEAU-DESVOIDY, André Jean Baptiste, (1799-1857)
RODLER, Alfred, (1861-1890)
ROEDER, Charles, (1848-1911)
ROFE, John, (1801-1878)
ROLLE, Friedrich, (1827-1887)
ROLLIER, Louis, (1859-1931)
ROMAN, Albert Frédéric, (1871-1943)
RÖMER, Carl Ferdinand von, (1818-1891)
ROMINGER, Carl Ludwig, (1820-1907)
ROTHPLETZ, August, (1853-1918)
ROWE, Arthur Walton, (1858-1926)
RUEDEMANN, Rudolf, (1864-1956)
RÜPPELL, Eduard, (1794-1884)
SACCO, Federico, (1864-1948)
SALTER, John William, (1820-1869)
SALWEY, Humphry, (1803-1877)
SANDBERGER, Fridolin Carl Ludwig von, (1826-1898)
SARDESON, Frederick William, (1866-1959)
SAVIN, Léon Héli, Lieutenant-Colonel, (1851-1907)
SCACCHI, Arcangelo, (1810-1893)
SCHALCH, Ferdinand, (1848-1918)
SCHELLWIEN, Ernst, (1866-1906)

INVERTEBRATE PALAEONTOLOGY 1850-1899 Cont'd.

SCHIMPER, Karl Friedrich, (1803-1867)
SCHLECHTENDAL, Dietrich von, (1834-1916)
SCHLOENBACH, Georg Justin Carl Urban, (1841-1870)
SCHLUETER, Clemens August, (1835-1906)
SCHMID, Ernst Erhard, (1815-1885)
SCHMIDT, Friedrich Bogdanovich "Bohdan", (1832-1908)
SCHRÖDER, Richard, (1853-1916)
SCHROEDER, Henry Carl, (1859-1927)
SCHUCHERT, Charles, (1858-1942)
SCHÜTZE, Ewald, (1873-1908)
SCHWERTSCHLAGER, Joseph, (1853-1924)
SCOULER, John, (-1874)
SCUDDER, Samuel Hubbard, (1837-1911)
SEDGWICK, Adam, Reverend, (1785-1873)
SEGUENZA, Luigi, (1873-1908)
SELLARDS, Elias Howard, (1875-1961)
SELYS-LONGCHAMPS, Michel Edmond de, Baron, (1813-1900)
SEMPER, Otto, (1830-1907)
SERRES, Marcel de, (1783-1862)
SEUNES, Jean, (1849-1920)
SHAKESPEAR, Ethel Mary Reader, (née Wood), Dame, (1871-?1946)
SHALER, Nathaniel Southgate, (1841-1906)
SHARMAN, George, (1832-1914)
SHIMEK, Bohumil, (1861-1937)
SHIMER, Hervey Woodburn, (1872-1965)
SHRUBSOLE, George William, (1827-1893)
SHUMARD, Benjamin Franklin, (1820-1869)
SIEMIRADZKI, Józef, (1858-1933)
SIMPSON, Martin, (1798-1892)
SISMONDA, Eugenio, (1815-1870)
SLADEN, Walter Percy, (1849-1900)
SMITH, Edgar Albert, (1847-1916)
SMITH, James Perrin, (1864-1931)
SMITH, John, (1845-1930)
SMITHE, Frederick, Reverend, (1822-1900)
SOENDEROP, Fritz Hermann Gustav, (1876-1914)
SOLLAS, William Johnson, (1849-1936)
SORDELLI, Ferdinando, (?1837-1916)
SOWERBY, George Brettingham, I, (1788-1854)
SPEYER, Oscar, (1827-1882)
SPRINGER, Frank, (1848-1924)
STABILE, Giuseppe, Abate, (1827-1869)
STANTON, Timothy William, (1860-1953)
STATUTI, Augusto, (1829-1911)
ŞTEFANESCU, Gregoriu, (1838-1911)
ŞTEFĂNESCU, Sabba, (1857-1931)
STEIN, Johannes Philip Emil Friedrich, (1814-1882)
STEININGER, Johann, (1794-1874)
STEPHENSON, Lloyd William, (1876-1962)
STOBBS, John Thomas, (?1866-1942)
STOKES, Charles, (1783-1853)
STOLICZKA, Ferdinand, (1838-1874)
STOLLEY, Ernst, (1869-1944)
STOPES, Henry, (1852-1902)

INVERTEBRATE PALAEONTOLOGY 1850-1899 Cont'd.

STOPPANI, Antonio, (1824-1891)
STORRIE, John, (1844-1901)
STRICKLAND, Hugh Edwin, (1811-1853)
SUESS, Eduard, (1831-1914)
SWINNERTON, Henry Hurd, (1875-1967)
SZABŐ, József, (1822-1894)
TARAMELLI, Torquato, (1845-1922)
TATE, Ralph, (1840-1901)
TAURER-GALLENSTEIN, Hans von, Ritter, (1846-1927)
TAUSCH VON GLÖCKELSTHURN, Leopold, (1858-1899)
TAWNEY, Edward Bernard, (1840-1882)
TCHERNYCHEV, Féodoss Nikolaievich, (1856-1914)
TERQUEM, Olry, (1797-1887)
THEVENIN, Armand, (1870-1918)
THOMAS, Abram Owen, (1876-1931)
THOMSON, James, (1823-1900)
TILTON, John Littlefield, (1863-1930)
TOMBLESON, Frederick Binns, (?1872-1938)
TÖRNQUIST, Sven Leonhard, (1840-1920)
TOUCAS, Joseph Aristide, (1843-1911)
TOULA, Franz, (1845-1920)
TOURNOÜER, Jacques Raoul, (1822-1882)
TRASK, John Boardman, (1824-1879)
TRAUTSCHOLD, Hermann Adolfovich, (1817-1902)
TRAXLER, Lászlo, (1868-1898)
TREACHER, Llewellyn, (1859-1943)
TRIBOLET, Georges de, (1830-1873)
TRIBOLET-HARDY, Frédéric Maurice de, (1852-1929)
TROSCHEL, Franz Hermann, (1810-1882)
TULLBERG, Sven Axel Theodor, (1852-1886)
TWENHOFEL, William Henry, (1875-1957)
UBAGHS, Johann Casimir, (1829-1894)
UHLIG, Victor Karl, (1857-1911)
ULRICH, Edward Oscar, (1857-1944)
UNGER, Franz Joseph Andreas Nicolas, (1800-1870)
VACEK, Michael, (1848-1925)
VALPY, Robert Harris, (1819-1904)
VAN INGEN, Gilbert, (1869-1928)
VAUGHAN, Arthur, (1868-1915)
VERNEUIL, Philippe Edouard Poulletier de, (1805-1873)
VERRILL, Addison Emery, (1839-1926)
VINCENT, Gérard, (?1824-1899)
VITÅLIS, István, (1871-1947)
VOGDES, Anthony Wayne, General, (1843-1923)
VOGT, Carl, (1817-1895)
VOLBORTH, Aleksandr Fedorovich, (1800-1876)
WAAGEN, Wilhelm Heinrich, (1841-1900)
WACHSMUTH, Charles, (1829-1896)
WAGNER, William, (1796-1885)
WÄHNER, Franz, (1856-1932)
WALCOTT, Charles Doolittle, (1850-1927)
WALKER, John Francis, (1839-1907)
WANNER, Johannes, (1878-1957)
WATELET, Adolphe, (1811-1880)

INVERTEBRATE PALAEONTOLOGY 1850-1899 Cont'd.

WEINKAUFF, H. C., (1817-1886)
WELLER, Stuart, (1870-1927)
WELTNER, Wilhelm, (1854-1917)
WESTWOOD, John Obadiah, (1805-1893)
WETHERELL, Nathaniel Thomas, (1800-1875)
WEYENBERGH, Hendrik, (1842-1885)
WHIDBORNE, George Ferris, Reverend, (1846-1910)
WHITE, Charles Abiathar, (1826-1910)
WHITEAVES, John Joseph Frederick, (1835-1909)
WHITFIELD, Robert Parr, (1828-1910)
WHYMPER, Edward, (1840-1911)
WIECHMANN, Carl Michael, (1828-1883)
WIMAN, Carl Johan Josef Ernst, (1867-1944)
WINCHELL, Alexander, (1824-1891)
WINCHELL, Newton Horace, (1829-1914)
WINGRAVE, Vitruvius Harold Wyatt, (1858-1938)
WINKLER, Tiberius Cornelius, (1822-1897)
WOOD, Searles Valentine, (1798-1880)
WOODS, Henry, (1868-1952)
WOODWARD, Bernard Barham, (1853-1930)
WOODWARD, Henry, (1832-1921)
WOODWARD, Horace Bolingbroke, (1848-1914)
WOODWARD, Samuel P., (1821-1865)
WORSFORD, Frederick H., (1870-1954)
WORTHEN, Amos Henry, (1812-1888)
WRIGHT, James, (1878-1957)
WRIGHT, Thomas, (1809-1884)
WYATT-EDGELL, Henry Adrian, (1847-1866)
YABE, Hisakatsu, (1878-1969)
YAKOVLEV, N. N., (1870-1966)
YOUNG, John, (1823-1900)
ZEISE, Oskar, (1860-1925)
ZIETZ, Amandus Christian Heinrich, (1839-1921)
ZITTEL, Karl Alfred von, (1839-1904)
ZUGMAYER, Heinrich, (1840-1917)

1900 to present

ABBOTT, William James Lewis, (1853-1933)
ABEL, Othenio, (1875-1946)
ADKINS, Walter Scott, (1890-1956)
AGASSIZ, Alexander, (1835-1910)
AIM, William Lennie, (?1872-1938)
ALDRICH, Truman Heminway, (1848-1932)
ALEXANDER, Frances Elizabeth Somerville, (1908-1958)
ALLAN, Robin Sutcliffe, (1900-1966)
ALLEN, Henry Atwool, (?1854-1934)
ALLOITEAU, James Henri Ernest, (1890-1969)
AMALITSKII, Vladimir Prochorovich, (1860-1917)
AMI, Henry Mark, (1858-1931)
AMMON, Ludwig von, (1850-1922)
ANDERSSON, Oscar Frithiof, (1867-1911)
ANDRÉE, Karl, (1880-1959)
ANDRUSOV, Dimitri, (1897-1976)
ANDRUSSOV, Nikolai Ivanovich, (1861-1924)

INVERTEBRATE PALAEONTOLOGY 1900 to present Cont'd.

APPLIN, Esther Richards, (1895-1972)
ARKELL, William Joscelyn, (1904-1958)
ARNOLD, Ralph, (1875-1961)
ASTRE, Gaston, (1896-)
AUGUSTA, Josef, (1903-1968)
BAKER, Frank Collins, (1867-1942)
BALDWIN, Walter, (?1878-1956)
BALL, John Rice, (1881-1953)
BANCROFT, John Bevis Beeston, (-1944)
BARNES, Jonathan, (-1928)
BARNETT, Peter, (1922-1946)
BARON, Richard, Reverend, (1847-1907)
BARRIS, Willis Hervey, (1820-1901)
BARROIS, Charles Eugène, (1851-1939)
BARTRUM, John Arthur, (1885-1949)
BASSE DE MÉNORVAL, Éliane Marguerite, (1899-)
BASSLER, Raymond Smith, (1878-1961)
BATHER, Francis Arthur, (1863-1934)
BAUMBERGER, Ernst, (1866-1935)
BAUR, Carl Theodor von, (1836-1911)
BEECHER, Charles Emerson, (1856-1904)
BEKKER, Hendrik, (1891-1925)
BELLINI, Raffaelo, (1874-1930)
BENEDEN, Edouard van, (1846-1910)
BENSON, William Noel, (1885-1957)
BERENDT, Gottlieb Michael, (1836-1920)
BERGERON, Jules, (1853-1919)
BERNARD, Henry Meyners, (1953-1909)
BERNOUILLI, Walter, (1885-1946)
BEUDANT, François-Sulpice, (1787-1850)
BEUSHAUSEN, Hermann Emil Louis, (1863-1904)
BIEBER, Charles Leonard, (1901-1965)
BIGOT, Alexandre, (1863-1953)
BILLINGS, Walter R., (?1849-1920)
BIRKMAIER, Anton, (-1926)
BIRLEY, Caroline, (1851-1907)
BISAT, William Sawney, (1886-1073)
BLACKMORE, Humphrey Purnell, (-1928)
BLAKE, John Frederick, Reverend, (1839-1906)
BLANCKENHORN, Max Ludwig Paul, (1861-1947)
BLASCHKE, Friedrich, (1883-1911)
BOBIES, Carl August, (1898-1958)
BODYLEVSKY, Vitaly Ivanovich, (1898-1968)
BOFILL Y POCH, Arturo, (1846-1929)
BÖHMERSHEIM, August Böhm von, Edler, (1858-1930)
BOLTON, Herbert, (1863-1936)
BÖSE, Emil, (1868-1972)
BOUČEK, Bedřich, (1904-1975)
BOUCOT, Arthur James, (1924-)
BOURGUIGNAT, Jules René, (1829-?1919)
BOUSSAC, Jean, (1885-1917)
BRADLEY, Wilmot Hyde, (1899-)
BRADY, Lionel Francis "Major", (1880-1963)
BRANDES, Theodor, (-1916)

INVERTEBRATE PALAEONTOLOGY 1900 to present Cont'd.

BRANDT, Alexander von, (1844-1932)
BRANSON, Carl Colton, (1906-1975)
BRAUER, Friedrich Moritz, (1832-1904)
BRØGGER, Waldemar Christofer, (1851-1940)
BROILI, Ferdinand, (1874-1946)
BROUWER, Hendrik Albertus, (1886-1973)
BROWN, Noel K., Jr., (1925-1968)
BROWN, Thomas Clachar, (1882-1934)
BRUSINA, Spiridion, (1845-1908)
BRYDONE, Reginald Marr, (-1943)
BUCKMAN, Sidney Savory, (1860-1929)
BULLEN, Robert Ashington, Reverend, (1850-1912)
BULMAN, Oliver Meredith Boone, (1902-1974)
BURCKHARDT, Carl Rudolf, (1866-1908)
BURNABY, Thomas Patrick, (1924-1968)
CALEY, John Fletcher, (1903-1971)
CALL, Richard Ellsworth, (1856-1917)
CALVIN, Samuel, (1840-1911)
CAMPBELL, Henry Donald, (1862-1934)
CAPELLINI, Giovanni, (1833-1922)
CARPENTER, Philip Herbert, (1852-1891)
CARRUTHERS, Robert George, (1880-1965)
CARRUTHERS, William, (1830-1922)
CATTOI, Noemí Violeta, (1911-1965)
CAZIOT, Eugène, (1844-1931)
CHADWICK, George Halcott, (1876-1953)
CHAO, Yatseng T., (1898-1929)
CHAPMAN, Frederick, (1864-1943)
CHARLESWORTH, Edward, (1813-1893)
CHELOT, Émile, (1862-1930)
CHOFFAT, Paul, (1849-1919)
CHUBB, Lawrence John, (?1888-1971)
CLARK, Bruce Lawrence, (1880-1945)
CLARK, William Bullock, (1860-1917)
CLARKE, John Mason, (1857-1925)
CLELAND, Herdman Fitzgerald, (1869-1935)
CLIFTON, Roland Leroy, (1884-1962)
CLINE, Lewis Manning, (1909-1971)
COCKERELL, Theodore Dru Alison, (1866-1948)
COLENUTT, George William, (?1862-1944)
COLLOT, Louis Marie François, (1846-1915)
COOKE, Charles Wythe, (1887-1971)
COOPER, Chalmer Lewis, (-)
COOPER, James Graham, (1830-1902)
CORNWALL, Ira Edmund, (1875-1962)
CORTI, Benedetto, (1868-1907)
CORYELL, Horace Noble, (1888-1965)
COSSMANN, Maurice, (1850-1924)
COTTREAU, Jean, (1877-1945)
COUPER, James Graham, (1830-1902)
COX, Leslie Reginald, (1897-1965)
CRANE, Edward, (1822-1901)
CRICK, George Charles, (1856-1917)
CRONEIS, Carey, (1901-1972)

INVERTEBRATE PALAEONTOLOGY 1900 to present Cont'd.

CUNNINGTON, William, (1813-1906)
CURRIE, Ethel Dobbie, (-1963)
CURRIE, Leslie D., (1904-1925)
D'ACHIARDI, Antonio, (1839-1902)
DALL, William Healey, (1845-1927)
DANGEARD, Louis, (-)
DARELL, Robert Darell Smythe, (1851-1936)
DAUTZENBERG, Philippe, (1849-1935)
DAVIES, Arthur Morley, (1869-1959)
DAVIES, John Henry, (1887-1971)
DAVIS, Arthur George, (1892-1957)
DECKER, Charles Elijah, (1868-1958)
DEECKE, Johannes Ernst Wilhelm, (1862-1934)
DE GREGORIO, Antonio, Marchese, (1855-1930)
DEHÉE, René, (1898-1928)
DEICHMÜLLER, Johannes, (1854-)
DEISS, Charles Frederick, (1903-1959)
DELHAES, Wilhelm, (1883-191)
DENCKMANN, August, (1860-1925)
DE NICOLIS, Enrico, (1841-1908)
DERBY, Orville Adelbert, (1851-1915)
DESTINEZ, Pierre, (-1911)
DEWALQUE, Gilles Joseph Gustave, (1828-1905)
DIBLEY, George Edward, (1857-1943)
DIENER, Carl, (1862-1928)
DIETRICH, Wilhelm Otto, (1881-1964)
DI STEFANO, Giovanni, (1856-1918)
DOLLÉ, Louis, (1878-1965)
DOLLFUS, Gustave Frédéric, (1850-1931)
DONCIEUX, Louis, (1874-1960)
DONNEZAN, Albert, (1846-1914)
DORMAL, Victor, (-1900)
DOUGLAS, James Archibald, (1884-1978)
DOUVILLÉ, Joseph Henri Ferdinand, (1846-1937)
DOUVILLÉ, Robert, (1881-1914)
DREVERMANN, Fritz, (1875-1932)
DUPONT, Édouard François, (1841-1911)
DVOŘÁK, Josef, (1904-)
DWERRYHOUSE, Arthur Richard, (1867-1931)
EDMONDS, Charles, (1886-1964)
EISEL, Robert, (1826-1917)
EISENACK, Alfred, (1891-)
ELLES, Gertrude Lilian, (1872-1960)
ELLISOR, Alva Christine, (1892-1964)
ENGLEHEART, Francis Henry Arnold, (-1963)
EPPS, Frank James, (?1891-1974)
ERNI, Arthur, (1885-1945)
ETHERIDGE, Robert, (1819-1903)
ETHERIDGE, Robert Jr., (1847-1920)
FABIANI, Ramiro, (1879-1954)
FAVRE, Ernest, (1845-1925)
FELIX, Johannes Paul, (1859-1941)
FENDRYCH, Otakar, (1860-1945)
FERASIN, Francesco, (1928-1958)

INVERTEBRATE PALAEONTOLOGY 1900 to present Cont'd.

FIEBELKORN, Max, (1869-1912)
FINLAY, Harold John, (1901-1951)
FISHER, Osmond, Reverend, (1817-1914)
FLEURY, Ernest, (1878-1958)
FOERSTE, August Frederick, (1862-1936)
FOORD, Arthur Humphreys, (1845-1934)
FORD, Trevor David, (1932-)
FORESTI, Lodovico, (1829-1913)
FORMA, Ernesto, (1869-1911)
FORNASINI, Carlo, (1854-1931)
FORTAU, René, (1867-1920)
FRAIPONT, Julien, (1857-1910)
FRAUSCHER, Karl, (1852-1914)
FRECH, Fritz, (1861-1917)
FRENGUELLI, Joaquîn, (1883-1958)
FRIEDBERG, Wilhelm, (1873-1941)
FRITSCH, Anton, (1832-1913)
FRITSCH, Karl von, Freiherr, (1838-1906)
FRIZZELL, Donald Leslie, (1906-1972)
FUCHS, Theodor, (1842-1925)
FUGGER, Friedrich Eberhard, (1842-1919)
GAGEL, Kurt, (1865-1927)
GALLOWAY, Jesse James, (1882-1962)
GARDNER, Julia Anna, (1882-1960)
GARRETSON, Mary Welleck, (1896-1971)
GARWOOD, Edmund Johnston, (1864-1949)
GASTER, Christopher Theodore Augustine, (1878-1963)
GAUTHIER, Victor Auguste, (1837-1911)
GEINITZ, Hanns Bruno, (1814-1900)
GEMMELLARO, Mariano, (1879-1921)
GÉRARD, Charles, Colonel, (1866-1953)
GERBER, Eduard, (1876-1956)
GEYER, David, (1855-1932)
GEYER, Georg, (1857-1938)
GIARD, Alfred, (1846-1908)
GIERTZ, Johannes, (1916-1943)
GILLARD, Pierre A., (1907-1947)
GIRTY, George Herbert, (1869-1939)
GOLDRING, Winifred, (1888-1971)
GOMES, Jacinto Pedro, (1844-1910)
GORDON, Charles Henry, (1857-1934)
GORJANOVIĆ-KRAMBERGER, Dragutin, (1856-1937)
GORSKY, I. I., (1893-1975)
GORTANI, Michele, (1883-1966)
GOSSELET, Jules Auguste, (1832-1916)
GOTTSCHE, Carl Christian, (1855-1909)
GOTTSCHICK, Franz, (1865-1927)
GRABAU, Amadeus William, (1870-1946)
GRAHAM, Joseph John, (1909-1967)
GRAHAM, William Armstrong Patterson, (1899-1934)
GREGORY, John Walter, (1864-1932)
GREPPIN, Eduard, (1856-1927)
GROOM, Theodore Thomas, (1863-)
GRZYBOWSKI, Józef, (1869-1922)

INVERTEBRATE PALAEONTOLOGY 1900 to present Cont'd.

GUILLAUME, Louis, (1894-1952)
GULICK, John Thomas, (1832-1923)
GUPPY, Robert John Lechmere, (1836-1916)
GURLEY, William Frank Eugene, (1854-1943)
HÄBERLE, Daniel, (1864-1934)
HAECKEL, Ernst Heinrich Philipp August, (1834-1919)
HAHN, Friedrich Felix (Fritz), (1885-1914)
HALAVÁTS, Gyula, (1853-1926)
HALL, Thomas Sergeant, (1858-1915)
HANDLIRSCH, Anton, (1865-1935)
HANDSCHIN, Eduard, (1894-1962)
HANNA, G. Dallas, (1887-1970)
HARBORT, Erich, (1879-1929)
HARMER, Frederic William, (1835-1923)
HARPER, John Childs, (1913-1977)
HARRINGTON, Horacio Jaime, (1910-1973)
HARRIS, George Frederick, (1862-1906)
HARRIS, Gilbert Dennison, (1864-1952)
HARRISON, Janet Mitchell Marr, (1893-1971)
HASSE, Carl, (1841-1922)
HAWKINS, Herbert Leader, (1887-1968)
HAYWARD, John, (-1958)
HENNIG, Edwin, (1882-)
HERITSCH, Franz, (1882-1945)
HERNÁNDEZ-PACHECO, Eduardo, (-1965)
HERON-ALLEN, Edward, (1861-1943)
HERRICK, Clarence Luther, (1858-1904)
HERRIES, Robert Stansfield, (?1860-1941)
HERRMANN, Fritz Heinrich Ulrich, (1883-1920)
HEYDEN, Lucas von, (1838-1915)
HICKLING, Henry George Albert, (1883-1954)
HILBER, Vincenz, (1853-1931)
HILGENDORF, Franz, (1839-1904)
HILL, Dorothy, (-)
HIND, Wheelton, (1859-1920)
HINDE, George Jennings, (1839-1918)
HOLCOMB, Leroy D., (1919-1974)
HOLM, Edvard Johann Gerhard, (1853-1926)
HOLMES, Francis S., (1815-1882)
HOLTEDAHL, Olaf, (1885-1975)
HOLUB, Emil, (1847-1902)
HOLWILL, Frederick James Winsor, (1925-1966)
HOLZAPFEL, Eduard, (1853-1913)
HOPKINSON, John, (1844-1919)
HORUSITZKY, Henrik, (1870-1945)
HOURCQ, Victor, (1901-1961)
HOWELL, Benjamin Franklin, (1890-1976)
HUDLESTON, Wilfrid Hudleston, (né Simpson), (1828-1909)
HUDSON, George Henry, (1855-1934)
HUDSON, Robert George Spencer, (1895-1965)
HUENE, Friedrich von, Freiherr, (1875-1969)
HUTTON, Frederick Wollaston, (1836-1905)
HUTTON, Harriet Mary, (1873-1937)
ISSEL, Arturo, (1842-1922)

INVERTEBRATE PALAEONTOLOGY 1900 to present Cont'd.

JACKSON, James Frederick, (1896-1966)
JAECKEL, Otto, (1863-1929)
JANENSCH, Werner, (1878-1969)
JAROSZ, Jan, (1877-1944)
JODOT, Paul, (1879-1961)
JOHNSON, Ralph Gordon, (1927-1976)
JOHNSTON, Mary Sophia, (1875-1955)
JOHNSTON, Robert Mackenzie, (-1918)
JOLEAUD, Léonce, (1880-1938)
JONES, Ernest Lloyd, (?1863-1942)
JONES, Thomas Rupert, (1819-1911)
JUKES-BROWNE, Alfred John, (1851-1914)
JULIEN, Pierre Alphonse, (1838-1905)
JURADO, Martín Doello, (1884-1948)
KAFKA, Josef, (1858-1929)
KARPINSKY, Alexander Petrovich, (1847-1936)
KARSCH, Ferdinand, (1853-1936)
KASTNER, Karl, (1847-1907)
KATZER, Bedřich, (1861-1925)
KAUFMAN, Rudolf, (1909-ca 1942)
KAY, George Marshall, (1904-1975)
KAYSER, Friedrich Heinrich Emanuel, (1845-1927)
KELUS, Aleksander, (1902-1946)
KENNARD, Alfred Santer, (1870-1948)
KESSLER, Paul, (?1882-1927)
KETTNEROVÁ, Marja (Marie), (1900-1933)
KEW, William Stephen Webster, (1890-1961)
KEYES, Charles Rollin, (1864-1942)
KEYTE, Ivy Allen, (1878-1931)
KIAER, Johan Aschehong, (1869-1931)
KINDLE, Edward Martin, (1869-1940)
KIRKBY, James Walker, (1834-1901)
KIRKPATRICK, Randolph, (-1950)
KITCHIN, Finlay Lorimer, (1870-1934)
KLÄHN, Hans, (1884-1933)
KLINE, Virginia Harriett, (1910-1959)
KLOUČEK, Celda, (1855-1935)
KOBAYASHI, Teiichi, (1901-)
KOBY, Frédéric Louis, (1852-1930)
KOCH, Gottlieb Karl David von, (1849-1914)
KOERT, Willi, (1875-1927)
KOKEN, Ernst Friedrich Rudolph Karl von, (1860-1912)
KOLBE, Hermann Julius, (1855-)
KOLOSVÁRY, Gábor, (1901-1968)
KONGIEL, Roman, (1904-1960)
KOZLOWSKI, Roman, (1889-1977)
KRUITZINGA, Pieter, (1885-1967)
KÜHN, Othmar, (1892-1969)
KUTASSY, Endre, (1898-1938)
LAKE, Philip, (1865-1949)
LAMBERT, Jules Mathieu, (1848-1940)
LAMONT, Archie, (-)
LANDOIS, Hermann, (1835-1905)
LANG, William Dickson, (1878-1966)

INVERTEBRATE PALAEONTOLOGY 1900 to present Cont'd.

LAPPARENT, Albert Auguste de, (1839-1908)
LAPWORTH, Charles, (1842-1920)
LAUBE, Gustav C., (1839-1923)
LAW, Robert, (1840-1907)
LEANZA, Armando F., (1919-1975)
LECOINTRE, Georges, (1888-1972)
LECOMPTE, Marius, (1902-1973)
LEITCH, Duncan, (1904-1956)
LEONHARD, Richard, (1870-1916)
LERICHE, Maurice, (1875-1948)
LEWIS, Herbert Price, (?1895-1947)
LIENENKLAUS, Ernst, (1849-1905)
LINDSTRÖM, Jonas Gustaf Oscar, (1829-1901)
LINSTOW, Otto von, (1872-1929)
LISSON, Carlos I., (1868?-1957)
LOCARD, Arnould, (1841-1904)
LONGSTAFF, Mary Jane, (née Donald), (1856-1935)
LOOMIS, Theodor, (1875-1909)
LÖRENTHEY, Imre, (1867-1917)
LORENZ, Theodor, (1875-1909)
LORIOL LE FORT, Charles Louis Perceval de, (1828-1908)
LOVISATO, Domenico, (1842-1916)
LUNIEWSKI, Adam, (1887-1945)
LÜTKEN, Christian Fredrik, (1827-1901)
LYELL, Charles, Sir, (1797-1875)
MAAS, Otto, (1867-1916)
MACPHERSON, José, Don, (1839-1902)
MAILLARD, Gustave, (1860-1891)
MAIRE, Julien Victor, (1856-1941)
MAKOWSKY, Alexander, (1833-1908)
MALAISE, Constantin H. L., (1834-1916)
MALDONADO-KOERDELL, Manuel, (1908-1972)
MANSFIELD, Wendell Clay, (1874-1939)
MARÈS, Paul, (1826-1900)
MARSHALL, Patrick, (1869-1950)
MARTELLI, Alessandro, (1877-1934)
MARTIN, Handel Tongue, (1862-1931)
MARTIN, Johann Karl Ludwig, (1851-1942)
MARWICK, John, (1891-1978)
MAYER-EYMAR, Karl David Wilhelm, (1826-1907)
MELI, Romolo, (1852-1920)
MENZEL, Karl Gustav Johannes, (1875-1914)
MERTIN, Hans, (1902-1941)
MEUNIER, Fernand, (1868-1926)
MICHAEL, Richard, (1869-1925)
MICKWITZ, Karl August von, (1849-1910)
MILNE-EDWARDS, Alphonse, (1835-1900)
MILTON, John Herbert, (?1862-1925)
MITZOPOULOS, Maximos K., (1896-1968)
MOBERG, Johan Christian, (1854-1915)
MOJSISOVICS, Johann August Georg Edmund, Edler von Mojsvàr, (1839-1907)
MONTEROSATO, Tommaso di Maria, Marchese di, (1841-1927)
MOORE, Edward William James, (1887-1970)

INVERTEBRATE PALAEONTOLOGY 1900 to present Cont'd.

MOORE, Raymond Cecil, (1892-1974)
MORELLET, Lucien, (1882-1945)
MORET, Léon, (1890-1972)
MUIR-WOOD, Helen Marguerite, (1896-1968)
MÜLLER, Karl Christian Gottfried Ludwig, (1862-1906)
MULLER, Siemon William, (1900-1970)
MUNIER-CHALMAS, Ernest Philippe Auguste, (1843-1903)
NELSON, Lloyd Alvino, (1895-1964)
NEWTON, Richard Bullen, (1850-1926)
OECHSLE, Richard, (1908-)
OEHLERT, Daniel Victor, (1849-1920)
OERTEL, Walter, (1889-1924)
OMBONI, Giovanni, (1829-1910)
OPPENHEIM, Leo Paul, (1863-1934)
OPPLIGER, Fritz, (1861-1932)
ORLOV, Uri Aleksandrovich, (1893-1966)
ORTMANN, Arnold Edward, (1863-1927)
OUSTALET, Émile, (1844-1905)
PACHUCKI, Czeslaw, (1898-1965)
PACKARD, Alpheus Spring, Jr., (1839-1905)
PALMER, Robert Hastings, (1882-1948)
PANTANELLI, Dante, (1844-1913)
PAQUIER, Victor Lucien, (1870-1911)
PARKER, William Albert, (1855-1918)
PARKINSON, Donald, (1891-1974)
PARKS, William Arthur, (1868-1936)
PARONA, Carlo Fabrizio, (1855-1939)
PARSONS, Leonard Miles, (1883-1956)
PATRUNKY, H., (1854-1930)
PAULCKE, Wilhelm, (1873-1949)
PAULUCCI, Marianna, Marchesa, (1835-1919)
PEREIRA DE LIMA, Wenceslau de Sousa, (1858-1920)
PERGENS, Eduard Willem Gerard, (1862-1917)
PERNER, Jaroslav, (1869-1947)
PERON, Alphonse, (1834-1908)
PERRINE, Irving, (1884-1955)
PERRY, Thomas Gregory, (1919-1972)
PERVINQUIÈRE, Léon, (1873-1913)
PETKOVIĆ, Vladamir K., (1873-1934)
PETRBOK, Jaroslav, (1881-1960)
PHILIPPI, Rudolph Amandus, (1808-1904)
PIETZCKER, Franz, (1885-1914)
PILSBRY, Henry Augustus, (1862-1957)
PLAS, Vladimír, (1905-)
POCOCK, Roy Woodhouse, (1887-1954)
POČTA, Filip, (1859-1924)
POMPECKJ, Josef Felix, (1867-1930)
POPOVICI-HAŢEG, Valeriu, (1868-1929)
POULSEN, Christian, (1896-1975)
PRADÁČOVÁ, Marcella, (1912-1972)
PŘIBYL, Alois, (1904-)
PRINGLE, John, (1877-1948)
PRUVOST, Pierre Eugéne Marie Joseph, (1890-1967)
QUITZOW, August Alfred Wilhelm, (1881-1914)

INVERTEBRATE PALAEONTOLOGY 1900 to present Cont'd.

RADOVANOVIĆ, Svetolik A., (1863-1928)
RAKUSZ, Gyula, (1896-1932)
RAMA RAO, Lakshmeswar, (1896-1974)
RATHBURN, Richard, (1852-1918)
RAVN, Jesper Peter Johansen, (1866-1951)
RAW, Frank, (?1875-1961)
REBEL, Hans, (1861-)
REED, Frederick Richard Cowper, (1869-1946)
REESIDE, John Bernard, Jr., (1889-1958)
REICHARDT, Wolfgang, (1909-1937)
REICHEL, Manfred, (1896-)
REIS, Otto Maria, (1862-1934)
REMEŠ, Mořic, (1867-1959)
RENEVIER, Eugène, (1831-1906)
REYNOLDS, Sidney Hugh, (1867-1949)
RIAZ, Auguste de, (-1920)
RICE, Elmer Michael, (1900-1971)
RICHARDSON, Linsdall, (1881-1967)
RICHTER, Emma, (1888-1956)
RICHTER, Rudolf, (1881-1957)
RIDLEY, Henry Nicholas, (1855-1956)
RIGAUX, Edmond, (1839-1915)
RISTORI, Giuseppe, (1856-1905)
ROBERTS, Thomas, (1856-1892)
RODIGHIERO, Andrea, (1892-1917)
ROEDER, Charles, (1848-1911)
ROLLIER, Louis, (1859-1931)
ROMAN, Albert Frédéric, (1871-1943)
ROMINGER, Carl Ludwig, (1820-1907)
ROSS, Frank Stenhouse, (1890-1975)
ROTHPLETZ, August, (1853-1918)
ROWE, Arthur Walton, (1858-1926)
RUEDEMANN, Rudolf, (1864-1956)
RUSSELL, Loris Shano, (1904-)
RUTTEN, Martin Gerard, (1910-1970)
SACCO, Federico, (1864-1948)
SALÉE, Achille, Chanoine, (1883-1932)
SALOPEK, Marijan, (1883-1967)
SARDESON, Frederick William, (1866-1959)
SÄVE-SÖDERBERGH, Gunnar, (1910-1948)
SAVIN, Léon Héli, Lieutenant-Colonel, (1851-1907)
SCHALCH, Ferdinand, (1848-1918)
SCHELLWIEN, Ernst, (1866-1906)
SCHENCK, Hubert Gregory, (1897-1960)
SCHINDEWOLF, Otto Heinrich, (1896-1972)
SCHLECHTENDAL, Dietrich von, (1834-1916)
SCHLUETER, Clemens August, (1835-1906)
SCHMIDT, Friedrich Bogdanovich "Bohdan", (1832-1908)
SCHMIDT, Hermann, (1892-1978)
SCHMIDT, Wilhelm Erich, (1882-1944)
SCHMIDT, Wolfgang, (1914-1972)
SCHRÉTER, Zóltan, (1882-1970)
SCHRÖDER, Richard, (1853-1916)
SCHROEDER, Henry Carl, (1859-1927)

INVERTEBRATE PALAEONTOLOGY 1900 to present Cont'd.

SCHUCHERT, Charles, (1858-1942)
SCHÜTZE, Ewald, (1873-1908)
SCHWERTSCHLAGER, Joseph, (1853-1924)
SCOTT, Gayle, (1894-1948)
SCUDDER, Samuel Hubbard, (1837-1911)
SEGUENZA, Luigi, (1873-1908)
SELLARDS, Elias Howard, (1875-1961)
SEMPER, Otto, (1830-1907)
SEUNES, Jean, (1849-1920)
SHAKESPEAR, Ethel Mary Reader, (née Wood), Dame, (1871-?1946)
SHALER, Nathaniel Southgate, (1841-1906)
SHANNON, William George St. John, (1889-1946)
SHARMAN, George, (1832-1914)
SHIDELER, William Henry, (1886-1958)
SHIMEK, Bohumil, (1861-1937)
SHIMER, Hervey Woodburn, (1872-1965)
SHROCK, Robert Rakes, (1904-)
SIEMIRADZKI, Józef, (1858-1933)
SINGLETON, F. A., (-1947)
SMITH, Edgar Albert, (1847-1916)
SMITH, James Perrin, (1864-1931)
SMITH, John, (1845-1930)
SMITH, Robert Hendee "Smokey", (1916-1946)
SMITH, Stanley, (1883-1955)
SMYTH, Louis Bouvier, (1933-?1953)
SOENDEROP, Fritz Hermann Gustav, (1876-1914)
SOLLAS, William Johnson, (1849-1936)
SORDELLI, Ferdinando, (?1837-1916)
SORGENFREI, Theodor, (1915-1972)
SPATH, Leonard Frank, (1882-1957)
SPRINGER, Frank, (1848-1924)
STAFF, Hans von, (1883-1915)
STAINBROOK, Merrill Addison, (1897-1956)
STANTON, Timothy William, (1860-1953)
ŞTEFANESCU, Gregoriu, (1838-1911)
ŞTEFĂNESCU, Sabba, (1857-1931)
STEPHENSON, Lloyd William, (1876-1962)
STEWART, Grace Anne, (1893-1970)
STOBBS, John Thomas, (?1866-1942)
STOLLEY, Ernst, (1869-1944)
STOPES, Henry, (1852-1902)
STORRIE, John, (1844-1901)
STRAELEN, Victor Emile van, (1889-1964)
STRAW, Stephen Henry, (?1892-1963)
STUMM, Erwin Charles, (1908-1969)
SUESS, Eduard, (1831-1914)
ŠUF, Jiří, (1891-1976)
SUTTON, Arle Herbert, (1896-1972)
SWINNERTON, Henry Hurd, (1875-1967)
TARAMELLI, Torquato, (1845-1922)
TATE, Ralph, (1840-1901)
TAURER-GALLENSTEIN, Hans von, Ritter, (1846-1927)
TAYLOR, Herbert, (-1963)
TCHERNYCHEV, Féodoss Nikolaievich, (1856-1914)

INVERTEBRATE PALAEONTOLOGY 1900 to present Cont'd.

TEICHERT, Curt, (-)
TEMPLEMAN, Andrew, (1887-1945)
TESCH, P., (1879-1961)
THALMANN, Hans E., (1899-1975)
THEVENIN, Armand, (1870-1918)
THIÉRY, Paul, (-1927)
THOMAS, Abram Owen, (1876-1931)
THOMAS, Henry Dighton, (1900-1966)
THOMAS, Horace David "Bill", (1905-1967)
THOMAS, Leo A., (1917-1965)
THOMAS, Norman Louis, (1897-1970)
THOMSON, James Allan, (1881-1928)
TILLYARD, Robin John, (1881-1937)
TILTON, John Littlefield, (1863-1930)
TING, Ven Kiang, (?1887-1936)
TIPPER, George Howlett, (1881-1947)
TOMBLESON, Frederick Binns, (?1872-1938)
TÖRNQUIST, Sven Leonhard, (1840-1920)
TOUCAS, Joseph Aristide, (1843-1911)
TOULA, Franz, (1845-1920)
TRASK, Parker Davies, (1899-1961)
TRAUTH, Friedrich, (1883-1967)
TRAUTSCHOLD, Hermann Adolfovich, (1817-1902)
TREACHER, Llewellyn, (1859-1943)
TRECHMANN, Charles Taylor, (1885-1964)
TRIBOLET-HARDY, Frédéric Maurice de, (1852-1929)
TWENHOFEL, William Henry, (1875-1957)
UHLIG, Victor Karl, (1857-1911)
ULRICH, Edward Oscar, (1857-1944)
UMBGROVE, Johannes Herman Frederik, (1899-1954)
VACEK, Michael, (1848-1925)
VALPY, Robert Harris, (1819-1904)
VAN INGEN, Gilbert, (1869-1928)
VAUGHAN, Arthur, (1868-1915)
VERRILL, Addison Emery, (1839-1926)
VÍGH, Gyula, (1889-1958)
VITÁLIS, István, (1871-1947)
VOGDES, Anthony Wayne, General, (1843-1923)
WAGNER, Walter Richard, (1928-1974)
WÄHNER, Franz, (1856-1932)
WALCOTT, Charles Doolittle, (1850-1927)
WALKER, John Francis, (1839-1907)
WANNER, Johannes, (1878-1957)
WEAVER, Charles Edwin, (1880-1958)
WEGNER, Theodor, (1880-1934)
WEIGELT, Johannes, (1890-1948)
WEINFURTER, Emil, (1904-1968)
WELLER, James Marvin, (1899-1976)
WELLER, Stuart, (1870-1927)
WELTNER, Wilhelm, (1854-1917)
WEPFER, Emil, (1882-1930)
WESTERGÅRD, Anton Hilmer, (1880-1968)
WETZEL, Otto Christian August, (1891-1971)
WHIDBORNE, George Ferris, Reverend, (1846-1910)

INVERTEBRATE PALAEONTOLOGY 1900 to present Cont'd.

WHITE, Charles Abiathar, (1826-1910)
WHITEAVES, John Joseph Frederick, (1835-1909)
WHITFIELD, Robert Parr, (1828-1910)
WHITTARD, Walter Frederick, (1902-1966)
WHYMPER, Edward, (1840-1911)
WILBERT, Louis Joseph "Slick", Jr., (1919-1953)
WILSON, Alice Evelyn, (1881-1964)
WILSON, Eugene Jack, (1921-1976)
WIMAN, Carl Johan Josef Ernst, (1867-1944)
WINCHELL, Newton Horace, (1829-1914)
WING, Monta Eldo, (1897-1976)
WINGRAVE, Vitruvius Harold Wyatt, (1858-1938)
WINTON, Will McClain, (1885-1960)
WIRZ, Albert Eduard, (1910-1973)
WITHERS, Thomas Henry, (1883-1953)
WOODS, Ethel Gertrude, (-1939)
WOODS, Henry, (1868-1952)
WOODWARD, Bernard Barham, (1853-1930)
WOODWARD, Henry, (1832-1921)
WOODWARD, Horace Bolingbroke, (1848-1914)
WORSFORD, Frederick H., (1870-1954)
WRIGHT, James, (1878-1957)
WRIGHT, Jean Davies, (-)
WRIGHT, "Put", (-1962)
WRIGHT, William, (-1936)
YABE, Hisakatsu, (1878-1969)
YAKOVLEV, N. N., (1870-1966)
ZÁZVORKA, Vlastislav, (1903-)
ZEALLEY, Arthur Edmund Victor, (1886-1918)
ZEISE, Oskar, (1860-1925)
ZEUNER, Frederick Everard, (1905-1963)
ZIETZ, Amandus Christian Heinrich, (1839-1921)
ZITTEL, Karl Alfred von, (1839-1904)
ZUGMAYER, Heinrich, (1840-1917)

Time unknown

DACQUÉ, Edgar, (-)
YANISHEVSKY, M. E., (-)

MARINE GEOLOGY

Up to 1800

TREMBLEY, Abraham, (1710-1784)

1800-1849

GUYOT, Arnold Henri, (1807-1884)
MAURY, Matthew Fontaine, (1806-1873)
POURTALÈS, Louis François de, (1823-1880)
WALLICH, George Charles, (1815-1899)

1850-1899

AGASSIZ, Alexander, (1835-1910)
ANDRUSSOV, Nikolai Ivanovich, (1861-1924)
FUCHS, Theodor, (1842-1925)
GUYOT, Arnold Henri, (1807-1884)
MAURY, Matthew Fontaine, (1806-1873)
MOLENGRAAFF, Gustaaf Adolf Frederik, (1860-1942)
POURTALÈS, Louis François de, (1823-1880)
WALLICH, George Charles, (1815-1899)
WALTHER, Johannes, (1860-1937)
WYVILLE THOMSON, Charles, Sir, (1830-1882)

1900 to present

AGASSIZ, Alexander, (1835-1910)
ANDRUSSOV, Nikolai Ivanovich, (1861-1924)
BRAMLETTE, Milton Nunn, (1896-1977)
CARRITT, Dayton Ernest, (1915-)
DANGEARD, Louis, (-)
DIETZ, Robert Sinclair, (1914-)
EWING, William Maurice, (1906-1974)
FUCHS, Theodor, (1842-1925)
GEBELEIN, Conrad Dennis, (1945-1978)
HÄNTZSCHEL, Walter, (1904-1972)
HEADLEE, Larry A., (1938-1970)
HEEZEN, Bruce Charles, (1924-1977)
HILL, Maurice Neville, (1919-1966)
HOLTEDAHL, Olaf, (1885-1975)
KUENEN, Phillip Henry, (1902-1976)
LIDZ, Louis, (1938-1967)
MOLENGRAAFF, Gustaaf Adolf Frederik, (1860-1942)
PICCARD, Auguste, (1884-1962)
PICCARD, Jacques Ernst, (1922-)
ROSSBY, Carl-Gustaf Arvid, (1898-1957)
SHEPARD, Francis P., (-)
SHIPEK, Carl Joseph, (1917-1969)
SOVINSKY, Viacheslav Nicholas, (1894-1962)
STETSON, Henry Crosby, (1900-1955)
TRASK, Parker Davies, (1899-1961)
VON ARX, William Stelling, (1916-)
WALTHER, Johannes, (1860-1937)
WHITTARD, Walter Frederick, (1902-1966)

MATHEMATICAL GEOLOGY

1800-1849

COLBY, Thomas F., Major-General, (?1783-1852)
HOPKINS, William, (1793-1866)

MATHEMATICAL GEOLOGY Cont'd.

1850-1899
BÖHMERSHEIM, August Böhm von, Edler, (1858-1930)
COLBY, Thomas F., Major-General, (?1783-1852)
GREEN, Alexander Henry, (1832-1896)
HOPKINS, William, (1793-1866)

1900 to present
BÖHMERSHEIM, August Böhm von, Edler, (1858-1930)
EARLE, Kenneth Wilson, (1891-1944)
RUTTEN, Martin Gerard, (1910-1970)
SIMPSON, Stephen Milton, Jr., (1929-)

METAMORPHIC PETROLOGY
Up to 1800
BEDDOES, Thomas, (1760-1808)
HUTTON, James, (1726-1797)
OTLEY, Jonathan, (1766-1856)

1800-1849
ALLPORT, Samuel, (1816-1897)
BEDDOES, Thomas, (1760-1808)
FORBES, David, (1828-1876)
HOFFMANN, Friedrich, (1797-1836)
HUNT, Thomas Sterry, (1826-1892)
JOKÉLY, Johann, (1826-1862)
NICOL, James, (1810-1879)
OTLEY, Jonathan, (1766-1856)
SORBY, Henry Clifton, (1826-1908)

1850-1899
ALLPORT, Samuel, (1816-1897)
BARROW, George, (1853-1932)
BASTIN, Edson Sunderland, (1878-1953)
BECKE, Friedrich Johann Karl, (1855-1931)
BECKER, George Ferdinand, (1847-1919)
BELYANKIN, Dmitri Stepanovich, (1876-1953)
BERGEAT, Alfred, Edmund, (1866-1924)
BOŘICKÝ, Emanuel, (1840-1881)
BOULE, Marcellin, (1861-1942)
BRUHNS, Franz Siegfried Willy, (1864-1929)
CROSBY, William Otis, (1850-1925)
DAVIES, Thomas, (1837-1892)
DRESSER, John Alexander, (1866-1954)
ELLES, Gertrude Lilian, (1872-1960)
EMERSON, Benjamin Kendall, (1843-1932)
ERDMANNSDÖFFER, Otto Heinrich, (1876-1955)
FORBES, David, (1828-1876)
FRANKS, Godfrey Firth, (1853-1917)
GARDINER, Charles Irving, (?1868-1940)
GORDON, Charles Henry, (1857-1934)
GREENLY, Edward, (1861-1951)
GRUBENMANN, Ulrich, (1850-1924)
HARKER, Alfred, (1859-1939)
HOBBS, William Herbert, (1864-1953)

METAMORPHIC PETROLOGY 1850-1899 Cont'd.

HÖGBOM, Arvid Gustaf, (1857-1940)
HORNE, John, (1848-1928)
HUNT, Thomas Sterry, (1826-1892)
JOKÉLY, Johann, (1826-1862)
KOLDERUP, Carl Fredrik, (1869-1942)
KORN, Johannes, (1862-1927)
KUPFFER, August Ernestovich, (1842-1918)
LACHAT, Hippolyte, (1829-1901)
MALLARD, François Ernest, (1833-1894)
MARSHALL, James Garth, (-1873)
MARTELLI, Alessandro, (1877-1934)
MIDDLEMISS, Charles Stewart, (1859-1945)
MILCH, Ludwig, (1867-1928)
MOORE, Elwood S., (1879-1966)
NICOL, James, (1810-1879)
OTLEY, Jonathan, (1766-1856)
ROCCATI, Alessandro, (1872-1928)
ROSENBUSCH, Karl Harry Ferdinand, (1836-1914)
RUTLEY, Frank, (1842-1904)
SORBY, Henry Clifton, (1826-1908)
SOUZA BRANDÃO, Vicente de, (1863-1916)
SUESS, Franz Eduard, (1867-1941)
TARASENKO, Vasily Efimovich, (1859-1926)
TEALL, Jethro Justinian Harris, Sir, (1849-1924)
VAN HISE, Charles Richard, (1857-1918)
WARREN, Charles Hyde, (1876-1950)
WEINSCHENK, Ernst Heinrich Oskar Kasimir, (1865-1921)
WILKMAN, Wanold Wrydon, (1872-?1938)
WILSON, J. S. Grant, (1855-1908)

1900 to present

ANDREATTA, Benedettino Ciro, (1906-1960)
BAILEY, Edward Battersby, Sir, (1881-1965)
BAKER, George, (1908-1975)
BARROW, George, (1853-1932)
BARTH, Thomas Frederik Weiby, (1899-1971)
BASTIN, Edson Sunderland, (1878-1953)
BECKE, Friedrich Johann Karl, (1855-1931)
BECKER, George Ferdinand, (1847-1919)
BELLAIR, Pierre, (1910-)
BELYANKIN, Dmitri Stepanovich, (1876-1953)
BERGEAT, Alfred, Edmund, (1866-1924)
BOULE, Marcellin, (1861-1942)
BOWEN, Norman Levi, (1887-1956)
BROWN, Peter Evans, (-)
BRUHNS, Franz Siegfried Willy, (1864-1929)
BURTON, Richard Cooksey, (1890-1916)
CAMPBELL, Robert, (1881-1957)
CHHIBBER, Harbans Lal, (1899-1955)
COOKE, Harold Caswell, (1884-1956)
CORNELIUS, Hans Peter, (1888-)
CRAIG, Robert Meldrum, (?1882-1956)
CROSBY, William Otis, (1850-1925)
DAVIES, Thomas, (1837-1892)

METAMORPHIC PETROLOGY 1900 to present Cont'd.

DRESSER, John Alexander, (1866-1954)
EDWARDS, Austin Burton, (1909-1960)
ELLES, Gertrude Lilian, (1872-1960)
EMERSON, Benjamin Kendall, (1843-1932)
ERDMANNSDÖFFER, Otto Heinrich, (1876-1955)
EYLES, Victor Ambrose, (1896-1978)
FABIAN, Rudolf, (1912-1943)
FAIRBAIRN, Harold Williams, (1906-)
FRANKS, Godfrey Firth, (1853-1917)
GARDINER, Charles Irving, (?1868-1940)
GILLSON, Joseph Lincoln, (1895-1964)
GLANGEAUD, Louis, (1903-)
GOLDSCHMIDT, Victor Moritz, (1888-1947)
GOODSPEED, George Edward, (1887-1974)
GORDON, Charles Henry, (1857-1934)
GREENLY, Edward, (1861-1951)
GROUT, Frank Fitch, (1880-1958)
GROVES, Arthur William, (1903-1956)
GRUBENMANN, Ulrich, (1850-1924)
HARKER, Alfred, (1859-1939)
HARRY, William Trevelyan, (?1922-1964)
HEJTMAN, Bohuslav, (1901-)
HOBBS, William Herbert, (1864-1953)
HÖGBOM, Arvid Gustaf, (1857-1940)
HOLLINGWORTH, Sydney Ewart, (1899-1966)
HOLMES, Arthur, (1890-1965)
HOOKER, Marjorie, (1908-1976)
HORNE, John, (1848-1928)
HUTCHINSON, William W., (-)
HUTTON, Colin Osborne, (1910-1971)
KOLDERUP, Carl Fredrik, (1869-1942)
KORN, Johannes, (1862-1927)
KUPFFER, August Ernestovich, (1842-1918)
LACHAT, Hippolyte, (1829-1901)
LEITH, Charles Kenneth, (1875-1956)
LOBJOIT, William Michael, (1932-1967)
LONGCHAMBON, Michel, (1886-1916)
MÄKINEN, Eero, (1886-1953)
MARIČ, Luka, (1899-)
MARTELLI, Alessandro, (1877-1934)
MIDDLEMISS, Charles Stewart, (1859-1945)
MILCH, Ludwig, (1867-1928)
MILLER, William John, (1880-1965)
MILLOT, Georges, (1917-)
MOORE, Elwood S., (1879-1966)
MOORHOUSE, Walter Wilson, (1913-1969)
MORRIS, Frederick Kuhne, (1885-1962)
NAIDU, Purama Rangashayee Jagapathi, (1903-1970)
NEALE, Ernest Richard Ward, (1923-)
NEWHOUSE, Walter Harry, (1897-1969)
ORCEL, Jean, (1896-)
PAVLOVIČ, Stojan, (1903-)
PAWLICA, Wladislaw, (1886-1919)
PERRIN, René, (1893-1966)

METAMORPHIC PETROLOGY 1900 to present Cont'd.

PIETZSCH, Kurt, (1884-1964)
POLDERVAART, Arie, (1918-1964)
PROKOP, František, (1898-)
RAMDOHR, Paul, (1890-)
RAVIER, Jean, (1922-)
READ, Herbert Harold, (1889-1970)
RICHEY, James Ernest, (1886-1968)
ROCCATI, Alessandro, (1872-1928)
ROSENBUSCH, Karl Harry Ferdinand, (1836-1914)
RUNNER, Joseph James, (1885-1970)
RUTLEY, Frank, (1842-1904)
SCHOLTZ, Douglas Leinster, (1902-1977)
SCHÜLLER, Arno, (1908-1968)
SEKANINA, Josef, (1900-)
SORBY, Henry Clifton, (1826-1908)
SOUZA BRANDÃO, Vicente de, (1863-1916)
STILLWELL, Frank Leslie, (1888-1963)
SUESS, Franz Eduard, (1867-1941)
SUTTON, John, (-)
SVOBODA, Josef, (1898-)
TARASENKO, Vasily Efimovich, (1859-1926)
TEALL, Jethro Justinian Harris, Sir, (1849-1924)
THOMSON, James Allan, (1881-1928)
TILLEY, Cecil Edgar, (1894-1973)
TOZER, Charles Fallows, (1918-1966)
URBAN, Karel, (1905-)
VACHTL, Josef, (1907-)
VANDENDRIESSCHE, Adrien, (1914-1940)
VAN HISE, Charles Richard, (1857-1918)
VOGT, Thorolf, (1888-1958)
VON ECKERMANN, Claes Walther Harry, (1886-1969)
WAGER, Lawrence Rickard, (1904-1965)
WALDBAUM, David Robert, (1937-1974)
WARREN, Charles Hyde, (1876-1950)
WEINSCHENK, Ernst Heinrich Oskar Kasimir, (1865-1921)
WESTERVELD, J., (1905-1962)
WHETTEN, John T., (-)
WILKMAN, Wanold Wrydon, (1872-?1938)
WILSON, J. S. Grant, (1855-1908)
ZAPLETAL, Karel, (1903-1972)
ZOUBEK, Vladimír, (1903-)

Time unknown

WOLFE, Caleb Wroe, (-)

METEORITICS

Up to 1800

AVICENNA, (980-1037)
BERTRAND DE DOUE, Jacques Mathieu, (1776-1862)
CHILDREN, John George, (1777-1852)
HUMBOLDT, Friedrich Wilhelm Heinrich Alexander von, Freiherr, (1769-1859)
LAVOISIER, Antoine Laurent, (1743-1794)
LOMONOSOV, Mikhail Vasilyevich, (1711-1765)

METEORITICS Up to 1800 Cont'd.

PLINY THE ELDER (CAIUS PLINIUS SECUNDUS), (?20 A.D.-79 A.D.)

1800-1849

BERTRAND DE DOUE, Jacques Mathieu, (1776-1862)
CHILDREN, John George, (1777-1852)
CRAHAY, Jacques Guillaume, (1789-1855)
DANA, James Dwight, (1813-1895)
DAUBRÉE, Gabriel Auguste, (1814-1896)
FERREL, William, (1817-1891)
GENTH, Friedrick August Ludwig Karl Wilhelm, later called Frederick Augustus, (1820-1893)
GREG, Robert Phillips, (1826-1906)
HAIDINGER, Wilhelm Karl von, Ritter, (1795-1871)
HOFF, Karl Ernst Adolf von, (1771-1837)
HUGGINS, William, Sir, (1824-1910)
HUMBOLDT, Friedrich Wilhelm Heinrich Alexander von, Freiherr, (1769-1859)
LA TOUCHE, James Digues, Reverend, (1824-1899)
MAURY, Matthew Fontaine, (1806-1873)
PERREY, Alexis, (1807-1882)
PHILLIPS, John, (1800-1874)
ROSE, Gustav, (1798-1873)
SAINTE-CLAIRE DEVILLE, Charles, (1814-1876)
SERPIERI, Alessandro, (1823-1885)
SIMONY, Friedrich, (1813-1896)
SORBY, Henry Clifton, (1826-1908)
SZABÓ, József, (1822-1894)

1850-1899

BALL, John, (-1941)
BARRINGER, Daniel Moreau, Sr., (1860-1929)
BERTRAND DE DOUE, Jacques Mathieu, (1776-1862)
BLANFORD, Henry Francis, (1834-1893)
BOWMAN, Herbert Lister, (1874-1942)
BREZINA, Maria Aristides, (1848-1909)
BUSZ, Karl Heinrich Emil Georg, (1863-1930)
COHEN, Emil Wilhelm, (1842-1905)
COLEMAN, Arthur Philemon, (1852-1939)
CRAHAY, Jacques Guillaume, (1789-1855)
DANA, Edward Salisbury, (1849-1935)
DANA, James Dwight, (1813-1895)
DAUBRÉE, Gabriel Auguste, (1814-1896)
DAVIS, William Morris, (1850-1934)
DUNN, Edward John, (1844-1937)
FARRINGTON, Oliver Cummings, (1864-1933)
FERREL, William, (1817-1891)
FLETCHER, Lazarus, Sir, (1854-1921)
FLIGHT, Walter, (1841-1885)
FOOT, Frederick J., (-1867)
FORT, Charles, (1874-1932)
GENTH, Friedrick August Ludwig Karl Wilhelm, later called Frederick Augustus, (1820-1893)
GREG, Robert Phillips, (1826-1906)
GREGORY, James Reynolds, (1832-1899)

METEORITICS 1850-1899 Cont'd.

GRISHCHINSKY, Petr Iuustinovich, (-1920)
HAIDINGER, Wilhelm Karl von, Ritter, (1795-1871)
HARMER, Frederic William, (1835-1923)
HIDDEN, William Earl, (1853-1918)
HINRICHS, Gustavus Detlef, (-1922)
HIORTDAHL, Thorstein Hallager, (1839-1925)
HOFF, Karl Ernst Adolf von, (1771-1837)
HÖGBOM, Arvid Gustaf, (1857-1940)
HOVEY, Edmund Otis, (1862-1924)
HUGGINS, William, Sir, (1824-1910)
HUMBOLDT, Friedrich Wilhelm Heinrich Alexander von, Freiherr, (1769-1859)
INOSTRANTZEV, Aleksandr Aleksandrovich, (1843-1920)
KLEIN, Johann Friedrich Carl, (1842-1907)
KROTOV, Petr Ivanovich, (1852-1914)
KUPFFER, August Ernestovich, (1842-1918)
KYLE, John Joseph Jolly, called Juan José, (1838-1922)
LACROIX, François Antoine Alfred, (1863-1948)
LAIS, Giuseppe, Padre, (1845-1921)
LA TOUCHE, James Digues, Reverend, (1824-1899)
LIVERSIDGE, Archibald, (1847-1927)
LOCKYER, Joseph Norman, Sir, (1836-1920)
LYONS, Henry George, (1864-?1943)
MAKOWSKY, Alexander, (1833-1908)
MARSHALL, James Garth, (-1873)
MAURY, Matthew Fontaine, (1806-1873)
MERCANTON, Paul Louis, (1876-1963)
MEUNIER, Étienne Stanislas, (1843-1925)
PALAZZO, Luigi, (1861-1933)
PERREY, Alexis, (1807-1882)
PHILLIPS, John, (1800-1874)
PRIOR, George Thurland, (1862-1936)
ROSE, Gustav, (1798-1873)
SAINTE-CLAIRE DEVILLE, Charles, (1814-1876)
SCOTT, Robert Henry, (1833-1916)
SERPIERI, Alessandro, (1823-1885)
SIMPSON, Edward Sydney, (1875-1939)
SMITH, Lawrence, (-1883)
SORBY, Henry Clifton, (1826-1908)
SPENCER, Leonard James, (1870-1959)
STREIFF-BECKER, Rudolf, (1873-1959)
SZABÓ, József, (1822-1894)
SZTERÉNYI, Hugo, (1857-1909)
TCHERNYCHEV, Féodoss Nikolaievich, (1856-1914)
TSCHERMAK VON SEYSENEGG, Gustav, (1836-1927)
WARD, Henry Augustus, (1834-1906)
WEINSCHENK, Ernst Heinrich Oskar Kasimir, (1865-1921)
WICHMANN, Carl Ernst Arthur, (1851-1927)
WILLS, Robert Glover, (1878-1950)
WINCHELL, Alexander Newton, (1874-1958)
WINWOOD, Henry Hoyte, (1830-1920)
WÜLFING, Ernst Anton, (1860-1930)

METEORITICS Cont'd.

1900 to present

AHRENS, Louis Herman, (1918-)
ALDERMAN, Arthur Richard, (1901-)
BAKER, George, (1908-1975)
BALL, John, (-1941)
BARRINGER, Daniel Moreau, Sr., (1860-1929)
BOWMAN, Herbert Lister, (1874-1942)
BRADY, Lionel Francis "Major", (1880-1963)
BREZINA, Maria Aristides, (1848-1909)
BUSZ, Karl Heinrich Emil Georg, (1863-1930)
COHEN, Emil Wilhelm, (1842-1905)
COLEMAN, Arthur Philemon, (1852-1939)
CUTTITTA, Frank, (1912-1974)
DANA, Edward Salisbury, (1849-1935)
DAVIS, William Morris, (1850-1934)
DITTLER, Emil, (1882-1945)
DUNN, Edward John, (1844-1937)
EDWARDS, Austin Burton, (1909-1960)
ELLSWORTH, Hardy Vincent, (1889-1952)
EYTHORSSON, Jon, (1895-1968)
FARRINGTON, Oliver Cummings, (1864-1933)
FERMOR, Lewis Leigh, Sir, (1880-1954)
FLETCHER, Lazarus, Sir, (1854-1921)
FORT, Charles, (1874-1932)
FOSHAG, William Frederick, (1894-1956)
GENTNER, Wolfgang, (1906-)
GLASS, Jewell Jeanette, (1888-1966)
GRISHCHINSKY, Petr Iuustinovich, (-1920)
HARMER, Frederic William, (1835-1923)
HIDDEN, William Earl, (1853-1918)
HIDE, Raymond, (1929-)
HINRICHS, Gustavus Detlef, (-1922)
HIORTDAHL, Thorstein Hallager, (1839-1925)
HÖGBOM, Alvar, (1894-1939)
HÖGBOM, Arvid Gustaf, (1857-1940)
HOINKES, Herfried, (1916-1973)
HOVEY, Edmund Otis, (1862-1924)
HUGGINS, William, Sir, (1824-1910)
HUTTENLOCHER, Heinrich, (1890-1954)
INOSTRANTZEV, Aleksandr Aleksandrovich, (1843-1920)
JÉRÉMINE, Elisabeth, (née Tschernaieff), (1879-1964)
KAVECKIS, Mykolas, (1889-)
KLEIN, Johann Friedrich Carl, (1842-1907)
KOKTA, Jaroslav, (1894-)
KROTOV, Petr Ivanovich, (1852-1914)
KULIK, Leonid Alekseevich, (1883-1942)
KUPFFER, August Ernestovich, (1842-1918)
KYLE, John Joseph Jolly, called Juan José, (1838-1922)
LACROIX, François Antoine Alfred, (1863-1948)
LAIS, Giuseppe, Padre, (1845-1921)
LEIGHTON, Henry, (1884-1963)
LIVERSIDGE, Archibald, (1847-1927)
LOCKYER, Joseph Norman, Sir, (1836-1920)
LOEWE, Fritz, (1895-1974)

METEORITICS 1900 to present Cont'd.

LONSDALE, John Tipton, (1895-1960)
LYONS, Henry George, (1864-?1943)
MAKOWSKY, Alexander, (1833-1908)
MAXIM, Ion Al., (1902-1968)
MERCANTON, Paul Louis, (1876-1963)
MEUNIER, Étienne Stanislas, (1843-1925)
NININGER, Harvey Harlow, (-)
OATES, Frank, (1890-1945)
PAČAK, Oldřich, (1882-)
PALAZZO, Luigi, (1861-1933)
PASTORE, Franco, (1885-1958)
PINSON, William Hamet, Jr., (1919-)
PRIOR, George Thurland, (1862-1936)
ROSSBY, Carl-Gustaf Arvid, (1898-1957)
SCOTT, Robert Henry, (1833-1916)
SEKANINA, Josef, (1900-)
SELGA, Miguel, Reverend Father, S. J., (1879-1956)
SHAND, Samuel James, (1882-1957)
SIMPSON, Edward Sydney, (1875-1939)
SLAVÍKOVÁ, Ludmila, (1890-1943)
SMITH, Thomas Hodge, (1894-1945)
SORBY, Henry Clifton, (1826-1908)
SPENCER, Leonard James, (1870-1959)
STEINOCHER, Vladimír, (1902-1972)
STILLWELL, Frank Leslie, (1888-1963)
STREIFF-BECKER, Rudolf, (1873-1959)
SZTERÉNYI, Hugo, (1857-1909)
TCHERNYCHEV, Féodoss Nikolaievich, (1856-1914)
THOMAS, Trevor Morgan, (?1919-1974)
TSCHERMAK VON SEYSENEGG, Gustav, (1836-1927)
UREY, Harold Clayton, (1893-)
VIGILEV, Boris, (1883-1924)
WARD, Henry Augustus, (1834-1906)
WEGENER, Alfred Lothar, (1880-1930)
WEINSCHENK, Ernst Heinrich Oskar Kasimir, (1865-1921)
WICHMANN, Carl Ernst Arthur, (1851-1927)
WILLS, Robert Glover, (1878-1950)
WINCHELL, Alexander Newton, (1874-1958)
WINWOOD, Henry Hoyte, (1830-1920)
WÜLFING, Ernst Anton, (1860-1930)

MICROPALAEONTOLOGY

Up to 1800

BATSCH, August Johann Georg Karl, (1761-1802)
BECCARI, Jacopo Bartolommeo, (1682-1766)
BIANCHI, Johannes-Janus Plancus, (1695-1775)
BRUGUIÈRES, Jean Guillaume, (1750-1799)
DEFRANCE, Jacques Louis Marie, (1758-1850)
FICHTEL, Leopold von, (fl.1798)
GUALTIERI, Nicolaus, (1688-1744)
SOLDANI, Ambrogio, Pater, (1736-1808)

1800-1849

ANGELIN, Nils Peter, (1805-1876)

MICROPALAEONTOLOGY 1800-1849 Cont'd.

ARCHIAC, Étienne Jules Adolphe Dexmier de Simon, Vicomte d', (1802-1868)
BAILEY, Jacob Whitman, (1811-1857)
BATSCH, August Johann Georg Karl, (1761-1802)
BOWERBANK, James Scott, (1797-1877)
CARPENTER, William Benjamin, (1813-1885)
COSTA, Oronzio Gabriele, (1787-1867)
DEFRANCE, Jacques Louis Marie, (1758-1850)
DUJARDIN, Félix, (1801-1860)
EHRENBERG, Christian Gottfried, (1795-1876)
FICHTEL, Leopold von, (fl.1798)
FREYER, Heinrich, (1802-1865)
GRUNOW, Albert, (1826-1914)
GÜMBEL, Carl Wilhelm von, Ritter, (1823-1898)
HAGENOW, Friedrich von, (1797-1865)
HAIDINGER, Wilhelm Karl von, Ritter, (1795-1871)
HAIME, Jules, (1824-1856)
HANTKEN, Miksa von Prudnik, (1821-1893)
HOSIUS, August, (1825-1896)
JONES, Thomas Rupert, (1819-1911)
KARRER, Felix, (1825-1903)
MANTELL, Gideon Algernon, (1790-1852)
MARSSON, Theodor, (1816-1892)
MOLL, Johann Paul Karl von, (fl.1798)
NEUGEBOREN, Johann Ludwig, (1806-1887)
PANDER, Christian Heinrich, (1794-1865)
PARKER, William Kitchen, (1823-1890)
PETERS, Karl, (1825-1881)
READE, Joseph Bancroft, Reverend, (1801-1870)
REUSS, August Emanuel von, Ritter, (1811-1873)
ROLLE, Friedrich, (1827-1887)
SCHACKO, Gustav, (1824-1917)
SCHLUMBERGER, Charles, (1825-1904)
SCHMIDT, Adolf, (1812-1899)
SISMONDA, Angelo, (1807-1878)
SLACK, Henry James, (1818-1896)
SOLDANI, Ambrogio, Pater, (1736-1808)
SPEYER, Oscar, (1827-1882)
TERQUEM, Olry, (1797-1887)
YOUNG, John, (1823-1900)
ZBORZEWSKI, Wojciech, (-)

1850-1899

ANDRUSSOV, Nikolai Ivanovich, (1861-1924)
ANGELIN, Nils Peter, (1805-1876)
ARCHIAC, Étienne Jules Adolphe Dexmier de Simon, Vicomte d', (1802-1868)
BAGG, Rufus Mather, (1869-1946)
BAILEY, Jacob Whitman, (1811-1857)
BARROIS, Charles Eugène, (1851-1939)
BASSLER, Raymond Smith, (1878-1961)
BERTHELIN, Jean Georges, (1840-1897)
BOWERBANK, James Scott, (1797-1877)
BRADY, Henry Bowman, (1835-1891)

MICROPALAEONTOLOGY 1850-1899 Cont'd.

BRANSON, Edwin Bayer, (1877-1950)
BURBACH, Otto Carl Heinrich, (1838-1888)
BURROWS, Henry William, (?1860-1939)
CARPENTER, William Benjamin, (1813-1885)
CAYEUX, Lucien, (1864-1944)
CHAPMAN, Frederick, (1864-1943)
CORTI, Benedetto, (1868-1907)
COSTA, Oronzio Gabriele, (1787-1867)
CRICK, Walter Drawbridge, (1857-1903)
DAVIES, Arthur Morley, (1869-1959)
DE LA HARPE, Philippe, (1830-1882)
DONCIEUX, Louis, (1874-1960)
DOUVILLÉ, Joseph Henri Ferdinand, (1846-1937)
DUJARDIN, Félix, (1801-1860)
EHRENBERG, Christian Gottfried, (1795-1876)
FISCHLI, Heinrich, (1857-1932)
FOORD, Arthur Humphreys, (1845-1934)
FORMA, Ernesto, (1869-1911)
FORNASINI, Carlo, (1854-1931)
FRANKE, Adolf, (1860-1942)
FRANZENAU, Agoston, (1856-1919)
FREYER, Heinrich, (1802-1865)
FUCHS, Theodor, (1842-1925)
GRUNOW, Albert, (1826-1914)
GRZYBOWSKI, Józef, (1869-1922)
GÜMBEL, Carl Wilhelm von, Ritter, (1823-1898)
GUPPY, Robert John Lechmere, (1836-1916)
HAECKEL, Ernst Heinrich Philipp August, (1834-1919)
HAGENOW, Friedrich von, (1797-1865)
HAIDINGER, Wilhelm Karl von, Ritter, (1795-1871)
HAIME, Jules, (1824-1856)
HANTKEN, Miksa von Prudnik, (1821-1893)
HERON-ALLEN, Edward, (1861-1943)
HERRICK, Clarence Luther, (1858-1904)
HINDE, George Jennings, (1839-1918)
HOSIUS, August, (1825-1896)
HOWCHIN, Walter, Reverend, (1845-1937)
JONES, Thomas Rupert, (1819-1911)
KARRER, Felix, (1825-1903)
KAUFMANN, Franz Joseph, (-1892)
KIRKBY, James Walker, (1834-1901)
KORN, Johannes, (1862-1927)
LIEBUS, Adalbert, (1876-1945)
LIENENKLAUS, Ernst, (1849-1905)
LÖRENTHEY, Imre, (1867-1917)
MADSEN, Victor Christian, (1865-1947)
MANTELL, Gideon Algernon, (1790-1852)
MARSSON, Theodor, (1816-1892)
MARTIN, Johann Karl Ludwig, (1851-1942)
MILLETT, Fortescue William, (1833-1915)
MÖCKLER, Frederick James, (?1845-1913)
MOON, Frederick William, (-1925)
MUNIER-CHALMAS, Ernest Philippe Auguste, (1843-1903)
NATHORST, Alfred Gabriel, (1850-1921)

MICROPALAEONTOLOGY 1850-1899 Cont'd.

NEUGEBOREN, Johann Ludwig, (1806-1887)
NEWTON, Edwin Tulley, (1840-1930)
NEWTON, Richard Bullen, (1850-1926)
PANDER, Christian Heinrich, (1794-1865)
PANTANELLI, Dante, (1844-1913)
PARKER, William Kitchen, (1823-1890)
PERNER, Jaroslav, (1869-1947)
PETERS, Karl, (1825-1881)
READE, Joseph Bancroft, Reverend, (1801-1870)
REINSCH, Paul Friedrich, (1836-1914)
REIS, Otto Maria, (1862-1934)
REUSS, August Emanuel von, Ritter, (1811-1873)
ROLLE, Friedrich, (1827-1887)
ROLLIER, Louis, (1859-1931)
RÜST, David, (1831-1916)
RZEHÁK, Anton, (1855-1923)
SCHACKO, Gustav, (1824-1917)
SCHELLWIEN, Ernst, (1866-1906)
SCHLUMBERGER, Charles, (1825-1904)
SCHMIDT, Adolf, (1812-1899)
SEGUENZA, Giuseppe, (1833-1889)
SHERBORN, Charles Davies, (1861-1942)
SHRUBSOLE, William Hobbs, (1837-1927)
SISMONDA, Angelo, (1807-1878)
SLACK, Henry James, (1818-1896)
SPEYER, Oscar, (1827-1882)
STOLTZ, Karl, (?1858-1927)
STÜRZENBAUM, József, (1845-1881)
TERQUEM, Olry, (1797-1887)
TERRIGI, Guglielmo, (1831-1892)
THOMAS, Abram Owen, (1876-1931)
TOBLER, August, (1872-1929)
UDDEN, Johan August, (1859-1932)
UHLIG, Victor Karl, (1857-1911)
ULRICH, Edward Oscar, (1857-1944)
VAUGHAN, Thomas Wayland, (1870-1952)
YOUNG, John, (1823-1900)
ZEISE, Oskar, (1860-1925)

1900 to present

ANDRUSSOV, Nikolai Ivanovich, (1861-1924)
APPLIN, Esther Richards, (1895-1972)
ASANO, Kiyoshi, (1910-)
BAGG, Rufus Mather, (1869-1946)
BAKER, George, (1908-1975)
BANDY, Orville Lee, (1917-1973)
BARROIS, Charles Eugène, (1851-1939)
BASSLER, Raymond Smith, (1878-1961)
BLACK, Maurice, (1904-1973)
BLOW, Walter Harry, (1924-1972)
BOUSSAC, Jean, (1885-1917)
BRADLEY, Wilmot Hyde, (1899-)
BRAMLETTE, Milton Nunn, (1896-1977)
BRANSON, Edwin Bayer, (1877-1950)

MICROPALAEONTOLOGY 1900 to present Cont'd.

BRANSON, Edwin Robert, (1908-1971)
BROTZEN, Fritz, (1902-1968)
BROWN, Noel K., Jr., (1925-1968)
BURNABY, Thomas Patrick, (1924-1968)
BURROWS, Henry William, (?1860-1939)
CARSON, Carlton Mann "Kit", (1898-1972)
CAYEUX, Lucien, (1864-1944)
CHAPMAN, Frederick, (1864-1943)
CHLEBOWSKI, Tadeusz, (1908-1961)
CLOUGH, Kenneth H., (1903-1966)
COOPER, Chalmer Lewis, (-)
CORTI, Benedetto, (1868-1907)
CORYELL, Horace Noble, (1888-1965)
COSIJN, Adrianus Johannes, (1890-1945)
CRESPIN, Irene, (-)
CRICK, Walter Drawbridge, (1857-1903)
CUSHMAN, Joseph Augustine, (1881-1949)
CUVILLIER, Jean, (1899-1969)
CUYLER, Robert Hamilton, (1908-1944)
DAVIES, Arthur Morley, (1869-1959)
DAVIES, Lewis Merson, Lieutenant-Colonel, (1882-1955)
DAVIS, Arthur George, (1892-1957)
DEFLANDRE, Georges Victor, (1897-1973)
DEFLANDRE-RIGAUD, Marthe, (-)
DONCIEUX, Louis, (1874-1960)
DOUVILLÉ, Joseph Henri Ferdinand, (1846-1937)
DOUVILLÉ, Robert, (1881-1914)
EARGLE, Dolan Hoye, (1905-1973)
ELEY, Hugh Moore, (1902-1942)
ELLIS, Brooks Fleming, (1897-1976)
ELLISOR, Alva Christine, (1892-1964)
FAULKNER, Joseph Stanislaus "Ginger", (1928-1963)
FINLAY, Harold John, (1901-1951)
FISCHLI, Heinrich, (1857-1932)
FLETCHER, Corbin Drummond, (1900-1931)
FOORD, Arthur Humphreys, (1845-1934)
FORMA, Ernesto, (1869-1911)
FORNASINI, Carlo, (1854-1931)
FRANKE, Adolf, (1860-1942)
FRANZENAU, Agoston, (1856-1919)
FRENGUELLI, Joaquín, (1883-1958)
FRIZZELL, Donald Leslie, (1906-1972)
FUCHS, Theodor, (1842-1925)
GALLOWAY, Jesse James, (1882-1962)
GOUDKOFF, Paul Pavel, (1880-1955)
GRAHAM, Joseph John, (1909-1967)
GRIMM, Kenneth Edgar, (1912-1961)
GROSS, Walter Robert, (1903-1974)
GRUNOW, Albert, (1826-1914)
GRZYBOWSKI, Józef, (1869-1922)
GUBLER, Jean, (1902-1940)
GUPPY, Robert John Lechmere, (1836-1916)
HAECKEL, Ernst Heinrich Philipp August, (1834-1919)
HANNA, G. Dallas, (1887-1970)

MICROPALAEONTOLOGY 1900 to present Cont'd.

HARPER, John Childs, (1913-1977)
HARRIS, Reginald W., (-)
HERON-ALLEN, Edward, (1861-1943)
HERRICK, Clarence Luther, (1858-1904)
HINDE, George Jennings, (1839-1918)
HOLCOMB, Leroy D., (1919-1974)
HORTON, Harold Merritt "Wog", (1897-1966)
HOWCHIN, Walter, Reverend, (1845-1937)
HUDDLE, John Warfield, (1907-1975)
HUGHES, Donald Dudley, (1893-1946)
HUTSON, Ezekiel Burney, (1897-1961)
IRELAND, Hubert Andrew, (1904-1973)
JARVIS, Daniel, (1920-1973)
JONES, Daniel J., (1889-1965)
JONES, Thomas Rupert, (1819-1911)
JORDAN, Louise, (1908-1966)
KAMPTNER, Erwin, (1889-1972)
KARRER, Felix, (1825-1903)
KIELAN-JAWOROWSKA, Zofia, (-)
KIRKBY, James Walker, (1834-1901)
KIRKPATRICK, Randolph, (-1950)
KLINE, Virginia Harriett, (1910-1959)
KOCHANSKY-DEVIDÉ, Vanda, (1915-)
KORN, Johannes, (1862-1927)
KOZLOWSKI, Roman, (1889-1977)
KUIPER, Willem Nicolaas, (1891-1952)
KÜPPER, Klaus, (1930-1957)
LAW, John Melvin, (1914-1967)
LE CALVEZ, Jean, (1908-1954)
LEWIS, Herbert Price, (?1895-1947)
LIEBUS, Adalbert, (1876-1945)
LIENENKLAUS, Ernst, (1849-1905)
LŐRENTHEY, Imre, (1867-1917)
LOWNDES, Ashley Gordon, (-1956)
MADSEN, Victor Christian, (1865-1947)
MAJZON, László, (1904-1973)
MARPLE, Mildred Fisher, (1898-1966)
MARTIN, Johann Karl Ludwig, (1851-1942)
MARTIN, Lewis, (1920-1973)
MARTINOTTI, Anna, (-1931)
MEHL, Maurice Goldsmith, (1887-1966)
MESSINA, Angelina Rose, (1910-1968)
MILLER, Guy E., "Doc", (1893-1977)
MILLER, John Cummins, (1895-1951)
MILLETT, Fortescue William, (1833-1915)
MITZOPOULOS, Maximos K., (1896-1968)
MÖCKLER, Frederick James, (?1845-1913)
MOON, Frederick William, (-1925)
MORELLET, Lucien, (1882-1945)
MORET, Léon, (1890-1972)
MUNIER-CHALMAS, Ernest Philippe Auguste, (1843-1903)
NAGAPPA, Yedatore, Shri, (1907-1960)
NATHORST, Alfred Gabriel, (1850-1921)
NEWTON, Edwin Tulley, (1840-1930)

MICROPALAEONTOLOGY 1900 to present Cont'd.

NEWTON, Richard Bullen, (1850-1926)
NØRVANG, Aksel, (1914-1968)
NOTH, Rudolf, (1884-1954)
NUTTALL, Winfred Laurence Falkiner, (1897-1969)
PANTANELLI, Dante, (1844-1913)
PERNER, Jaroslav, (1869-1947)
PLUMMER, Helen Jeanne (née Skewes), (1891-1951)
PŘIBYL, Alois, (1904-)
RAMA RAO, Lakshmeswar, (1896-1974)
REICHEL, Manfred, (1896-)
REINHOLD, Thomas, (1890-1955)
REINSCH, Paul Friedrich, (1836-1914)
REIS, Otto Maria, (1862-1934)
ROGATZ, Henry, (1902-1974)
ROLLIER, Louis, (1859-1931)
ROUNDY, Paul Vere, (1884-1937)
RUIZ DE GAONA, Máximo, Padre, (-1971)
RÜST, David, (1831-1916)
RUTTEN, Louis Martin Robert, (1884-1946)
RUTTEN, Martin Gerard, (1910-1970)
RYDE, John Walter, (1898-1961)
RZEHÂK, Anton, (1855-1923)
SANDIDGE, John Roy, (1897-1976)
SCHACKO, Gustav, (1824-1917)
SCHELLWIEN, Ernst, (1866-1906)
SCHENCK, Hubert Gregory, (1897-1960)
SCHLUMBERGER, Charles, (1825-1904)
SCHMID, Kurt, (1906-1957)
SCHUBERT, Richard Johann, (1876-1915)
SENN, Alfred, (1899-1949)
SHERBORN, Charles Davies, (1861-1942)
SHRUBSOLE, William Hobbs, (1837-1927)
SMITH, Homer James, (1901-1954)
SMITH, Robert Hendee "Smokey", (1916-1946)
SPARKS, Dale Darrell Dolsun, (1894-1938)
STAFF, Hans von, (1883-1915)
STEWART, Grace Anne, (1893-1970)
STOLTZ, Karl, (?1858-1927)
STRAW, Stephen Henry, (?1892-1963)
SYLVESTER-BRADLEY, Peter Colley, (1914?-1978)
SYNIEWSKA, Janina, (1893-1951)
TEVES, Juan S., "Steve", (1902-1958)
THALMANN, Hans E., (1899-1975)
THOMAS, Abram Owen, (1876-1931)
THOMAS, Norman Louis, (1897-1970)
TOBLER, August, (1872-1929)
TRAUTH, Friedrich, (1883-1967)
TRIEBEL, Erich, (1894-1971)
UDDEN, Johan August, (1859-1932)
UHLIG, Victor Karl, (1857-1911)
ULRICH, Edward Oscar, (1857-1944)
UMBGROVE, Johannes Herman Frederik, (1899-1954)
VADÁSZ, Elemér, (1885-1970)
VAUGHAN, Thomas Wayland, (1870-1952)

MICROPALAEONTOLOGY 1900 to present Cont'd.

VLERK, Isaak Martinus van der, (?1892-1974)
WEINFURTER, Emil, (1904-1968)
WEINZIERL, Laura Lee, (née Lane), (1900-1928)
WETZEL, Otto Christian August, (1891-1971)
WETZEL, Walter, (1887-)
WHITE, Maynard Pressley, (1896-1961)
WICHER, Carl A., (1901-1957)
WILSON, Eugene Jack, (1921-1976)
ZEISE, Oskar, (1860-1925)
ŽYLA, Antoni, (1902-1966)

MINERAL COLLECTING

Up to 1800

BOURNON, Jacques Louis de, Comte, (1751-1825)
BRÜNNICH, Morten Thrane, (1737-1827)
CEDERBAUM, Peter Christopher, (1733-1795)
COLLING, Olof, (1677-1754)
FORSTER, Adolarius Jacob, (1739-1806)
GERSDORF, Adolf Traugott von, (1744-1807)
GIBBS, George, Colonel, (1776-1833)
LETTSOM, John Coakley, (1744-1815)
MOLTKE, Adam Gottlob, Grev, (1710-1762)
PERKINS, Benjamin Douglas, (1774-1810)
PETIVER, James, (1658-1718)
RASHLEIGH, Philip, (1729-1811)
TRADESCANT, John, The Elder, (1587-1638)
TU WAN, (fl.1115-1133)
VALLERIUS, Göran, (1683-1744)
WORM, Olle, (1588-1654)

1800-1849

ADAM, William, (1794?-1873)
BOURNON, Jacques Louis de, Comte, (1751-1825)
BRÜNNICH, Morten Thrane, (1737-1827)
CHALUBIŃSKI, Tytus, (1820-1889)
DAMON, Robert, (1814-1889)
FONTAINE, Charles Aloyse, Chanoine, (1754-1834)
FORSTER, Adolarius Jacob, (1739-1806)
GERSDORF, Adolf Traugott von, (1744-1807)
GIBBS, George, Colonel, (1776-1833)
GLEN, David Corse, (1824-)
HUTTON, Robert, (1784-1870)
LETTSOM, John Coakley, (1744-1815)
MANDELSLOH, Friedrich von, Graf, (1795-1870)
MEADE, William, (1789-1862)
MÜLLER, Moritz, (1791-1858)
PERKINS, Benjamin Douglas, (1774-1810)
RANDALL, John, (1810-1910)
RASHLEIGH, Philip, (1729-1811)
ROSALES, Henry, (?1822-1916)
WAPPLER, Rudolph Bruno, (1828-1888)

1850-1899

BALLY-PRIOR, Peter Eduard (né Bally), (1847-1926)

MINERAL COLLECTING 1850-1899 Cont'd.

BEALE, William Phipson, Sir, (1839-1922)
BELL, Robert, (1864-1934)
BEMENT, Clarence S., (1843-1923)
BOSCH, Carl, (1874-1940)
BRINDLEY, William, (1832-1919)
CHAŁUBIŃSKI, Tytus, (1820-1889)
COLBURN, Burnham Standish, (1872-1959)
DAMON, Robert, (1814-1889)
FONTAINE, Charles Aloyse, Chanoine, (1754-1834)
FOOTE, Albert E., (1846-1895)
GLEN, David Corse, (1824-)
GORDON, John More, Reverend, (1849-1922)
HUNTER-SELKIRK, John R. S., (1835-1898)
HUTTON, Robert, (1784-1870)
LEVISON, Wallace Goold, (1846-1924)
LOEHR, August von, Ritter, (1847-1917)
MANDELSLOH, Friedrich von, Graf, (1795-1870)
MÜLLER, Moritz, (1791-1858)
ORSINI, Antonio, (-1870)
RACHOY, Josef, (-1902)
RANDALL, John, (1810-1910)
RICHTER, Ernst Julius, (-1868)
RINDELL, Artur, (1852-1936)
ROEBLING, Washington Augustus, Colonel, (1837-1926)
ROSALES, Henry, (?1822-1916)
SCHALCH, Ferdinand, (1848-1918)
SELIGMANN, Gustav, (1849-1920)
THOMPSON, Beeby, (1848-1931)
VONSEN, Magnus, (1879-1954)
WAPPLER, Rudolph Bruno, (1828-1888)

1900 to present
BALLY-PRIOR, Peter Eduard (né Bally), (1847-1926)
BEALE, William Phipson, Sir, (1839-1922)
BELL, Robert, (1864-1934)
BEMENT, Clarence S., (1843-1923)
BJAREBY, Alfred Gunnar, (1899-1967)
BOSCH, Carl, (1874-1940)
BRINDLEY, William, (1832-1919)
COIL, Clarence, (-)
COLBURN, Burnham Standish, (1872-1959)
FINCKH, Alfred, (-1914)
FINNEMAN, Karl Johan, (1880-1953)
FOOTE, Warren, (-1936)
FORD, Hugh Alexander, (1885-1966)
GORDON, John More, Reverend, (1849-1922)
JENNI, Clarence M., Colonel, (1896-)
LEVISON, Wallace Goold, (1846-1924)
LOEHR, August von, Ritter, (1847-1917)
RACHOY, Josef, (-1902)
RANDALL, John, (1810-1910)
RINDELL, Artur, (1852-1936)
ROEBLING, Washington Augustus, Colonel, (1837-1926)
ROSALES, Henry, (?1822-1916)

MINERAL COLLECTING 1900 to present Cont'd.

RUTHERFORD, Hugo, Sir, (1887-1942)
SCHALCH, Ferdinand, (1848-1918)
SELIGMANN, Gustav, (1849-1920)
THOMPSON, Beeby, (1848-1931)
VONSEN, Magnus, (1879-1954)
WATTISON, John Thomas, (1884-1974)
ZODAC, Peter, (1894-1967)

Time unknown
JENSEN, David E., (-)
PERHAM, Frank, (-)
SCHORTMANN, Alvin, (-)
SCHORTMANN, Raymond, (-)
YEDLIN, Neal, (-)

MINERALOGY
Up to 1800
AGATHARCHIDES of Cnidus, (170-100 B.C.)
AGRICOLA, Georgius, (1494-1555)
AIKIN, Arthur, (1773-1854)
ALBERTUS MAGNUS, (1193-1280)
ALBINUS, Petrus, (1543-1598)
ALDROVANDI, Ulisse, (1522-1605)
ALLAN, Thomas, (1777-1833)
ALLEN, William, (1770-1843)
ANDRADA E SILVA, José Bonifácio de, (1762-1838)
ANDRÉ, Kristián Karel, (1763-1831)
ARISTOTLE, (384-322 B.C.)
AUBUISSON DE VOISINS, Jean François d', (1769-1841)
AVICENNA, (980-1037)
BABINGTON, William, (1756-1833)
BARTHOLIN, Thomas, (1616-1680)
BAUME, Antoine, (1728-1804)
BECHER, Johann Joachim, (1635-1682)
BENKÖ, Ferencz, (fl.1786)
BERGMAN, Torbern Olof, (1735-1784)
BIRINGUCCIO, Vannoccio, (1480-1538)
BIRUNI, (pseud. of Abu-al-Rayán Muhammad ibn-Ahmad al-Biruni), (ca.972-1048)
BOODTS, Anselmus Boëtius de, (1550-1634)
BORCH, Michael Jan, (1751-1810)
BORN, Ignaz von, Baron, (1742-1791)
BOURGUET, Louis, (1678-1742)
BRADBURY, John, (1765-1823)
BROCHANT DE VILLIERS, André Jean Marie, (1772-1840)
BRONGNIART, Alexandre, (1778-1847)
BROOKE, Henry James, (1771-1857)
BUCH, Christian Leopold von, Freiherr, (1774-1853)
CHARPENTIER, Johann Friedrich Wilhelm Toussaint von, (1728-1805)
CHILDREN, John George, (1777-1852)
CLARKE, Edward Daniel, (1769-1822)
CORDIER, Pierre Louis Antoine, (1777-1861)
CRONSTEDT, Axel Fredrik, (1722-1765)
DA COSTA, Emanuel Mendes, (1717-1791)

MINERALOGY Up to 1800 Cont'd.

DAUBENTON, Louis Jean Marie, (1716-1799)
DINWIDDIE, James, (-)
EATON, Amos, (1776-1842)
EMMERLING, Ludwig August, (1765-1841)
ESMARCH, Jens, (1762-1839)
ESPLING, Olof, (1751-1821)
FABRICIUS, Georg, (1516-1571)
FAREY, John, (1766-1826)
FAUJAS DE SAINT FOND, Barthélémy, (1741-1819)
FERBER, Johann Jacob, (1743-1790)
FICHTEL, Johann Ehrenreich von, (1732-1795)
FORSIUS (HELSINGFORSIUS), Sigfrid Aronius, (ca.1550-1624)
FORSTER, Westgarth, (1772-1835)
FREIESLEBEN, Johann Carl, (1774-1846)
FRIDVALSKY, János, (1730-1784)
FUCHS, Johann Nepomuk von, (1774-1856)
GADD, Pehr Adrian, (1727-1797)
GESSNER, Conrad, (1516-1565)
GIESECKE, Charles Lewis, Sir, (né Metzle, Karl Ludwig), (1761-1833)
GREGOR, William, Reverend, (1761-1817)
GREVILLE, Charles Francis, Right Honourable, (1749-1809)
GUTHRIE, Matthew, (1743-1807)
HARIOT, Thomas, (1560-1621)
HAÜY, René-Just, Abbé, (1743-1822)
HELMONT, Johann Baptista van, (1578-1644)
HENCKEL, Johann Friedrich, (1679-1744)
HENNAH, Richard, Reverend, (?1765-1846)
HEULAND, John Henry, (1778-1856)
HILL, John, Sir, (1706-1775)
HOFFMANN, Jan Leonard, (1710-1782)
HUTTON, James, (1726-1797)
IMPERATO, Ferrante, (1550-1631)
JAMESON, Robert, (1774-1854)
JAŚKIEWICZ, Jan D. P., (1749-1809)
JORDEN, Edward, (1569-1632)
JUNDZILL, Stanislaw Bonifacy, (1761-1847)
KALM, Pehr, (1716-1779)
KARSTEN, Dietrich Ludwig Gustav, (1768-1810)
KENTMANN, Johannes, (1518-1574)
KIRWAN, Richard, (1733-1812)
KLUK, Krzysztof, (1739-1796)
KOLLĄTAJ, Hugon, (1750-1812)
KOMENSKÝ, Jan Ámos, (1592-1670)
KONIG, Karl Dietrich Eberhardt, called Charles, (1774-1851)
LAVOISIER, Antoine Laurent, (1743-1794)
LAXMAN, Erik, (1737-1796)
LEHMANN, Johann Gottlob, (1719-1767)
LENZ, Johann Georg, (1748-1832)
LESKE, Nathanael Gottfried, (1751-1786)
LINDACKER, Jan Tadeáš, (1768-1816)
LINNE, Carl von, (1707-1778)
MARSIGLI, Luigi Ferdinando, (ca.1330-1394)
MATHESIUS, Johannes, (1504-1565)

MINERALOGY Up to 1800 Cont'd.

MAWE, John, (1764-1829)
MERRET, Christopher, (1614-)
MILLER, Samuel, (1769-1850)
MOHS, Carl Friedrich Christian, (1773-1839)
MONNET, Antoine Grimoald, (1734-1817)
MONTICELLI, Teodoro, Abate, (1757-1845)
MONTLOSIER, Fr. Dominique Reynaud, comte de, (1755-1838)
NASIREDDIN, Mukhamed (called Tusi), (1201-1274)
NAU, Bernhard Sebastian von, (1766-1845)
NICOL, William, (ca.1768-1851)
NICOLS, Thomas, (fl. 1652)
OTLEY, Jonathan, (1766-1856)
PARACELSUS, (pseud. of HOHENHEIM, Philippus Theophrastus Bombastus von), (1493-1541)
PETTKO, Ján, (1812-1890)
PLINY THE ELDER (CAIUS PLINIUS SECUNDUS), (?20 A.D.-79 A.D.)
POTT, Johann Heinrich, (1692-1777)
PRIESTLEY, Joseph, (1733-1804)
REUSS, Franz Ambros, (1761-1830)
RIO, Andrés Manuel del, (1764-1849)
SAPIEHA, Aleksander, (1773-1812)
SAUSSURE, Horace Bénédict de, (1740-1799)
SAUSSURE, Nicolas Théodore de, (1765-1845)
SCHEELE, Karl Wilhelm, (1742-1786)
SCHEUCHZER, Johann Jakob, (1672-1733)
SCHLOTHEIM, Ernst Friedrich von, Freiherr, (1764-1832)
SCHMEISSER, Johann August, (1767-1837)
SCHOEPF, Johann David, (1752-1800)
SENDIVOGIUS, Michael, (1556-1636 or 1646)
SICULUS, Diodorus, (ca.85-21 B.C.)
STASZIC, Stanislaw, (1755-1826)
ŠTERNBERK, Kašpar Mariá, Hrabe, (1761-1838)
SWEDENBORG, Emanuel, (1688-1772)
TELEKI, Domokos, Graf, (-1803)
THEOPHRASTUS, (371-ca.286 B.C.)
THOMSON, Thomas, (1773-1852)
TIDSTRÖM, Andreas Philip, (1723-1779)
TONDI, Matteo, (1762-1835)
TOWNSON, Robert, (1763-1827)
TREBRA, Friedrich Wilhelm Heinrich von, (1740-1819)
VOGT, Johann Georg Mauriz, (1669-1731)
VOIGT, Johann Karl Wilhelm, (1752-1821)
VON ENGESTRÖM, Gustaf, (1738-1813)
WAD, Gregers, (1755-1832)
WALKER, John, Reverend, (1731-1803)
WALLERIUS, Johan Gottschalk, (1709-1785)
WATERHOUSE, Benjamin, (1753-1846)
WEAVER, Thomas, (1773-1855)
WEBSTER, John, (1610-1682)
WERNER, Abraham Gottlob, (1749-1817)
WOLLASTON, William Hyde, (1766-1828)
ZIMMERMANN, Christian, (1765-1853)

MINERALOGY Cont'd.

1800-1849

ABICH, Hermann Wilhelm, (1806-1886)
ADAM, William, (1794?-1873)
AIKIN, Arthur, (1773-1854)
ALGER, Francis, (1807-1863)
ALLAN, Thomas, (1777-1833)
ALLEN, William, (1770-1843)
ANDRADA E SILVA, José Bonifácio de, (1762-1838)
ANDRÉ, Kristián Karel, (1763-1831)
ARPPE, Adolf Edvard, (1818-1894)
ATHERSTONE, William Guybon, Honourable, (-1898)
ATTWOOD, Melville, (1812-1898)
AUBUISSON DE VOISINS, Jean François d', (1769-1841)
BABINGTON, William, (1756-1833)
BARNES, Daniel Henry, (1785-1828)
BECK, Theodoric Romeyn, (1791-1855)
BERZELIUS, Jöns Jakob, (1779-1848)
BEYRICH, August Heinrich Ernst, (1815-1896)
BISCHOFF, Karl Gustav, (1792-1870)
BLOMSTRAND, Christian Wilhelm, (1826-1897)
BLUM, Johann Reinhard, (1802-1883)
BOLL, Jacob, (1828-1880)
BONSDORFF, Pehr Adolf von, (1791-1839)
BORCH, Michael Jan, (1751-1810)
BOUSSINGAULT, Joseph Dieudonné Jean-Baptiste, (1807-1887)
BOUVÉ, Thomas Tracy, (1815-1896)
BOWEN, George Thomas, (1803-1828)
BRADBURY, John, (1765-1823)
BRAVAIS, Auguste, (1811-1863)
BREITHAUPT, Johann Friedrich August, (1791-1873)
BROCHANT DE VILLIERS, André Jean Marie, (1772-1840)
BRONGNIART, Alexandre, (1778-1847)
BROOKE, Henry James, (1771-1857)
BRUMBY, Richard T., (1804-1856)
BRYSON, Alexander, (1816-1866)
BUCH, Christian Leopold von, Freiherr, (1774-1853)
BUCKLAND, William, Reverend, (1784-1856)
BUNSEN, Robert Wilhelm, (1811-1899)
BURAT, Amédée, (1809-1883)
CATULLO, Tommaso Antonio, (1782-1869)
CAUCHY, F. P., (-?1842)
CHAPMAN, Edward John, (1821-1904)
CHARPENTIER, Johann Friedrich Wilhelm Toussaint von, (1728-1805)
CHILDREN, John George, (1777-1852)
CLARKE, Edward Daniel, (1769-1822)
CONYBEARE, John Josiah, Reverend, (1779-1824)
CORDIER, Pierre Louis Antoine, (1777-1861)
COTTA, Carl Bernhard von, (1808-1879)
CREDNER, Carl Friedrich Heinrich, (1809-1876)
ČŽJŽEK, Johann Baptist Anton Karl, (1806-1855)
DAMOUR, Alexis, (1808-1902)
DANA, James Dwight, (1813-1895)
DANA, James Freeman, (1793-1827)
DAUBRÉE, Gabriel Auguste, (1814-1896)

MINERALOGY 1800-1849 Cont'd.

DAVIES, William, (1814-1891)
DELAFOSSE, Gabriel, (1796-1878)
DELESSE, Achille Ernest Oscar Joseph, (1817-1881)
DES CLOIZEAUX, Alfred Louis Olivier Legrand, (1817-1897)
DEWALQUE, Gilles Joseph Gustave, (1828-1905)
DEWEY, Chester, (1784-1867)
DINWIDDIE, James, (-)
DOMEYKO, Ignacy, (1802-1889)
DUCATEL, Julius Timoleon, (1796-1849)
DUFRÉNOY, Pierre Armand, (1792-1857)
DUNKER, Wilhelm Bernhard Rudolph Hadrian, (1809-1885)
EATON, Amos, (1776-1842)
EMMERLING, Ludwig August, (1765-1841)
ERNI, Henri, (1822-1885)
ESMARCH, Jens, (1762-1839)
ESTREICHER, Alojzy, (-?1852)
FAREY, John, (1766-1826)
FAUJAS DE SAINT FOND, Barthélémy, (1741-1819)
FEISTMANTEL, Karel, (1819-1885)
FISCHER, Leopold Heinrich, (1817-1886)
FORCHHAMMER, Johan Georg, (1794-1865)
FORSTER, Westgarth, (1772-1835)
FOUQUÉ, Ferdinand André, (1828-1904)
FOURNET, Joseph Jean Baptiste Xavier, (1801-1869)
FREIESLEBEN, Johann Carl, (1774-1846)
FUCHS, Johann Nepomuk von, (1774-1856)
FUCHS, Wilhelm, (1802-1853)
FURUHJELM, Enoch Hjalmar, (1823-1886)
GADOLIN, Axel Wilhelm, General, (1828-1892)
GAY, Martin, (1803-1850)
GENTH, Friedrick August Ludwig Karl Wilhelm, later called Frederick Augustus, (1820-1893)
GERMAR, Ernst Friedrich, (1786-1853)
GIESECKE, Charles Lewis, Sir, (né Metzle, Karl Ludwig), (1761-1833)
GLOCKER, Ernst Friedrich, (1793-1858)
GREG, Robert Phillips, (1826-1906)
GREGOR, William, Reverend, (1761-1817)
GREVILLE, Charles Francis, Right Honourable, (1749-1809)
GUEYMARD, François Émile, (1788-1870)
GUISCARDI, Guglielmo, (1821-1885)
GUTHRIE, Matthew, (1743-1807)
HAIDINGER, Wilhelm Karl von, Ritter, (1795-1871)
HANUŠ, František, (1860-1937)
HAÜY, René-Just, Abbé, (1743-1822)
HAYDEN, Ferdinand Vandiveer, (1829-1887)
HEDDLE, Matthew Forster, (1828-1897)
HENNAH, Richard, Reverend, (?1765-1846)
HENSLOW, John Stevens, Reverend, (1796-1861)
HEULAND, John Henry, (1778-1856)
HITCHCOCK, Edward, (1793-1864)
HOERNES, Moritz, (1815-1868)
HOFF, Thomas Christian Frederik August, (1822-1882)
HOHENEGGER, Ludwig, (1807-1864)

MINERALOGY 1800-1849 Cont'd.

HORNER, Leonard, (1785-1864)
HOW, Henry, (1828-1879)
HUNT, Thomas Sterry, (1826-1892)
HUNTER, Robert, Reverend, (1823-1897)
IGELSTRÖM, Lars Johan, (1822-1897)
JACKSON, Charles Thomas, (1805-1880)
JAMESON, Robert, (1774-1854)
JAŚKIEWICZ, Jan D. P., (1749-1809)
JOHNSTON, James Finlay Weir, (1796-1855)
JOKÉLY, Johann, (1826-1862)
JUNDZILL, Stanislaw Bonifacy, (1761-1847)
KARSTEN, Dietrich Ludwig Gustav, (1768-1810)
KARSTEN, Gustav, (1820-1900)
KARSTEN, Hermann, (1809-1877)
KEFERSTEIN, Christian, (1784-1866)
KENNGOTT, Gustav Adolf, (1818-1897)
KIRWAN, Richard, (1733-1812)
KJERULF, Theodor, (1825-1888)
KOBELL, Franz von, (1803-1882)
KOKSHAROV, Nikola J. von, General, (1818-1893)
KOLENATÝ, Bedřich, (1812-1864)
KOLLĄTAJ, Hugon, (1750-1812)
KONIG, Karl Dietrich Eberhardt, called Charles, (1774-1851)
KREJČÍ, Jan, (1825-1887)
LEA, Isaac, (1792-1886)
LEIDY, Joseph, (1823-1891)
LENZ, Johann Georg, (1748-1832)
LEONHARD, Gustav von, (1816-1878)
LEONHARD, Karl Cäsar von, (1779-1862)
LÉVY, Armand, (1795-1861)
LINDACKER, Jan Tadeáš, (1768-1816)
LIPOLD, Markus Vincenc, (1816-1883)
LLOYD, J. A., Colonel, (-1856?)
LOCKE, John, (1792-1856)
LUDLAM, Henry, (1822-1880)
MACLEAY, William Sharp, (1792-1865)
MARCOU, Jules, (1824-1898)
MAWE, John, (1764-1829)
McCOY, Frederick, Sir, (1823-1899)
MELION, Josef Vincenc, (1813-1905)
MENGE, Johann, (-)
MERIAN, Peter, (1795-1883)
MEYN, Ludwig, (1820-1878)
MIERS, John, (1789-1879)
MILLER, Samuel, (1769-1850)
MILLER, William Hallowes, (1801-1880)
MITCHELL, Elisha, Reverend, (1793-1857)
MOHS, Carl Friedrich Christian, (1773-1839)
MOLYNEUX, William, (1824-1882)
MONNET, Antoine Grimoald, (1734-1817)
MONTICELLI, Teodoro, Abate, (1757-1845)
MONTLOSIER, Fr. Dominique Reynaud, comte de, (1755-1838)
MOORE, Nathaniel Fish, (1782-1872)
MUDGE, Benjamin Franklin, (1817-1879)

MINERALOGY 1800-1849 Cont'd.

MURRAY, Alexander, (1810-1884)
NAU, Bernhard Sebastian von, (1766-1845)
NAUMANN, Carl Friedrich, (1797-1873)
NECKER DE SAUSSURE, Louis Albert, (1786-1861)
NENDTVICH, Károly, (1811-?1892)
NEUMANN, Franz Ernest, (1798-1895)
NICOL, James, (1810-1879)
NICOL, William, (ca.1768-1851)
NÖGGERATH, Johann Jakob, (1788-1877)
NORDENSKIÖLD, Nils Gustaf, (1792-1866)
OLMSTED, Denison, (1791-1859)
OTLEY, Jonathan, (1766-1856)
PARTSCH, Paul, (1791-1856)
PERCY, John, (1817-1889)
PEREIRA DA COSTA, Francisco Antonio, (1809-1889)
PETERS, Karl, (1825-1881)
PETTKO, Ján, (1812-1890)
PFAFF, Friedrich, (1825-1886)
PHILLIPS, William, (1800-1834)
PILLA, Leopoldo, (1805-1848)
PÖSEPNÝ, František, (1836-1895)
PRESL, Jan Svatopluk, (1791-1849)
PRIESTLEY, Joseph, (1733-1804)
PUSCH, Georg Gottlieb, (1790-1846)
RAMMELSBERG, Karl Friedrich, (1813-1899)
RAULIN, Félix Victor, (1815-1905)
RAUMER, Karl Georg von, (1783-1865)
REUSS, August Emanuel von, Ritter, (1811-1873)
REUSS, Franz Ambros, (1761-1830)
RIO, Andrés Manuel del, (1764-1849)
RÖMER, Carl Ferdinand von, (1818-1891)
RÖMER, Hermann, (1816-1894)
ROSE, Alexander, (1781-1860)
ROSE, Gustav, (1798-1873)
RUMPF, Ludwig, (1793-1862)
RUSKIN, John, (1819-1900)
SAINTE-CLAIRE DEVILLE, Charles, (1814-1876)
SANDBERGER, Fridolin Carl Ludwig von, (1826-1898)
SAPIEHA, Aleksander, (1773-1812)
SARTORIUS VON WALTERSHAUSEN, Wolfgang, (1809-1876)
SAUSSURE, Nicolas Théodore de, (1765-1845)
SCACCHI, Arcangelo, (1810-1893)
SCHENKENBERG, F. C. A., (-)
SCHLOTHEIM, Ernst Friedrich von, Freiherr, (1764-1832)
SCHMEISSER, Johann August, (1767-1837)
SCHMID, Ernst Erhard, (1815-1885)
SCHOEPF, Johann David, (1752-1800)
SCHOOLCRAFT, Henry Rowe, (1793-1864)
SCHRAUF, Albrecht, (1837-1897)
SELLA, Quintino, (1827-1884)
SILLIMAN, Benjamin, Sr., (1779-1864)
SILLIMAN, Benjamin Jr., (1816-1885)
SJÖGREN, Carl Anton Hjalmar, (1822-1893)
SOLTYK, Michal, (-)

MINERALOGY 1800-1849 Cont'd.

SOWERBY, Henry, (1825-1891)
STASZIC, Stanislaw, (1755-1826)
ŠTERNBERK, Kašpar Mariá, Hrabe, (1761-1838)
STORY-MASKELYNE, Mervin Herbert Nevil, (1823-1911)
STROBEL, Pellegrino, (1821-1895)
SWALLOW, George Clinton, (1817-1899)
SZABÓ, József, (1822-1894)
TELEKI, Domokos, Graf, (-1803)
TENNANT, James, (1808-1881)
TENORE, Gaetano, (1826-1903)
THOMSON, Thomas, (1773-1852)
TONDI, Matteo, (1762-1835)
TOROSIEWICZ, Teodor, (-)
TOWNSON, Robert, (1763-1827)
TRAUTSCHOLD, Hermann Adolfovich, (1817-1902)
TREBRA, Friedrich Wilhelm Heinrich von, (1740-1819)
VIVIAN, William, (1817-1879)
VOIGT, Johann Karl Wilhelm, (1752-1821)
VON ENGESTRÖM, Gustaf, (1738-1813)
WAD, Gregers, (1755-1832)
WALKER, John, Reverend, (1731-1803)
WALMSTEDT, Lars Edward, (1819-1892)
WATERHOUSE, Benjamin, (1753-1846)
WEAVER, Thomas, (1773-1855)
WEBSKY, Christian Friedrich Martin, (1824-1886)
WEBSTER, John White, (1793-1850)
WEISS, Christian Samuel, (1780-1856)
WERNER, Abraham Gottlob, (1749-1817)
WHEWELL, William, Reverend, (1794-1866)
WÖHLER, Friedrich, (1800-1883)
WOLLASTON, William Hyde, (1766-1828)
WOODROW, James, (1828-1907)
ZIMMERMANN, Christian, (1765-1853)
ZIPPE, František Xaver Maximilian, (1791-1863)
ZIPSER, Christian Andreas, (1783-1864)

1850-1899

ABICH, Hermann Wilhelm, (1806-1886)
ADAMS, Frank Dawson, (1859-1942)
AF SCHULTÉN, August Benjamin, (1856-1913)
AIKIN, Arthur, (1773-1854)
ALGER, Francis, (1807-1863)
AMINOFF, Gregori, (1883-1947)
ANDERSON, Charles, (1876-1944)
ANDERSSON, Axel R., (1858-1952)
ARPPE, Adolf Edvard, (1818-1894)
ARTINI, Ettore, (1866-1928)
ARZRUNI, Andreas, (1847-1898)
ASHCROFT, Frederick Noel, (1878-1949)
ATHERSTONE, William Guybon, Honourable, (-1898)
ATTWOOD, Melville, (1812-1898)
AZÉMA, Léon, Colonel, (1852-1931)
BABÁNEK, František, (1836-1910)
BACKLUND, Helge Götrik, (1878-1958)

MINERALOGY 1850-1899 Cont'd.

BÄCKSTROM, Helge Mattias, (1865-1932)
BALDACCI, Luigi, (1850-1927)
BALDAUF, Richard Julius, (1848-1931)
BARET, Charles, (1831-1910)
BARNETT, Andrew Ketchen, (1852-1914)
BARRON, Thomas, (1867-1906)
BARVÍŘ, Jindřich Ladislav, (1863-1952)
BASCOM, Florence, (1862-1945)
BASTIN, Edson Sunderland, (1878-1953)
BAUER, Heinrich E., (1840-1896)
BAUER, Max Hermann, (1844-1917)
BAUERMAN, Hilary, (1833-1909)
BAUMHAUER, Heinrich Adolf, (1848-1926)
BAYLEY, William Shirley, (1861-1943)
BECK, Theodoric Romeyn, (1791-1855)
BECKE, Friedrich Johann Karl, (1855-1931)
BECKENKAMP, Jakob, (1855-1931)
BECKER, Ewald, (-1873)
BELYANKIN, Dmitri Stepanovich, (1876-1953)
BEMMELEN, Jacob Maarten van, (1830-1911)
BERGT, Walther Adolf, (1864-1941)
BERKEY, Charles Peter, (1867-1955)
BERTOLIO, Leopoldo Sollman, (-?1922)
BERTRAND, Émile, (1844-1909)
BERWERTH, Friedrich Martin, (1850-1918)
BEUTELL, Albert, (1859-1921)
BEYRICH, August Heinrich Ernst, (1815-1896)
BISCHOFF, Karl Gustav, (1792-1870)
BLAKE, John Marcus, (1838-1920)
BLOMSTRAND, Christian Wilhelm, (1826-1897)
BLUM, Johann Reinhard, (1802-1883)
BODENBERGER, Wilhelm, (1857-1941)
BOERIS, Giovanni, (1867-1946)
BØGGILD, Ove Balthazar, (1872-1956)
BOHDANOWICZ, Karol, (1864-1947)
BOLL, Jacob, (1828-1880)
BOŘICKÝ, Emanuel, (1840-1881)
BOUSSINGAULT, Joseph Dieudonné Jean-Baptiste, (1807-1887)
BOUVÉ, Thomas Tracy, (1815-1896)
BOWLES, Oliver, (1877-1958)
BOWMAN, Herbert Lister, (1874-1942)
BRALY, Adrien, (1864-1927)
BRAUNS, Reinhard Anton, (1861-1937)
BRAVAIS, Auguste, (1811-1863)
BRAVO, José J., (1874-1927)
BREITHAUPT, Johann Friedrich August, (1791-1873)
BREZINA, Maria Aristides, (1848-1909)
BROCK, Reginald Walter, (1874-1935)
BRØGGER, Waldemar Christofer, (1851-1940)
BROWN, Amos Peaslee, (1864-1917)
BRUGNATELLI, Luigi, (1859-1928)
BRUHNS, Franz Siegfried Willy, (1864-1929)
BRUMBY, Richard T., (1804-1856)
BRUSH, George Jarvis, (1831-1912)

MINERALOGY 1850-1899 Cont'd.

BRYSON, Alexander, (1816-1866)
BUCCA, Lorenzo, (1857-1930)
BUCH, Christian Leopold von, Freiherr, (1774-1853)
BÜCKING, Ferdinand Carl Bertram Hugo, (1851-1932)
BUCKLAND, William, Reverend, (1784-1856)
BUEHLER, Henry Andrew, (1876-1944)
BUNSEN, Robert Wilhelm, (1811-1899)
BURAT, Amédée, (1809-1883)
BURKART, Eduard, (1865-1941)
BUSZ, Karl Heinrich Emil Georg, (1863-1930)
BUTLER, Francis Henry, (1849-1935)
BUTTGENBACH, Henri, (1874-1964)
CĂDERE, Dimitrie M., (1876-1941)
CAHN, Lazard, (1865-1940)
CANFIELD, Frederick Alexander, (1849-1926)
CARNOT, Marie Adolphe, (1839-1920)
CASWELL, John Henry, (1846-1909)
CATULLO, Tommaso Antonio, (1782-1869)
CESÀRO, Guiseppe Raimondo Pio, (1849-1939)
CHAMBERLAIN, Arthur, (-)
CHANCOURTOIS, A. E. Béguyer de, (-1886?)
CHANDLER, Charles Frederick, (1836-1925)
CHAPMAN, Edward John, (1821-1904)
CHURCH, Arthur Herbert, Sir, (1834-1915)
CHYDENIUS, Johan Jakob "Hannes", (1836-1890)
CLARAZ, Georges, (1832-1930)
CLARKE, Frank Wigglesworth, (1847-1931)
CLERICI, Enrico, (1862-1938)
COBÎLCESCU, Grigore, (1826-1892)
COCKIN, George Marmaduke, (1852-1918)
COLEMAN, Arthur Philemon, (1852-1939)
COLLINS, Arthur L., (1868-1902)
COLLINS, Henry Francis, (?1864-1924)
COLLINS, Joseph Henry, (1841-1916)
COOMARASWAMY, Ananda K., (1877-1947)
CORDIER, Pierre Louis Antoine, (1777-1861)
CORNET, François Léopold, (1834-1887)
COSTA SENA, Joaquim Candido da, (1852-1919)
COTTA, Carl Bernhard von, (1808-1879)
CRAWFORD, Ralph Dixon, (1873-1950)
CREDNER, Carl Friedrich Heinrich, (1809-1876)
CROOK, Alja Robinson, (1864-1930)
CROOK, Thomas, (1876-1937)
CROSBY, William Otis, (1850-1925)
CULLIS, Charles Gilbert, (1871-1941)
CURRIE, James, (1863-1930)
ČŽJŽEK, Johann Baptist Anton Karl, (1806-1855)
D'ACHIARDI, Giovanni, (1872-1944)
DAMOUR, Alexis, (1808-1902)
DANA, Edward Salisbury, (1849-1935)
DANA, James Dwight, (1813-1895)
DARTON, Nelson Horatio, (1865-1948)
DAUBRÉE, Gabriel Auguste, (1814-1896)
DAVIES, Thomas, (1837-1892)

MINERALOGY 1850-1899 Cont'd.

DAVIES, William, (1814-1891)
DAVISON, Ernest Henry, (1877-1944)
DELAFOSSE, Gabriel, (1796-1878)
DELESSE, Achille Ernest Oscar Joseph, (1817-1881)
DES CLOIZEAUX, Alfred Louis Olivier Legrand, (1817-1897)
DEWALQUE, Gilles Joseph Gustave, (1828-1905)
DEWEY, Chester, (1784-1867)
DEWEY, Henry, (1876-1965)
DICK, Allan Brugh, (1833-1926)
DOELTER y CISTERLICH, Cornelio August, (1850-1930)
DOMEYKO, Ignacy, (1802-1889)
DRUGMAN, Julien Léon Eugène, (1875-1950)
DUBICKI, Józef, (-?1890)
DUDGEON, Patrick, (-1895)
DUFFOUR, Alexis, (1877-1948)
DUFRÉNOY, Pierre Armand, (1792-1857)
DUNIN-BORKOWSKI, Stanislaw von, Graf, (1864-1934)
DUNKER, Wilhelm Bernhard Rudolph Hadrian, (1809-1885)
DUPARC, Louis Claude, (1866-1932)
EAKLE, Arthur Starr, (1862-1931)
EGLESTON, Thomas, (1832-1900)
EMMONS, Samuel Franklin, (1841-1911)
ENDLICH, Frederic Miller, (1851-1899)
ENGLISH, George Letchworth, (1864-1944)
ERNI, Henri, (1822-1885)
ESTREICHER, Alojzy, (-?1852)
ETHERIDGE, Robert Jr., (1847-1920)
EVANS, John William, (1857-1930)
FABER, Harald Nicolai, (1856-1943)
FARRINGTON, Oliver Cummings, (1864-1933)
FEDEROV, Evgraf Stepanovich, (1853-1919)
FEISTMANTEL, Karel, (1819-1885)
FENNER, Clarence Norman, (1870-1949)
FERRIER, Walter F., (1865-1950)
FIRKET, Adolphe, (1837-1905)
FIRTSCH, Georg, (1859-1914)
FISCHER, Leopold Heinrich, (1817-1886)
FLETCHER, Lazarus, Sir, (1854-1921)
FLETCHER, Mark, Reverend, (1870-1956)
FLIGHT, Walter, (1841-1885)
FLINK, Gustaf, (1849-1931)
FONTAINE, William Morris, (1835-1913)
FOOTE, Harry Ward, (1875-1942)
FORCHHAMMER, Johan Georg, (1794-1865)
FORD, William Ebenezer, (1878-1939)
FORIR, Henri, (1856-1907)
FOSTER, Clement Le Neve, Sir, (1841-1904)
FOUQUÉ, Ferdinand André, (1828-1904)
FOURNET, Joseph Jean Baptiste Xavier, (1801-1869)
FRANZENAU, Agoston, (1856-1919)
FRAZER, Persifor, (1844-1909)
FRAZIER, Benjamin West, (1841-1905)
FRENZEL, Friedrich August, (1842-1902)
FRIEDEL, Charles, (1832-1899)

MINERALOGY 1850-1899 Cont'd.

FRIEDEL, Georges, (1865-1933)
FUCHS, Johann Nepomuk von, (1774-1856)
FUCHS, Wilhelm, (1802-1853)
FUGGER, Friedrich Eberhard, (1842-1919)
FUKUCHI, Nobuyo, (1877-1934)
FURUHJELM, Enoch Hjalmar, (1823-1886)
GADOLIN, Axel Wilhelm, General, (1828-1892)
GARRIGOU, Félix, (1835-1920)
GAUBERT, Marie Benoit Paul, (1865-)
GAUDEFROY, Christophe Léon, Chanoine, (1878-1971)
GAUTIER, Emile Justin Armand, (1837-1920)
GAVELIN, Axel Olof, (1875-1947)
GENTH, Friedrick August Ludwig Karl Wilhelm, later called Frederick Augustus, (1820-1893)
GERBER, Eduard, (1876-1956)
GERMAR, Ernst Friedrich, (1786-1853)
GLINKA, Sergei Fedorovich, (1855-1933)
GLOCKER, Ernst Friedrich, (1793-1858)
GOLDSCHMIDT, Victor, (1853-1933)
GONÇALVES GUIMARÃES, Antonio José, (1875-1919)
GOODCHILD, John George, (1844-1906)
GOSSNER, Balthasar, (1877-1937)
GRÄNZER, Josef, (1857-1934)
GRATACAP, Louis Pope, (1850-1917)
GREG, Robert Phillips, (1826-1906)
GREGORY, James Reynolds, (1832-1899)
GRISHCHINSKY, Petr Iuustinovich, (-1920)
GRÖNWALL, Karl Anders Axel, (1869-1944)
GROSSPIETSCH, Oskar, (-1920)
GROTH, Paul Heinrich von, Ritter, (1843-1927)
GRÜNLING, Friedrich, (1857-1919)
GUEYMARD, François Émile, (1788-1870)
GUILD, Frank Nelson, (1870-1939)
GUISCARDI, Guglielmo, (1821-1885)
GURLDT, Adolf Friedrich, (1829-1902)
GURNEY, Henry Palin, Reverend, (1847-1904)
HAGUE, Arnold, (1840-1917)
HAIDINGER, Wilhelm Karl von, Ritter, (1795-1871)
HALL, Marshall, Captain, (1831-1896)
HAMBERG, Axel, (1863-1933)
HAMMER, Wilhelm, (1875-1942)
HANUŠ, František, (1860-1937)
HARDMAN, Edward Townley, (1845-1887)
HARRINGTON, Bernard J., (1848-1907)
HARRISON, John Burchmore, Sir, (1856-1926)
HAUTEFEUILLE, Paul Gabriel, (1836-1902)
HAYDEN, Ferdinand Vandiveer, (1829-1887)
HECTOR, James, Sir, (1834-1907)
HEDDLE, Matthew Forster, (1828-1897)
HELLAND, Amund Theodor, (1846-1918)
HELMHACKER, Rudolf Václav, (1840-1915)
HENRICH, Ferdinand, (1837-1915)
HENSLOW, John Stevens, Reverend, (1796-1861)
HESNARD, Emile E., (1867-1929)

MINERALOGY 1850-1899 Cont'd.

HESS, Frank L., (1871-1955)
HIBSCH, Josef Emanuel, (1852-1940)
HIDDEN, William Earl, (1853-1918)
HIKI, Tadasu, (1866-1927)
HILLEBRAND, William Francis, (1853-1925)
HINRICHS, Gustavus Detlef, (-1922)
HINTERLECHNER, Karl, (1874-1932)
HINTZE, Carl Adolf Ferdinand, (1851-1916)
HIORTDAHL, Thorstein Hallager, (1839-1925)
HIRSCHWALD, Julius, (1845-1928)
HITCHCOCK, Edward, (1793-1864)
HOBLEY, Charles William, (1867-1947)
HOERNES, Moritz, (1815-1868)
HOFF, Thomas Christian Frederik August, (1822-1882)
HOFFMANN, George Christian, (1837-1917)
HOFMANN, Adolf, (1853-1913)
HOFMANN, Károly, (1839-1891)
HÖGBOM, Arvid Gustaf, (1857-1940)
HOHENEGGER, Ludwig, (1807-1864)
HOLLAND, Thomas Henry, Sir, (1868-1947)
HOLMQUIST, Per Johan, (1866-1946)
HORNER, Leonard, (1785-1864)
HOVEY, Edmund Otis, (1862-1924)
HOW, Henry, (1828-1879)
HUBBARD, Lucius Lee, (1849-1933)
HUGI, Emil, (1873-1937)
HUNT, Thomas Sterry, (1826-1892)
HUNTER, Robert, Reverend, (1823-1897)
HUSSAK, Franz Eugen, (1856-1911)
HUTCHINSON, Arthur, (1866-1937)
HYLAND, John Shearson, (1866-1898)
IGELSTRÖM, Lars Johan, (1822-1897)
INOSTRANTZEV, Aleksandr Aleksandrovich, (1843-1920)
ISSEL, Arturo, (1842-1922)
JACKSON, Charles Thomas, (1805-1880)
JAHN, Jaroslav Jiljí, (1865-1934)
JAMESON, Robert, (1774-1854)
JANNETTAZ, Edouard, (-1900)
JASTRZĘBOWSKI, Wojciech, (-)
JELSKI, Konstanty, (1837-1896)
JERVIS, William Paget, (1831-1906)
JESPERSEN, Johan Magnus Peder, (1833-1917)
JIMBO, Kotora, (1867-1924)
JOHANSSON, Karl Fredrick, (1866-1933)
JOHNSEN, Arrien, (1877-1934)
JOHNSTON, James Finlay Weir, (1796-1855)
JOHN VON JOHNESBERG, Conrad Heinrich, Edler, (1852-1918)
JOKÉLY, Johann, (1826-1862)
JOLY, John, (1857-1933)
JUDD, John Wesley, (1840-1916)
JULIEN, Alexis Anastay, (1840-1919)
KAISER, Friedrich Wilhelm Erich, (1871-1934)
KALECSINSZKY, Sandór, (1857-1911)
KALKOWSKY, Ernst Louis, (1851-1938)

MINERALOGY 1850-1899 Cont'd.

KARANDYEEV, Vissarion Vissarionovich, (-1916)
KARSTEN, Gustav, (1820-1900)
KARSTEN, Hermann, (1809-1877)
KEFERSTEIN, Christian, (1784-1866)
KEMP, James Furman, (1859-1926)
KENNGOTT, Gustav Adolf, (1818-1897)
KHRUSHCHOV, Konstantin Dmitrievich, (? -1912)
KINCH, Edward, (1848-1920)
KIŠPATIĆ, Mišo, (1851-1926)
KITSON, Albert Ernest, Sir, (1868-1937)
KITTO, Benedict, (1841-1928)
KJERULF, Theodor, (1825-1888)
KLEIN, Johann Friedrich Carl, (1842-1907)
KLOCKE, Friedrich, (1847-1885)
KLOCKMANN, Friedrich Ferdinand Hermann, (1858-1937)
KLVAŇA, Josef, (1857-1919)
KOBELL, Franz von, (1803-1882)
KOECHLIN, Rudolf Ignatz, (1862-1939)
KOKSHAROV, Nikola J. von, General, (1818-1893)
KOLBECK, Friedrich Ludwig Wilhelm, (1860-1943)
KOLDERUP, Carl Fredrik, (1869-1942)
KOLENATÝ, Bedřich, (1812-1864)
KONIG, Karl Dietrich Eberhardt, called Charles, (1774-1851)
KOVÁŘ, František, (1862-1941)
KRANTZ, Friedrich Ludwig Robert, (1859-1926)
KRATOCHVÍL, Josef, (1878-1958)
KRAUS, Edward Henry, (1875-1973)
KREJČÍ, Jan, (1825-1887)
KRENNER, József Sándor, (1839-1920)
KREUTZ, Felix Franz Xaver, (1844-1910)
KREUTZ, Szczęsny, (-?1887)
KROTOV, Petr Ivanovich, (1852-1914)
KUČERA, Bruno, (1863-1934)
KUNZ, George Frederick, (1856-1932)
KUPFFER, August Ernestovich, (1842-1918)
KURNAKOV, Nikolai Semenovich, (1860-1941)
LACROIX, François Antoine Alfred, (1863-1948)
LALOY, Roger, (1851-1909)
LAPPARENT, Albert Auguste de, (1839-1908)
LASAULX, Arnold von, (1839-1886)
LAUNAY, Louis Alphonse Auguste de, (1860-1938)
LEA, Isaac, (1792-1886)
LEATHERS, Walter S., (1874-)
LEHMANN, Johannes Georg, (1851-1925)
LEIDY, Joseph, (1823-1891)
LEITZ, Ernst, Senior, (1871-1956)
LEONHARD, Gustav von, (1816-1878)
LEONHARD, Karl Cäsar von, (1779-1862)
LEUZE, Alfred, (1845-1899)
LÉVY, Armand, (1795-1861)
LEWIS, A. E., (-1906)
LEWIS, Henry Carvill, (1853-1888)
LIEBER, Oscar M., (-1862)
LIEBISCH, Theodor, (1852-1922)

MINERALOGY 1850-1899 Cont'd.

LIFFA, Aurél, (1872-1957)
LINCIO, Gabrielle, (1874-1938)
LINDGREN, Waldemar, (1860-1939)
LINDON, Edward Byron, (1860-1891)
LIPOLD, Markus Vincenc, (1816-1883)
LIVERSIDGE, Archibald, (1847-1927)
LLOYD, J. A., Colonel, (-1856?)
LOEWINSSON-LESSING, Franc Julyevich, (1859-1939)
LOOMIS, Frederic Brewster, (1873-1937)
LOUDERBACK, George Davis, (1874-1957)
LOVISATO, Domenico, (1842-1916)
LUDLAM, Henry, (1822-1880)
LUDWIG, Ernst, (1842-1915)
LUQUER, Lea McIlvaine, (1864-1930)
MACLEAY, William Sharp, (1792-1865)
MAKOWSKY, Alexander, (1833-1908)
MALLADRA, Alessandro, (1865-1945)
MALLARD, François Ernest, (1833-1894)
MALLETT, Frederic Richard, (1841-1921)
MANASSE, Ernesto, (1875-1922)
MANCHESTER, James Greenfield, (1871-1948)
MARCOU, Jules, (1824-1898)
MATHEWS, Edward Bennett, (1869-1944)
MATTHEWS, Edward Bennett, (1869-1944)
MAUGIN, Charles, (1878-1958)
McCAUGHEY, William John, (1882-1962)
McCOY, Frederick, Sir, (1823-1899)
McMAHON, Charles Alexander, Lieutenant-General, (1830-1904)
MELCZER, Gusztáv, (1869-1907)
MELION, Josef Vincenc, (1813-1905)
MELLO, John Magens, Reverend, (1836-1914)
MERENSKY, Hans, (1871-1952)
MERIAN, Peter, (1795-1883)
MERRILL, George Perkins, (1854-1929)
MERRILL, Loren B., (?1853-1930)
MEUNIER, Étienne Stanislas, (1843-1925)
MEYN, Ludwig, (1820-1878)
MICHEL, Léopold Jean, (1846-1919)
MICHEL-LÉVY, Albert, (1877-1955)
MICHEL-LÉVY, Auguste, (1844-1911)
MIELEITNER, Karl, (1890-1923)
MIERS, Henry Alexander, Sir, (1858-1942)
MIERS, John, (1789-1879)
MILCH, Ludwig, (1867-1928)
MILLER, Benjamin Le Roy, (1874-1944)
MILLER, Samuel, (1769-1850)
MILLER, Willet Green, (1839-1925)
MILLER, William Hallowes, (1801-1880)
MILLOSEVICH, Federico, (1875-1942)
MITCHELL, Elisha, Reverend, (1793-1857)
MOISSAN, Henri, (1852-1907)
MOLYNEUX, William, (1824-1882)
MONETTI, Luigi, (1864-1926)
MOORE, Nathaniel Fish, (1782-1872)

MINERALOGY 1850-1899 Cont'd.

MORGAN, Glenn Beckley, (1876-1925)
MOROZEWICZ, Józef Marian, (1865-1941)
MORTON, James F., (1870-1941)
MOSES, Alfred Joseph, (1859-1920)
MOSS, Richard Jackson, (1847-1934)
MRAZEC, Lodovic, (1867-1944)
MUDGE, Benjamin Franklin, (1817-1879)
MÜGGE, Johannes Otto Konrad, (1858-1932)
MÜLLER, Hugo Heinrich Wilhelm, (1833-1915)
MURRAY, Alexander, (1810-1884)
NASON, Henry Bradford, (1831-1895)
NAUMANN, Carl Friedrich, (1797-1873)
NECHAEV, Aleksyei Vasil'evich, (1864-1915)
NECKER DE SAUSSURE, Louis Albert, (1786-1861)
NENDTVICH, Károly, (1811-?1892)
NEUMANN, Franz Ernest, (1798-1895)
NICOL, James, (1810-1879)
NICOL, William, (ca.1768-1851)
NIES, Friedrich, (1839-1895)
NIKITIN, Vasily Vasiliivich, (1867-1942)
NÖGGERATH, Johann Jakob, (1788-1877)
NORDENSKIÖLD, Nils Adolf Erik, (1832-1901)
NORDENSKIÖLD, Nils Gustaf, (1792-1866)
NOVARESE, Vittorio, (1861-1948)
OEBBEKE, Konrad, (1853-1932)
OLMSTED, Denison, (1791-1859)
O'REILLY, Joseph P., (1829-1905)
ORLOV, Nikolai Arsenevich, (-1919)
OSANN, Carl Alfred, (1859-1923)
OTLEY, Jonathan, (1766-1856)
OTTMER, Eduard Julius Otto, (1846-1886)
PANEBIANCO, Ruggero, (1848-1930)
PARK, James, (1857-1946)
PARSONS, Arthur Leonard, (1873-1957)
PARSONS, James, (1876-1908)
PARTSCH, Paul, (1791-1856)
PEARCE, Richard, (1837-1927)
PENFIELD, Samuel Lewis, (1856-1906)
PENROSE, Richard Alexander Fullerton, Jr., (1863-1931)
PERCEVAL, Spencer George, (1838-1922)
PERCY, John, (1817-1889)
PEREIRA DA COSTA, Francisco Antonio, (1809-1889)
PETERS, Karl, (1825-1881)
PETTERD, William Frederick, (1849-1910)
PFAFF, Friedrich, (1825-1886)
PHILLIPS, Alexander Hamilton, (1866-1937)
PHILLIPS, William, (1800-1834)
PHILLIPS, William Battle, (1857-1918)
PICCINELLI, Piero, Conte, (1865-1937)
PIRSSON, Louis Valentine, (1860-1919)
PISANI DE SERRES, Felix, (1831-1920)
PLAGEMANN, A., (1858-1908)
PLATNAUER, Henry Maurice, (1857-1939)
POLLARD, William, (1870-1946)

MINERALOGY 1850-1899 Cont'd.

PONI, Petru, (1841-1925)
POPOFF, Boris Anfirovich, (1871-1950)
PORTH, Emil, (1832-1858)
POŠEPNÝ, František, (1836-1895)
PRATT, Joseph Hyde, (1870-1942)
PRIMICS, György, (1849-1893)
PRIOR, George Thurland, (1862-1936)
PROCHÁZKA, Vladimír Josef, (1862-1913)
PUFAHL, Otto Hermann August, (1855-1924)
RAMMELSBERG, Karl Friedrich, (1813-1899)
RAMSAY, Wilhelm, (1865-1928)
RANSOME, Frederick Leslie, (1868-1935)
RATH, Gerhard vom, (1830-1888)
RAULIN, Félix Victor, (1815-1905)
RAUMER, Karl Georg von, (1783-1865)
REDLICH, Karl A., (1869-1942)
REPOSSI, Emilio, (1876-1931)
REUSS, August Emanuel von, Ritter, (1811-1873)
REYNOLDS, James Emerson, (1844-1920)
RICHARDSON, Charles Henry, (1862-1935)
RIDLEY, Henry Nicholas, (1855-1956)
RIES, Heinrich, (1871-1951)
RINNE, Friedrich Wilhelm Berthold, (1863-1933)
ROCCATI, Alessandro, (1872-1928)
RÖMER, Carl Ferdinand von, (1818-1891)
RÖMER, Hermann, (1816-1894)
ROMEU, Albert de, (1875-1915)
ROSE, Alexander, (1781-1860)
ROSE, Gustav, (1798-1873)
ROSENBUSCH, Karl Harry Ferdinand, (1836-1914)
ROSIWAL, August Karl, (1860-1923)
RUDLER, Frederick William, (1840-1915)
RUMPF, Johann, (1841-1923)
RUMPF, Ludwig, (1793-1862)
RUSKIN, John, (1819-1900)
RUSSELL, Arthur Edward Ian Montagu, Sir, (1878-1964)
RUTLEY, Frank, (1842-1904)
RYBA, Frantisek, (1867-1918)
RZEHÁK, Anton, (1855-1923)
SADEBECK, Alexander, (1843-1879)
SAINTE-CLAIRE DEVILLE, Charles, (1814-1876)
SAINTE-CLAIRE DEVILLE, Henri, (-1881)
SALMOJRAGHI, Francesco, (1837-?1910)
SAMOILOV, Yakov Vladimirovich, (1870-1925)
SANDBERGER, Fridolin Carl Ludwig von, (1826-1898)
SARGENT, Henry Crunden, (1851-1931)
SARTORIUS VON WALTERSHAUSEN, Wolfgang, (1809-1876)
SAUER, Gustav Adolf, (1825-1932)
SCACCHI, Arcangelo, (1810-1893)
SCACCHI, Eugenio, (1854-1929)
SCHARIZER, Rudolf, (1859-1935)
SCHENKENBERG, F. C. A., (-)
SCHETELIG, Jakob Grubbe Cock, (1875-1935)
SCHMID, Ernst Erhard, (1815-1885)

MINERALOGY 1850-1899 Cont'd.

SCHMIDT, Carl, (1862-1923)
SCHMIDT, Sándor, (1855-?1905)
SCHOOLCRAFT, Henry Rowe, (1793-1864)
SCHRADER, Frank Charles, (1860-1944)
SCHRAUF, Albrecht, (1837-1897)
SCHUSTER, Max, (1856-1887)
SCOTT, Robert Henry, (1833-1916)
SELLA, Quintino, (1827-1884)
SEMSEY UR SEMSEI, Andor, (-1923)
SILLIMAN, Benjamin, Sr., (1779-1864)
SILLIMAN, Benjamin Jr., (1816-1885)
SIMPSON, Edward Sydney, (1875-1939)
SJÖGREN, Carl Anton Hjalmar, (1822-1893)
SJÖGREN, Sten Anders Hjalmar, (1856-1922)
SLAVÍK, František, (1876-1957)
SLOAN, Earle, (1858-1926)
SMITH, Lawrence, (-1883)
SOLLAS, William Johnson, (1849-1936)
SOLLY, Richard Harrison, (1851-1925)
SOUZA BRANDÃO, Vicente de, (1863-1916)
SOWERBY, Henry, (1825-1891)
SPENCER, Leonard James, (1870-1959)
SPIREK, Vincenz, (1852-1907)
STATUTI, Augusto, (1829-1911)
STEENSTRUP, Knud Johannes Vogelius, (1842-1913)
STEIGER, George, (1869-1944)
STELLA, Augusto, (1863-1944)
STELZNER, Alfred Wilhelm, (1840-1895)
ŠTĚRBA-BÖHM, Jan Stanislav, (1874-1938)
STORY-MASKELYNE, Mervin Herbert Nevil, (1823-1911)
STRENG, August, (1830-1897)
STROBEL, Pellegrino, (1821-1895)
STRÜVER, Johannes Karl Theodor, called Giovanni, (1842-1915)
SUESS, Franz Eduard, (1867-1941)
SUTTON, John Richard, (1864-1937)
SWALLOW, George Clinton, (1817-1899)
SZABÓ, József, (1822-1894)
SZÁDECZKY, Gyula, Baron Szádeczky-Kardoss of Szádecsne and Kafdosfalva, (1860-1935)
SZONTAGH, Tamás, (1851-1937)
SZTERÉNYI, Hugo, (1857-1909)
TARASENKO, Vasily Efimovich, (1859-1926)
TAYLER, Joseph Walter, (-?1896)
TENNANT, James, (1808-1881)
TENORE, Gaetano, (1826-1903)
THOMAS, Herbert Henry, (1876-1935)
THOMSON, Thomas, (1773-1852)
THUGUTT, Stanislaw Józef, (1862-1956)
TOOTHAKER, Charles Robinson, (1873-1952)
TÖRNEBOHM, Alfred Elis, (1838-1911)
TOROSIEWICZ, Teodor, (-)
TRAUTSCHOLD, Hermann Adolfovich, (1817-1902)
TRAVERSO, Giovanni Battista, (1843-1914)
TRECHMANN, Charles Otto, (1851?-1917)

MINERALOGY 1850-1899 Cont'd.

TRIBOLET-HARDY, Frédéric Maurice de, (1852-1929)
TSCHERMAK VON SEYSENEGG, Gustav, (1836-1927)
TUČAN, Fran, (1878-1954)
TWITCHELL, Mayville William, (1866-1927)
UDDEN, Johan August, (1859-1932)
USSING, Niels Viggo, (1864-1911)
VALENTIN, Jean, (1877-1898)
VALMONT DE BOMARE, Jacques Christophe, (-)
VAN HISE, Charles Richard, (1857-1918)
VAN HORN, Frank Robertson, (1872-1933)
VAN'T HOFF, Jacobus Hendricus, (1852-1911)
VATER, Heinrich, (1859-1930)
VAUX, George, Jr., (1863-1927)
VERNADSKI, Vladimir Ivanovich, (1863-1945)
VÉSIGNIÉ, Louis, Colonel, (1870-1954)
VIOLA, Carlo Maria, (1855-1925)
VIVIAN, William, (1817-1879)
VOGELSANG, Hermann, (1838-1874)
VOIGT, Waldemar, (1850-1919)
VRBA, Karel, (1845-1922)
WAAGEN, Wilhelm Heinrich, (1841-1900)
WADA, Tsunashirô, (1856-1920)
WALKER, Thomas Leonard, (1867-1942)
WALLERANT, Frédéric Félix Auguste, (1858-1936)
WALMSTEDT, Lars Edward, (1819-1892)
WARD, Henry Augustus, (1834-1906)
WARD, James Clifton, Reverend, (1843-1880)
WARREN, Charles Hyde, (1876-1950)
WASHINGTON, Henry Stephens, (1867-1934)
WATSON, Thomas Leonard, (1871-1924)
WEBSKY, Christian Friedrich Martin, (1824-1886)
WEBSTER, John White, (1793-1850)
WEIBULL, Kristian Oskar Mats, (1856-1923)
WEIDMAN, Samuel, (1870-1945)
WEINBERGER, Isidor, (1838-1915)
WEINSCHENK, Ernst Heinrich Oskar Kasimir, (1865-1921)
WEISBACH, Albin, (1833-1901)
WEISS, Christian Ernst, (1833-1890)
WEISS, Christian Samuel, (1780-1856)
WELLS, Horace Lemuel, (1855-1924)
WELLS, Roger Clark, (1877-1944)
WELSCH, Jules, (1858-1929)
WEYBERG, Zygmunt, (1872-1944)
WHITLOCK, Herbert Percy, (1868-1948)
WICHMANN, Carl Ernst Arthur, (1851-1927)
WIIK, Fredrik Johan, (1839-1909)
WILDER, Frank Alonzo, (1870-1930)
WILLIAMS, George Huntington, (1856-1894)
WILLIAMS, John Francis, (1862-1891)
WILSON, J. S. Grant, (1855-1908)
WINCHELL, Alexander Newton, (1874-1958)
WÖHLER, Friedrich, (1800-1883)
WOLF, Theodor, (1841-1924)
WOLFF, John Elict, (1857-1940)

MINERALOGY 1850-1899 Cont'd.

WOODROW, James, (1828-1907)
WOODRUFF, Elmer Grant, (1872-1952)
WOODWARD, Charles Josiah, (1838-1932)
WRIGHT, Frederick Eugene, (1877-1953)
WÜLFING, Ernst Anton, (1860-1930)
WYNDHAM, T. Heathcote, (?1842-1876)
WYROUBOFF, Gr., (1843-1913)
YACHEVSKY, Leonard Antonovich, (1858-1916)
YEATES, William Smith, (1856-1908)
YOUNG, George Albert, (1898-1947)
ZEMYATCHENSKY, Petr Andreevich, (1856-1942)
ZEPHAROVICH, Victor Leopold von, Ritter, (1830-1890)
ZHEMCHUZHNYI, Sergei Fedorovitch, (1873-1929)
ZIMÁNYI, Károly, (1862-1941)
ZIMMERMANN, Christian, (1765-1853)
ZIPPE, František Xaver Maximilian, (1791-1863)
ZIPSER, Christian Andreas, (1783-1864)
ZIRKEL, Ferdinand, (1838-1912)
ZYNDEL, Fortunat, (1882-1917)

1900 to present

ADAMS, Frank Dawson, (1859-1942)
AF SCHULTÉN, August Benjamin, (1856-1913)
AHRENS, Louis Herman, (1918-)
AIERS, Arthur Joseph, (1896-1965)
AIROLDI, Marco, (1900-1937)
ALDERMAN, Arthur Richard, (1901-)
ALLING, Harold Lattimore, (1888-1960)
ALOISI, Piero, (1881-1938)
AMINOFF, Gregori, (1883-1947)
ANDERSEN, Olaf, (1884-1941)
ANDERSON, Charles, (1876-1944)
ANDERSSON, Axel R., (1858-1952)
ANDREATTA, Benedettino Ciro, (1906-1960)
ARMASHEVSKY, Peter Jakovlevich, (-1919)
ARTINI, Ettore, (1866-1928)
ASHCROFT, Frederick Noel, (1878-1949)
AZÉMA, Léon, Colonel, (1852-1931)
BABÁNEK, František, (1836-1910)
BABIČ, Bogdan, (1891-1961)
BACKLUND, Helge Götrik, (1878-1958)
BÄCKSTROM, Helge Mattias, (1865-1932)
BAKER, George, (1908-1975)
BALDACCI, Luigi, (1850-1927)
BALDAUF, Richard Julius, (1848-1931)
BALOGH, Ernö, (1882-1969)
BANDY, Mark Chance, (1900-1963)
BÁNYAI, János, (1886-1971)
BAREN, F. H. van, (-1975)
BARET, Charles, (1831-1910)
BARNETT, Andrew Ketchen, (1852-1914)
BARRON, Thomas, (1867-1906)
BARSONOV, Georgyi Pavlovich, (1897-)
BARTH, Thomas Frederik Weiby, (1899-1971)

MINERALOGY 1900 to present Cont'd.

BARTLETT, Helen Blair, (1901-1969)
BARVÍŘ, Jindřich Ladislav, (1863-1952)
BASCOM, Florence, (1862-1945)
BASTIN, Edson Sunderland, (1878-1953)
BATEMAN, Alan Mara, (1889-1971)
BAUER, Lawson H., (1889-1954)
BAUER, Max Hermann, (1844-1917)
BAUERMAN, Hilary, (1833-1909)
BAUMHAUER, Heinrich Adolf, (1848-1926)
BAYLEY, William Shirley, (1861-1943)
BECKE, Friedrich Johann Karl, (1855-1931)
BECKENKAMP, Jakob, (1855-1931)
BEDER, Robert, (1885-1930)
BELYANKIN, Dmitri Stepanovich, (1876-1953)
BEMMELEN, Jacob Maarten van, (1830-1911)
BERGT, Walther Adolf, (1864-1941)
BERKEY, Charles Peter, (1867-1955)
BERMAN, Harry, (1902-1944)
BERTOLIO, Leopoldo Sollman, (-?1922)
BERTRAND, Émile, (1844-1909)
BERWERTH, Friedrich Martin, (1850-1918)
BETECHTIN, Anatolii Georgievich, (1897-1962)
BEUDANT, François-Sulpice, (1787-1850)
BEUTELL, Albert, (1859-1921)
BIANCHI, Angelo, (1892-1970)
BILLIET, Valère Louis, (1903-1945)
BLAKE, George Stanfield, (-1940)
BLAKE, John Marcus, (1838-1920)
BODENBERGER, Wilhelm, (1857-1941)
BOEKE, Hendrik Enno, (1881-1918)
BOERIS, Giovanni, (1867-1946)
BØGGILD, Ove Balthazar, (1872-1956)
BØGVAD, Richard, (1897-1952)
BOHDANOWICZ, Karol, (1864-1947)
BOLDYREV, Anatolii Kapitonovich, (1883-1946)
BORGES, Alexandre, (1898-1955)
BORN, Kendall Eugene, (1908-1947)
BOUBÉE, Nérée, (1893-1967)
BOWLES, Oliver, (1877-1958)
BOWMAN, Herbert Lister, (1874-1942)
BOYD, Alexander John, (1891-1969)
BRADLEY, William Frank, (1908-1973)
BRAGG, William Lawrence, Sir, (1890-1971)
BRALY, Adrien, (1864-1927)
BRAMMALL, Alfred, (?1879-1954)
BRANNOCK, Kent Combs "Casey", (1923-1973)
BRAUNS, Reinhard Anton, (1861-1937)
BRAVO, José J., (1874-1927)
BREZINA, Maria Aristides, (1848-1909)
BROCK, Reginald Walter, (1874-1935)
BRØGGER, Waldemar Christofer, (1851-1940)
BROGGI, Jorge Albert, (1888-1966)
BROWN, Amos Peaslee, (1864-1917)
BROWN, John Coggin, (1884-1962)

MINERALOGY 1900 to present Cont'd.

BRUET, Edmond, (1890-)
BRUGNATELLI, Luigi, (1859-1928)
BRUHNS, Franz Siegfried Willy, (1864-1929)
BRUSH, George Jarvis, (1831-1912)
BUCCA, Lorenzo, (1857-1930)
BÜCKING, Ferdinand Carl Bertram Hugo, (1851-1932)
BUEHLER, Henry Andrew, (1876-1944)
BUERGER, Martin Julian, (-)
BURFOOT, James Dabney "Dan", Jr., (1896-1966)
BURKART, Eduard, (1865-1941)
BUSZ, Karl Heinrich Emil Georg, (1863-1930)
BUTLER, Francis Henry, (1849-1935)
BUTLER, Gurdon Montague, (1881-1961)
BUTTGENBACH, Henri, (1874-1964)
CǍDERE, Dimitrie M., (1876-1941)
CAHN, Lazard, (1865-1940)
CAMPBELL, Ian, (1900-1978)
CANFIELD, Frederick Alexander, (1849-1926)
CARNOT, Marie Adolphe, (1839-1920)
CARPANESE, Tullio, (-?1940)
CARROLL, Dorothy, (1907-1970)
CASWELL, John Henry, (1846-1909)
CESÀRO, Guiseppe Raimondo Pio, (1849-1939)
CHAMBERLAIN, Arthur, (-)
CHANDLER, Charles Frederick, (1836-1925)
CHAPMAN, Edward John, (1821-1904)
CHERVET, Jean, (1904-1962)
CHEVENARD, Pierre, (1888-1960)
CHUDOBA, Karl F., (1898-1976)
CHURCH, Arthur Herbert, Sir, (1834-1915)
CHYDENIUS, Johan Jakob "Hannes", (1836-1890)
CLARAZ, Georges, (1832-1930)
CLARK, Robert Watson, (1884-1948)
CLARKE, Frank Wigglesworth, (1847-1931)
CLAUSEN, Hans Carl Eisenreich, (1898-1971)
CLERICI, Enrico, (1862-1938)
COBÎLCESCU, Grigore, (1826-1892)
COCHRAN, Kenneth L., (1910-1973)
COCKIN, George Marmaduke, (1852-1918)
COLEMAN, Arthur Philemon, (1852-1939)
COLLINS, Arthur L., (1868-1902)
COLLINS, Henry Francis, (?1864-1924)
COLLINS, Joseph Henry, (1841-1916)
COMBE, A. D., (1893-1949)
CONFORTO, Bruno, (1918-1963)
CONNOLLY, Joseph Peter, (1890-1947)
COOK, Charles Wilford, (1882-1933)
COOMARASWAMY, Ananda K., (1877-1947)
COOPER, Byron Nelson, (1912-1971)
CORNELIUS, Hans Peter, (1888-)
CORNU, Felix, (1882-1909)
COSTA SENA, Joaquim Candido da, (1852-1919)
CRAWFORD, Ralph Dixon, (1873-1950)
CROOK, Alja Robinson, (1864-1930)

MINERALOGY 1900 to present Cont'd.

CROOK, Thomas, (1876-1937)
CROSBY, William Otis, (1850-1925)
CULLIS, Charles Gilbert, (1871-1941)
CURRIE, James, (1863-1930)
CUTTITTA, Frank, (1912-1974)
D'ACHIARDI, Giovanni, (1872-1944)
DAMOUR, Alexis, (1808-1902)
DANA, Edward Salisbury, (1849-1935)
DARTON, Nelson Horatio, (1865-1948)
DAVIES, George Macdonald, (1885-1973)
DAVIES, Thomas, (1837-1892)
DAVISON, Ernest Henry, (1877-1944)
DE ARAUJO FERRAZ, Jorge Palmiro, (1883-1926)
DENNEN, William Henry, (1920-)
DENNING, Reynolds McConnell, (1916-1967)
DEWALQUE, Gilles Joseph Gustave, (1828-1905)
DEWEY, Henry, (1876-1965)
DICK, Allan Brugh, (1833-1926)
DITTLER, Emil, (1882-1945)
DOELTER y CISTERLICH, Cornelio August, (1850-1930)
DRUGMAN, Julien Léon Eugène, (1875-1950)
DUBYANSKY, Viktor Viktorovich, (1880-1925)
DUFFOUR, Alexis, (1877-1948)
DUNIN-BORKOWSKI, Stanislaw von, Graf, (1864-1934)
DUPARC, Louis Claude, (1866-1932)
EAKLE, Arthur Starr, (1862-1931)
EDELMANN, Cornelis Hendrik, (1901-1964)
EDWARDS, Austin Burton, (1909-1960)
EGLESTON, Thomas, (1832-1900)
EHRMANN, Martin L., (1903-1972)
ELLSWORTH, Hardy Vincent, (1889-1952)
EMMONS, Samuel Franklin, (1841-1911)
ENGLISH, George Letchworth, (1864-1944)
ESCHER, Berend George, (-1967)
ETHERIDGE, Robert Jr., (1847-1920)
EVANS, John William, (1857-1930)
FABER, Harald Nicolai, (1856-1943)
FAESSLER, Carl, (1895-1957)
FAIRCHILD, John Gifford, (1882-1965)
FALINI, Filippo, (1917-1963)
FANKUCHEN, Isidor, (1905-1964)
FARRINGTON, Oliver Cummings, (1864-1933)
FEDEROV, Evgraf Stepanovich, (1853-1919)
FENNER, Clarence Norman, (1870-1949)
FERMOR, Lewis Leigh, Sir, (1880-1954)
FERRIER, Walter F., (1865-1950)
FERSMAN, Aleksander Yevgenyevich, (1883-1945)
FIALA, František, (1903-)
FIRKET, Adolphe, (1837-1905)
FIRTSCH, Georg, (1859-1914)
FLEET, William Frederick, (1890-1966)
FLETCHER, Lazarus, Sir, (1854-1921)
FLETCHER, Mark, Reverend, (1870-1956)
FLINK, Gustaf, (1849-1931)

MINERALOGY 1900 to present Cont'd.

FÖLDVÁRI, Aladár, (1906-1973)
FONTAINE, William Morris, (1835-1913)
FOOTE, Harry Ward, (1875-1942)
FORD, Trevor David, (1932-)
FORD, William Ebenezer, (1878-1939)
FORIR, Henri, (1856-1907)
FOSHAG, William Frederick, (1894-1956)
FOSTER, Clement Le Neve, Sir, (1841-1904)
FOUQUÉ, Ferdinand André, (1828-1904)
FRANZENAU, Agoston, (1856-1919)
FRAZER, Persifor, (1844-1909)
FRAZIER, Benjamin West, (1841-1905)
FREIRE DE ANDRADE, Carlos Benot, (1893-1956)
FRENZEL, Friedrich August, (1842-1902)
FRIEDEL, Edmond, (1895-1972)
FRIEDEL, Georges, (1865-1933)
FROHBERG, Max Hans, (1901-1970)
FUGGER, Friedrich Eberhard, (1842-1919)
FUKUCHI, Nobuyo, (1877-1934)
FURCRON, Aurelius Sydney, (1899-1971)
GABRIEL, Alton, (1902-1954)
GALLAGHER, David, (1906-1968)
GALPIN, Sidney Longman, (1886-1962)
GARRIGOU, Félix, (1835-1920)
GAUDEFROY, Christophe Léon, Chanoine, (1878-1971)
GAULT, Hugh Richard, (1915-1961)
GAUTIER, Emile Justin Armand, (1837-1920)
GAVELIN, Axel Olof, (1875-1947)
GERBER, Eduard, (1876-1956)
GESTER, Stephen Howell, (1889-1967)
GIANNOTTI, Carlo, (1897-1929)
GILL, James Rogers, (1922-1972)
GILLERMAN, Elliot "Bud", (1913-1974)
GILLSON, Joseph Lincoln, (1895-1964)
GILLULY, James, (-)
GLASS, Jewell Jeanette, (1888-1966)
GLINKA, Sergei Fedorovich, (1855-1933)
GOLDSCHLAG, Maurycy, (1891-1917)
GOLDSCHMIDT, Victor, (1853-1933)
GOLDSCHMIDT, Victor Moritz, (1888-1947)
GONÇALVES GUIMARÃES, Antonio José, (1875-1919)
GOODCHILD, John George, (1844-1906)
GOODSPEED, George Edward, (1887-1974)
GORDON, Samuel George, (1897-1952)
GÖRGEY VON GÖRGÖ, Rudolf, (1886-1915)
GOSSNER, Balthasar, (1877-1937)
GOWER, John A., (1921-1972)
GRAHAM, Richard Percival Devereux, (1880-1965)
GRAMONT, Antoine Alfred Arnaud Xavier Louis de, Comte, (1861-1923)
GRANQUIST, William T., (1923-1974)
GRÄNZER, Josef, (1857-1934)
GRATACAP, Louis Pope, (1850-1917)
GRATON, Louis Caryl, (1880-1970)
GRAWE, Oliver Rudolph, (1901-1965)

MINERALOGY 1900 to present Cont'd.

GRISHCHINSKY, Petr Iuustinovich, (-1920)
GRÖNWALL, Karl Anders Axel, (1869-1944)
GROSSPIETSCH, Oskar, (-1920)
GROTH, Paul Heinrich von, Ritter, (1843-1927)
GROUT, Frank Fitch, (1880-1958)
GRÜNLING, Friedrich, (1857-1919)
GUILD, Frank Nelson, (1870-1939)
GUNTER, Herman, (1885-1972)
GURLDT, Adolf Friedrich, (1829-1902)
GURNEY, Henry Palin, Reverend, (1847-1904)
HAGUE, Arnold, (1840-1917)
HALL, George Martin, (1891-1941)
HALL, Thomas Clifford Fitzwilliam, (1886-1956)
HALLIMOND, Arthur Francis, (1890-1968)
HAM, William Eugene, (1916-1970)
HAMBERG, Axel, (1863-1933)
HAMILTON, Peggy-Kay, (1922-1959)
HAMMER, Wilhelm, (1875-1942)
HARRINGTON, Bernard J., (1848-1907)
HARRISON, John Burchmore, Sir, (1856-1926)
HAUTEFEUILLE, Paul Gabriel, (1836-1902)
HAWKINS, Alfred Cary, (1887-1954)
HAYES, George Gibbons, (1889-)
HECTOR, James, Sir, (1834-1907)
HEIDE, Friedrich, (1891-)
HELLAND, Amund Theodor, (1846-1918)
HELMHACKER, Rudolf Václav, (1840-1915)
HENRICH, Ferdinand, (1837-1915)
HERPERS, Henry F., Jr., (1915-1952)
HERRMANN, Margit, (-?1957)
HERZENBERG, Robert, (1885-1955)
HESLOP, Mary Kingdon, (?1884-?1954)
HESNARD, Emile E., (1867-1929)
HESS, Frank L., (1871-1955)
HIBSCH, Josef Emanuel, (1852-1940)
HIDDEN, William Earl, (1853-1918)
HIGAZY, Riad Abdel-Megid, (1919-1967)
HIKI, Tadasu, (1866-1927)
HILL, James Wilcott, (1920-1954)
HILLEBRAND, William Francis, (1853-1925)
HIMMELBAUER, Alfred, (1884-1943)
HINRICHS, Gustavus Detlef, (-1922)
HINTERLECHNER, Karl, (1874-1932)
HINTZE, Carl Adolf Ferdinand, (1851-1916)
HIORTDAHL, Thorstein Hallager, (1839-1925)
HIRSCHWALD, Julius, (1845-1928)
HLÁVKA, Karel, (1905-)
HOBLEY, Charles William, (1867-1947)
HODGE, Edwin Thomas, (1887-1970)
HOERNES, Dieter, (1912-1955)
HOFFMANN, George Christian, (1837-1917)
HOFMANN, Adolf, (1853-1913)
HÖGBOM, Arvid Gustaf, (1857-1940)
HOLDEN, Edward Fuller, (1901-1925)

MINERALOGY 1900 to present Cont'd.

HOLLAND, Thomas Henry, Sir, (1868-1947)
HOLMQUIST, Per Johan, (1866-1946)
HONESS, Arthur Pharaoh, (1887-1942)
HOOKER, Marjorie, (1908-1976)
HOVEY, Edmund Otis, (1862-1924)
HOWER, John, Jr., (1927-)
HUBBARD, Lucius Lee, (1849-1933)
HUGI, Emil, (1873-1937)
HULL, Joseph Poyer Deyo, (1889-1967)
HUNT, Walter Frederick, (1882-1975)
HUSSAK, Franz Eugen, (1856-1911)
HUTCHINSON, Arthur, (1866-1937)
HUTTENLOCHER, Heinrich, (1890-1954)
HUTTON, Colin Osborne, (1910-1971)
INEICHEN, Giovanni, (-1952)
INOSTRANTZEV, Aleksandr Aleksandrovich, (1843-1920)
ISSEL, Arturo, (1842-1922)
JAHN, Jaroslav Jiljí, (1865-1934)
JANNETTAZ, Edouard, (-1900)
JERVIS, William Paget, (1831-1906)
JESPERSEN, Johan Magnus Peder, (1833-1917)
JESUS, Amílcar Mário de, (1895-1960)
JIMBŌ, Kotora, (1867-1924)
JOESTING, Henry Rochambeau, (1903-1965)
JOHANSSON, Karl Fredrick, (1866-1933)
JOHNSEN, Arrien, (1877-1934)
JOHNSON, J. P., (1880-1918)
JOHNSTON, Ashton William, (1910-)
JOHN VON JOHNESBERG, Conrad Heinrich, Edler, (1852-1918)
JOLY, John, (1857-1933)
JONES, Daniel J., (1889-1965)
JOURDAIN, André, (1890-1967)
JUDD, John Wesley, (1840-1916)
JULIEN, Alexis Anastay, (1840-1919)
JURKOVIČ, Ivan, (1917-)
KAISER, Friedrich Wilhelm Erich, (1871-1934)
KALECSINSZKY, Sandór, (1857-1911)
KALKOWSKY, Ernst Louis, (1851-1938)
KARANDYEEV, Vissarion Vissarionovich, (-1916)
KARL, Franz, (1918-1972)
KAŠPAR, Jan, (1898-)
KAVECKIS, Mykolas, (1889-)
KEMP, James Furman, (1859-1926)
KERCKHOFF, P. J. van, (-)
KHRUSHCHOV, Konstantin Dmitrievich, (? -1912)
KINCH, Edward, (1848-1920)
KINGSBURY, Arthur William Gerald, (1906-1968)
KIŠPATIĆ, Mišo, (1851-1926)
KITAJEWSKI, Adam Maksymilian, (-)
KITSON, Albert Ernest, Sir, (1868-1937)
KITTO, Benedict, (1841-1928)
KLEBER, Will, (-1970)
KLEIN, A. Albert, (1889-1953)
KLEIN, Johann Friedrich Carl, (1842-1907)

MINERALOGY 1900 to present Cont'd.

KLOCKMANN, Friedrich Ferdinand Hermann, (1858-1937)
KLVAŇA, Josef, (1857-1919)
KOCH, Sándor, (1906-)
KODYM, Odolen, (1898-1963)
KOECHLIN, Rudolf Ignatz, (1862-1939)
KÖHLER, Alexander, (1893-1955)
KOKTA, Jaroslav, (1894-)
KOLBECK, Friedrich Ludwig Wilhelm, (1860-1943)
KOLDERUP, Carl Fredrik, (1869-1942)
KOLDERUP, Niels-Henrik, (1898-1971)
KOUTEK, Jaromír, (1902-)
KOVÁŘ, František, (1862-1941)
KRANTZ, Friedrich Ludwig Robert, (1859-1926)
KRATOCHVÍL, Josef, (1878-1958)
KRAUS, Edward Henry, (1875-1973)
KRENNER, József Sándor, (1839-1920)
KREUTZ, Felix Franz Xaver, (1844-1910)
KREUTZ, Stefan, (1883-1941)
KRIEGER, Philip, (1900-1940)
KRISHNAN, Maharajapuram Sitaram, Padma Bhusan, (1898-1970)
KROKSTRÖM, Torsten Hugo, (1905-1950)
KROTOV, Petr Ivanovich, (1852-1914)
KRUTA, Tomáš, (1906-)
KUČERA, Bruno, (1863-1934)
KULIK, Leonid Alekseevich, (1883-1942)
KUNZ, George Frederick, (1856-1932)
KUPFFER, August Ernestovich, (1842-1918)
KURNAKOV, Nikolai Semenovich, (1860-1941)
KUŹNIAR, Czeslaw Faustyn, (1889-1955)
LACROIX, François Antoine Alfred, (1863-1948)
LALOY, Roger, (1851-1909)
LANDWEHR, Walter Richard, (1893-1972)
LAPHAM, Davis M., (1931-1974)
LAPPARENT, Albert Auguste de, (1839-1908)
LAPPARENT, Jacques de, (-?1949)
LARSEN, Esper Signius, III, (1912-1961)
LAUNAY, Louis Alphonse Auguste de, (1860-1938)
LEATHERS, Walter S., (1874-)
LEDOUX, Auguste, (1887-1918)
LEE, Olan Ivan, (1888-1952)
LEHMANN, Johannes Georg, (1851-1925)
LEITZ, Ernst, Senior, (1871-1956)
LEONHARDT, Johannes, (1893-1959)
LEUTWEIN, Friedrich, (1911-1974)
LEWIS, A. E., (-1906)
LIEBISCH, Theodor, (1852-1922)
LIFFA, Aurél, (1872-1957)
LINCIO, Gabrielle, (1874-1938)
LINDGREN, Waldemar, (1860-1939)
LIVERSIDGE, Archibald, (1847-1927)
LOEWINSSON-LESSING, Franc Julyevich, (1859-1939)
LONSDALE, John Tipton, (1895-1960)
LOOMIS, Frederic Brewster, (1873-1937)
LOUDERBACK, George Davis, (1874-1957)

MINERALOGY 1900 to present Cont'd.

LOUGHLIN, Gerald Francis, (1880-1946)
LOVISATO, Domenico, (1842-1916)
LÖW, Márton, (1885-1943)
LUCAS, Elmer Lawrence, (1898-1966)
LUDWIG, Ernst, (1842-1915)
LUQUER, Lea McIlvaine, (1864-1930)
MACDONALD, Simon, (1927-1971)
MACHATSCHKI, Felix Karl Ludwig, (1895-1970)
MACKENZIE, Graham Stewart, (1905-1960)
MacROBERT OF DOUNESIDE AND CROMAR, Rachel, Lady (née Workman), (-1954)
MAGISTRETTI, Luigi, (-?1959)
MAKOWSKY, Alexander, (1833-1908)
MALKOWSKI, Stanislaw, (1889-1962)
MALLADRA, Alessandro, (1865-1945)
MALLETT, Frederic Richard, (1841-1921)
MANASSE, Ernesto, (1875-1922)
MANCHESTER, James Greenfield, (1871-1948)
MANDARINO, Joseph Anthony, (-)
MARBLE, John Putnam, (1897-1955)
MARIČ, Luka, (1899-)
MARMO, Vladi, (1914-1969)
MATHEWS, Edward Bennett, (1869-1944)
MATTHEWS, Edward Bennett, (1869-1944)
MAUGIN, Charles, (1878-1958)
MAURITZ, Béla, (1881-1971)
McCAUGHEY, William John, (1882-1962)
McDONALD, Stanley Matthews, (1915-1950)
McMAHON, Charles Alexander, Lieutenant-General, (1830-1904)
McVAY, Thomas Newkirk, (1891-1971)
MEEN, Victor Ben, (1910-1971)
MELCZER, Gusztáv, (1869-1907)
MELION, Josef Vincenc, (1813-1905)
MELLIS, Otto, (1906-1970)
MELLO, John Magens, Reverend, (1836-1914)
MENNELL, Frederick Philip, (1880-1966)
MERENSKY, Hans, (1871-1952)
MERRILL, George Perkins, (1854-1929)
MERRILL, Loren B., (?1853-1930)
MEUNIER, Étienne Stanislas, (1843-1925)
MICHEL, Léopold Jean, (1846-1919)
MICHEL-LÉVY, Albert, (1877-1955)
MICHEL-LÉVY, Auguste, (1844-1911)
MIERS, Henry Alexander, Sir, (1858-1942)
MIHOLIČ, Stanko, (1891-1960)
MILCH, Ludwig, (1867-1928)
MILLER, Benjamin Le Roy, (1874-1944)
MILLER, Willet Green, (1839-1925)
MILLER, William John, (1880-1965)
MILLOSEVICH, Federico, (1875-1942)
MILLOT, Georges, (1917-)
MITSCHERLICH, Eilhard, (1794-1863)
MOISSAN, Henri, (1852-1907)
MONETTI, Luigi, (1864-1926)

MINERALOGY 1900 to present Cont'd.

MOORE, Paul Brian, (1940-)
MORGAN, Glenn Beckley, (1876-1925)
MOROZEWICZ, Józef Marian, (1865-1941)
MORRIS, Frederick Kuhne, (1885-1962)
MORTOLA, Edelmira Ines, (-1973)
MORTON, James F., (1870-1941)
MOSES, Alfred Joseph, (1859-1920)
MOSS, Richard Jackson, (1847-1934)
MRAZEC, Lodovic, (1867-1944)
MÜGGE, Johannes Otto Konrad, (1858-1932)
MÜLLER, Hugo Heinrich Wilhelm, (1833-1915)
MURDOCK, Joseph, (1890-1973)
MURRAY, Albert Nelson, (1894-1961)
MYERS, William Marsh, (1892-1951)
NAIDU, Purama Rangashayee Jagapathi, (1903-1970)
NÁRAY-SZABÓ, István, (1899-1972)
NECHAEV, Aleksyei Vasil'evich, (1864-1915)
NECHAY, Wiktor, (1895-1940)
NĚMEC, František, (1905-1972)
NEUHAUS, Alfred, (1903-1975)
NEWHOUSE, Walter Harry, (1897-1969)
NICOLAS, Jean, (1925-)
NIGGLI, Paul, (1888-1953)
NIKITIN, Vasily Vasiliivich, (1867-1942)
NOÉ, Emilio, (1881-1940)
NORDENSKIÖLD, Nils Adolf Erik, (1832-1901)
NØRREGAARD, Erik Martin, (1880-1959)
NOVÁČK, Radim, (1905-1942)
NOVÁK, Jiří, (?1902-1971)
NOVARESE, Vittorio, (1861-1948)
NOVOTNÝ, Miroslav, (1921-1971)
OATES, Frank, (1890-1945)
OEBBEKE, Konrad, (1853-1932)
ONORATO, Ettore, (-)
ORCEL, Jean, (1896-)
O'REILLY, Joseph P., (1829-1905)
ORLOV, Nikolai Arsenevich, (-1919)
OSANN, Carl Alfred, (1859-1923)
PAČAK, Oldřich, (1882-)
PALACHE, Charles, (1869-1954)
PANEBIANCO, Ruggero, (1848-1930)
PANICHI, Ugo, (-?1967)
PARK, James, (1857-1946)
PARKER, Robert Luling, (1893-1973)
PARSONS, Arthur Leonard, (1873-1957)
PARSONS, James, (1876-1908)
PARTRIDGE, Francis Chamberlain, (1903-1939)
PAVLOVIĆ, Stojan, (1903-)
PAWLICA, Wladislaw, (1886-1919)
PEACOCK, Martin Alfred, (1898-1950)
PEARCE, Richard, (1837-1927)
PECK, Arthur Becker, (1892-1943)
PECORA, William Thomas, (1913-1972)
PEGAU, Albert August, (1894-1963)

MINERALOGY 1900 to present Cont'd.

PENFIELD, Samuel Lewis, (1856-1906)
PENROSE, Richard Alexander Fullerton, Jr., (1863-1931)
PENTA, Francesco, (1899-1965)
PERCEVAL, Spencer George, (1838-1922)
PEREIRA DE SAMPAIO FORJAZ PIMENTEL, António Augusto Álvares, (1893-)
PETTERD, William Frederick, (1849-1910)
PEYTON, Garland, (1892-1964)
PHILLIPS, Alexander Hamilton, (1866-1937)
PHILLIPS, William Battle, (1857-1918)
PICCINELLI, Piero, Conte, (1865-1937)
PIOTROWSKI, Hugon Ludwik, (1897-1962)
PIRSSON, Louis Valentine, (1860-1919)
PISANI DE SERRES, Felix, (1831-1920)
PLAGEMANN, A., (1858-1908)
PLATNAUER, Henry Maurice, (1857-1939)
POCOCK, Roy Woodhouse, (1887-1954)
POLLARD, William, (1870-1946)
PONI, Petru, (1841-1925)
POOLE, Henry Skeffington, (1884-1917)
POOR, Russell Spurgeon, (1899-1972)
POPOFF, Boris Anfirovich, (1871-1950)
PRATT, Joseph Hyde, (1870-1942)
PRIOR, George Thurland, (1862-1936)
PROCHÁZKA, Vladimír Josef, (1862-1913)
PUEL, H. E., (-1964)
PUFAHL, Otto Hermann August, (1855-1924)
QUENSEL, Percy Dudgeon, (1881-1966)
QUERCIGH, Emanuele, (1885-1939)
RABBITT, John Charles, (1907-1957)
RADLEY, Ernest George, (1880-1945)
RAGGATT, Harold George, Sir, (1900-1968)
RAMAN, Chandrasekhara Venkata, Sir, (1888-1970)
RAMDOHR, Paul, (1890-)
RAMSAY, Wilhelm, (1865-1928)
RAMSDELL, Lewis Stephen, (1895-1975)
RANSOME, Frederick Leslie, (1868-1935)
RATH, Charles Martin, (1881-1959)
RAULIN, Félix Victor, (1815-1905)
READ, Herbert Harold, (1889-1970)
REDLICH, Karl A., (1869-1942)
REICHERT, Robert, (1901-1937)
REICHERT, Stanley Orville, (1912-1970)
REPOSSI, Emilio, (1876-1931)
REYNOLDS, James Emerson, (1844-1920)
RICH, Charles Irvin, (1918-1975)
RICHARDSON, Charles Henry, (1862-1935)
RIDLEY, Henry Nicholas, (1855-1956)
RIEDEL, Josef, (1891-1966)
RIES, Heinrich, (1871-1951)
RIMANN, Eberhard, (1882-1944)
RINNE, Friedrich Wilhelm Berthold, (1863-1933)
ROCCATI, Alessandro, (1872-1928)
ROGERS, Austin Flint, (1887-1957)

MINERALOGY 1900 to present Cont'd.

ROMAN, David, (1884-1927)
ROMEU, Albert de, (1875-1915)
ROSE, Hermann, (1883-1976)
ROSENBUSCH, Karl Harry Ferdinand, (1836-1914)
ROSENHOLTZ, Joseph Leon, (1899-1963)
ROSICKÝ, Vojtěch, (1880-1942)
ROSIWAL, August Karl, (1860-1923)
ROSS, Clarence Samuel, (1880-1975)
ROST, Rudolf, (1902-)
ROZEN, Zygmunt, (-1936)
RUDLER, Frederick William, (1840-1915)
RUMPF, Johann, (1841-1923)
RUSSELL, Arthur Edward Ian Montagu, Sir, (1878-1964)
RUTLEY, Frank, (1842-1904)
RYBA, Frantisek, (1867-1918)
RYŠ, Josef, (1879-1960)
RZACZYŃSKI, Gabriel, (-)
RZEHÁK, Anton, (1855-1923)
SALMOJRAGHI, Francesco, (1837-?1910)
SAMOILOV, Yakov Vladimirovich, (1870-1925)
SAMPSON, David Noel, (1917-1971)
SAPPINGTON, Chester, (1905-1968)
SARGENT, Henry Crunden, (1851-1931)
SARJEANT, William Antony Swithin, (1935-)
SAUER, Gustav Adolf, (1825-1932)
SCACCHI, Eugenio, (1854-1929)
SCHAIRER, John Frank, (1904-1970)
SCHALLER, Waldemar Theodore, (1882-1967)
SCHARIZER, Rudolf, (1859-1935)
SCHEIBE, Robert, (1859-1923)
SCHETELIG, Jakob Grubbe Cock, (1875-1935)
SCHEUMANN, Karl Hermann, (1881-1964)
SCHLICHTEN, Otto Charles von, (1886-1950)
SCHMIDT, Carl, (1862-1923)
SCHMIDT, Sándor, (1855-?1905)
SCHMIDT, Walter, (1885-1945)
SCHNEIDERHÖHN, Hans, (1887-1962)
SCHOEP, Alfred Ferdinandus Maria Ghislenus, (1881-1966)
SCHOLTZ, Douglas Leinster, (1902-1977)
SCHOUTEN, C., (1891-1963)
SCHRADER, Frank Charles, (1860-1944)
SCHÜLLER, Arno, (1908-1968)
SCHULZ, Karl, (1885-1928)
SCOTT, Alexander, (1890-1951)
SCOTT, Robert Henry, (1833-1916)
SEKANINA, Josef, (1900-)
SEMSEY UR SEMSEI, Andor, (-1923)
SHAND, Samuel James, (1882-1957)
SHORT, Maxwell Naylor, (1889-1952)
SIMPSON, Edward Sydney, (1875-1939)
SJÖGREN, Sten Anders Hjalmar, (1856-1922)
SLAVÍK, František, (1876-1957)
SLAVÍKOVÁ, Ludmila, (1890-1943)
SLAWSON, Chester Baker, (1898-1964)

MINERALOGY 1900 to present Cont'd.

SLOAN, Earle, (1858-1926)
SMIRNOV, Sergei Sergei'evich, (1895-1947)
SMITH, Denis Raymond James, (1937-1964)
SMITH, Herbert Gladstone, (-1950)
SMITH, James Robert, (1925-1972)
SMITH, Thomas Hodge, (1894-1945)
SMITH, Warren D., (1880-1950)
SOLLAS, William Johnson, (1849-1936)
SOLLY, Richard Harrison, (1851-1925)
SOUZA BRANDÃO, Vicente de, (1863-1916)
SOWERBY, Arthur de Carle, (1885-1954)
SPANGENBERG, Kurt, (1889-1957)
SPENCER, Edmondson, (?1885-1970)
SPENCER, Leonard James, (1870-1959)
SPIREK, Vincenz, (1852-1907)
STEENSTRUP, Knud Johannes Vogelius, (1842-1913)
STEIGER, George, (1869-1944)
STEINMETZ, Hermann, (1879-1964)
STEINOCHER, Vladimír, (1902-1972)
STEJSKAL, Jan, (1902-1972)
STELLA, Augusto, (1863-1944)
ŠTĚRBA-BÖHM, Jan Stanislav, (1874-1938)
STILLWELL, Frank Leslie, (1888-1963)
STONE, Robert L., (1912-1969)
STORY-MASKELYNE, Mervin Herbert Nevil, (1823-1911)
STRINGHAM, Bronson F., (1907-1968)
STRÜVER, Johannes Karl Theodor, called Giovanni, (1842-1915)
STUART, Duncan, VII, (1905-1969)
SUBRAMANIAM, Anantharama Parameshwara, (1921-1973)
SUESS, Franz Eduard, (1867-1941)
SUTTON, John Richard, (1864-1937)
SVOBODA, Josef, (1898-)
SZÁDECZKY, Gyula, Baron Szádeczky-Kardoss of Szádecsne and Kafdosfalva, (1860-1935)
SZONTAGH, Tamás, (1851-1937)
SZTERÉNYI, Hugo, (1857-1909)
TABER, Stephen, (1882-1963)
TAJDER, Miroslav, (1909-)
TANTON, Thomas Leslie, (1890-1971)
TARASENKO, Vasily Efimovich, (1859-1926)
TARR, William Arthur, (1881-1939)
TENORE, Gaetano, (1826-1903)
TERTSCH, Hermann, (1880-1963)
THOMAS, Herbert Henry, (1876-1935)
THOMAS, Trevor Morgan, (?1919-1974)
THOMPSON, Robert Mitchell, (1918-1967)
THOMSON, Joseph Ellis, (1882-1944)
THOMSON, Robert, (1902-1970)
THUGUTT, Stanislaw Józef, (1862-1956)
TILLEY, Cecil Edgar, (1894-1973)
TOBORFFY, Zoltán, (1883-1927)
TOKARSKI, Julian, (1883-1961)
TOKODY, László, (1893-1964)
TOMKEIEFF, Sergei Ivanovich, (1892-1968)

MINERALOGY 1900 to present Cont'd.

TOOTHAKER, Charles Robinson, (1873-1952)
TÖRNEBOHM, Alfred Elis, (1838-1911)
TRAINER, David Woolsey, Jr., (1899-1965)
TRAUTSCHOLD, Hermann Adolfovich, (1817-1902)
TRAVERSO, Giovanni Battista, (1843-1914)
TRECHMANN, Charles Otto, (1851?-1917)
TRECHMANN, Charles Taylor, (1885-1964)
TRIBOLET-HARDY, Frédéric Maurice de, (1852-1929)
TRÖGER, Walter Ehrenreich, (1901-1963)
TRONQUOY, René, (1884-1915)
TRYGGVASON, Tómas, (1907-1965)
TSCHERMAK VON SEYSENEGG, Gustav, (1836-1927)
TUČAN, Fran, (1878-1954)
TUČEK, Karel, (1906-)
TWITCHELL, Mayville William, (1866-1927)
UDDEN, Johan August, (1859-1932)
UHLIG, Johannes, (1883-1918)
ULRICH, František, (1899-1941)
UNGEMACH, Henri Léon, (1879-1936)
URBAIN, Georges, (-1938)
USSING, Niels Viggo, (1864-1911)
VACHTL, Josef, (1907-)
VAN AMRINGE, Edwin Verne, (1899-1956)
VANDENDRIESSCHE, Adrien, (1914-1940)
VAN HISE, Charles Richard, (1857-1918)
VAN HORN, Frank Robertson, (1872-1933)
VAN'T HOFF, Jacobus Hendricus, (1852-1911)
VATER, Heinrich, (1859-1930)
VAUX, George, Jr., (1863-1927)
VENDEL, Aladár, (1886-1971)
VENDL, Maria, (1890-1945)
VERNADSKI, Vladimir Ivanovich, (1863-1945)
VESELÝ, Václav, (1896-1956)
VÉSIGNIÉ, Louis, Colonel, (1870-1954)
VICKERS, Arthur Eric Jarvis, (1897-1958)
VIDACS, Aladár, (1909-?1965)
VÍGH, Gyula, (1889-1958)
VINCIENNE, Henri, (1898-1965)
VIOLA, Carlo Maria, (1855-1925)
VLASOV, Kuz'ma Alekseevich, (1905-1964)
VOGT, Thorolf, (1888-1958)
VOIGT, Waldemar, (1850-1919)
VON ECKERMANN, Claes Walther Harry, (1886-1969)
VRABKA, František, (1924-1957)
VRBA, Karel, (1845-1922)
WADA, Tsunashirô, (1856-1920)
WALDBAUM, David Robert, (1937-1974)
WALKER, George Forbes, (1916-1970)
WALKER, Thomas Leonard, (1867-1942)
WALLACE, Robert Charles, (1881-1955)
WALLERANT, Frédéric Félix Auguste, (1858-1936)
WARD, Henry Augustus, (1834-1906)
WARREN, Charles Hyde, (1876-1950)
WARREN, Harry V., (?1904-)

MINERALOGY 1900 to present Cont'd.

WASHINGTON, Henry Stephens, (1867-1934)
WASSERSTEIN, Benno, (1906-1956)
WATSON, Thomas Leonard, (1871-1924)
WAWRYK, Wlodzimierz, (1902-1963)
WEBB, David Knowlton, Jr., (1928-1973)
WEIBULL, Kristian Oskar Mats, (1856-1923)
WEIDMAN, Samuel, (1870-1945)
WEINBERGER, Isidor, (1838-1915)
WEINSCHENK, Ernst Heinrich Oskar Kasimir, (1865-1921)
WEISBACH, Albin, (1833-1901)
WEISS, Eriol Joseph, (1918-1967)
WELLS, Horace Lemuel, (1855-1924)
WELLS, Roger Clark, (1877-1944)
WELSCH, Jules, (1858-1929)
WEYBERG, Zygmunt, (1872-1944)
WHITLOCK, Herbert Percy, (1868-1948)
WICHMANN, Carl Ernst Arthur, (1851-1927)
WIGGLESWORTH, Edward, (1885-1945)
WIIK, Fredrik Johan, (1839-1909)
WILDER, Frank Alonzo, (1870-1930)
WILDER, Newell Morris "Newt", (1908-1946)
WILKERSON, Albert Samuel, (1897-1963)
WILSON, J. S. Grant, (1855-1908)
WINCHELL, Alexander Newton, (1874-1958)
WINKLER-HERMADEN, Artur, (1890-1963)
WISSER, Edward Hollister, (1895-1970)
WOJNO, Tadeusz Jerzy, (1884-1971)
WOLF, Theodor, (1841-1924)
WOLFF, John Elict, (1857-1940)
WOODROW, James, (1828-1907)
WOODRUFF, Elmer Grant, (1872-1952)
WOODWARD, Charles Josiah, (1838-1932)
WRIGHT, Frederick Eugene, (1877-1953)
WRIGHT, Harold Douglas, (1921-1969)
WROOST, Volkmar, (1912-1936)
WÜLFING, Ernst Anton, (1860-1930)
WYROUBOFF, Gr., (1843-1913)
YACHEVSKY, Leonard Antonovich, (1858-1916)
YAGODA, Herman, (1908-1964)
YEATES, William Smith, (1856-1908)
YOUNG, George Albert, (1898-1947)
ZAMBONINI, Ferruccio, (1880-1932)
ZEMYATCHENSKY, Petr Andreevich, (1856-1942)
ZENZÉN, Nils, (1883-1959)
ZERFOSS, Samuel, (1912-1958)
ZHEMCHUZHNYI, Sergei Fedorovitch, (1873-1929)
ZIJLSTRA, Gerrit, (-1962)
ZIMÁNYI, Károly, (1862-1941)
ZIRKEL, Ferdinand, (1838-1912)
ZPALETAL, Karel, (1903-1972)
ZYNDEL, Fortunat, (1882-1917)

Time unknown
FONBERG, Ignacy, (-)

MINERALOGY Time unknown Cont'd.

GRIGORIEV, Dmitri Pavlovitch, (-)
KAHLMETER, Henrik, (-)
LIPPI, Carmine Antonio, (-)
WOLFE, Caleb Wroe, (-)

MINERALOGY <u>See also</u> METEORITICS, MINERAL COLLECTING

MINING ENGINEERING

Up to 1800

BORDA, José de la, Don, (1699-1778) [PROSPECTORS, DIVINERS AND MINING ENGINEERS]
HELMS, Anton Zacharias, (1751-1803) [PROSPECTORS, DIVINERS AND MINING ENGINEERS]
TREVITHICK, Richard, (1771-1833) [PROSPECTORS, DIVINERS AND MINING ENGINEERS]
WOOLF, Arthur, (1766-1837) [PROSPECTORS, DIVINERS AND MINING ENGINEERS]

1800-1849

HELMS, Anton Zacharias, (1751-1803) [PROSPECTORS, DIVINERS AND MINING ENGINEERS]
TREVITHICK, Richard, (1771-1833) [PROSPECTORS, DIVINERS AND MINING ENGINEERS]
WILLIAMS, Thomas, (1737-1802) [PROSPECTORS, DIVINERS AND MINING ENGINEERS]
WOOLF, Arthur, (1766-1837) [PROSPECTORS, DIVINERS AND MINING ENGINEERS]

1850-1899

HELMS, Anton Zacharias, (1751-1803) [PROSPECTORS, DIVINERS AND MINING ENGINEERS]
JULIAN, Henry Forbes, (1861-1912) [PROSPECTORS, DIVINERS AND MINING ENGINEERS]
LEDOUX, Albert Reid, (1852-1923) [PROSPECTORS, DIVINERS AND MINING ENGINEERS]
LONGYEAR, Edmund Joseph, (1864-1954) [PROSPECTORS, DIVINERS AND MINING ENGINEERS]
MACKAY, John William, (1831-1902) [PROSPECTORS, DIVINERS AND MINING ENGINEERS]
McCARTHY, Edward Thomas, (-) [PROSPECTORS, DIVINERS AND MINING ENGINEERS]
TAYLOR, William P., (1859-) [PROSPECTORS, DIVINERS AND MINING ENGINEERS]
TRELAWNEY-ANSELL, E. C., (-) [PROSPECTORS, DIVINERS AND MINING ENGINEERS]
WAY, Herbert W. L., (-) [PROSPECTORS, DIVINERS AND MINING ENGINEERS]

1900 to present

ANDERTON, Russell, (-) [PROSPECTORS, DIVINERS AND MINING ENGINEERS]
BERTHELOT, Charles, (-) [PROSPECTORS, DIVINERS AND MINING ENGINEERS]

MINING ENGINEERING 1900 to present Cont'd.

BOYLEN, Matthew James, (1911?-) [PROSPECTORS, DIVINERS AND MINING ENGINEERS]
BROWN, Eldon L., (-) [PROSPECTORS, DIVINERS AND MINING ENGINEERS]
HELLENS, Alexander Daniel, (1911-) [PROSPECTORS, DIVINERS AND MINING ENGINEERS]
HUSTON, Charles, (1908-) [PROSPECTORS, DIVINERS AND MINING ENGINEERS]
JULIAN, Henry Forbes, (1861-1912) [PROSPECTORS, DIVINERS AND MINING ENGINEERS]
LaBINE, Charles L., (1888-) [PROSPECTORS, DIVINERS AND MINING ENGINEERS]
LEDOUX, Albert Reid, (1852-1923) [PROSPECTORS, DIVINERS AND MINING ENGINEERS]
LINDSLEY, Thayer, (1882-) [PROSPECTORS, DIVINERS AND MINING ENGINEERS]
LONGYEAR, Edmund Joseph, (1864-1954) [PROSPECTORS, DIVINERS AND MINING ENGINEERS]
McCARTHY, Edward Thomas, (-) [PROSPECTORS, DIVINERS AND MINING ENGINEERS]
MOSHER, Alex, (1900-) [PROSPECTORS, DIVINERS AND MINING ENGINEERS]
MOSHER, Murdock C., (-) [PROSPECTORS, DIVINERS AND MINING ENGINEERS]
PARKER, Ralph D., (1898-) [PROSPECTORS, DIVINERS AND MINING ENGINEERS]
RAINIER, Peter W., (-) [PROSPECTORS, DIVINERS AND MINING ENGINEERS]
REDPATH, James B., (-) [PROSPECTORS, DIVINERS AND MINING ENGINEERS]
ROBERTSON, Andrew, (1903?-) [PROSPECTORS, DIVINERS AND MINING ENGINEERS]
TAYLOR, William P., (1859-) [PROSPECTORS, DIVINERS AND MINING ENGINEERS]
TRELAWNEY-ANSELL, E. C., (-) [PROSPECTORS, DIVINERS AND MINING ENGINEERS]
TREMLETT, Rex, (-) [PROSPECTORS, DIVINERS AND MINING ENGINEERS]
WATTS, Murray Edmund, (1909-) [PROSPECTORS, DIVINERS AND MINING ENGINEERS]

MINING GEOLOGY

Up to 1800

AUBUISSON DE VOISINS, Jean François d', (1769-1841)
BALD, Robert, (1776-1861)
BARROW, John, Sir, (1764-1848)
BEAUSOLEIL, Jean du Chastelet, Baron de .... et d'Auffembach, (ca 1578-1620)
BEAUSOLEIL, Martine de Bertereau, Baronne de .... et d'Auffembach, (fl.1600-1630)
BUSHELL, Thomas, (?1594-1674) [PROSPECTORS, DIVINERS AND MINING ENGINEERS]
CRONSTEDT, Axel Fredrik, (1722-1765)
DUHAMEL, Jean Pierre François Guillaume, (1730-1816)

MINING GEOLOGY Up to 1800 Cont'd.

ELHUYAR Y DE ZUBICE, Fausto de, (1755-1833)
EMMERLING, Ludwig August, (1765-1841)
ERCKER, Lazarus, (-1593)
ESMARCH, Jens, (1762-1839)
FERBER, Johann Jacob, (1743-1790)
FORSTER, Westgarth, (1772-1835)
HENCKEL, Johann Friedrich, (1679-1744)
HERMELIN, Samuel Gustaf, (1744-1820)
JENKIN, William, (1738-1820) [PROSPECTORS, DIVINERS AND MINING ENGINEERS]
KARSTEN, Dietrich Ludwig Gustav, (1768-1810)
MATHESIUS, Johannes, (1504-1565)
MAWE, John, (1764-1829)
RASPE, Rudolf Erich, (1737-1794)
RIO, Andrés Manuel del, (1764-1849)
RÜLEIN VON CALW, Ulrich, (ca.1467-1523)
STASZIC, Stanislaw, (1755-1826)
STRACHEY, John, (1671-1743)
VOGT, Johann Georg Mauriz, (1669-1731)
VOIGT, Johann Karl Wilhelm, (1752-1821)
WERNER, Abraham Gottlob, (1749-1817)
ZIMMERMANN, Christian, (1765-1853)

1800-1849
ATTWOOD, Melville, (1812-1898)
AUBUISSON DE VOISINS, Jean François d', (1769-1841)
BALD, Robert, (1776-1861)
BARROW, John, Sir, (1764-1848)
BOUSSINGAULT, Joseph Dieudonné Jean-Baptiste, (1807-1887)
BRARD, Cyprien Prosper, (1788-1838)
BROWN, Richard, (1805-1882)
BURAT, Amédée, (1809-1883)
BURKART, Hermann Josef, (1798-1874)
CHARPENTIER, Jean de, (1786-1855)
CLARKE, William Branwhite, Reverend, (1798-1878)
COSSHAM, Handel, (1824-?1890)
CREDNER, Carl Friedrich Heinrich, (1809-1876)
DE PRADO, Casiano, Don, (1797-1866?)
DOMEYKO, Ignacy, (1802-1889)
DUHAMEL, Jean Pierre François Guillaume, (1730-1816)
ELHUYAR Y DE ZUBICE, Fausto de, (1755-1833)
ELST, E. van de, (1826-1897)
EMMERLING, Ludwig August, (1765-1841)
ENYS, John Samuel, (1797-1872)
ESMARCH, Jens, (1762-1839)
FORSTER, Westgarth, (1772-1835)
FUCHS, Wilhelm, (1802-1853)
GÄTZSCHMANN, Moritz Ferdinand, (1800-1895)
GENTH, Friedrick August Ludwig Karl Wilhelm, later called Frederick Augustus, (1820-1893)
GOBANZ, Alexander, (1826-1895)
GREENWELL, George Clementson, (1821-1900)
HARGRAVES, Edward Hammond, (1816-1891)
HAUCHECORNE, Heinrich Lambert Wilhelm, (1828-1900)

MINING GEOLOGY 1800-1849 Cont'd.

HELMERSEN, Gregor von, (1803-1885)
HENWOOD, William Jory, (1805-1875)
HERMELIN, Samuel Gustaf, (1744-1820)
HIRIAKOFF, Modeste, (1812-1894)
HOPKINS, Evan, (?1810-1867)
HURST, Thomas Grainge, (1824-1890)
HUXLEY, Thomas Henry, (1825-1895)
JENKIN, William, (1738-1820) [PROSPECTORS, DIVINERS AND MINING ENGINEERS]
KARSTEN, Dietrich Ludwig Gustav, (1768-1810)
KÜHN, Karl Amadeus, (1783-1848)
LOGAN, William Edmond, Sir, (1798-1875)
MACFARLANE, James, (1819-1885)
MAWE, John, (1764-1829)
MEADE, Richard, (1827-1894)
MURCHISON, Roderick Impey, Sir, (1792-1871)
NÖGGERATH, Johann Jakob, (1788-1877)
PERCIVAL, James Gates, (1795-1856)
PHILLIPS, John Arthur, (1822-1887)
QUIQUEREZ, Auguste, (1801-1882)
RAMSAY, Andrew Crombie, Sir, (1814-1891)
RICKARD, Francis Ignacio, (-)
RIO, Andrés Manuel del, (1764-1849)
ROSALES, Henry, (?1822-1916)
RUBIDGE, Richard Nathaniel, (1821?-1869)
SALMON, Henry Curwen, (1827-1873)
SCHAFHÄUTL, Karl Franz Emil, (1803-1890)
SCOTT, Marcus Watson Tate, (1814-1875)
SJÖGREN, Carl Anton Hjalmar, (1822-1893)
SMYTH, Warington Wilkinson, Sir, (1817-1890)
STASZIC, Stanislaw, (1755-1826)
STROMBECK, August von, (1808-1900)
SWALLOW, George Clinton, (1817-1899)
TAYLOR, John, (1780-1864)
TAYLOR, Richard Cowling, (1789-1851)
VANUXEM, Lardner, (1792-1848)
VOIGT, Johann Karl Wilhelm, (1752-1821)
WEISSENBACH, Karl Gustav Adalbert von, (1797-1848)
WERNER, Abraham Gottlob, (1749-1817)
WOOD, Nicholas, (1795-1865)
XÁNTUS, János, (1825-1894)
ZIMMERMANN, Christian, (1765-1853)

1850-1899

ADAMS, George Whitney, (1872-1930)
ANSTIE, John, (-1900)
ASHBURNER, Charles Albert, (1854-1889)
ATTWOOD, George, (1845-1912)
ATTWOOD, Melville, (1812-1898)
AVERY, Charles Dwight, (1877-1947)
BALD, Robert, (1776-1861)
BALDAUF, Richard Julius, (1848-1931)
BALL, Francis Matthew, (1877-1956)
BARAFFAEL, Angelo, (1866-1928)

MINING GEOLOGY 1850-1899 Cont'd.

BARNETT, Andrew Ketchen, (1852-1914)
BARRINGER, Daniel Moreau, Sr., (1860-1929)
BAUER, Heinrich E., (1840-1896)
BAWDEN, Samuel Francis, (1854-1884)
BECHER, Harry Macdonald, (1855-1893)
BECKER, George Ferdinand, (1847-1919)
BELL, James Macintosh, (1877-1934)
BOUSSINGAULT, Joseph Dieudonné Jean-Baptiste, (1807-1887)
BOWLES, Oliver, (1877-1958)
BRAVO, José J., (1874-1927)
BROOKS, Thomas Benton, (1836-1900)
BROWN, Charles Barrington, (1839-1917)
BROWN, Richard, (1805-1882)
BUCKLEY, Ernest Robertson, (1872-1912)
BURAT, Amédée, (1809-1883)
BURKART, Hermann Josef, (1798-1874)
CAMSELL, Charles, (1876-1958)
CARPENTER, Franklin Reuben, (1848-1910)
CARSTENS, Carl Wilhelm, (1877-?1950)
CHARPENTIER, Jean de, (1786-1855)
CHEVALIER, Jerome Archibald, (1877-1940)
CLARKE, William Branwhite, Reverend, (1798-1878)
COCKIN, George Marmaduke, (1852-1918)
COKE, George Elmsley, (?   -1911)
COLLINS, Arthur L., (1868-1902)
COLLINS, Joseph Henry, (1841-1916)
COLLINS, William Henry, (1878-1937)
CORTESE, Emilio, (1856-1938)
COSSHAM, Handel, (1824-?1890)
COSTA SENA, Joaquim Candido da, (1852-1919)
CREDNER, Carl Friedrich Heinrich, (1809-1876)
CULLIS, Charles Gilbert, (1871-1941)
CUMINGS, Willard Lawsen, (1874-1945)
CZARNOCKI, Stefan, (1878-1947)
DAINTREE, Richard, (1831-1878)
DAMM, Carl Oluf Bernhard, (1869-1949)
DAVID, Tannatt William Edgeworth, Sir, (1858-1934)
DAVIES, David Christopher, (1827-1885)
DAVIES, David Farr, (1871-1935)
DE PRADO, Casiano, Don, (1797-1866?)
DOMEYKO, Ignacy, (1802-1889)
DOUGLAS, James, (1837-   )
DRAPER, David, (1849-1929)
DRESSER, John Alexander, (1866-1954)
DU BOIS, Georges C., (1874-   )
DUNN, Edward John, (1844-1937)
ELST, E. van de, (1826-1897)
EMMONS, Samuel Franklin, (1841-1911)
ENDLICH, Frederic Miller, (1851-1899)
ENYS, John Samuel, (1797-1872)
FALCONER, John Downie, (1876-1947)
FERRIER, Walter F., (1865-1950)
FINLAYSON, Alexander Moncreiff, (-1917)
FLORES REYES, Teodoro, (1873-1955)

MINING GEOLOGY 1850-1899 Cont'd.

FOSTER, Clement Le Neve, Sir, (1841-1904)
FRAZIER, Benjamin West, (1841-1905)
FUCHS, Wilhelm, (1802-1853)
GATTNAR, Josef, (1854-1928)
GÄTZSCHMANN, Moritz Ferdinand, (1800-1895)
GENTH, Friedrick August Ludwig Karl Wilhelm, later called
Frederick Augustus, (1820-1893)
GIBBONS, Arthur, (1867-1938)
GOBANZ, Alexander, (1826-1895)
GRANSTRÖM, Gustaf Abraham, (1851-1941)
GREENWELL, George Clementson, (1821-1900)
GÜRICH, Georg, (1859-1938)
HABERFELNER, Josef, (1830-1913)
HAMMOND, John Hays, (1855-1936)
HANNAY, Harry Campbell, (1874-1961)
HARGRAVES, Edward Hammond, (1816-1891)
HARRINGTON, Bernard J., (1848-1907)
HATCH, Frederick Henry, (?1864-1932)
HAUCHECORNE, Heinrich Lambert Wilhelm, (1828-1900)
HEDSTROM, Herman Oskar, (1867-?1948)
HELMERSEN, Gregor von, (1803-1885)
HELMHACKER, Rudolf Václav, (1840-1915)
HENDERSON, James Alexander Leo, (?1874-1940)
HENWOOD, William Jory, (1805-1875)
HERRICK, Clarence Luther, (1858-1904)
HESLOP, Septimus, (?1860-1945)
HESS, Frank L., (1871-1955)
HIRIAKOFF, Modeste, (1812-1894)
HOLMES, Joseph Austin, (1859-1915)
HOLMSEN, Andreas, (1869-1955)
HOPKINS, Evan, (?1810-1867)
HORNEMAN, Hans Henrik, (1878-1945)
HURST, Thomas Grainge, (1824-1890)
HUTCHINSON, Lon Lewis, (1877-1947)
HUXLEY, Thomas Henry, (1825-1895)
HYLAND, John Shearson, (1866-1898)
IRVING, John Duer, (1874-1918)
IRVING, Roland Duer, (1847-1888)
JACK, Robert Lockhart, (1878-1964)
JAHN, Jaroslav Jiljí, (1865-1934)
JOHANSSON, Karl Fredrick, (1866-1933)
JUDD, John Wesley, (1840-1916)
KARPINSKY, Alexander Petrovich, (1847-1936)
KEYES, Charles Rollin, (1864-1942)
KITSON, Albert Ernest, Sir, (1868-1937)
KOCH, Walter Edward, (1848-1916)
KONTKIEWICZ, Stanislaw, Sr., (1849-1924)
LAMBERT, Paul, (1869-1950)
LEBOUR, George Alexander Louis, (1847-1918)
LINDON, Edward Byron, (1860-1891)
LIVERSIDGE, Archibald, (1847-1927)
LOGAN, William Edmond, Sir, (1798-1875)
LOMAX, James, (1857-?1934)
LONGBOTTOM, Arthur, (? -1910)

MINING GEOLOGY 1850-1899 Cont'd.

LOUIS, Henry, (1855-1939)
MACFARLANE, James, (1819-1885)
MALLADA, Lucas, (1841-1921)
MARKS, Edward Seaborn, (?1875-1950)
MAY DE MADIIS, Alexis Freiherr von, (1852-1911)
McCONNELL, Richard George, (1857-1942)
McDONALD, William, (1876-1943)
McINNES, William, (1858-1925)
McKIM, John Walter, (1877-1927)
MEADE, Richard, (1827-1894)
MELL, Patrick Hues, (1850-1918)
MERRICKS, Frank, (?1863-1936)
MICHALSKI, Aleksander Octavianovich, (1855-1904)
MILLS, James Ellison, (1834-1901)
MORGAN, Peter (Percy) Gates, (1867-1927)
MÜNSTER, Thomas Georg, (1855-1938)
MURCHISON, Roderick Impey, Sir, (1792-1871)
NÖGGERATH, Johann Jakob, (1788-1877)
NORELIUS, Carl Oskar, (1861-1941)
O'REILLY, Joseph P., (1829-1905)
PARK, James, (1857-1946)
PEARCE, Richard, (1837-1927)
PECK, Frederick Burritt, (1860-1925)
PELLATI, Nicola, (1835-1907)
PENNING, William Henry, (1838-1902)
PENROSE, Richard Alexander Fullerton, Jr., (1863-1931)
PERCIVAL, James Gates, (1795-1856)
PETTEE, William Henry, (1836-1904)
PHILLIPS, John Arthur, (1822-1887)
PICKERING, William Henry, (1858-1912)
PIKE, John William, (1839-1874)
PLANT, Nathaniel, (1832-1875)
PUMPELLY, Raphael, (1837-1923)
PURDUE, Albert Homer, (1861-1917)
QUIQUEREZ, Auguste, (1801-1882)
RAMSAY, Andrew Crombie, Sir, (1814-1891)
RAYMOND, Rossiter Worthington, (1840-1918)
RICKARD, Francis Ignacio, (-)
RICKARD, Thomas Arthur, (1864-1953)
ROSALES, Henry, (?1822-1916)
ROTHWELL, Richard Pennefeather, (1836-1901)
RUBIDGE, Richard Nathaniel, (1821?-1869)
RUTOT, Aimé Louis, (1847-1933)
SALMON, Henry Curwen, (1827-1873)
SARGENT, Harry Lionel, (1871-1955)
SAUER, Julius, (1849-1911)
SCHAFHÄUTL, Karl Franz Emil, (1803-1890)
SCHLUMBERGER, Conrad, (1878-1936)
SCOTT, Marcus Watson Tate, (1814-1875)
SJÖGREN, Carl Anton Hjalmar, (1822-1893)
SMYTH, Warington Wilkinson, Sir, (1817-1890)
SOLITANDER, Carl Probus, (1843-1918)
SOPWITH, Arthur, (1843-1920)
ST. JOHN, Orestes Hawley, (1841-1921)

MINING GEOLOGY 1850-1899 Cont'd.

STOKES, Arthur Henry, (1842-1910)
STROMBECK, August von, (1808-1900)
SUTTON, William John, (1859-1914)
SWALLOW, George Clinton, (1817-1899)
TAYLOR, John, (1780-1864)
TEALE, Edmund Oswald, Sir, (1874-1971)
TERMIER, Pierre, (1859-1930)
TRUSCOTT, Samuel John, (?1870-1950)
TWELVETREES, William Harper, (1848-1919)
ULFFERS, Ernst, (1837-1919)
VAN HORN, Frank Robertson, (1872-1933)
VEATCH, Arthur Clifford, (1878-1938)
WADA, Tsunashirô, (1856-1920)
WATERSCHOOT VAN DER GRACHT, Willem Anton Josef Maria van, (1873-1943)
WENGER, Marian, (1864-1939)
WESTLUND, Erik Wiktor, (-?1954)
WILKINSON, Charles Smith, (1843-1891)
WILLIAMS, Gardner Fred, (1842-1922)
WILSON, Alfred William Gunning, (1873-1954)
WINCHELL, Horace Vaughn, (1865-1923)
WINDIAKIEWICZ, Edward, (1858-1942)
WOOD, Nicholas, (1795-1865)
WRIGHT, Charles E., (1843-1888)
XÁNTUS, János, (1825-1894)
ZIMMERMANN, Christian, (1765-1853)

1900 to present

ABARQUEZ, Ramon Fabella, (1891-1959)
ABERNATHY, George Elmer, (1888-1948)
ADAMS, George Whitney, (1872-1930)
AHO, Aaro E., (-1977)
ALCOCK, Frederick James, (1888-1972)
ANDERSON, Robert Beattie, (1904-1970)
ANDREW, Arthur Robert, (1881-1941)
ARMSTRONG, Harold Kerr, (1893-1962)
ATTWOOD, George, (1845-1912)
AVERY, Charles Dwight, (1877-1947)
BAKELS, P. S., (-1960)
BALDAUF, Richard Julius, (1848-1931)
BALL, Francis Matthew, (1877-1956)
BALL, Harry Standish, (?1888-1941)
BARAFFAEL, Angelo, (1866-1928)
BARNETT, Andrew Ketchen, (1852-1914)
BARRETT, Leslie Park, (1887-1972)
BARRINGER, Daniel Moreau, Sr., (1860-1929)
BARRON, William Ernest, (1880-1935)
BECKER, Clyde M., (1882-1938)
BECKER, George Ferdinand, (1847-1919)
BEDER, Robert, (1885-1930)
BELL, James Macintosh, (1877-1934)
BELL, John Arnold, (1939-1970)
BENNETT, Walter Hellyer, (1892-1971)
BILLINGSLEY, Paul, (1887-1962)

MINING GEOLOGY 1900 to present Cont'd.

BISHOPP, Douglas Wallace, (1900-1977)
BOWLES, Oliver, (1877-1958)
BRACE, Orval L., (1893-1975)
BRAVO, José J., (1874-1927)
BRINKERHOFF, Ira A., (1904-1969)
BROCK, Byron Britton, (1904-1972)
BROGGI, Jorge Albert, (1888-1966)
BROWN, Bahngrell Walter "Bill", (1916-)
BROWN, Charles Barrington, (1839-1917)
BRYAN, Joseph Jefferson, (1905-1951)
BUCKLEY, Ernest Robertson, (1872-1912)
BURTON, George Edgar, (1880-1953)
BURWELL, Edward Bouldin, Jr., (1894-1964)
BUTLER, Gurdon Montague, (1881-1961)
BUTTERLEY, Archibald Donald, (1896-1957)
BYRNE, Jerome Cotter, (1910-)
CAMSELL, Charles, (1876-1958)
CARPENTER, Franklin Reuben, (1848-1910)
CARSTENS, Carl Wilhelm, (1877-?1950)
CHEVALIER, Jerome Archibald, (1877-1940)
CHLEBOWSKI, Tadeusz, (1908-1961)
COCHRAN, Kenneth L., (1910-1973)
COCKIN, George Marmaduke, (1852-1918)
COFFIN, Reuben Clare, (1886-1972)
COLLINS, Arthur L., (1868-1902)
COLLINS, Joseph Henry, (1841-1916)
COLLINS, William Henry, (1878-1937)
COLLOM, Roy Edward, (1881-1948)
CONDIT, Daniel Dale, (1886-1955)
CONNOR, James, (1894-1977)
CONWAY, William Palen "Tony", Jr., (1914-1944)
CORTESE, Emilio, (1856-1938)
COSTA SENA, Joaquim Candido da, (1852-1919)
COX, Guy Henry, (1881-1922)
CRICKMAY, Geoffrey W., (1905-1971)
CROOKSHANK, Henry, (1893-1972)
CROSS, Julian Gifford, (1889-)
CROWN, Walter James, (1901-1951)
CULLIS, Charles Gilbert, (1871-1941)
CUMINGS, Willard Lawsen, (1874-1945)
CURRIER, Louis Wade, (1890-1970)
CZARNOCKI, Stefan, (1878-1947)
CZERMAK, Friedrich, (1890-1960)
DAMM, Carl Oluf Bernhard, (1869-1949)
DANIELLSON, Johan Patrik, (-?1954)
DAVID, Tannatt William Edgeworth, Sir, (1858-1934)
DAVIES, David Farr, (1871-1935)
DAVIS, Elmer Fred, (1887-1974)
DAVIS, Ralph Emerson, (1884-1968)
DE LA SAUCE, Wilhelm, (1882-1955)
DENNING, Reynolds McConnell, (1916-1967)
DENNIS, Frederick William, (1883-1939)
DERRY, Duncan Ramsay, (?1908-)
DIETZ, Curt, (1882-1965)

MINING GEOLOGY 1900 to present Cont'd.

DIXON, Henry George Dacres, (1887-1947)
DONOVAN, Percy Williams, (1879-1973)
DOORNINCK, Nico H. van, (1893-1956)
DRAPER, David, (1849-1929)
DRESSER, John Alexander, (1866-1954)
DRYSDALE, Charles Wales, (1885-1917)
DU BOIS, Georges C., (1874-)
DUGGLEBY, Arthur Francis, (1893-1945)
DUNN, Edward John, (1844-1937)
EDWARDS, Austin Burton, (1909-1960)
EDWARDS, Everett C., (1897-1972)
EDWARDS, Merwin Guy, (1889-1969)
ELLIOTT, Harold Charles, (1910-1972)
EMMONS, Samuel Franklin, (1841-1911)
FAESSLER, Carl, (1895-1957)
FALCONER, John Downie, (1876-1947)
FERRIER, Walter F., (1865-1950)
FINLAYSON, Alexander Moncreiff, (-1917)
FLORES REYES, Teodoro, (1873-1955)
FOHS, Ferdinand Julius, (1884-1965)
FOSLIE, Steinar, (1887-1951)
FOSTER, Clement Le Neve, Sir, (1841-1904)
FOWLER, George Malcolm, (1885-1972)
FOX, Cyril Sankey, Sir, (1886-1951)
FOXWORTH, Wyckliff Riley, (1936-1965)
FRASER, Horace John "Dutch", (1905-1969)
FRAZIER, Benjamin West, (1841-1905)
FRITTS, Crawford Ellsworth, (1927-1972)
GABRIEL, Alton, (1902-1954)
GALLAGHER, David, (1906-1968)
GARRETSON, Mary Welleck, (1896-1971)
GASKIN, Alan Robert James, (1927-1974)
GATTNAR, Josef, (1854-1928)
GIBBONS, Arthur, (1867-1938)
GILLANDERS, Earl B., (-)
GILLES, Verner Arthur, (1886-1954)
GOLDKUTH, Algot, (-1957)
GOLDMAN, Marcus Isaac, (1881-1965)
GORDON, Dugald, (1886-1958)
GRANSTRÖM, Gustaf Abraham, (1851-1941)
GRATON, Louis Caryl, (1880-1970)
GROND, Gerard Joseph Alfons, (1890-1970)
GROOTHOFF, Christiaan Theodor, (1887-1969)
GRUTTERINK, Jan Adolf, (1879-1949)
GÜRICH, Georg, (1859-1938)
HABERFELNER, Josef, (1830-1913)
HALL, Thomas Clifford Fitzwilliam, (1886-1956)
HAMILTON, Henry L., (1882-1925)
HAMMOND, John Hays, (1855-1936)
HANNAY, Harry Campbell, (1874-1961)
HARQUAIL, James A., (-)
HARRINGTON, Bernard J., (1848-1907)
HATCH, Frederick Henry, (?1864-1932)
HATFIELD, Willis Charles, "Luke", (1898-1970)

MINING GEOLOGY 1900 to present Cont'd.

HAUCHECORNE, Heinrich Lambert Wilhelm, (1828-1900)
HAWLEY, James Edwin, (1897-1965)
HEDSTROM, Herman Oskar, (1867-?1948)
HELMHACKER, Rudolf Václav, (1840-1915)
HENDERSON, James Alexander Leo, (?1874-1940)
HENLEY, Arthur Sidney, (1879-1936)
HEROLD, Stanley Carrollton, (1883-1958)
HERON, Alexander Macmillan, (1884-1971)
HERRICK, Clarence Luther, (1858-1904)
HESLOP, Septimus, (?1860-1945)
HESS, Frank L., (1871-1955)
HIBBERT, Frank, Colonel, (?1889-1951)
HILL, Robert G., (1914-1973)
HÖGBOM, Bertil, (-1962)
HOLMES, Joseph Austin, (1859-1915)
HOLMSEN, Andreas, (1869-1955)
HORNEMAN, Hans Henrik, (1878-1945)
HOWELLS, Jenkyn Owen, (-1950)
HOYT, John Harger, (1928-1970)
HUDSON, Frank Samuel, (1890-1967)
HUGHES, William John, (?1899-1961)
HUMMEL-GUMAELIUS, Torsten, (-?1949)
HUTCHINSON, Lon Lewis, (1877-1947)
IRVING, John Duer, (1874-1918)
ISLAMOV, Oskar Islamovich, (1910-1978)
ITERSON, F. K. Th. van, (-1957)
JACK, Robert Lockhart, (1878-1964)
JACOB, Charles Edward, (1914-1970)
JAHN, Jaroslav Jiljí, (1865-1934)
JOESTING, Henry Rochambeau, (1903-1965)
JOHANSSON, Karl Fredrick, (1866-1933)
JOHNSON, J. P., (1880-1918)
JOHNSTON, Ashton William, (1910-)
JONES, Verner E., (1905-1977)
JONES, William Francis, (1885-1941)
JONES, William Richard, (1890-1970)
JONSON, Per Abraham, (-?1957)
JOUBIN, Francis Renault, (1911-)
JOYCE, John Raymond Foggan, (?1907-1963)
JUDD, John Wesley, (1840-1916)
JUDSON, Wilber, (1880-1951)
JUNGNER, Jonas Gustaf, (-?1942)
KARPINSKY, Alexander Petrovich, (1847-1936)
KENNEDY, Horas Tristram, Lieutenant, (1889-1917)
KENNEDY, William, (-1926)
KEYES, Charles Rollin, (1864-1942)
KEYES, Wilson, (1901-1936)
KIHLSTEDT, Folke H., (1901?-1956)
KINNEY, Harry Dill, (1886-1961)
KITSON, Albert Ernest, Sir, (1868-1937)
KOCH, Walter Edward, (1848-1916)
KONTKIEWICZ, Stanislaw, Sr., (1849-1924)
KONTKIEWICZ, Stanislaw Jr., (1883-1951)
KROENLEIN, George Alfred, (1898-1940)

MINING GEOLOGY 1900 to present Cont'd.

LADNER, Albert L., (1905-1975)
LAMBERT, Paul, (1869-1950)
LANDWEHR, Walter Richard, (1893-1972)
LARSH, Napoleon Bonaparte "Po", (1897-1967)
LEBOUR, George Alexander Louis, (1847-1918)
LECHNER, Karl, (1899-1958)
LECKWIJCK, Willem Peter van, (1902-1975)
LIVERSIDGE, Archibald, (1847-1927)
LOMAX, James, (1857-?1934)
LONGBOTTOM, Arthur, (?   -1910)
LOUGHLIN, Gerald Francis, (1880-1946)
LOUIS, Henry, (1855-1939)
LUNDBERG, Sven, (1900-?1948)
MACKAY, Robert Alexander, (?1904-1961)
MACKENZIE, Graham Stewart, (1905-1960)
MALLADA, Lucas, (1841-1921)
MALOUF, S. E., (1914-   )
MARKS, Edward Seaborn, (?1875-1950)
MARSMAN, J. H., (1892-1956)
MARTYN, Phil F., (1903-1974)
MAUFE, Herbert Brantwood, (1879-1946)
MAY DE MADIIS, Alexis Freiherr von, (1852-1911)
McBEAN, John Williams, (1913-   )
McCONNEL, Roger H., (1908-1971)
McCONNELL, Richard George, (1857-1942)
McDONALD, William, (1876-1943)
McINNES, William, (1858-1925)
McKIM, John Walter, (1877-1927)
McKINSTRY, Hugh Exton, (1896-1961)
McNUTT, Vachel Harry, (1888-1936)
MELL, Patrick Hues, (1850-1918)
MERRICKS, Frank, (?1863-1936)
MICHALSKI, Aleksander Octavianovich, (1855-1904)
MICHENER, Charles, (-   )
MILLS, James Ellison, (1834-1901)
MORGAN, Peter (Percy) Gates, (1867-1927)
MUNCK, Victor C. E. A. de, (1920-1956)
MÜNSTER, Thomas Georg, (1855-1938)
MYERS, William Marsh, (1892-1951)
MYLIUS, Louis Aubrey, (1890-1965)
NARAMORE, Chester "Chet", (1879-1961)
NEEDHAM, Claude Ervin, (1894-1950)
NELSON, Lloyd Alvino, (1895-1964)
NES, C. L. van, (?1878-1963)
NOBLE, Earl Bart, (1894-1967)
NOLAN, Philip Edward, (1887-1958)
NORELIUS, Carl Oskar, (1861-1941)
OFFICER, Herbert George, (1889-1937)
O'REILLY, Joseph P., (1829-1905)
OWEN, William Joseph, (1929-1959)
PACKARD, Henry Jessup, (1883-1930)
PAPP, Simon, (1886-1970)
PARK, James, (1857-1946)
PARKS, Roland Dane, (1900-1972)

MINING GEOLOGY 1900 to present Cont'd.

PEARCE, Richard, (1837-1927)
PECK, Frederick Burritt, (1860-1925)
PELLATI, Nicola, (1835-1907)
PENNING, William Henry, (1838-1902)
PENNY, Frederick Willoughby, (1887-1950)
PENROSE, Richard Alexander Fullerton, Jr., (1863-1931)
PERRY, Thomas Gregory, (1919-1972)
PETTEE, William Henry, (1836-1904)
PETTY, Thomas Franklin, (1900-1962)
PEYTON, Garland, (1892-1964)
PHILLIPS, David Watkins, (1899-1962)
PICKERING, William Henry, (1858-1912)
POBORSKI, Czeslaw, (1916-1964)
POLLARD, Ernest Rex, (1922-1966)
POND, Walter Franklin, (1885-1974)
POOLE, Henry Skeffington, (1884-1917)
PORCH, Edwin Leight, Jr., (1889-1948)
PUMPELLY, Raphael, (1837-1923)
PURDUE, Albert Homer, (1861-1917)
QUIRK, Roger Nathaniel, (?1909-1964)
RAMDOHR, Paul, (1890-)
RAYMOND, Rossiter Worthington, (1840-1918)
REBER, Louis Ehrhart, Jr., (1889-1966)
RICKARD, Thomas Arthur, (1864-1953)
RIGGS, Robert Jennings, (1888-1965)
ROBERTS, Hugh Marine, (1885-1965)
ROCH, Edouard, (1901-)
ROSALES, Henry, (?1822-1916)
ROTHWELL, Richard Pennefeather, (1836-1901)
RUDOLPH, Joseph, (1897-1953)
RUNNER, Joseph James, (1885-1970)
RUTBERG, Karl, (-1961)
RUTOT, Aimé Louis, (1847-1933)
SAHA, Baidya Nath, Babu, (-1929)
SANDERSON, James Owen Gresham "Pete", (1898-1963)
SANDS, John Melville, (1883-1962)
SARGENT, Harry Lionel, (1871-1955)
SATIJN, P. J. M., (?1907-1962)
SAUER, Julius, (1849-1911)
SCHLAIKJER, Erich Maren, (1905-1972)
SCHLUMBERGER, Conrad, (1878-1936)
SCHMITT, Harrison Ashley, (1896-1966)
SCHOLL, Louis August, Jr., (1887-1953)
SCHOLS, Hendrikus, (1901-1957)
SCHUMACHER, Friedrich, (1884-)
SCOFIELD, Lloyd M., (1897-1969)
SERVISS, Frederick LeVerne, (1895-1954)
SEVERY, Clarence Luther, (1885-1975)
SHALER, Millard King, (1880-1942)
SHAW, Everett Simeon, (1885-1955)
SHEARER, Harold Kurtz, (1889-1946)
SHERRILL, Richard Ellis, (1899-1953)
SHORT, Maxwell Naylor, (1889-1952)
SIMONS, Herbert Samuel, (1891-1956)

MINING GEOLOGY 1900 to present Cont'd.

SINGEWALD, Joseph Theophilus, Jr., (1884-1963)
SMITH, Warren D., (1880-1950)
SOLITANDER, Carl Probus, (1843-1918)
SOPWITH, Arthur, (1843-1920)
SPIKE, Wilfred Ivan, (1925-1969)
SPOFFORD, Howard Nelson, (1881-1950)
STAPLES, Ernest Henry, (1883-1957)
STEUBING, William C., (1881-1943)
ST. JOHN, Orestes Hawley, (1841-1921)
STOČES, Bohuslav, (1890-)
STOKES, Arthur Henry, (1842-1910)
STRANGWAY, David William, (1934-)
SULLIVAN, C. J., (1914?-)
SUNDBERG, Karl, (1891-1938)
SUTTON, Arle Herbert, (1896-1972)
SUTTON, William John, (1859-1914)
SWAMY, Kolagotla Bhimeswara, Sri, (-1950)
SWARTZ, Joel Howard, (1893-1971)
ŚWIDZINSKI, Henryk, (1904-1969)
TASHMAN, Djavad Eyoub, (1893-1956)
TAYLOR, James Carlton, (1924-1976)
TAZIEFF, Haroun, (1914-)
TEALE, Edmund Oswald, Sir, (1874-1971)
TEMPLETON, Eugene Crittenden, (1889-1932)
TERMIER, Pierre, (1859-1930)
TEVES, Juan S., "Steve", (1902-1958)
THOMAS, David, (1880-1934)
THOMAS, George Dewey, (1898-1975)
THOMAS, Ronald Hawksby, Major, (1885-?1940)
THOMSON, Joseph Ellis, (1882-1944)
TIPPER, George Howlett, (1881-1947)
TOKARSKI, Julian, (1883-1961)
TRÜMPY, Daniel, (1893-1971)
TRUSCOTT, Samuel John, (?1870-1950)
TSCHERNIG, Emil, (1896-1970)
TWELVETREES, William Harper, (1848-1919)
UGLOW, William Lawrence, (1884-1926)
ULFFERS, Ernst, (1837-1919)
VANDER LECK, Lawrence, (1890-1954)
VAN HORN, Frank Robertson, (1872-1933)
VEATCH, Arthur Clifford, (1878-1938)
VERSLUYS, Jan, (1880-1935)
VINCIENNE, Henri, (1898-1965)
WADA, Tsunashirô, (1856-1920)
WADIA, Darashaw Nosherwan, (1883-1969)
WAGNER, Percy Albert, (1885-1929)
WALLROTH, Gunnar, (-?1949)
WANDKE, Alfred, (1887-1941)
WASHBURNE, Chester Wesley, (1883-1971)
WASSERSTEIN, Benno, (1906-1956)
WATERSCHOOT VAN DER GRACHT, Willem Anton Josef Maria van, (1873-1943)
WATTS, Murray Edmund, (1909-)
WEBB, John Benwell, (1901-1975)

MINING GEOLOGY 1900 to present Cont'd.

WENGER, Marian, (1864-1939)
WESTLUND, Erik Wiktor, (-?1954)
WHEALTON, Rowland Gardiner "Peg", (1896-1956)
WHEELER, Charles Douglas, (?1922-1944)
WHITE, William Harrison, (1913-1972)
WHITEHEAD, Walter Lucius, (1891-1969)
WHITTIER, William Harrison, (1889-1951)
WILD, Robert Powley, (1882-1946)
WILDER, Newell Morris "Newt", (1908-1946)
WILLIAMS, Gardner Fred, (1842-1922)
WILLIAMS, George O. "Skipper", (1897-1967)
WILLIAMS, Mervyn, (1934-1970)
WILMOTH, Leslie James, (1881-1963)
WILSON, Alfred William Gunning, (1873-1954)
WILSON, James W., (?1925-1964)
WILSON, Thomas Carroll, (1906-1969)
WINCHELL, Horace Vaughn, (1865-1923)
WINDIAKIEWICZ, Edward, (1858-1942)
WINGE, Knut Ingel, (-1958)
WIRZ, Albert Eduard, (1910-1973)
WOLF, Albert G., (1885-1964)
WOOD, Frederick Wilfred, (?1904-1961)
WOOD, James George, (1888-1928)
WRIGHT, Charles Will, (1879-)
YEAGER, Lloyd Irvin, (1900-1938)
ZEIJLMANS VAN EMMICHOVEN, Cornelis Pieter Adriaan, (1897-1952)
ZEZI, Pietro, (1884-1914)

Time unknown
ELZVIK, Erik, (-)
RANDS, William Henry, (-)

PALAEOBOTANY

Up to 1800

ALBERTUS MAGNUS, (1193-1280)
AYCKE, Johann Christian, (1766-1854)
BROWN, Robert, (1773-1858)
FAUJAS DE SAINT FOND, Barthélémy, (1741-1819)
KIRCHER, Athanasius, (1601-1680)
MONTI, Giuseppe, (1682-1760)
NAU, Bernhard Sebastian von, (1766-1845)
PARKINSON, James, (1755-1824)
SCHLOTHEIM, Ernst Friedrich von, Freiherr, (1764-1832)
SCHULZE, Christian Friedrich, (1730-1775)
SERRES, Marcel de, (1783-1862)
ŠTERNBERK, Kašpar Marià, Hrabe, (1761-1838)

1800-1849

ANDRAE, Carl Justus, (1816-1885)
ARGYLL, George Douglas, Duke of, (1823-1900)
ARTIS, Edmund Tyrell, (1789-1847)
AYCKE, Johann Christian, (1766-1854)
BOWERBANK, James Scott, (1797-1877)
BRAUN, Heinrich Alexander, (1805-1877)
BRAUN, Karl Friedrich Wilhelm, (1800-1864)
BRAUNS, David August, (1827-1893)
BRONGNIART, Adolphe Théodore, (1801-1876)
BROWN, Robert, (1773-1858)
BUNBURY, Charles James Fox, Sir, (1809-1886)
CANDOLLE, Alphonse de, (1806-1893)
CASPARY, Robert, (1818-1887)
COEMANS, Eugène H. L. G., (1825-1871)
CRÉPIN, François, (1830-1903)
DAWES, John Samuel, (1802-1878)
DAWSON, John William, Sir, (1820-1899)
DE VISIANI, Roberto, (1800-1878)
DE ZIGNO, Achille, Barone, (1813-1888)
DICKSON, Oscar, (1823-1897)
ENGELMANN, George, (1809-1884)
ESCHER VON DER LINTH, Arnold, (1807-1872)
ETTINGSHAUSEN, Constantin von, Freiherr, (1826-1897)
FAUJAS DE SAINT FOND, Barthélémy, (1741-1819)
FEISTMANTEL, Karel, (1819-1885)
FISCHER-OOSTER, Carl von, (1807-1875)
FRESENIUS, Georg, (1808-1866)
FREYER, Heinrich, (1802-1865)
GAUDIN, Charles Théophile, (1822-1866)
GERMAR, Ernst Friedrich, (1786-1853)
GEYLER, Hermann Theodor, (1834-1889)
GOLDENBERG, Friedrich, (1799-1881)
GÖPPERT, Heinrich Robert, (1801-1884)
GRAY, Asa, (1810-1888)
GÜMBEL, Carl Wilhelm von, Ritter, (1823-1898)
GUTBIER, Christian August von, (1798-1866)
HASSENCAMP, Ernst, (1824-1881)
HAZSLINSZKY, Hazslini Frigyes Ágost, (1818-1896)
HEER, Oswald, (1809-1883)
HISLOP, Stephen, Reverend, (1817-1863)

PALAEOBOTANY 1800-1849 Cont'd.

HOFFMANN, Friedrich, (1797-1836)
HOFMEISTER, Wilhelm Friedrich Benedict, (1824-1877)
HOOKER, Joseph Dalton, Sir, (1817-1911)
HOSIUS, August, (1825-1896)
HUNTER, Robert, Reverend, (1823-1897)
HUTTON, William, (-1861?)
JÄGER, Georg Friedrich, (1785-1866)
KLIPSTEIN, August Wilhelm, (1801-1894)
KURR, Johann Gottlob von, (1798-1870)
LESQUEREUX, Charles Leo, (1806-1889)
LUDWIG, Rudolph August Birminhold Sebastian, (1812-1880)
MARÈS, Paul, (1826-1900)
MARTIUS, Karl Friedrich Philipp von, (1794-1868)
MASSALONGO, Abramo Bartolomeo, (1824-1860)
MENEGHINI, Giuseppe Giovanni Antonio, (1811-1889)
MENGE, Anton, (1808-1880)
MORIÈRE, Jules Pierre Gilles, (1817-1888)
MÜLLER, Ferdinand von, (1825-1896)
NAU, Bernhard Sebastian von, (1766-1845)
NÖLDEKE, Karl, (1815-1898)
NOULET, Jean Baptiste, (1802-1890)
OLDHAM, Thomas, (1816-1878)
OTTO, Ernst von, (1799-1863)
PARKINSON, James, (1755-1824)
PEACH, Charles William, (1800-1886)
PLANCHON, Jules Émile, (1823-1888)
PRESL, Karl Borivoj, (1794-1852)
PROBST, Joseph, (1823-1905)
REUSS, August Emanuel von, Ritter, (1811-1873)
ROLLE, Friedrich, (1827-1887)
ROSSMÄSSLER, Emil Adolf, (1806-1867)
ROYLE, John Forbes, (-1857?)
RUMPF, Ludwig, (1793-1862)
SAPORTA, Louis Charles Joseph Gaston de, Marquis, (1823-1895)
SAUVEUR, Dieudonné Jean Joseph, (1797-1862)
SCHENK, Joseph August, (1815-1891)
SCHIMPER, Philipp Wilhelm, (1808-1880)
SCHLEIDEN, Matthias Jakob, (1804-1881)
SCHLOTHEIM, Ernst Friedrich von, Freiherr, (1764-1832)
SERRES, Marcel de, (1783-1862)
SISMONDA, Eugenio, (1815-1870)
STENZEL, Karl Gustav, (1826-1905)
ŠTERNBERK, Kašpar Marià, Hrabe, (1761-1838)
STOKES, Charles, (1783-1853)
STUR, Dénes, (1827-1893)
TORREY, John, (1796-1873)
WATELET, Adolphe, (1811-1880)
WILLIAMSON, William Crawford, (1816-1895)
WIRTGEN, Philipp Wilhelm, (1806-1870)

1850-1899

AMALITSKII, Vladimir Prochorovich, (1860-1917)
ANDRAE, Carl Justus, (1816-1885)
ARGYLL, George Douglas, Duke of, (1823-1900)

PALAEOBOTANY 1850-1899 Cont'd.

AYCKE, Johann Christian, (1766-1854)
BAYER, Edvin Jaroslav, (1862-1927)
BECK, Carl Richard, (1858-1919)
BERTRAND, C. Eg., (1851-1920)
BEUST, Fritz von, (1856-1908)
BLYTT, Axel, (1843-1898)
BOWER, Fredrik Orpen, (1855-1948)
BOWERBANK, James Scott, (1797-1877)
BRAUN, Heinrich Alexander, (1805-1877)
BRAUN, Karl Friedrich Wilhelm, (1800-1864)
BRAUNS, David August, (1827-1893)
BRONGNIART, Adolphe Théodore, (1801-1876)
BROWN, Robert, (1773-1858)
BUNBURY, Charles James Fox, Sir, (1809-1886)
BUREAU, Edouard, (1830-1918)
CAMPBELL, Douglas Houghton, (1859-1953)
CANDOLLE, Alphonse de, (1806-1893)
CARPENTIER, Alfred, Chanoine, (1878-1952)
CARRUTHERS, William, (1830-1922)
CASPARY, Robert, (1818-1887)
CAYEUX, Lucien, (1864-1944)
CLAYPOLE, Edward Waller, (1835-1901)
COEMANS, Eugène H. L. G., (1825-1871)
COMPTNER, Gustav, (1831-1922)
CONWENTZ, Hugo, (1855-1922)
CRAMER, Carl Edward, (1831-1901)
CRÉPIN, François, (1830-1903)
DAVIES, David, (1870-1931)
DAWES, John Samuel, (1802-1878)
DAWSON, John William, Sir, (1820-1899)
DEANE, Henry Jr., (-1924)
DE VISIANI, Roberto, (1800-1878)
DE ZIGNO, Achille, Barone, (1813-1888)
DICKSON, Oscar, (1823-1897)
DUSÉN, Per, (1855-1926)
ENGELMANN, George, (1809-1884)
ESCHER VON DER LINTH, Arnold, (1807-1872)
ETTINGSHAUSEN, Constantin von, Freiherr, (1826-1897)
FEISTMANTEL, Karel, (1819-1885)
FEISTMANTEL, Otokar, (1848-1891)
FELIX, Johannes Paul, (1859-1941)
FISCHER-OOSTER, Carl von, (1807-1875)
FLICHE, Paul, (1836-1908)
FONTAINE, William Morris, (1835-1913)
FRAUSCHER, Karl, (1852-1914)
FRESENIUS, Georg, (1808-1866)
FREYER, Heinrich, (1802-1865)
FRIEDRICH, Paul, (1856-1918)
FRITSCH, Karl von, Freiherr, (1838-1906)
GAUDIN, Charles Théophile, (1822-1866)
GEINITZ, Eugen, (1854-1925)
GERBER, Eduard, (1876-1956)
GERMAR, Ernst Friedrich, (1786-1853)
GEYLER, Hermann Theodor, (1834-1889)

PALAEOBOTANY 1850-1899 Cont'd.

GOLDENBERG, Friedrich, (1799-1881)
GÖPPERT, Heinrich Robert, (1801-1884)
GOULD, Charles Newton, (1868-1949)
GRAND'EURY, François Cyrille, (1839-1917)
GRAY, Asa, (1810-1888)
GÜMBEL, Carl Wilhelm von, Ritter, (1823-1898)
GUTBIER, Christian August von, (1798-1866)
HARTZ, Nikolaj Eeg Kruse, (1867-1937)
HASSENCAMP, Ernst, (1824-1881)
HAZSLINSZKY, Hazslini Frigyes Ágost, (1818-1896)
HEER, Oswald, (1809-1883)
HISLOP, Stephen, Reverend, (1817-1863)
HOFMANN, Adolf, (1853-1913)
HOFMEISTER, Wilhelm Friedrich Benedict, (1824-1877)
HOLLENDONNER, Ferenc, (1882-1935)
HOLLICK, Arthur, (1857-1933)
HOOKER, Joseph Dalton, Sir, (1817-1911)
HOSIUS, August, (1825-1896)
HOWE, Marshall Avery, (1867-1936)
HUNTER, Robert, Reverend, (1823-1897)
HUTTON, William, (-1861?)
JÄGER, Georg Friedrich, (1785-1866)
JENTZSCH, Alfred, (1850-1925)
JOHNSTON, Robert Mackenzie, (-1918)
JONGMANS, Wilhelmus Josephus, (1878-1957)
KARPINSKY, Alexander Petrovich, (1847-1936)
KATZER, Bedřich, (1861-1925)
KIDSTON, Robert, (1852-1924)
KLIPSTEIN, August Wilhelm, (1801-1894)
KRAŠAN, Franz, (1840-1907)
KRASSER, Fridolin, (1863-1922)
KRAUS, Gregor, (1841-1915)
KURR, Johann Gottlob von, (1798-1870)
KURTZ, Fritz, (1854-1920)
LANG, William Henry, (1874-1960)
LESQUEREUX, Charles Leo, (1806-1889)
LINGELSHEIM, Alexander von, (1874-1937)
LOOMIS, Theodor, (1875-1909)
LORENZ, Theodor, (1875-1909)
LUDWIG, Rudolph August Birminhold Sebastian, (1812-1880)
MARÈS, Paul, (1826-1900)
MARION, Antoine Fortuné, (1846-1900)
MARTIUS, Karl Friedrich Philipp von, (1794-1868)
MASLEN, Arthur John, (1871-1954)
MASSALONGO, Abramo Bartolomeo, (1824-1860)
McMURTRIE, James, (1840-1914)
MENEGHINI, Giuseppe Giovanni Antonio, (1811-1889)
MENGE, Anton, (1808-1880)
MENZEL, Paul, (1864-1927)
MICHAEL, Richard, (1869-1925)
MÖLLER, Hjalmar August, (1866-1941)
MORIÈRE, Jules Pierre Gilles, (1817-1888)
MÜLLER, Ferdinand von, (1825-1896)
NATHORST, Alfred Gabriel, (1850-1921)

PALAEOBOTANY 1850-1899 Cont'd.

NEWELL ARBER, Edward Alexander, (1870-1918)
NOÉ VON ARCHENEGG, Adolph Carl, (1873-1939)
NÖLDEKE, Karl, (1815-1898)
NOULET, Jean Baptiste, (1802-1890)
OLDHAM, Thomas, (1816-1878)
OTTO, Ernst von, (1799-1863)
PALLA, Eduard, (1864-1922)
PEACH, Charles William, (1800-1886)
PEREIRA DE LIMA, Wenceslau de Sousa, (1858-1920)
PILAR, Georg, (1847-1893)
PLANCHON, Jules Émile, (1823-1888)
POTONIÉ, Henry Edmond, (1857-1913)
PRESL, Karl Borivoj, (1794-1852)
PROBST, Joseph, (1823-1905)
PROSSER, Charles Smith, (1860-1916)
PURKYNĚ, Cyril, (1862-1937)
RACIBORSKI, Maryan, (1863-1917)
REED, Frederick Richard Cowper, (1869-1946)
REHM, Heinrich, (1828-1916)
REID, Clement, (1853-1916)
REID, Eleanor Mary, (née Wynne Edwards), (1860-1953)
REINSCH, Paul Friedrich, (1836-1914)
REIS, Otto Maria, (1862-1934)
RENAULT, Bernard, (1836-1904)
REUSS, August Emanuel von, Ritter, (1811-1873)
RICHTER, Paul Bogoslav, (1854-1911)
RIDLEY, Henry Nicholas, (1855-1956)
RISTORI, Giuseppe, (1856-1905)
ROLLE, Friedrich, (1827-1887)
ROSSMÄSSLER, Emil Adolf, (1806-1867)
ROYLE, John Forbes, (-1857?)
RUMPF, Ludwig, (1793-1862)
RYBA, Frantisek, (1867-1918)
SAPORTA, Louis Charles Joseph Gaston de, Marquis, (1823-1895)
SAUVEUR, Dieudonné Jean Joseph, (1797-1862)
SCHENK, Joseph August, (1815-1891)
SCHIMPER, Philipp Wilhelm, (1808-1880)
SCHLECHTENDAL, Dietrich von, (1834-1916)
SCHLEIDEN, Matthias Jakob, (1804-1881)
SCHMALHAUSEN, Johannes Theodor, (1849-1894)
SCHÖNFELD, Georg, (1878-1926)
SCHRÖTER, Carl, (1855-?1925)
SCHUMANN, Karl, (1851-1904)
SCHÜTZE, Ewald, (1873-1908)
SCOTT, Dukinfield Henry, (1854-1934)
SELLARDS, Elias Howard, (1875-1961)
SERRES, Marcel de, (1783-1862)
SEWARD, Albert Charles, Sir, (1863-1941)
SISMONDA, Eugenio, (1815-1870)
SOLMS-LAUBACH, Hermann zu, Graf, (1842-1915)
SORDELLI, Ferdinando, (?1837-1916)
STACHE, Guido, (1833-1921)
STAINER, Xavier, (1865-)
STAUB, Morić, (1842-1904)

PALAEOBOTANY 1850-1899 Cont'd.

STENZEL, Karl Gustav, (1826-1905)
STERZEL, Johann Traugott, (1841-1914)
STOKES, Charles, (1783-1853)
STOLLER, Jakob, (1873-1930)
STRASBURGER, Eduard, (1844-1912)
STRENG, August, (1830-1897)
STUDER, Theophil, (1845-1922)
STUR, Dénes, (1827-1893)
SUTCLIFFE, William Henry, (1855-1913)
TATE, Ralph, (1840-1901)
TENISON-WOODS, Julian Edmund, Reverend, (1832-1889)
TORREY, John, (1796-1873)
VATER, Heinrich, (1859-1930)
VELENOVSKÝ, Josef, (1858-1949)
WARD, Lester Frank, (1841-1913)
WATELET, Adolphe, (1811-1880)
WEBER, Carl Albert, (1856-1931)
WETTSTEIN, Richard, (1863-1931)
WHITE, Charles David, (1862-1935)
WHITLA, Francis, (fl.1830-1853)
WIELAND, George Reber, (1865-1953)
WILLIAMSON, William Crawford, (1816-1895)
WIRTGEN, Philipp Wilhelm, (1806-1870)
WÜNSCH, Edward A., (1822-1895)
YABE, Hisakatsu, (1878-1969)
ZEILLER, Charles René, (1847-1915)
ZWANZIGER, Gustav Adolf, (1837-1893)

1900 to present

AMALITSKII, Vladimir Prochorovich, (1860-1917)
ANDRÉANSZKY, Gábor, (1895-1967)
BARTRUM, John Arthur, (1885-1949)
BAYER, Edvin Jaroslav, (1862-1927)
BECK, Carl Richard, (1858-1919)
BERRY, Edward Willard, (1900-1968)
BERTRAND, Paul, (1879-1944)
BEUST, Fritz von, (1856-1908)
BOCHEŃSKI, Tadeusz Antoni, (1901-1958)
BOUREAU, Edouard Léon François, (1913-)
BOWER, Fredrik Orpen, (1855-1948)
BUREAU, Edouard, (1830-1918)
CAMPBELL, Douglas Houghton, (1859-1953)
CARPENTIER, Alfred, Chanoine, (1878-1952)
CARRUTHERS, William, (1830-1922)
CASTELLANOS, Alberto, (1896-1968)
CAYEUX, Lucien, (1864-1944)
CHALONER, William G., (-)
CHANEY, Ralph Works, (1890-1971)
CLAYPOLE, Edward Waller, (1835-1901)
COMPTNER, Gustav, (1831-1922)
CONWENTZ, Hugo, (1855-1922)
CRAMER, Carl Edward, (1831-1901)
CROFT, William Noble, (1915-1953)
DAVIES, David, (1870-1931)

PALAEOBOTANY 1900 to present Cont'd.

DEANE, Henry Jr., (-1924)
DIX, Emily, (1904-1973)
DON, Archibald William Robertson, (1891-1916)
DORLODOT, Henry de, Chanoîne, (1855-1929)
DUSÉN, Per, (1855-1926)
EDWARDS, Wilfred Norman, (1890-1956)
FELIX, Johannes Paul, (1859-1941)
FLICHE, Paul, (1836-1908)
FLORSCHÜTZ, Frans, (1887-1965)
FONTAINE, William Morris, (1835-1913)
FRAUSCHER, Karl, (1852-1914)
FRIEDRICH, Paul, (1856-1918)
FRITSCH, Karl von, Freiherr, (1838-1906)
GEINITZ, Eugen, (1854-1925)
GERBER, Eduard, (1876-1956)
GOODE, R. H., Reverend, (1888-1969)
GORDON, William Thomas, (1884-1950)
GOTHAN, Walther, (1879-1954)
GOULD, Charles Newton, (1868-1949)
GRAHAM, Roy, (1908-1939)
GRAMBAST, Louis, (-1976)
GRAND'EURY, François Cyrille, (1839-1917)
HARTZ, Nikolaj Eeg Kruse, (1867-1937)
HEARD, Albert, (1894-1944)
HERAK, Milan, (1917-)
HOFMANN, Adolf, (1853-1913)
HOLDEN, Henry Smith, (1887-1963)
HOLLENDONNER, Ferenc, (1882-1935)
HOLLICK, Arthur, (1857-1933)
HOOKER, Joseph Dalton, Sir, (1817-1911)
HORN AF RANTZIEN, Henning, (1922-1960)
HOSKINS, John Hobart, (1896-1957)
HOWE, Marshall Avery, (1867-1936)
JAROŃ, Bronislaw, (1905-1942)
JENTZSCH, Alfred, (1850-1925)
JOHNSTON, Robert Mackenzie, (-1918)
JONGMANS, Wilhelmus Josephus, (1878-1957)
JURASKY, Karl Alfons, (1903-1945)
JUST, Theodor Karl, (1904-1959)
KARPINSKY, Alexander Petrovich, (1847-1936)
KATZER, Bedřich, (1861-1925)
KIDSTON, Robert, (1852-1924)
KOLBE, Robert Wilhelm, (1882-1960)
KRAŠAN, Franz, (1840-1907)
KRASSER, Fridolin, (1863-1922)
KRAUS, Gregor, (1841-1915)
KRÄUSEL, Richard, (1890-1966)
KRYSHTOFOVICH, Afrikan Nikolaevich, (1885-1953)
KURTZ, Fritz, (1854-1920)
LANG, William Henry, (1874-1960)
LESLIE, Thomas Nicholas, (-1942)
LILPOP, Jerzy, (1888-1945)
LINGELSHEIM, Alexander von, (1874-1937)
LOOMIS, Theodor, (1875-1909)

PALAEOBOTANY 1900 to present Cont'd.

LORENZ, Theodor, (1875-1909)
LOUBIÈRE, A., (-)
MÄGDEFRAU, Karl, (1907-)
MALDONADO-KOERDELL, Manuel, (1908-1972)
MARÈS, Paul, (1826-1900)
MASLEN, Arthur John, (1871-1954)
McMURTRIE, James, (1840-1914)
MENÉNDEZ, Carlos Alberto, (1921-1976)
MENZEL, Paul, (1864-1927)
MICHAEL, Richard, (1869-1925)
MÖLLER, Hjalmar August, (1866-1941)
MORELLET, Lucien, (1882-1945)
MUKERJI, Sushil Kumar, (1896-1934)
MURR, Josef, (-1932)
NATHORST, Alfred Gabriel, (1850-1921)
NEUBURG, Maria Feodorovna, (1894-1962)
NEWELL ARBER, Edward Alexander, (1870-1918)
NIKLEWSKI, Jerzy, (1937-1971)
NOÉ VON ARCHENEGG, Adolph Carl, (1873-1939)
PABST, Marie B., (1909-1963)
PALLA, Eduard, (1864-1922)
PEREIRA DE LIMA, Wenceslau de Sousa, (1858-1920)
PEVALEK, Ivo, (1893-)
PIA, Julius von, (1887-1943)
PIECH, Kazimierz, (1893-1944)
POTONIÉ, Henry Edmond, (1857-1913)
PROBST, Joseph, (1823-1905)
PROSSER, Charles Smith, (1860-1916)
PURKYNĚ, Cyril, (1862-1937)
RACIBORSKI, Maryan, (1863-1917)
REED, Frederick Richard Cowper, (1869-1946)
REHM, Heinrich, (1828-1916)
REID, Clement, (1853-1916)
REID, Eleanor Mary, (née Wynne Edwards), (1860-1953)
REINSCH, Paul Friedrich, (1836-1914)
REIS, Otto Maria, (1862-1934)
RENAULT, Bernard, (1836-1904)
RICHTER, Paul Bogoslav, (1854-1911)
RIDLEY, Henry Nicholas, (1855-1956)
RISTORI, Giuseppe, (1856-1905)
RUDOLPH, Karl, (1881-1937)
RYBA, Frantisek, (1867-1918)
SAHNI, Birbal, (1891-1949)
SAMUELSSON, Gunnar, (-?1945)
SCHLECHTENDAL, Dietrich von, (1834-1916)
SCHÖNFELD, Georg, (1878-1926)
SCHRÖTER, Carl, (1855-?1925)
SCHUMANN, Karl, (1851-1904)
SCHÜTZE, Ewald, (1873-1908)
SCOTT, Dukinfield Henry, (1854-1934)
SELLARDS, Elias Howard, (1875-1961)
SEMAKA, Alexandru, (1924-1970)
ŠETLÍK, Jaromír, (1896-)
SEWARD, Albert Charles, Sir, (1863-1941)

PALAEOBOTANY 1900 to present Cont'd.

SOLMS-LAUBACH, Hermann zu, Graf, (1842-1915)
SORDELLI, Ferdinando, (?1837-1916)
STACHE, Guido, (1833-1921)
STAINER, Xavier, (1865-)
STARK, Peter, (1888-1932)
STAUB, Morić, (1842-1904)
STENZEL, Karl Gustav, (1826-1905)
STERZEL, Johann Traugott, (1841-1914)
STOLLER, Jakob, (1873-1930)
STOPES, Marie Charlotte Carmichael, (1880-1958)
STRASBURGER, Eduard, (1844-1912)
STUDER, Theophil, (1845-1922)
SUTCLIFFE, William Henry, (1855-1913)
SZAFER, Wladyslaw, (1886-1970)
TATE, Ralph, (1840-1901)
THOMAS, Hugh Hamshaw, (1885-1962)
VATER, Heinrich, (1859-1930)
VELENOVSKÝ, Josef, (1858-1949)
WALTON, John, (-1971)
WARD, Lester Frank, (1841-1913)
WATSON, David Meredith Seares, (1886-1973)
WEBER, Carl Albert, (1856-1931)
WEIGELT, Johannes, (1890-1948)
WETTSTEIN, Richard, (1863-1931)
WEYLAND, Hermann, (1888-)
WHITE, Charles David, (1862-1935)
WIELAND, George Reber, (1865-1953)
WILSON, Leonard Richard, (1906-)
YABE, Hisakatsu, (1878-1969)
ZEILLER, Charles René, (1847-1915)

Time unknown
KERNER, Fritz, (-)

PALAEOCLIMATOLOGY

Up to 1800
CRICHTON, Alexander, Sir, (1763-1856)

1800-1849
CRICHTON, Alexander, Sir, (1763-1856)
CROLL, James, (1821-1890)
WALLACE, Alfred Russel, (1823-1913)

1850-1899
BARRELL, Joseph, (1869-1919)
BLYTT, Axel, (1843-1898)
CRICHTON, Alexander, Sir, (1763-1856)
CROLL, James, (1821-1890)
KAYSER, Friedrich Heinrich Emanuel, (1845-1927)
SCHUCHERT, Charles, (1858-1942)
WALLACE, Alfred Russel, (1823-1913)
WOEIKOF, Alexander, (1842-1916)

PALAEOCLIMATOLOGY Cont'd.

1900 to present
BANDY, Orville Lee, (1917-1973)
BÁRDHARSON, Gudhmundur G., (1880-1933)
BARRELL, Joseph, (1869-1919)
CAYEUX DE SENARPONT, André, called CAILLEUX, André, (1907-)
KAYSER, Friedrich Heinrich Emanuel, (1845-1927)
PIA, Julius von, (1887-1943)
SCHUCHERT, Charles, (1858-1942)
STOCK, Chester, (1892-1950)
TODD, Thomas W., (1927-1969)
WALLACE, Alfred Russel, (1823-1913)
WOEIKOF, Alexander, (1842-1916)

PALAEOECOLOGY
1800-1849
FORBES, Edward, (1815-1854)
HEER, Oswald, (1809-1883)

1850-1899
DAVIES, David, (1870-1931)
FORBES, Edward, (1815-1854)
HEER, Oswald, (1809-1883)
KOVALEVSKY, Vladimir Onufrievich, (1843-1883)

1900 to present
ALLAN, Robin Sutcliffe, (1900-1966)
BANDY, Orville Lee, (1917-1973)
BLANC, Alberto Carlo, (1906-1960)
CHANEY, Ralph Works, (1890-1971)
CUSHMAN, Joseph Augustine, (1881-1949)
DAVIES, David, (1870-1931)
LAMONT, Archie, (-)
SCHMIDT, Hermann, (1892-1978)
STOCK, Chester, (1892-1950)
TRUEMAN, Arthur Elijah, Sir, (1894-1956)
WASMUND, Erich, (1902-1945)

PALAEOICHNOLOGY
Up to 1800
GIMBERNAT, Carlos de, Don, (1765-1839)

1800-1849
BARKAS, Thomas Pallister, (1819-1891)
BARRATT, Joseph, (1796-1882)
BINNEY, Edward William, (1812-1881)
BRODIE, Peter Bellinger, Reverend, (1815-1897)
DAWSON, John William, Sir, (1820-1899)
DEANE, James, (1801-1858)
FISCHER-OOSTER, Carl von, (1807-1875)
GIMBERNAT, Carlos de, Don, (1765-1839)
GIRARD, Carl Adolph, (1814-1878)
GRANT, Robert Edmond, (1793-1874)
HAIDINGER, Wilhelm Karl von, Ritter, (1795-1871)
HITCHCOCK, Edward, (1793-1864)

PALAEOICHNOLOGY 1800-1849 Cont'd.

HUXLEY, Thomas Henry, (1825-1895)
LOGAN, William Edmond, Sir, (1798-1875)
MORTON, George Highfield, (1826-1900)
RUMPF, Ludwig, (1793-1862)
SCHMID, Ernst Erhard, (1815-1885)
SELWYN, Alfred Richard Cecil, (1824-1902)
TYLOR, Alfred, (1823-1884)

1850-1899

ABEL, Othenio, (1875-1946)
BARBOUR, Erwin Hinckley, (1856-1947)
BARKAS, Thomas Pallister, (1819-1891)
BARRATT, Joseph, (1796-1882)
BEASLEY, Henry Charles, (1836-1919)
BECKLES, Samuel H., (-1890)
BERNHARDI, Reinhard, (fl.1834-1841)
BINNEY, Edward William, (1812-1881)
BRODIE, Peter Bellinger, Reverend, (1815-1897)
BRODRICK, Harold, (1872-1946)
CLAYDEN, Arthur William, (1855-1944)
CORFIELD, William, (1843-1903)
DAMES, Wilhelm Barnim, (1843-1898)
DAWSON, John William, Sir, (1820-1899)
DEANE, James, (1801-1858)
FISCHER-OOSTER, Carl von, (1807-1875)
GILMORE, Charles Whitney, (1874-1945)
GIRARD, Carl Adolph, (1814-1878)
GRANT, Robert Edmond, (1793-1874)
HAIDINGER, Wilhelm Karl von, Ritter, (1795-1871)
HARKNESS, Robert, (1816-1878)
HITCHCOCK, Edward, (1793-1864)
HUXLEY, Thomas Henry, (1825-1895)
JAMES, Joseph Francis, (1857-1897)
JORDAN, David Starr, (1851-1931)
LOGAN, William Edmond, Sir, (1798-1875)
LULL, Richard Swann, (1867-1957)
MAIDWELL, Frederic Thomas, (1872-1921)
MATTHEW, George Frederic, (1837-1923)
MORTON, George Highfield, (1826-1900)
NOPCSA VON FELSŐ-SZILVÁS, Ferenc, baron, (1877-1933)
RUMPF, Ludwig, (1793-1862)
SCHMID, Ernst Erhard, (1815-1885)
SELWYN, Alfred Richard Cecil, (1824-1902)
SHIPMAN, James, (1848-1901)
SMITH, John, (1845-1930)
SNOW, Francis Huntington, (1840-1908)
STEINDACHNER, Franz, (1834-1919)
TYLOR, Alfred, (1823-1884)
VARTY SMITH, George, (-1924)

1900 to present

ABEL, Othenio, (1875-1946)
BARBOUR, Erwin Hinckley, (1856-1947)
BEASLEY, Henry Charles, (1836-1919)

PALAEOICHNOLOGY 1900 to present Cont'd.

BRADLEY, Wilmot Hyde, (1899-)
BRADY, Lionel Francis "Major", (1880-1963)
BRODRICK, Harold, (1872-1946)
CARMAN, Joel Ernest, (1882-1966)
CLAYDEN, Arthur William, (1855-1944)
CORFIELD, William, (1843-1903)
CURRIE, Philip John, (1949-)
GILMORE, Charles Whitney, (1874-1945)
HÄNTZSCHEL, Walter, (1904-1972)
HICKLING, Henry George Albert, (1883-1954)
HOUSTON, Samuel Henry, (1905-1947)
HOYT, John Harger, (1928-1970)
JORDAN, David Starr, (1851-1931)
LULL, Richard Swann, (1867-1957)
MAIDWELL, Frederic Thomas, (1872-1921)
MATTHEW, George Frederic, (1837-1923)
MOODIE, Roy Lee, (1880-1934)
MORTON, George Highfield, (1826-1900)
NOPCSA VON FELSÖ-SZILVÁS, Ferenc, baron, (1877-1933)
PEABODY, Frank Elmer, (1914-1958)
SARJEANT, William Antony Swithin, (1935-)
SELWYN, Alfred Richard Cecil, (1824-1902)
SHIPMAN, James, (1848-1901)
SHULER, Ellis William, (1881-1954)
SMITH, John, (1845-1930)
SNOW, Francis Huntington, (1840-1908)
STEINDACHNER, Franz, (1834-1919)
SZMUC, Eugene Joseph, (1927-1976)
VARTY SMITH, George, (-1924)

PALAEONTOLOGY (undivided)

Up to 1800

ALBERTI, Leon Battista, (1404-1472)
ARDUINO, Giovanni, (1714-1795)
BAUHIN, Jean, (1541-1613)
BERGMAN, Torbern Olof, (1735-1784)
BERINGER, Johann Bartholomaeus Adam, (1667-1740)
BOCCONE, Paolo, (1633-1704)
CARDANO, Girolamo, (1501-1576)
CAROSI, Jan Filip, (1744-1799)
CLARK, William, (1770-1838)
CROUSAZ, Jean-Pierre de, (1663-1750)
DA COSTA, Emanuel Mendes, (1717-1791)
DEZALLIER D'ARGENVILLE, Antoine Joseph, (1680-1765)
FRACASTORO, Girolamo Hieronymus, (1483-1553)
FÜCHSEL, Georg Christian, (1722-1773)
GALEAZZI, Domenico Mario Gusmano, (1686-ca.1775)
GESSNER, Conrad, (1516-1565)
GOETHE, Johann Wolfgang von, (1749-1832)
HALL, Elias, (1764-1853)
HOOKE, Robert, (1635-1703)
KAPPELER, Moritz, (1685-1769)
KIRBY, William, Reverend, (1759-1851?)
KLEIN, Jakob Theodor, (1685-1759)

PALAEONTOLOGY (undivided) Up to 1800 Cont'd.

McDONNELL, James, (?1763-1845)
MURALT, Johannes von, (1645-1733)
PALISSY, Bernard, (1510-1589)
PARROT, Georg Friedrich, (1767-1852)
PLOT, Robert, (1640-1696)
RAY, John, (1627-1705)
RÉAUMUR, René Antoine Ferchault de, (1683-1757)
SCILLA, Agostino, (1639-1700)
SLOANE, Hans, Sir, (1660-1753)
SOWERBY, James, (1757-1822)
STUKELEY, William, (1687-1765)
VALENTINI, Michael Bernhard, (1657-1729)
VALLISNERI, Antonio, (1661-1730)
VINCI, Leonardo da, (1452-1519)
WEAVER, Thomas, (1773-1855)
WOODWARD, John, (1665-1728)

1800-1849

ALTH, Alois von, Edler, (1819-1886)
ARGYLL, George Douglas, Duke of, (1823-1900)
BARRATT, Joseph, (1796-1882)
BIEDERMANN-IMHOOF, Adolf W. G., (1829-1900)
BIELZ, Eduard Albert, (1827-1898)
BRAUN, Heinrich Alexander, (1805-1877)
BRODERIP, William John, (-1859)
BRODIE, Peter Bellinger, Reverend, (1815-1897)
BRUMBY, Richard T., (1804-1856)
BUCKLAND, Francis Trevelyan, (1826-1880)
CARUS, Carl Gustav, (1789-1869)
CASH, William, (1843-1914)
CHAMBERS, Robert, (1802-1871)
CHAPUIS, Félicien, (1824-1879)
CIOFALO, Saverio, (1825-1925)
CLARK, William, (1770-1838)
CONDON, Thomas, (1822-1907)
CONYBEARE, John Josiah, Reverend, (1779-1824)
COSTA, Oronzio Gabriele, (1787-1867)
COTTA, Carl Bernhard von, (1808-1879)
DARWIN, Charles Robert, (1809-1882)
DICK, Robert, (1811-1866)
DUNKER, Wilhelm Bernhard Rudolph Hadrian, (1809-1885)
EBRAY, Charles Henri Théophile, (1823-1879)
EIGHTS, James, (1798-1882)
ERMAN, Georg Adolf, (1806-1877)
GAUDRY, Jean Albert, (1827-1908)
GEOFFROY SAINT-HILAIRE, Isidore, (1805-1861)
GOETHE, Johann Wolfgang von, (1749-1832)
GUISCARDI, Guglielmo, (1821-1885)
GÜNTHER, A. F., (?1805-1871)
HALL, Elias, (1764-1853)
HALL, James, (1811-1898)
HANTKEN, Miksa von Prudnik, (1821-1893)
HAWN, Frederick, (1810-1898)
HEYMANN, Hermann, (-1871)

PALAEONTOLOGY (undivided) 1800-1849 Cont'd.

HISLOP, Stephen, Reverend, (1817-1863)
HOWSE, Richard, (1821-1901)
HUGI, Franz Joseph, (1796-1855)
JACKSON, Charles Thomas, (1805-1880)
KARSTEN, Hermann, (1809-1877)
KAYE, Charles Turton, (1812-1846)
KIRBY, William, Reverend, (1759-1851?)
KLIPSTEIN, August Wilhelm, (1801-1894)
KOENEN, Adolf von, (1837-1915)
LECOQ, Henri, (1802-1871)
LEE, John Edward, (1808-1887)
LEUNIS, Johann, (1802-1873)
McDONNELL, James, (?1763-1845)
MENDEL, Johann Gregor, (1822-1884)
MEYN, Ludwig, (1820-1878)
MOESCH, Casimir, (1827-1898)
MORRIS, John, (1810-1886)
MUDGE, Benjamin Franklin, (1817-1879)
NILSSON, Sven, (1787-1883)
ORBIGNY, Alcide Charles Victor Dessalines d', (1802-1857)
OTTO, Ernst von, (1799-1863)
PARROT, Georg Friedrich, (1767-1852)
PIETTE, Louis Edouard Stanislas, (1827-1906)
PUSCH, Georg Gottlieb, (1790-1846)
QUENSTEDT, Friedrich August von, (1809-1889)
RAFINESQUE-SCHMALTZ, Constantine Samuel, (1783-1840)
RIVIÈRE, Auguste, (1809-1877)
RÖMER, Friedrich Adolph, (1809-1869)
RŐNAY, Jácint, Bishop, (1814-1889)
ROSE, Alexander, (1781-1860)
SCHAFHÄUTL, Karl Franz Emil, (1803-1890)
SCHOTT, Heinrich, (-1865)
SMITH, Joshua Toulmin, (1816-1869)
SOWERBY, George Brettingham, II, (1812-1884)
SOWERBY, James, (1757-1822)
SOWERBY, James de Carle, (1787-1871)
SPENCER, Herbert, (1820-1903)
STEENSTRUP, Johannes Japetus Smith, (1813-1897)
STROMBECK, August von, (1808-1900)
SYMONDS, William Samuel, Reverend, (1818-1887)
TOURNAL, Paul, (1805-1872)
VERNEUIL, Philippe Edouard Poulletier de, (1805-1873)
VILANOVA Y PIERA, Juan, (1822-1893)
WALLACE, Alfred Russel, (1823-1913)
WEAVER, Thomas, (1773-1855)
WILTSHIRE, Thomas, Reverend, (1826-1902)
WOOD, John George, Reverend, (1827-1889)
WYMAN, Jeffries, (1814-1874)
ZIPSER, Christian Andreas, (1783-1864)

1850-1899

ALTH, Alois von, Edler, (1819-1886)
AREITIO Y LARRINAGA, Alfonso de, (-1884)
ARGYLL, George Douglas, Duke of, (1823-1900)

PALAEONTOLOGY (undivided) 1850-1899 Cont'd.

ARTHABER, Gustav von, Edler, (1864-1943)
BĄKOWSKI, Józef, (-1888)
BARRATT, Joseph, (1796-1882)
BARRELL, Joseph, (1869-1919)
BASSANI, Francesco, (1853-1916)
BECKER, Ewald, (-1873)
BENECKE, Ernst Wilhelm, (1838-1917)
BIEDERMANN-IMHOOF, Adolf W. G., (1829-1900)
BIELZ, Eduard Albert, (1827-1898)
BLAYAC, Joseph, (1865-1936)
BLEEKRODE, S., (-)
BÖHM, Georg, (1854-1913)
BONARELLI, Guido, (1871-1951)
BOSNIASKI, Sigismund Grzymala, (1831-)
BRANCA, Wilhelm von, (1844-1928)
BRAUN, Heinrich Alexander, (1805-1877)
BRODERIP, William John, (-1859)
BRODIE, Peter Bellinger, Reverend, (1815-1897)
BROWN, Amos Peaslee, (1864-1917)
BRUDER, George, (1856-1916)
BRUMBY, Richard T., (1804-1856)
BUCKLAND, Francis Trevelyan, (1826-1880)
CACCIAMALI, Giambattista, (1856-1934)
CANAVARI, Mario, (1855-1928)
CARAPEZZA, E., (-1915)
CARUS, Carl Gustav, (1789-1869)
CASH, William, (1843-1914)
CERULLI-IRELLI, Serafino, (1873-1946)
CHAMBERS, Robert, (1802-1871)
CHAPUIS, Félicien, (1824-1879)
CHECCHIA-RISPOLI, Giuseppe, (1877-1947)
CIOFALO, Saverio, (1825-1925)
CLERICI, Enrico, (1862-1938)
CLIFFORD, William Hugh, Right Honourable, (1858-1943)
COBBOLD, Edgar Sterling, (1851-1936)
COBÎLCESCU, Grigore, (1826-1892)
CONDON, Thomas, (1822-1907)
COSTA, Oronzio Gabriele, (1787-1867)
COTTA, Carl Bernhard von, (1808-1879)
CROOK, Alja Robinson, (1864-1930)
DAL PIAZ, Giorgio, (1872-1962)
DARWIN, Charles Robert, (1809-1882)
DE ALESSANDRI, Giulio, (-?1920)
DENINGER, Karl, (1878-1917)
DE REGNY, Paolo Vinassa, (1871-1957)
DICK, Robert, (1811-1866)
DÖDERLEIN, Ludwig, (1855-1936)
DOSS, Karl Bruno, (1861-1919)
DUNKER, Wilhelm Bernhard Rudolph Hadrian, (1809-1885)
DYBOWSKI, Wladyslaw, (1838-1910)
EBRAY, Charles Henri Théophile, (1823-1879)
ECK, Heinrich von, (1837-1925)
EIGHTS, James, (1798-1882)
ERMAN, Georg Adolf, (1806-1877)

PALAEONTOLOGY (undivided) 1850-1899 Cont'd.

FABER, Harald Nicolai, (1856-1943)
FALLAUX, Cornelius, (-1885)
FRANKS, Godfrey Firth, (1853-1917)
FRIČ, Antonín, (1832-1913)
FUCINI, Alberto, (1864-1940)
GAUDRY, Jean Albert, (1827-1908)
GEOFFROY SAINT-HILAIRE, Isidore, (1805-1861)
GRINNELL, George Bird, (1849-1938)
GUISCARDI, Guglielmo, (1821-1885)
GÜNTHER, A. F., (?1805-1871)
GÜRICH, Georg, (1859-1938)
HAAS, Hippolyt Julius, (1855-1913)
HABERFELNER, Josef, (1830-1913)
HAGEN, Max, (1831-1893)
HALL, Elias, (1764-1853)
HALL, James, (1811-1898)
HANTKEN, Miksa von Prudnik, (1821-1893)
HAUG, Gustave Émile, (1861-1927)
HAWN, Frederick, (1810-1898)
HENDERSON, Junius, (1865-1937)
HEYMANN, Hermann, (-1871)
HISLOP, Stephen, Reverend, (1817-1863)
HOERNES, Rudolf, (1850-1912)
HOWSE, Richard, (1821-1901)
HUGI, Franz Joseph, (1796-1855)
HYATT, Alpheus, (1838-1902)
JACKSON, Charles Thomas, (1805-1880)
JAMES, Joseph Francis, (1857-1897)
JESPERSEN, Johan Magnus Peder, (1833-1917)
KARSTEN, Hermann, (1809-1877)
KAYE, Charles Turton, (1812-1846)
KERR, John Graham, (1869-1957)
KINKELIN, Georg Friedrich, (1836-1913)
KISSLING, Ernst, (1865-1936)
KLIPSTEIN, August Wilhelm, (1801-1894)
KNOWLTON, Frank Hall, (1860-1926)
KOENEN, Adolf von, (1837-1915)
KOSSMAT, Franz, (1871-1938)
KRAFFT VON DELLMENSINGEN, Albrecht, (1871-1901)
KURCK, Clas, (1849-1937)
LECOQ, Henri, (1802-1871)
LEE, John Edward, (1808-1887)
LENNIER, Gustave, (-1905)
LEUNIS, Johann, (1802-1873)
LINNARSSON, Jonas Gustaf Oscar, (1841-1881)
LUNDGREN, Sven Anders Bernhard, (1843-1897)
MacHENRY, Alexander, (1843-1919)
MALLADA, Lucas, (1841-1921)
MALLING, Carl Peter, (1862-1927)
MATLEY, Charles Alfred, (1866-1947)
McCLUNG, Clarence Erwin, (1870-1946)
MENDEL, Johann Gregor, (1822-1884)
MENIN, Aurelio, Padre, (-)
MERRILL, George Perkins, (1854-1929)

PALAEONTOLOGY (undivided) 1850-1899 Cont'd.

MEUNIER, Étienne Stanislaw, (1843-1925)
MEYN, Ludwig, (1820-1878)
MILLER, Samuel A., (1837-1897)
MOESCH, Casimir, (1827-1898)
MOREY, Frank, (?1858-1925)
MORGAN, Jacques Jean Marie de, (1857-1924)
MORRIS, John, (1810-1886)
MORTENSEN, Theodor, (1868-1952)
MUDGE, Benjamin Franklin, (1817-1879)
NEGRI, Arturo, (-1896)
NERY DA ENCARNAÇAO DELGADO, Joaquim Filipe, (1835-1908)
NIEVO, Ippolito, (1867-1916)
NILSSON, Sven, (1787-1883)
ORBIGNY, Alcide Charles Victor Dessalins d', (1802-1857)
ORCUTT, William Warren, (1869-1942)
OSWALD, Felix, (1866-1958)
OTTMER, Eduard Julius Otto, (1846-1886)
OTTO, Ernst von, (1799-1863)
PAHLEN, Alexis von der Baron, (1850-1925)
PARKER, Thomas Jeffrey, (1850-1897)
PARKINSON, Cyril, (1854?-1919)
PARROT, Georg Friedrich, (1767-1852)
PENHALLOW, David Pearce, (1854-1910)
PIETTE, Louis Edouard Stanislas, (1827-1906)
PRIEST, Simeon, (1876-1954)
QUENSTEDT, Friedrich August von, (1809-1889)
RAMÈS, Jean-Baptiste, (1832-1894)
REDTENBACHER, Anton, (1841-1911)
RIVIÈRE, Auguste, (1809-1877)
RÖMER, Friedrich Adolph, (1809-1869)
RÓNAY, Jacint, Bishop, (1814-1889)
ROSE, Alexander, (1781-1860)
RUTOT, Aime Louis, (1847-1933)
SAYN, Gustave, (1862-1933)
SCALIA, Salvatore, (1874-1923)
SCHAFHÄUTL, Karl Franz Emil, (1803-1890)
SCHOTT, Heinrich, (-1865)
SEEBACH, Karl Albert Ludwig von, (1839-1880)
SEGUENZA, Giuseppe, (1833-1889)
SMITH, Joshua Toulmin, (1816-1869)
SOWERBY, George Brettingham, II, (1812-1884)
SOWERBY, James de Carle, (1787-1871)
SPANDEL, Erich, (1855-1909)
SPENCER, Herbert, (1820-1903)
STACHE, Giudo, (1833-1921)
STAINER, Xavier, (1865-)
STEENSTRUP, Johannes Japetus Smith, (1813-1897)
STROMBECK, August von, (1808-1900)
STRÜBIN, Karl, (1876-1916)
STRUCKMANN, Carl Eberhard Friedrich, (1833-1898)
SYMONDS, William Samuel, Reverend, (1818-1887)
SZAJNOCHA, Wladyslaw, (1857-1928)
TEISSEYRE, Wawrzyniec, (1860-1939)
TELLER, Friedrich, (1852-1913)
TENISON-WOODS, Julian Edmund, Reverend, (1832-1889)

PALAEONTOLOGY (undivided) 1850-1899 Cont'd.

THURMANN, Jules, (1804-1855)
TOMMASI, Annibale, (1858-1921)
TOURNAL, Paul, (1805-1872)
TUCCIMEI, Giuseppe, (1851-1915)
VASSEUR, Gaston Casimir, (1855-1915)
VERNEUIL, Philippe Edouard Poulletier de, (1805-1873)
VIDAL I CARRERAS, Lluis Mariano, (1842-1922)
VILANOVA Y PIERA, Juan, (1822-1893)
WAGA, Antoni Stanislaw Floryan, (-?1890)
WAHNSCHAFFE, Gustav Albert Bruno Felix, (1851-1914)
WALLACE, Alfred Russel, (1823-1913)
WALTHER, Johannes, (1860-1937)
WILCKENS, Otto Rudolf, (1876-1943)
WILLIAMS, Henry Shaler, (1847-1918)
WILTSHIRE, Thomas, Reverend, (1826-1902)
WIŚNIOWSKI, Tadeusz, (1865-1933)
WOOD, John George, Reverend, (1827-1889)
WÜST, Ewald, (1875-1934)
WYMAN, Jeffries, (1814-1874)
YOUNG, John, (1835-1902)
ZIPSER, Christian Andreas, (1783-1864)

1900 to present

ABRARD, René, (1892-1973)
ALIMEN, Henriette, (1900-)
ANIČ, Dragutin, (1898-1961)
ARTHABER, Gustav von, Edler, (1864-1943)
BÁRDHARSON, Gudhmundur G., (1880-1933)
BARRELL, Joseph, (1869-1919)
BASSANI, Francesco, (1853-1916)
BENECKE, Ernst Wilhelm, (1838-1917)
BLAYAC, Joseph, (1865-1936)
BÖHM, Georg, (1854-1913)
BONARELLI, Guido, (1871-1951)
BRANCA, Wilhelm von, (1844-1928)
BRAVINDER, Kenneth Mason, (1907-1946)
BROWN, Amos Peaslee, (1864-1917)
BRUDER, George, (1856-1916)
BÜRGL, Hans, (1907-1966)
CACCIAMALI, Giambattista, (1856-1934)
CANAVARI, Mario, (1855-1928)
CARAPEZZA, E., (-1915)
CASTELL, Cyril Philip, (1907-1972)
CERULLI-IRELLI, Serafino, (1873-1946)
CHECCHIA-RISPOLI, Giuseppe, (1877-1947)
CIOFALO, Saverio, (1825-1925)
CLERICI, Enrico, (1862-1938)
CLIFFORD, William Hugh, Right Honourable, (1858-1943)
CLOUD, Preston Ercelle, Jr., (1912-)
COBBOLD, Edgar Sterling, (1851-1936)
COBÎLCESCU, Grigore, (1826-1892)
CONDON, Thomas, (1822-1907)
COOK, Harold James, (1887-?1962)
CROOK, Alja Robinson, (1864-1930)

PALAEONTOLOGY (undivided) 1900 to present Cont'd.

DAL PIAZ, Giambattista, (1927-)
DAL PIAZ, Giorgio, (1872-1962)
DE ALESSANDRI, Giulio, (-?1920)
DENINGER, Karl, (1878-1917)
DE REGNY, Paolo Vinassa, (1871-1957)
DÖDERLEIN, Ludwig, (1855-1936)
DOSS, Karl Bruno, (1861-1919)
DUDICH, Endre, (1895-1971)
DYBOWSKI, Wladyslaw, (1838-1910)
ECK, Heinrich von, (1837-1925)
EVANS, Emile, (1883-1966)
FABER, Harald Nicolai, (1856-1943)
FABIANI, Ramiro, (1879-1954)
FRANKS, Godfrey Firth, (1853-1917)
FRIČ, Antonín, (1832-1913)
FUCINI, Alberto, (1864-1940)
FURON, Raymond, (1898-)
GAUDRY, Jean Albert, (1827-1908)
GERMANN, John Christian, (1903-1954)
GORIZDRO-KULCZYCKA, Zinaida, (1884-1949)
GRIMM, Kenneth Edgar, (1912-1961)
GRINNELL, George Bird, (1849-1938)
GÜRICH, Georg, (1859-1938)
HAAS, Hippolyt Julius, (1855-1913)
HABERFELNER, Josef, (1830-1913)
HALDANE, John Burdon Sanderson, (1892-1964)
HÄNTZSCHEL, Walter, (1904-1972)
HAUG, Gustave Émile, (1861-1927)
HENDERSON, Junius, (1865-1937)
HIBBERT-WARE, Samuel, (1782-1848)
HOERNES, Rudolf, (1850-1912)
HOFMANN, Elise, (1889-1955)
HOWSE, Richard, (1821-1901)
HUDDLE, John Warfield, (1907-1975)
HUPÉ, Pierre, (1907-)
HUSSEY, Russell Claudius, (1888-1970)
HYATT, Alpheus, (1838-1902)
JESPERSEN, Johan Magnus Peder, (1833-1917)
KAUTSKY, Fritz, (1890-1963)
KELLOW, Moses, (-1943)
KERR, John Graham, (1869-1957)
KING, William Bernard Robinson, (1899-1963)
KINKELIN, Georg Friedrich, (1836-1913)
KISSLING, Ernst, (1865-1936)
KNOWLTON, Frank Hall, (1860-1926)
KOENEN, Adolf von, (1837-1915)
KOSSMAT, Franz, (1871-1938)
KRAFFT VON DELLMENSINGEN, Albrecht, (1871-1901)
KURCK, Clas, (1849-1937)
LECKWIJCK, Willem Peter van, (1902-1975)
LENNIER, Gustave, (-1905)
LOTZE, Franz, (1903-1971)
MacHENRY, Alexander, (1843-1919)
MALLADA, Lucas, (1841-1921)

PALAEONTOLOGY (undivided) 1900 to present Cont'd.

MALLING, Carl Peter, (1862-1927)
MATLEY, Charles Alfred, (1866-1947)
MAXIM, Ion Al., (1902-1968)
McCLUNG, Clarence Erwin, (1870-1946)
McLEARN, Frank Harris, (1885-1964)
MERCIAI, Giuseppe, (1877-1939)
MERRILL, George Perkins, (1854-1929)
MEUNIER, Étienne Stanislas, (1843-1925)
MONGIN, Denise, (-)
MOODIE, Roy Lee, (1880-1934)
MOREY, Frank, (?1858-1925)
MORGAN, Jacques Jean Marie de, (1857-1924)
MORTENSEN, Theodor, (1868-1952)
NERY DA ENCARNAÇAO DELGADO, Joaquim Filipe, (1835-1908)
NICOLAS, Jean Paul, (1921-1967)
NIEVO, Ippolito, (1867-1916)
NOWAK, Jan, (1880-1940)
ORCUTT, William Warren, (1869-1942)
OSWALD, Felix, (1866-1958)
PAECKELMANN, Werner, (1890-1952)
PAHLEN, Alexis von der Baron, (1850-1925)
PANOW, Eugeniusz, (1890-1958)
PARKINSON, Cyril, (1854?-1919)
PENHALLOW, David Pearce, (1854-1910)
PETROCCHI, Carlo, (1898-1956)
PIA, Julius von, (1887-1943)
PIETTE, Louis Edouard Stanislas, (1827-1906)
PRIEST, Simeon, (1876-1954)
RADLEY, Ernest George, (1880-1945)
RAJNATH, Shri, (1904-1972)
REDTENBACHER, Anton, (1841-1911)
ROUNDY, Paul Vere, (1884-1937)
RUTOT, Aimé Louis, (1847-1933)
SAYN, Gustave, (1862-1933)
SCALIA, Salvatore, (1874-1923)
SCHOLZ, Erich, (1884-1914)
SOMMERMEIER, Leopold, (1885-1963)
SPANDEL, Erich, (1855-1909)
SPENCER, Herbert, (1820-1903)
SPENGLER, Erich, (1886-1962)
STACHE, Guido, (1833-1921)
STAESCHE, Karl, (1902-)
STAINER, Xavier, (1865-)
STRÜBIN, Karl, (1876-1916)
SUGDEN, Walter "Bill", (1923-1977)
SZAJNOCHA, Wladyslaw, (1857-1928)
TEISSEYRE, Wawrzyniec, (1860-1939)
TELLER, Friedrich, (1852-1913)
THOMASSON, Harald, (1893-1937)
THURMANN, Jules, (1804-1855)
TIEJE, Arthur Jerrold, (1891-1944)
TOMMASI, Annibale, (1858-1921)
TROLL-OBERGFELL, Oskar von, (1882-1972)
TRUEMAN, Arthur Elijah, Sir, (1894-1956)

PALAEONTOLOGY (undivided) 1900 to present Cont'd.

TUCCIMEI, Giuseppe, (1851-1915)
UNTERMANN, Billie R., (1906-1973)
VASSEUR, Gaston Casimir, (1855-1915)
VIDAL I CARRERAS, Lluis Mariano, (1842-1922)
WAHNSCHAFFE, Gustav Albert Bruno Felix, (1851-1914)
WALLACE, Alfred Russel, (1823-1913)
WALTHER, Johannes, (1860-1937)
WARREN, Percival Sidney, (1890-1970)
WETZEL, Walter, (1887-)
WHITESIDE, Robert Massie, (1895-1936)
WILCKENS, Otto Rudolf, (1876-1943)
WILLIAMS, Henry Shaler, (1847-1918)
WILSON, Vernon, (1907-1971)
WILTSHIRE, Thomas, Reverend, (1826-1902)
WINKLER-HERMADEN, Artur, (1890-1963)
WIŚNIOWSKI, Tadeusz, (1865-1933)
WRIGLEY, Arthur George, (1885-1953)
WÜST, Ewald, (1875-1934)
YOUNG, John, (1835-1902)
ZUFFARDI, Pietro, (1885-1915)

PALAEONTOLOGY <u>See also</u> INVERTEBRATE PALAEONTOLOGY, MICROPALAEONTOLOGY, PALAEOECOLOGY, PALAEOICHNOLOGY, PALYNOLOGY, VERTEBRATE PALAEONTOLOGY

PALYNOLOGY

1800-1849

DEANE, Henry, (1807-1874)
EHRENBERG, Christian Gottfried, (1795-1876)
READE, Joseph Bancroft, Reverend, (1801-1870)
SCHAUROTH, Karl von, Freiherr, (1818-1893)

1850-1899

DEANE, Henry, (1807-1874)
DE GEER, Gerard Jacob, Baron, (1858-1943)
EHRENBERG, Christian Gottfried, (1795-1876)
READE, Joseph Bancroft, Reverend, (1801-1870)
REINSCH, Paul Friedrich, (1836-1914)
SCHAUROTH, Karl von, Freiherr, (1818-1893)
WHITE, Henry Hopley, (-?1876)

1900 to present

BERRY, Edward Willard, (1900-1968)
CHALONER, William G., (-)
CONRAD, Walter, (1888-1943)
DEFLANDRE, Georges Victor, (1897-1973)
DEFLANDRE-RIGAUD, Marthe, (-)
DE GEER, Gerard Jacob, Baron, (1858-1943)
DUBOIS, Georges, (1890-1953)
EISENACK, Alfred, (1891-)
ERDTMAN, Otto Gunnar Elias, (1897-1973)

PALYNOLOGY 1900 to present Cont'd.

FLORSCHÜTZ, Frans, (1887-1965)
GRANLUND, Erik, (1892-1938)
HOFFMEISTER, William S., (-)
HOFMANN, Elise, (1889-1955)
JARÓN, Bronislaw, (1905-1942)
KRÄUSEL, Richard, (1890-1966)
MARSHALL, Alan Edward, (1938-1970)
MENÉNDEZ, Carlos Alberto, (1921-1976)
MOORE, Leslie Rowsell, (1914-)
POTONIÉ, Robert, (1889-1974)
RASHID, Muhammad Abdur, (1938-1969)
REINSCH, Paul Friedrich, (1836-1914)
RUDOLPH, Karl, (1881-1937)
SARJEANT, William Antony Swithin, (1935-)
STARK, Peter, (1888-1932)
THOMSON, Paul William, (1891-1957)
WETZEL, Otto Christian August, (1891-1971)
WILSON, Leonard Richard, (1906-)

PEDOLOGY

1800-1849

DARWIN, Charles Robert, (1809-1882)
SZABÓ, József, (1822-1894)

1850-1899

AGAFONOFF, Valerian K., (1863-1955)
DARWIN, Charles Robert, (1809-1882)
GLINKA, Konstantin Dmitrievich, (1867-1927)
GÜLL, Vilmos, (1876-1909)
HARRISON, John Burchmore, Sir, (1856-1926)
HIBSCH, Josef Emanuel, (1852-1940)
HILGARD, Eugene Woldemar, (1833-1916)
MARBUT, Curtis Fletcher, (1863-1935)
ORTH, Albert, (1835-1915)
SZABÓ, József, (1822-1894)
WEIBULL, Kristian Oskar Mats, (1856-1923)

1900 to present

AGAFONOFF, Valerian K., (1863-1955)
BAREN, F. H. van, (-1975)
BOOBERG, Gunnar, (1892-1944)
CERNESCU, Nicolae C., (1904-1967)
DINES, Henry George, (1891-?1964)
EDELMANN, Cornelis Hendrik, (1901-1964)
EDMONDS, Frederic Harrison, (1898-1965)
ERHART, Henri, (-)
GALPIN, Sidney Longman, (1886-1962)
GLINKA, Konstantin Dmitrievich, (1867-1927)
GÜLL, Vilmos, (1876-1909)
HARRISON, John Burchmore, Sir, (1856-1926)
HIBSCH, Josef Emanuel, (1852-1940)
HILGARD, Eugene Woldemar, (1833-1916)
HULL, Joseph Poyer Deyo, (1889-1967)
IRELAND, Hubert Andrew, (1904-1973)

PEDOLOGY 1900 to present Cont'd.

JEFFRIES, Charles Davis, (1896-1965)
JONES, Brynmor, (1902-1941)
KUBIËNA, Walter L., (1897-1970)
LOZINSKI, Walery, (1880-1944)
MARBUT, Curtis Fletcher, (1863-1935)
McDOUGALL, John Edward, (1914-1964)
MILTHERS, Keld, (1907-1960)
MULLER, Siemon William, (1900-1970)
NØRVANG, Aksel, (1914-1968)
ORTH, Albert, (1835-1915)
RITTENHOUSE, Gordon, (1910-1974)
RUSSOM, Vaughn Walter, (1891-1955)
SCHERF, Emil, (1889-1967)
SMITH, Jeremy, (1929-1959)
SMOLÍK, Ladislav, (1895-1960)
URBÁNEK, Ladislav, (1901-)
WEIBULL, Kristian Oskar Mats, (1856-1923)
ŽEBERA, Karel, (1901-)

PETROLEUM GEOLOGY

1800-1849

NEFF, Peter, (1827-1903)
NICOL, Germain Louis, (1922-1973)
SILLIMAN, Benjamin Jr., (1816-1885)

1850-1899

ADAMS, George Whitney, (1872-1930)
ADDA, Kálmán, (1862-1901)
ANDERSON, Frank Marion, (1863-1945)
ARNOLD, Ralph, (1875-1961)
ASHBURNER, Charles Albert, (1854-1889)
ASHLEY, George Hall, (1866-1951)
BAILEY, Tom Esmond Geoffrey, Captain, (-1919)
BATES, Mowry, (1875-1926)
BEADNELL, Hugh John Llewellyn, Major, (1874-1944)
BEEBY-THOMPSON, Arthur, (1873-1968)
BEEDE, Joshua William, (1871-1940)
CAMERON, A. E., (-)
CHEVALIER, Jerome Archibald, (1877-1940)
CRIDER, Albert Foster, (1873-1945)
CZARNOCKI, Stefan, (1878-1947)
DICKERSON, Roy Ernest, (1877-1944)
DIXON, Ernest Edward Leslie, (1876-1963)
ELDRIDGE, George Homans, (1854-1905)
ERB, Josef Theodor, (1874-1934)
ETHERIDGE, Robert Jr., (1847-1920)
FISHER, Cassius Asa, (1872-1930)
GETZENDANER, Frank Marshall, (1876-1957)
GOUBKIN, Ivan Mikhailovitch, (1871-1939)
GUBKIN, Ivan Mikhailovich, (1874?-)
HAGER, Lee, (1874-1944)
HARRIS, Gilbert Dennison, (1864-1952)
HAWORTH, Erasmus, (1855-1932)
HAYES, Charles Willard, (1858-1916)

PETROLEUM GEOLOGY 1850-1899 Cont'd.

HEATH, Daisy Winifred, (1875-1954)
HENDERSON, James Alexander Leo, (?1874-1940)
HENNEN, Ray Vernon, (1875-1958)
HUTCHINSON, Lon Lewis, (1877-1947)
KARPINSKY, Alexander Petrovich, (1847-1936)
KHARLICHKOV, Konstantin Vasil'evich, (-1921)
KISSLING, Ernst, (1865-1936)
LUPTON, Charles Thomas, (1878-1935)
MACLAREN, James Malcolm, (1873-1935)
MARTIN, George C., (1875-1943)
MATSON, George Charlton, (1873-1940)
McKIM, John Walter, (1877-1927)
MEAD, Roy Gibbons, (1881-1956)
MOLENGRAAFF, Gustaaf Adolf Frederik, (1860-1942)
MOON, Frederick William, (-1925)
MRAZEC, Lodovic, (1867-1944)
MUNN, Malcolm John, (1874-1949)
NAGYSUR, Hugo Böckh von, (1874-1931)
NEFF, Peter, (1827-1903)
NICOL, Germain Louis, (1922-1973)
OCKENDEN, Maurice Albion, (1874-1954)
OHERN, Daniel Webster, (1870-1953)
ORCUTT, William Warren, (1869-1942)
ORDOÑEZ, Ezequiel, (1867-1950)
POLLOK, Rudolf R., (1876-1929)
PORRIO, Cesare, (1865-1940)
REDWOOD, Boverton, Sir, (1846-1919)
RICHARDSON, George Burr, (1872-1949)
SCHLUMBERGER, Conrad, (1878-1936)
SILLIMAN, Benjamin Jr., (1816-1885)
SJÖGREN, Sten Anders Hjalmar, (1856-1922)
SNYDER, John Young, (1872-1939)
SONNEVELD, Jan, (1877-195)
STEWART, Philip Charteris Anstruther, (1874-1923)
SZAJNOCHA, Wladyslaw, (1857-1928)
TAYLOR, Charles Henry, (1877-1964)
TEISSEYRE, Wawrzyniec, (1860-1939)
TOBLER, August, (1872-1929)
TOLWIŃSKI, Konstanty, (1876-1961)
VALERIUS, Claude N., (-)
VALERIUS, Michael Marcus, (1872-1947)
VEATCH, Arthur Clifford, (1878-1938)
WADE, Arthur, (1878-1951)
WALTMAN, William Dewitt, (1875-1955)
WANNER, Johannes, (1878-1957)
WATERSCHOOT VAN DER GRACHT, Willem Anton Josef Maria van, (1873-1943)
WHITE, Israel Charles, (1848-1927)
WOODRUFF, Elmer Grant, (1872-1952)
WOOLNOUGH, Walter George, (1876-1958)
ZUBER, Rudolf, (1858-1920)
ZYNDEL, Fortunat, (1882-1917)

PETROLEUM GEOLOGY Cont'd.

1900 to present

ADAMS, George Whitney, (1872-1930)
ADAMS, John Emery, (1899-1971)
ADAMS, Noah C., (1891-1927)
ADAMS, William Clyde, (1892-1957)
ADKINS, Walter Scott, (1890-1956)
AID, Harry, (1897-1927)
AIMER, James David, (1899-1973)
AINSWORTH, David, (1898-1967)
ALCORN, Avary Hunt, (1902-1967)
ALLEN, Donald Meredith, (1899-1971)
ALLEN, Walter J., (1890-1961)
AL-NAQIB, Khorshid Mohammed, (?1927-1972)
AMBROSE, Arthur Warren, (1889-1952)
AMBROSE, John Willis, (-1974)
ANDERSON, Frank Marion, (1863-1945)
ANDERSON, Robert Lee, (1912-1959)
ANDRAU, Evert Willem Karel, (1900-1951)
ANDREW, Arthur Robert, (1881-1941)
ANSBERRY, Leroy Eugene, (1923-1961)
ARGABRITE, William Greame, (1885-1963)
ARMSTRONG, Harold Kerr, (1893-1962)
ARMSTRONG, William Burrus, (1917-1954)
ARNOLD, Ralph, (1875-1961)
ASHLEY, George Hall, (1866-1951)
ASHMORE, Herman T., (1916-1976)
ASHMORE, Leland Wayne, (1916-1964)
ATHY, Lawrence Ferdinand, (1898-1955)
ATKINSON, William H., (1898-1972)
AUGUR, Irving V., (1887-1930)
AURIN, Fritz Love, (1891-1969)
BAILEY, Ralph Fraser, (1920-1953)
BAILEY, Tom Esmond Geoffrey, Captain, (-1919)
BAIRD, Chester Alvan, (1896-1959)
BALDWIN, William Felbert Jackson, Jr., (1927-1972)
BALE, Hubert Eugene, (1898-1969)
BALLARD, James Logan, (1902-1974)
BALLARD, William Norval, (1906-1976)
BANDY, Mark Chance, (1900-1963)
BANDY, Orville Lee, (1917-1973)
BANKS, Thomas Russell, (1900-1948)
BARBER, Cecil Thomas, (-)
BARD, Richards J., (1900-1977)
BARLOW, Maerlant M. "Pete", (1906-1950)
BARNES, Roy Merrill, (1891-1960)
BARRINGTON, Thomas W., (-)
BARROW, Geoffrey, (1899-1963)
BARROW, Leonidas Theodore "Slim", (1895-1978)
BARTELL, Lawrence Denny, (1902-1955)
BARTON, Donald Clinton, (1889-1939)
BARTRAM, John Greer, (1893-1955)
BASSINGER, William E., (1928-1977)
BATES, Mowry, (1875-1926)
BEACH, Floyd Kellogg, (1884-1964)

PETROLEUM GEOLOGY 1900 to present Cont'd.

BEACH, Hugh Hamilton, (1909-1973)
BEADNELL, Hugh John Llewellyn, Major, (1874-1944)
BECKER, Clyde M., (1882-1938)
BECKER, Robert M., (1919-1975)
BECKER, Ted G., (1921-1975)
BECKMANN, Lawrence John, (1912-1976)
BEEBY-THOMPSON, Arthur, (1873-1968)
BEEDE, Joshua William, (1871-1940)
BEERY, Roy F., Jr., (1917-1964)
BELL, Alfred Hannam, (1895-1977)
BELL, Olin Grogan, (1893-1970)
BELTZ, Edward W., (1891-1970)
BENNETT, Clyde M., (1881-1935)
BENSON, Dale L., (1899-1968)
BENTZ, Alfred, (1897-1964)
BENTZ, Ivan Vincent, (1896-1938)
BERG, Ernest Lyle, (1910-1953)
BERNARD, Hugh Allen, (1915-1975)
BERNOUILLI, Walter, (1885-1946)
BEVIER, George Means, (1888-1972)
BIGGS, Charles L. "Tib", (1919-1971)
BINNEY, Edwin, Jr., (1899-1928)
BIRAUD, François, (1899-1936)
BLANCHARD, William Grant, Jr., (1894-1952)
BLANKENSHIP, Joseph Croxton, (1930-1964)
BLANPIED, Bernerd William, (1897-1975)
BLOOMER, Philip A., Jr., (1922-1967)
BLOW, Walter Harry, (1924-1972)
BLUMENTHAL, Milton Diez, (1922-1964)
BOCKELMAN, Frederick H., Jr., (1923-1968)
BOGARDUS, Robert Stephen "Bud", (1914-1953)
BÖHM, Ferenc, (1881-1940)
BOLYARD, Garrett Lincoln, (1897-1963)
BOND, Lewis Alexander, (1896-1972)
BONTZ, Conrad K., (1894-1927)
BOSS, Richard F., (1923-1976)
BOSWORTH, Thomas Owen, (1882-1928)
BOURCHIER, John Rickman, (1896-1927)
BOWEN, James P., (1889-1960)
BOWLING, Leslie, (1908-1977)
BOYD, Harold Edward, (1887-1952)
BOYLE, Walter J., (1897-1973)
BRADLEY, Everett L., (1897-1969)
BRAMKAMP, Richard Allen, (1910-1958)
BRAMWYCH, Michael, (1911-1974)
BRANSON, Edwin Robert, (1908-1971)
BRANT, Ralph Allen, (1899-1958)
BRAUCHLI, R. W., (1895-1975)
BRAVINDER, Kenneth Mason, (1907-1946)
BREMER, B. E. "Boots", (1904-1974)
BREMNER, Carl St. John, (1895-1944)
BREWER, Charles, Jr., (1900-1944)
BRILL, Virgil August, (1905-1972)
BRINKERHOFF, Ira A., (1904-1969)

PETROLEUM GEOLOGY 1900 to present Cont'd.

BRISCOE, Glenn Oliver, (1898-1974)
BROWN, Alton Roe, (1930-1975)
BROWN, John Albert, (1901-1954)
BROWN, Noel K., Jr., (1925-1968)
BROWN, Otto E., (1896-1973)
BRYAN, Joseph Jefferson, (1905-1951)
BRYAN, Roger Wesley, (1927-1971)
BUJALSKI, Boleslaw, (1888-1945)
BUNTE, Arnold S., (1902-1968)
BURGER, Robert W. "Bush", (1906-1945)
BÜRGL, Hans, (1907-1966)
BURNS, Russell Warren, (1912-1967)
BURRESS, Walter March, (1893-1950)
BURTON, George Edgar, (1880-1953)
BUSK, Henry Gould, (?1891-1956)
BUTCHER, Cary Preston, (1898-1950)
BUTCHER, Walter William, (1916-1966)
BUTTERWORTH, Emerson McMillin, (1894-1961)
BUTTRAM, George Franklin, (1886-1966)
BYBEE, Halbert Pleasant, (1888-1957)
BYRON, Clifford Winkler, (1901-1970)
CALAME, Jean Jacques, (1923-1973)
CALEY, John Fletcher, (1903-1971)
CAMERON, A. E., (-)
CAMP, John F. Jr., (1915-1976)
CAMPBELL, Harry A., (1893-1951)
CAMPBELL, Keir Arthur, (-1955)
CAMPBELL, Robert Burns, (1892-1955)
CANADA, William Richards, (1909-1962)
CANFIELD, Charles Reiter, (1906-1964)
CANNON, Robert Lee, (1899-1961)
CARLSON, Charles Gordon, (1895-1977)
CARLTON, Dave Pierce, (1890-1976)
CARPENTER, Everett, (1884-1968)
CARR, William L., (1919-1975)
CARROLL, Don Llewellyn, (1902-1954)
CARSON, Carlton Mann "Kit", (1898-1972)
CARTER, Frank B., (1898-1968)
CARTER, Thomas Ray, (1927-1959)
CASEBEER, Charles Theodore, (1905-1941)
CASEY, John Hestwood, (1923-1967)
CASSEL, Chester A. "Chet", (1897-1971)
CASTEEL, Sanford Henry, (1898-1952)
CASTER, Ermil Leslie, (1896-1955)
CASTON, John E. F., (1925-1972)
CHARLES, Homer Herbert, (1900-1963)
CHATMAN, Cecil La Mar, (1902-1938)
CHENEY, Charles Albert, (1885-1937)
CHENEY, George Monroe, (1893-1952)
CHEVALIER, Jerome Archibald, (1877-1940)
CHISHOLM, William Francis, (1886-1944)
CHLEBOWSKI, Tadeusz, (1908-1961)
CLAPP, Frederick Gardner, (1879-1944)
CLARK, Chester Charles, (1885-1953)

PETROLEUM GEOLOGY 1900 to present Cont'd.

CLARK, Frank Rinker, (1881-1974)
CLARK, Howard, (1885-1972)
CLARK, Leslie Madison, (1898-1976)
CLARK, Paul Henry, (1909-1951)
CLARKE, Carl W., (1896-1928)
CLIFTON, Roland Leroy, (1884-1962)
CLINE, Lewis Manning, (1909-1971)
CLINKSCALES, Albert Sidney, (1894-1969)
CLOOS, Hans, (1885-1951)
COCHRAN, Kenneth L., (1910-1973)
COFFIN, Reuben Clare, (1886-1972)
COLLINS, Melvin J., (1897-1936)
COLLOM, Roy Edward, (1881-1948)
CONDIT, Daniel Dale, (1886-1955)
CONKLING, Richard A., (1885-1952)
CONKLING, Russell C., (1897-1955)
CONWAY, William Palen "Tony", Jr., (1914-1944)
COOKE, Charles Wythe, (1887-1971)
COOPER, Herschel H., (1894-1954)
COPASS, Jack Malcolm, (1906-1963)
CORTES, Henry Cornelius, (1892-1957)
CORYELL, Horace Noble, (1888-1965)
COSIJN, Adrianus Johannes, (1890-1945)
COTTER, Gerald de Purcell, (1881-1941)
COTTINGHAM, Virgil Elisha, (1891-1958)
COX, Guy Henry, (1881-1922)
COX, William Edgerton, (1912-1957)
CRAIG, William Henry, (1897-1951)
CRANDALL, Richard Robbins, (1902-1949)
CRAWFORD, David Johnston, (1899-1960)
CREWS, George Curtis, (1837-1977)
CRICKMAY, Geoffrey W., (1905-1971)
CRIDER, Albert Foster, (1873-1945)
CROSS, Rodman K., (1911-1970)
CROWN, Walter James, (1901-1951)
CRUM, Harry Edwin, (1892-1955)
CUNNINGHAM, George Magee, (1895-1963)
CUSHMAN, Joseph Augustine, (1881-1949)
CZARNOCKI, Stefan, (1878-1947)
DACEY, Reginald Lynn, (1908-1951)
DAMON, Fred T., (1912-1977)
DANIELS, James Ira, (1896-1967)
DARBY, Raymond Vern, (1896-1953)
DAVIE, Thomas Early, (1906-1959)
DAVILA, Salvador Ortiz, (1902-1931)
DAVIS, Arthur George, (1892-1957)
DAVIS, Ralph Emerson, (1884-1968)
DAVIS, Thornton, (1892-1969)
DAWSON, Joseph Meadows, (1895-1946)
DAWSON, Thomas Albert, (1916-1974)
DECKER, Charles LaVerne, (1898-1955)
DeGOLYER, Everette Lee, (1886-1956)
DE KOCH, William Gaspar, (1898-1954)
DENISON, Albert Rodger, (1897-1962)

PETROLEUM GEOLOGY 1900 to present Cont'd.

DERDEN, Jesse Homer, (1890-1944)
DeRUDDER, Ronald D., (1932-1975)
DEUSSEN, Alexander, (1882-1959)
DEWHURST, Thomas, (1881-1973)
DeWOLF, Frank Walbridge, (1881-1957)
DICKERSON, Roy Ernest, (1877-1944)
DICKSON, Fielding Breeden, (1912-1972)
DILLÉ, Glenn Scott, (1896-1963)
DIXON, Ernest Edward Leslie, (1876-1963)
DIXON, Henry George Dacres, (1887-1947)
DODSON, Floyd Cummings, (1895-1949)
DONAVAN, Jack Hilder, (1921-1953)
DONOGHUE, David Patrick, (1891-1958)
DOORNINCK, Nico H. van, (1893-1956)
DORSEY, George Edwin, (1892-1953)
DOUGLAS, James Mainwaring, (1887-1952)
DOUGLAS, John Gray, (1900-1974)
DOYLE, Robert, (1913-1963)
DRAKE, Cecil, (1899-1949)
DREHER, Otto, (1888-1945)
DRESSER, Myron Amos, (1894-1973)
DUCOTE, James Wardy, (1919-1964)
DURHAM, Charles Albert, (1910-1954)
DUSTON, Arthur W., (1886-1938)
DUTY, Richard Emerson, (1908-1966)
EARL, Eugene Leonard "Red", (1906-1956)
EATON, Arthur, (1886-1923)
EATON, Joseph Edmund, (1895-1958)
EBY, James Brian, (1896-)
EDMONDS, Frederic Harrison, (1898-1965)
EDSON, Dwight J., (1896-1976)
EDSON, Fanny Carter, (1887-1952)
EDWARDS, Everett C., (1897-1972)
EDWARDS, Merwin Guy, (1889-1969)
ELDER, Stanley Gordon, (1904-1957)
ELDRIDGE, George Homans, (1854-1905)
ELLEDGE, Emmett Russell, (1898-1960)
ELLISON, Charles Walker, (1897-1953)
ELLISON, Kenneth Alfred, (1902-1963)
ELLISOR, Alva Christine, (1892-1964)
ELSON, William J., (1895-1965)
ELY, Fred Burchard, (1883-1946)
ERB, Josef Theodor, (1874-1934)
ERDMANN, Charles Edgar, (1897-1971)
ESAREY, Ralph Emerson, (1901-1975)
ESPACH, Ralph Homeward, (1900-1956)
ETHERIDGE, Robert Jr., (1847-1920)
EVANS, Charles Sparling, (1894-1962)
EVANS, Percy, (1892-1974)
EVANS, Vincent, (1907-1951)
EYSSELL, Alfred Raab, (1898-1944)
FAHLE, Paul Bartlett, (1910-1954)
FALCON, Norman Leslie, (-)
FANGUAY, Nolan Anthony, (1916-1942)

PETROLEUM GEOLOGY 1900 to present Cont'd.

FARISH, Linn Markley, (1901-1944)
FARLEE, Richard Wilson, (1913-1961)
FATH, Arthur Earl, (1887-1976)
FAULKNER, Joseph Stanislaus "Ginger", (1928-1963)
FAUSSONE, Luigi, (1880-1925)
FERGUSON, John Logan, (1893-1973)
FERGUSON, Roy Noble, (1883-1961)
FERRERO, Clarence de Labretonne, (1903-1954)
FICHTER, Hans Jakob, (1908-1977)
FINCH, Elmer Harrison, (1883-1953)
FISHER, Cassius Asa, (1872-1930)
FISK, Harold Norman, (1908-1964)
FITTS, John, (1879-1942)
FITZGERALD, James, Jr., (1903-1951)
FITZGERALD, Paul Eugene, (1903-1961)
FITZGERALD, Thomas John, (1903-1973)
FLESCH, Gary Allen, (1949-1976)
FLY, Paul J., (1907-1967)
FOHS, Ferdinand Julius, (1884-1965)
FOLEY, Edward J., (1908-1971)
FOLGER, James Anthony, (1896-1956)
FORTINE, Dwight Hugh, (1904-1949)
FOSSA-MANCINI, Enrico, (1884-1950)
FOSSHAGE, Ernest Willard, (1908-1955)
FOWLER, Charles Worth, Jr., (1899-1948)
FOXWORTH, Wyckliff Riley, (1936-1965)
FRALICH, Charles Eugene, (1902-1964)
FRAZIER, David Edward, (1931-1976)
FREED, Richard, (1908-1967)
FREI, Ernst, (1894-1974)
FREY, Alfred Paul, (1893-1963)
FREY, M. Gordon, (1913-)
FRIEDL, Karl, (1898-1966)
FRIZZELL, Donald Leslie, (1906-1972)
FROST, Jack, (1900-1970)
FUELLHART, Donald Egbert, (1899-1945)
FULTON, Loris John, (1906-1974)
GAITHER, Benjamin Mason, (1919-1960)
GALLAGHER, William G., Jr., (1898-1932)
GALLIE, John Francis, III, (1913-1951)
GARDINER, Thomas Moir, Jr., (1898-1930)
GARDNER, Frank J., (1915-1975)
GARDNER, James Henry, (1883-1964)
GARFIAS, Valentin R., (-)
GARIEPY, Glen Bernard, (1904-1962)
GARRETT, Lovic Pierce, (1880-1943)
GARRETT, Melvin Miller "Dutch", (1894-1969)
GARRETT, Robert Ernest, (1888-1947)
GARRETT, T. R. H., Major, (-1938)
GEIS, Wilfrid Hoy "Bill", (1892-1961)
GEORGE, William Owsley, (1892-1966)
GESTER, George Clark, (1884-1959)
GESTER, Stephen Howell, (1889-1967)
GETZENDANER, Avalette Edmond, (1902-1962)

PETROLEUM GEOLOGY 1900 to present Cont'd.

GETZENDANER, Frank Marshall, (1876-1957)
GIERHART, Guy B. "Joe", (1903-1968)
GINTER, Roy Lamont, (1892-1958)
GOERNER, Frederic Stephen, (1915-1969)
GOLLNICK, Robert Louis, (1918-1963)
GOODELL, Horace N. "Hank", (1915-1971)
GOODSON, Joseph Arthur, (1910-1946)
GORDON, Dugald, (1886-1958)
GORMAN, Joseph M., (1910-1967)
GOUBKIN, Ivan Mikhailovitch, (1871-1939)
GOUDKOFF, Paul Pavel, (1880-1955)
GOYER, Sidney Carroll, (1934-1965)
GRAHAM, Daniel Willard, (1921-1954)
GRAHAM, David H., (1910-1944)
GRAHAM, Joseph John, (1909-1967)
GREEN, Guy Emmett "Squire", (1896-1965)
GREENE, Frank Cook, (1886-1975)
GRIFFITH, Glenn Rodgers Vickery, (1891-1964)
GRIGORAŞ, Nicolae, (1913-1969)
GRIMES, Russell Winters, (1900-1969)
GROSS, Arthur Bunert "Abie", (1904-1965)
GUBKIN, Ivan Mikhailovich, (1874?-)
GUINN, Delmar Richard, (1902-1955)
HAGER, Lee, (1874-1944)
HAKE, Benjamin Franklin, (1895-1973)
HALEPESKA, Dennis Leonard, (1920-1958)
HALL, Elwin B., (1891-1949)
HALSE, George Whitfield, (1885-1968)
HAMILTON, Henry L., (1882-1925)
HAMMILL, Chester A., (1889-1941)
HAMMOND, Weldon Woolf, (1908-1964)
HAMMONDS, G. Scott, (1910-1965)
HANDLEY, Howard Walter, (1905-1940)
HANNA, G. Dallas, (1887-1970)
HANSELL, James Myron, (1902-1970)
HANSON, Ernest Albin, (1899-1973)
HANSON, Wallace W., (1921-1971)
HARDISON, Harvey, (1895-1949)
HARES, Charles Joseph, (1881-1970)
HARPER, O. C., (1897-1973)
HARRELL, Marshall Allen, (1900-1959)
HARRINGTON, George Leavitt, (1883-1972)
HARRINGTON, Horacio Jaime, (1910-1973)
HARRIS, Gilbert Dennison, (1864-1952)
HARRISON, George Haymond, (1903-1949)
HARRISON, John Vernon, (1892-1972)
HARRISON, John Wiley, (1916-1952)
HARRISON, Thomas Samuel, (1881-1964)
HARTWIG, Albert Ernest, Jr., (1922-1966)
HATFIELD, Willis Charles, "Luke", (1898-1970)
HAUPTMAN, Charles R., (1891-1971)
HAWKINS, Glenn Dewayne, (1897-1970)
HAWLEY, James Edwin, (1897-1965)
HAWORTH, Erasmus, (1855-1932)

PETROLEUM GEOLOGY 1900 to present Cont'd.

HAYES, Charles Willard, (1858-1916)
HAYWARD, Raymond Edward, (1932-1964)
HAZELTINE, Roy S., (1890-1972)
HAZZARD, John Charles, (1903-1973)
HAZZARD, Roy Thorpe, (1893-1976)
HEADLEE, Larry A., (1938-1970)
HEALD, Kenneth Conrad, (1888-1971)
HEATH, Daisy Winifred, (1875-1954)
HEDRICK, Oley Foster, (1891-1950)
HEMPEL, Joachim, (1884-1955)
HENDERSON, Harold H. "Heinie", (1905-1949)
HENDERSON, James Alexander Leo, (?1874-1940)
HENDON, Bryan, (1896-1926)
HENDRICKSON, Victor James, (1892-1953)
HENLEY, Arthur Sidney, (1879-1936)
HENNEN, Ray Vernon, (1875-1958)
HENNY, Gerhard, (1890-1944)
HERALD, Frank Edward, (1887-1965)
HEROLD, Stanley Carrollton, (1883-1958)
HEROY, William Bayard, Sr., (1883-1971)
HERRERA, Mariano B., (1918-1951)
HERRING, Donald Grant, Jr., (1918-1962)
HEWITT, Robert Leigh, (1911-1956)
HIGGINS, Daniel Franklin, Jr., (1884-1930)
HIGGS, Morton Thomas, (1896-1956)
HILL, Robert G., (1914-1973)
HILL, Robert Luce, (1937-1964)
HINDES, Earle P., (1888-1962)
HITCHEN, Charles Stansfield, (1904-1971)
HLAUSCHEK, Hans, (1896-1971)
HOGAN, Dana, (1890-1945)
HOLCOMB, Leroy D., (1919-1974)
HOLLISTER, Joseph Steffens, (1903-1953)
HOLMAN, Eugene, (1895-1962)
HOLST PELLEKAAN, William van, (1880-1949)
HONESS, Arthur Pharaoh, (1887-1942)
HONESS, Charles William, (1885-1949)
HOOD, Forrest Wilson, (1897-1963)
HOOKWAY, Lozell Charles, (1905-1938)
HOOVER, James Barr, (1902-1953)
HOPKINS, Edwin Butcher, (1882-1940)
HOPPER, Walter Everett, (1886-1971)
HORTON, Harold Merritt "Wog", (1897-1966)
HOSKINS, Baker, Jr., (1900-1941)
HOTZ, Walter, (1884-1958)
HOURCQ, Victor, (1901-1961)
HOUSTON, Samuel Henry, (1905-1947)
HOWELL, Jesse V., (1891-1971)
HOWELL, William Frederick, (1895-1965)
HOYLE, Charles Robert, Sr., (1891-1954)
HOYT, John Harger, (1928-1970)
HUBBARD, Bela, (1890-1959)
HUBBARD, George Hulme, (1899-1969)
HUBMAN, Carl Wesley, (1905-1947)

PETROLEUM GEOLOGY 1900 to present Cont'd.

HUDDLE, John Warfield, (1907-1975)
HUDSON, Frank Samuel, (1890-1967)
HUFFMAN, Edward Archibald, (1912-1954)
HUGHES, Donald Dudley, (1893-1946)
HUGHES, Richard, (1889-1969)
HUGHES, Richard John, Jr., (1915-1968)
HULL, Joseph Poyer Deyo, (1889-1967)
HUME, George Sherwood, (1893-1965)
HUMMEL, Carl Bernard, (1896-1923)
HUNT, Edwin Harold, (1897-1949)
HUNTLEY, Charles Stirling, (1895-1926)
HURNDALL, John P., (1902-1966)
HUTCHINS, Philip Frank, (1927-1955)
HUTCHINSON, Frederick McAlpine, (1884-1952)
HUTCHINSON, Lon Lewis, (1877-1947)
HUTCHISON, Arthur Gilbertson, (1902-1949)
HUTSON, Ezekiel Burney, (1897-1961)
HYDE, Clarence Edgar, (1894-1954)
ICKES, Eugene Law, (1885-1941)
IDDINGS, Arthur, (1891-1957)
ILLING, Vincent Charles, (1890-1969)
INEICHEN, Giovanni, (-1952)
IRELAND, Hubert Andrew, (1904-1973)
IRWIN, Marvin Leroy "Mel", (1915-1964)
ISENBERGER, Nathan Perry, (1896-1956)
JABLONSKI, Eugene, (1892-1975)
JAMES, John William, (1921-1977)
JAMISON, Claude E., (1879-1939)
JANES, Charles Arthur, (1928-1966)
JANOSKY, Robert A., (1923-1955)
JEFFREYS, Geoffrey, (1885-1953)
JENNINGS, Charles I. "Al", (1894-1968)
JENNINGS, Philip H., (1899-1974)
JENSEN, Christian "Cope", (1905-1961)
JEWETT, John Mark, (1896-1970)
JILLSON, Willard Rouse, (1890-)
JOHNS, Walter Wendell Schell, (1911-1963)
JOHNSON, Clifton Wood, (1905-1954)
JOHNSON, Earle Winston, (1902-1953)
JOHNSON, Joseph Wright, (1917-1957)
JOHNSON, Norman Lee, (1926-1970)
JONES, Daniel J., (1889-1965)
JONES, David Glynn, (1904-1974)
JONES, Ogden Sherman, (1891-1952)
JONES, Verner E., (1905-1977)
JONES, William Francis, (1885-1941)
JORDAN, Louise, (1908-1966)
JUDSON, Sidney Arthur, (1887-1971)
JUNG, Jean, (1896-)
KARPINSKY, Alexander Petrovich, (1847-1936)
KAUENHOWEN, Walter, (1900-1942)
KAY, Fred Hall, (1885-1943)
KEELER, William Weaver, (1897-1943)
KELLY, Donald, (1898-1949)

PETROLEUM GEOLOGY 1900 to present Cont'd.

KEMMERLING, Georges Laure Louis, (1888-1932)
KENNEDY, Luther Eugene, (1891-1966)
KENNEDY, William, (-1926)
KEPPLER, Leo George, (1890-1971)
KERTAI, György, (1912-1968)
KEW, William Stephen Webster, (1890-1961)
KEYES, Wilson, (1901-1936)
KHARLICHKOV, Konstantin Vasil'evich, (-1921)
KIDD, Major Gentry, (1904-1974)
KIMBALL, Kent K., (1892-1953)
KING, Jack Joseph, (1902-1929)
KING, Robert Bruce, (1923-1977)
KINKEL, John Franklin, (1899-1936)
KIRBY, Grady, (1889-1969)
KIRBY, Maurice E. "Bud", (1909-1963)
KISLING, James William, Jr., (1898-1973)
KISSLING, Ernst, (1865-1936)
KITSON, Howard Walde, (1883-1931)
KLEIBER, Karl Ernst, (1912-1972)
KLEINPELL, William Darwin, (1898-1959)
KLINE, Virginia Harriett, (1910-1959)
KLINGER, Edgar Dewitt, (1899-1973)
KLOMPÉ, Theodorus Hendricus Franciscus, (?1904-1963)
KNEBEL, G. Moses, (1899-1974)
KNIGHT, Garold Louis "Jerry", (1894-1968)
KNOWLES, Gerald Elgin, (1923-1976)
KNOX, Thomas Knight, (1896-1949)
KOCH, Harold Kenneth, (1927-1973)
KOZÁRY, Myron Theodore "Mirko", (1918-1966)
KRAJEWSKI, Stanislaw, (1890-1968)
KRALIS, Dan F., Jr., (1919-1962)
KRAMER, William Baltser, III, (1896-1963)
KREIDLER, William Lynn, (1911-1975)
KREPPS, William J., (1924-1976)
KRIDLER, George Matthew, (1889-1953)
KROENLEIN, George Alfred, (1898-1940)
KRYNINE, Paul Dimitri, (1902-1964)
KUBELLA, Karl, (1922-1973)
KUIPER, Willem Nicolaas, (1891-1952)
KURTZ, Robert Gourdin, (1898-1958)
KWOKEK, Stefan, (1923-1970)
LADNER, Albert L., (1905-1975)
LAHEE, Frederic Henry, (1884-1968)
LAKE, Merrill Evans, (1902-1938)
LAMMERS, Edward Chauncey Hinman, (1910-1971)
LANCASTER, L. Kenneth, (1906-1971)
LANDES, Robert William, (1910-1969)
LARSH, Napoleon Bonaparte "Po", (1897-1967)
LARSH, Walter W., (1893-1951)
LATIMER, Frank Parkes, (1903-1928)
LATTA, Albert Lloyd, (1907-1960)
LAW, John Melvin, (1914-1967)
LEACH, Thomas Witt, (1896-1966)
LEATHEROCK, Otto, (1901-1941)

PETROLEUM GEOLOGY 1900 to present Cont'd.

LEDBETTER, William Robert, (1921-1971)
LEDINGHAM, Glen Wallace, (1911-1978)
LEE, Marvin, (1889-1953)
LEES, George Martin, (1898-1955)
LEHNER, Ernst J., (1895-1968)
LEPPER, Gordon Willis, (1889-1950)
LESTER, John Lawrence, (1914-1977)
LEUTENEGGER, Willy Otto, (1903-1972)
LEVORSEN, Arville Irving, (1894-1965)
LEWIS, Francis Leonard, Jr., (1921-1953)
LEYDS, Louis Willem, (1887-1963)
LIDDLE, Ralph Alexander, (1896-1963)
LILIENBORG, Burton Armand, (1900-1949)
LILLIGREN, Johnathan Martin, (1880-1956)
LIND, Charles McClelland, (1911-1957)
LINDEN, Bernard Hugo "Ben" van der, (-1958)
LINEHAN, Charles M., (1913-1973)
LINES, Frederic George, (1921-1963)
LLOYD, Edwin Russell, (1882-1955)
LLOYD, Edwin Russell "Rusty", Jr., (1924-1951)
LOPEZ, Abdon Joseph, (1924-1972)
LÓPEZ VÁZQUEZ, Andrés, (1903-1954)
LOSKAMP, Alvin Powell, (1900-1954)
LOWMAN, Shepard Wetmore, (1899-1968)
LUCKY, Maurice Cecil, (1895-1959)
LUPTON, Benjamin Charles, (1908-1964)
LUPTON, Charles Thomas, (1878-1935)
MACKENZIE, Andrew Nicholls "Kelley", (1888-1951)
MACLAREN, James Malcolm, (1873-1935)
MacNAUGHTON, Lewis Winslow, (1902-1969)
MACOVEI, Gheorghe, (1880-1969)
MACREADY, George Alexander, (1885-1955)
MAEBIUS, Jed B., (1907-1969)
MALOTT, Clyde A., (-1950)
MANN, Robert Scott, (1915-1956)
MARCHESINI, Enrico, (1911-1968)
MARKHAM, Edmund Olis, (1893-1960)
MARR, Francis Alleyne, (1893-1942)
MARSHALL, Alan Edward, (1938-1970)
MARTI, Ernest John, (1913-1964)
MARTIN, Francis Ivan "Shorty", (1896-1952)
MARTIN, George C., (1875-1943)
MARTIN, Lewis, (1920-1973)
MARTYN, Phil F., (1903-1974)
MASON, Shirley Lowell, (1893-1975)
MATSON, George Charlton, (1873-1940)
MAUCINI, Joseph John, (1893-1955)
MAXSON, John Haviland, (1906-1966)
MAY, Arthur Raymond, (1893-1948)
MAYFIELD, William I., (1912-1972)
MAYO, Herbert Theodore, (1889-1965)
McARTHUR, Donald, (1895-1960)
McCAULEY, Dan Wesley, (1916-1967)
McCLAIN, Alan Hartman, (1899-1960)

PETROLEUM GEOLOGY 1900 to present Cont'd.

McCLOY, Walter Lowrie, Jr., (1912-1954)
McCLURE, Drexel Eugene, (1928-1977)
McCOLLOUGH, Edward Lamb, (1931-1955)
McCOY, Alexander Watts, (1889-1944)
McDONALD, Stanley Matthews, (1915-1950)
McDOUGALL, John Edward, (1914-1964)
McFARLAND, Leland R., (1904-1973)
McFARLAND, Russell Scott, (1893-1968)
McFERRON, George Irwin, (1895-1969)
McGEHEE, Julius Rex, (1905-1963)
McGLOTHLIN, William Claude "Big Mack", (1895-1942)
McINTYRE, Paul Joseph, (1893-1963)
McKIM, John Walter, (1877-1927)
McLAREN, Robert Lee, (1901-1953)
McMANNIS, William J., (1924-1973)
McMINN, Paul Meloy, (1919-1974)
McNEELY, Robert Barnes, (1902-1966)
McNEIL, Harold E., (1910-1965)
McNUTT, Vachel Harry, (1888-1936)
MEAD, Roy Gibbons, (1881-1956)
MEHL, Maurice Goldsmith, (1887-1966)
MEKEL, Johannes Antonius Alphonsus, (1891-1942)
MENCHER, Ely, (1913-)
MENKEN, Frederick Andrew, (1905-1965)
MERRITT, John Wesley, (1885-1958)
METCALF, Roy J., (1889-1941)
METRE, Waman Bapu, (1906-1970)
MEYER, Edouard, (1899-1974)
MIGAUX, Léon, (1897-1914)
MILLER, Buford, (1906-1962)
MILLER, Carroll C., (1899-1974)
MILLER, Guy E., "Doc", (1893-1977)
MILLER, John Cummins, (1895-1951)
MILLER, Wendell Zerbe, (1892-1958)
MILLER, Willard L., (1896-1973)
MILLER, William F., (1929-1976)
MILNER, Charles Albert, Jr., (1900-1971)
MITCHELL, James Garland, (1908-1972)
MITCHELL, Kingsley Camden, (1903-1931)
MIX, Charles Alexander, (1898-1954)
MOLENGRAAFF, Gerard Johan Hendrik, (1887-1961)
MOLENGRAAFF, Gustaaf Adolf Frederik, (1860-1942)
MONNETT, Victor Elvert, (1889-1972)
MONTGOMERY, Curtis Hall, (1897-1944)
MOODY, Clarence Lemuel, (1888-1963)
MOODY, Graham Blair, (1889-1975)
MOON, Frederick William, (-1925)
MOORE, Carl Allphin, (1911-1973)
MOORE, Gilbert Parvin, (1898-1970)
MOORE, James Augustus, (1910-1955)
MOORE, Prentiss D., (1895-1955)
MORAN, Robert Breck, (1879-1961)
MORGAN, Frank Albert, (1898-1976)
MORGAN, George Dillon, (1894-1950)

PETROLEUM GEOLOGY 1900 to present Cont'd.

MORLEY, Harold Thompson "Midge", (1897-1976)
MORRILL, Junius Hughes, (1913-1961)
MORRIS, Albert Ferd, (1898-1957)
MORRIS, Harry Newton, Jr., (1920-1952)
MORSE, Paul Franklin, (1897-1929)
MRAZEC, Lodovic, (1867-1944)
MUIR, John Malcolm, (1885-1938)
MUNN, Malcolm John, (1874-1949)
MURCHISON, Eugene Alexander, III, (1931-1953)
MURPHY, James Kenneth, (1900-1955)
MURRAY, George H., Jr., (1928-1971)
MYERS, John Clark, (1894-1970)
MYERS, Julian Q., (1889-1950)
MYERS, Thurman H., (1890-1940)
MYLIUS, Louis Aubrey, (1890-1965)
NAGAPPA, Yedatore, Shri, (1907-1960)
NAGYSUR, Hugo Böckh von, (1874-1931)
NANCE, James W., (1908-1977)
NARAMORE, Chester "Chet", (1879-1961)
NEARHOOD, Clinton Elsworth, (1915-1957)
NEDOM, Henry Arthur, (1895-1940)
NEFF, Peter, (1827-1903)
NEIGHBOURS, Glen Reed, (1917-1961)
NELSON, Richard Newman, (1897-1964)
NETTERSTROM, Paul Wagner, (1914-1960)
NEUMAN, Leonard Joseph, (1900-1953)
NEWBY, Warner Wilson, (1898-1926)
NIGHTINGALE, William Thomas, (1897-1964)
NISBET, John McHenry, (1894-1962)
NOBLE, Earl Bart, (1894-1967)
NOLAN, Edward Dana, (1890-1926)
NOLAN, Philip Edward, (1887-1958)
NOMLAND, Jorgen O., (1885-1943)
NORDEN, John Alexander Erdelyi-Fazekas, (né Erdélyi-Fazekas), (1912-1977)
NORVILLE, Glen Smith, (1899-1963)
NOTESTEIN, Frank B., (1885-1973)
NOWLAN, Harry Hackett, (1888-1958)
NUTTALL, Winfred Laurence Falkiner, (1897-1969)
NYGREN, Walter Eric, (1907-1971))
O'BRIEN, Francis Eugene, (1916-1968)
OCKENDEN, Maurice Albion, (1874-1954)
OFFICER, Herbert George, (1889-1937)
OHERN, Daniel Webster, (1870-1953)
OLCUTT, David Perry, (1896-1958)
OLDS, Thomas Hartman, (1883-1932)
ORCUTT, William Warren, (1869-1942)
ORDOÑEZ, Ezequiel, (1867-1950)
OWEN, William Joseph, (1929-1959)
OWENS, Frith Cravens, (1898-1977)
OWENS, John Edward, (1912-1963)
PACKARD, Henry Jessup, (1883-1930)
PAINE, Paul, (1881-1957)
PAINE, William R., (-)

PETROLEUM GEOLOGY 1900 to present Cont'd.

PALMER, Leland L., (1906-1978)
PALMER, Robert Hastings, (1882-1948)
PANYITY, Louis Samuel, (1890-1943)
PAPP, Simon, (1886-1970)
PARGETER, Reginald Clive, (1919-1972)
PARKER, Ben Hutchinson, (1902-1969)
PARSONS, Ernest, (?1887-1963)
PEABODY, Harlan William, (1881-1960)
PEATTIE, Roderick, (1891-)
PEMBERTON, John Roy "Bill", (1884-1968)
PENNEL, Harry David, (1907-1971)
PENNY, Frederick Willoughby, (1887-1950)
PEPPERBERG, Leon J., (1883-1937)
PERCIVAL, Frederick George, (1887-1973)
PERINI, Vincent Charles "Bud", Jr., (1895-1965)
PERRINE, Irving, (1884-1955)
PETERSEN, Georg, (1898-)
PETERSON, Arthur Frederick, (1911-1967)
PETERSON, Clarence J., (1885-1948)
PETSCH, Arthur Henry, (1895-1969)
PETTY, Dabney E., (1891-1963)
PETTY, Thomas Franklin, (1900-1962)
PHELPS, Frederick George, (1913-1964)
PHILBRICK, Eugene Patrick, (1897-1951)
PHILLIPS, Eldridge Douglas, (1894-1931)
PICARD, Meredith Dane, (1927-)
PICKLESIMER, Harold Greeley, (1916-1952)
PILCHER, Benjamin Luther, Jr., (1906-1946)
PINFOLD, Ernest Sheppard, (1888-1969)
PLESMAN, Albert, (1889-1953)
PLUMMER, Frederick Byron, (1886-1947)
POLLOCK, Charles Murray, (1898-1969)
POLLOK, Rudolf R., (1876-1929)
PORCH, Edwin Leight, Jr., (1889-1948)
PORRIO, Cesare, (1865-1940)
POTEET, James Harold, (1896-1965)
POTTER, Grover Cleveland, (1892-1975)
POTTER, Nelson Bryan, (1897-1965)
POWELL, Ralph Sterling, (1901-1951)
POWERS, Sidney, (1890-1932)
PRATLEY, Henry Hart, (1893-1941)
PRATT, Wallace Everette, (1885-)
PREECE, Rae, (1897-1961)
PRETTYMAN, Thomas Mann, (1888-1940)
PRICE, Arthur Sebert, (1904-1948)
PROUT, Franklin S. "Shorty", (1889-1940)
PUTNAM, L. G., (1898-1936)
RAINWATER, Edward Harriman "Rainey", (1909-1972)
RAMSEY, Thomas Yancey, (1910-1968)
RATH, Charles Martin, (1881-1959)
RAY, Robert Hillyer, (1902-1968)
REA, Henry Carter, (1900-1963)
REDWOOD, Boverton, Sir, (1846-1919)
REED, Eugene C., (1901-1976)

PETROLEUM GEOLOGY 1900 to present Cont'd.

REED, Lyman Coleman, (1900-1970)
REED, Ralph Daniel, (1889-1940)
REESE, Richard Griffith, (1902-1964)
REID, William McCormick, (1911-1943)
REIFF, Allan Eugene, (1908-1933)
REISS, John W., (1920-1943)
RENAUD, Charles Louis, (1898-1964)
RENICK, Brink Coleman, (1897-1971)
REYES, Felix Celso, (1924-1972)
REYNOLDS, Karl William, (1894-1970)
REYNOLDS, Roy A., (1889-1943)
RHOADES, Ralph Omer, (1895-1961)
RHOADES, Roger F., (1905-1972)
RICE, Elmer Michael, (1900-1971)
RICH, John Lyon, (1884-1956)
RICHARDS, James Taylor "Dick", (1892-1953)
RICHARDSON, F. D. S. "Ricardo", (-1966)
RICHARDSON, George Burr, (1872-1949)
RIDGEWAY, Bertrand S., (1899-1942)
RIFE, Byron, (1892-1974)
RIGGS, Robert Jennings, (1888-1965)
RITER, Samuel W., (1896-1941)
RITTENHOUSE, Gordon, (1910-1974)
ROARK, Louis, (1890-1950)
ROBERTS, Marion Stanton, (1909-1963)
ROBERTS, Morgan Evan, (1894-1941)
ROBINSON, Bryan Franklin, (1897-1949)
ROBINSON, Ernest Guy, (1898-1973)
ROGALA, Wojciech, (1884-1947)
ROGATZ, Henry, (1902-1974)
ROGERS, Gaillard Sherburne, (1889-1919)
ROGERS, Jesse Armstead, (1913-1962)
ROMANES, James, (1886-1942)
RONES, Morris, (1922-1965)
ROSEN, George D., (1914-1978)
ROW, Charles Herbert, (1895-1970)
ROWLEY, Alden Bruce, (1888-1956)
RUBEY, William Walden, (1898-1974)
RUDOLPH, Joseph, (1897-1953)
RUKAS, Justin Matthew, (1910-1963)
RUSS, Leon Fair, (1886-1953)
RUSSELL, Philip G., (1891-1971)
RUSSELL, William Low, (1897-1976)
RUSSOM, Vaughn Walter, (1891-1955)
RUTSCH, Rolf F., (1902-1975)
RUTTEN, Martin Gerard, (1910-1970)
RYAN, Russell Fossler, (1897-1935)
RYCHLICKI, Jan, (-1942)
SANDERSON, James Owen Gresham "Pete", (1898-1963)
SANDIDGE, John Roy, (1897-1976)
SANDS, John Melville, (1883-1962)
SANFORD, Wendell Glenn, (1914-1958)
SANSOM, Charles Alan, (1902-1945)
SAPPINGTON, Chester, (1905-1968)

PETROLEUM GEOLOGY 1900 to present Cont'd.

SAUNDERS, Lowell Waller, (1901-1954)
SAWYER, Roger W., (1895-1941)
SCHAUB, Hans Peter, (1912-1965)
SCHILLING, Karl Howard, (1894-1945)
SCHLAIKJER, Erich Maren, (1905-1972)
SCHLUMBERGER, Conrad, (1878-1936)
SCHLUMBERGER, Marcel, (1884-1953)
SCHMIDT, Karl A. "Gus", (1900-1965)
SCHMITTOU, Maurice B., (1899-1929)
SCHNEEGANS, Daniel, (1907-1949)
SCHNEIDER, George William, (1899-1963)
SCHOLL, Guy Joseph, (1900-1968)
SCHOLL, Louis August, Jr., (1887-1953)
SCHROYER, Charles Ross, (1887-1932)
SCHUSTER, Rudolf, (1931-1968)
SCHWABENLAND, James Richard, (1926-1968)
SCHWARZ, Melbert Edgar, (1898-1957)
SCHWENNESEN, Alvin Theodore, (1883-1962)
SCOPES, John Thomas, (1900-)
SCOTT, Horace Lafayette, (1901-1976)
SCOTT, Walter Winthrop, (1893-1939)
SCOTTI, Ariberto, (1885-1925)
SCRUTON, Philip Challacombe, (1922-1962)
SEASHORE, Paul Theodore, (1898-1954)
SEGALL, Julius, (1884-1928)
SENN, Alfred, (1899-1949)
SEVERY, Clarence Luther, (1885-1975)
SHAW, Eugene Wesley, (1881-1935)
SHAW, Everett Simeon, (1885-1955)
SHEARER, Harold Kurtz, (1889-1946)
SHERBORNE, John Elliott, Sr., (1911-1974)
SHERMAN, Roger Hiram, (1904-1954)
SHERRILL, Richard Ellis, (1899-1953)
SHERWOOD, Theodore C., Jr., (1896-1928)
SICKLER, John Miller, Jr., (1891-1970)
SIMMONS, Jesse Elmore, (1907-1941)
SIMMONS, Kenneth Albert, (1906-1971)
SIMPSON, Stephen Milton, Jr., (1929-)
SJÖGREN, Sten Anders Hjalmar, (1856-1922)
SMITH, G. Wendell, (1917-1974)
SMITH, Gerald Julian, (1903-1976)
SMITH, Homer James, (1901-1954)
SMITH, Neal Johnstone, (1912-1971)
SMITH, Robert Hendee "Smokey", (1916-1946)
SNIDER, Luther Crocker, (1882-1947)
SNIVELEY, H. Norman, (1879-1959)
SNYDER, John Young, (1872-1939)
SOMMERMEIER, Leopold, (1885-1963)
SONNEVELD, Jan, (1877-195)
SOUTHER, John Barron, (1894-1964)
SPALDING, George M., (1930-1977)
SPARKS, Dale Darrell Dolsun, (1894-1938)
SPEED, Carleton D., Jr., (1903-1970)
SPOFFORD, Howard Nelson, (1881-1950)

PETROLEUM GEOLOGY 1900 to present Cont'd.

SPRAGUE, Harry Alvah, (1913-1950)
SPRATT, Joseph Grant, (1904-1978)
SPROULE, John Campbell, (1905-1970)
STALDER, Walter "Cap", (1881-1949)
STAMP, Laurence Dudley, Sir, (1898-1966)
STANLEY, M. Wood, (1893-1929)
STEBINGER, Eugene, (1883-1951)
STEELE, Horace C., (1930-1977)
STEHR, Raymond Albert, (1903-1967)
STEIG, Maynard Herman, (1904-1965)
STEINY, Homer Julius, (1892-1974)
STELFOX, Reichie S., (1904-1932)
STEUBING, William C., (1881-1943)
STEVENS, George D., (1898-1965)
STEWART, Irvine E., (1893-1965)
STEWART, Philip Charteris Anstruther, (1874-1923)
STONE, A. John F., (1917-1969)
STONER, Oscar E., Sr., (1892-1965)
STREIFF, Viktor, (1909-1973)
STUDER, Joseph Paul, (1916-1951)
STUTZER, Otto, (1881-1936)
SUFFIELD, Harry F., (1916-1966)
SUGDEN, Walter "Bill", (1923-1977)
SULLINS, M. Joseph, (1930-1972)
SUMAN, John Robert, (1890-1972)
SUTTON, Frederick A., (1894-1950)
SWAIN, James Fulton, (1903-1948)
SWANN, David H., (1915-1968)
SWEENEY, Henry Nunn, (1914-1972)
SWIGER, Rual Bower, (1897-1950)
SZAJNOCHA, Wladyslaw, (1857-1928)
TAGUE, Glenn Charles, (1905-1964)
TALLANT, Roy Leon, (1912-1961)
TANDY, James Hiram "Spot", (1892-1962)
TANNER, Hugh Alexander, (1904-1954)
TASHMAN, Djavad Eyoub, (1893-1956)
TAYLOR, Charles Henry, (1877-1964)
TAYLOR, Garvin Lawrence, (1908-1974)
TAYLOR, James Carlton, (1924-1976)
TAYLOR, W. Harlan, (1908-1972)
TEAS, Livingstone Pierson, (1894-1970)
TEISSEYRE, Wawrzyniec, (1860-1939)
TELEGDI ROTH, Károly, (1886-?1956)
TEMPLETON, Eugene Crittenden, (1889-1932)
TETRICK, Paul Roderick, (1920-1956)
THAEMAR, James T., Sr., (1911-1972)
THALMANN, Hans E., (1899-1975)
THIES, Roland K., (1905-1973)
THOMAS, George Dewey, (1898-1975)
THOMAS, Harry George, (1929-1976)
THOMAS, Horace David "Bill", (1905-1967)
THOMAS, John Elmer, (1892-1949)
THOMAS, Norman Louis, (1897-1970)
THOMPSON, E. G., (1898-1973)

PETROLEUM GEOLOGY 1900 to present Cont'd.

THOMPSON, Jean Cleveland, (1887-1929)
THOMPSON, Robert Raymond, (1893-1954)
THOMPSON, Warren O., (1898-1968)
THORNBURG, Dwight Holbrook "Duke", (1895-1965)
THORNTON, Owen Fuller, (1916-1953)
THORSEN, Carl Peter, (1927-1964)
TIDY, Anthony John Lintott, (1928-1969)
TILLOTSON, Allen Walter, (1905-1964)
TOBLER, August, (1872-1929)
TODD, Thomas W., (1927-1969)
TOLER, Henry N., (1901-1970)
TOLWIŃSKI, Konstanty, (1876-1961)
TOMLINSON, Charles Weldon, (1892-1960)
TOMPKINS, Joseph D., (1905-1974)
TOOMEY, Charles Calvin, (1892-1957)
TOWSE, Donald Frederick, (1924-)
TRASK, Parker Davies, (1899-1961)
TRICE, Edwin Leslie, Jr., (1929-1967)
TROXELL, John N., (1895-1970)
TRÜMPY, Daniel, (1893-1971)
TURNER, Joseph Harold, (1900-1950)
UREN, Lester Charles, (1888-1960)
VALERIUS, Michael Marcus, (1872-1947)
VAN COUVERING, Martin, (1888-1976)
VANDER LECK, Lawrence, (1890-1954)
VEATCH, Arthur Clifford, (1878-1938)
VEIGEL, Edward C., (1920-1949)
VERNON, Jesse, (1894-1972)
VER WIEBE, Walter August, (1887-1961)
VETTERS, Hermann, (1880-1941)
VIENNOT, Pierre, (1891-1931)
WADE, Arthur, (1878-1951)
WAGENER, Charles Henry "Chick", (1897-1950)
WAGNER, Carroll M., (1893-1965)
WAGNER, Cary Richard, Jr., (1917-1962)
WALKER, Lucian Hickman, (1894-1970)
WALLACE, Maurice Hudiburg, (1917-1961)
WALTERS, Mason G., (1903-1967)
WALTMAN, William Dewitt, (1875-1955)
WANGSNESS, Orrin Jerold, (1917-1967)
WANNER, Johannes, (1878-1957)
WAPPLER, Ed Dickinson, (1908-1937)
WARE, John McKee, (1905-1973)
WARE, Thomas Preston, (1909-1972)
WARING, Gerald Ashley, (1883-1971)
WARING, Waldo W., (1903-1955)
WARREN, Percival Sidney, (1890-1970)
WASHBURNE, Chester Wesley, (1883-1971)
WASHINGTON-GRAY, Kenneth, (?1899-1963)
WASSON, Theron, (1887-1970)
WATERMAN, Willis D., (1925-1968)
WATERSCHOOT VAN DER GRACHT, Willem Anton Josef Maria van, (1873-1943)
WDOWIARZ, Jan, (1903-1956)

PETROLEUM GEOLOGY 1900 to present Cont'd.

WEAVER, Paul, (1888-1964)
WEBB, John Benwell, (1901-1975)
WEDEL, Arthur Albert, (1898-1941)
WEEKS, Herbert John, (1897-1960)
WEICHBRODT, Harold Tague, (1921-1966)
WEIGNER, Stanislaw Wiktor, (1886-1935)
WESLEY, George Rutherford, (1905-1965)
WHEALTON, Rowland Gardiner "Peg", (1896-1956)
WHEELER, Orby Clinton, (1898-1970)
WHITAKER, Harvey, (1906-1964)
WHITCOMB, Bruce, (1900-1950)
WHITE, Israel Charles, (1848-1927)
WHITE, Maynard Pressley, (1896-1961)
WHITE, Stanley Battershill, (1899-1966)
WHITEHEAD, Robert Brooks, (1889-1936)
WHITEHEAD, Walter Lucius, (1891-1969)
WHITESIDE, Charles Edwin, (1897-1927)
WHITESIDE, Robert Massie, (1895-1936)
WHITTIER, William Harrison, (1889-1951)
WHITWELL, Earl Van Dorn, (1892-1963)
WIEDENMAYER, Carl, (1897-1951)
WIER, Harvey James, Jr., (1924-1950)
WILBERT, Louis Joseph "Slick", Jr., (1919-1953)
WILDER, Newell Morris "Newt", (1908-1946)
WILHELM, Oscar, (1897-1975)
WILLIAMS, George O. "Skipper", (1897-1967)
WILMOTH, Leslie James, (1881-1963)
WILSEY, James Arthur, Jr., (1921-1949)
WILSON, Bernard Wright, (1922-1976)
WILSON, Harry H., (1902-1929)
WILSON, Joseph Garland, (1902-1955)
WILSON, Malcolm Earl, Sr., (1885-1960)
WILSON, Roy Arthur, (1891-1970)
WILSON, Thomas Carroll, (1906-1969)
WILSON, Walter Byron "Shorty", (1885-1951)
WILSON, William Jay, (1922-1971)
WIMBISH, Forrest E., (1900-1972)
WINCHESTER, Dean Eddy, (1883-1936)
WINES, Donald Bradford, (1902-1958)
WINTER, Claud Victor, Jr., (1925-1968)
WINTERER, Edward Virgil "Tubby", (1897-1944)
WIRZ, Albert Eduard, (1910-1973)
WITZ, Albert Eduard, (1910-1973)
WOLF, Albert G., (1885-1964)
WOOD, Robert Harrison, (1887-1956)
WOOD, Virgil O., (1892-1954)
WOODRUFF, Elmer Grant, (1872-1952)
WOODS, Earl Hazen, (1902-1950)
WOOLNOUGH, Walter George, (1876-1958)
WRATHER, William Embry, (1883-1963)
WRIGHT, Andrew Clemmons, (1893-1968)
WRIGHT, Harry Favill, (1888-1944)
YEAGER, Lloyd Irvin, (1900-1938)
YEE, Bing Q., (1924-1973)

PETROLEUM GEOLOGY 1900 to present Cont'd.

YOUNG, Karl Etienne "Happy", (1903-1936)
YOUNGQUIST, Walter, (-)
YULE, Frank Ingalls, (1926-1961)
ZAVOICO, Basil B., (1899-1975)
ZIMMERMAN, Charles Collingwood, (1905-1961)
ZOLLER, Lawrence J., (1895-1973)
ZUBER, Rudolf, (1858-1920)
ZUBER, Stanislaw, (1893-1947)
ZWIERZYCKI, Józef, (1888-1961)
ZYNDEL, Fortunat, (1882-1917)

Time unknown
ANGELL, Cyrus D., (-)

PETROLOGY (undivided)

Up to 1800
CAROSI, Jan Filip, (1744-1799)
CORDIER, Pierre Louis Antoine, (1777-1861)
HOOKE, Robert, (1635-1703)
KALM, Pehr, (1716-1779)
MACCULLOCH, John, (1773-1835)
MAILLET, Benoît de, (1656-1738)
MERRET, Christopher, (1614-)
NOSSENI, Johann Maria, (1544-1620)
ROUELLE, Guillaume François, (1703-1770)
SOULAVIE, Jean Louis Giraud, Abbé, (1752-1813)
STUKELEY, William, (1687-1765)
THEOPHRASTUS, (371-ca.286 B.C.)
WERNER, Abraham Gottlob, (1749-1817)

1800-1849
BISCHOFF, Karl Gustav, (1792-1870)
CLARKE, William Branwhite, Reverend, (1798-1878)
CORDIER, Pierre Louis Antoine, (1777-1861)
DAUBRÉE, Gabriel Auguste, (1814-1896)
DE LA VALLÉE POUSSIN, Charles Louis Joseph Xavier, (1827-1903)
DELESSE, Achille Ernest Oscar Joseph, (1817-1881)
ESCHER VON DER LINTH, Arnold, (1807-1872)
FOUQUÉ, Ferdinand André, (1828-1904)
HOLL, Harvey Buchanan, (1820-1886)
HORNER, Leonard, (1785-1864)
HUDLESTON, Wilfrid Hudleston, (né Simpson), (1828-1909)
KEILHAU, Baltazar Mathias, (1797-1858)
KENNGOTT, Gustav Adolf, (1818-1897)
LEONHARD, Gustav von, (1816-1878)
MACCULLOCH, John, (1773-1835)
MARTIN, Peter John, (1786?-1860)
MAYER-EYMAR, Karl David Wilhelm, (1826-1907)
NECKER DE SAUSSURE, Louis Albert, (1786-1861)
PARRAN, Alphonse, (1827-1903)
PETTERSEN, Karl, (1826-1890)
PFAFF, Friedrich, (1825-1886)
POŠEPNÝ, František, (1836-1895)
RAUMER, Karl Georg von, (1783-1865)

PETROLOGY (undivided) 1800-1849 Cont'd.

ROTH, Justus Ludwig Adolph, (1818-1892)
SARTORIUS VON WALTERSHAUSEN, Wolfgang, (1809-1876)
SOULAVIE, Jean Louis Giraud, Abbé, (1752-1813)
SYMONDS, William Samuel, Reverend, (1818-1887)
WERNER, Abraham Gottlob, (1749-1817)

1850-1899

ADYE, Ernest Howard, (1857-1918)
AMI, Henry Mark, (1858-1931)
ANDRIAN-WERBURG, Ferdinand von, Freiherr, (1835-1914)
BACKLUND, Helge Götrik, (1878-1958)
BALTZER, Richard Armin, (1842-1913)
BARRELL, Joseph, (1869-1919)
BASCOM, Florence, (1862-1945)
BAYLEY, William Shirley, (1861-1943)
BENNETT, Frederick William, (1860-1931)
BERTRAND, Léon, (1869-1947)
BERWERTH, Friedrich Martin, (1850-1918)
BISCHOFF, Karl Gustav, (1792-1870)
BLAKE, John Frederick, Reverend, (1839-1906)
BLANFORD, William Thomas, (1832-1905)
BLEICHER, Marie-Gustave, (1838-1901)
BONNEY, Thomas George, Reverend, (1833-1923)
BRAUNS, Reinhard Anton, (1861-1937)
BRAVO, José J., (1874-1927)
BRØGGER, Waldemar Christofer, (1851-1940)
BUCCA, Lorenzo, (1857-1930)
BÜCKING, Ferdinand Carl Bertram Hugo, (1851-1932)
BUSZ, Karl Heinrich Emil Georg, (1863-1930)
CALLAWAY, Charles, (1838-1915)
CHRISTA, Emanuel, (1874-1948)
CLARKE, William Branwhite, Reverend, (1798-1878)
CLELAND, Alexander M'Ivor, (1862-)
COHEN, Emil Wilhelm, (1842-1905)
COLE, Grenville Arthur James, (1859-1924)
COLONY, Roy Jed, (1870-1936)
CORDIER, Pierre Louis Antoine, (1777-1861)
CROOK, Alja Robinson, (1864-1930)
CROOK, Thomas, (1876-1937)
CROSBY, William Otis, (1850-1925)
CUSHING, Henry Platt, (1860-1921)
D'ACHIARDI, Giovanni, (1872-1944)
DALTON, William H., (?1848-1929)
DANZIG, Emil, (1855-1943)
DAUBRÉE, Gabriel Auguste, (1814-1896)
DAVISON, Ernest Henry, (1877-1944)
DE LA VALLÉE POUSSIN, Charles Louis Joseph Xavier, (1827-1903)
DELESSE, Achille Ernest Oscar Joseph, (1817-1881)
DEWEY, Henry, (1876-1965)
DUPARC, Louis Claude, (1866-1932)
DWERRYHOUSE, Arthur Richard, (1867-1931)
ESCHER VON DER LINTH, Arnold, (1807-1872)
EVANS, John William, (1857-1930)
FEDEROV, Evgraf Stepanovich, (1853-1919)

PETROLOGY (undivided) 1850-1899 Cont'd.

FENNER, Clarence Norman, (1870-1949)
FLETT, John Smith, Sir, (1869-1947)
FOUQUÉ, Ferdinand André, (1828-1904)
FOX-STRANGWAYS, Charles Edward, (1844-1910)
GAVELIN, Axel Olof, (1875-1947)
GONÇALVES GUIMARÃES, Antonio José, (1875-1919)
HAGUE, Arnold, (1840-1917)
HAMMER, Wilhelm, (1875-1942)
HARKER, Allen, (1848-1894)
HICKS, Henry, (1837-1899)
HILL, James Bastian, (1861-1927)
HINTERLECHNER, Karl, (1874-1932)
HIRSCHWALD, Julius, (1845-1928)
HOBSON, Bernard, (1860-1933)
HOFMANN, Károly, (1839-1891)
HOLL, Harvey Buchanan, (1820-1886)
HOLMQUIST, Per Johan, (1866-1946)
HORNER, Leonard, (1785-1864)
HOVEY, Edmund Otis, (1862-1924)
HUDLESTON, Wilfrid Hudleston, (né Simpson), (1828-1909)
HUGI, Emil, (1873-1937)
HUSSAK, Franz Eugen, (1856-1911)
JOHNSEN, Arrien, (1877-1934)
JOLY, John, (1857-1933)
JUDD, John Wesley, (1840-1916)
KALECSINSZKY, Sandór, (1857-1911)
KATZER, Bedřich, (1861-1925)
KEILHAU, Baltazar Mathias, (1797-1858)
KENNGOTT, Gustav Adolf, (1818-1897)
KHRUSHCHOV, Konstantin Dmitrievich, (? -1912)
KITCHIN, Finlay Lorimer, (1870-1934)
KLVAŇA, Josef, (1857-1919)
KOCH, Antal von Bodrog, (1843-1927)
KOLBECK, Friedrich Ludwig Wilhelm, (1860-1943)
KOTÔ, Bundjirô, (1856-1935)
KRATOCHVÍL, Josef, (1878-1958)
KREUTZ, Felix Franz Xaver, (1844-1910)
KUSAKABE, Shiroka, (1872-1924)
LACROIX, François Antoine Alfred, (1863-1948)
LANE, Alfred Church, (1863-1948)
LASAULX, Arnold von, (1839-1886)
LEHMANN, Johannes Georg, (1851-1925)
LEITZ, Ernst, Senior, (1871-1956)
LENK, Hans, (1863-1938)
LEONHARD, Gustav von, (1816-1878)
LIEBISCH, Theodor, (1852-1922)
LOSSEN, Karl August, (1841-1893)
MALAISE, Constantin H. L., (1834-1916)
MALLARD, François Ernest, (1833-1894)
MANASSE, Ernesto, (1875-1922)
MARTIN, Peter John, (1786?-1860)
MATHEWS, Edward Bennett, (1869-1944)
MATTHEWS, Edward Bennett, (1869-1944)
MAYER-EYMAR, Karl David Wilhelm, (1826-1907)

PETROLOGY (undivided) 1850-1899 Cont'd.

McMAHON, Charles Alexander, Lieutenant-General, (1830-1904)
MICHEL-LÉVY, Albert, (1877-1955)
MICHEL-LÉVY, Auguste, (1844-1911)
MILLER, Willet Green, (1839-1925)
MORGAN, Conwy Lloyd, (?1852-1936)
MRAZEC, Lodovic, (1867-1944)
MUNTEANI-MURGOCI, Gheorghe, (1872-1925)
NECKER DE SAUSSURE, Louis Albert, (1786-1861)
NIKITIN, Vasily Vasiliivich, (1867-1942)
NYHOLM, Emil Teodor, (1869-1908)
PARRAN, Alphonse, (1827-1903)
PETTERSEN, Karl, (1826-1890)
PFAFF, Friedrich, (1825-1886)
POLLARD, William, (1870-1946)
POPOFF, Boris Anfirovich, (1871-1950)
POŠEPNÝ, František, (1836-1895)
PRIOR, George Thurland, (1862-1936)
RAMSAY, Wilhelm, (1865-1928)
RANSOME, Frederick Leslie, (1868-1935)
RATH, Gerhard vom, (1830-1888)
RAUMER, Karl Georg von, (1783-1865)
REPOSSI, Emilio, (1876-1931)
REYER, Eduard, (1849-1914)
RINNE, Friedrich Wilhelm Berthold, (1863-1933)
ROTH, Justus Ludwig Adolph, (1818-1892)
ROTH, Samu, (?1854-?1888)
SARTORIUS VON WALTERSHAUSEN, Wolfgang, (1809-1876)
SCHAFARZIK, Ferenc, (1854-1927)
SCHETELIG, Jakob Grubbe Cock, (1875-1935)
SCOTT, Robert Henry, (1833-1916)
SEYLER, Clarence Arthur, (1866-1959)
SLAVÍK, František, (1876-1957)
STACHE, Guido, (1833-1921)
STEENSTRUP, Knud Johannes Vogelius, (1842-1913)
STEINMANN, Gustav, (1856-1929)
STELZNER, Alfred Wilhelm, (1840-1895)
SYMONDS, William Samuel, Reverend, (1818-1887)
SZÁDECZKY, Gyula, Baron Szádeczky-Kardoss of Szádecsne and Kafdosfalva, (1860-1935)
SZONTAGH, Tamás, (1851-1937)
TERMIER, Pierre, (1859-1930)
TIDDEMAN, Richard Hill, (1842-1917)
TODD, James Edward, (1846-1922)
TOIT, Alexander Logie du, (1878-1948)
TÖRNEBOHM, Alfred Elis, (1838-1911)
TUČAN, Fran, (1878-1954)
TURTON, William Henry, (1862-1942)
TUTCHER, John William, (1858-1951)
UDDEN, Johan August, (1859-1932)
USSING, Niels Viggo, (1864-1911)
VAN HORN, Frank Robertson, (1872-1933)
VOGT, Johan Herman Lie, (1858-1932)
VRBA, Karel, (1845-1922)
WALCOTT, Charles Doolittle, (1850-1927)

PETROLOGY (undivided) 1850-1899 Cont'd.

WALKER, Thomas Leonard, (1867-1942)
WARD, James Clifton, Reverend, (1843-1880)
WATSON, Thomas Leonard, (1871-1924)
WEIBULL, Kristian Oskar Mats, (1856-1923)
WEYBERG, Zygmunt, (1872-1944)
WHITE, Theodore Greely, (1872-1901)
WIIK, Fredrik Johan, (1839-1909)
WOLFF, John Elict, (1857-1940)
WOODWARD, Horace Bolingbroke, (1848-1914)
WRIGHT, Frederick Eugene, (1877-1953)
WÜLFING, Ernst Anton, (1860-1930)

1900 to present

ADYE, Ernest Howard, (1857-1918)
AHRENS, Louis Herman, (1918-)
ALDERMAN, Arthur Richard, (1901-)
ALOISI, Piero, (1881-1938)
AMBROŽ, Vojtěch, (1905-1962)
AMI, Henry Mark, (1858-1931)
ANDERSEN, Olaf, (1884-1941)
ANDERSON, George Harold, (1893-1956)
ANDRIAN-WERBURG, Ferdinand von, Freiherr, (1835-1914)
BACKLUND, Helge Götrik, (1878-1958)
BAJPAI, Mahesh Prasad, (1907-1938)
BALTZER, Richard Armin, (1842-1913)
BÁNYAI, János, (1886-1971)
BARNES, Jonathan, (-1928)
BARRABÉ, Louis, (1895-1961)
BARRELL, Joseph, (1869-1919)
BASCOM, Florence, (1862-1945)
BAYLEY, William Shirley, (1861-1943)
BEDER, Robert, (1885-1930)
BENNETT, Frederick William, (1860-1931)
BERTRAND, Léon, (1869-1947)
BERWERTH, Friedrich Martin, (1850-1918)
BETECHTIN, Anatolii Georgievich, (1897-1962)
BIANCHI, Angelo, (1892-1970)
BLAKE, John Frederick, Reverend, (1839-1906)
BLANFORD, William Thomas, (1832-1905)
BLEICHER, Marie-Gustave, (1838-1901)
BOLDYREV, Anatolii Kapitonovich, (1883-1946)
BONNEY, Thomas George, Reverend, (1833-1923)
BOSWELL, Percy George Hamnall, (1886-1960)
BOUČEK, Bedřich, (1904-1975)
BRAUNS, Reinhard Anton, (1861-1937)
BRAVO, José J., (1874-1927)
BRØGGER, Waldemar Christofer, (1851-1940)
BRUET, Edmond, (1890-)
BUCCA, Lorenzo, (1857-1930)
BÜCKING, Ferdinand Carl Bertram Hugo, (1851-1932)
BURFOOT, James Dabney "Dan", Jr., (1896-1966)
BUSZ, Karl Heinrich Emil Georg, (1863-1930)
CADY, Gilbert Haven, (1882-1970)
CALLAWAY, Charles, (1838-1915)

PETROLOGY (undivided) 1900 to present Cont'd.

ČECHOVIČ, Vsevolod, (1900-1961)
CHRISTA, Emanuel, (1874-1948)
CLARK, Robert Watson, (1884-1948)
CLOOS, Ernst, (1898-1974)
COHEN, Emil Wilhelm, (1842-1905)
COLE, Grenville Arthur James, (1859-1924)
COLONY, Roy Jed, (1870-1936)
CONNOLLY, Joseph Peter, (1890-1947)
CROOK, Alja Robinson, (1864-1930)
CROOK, Thomas, (1876-1937)
CROSBY, William Otis, (1850-1925)
CUSHING, Henry Platt, (1860-1921)
D'ACHIARDI, Giovanni, (1872-1944)
DALTON, William H., (?1848-1929)
DANZIG, Emil, (1855-1943)
DAVIDSON, Charles Findlay, (1911-1967)
DAVISON, Ernest Henry, (1877-1944)
DE LA VALLÉE POUSSIN, Charles Louis Joseph Xavier, (1827-1903)
DEMAY, André, (1890-1964)
DESIO, Ardito, (1897-)
DEUBEL, Fritz, (1898-1966)
DEWEY, Henry, (1876-1965)
DITTLER, Emil, (1882-1945)
DUPARC, Louis Claude, (1866-1932)
DUPARQUE, André, (1892-1960)
DWERRYHOUSE, Arthur Richard, (1867-1931)
ESKOLA, Pentti Eelis, (1883-1964)
EVANS, Emile, (1883-1966)
EVANS, John William, (1857-1930)
EVANS, Percy, (1892-1974)
FALINI, Filippo, (1917-1963)
FEDEROV, Evgraf Stepanovich, (1853-1919)
FENNER, Clarence Norman, (1870-1949)
FERMOR, Lewis Leigh, Sir, (1880-1954)
FLETT, John Smith, Sir, (1869-1947)
FÖLDVÁRI, Aladár, (1906-1973)
FOUQUÉ, Ferdinand André, (1828-1904)
FOX-STRANGWAYS, Charles Edward, (1844-1910)
FRITSCH, Wolfgang, (1928-1970)
GARTNER, Otokar, (1889-1961)
GAVELIN, Axel Olof, (1875-1947)
GILBERT, Charles J., (-?1934)
GOLDMAN, Marcus Isaac, (1881-1965)
GOLDSCHMIDT, Victor Moritz, (1888-1947)
GONÇALVES GUIMARÃES, Antonio José, (1875-1919)
GRAHAM, William Armstrong Patterson, (1899-1934)
GRANTHAM, Donald Robert, (1896-1969)
GRČEV, Kočo, (1929-1963)
GREGORY, Hubert Harold, (-?1950)
GRENGG, Roman, (1884-1972)
HAGUE, Arnold, (1840-1917)
HAMMER, Wilhelm, (1875-1942)
HAUSER, Alois, (1899-1955)
HEIDE, Friedrich, (1891-)

PETROLOGY (undivided) 1900 to present Cont'd.

HERRMANN, Margit, (-?1957)
HESS, Harry Hammond, (1906-1969)
HILL, James Bastian, (1861-1927)
HIMMELBAUER, Alfred, (1884-1943)
HINTERLECHNER, Karl, (1874-1932)
HIRSCHWALD, Julius, (1845-1928)
HOBSON, Bernard, (1860-1933)
HÖGBOM, Alvar, (1894-1939)
HOLMQUIST, Per Johan, (1866-1946)
HOVEY, Edmund Otis, (1862-1924)
HUDDLE, John Warfield, (1907-1975)
HUDLESTON, Wilfrid Hudleston, (né Simpson), (1828-1909)
HUGI, Emil, (1873-1937)
HUSSAK, Franz Eugen, (1856-1911)
HUTTENLOCHER, Heinrich, (1890-1954)
HUTTON, Colin Osborne, (1910-1971)
JÉRÉMINE, Elisabeth, (née Tschernaieff), (1879-1964)
JESUS, Amílcar Mário de, (1895-1960)
JEWETT, John Mark, (1896-1970)
JOHANSSON, Harald Elias, (1880-1931)
JOHNSEN, Arrien, (1877-1934)
JOHNSON, Clifton Wood, (1905-1954)
JOLY, John, (1857-1933)
JUDD, John Wesley, (1840-1916)
KALÁŠEK, Jan, (1903-)
KALECSINSZKY, Sandór, (1857-1911)
KARL, Franz, (1918-1972)
KATZER, Bedřich, (1861-1925)
KHRUSHCHOV, Konstantin Dmitrievich, (? -1912)
KITCHIN, Finlay Lorimer, (1870-1934)
KIVI, Wilho John, (1918-1950)
KLEIN, A. Albert, (1889-1953)
KLVAŇA, Josef, (1857-1919)
KOCH, Antal von Bodrog, (1843-1927)
KÖHLER, Alexander, (1893-1955)
KOLBECK, Friedrich Ludwig Wilhelm, (1860-1943)
KOLDERUP, Niels-Henrik, (1898-1971)
KOTÔ, Bundjirô, (1856-1935)
KRATOCHVÍL, Josef, (1878-1958)
KREUTZ, Felix Franz Xaver, (1844-1910)
KREUTZ, Stefan, (1883-1941)
KRISHNAN, Maharajapuram Sitaram, Padma Bhusan, (1898-1970)
KROKSTRÖM, Torsten Hugo, (1905-1950)
KRUTA, Tomáš, (1906-)
KUSAKABE, Shiroka, (1872-1924)
KUŹNIAR, Czeslaw Faustyn, (1889-1955)
LACROIX, François Antoine Alfred, (1863-1948)
LANE, Alfred Church, (1863-1948)
LAPPARENT, Jacques de, (-?1949)
LEECH, John George Chambré, (1890-1966)
LEHMANN, Johannes Georg, (1851-1925)
LEITCH, Patrick Arthur, (-1943)
LEITZ, Ernst, Senior, (1871-1956)
LELUBRE, Maurice, (1916-)

PETROLOGY (undivided) 1900 to present Cont'd.

LENK, Hans, (1863-1938)
LIEBISCH, Theodor, (1852-1922)
LONSDALE, John Tipton, (1895-1960)
MALAISE, Constantin H. L., (1834-1916)
MANASSE, Ernesto, (1875-1922)
MARMO, Vladi, (1914-1969)
MATHEWS, Edward Bennett, (1869-1944)
MATTHEWS, Edward Bennett, (1869-1944)
MAURITZ, Béla, (1881-1971)
MAYER-EYMAR, Karl David Wilhelm, (1826-1907)
McMAHON, Charles Alexander, Lieutenant-General, (1830-1904)
MICHEL-LÉVY, Albert, (1877-1955)
MICHEL-LÉVY, Auguste, (1844-1911)
MILES, Keith Rodney, (1915-1972)
MILLER, Don John, (1919-1961)
MILLER, Willet Green, (1839-1925)
MOORE, Edward William James, (1887-1970)
MORGAN, Conwy Lloyd, (?1852-1936)
MOSELEY, John Reed, (1900-1959)
MRAZEC, Lodovic, (1867-1944)
MUNCK, Victor C. E. A. de, (1920-1956)
MUNTEANI-MURGOCI, Gheorghe, (1872-1925)
NECHAY, Wiktor, (1895-1940)
NETOLICKÝ, Jan, (1934-1963)
NIGGLI, Paul, (1888-1953)
NIKITIN, Vasily Vasiliivich, (1867-1942)
NYHOLM, Emil Teodor, (1869-1908)
OSTADAL, Rudolf, (1897-1970)
PANTÓ, Gábor, (1917-1972)
PARRAN, Alphonse, (1827-1903)
PATTON, Leroy Thompson, (1880-1957)
PIA, Julius von, (1887-1943)
POLLARD, William, (1870-1946)
POPOFF, Boris Anfirovich, (1871-1950)
POTONIÉ, Robert, (1889-1974)
PRADÁČOVÁ, Marcella, (1912-1972)
PRIOR, George Thurland, (1862-1936)
RADLEY, Ernest George, (1880-1945)
RAMA RAO, Bellu, (1890-1970)
RAMA RAO, Lakshmeswar, (1896-1974)
RAMSAY, Wilhelm, (1865-1928)
RANSOME, Frederick Leslie, (1868-1935)
RATH, Charles Martin, (1881-1959)
REICHERT, Robert, (1901-1937)
REPOSSI, Emilio, (1876-1931)
REYER, Eduard, (1849-1914)
RHODES, James Eric Wynfield, (?1888-1976)
RICHARDSON, William Alfred, (1887-?1964)
RINNE, Friedrich Wilhelm Berthold, (1863-1933)
ROGERS, Gaillard Sherburne, (1889-1919)
ROSENHOLTZ, Joseph Leon, (1899-1963)
ROUBAULT, Marcel Edouard, (1905-1974)
ROZEN, Zygmunt, (-1936)
RYDZEWSKI, Bronislaw, (1884-1945)

PETROLOGY (undivided) 1900 to present Cont'd.

SAHA, Baidya Nath, Babu, (-1929)
SCHAFARZIK, Ferenc, (1854-1927)
SCHETELIG, Jakob Grubbe Cock, (1875-1935)
SCHEUMANN, Karl Hermann, (1881-1964)
SCHNEIDERHÖHN, Hans, (1887-1962)
SCOTT, Robert Henry, (1833-1916)
SEYLER, Clarence Arthur, (1866-1959)
SHANNON, William George St. John, (1889-1946)
SHROCK, Robert Rakes, (1904-)
SLAVÍK, František, (1876-1957)
SMITH, James Robert, (1925-1972)
SMOLARSKA, Irena, (1928-1972)
SMYTH, Louis Bouvier, (1933-?1953)
STACHE, Guido, (1833-1921)
STEENSTRUP, Knud Johannes Vogelius, (1842-1913)
STEINMANN, Gustav, (1856-1929)
STEJSKAL, Jan, (1902-1972)
STEWART, Duncan, (1905-1969)
STILLE, Hans, (1876-1966)
SUBRAMANIAM, Anantharama Parameshwara, (1921-1973)
SZÁDECZKY, Gyula, Baron Szádeczky-Kardoss of Szádecsne and Kafdosfalva, (1860-1935)
SZONTAGH, Tamás, (1851-1937)
TER MEULEN, J. A. C., (1902-1960)
TERMIER, Pierre, (1859-1930)
THOMSON, Paul William, (1891-1957)
TIDDEMAN, Richard Hill, (1842-1917)
TIPPER, George Howlett, (1881-1947)
TODD, James Edward, (1846-1922)
TOIT, Alexander Logie du, (1878-1948)
TOKARSKI, Julian, (1883-1961)
TÖRNEBOHM, Alfred Elis, (1838-1911)
TÖRÖK, Zoltán, (1893-1963)
TORRE DE ASSUNÇÃO, Carlos Fernando, (1901-)
TRÖGER, Walter Ehrenreich, (1901-1963)
TUČAN, Fran, (1878-1954)
TUČEK, Karel, (1906-)
TURTON, William Henry, (1862-1942)
TUTCHER, John William, (1858-1951)
UDDEN, Johan August, (1859-1932)
URBAIN, Pierre, (1895-1968)
USSING, Niels Viggo, (1864-1911)
VAN HORN, Frank Robertson, (1872-1933)
VENDEL, Aladár, (1886-1971)
VINCIENNE, Henri, (1898-1965)
VLASOV, Kuz'ma Alekseevich, (1905-1964)
VOGT, Johan Herman Lie, (1858-1932)
VRBA, Karel, (1845-1922)
WALCOTT, Charles Doolittle, (1850-1927)
WALKER, Frederick, (1898-1968)
WALKER, Thomas Leonard, (1867-1942)
WATSON, Thomas Leonard, (1871-1924)
WAWRYK, Wlodzimierz, (1902-1963)
WEIBULL, Kristian Oskar Mats, (1856-1923)

PETROLOGY (undivided) 1900 to present Cont'd.

WEYBERG, Zygmunt, (1872-1944)
WHITE, Theodore Greely, (1872-1901)
WIIK, Fredrik Johan, (1839-1909)
WILCOCKSON, William Howson, (1891-1976)
WINKLER-HERMADEN, Artur, (1890-1963)
WOLFF, John Elict, (1857-1940)
WOODWARD, Horace Bolingbroke, (1848-1914)
WORDIE, James Mann, Sir, (?1890-1962)
WRIGHT, Frederick Eugene, (1877-1953)
WÜLFING, Ernst Anton, (1860-1930)
WUNDERLICH, Hans-Georg, (1928-1974)
YEMTCHUKHNIKOV, G. A., (1885-1957)

Time unknown
MALECKI, J., (-)

PETROLOGY *See also* IGNEOUS PETROLOGY, METAMORPHIC PETROLOGY, METEORITICS, SEDIMENTOLOGY

PHOTOGEOLOGY
1800-1849
NEFF, Peter, (1827-1903)

1850-1899
ARMSTRONG, Henry Edward, (1848-1937)
ASHCROFT, Frederick Noel, (1878-1949)
DE GEER, Gerard Jacob, Baron, (1858-1943)
HELBLING, Robert, (1874-1954)
HELBRONNER, Paul, (1871-1938)
KJELLSTRÖM, Carl Johan Otto, (1855-1913)
NEFF, Peter, (1827-1903)
TUTCHER, John William, (1858-1951)

1900 to present
AIERS, Arthur Joseph, (1896-1965)
ANDERSON, Robert Lee, (1912-1959)
ANSBERRY, Leroy Eugene, (1923-1961)
ARMSTRONG, Henry Edward, (1848-1937)
ASHCROFT, Frederick Noel, (1878-1949)
BLOODWORTH, Bill Lloyd, (1916-1971)
DE GEER, Gerard Jacob, Baron, (1858-1943)
EARDLEY, Armand John, (1901-1972)
FINSTERWALDER, Richard, (1899-1963)
GUZIK, Kazimierz, (1911-1970)
HELBLING, Robert, (1874-1954)
HELBRONNER, Paul, (1871-1938)
KJELLSTRÖM, Carl Johan Otto, (1855-1913)
LAMSENS, O. Robert, (1922-1971)
MARCHESINI, Enrico, (1911-1968)
MELTON, Frank A., (-)
NEFF, Peter, (1827-1903)
RADLEY, Ernest George, (1880-1945)
REA, Henry Carter, (1900-1963)
READER, Thomas William, (-?1923)

PHOTOGEOLOGY 1900 to present Cont'd.

RICHARDS, George L., Jr., (1908-1944)
SCHERMERHORN, W., (1894-)
SHIPEK, Carl Joseph, (1917-1969)
TUTCHER, John William, (1858-1951)

PROSPECTING

Up to 1800

NISSER, Pedro, (1799-1878) [PROSPECTORS, DIVINERS AND MINING ENGINEERS]

1800-1849

NISSER, Pedro, (1799-1878) [PROSPECTORS, DIVINERS AND MINING ENGINEERS]

1850-1899

BOND, Marshall, (1867-1941) [PROSPECTORS, DIVINERS AND MINING ENGINEERS]
GARDINER, Howard Calhoun, (1826-1917) [PROSPECTORS, DIVINERS AND MINING ENGINEERS]
HANNAN, Patrick, (1842-1925) [PROSPECTORS, DIVINERS AND MINING ENGINEERS]
JEBB, John Gladwyn, (1841-1893) [PROSPECTORS, DIVINERS AND MINING ENGINEERS]
MACDONALD, Alexander, (-) [PROSPECTORS, DIVINERS AND MINING ENGINEERS]
MITCHELL, George M., (-) [PROSPECTORS, DIVINERS AND MINING ENGINEERS]
PEDRICK, Howard A. "Dad", (1863-?1930) [PROSPECTORS, DIVINERS AND MINING ENGINEERS]
REINHART, Herman Francis, (1832-1889) [PROSPECTORS, DIVINERS AND MINING ENGINEERS]
TRELAWNEY-ANSELL, E. C., (-) [PROSPECTORS, DIVINERS AND MINING ENGINEERS]
VANDERLIP, Washington B., (-) [PROSPECTORS, DIVINERS AND MINING ENGINEERS]
WAY, Herbert W. L., (-) [PROSPECTORS, DIVINERS AND MINING ENGINEERS]

1900 to present

ALBERT, Herman W., (1882-) [PROSPECTORS, DIVINERS AND MINING ENGINEERS]
BEAUCHEMIN, Pierre, (-) [PROSPECTORS, DIVINERS AND MINING ENGINEERS]
BERRINGTON, Roger, (-) [PROSPECTORS, DIVINERS AND MINING ENGINEERS]
BOND, Marshall, (1867-1941) [PROSPECTORS, DIVINERS AND MINING ENGINEERS]
BOYLEN, Matthew James, (1911?-) [PROSPECTORS, DIVINERS AND MINING ENGINEERS]
CLUNE, Frank, (-) [PROSPECTORS, DIVINERS AND MINING ENGINEERS]
CONNELL, Frederick M., (1885-) [PROSPECTORS, DIVINERS AND MINING ENGINEERS]

PROSPECTING 1900 to present Cont'd.

FAIRBRIDGE, Kingsley, (1885-1924) [PROSPECTORS, DIVINERS AND MINING ENGINEERS]
FRASER, John A., (-) [PROSPECTORS, DIVINERS AND MINING ENGINEERS]
GARDINER, Howard Calhoun, (1826-1917) [PROSPECTORS, DIVINERS AND MINING ENGINEERS]
HANNAN, Patrick, (1842-1925) [PROSPECTORS, DIVINERS AND MINING ENGINEERS]
HORNBY, John, (1880-1927) [PROSPECTORS, DIVINERS AND MINING ENGINEERS]
HUESTIS, Herman H., (1907-) [PROSPECTORS, DIVINERS AND MINING ENGINEERS]
JOWSEY, Robert J., (1905-) [PROSPECTORS, DIVINERS AND MINING ENGINEERS]
LaBINE, Charles L., (1888-) [PROSPECTORS, DIVINERS AND MINING ENGINEERS]
LABINE, Gilbert Adelore, (1890-) [PROSPECTORS, DIVINERS AND MINING ENGINEERS]
LASSETER, Lewis Harold Bell "Das", (1883-1930) [PROSPECTORS, DIVINERS AND MINING ENGINEERS]
LEAHY, Michael, (-) [PROSPECTORS, DIVINERS AND MINING ENGINEERS]
MADSEN, Marius Kristian, (1899-) [PROSPECTORS, DIVINERS AND MINING ENGINEERS]
MASSON, Charles E., (-) [PROSPECTORS, DIVINERS AND MINING ENGINEERS]
MITCHELL, George M., (-) [PROSPECTORS, DIVINERS AND MINING ENGINEERS]
MOSHER, Alex, (1900-) [PROSPECTORS, DIVINERS AND MINING ENGINEERS]
MOSHER, Murdock C., (-) [PROSPECTORS, DIVINERS AND MINING ENGINEERS]
MUNSTERHJELM, Erik, (-) [PROSPECTORS, DIVINERS AND MINING ENGINEERS]
PEDRICK, Howard A. "Dad", (1863-?1930) [PROSPECTORS, DIVINERS AND MINING ENGINEERS]
RAINIER, Peter W., (-) [PROSPECTORS, DIVINERS AND MINING ENGINEERS]
RICHARDSON, Gwen, (-) [PROSPECTORS, DIVINERS AND MINING ENGINEERS]
ROBERTSON, Andrew, (1903?-) [PROSPECTORS, DIVINERS AND MINING ENGINEERS]
RUTZEBECK, Hjalmar, (-) [PROSPECTORS, DIVINERS AND MINING ENGINEERS]
SPRINGER, Karl J., (-) [PROSPECTORS, DIVINERS AND MINING ENGINEERS]
TRELAWNEY-ANSELL, E. C., (-) [PROSPECTORS, DIVINERS AND MINING ENGINEERS]
TREMLETT, Rex, (-) [PROSPECTORS, DIVINERS AND MINING ENGINEERS]
VANDERLIP, Washington B., (-) [PROSPECTORS, DIVINERS AND MINING ENGINEERS]
WATTS, Murray Edmund, (1909-) [PROSPECTORS, DIVINERS AND MINING ENGINEERS]

QUATERNARY GEOLOGY

Up to 1800

GYLLENHAAL, Johan Gabriel, (1750-1788)
MACKENZIE, Alexander, Sir, (ca.1767-1830)
OWEN, George, (1552-1613)

1800-1849

AGASSIZ, Louis Jean Rodolphe, (1807-1873)
BAYFIELD, Thomas Gabriel, (1817-1893)
BELL, Dugald, (1827-1892)
BOLL, Ernst Friedrich August, (1817-1868)
BOUCHER DE CRÈVECOEUR DE PERTHES, Jacques, (1788-1868)
BOUVÉ, Thomas Tracy, (1815-1896)
BRAVAIS, Auguste, (1811-1863)
CAMPBELL, John Francis, (1821-1885)
CLOSE, Maxwell Henry, Reverend, (1822-1903)
COOK, George Hammell, (1818-1889)
CRAIG, Robert, (?1821-1901)
EVANS, John, Sir, (1823-1908)
FLOWER, John Wickham, (1807-1873)
GUYOT, Arnold Henri, (1807-1884)
HAAST, John Francis Julius von, Sir, (1824-1887)
HIND, Henry Youle, (1823-1908)
JAMIESON, Thomas Francis, (1829-1913)
KING, Samuel William, (1821-1868)
KOENEN, Adolf von, (1837-1915)
LUCY, William Charles, (1822-1898)
MACKENZIE, Alexander, Sir, (ca.1767-1830)
MACKINTOSH, Daniel, (1815-1891)
MARTIN, Peter John, (1786?-1860)
MÖBIUS, Karl August, (1825-1908)
MOLYNEUX, William, (1824-1882)
MORLOT, Adolf von, (1820-1867)
MORTIMER, John Robert, (1825-1911)
MORTON, George Highfield, (1826-1900)
NORTON, Henry, (?1812-1892)
OMBONI, Giovanni, (1829-1910)
PRESTWICH, Joseph, Sir, (1812-1896)
RICKETTS, Charles, (1818-1904)
RÖMER, Carl Ferdinand von, (1818-1891)
SMITH, Gilbert Nicholas, Reverend, (1796-1877)
SMITH, James, (1782-1867)
STEENSTRUP, Johannes Japetus Smith, (1813-1897)
TAYLOR, Richard Cowling, (1789-1851)
TORELL, Otto Martin, (1828-1900)
TYLOR, Alfred, (1823-1884)
WATSON, Robert Boog, Reverend, (?1824-1910)
YOUNG, Robert, Right Honourable, (1822-1917)

1850-1899

ABBOTT, William James Lewis, (1853-1933)
ADAMSON, Samuel A., (-?1890)
AGASSIZ, Louis Jean Rodolphe, (1807-1873)
AILIO, Julius, (1872-1933)
BAIN, Francis, (1842-1894)
BALL, Robert Stawell, Sir, (1840-1913)

QUATERNARY GEOLOGY 1850-1899 Cont'd.

BARTON, George Hunt, (1852-1933)
BAYFIELD, Thomas Gabriel, (1817-1893)
BELL, Dugald, (1827-1892)
BERENDT, Gottlieb Michael, (1836-1920)
BERNHARDI, Reinhard, (fl.1834-1841)
BÖHMERSHEIM, August Böhm von, Edler, (1858-1930)
BOLL, Ernst Friedrich August, (1817-1868)
BOUCHER DE CRÈVECOEUR DE PERTHES, Jacques, (1788-1868)
BOUVÉ, Thomas Tracy, (1815-1896)
BOYER, Georges, (-1892)
BRAVAIS, Auguste, (1811-1863)
BROWN, John Allen, (1831-1903)
BRÜCKNER, Eduard, (1862-1927)
BULL, Alfred Joseph, (1875-1950)
BULLOWS, Wilfred Lowson, (?1869-1943)
BURY, Henry, (?1862-1958)
CALKER, Friedrich Julius Peter von, (1841-1913)
CALL, Richard Ellsworth, (1856-1917)
CALVIN, Samuel, (1840-1911)
CAMPBELL, John Francis, (1821-1885)
CARNEY, Frank, (1868-1934)
CHADWICK, George Halcott, (1876-1953)
CHAMBERLIN, Thomas Chrowder, (1843-1928)
CHRISTEN, Sydney Mary, (née THOMPSON), (-1923)
CHUDEAU, René, (1864-1921)
CLARK, James Edmund, (1850-1944)
CLOSE, Maxwell Henry, Reverend, (1822-1903)
CODRINGTON, Thomas, (1829-1918)
COMMONT, Victor, (1866-1918)
COOK, George Hammell, (1818-1889)
CORNER, Frank, (1861-1939)
CRAIG, Robert, (?1821-1901)
CROFTS, William Hastings, (1861-1935)
DALTON, William H., (?1848-1929)
DAWKINS, William Boyd, Sir, (1837-1929)
DEECKE, Johannes Ernst Wilhelm, (1862-1934)
DEELEY, Richard Mountford, (1855-1944)
DELVAUX, Émile, (1837-1901)
DEPÉRET, Charles Jean Julien, (1854-1929)
DE RANCE, Charles Eugène, (1847-1906)
DEWEY, Henry, (1876-1965)
DONNELLY, Ignatius, (1831-1901)
DOSS, Karl Bruno, (1861-1919)
DU PASQUIER, Léon, (1864-1897)
DWERRYHOUSE, Arthur Richard, (1867-1931)
EVANS, John, Sir, (1823-1908)
FEILDEN, Henry Wemyss, Colonel, (1838-1921)
FLEURY, Ernest, (1878-1958)
FLOWER, John Wickham, (1807-1873)
FOOT, Frederick J., (-1867)
GAÅL, István, (1877-1957)
GEIKIE, James, (1839-1915)
GELLERSTEDT, Göran, (1834-1913)
GÖLLER, August, (1878-1965)

QUATERNARY GEOLOGY 1850-1899 Cont'd.

GOODCHILD, John George, (1844-1906)
GORDON, Charles Henry, (1857-1934)
GRØNLIE, Ole Tobias, (1871-1954)
GUYOT, Arnold Henri, (1807-1884)
HAAST, John Francis Julius von, Sir, (1824-1887)
HALL, Charles Monroe, (1870-1903)
HARMER, Frederic William, (1835-1923)
HARRISON, Edward Richard, Sir, (1872-1960)
HARRISON, William Jerome, (1845-1908)
HAWARD, Frederick James Nairn, (1871-1953)
HICKS, Henry, (1837-1899)
HINCH, John de Witt, (?1875-1931)
HIND, Henry Youle, (1823-1908)
HUBBARD, George David, (1871-1958)
JAMES, Joseph Francis, (1857-1897)
JAMIESON, Thomas Francis, (1829-1913)
JEHU, Thomas John, (1871-?1943)
JESSEN, Axel Hans, (1868-1952)
JOHNSON, Walter, (1867-1950)
JONES, J. Claude, (1877-1932)
JONES, Owen Thomas, (1878-1967)
KADIČ, Ottokár, (1876-1957)
KEILHACK, Konrad, (1858-1944)
KENDALL, Percy Fry, (1856-1936)
KENNARD, Alfred Santer, (1870-1948)
KIDNER, Henry, (?1851-1925)
KILROE, James Robinson, (1848-1927)
KING, Samuel William, (1821-1868)
KINKELIN, Georg Friedrich, (1836-1913)
KISSLING, Ernst, (1865-1936)
KOENEN, Adolf von, (1837-1915)
KŘIŽ, Martin, (1841-1916)
LAMPLUGH, George William, (1859-1926)
LAUFER, Ernst, (1850-1893)
LEVERETT, Frank, (1859-1943)
LEWIŃSKI, Jan Piotr, (1876-1939)
LEWIS, Henry Carvill, (1853-1888)
LLOYD, T. G. B., (1829-1876)
LORY, Pierre, (1866-1956)
LUCY, William Charles, (1822-1898)
LUKASZEWICZ, Józef, (1863-1928)
LYELL, Charles, Sir, (1797-1875)
MACBRIDE, Thomas Huston, (1848-1934)
MACKINTOSH, Daniel, (1815-1891)
MANTLE, Harry G., (ca.1858-1937)
MARINONI, Camillo, (1845-1883)
MARTIN, Peter John, (1786?-1860)
MAŠKA, Karel Jaroslov, (1851-1915)
MAW, George, (1832-1912)
McCABE, Joseph, (1867-1955)
MELLO, John Magens, Reverend, (1836-1914)
MILLER, Hugh Jr., (1850-1896)
MÖBIUS, Karl August, (1825-1908)
MOLYNEUX, William, (1824-1882)

QUATERNARY GEOLOGY 1850-1899 Cont'd.

MONTAG, Emil, (1875-1960)
MORLOT, Adolf von, (1820-1867)
MORTIMER, John Robert, (1825-1911)
MORTON, George Highfield, (1826-1900)
NORDENSKIÖLD, Nils Otto Gustaf, (1869-1928)
NORDMANN, Valdemar Johan Heinrich, (1872-1962)
NORTON, Henry, (?1812-1892)
NORTON, William Harmon, (-)
OGILVIE, Ida Helen, (1874-1963)
OMBONI, Giovanni, (1829-1910)
ØYEN, Peter Arnaeus, (1863-1932)
PARSONS, Henry Franklin, (1846-1913)
PENNING, William Henry, (1838-1902)
POPOFF, Boris Anfirovich, (1871-1950)
PRESTWICH, Joseph, Sir, (1812-1896)
PURKYNĚ, Cyril, (1862-1937)
RAMSAY, Wilhelm, (1865-1928)
READE, Thomas Mellard, (1832-1909)
REID, Clement, (1853-1916)
REMELÉ, Adolf Karl, (1839-1915)
RICKETTS, Charles, (1818-1904)
RÖMER, Carl Ferdinand von, (1818-1891)
ROMER, Eugeniusz, (1871-1954)
RUTKOWSKI, Feliks, (1878-1964)
RUTLEY, Frank, (1842-1904)
RUTOT, Aimé Louis, (1847-1933)
SALISBURY, Rollin D., (1858-1922)
SALTER, Alfred Edward, (?1863-1926)
SARDESON, Frederick William, (1866-1959)
SEDERHOLM, Jakob Johannes, (1863-1934)
SHIMEK, Bohumil, (1861-1937)
SIEGERT, Leo, (1872-1917)
SIMPSON, George Clarke, Sir, (1878-1965)
SMITH, Gilbert Nicholas, Reverend, (1796-1877)
SMITH, James, (1782-1867)
SMITH, Reginald Allender, (?1847-1940)
STEENSTRUP, Johannes Japetus Smith, (1813-1897)
STEINMANN, Gustav, (1856-1929)
STIRRUP, Mark, (1831-1907)
STOPES, Henry, (1852-1902)
SWEET, George, (-?1920)
TENISON-WOODS, Julian Edmund, Reverend, (1832-1889)
TIDDEMAN, Richard Hill, (1842-1917)
TIMMINS, Arthur, Major, (?1858-1940)
TODD, James Edward, (1846-1922)
TORELL, Otto Martin, (1828-1900)
TÖRNEBOHM, Alfred Elis, (1838-1911)
TURTON, William Henry, (1862-1942)
TYLOR, Alfred, (1823-1884)
UPHAM, Warren, (1850-1934)
VERRI, Antonio, (1839-1925)
WAHNSCHAFFE, Gustav Albert Bruno Felix, (1851-1914)
WARREN, Samuel Hazzledine, (1872-1958)
WATSON, Robert Boog, Reverend, (?1824-1910)

QUATERNARY GEOLOGY 1850-1899 Cont'd.

WILLIAMS, Arthur Herbert, (?1865-1935)
WINCHELL, Newton Horace, (1829-1914)
WINKLER, Tiberius Cornelius, (1822-1897)
WOLDŘICH, Jan Nepomuk, (1834-1906)
WOOD, Searles Valentine Jr., (1830-1884)
WOODWARD, Henry, (1832-1921)
WOOLACOTT, David, (1872-1924)
WRIGHT, Albert Allen, (1846-1905)
YOUNG, Alfred Collett, (1862-1932)
YOUNG, John, (1835-1902)
YOUNG, Robert, Right Honourable, (1822-1917)
ŽELÍZKO, Jan Vratislav, (1874-1938)

1900 to present

ABBOTT, William James Lewis, (1853-1933)
AHLMANN, Hans Wilhelmsson, (1889-1974)
AILIO, Julius, (1872-1933)
ALLING, Harold Lattimore, (1888-1960)
AMBROŽ, Vojtěch, (1905-1962)
ANDREWS, Richard Wallis, (?1927-1967)
ARKELL, William Joscelyn, (1904-1958)
BALL, Robert Stawell, Sir, (1840-1913)
BÁRDHARSON, Gudhmundur G., (1880-1933)
BARTON, George Hunt, (1852-1933)
BAYER, Josef, (1882-1931)
BECK, Paul, (1882-1958)
BENNETT, Walter Hellyer, (1892-1971)
BERENDT, Gottlieb Michael, (1836-1920)
BESCHEL, Roland, (1928-1971)
BIEBER, Charles Leonard, (1901-1965)
BISAT, William Sawney, (1886-1973)
BLANC, Alberto Carlo, (1906-1960)
BLANC, Gian Alberto, Barone, (1879-1966)
BÖHMERSHEIM, August Böhm von, Edler, (1858-1930)
BOSWORTH, Thomas Owen, (1882-1928)
BOURCART, Jacques, (1891-1965)
BRANDER, Gunnar, (1903-1940)
BROWN, John Allen, (1831-1903)
BRÜCKNER, Eduard, (1862-1927)
BULL, Alfred Joseph, (1875-1950)
BULLOWS, Wilfred Lowson, (?1869-1943)
BURY, Henry, (?1862-1958)
CADY, Richard Carlysle, (1908-1943)
CALKER, Friedrich Julius Peter von, (1841-1913)
CALL, Richard Ellsworth, (1856-1917)
CALVIN, Samuel, (1840-1911)
CARMAN, Joel Ernest, (1882-1966)
CARNEY, Frank, (1868-1934)
CARRUTHERS, Robert George, (1880-1965)
CHADWICK, George Halcott, (1876-1953)
CHAMBERLIN, Thomas Chrowder, (1843-1928)
CHARLESWORTH, John Kaye, (1889-   )
CHRISTEN, Sydney Mary, (née THOMPSON), (-1923)
CHUDEAU, René, (1864-1921)

QUATERNARY GEOLOGY 1900 to present Cont'd.

CLARK, James Edmund, (1850-1944)
CLOSE, Maxwell Henry, Reverend, (1822-1903)
CODRINGTON, Thomas, (1829-1918)
COMBAZ, Paul, Chanoine, (-1930)
COMMONT, Victor, (1866-1918)
CORNER, Frank, (1861-1939)
CRAIG, Robert, (?1821-1901)
CROFTS, William Hastings, (1861-1935)
DALTON, William H., (?1848-1929)
DAWKINS, William Boyd, Sir, (1837-1929)
DEECKE, Johannes Ernst Wilhelm, (1862-1934)
DEELEY, Richard Mountford, (1855-1944)
DELL, John Alexander, (1882-1964)
DELVAUX, Émile, (1837-1901)
DEPÉRET, Charles Jean Julien, (1854-1929)
DE RANCE, Charles Eugène, (1847-1906)
DEWEY, Henry, (1876-1965)
DOBSON, Peter, (-)
DONNELLY, Ignatius, (1831-1901)
DOSS, Karl Bruno, (1861-1919)
DRAGO, Ciro, (1895-1960)
DUBOIS, Georges, (1890-1953)
DÜCKER, Alfred, (1908-1969)
DWERRYHOUSE, Arthur Richard, (1867-1931)
EDELMANN, Cornelis Hendrik, (1901-1964)
EDMONDS, Frederic Harrison, (1898-1965)
EVANS, John, Sir, (1823-1908)
FEILDEN, Henry Wemyss, Colonel, (1838-1921)
FLEURY, Ernest, (1878-1958)
FLINT, Richard Foster, (1902-1976)
FLORIN, Sten, (-)
FLORSCHÜTZ, Frans, (1887-1965)
FRYXELL, Roald Hilding, (1934-1974)
GAÁL, István, (1877-1957)
GEIKIE, James, (1839-1915)
GELLERSTEDT, Göran, (1834-1913)
GILBERT, Charles J., (-?1934)
GOLDTHWAIT, James Walter, (1880-1947)
GÖLLER, August, (1878-1965)
GOODCHILD, John George, (1844-1906)
GOODING, Ansel Miller, (1924-1976)
GOODLET, George Anderson, (1910-1970)
GORDON, Charles Henry, (1857-1934)
GRANLUND, Erik, (1892-1938)
GRAY, Joseph William, (-1935)
GRØNLIE, Ole Tobias, (1871-1954)
GWYNNE, Charles Sumner, (1885-1972)
HALICKA, Antonina, (1908-1973)
HALL, Charles Monroe, (1870-1903)
HARMER, Frederic William, (1835-1923)
HARRISON, Edward Richard, Sir, (1872-1960)
HARRISON, William Jerome, (1845-1908)
HAWARD, Frederick James Nairn, (1871-1953)
HAYWARD, John, (-1958)

QUATERNARY GEOLOGY 1900 to present Cont'd.

HINCH, John de Witt, (?1875-1931)
HIND, Henry Youle, (1823-1908)
HINTON, Martin Alister Campbell, (1883-1961)
HOLLINGWORTH, Sydney Ewart, (1899-1966)
HOLMES, George William, (1922-1970)
HOLMSEN, Gunnar, (1880-1976)
HOLTEDAHL, Olaf, (1885-1975)
HORBERG, Carl Leland, (1910-1955)
HUBBARD, George David, (1871-1958)
JAMIESON, Thomas Francis, (1829-1913)
JEHU, Thomas John, (1871-?1943)
JESSEN, Axel Hans, (1868-1952)
JOHNSON, J. P., (1880-1918)
JOHNSON, Walter, (1867-1950)
JOLEAUD, Léonce, (1880-1938)
JONES, J. Claude, (1877-1932)
JONES, Owen Thomas, (1878-1967)
KADIČ, Ottokár, (1876-1957)
KAVECKIS, Mykolas, (1889-)
KEILHACK, Konrad, (1858-1944)
KENDALL, Percy Fry, (1856-1936)
KENNARD, Alfred Santer, (1870-1948)
KIDNER, Henry, (?1851-1925)
KILROE, James Robinson, (1848-1927)
KINKELIN, Georg Friedrich, (1836-1913)
KISSLING, Ernst, (1865-1936)
KLEBELSBERG, Raimund von, (1886-1957)
KOENEN, Adolf von, (1837-1915)
KORMOS, Tivadar, (1881-1948)
KŘIŽ, Martin, (1841-1916)
KRYNINE, Paul Dimitri, (1902-1964)
KYRLE, Georg, (1887-1937)
LAHEE, Frederic Henry, (1884-1968)
LAMPLUGH, George William, (1859-1926)
LEIGHTON, Morris Morgan, (1887-1971)
LENCEWICZ, Stanislaw, (1899-1944)
LEVERETT, Frank, (1859-1943)
LEWIŃSKI, Jan Piotr, (1876-1939)
LORY, Pierre, (1866-1956)
LOZINSKI, Walery, (1880-1944)
LUKASZEWICZ, Józef, (1863-1928)
LUNDQVIST, Gösta, (1894-1967)
LYELL, Charles, Sir, (1797-1875)
MACBRIDE, Thomas Huston, (1848-1934)
MacCLINTOCK, Paul, (1891-1970)
MANTLE, Harry G., (ca. 1858-1937)
MARSTON, Alvan Theophilus, (1889-1971)
MAŠKA, Karel Jaroslov, (1851-1915)
MAUFE, Herbert Brantwood, (1879-1946)
MAW, George, (1832-1912)
MELLO, John Magens, Reverend, (1836-1914)
MIHÁLTZ, István, (1897-1964)
MILTHERS, Keld, (1907-1960)
MONTAG, Emil, (1875-1960)

QUATERNARY GEOLOGY 1900 to present Cont'd.

MORTIMER, John Robert, (1825-1911)
MORTON, George Highfield, (1826-1900)
NEUMANN, Hans, (1910-1941)
NIKLEWSKI, Jerzy, (1937-1971)
NORDENSKIÖLD, Nils Otto Gustaf, (1869-1928)
NORDMANN, Valdemar Johan Heinrich, (1872-1962)
NORTON, William Harmon, (-)
OGILVIE, Ida Helen, (1874-1963)
OMBONI, Giovanni, (1829-1910)
ÖSTER, Johannes, (1902-1975)
OVERY, Charles, Reverend, (1882-)
ØYEN, Peter Arnaeus, (1863-1932)
PARSONS, Henry Franklin, (1846-1913)
PENNING, William Henry, (1838-1902)
PETRBOK, Jaroslav, (1881-1960)
PFANNENSTIEL, Max, (1902-1976)
POLANSKI, Jorge, (1892-1975)
PONTIER, G., (-1933)
POPOFF, Boris Anfirovich, (1871-1950)
PURKYNĚ, Cyril, (1862-1937)
RAMSAY, Wilhelm, (1865-1928)
READE, Thomas Mellard, (1832-1909)
REED, Eugene C., (1901-1976)
REID, Clement, (1853-1916)
REMELÉ, Adolf Karl, (1839-1915)
RICKETTS, Charles, (1818-1904)
ROMER, Eugeniusz, (1871-1954)
ROSENDAHL, Halvor, (1890-1963)
RUTKOWSKI, Feliks, (1878-1964)
RUTLEY, Frank, (1842-1904)
RUTOT, Aimé Louis, (1847-1933)
SAINTY, James Edward, (?1882-1967)
SALISBURY, Rollin D., (1858-1922)
SALTER, Alfred Edward, (?1863-1926)
SANDEGREN, Ragnar, (1887-1966)
SARDESON, Frederick William, (1866-1959)
SAWICKI, Ludwik, (1893-1972)
SCHERF, Emil, (1889-1967)
SCHOEWE, Walter Henry, (1891-1966)
SEDERHOLM, Jakob Johannes, (1863-1934)
SHEPPS, Vincent Chester, (1928-1967)
SHIMEK, Bohumil, (1861-1937)
SIEGERT, Leo, (1872-1917)
SIMPSON, George Clarke, Sir, (1878-1965)
SLATER, George, (-1956)
SMITH, Reginald Allender, (?1847-1940)
SOUTO, Alberto, (1888-1961)
SPREITZER, Hans, (1897-1973)
STEENHUIS, Jacob Frederik, (1883-1958)
STEINMANN, Gustav, (1856-1929)
STIRRUP, Mark, (1831-1907)
STOPES, Henry, (1852-1902)
ŠUF, Jiří, (1891-1976)
SUNDELIN, Uno, (-1926)

QUATERNARY GEOLOGY 1900 to present Cont'd.

SWEET, George, (-?1920)
ŚWIDERSKI, Bohdan, (1892-1943)
TESCH, P., (1879-1961)
THOMASSON, Harald, (1893-1937)
TIDDEMAN, Richard Hill, (1842-1917)
TIMMINS, Arthur, Major, (?1858-1940)
TODD, James Edward, (1846-1922)
TÖRNEBOHM, Alfred Elis, (1838-1911)
TROLL, Carl, (1899-)
TROTTER, Frederick Murray, (1897-1968)
TRZASKA-JURKIEWICZOWA, Irena, (1912-1966)
TURTON, William Henry, (1862-1942)
UPHAM, Warren, (1850-1934)
URBÁNEK, Ladislav, (1901-)
VERRI, Antonio, (1839-1925)
VLERK, Isaak Martinus van der, (?1892-1974)
VON POST, Ernst Jakob Lennart, (1884-1951)
WAHNSCHAFFE, Gustav Albert Bruno Felix, (1851-1914)
WALKER, Paul, (1934-1959)
WANLESS, Harold Rollin, (1898-1970)
WARD, Freeman, (1879-1943)
WARREN, Samuel Hazzledine, (1872-1958)
WASMUND, Erich, (1902-1945)
WATSON, Robert Boog, Reverend, (?1824-1910)
WEILER, Wilhelm, (1890-1972)
WEINBERGER, Ludwig, (1911-1966)
WETZEL, Walter, (1887-)
WHITE, George Willard, (1903-)
WILLIAMS, Arthur Herbert, (?1865-1935)
WINCHELL, Newton Horace, (1829-1914)
WOLDŘICH, Jan Nepomuk, (1834-1906)
WOLDSTEDT, Paul, (1888-1973)
WOODWARD, Henry, (1832-1921)
WOOLACOTT, David, (1872-1924)
WORCESTER, Philip George, (1884-1970)
WRIGHT, Albert Allen, (1846-1905)
YOUNG, Alfred Collett, (1862-1932)
YOUNG, John, (1835-1902)
YOUNG, Robert, Right Honourable, (1822-1917)
ZANS, Verners Aleksandrs, (1904-1961)
ŽEBERA, Karel, (1901-)
ŽELÍZKO, Jan Vratislav, (1874-1938)
ZEUNER, Frederick Everard, (1905-1963)

SEDIMENTOLOGY

Up to 1800

BOYLE, Robert, Honourable, (1627-1691)
CARDANO, Girolamo, (1501-1576)
HUTTON, James, (1726-1797)
MITCHILL, Samuel Latham, (1764-1831)
PALISSY, Bernard, (1510-1589)
SICULUS, Diodorus, (ca.85-21 B.C.)
SWEDENBORG, Emanuel, (1688-1772)
TREMBLEY, Abraham, (1710-1784)
VINCI, Leonardo da, (1452-1519)

1800-1849

BAER, Karl Ernst von, (1792-1876)
LA TOUCHE, James Digues, Reverend, (1824-1899)
MITCHILL, Samuel Latham, (1764-1831)
RIDDELL, John Leonard, (1807-1865)
SORBY, Henry Clifton, (1826-1908)
STOKES, George Gabriel, Sir, (1819-1903)

1850-1899

ABBOTT, George, (1844-1925)
BAER, Karl Ernst von, (1792-1876)
BARRELL, Joseph, (1869-1919)
BROWN, Amos Peaslee, (1864-1917)
CAYEUX, Lucien, (1864-1944)
DAVIS, Charles Albert, (1861-1916)
ECK, Heinrich von, (1837-1925)
GOSSLING, Frank Newbery, (1888-1946)
HUNT, Arthur Roope, (1843-1914)
JONES, Owen Thomas, (1878-1967)
LA TOUCHE, James Digues, Reverend, (1824-1899)
LEWIŃSKI, Jan Piotr, (1876-1939)
LOMAS, Joseph, (1860-1908)
LYELL, Charles, Sir, (1797-1875)
MACKIE, William, (1856-1932)
MANKIN, Charles John, (1932-)
MARR, John Edward, (1857-1933)
MOORE, Elwood S., (1879-1966)
MURRAY, John, Sir, (1841-1914)
NEUMAYR, Melchior, (1845-1890)
PARSONS, Henry Franklin, (1846-1913)
PENCK, Friedrich Carl Albrecht, (1858-1945)
RAISIN, Catherine Alice, (1855-1945)
RENARD, Alphonse François, (1842-1903)
RIDDELL, John Leonard, (1807-1865)
SKEATS, Ernest Willington, (1875-1953)
SOLLAS, William Johnson, (1849-1936)
SORBY, Henry Clifton, (1826-1908)
SPRING, Walthère, (-)
STOKES, George Gabriel, Sir, (1819-1903)
TWENHOFEL, William Henry, (1875-1957)
UDDEN, Johan August, (1859-1932)
VAN'T HOFF, Jacobus Hendricus, (1852-1911)

SEDIMENTOLOGY Cont'd.

1900 to present

ABBOTT, George, (1844-1925)
AHRENS, Louis Herman, (1918-)
ANDERSON, Gustavus Edwin, (1879-1940)
ANDRÉE, Karl, (1880-1959)
ANDREWS, Richard Wallis, (?1927-1967)
ANDRUSOV, Dimitri, (1897-1976)
ATHY, Lawrence Ferdinand, (1898-1955)
BAKER, Herbert Arthur, (1885-1954)
BARNES, Jonathan, (-1928)
BARRELL, Joseph, (1869-1919)
BAUDOIN, René, (-)
BELLAIR, Pierre, (1910-)
BERNARD, Hugh Allen, (1915-1975)
BLACK, Maurice, (1904-1973)
BOSWELL, Percy George Hamnall, (1886-1960)
BRAMLETTE, Milton Nunn, (1896-1977)
BROWN, Amos Peaslee, (1864-1917)
BUCHER, Walter Herman, (1888-1965)
CAMPBELL, Robert, (1881-1957)
CARROLL, Dorothy, (1907-1970)
CAYEUX, Lucien, (1864-1944)
CAYEUX DE SENARPONT, André, called CAILLEUX, André, (1907-)
COOPER, Byron Nelson, (1912-1971)
DAMON, Henry Gordon, (1883-1960)
DAVIS, Charles Albert, (1861-1916)
DAVIS, Elmer Fred, (1887-1974)
DIETZ, Robert Sinclair, (1914-)
ECK, Heinrich von, (1837-1925)
EDELMANN, Cornelis Hendrik, (1901-1964)
EDWARDS, Austin Burton, (1909-1960)
ELLIS, Charles W., Jr., (1934-1970)
ESAREY, Ralph Emerson, (1901-1975)
FISK, Harold Norman, (1908-1964)
FLEET, William Frederick, (1890-1966)
FRAZIER, David Edward, (1931-1976)
GEBELEIN, Conrad Dennis, (1945-1978)
GILLULY, James, (-)
GOLDMAN, Marcus Isaac, (1881-1965)
GOREAU, Thomas Fritz, (1924-1970)
GOSSLING, Frank Newbery, (1888-1946)
GRAWE, Oliver Rudolph, (1901-1965)
GRIPP, Karl, (1891-)
GROSSZ, Ádám, (1925-1973)
GROUT, Frank Fitch, (1880-1958)
HALLIMOND, Arthur Francis, (1890-1968)
HAM, William Eugene, (1916-1970)
HARRINGTON, Horacio Jaime, (1910-1973)
HAYWARD, Henry Arthur, (-1964)
HEARD, Albert, (1894-1944)
HODGE, Matthew Burnop, (-1951)
HOYT, John Harger, (1928-1970)
HUNT, Arthur Roope, (1843-1914)
HUTCHINS, Philip Frank, (1927-1955)

SEDIMENTOLOGY 1900 to present Cont'd.

ILLING, Vincent Charles, (1890-1969)
JONES, Owen Thomas, (1878-1967)
KLEIBER, Karl Ernst, (1912-1972)
KOKOSZYŃSKA, Bronislawa, (1897-1959)
KRYNINE, Paul Dimitri, (1902-1964)
KUENEN, Phillip Henry, (1902-1976)
LEWIŃSKI, Jan Piotr, (1876-1939)
LOMAS, Joseph, (1860-1908)
LYELL, Charles, Sir, (1797-1875)
MACKIE, William, (1856-1932)
MARR, John Edward, (1857-1933)
MARTIN, Edgar Charles, (1884-1955)
MEHL, Maurice Goldsmith, (1887-1966)
MOLENGRAAFF, Gerard Johan Hendrik, (1887-1961)
MOORE, Elwood S., (1879-1966)
MOSELEY, John Reed, (1900-1959)
MURRAY, John, Sir, (1841-1914)
PARSONS, Henry Franklin, (1846-1913)
PEET, Geoffrey Alan, (1922-1959)
PENCK, Friedrich Carl Albrecht, (1858-1945)
PICARD, Meredith Dane, (1927-)
POLDERVAART, Arie, (1918-1964)
PRATJE, Otto, (1890-1925)
PRUVOST, Pierre Eugéne Marie Joseph, (1890-1967)
RAISIN, Catherine Alice, (1855-1945)
RENARD, Alphonse François, (1842-1903)
RICHARDSON, William Alfred, (1887-?1964)
RITTENHOUSE, Gordon, (1910-1974)
SCHRÖDER, Hans Adolf Karl Eckart, (1899-1968)
SCRUTON, Philip Challacombe, (1922-1962)
SHEPARD, Francis P., (-)
SHROCK, Robert Rakes, (1904-)
SKEATS, Ernest Willington, (1875-1953)
SMITH, Homer James, (1901-1954)
SMITH, Ned Myron, (1923-1976)
SOLLAS, William Johnson, (1849-1936)
SORBY, Henry Clifton, (1826-1908)
STAMP, Laurence Dudley, Sir, (1898-1966)
STETSON, Henry Crosby, (1900-1955)
STOKES, George Gabriel, Sir, (1819-1903)
SUGDEN, Walter "Bill", (1923-1977)
SVOBODA, J. V., (1921-1972)
TARR, William Arthur, (1881-1939)
TAYLOR, Garvin Lawrence, (1908-1974)
TAYLOR, James Haward, (1909-1968)
THOMASSON, Harald, (1893-1937)
THOMPSON, Warren O., (1898-1968)
TODD, Thomas W., (1927-1969)
TRAINER, David Woolsey, Jr., (1899-1965)
TRASK, Parker Davies, (1899-1961)
TRÜMPY, Rudolf D., (-)
TWENHOFEL, William Henry, (1875-1957)
TYLER, Stanley Allen, (1906-1963)
UDDEN, Johan August, (1859-1932)

SEDIMENTOLOGY 1900 to present Cont'd.

VAN'T HOFF, Jacobus Hendricus, (1852-1911)
WADELL, Hakon, (1895-1962)
WANLESS, Harold Rollin, (1898-1970)
WELLER, James Marvin, (1899-1976)
WHETTEN, John T., (-)

SEISMOLOGY

Up to 1800

BERTRAND, Élie, (1712-1797)
BEVIS (or BEVANS), John, (1693-1771)
CASSIUS, Dio Cocceianus, ([155-235] A.D.)
DOLOMIEU, Dieudonné Sylvain Guy Tancrède, called Déodat, Gratet de, (1750-1801)
FÜCHSEL, Georg Christian, (1722-1773)
GRIMALDI, Francesco Antonio, (1740-1784)
HAMILTON, William, Sir, (1730-1803)
HEBERDEN, Thomas, (1703-1769)
HOOKE, Robert, (1635-1703)
HUMBOLDT, Friedrich Wilhelm Heinrich Alexander von, Freiherr, (1769-1859)
KIRCHER, Athanasius, (1601-1680)
LOMONOSOV, Mikhail Vasilyevich, (1711-1765)
MICHELL, John, Reverend, (1724-1793)
PALISSY, Bernard, (1510-1589)
PAUSANIAS, (ca.130-180 A.D.)
PLINY THE ELDER (CAIUS PLINIUS SECUNDUS), (?20 A.D.-79 A.D.)
RAY, John, (1627-1705)
SICULUS, Diodorus, (ca.85-21 B.C.)
STUKELEY, William, (1687-1765)
VIVENZIO, Giovanni, (-)
WINTHROP, John, (1714-1779)

1800-1849

BERTELLI, Petro Timoteo, (1826-1905)
BRYCE, James, (1806-1877)
CECCHI, Filippo, (1822-1887)
DOLOMIEU, Dieudonné Sylvain Guy Tancrède, called Déodat, Gratet de, (1750-1801)
EGEN, P. N. C., (1793-1849)
GRAHAM, Maria, Mrs., (1785-1842)
HAMILTON, William, Sir, (1730-1803)
HOCHSTETTER, Ferdinand von, Ritter, (1829-1884)
HOFF, Karl Ernst Adolf von, (1771-1837)
HOME, David Milne, (1805-1890)
HOPKINS, William, (1793-1866)
HUMBOLDT, Friedrich Wilhelm Heinrich Alexander von, Freiherr, (1769-1859)
MACFARLANE, James, (1819-1885)
MALLET, Robert, (1810-1881)
MERIAN, Peter, (1795-1883)
NÖGGERATH, Johann Jakob, (1788-1877)
OLDHAM, Thomas, (1816-1878)
PALMIERI, Luigi, (1807-1896)
PERREY, Alexis, (1807-1882)

SEISMOLOGY 1800-1849 Cont'd.

RICKETTS, Charles, (1818-1904)
ROGERS, Henry Darwin, (1808-1866)
SCHMIDT, Johann Friedrich Julius, (1825-1884)
SCROPE, George Poulett, (né Thomson), (1797-1876)
SERPIERI, Alessandro, (1823-1885)
TACCHINI, Pietro, (1838-1905)
VOLGER, Georg Heinrich Otto, (ca 1822-1897)

1850-1899
ABBE, Cleveland, Jr., (1872-1934)
ABBOT, Henry Larcom, (1831-1927)
BAIRD SMITH, R., Colonel, (-)
BARATTA, Mario, (1868-1935)
BERTELLI, Petro Timoteo, (1826-1905)
BERTRAND, Marcel Alexandre, (1847-1907)
BONETTI, Filippo, (1854-1911)
BRANNER, John Casper, (1850-1922)
BRYCE, James, (1806-1877)
CECCHI, Filippo, (1822-1887)
CHANCOURTOIS, A. E. Béguyer de, (-1886?)
CHOFFAT, Paul, (1849-1919)
DAVIS, William Morris, (1850-1934)
DAVISON, Charles, (1858-1940)
DAY, Arthur Louis, (1869-1960)
DE ROSSI, Michele Stefano, (1834-1898)
DRĂGHICEANU, Mathei M., (1844-1939)
DUTTON, Clarence Edward, Major, (1841-1912)
ETZOLD, Franz, (1859-1928)
FALB, Rudolf, (-)
FLORES REYES, Teodoro, (1873-1955)
FOREL, François Alphonse, (1841-1912)
GEINITZ, Eugen, (1854-1925)
GERLAND, Georg Karl Cornelius, (1833-1919)
GORDON, Charles Henry, (1857-1934)
HARBOE, Edouard Georg, (1845-1919)
HOBBS, William Herbert, (1864-1953)
HOCHSTETTER, Ferdinand von, Ritter, (1829-1884)
HOERNES, Rudolf, (1850-1912)
HOFF, Karl Ernst Adolf von, (1771-1837)
HOGBEN, George, (1853-1920)
HOLDEN, Edward Singleton, (1846-1914)
HOME, David Milne, (1805-1890)
HOPKINS, William, (1793-1866)
HUMBOLDT, Friedrich Wilhelm Heinrich Alexander von, Freiherr, (1769-1859)
JOHNSTON-LAVIS, Henry James, (1856-1914)
KIKUCHI, Dairoko, Baron, (1855-1917)
KOLDERUP, Carl Fredrik, (1869-1942)
KOTÔ, Bundjirô, (1856-1935)
KUSAKABE, Shiroka, (1872-1924)
LANCASTER, A., (-?1908)
LASAULX, Arnold von, (1839-1886)
LENOX-CONYNGHAM, Gerald Ponsonby, Colonel Sir, (1866-1956)
LIBBEY, William, (1855-1927)

SEISMOLOGY 1850-1899 Cont'd.

MACFARLANE, James, (1819-1885)
MALLADRA, Alessandro, (1865-1945)
MALLET, Robert, (1810-1881)
MENDENHALL, Thomas Corwin, (1841-1924)
MERCALLI, Giuseppe, (1850-1914)
MERIAN, Peter, (1795-1883)
MIDDLEMISS, Charles Stewart, (1859-1945)
MILNE, John, (1850-1913)
MOHOROVIČIĆ, Andrya, (1857-1936)
MONTESSUS DE BALLORE, Fernand Jean Baptiste Marie Bernard, Comte de, (1851-1923)
NAUMANN, Edmund, (1854-1927)
NÖGGERATH, Johann Jakob, (1788-1877)
ODENBACH, Frederick L., Father, (1857-1933)
OLDHAM, Richard Dixon, (1858-1936)
OLDHAM, Thomas, (1816-1878)
OMORI, Fusakichi, (1868-1923)
O'REILLY, Joseph P., (1829-1905)
PALAZZO, Luigi, (1861-1933)
PALMIERI, Luigi, (1807-1896)
PERREY, Alexis, (1807-1882)
PILAR, Georg, (1847-1893)
PURDUE, Albert Homer, (1861-1917)
REBEUR-PASCHWITZ, Ernst von, (1861-1895)
RICKETTS, Charles, (1818-1904)
RIDLEY, Henry Nicholas, (1855-1956)
ROCKWOOD, Charles Greene, (1843-1913)
RODOLPH, Emil, (1853-1915)
ROGERS, Henry Darwin, (1808-1866)
RUDOLPH, Emil, (1853-1915)
SAPPER, Karl Theodor, (1866-1945)
SCHMIDT, Johann Friedrich Julius, (1825-1884)
SCROPE, George Poulett, (né Thomson), (1797-1876)
SEE, Thomas Jefferson Jackson, (1866-)
SEEBACH, Karl Albert Ludwig von, (1839-1880)
SEIKEI, Sekiya, (1855-1896)
SERPIERI, Alessandro, (1823-1885)
SHIDA, Toshi, (1876-1936)
STANLEY, William Ford Robinson, (1828-1909)
STONE, Ralph Walter, (1876-1964)
SUESS, Eduard, (1831-1914)
TACCHINI, Pietro, (1838-1905)
TARR, Ralph Stockman, (1864-1912)
VOLGER, Georg Heinrich Otto, (ca 1822-1897)
WILLIS, Bailey, (1857-1949)
YAMASAKI, Naomasa, (1870-1929)

1900 to present

ABBE, Cleveland, Jr., (1872-1934)
ADKINS, John Nathaniel, (1911-)
BALLOU, Albert Lorenzo, Jr., (1917-1962)
BARATTA, Mario, (1868-1935)
BERTELLI, Petro Timoteo, (1826-1905)
BERTRAND, Marcel Alexandre, (1847-1907)

SEISMOLOGY 1900 to present Cont'd.

BOND, Lewis Alexander, (1896-1972)
BONETTI, Filippo, (1854-1911)
BRACE, William Francis, (1926-)
BRANNER, John Casper, (1850-1922)
BULLEN, Keith Edward, (1906-1976)
BULLERWELL, William, (1916-1977)
BUWALDA, John Peter, (1886-1954)
CHOFFAT, Paul, (1849-1919)
DAVIS, Elmer Fred, (1887-1974)
DAVIS, William Morris, (1850-1934)
DAVISON, Charles, (1858-1940)
DAY, Arthur Louis, (1869-1960)
DRĂGHICEANU, Mathei M., (1844-1939)
DUTTON, Clarence Edward, Major, (1841-1912)
DVOŘÁK, Arnošt, (1904-)
ETZOLD, Franz, (1859-1928)
FLORES REYES, Teodoro, (1873-1955)
FOREL, François Alphonse, (1841-1912)
GANGI, Anthony Frank, (1929-)
GEINITZ, Eugen, (1854-1925)
GERLAND, Georg Karl Cornelius, (1833-1919)
GOLDEN, John Muro, (1900-1945)
GORDON, Charles Henry, (1857-1934)
GUTENBERG, Beno, (1889-1960)
HARBOE, Edouard Georg, (1845-1919)
HECK, Nicholas Hunter, (1882-1953)
HEILAND, Carl August, (1899-1956)
HILLER, Wilhelm, (1899-)
HOBBS, William Herbert, (1864-1953)
HOERNES, Rudolf, (1850-1912)
HOGBEN, George, (1853-1920)
HOLDEN, Edward Singleton, (1846-1914)
INESON, Jack, (1917-1970)
ISHIMOTO, Mishio, (1893-1940)
JANCZEWSKI, Edward Walery, (1887-1959)
JOHNSTON-LAVIS, Henry James, (1856-1914)
KEMMERLING, Georges Laure Louis, (1888-1932)
KIKUCHI, Dairoko, Baron, (1855-1917)
KOLDERUP, Carl Fredrik, (1869-1942)
KOLDERUP, Niels-Henrik, (1898-1971)
KOTÔ, Bundjirô, (1856-1935)
KUSAKABE, Shiroka, (1872-1924)
LANCASTER, A., (-?1908)
LENOX-CONYNGHAM, Gerald Ponsonby, Colonel Sir, (1866-1956)
MACDONALD, Gordon James Fraser, (1929-)
MACELWANE, James Bernard, Father, S.J., (1883-1956)
MALLADRA, Alessandro, (1865-1945)
MARSDEN, Ernest, (1889-1970)
MENDENHALL, Thomas Corwin, (1841-1924)
MERCALLI, Giuseppe, (1850-1914)
MIDDLEMISS, Charles Stewart, (1859-1945)
MILLER, Don John, (1919-1961)
MILNE, John, (1850-1913)
MOHOROVIČIĆ, Andrya, (1857-1936)

SEISMOLOGY 1900 to present Cont'd.

MONTESSUS DE BALLORE, Fernand Jean Baptiste Marie Bernard, Comte de, (1851-1923)
MOODY, Graham Blair, (1889-1975)
NAGATA, Takesi, (1913-)
NAUMANN, Edmund, (1854-1927)
O'CONNELL, Daniel Trugott, (1901-1971)
ODENBACH, Frederick L., Father, (1857-1933)
OLDHAM, Richard Dixon, (1858-1936)
OMORI, Fusakichi, (1868-1923)
O'REILLY, Joseph P., (1829-1905)
PALAZZO, Luigi, (1861-1933)
PECORA, William Thomas, (1913-1972)
PETTY, Dabney E., (1891-1963)
PRATLEY, Henry Hart, (1893-1941)
PRESS, Frank, (1924-)
PURDUE, Albert Homer, (1861-1917)
RICKETTS, Charles, (1818-1904)
RIDLEY, Henry Nicholas, (1855-1956)
RIEBER, Frank, (1891-1948)
ROCKWOOD, Charles Greene, (1843-1913)
RODOLPH, Emil, (1853-1915)
RUDOLPH, Emil, (1853-1915)
SAPPER, Karl Theodor, (1866-1945)
SEE, Thomas Jefferson Jackson, (1866-)
SHIDA, Toshi, (1876-1936)
SIMMONS, Marvin Gene, (1929-)
SLICHTER, Louis Byrne, (1896-)
SMITH, Aylwin L., (1898-1965)
SMIT SIBINGA, Gerard Leonard, (1895-1963)
STANLEY, William Ford Robinson, (1828-1909)
STONE, Ralph Walter, (1876-1964)
STONELEY, Robert, (1894-1976)
SUESS, Eduard, (1831-1914)
TABER, Stephen, (1882-1963)
TACCHINI, Pietro, (1838-1905)
TARR, Ralph Stockman, (1864-1912)
TOKSÖZ, M. Nafi, (1934-)
TRISTÁN, José Fidel, (-1932)
WILLIS, Bailey, (1857-1949)
YAMASAKI, Naomasa, (1870-1929)

SELENOLOGY

Up to 1800

HERSCHEL, Frederick William, Sir, (1738-1822)

1800-1849

DANA, James Dwight, (1813-1895)
HERSCHEL, Frederick William, Sir, (1738-1822)
SLACK, Henry James, (1818-1896)

1850-1899

DANA, James Dwight, (1813-1895)
SLACK, Henry James, (1818-1896)

SELENOLOGY Cont'd.

1900 to present
FRYXELL, Roald Hilding, (1934-1974)
HESS, Harry Hammond, (1906-1969)
MASON, Arnold Caverly, (1906-1961)
PRESS, Frank, (1924-)
TOLANSKY, Samuel, (?1908-1973)

SPELAEOLOGY
Up to 1800
AVICENNA, (980-1037)

1800-1849
AICHHORN, Sigmund, (1814-1892)
BUCKLAND, William, Reverend, (1784-1856)
CHRISTY, Henry, (1810-1865)
DUBOIS, Auguste, (1862-1923)
FILHOL, Edouard, (1814-1883)
FRAAS, Oskar Friedrich von, (1824-1897)
LUND, Peter Wilhelm, (1801-1880)
LUXMOORE, Edward Bouverie, (1829-1893)
PENGELLY, William, (1812-1894)
SCHAAFHAUSEN, Hermann, (1816-1893)
SCHMERLING, Philippe Charles, (1791-1836)
SCHMIDL, Adolph A., (1802-1863)
TOURNAL, Paul, (1805-1872)
WAGNER, Johann Andreas, (1797-1861)

1850-1899
ABSOLON, Karel, (1877-1960)
AICHHORN, Sigmund, (1814-1892)
BOEGAN, Eugenio, (1875-1939)
BORSARI, Ferdinando, (1858-1891)
BOTTI, Ulderigo, (1822-1905)
BREUIL, Henri, Abbé, (1877-1961)
BRODRICK, Harold, (1872-1946)
BRUNO, Carlo, (1831-1916)
BUCKLAND, William, Reverend, (1784-1856)
CHRISTY, Henry, (1810-1865)
DAVIES, Henry Nathaniel, (-1920)
DUBOIS, Auguste, (1862-1923)
FILHOL, Edouard, (1814-1883)
FOURNIER, Eugène, (1871-1941)
FRAAS, Oskar Friedrich von, (1824-1897)
GARRIGOU, Félix, (1835-1920)
GORDON, Charles Henry, (1857-1934)
HAUTHAL, Rudolf, (1854-1928)
HICKS, Henry, (1837-1899)
HOVEY, Horace Carter, Reverend, (1833-1914)
HUGHES, Thomas McKenny, (1832-1917)
ISSEL, Arturo, (1842-1922)
JONES, Ernest Lloyd, (?1863-1942)
KADIČ, Ottokár, (1876-1957)
KŘIŽ, Martin, (1841-1916)
LUND, Peter Wilhelm, (1801-1880)

SPELAEOLOGY 1850-1899 Cont'd.

LUXMOORE, Edward Bouverie, (1829-1893)
PENGELLY, William, (1812-1894)
PENNINGTON, Rooke, (-?1887)
SCHAAFHAUSEN, Hermann, (1816-1893)
SCHMIDL, Adolph A., (1802-1863)
STONE, Ralph Walter, (1876-1964)
TOURNAL, Paul, (1805-1872)
WAGNER, Johann Andreas, (1797-1861)

1900 to present

ABSOLON, Karel, (1877-1960)
ACLAND, Henry Dyke, (-1936)
BŎCEK, Antonín, (1880-1956)
BOEGAN, Eugenio, (1875-1939)
BOTTI, Ulderigo, (1822-1905)
BREUIL, Henri, Abbé, (1877-1961)
BRODRICK, Harold, (1872-1946)
BRUNO, Carlo, (1831-1916)
CASTERET, Norbert, (1897-)
DAVIES, Henry Nathaniel, (-1920)
DE SOUSA, Vasco Mendes, (1917-1962)
DUDICH, Endre, (1895-1971)
FOURNIER, Eugène, (1871-1941)
GARRIGOU, Félix, (1835-1920)
GLORY, André, Abbé, (-)
GORDON, Charles Henry, (1857-1934)
HAUTHAL, Rudolf, (1854-1928)
HOVEY, Horace Carter, Reverend, (1833-1914)
HUGHES, Thomas McKenny, (1832-1917)
ISSEL, Arturo, (1842-1922)
JONES, Ernest Lloyd, (?1863-1942)
KADIČ, Ottokár, (1876-1957)
KŘIŽ, Martin, (1841-1916)
KYRLE, Georg, (1887-1937)
MacENERY, John, Reverend, (1796-1841)
PALMER, L. S., (1890-1962)
PATRIZI, Saverio, (1902-1957)
PEET, Geoffrey Alan, (1922-1959)
PEGG, E. J., (?1899-1948)
RENAULT, Philippe, (-)
STONE, Ralph Walter, (1876-1964)
TASNÁDI-KUBACSKA, András, (1902-1977)
VIRET, Jean, (1894-1970)

STRATIGRAPHY

Up to 1800

AIKIN, Arthur, (1773-1854)
ANKER, Matthias Joseph, (1771-1843)
ARDUINO, Giovanni, (1714-1795)
AVICENNA, (980-1037)
BAIN, Andrew Geddes, (1797-1864)
BAKEWELL, Robert, (1768-1843)
BERGMAN, Torbern Olof, (1735-1784)
BRADBURY, John, (1765-1823)

STRATIGRAPHY Up to 1800 Cont'd.

BROCCHI, Giovanni Battista (Giambattista), (1772-1826)
BROCHANT DE VILLIERS, André Jean Marie, (1772-1840)
BRONGNIART, Alexandre, (1778-1847)
BUCH, Christian Leopold von, Freiherr, (1774-1853)
CLARK, William, (1770-1838)
CUVIER, Jean Léopold Nicolas Fréderic, (1769-1832)
DOLOMIEU, Dieudonné Sylvain Guy Tancrède, called Déodat, Gratet de, (1750-1801)
EATON, Amos, (1776-1842)
ESCHER VON DER LINTH, Hans Conrad, (1767-1823)
ESCHWEGE, Wilhelm Ludwig, Baron, (1777-1855)
ESMARCH, Jens, (1762-1839)
EVANS, Lewis, (1700-1756)
FAREY, John, (1766-1826)
FAUJAS DE SAINT FOND, Barthélémy, (1741-1819)
FISCHER VON WALDHEIM, Johann Gotthelf Friedrich, (1771-1853)
FLURL, Mathias von, (1756-1823)
FÜCHSEL, Georg Christian, (1722-1773)
GREENOUGH, George Bellas, (1778-1855)
GUETTARD, Jean Étienne, (1715-1786)
HACQUET, Belsazar, (1739-1815)
HAILSTONE, John, Reverend, (1759-1847)
HALL, Elias, (1764-1853)
HARIOT, Thomas, (1560-1621)
HEHL, Johann Carl Ludwig, (1774-1853)
HISINGER, Wilhelm, (1766-1852)
HUMBOLDT, Friedrich Wilhelm Heinrich Alexander von, Freiherr, (1769-1859)
HUNTER, John, (1728-1793)
HUTCHINS, Thomas, (1730-1789)
IMPERATO, Ferrante, (1550-1631)
JAMESON, Robert, (1774-1854)
LAVOISIER, Antoine Laurent, (1743-1794)
LEHMANN, Johann Gottlob, (1719-1767)
LESUEUR, (1778-1846)
LISTER, Martin, (1638-1712)
LUC, Jean André de, (1727-1817)
MACCULLOCH, John, (1773-1835)
MACLURE, William, (1763-1840)
MARASCHINI, Pietro, (1724-1825)
MARSIGLI, Luigi Ferdinando, (ca.1330-1394)
MATON, William George, (1774-1835)
MÉTHERIE, Jean Claude de la, (1743-1817)
MONNET, Antoine Grimoald, (1734-1817)
OTLEY, Jonathan, (1766-1856)
PETTKO, Ján, (1812-1890)
PINGEL, Christian, (1793-1852)
RENNELL, James, Major, (1742-1830)
SAPIEHA, Aleksander, (1773-1812)
SARTORIUS, Georg Christian, (1774-1838)
SAUSSURE, Horace Bénédict de, (1740-1799)
SCHLOTHEIM, Ernst Friedrich von, Freiherr, (1764-1832)
SMITH, William, (1769-1839)
SOULAVIE, Jean Louis Giraud, Abbé, (1752-1813)

STRATIGRAPHY Up to 1800 Cont'd.

STENO, Nicolaus, (1631-1687)
STRACHEY, John, (1671-1743)
SWEDENBORG, Emanuel, (1688-1772)
THOMAS, David, (1776-1859)
TOWNSEND, Joseph, Reverend, (1739-1816)
TROOST, Gerard, (1776-1850)
WALKER, John, Reverend, (1731-1803)
WATSON, White, (1760-1835)
WEBSTER, Thomas, (1773-1844)
WHITEHURST, John, (1713-1788)
WOODWARD, John, (1665-1728)

1800-1849

ABERLE, Carl, (1818-1892)
ABICH, Hermann Wilhelm, (1806-1886)
ADAM, William, (1794?-1873)
ADAMS, Charles Baker, (1814-1853)
AIKIN, Arthur, (1773-1854)
ALBERTI, Friedrich August von, (1795-1878)
ALGER, Francis, (1807-1863)
ALTH, Alois von, Edler, (1819-1886)
ANGELIN, Nils Peter, (1805-1876)
ANKER, Matthias Joseph, (1771-1843)
APLIN, Christopher d'Oyley Hale, (1819-1875)
ARCHIAC, Étienne Jules Adolphe Dexmier de Simon, Vicomte d', (1802-1868)
ARGYLL, George Douglas, Duke of, (1823-1900)
ARTIS, Edmund Tyrell, (1789-1847)
ATHERSTONE, William Guybon, Honourable, (-1898)
AUSTIN, Thomas, Fort-Major, (1794-1881)
AVELINE, William Talbot, (1822-1903)
BAILY, William Hellier, (1819-1888)
BAIN, Andrew Geddes, (1797-1864)
BAKEWELL, Robert, (1768-1843)
BAPTISTA, Isidoro Emilio, (1815-1863)
BARKAS, Thomas Pallister, (1819-1891)
BARNES, Daniel Henry, (1785-1828)
BARRANDE, Joachim, (1799-1883)
BATES, Henry Walter, (1825-1892)
BAUZÁ, Felipe, (1802-1875)
BAYLE, Émile, (1819-1895)
BEAUMONT, Jean Baptiste Armand Louis Léonce Élie de, (1798-1874)
BEESLEY, Thomas, (1818-1896)
BELL, Dugald, (1827-1892)
BENSTED, William Harding, (1802-1873)
BESSER, Wilibald Swibert Joseph Gottlieb von, (1784-1842)
BEST, Edward, (1824-1906)
BEYRICH, August Heinrich Ernst, (1815-1896)
BIANCONI, Gian Giuseppe, (1809-1878)
BIELZ, Eduard Albert, (1827-1898)
BIGSBY, John Jeremiah, (1792-1881)
BILLINGS, Elkanah, (1820-1876)
BINKHOORST TOT DEN BINKHOORST, Johan Theodoor, (1810-1876)
BINNEY, Edward William, (1812-1881)

STRATIGRAPHY 1800-1849 Cont'd.

BOLL, Ernst Friedrich August, (1817-1868)
BOLTON, John, (1788-1873)
BONNARD, Augustin Henri de, (1781-1857)
BOSQUET, Joseph Augustin Hubert, (1814-1880)
BOURGUIGNAT, Jules René, (1829-?1919)
BRADBURY, John, (1765-1823)
BRARD, Cyprien Prosper, (1788-1838)
BRAUNS, David August, (1827-1893)
BRIART, Alphonse, (1825-1898)
BRIGHT, Richard, (1789-1858)
BRISTOW, Henry William, (1817-1889)
BROADHEAD, Garland Carr, (1827-1912)
BROCCHI, Giovanni Battista (Giambattista), (1772-1826)
BROCHANT DE VILLIERS, André Jean Marie, (1772-1840)
BROCKBANK, Richard Bowman, (1824-1912)
BRODIE, Peter Bellinger, Reverend, (1815-1897)
BRONGNIART, Alexandre, (1778-1847)
BRYCE, James, (1806-1877)
BUCH, Christian Leopold von, Freiherr, (1774-1853)
BUCKLAND, William, Reverend, (1784-1856)
BUCKMAN, James, (1814-1884)
CAMPICHE, Gustave, (1809-1870)
CANAVAL, Josef Leodegar, (1820-1898)
CARNALL, Rudolf von, (1804-1874)
CARTER, Henry John, (1813-1895)
CASTELLI, Albin Benedict, (1822-1892)
CATLIN, George, (1796-1872)
CATULLO, Tommaso Antonio, (1782-1869)
CAUMONT, Arcisse de, (1801-1872)
CAUTLEY, Proby T., Colonel Sir, (1802-1871)
CIOFALO, Saverio, (1825-1925)
CLARK, George Thomas, (1811-1898)
CLARK, William, (1770-1838)
CLARKE, William Branwhite, Reverend, (1798-1878)
COCCHI, Igino, (1827-1913)
COLLENOT, Jean Jacques, (1814-1893)
CONDON, Thomas, (1822-1907)
CONRAD, Timothy Abbott, (1803-1877)
CONTEJEAN, Charles Louis, (1824-1907)
CONYBEARE, John Josiah, Reverend, (1779-1824)
CONYBEARE, William Daniel, Reverend, (1787-1857)
COOK, George Hammell, (1818-1889)
COQUAND, Henri, (1813-1881)
CORDA, August Josef, (1811-1849)
COSTA, Pereira da, (1809-1889)
COTTA, Carl Bernhard von, (1808-1879)
COX, Edward Travers, (1821-1907)
COZZENS, Issachar, Jr., (1780-1865)
CRAIG, Robert, (?1821-1901)
CREDNER, Carl Friedrich Heinrich, (1809-1876)
CROLL, James, (1821-1890)
CROSS, John Edward, Reverend, (1821-1897)
CUMMING, Joseph G., Reverend, (1812-1868)
CUNNINGTON, William, (1813-1906)

STRATIGRAPHY 1800-1849 Cont'd.

CURIONI, Giulio, (1796-1878)
CUVIER, Jean Léopold Nicolas Frédéric, (1769-1832)
DANA, James Freeman, (1793-1827)
DARWIN, Charles Robert, (1809-1882)
DAWSON, John William, Sir, (1820-1899)
DEANE, James, (1801-1858)
DECHEN, Ernst Heinrich Carl von, (1800-1889)
DE LA BECHE, Henry Thomas, Sir, (1796-1855)
DE LA CONDAMINE, H. M., Reverend, (-1854)
DE LA VALLÉE POUSSIN, Charles Louis Joseph Xavier, (1827-1903)
DELESSE, Achille Ernest Oscar Joseph, (1817-1881)
DE PRADO, Casiano, Don, (1797-1866?)
DESNOYERS, Jules, (1801-1887)
DESOR, Pierre Jean Edouard, (1811-1882)
DEWALQUE, Gilles Joseph Gustave, (1828-1905)
DE ZIGNO, Achille, Barone, (1813-1888)
DICK, Robert, (1811-1866)
DIXON, Frederick, (1799-1849)
DODERLEIN, Pietro, (1810-1895)
DOLOMIEU, Dieudonné Sylvain Guy Tancrède, called Déodat, Gratet de, (1750-1801)
DOROSZHIN, Peter P., (-)
DOWKER, George, (1828-1899)
DREW, Frederick, (1836-1891)
DUBOIS, Auguste, (1862-1923)
DUCATEL, Julius Timoleon, (1796-1849)
DUFRÉNOY, Pierre Armand, (1792-1857)
DUMAS, Émilien, (1804-1870)
DUNCAN, Peter Martin, (1824-1891)
DUNKER, Wilhelm Bernhard Rudolph Hadrian, (1809-1885)
DU NOYER, George Victor, (-1869)
DUNSTAN, Benjamin, (1864-1933)
EATON, Amos, (1776-1842)
EBRAY, Charles Henri Théophile, (1823-1879)
EGERTON, William Henry, Reverend, (1811-1910)
EHRENBERG, Christian Gottfried, (1795-1876)
EHRLICH, Franz Carl, (1808-1886)
EICHWALD, Eduard Iwanowitsch von, (1795-1876)
EISEL, Robert, (1826-1917)
EMMONS, Ebenezer, (1800-1863)
EMMRICH, Hermann Friedrich, (1815-1879)
EMORY, William Hemsley, (1811-1887)
ENGELSPACH, Auguste, (LARIVIÈRE, pseud.), (1799-1831)
ERDMANN, Axel Joachim, (-)
ESCHER VON DER LINTH, Arnold, (1807-1872)
ESCHER VON DER LINTH, Hans Conrad, (1767-1823)
ESCHWEGE, Wilhelm Ludwig, Baron, (1777-1855)
ESER, Anton Friedrich, (1798-1873)
ESMARCH, Jens, (1762-1839)
ETHERIDGE, Robert, (1819-1903)
EWALD, Julius Wilhelm, (1811-1891)
FANNING, Tolbert, (1810-1874)
FAREY, John, (1766-1826)
FAUJAS DE SAINT FOND, Barthélemy, (1741-1819)

STRATIGRAPHY 1800-1849 Cont'd.

FAVRE, Jean Alphonse, (1815-1890)
FEATHERSTONHAUGH, George William, (1780-1866)
FEISTMANTEL, Karel, (1819-1885)
FINCH, John, (fl.1823-1835)
FISCHER-OOSTER, Carl von, (1807-1875)
FISCHER VON WALDHEIM, Johann Gotthelf Friedrich, (1771-1853)
FISHER, Osmond, Reverend, (1817-1914)
FITTON, William Henry, (1780-1861)
FLURL, Mathias von, (1756-1823)
FOETTERLE, František, (1823-1876)
FONTAINE, Charles Aloyse, Chanoine, (1754-1834)
FORBES, Edward, (1815-1854)
FORCHHAMMER, Johan Georg, (1794-1865)
FOUQUÉ, Ferdinand André, (1828-1904)
FOURNET, Joseph Jean Baptiste Xavier, (1801-1869)
FRÉMONT, John Charles, (1813-1890)
FROSSARD, Émilien, (1802-1881)
FURUHJELM, Enoch Hjalmar, (1823-1886)
GALEOTTI, Henri Guillaume, (1814-1858)
GANLY, Patrick, (1809-1899)
GASTALDI, Bartolomeo, (1818-1879)
GEINITZ, Hanns Bruno, (1814-1900)
GEMMELLARO, Carlo, (1787-1866)
GERLACH, Heinrich, (1822-1871)
GESNER, Abraham, (1797-1864)
GILLIÉRON, Victor, (1826-1890)
GIORDANO, Felice, (1825-1892)
GIRARD, Carl Adolph, (1814-1878)
GLOCKER, Ernst Friedrich, (1793-1858)
GODWIN-AUSTEN, Robert Alfred Cloyne, (1808-1884)
GORDON, George, Reverend, (1801-1893)
GRAVES, Louis, (1791-1857)
GREENOUGH, George Bellas, (1778-1855)
GREPPIN, Jean Baptiste, (-1881)
GRESSLY, Amanz, (1814-1865)
GREWINGK, Konstantin, (1819-1887)
GRIFFITH, Richard John, Sir, (1784-1878)
GRUNER, Louis, (1809-1883)
GUIDONI, Gerolamo, (1794-1870)
GÜMBEL, Carl Wilhelm von, Ritter, (1823-1898)
GUNN, John, Reverend, (1801-1890)
HAAST, John Francis Julius von, Sir, (1824-1887)
HACQUET, Belsazar, (1739-1815)
HAIDINGER, Wilhelm Karl von, Ritter, (1795-1871)
HAILSTONE, John, Reverend, (1759-1847)
HALL, Elias, (1764-1853)
HALL, James, (1811-1898)
HAMILTON, William John, (1805-1867)
HANTKEN, Miksa von Prudnik, (1821-1893)
HANUŠ, František, (1860-1937)
HARRIS, William, (1797-1877)
HASSENCAMP, Ernst, (1824-1881)
HASTINGS, Barbara, Marchioness of, (1810-1858)
HAUER, Franz von, Ritter, (1822-1899)

STRATIGRAPHY 1800-1849 Cont'd.

HAUSMANN, Johann Friedrich Ludwig, (1782-1859)
HAWN, Frederick, (1810-1898)
HAYDEN, Ferdinand Vandiveer, (1829-1887)
HÉBERT, Edmond, (1812-1890)
HEHL, Johann Carl Ludwig, (1774-1853)
HELMERSEN, Gregor von, (1803-1885)
HENWOOD, William Jory, (1805-1875)
HERBICH, Franz, (1821-1887)
HIND, Henry Youle, (1823-1908)
HISINGER, Wilhelm, (1766-1852)
HISLOP, Stephen, Reverend, (1817-1863)
HOCHSTETTER, Ferdinand von, Ritter, (1829-1884)
HOFF, Karl Ernst Adolf von, (1771-1837)
HOFFMANN, Friedrich, (1797-1836)
HOHENEGGER, Ludwig, (1807-1864)
HOLL, Harvey Buchanan, (1820-1886)
HOME, David Milne, (1805-1890)
HOMFRAY, David, (1822-1893)
HORNER, Leonard, (1785-1864)
HOSIUS, August, (1825-1896)
HOUGHTON, Douglass, (1809-1845)
HOWSE, Richard, (1821-1901)
HUBBARD, Bela, (1814-1896)
HUDLESTON, Wilfrid Hudleston, (né Simpson), (1828-1909)
HUMBOLDT, Friedrich Wilhelm Heinrich Alexander von, Freiherr, (1769-1859)
HUNTER, Robert, Reverend, (1823-1897)
HUTTON, William, (-1861?)
HUXLEY, Thomas Henry, (1825-1895)
JACQUEMONT, Victor, (1801-1832)
JACQUOT, Eugène, (1817-1903)
JAMESON, Robert, (1774-1854)
JEANJEAN, Adrien, (1821-1897)
JOHNSON, Henry, (1823-1885)
JOHNSTRUP, Johannes Frederik, (1818-1894)
JONES, Thomas Rupert, (1819-1911)
JUKES, Joseph Beete, (1811-1869)
KARRER, Felix, (1825-1903)
KEENE, William, (1798-1872)
KEFERSTEIN, Christian, (1784-1866)
KEILHAU, Baltazar Mathias, (1797-1858)
KERR, Washington Caruthers, (1827-1885)
KEYSERLING, Alexander Andreevich, Count, (1815-1891)
KINAHAN, George Henry, (1829-1908)
KING, Samuel William, (1821-1868)
KING, William, (1809-1886)
KJERULF, Theodor, (1825-1888)
KLAASSEN, Hendericus Martinus, (1828-1910)
KLIPSTEIN, August Wilhelm, (1801-1894)
KOCH, Carl, (1827-1882)
KOCH, Friedrich Eduard, (1817-1894)
KOENEN, Adolf von, (1837-1915)
KREJČÍ, Jan, (1825-1887)
KUDERNATSCH, Johann, (1819-1856)

STRATIGRAPHY 1800-1849 Cont'd.

KUTORGA, Stepan Semenovich, (1805-1861)
LAMOUROUX, Jean Vincent Félix, (1779-1825)
LAPHAM, Increase Allen, (1811-1875)
LARTET, Edouard Armand Isidore Hippolyte, (1801-1871)
LA TOUCHE, James Digues, Reverend, (1824-1899)
LECKENBY, John, (1814-1877)
LEICHHARDT, Friedrich Wilhelm Ludwig, (1813-1848)
LEONHARD, Gustav von, (1816-1878)
LESLEY, J. Peter, (1819-1903)
LESQUEREUX, Charles Leo, (1806-1889)
LESUEUR, (1778-1846)
LEUBE, Gustav, (1808-1881)
LEYMERIE, Alexander, (1801-1878)
LILL VON LILIENBACH, Karl, (1798-1831)
LINDSTRÖM, Jonas Gustaf Oscar, (1829-1901)
LIPOLD, Markus Vincenc, (1816-1883)
LOCKE, John, (1792-1856)
LOFTUS, William Kennett, (1822-1859)
LOGAN, William Edmond, Sir, (1798-1875)
LONG, Stephen Harriman, (1784-1864)
LORIOL LE FORT, Charles Louis Perceval de, (1828-1908)
LORY, Charles, (1823-1889)
LUC, Jean André de, (1727-1817)
LUCY, William Charles, (1822-1898)
LUDWIG, Rudolph August Birminhold Sebastian, (1812-1880)
MacADAM, James, (1801-1861)
MACCULLOCH, John, (1773-1835)
MACLAREN, Charles, (1782-1866)
MACLURE, William, (1763-1840)
MAJENDIE, Ashurst, (?1784-1867)
MANSEL-PLEYDELL, John Clavell, (1817-1902)
MANTELL, Gideon Algernon, (1790-1852)
MARASCHINI, Pietro, (1724-1825)
MARCOU, Jules, (1824-1898)
MARCY, Oliver, (1820-1899)
MARÈS, Paul, (1826-1900)
MARTIN, Peter John, (1786?-1860)
MARZARI PENCATI, Giuseppe Conte, (1779-1836)
MATHER, William Williams, (1804-1859)
MATHERON, Pierre Émile Philippe, (1807-1899)
MATON, William George, (1774-1835)
MAYER-EYMAR, Karl David Wilhelm, (1826-1907)
McCOY, Frederick, Sir, (1823-1899)
MEDLICOTT, Henry Benedict, (1829-1905)
MEEK, Fielding Bradford, (1817-1876)
MENEGHINI, Giuseppe Giovanni Antonio, (1811-1889)
MERIAN, Peter, (1795-1883)
MÉTHERIE, Jean Claude de la, (1743-1817)
MEYN, Ludwig, (1820-1878)
MICHELOTTI, Giovanni, (1814-1898)
MILLER, Hugh, (1802-1856)
MILLINGTON, John, (1779-1886)
MOESCH, Casimir, (1827-1898)
MONNET, Antoine Grimoald, (1734-1817)

STRATIGRAPHY 1800-1849 Cont'd.

MONTMOLLIN, Auguste de, (1808-1898)
MOORE, Charles, (1814-1881)
MOORE, Francis, Jr., (1808-1864)
MOORE, John Carrick, (1804-1898)
MORIÈRE, Jules Pierre Gilles, (1817-1888)
MORLOT, Adolf von, (1820-1867)
MORRIS, John, (1810-1886)
MORTIMER, John Robert, (1825-1911)
MORTON, George Highfield, (1826-1900)
MUDGE, Benjamin Franklin, (1817-1879)
MÜLLER, Albrecht, (1819-1890)
MURCHISON, Roderick Impey, Sir, (1792-1871)
MURRAY, Alexander, (1810-1884)
NAUMANN, Carl Friedrich, (1797-1873)
NEFF, Peter, (1827-1903)
NEWBERRY, John Strong, (1822-1892)
NICOL, Germain Louis, (1922-1973)
NICOL, James, (1810-1879)
NICOLET, Adolphe Célestin, (1803-1871)
NICOLLET, Joseph Nicolas, (1786-1843)
NORTHAMPTON, Joshua Alwyne Spencer, Marquis of, (1790-1851)
NORTON, Henry, (?1812-1892)
NORWOOD, Joseph Granville, (1807-1895)
OEYNHAUSEN, Karl August Ludwig, (1795-1865)
OLDHAM, Thomas, (1816-1878)
OLMSTED, Denison, (1791-1859)
OMALIUS D'HALLOY, Jean-Baptist Julien d', (1783-1875)
ORBIGNY, Alcide Charles Victor Dessalines d', (1802-1857)
ORMEROD, George Wareing, (1810-1891)
OTLEY, Jonathan, (1766-1856)
OWEN, David Dale, (1807-1860)
OWEN, Richard Dale, (1810-1890)
PAGLIA, Enrico, (-1889)
PAILLETTE, Antoine Adrien, (-1858)
PANČIČ, Joseph, (-1888)
PANDER, Christian Heinrich, (1794-1865)
PANHUYS, George Emile Auguste van, (1796-1871)
PARETO, Lorenzo, (1800-1865)
PARRAN, Alphonse, (1827-1903)
PASINI, Lodovico, (1804-1870)
PASSY, Antoine François, (1792-1873)
PATTISON, Samuel Rowles, (1809-1901)
PENGELLY, William, (1812-1894)
PERCIVAL, James Gates, (1795-1856)
PEREIRA DA COSTA, Francisco Antonio, (1809-1889)
PETERS, Karl, (1825-1881)
PETTKO, Ján, (1812-1890)
PHILIPPI, Rudolph Amandus, (1808-1904)
PHILLIPS, John, (1800-1874)
PHILLIPS, William, (1800-1834)
PICHLER, Adolf, (1819-1900)
PIDANCET, Pierre Marie Just, (1823-1871)
PIETTE, Louis Edouard Stanislas, (1827-1906)
PILLA, Leopoldo, (1805-1848)

STRATIGRAPHY 1800-1849 Cont'd.

PILLET, Louis, (1819-1894)
PINGEL, Christian, (1793-1852)
PIPER, George Harry, (1819-1897)
PIRONA, Giulio Andrea, (1822-1895)
PISSIS, Aimé, (1812-1889)
PLANT, Major John, (1819-?1892)
PLATZ, Philipp, (1827-1900)
POMEL, Nicolas Auguste, (1821-1898)
PORTLOCK, Joseph Ellison, Major-General, (1794-1864)
POTIER, Alfred, (1840-1905)
PRESL, Jan Svatopluk, (1791-1849)
PRESTWICH, Joseph, Sir, (1812-1896)
PRÉVOST, Louis Constant, (1787-1856)
PRINZINGER, Heinrich, (1822-1908)
PUSCH, Georg Gottlieb, (1790-1846)
QUENSTEDT, Friedrich August von, (1809-1889)
RAMSAY, Andrew Crombie, Sir, (1814-1891)
RANDALL, John, (1810-1910)
RAULIN, Félix Victor, (1815-1905)
REEVE, James, (1833-1920)
RENNELL, James, Major, (1742-1830)
REUSS, August Emanuel von, Ritter, (1811-1873)
RHEA, Matthew, (1795-1879)
RIBEIRO, Carlos, General, (1813-1882)
RICHARDSON, Benjamin, Reverend, (-1832)
RICHTER, Reinhard, (1813-1884)
RICKETTS, Charles, (1818-1904)
RIDDELL, John Leonard, (1807-1865)
RIEMSDIJK, A. W. G. van, (1803-1871)
RIEPL, Franz Xaver, (1790-1857)
ROGERS, Henry Darwin, (1808-1866)
ROGERS, William Barton, (1804-1882)
ROLLE, Friedrich, (1827-1887)
RÖMER, Carl Ferdinand von, (1818-1891)
RÖMER, Friedrich Adolph, (1809-1869)
RÖMER, Hermann, (1816-1894)
ROMINGER, Carl Ludwig, (1820-1907)
ROSE, Alexander, (1781-1860)
ROSE, Caleb Burrell, (1790-1872)
ROSTHORN, Franz von, Edler, (1795-1877)
ROUAULT, Marie, (1813-1881)
ROUVILLE, Paul Gervais de, (1823-1907)
ROZET, Antoine, Commandant, (1798-1858)
RUBIDGE, Richard Nathaniel, (1821?-1869)
RUSSEGGER, Josef, (-)
RÜTIMEYER, Karl Ludwig, (1825-1895)
SAEMANN, Louis, (1821-1866)
SALTER, John William, (1820-1869)
SALWEY, Humphry, (1803-1877)
SANDBERGER, Fridolin Carl Ludwig von, (1826-1898)
SANDERS, William, (1799-1875)
SAPIEHA, Aleksander, (1773-1812)
SARTORIUS, Georg Christian, (1774-1838)
SAUVAGE, François Clément, (1814-1872)

STRATIGRAPHY 1800-1849 Cont'd.

SAVI, Paolo, (1798-1871)
SAWKINS, James Gay, (1806-1878)
SCARABELLI GOMMI FLAMINI, Giuseppe, (1820-1905)
SCHAFHÄUTL, Karl Franz Emil, (1803-1890)
SCHAUROTH, Karl von, Freiherr, (1818-1893)
SCHIEL, Jacob Heinrich, (1813-)
SCHIMPER, Philipp Wilhelm, (1808-1880)
SCHLOTHEIM, Ernst Friedrich von, Freiherr, (1764-1832)
SCHLUMBERGER, Charles, (1825-1904)
SCHMID, Ernst Erhard, (1815-1885)
SCHOOLCRAFT, Henry Rowe, (1793-1864)
SCHWIPPEL, Karl, (1821-1911)
SEDGWICK, Adam, Reverend, (1785-1873)
SEELAND, Ferdinand, (1822-1901)
SHARP, Samuel, (1814-1882)
SHARPE, Daniel, (1806-1856)
SHUMARD, Benjamin Franklin, (1820-1869)
SHUMARD, George Getz, (1826-1867)
SIMPSON, Martin, (1798-1892)
SISMONDA, Angelo, (1807-1878)
SMITH, James, (1782-1867)
SMITH, William, (1769-1839)
SMYTH, Warington Wilkinson, Sir, (1817-1890)
SOULAVIE, Jean Louis Giraud, Abbé, (1752-1813)
SPEYER, Oscar, (1827-1882)
STARING, W. C. H., (1808-1874)
STEININGER, Johann, (1794-1874)
STEPHENS, William John, (1829-1890)
STOW, George B. William, (1822-1882)
STRICKLAND, Hugh Edwin, (1811-1853)
STROMBECK, August von, (1808-1900)
STUDER, Bernhard Rudolf, (1794-1887)
STUR, Dénes, (1827-1893)
STUTCHBURY, Samuel, (1798-1859)
SÜSSMILCH, Carl Adolf, (?1874-1946)
SWALLOW, George Clinton, (1817-1899)
SZABÓ, József, (1822-1894)
TATE, George, (1805-1871)
TENORE, Michele, (-)
TERQUEM, Olry, (1797-1887)
THOMAS, David, (1776-1859)
THOMPSON, Zadock, (1796-1856)
THOMSON, James, (1823-1900)
TOPLEY, William, (1841-1894)
TOURNOÜER, Jacques Raoul, (1822-1882)
TOWNSEND, Joseph, Reverend, (1739-1816)
TRAUTSCHOLD, Hermann Adolfovich, (1817-1902)
TREVELYAN, Walter Calverley, Sir, (1797-1879)
TRIGER, Jacques Jules, (1801-1867)
TROOST, Gerard, (1776-1850)
TUOMEY, Michael, (1805-1857)
UBAGHS, Johann Casimir, (1829-1894)
VALPY, Robert Harris, (1819-1904)
VAN DER MAELEN, Philippe Marie Guillaume, (1795-1869)

STRATIGRAPHY 1800-1849 Cont'd.

VANUXEM, Lardner, (1792-1848)
VASCONCELLOS DE PEREIRA CABRAL, Frederico Augusto de, (-)
VERNEUIL, Philippe Edouard Poulletier de, (1805-1873)
VICARY, William, (1811-1903)
VILANOVA Y PIERA, Juan, (1822-1893)
VINCENT, Gérard, (?1824-1899)
VIQUESNEL, Auguste, (1800-1867)
VOLBORTH, Aleksandr Fedorovich, (1800-1876)
WAEL, Norbert Charles Louis de, (1817-1901)
WALKER, John, Reverend, (1731-1803)
WALLACE, Alfred Russel, (1823-1913)
WATELET, Adolphe, (1811-1880)
WATSON, White, (1760-1835)
WEBSTER, Thomas, (1773-1844)
WETHERELL, Nathaniel Thomas, (1800-1875)
WHITE, Charles Abiathar, (1826-1910)
WHITNEY, Josiah Dwight, (1819-1896)
WHITTLESEY, Charles, Colonel, (1808-1886)
WILTSHIRE, Thomas, Reverend, (1826-1902)
WINCHELL, Alexander, (1824-1891)
WING, Augustus, Reverend, (1808-1876)
WITCHELL, Edwin, (1823-1887)
WOLF, Heinrich, (1825-1882)
WOOD, Nicholas, (1795-1865)
WOODWARD, Samuel, (1790-1838)
WORTHEN, Amos Henry, (1812-1888)
WRIGHT, Thomas, (1809-1884)
YOUNG, John, (1823-1900)
ZIEGLER, Jakob Melchior, (1801-1883)
ZITTEL, Karl Alfred von, (1839-1904)

1850-1899

ABBOTT, William James Lewis, (1853-1933)
ABEL, Othenio, (1875-1946)
ABERLE, Carl, (1818-1892)
ABICH, Hermann Wilhelm, (1806-1886)
ADAMS, Charles Baker, (1814-1853)
ADAMS, George Irving, (1870-1932)
ADAMSON, Samuel A., (-?1890)
AHLBURG, Johannes Heinrich Wilhelm, (1833-1919)
AIKIN, Arthur, (1773-1854)
ALBERTI, Friedrich August von, (1795-1878)
ALGER, Francis, (1807-1863)
ALLEN, Henry Atwool, (?1854-1934)
ALTH, Alois von, Edler, (1819-1886)
AMALITSKII, Vladimir Prochorovich, (1860-1917)
AMEGHINO, Florentino, (1854-1911)
AMI, Henry Mark, (1858-1931)
AMMON, Ludwig von, (1850-1922)
AMPFERER, Otto, (1875-1947)
ANDERSON, William, (1860-1915)
ANDERSSON, Johann Gunnar, (1874-1960)
ANDREWS, Ernest Clayton, (1870-1948)
ANDRUSSOV, Nikolai Ivanovich, (1861-1924)

STRATIGRAPHY 1850-1899 Cont'd.

ANGELIN, Nils Peter, (1805-1876)
APLIN, Christopher d'Oyley Hale, (1819-1875)
ARCHIAC, Étienne Jules Adolphe Dexmier de Simon, Vicomte d', (1802-1868)
AREITIO Y LARRINAGA, Alfonso de, (-1884)
ARGYLL, George Douglas, Duke of, (1823-1900)
ARMSTRONG, Henry Edward, (1848-1937)
ARNAUD, Hilaire, (1827-1908)
ARNOLD, Ralph, (1875-1961)
ARTHABER, Gustav von, Edler, (1864-1943)
ASHBURNER, Charles Albert, (1854-1889)
ASHLEY, George Hall, (1866-1951)
ATHERSTONE, William Guybon, Honourable, (-1898)
AUSTIN, Thomas, Fort-Major, (1794-1881)
AUTHELIN, Charles, (1871-1903)
AVELINE, William Talbot, (1822-1903)
BAGG, Rufus Mather, (1869-1946)
BAILEY, Loring Woart, (1839-1925)
BAILEY, Tom Esmond Geoffrey, Captain, (-1919)
BAILY, William Hellier, (1819-1888)
BAIN, Francis, (1842-1894)
BAIRD SMITH, R., Colonel, (-)
BAKER, Benjamin Alfred, (1859-1939)
BALL, John, (-1941)
BALL, Valentine, (1843-1895)
BAPTISTA, Isidoro Emilio, (1815-1863)
BARKAS, Thomas Pallister, (1819-1891)
BARLOW, Alfred Ernest, (1861-1914)
BARON, Richard, Reverend, (1847-1907)
BARRANDE, Joachim, (1799-1883)
BARRELL, Joseph, (1869-1919)
BARROIS, Charles Eugène, (1851-1939)
BARRON, Thomas, (1867-1906)
BARROW, George, (1853-1932)
BARTON, George Hunt, (1852-1933)
BASSANI, Francesco, (1853-1916)
BASSLER, Raymond Smith, (1878-1961)
BATES, Henry Walter, (1825-1892)
BAUER, Heinrich E., (1840-1896)
BAUMBERGER, Ernst, (1866-1935)
BAUR, Carl Theodor von, (1836-1911)
BAUZÁ, Felipe, (1802-1875)
BAYAN, Joseph Félix Ferdinand, (1845-1874)
BAYLE, Émile, (1819-1895)
BEACHLER, Charles S., (1870-1894)
BEADNELL, Hugh John Llewellyn, Major, (1874-1944)
BEAUGÉ, Alfred, (1878-1935)
BEAUMONT, Jean Baptiste Armand Louis Léonce Élie de, (1798-1874)
BECK, Carl Richard, (1858-1919)
BECKER, Ewald, (-1873)
BECKLES, Samuel H., (-1890)
BEEDE, Joshua William, (1871-1940)
BEESLEY, Thomas, (1818-1896)
BEHR, Johannes, (1875-1973)

STRATIGRAPHY 1850-1899 Cont'd.

BELL, Alfred Graham, (?1875-1964)
BELL, Dugald, (1827-1892)
BELL, James Macintosh, (1877-1934)
BELL, Robert, (1841-1917)
BEMROSE, Henry Howe, (1857-1939)
BENECKE, Ernst Wilhelm, (1838-1917)
BENNETT, Francis James, (1845-1920)
BENSTED, William Harding, (1802-1873)
BERENDT, Gottlieb Michael, (1836-1920)
BERGERON, Jules, (1853-1919)
BERGHELL, Hugo, (1864-1922)
BERTRAND, Léon, (1869-1947)
BERTRAND, Marcel Alexandre, (1847-1907)
BEST, Edward, (1824-1906)
BEUSHAUSEN, Hermann Emil Louis, (1863-1904)
BEYER, Samuel Walker, (1865-1931)
BEYRICH, August Heinrich Ernst, (1815-1896)
BIANCONI, Gian Giuseppe, (1809-1878)
BIELZ, Eduard Albert, (1827-1898)
BIGOT, Alexandre, (1863-1953)
BIGSBY, John Jeremiah, (1792-1881)
BILLINGS, Elkanah, (1820-1876)
BINKHOORST TOT DEN BINKHOORST, Johan Theodoor, (1810-1876)
BINNEY, Edward William, (1812-1881)
BIOCHE, Alphonse, (1844-1918)
BIRD, Charles, (1843-1910)
BIRDS, James Adey, (1831-1894)
BIRKMAIER, Anton, (-1926)
BLAKE, John Frederick, Reverend, (1839-1906)
BLAKE, John Hopwood, (1843-1901)
BLANCKENHORN, Max Ludwig Paul, (1861-1947)
BLANFORD, Henry Francis, (1834-1893)
BLANFORD, William Thomas, (1832-1905)
BLAYAC, Joseph, (1865-1936)
BLEICHER, Marie-Gustave, (1838-1901)
BLOMBERG, Karl Albert, (1844-1922)
BÖHM, Georg, (1854-1913)
BÖHMERSHEIM, August Böhm von, Edler, (1858-1930)
BOLL, Ernst Friedrich August, (1817-1868)
BOLTON, Herbert, (1863-1936)
BOLTON, John, (1788-1873)
BONARELLI, Guido, (1871-1951),
BONNEY, Thomas George, Reverend, (1833-1923)
BORSARI, Ferdinando, (1858-1891)
BÖSE, Emil, (1868-1972)
BOSQUET, Joseph Augustin Hubert, (1814-1880)
BOULE, Marcellin, (1861-1942)
BOURGUIGNAT, Jules René, (1829-?1919)
BRADLEY, Frank Howe, (1838-1879)
BRANDES, Theodor, (-1916)
BRANNER, John Caspar, (1850-1922)
BRANSON, Edwin Bayer, (1877-1950)
BRAUNS, David August, (1827-1893)
BRESSON, Arthur, (1873-1946)
BREWER, William Henry, (1828-1910)

STRATIGRAPHY 1850-1899 Cont'd.

BRIART, Alphonse, (1825-1898)
BRIGHT, Richard, (1789-1858)
BRISTOW, Henry William, (1817-1889)
BROADHEAD, Garland Carr, (1827-1912)
BROCK, Reginald Walter, (1874-1935)
BROCKBANK, Richard Bowman, (1824-1912)
BRODIE, Peter Bellinger, Reverend, (1815-1897)
BRØGGER, Waldemar Christofer, (1851-1940)
BROOKS, Alfred Hulse, (1871-1924)
BROOKS, Thomas Benton, (1836-1900)
BROWN, Charles Barrington, (1839-1917)
BROWN, Henry Yorke Lyell "Stoney", (1844-1928)
BRUDER, George, (1856-1916)
BRUNNER VON WATTENWYL, Karl, (1823-1914)
BRUSINA, Spiridion, (1845-1908)
BRYCE, James, (1806-1877)
BUCH, Christian Leopold von, Freiherr, (1774-1853)
BUCKLAND, William, Reverend, (1784-1856)
BUCKMAN, James, (1814-1884)
BUCKMAN, Sidney Savory, (1860-1929)
BUEHLER, Henry Andrew, (1876-1944)
BUKOWSKI, Gejza, (1858-1937)
BULL, Alfred Joseph, (1875-1950)
BURCKHARDT, Carl, (1869-1935)
BUREAU, Edouard, (1830-1918)
BURROUGHS, John, (1837-1921)
BURROWS, Henry William, (?1860-1939)
BUTTS, Charles, (1863-1946)
BUXTORF, August, (1877-)
CAIRNS, Robert, (1854-1911)
CALDERON Y ARANA, Salvador, (1851-1911)
CALL, Richard Ellsworth, (1856-1917)
CALLAWAY, Charles, (1838-1915)
CALVIN, Samuel, (1840-1911)
CAMERON, A. E., (-)
CAMPBELL, Guy, (1877-1970)
CAMPBELL, Henry Donald, (1862-1934)
CAMPBELL, Marius Robinson, (1858-1940)
CAMPICHE, Gustave, (1809-1870)
CANAVAL, Josef Leodegar, (1820-1898)
CANAVARI, Mario, (1855-1928)
CAPELLINI, Giovanni, (1833-1922)
CARNALL, Rudolf von, (1804-1874)
CARPENTIER, Alfred, Chanoine, (1878-1952)
CARR, John Wesley, (1862-[1939])
CARTER, Henry John, (1813-1895)
CARUS-WILSON, C., (1857-1934)
CARVALHO, Anselmo Ferraz de, (1878-1955)
CASTELLI, Albin Benedict, (1822-1892)
CASTRO Y SUERO, Manuel Fernandez de, (1825-1895)
CATLIN, George, (1796-1872)
CATULLO, Tommaso Antonio, (1782-1869)
CAUMONT, Arcisse de, (1801-1872)
CAUTLEY, Proby T., Colonel Sir, (1802-1871)

STRATIGRAPHY 1850-1899 Cont'd.

CAYEUX, Lucien, (1864-1944)
CHADWICK, George Halcott, (1876-1953)
CHAMBERLIN, Thomas Chrowder, (1843-1928)
CHAMPERKNOWNE, Arthur, (1839-1887)
CHANCOURTOIS, A. E. Béguyer de, (-1886?)
CHAPER, Maurice Armand, (1834-1896)
CHARLESWORTH, Edward, (1813-1893)
CHECCHIA-RISPOLI, Giuseppe, (1877-1947)
CHELLONNEIX, Émile, (-1885)
CHOFFAT, Paul, (1849-1919)
CHUDEAU, René, (1864-1921)
CIOFALO, Saverio, (1825-1925)
CLARK, George Thomas, (1811-1898)
CLARK, William Bullock, (1860-1917)
CLARKE, John Mason, (1857-1925)
CLARKE, William Branwhite, Reverend, (1798-1878)
CLAYPOLE, Edward Waller, (1835-1901)
CLELAND, Herdman Fitzgerald, (1869-1935)
CLOUGH, Charles Thomas, (1852-1916)
CLUTTERBUCK, Robert, (-1879)
COBBOLD, Edgar Sterling, (1851-1936)
COCCHI, Igino, (1827-1913)
COCKIN, George Marmaduke, (1852-1918)
COHEN, Emil Wilhelm, (1842-1905)
COLE, Grenville Arthur James, (1859-1924)
COLLENOT, Jean Jacques, (1814-1893)
COLLOT, Louis Marie François, (1846-1915)
COLONY, Roy Jed, (1870-1936)
COMPTNER, Gustav, (1831-1922)
CONDON, Thomas, (1822-1907)
CONRAD, Timothy Abbott, (1803-1877)
CONTEJEAN, Charles Louis, (1824-1907)
CONYBEARE, William Daniel, Reverend, (1787-1857)
COOK, George Hammell, (1818-1889)
COOMARASWAMY, Ananda K., (1877-1947)
COOPER-KING, Charles, Lieutenant-Colonel, (1843-1898)
COQUAND, Henri, (1813-1881)
CORBIÈRE, Louis, (1850-1941)
CORNET, François Léopold, (1834-1887)
CORNET, René Jules, (1865-1929)
CORSTORPHINE, George Steuart, (1865-1919)
CORTESE, Emilio, (1856-1938)
COSSMANN, Maurice, (1850-1924)
COSTA, Pereira da, (1809-1889)
COTTA, Carl Bernhard von, (1808-1879)
COTTEAU, Gustave Honoré, (1818-1894)
COURTY, Georges, (1875-1953)
COX, Edward Travers, (1821-1907)
COZZENS, Issachar, Jr., (1780-1865)
CRAIG, Robert, (?1821-1901)
CRAWFORD, Ralph Dixon, (1873-1950)
CREDNER, Carl Friedrich Heinrich, (1809-1876)
CREDNER, Carl Hermann Georg, (1841-1913)
CRICK, Walter Drawbridge, (1857-1903)

STRATIGRAPHY 1850-1899 Cont'd.

CRIDER, Albert Foster, (1873-1945)
CROLL, James, (1821-1890)
CROSFIELD, Margaret Chorley, (1859-1952)
CROSS, John Edward, Reverend, (1821-1897)
CUMMING, Joseph G., Reverend, (1812-1868)
CUNNINGTON, William, (1813-1906)
CURIONI, Giulio, (1796-1878)
CUSHING, Henry Platt, (1860-1921)
D'ACHIARDI, Antonio, (1839-1902)
DAINTREE, Richard, (1831-1878)
DAKYNS, John Roche, (1836-1910)
DALIMIER, Paul Louis Marie, (1835-1863)
DAL PIAZ, Giorgio, (1872-1962)
DALTON, William H., (?1848-1929)
DAMES, Wilhelm Barnim, (1843-1898)
DANZIG, Emil, (1855-1943)
DARELL, Robert Darell Smythe, (1851-1936)
DARTON, Nelson Horatio, (1865-1948)
DARWIN, Charles Robert, (1809-1882)
DATHE, Johann Friedrich Ernst, (1845-1917)
DAVID, Tannatt William Edgeworth, Sir, (1858-1934)
DAVIES, Arthur Morley, (1869-1959)
DAVIES, David Christopher, (1827-1885)
DAVIS, James William, (1846-1893)
DAWSON, George Mercer, (1849-1901)
DAWSON, John William, Sir, (1820-1899)
DE ALESSANDRI, Giulio, (-?1920)
DEANE, James, (1801-1858)
DECHEN, Ernst Heinrich Carl von, (1800-1889)
DECKER, Charles Elijah, (1868-1958)
DEECKE, Johannes Ernst Wilhelm, (1862-1934)
DE LA BECHE, Henry Thomas, Sir, (1796-1855)
DE LA CONDAMINE, H. M., Reverend, (-1854)
DE LA VALLÉE POUSSIN, Charles Louis Joseph Xavier, (1827-1903)
DELÉPINE, Gaston-Georges, Monseigneur, (1878-1963)
DELESSE, Achille Ernest Oscar Joseph, (1817-1881)
DELVAUX, Émile, (1837-1901)
DENCKMANN, August, (1860-1925)
DE NICOLIS, Enrico, (1841-1908)
DENINGER, Karl, (1878-1917)
DEPÉRET, Charles Jean Julien, (1854-1929)
DE PRADO, Casiano, Don, (1797-1866?)
DE RANCE, Charles Eugène, (1847-1906)
DERBY, Orville Adelbert, (1851-1915)
DE REGNY, Paolo Vinassa, (1871-1957)
DESNOYERS, Jules, (1801-1887)
DESOR, Pierre Jean Edouard, (1811-1882)
DEWALQUE, Gilles Joseph Gustave, (1828-1905)
DEWEY, Henry, (1876-1965)
DE ZIGNO, Achille, Barone, (1813-1888)
DICK, Robert, (1811-1866)
DICKERSON, Roy Ernest, (1877-1944)
DIENER, Carl, (1862-1928)
DI STEFANO, Giovanni, (1856-1918)

STRATIGRAPHY 1850-1899 Cont'd.

DIXON, Ernest Edward Leslie, (1876-1963)
DODERLEIN, Pietro, (1810-1895)
DOLLÉ, Louis, (1878-1965)
DOLLFUS, Gustave Frédéric, (1850-1931)
DOLLOT, Auguste, (1841-1924)
DONCIEUX, Louis, (1874-1960)
DORMAL, Victor, (-1900)
DOUVILLÉ, Joseph Henri Ferdinand, (1846-1937)
DOUXAMI, Henri, (1871-1913)
DOWKER, George, (1828-1899)
DOWLING, Donaldson Bogart, (1858-1925)
DRĂGHICEANU, Mathei M., (1844-1939)
DRAKE, Noah Fields, (1864-1945)
DRESSER, John Alexander, (1866-1954)
DREVERMANN, Fritz, (1875-1932)
DREW, Frederick, (1836-1891)
DUBOIS, Auguste, (1862-1923)
DUFRÉNOY, Pierre Armand, (1792-1857)
DUMAS, Émilien, (1804-1870)
DUMBLE, Edwin Theodore, (1852-1927)
DUNCAN, Peter Martin, (1824-1891)
DUNKER, Wilhelm Bernhard Rudolph Hadrian, (1809-1885)
DUNLOP, Andrew, (1845-1915)
DUNN, Edward John, (1844-1937)
DU NOYER, George Victor, (-1869)
DUNSTAN, Benjamin, (1864-1933)
DUPONT, Édouard François, (1841-1911)
DUTTON, Clarence Edward, Major, (1841-1912)
DWIGHT, William Buck, (1833-1906)
EBEL, Johann Gottfried, (1764-1830)
EBRAY, Charles Henri Théophile, (1823-1879)
ECK, Heinrich von, (1837-1925)
EGAN, Frederick William, (1836-1901)
EGERTON, William Henry, Reverend, (1811-1910)
EHRENBERG, Christian Gottfried, (1795-1876)
EHRLICH, Franz Carl, (1808-1886)
EICHWALD, Eduard Iwanowitsch von, (1795-1876)
EISEL, Robert, (1826-1917)
ELDRIDGE, George Homans, (1854-1905)
ELLES, Gertrude Lilian, (1872-1960)
ELLS, Robert Wheelock, (1845-1911)
ELSDEN, James Vincent, (?1856-1930)
EMERSON, Benjamin Kendall, (1843-1932)
EMMONS, Ebenezer, (1800-1863)
EMMRICH, Hermann Friedrich, (1815-1879)
EMORY, William Hemsley, (1811-1887)
ENDLICH, Frederic Miller, (1851-1899)
ENGEL, Theodor, (1842-1933)
ENGELMANN, August Heinrich, (1831-1899)
ERDMANN, Axel Joachim, (-)
ERDMANN, Edvard, (1840-1923)
ESCHER VON DER LINTH, Arnold, (1807-1872)
ESCHWEGE, Wilhelm Ludwig, Baron, (1777-1855)
ESER, Anton Friedrich, (1798-1873)

STRATIGRAPHY 1850-1899 Cont'd.

ÉTALLON, Claude Auguste, (1826-1862)
ETHERIDGE, Robert, (1819-1903)
ETHERIDGE, Robert Jr., (1847-1920)
EUDES-DESLONGCHAMPS, Eugène, (1830-1889)
EVANS, Caleb, (1831-1886)
EWALD, Julius Wilhelm, (1811-1891)
EXTON, Hugh, (1833-1903)
FABRE, Georges, (-1911)
FALCONER, John Downie, (1876-1947)
FALLAUX, Cornelius, (-1885)
FALSAN, Albert, (1833-1902)
FANNING, Tolbert, (1810-1874)
FARMERY, Joseph Richard, (1872-1928)
FAVRE, Ernest, (1845-1925)
FAVRE, Jean Alphonse, (1815-1890)
FAYOL, Henri, (1841-1925)
FEARNSIDES, William George, (1879-1968)
FEATHERSTONHAUGH, George William, (1780-1866)
FEISTMANTEL, Karel, (1819-1885)
FENDRYCH, Otakar, (1860-1945)
FERRONNIÈRE, Georges, (1875-1922)
FIEBELKORN, Max, (1869-1912)
FINCKH, Ludwig, (1871-1930)
FIRKET, Adolphe, (1837-1905)
FISCHER, Hugo, (1864-1926)
FISCHER-OOSTER, Carl von, (1807-1875)
FISCHER VON WALDHEIM, Johann Gotthelf Friedrich, (1771-1853)
FISHER, Cassius Asa, (1872-1930)
FISHER, Osmond, Reverend, (1817-1914)
FITTON, William Henry, (1780-1861)
FLETCHER, Hugh, (1848-1909)
FLETT, John Smith, Sir, (1869-1947)
FLEURY, Ernest, (1878-1958)
FLIEGEL, Gotthard, (1873-1947)
FLORES REYES, Teodoro, (1873-1955)
FOETTERLE, František, (1823-1876)
FOLLMANN, Otto, (1856-1926)
FONTAINE, Charles Aloyse, Chanoine, (1754-1834)
FONTANNES, Charles François, (1839-1886)
FOOT, Frederick J., (-1867)
FOOTE, R. Bruce, (-1912)
FORBES, Edward, (1815-1854)
FORCHHAMMER, Johan Georg, (1794-1865)
FORIR, Henri, (1856-1907)
FORTAU, René, (1867-1920)
FOSTER, Clement Le Neve, Sir, (1841-1904)
FOUQUÉ, Ferdinand André, (1828-1904)
FOURMARIER, Paul Frédéric Joseph Lucien Lauren, (1877-1970)
FOURNET, Joseph Jean Baptiste Xavier, (1801-1869)
FOURNIER, Eugène, (1871-1941)
FOX-STRANGWAYS, Charles Edward, (1844-1910)
FRAAS, Eberhard, (1862-1915)
FRAIPONT, Julien, (1857-1910)
FRANKE, Adolf, (1860-1942)

STRATIGRAPHY 1850-1899 Cont'd.

FRASER, James Pinkerton, (-)
FRAZER, Persifor, (1844-1909)
FRECH, Fritz, (1861-1917)
FRÉMONT, John Charles, (1813-1890)
FRIEDBERG, Wilhelm, (1873-1941)
FRIEDEL, Georges, (1865-1933)
FRITSCH, Karl von, Freiherr, (1838-1906)
FROSSARD, Émilien, (1802-1881)
FROSTERUS, Gustaf Benjamin, (1866-1931)
FRÜH, Jakob, (1852-)
FUCHS, Alexander, (1874-1935)
FUCHS, Philippe Jacques Edmond, (1837-1889)
FUCHS, Theodor, (1842-1925)
FUGGER, Friedrich Eberhard, (1842-1919)
FURUHJELM, Enoch Hjalmar, (1823-1886)
GAÁL, István, (1877-1957)
GABB, William More, (1839-1878)
GALE, Hoyt Stoddard, (1876-1952)
GALEOTTI, Henri Guillaume, (1814-1858)
GANLY, Patrick, (1809-1899)
GARDINER, Charles Irving, (?1868-1940)
GARWOOD, Edmund Johnston, (1864-1949)
GASTALDI, Bartolomeo, (1818-1879)
GASTER, Christopher Theodore Augustine, (1878-1963)
GAVELIN, Axel Olof, (1875-1947)
GEIKIE, Archibald, Sir, (1835-1924)
GEIKIE, James, (1839-1915)
GEINITZ, Eugen, (1854-1925)
GEINITZ, Hanns Bruno, (1814-1900)
GEMMELLARO, Carlo, (1787-1866)
GENTIL, Louis, (1868-1925)
GERBER, Eduard, (1876-1956)
GERLACH, Heinrich, (1822-1871)
GESNER, Abraham, (1797-1864)
GEYER, David, (1855-1932)
GEYER, Georg, (1857-1938)
GIBBONS, Arthur, (1867-1938)
GIBSON, Walcot, (1864-1941)
GIDLEY, James Williams, (1866-1931)
GILBERT, Grove Karl, (1843-1918)
GILLIÉRON, Victor, (1826-1890)
GIORDANO, Felice, (1825-1892)
GIRARD, Carl Adolph, (1814-1878)
GIRARDOT, Charles Joseph Albert Julien, (1848-1930)
GIROUX, N. J., (1859-1896)
GIRTY, George Herbert, (1869-1939)
GLASS, Norman, Reverend, (1832-1895)
GLENN, Leonidas Chalmers, (1871-1951)
GLOCKER, Ernst Friedrich, (1793-1858)
GODWIN-AUSTEN, Robert Alfred Cloyne, (1808-1884)
GOODCHILD, John George, (1844-1906)
GOODYEAR, Edith, (1879-1959)
GORDON, Charles Henry, (1857-1934)
GORDON, George, Reverend, (1801-1893)

STRATIGRAPHY 1850-1899 Cont'd.

GORJANOVIĆ-KRAMBERGER, Dragutin, (1856-1937)
GOSSELET, Jules Auguste, (1832-1916)
GOSSLING, Frank Newbery, (1888-1946)
GOTTSCHE, Carl Christian, (1855-1909)
GOULD, Charles, (-1895)
GOULD, Charles Newton, (1868-1949)
GRABAU, Amadeus William, (1870-1946)
GRAND'EURY, François Cyrille, (1839-1917)
GRANT, Ulysses Sherman, (1867-1932)
GRAVES, Louis, (1791-1857)
GREBE, Heinrich, (1831-1903)
GREEN, Alexander Henry, (1832-1896)
GREEN, John Frederick Norman, (?1873-1949)
GREEN, Upfield, (1834-1917)
GREENE, Alfred Ernest, (1878-1940)
GREENOUGH, George Bellas, (1778-1855)
GREGORY, Herbert E., (1869-1952)
GREGORY, John Walter, (1864-1932)
GREPPIN, Eduard, (1856-1927)
GREPPIN, Jean Baptiste, (-1881)
GRESSLY, Amanz, (1814-1865)
GREWINGK, Konstantin, (1819-1887)
GRIESBACH, Karl Ludolf, (1847-1907)
GRIFFITH, Richard John, Sir, (1784-1878)
GRODDECK, Albrecht Ludwig von, (1837-1887)
GROEBER, Paul Friedrich Carl, (1885-1968)
GRÖNWALL, Karl Anders Axel, (1869-1944)
GROOM, Theodore Thomas, (1863-)
GRUNER, Louis, (1809-1883)
GRZYBOWSKI, Józef, (1869-1922)
GUÉBHARD, Adrien, (1849-1924)
GUIDONI, Gerolamo, (1794-1870)
GUILLIER, Albert, (1839-1885)
GUMAELIUS, Otto Joël, (1835-1899)
GÜMBEL, Carl Wilhelm von, Ritter, (1823-1898)
GUNN, John, Reverend, (1801-1890)
GUNN, William, (1837-1902)
GÜRICH, Georg, (1859-1938)
GURLDT, Adolf Friedrich, (1829-1902)
GUTZWILLER, Andreas, (1845-1917)
HAAST, John Francis Julius von, Sir, (1824-1887)
HÄBERLE, Daniel, (1864-1934)
HABETS, Alfred, (1839-1908)
HAIDINGER, Wilhelm Karl von, Ritter, (1795-1871)
HALFAR, Anton, (1836-1893)
HALL, Arthur Lewis, (1872-1955)
HALL, Charles Monroe, (1870-1903)
HALL, Christopher Webber, (1845-1911)
HALL, Elias, (1764-1853)
HALL, James, (1811-1898)
HALL, Thomas Sergeant, (1858-1915)
HALL, Townshend Monckton, (1845-1899)
HAMILTON, William John, (1805-1867)
HANTKEN, Miksa von Prudnik, (1821-1893)

STRATIGRAPHY 1850-1899 Cont'd.

HANUŠ, František, (1860-1937)
HARADA, Toyokitsi, (-1894)
HARDMAN, Edward Townley, (1845-1887)
HARKER, Allen, (1848-1894)
HARPER, Lewis, (-)
HARRINGTON, Bernard J., (1848-1907)
HARRIS, George Frederick, (1862-1906)
HARRIS, Gilbert Dennison, (1864-1952)
HARRIS, William, (1797-1877)
HARRISON, William Jerome, (1845-1908)
HARTT, Charles Frederic, (1840-1878)
HASSENCAMP, Ernst, (1824-1881)
HASTINGS, Barbara, Marchioness of, (1810-1858)
HATCH, Frederick Henry, (?1864-1932)
HAUER, Franz von, Ritter, (1822-1899)
HAUG, Gustave Émile, (1861-1927)
HAUSMANN, Johann Friedrich Ludwig, (1782-1859)
HAWN, Frederick, (1810-1898)
HAWORTH, Erasmus, (1855-1932)
HAY, Robert, (1835-1895)
HAYDEN, Ferdinand Vandiveer, (1829-1887)
HAYES, Charles Willard, (1858-1916)
HÉBERT, Edmond, (1812-1890)
HECTOR, James, Sir, (1834-1907)
HEIM, Albert, (1849-1937)
HELBRONNER, Paul, (1871-1938)
HELMERSEN, Gregor von, (1803-1885)
HENDERSON, Junius, (1865-1937)
HENNEN, Ray Vernon, (1875-1958)
HENWOOD, William Jory, (1805-1875)
HERBICH, Franz, (1821-1887)
HERRICK, Clarence Luther, (1858-1904)
HERRIES, Robert Stansfield, (?1860-1941)
HESS, Frank L., (1871-1955)
HIBSCH, Josef Emanuel, (1852-1940)
HICKS, Henry, (1837-1899)
HILBER, Vincenz, (1853-1931)
HILGARD, Eugene Woldemar, (1833-1916)
HILL, Frank A., (1858-1915)
HILL, James Bastian, (1861-1927)
HILL, Robert Thomas, (1858-1941)
HILL, William, (1849-1912)
HILLS, Richard Charles, (1848-1923)
HINCH, John de Witt, (?1875-1931)
HIND, Henry Youle, (1823-1908)
HIND, Wheelton, (1859-1920)
HINDE, George Jennings, (1839-1918)
HINTERLECHNER, Karl, (1874-1932)
HISINGER, Wilhelm, (1766-1852)
HISLOP, Stephen, Reverend, (1817-1863)
HITCHCOCK, Charles Henry, (1836-1919)
HOBSON, Bernard, (1860-1933)
HOCHSTETTER, Ferdinand von, Ritter, (1829-1884)
HOERNES, Rudolf, (1850-1912)

STRATIGRAPHY 1850-1899 Cont'd.

HOFF, Karl Ernst Adolf von, (1771-1837)
HOFFMANN, George Christian, (1837-1917)
HOFMANN, Adolf, (1853-1913)
HOFMANN, Károly, (1839-1891)
HOHENEGGER, Ludwig, (1807-1864)
HOLDEN, Roy Jay, (1870-1945)
HOLGATE, Benjamin, (1838-1915)
HOLL, Harvey Buchanan, (1820-1886)
HOLLAND, Thomas Henry, Sir, (1868-1947)
HOLM, Edvard Johann Gerhard, (1853-1926)
HOLMES, Francis S., (1815-1882)
HOLMES, T. Vincent, (?1841-1923)
HOLZAPFEL, Eduard, (1853-1913)
HOME, David Milne, (1805-1890)
HOMFRAY, David, (1822-1893)
HONY, W. Edward, Archdeacon, (-)
HOOLEY, Reginald Walter, (1866-1923)
HORNE, John, (1848-1928)
HORNER, Leonard, (1785-1864)
HORWITZ, Ludwik, (1875-1943)
HOSIUS, August, (1825-1896)
HOWCHIN, Walter, Reverend, (1845-1937)
HOWE, Ernest, (1875-1932)
HOWELL, Edwin Eugene, (1845-1911)
HOWELL, Henry Hyatt, (1834-1915)
HOWSE, Richard, (1821-1901)
HUBBARD, Bela, (1814-1896)
HUDLESTON, Wilfrid Hudleston, (né Simpson), (1828-1909)
HUENE, Friedrich von, Freiherr, (1875-1969)
HUGHES, Theodore W. H., (?1843-1907)
HUGHES, Thomas McKenny, (1832-1917)
HULL, Edward, (1829-1917)
HUMBOLDT, Friedrich Wilhelm Heinrich Alexander von, Freiherr, (1769-1859)
HUME, William Fraser, (1867-1949)
HUNT, Arthur Roope, (1843-1914)
HUNTER, Robert, Reverend, (1823-1897)
HUNTER-SELKIRK, John R. S., (1835-1898)
HUTCHINSON, Lon Lewis, (1877-1947)
HUTTON, Frederick Wollaston, (1836-1905)
HUTTON, Harriet Mary, (1873-1937)
HUTTON, William, (-1861?)
HUXLEY, Thomas Henry, (1825-1895)
IBBETSON, L. L. Boscawen, Captain, (-1869)
IDDINGS, Joseph Paxson, (1857-1920)
INBERG, Isak Johan, (1853-1893)
INVILLIERS, Edward Vincent d', (1857-1928)
IRVING, John Duer, (1874-1918)
IRVING, Roland Duer, (1847-1888)
JACCARD, Samuel Auguste, (1833-1895)
JACK, Robert Lockhart, (1878-1964)
JACOB, Charles François Étienne, (1878-1962)
JACOBS, Elbridge Churchill, (1873-1957)
JACQUOT, Eugène, (1817-1903)

STRATIGRAPHY 1850-1899 Cont'd.

JAGGAR, Thomas Augustus, (1871-1953)
JAHN, Jaroslav Jiljí, (1865-1934)
JAMESON, Robert, (1774-1854)
JANET, Léon, (1861-1909)
JAROSZ, Jan, (1877-1944)
JEANJEAN, Adrien, (1821-1897)
JEHU, Thomas John, (1871-?1943)
JENNINGS, Albert Vaughan, (1864-1903)
JENNY, Fridolin, (1862-1933)
JENTZSCH, Alfred, (1850-1925)
JESPERSEN, Johan Magnus Peder, (1833-1917)
JESSEN, Axel Hans, (1868-1952)
JIMBO, Kotora, (1867-1924)
JOHANNSEN, Albert, (1871-1962)
JOHNS, Cosmo, (?1866-1951)
JOHNSON, Henry, (1823-1885)
JOHNSTON, Robert Mackenzie, (-1918)
JOHNSTRUP, Johannes Frederik, (1818-1894)
JONES, J. Claude, (1877-1932)
JONES, Owen Thomas, (1878-1967)
JONES, Thomas Rupert, (1819-1911)
JONGMANS, Wilhelmus Josephus, (1878-1957)
JÖNSSON, Jöns, (1848-1894)
JOWETT, Albert, (1875-?1958)
JUDD, John Wesley, (1840-1916)
JUKES, Joseph Beete, (1811-1869)
JUKES-BROWNE, Alfred John, (1851-1914)
JULIEN, Pierre Alphonse, (1838-1905)
KAISER, Friedrich Wilhelm Erich, (1871-1934)
KARPINSKY, Alexander Petrovich, (1847-1936)
KARRER, Felix, (1825-1903)
KASTNER, Karl, (1847-1907)
KATZER, Bedřich, (1861-1925)
KAUFMANN, Franz Joseph, (-1892)
KAYSER, Friedrich Heinrich Emanuel, (1845-1927)
KEENE, William, (1798-1872)
KEEPING, Walter, (1854-1888)
KEFERSTEIN, Christian, (1784-1866)
KEILHAU, Baltazar Mathias, (1797-1858)
KEITH, Arthur, (1864-1944)
KENDALL, Percy Fry, (1856-1936)
KERFORNE, Fernand, (1864-1927)
KERR, Washington Caruthers, (1827-1885)
KEYES, Charles Rollin, (1864-1942)
KEYSERLING, Alexander Andreevich, Count, (1815-1891)
KEYTE, Ivy Allen, (1878-1931)
KIAER, Johan Aschehong, (1869-1931)
KILIAN, Charles Constant Wilfrid, (1862-1925)
KILROE, James Robinson, (1848-1927)
KINAHAN, George Henry, (1829-1908)
KINDLE, Edward Martin, (1869-1940)
KING, Clarence Rivers, (1842-1901)
KING, Samuel William, (1821-1868)
KING, William, (1809-1886)

STRATIGRAPHY 1850-1899 Cont'd.

KIRKBY, James Walker, (1834-1901)
KISSLING, Ernst, (1865-1936)
KITCHIN, Finlay Lorimer, (1870-1934)
KITSON, Albert Ernest, Sir, (1868-1937)
KITTL, Ernst Anton Leopold, (1854-1913)
KJELLSTRÖM, Carl Johan Otto, (1855-1913)
KJERULF, Theodor, (1825-1888)
KLAASSEN, Hendericus Martinus, (1828-1910)
KLAUTZSCH, Adolf, (1869-1927)
KLEBS, Richard, (1850-1911)
KLEMM, Gustav, (1858-1938)
KLIPSTEIN, August Wilhelm, (1801-1894)
KLOUČEK, Celda, (1855-1935)
KNIGHT, Wilbur Clinton, (1858-1903)
KNIPE, Henry Robert, (1855-1918)
KOBY, Frédéric Louis, (1852-1930)
KOCH, Antal von Bodrog, (1843-1927)
KOCH, Carl, (1827-1882)
KOCH, Ferdo, (1874-?1936)
KOCH, Friedrich Eduard, (1817-1894)
KOCH, Gustav Adolf, (?1846-1921)
KOENEN, Adolf von, (1837-1915)
KOERT, Willi, (1875-1927)
KOKEN, Ernst Friedrich Rudolph Karl von, (1860-1912)
KONTKIEWICZ, Stanislaw, Sr., (1849-1924)
KORN, Johannes, (1862-1927)
KORNERUP, Andreas, (?1856-1880)
KRAFFT VON DELLMENSINGEN, Albrecht, (1871-1901)
KREJČÍ, Jan, (1825-1887)
KREUTZ, Felix Franz Xaver, (1844-1910)
KŘIŽ, Martin, (1841-1916)
KUDERNATSCH, Johann, (1819-1856)
KUMMEL, Henry Barnard, (1867-1945)
KUNTH, Karl Ernst Albrecht, (1842-1871)
KUSTA, Johann, (1845-1900)
KUTORGA, Stepan Semenovich, (1805-1861)
KUŹNIAR, Wiktor, ([1871]-1935)
KYNASTON, Herbert, (1868-1915)
LAFLAMME, Joseph Clovis Kemner, Monseigneur, (1849-1910)
LAKE, Philip, (1865-1949)
LAMBERT, Alan, (1837-1927)
LAMBERT, Jules Mathieu, (1848-1940)
LAMBERT, Paul, (1869-1950)
LAMPLUGH, George William, (1859-1926)
LANDES, Henry, (1867-1936)
LANDON, Joseph, (-1902)
LANE, Alfred Church, (1863-1948)
LANG, William Dickson, (1878-1966)
LANGTON, David W., Jr, (1864-?1908)
LAPHAM, Increase Allen, (1811-1875)
LAPPARENT, Albert Auguste de, (1839-1908)
LAPWORTH, Charles, (1842-1920)
LAPWORTH, Herbert, (1875-1933)
LARTET, Edouard Armand Isidore Hippolyte, (1801-1871)

STRATIGRAPHY 1850-1899 Cont'd.

LARTET, Louis, (1840-1899)
LASAULX, Arnold von, (1839-1886)
LA TOUCHE, James Digues, Reverend, (1824-1899)
LAUFER, Ernst, (1850-1893)
LAUNAY, Louis Alphonse Auguste de, (1860-1938)
LEACH, Arthur Leonard, (1869-1957)
LEBOUR, George Alexander Louis, (1847-1918)
LECKENBY, John, (1814-1877)
LEE, Willis Thomas, (1864-1926)
LEIGHTON, Thomas, (1858-1906)
LE MESLE, George, (-1895)
LEMOINE, Paul, (1878-1940)
LENNIER, Gustave, (-1905)
LENZ, Oskar, (1848-1925)
LEONARD, Arthur Gray, (1865-1932)
LEONARD, Hugh, (1841-1909)
LEONHARD, Gustav von, (1816-1878)
LEONHARD, Richard, (1870-1916)
LEPPLA, August, (1859-1924)
LEPSIUS, Karl Georg Richard, (1851-1915)
LERICHE, Maurice, (1875-1948)
LESLEY, J. Peter, (1819-1903)
LESQUEREUX, Charles Leo, (1806-1889)
LEUBE, Gustav, (1808-1881)
LEWIŃSKI, Jan Piotr, (1876-1939)
LEWIS, Henry Carvill, (1853-1888)
LEWIS, Joseph Volney, (1869-1969)
LEYMERIE, Alexander, (1801-1878)
LIEBER, Oscar M., (-1862)
LIEBUS, Adalbert, (1876-1945)
LIMANOWSKI, Mieczyslaw, (1876-1947)
LINDGREN, Waldemar, (1860-1939)
LINDSTRÖM, Axel Fredrik, (1839-1911)
LINDSTRÖM, Jonas Gustaf Oscar, (1829-1901)
LINSTOW, Otto von, (1872-1929)
LIPOLD, Markus Vincenc, (1816-1883)
LISSON, Carlos I., (1868?-1957)
LLOYD, T. G. B., (1829-1876)
LOBLEY, James Logan, (1833-1913)
LÓCZY, Lajos, (1849-1920)
LOFTUS, William Kennett, (1822-1859)
LOGAN, William Edmond, Sir, (1798-1875)
LOHEST, Max, (1857-1926)
LOMAS, Joseph, (1860-1908)
LOMNICKI, Marian Alojzy R., (1845-1915)
LONG, Stephen Harriman, (1784-1864)
LOOMIS, Theodor, (1875-1909)
LÖRENTHEY, Imre, (1867-1917)
LORENZ, Theodor, (1875-1909)
LORETZ, Hermann, (1836-1917)
LORIOL LE FORT, Charles Louis Perceval de, (1828-1908)
LORY, Charles, (1823-1889)
LORY, Pierre, (1866-1956)
LOUGHRIDGE, Robert Hills, (1843-1917)

STRATIGRAPHY 1850-1899 Cont'd.

LOW, Albert Peter, (1861-1952)
LUCIUS, Michel, (1876-1961)
LUCY, William Charles, (1822-1898)
LUDWIG, Rudolph August Birminhold Sebastian, (1812-1880)
LUGEON, Maurice, (1870-1953)
LUKASZEWICZ, Józef, (1863-1928)
LUNDGREN, Sven Anders Bernhard, (1843-1897)
LUNIEWSKI, Adam, (1887-1945)
LUPTON, Charles Thomas, (1878-1935)
LYDEKKER, Richard, (1849-1915)
LYELL, Charles, Sir, (1797-1875)
LYMAN, Benjamin Smith, (1835-1920)
MAAS, Günther Heinrich Julius Max, (1871-1905)
MacADAM, James, (1801-1861)
MACBRIDE, Thomas Huston, (1848-1934)
MacHENRY, Alexander, (1843-1919)
MACLAREN, Charles, (1782-1866)
MACLAREN, James Malcolm, (1873-1935)
MACPHERSON, José, Don, (1839-1902)
MAILLARD, Gustave, (1860-1891)
MAIRE, Julien Victor, (1856-1941)
MAJENDIE, Ashurst, (?1784-1867)
MAKOWSKY, Alexander, (1833-1908)
MALAISE, Constantin H. L., (1834-1916)
MALLADA, Lucas, (1841-1921)
MALLING, Carl Peter, (1862-1927)
MANKIN, Charles John, (1932-)
MANSEL-PLEYDELL, John Clavell, (1817-1902)
MANSFIELD, Wendell Clay, (1874-1939)
MANTELL, Gideon Algernon, (1790-1852)
MARCOU, Jules, (1824-1898)
MARCY, Oliver, (1820-1899)
MARÈS, Paul, (1826-1900)
MARGERIE, Emmanuel Marie Pierre Martin Jacquin de, (1862-1953)
MARINELLI, Olinto, (1874-1926)
MARKHAM, Clements Robert, Sir, (1830-1916)
MARR, John Edward, (1857-1933)
MARSHALL, Patrick, (1869-1950)
MARSHAM-TOWNSHEND, Robert Honourable, (1834-1915)
MARTELLI, Alessandro, (1877-1934)
MARTIN, David, (1842-1918)
MARTIN, George C., (1875-1943)
MARTIN, Johann Karl Ludwig, (1851-1942)
MARTIN, Peter John, (1786?-1860)
MARVINE, Archibald Robertson, (1848-1876)
MATHER, William Williams, (1804-1859)
MATHERON, Pierre Émile Philippe, (1807-1899)
MATHEWS, Edward Bennett, (1869-1944)
MATLEY, Charles Alfred, (1866-1947)
MATSON, George Charlton, (1873-1940)
MATTHES, François Emile, (1874-1948)
MATTIROLO, Ettore, (1853-1923)
MAWSON, Joseph, (1830-1927)
MAYER-EYMAR, Karl David Wilhelm, (1826-1907)

STRATIGRAPHY 1850-1899 Cont'd.

McCALLEY, Henry, (1852-1904)
McCONNELL, Richard George, (1857-1942)
McCOY, Frederick, Sir, (1823-1899)
McGEE, William John, (1853-1912)
McHENRY, Alexander, (1843-1919)
McINTYRE, Peter, (1863-1940)
McKAY, Alexander, (1841-1917)
McMURTRIE, James, (1840-1914)
McPHERSON, William, (-1928)
MEADOWS, Walter Hennesy, (?1864-1937)
MEDLICOTT, Henry Benedict, (1829-1905)
MEEK, Fielding Bradford, (1817-1876)
MEINZER, Oscar Edward, (1876-1948)
MELI, Romolo, (1852-1920)
MELZI, Gilberto, Conte, (1868-1899)
MENDENHALL, Walter Curran, (1871-1957)
MENEGHINI, Giuseppe Giovanni Antonio, (1811-1889)
MENZEL, Karl Gustav Johannes, (1875-1914)
MERIAN, Peter, (1795-1883)
MEUNIER, Étienne Stanislas, (1843-1925)
MEŸER, Charles John Adrian, (1832-1900)
MEYER, Erich, (1874-1915)
MEYN, Ludwig, (1820-1878)
MICHAEL, Richard, (1869-1925)
MICHALSKI, Aleksander Octavianovich, (1855-1904)
MICHEL-LÉVY, Auguste, (1844-1911)
MICHELOTTI, Giovanni, (1814-1898)
MIDDLEMISS, Charles Stewart, (1859-1945)
MIEG-KROH, Mathieu, (1849-1911)
MILLER, Benjamin Le Roy, (1874-1944)
MILLER, Hugh, (1802-1856)
MILLER, Hugh Jr., (1850-1896)
MILLINGTON, John, (1779-1886)
MILLS, James Ellison, (1834-1901)
MILNE, John, (1850-1913)
MILTHERS, Paul Christian Vilhelm Madsen, (1865-1962)
MILTON, John Herbert, (?1862-1925)
MOBERG, Johan Christian, (1854-1915)
MOBERG, Karl Adolf, (1840-1901)
MOESCH, Casimir, (1827-1898)
MOJSISOVICS, Johann August Georg Edmund, Edler von Mojsvàr, (1839-1907)
MOLENGRAAFF, Gustaaf Adolf Frederik, (1860-1942)
MOLYNEUX, Arthur John Charles, (?1866-1921)
MONTMOLLIN, Auguste de, (1808-1898)
MOON, Frederick William, (-1925)
MOORE, Charles, (1814-1881)
MOORE, Francis, Jr., (1808-1864)
MOORE, John Carrick, (1804-1898)
MORAN, Thomas, (1837-1926)
MORGAN, Conwy Lloyd, (?1852-1936)
MORGAN, Jacques Jean Marie de, (1857-1924)
MORGAN, Joseph Bickerton, (1859-1894)
MORGAN, Peter (Percy) Gates, (1867-1927)

STRATIGRAPHY 1850-1899 Cont'd.

MORIÈRE, Jules Pierre Gilles, (1817-1888)
MORLOT, Adolf von, (1820-1867)
MORRIS, John, (1810-1886)
MORSE, William Clifford, (1874-1962)
MORTIMER, John Robert, (1825-1911)
MORTON, George Highfield, (1826-1900)
MOURLON, Michel Félix, (1845-1915)
MUDGE, Benjamin Franklin, (1817-1879)
MÜHLBERG, Friedrich, (1840-1915)
MÜLLER, Albrecht, (1819-1890)
MÜLLER, Karl Christian Gottfried Ludwig, (1862-1906)
MUNIER-CHALMAS, Ernest Philippe Auguste, (1843-1903)
MUNN, Malcolm John, (1874-1949)
MUNTHE, Henrik Vilhelm, (1860-1958)
MURCHISON, Roderick Impey, Sir, (1792-1871)
MURRAY, Alexander, (1810-1884)
MUSHET, David, (-?1859)
MYLIUS, Hugo, (1876-1918)
NAGYSUR, János Böckh von, (1840-1909)
NAUMANN, Carl Friedrich, (1797-1873)
NAUMANN, Edmund, (1854-1927)
NEFF, Peter, (1827-1903)
NERY DA ENCARNAÇAO DELGADO, Joaquim Filipe, (1835-1908)
NEUMAYR, Melchior, (1845-1890)
NEWBERRY, John Strong, (1822-1892)
NEWLAND, David Hale, (1872-1943)
NICHOLSON, Henry Alleyne, (1844-1899)
NICKLÈS, René, (1859-1917)
NICOL, Germain Louis, (1922-1973)
NICOL, James, (1810-1879)
NICOLET, Adolphe Célestin, (1803-1871)
NIES, Friedrich, (1839-1895)
NOLAN, Joseph, (1841-1902)
NOPCSA VON FELSÖ-SZILVÁS, Ferenc, baron, (1877-1933)
NORTHAMPTON, Joshua Alwyne Spencer, Marquis of, (1790-1851)
NORTON, Henry, (?1812-1892)
NORTON, William Harmon, (-)
NORWOOD, Charles Joseph, (1853-1927)
NORWOOD, Joseph Granville, (1807-1895)
NOVÁK, Otomar Pravoslav, (1851-1892)
OEHLERT, Daniel Victor, (1849-1920)
OEYNHAUSEN, Karl August Ludwig, (1795-1865)
OGILVIE, Francis Grant, Sir, (1858-1930)
OGILVIE, Ida Helen, (1874-1963)
OGILVIE, William, (1846-1912)
OHERN, Daniel Webster, (1870-1953)
O'KELLY, Joseph, (1832-1883)
OLDHAM, Charles Aemilius, (1831?-1869)
OLDHAM, Richard Dixon, (1858-1936)
OLDHAM, Thomas, (1816-1878)
OLMSTED, Denison, (1791-1859)
OLRY, Albert, (1847-1913)
OMALIUS D'HALLOY, Jean-Baptist Julien d', (1783-1875)
OPPEL, Albert, (1831-1865)

STRATIGRAPHY 1850-1899 Cont'd.

ORBIGNY, Alcide Charles Victor Dessalines d', (1802-1857)
ORDOÑEZ, Ezequiel, (1867-1950)
ORMEROD, George Wareing, (1810-1891)
ORTLIEB, Jean, (1839-1890)
ORTON, Edward Francis Baxter, (1829-1899)
OSMAN, Charles William, (1855-1932)
OSWALD, Felix, (1866-1958)
OTLEY, Jonathan, (1766-1856)
OWEN, David Dale, (1807-1860)
OWEN, Richard Dale, (1810-1890)
PAGLIA, Enrico, (-1889)
PAILLETTE, Antoine Adrien, (-1858)
PANČIČ, Joseph, (-1888)
PANDER, Christian Heinrich, (1794-1865)
PANHUYS, George Emile Auguste van, (1796-1871)
PANTANELLI, Dante, (1844-1913)
PAPP, Károly, (1873-1963)
PAQUIER, Victor Lucien, (1870-1911)
PARETO, Lorenzo, (1800-1865)
PARK, James, (1857-1946)
PARKINSON, Cyril, (1854?-1919)
PARKS, William Arthur, (1868-1936)
PARONA, Carlo Fabrizio, (1855-1939)
PARRAN, Alphonse, (1827-1903)
PARSONS, Henry Franklin, (1846-1913)
PASINI, Lodovico, (1804-1870)
PASSY, Antoine François, (1792-1873)
PATTISON, Samuel Rowles, (1809-1901)
PAUL, Carl Maria, (1838-1900)
PEACH, Benjamin Neeve, (1842-1926)
PECK, Frederick Burritt, (1860-1925)
PELLATI, Nicola, (1835-1907)
PENGELLY, William, (1812-1894)
PENNING, William Henry, (1838-1902)
PERCIVAL, James Gates, (1795-1856)
PEREIRA DA COSTA, Francisco Antonio, (1809-1889)
PEREIRA DE LIMA, Wenceslau de Sousa, (1858-1920)
PERKINS, George Henry, (1844-1933)
PERNER, Jaroslav, (1869-1947)
PERVINQUIÈRE, Léon, (1873-1913)
PETERS, Karl, (1825-1881)
PETHŐ, Gyula, (1848-1902)
PETKOVIČ, Vladamir K., (1873-1934)
PETRASCHEK, Wilhelm, (1876-1967)
PHILIPPI, Rudolph Amandus, (1808-1904)
PHILLIPS, John, (1800-1874)
PHILLIPS, William, (1800-1834)
PICHLER, Adolf, (1819-1900)
PIDANCET, Pierre Marie Just, (1823-1871)
PIETTE, Louis Edouard Stanislas, (1827-1906)
PILAR, Georg, (1847-1893)
PILGRIM, Henry Guy Ellcock, (1875-1943)
PILLET, Louis, (1819-1894)
PINGEL, Christian, (1793-1852)

STRATIGRAPHY 1850-1899 Cont'd.

PINGEL, Johann Victorinus, (1834-1919)
PIPER, George Harry, (1819-1897)
PIRONA, Giulio Andrea, (1822-1895)
PISSIS, Aimé, (1812-1889)
PLANT, Major John, (1819-?1892)
PLATT, Franklin, (1844-1900)
PLATZ, Philipp, (1827-1900)
POČTA, Filip, (1859-1924)
POLLARD, William, (1870-1946)
POLWHELE, Thomas Roxburgh, (1831-1909)
POMEL, Nicolas Auguste, (1821-1898)
POMPECKJ, Josef Felix, (1867-1930)
POPOVICI-HAŢEG, Valeriu, (1868-1929)
PORTER, Henry, (1832-1868)
PORTLOCK, Joseph Ellison, Major-General, (1794-1864)
POSEWITZ, Tivadar, (1850-1917)
POTIER, Alfred, (1840-1905)
POWELL, John Wesley, (1834-1902)
PREISWERK, Heinrich, (1876-1940)
PRELLER, Charles du Riche, (1844-)
PRESTON, Henry, (1852-1940)
PRESTWICH, Joseph, Sir, (1812-1896)
PRÉVOST, Louis Constant, (1787-1856)
PRIEM, Fernand, (1857-1919)
PRIEST, Simeon, (1876-1954)
PRINGLE, John, (1877-1948)
PRINZINGER, Heinrich, (1822-1908)
PROCHÁZKA, Vladimír Josef, (1862-1913)
PROSSER, Charles Smith, (1860-1916)
PUMPELLY, Raphael, (1837-1923)
PURDUE, Albert Homer, (1861-1917)
PURKYNĚ, Cyril, (1862-1937)
QUENSTEDT, Friedrich August von, (1809-1889)
RADIMSKY, Wenzel, (-1895)
RAMSAY, Andrew Crombie, Sir, (1814-1891)
RANDALL, John, (1810-1910)
RANSOME, Frederick Leslie, (1868-1935)
RATHBURN, Richard, (1852-1918)
RAULIN, Félix Victor, (1815-1905)
RAVN, Jesper Peter Johansen, (1866-1951)
REDTENBACHER, Anton, (1841-1911)
REED, Frederick Richard Cowper, (1869-1946)
REEVE, James, (1833-1920)
REID, Clement, (1853-1916)
REIS, Otto Maria, (1862-1934)
REISER, Karl August, (1853-1922)
RENEVIER, Eugène, (1831-1906)
RENIER, Armand Marie Vincent Joseph, (1876-1951)
REUSCH, Hans Henrik, (1852-1922)
REUSS, August Emanuel von, Ritter, (1811-1873)
RÉVIL, Joseph, (1849-1931)
REYNOLDS, Sidney Hugh, (1867-1949)
RHEA, Matthew, (1795-1879)
RIAZ, Auguste de, (-1920)

STRATIGRAPHY 1850-1899 Cont'd.

RIBEIRO, Carlos, General, (1813-1882)
RICE, William North, (1845-1928)
RICHARDSON, Charles Henry, (1862-1935)
RICHARDSON, George Burr, (1872-1949)
RICHARZ, Stephan, Pater, (1874-1934)
RICHTER, Reinhard, (1813-1884)
RICHTHOFEN, Paul Wilhelm Ferdinand von, Freiherr, (1833-1905)
RICKETTS, Charles, (1818-1904)
RIDDELL, John Leonard, (1807-1865)
RIEMSDIJK, A. W. G. van, (1803-1871)
RIEPL, Franz Xaver, (1790-1857)
RIES, Heinrich, (1871-1951)
RIGAUX, Edmond, (1839-1915)
RISTORI, Giuseppe, (1856-1905)
ROBERTS, Thomas, (1856-1892)
RODLER, Alfred, (1861-1890)
ROEDER, Charles, (1848-1911)
ROGERS, Arthur William, (1872-1946)
ROGERS, Henry Darwin, (1808-1866)
ROGERS, William Barton, (1804-1882)
ROLLE, Friedrich, (1827-1887)
ROLLIER, Louis, (1859-1931)
ROMAN, Albert Frédéric, (1871-1943)
RÖMER, Carl Ferdinand von, (1818-1891)
RÖMER, Friedrich Adolph, (1809-1869)
RÖMER, Hermann, (1816-1894)
ROMINGER, Carl Ludwig, (1820-1907)
RØRDAM, Kristian, (1860-1939)
ROSE, Alexander, (1781-1860)
ROSE, Caleb Burrell, (1790-1872)
ROSENBUSCH, Karl Harry Ferdinand, (1836-1914)
ROSSI, Arturo, (1860-1892)
ROSTHORN, Franz von, Edler, (1795-1877)
ROTHPLETZ, August, (1853-1918)
ROUAULT, Marie, (1813-1881)
ROUVILLE, Paul Gervais de, (1823-1907)
ROWE, Richard Burton, (1872-1902)
ROZET, Antoine, Commandant, (1798-1858)
RUBIDGE, Richard Nathaniel, (1821?-1869)
RUSSEGGER, Josef, (-)
RUSSELL, Israel Cook, (1852-1906)
RÜTIMEYER, Karl Ludwig, (1825-1895)
RUTKOWSKI, Feliks, (1878-1964)
RUTLEY, Frank, (1842-1904)
RUTOT, Aimé Louis, (1847-1933)
RYBA, Frantisek, (1867-1918)
RZEHÁK, Anton, (1855-1923)
SACCO, Federico, (1864-1948)
SADEBECK, Alexander, (1843-1879)
SAEMANN, Louis, (1821-1866)
SAFFORD, James Merrill, (1822-1907)
SALISBURY, Rollin D., (1858-1922)
SALTER, John William, (1820-1869)
SALWEY, Humphry, (1803-1877)

STRATIGRAPHY 1850-1899 Cont'd.

SANDBERGER, Fridolin Carl Ludwig von, (1826-1898)
SANDERS, William, (1799-1875)
SANGER, Frederick Robert, (1874-1942)
SARDESON, Frederick William, (1866-1959)
SARGENT, Henry Crunden, (1851-1931)
SAUER, Gustav Adolf, (1825-1932)
SAUVAGE, François Clément, (1814-1872)
SAUVAGE, Henri Émile, (1842-1917)
SAVI, Paolo, (1798-1871)
SAWKINS, James Gay, (1806-1878)
SCARABELLI GOMMI FLAMINI, Giuseppe, (1820-1905)
SCHAFHÄUTL, Karl Franz Emil, (1803-1890)
SCHALCH, Ferdinand, (1848-1918)
SCHARDT, W. Hans, (1858-1931)
SCHAUROTH, Karl von, Freiherr, (1818-1893)
SCHELLWIEN, Ernst, (1866-1906)
SCHETELIG, Jakob Grubbe Cock, (1875-1935)
SCHIMPER, Philipp Wilhelm, (1808-1880)
SCHLOENBACH, Georg Justin Carl Urban, (1841-1870)
SCHLOSSER, Max, (1854-1933)
SCHLUETER, Clemens August, (1835-1906)
SCHLUMBERGER, Charles, (1825-1904)
SCHLUNCK, Johannes, (1876-1915)
SCHMID, Ernst Erhard, (1815-1885)
SCHMIDT, Carl, (1862-1923)
SCHMIDT, Friedrich Bogdanovich "Bohdan", (1832-1908)
SCHOLZ, Max, (1832-1892)
SCHOOLCRAFT, Henry Rowe, (1793-1864)
SCHOTTLER, Wilhelm, (1869-1932)
SCHRADER, Frank Charles, (1860-1944)
SCHUCHERT, Charles, (1858-1942)
SCHUMACHER, Eugen, (1851-1922)
SCHWARZ, Ernest Hubert Lewis, (1873-1928)
SCHWEINFURTH, Georg, (1836-1925)
SCHWIPPEL, Karl, (1821-1911)
SCOTT, William Berryman, (1858-1947)
SECCO, Andrea, (-1889)
SEDGWICK, Adam, Reverend, (1785-1873)
SEEBACH, Karl Albert Ludwig von, (1839-1880)
SEELAND, Ferdinand, (1822-1901)
SEELEY, Harry Govier, (1839-1909)
SEGUENZA, Giuseppe, (1833-1889)
SEGUENZA, Luigi, (1873-1908)
SELLARDS, Elias Howard, (1875-1961)
SEMPER, Otto, (1830-1907)
SEUNES, Jean, (1849-1920)
SHALER, Nathaniel Southgate, (1841-1906)
SHARMAN, George, (1832-1914)
SHARP, Samuel, (1814-1882)
SHARPE, Daniel, (1806-1856)
SHATTUCK, George Burbank, (1869-)
SHEPARD, Edward Martin, (1854-1934)
SHERBORN, Charles Davies, (1861-1942)
SHERLOCK, R. L., (1875-1948)

STRATIGRAPHY 1850-1899 Cont'd.

SHIMER, Hervey Woodburn, (1872-1965)
SHIPMAN, James, (1848-1901)
SHRUBSOLE, William Hobbs, (1837-1927)
SHUMARD, Benjamin Franklin, (1820-1869)
SHUMARD, George Getz, (1826-1867)
SIEGERT, Leo, (1872-1917)
SIEMIRADZKI, Józef, (1858-1933)
SIMPSON, Martin, (1798-1892)
SISMONDA, Angelo, (1807-1878)
SLATER, Herbert Kelsall, (1875-1913)
SMALL, Evan William, (1859-1933)
SMITH, Eugene Allen, (1841-1927)
SMITH, George Otis, (1871-1944)
SMITH, James, (1782-1867)
SMITH, John, (1845-1930)
SMITH, William Richard Tangier, (1869-1972)
SMOCK, John Conover, (1842-1925)
SMYTH, Warington Wilkinson, Sir, (1817-1890)
SOENDEROP, Fritz Hermann Gustav, (1876-1914)
SOLLAS, William Johnson, (1849-1936)
SOMERVAIL, Alexander, (1841-1910)
SOREIL, Gustave Joseph, (1842-1907)
SPEYER, Oscar, (1827-1882)
STACHE, Guido, (1833-1921)
STAINER, Xavier, (1865-)
STANTON, Timothy William, (1860-1953)
STAPFF, Friedrich Moritz, (1836-1895)
STARING, W. C. H., (1808-1874)
STEBBING, William Pinchard Delane, (1873-1961)
ŞTEFANESCU, Gregoriu, (1838-1911)
ŞTEFĂNESCU, Sabba, (1857-1931)
STEIGER, George, (1869-1944)
STEININGER, Johann, (1794-1874)
STEINMANN, Gustav, (1856-1929)
STELZNER, Alfred Wilhelm, (1840-1895)
STEPHENS, William John, (1829-1890)
STEPHENSON, Lloyd William, (1876-1962)
STEVENSON, John James, (1841-1924)
STILES, Harold, Sir, (?1862-1946)
ST. JOHN, Orestes Hawley, (1841-1921)
STOBBS, John Thomas, (?1866-1942)
STOLICZKA, Ferdinand, (1838-1874)
STOLLEY, Ernst, (1869-1944)
STORRIE, John, (1844-1901)
STOW, George B. William, (1822-1882)
STRAHAN, Aubrey, Sir, (1852-1928)
STRICKLAND, Hugh Edwin, (1811-1853)
STROMBECK, August von, (1808-1900)
STRÜBIN, Karl, (1876-1916)
STRUCKMANN, Carl Eberhard Friedrich, (1833-1898)
STUDER, Bernhard Rudolf, (1794-1887)
STUR, Dénes, (1827-1893)
STÜRZENBAUM, József, (1845-1881)
STUTCHBURY, Samuel, (1798-1859)

STRATIGRAPHY 1850-1899 Cont'd.

SUESS, Eduard, (1831-1914)
SUESS, Franz Eduard, (1867-1941)
SÜSSMILCH, Carl Adolf, (?1874-1946)
SUTCLIFFE, William Henry, (1855-1913)
SVEDMARK, Lennart Eugène, (1847-1922)
SWALLOW, George Clinton, (1817-1899)
SWINNERTON, Henry Hurd, (1875-1967)
SYMES, Richard Glascott, (1840-1906)
SZABÓ, József, (1822-1894)
SZÁDECZKY, Gyula, Baron Szádeczky-Kardoss of Szádecsne and Kafdosfalva, (1860-1935)
SZAJNOCHA, Wladyslaw, (1857-1928)
TAFF, Joseph Alexander, (1862-1944)
TARAMELLI, Torquato, (1845-1922)
TATE, George, (1805-1871)
TATE, Ralph, (1840-1901)
TAURER-GALLENSTEIN, Hans von, Ritter, (1846-1927)
TAUSCH VON GLÖCKELSTHURN, Leopold, (1858-1899)
TAWNEY, Edward Bernard, (1840-1882)
TAYLOR, John Ellor, (1835-1895)
TCHERNYCHEV, Féodoss Nikolaievich, (1856-1914)
TEALE, Edmund Oswald, Sir, (1874-1971)
TEALL, Jethro Justinian Harris, Sir, (1849-1924)
TEISSEYRE, Wawrzyniec, (1860-1939)
TELLER, Friedrich, (1852-1913)
TENISON-WOODS, Julian Edmund, Reverend, (1832-1889)
TERMIER, Pierre, (1859-1930)
TERQUEM, Olry, (1797-1887)
THEVENIN, Armand, (1870-1918)
THOMAS, Abram Owen, (1876-1931)
THOMAS, David, (1776-1859)
THOMAS, Herbert Henry, (1876-1935)
THOMAS, Philippe, (1843-1910)
THOMPSON, Beeby, (1848-1931)
THOMPSON, Zadock, (1796-1856)
THOMSON, Alexander Gordon Milne, (1866-1919)
THOMSON, James, (1823-1900)
THORNTON, Richard, (1838-1863)
THURMANN, Jules, (1804-1855)
TIDDEMAN, Richard Hill, (1842-1917)
TIETZE, Emil, (1845-1931)
TIETZE, Oskar Adolf Eduard, (1874-1920)
TILTON, John Littlefield, (1863-1930)
TIMMINS, Arthur, Major, (?1858-1940)
TOBLER, August, (1872-1929)
TODD, James Edward, (1846-1922)
TOIT, Alexander Logie du, (1878-1948)
TOLL, Eduard von, Baron, (1858-1902)
TOLWIŃSKI, Konstanty, (1876-1961)
TOMBLESON, Frederick Binns, (?1872-1938)
TOMMASI, Annibale, (1858-1921)
TOPLEY, William, (1841-1894)
TÖRNQUIST, Sven Leonhard, (1840-1920)
TOUCAS, Joseph Aristide, (1843-1911)

STRATIGRAPHY 1850-1899 Cont'd.

TOULA, Franz, (1845-1920)
TOURNOÜER, Jacques Raoul, (1822-1882)
TRABUCCO, Giacomo, (1845-1925)
TRASK, John Boardman, (1824-1879)
TRAUTSCHOLD, Hermann Adolfovich, (1817-1902)
TREACHER, Llewellyn, (1859-1943)
TRENCH, Richard, (1836-1861)
TRENTANOVE, Giorgio Morando, (1874-1914)
TREVELYAN, Walter Calverley, Sir, (1797-1879)
TRIBOLET, Georges de, (1830-1873)
TRIBOLET-HARDY, Frédéric Maurice de, (1852-1929)
TRIGER, Jacques Jules, (1801-1867)
TUCCIMEI, Giuseppe, (1851-1915)
TULLBERG, Sven Axel Theodor, (1852-1886)
TUOMEY, Michael, (1805-1857)
TWENHOFEL, William Henry, (1875-1957)
TWITCHELL, Mayville William, (1866-1927)
TYLDEN-WRIGHT, Charles, (1832-1900)
TYRRELL, Joseph Burr, (1858-1957)
UBAGHS, Johann Casimir, (1829-1894)
UDALL, John, (1841-1920)
UDDEN, Johan August, (1859-1932)
UHLIG, Victor Karl, (1857-1911)
ULRICH, Edward Oscar, (1857-1944)
USSHER, William Augustus Edmond, (1849-1920)
VACEK, Michael, (1848-1925)
VAGHI, Vittorio, (-1921)
VALLE DE LERSUNDI, Alfonso del, (1875-1944)
VALPY, Robert Harris, (1819-1904)
VAN DER MAELEN, Philippe Marie Guillaume, (1795-1869)
VAN HISE, Charles Richard, (1857-1918)
VAN INGEN, Gilbert, (1869-1928)
VARTY SMITH, George, (-1924)
VASCONCELLOS DE PEREIRA CABRAL, Frederico Augusto de, (-)
VASSEUR, Gaston Casimir, (1855-1915)
VATER, Heinrich, (1859-1930)
VAUGHAN, Arthur, (1868-1915)
VAUGHAN, Thomas Wayland, (1870-1952)
VAUX, Adolphe de, (-1899)
VEATCH, Arthur Clifford, (1878-1938)
VERBEEK, Rogier Diederick Marius, (1845-1926)
VERNEUIL, Philippe Edouard Poulletier de, (1805-1873)
VICARY, William, (1811-1903)
VILANOVA Y PIERA, Juan, (1822-1893)
VILLA, Antonio, (-1885)
VILLA, Giovanni Battista, (-1887)
VINCENT, Gérard, (?1824-1899)
VIQUESNEL, Auguste, (1800-1867)
VITÁLIS, István, (1871-1947)
VIZER, Vilmos, (1873-1941)
VOGT, Johan Herman Lie, (1858-1932)
VOLBORTH, Aleksandr Fedorovich, (1800-1876)
WAAGEN, Lukas, (1877-1959)
WAAGEN, Wilhelm Heinrich, (1841-1900)

STRATIGRAPHY 1850-1899 Cont'd.

WAEL, Norbert Charles Louis de, (1817-1901)
WÄHNER, Franz, (1856-1932)
WAHNSCHAFFE, Gustav Albert Bruno Felix, (1851-1914)
WALCOTT, Charles Doolittle, (1850-1927)
WALKER, Edward Eaton, (-1903)
WALKER, John Francis, (1839-1907)
WALKER, Thomas Leonard, (1867-1942)
WALL, George Parkes, (1832-1912)
WALLACE, Alfred Russel, (1823-1913)
WALLERANT, Frédéric Félix Auguste, (1858-1936)
WANNER, Johannes, (1878-1957)
WARD, James Clifton, Reverend, (1843-1880)
WARD, John, (1837-1906)
WARD, Leonard Keith, (1879-1964)
WARDLE, Thomas, Sir, (1831-1909)
WATELET, Adolphe, (1811-1880)
WATSON, Thomas Leonard, (1871-1924)
WATTS, William Whitehead, (1860-1947)
WEBER, Carl Albert, (1856-1931)
WEBER, Julius, (1864-1924)
WEIDMAN, Samuel, (1870-1945)
WEISE, Ernst, (1843-1941)
WEISS, Christian Ernst, (1833-1890)
WELLER, Stuart, (1870-1927)
WELLS, Roger Clark, (1877-1944)
WELSCH, Jules, (1858-1929)
WESTGATE, Lewis Gardner, (1868-1948)
WESTON, Thomas Chesmer, (1832-1910)
WETHERELL, Nathaniel Thomas, (1800-1875)
WETTSTEIN, Alexander, (1861-1887)
WHITAKER, William, (1836-1925)
WHITE, Charles Abiathar, (1826-1910)
WHITE, Charles David, (1862-1935)
WHITE, Israel Charles, (1848-1927)
WHITEAVES, John Joseph Frederick, (1835-1909)
WHITNEY, Josiah Dwight, (1819-1896)
WHITTLESEY, Charles, Colonel, (1808-1886)
WILDER, Frank Alonzo, (1870-1930)
WILKINSON, Charles Smith, (1843-1891)
WILKMAN, Wanold Wrydon, (1872-?1938)
WILKS, Mark, (?1862-1945)
WILLARD, Daniel Everett, (1862-1947)
WILLIAMS, Arthur Herbert, (?1865-1935)
WILLIAMS, George Huntington, (1856-1894)
WILLIAMS, Henry Shaler, (1847-1918)
WILLIS, Bailey, (1857-1949)
WILSON, Charles Henry, (1851-1887)
WILSON, Edward, (?1849-1898)
WILSON, J. S. Grant, (1855-1908)
WILTSHIRE, Thomas, Reverend, (1826-1902)
WIMAN, Carl Johan Josef Ernst, (1867-1944)
WINCHELL, Alexander, (1824-1891)
WINCHELL, Horace Vaughn, (1865-1923)
WINCHELL, Newton Horace, (1829-1914)

STRATIGRAPHY 1850-1899 Cont'd.

WINDHAUSEN, Anselm, (1882-1932)
WING, Augustus, Reverend, (1808-1876)
WINTER, William Percy, (?1870-1950)
WINWOOD, Henry Hoyte, (1830-1920)
WIŚNIOWSKI, Tadeusz, (1865-1933)
WITCHELL, Edwin, (1823-1887)
WOHLGEMUTH, Jules, (-1893)
WOLF, Heinrich, (1825-1882)
WOLFF, Wilhelm, (1872-1951)
WOOD, Nicholas, (1795-1865)
WOODRUFF, Elmer Grant, (1872-1952)
WOODWARD, Harry Page, (1858-1917)
WOODWARD, Horace Bolingbroke, (1848-1914)
WOODWARD, Robert Simpson, (1849-1924)
WOOLACOTT, David, (1872-1924)
WOOLNOUGH, Walter George, (1876-1958)
WORTH, Richard Nicholas, (-1896)
WORTHEN, Amos Henry, (1812-1888)
WRIGHT, Frederick Eugene, (1877-1953)
WRIGHT, George Frederick, (1838-1921)
WRIGHT, James, (1878-1957)
WRIGHT, Thomas, (1809-1884)
WRIGHT, William Bourke, (1876-1939)
WUNDT, Georg von, (1845-1929)
WYATT-EDGELL, Henry Adrian, (1847-1866)
WYNNE, Arthur Beavor, (1835-1906)
YABE, Hisakatsu, (1878-1969)
YACHEVSKY, Leonard Antonovich, (1858-1916)
YOUNG, Alfred Prentice, (1841-1919)
YOUNG, George Albert, (1898-1947)
YOUNG, John, (1823-1900)
YOUNG, John, (1835-1902)
ZARĘCZNY, Stanislaw Wawrzyniec, (1848-1909)
ZEISE, Oskar, (1860-1925)
ZIEGLER, Jakob Melchior, (1801-1883)
ZIMMERMANN, Ernst, (1860-1944)
ZITTEL, Karl Alfred von, (1839-1904)
ZLATARSKI, Georgi Nikolov, (1854-1909)
ZUGMAYER, Heinrich, (1840-1917)

1900 to present

ABBOTT, William James Lewis, (1853-1933)
ABEL, Othenio, (1875-1946)
ABERNATHY, George Elmer, (1888-1948)
ABRARD, René, (1892-1973)
ADAMS, George Irving, (1870-1932)
ADKINS, Walter Scott, (1890-1956)
AHLBURG, Johannes Heinrich Wilhelm, (1833-1919)
AHO, Aaro E., (-1977)
ALCOCK, Frederick James, (1888-1972)
ALEXANDER, Frances Elizabeth Somerville, (1908-1958)
ALIMEN, Henriette, (1900-)
ALLAN, John Andrew, (1884-1955)
ALLAN, Robin Sutcliffe, (1900-1966)

STRATIGRAPHY 1900 to present Cont'd.

ALLEN, Henry Atwool, (?1854-1934)
AMALITSKII, Vladimir Prochorovich, (1860-1917)
AMEGHINO, Florentino, (1854-1911)
AMI, Henry Mark, (1858-1931)
AMMON, Ludwig von, (1850-1922)
AMPFERER, Otto, (1875-1947)
ANDERSON, Alfred Leonard, (1900-1964)
ANDERSON, William, (1860-1915)
ANDERSSON, Johann Gunnar, (1874-1960)
ANDREWS, Ernest Clayton, (1870-1948)
ANDREWS, Richard Wallis, (?1927-1967)
ANDRUSOV, Dimitri, (1897-1976)
ANDRUSSOV, Nikolai Ivanovich, (1861-1924)
ANELLI, Mario, (1882-1953)
ANIČ, Dragutin, (1898-1961)
ARAMBOURG, Camille, (1885-1969)
ARKELL, William Joscelyn, (1904-1958)
ARMSTRONG, Henry Edward, (1848-1937)
ARNAUD, Hilaire, (1827-1908)
ARNOLD, Ralph, (1875-1961)
ARTHABER, Gustav von, Edler, (1864-1943)
ASANO, Kiyoshi, (1910-)
ASHLEY, George Hall, (1866-1951)
ASTRE, Gaston, (1896-)
AUGUSTA, Josef, (1903-1968)
AUTHELIN, Charles, (1871-1903)
AVELINE, William Talbot, (1822-1903)
BADEN-POWELL, Donald Ferlys Wilson, (1897-1973)
BAGG, Rufus Mather, (1869-1946)
BAILEY, Loring Woart, (1839-1925)
BAILEY, Tom Esmond Geoffrey, Captain, (-1919)
BAKER, Benjamin Alfred, (1859-1939)
BAKER, Herbert Arthur, (1885-1954)
BALK, Robert, (1899-1955)
BALL, John, (-1941)
BALL, John Rice, (1881-1953)
BALLARD, William Norval, (1906-1976)
BANCROFT, John Bevis Beeston, (-1944)
BANDY, Orville Lee, (1917-1973)
BARBER, Cecil Thomas, (-)
BARBOUR, George Brown, (1891-1977)
BARLOW, Alfred Ernest, (1861-1914)
BARON, Richard, Reverend, (1847-1907)
BARRABÉ, Louis, (1895-1961)
BARRELL, Joseph, (1869-1919)
BARRETT, Leslie Park, (1887-1972)
BARRETT, Lucas, (1837-1862)
BARROIS, Charles Eugène, (1851-1939)
BARRON, Thomas, (1867-1906)
BARROW, George, (1853-1932)
BARTLETT, Benjamin Pope, (-1928)
BARTON, George Hunt, (1852-1933)
BARTRUM, John Arthur, (1885-1949)
BASSANI, Francesco, (1853-1916)

STRATIGRAPHY 1900 to present Cont'd.

BASSE DE MÉNORVAL, Éliane Marguerite, (1899-)
BASSLER, Raymond Smith, (1878-1961)
BATEMAN, Alan Mara, (1889-1971)
BATES, Robert Glenn, (1923-1972)
BAUMBERGER, Ernst, (1866-1935)
BAUR, Carl Theodor von, (1836-1911)
BEACH, Hugh Hamilton, (1909-1973)
BEADNELL, Hugh John Llewellyn, Major, (1874-1944)
BEAUGÉ, Alfred, (1878-1935)
BECK, Carl Richard, (1858-1919)
BEEDE, Joshua William, (1871-1940)
BEEKLEY, Albert Leon, (1883-1952)
BEHR, Johannes, (1875-1973)
BEKKER, Hendrik, (1891-1925)
BELL, Alfred Graham, (?1875-1964)
BELL, Alfred Hannam, (1895-1977)
BELL, James Macintosh, (1877-1934)
BELL, John Arnold, (1939-1970)
BELL, Olin Grogan, (1893-1970)
BELL, Robert, (1841-1917)
BEM-BÖHM, Boleszláv, (1900-1964)
BEMROSE, Henry Howe, (1857-1939)
BENECKE, Ernst Wilhelm, (1838-1917)
BENNETT, Francis James, (1845-1920)
BENSON, William Noel, (1885-1957)
BENTZ, Alfred, (1897-1964)
BERENDT, Gottlieb Michael, (1836-1920)
BERGERON, Jules, (1853-1919)
BERGHELL, Hugo, (1864-1922)
BERNARD, Hugh Allen, (1915-1975)
BERNOUILLI, Walter, (1885-1946)
BERRY, Edward Willard, (1900-1968)
BERTRAND, Léon, (1869-1947)
BERTRAND, Marcel Alexandre, (1847-1907)
BEST, Edward, (1824-1906)
BEUDANT, François-Sulpice, (1787-1850)
BEUSHAUSEN, Hermann Emil Louis, (1863-1904)
BEYER, Samuel Walker, (1865-1931)
BEYSCHLAG, Franz, (-?1935)
BIEBER, Charles Leonard, (1901-1965)
BIGOT, Alexandre, (1863-1953)
BIOCHE, Alphonse, (1844-1918)
BION, Herbert Stanley, (1888-1915)
BIRD, Charles, (1843-1910)
BIRKMAIER, Anton, (-1926)
BISAT, William Sawney, (1886-1973)
BISHOP, Walter William, (1931-1977)
BISSET, Charles Barron, (-?1966)
BLAKE, George Stanfield, (-1940)
BLAKE, John Frederick, Reverend, (1839-1906)
BLAKE, John Hopwood, (1843-1901)
BLANC, Alberto Carlo, (1906-1960)
BLANC, Gian Alberto, Barone, (1879-1966)
BLANCKENHORN, Max Ludwig Paul, (1861-1947)

STRATIGRAPHY 1900 to present Cont'd.

BLANFORD, William Thomas, (1832-1905)
BLAYAC, Joseph, (1865-1936)
BLEICHER, Marie-Gustave, (1838-1901)
BLOMBERG, Karl Albert, (1844-1922)
BLONDEL, Fernand, (1894-1968)
BLOOM, Edward Francis Dandridge, (-1942)
BLOW, Walter Harry, (1924-1972)
BOBIES, Carl August, (1898-1958)
BODYLEVSKY, Vitaly Ivanovich, (1898-1968)
BOGDANOV, Alexei Alexeievitch, (1906-1971)
BÖHM, Georg, (1854-1913)
BÖHMERSHEIM, August Böhm von, Edler, (1858-1930)
BOLTON, Herbert, (1863-1936)
BONARELLI, Guido, (1871-1951)
BONNEY, Thomas George, Reverend, (1833-1923)
BONTE, Antoine Jean Marie Joseph, (1908-)
BOOCOCK, Colin, (1925-1973)
BORGES, Alexandre, (1898-1955)
BORN, Kendall Eugene, (1908-1947)
BORRELLO, Angel V., (1918-1971)
BÖSE, Emil, (1868-1972)
BOSWELL, Percy George Hamnall, (1886-1960)
BOUČEK, Bedřich, (1904-1975)
BOUCOT, Arthur James, (1924-)
BOULE, Marcellin, (1861-1942)
BOURCART, Jacques, (1891-1965)
BOURGUIGNAT, Jules René, (1829-?1919)
BOUSSAC, Jean, (1885-1917)
BRADLEY, William Frank, (1908-1973)
BRAMLETTE, Milton Nunn, (1896-1977)
BRANDES, Theodor, (-1916)
BRANNER, John Casper, (1850-1922)
BRANSON, Carl Colton, (1906-1975)
BRANSON, Edwin Bayer, (1877-1950)
BRANSON, Edwin Robert, (1908-1971)
BRANT, Ralph Allen, (1899-1958)
BRESSON, Arthur, (1873-1946)
BREWER, William Henry, (1828-1910)
BROCK, Byron Britton, (1904-1972)
BROCK, Reginald Walter, (1874-1935)
BROCKBANK, Richard Bowman, (1824-1912)
BRØGGER, Waldemar Christofer, (1851-1940)
BROGGI, Jorge Albert, (1888-1966)
BROMEHEAD, Cyril Edward Nowill, (1885-1952)
BROOKS, Alfred Hulse, (1871-1924)
BRÖSAMLEN, Otto, (1885-1966)
BROTZEN, Fritz, (1902-1968)
BROUWER, Hendrik Albertus, (1886-1973)
BROWN, Charles Barrington, (1839-1917)
BROWN, Henry Yorke Lyell "Stoney", (1844-1928)
BROWN, James Macmillan, (1921-1970)
BROWN, John Coggin, (1884-1962)
BROWN, Noel K., Jr., (1925-1968)
BRUDER, George, (1856-1916)

STRATIGRAPHY 1900 to present Cont'd.

BRUNNER VON WATTENWYL, Karl, (1823-1914)
BRUSINA, Spiridion, (1845-1908)
BRYAN, Walter Heywood, (1891-1966)
BRYDONE, Reginald Marr, (-1943)
BUBNOFF, Serge von, (1888-1957)
BUCKMAN, Sidney Savory, (1860-1929)
BUEHLER, Henry Andrew, (1876-1944)
BUKOWSKI, Gejza, (1858-1937)
BULL, Alfred Joseph, (1875-1950)
BURCKHARDT, Carl, (1869-1935)
BUREAU, Edouard, (1830-1918)
BURFOOT, James Dabney "Dan", Jr., (1896-1966)
BÜRGL, Hans, (1907-1966)
BURNETT, George Alexander, (1892-1952)
BURROUGHS, John, (1837-1921)
BURROWS, Henry William, (?1860-1939)
BURTON, Richard Cooksey, (1890-1916)
BUTTS, Charles, (1863-1946)
BUWALDA, John Peter, (1886-1954)
BUXTORF, August, (1877-)
CADISCH, Joos, (1895-)
CADY, Gilbert Haven, (1882-1970)
CAIRE, André, (1923-)
CAIRNES, Delorme Donaldson, (1879-1917)
CAIRNS, Robert, (1854-1911)
CALDERON Y ARANA, Salvador, (1851-1911)
CALEY, John Fletcher, (1903-1971)
CALL, Richard Ellsworth, (1856-1917)
CALLAWAY, Charles, (1838-1915)
CALVIN, Samuel, (1840-1911)
CAMERON, A. E., (-)
CAMPBELL, Arthur Byron, (1924-1971)
CAMPBELL, Guy, (1877-1970)
CAMPBELL, Henry Donald, (1862-1934)
CAMPBELL, Marius Robinson, (1858-1940)
CAMPBELL, Robert, (1881-1957)
CAMPBELL, Robert Burns, (1892-1955)
CANAVARI, Mario, (1855-1928)
CAPELLINI, Giovanni, (1833-1922)
CARMAN, Joel Ernest, (1882-1966)
CARPENTIER, Alfred, Chanoine, (1878-1952)
CARR, John Wesley, (1862-[1939])
CARROLL, Dorothy, (1907-1970)
CARRUTHERS, Robert George, (1880-1965)
CARUS-WILSON, C., (1857-1934)
CARVALHO, Anselmo Ferraz de, (1878-1955)
CASTRO Y SUERO, Manuel Fernandez de, (1825-1895)
CATHCART, Stanley Holman, (1889-1953)
CAYEUX, Lucien, (1864-1944)
ČECH, Vladimír, (1901-)
ČELEBRIČ, Dorde, (1929-1970)
ČEPEK, Ladislav, (1899-1974)
CHADWICK, George Halcott, (1876-1953)
CHAMBERLIN, Thomas Chrowder, (1843-1928)

STRATIGRAPHY 1900 to present Cont'd.

CHARLESWORTH, Edward, (1813-1893)
CHAUVE, Pierre, (1930-)
CHECCHIA-RISPOLI, Giuseppe, (1877-1947)
CHHIBBER, Harbans Lal, (1899-1955)
CHIPMAN, Kenneth G., (-)
CHLEBOWSKI, Tadeusz, (1908-1961)
CHOFFAT, Paul, (1849-1919)
CHUBB, Lawrence John, (?1888-1971)
CHUDEAU, René, (1864-1921)
CIOFALO, Saverio, (1825-1925)
CLAR, Eberhard, (1902-)
CLARK, William Bullock, (1860-1917)
CLARKE, John Mason, (1857-1925)
CLAYPOLE, Edward Waller, (1835-1901)
CLELAND, Herdman Fitzgerald, (1869-1935)
CLIFTON, Roland Leroy, (1884-1962)
CLINE, Lewis Manning, (1909-1971)
CLOOS, Hans, (1885-1951)
CLOUD, Preston Ercelle, Jr., (1912-)
CLOUGH, Charles Thomas, (1852-1916)
COBBOLD, Edgar Sterling, (1851-1936)
COCKBURN, Alexander Murray, (1901-1959)
COCKIN, George Marmaduke, (1852-1918)
COHEN, Emil Wilhelm, (1842-1905)
COLE, Grenville Arthur James, (1859-1924)
COLLINS, William Ion, (1897-1964)
COLLOT, Louis Marie François, (1846-1915)
COLONY, Roy Jed, (1870-1936)
COMBE, A. D., (1893-1949)
COMPTNER, Gustav, (1831-1922)
CONDON, Thomas, (1822-1907)
CONTEJEAN, Charles Louis, (1824-1907)
COOKE, Charles Wythe, (1887-1971)
COOMARASWAMY, Ananda K., (1877-1947)
COOPER, Byron Nelson, (1912-1971)
COOPER, Chalmer Lewis, (-)
CORNET, René Jules, (1865-1929)
CORSTORPHINE, George Steuart, (1865-1919)
CORTESE, Emilio, (1856-1938)
COSSMANN, Maurice, (1850-1924)
COTTER, Gerald de Purcell, (1881-1941)
COURTY, Georges, (1875-1953)
COX, Edward Travers, (1821-1907)
CRAIG, Robert, (?1821-1901)
CRAIG, Robert Meldrum, (?1882-1956)
CRAWFORD, Ralph Dixon, (1873-1950)
CREDNER, Carl Hermann Georg, (1841-1913)
CRICK, Walter Drawbridge, (1857-1903)
CRICKMAY, Geoffrey W., (1905-1971)
CRIDER, Albert Foster, (1873-1945)
CROFT, William Noble, (1915-1953)
CRONEIS, Carey, (1901-1972)
CROOKSHANK, Henry, (1893-1972)
CROSFIELD, Margaret Chorley, (1859-1952)

STRATIGRAPHY 1900 to present Cont'd.

CROWTHER, Arthur Frederick, (1929-1954)
CULEK, Antonin, (1904-1944)
CULVER, Harold Eugene, (1883-1970)
CUNNINGTON, C. H., Lieutenant, (1889-1918)
CUNNINGTON, William, (1813-1906)
CURRIE, Ethel Dobbie, (-1963)
CURRIER, Louis Wade, (1890-1970)
CUSHING, Henry Platt, (1860-1921)
CUSHMAN, Joseph Augustine, (1881-1949)
CUYLER, Robert Hamilton, (1908-1944)
CZARNOCKI, Jan, (1889-1951)
D'ACHIARDI, Antonio, (1839-1902)
DAGUIN, Fernand, (1889-1948)
DAHLGRÜN, Fritz, (1894-1954)
DAKE, Charles Laurence, (1883-1934)
DAKYNS, John Roche, (1836-1910)
DALLONI, Marius Gustave, (1880-1959)
DAL PIAZ, Giambattista, (1927-)
DAL PIAZ, Giorgio, (1872-1962)
DALTON, William H., (?1848-1929)
DAMON, Henry Gordon, (1883-1960)
DANE, Carle Hamilton, (1900-1968)
DANGEARD, Louis, (-)
DANZIG, Emil, (1855-1943)
DARELL, Robert Darell Smythe, (1851-1936)
DARNAY-DORNYAY, Béla, (1887-1965)
DARTON, Nelson Horatio, (1865-1948)
DATHE, Johann Friedrich Ernst, (1845-1917)
DAVID, Tannatt William Edgeworth, Sir, (1858-1934)
DAVIDSON, Charles Findlay, (1911-1967)
DAVIES, Arthur Morley, (1869-1959)
DAVIES, George Macdonald, (1885-1973)
DAVIES, John Henry, (1887-1971)
DAVIES, Lewis Merson, Lieutenant-Colonel, (1882-1955)
DAVIS, Arthur George, (1892-1957)
DAWSON, George Mercer, (1849-1901)
DAWSON, Thomas Albert, (1916-1974)
DE ALESSANDRI, Giulio, (-?1920)
DECKER, Charles Elijah, (1868-1958)
DE DYCKER, Raymond, (1903-1947)
DEECKE, Johannes Ernst Wilhelm, (1862-1934)
DEHÉE, René, (1898-1928)
DEISS, Charles Frederick, (1903-1959)
DE LA VALLÉE POUSSIN, Charles Louis Joseph Xavier, (1827-1903)
DELEAU, Paul Camille, (1904-)
DELÉPINE, Gaston-Georges, Monseigneur, (1878-1963)
DELVAUX, Émile, (1837-1901)
DEMAY, André, (1890-1964)
DENCKMANN, August, (1860-1925)
DE NICOLIS, Enrico, (1841-1908)
DENINGER, Karl, (1878-1917)
DENIZOT, Georges, (-)
DEPÉRET, Charles Jean Julien, (1854-1929)
DE RANCE, Charles Eugène, (1847-1906)

STRATIGRAPHY 1900 to present Cont'd.

DERBY, Orville Adelbert, (1851-1915)
DE REGNY, Paolo Vinassa, (1871-1957)
DESIO, Ardito, (1897-)
DESTINEZ, Pierre, (-1911)
DEUBEL, Fritz, (1898-1966)
DEUSSEN, Alexander, (1882-1959)
DEWALQUE, Gilles Joseph Gustave, (1828-1905)
DEWEY, Henry, (1876-1965)
DEWHURST, Thomas, (1881-1973)
DeWOLF, Frank Walbridge, (1881-1957)
DICKERSON, Roy Ernest, (1877-1944)
DIENER, Carl, (1862-1928)
DIETRICH, Wilhelm Otto, (1881-1964)
DINES, Henry George, (1891-?1964)
DINHAM, Charles Hawker, (1883-1955)
DI STEFANO, Giovanni, (1856-1918)
DIX, Emily, (1904-1973)
DIXON, Ernest Edward Leslie, (1876-1963)
DOCWRA, Colin, (-1942)
DOELL, Richard Rayman, (1923-)
DOLLÉ, Louis, (1878-1965)
DOLLFUS, Gustave Fréderic, (1850-1931)
DOLLOT, Auguste, (1841-1924)
DONCIEUX, Louis, (1874-1960)
DORLODOT, Henry de, Chanoîne, (1855-1929)
DORMAL, Victor, (-1900)
DORSEY, George Edwin, (1892-1953)
DOUGLAS, James Archibald, (1884-1978)
DOUVILLÉ, Joseph Henri Ferdinand, (1846-1937)
DOUVILLÉ, Robert, (1881-1914)
DOUXAMI, Henri, (1871-1913)
DOWLING, Donaldson Bogart, (1858-1925)
DRĂGHICEANU, Mathei M., (1844-1939)
DRAKE, Noah Fields, (1864-1945)
DRESSER, John Alexander, (1866-1954)
DREVERMANN, Fritz, (1875-1932)
DUBOIS, Georges, (1890-1953)
DÜCKER, Alfred, (1908-1969)
DUMBLE, Edwin Theodore, (1852-1927)
DUNLOP, Andrew, (1845-1915)
DUNN, Edward John, (1844-1937)
DUPONT, Édouard François, (1841-1911)
DURAND DELGA, Michel, (1923-)
DUTERTRE, August Pierre, (1890-1940)
DUTTON, Clarence Edward, Major, (1841-1912)
DUTY, Richard Emerson, (1908-1966)
DVOŘÁK, Josef, (1904-)
DYER, William Spafford, (1894-1941)
EARDLEY, Armand John, (1901-1972)
EARGLE, Dolan Hoye, (1905-1973)
EARLE, Kenneth Wilson, (1891-1944)
EASTWOOD, Tom, (1889?-1970)
EBEL, Johann Gottfried, (1764-1830)
EBY, James Brian, (1896-)

STRATIGRAPHY 1900 to present Cont'd.

ECK, Heinrich von, (1837-1925)
EDMONDS, Charles, (1886-1964)
EDMONDS, Frederic Harrison, (1898-1965)
EDMUNDS, Francis Hereward, (1893-1960)
EDSON, Fanny Carter, (1887-1952)
EDWARDS, Merwin Guy, (1889-1969)
EGAN, Frederick William, (1836-1901)
EGERTON, William Henry, Reverend, (1811-1910)
EISEL, Robert, (1826-1917)
EKBLAW, George Elbert, (1895-1972)
ELDER, Stanley Gordon, (1904-1957)
ELDRIDGE, George Homans, (1854-1905)
ELLES, Gertrude Lilian, (1872-1960)
ELLIS, Charles W., Jr., (1934-1970)
ELLISOR, Alva Christine, (1892-1964)
ELLS, Robert Wheelock, (1845-1911)
ELLSWORTH, Hardy Vincent, (1889-1952)
ELSDEN, James Vincent, (?1856-1930)
EMERSON, Benjamin Kendall, (1843-1932)
EMMONS, William Harvey, (1876-1948)
ENGEL, Theodor, (1842-1933)
EPPS, Frank James, (?1891-1974)
ERDMANN, Charles Edgar, (1897-1971)
ERDMANN, Edvard, (1840-1923)
ERNI, Arthur, (1885-1945)
ESAREY, Ralph Emerson, (1901-1975)
ESKOLA, Pentti Eelis, (1883-1964)
ETHERIDGE, Robert, (1819-1903)
ETHERIDGE, Robert Jr., (1847-1920)
EVANS, Charles Sparling, (1894-1962)
EVANS, Percy, (1892-1974)
EWING, William Maurice, (1906-1974)
EXTON, Hugh, (1833-1903)
FABIANI, Ramiro, (1879-1954)
FABRE, Georges, (-1911)
FAGG, Christopher Charles, (1883-1965)
FAIRCHILD, John Gifford, (1882-1965)
FALCONER, John Downie, (1876-1947)
FALLOT, Paul, (1889-1960)
FALSAN, Albert, (1833-1902)
FARMERY, Joseph Richard, (1872-1928)
FATH, Arthur Earl, (1887-1976)
FAVRE, Ernest, (1845-1925)
FAYOL, Henri, (1841-1925)
FEARNSIDES, William George, (1879-1968)
FENDRYCH, Otakar, (1860-1945)
FERASIN, Francesco, (1928-1958)
FERGUSON, John Logan, (1893-1973)
FERMOR, Lewis Leigh, Sir, (1880-1954)
FERRONNIÈRE, Georges, (1875-1922)
FIEBELKORN, Max, (1869-1912)
FIELD, Richard Montgomery, (1885-1961)
FINCH, Elmer Harrison, (1883-1953)
FINCKH, Ludwig, (1871-1930)

STRATIGRAPHY 1900 to present Cont'd.

FINLAY, Harold John, (1901-1951)
FIRKET, Adolphe, (1837-1905)
FISCHER, Hugo, (1864-1926)
FISHER, Cassius Asa, (1872-1930)
FISHER, Osmond, Reverend, (1817-1914)
FISK, Harold Norman, (1908-1964)
FLETCHER, Hugh, (1848-1909)
FLETT, John Smith, Sir, (1869-1947)
FLEURY, Ernest, (1878-1958)
FLIEGEL, Gotthard, (1873-1947)
FLINT, Richard Foster, (1902-1976)
FLORES REYES, Teodoro, (1873-1955)
FOLEY, Edward J., (1908-1971)
FOLGER, James Anthony, (1896-1956)
FOLGNER, Raimund, (1882-1916)
FOLLMANN, Otto, (1856-1926)
FOOTE, R. Bruce, (-1912)
FORD, Trevor David, (1932-)
FORIR, Henri, (1856-1907)
FORSTER-COOPER, Clive, (1880-1947)
FORTAU, René, (1867-1920)
FOSLIE, Steinar, (1887-1951)
FOSTER, Clement Le Neve, Sir, (1841-1904)
FOSTER, Margaret Dorothy, (1895-1970)
FOUQUÉ, Ferdinand André, (1828-1904)
FOURMARIER, Paul Frédéric Joseph Lucien Lauren, (1877-1970)
FOURNIER, Eugène, (1871-1941)
FOX, Cyril Sankey, Sir, (1886-1951)
FOX-STRANGWAYS, Charles Edward, (1844-1910)
FRAAS, Eberhard, (1862-1915)
FRAIPONT, Julien, (1857-1910)
FRANKE, Adolf, (1860-1942)
FRAZER, Persifor, (1844-1909)
FRECH, Fritz, (1861-1917)
FREI, Ernst, (1894-1974)
FRENGUELLI, Joaquín, (1883-1958)
FRIEDBERG, Wilhelm, (1873-1941)
FRIEDEL, Georges, (1865-1933)
FRIES, Carl, Jr., (1910-1965)
FRITSCH, Karl von, Freiherr, (1838-1906)
FRITTS, Crawford Ellsworth, (1927-1972)
FRIZZELL, Donald Leslie, (1906-1972)
FROSTERUS, Gustaf Benjamin, (1866-1931)
FRÜH, Jakob, (1852-)
FUCHS, Alexander, (1874-1935)
FUCHS, Theodor, (1842-1925)
FUCHS, Vivian Ernest, Sir, (1908-)
FUGGER, Friedrich Eberhard, (1842-1919)
FURCRON, Aurelius Sydney, (1899-1971)
FURON, Raymond, (1898-)
GAÁL, István, (1877-1957)
GAGNEBIN, Élie, (1891-1949)
GALE, Hoyt Stoddard, (1876-1952)
GALLAGHER, David, (1906-1968)

STRATIGRAPHY 1900 to present Cont'd.

GALLWITZ, Hans, (1896-1958)
GARDINER, Charles Irving, (?1868-1940)
GARDNER, James Henry, (1883-1964)
GARDNER, Julia Anna, (1882-1960)
GARWOOD, Edmund Johnston, (1864-1949)
GASTER, Christopher Theodore Augustine, (1878-1963)
GAULT, Hugh Richard, (1915-1961)
GAVELIN, Axel Olof, (1875-1947)
GEIJER, Per Adolf, (1886-1976)
GEIKIE, Archibald, Sir, (1835-1924)
GEIKIE, James, (1839-1915)
GEINITZ, Eugen, (1854-1925)
GEINITZ, Hanns Bruno, (1814-1900)
GEMMELLARO, Mariano, (1879-1921)
GENTIL, Louis, (1868-1925)
GEORGE, William Owsley, (1892-1966)
GERBER, Eduard, (1876-1956)
GEYER, David, (1855-1932)
GEYER, Georg, (1857-1938)
GIBBONS, Arthur, (1867-1938)
GIBSON, Walcot, (1864-1941)
GIDLEY, James Williams, (1866-1931)
GIGNOUX, Maurice Irénée Marie, (1881-1955)
GILBERT, Grove Karl, (1843-1918)
GILL, James Rogers, (1922-1972)
GILLARD, Pierre A., (1907-1947)
GILLERMAN, Elliot "Bud", (1913-1974)
GILLULY, James, (-)
GINTER, Roy Lamont, (1892-1958)
GIRARDOT, Charles Joseph Albert Julien, (1848-1930)
GIRTY, George Herbert, (1869-1939)
GLANGEAUD, Louis, (1903-)
GLASER, Theodor, (1894-1922)
GLAUERT, Ludwig, (1879-1963)
GLENN, Leonidas Chalmers, (1871-1951)
GODON, Joseph, Chanoine, (-1932)
GOLDMAN, Marcus Isaac, (1881-1965)
GOLDRING, Winifred, (1888-1971)
GOODCHILD, John George, (1844-1906)
GOODLET, George Anderson, (1910-1970)
GOODYEAR, Edith, (1879-1959)
GORDON, Charles Henry, (1857-1934)
GORIZDRO-KULCZYCKA, Zinaida, (1884-1949)
GORJANOVIĆ-KRAMBERGER, Dragutin, (1856-1937)
GORMAN, Joseph M., (1910-1967)
GORSKY, I. I., (1893-1975)
GORTANI, Michele, (1883-1966)
GOSSELET, Jules Auguste, (1832-1916)
GOSSLING, Frank Newbery, (1888-1946)
GOTTSCHE, Carl Christian, (1855-1909)
GOULD, Charles Newton, (1868-1949)
GRABAU, Amadeus William, (1870-1946)
GRAHAM, Joseph John, (1909-1967)
GRAHAM, William Armstrong Patterson, (1899-1934)

STRATIGRAPHY 1900 to present Cont'd.

GRAMBAST, Louis, (-1976)
GRAND'EURY, François Cyrille, (1839-1917)
GRANT, Ulysses Sherman, (1867-1932)
GRANTHAM, Donald Robert, (1896-1969)
GRATON, Louis Caryl, (1880-1970)
GRAY, Joseph William, (-1935)
GREBE, Heinrich, (1831-1903)
GREEN, John Frederick Norman, (?1873-1949)
GREEN, Upfield, (1834-1917)
GREENE, Alfred Ernest, (1878-1940)
GREENE, Frank Cook, (1886-1975)
GREGORY, Herbert E., (1869-1952)
GREGORY, Hubert Harold, (-?1950)
GREGORY, John Walter, (1864-1932)
GREPPIN, Eduard, (1856-1927)
GRIESBACH, Karl Ludolf, (1847-1907)
GRIFFITH, Glenn Rodgers Vickery, (1891-1964)
GRIGORAŞ, Nicolae, (1913-1969)
GRIPP, Karl, (1891-)
GROEBER, Paul Friedrich Carl, (1885-1968)
GRÖNWALL, Karl Anders Axel, (1869-1944)
GROOM, Theodore Thomas, (1863-)
GROTH, Jean, (1899-1916)
GROUT, Frank Fitch, (1880-1958)
GRUTTERINK, Jan Adolf, (1879-1949)
GRZYBOWSKI, Józef, (1869-1922)
GUBLER, Jean, (1902-1940)
GUÉBHARD, Adrien, (1849-1924)
GUILLAUME, Louis, (1894-1952)
GUILLAUME, Solange, (1932-)
GUILLOU, Robert Barton, (1923-1965)
GUNN, William, (1837-1902)
GUNTER, Herman, (1885-1972)
GÜNZLER-SEIFFERT, Hans, (1883-1957)
GÜRICH, Georg, (1859-1938)
GURLDT, Adolf Friedrich, (1829-1902)
GUTZWILLER, Andreas, (1845-1917)
GUZIK, Kazimierz, (1911-1970)
HÄBERLE, Daniel, (1864-1934)
HABETS, Alfred, (1839-1908)
HAHN, Friedrich Felix (Fritz), (1885-1914)
HALL, Arthur Lewis, (1872-1955)
HALL, Charles Monroe, (1870-1903)
HALL, Christopher Webber, (1845-1911)
HALL, George Martin, (1891-1941)
HALL, Thomas Clifford Fitzwilliam, (1886-1956)
HALL, Thomas Sergeant, (1858-1915)
HALLIMOND, Arthur Francis, (1890-1968)
HAM, William Eugene, (1916-1970)
HAMLING, Joseph Green, (-?1935)
HANNESSON, Pálmi, (1898-1956)
HANSELL, James Myron, (1902-1970)
HANSON, Ernest Albin, (1899-1973)
HARBORT, Erich, (1879-1929)

STRATIGRAPHY 1900 to present Cont'd.

HARES, Charles Joseph, (1881-1970)
HARPER, John Childs, (1913-1977)
HARRINGTON, Bernard J., (1848-1907)
HARRINGTON, George Leavitt, (1883-1972)
HARRINGTON, Horacio Jaime, (1910-1973)
HARRIS, George Frederick, (1862-1906)
HARRIS, Gilbert Dennison, (1864-1952)
HARRIS, Reginald W., (-)
HARRISON, John Vernon, (1892-1972)
HARRISON, William Jerome, (1845-1908)
HARTLEY, John Jerom, (1886-)
HATCH, Frederick Henry, (?1864-1932)
HAUG, Gustave Émile, (1861-1927)
HAUPTMAN, Charles R., (1891-1971)
HAVARD, Charles Gentry, (1921-1950)
HAWKINS, Alfred Cary, (1887-1954)
HAWKINS, Herbert Leader, (1887-1968)
HAWORTH, Erasmus, (1855-1932)
HAYES, Albert Orion, (1882-1965)
HAYES, Charles Willard, (1858-1916)
HAZZARD, Roy Thorpe, (1893-1976)
HEADLEE, Larry A., (1938-1970)
HEARD, Albert, (1894-1944)
HECTOR, James, Sir, (1834-1907)
HEIM, Albert, (1849-1937)
HEIM, Arnold, (?1882-1965)
HEINZELMANN, Karl, (1911-1966)
HELBRONNER, Paul, (1871-1938)
HENDERSON, Junius, (1865-1937)
HENDRICKSON, Victor James, (1892-1953)
HENNEN, Ray Vernon, (1875-1958)
HENNIG, Edwin, (1882-)
HENRY, J., (-1924)
HERAK, Milan, (1917-)
HERALD, Frank Edward, (1887-1965)
HERITSCH, Franz, (1882-1945)
HERNÂNDEZ-PACHECO, Eduardo, (-1965)
HERNÂNDEZ-PACHECO, Francisco, (1899-1976)
HERON, Alexander Macmillan, (1884-1971)
HEROY, William Bayard, Sr., (1883-1971)
HERPERS, Henry F., Jr., (1915-1952)
HERRICK, Clarence Luther, (1858-1904)
HERRIES, Robert Stansfield, (?1860-1941)
HERRMANN, Fritz Heinrich Ulrich, (1883-1920)
HESS, Frank L., (1871-1955)
HEWETT, Donnel Foster, (1881-1971)
HEWITT, Henry Dixon, (-?1966)
HEWITT, Robert Leigh, (1911-1956)
HIBBARD, Claude W., (1905-1973)
HIBBERT-WARE, Samuel, (1782-1848)
HIBSCH, Josef Emanuel, (1852-1940)
HIGAZY, Riad Abdel-Megid, (1919-1967)
HILBER, Vincenz, (1853-1931)
HILGARD, Eugene Woldemar, (1833-1916)

STRATIGRAPHY 1900 to present Cont'd.

HILL, Dorothy, (-)
HILL, Frank A., (1858-1915)
HILL, James Bastian, (1861-1927)
HILL, James Wilcott, (1920-1954)
HILL, Robert Thomas, (1858-1941)
HILL, William, (1849-1912)
HILLS, Richard Charles, (1848-1923)
HIMUS, Godfrey Wilfred, (1894-1964)
HINCH, John de Witt, (?1875-1931)
HIND, Henry Youle, (1823-1908)
HIND, Wheelton, (1859-1920)
HINDE, George Jennings, (1839-1918)
HINTERLECHNER, Karl, (1874-1932)
HITCHCOCK, Charles Henry, (1836-1919)
HITCHON, Hubert, (1899-1962)
HLÁVKA, Karel, (1905-)
HOBSON, Bernard, (1860-1933)
HODGE, Edwin Thomas, (1887-1970)
HODGE, Matthew Burnop, (-1951)
HOERNES, Rudolf, (1850-1912)
HOFFMANN, George Christian, (1837-1917)
HOFMANN, Adolf, (1853-1913)
HOLDEN, Roy Jay, (1870-1945)
HOLGATE, Benjamin, (1838-1915)
HOLLAND, Thomas Henry, Sir, (1868-1947)
HOLLINGWORTH, Sydney Ewart, (1899-1966)
HOLM, Edvard Johann Gerhard, (1853-1926)
HOLMES, Francis S., (1815-1882)
HOLMES, T. Vincent, (?1841-1923)
HOLMSEN, Gunnar, (1880-1976)
HOLTEDAHL, Olaf, (1885-1975)
HOLZAPFEL, Eduard, (1853-1913)
HOOLEY, Reginald Walter, (1866-1923)
HOPPER, Walter Everett, (1886-1971)
HÓRAK, Lubomír, (1935-1962)
HORBERG, Carl Leland, (1910-1955)
HORNE, John, (1848-1928)
HORUSITZKY, Ferenc, (1901-1971)
HORWITZ, Ludwik, (1875-1943)
HOURCQ, Victor, (1901-1961)
HOWCHIN, Walter, Reverend, (1845-1937)
HOWE, Ernest, (1875-1932)
HOWELL, Benjamin Franklin, (1890-1976)
HOWELL, Edwin Eugene, (1845-1911)
HOWELL, Henry Hyatt, (1834-1915)
HOWSE, Richard, (1821-1901)
HOYT, John Harger, (1928-1970)
HUDDLE, John Warfield, (1907-1975)
HUDLESTON, Wilfrid Hudleston, (né Simpson), (1828-1909)
HUDSON, Frank Samuel, (1890-1967)
HUDSON, Robert George Spencer, (1895-1965)
HUENE, Friedrich von, Freiherr, (1875-1969)
HUFFMAN, Edward Archibald, (1912-1954)
HUGHES, Donald Dudley, (1893-1946)

STRATIGRAPHY 1900 to present Cont'd.

HUGHES, Richard John, Jr., (1915-1968)
HUGHES, Theodore W. H., (?1843-1907)
HUGHES, Thomas McKenny, (1832-1917)
HULL, Edward, (1829-1917)
HULL, Joseph Poyer Deyo, (1889-1967)
HUME, George Sherwood, (1893-1965)
HUME, William Fraser, (1867-1949)
HUNT, Arthur Roope, (1843-1914)
HUPÉ, Pierre, (1907-)
HUSSEY, Russell Claudius, (1888-1970)
HUTCHINSON, Lon Lewis, (1877-1947)
HUTTON, Colin Osborne, (1910-1971)
HUTTON, Frederick Wollaston, (1836-1905)
HUTTON, Harriet Mary, (1873-1937)
IDDINGS, Joseph Paxson, (1857-1920)
ILLING, Vincent Charles, (1890-1969)
INVILLIERS, Edward Vincent d', (1857-1928)
IRELAND, Hubert Andrew, (1904-1973)
IRVING, John Duer, (1874-1918)
JACK, Robert Lockhart, (1878-1964)
JACKSON, James Frederick, (1896-1966)
JACOB, Charles Edward, (1914-1970)
JACOB, Charles François Étienne, (1878-1962)
JACOBS, Elbridge Churchill, (1873-1957)
JACQUET, Fernand, (1908-1937)
JACQUOT, Eugène, (1817-1903)
JAGGAR, Thomas Augustus, (1871-1953)
JAHN, Jaroslav Jiljí, (1865-1934)
JANET, Léon, (1861-1909)
JAROSZ, Jan, (1877-1944)
JARVIS, Daniel, (1920-1973)
JEHU, Thomas John, (1871-?1943)
JENNINGS, Albert Vaughan, (1864-1903)
JENNY, Fridolin, (1862-1933)
JENTZSCH, Alfred, (1850-1925)
JESPERSEN, Johan Magnus Peder, (1833-1917)
JESSEN, Axel Hans, (1868-1952)
JEWETT, John Mark, (1896-1970)
JILLSON, Willard Rouse, (1890-)
JIMBŌ, Kotora, (1867-1924)
JODOT, Paul, (1879-1961)
JOESTING, Henry Rochambeau, (1903-1965)
JOHANNSEN, Albert, (1871-1962)
JOHNS, Cosmo, (?1866-1951)
JOHNSON, Bertrand Leroy, (1882-1962)
JOHNSON, Charles George, (1914-1969)
JOHNSTON, Ashton William, (1910-)
JOHNSTON, Robert Mackenzie, (-1918)
JOHNSTON, William Drumm, Jr., (1899-1972)
JOLEAUD, Léonce, (1880-1938)
JONES, Brynmor, (1902-1941)
JONES, Daniel J., (1889-1965)
JONES, Hubert Cecil, (1881-1963)
JONES, J. Claude, (1877-1932)

STRATIGRAPHY 1900 to present Cont'd.

JONES, Owen Thomas, (1878-1967)
JONES, Thomas Rupert, (1819-1911)
JONES, William Richard, (1890-1970)
JONGMANS, Wilhelmus Josephus, (1878-1957)
JORDAN, Louise, (1908-1966)
JOSSÉ, Kazimierz, (1892-1942)
JOWETT, Albert, (1875-?1958)
JUDD, John Wesley, (1840-1916)
JUDSON, Sidney Arthur, (1887-1971)
JUKES-BROWNE, Alfred John, (1851-1914)
JULIEN, Pierre Alphonse, (1838-1905)
JUNNER, Norman Ross, (1891-1973)
KAISER, Friedrich Wilhelm Erich, (1871-1934)
KALÁŠEK, Jan, (1903-)
KARPINSKY, Alexander Petrovich, (1847-1936)
KARRER, Felix, (1825-1903)
KASTNER, Karl, (1847-1907)
KATZER, Bedřich, (1861-1925)
KAUFMAN, Rudolf, (1909-ca 1942)
KAUTSKY, Fritz, (1890-1963)
KAY, Fred Hall, (1885-1943)
KAY, George Marshall, (1904-1975)
KAYSER, Friedrich Heinrich Emanuel, (1845-1927)
KEITH, Arthur, (1864-1944)
KENDALL, Percy Fry, (1856-1936)
KENNEDY, Horas Tristram, Lieutenant, (1889-1917)
KENNEDY, Luther Eugene, (1891-1966)
KENNEDY, William, (-1926)
KENNERLEY, John Brian, (1937-1976)
KERFORNE, Fernand, (1864-1927)
KETTNER, Radim, (1891-1967)
KEW, William Stephen Webster, (1890-1961)
KEYES, Charles Rollin, (1864-1942)
KEYTE, Ivy Allen, (1878-1931)
KIAER, Johan Aschehong, (1869-1931)
KILIAN, Charles Constant Wilfrid, (1862-1925)
KILIAN, François Théodore Conrad, (1898-1950)
KILROE, James Robinson, (1848-1927)
KIMBALL, Kent K., (1892-1953)
KINAHAN, George Henry, (1829-1908)
KINDLE, Edward Martin, (1869-1940)
KING, Clarence Rivers, (1842-1901)
KING, William Bernard Robinson, (1899-1963)
KIRBY, Maurice E. "Bud", (1909-1963)
KIRKBY, James Walker, (1834-1901)
KISSLING, Ernst, (1865-1936)
KITCHIN, Finlay Lorimer, (1870-1934)
KITSON, Albert Ernest, Sir, (1868-1937)
KITTL, Ernst Anton Leopold, (1854-1913)
KJELLSTRÖM, Carl Johan Otto, (1855-1913)
KLAASSEN, Hendericus Martinus, (1828-1910)
KLAUTZSCH, Adolf, (1869-1927)
KLEBS, Richard, (1850-1911)
KLEIBER, Karl Ernst, (1912-1972)

STRATIGRAPHY 1900 to present Cont'd.

KLEINPELL, William Darwin, (1898-1959)
KLEMM, Gustav, (1858-1938)
KLINE, Virginia Harriett, (1910-1959)
KLOMPÉ, Theodorus Hendricus Franciscus, (?1904-1963)
KLOUČEK, Celda, (1855-1935)
KNIGHT, Garold Louis "Jerry", (1894-1968)
KNIGHT, Samuel Howell, (1892-1975)
KNIGHT, Wilbur Clinton, (1858-1903)
KNIPE, Henry Robert, (1855-1918)
KNOPF, Adolph, (1882-1966)
KOBAYASHI, Teiichi, (1901-)
KOBY, Frédéric Louis, (1852-1930)
KOBYLECKI, Mieczyslaw, (1909-1944)
KOCH, Antal von Bodrog, (1843-1927)
KOCH, Ferdo, (1874-?1936)
KOCH, Gustav Adolf, (?1846-1921)
KOCH, Lauge, (1892-1964)
KOCKEL, Carl Walter, (1898-1966)
KODYM, Odolen, (1898-1963)
KOENEN, Adolf von, (1837-1915)
KOERT, Willi, (1875-1927)
KOKEN, Ernst Friedrich Rudolph Karl von, (1860-1912)
KOKOSZYŃSKA, Bronislawa, (1897-1959)
KÖLBL, Leopold, (1895-1970)
KOMOROWICZ, Maurycy, (1881-1923)
KONTKIEWICZ, Stanislaw, Sr., (1849-1924)
KORMOS, Tivadar, (1881-1948)
KORN, Johannes, (1862-1927)
KOSCHMANN, Albert Herbert, (1896-1962)
KOUTEK, Jaromír, (1902-)
KOZLOWSKI, Roman, (1889-1977)
KRAFFT VON DELLMENSINGEN, Albrecht, (1871-1901)
KRAMER, William Baltser, III, (1896-1963)
KREIDLER, William Lynn, (1911-1975)
KREUTZ, Felix Franz Xaver, (1844-1910)
KRISHNAN, Maharajapuram Sitaram, Padma Bhusan, (1898-1970)
KŘIŽ, Martin, (1841-1916)
KRUITZINGA, Pieter, (1885-1967)
KÜHN, Othmar, (1892-1969)
KUMMEL, Henry Barnard, (1867-1945)
KUŹNIAR, Wiktor, ([1871]-1935)
KYNASTON, Herbert, (1868-1915)
LACHMANN, Richard, (1885-1916)
LACOSTE, Jean Pierre Victor, (1901-1942)
LAFLAMME, Joseph Clovis Kemner, Monseigneur, (1849-1910)
LAKE, Philip, (1865-1949)
LAMARE, Pierre, (1894-1967)
LAMBERT, Alan, (1837-1927)
LAMBERT, Jules Mathieu, (1848-1940)
LAMBERT, Paul, (1869-1950)
LAMONT, Archie, (-)
LAMPLUGH, George William, (1859-1926)
LANCASTER, L. Kenneth, (1906-1971)
LANDES, Henry, (1867-1936)

STRATIGRAPHY 1900 to present Cont'd.

LANDES, Robert William, (1910-1969)
LANDON, Joseph, (-1902)
LANE, Alfred Church, (1863-1948)
LANG, William Dickson, (1878-1966)
LANQUINE, Antonin, (-)
LAPPARENT, Albert Auguste de, (1839-1908)
LAPPARENT, Albert Félix de, (1905-1975)
LAPPARENT, Jacques de, (-?1949)
LAPWORTH, Charles, (1842-1920)
LAPWORTH, Herbert, (1875-1933)
LARSEN, Esper Signius, III, (1912-1961)
LAUNAY, Louis Alphonse Auguste de, (1860-1938)
LAWRIE, William Henry, (?1884-1943)
LEACH, Arthur Leonard, (1869-1957)
LEANZA, Armando F., (1919-1975)
LEATHEROCK, Otto, (1901-1941)
LEBOUR, George Alexander Louis, (1847-1918)
LECKWIJCK, Willem Peter van, (1902-1975)
LECOINTRE, Georges, (1888-1972)
LECOMPTE, Marius, (1902-1973)
LEE, Willis Thomas, (1864-1926)
LEES, George Martin, (1898-1955)
LEIGHTON, Henry, (1884-1963)
LEIGHTON, Morris Morgan, (1887-1971)
LEIGHTON, Thomas, (1858-1906)
LEITCH, Duncan, (1904-1956)
LEMOINE, Paul, (1878-1940)
LENNIER, Gustave, (-1905)
LENZ, Oskar, (1848-1925)
LEONARD, Arthur Gray, (1865-1932)
LEONARD, Hugh, (1841-1909)
LEONHARD, Richard, (1870-1916)
LEPPLA, August, (1859-1924)
LEPSIUS, Karl Georg Richard, (1851-1915)
LERICHE, Maurice, (1875-1948)
LESLEY, J. Peter, (1819-1903)
LEUZE, Johannes, (1883-1915)
LEVORSEN, Arville Irving, (1894-1965)
LEWIŃSKI, Jan Piotr, (1876-1939)
LEWIS, G. Edward, (-)
LEWIS, Herbert Price, (?1895-1947)
LEWIS, Joseph Volney, (1869-1969)
LIDDLE, Ralph Alexander, (1896-1963)
LIEBUS, Adalbert, (1876-1945)
LIGHTFOOT, Ben, (1888-1966)
LIMANOWSKI, Mieczyslaw, (1876-1947)
LINDGREN, Waldemar, (1860-1939)
LINDSTRÖM, Axel Fredrik, (1839-1911)
LINDSTRÖM, Jonas Gustaf Oscar, (1829-1901)
LINSTOW, Otto von, (1872-1929)
LISSON, Carlos I., (1868?-1957)
LITTLE, Otway Henry, (-1956)
LOBJOIT, William Michael, (1932-1967)
LOBLEY, James Logan, (1833-1913)

STRATIGRAPHY 1900 to present Cont'd.

LÓCZY, Lajos, (1849-1920)
LOHEST, Max, (1857-1926)
LOMAS, Joseph, (1860-1908)
LOMNICKI, Marian Alojzy R., (1845-1915)
LONSDALE, John Tipton, (1895-1960)
LOOMIS, Theodor, (1875-1909)
LÖRENTHEY, Imre, (1867-1917)
LORENZ, Theodor, (1875-1909)
LORETZ, Hermann, (1836-1917)
LORIOL LE FORT, Charles Louis Perceval de, (1828-1908)
LORY, Pierre, (1866-1956)
LOTZE, Franz, (1903-1971)
LOUGHLIN, Gerald Francis, (1880-1946)
LOUGHRIDGE, Robert Hills, (1843-1917)
LOW, Albert Peter, (1861-1952)
LOW, Julian W., (1904-1968)
LOWMAN, Shepard Wetmore, (1899-1968)
LUCIUS, Michel, (1876-1961)
LUGEON, Maurice, (1870-1953)
LUKASZÉWICZ, Józef, (1863-1928)
LUKOVIČ, Milan, (1889-)
LUNDQVIST, Gösta, (1894-1967)
LUNIEWSKI, Adam, (1887-1945)
LUPTON, Charles Thomas, (1878-1935)
LYDEKKER, Richard, (1849-1915)
LYELL, Charles, Sir, (1797-1875)
LYMAN, Benjamin Smith, (1835-1920)
MAACK, Reinhard, (1892-1969)
MAAS, Günther Heinrich Julius Max, (1871-1905)
MACBRIDE, Thomas Huston, (1848-1934)
MACDONALD, Simon, (1927-1971)
MACGREGOR, Murray, (1884-1966)
MacHENRY, Alexander, (1843-1919)
MACKENZIE, Graham Stewart, (1905-1960)
MACLAREN, James Malcolm, (1873-1935)
MACMILLAN, William Edward Frank, (-1954)
MACOVEI, Gheorghe, (1880-1969)
MACPHERSON, José, Don, (1839-1902)
MADIGAN, Cecil Thomas, (1889-1946)
MAGNUSSON, Nils Harald, (1890-1976)
MAILLARD, Gustave, (1860-1891)
MAIRE, Julien Victor, (1856-1941)
MAKOWSKY, Alexander, (1833-1908)
MALAISE, Constantin H. L., (1834-1916)
MALAVOY, Jean, (1903-1945)
MALDONADO-KOERDELL, Manuel, (1908-1972)
MALLADA, Lucas, (1841-1921)
MALLING, Carl Peter, (1862-1927)
MALOTT, Clyde A., (-1950)
MANSEL-PLEYDELL, John Clavell, (1817-1902)
MANSFIELD, Wendell Clay, (1874-1939)
MANSON, William, (1891-1973)
MARÈS, Paul, (1826-1900)
MARGERIE, Emmanuel Marie Pierre Martin Jacquin de, (1862-1953)

STRATIGRAPHY 1900 to present Cont'd.

MARGETIČ, Marko, (1891-1955)
MARINELLI, Olinto, (1874-1926)
MARKHAM, Clements Robert, Sir, (1830-1916)
MARPLE, Mildred Fisher, (1898-1966)
MARR, John Edward, (1857-1933)
MARSELL, Ray Everett, (1893-1971)
MARSHALL, Patrick, (1869-1950)
MARSHAM-TOWNSHEND, Robert Honourable, (1834-1915)
MARTELLI, Alessandro, (1877-1934)
MARTIN, David, (1842-1918)
MARTIN, Edgar Charles, (1884-1955)
MARTIN, Francis Ivan "Shorty", (1896-1952)
MARTIN, George C., (1875-1943)
MARTIN, Johann Karl Ludwig, (1851-1942)
MARWICK, John, (1891-1978)
MASON, Arnold Caverly, (1906-1961)
MATĚJKA, Alois, (1898-)
MATHER, Kirtley Fletcher, (1888-1978)
MATHEWS, Edward Bennett, (1869-1944)
MATLEY, Charles Alfred, (1866-1947)
MATSON, George Charlton, (1873-1940)
MATTHES, François Emile, (1874-1948)
MATTIROLO, Ettore, (1853-1923)
MAUFE, Herbert Brantwood, (1879-1946)
MAWSON, Douglas, Sir, (1882-1958)
MAWSON, Joseph, (1830-1927)
MAXSON, John Haviland, (1906-1966)
MAYER-EYMAR, Karl David Wilhelm, (1826-1907)
MAYFIELD, Samuel Martin, (1888?-1963)
MAZUREK, Alojzy, (1885-1944)
McCALLEY, Henry, (1852-1904)
McCONNEL, Roger H., (1908-1971)
McCONNELL, Richard George, (1857-1942)
McDONALD, Stanley Matthews, (1915-1950)
McGEE, William John, (1853-1912)
McGEHEE, Julius Rex, (1905-1963)
McGUINNESS, Charles Lee, (1914-1971)
McHENRY, Alexander, (1843-1919)
McINTYRE, Peter, (1863-1940)
McKAY, Alexander, (1841-1917)
McLAUGHIN, Dean Benjamin, (1901-1965)
McMANNIS, William J., (1924-1973)
McMURTRIE, James, (1840-1914)
McPHERSON, William, (-1928)
MEADOWS, Walter Hennesy, (?1864-1937)
MEDLICOTT, Henry Benedict, (1829-1905)
MEFFERD, Ralph L., (1909-1964)
MEHL, Maurice Goldsmith, (1887-1966)
MEINZER, Oscar Edward, (1876-1948)
MELI, Romolo, (1852-1920)
MENCHER, Ely, (1913-)
MENDENHALL, Walter Curran, (1871-1957)
MENZEL, Karl Gustav Johannes, (1875-1914)
MERRITT, Clifford A., (-)

STRATIGRAPHY 1900 to present Cont'd.

MERTIN, Hans, (1902-1941)
MEUNIER, Étienne Stanislas, (1843-1925)
MEYENDORFF, André de, (1918-1942)
MEYER, Erich, (1874-1915)
MICHAEL, Richard, (1869-1925)
MICHALSKI, Aleksander Octavianovich, (1855-1904)
MICHEL-LÉVY, Auguste, (1844-1911)
MIDDLEMISS, Charles Stewart, (1859-1945)
MIEG-KROH, Mathieu, (1849-1911)
MILES, Keith Rodney, (1915-1972)
MILLER, Benjamin Le Roy, (1874-1944)
MILLER, Don John, (1919-1961)
MILLER, Wendell Zerbe, (1892-1958)
MILLER, Willard L., (1896-1973)
MILLER, William John, (1880-1965)
MILLS, James Ellison, (1834-1901)
MILNE, John, (1850-1913)
MILTHERS, Paul Christian Vilhelm Madsen, (1865-1962)
MILTON, John Herbert, (?1862-1925)
MIRĂUŢĂ, Orest, (1931-1967)
MISER, Hugh Dinsmore, (1884-1969)
MITCHELL, George Hoole "Mick", (1902-1976)
MITZOPOULOS, Maximos K., (1896-1968)
MOBERG, Johan Christian, (1854-1915)
MOBERG, Karl Adolf, (1840-1901)
MOJSISOVICS, Johann August Georg Edmund, Edler von Mojsvàr, (1839-1907)
MOLENGRAAFF, Gustaaf Adolf Frederik, (1860-1942)
MOLYNEUX, Arthur John Charles, (?1866-1921)
MONCKTON, Horace Woolaston, (1857-1931)
MONGIN, Denise, (-)
MOON, Frederick William, (-1925)
MOORE, Carl Allphin, (1911-1973)
MOORE, Edward William James, (1887-1970)
MOORE, Leslie Rowsell, (1914-)
MOORE, Raymond Cecil, (1892-1974)
MORAN, Thomas, (1837-1926)
MORELLET, Lucien, (1882-1945)
MORET, Léon, (1890-1972)
MORGAN, Conwy Lloyd, (?1852-1936)
MORGAN, George Dillon, (1894-1950)
MORGAN, Jacques Jean Marie de, (1857-1924)
MORGAN, Peter (Percy) Gates, (1867-1927)
MORSE, Paul Franklin, (1897-1929)
MORSE, William Clifford, (1874-1962)
MORTIMER, John Robert, (1825-1911)
MORTON, George Highfield, (1826-1900)
MOSELEY, John Reed, (1900-1959)
MOSSOCZY, Zbigniew, (1907-1963)
MOURLON, Michel Félix, (1845-1915)
MÜHLBERG, Friedrich, (1840-1915)
MUIR, John Malcolm, (1885-1938)
MULLEN, John Timothy, Jr., (1906-1959)
MÜLLER, Heinrich, (1887-1914)

STRATIGRAPHY 1900 to present Cont'd.

MÜLLER, Karl Christian Gottfried Ludwig, (1862-1906)
MULLER, Siemon William, (1900-1970)
MUNIER-CHALMAS, Ernest Philippe Auguste, (1843-1903)
MUNN, Malcolm John, (1874-1949)
MUNTHE, Henrik Vilhelm, (1860-1958)
MURRAY, Albert Nelson, (1894-1961)
MYERS, John Clark, (1894-1970)
MYLIUS, Hugo, (1876-1918)
MYLIUS, Louis Aubrey, (1890-1965)
NAGAPPA, Yedatore, Shri, (1907-1960)
NAGYSUR, János Böckh von, (1840-1909)
NARAMORE, Chester "Chet", (1879-1961)
NAUMANN, Edmund, (1854-1927)
NEALE, Ernest Richard Ward, (1923-)
NEFF, Peter, (1827-1903)
NELSON, Lloyd Alvino, (1895-1964)
NĚMEJC, František, (1901-)
NERY DA ENCARNAÇAO DELGADO, Joaquim Filipe, (1835-1908)
NEUBURG, Maria Feodorovna, (1894-1962)
NEWLAND, David Hale, (1872-1943)
NICKLÈS, René, (1859-1917)
NININGER, Robert D., (-)
NOBLE, Levi Fatzinger, (1882-1965)
NOLAN, Joseph, (1841-1902)
NOPCSA VON FELSÖ-SZILVÁS, Ferenc, baron, (1877-1933)
NORBERG, Hans L., (-?1918)
NORTON, William Harmon, (-)
NORWOOD, Charles Joseph, (1853-1927)
NOSZKY, Jenö, Sr., (1880-1951)
NOSZKY, Jenö Jr., (1909-1970)
NOWAK, Jan, (1880-1940)
OAKES, Malcolm Christie, (1890-1977)
OATES, Frank, (1890-1945)
ODLING, Marmaduke, (?1888-1956)
OEHLERT, Daniel Victor, (1849-1920)
OERTEL, Walter, (1889-1924)
OGILVIE, Francis Grant, Sir, (1858-1930)
OGILVIE, Ida Helen, (1874-1963)
OGILVIE, William, (1846-1912)
OHERN, Daniel Webster, (1870-1953)
OLDHAM, Richard Dixon, (1858-1936)
OLLERENSHAW, Neil, (-)
OLRY, Albert, (1847-1913)
OPPENHEIMER, Josef, (1883-1933)
ORDOÑEZ, Ezequiel, (1867-1950)
ORIEL, Benjamin, Reverend, (-1936)
ORTON, Edward Francis Baxter, (1829-1899)
OSBORNE, George Davenport, (1899-1955)
OSBORNE WHITE, Harold J., (-1954)
OSMAN, Charles William, (1855-1932)
OSWALD, Felix, (1866-1958)
PACHUCKI, Czeslaw, (1898-1965)
PAECKELMANN, Werner, (1890-1952)
PAINE, William R., (-)

STRATIGRAPHY 1900 to present Cont'd.

PALMER, Robert Hastings, (1882-1948)
PALMER, Rupert William, (1890-1922)
PANOW, Eugeniusz, (1890-1958)
PANTANELLI, Dante, (1844-1913)
PAPP, Károly, (1873-1963)
PAPP, Simon, (1886-1970)
PAQUIER, Victor Lucien, (1870-1911)
PARGETER, Reginald Clive, (1919-1972)
PARK, James, (1857-1946)
PARKINSON, Cyril, (1854?-1919)
PARKINSON, Donald, (1891-1974)
PARKS, William Arthur, (1868-1936)
PARONA, Carlo Fabrizio, (1855-1939)
PARRAN, Alphonse, (1827-1903)
PARSONS, Henry Franklin, (1846-1913)
PARSONS, Leonard Miles, (1883-1956)
PASTORE, Franco, (1885-1958)
PATTISON, Samuel Rowles, (1809-1901)
PATTON, Leroy Thompson, (1880-1957)
PÁVAI, Vajna Ferenc, (1886-1964)
PEABODY, Harlan William, (1881-1960)
PEACH, Benjamin Neeve, (1842-1926)
PECK, Frederick Burritt, (1860-1925)
PECORA, William Thomas, (1913-1972)
PEGAU, Albert August, (1894-1963)
PEGG, E. J., (?1899-1948)
PELLATI, Nicola, (1835-1907)
PENNEL, Harry David, (1907-1971)
PENNING, William Henry, (1838-1902)
PEPPER, James Franklin, (1898-1963)
PEPPERBERG, Leon J., (1883-1937)
PEREIRA DE LIMA, Wenceslau de Sousa, (1858-1920)
PERKINS, George Henry, (1844-1933)
PERNER, Jaroslav, (1869-1947)
PERRINE, Irving, (1884-1955)
PERRY, Thomas Gregory, (1919-1972)
PERVINQUIÈRE, Léon, (1873-1913)
PETHÖ, Gyula, (1848-1902)
PETKOVIĆ, Kosta, (1903-)
PETKOVIĆ, Vladamir K., (1873-1934)
PETRASCHEK, Wilhelm, (1876-1967)
PETROCCHI, Carlo, (1898-1956)
PHILIPPI, Rudolph Amandus, (1808-1904)
PHILLIPS, David Watkins, (1899-1962)
PIA, Julius von, (1887-1943)
PICKLESIMER, Harold Greeley, (1916-1952)
PIETTE, Louis Edouard Stanislas, (1827-1906)
PIETZCKER, Franz, (1885-1914)
PIETZSCH, Kurt, (1884-1964)
PILAR, Ludgero da Piedade, (1920-1967)
PILGRIM, Henry Guy Ellcock, (1875-1943)
PINFOLD, Ernest Sheppard, (1888-1969)
PINGEL, Johann Victorinus, (1834-1919)
PLAS, Vladimír, (1905-)

STRATIGRAPHY 1900 to present Cont'd.

PLATT, Franklin, (1844-1900)
PLUMMER, Frederick Byron, (1886-1947)
PLUMMER, Helen Jeanne (née Skewes), (1891-1951)
POCOCK, Roy Woodhouse, (1887-1954)
POČTA, Filip, (1859-1924)
POLANSKI, Jorge, (1892-1975)
POLDERVAART, Arie, (1918-1964)
POLLARD, Ernest Rex, (1922-1966)
POLLARD, William, (1870-1946)
POLWHELE, Thomas Roxburgh, (1831-1909)
POMPECKJ, Josef Felix, (1867-1930)
POND, Walter Franklin, (1885-1974)
PONTIER, G., (-1933)
POOR, Russell Spurgeon, (1899-1972)
POPOVICI-HAŢEG, Valeriu, (1868-1929)
POSEWITZ, Tivadar, (1850-1917)
POSTEL, Albert Williams, (1909-1966)
POTEET, James Harold, (1896-1965)
POWELL, John Wesley, (1834-1902)
POWERS, Sidney, (1890-1932)
PRATJE, Otto, (1890-1925)
PRATT, Wallace Everette, (1885-)
PREISWERK, Heinrich, (1876-1940)
PREMIK, Józef, (1890-1963)
PRESTON, Henry, (1852-1940)
PRETTYMAN, Thomas Mann, (1888-1940)
PŘIBYL, Alois, (1904-)
PRICE, Paul Holland, (1898-)
PRIEM, Fernand, (1857-1919)
PRIEST, Simeon, (1876-1954)
PRIESTLEY, Raymond Edward, Sir, (1886-1974)
PRINGLE, John, (1877-1948)
PRINZINGER, Heinrich, (1822-1908)
PROCHÁZKA, Milóš, (1923-1956)
PROCHÁZKA, Vladimír Josef, (1862-1913)
PROSSER, Charles Smith, (1860-1916)
PRUVOST, Pierre Eugéne Marie Joseph, (1890-1967)
PUGH, William John, Sir, (1892-1974)
PUMPELLY, Raphael, (1837-1923)
PURDUE, Albert Homer, (1861-1917)
PURKYNĚ, Cyril, (1862-1937)
PUTNAM, William Clement, (1908-1963)
QUIRKE, Terence Thomas, (1886-1947)
QUITZOW, August Alfred Wilhelm, (1881-1914)
RADLEY, Ernest George, (1880-1945)
RAEBURN, Colin, (1894-1970)
RAGGATT, Harold George, Sir, (1900-1968)
RAINWATER, Edward Harriman "Rainey", (1909-1972)
RAJNATH, Shri, (1904-1972)
RAKOVEC, Ivan, (1899-)
RAKUSZ, Gyula, (1896-1932)
RAMA RAO, Lakshmeswar, (1896-1974)
RANDALL, John, (1810-1910)
RANSOME, Frederick Leslie, (1868-1935)

STRATIGRAPHY 1900 to present Cont'd.

RASHID, Muhammad Abdur, (1938-1969)
RATHBURN, Richard, (1852-1918)
RAULIN, Félix Victor, (1815-1905)
RAVIER, Jean, (1922-)
RAVN, Jesper Peter Johansen, (1866-1951)
READ, Herbert Harold, (1889-1970)
READER, Thomas William, (-?1923)
REDTENBACHER, Anton, (1841-1911)
REECE, Alan W., (1921-1960)
REED, Frederick Richard Cowper, (1869-1946)
REESIDE, John Bernard, Jr., (1889-1958)
REICHARDT, Wolfgang, (1909-1937)
REID, Clement, (1853-1916)
REINHOLD, Thomas, (1890-1955)
REIS, Otto Maria, (1862-1934)
REISER, Karl August, (1853-1922)
REITHOFER, Otto, (1902-1965)
RENEVIER, Eugène, (1831-1906)
RENICK, Brink Coleman, (1897-1971)
RENIER, Armand Marie Vincent Joseph, (1876-1951)
REUSCH, Hans Henrik, (1852-1922)
RÉVIL, Joseph, (1849-1931)
REYNOLDS, Sidney Hugh, (1867-1949)
RHEDEN, J. J. Pannekoek van, (-1945)
RHODES, James Eric Wynfield, (?1888-1976)
RIAZ, Auguste de, (-1920)
RICE, William North, (1845-1928)
RICH, John Lyon, (1884-1956)
RICHARDS, James Taylor "Dick", (1892-1953)
RICHARDSON, Charles Henry, (1862-1935)
RICHARDSON, George Burr, (1872-1949)
RICHARDSON, Linsdall, (1881-1967)
RICHARZ, Stephan, Pater, (1874-1934)
RICHEY, James Ernest, (1886-1968)
RICHTHOFEN, Paul Wilhelm Ferdinand von, Freiherr, (1833-1905)
RICKETTS, Charles, (1818-1904)
RIES, Heinrich, (1871-1951)
RIGAUX, Edmond, (1839-1915)
RIGGS, Robert Jennings, (1888-1965)
RISTORI, Giuseppe, (1856-1905)
RITTENHOUSE, Gordon, (1910-1974)
ROARK, Louis, (1890-1950)
ROBBIE, James Andrew, (1910-1977)
ROBERTS, Morgan Evan, (1894-1941)
ROBERTS, Thomas, (1856-1892)
ROCH, Edouard, (1901-)
RODIGHIERO, Andrea, (1892-1917)
ROE, Frank Walton "Peter", (?1914-1965)
ROEDER, Charles, (1848-1911)
ROGALA, Wojciech, (1884-1947)
ROGERS, Arthur William, (1872-1946)
ROGERS, Gaillard Sherburne, (1889-1919)
ROLLIER, Louis, (1859-1931)
ROMAN, Albert Fréderic, (1871-1943)

STRATIGRAPHY 1900 to present Cont'd.

ROMINGER, Carl Ludwig, (1820-1907)
RONES, Morris, (1922-1965)
RØRDAM, Kristian, (1860-1939)
ROSENBURG, Georg, (1897-1969)
ROSENBUSCH, Karl Harry Ferdinand, (1836-1914)
ROSS, Clarence Samuel, (1880-1975)
ROTHPLETZ, August, (1853-1918)
ROUNDY, Paul Vere, (1884-1937)
ROUVILLE, Paul Gervais de, (1823-1907)
ROWE, Richard Burton, (1872-1902)
ROYO Y GOMEZ, José, (1895-1958)
ROZLOZSNIK, Pál, (1880-1940)
RUBEY, William Walden, (1898-1974)
RUNNER, Joseph James, (1885-1970)
RUSS, Leon Fair, (1886-1953)
RUSSELL, Israel Cook, (1852-1906)
RUSSELL, Loris Shano, (1904-)
RUSSELL, William Low, (1897-1976)
RUTKOWSKI, Feliks, (1878-1964)
RUTLEY, Frank, (1842-1904)
RUTOT, Aimé Louis, (1847-1933)
RUTTEN, Louis Martin Robert, (1884-1946)
RUTTEN, Martin Gerard, (1910-1970)
RYBA, Frantisek, (1867-1918)
RYDZEWSKI, Bronislaw, (1884-1945)
RZEHÂK, Anton, (1855-1923)
SACCO, Federico, (1864-1948)
SAFFORD, James Merrill, (1822-1907)
SAINTY, James Edward, (?1882-1967)
SALÉE, Achille, Chanoine, (1883-1932)
SALISBURY, Rollin D., (1858-1922)
SALOPEK, Marijan, (1883-1967)
SAMBASIVA IYER, Valiathur S., (-?1943)
SAMPSON, David Noel, (1917-1971)
SAMSONOWICZ, Jan, (1888-1959)
SANDFORD, Kenneth Stuart, (1899-1971)
SANGER, Frederick Robert, (1874-1942)
SARDESON, Frederick William, (1866-1959)
SARGENT, Henry Crunden, (1851-1931)
SARJEANT, William Antony Swithin, (1935-)
SAUER, Gustav Adolf, (1825-1932)
SAUVAGE, Henri Émile, (1842-1917)
SÄVE-SÖDERBERGH, Gunnar, (1910-1948)
SAWYER, Roger W., (1895-1941)
SCARABELLI GOMMI FLAMINI, Giuseppe, (1820-1905)
SCHALCH, Ferdinand, (1848-1918)
SCHALLER, Waldemar Theodore, (1882-1967)
SCHARDT, W. Hans, (1858-1931)
SCHEIBE, Robert, (1859-1923)
SCHELLWIEN, Ernst, (1866-1906)
SCHENCK, Hubert Gregory, (1897-1960)
SCHETELIG, Jakob Grubbe Cock, (1875-1935)
SCHINDEWOLF, Otto Heinrich, (1896-1972)
SCHLAGER, Max, (1903-1972)

STRATIGRAPHY 1900 to present Cont'd.

SCHLAIKJER, Erich Maren, (1905-1972)
SCHLOSSER, Max, (1854-1933)
SCHLUETER, Clemens August, (1835-1906)
SCHLUMBERGER, Charles, (1825-1904)
SCHLUNCK, Johannes, (1876-1915)
SCHMID, Kurt, (1906-1957)
SCHMIDT, Carl, (1862-1923)
SCHMIDT, Friedrich Bogdanovich "Bohdan", (1832-1908)
SCHMIDT, Hermann, (1892-1978)
SCHMIDT, Wolfgang, (1914-1972)
SCHOFIELD, Eric, (1911-1944)
SCHOLS, Hendrikus, (1901-1957)
SCHOTTLER, Wilhelm, (1869-1932)
SCHRADER, Frank Charles, (1860-1944)
SCHRÉTER, Zóltan, (1882-1970)
SCHRÖDER, Hans Adolf Karl Eckart, (1899-1968)
SCHUCHERT, Charles, (1858-1942)
SCHUMACHER, Eugen, (1851-1922)
SCHWARZ, Ernest Hubert Lewis, (1873-1928)
SCHWEINFURTH, Georg, (1836-1925)
SCHWENNESEN, Alvin Theodore, (1883-1962)
SCHWIPPEL, Karl, (1821-1911)
SCOTT, Gayle, (1894-1948)
SCOTT, William Berryman, (1858-1947)
SCRUTON, Philip Challacombe, (1922-1962)
SEARS, Julian Ducker, (1891-1970)
SEELAND, Ferdinand, (1822-1901)
SEELEY, Harry Govier, (1839-1909)
SEGUENZA, Luigi, (1873-1908)
SELLARDS, Elias Howard, (1875-1961)
SEMPER, Otto, (1830-1907)
SENN, Alfred, (1899-1949)
ŠETLÍK, Jaromír, (1896-)
SEUNES, Jean, (1849-1920)
SHALER, Nathaniel Southgate, (1841-1906)
SHARMAN, George, (1832-1914)
SHATTUCK, George Burbank, (1869-)
SHAW, Eugene Wesley, (1881-1935)
SHEARER, Harold Kurtz, (1889-1946)
SHEPARD, Edward Martin, (1854-1934)
SHERBORN, Charles Davies, (1861-1942)
SHERLOCK, R. L., (1875-1948)
SHERRILL, Richard Ellis, (1899-1953)
SHIDELER, William Henry, (1886-1958)
SHIMER, Hervey Woodburn, (1872-1965)
SHIPMAN, James, (1848-1901)
SHORT, Maxwell Naylor, (1889-1952)
SHROCK, Robert Rakes, (1904-)
SHRUBSOLE, William Hobbs, (1837-1927)
SHULER, Ellis William, (1881-1954)
SIBLY, Thomas Franklin, Sir, (1883-1948)
SIEGERT, Leo, (1872-1917)
SIEMIRADZKI, Józef, (1858-1933)
SIMMONS, William Charles, (1888-1951)

STRATIGRAPHY 1900 to present Cont'd.

SINCLAIR, Ewart Gladstone, (1884-1927)
SINGLETON, F. A., (-1947)
SKOCYLAS-CISZEWSKA, Kamila, (1902-1971)
SLATER, George, (-1956)
SLATER, Herbert Kelsall, (1875-1913)
SMALL, Evan William, (1859-1933)
SMITH, Bernard, (1881-1936)
SMITH, Eugene Allen, (1841-1927)
SMITH, George Otis, (1871-1944)
SMITH, John, (1845-1930)
SMITH, Ned Myron, (1923-1976)
SMITH, Robert Hendee "Smokey", (1916-1946)
SMITH, Stanley, (1883-1955)
SMITH, William Richard Tangier, (1869-1972)
SMOCK, John Conover, (1842-1925)
SMYTH, Louis Bouvier, (1933-?1953)
SNIDER, Luther Crocker, (1882-1947)
SOBRAL, José María, (-1961)
SOENDEROP, Fritz Hermann Gustav, (1876-1914)
SOLLAS, William Johnson, (1849-1936)
SOMERVAIL, Alexander, (1841-1910)
SOMMERMEIER, Leopold, (1885-1963)
SOREIL, Gustave Joseph, (1842-1907)
SORGENFREI, Theodor, (1915-1972)
SOVINSKY, Viacheslav Nicholas, (1894-1962)
SPATH, Leonard Frank, (1882-1957)
SPENGLER, Erich, (1886-1962)
ŠPINAR, Zdeněk Vlastimil, (1906-)
SPITZ, Albrecht, (1883-1918)
SPROULE, John Campbell, (1905-1970)
STACHE, Guido, (1833-1921)
STAESCHE, Karl, (1902-)
STAFF, Hans von, (1883-1915)
STAINBROOK, Merrill Addison, (1897-1956)
STAINER, Xavier, (1865-)
STAMP, Laurence Dudley, Sir, (1898-1966)
STANTON, Timothy William, (1860-1953)
STEBBING, William Pinchard Delane, (1873-1961)
STEBINGER, Eugene, (1883-1951)
STEENHUIS, Jacob Frederik, (1883-1958)
ŞTEFANESCU, Gregoriu, (1838-1911)
ŞTEFĂNESCU, Sabba, (1857-1931)
STEHR, Raymond Albert, (1903-1967)
STEIGER, George, (1869-1944)
STEINMANN, Gustav, (1856-1929)
STEJSKAL, Jan, (1902-1972)
STEPHENSON, Lloyd William, (1876-1962)
STEVENSON, John James, (1841-1924)
STEWART, Grace Anne, (1893-1970)
STEWART, Irvine E., (1893-1965)
STILES, Harold, Sir, (?1862-1946)
ST. JOHN, Orestes Hawley, (1841-1921)
STOBBS, John Thomas, (?1866-1942)
STOCK, Chester, (1892-1950)

STRATIGRAPHY 1900 to present Cont'd.

STOCKDALE, Paris Buell, (1896-1962)
STOLLEY, Ernst, (1869-1944)
STORRIE, John, (1844-1901)
STOW, Marcellus Henry, (1902-1957)
STRAELEN, Victor Emile van, (1889-1964)
STRAHAN, Aubrey, Sir, (1852-1928)
STRAW, Stephen Henry, (?1892-1963)
STRINGHAM, Bronson F., (1907-1968)
STRÜBIN, Karl, (1876-1916)
STUTZER, Otto, (1881-1936)
SUBRAMANIAM, Anantharama Parameshwara, (1921-1973)
SUESS, Eduard, (1831-1914)
SUESS, Franz Eduard, (1867-1941)
ŠUF, Jiří, (1891-1976)
SUGDEN, Walter "Bill", (1923-1977)
SUJKOWSKI, Zbigniew Leliwa, (1898-1954)
SÜMEGHY, József, (1892-1955)
SUTCLIFFE, William Henry, (1855-1913)
SUTTON, Arle Herbert, (1896-1972)
SUTTON, Frederick A., (1894-1950)
SVEDMARK, Lennart Eugène, (1847-1922)
SVOBODA, Josef, (1898-)
SWANN, David H., (1915-1968)
SWARTZ, Joel Howard, (1893-1971)
SWEETING, George Scotland, (1889-1977)
ŚWIDERSKI, Bohdan, (1892-1943)
ŚWIDZINSKI, Henryk, (1904-1969)
SWINNERTON, Henry Hurd, (1875-1967)
SYLVESTER-BRADLEY, Peter Colley, (1914?-1978)
SYMES, Richard Glascott, (1840-1906)
SZÁDECZKY, Gyula, Baron Szádeczky-Kardoss of Szádecsne and Kafdosfalva, (1860-1935)
SZAJNOCHA, Wladyslaw, (1857-1928)
SZMUC, Eugene Joseph, (1927-1976)
TAEGER, Henrik, (1881-1939)
TAFF, Joseph Alexander, (1862-1944)
TAGUE, Glenn Charles, (1905-1964)
TANDY, James Hiram "Spot", (1892-1962)
TANTON, Thomas Leslie, (1890-1971)
TARAMELLI, Torquato, (1845-1922)
TATE, Ralph, (1840-1901)
TAURER-GALLENSTEIN, Hans von, Ritter, (1846-1927)
TAYLOR, Griffith, (1881-1963)
TAYLOR, Herbert, (-1963)
TAYLOR, James Carlton, (1924-1976)
TAYLOR, James Haward, (1909-1968)
TCHERNYCHEV, Féodoss Nikolaievich, (1856-1914)
TEALE, Edmund Oswald, Sir, (1874-1971)
TEALL, Jethro Justinian Harris, Sir, (1849-1924)
TEICHERT, Curt, (-)
TEISSEYRE, Henryk, (1903-1975)
TEISSEYRE, Wawrzyniec, (1860-1939)
TELEGDI ROTH, Károly, (1886-?1956)
TELLER, Friedrich, (1852-1913)

STRATIGRAPHY 1900 to present Cont'd.

TEMPLEMAN, Andrew, (1887-1945)
TERMIER, Henri François Emile, (1897-)
TERMIER, Pierre, (1859-1930)
TESCH, P., (1879-1961)
THALMANN, Hans E., (1899-1975)
THEVENIN, Armand, (1870-1918)
THIÉRY, Paul, (-1927)
THIES, Roland K., (1905-1973)
THOMAS, Abram Owen, (1876-1931)
THOMAS, Herbert Henry, (1876-1935)
THOMAS, Horace David "Bill", (1905-1967)
THOMAS, Leo A., (1917-1965)
THOMAS, Philippe, (1843-1910)
THOMPSON, Beeby, (1848-1931)
THOMPSON, David Grosh, (1888-1943)
THOMPSON, Jean Cleveland, (1887-1929)
THOMSON, Alexander Gordon Milne, (1866-1919)
THOMSON, James Allan, (1881-1928)
THOMSON, Joseph Ellis, (1882-1944)
THORAL, Marcel, (1900-1956)
THURMANN, Jules, (1804-1855)
TIDDEMAN, Richard Hill, (1842-1917)
TIDY, Anthony John Lintott, (1928-1969)
TIEJE, Arthur Jerrold, (1891-1944)
TIETZE, Emil, (1845-1931)
TIETZE, Oskar Adolf Eduard, (1874-1920)
TILTON, John Littlefield, (1863-1930)
TIMMINS, Arthur, Major, (?1858-1940)
TING, Ven Kiang, (?1887-1936)
TIPPER, George Howlett, (1881-1947)
TOBLER, August, (1872-1929)
TODD, James Edward, (1846-1922)
TOIT, Alexander Logie du, (1878-1948)
TOLL, Eduard von, Baron, (1858-1902)
TOLWIŃSKI, Konstanty, (1876-1961)
TOMBLESON, Frederick Binns, (?1872-1938)
TOMLINSON, Charles Weldon, (1892-1960)
TOMMASI, Annibale, (1858-1921)
TONSO, Alessandro, (-)
TÖRNQUIST, Sven Leonhard, (1840-1920)
TOUCAS, Joseph Aristide, (1843-1911)
TOULA, Franz, (1845-1920)
TOWSE, Donald Frederick, (1924-)
TRABUCCO, Giacomo, (1845-1925)
TRASK, Parker Davies, (1899-1961)
TRAUTH, Friedrich, (1883-1967)
TRAUTSCHOLD, Hermann Adolfovich, (1817-1902)
TREACHER, Llewellyn, (1859-1943)
TRECHMANN, Charles Taylor, (1885-1964)
TRENTANOVE, Giorgio Morando, (1874-1914)
TRIBOLET-HARDY, Frédéric Maurice de, (1852-1929)
TROEDSSON, Gustaf T., (1891-1954)
TROTTER, Frederick Murray, (1897-1968)
TRUEMAN, Arthur Elijah, Sir, (1894-1956)

STRATIGRAPHY 1900 to present Cont'd.

TRZASKA-JURKIEWICZOWA, Irena, (1912-1966)
TSCHIKHACHEV, Peter A., (-)
TUCCIMEI, Giuseppe, (1851-1915)
TUČEK, Karel, (1906-)
TWENHOFEL, William Henry, (1875-1957)
TWITCHELL, Mayville William, (1866-1927)
TYRRELL, Joseph Burr, (1858-1957)
UDALL, John, (1841-1920)
UDDEN, Johan August, (1859-1932)
UGLOW, William Lawrence, (1884-1926)
UHLIG, Victor Karl, (1857-1911)
ULRICH, Edward Oscar, (1857-1944)
UMBGROVE, Johannes Herman Frederik, (1899-1954)
UNTERMANN, Billie R., (1906-1973)
UNTERMANN, Gerhard Ernest, Jr., (1898-1975)
URBAN, Karel, (1905-)
URBÁNEK, Ladislav, (1901-)
USSHER, William Augustus Edmond, (1849-1920)
VACEK, Michael, (1848-1925)
VADÁSZ, Elemér, (1885-1970)
VAGHI, Vittorio, (-1921)
VALENTINE, R. L., Lieutenant, (1890-1916)
VALLE DE LERSUNDI, Alfonso del, (1875-1944)
VALPY, Robert Harris, (1819-1904)
VAN AMRINGE, Edwin Verne, (1899-1956)
VAN HISE, Charles Richard, (1857-1918)
VAN INGEN, Gilbert, (1869-1928)
VARDABASSO, Silvio, (1891-1966)
VARTY SMITH, George, (-1924)
VASSEUR, Gaston Casimir, (1855-1915)
VATER, Heinrich, (1859-1930)
VAUGHAN, Arthur, (1868-1915)
VAUGHAN, Thomas Wayland, (1870-1952)
VEATCH, Arthur Clifford, (1878-1938)
VERBEEK, Rogier Diederick Marius, (1845-1926)
VER WIEBE, Walter August, (1887-1961)
VESTAL, Franklin Earl, (1884-1969)
VETTERS, Hermann, (1880-1941)
VIANNA, António, (1887-1949)
VICARY, William, (1811-1903)
VIENNOT, Pierre, (1891-1931)
VÍGH, Gyula, (1889-1958)
VIGILEV, Boris, (1883-1924)
VINCIENNE, Henri, (1898-1965)
VITÁLIS, István, (1871-1947)
VIZER, Vilmos, (1873-1941)
VLERK, Isaak Martinus van der, (?1892-1974)
VOGT, Johan Herman Lie, (1858-1932)
VOGT, Thorolf, (1888-1958)
VOLLRATH, Paul, (-1929)
VON ECKERMANN, Claes Walther Harry, (1886-1969)
WAAGEN, Lukas, (1877-1959)
WADIA, Darashaw Nosherwan, (1883-1969)
WAEL, Norbert Charles Louis de, (1817-1901)

STRATIGRAPHY 1900 to present Cont'd.

WAGNER, Walter Richard, (1928-1974)
WÄHNER, Franz, (1856-1932)
WAHNSCHAFFE, Gustav Albert Bruno Felix, (1851-1914)
WALCOTT, Charles Doolittle, (1850-1927)
WALCOTT, Mary Vaux, née Vaux, (-1940)
WALDBAUM, David Robert, (1937-1974)
WALKER, Edward Eaton, (-1903)
WALKER, John Francis, (1839-1907)
WALKER, Thomas Leonard, (1867-1942)
WALL, George Parkes, (1832-1912)
WALLACE, Alfred Russel, (1823-1913)
WALLACE, Robert Charles, (1881-1955)
WALLERANT, Frédéric Félix Auguste, (1858-1936)
WALLS, James G., (1907-)
WANLESS, Harold Rollin, (1898-1970)
WANNER, Johannes, (1878-1957)
WARD, Freeman, (1879-1943)
WARD, John, (1837-1906)
WARD, Leonard Keith, (1879-1964)
WARDLE, Thomas, Sir, (1831-1909)
WARING, Gerald Ashley, (1883-1971)
WARREN, Percival Sidney, (1890-1970)
WASHBURNE, Chester Wesley, (1883-1971)
WASSON, Theron, (1887-1970)
WATSON, Thomas Leonard, (1871-1924)
WATTS, William Whitehead, (1860-1947)
WDOWIARZ, Jan, (1903-1956)
WEAVER, Charles Edwin, (1880-1958)
WEAVER, Paul, (1888-1964)
WEBB, John Benwell, (1901-1975)
WEBER, Carl Albert, (1856-1931)
WEBER, Julius, (1864-1924)
WEDEL, Arthur Albert, (1898-1941)
WEEKS, Herbert John, (1897-1960)
WEEKS, Ludlow Jackson, (1899-1964)
WEGNER, Theodor, (1880-1934)
WEIDMAN, Samuel, (1870-1945)
WEIGELT, Johannes, (1890-1948)
WEISE, Ernst, (1843-1941)
WELLER, James Marvin, (1899-1976)
WELLER, Stuart, (1870-1927)
WELLS, Francis Gerrit, (1898-1969)
WELLS, Roger Clark, (1877-1944)
WELSCH, Jules, (1858-1929)
WEPFER, Emil, (1882-1930)
WESLEY, George Rutherford, (1905-1965)
WESTGATE, Lewis Gardner, (1868-1948)
WESTON, Thomas Chesmer, (1832-1910)
WETZEL, Walter, (1887-)
WHEELER, Orby Clinton, (1898-1970)
WHITE, Charles Abiathar, (1826-1910)
WHITE, Charles David, (1862-1935)
WHITE, George Willard, (1903-)
WHITE, Israel Charles, (1848-1927)

STRATIGRAPHY 1900 to present Cont'd.

WHITEAVES, John Joseph Frederick, (1835-1909)
WHITEHEAD, Talbot Haes, (1890-1966)
WHITESIDE, Robert Massie, (1895-1936)
WHITTARD, Walter Frederick, (1902-1966)
WHITTIER, William Harrison, (1889-1951)
WICHER, Carl A., (1901-1957)
WILBERT, Louis Joseph "Slick", Jr., (1919-1953)
WILDER, Frank Alonzo, (1870-1930)
WILHELM, Oscar, (1897-1975)
WILKINSON, William Donald, (1901-1969)
WILKMAN, Wanold Wrydon, (1872-?1938)
WILKS, Mark, (?1862-1945)
WILLARD, Daniel Everett, (1862-1947)
WILLETT, Richard Wright, (1912-1974)
WILLIAMS, Arthur Herbert, (?1865-1935)
WILLIAMS, Henry Shaler, (1847-1918)
WILLIAMS, Mervyn, (1934-1970)
WILLIS, Bailey, (1857-1949)
WILSON, Alice Evelyn, (1881-1964)
WILSON, Bernard Wright, (1922-1976)
WILSON, Eugene Jack, (1921-1976)
WILSON, J. S. Grant, (1855-1908)
WILSON, Malcolm Earl, Sr., (1885-1960)
WILSON, Reginald Charles, (1888-1943)
WILSON, Roy Arthur, (1891-1970)
WILSON, Vernon, (1907-1971)
WILTSHIRE, Thomas, Reverend, (1826-1902)
WIMAN, Carl Johan Josef Ernst, (1867-1944)
WINCHELL, Horace Vaughn, (1865-1923)
WINCHELL, Newton Horace, (1829-1914)
WINCHESTER, Dean Eddy, (1883-1936)
WINDHAUSEN, Anselm, (1882-1932)
WING, Monta Eldo, (1897-1976)
WINTER, William Percy, (?1870-1950)
WINTON, Will McClain, (1885-1960)
WINWOOD, Henry Hoyte, (1830-1920)
WIŚNIOWSKI, Tadeusz, (1865-1933)
WOLDSTEDT, Paul, (1888-1973)
WOLFF, Wilhelm, (1872-1951)
WOOD, Robert Harrison, (1887-1956)
WOOD, Virgil O., (1892-1954)
WOODRUFF, Elmer Grant, (1872-1952)
WOODRUFF, John Grant, (1898-1971)
WOODS, Earl Hazen, (1902-1950)
WOODS, Ethel Gertrude, (-1939)
WOODWARD, Harry Page, (1858-1917)
WOODWARD, Herbert Preston, (1899-1968)
WOODWARD, Horace Bolingbroke, (1848-1914)
WOODWARD, Robert Simpson, (1849-1924)
WOOLACOTT, David, (1872-1924)
WOOLDRIDGE, Sidney William, (1900-1963)
WOOLNOUGH, Walter George, (1876-1958)
WORCESTER, Philip George, (1884-1970)
WORLEY, W. F., (-1923)

STRATIGRAPHY 1900 to present Cont'd.

WRATHER, William Embry, (1883-1963)
WRIGHT, Frederick Eugene, (1877-1953)
WRIGHT, George Frederick, (1838-1921)
WRIGHT, James, (1878-1957)
WRIGHT, William, (-1936)
WRIGHT, William Bourke, (1876-1939)
WUNDT, Georg von, (1845-1929)
WYNNE, Arthur Beavor, (1835-1906)
YABE, Hisakatsu, (1878-1969)
YACHEVSKY, Leonard Antonovich, (1858-1916)
YOUNG, Alfred Prentice, (1841-1919)
YOUNG, George Albert, (1898-1947)
YOUNG, John, (1835-1902)
YOUNGQUIST, Walter, (-)
ZANS, Verners Aleksandrs, (1904-1961)
ZAPP, Alfred Dexter, (1916-1962)
ZARĘCZNY, Stanislaw Wawrzyniec, (1848-1909)
ZÁZVORKA, Vlastislav, (1903-)
ZEALLEY, Arthur Edmund Victor, (1886-1918)
ZEIJLMANS VAN EMMICHOVEN, Cornelis Pieter Adriaan, (1897-1952)
ZEISE, Oskar, (1860-1925)
ZELENKA, Ladislav, (1900-1931)
ZIMMERMANN, Ernst, (1860-1944)
ZITTEL, Karl Alfred von, (1839-1904)
ZLATARSKI, Georgi Nikolov, (1854-1909)
ZPALETAL, Karel, (1903-1972)
ZUGMAYER, Heinrich, (1840-1917)
ZWIERZYCKI, Józef, (1888-1961)
ŽYLA, Antoni, (1902-1966)

Time unknown
D'ISENGARD, Luigi, Barone, (-)
MACGREGOR, A. M., (-)
RANDS, William Henry, (-)
TEGENGREN, Felix, (-)

STRATIGRAPHY See also GEOCHRONOLOGY, PHOTOGEOLOGY, QUATERNARY GEOLOGY

STRUCTURAL GEOLOGY
Up to 1800
ALBERTUS MAGNUS, (1193-1280)
ARDUINO, Giovanni, (1714-1795)
BAKEWELL, Robert, (1768-1843)
BONNET, Charles, (1720-1793)
BOURGUET, Louis, (1678-1742)
BROCHANT DE VILLIERS, André Jean Marie, (1772-1840)
BRUNO, Filippo, called Giordano, (1548-1600)
BUCH, Christian Leopold von, Freiherr, (1774-1853)
CRICHTON, Alexander, Sir, (1763-1856)
DAVY, Humphry, Sir, (1778-1829)
ESMARCH, Jens, (1762-1839)
FAUJAS DE SAINT FOND, Barthélémy, (1741-1819)
FORSTER, Johann Reinhold, (1729-1798)

STRUCTURAL GEOLOGY Up to 1800 Cont'd.

GAUTIER, Henri, (1660-1734)
GREENOUGH, George Bellas, (1778-1855)
HACQUET, Belsazar, (1739-1815)
HOBBS, William, (fl. 1700-1715)
HUMBOLDT, Friedrich Wilhelm Heinrich Alexander von, Freiherr, (1769-1859)
HUTCHINS, Thomas, (1730-1789)
HUTTON, James, (1726-1797)
JAŚKIEWICZ, Jan D. P., (1749-1809)
LOMONOSOV, Mikhail Vasilyevich, (1711-1765)
MÉTHERIE, Jean Claude de la, (1743-1817)
OTLEY, Jonathan, (1766-1856)
PALISSY, Bernard, (1510-1589)
PALLAS, Peter Simon, (1741-1811)
PLAYFAIR, John, (1748-1819)
RASPE, Rudolf Erich, (1737-1794)
ROUELLE, Guillaume François, (1703-1770)
SAUSSURE, Horace Bénédict de, (1740-1799)
SENECA, Lucius Annaeus, (ca.4 B.C. - 65 A.D.)
SMITH, William, (1769-1839)
STASZIC, Stanislaw, (1755-1826)
STENO, Nicolaus, (1631-1687)
STRANGE, John, (1732-1799)
SWEDENBORG, Emanuel, (1688-1772)
VALLISNERI, Antonio, (1661-1730)
VINCI, Leonardo da, (1452-1519)
VOGT, Johann Georg Mauriz, (1669-1731)
WATSON, White, (1760-1835)
WERNER, Abraham Gottlob, (1749-1817)

1800-1849

ABICH, Hermann Wilhelm, (1806-1886)
ANSTED, David Thomas, (1814-1880)
APLIN, Christopher d'Oyley Hale, (1819-1875)
AVELINE, William Talbot, (1822-1903)
BABBAGE, Charles, (1792-1871)
BAILY, William Hellier, (1819-1888)
BAINBRIDGE, William, (1812-1860)
BAKEWELL, Robert, (1768-1843)
BEAUMONT, Jean Baptiste Armand Louis Léonce Élie de, (1798-1874)
BEST, Edward, (1824-1906)
BILLINGS, Elkanah, (1820-1876)
BOUÉ, Ami, (1794-1881)
BROCHANT DE VILLIERS, André Jean Marie, (1772-1840)
BUCH, Christian Leopold von, Freiherr, (1774-1853)
BURKART, Hermann Josef, (1798-1874)
CLARKE, William Branwhite, Reverend, (1798-1878)
CONDON, Thomas, (1822-1907)
CONRAD, Timothy Abbott, (1803-1877)
COX, Edward Travers, (1821-1907)
CREDNER, Carl Friedrich Heinrich, (1809-1876)
CRICHTON, Alexander, Sir, (1763-1856)
CROLL, James, (1821-1890)
DANA, James Dwight, (1813-1895)

STRUCTURAL GEOLOGY 1800-1849 Cont'd.

DARWIN, Charles Robert, (1809-1882)
DAVY, Humphry, Sir, (1778-1829)
DAWSON, John William, Sir, (1820-1899)
DE LA BECHE, Henry Thomas, Sir, (1796-1855)
DE LA CONDAMINE, H. M., Reverend, (-1854)
DE PRADO, Casiano, Don, (1797-1866?)
DEWALQUE, Gilles Joseph Gustave, (1828-1905)
DREW, Frederick, (1836-1891)
DU NOYER, George Victor, (-1869)
DUNSTAN, Benjamin, (1864-1933)
EMMONS, Ebenezer, (1800-1863)
ENGELSPACH, Auguste, (LARIVIÈRE, pseud.), (1799-1831)
ERDMANN, Axel Joachim, (-)
ESCHER VON DER LINTH, Arnold, (1807-1872)
ESMARCH, Jens, (1762-1839)
ETHERIDGE, Robert, (1819-1903)
FAUJAS DE SAINT FOND, Barthélémy, (1741-1819)
FAVRE, Jean Alphonse, (1815-1890)
FOETTERLE, František, (1823-1876)
FORBES, David, (1828-1876)
FORBES, Edward, (1815-1854)
FURUHJELM, Enoch Hjalmar, (1823-1886)
GANLY, Patrick, (1809-1899)
GIORDANO, Felice, (1825-1892)
GREEN, William Lowthian, (1819-1890)
GREENOUGH, George Bellas, (1778-1855)
GRESSLY, Amanz, (1814-1865)
GUEYMARD, François Émile, (1788-1870)
GUIDONI, Gerolamo, (1794-1870)
GÜMBEL, Carl Wilhelm von, Ritter, (1823-1898)
HAAST, John Francis Julius von, Sir, (1824-1887)
HACQUET, Belsazar, (1739-1815)
HALL, James, (1811-1898)
HANTKEN, Miksa von Prudnik, (1821-1893)
HAYDEN, Ferdinand Vandiveer, (1829-1887)
HELMERSEN, Gregor von, (1803-1885)
HOLL, Harvey Buchanan, (1820-1886)
HOPKINS, Evan, (?1810-1867)
HUBBARD, Bela, (1814-1896)
HUMBOLDT, Friedrich Wilhelm Heinrich Alexander von, Freiherr, (1769-1859)
HUXLEY, Thomas Henry, (1825-1895)
JACQUEMONT, Victor, (1801-1832)
JAŚKIEWICZ, Jan D. P., (1749-1809)
JUKES, Joseph Beete, (1811-1869)
KEYSERLING, Alexander Andreevich, Count, (1815-1891)
KINAHAN, George Henry, (1829-1908)
KJERULF, Theodor, (1825-1888)
KOCH, Carl, (1827-1882)
KREJČÍ, Jan, (1825-1887)
LAPHAM, Increase Allen, (1811-1875)
LESQUEREUX, Charles Leo, (1806-1889)
LILL VON LILIENBACH, Karl, (1798-1831)
LOCKE, John, (1792-1856)

STRUCTURAL GEOLOGY 1800-1849 Cont'd.

LOFTUS, William Kennett, (1822-1859)
LOGAN, William Edmond, Sir, (1798-1875)
LORY, Charles, (1823-1889)
MacADAM, James, (1801-1861)
MACLAREN, Charles, (1782-1866)
MAJENDIE, Ashurst, (?1784-1867)
MARCOU, Jules, (1824-1898)
MATHER, William Williams, (1804-1859)
MEDLICOTT, Henry Benedict, (1829-1905)
MEEK, Fielding Bradford, (1817-1876)
MERIAN, Peter, (1795-1883)
MÉTHERIE, Jean Claude de la, (1743-1817)
MILLER, Hugh, (1802-1856)
MILLINGTON, John, (1779-1886)
MOORE, Francis, Jr., (1808-1864)
MURCHISON, Roderick Impey, Sir, (1792-1871)
MURRAY, Alexander, (1810-1884)
NEWBERRY, John Strong, (1822-1892)
NORWOOD, Joseph Granville, (1807-1895)
NUTTALL, Thomas, (1786-1859)
OLDHAM, Thomas, (1816-1878)
OMALIUS D'HALLOY, Jean-Baptist Julien d', (1783-1875)
OTLEY, Jonathan, (1766-1856)
OWEN, David Dale, (1807-1860)
PALLAS, Peter Simon, (1741-1811)
PERCIVAL, James Gates, (1795-1856)
PETTERSEN, Karl, (1826-1890)
PLAYFAIR, John, (1748-1819)
PORTLOCK, Joseph Ellison, Major-General, (1794-1864)
POŠEPNÝ, František, (1836-1895)
POTIER, Alfred, (1840-1905)
PUSCH, Georg Gottlieb, (1790-1846)
QUIQUEREZ, Auguste, (1801-1882)
RAMSAY, Andrew Crombie, Sir, (1814-1891)
RAUMER, Karl Georg von, (1783-1865)
RIBEIRO, Carlos, General, (1813-1882)
ROFE, John, (1801-1878)
ROGERS, Henry Darwin, (1808-1866)
ROGERS, William Barton, (1804-1882)
ROMINGER, Carl Ludwig, (1820-1907)
ROYLE, John Forbes, (-1857?)
RUBIDGE, Richard Nathaniel, (1821?-1869)
RÜTIMEYER, Karl Ludwig, (1825-1895)
SALTER, John William, (1820-1869)
SANDERS, William, (1799-1875)
SCROPE, George Poulett, (né Thomson), (1797-1876)
SEDGWICK, Adam, Reverend, (1785-1873)
SELWYN, Alfred Richard Cecil, (1824-1902)
SHARPE, Daniel, (1806-1856)
SHUMARD, George Getz, (1826-1867)
SIMONY, Friedrich, (1813-1896)
SMITH, William, (1769-1839)
SMYTH, Warington Wilkinson, Sir, (1817-1890)
STASZIC, Stanislaw, (1755-1826)

STRUCTURAL GEOLOGY 1800-1849 Cont'd.

STOW, George B. William, (1822-1882)
STUDER, Bernhard Rudolf, (1794-1887)
STUTCHBURY, Samuel, (1798-1859)
SÜSSMILCH, Carl Adolf, (?1874-1946)
SWALLOW, George Clinton, (1817-1899)
SZABÓ, József, (1822-1894)
TAUNTON, John Hooke, (1820-1893)
THEOBALD, Gottfried Ludwig, (1810-1869)
TOPLEY, William, (1841-1894)
TROILIUS, Carl Oscar, (1813-[1899])
TSCHUDI, Johan Jacob von, (1818-1889)
VANUXEM, Lardner, (1792-1848)
VERNEUIL, Philippe Edouard Poulletier de, (1805-1873)
WAILES, Benjamin Leonard Couington, (1797-1862)
WATSON, White, (1760-1835)
WERNER, Abraham Gottlob, (1749-1817)
WHITE, Charles Abiathar, (1826-1910)
WHITNEY, Josiah Dwight, (1819-1896)
WHITTLESEY, Charles, Colonel, (1808-1886)
WINCHELL, Alexander, (1824-1891)
WOLF, Heinrich, (1825-1882)
WORTHEN, Amos Henry, (1812-1888)
ZEJSZNER, Ludwik, (1805-1871)
ZITTEL, Karl Alfred von, (1839-1904)

1850-1899

ABICH, Hermann Wilhelm, (1806-1886)
ALLEN, Henry Atwool, (?1854-1934)
AMI, Henry Mark, (1858-1931)
AMPFERER, Otto, (1875-1947)
ANDERSON, William, (1860-1915)
ANDRUSSOV, Nikolai Ivanovich, (1861-1924)
ANSTED, David Thomas, (1814-1880)
APLIN, Christopher d'Oyley Hale, (1819-1875)
ASHLEY, George Hall, (1866-1951)
AVELINE, William Talbot, (1822-1903)
BABBAGE, Charles, (1792-1871)
BAILY, William Hellier, (1819-1888)
BAINBRIDGE, William, (1812-1860)
BAKEWELL, Frederick Collier, (-1869)
BALL, Valentine, (1843-1895)
BALTZER, Richard Armin, (1842-1913)
BARETTI, Martino, (1841-1905)
BARLOW, Alfred Ernest, (1861-1914)
BARRELL, Joseph, (1869-1919)
BARROIS, Charles Eugène, (1851-1939)
BARRON, Thomas, (1867-1906)
BARROW, George, (1853-1932)
BARTON, George Hunt, (1852-1933)
BAYLEY, William Shirley, (1861-1943)
BEACHLER, Charles S., (1870-1894)
BEADNELL, Hugh John Llewellyn, Major, (1874-1944)
BEAUGÉ, Alfred, (1878-1935)
BEAUMONT, Jean Baptiste Armand Louis Léonce Élie de, (1798-1874)

STRUCTURAL GEOLOGY 1850-1899 Cont'd.

BECK, Carl Richard, (1858-1919)
BECKER, George Ferdinand, (1847-1919)
BEHR, Johannes, (1875-1973)
BELL, James Macintosh, (1877-1934)
BELL, Robert, (1841-1917)
BEMROSE, Henry Howe, (1857-1939)
BENNETT, Arthur, (1873-1941)
BENNETT, Francis James, (1845-1920)
BERGERON, Jules, (1853-1919)
BERTRAND, Léon, (1869-1947)
BERTRAND, Marcel Alexandre, (1847-1907)
BEST, Edward, (1824-1906)
BILLINGS, Elkanah, (1820-1876)
BLAKE, John Hopwood, (1843-1901)
BLANCKENHORN, Max Ludwig Paul, (1861-1947)
BLANFORD, William Thomas, (1832-1905)
BLOMBERG, Karl Albert, (1844-1922)
BÖHMERSHEIM, August Böhm von, Edler, (1858-1930)
BÖSE, Emil, (1868-1972)
BOUÉ, Ami, (1794-1881)
BOULE, Marcellin, (1861-1942)
BRESSON, Arthur, (1873-1946)
BROCK, Reginald Walter, (1874-1935)
BRØGGER, Waldemar Christofer, (1851-1940)
BROOKS, Alfred Hulse, (1871-1924)
BROOKS, Thomas Benton, (1836-1900)
BROWN, Charles Barrington, (1839-1917)
BROWN, Henry Yorke Lyell "Stoney", (1844-1928)
BRUNO, Carlo, (1831-1916)
BUCH, Christian Leopold von, Freiherr, (1774-1853)
BUEHLER, Henry Andrew, (1876-1944)
BULL, Alfred Joseph, (1875-1950)
BURCKHARDT, Carl, (1869-1935)
BURKART, Hermann Josef, (1798-1874)
BUTTS, Charles, (1863-1946)
BUXTORF, August, (1877-)
CACCIAMALI, Giambattista, (1856-1934)
CALL, Richard Ellsworth, (1856-1917)
CALLAWAY, Charles, (1838-1915)
CALVIN, Samuel, (1840-1911)
CAMPBELL, Henry Donald, (1862-1934)
CAPELLINI, Giovanni, (1833-1922)
CERMENATI, Mario, (1868-1924)
CHADWICK, George Halcott, (1876-1953)
CHOFFAT, Paul, (1849-1919)
CHRISTA, Emanuel, (1874-1948)
CLARKE, William Branwhite, Reverend, (1798-1878)
CLOUGH, Charles Thomas, (1852-1916)
COLE, Grenville Arthur James, (1859-1924)
CONDON, Thomas, (1822-1907)
CONRAD, Timothy Abbott, (1803-1877)
COOMARASWAMY, Ananda K., (1877-1947)
CORNET, René Jules, (1865-1929)
CORSTORPHINE, George Steuart, (1865-1919)

STRUCTURAL GEOLOGY 1850-1899 Cont'd.

COURTY, Georges, (1875-1953)
COX, Edward Travers, (1821-1907)
CRAWFORD, Ralph Dixon, (1873-1950)
CREDNER, Carl Friedrich Heinrich, (1809-1876)
CRICHTON, Alexander, Sir, (1763-1856)
CROLL, James, (1821-1890)
CROSBY, William Otis, (1850-1925)
DAKYNS, John Roche, (1836-1910)
DAL PIAZ, Giorgio, (1872-1962)
DALTON, William H., (?1848-1929)
DALY, Reginald Aldworth, (1871-1957)
DANA, James Dwight, (1813-1895)
DANZIG, Emil, (1855-1943)
DARTON, Nelson Horatio, (1865-1948)
DARWIN, Charles Robert, (1809-1882)
DATHE, Johann Friedrich Ernst, (1845-1917)
DAVID, Tannatt William Edgeworth, Sir, (1858-1934)
DAVIS, William Morris, (1850-1934)
DAWSON, George Mercer, (1849-1901)
DAWSON, John William, Sir, (1820-1899)
DE LA BECHE, Henry Thomas, Sir, (1796-1855)
DE LA CONDAMINE, H. M., Reverend, (-1854)
DE LORENZO, Giuseppe, (1871-1957)
DENCKMANN, August, (1860-1925)
DEPÉRET, Charles Jean Julien, (1854-1929)
DE PRADO, Casiano, Don, (1797-1866?)
DE RANCE, Charles Eugène, (1847-1906)
DERBY, Orville Adelbert, (1851-1915)
DE STEFANI, Carlo, (1851-1924)
DEWALQUE, Gilles Joseph Gustave, (1828-1905)
DEWEY, Henry, (1876-1965)
DIXON, Ernest Edward Leslie, (1876-1963)
DONNELLY, Ignatius, (1831-1901)
DOWLING, Donaldson Bogart, (1858-1925)
DRĂGHICEANU, Mathei M., (1844-1939)
DRESSER, John Alexander, (1866-1954)
DREW, Frederick, (1836-1891)
DUMBLE, Edwin Theodore, (1852-1927)
DU NOYER, George Victor, (-1869)
DUNSTAN, Benjamin, (1864-1933)
DUTTON, Clarence Edward, Major, (1841-1912)
EGAN, Frederick William, (1836-1901)
ELDRIDGE, George Homans, (1854-1905)
ELLES, Gertrude Lilian, (1872-1960)
ELLS, Robert Wheelock, (1845-1911)
EMERSON, Benjamin Kendall, (1843-1932)
EMMONS, Ebenezer, (1800-1863)
ENDLICH, Frederic Miller, (1851-1899)
ENGELMANN, August Heinrich, (1831-1899)
ERDMANN, Axel Joachim, (-)
ERDMANN, Edvard, (1840-1923)
ESCHER VON DER LINTH, Arnold, (1807-1872)
ETHERIDGE, Robert, (1819-1903)
ETHERIDGE, Robert Jr., (1847-1920)

STRUCTURAL GEOLOGY 1850-1899 Cont'd.

FALCONER, John Downie, (1876-1947)
FAVRE, Jean Alphonse, (1815-1890)
FEARNSIDES, William George, (1879-1968)
FERRONNIÈRE, Georges, (1875-1922)
FISHER, Cassius Asa, (1872-1930)
FLETT, John Smith, Sir, (1869-1947)
FOETTERLE, František, (1823-1876)
FOOT, Frederick J., (-1867)
FOOTE, R. Bruce, (-1912)
FORBES, David, (1828-1876)
FORBES, Edward, (1815-1854)
FORTAU, René, (1867-1920)
FOSTER, Clement Le Neve, Sir, (1841-1904)
FOURMARIER, Paul Fréderic Joseph Lucien Lauren, (1877-1970)
FOURNIER, Eugène, (1871-1941)
FOX-STRANGWAYS, Charles Edward, (1844-1910)
FRAZER, Persifor, (1844-1909)
FRECH, Fritz, (1861-1917)
FRIEDEL, Georges, (1865-1933)
FUCHS, Alexander, (1874-1935)
FURUHJELM, Enoch Hjalmar, (1823-1886)
GANLY, Patrick, (1809-1899)
GARDINER, Charles Irving, (?1868-1940)
GEIKIE, Archibald, Sir, (1835-1924)
GEIKIE, James, (1839-1915)
GENTIL, Louis, (1868-1925)
GERBER, Eduard, (1876-1956)
GIBSON, Walcot, (1864-1941)
GILBERT, Grove Karl, (1843-1918)
GIORDANO, Felice, (1825-1892)
GIROUX, N. J., (1859-1896)
GLENN, Leonidas Chalmers, (1871-1951)
GOODCHILD, John George, (1844-1906)
GORJANOVIĆ-KRAMBERGER, Dragutin, (1856-1937)
GOTTSCHE, Carl Christian, (1855-1909)
GOULD, Charles, (-1895)
GOULD, Charles Newton, (1868-1949)
GRANT, Ulysses Sherman, (1867-1932)
GREEN, Alexander Henry, (1832-1896)
GREEN, John Frederick Norman, (?1873-1949)
GREEN, Upfield, (1834-1917)
GREEN, William Lowthian, (1819-1890)
GREENE, Alfred Ernest, (1878-1940)
GREENLY, Edward, (1861-1951)
GREENOUGH, George Bellas, (1778-1855)
GREGORY, Herbert E., (1869-1952)
GREGORY, John Walter, (1864-1932)
GRESSLY, Amanz, (1814-1865)
GRIESBACH, Karl Ludolf, (1847-1907)
GRODDECK, Albrecht Ludwig von, (1837-1887)
GROEBER, Paul Friedrich Carl, (1885-1968)
GRZYBOWSKI, Józef, (1869-1922)
GUÉBHARD, Adrien, (1849-1924)
GUEYMARD, François Émile, (1788-1870)

STRUCTURAL GEOLOGY 1850-1899 Cont'd.

GUIDONI, Gerolamo, (1794-1870)
GÜMBEL, Carl Wilhelm von, Ritter, (1823-1898)
GUNN, William, (1837-1902)
GUTZWILLER, Andreas, (1845-1917)
HAAST, John Francis Julius von, Sir, (1824-1887)
HALFAR, Anton, (1836-1893)
HALL, Arthur Lewis, (1872-1955)
HALL, Christopher Webber, (1845-1911)
HALL, James, (1811-1898)
HANTKEN, Miksa von Prudnik, (1821-1893)
HARADA, Toyokitsi, (-1894)
HARDMAN, Edward Townley, (1845-1887)
HARKER, Alfred, (1859-1939)
HARRINGTON, Bernard J., (1848-1907)
HARRIS, Gilbert Dennison, (1864-1952)
HARTNAGEL, Chris Andrew, (1874-1962)
HATCH, Frederick Henry, (?1864-1932)
HAYDEN, Ferdinand Vandiveer, (1829-1887)
HAYES, Charles Willard, (1858-1916)
HECTOR, James, Sir, (1834-1907)
HEIM, Albert, (1849-1937)
HELMERSEN, Gregor von, (1803-1885)
HELMHACKER, Rudolf Václav, (1840-1915)
HENNEN, Ray Vernon, (1875-1958)
HERRICK, Clarence Luther, (1858-1904)
HESS, Frank L., (1871-1955)
HIBSCH, Josef Emanuel, (1852-1940)
HILL, Frank A., (1858-1915)
HILL, James Bastian, (1861-1927)
HILL, Robert Thomas, (1858-1941)
HINCH, John de Witt, (?1875-1931)
HINTERLECHNER, Karl, (1874-1932)
HOBBS, William Herbert, (1864-1953)
HOERNES, Rudolf, (1850-1912)
HOFFMANN, George Christian, (1837-1917)
HOLDEN, Roy Jay, (1870-1945)
HOLL, Harvey Buchanan, (1820-1886)
HOLLAND, Thomas Henry, Sir, (1868-1947)
HOLMES, T. Vincent, (?1841-1923)
HOLMQUIST, Per Johan, (1866-1946)
HOPKINS, Evan, (?1810-1867)
HORNE, John, (1848-1928)
HOWE, Ernest, (1875-1932)
HOWELL, Edwin Eugene, (1845-1911)
HOWELL, Henry Hyatt, (1834-1915)
HUBBARD, Bela, (1814-1896)
HUDSON, George Henry, (1855-1934)
HUGHES, Thomas McKenny, (1832-1917)
HULL, Edward, (1829-1917)
HUMBOLDT, Friedrich Wilhelm Heinrich Alexander von, Freiherr, (1769-1859)
HUME, William Fraser, (1867-1949)
HUTCHINSON, Lon Lewis, (1877-1947)
HUTTON, Frederick Wollaston, (1836-1905)

STRUCTURAL GEOLOGY 1850-1899 Cont'd.

HUXLEY, Thomas Henry, (1825-1895)
IDDINGS, Joseph Paxson, (1857-1920)
IRVING, John Duer, (1874-1918)
IRVING, Roland Duer, (1847-1888)
JACOB, Charles François Étienne, (1878-1962)
JACOBS, Elbridge Churchill, (1873-1957)
JAGGAR, Thomas Augustus, (1871-1953)
JAHN, Jaroslav Jiljí, (1865-1934)
JEHU, Thomas John, (1871-?1943)
JOHANNSEN, Albert, (1871-1962)
JOHNSEN, Arrien, (1877-1934)
JONES, Owen Thomas, (1878-1967)
JUKES, Joseph Beete, (1811-1869)
JUKES-BROWNE, Alfred John, (1851-1914)
KAISER, Friedrich Wilhelm Erich, (1871-1934)
KARPINSKY, Alexander Petrovich, (1847-1936)
KATZER, Bedřich, (1861-1925)
KEITH, Arthur, (1864-1944)
KEMP, James Furman, (1859-1926)
KERFORNE, Fernand, (1864-1927)
KEYES, Charles Rollin, (1864-1942)
KEYSERLING, Alexander Andreevich, Count, (1815-1891)
KILIAN, Charles Constant Wilfrid, (1862-1925)
KILROE, James Robinson, (1848-1927)
KINAHAN, George Henry, (1829-1908)
KING, Clarence Rivers, (1842-1901)
KITSON, Albert Ernest, Sir, (1868-1937)
KJERULF, Theodor, (1825-1888)
KLEMM, Gustav, (1858-1938)
KOCH, Carl, (1827-1882)
KOCH, Ferdo, (1874-?1936)
KORN, Johannes, (1862-1927)
KORNERUP, Andreas, (?1856-1880)
KOSSMAT, Franz, (1871-1938)
KRAFFT VON DELLMENSINGEN, Albrecht, (1871-1901)
KREJČÍ, Jan, (1825-1887)
KREUTZ, Felix Franz Xaver, (1844-1910)
KUMMEL, Henry Barnard, (1867-1945)
KYNASTON, Herbert, (1868-1915)
LACROIX, François Antoine Alfred, (1863-1948)
LAFLAMME, Joseph Clovis Kemner, Monseigneur, (1849-1910)
LAKE, Philip, (1865-1949)
LAMBE, Lawrence Morris, (1863-1919)
LAMBERT, Alan, (1837-1927)
LAMPLUGH, George William, (1859-1926)
LANDES, Henry, (1867-1936)
LANE, Alfred Church, (1863-1948)
LAPHAM, Increase Allen, (1811-1875)
LAPPARENT, Albert Auguste de, (1839-1908)
LAPWORTH, Charles, (1842-1920)
LAWSON, Andrew Cowper, (1861-1952)
LEE, Willis Thomas, (1864-1926)
LE MESLE, George, (-1895)
LEMOINE, Paul, (1878-1940)

STRUCTURAL GEOLOGY 1850-1899 Cont'd.

LEONARD, Hugh, (1841-1909)
LEPSIUS, Karl Georg Richard, (1851-1915)
LESQUEREUX, Charles Leo, (1806-1889)
LEWIS, Henry Carvill, (1853-1888)
LEWIS, Joseph Volney, (1869-1969)
LIEBER, Oscar M., (-1862)
LIEBUS, Adalbert, (1876-1945)
LIMANOWSKI, Mieczyslaw, (1876-1947)
LINDGREN, Waldemar, (1860-1939)
LÓCZY, Lajos, (1849-1920)
LOFTUS, William Kennett, (1822-1859)
LOGAN, William Edmond, Sir, (1798-1875)
LOHEST, Max, (1857-1926)
LORY, Charles, (1823-1889)
LORY, Pierre, (1866-1956)
LOSSEN, Karl August, (1841-1893)
LOUGHRIDGE, Robert Hills, (1843-1917)
LOW, Albert Peter, (1861-1952)
LÖWL, Ferdinand, (1856-1908)
LUBBOCK, John, Lord Avebury, (1834-1913)
LUCIUS, Michel, (1876-1961)
LUGEON, Maurice, (1870-1953)
LUKASZEWICZ, Józef, (1863-1928)
LUPTON, Charles Thomas, (1878-1935)
LYDEKKER, Richard, (1849-1915)
LYELL, Charles, Sir, (1797-1875)
LYMAN, Benjamin Smith, (1835-1920)
MacADAM, James, (1801-1861)
MACBRIDE, Thomas Huston, (1848-1934)
MACHADE E COSTA, Alfredo Augusto d'Olveira, (1870-?1952)
MacHENRY, Alexander, (1843-1919)
MACLAREN, Charles, (1782-1866)
MACLAREN, James Malcolm, (1873-1935)
MACPHERSON, José, Don, (1839-1902)
MAJENDIE, Ashurst, (?1784-1867)
MANKIN, Charles John, (1932-)
MARCOU, Jules, (1824-1898)
MARSHALL, Patrick, (1869-1950)
MARTIN, David, (1842-1918)
MARTIN, George C., (1875-1943)
MARVINE, Archibald Robertson, (1848-1876)
MATHER, William Williams, (1804-1859)
MATHEWS, Edward Bennett, (1869-1944)
MATSON, George Charlton, (1873-1940)
MATTHES, François Emile, (1874-1948)
McCALLEY, Henry, (1852-1904)
McCONNELL, Richard George, (1857-1942)
McGEE, William John, (1853-1912)
McHENRY, Alexander, (1843-1919)
McINTYRE, Peter, (1863-1940)
McKAY, Alexander, (1841-1917)
MEDLICOTT, Henry Benedict, (1829-1905)
MEEK, Fielding Bradford, (1817-1876)
MEINZER, Oscar Edward, (1876-1948)

STRUCTURAL GEOLOGY 1850-1899 Cont'd.

MENDENHALL, Walter Curran, (1871-1957)
MERIAN, Peter, (1795-1883)
MICHALSKI, Aleksander Octavianovich, (1855-1904)
MICHEL-LÉVY, Auguste, (1844-1911)
MIDDLEMISS, Charles Stewart, (1859-1945)
MILLER, Benjamin Le Roy, (1874-1944)
MILLER, Hugh, (1802-1856)
MILLER, Hugh Jr., (1850-1896)
MILLINGTON, John, (1779-1886)
MILTHERS, Paul Christian Vilhelm Madsen, (1865-1962)
MOHOROVIČIĆ, Andrya, (1857-1936)
MOJSISOVICS, Johann August Georg Edmund, Edler von Mojsvàr, (1839-1907)
MOLENGRAAFF, Gustaaf Adolf Frederik, (1860-1942)
MOLYNEUX, Arthur John Charles, (?1866-1921)
MOORE, Francis, Jr., (1808-1864)
MORAN, Thomas, (1837-1926)
MORGAN, Conwy Lloyd, (?1852-1936)
MORGAN, Peter (Percy) Gates, (1867-1927)
MOURLON, Michel Félix, (1845-1915)
MUIR, John, (1838-1914)
MUNN, Malcolm John, (1874-1949)
MUNTEANI-MURGOCI, Gheorghe, (1872-1925)
MUNTHE, Henrik Vilhelm, (1860-1958)
MURCHISON, Roderick Impey, Sir, (1792-1871)
MURRAY, Alexander, (1810-1884)
NAUMANN, Edmund, (1854-1927)
NEWBERRY, John Strong, (1822-1892)
NEWLAND, David Hale, (1872-1943)
NOLAN, Joseph, (1841-1902)
NORDENSKIÖLD, Nils Adolf Erik, (1832-1901)
NORTON, William Harmon, (-)
NORWOOD, Charles Joseph, (1853-1927)
NORWOOD, Joseph Granville, (1807-1895)
NOVARESE, Vittorio, (1861-1948)
NUTTALL, Thomas, (1786-1859)
OGILVIE, William, (1846-1912)
OGILVIE GORDON, Maria Matilda, Dame, (1864-1939)
O'KELLY, Joseph, (1832-1883)
OLDHAM, Charles Aemilius, (1831?-1869)
OLDHAM, Richard Dixon, (1858-1936)
OLDHAM, Thomas, (1816-1878)
OMALIUS D'HALLOY, Jean-Baptist Julien d', (1783-1875)
ORDOÑEZ, Ezequiel, (1867-1950)
ORTON, Edward Francis Baxter, (1829-1899)
OTLEY, Jonathan, (1766-1856)
OWEN, David Dale, (1807-1860)
PARK, James, (1857-1946)
PAUL, Carl Maria, (1838-1900)
PAULCKE, Wilhelm, (1873-1949)
PEACH, Benjamin Neeve, (1842-1926)
PENCK, Friedrich Carl Albrecht, (1858-1945)
PENNING, William Henry, (1838-1902)
PERCIVAL, James Gates, (1795-1856)

STRUCTURAL GEOLOGY 1850-1899 Cont'd.

PERKINS, George Henry, (1844-1933)
PETHŐ, Gyula, (1848-1902)
PETKOVIĆ, Vladamir K., (1873-1934)
PETRASCHEK, Wilhelm, (1876-1967)
PETTERSEN, Karl, (1826-1890)
PILAR, Georg, (1847-1893)
PILGRIM, Henry Guy Ellcock, (1875-1943)
PIRSSON, Louis Valentine, (1860-1919)
POLLARD, William, (1870-1946)
POLWHELE, Thomas Roxburgh, (1831-1909)
PORRIO, Cesare, (1865-1940)
PORTLOCK, Joseph Ellison, Major-General, (1794-1864)
PŎSEPNÝ, František, (1836-1895)
POTIER, Alfred, (1840-1905)
POWELL, John Wesley, (1834-1902)
PREISWERK, Heinrich, (1876-1940)
PRELLER, Charles du Riche, (1844-)
PRINGLE, John, (1877-1948)
PROSSER, Charles Smith, (1860-1916)
PUMPELLY, Raphael, (1837-1923)
QUIQUEREZ, Auguste, (1801-1882)
RAMSAY, Andrew Crombie, Sir, (1814-1891)
RAMSAY, Wilhelm, (1865-1928)
RANSOME, Frederick Leslie, (1868-1935)
RAUMER, Karl Georg von, (1783-1865)
READE, Thomas Mellard, (1832-1909)
REED, Frederick Richard Cowper, (1869-1946)
RÉVIL, Joseph, (1849-1931)
RIBEIRO, Carlos, General, (1813-1882)
RICHARDSON, Charles Henry, (1862-1935)
RICHARDSON, George Burr, (1872-1949)
RICHTHOFEN, Paul Wilhelm Ferdinand von, Freiherr, (1833-1905)
RIES, Heinrich, (1871-1951)
ROFE, John, (1801-1878)
ROGERS, Arthur William, (1872-1946)
ROGERS, Henry Darwin, (1808-1866)
ROGERS, William Barton, (1804-1882)
ROMAN, Albert Frédéric, (1871-1943)
ROMER, Eugeniusz, (1871-1954)
ROMINGER, Carl Ludwig, (1820-1907)
RØRDAM, Kristian, (1860-1939)
ROWE, Richard Burton, (1872-1902)
ROYLE, John Forbes, (-1857?)
RUBIDGE, Richard Nathaniel, (1821?-1869)
RÜTIMEYER, Karl Ludwig, (1825-1895)
RUTKOWSKI, Feliks, (1878-1964)
RUTLEY, Frank, (1842-1904)
SAFFORD, James Merrill, (1822-1907)
SALISBURY, Rollin D., (1858-1922)
SALOMON-CALVI, Wilhelm, (1868-1941)
SALTER, John William, (1820-1869)
SANDERS, William, (1799-1875)
SARDESON, Frederick William, (1866-1959)
SAUER, Gustav Adolf, (1825-1932)

STRUCTURAL GEOLOGY 1850-1899 Cont'd.

SCHAFARZIK, Ferenc, (1854-1927)
SCHAFFER, Franz Xavier, (1876-1954)
SCHALCH, Ferdinand, (1848-1918)
SCHARDT, W. Hans, (1858-1931)
SCHMIDT, Carl, (1862-1923)
SCHOTTLER, Wilhelm, (1869-1932)
SCHRADER, François Jean Daniel "Franz", (1844-1924)
SCHRADER, Frank Charles, (1860-1944)
SCHWARZ, Ernest Hubert Lewis, (1873-1928)
SCHWINNER, Robert Gangolf, (1878-1953)
SCROPE, George Poulett, (né Thomson), (1797-1876)
SEDGWICK, Adam, Reverend, (1785-1873)
SEE, Thomas Jefferson Jackson, (1866-)
SELWYN, Alfred Richard Cecil, (1824-1902)
SHALER, Nathaniel Southgate, (1841-1906)
SHARMAN, George, (1832-1914)
SHARPE, Daniel, (1806-1856)
SHEPARD, Edward Martin, (1854-1934)
SHERLOCK, R. L., (1875-1948)
SHUMARD, George Getz, (1826-1867)
SMITH, Eugene Allen, (1841-1927)
SMITH, George Otis, (1871-1944)
SMOCK, John Conover, (1842-1925)
SMYTH, Warington Wilkinson, Sir, (1817-1890)
SPENCER, Joseph William Winthrop, (1851-1921)
STEIGER, George, (1869-1944)
STEVENSON, John James, (1841-1924)
ST. JOHN, Orestes Hawley, (1841-1921)
STOLICZKA, Ferdinand, (1838-1874)
STOW, George B. William, (1822-1882)
STRAHAN, Aubrey, Sir, (1852-1928)
STUDER, Bernhard Rudolf, (1794-1887)
STUTCHBURY, Samuel, (1798-1859)
SUESS, Eduard, (1831-1914)
SUESS, Franz Eduard, (1867-1941)
SÜSSMILCH, Carl Adolf, (?1874-1946)
SWALLOW, George Clinton, (1817-1899)
SYMES, Richard Glascott, (1840-1906)
SZABÓ, József, (1822-1894)
SZÁDECZKY, Gyula, Baron Szádeczky-Kardoss of Szádecsne and Kafdosfalva, (1860-1935)
TAFF, Joseph Alexander, (1862-1944)
TAUNTON, John Hooke, (1820-1893)
TCHERNYCHEV, Féodoss Nikolaievich, (1856-1914)
TEALE, Edmund Oswald, Sir, (1874-1971)
TEALL, Jethro Justinian Harris, Sir, (1849-1924)
TEISSEYRE, Wawrzyniec, (1860-1939)
TELLER, Friedrich, (1852-1913)
TERMIER, Pierre, (1859-1930)
THEOBALD, Gottfried Ludwig, (1810-1869)
THOMAS, Herbert Henry, (1876-1935)
TIDDEMAN, Richard Hill, (1842-1917)
TOIT, Alexander Logie du, (1878-1948)
TOPLEY, William, (1841-1894)

STRUCTURAL GEOLOGY 1850-1899 Cont'd.

TOULA, Franz, (1845-1920)
TRASK, John Boardman, (1824-1879)
TRIBOLET-HARDY, Frédéric Maurice de, (1852-1929)
TROILIUS, Carl Oscar, (1813-[1899])
TSCHUDI, Johan Jacob von, (1818-1889)
TWITCHELL, Mayville William, (1866-1927)
TYRRELL, Joseph Burr, (1858-1957)
UDDEN, Johan August, (1859-1932)
UHLIG, Victor Karl, (1857-1911)
ULRICH, Edward Oscar, (1857-1944)
USSHER, William Augustus Edmond, (1849-1920)
USSING, Niels Viggo, (1864-1911)
VAN HISE, Charles Richard, (1857-1918)
VAN HORN, Frank Robertson, (1872-1933)
VAUGHAN, Thomas Wayland, (1870-1952)
VEATCH, Arthur Clifford, (1878-1938)
VERNEUIL, Philippe Edouard Poulletier de, (1805-1873)
VERRI, Antonio, (1839-1925)
WÄHNER, Franz, (1856-1932)
WAILES, Benjamin Leonard Couington, (1797-1862)
WALCOTT, Charles Doolittle, (1850-1927)
WALKER, Thomas Leonard, (1867-1942)
WALL, George Parkes, (1832-1912)
WARD, James Clifton, Reverend, (1843-1880)
WARD, Leonard Keith, (1879-1964)
WATSON, Thomas Leonard, (1871-1924)
WEIDMAN, Samuel, (1870-1945)
WELLER, Stuart, (1870-1927)
WELLS, Roger Clark, (1877-1944)
WESTGATE, Lewis Gardner, (1868-1948)
WESTON, Thomas Chesmer, (1832-1910)
WETTSTEIN, Alexander, (1861-1887)
WHITAKER, William, (1836-1925)
WHITE, Charles Abiathar, (1826-1910)
WHITE, Israel Charles, (1848-1927)
WHITEAVES, John Joseph Frederick, (1835-1909)
WHITNEY, Josiah Dwight, (1819-1896)
WHITTLESEY, Charles, Colonel, (1808-1886)
WHYMPER, Edward, (1840-1911)
WIIK, Fredrik Johan, (1839-1909)
WILCKENS, Otto Rudolf, (1876-1943)
WILDER, Frank Alonzo, (1870-1930)
WILKINSON, Charles Smith, (1843-1891)
WILKMAN, Wanold Wrydon, (1872-?1938)
WILLIAMS, George Huntington, (1856-1894)
WILLIS, Bailey, (1857-1949)
WILSON, Charles Henry, (1851-1887)
WILSON, J. S. Grant, (1855-1908)
WINCHELL, Alexander, (1824-1891)
WINCHELL, Horace Vaughn, (1865-1923)
WINDHAUSEN, Anselm, (1882-1932)
WIŚNIOWSKI, Tadeusz, (1865-1933)
WOLF, Heinrich, (1825-1882)
WOODRUFF, Elmer Grant, (1872-1952)

STRUCTURAL GEOLOGY 1850-1899 Cont'd.

WOODWARD, Harry Page, (1858-1917)
WOODWARD, Horace Bolingbroke, (1848-1914)
WOODWARD, Robert Simpson, (1849-1924)
WOODWORTH, Jay Backus, (1865-1925)
WOOLNOUGH, Walter George, (1876-1958)
WORTHEN, Amos Henry, (1812-1888)
WRIGHT, Frederick Eugene, (1877-1953)
WRIGHT, William Bourke, (1876-1939)
WYNNE, Arthur Beavor, (1835-1906)
WYVILLE THOMSON, Charles, Sir, (1830-1882)
YABE, Hisakatsu, (1878-1969)
YACHEVSKY, Leonard Antonovich, (1858-1916)
YOUNG, Alfred Prentice, (1841-1919)
YOUNG, George Albert, (1898-1947)
YOUNG, John, (1835-1902)
ZEJSZNER, Ludwik, (1805-1871)
ZITTEL, Karl Alfred von, (1839-1904)

1900 to present
ABERER, Ferdinand, (1913-1965)
AHO, Aaro E., (-1977)
ALCOCK, Frederick James, (1888-1972)
ALLAN, John Andrew, (1884-1955)
ALLEN, Henry Atwool, (?1854-1934)
AMI, Henry Mark, (1858-1931)
AMPFERER, Otto, (1875-1947)
ANDERSON, Alfred Leonard, (1900-1964)
ANDERSON, William, (1860-1915)
ANDREATTA, Benedettino Ciro, (1906-1960)
ANDREWS, Richard Wallis, (?1927-1967)
ANDRUSOV, Dimitri, (1897-1976)
ANDRUSSOV, Nikolai Ivanovich, (1861-1924)
ANELLI, Mario, (1882-1953)
ARBENZ, Paul, (1880-1943)
ARGAND, Émile, (1879-1940)
ASHLEY, George Hall, (1866-1951)
AUBOUIN, Jean, (1928-)
AVELINE, William Talbot, (1822-1903)
BAILEY, Edward Battersby, Sir, (1881-1965)
BAKER, Herbert Arthur, (1885-1954)
BALK, Robert, (1899-1955)
BALL, John Rice, (1881-1953)
BALLARD, William Norval, (1906-1976)
BALTZER, Richard Armin, (1842-1913)
BARBOUR, George Brown, (1891-1977)
BARETTI, Martino, (1841-1905)
BARLOW, Alfred Ernest, (1861-1914)
BARRELL, Joseph, (1869-1919)
BARRETT, Leslie Park, (1887-1972)
BARRETT, Lucas, (1837-1862)
BARROIS, Charles Eugène, (1851-1939)
BARRON, Thomas, (1867-1906)
BARROW, George, (1853-1932)
BARTON, George Hunt, (1852-1933)

STRUCTURAL GEOLOGY 1900 to present Cont'd.

BATEMAN, Alan Mara, (1889-1971)
BATES, Robert Glenn, (1923-1972)
BAYLEY, William Shirley, (1861-1943)
BEADNELL, Hugh John Llewellyn, Major, (1874-1944)
BEAUGÉ, Alfred, (1878-1935)
BECK, Carl Richard, (1858-1919)
BECKER, George Ferdinand, (1847-1919)
BECKWITH, Radcliffe Harold, (1900-1964)
BEEKLEY, Albert Leon, (1883-1952)
BEHR, Johannes, (1875-1973)
BELL, James Macintosh, (1877-1934)
BELL, John Arnold, (1939-1970)
BELL, Olin Grogan, (1893-1970)
BELL, Robert, (1841-1917)
BELLAIR, Pierre, (1910-)
BELOUSSOV, Vladimir Vladimirovitch, (-)
BEMROSE, Henry Howe, (1857-1939)
BENNETT, Arthur, (1873-1941)
BENNETT, Francis James, (1845-1920)
BENSON, William Noel, (1885-1957)
BERGERON, Jules, (1853-1919)
BERNOUILLI, Walter, (1885-1946)
BERTRAND, Léon, (1869-1947)
BERTRAND, Marcel Alexandre, (1847-1907)
BEST, Edward, (1824-1906)
BEYSCHLAG, Franz, (-?1935)
BIEBER, Charles Leonard, (1901-1965)
BION, Herbert Stanley, (1888-1915)
BISHOPP, Douglas Wallace, (1900-1977)
BISSET, Charles Barron, (-?1966)
BLAKE, John Hopwood, (1843-1901)
BLANCKENHORN, Max Ludwig Paul, (1861-1947)
BLANFORD, William Thomas, (1832-1905)
BLOMBERG, Karl Albert, (1844-1922)
BLONDEL, Fernand, (1894-1968)
BOGDANOV, Alexei Alexeievitch, (1906-1971)
BÖHMERSHEIM, August Böhm von, Edler, (1858-1930)
BOLDYREV, Anatolii Kapitonovich, (1883-1946)
BOOCOCK, Colin, (1925-1973)
BÖSE, Emil, (1868-1972)
BOUCOT, Arthur James, (1924-)
BOULE, Marcellin, (1861-1942)
BOUSSAC, Jean, (1885-1917)
BRADLEY, William Frank, (1908-1973)
BRAMMALL, Alfred, (?1879-1954)
BRANSON, Edwin Robert, (1908-1971)
BRESSON, Arthur, (1873-1946)
BROCK, Byron Britton, (1904-1972)
BROCK, Reginald Walter, (1874-1935)
BRØGGER, Waldemar Christofer, (1851-1940)
BROGGI, Jorge Albert, (1888-1966)
BROMEHEAD, Cyril Edward Nowill, (1885-1952)
BROOKS, Alfred Hulse, (1871-1924)
BROTZEN, Fritz, (1902-1968)

STRUCTURAL GEOLOGY 1900 to present Cont'd.

BROUWER, Hendrik Albertus, (1886-1973)
BROWN, Charles Barrington, (1839-1917)
BROWN, Henry Yorke Lyell "Stoney", (1844-1928)
BROWN, James Macmillan, (1921-1970)
BROWN, John Coggin, (1884-1962)
BROWN, Peter Evans, (-)
BRUCE, Everend Lester "Louie", (1884-1949)
BRUNO, Carlo, (1831-1916)
BRYAN, Walter Heywood, (1891-1966)
BUBNOFF, Serge von, (1888-1957)
BUCHER, Walter Herman, (1888-1965)
BUEHLER, Henry Andrew, (1876-1944)
BUJALSKI, Boleslaw, (1888-1945)
BULL, Alfred Joseph, (1875-1950)
BURCKHARDT, Carl, (1869-1935)
BURFOOT, James Dabney "Dan", Jr., (1896-1966)
BURNETT, George Alexander, (1892-1952)
BURTON, Richard Cooksey, (1890-1916)
BUTTS, Charles, (1863-1946)
BUWALDA, John Peter, (1886-1954)
BUXTORF, August, (1877-)
BYBEE, Halbert Pleasant, (1888-1957)
CACCIAMALI, Giambattista, (1856-1934)
CADISCH, Joos, (1895-)
CADY, Gilbert Haven, (1882-1970)
CAIRE, André, (1923-)
CAIRNES, Delorme Donaldson, (1879-1917)
CALL, Richard Ellsworth, (1856-1917)
CALLAWAY, Charles, (1838-1915)
CALVIN, Samuel, (1840-1911)
CAMPBELL, Arthur Byron, (1924-1971)
CAMPBELL, Henry Donald, (1862-1934)
CAMPBELL, Robert, (1881-1957)
CAPELLINI, Giovanni, (1833-1922)
CARROLL, Dorothy, (1907-1970)
CARRUTHERS, Robert George, (1880-1965)
CATHCART, Stanley Holman, (1889-1953)
ČECH, Vladimîr, (1901-)
ČECHOVIČ, Vsevolod, (1900-1961)
ČEPEK, Ladislav, (1899-1974)
CERMENATI, Mario, (1868-1924)
CHADWICK, George Halcott, (1876-1953)
CHAMBERLIN, Rollin Thomas, (1881-1948)
CHAPUT, Jules Ernest, (1882-1943)
CHHIBBER, Harbans Lal, (1899-1955)
CHIPMAN, Kenneth G., (-)
CHOFFAT, Paul, (1849-1919)
CHRISTA, Emanuel, (1874-1948)
CHUBB, Lawrence John, (?1888-1971)
CLAR, Eberhard, (1902-)
CLOOS, Ernst, (1898-1974)
CLOOS, Hans, (1885-1951)
CLOUGH, Charles Thomas, (1852-1916)
COCKBURN, Alexander Murray, (1901-1959)

STRUCTURAL GEOLOGY 1900 to present Cont'd.

COLE, Grenville Arthur James, (1859-1924)
COLLET, Léon William, (1880-1957)
CONDON, Thomas, (1822-1907)
COOKE, Charles Wythe, (1887-1971)
COOKE, Harold Caswell, (1884-1956)
COOMARASWAMY, Ananda K., (1877-1947)
COOMBS, Howard A., (-)
COOPER, Byron Nelson, (1912-1971)
CORNELIUS, Hans Peter, (1888-)
CORNET, René Jules, (1865-1929)
CORSTORPHINE, George Steuart, (1865-1919)
COTTER, Gerald de Purcell, (1881-1941)
COTTON, Leo Arthur, (1883-1963)
COURTY, Georges, (1875-1953)
COX, Edward Travers, (1821-1907)
CRAIG, Robert Meldrum, (?1882-1956)
CRAWFORD, Ralph Dixon, (1873-1950)
CRICKMAY, Geoffrey W., (1905-1971)
CROOKSHANK, Henry, (1893-1972)
CROSBY, William Otis, (1850-1925)
CROWTHER, Arthur Frederick, (1929-1954)
CRUM, Harry Edwin, (1892-1955)
CULEK, Antonin, (1904-1944)
CUNNINGTON, C. H., Lieutenant, (1889-1918)
CURRIER, Louis Wade, (1890-1970)
CZARNOCKI, Jan, (1889-1951)
DAGUIN, Fernand, (1889-1948)
DAHLGRÜN, Fritz, (1894-1954)
DAKE, Charles Laurence, (1883-1934)
DAKYNS, John Roche, (1836-1910)
DALE, Nelson Clark, (1880-1962)
DAL PIAZ, Giorgio, (1872-1962)
DALTON, William H., (?1848-1929)
DALY, Reginald Aldworth, (1871-1957)
DANE, Carle Hamilton, (1900-1968)
DANZIG, Emil, (1855-1943)
DARTON, Nelson Horatio, (1865-1948)
DATHE, Johann Friedrich Ernst, (1845-1917)
DAVID, Tannatt William Edgeworth, Sir, (1858-1934)
DAVIDSON, Charles Findlay, (1911-1967)
DAVIS, William Morris, (1850-1934)
DAWSON, George Mercer, (1849-1901)
DE DYCKER, Raymond, (1903-1947)
DEISS, Charles Frederick, (1903-1959)
DELEAU, Paul Camille, (1904-)
DE LORENZO, Giuseppe, (1871-1957)
DEMAY, André, (1890-1964)
DENCKMANN, August, (1860-1925)
DENNEN, William Henry, (1920-)
DEPÉRET, Charles Jean Julien, (1854-1929)
DE RANCE, Charles Eugène, (1847-1906)
DERBY, Orville Adelbert, (1851-1915)
DESIO, Ardito, (1897-)
DE STEFANI, Carlo, (1851-1924)

STRUCTURAL GEOLOGY 1900 to present Cont'd.

DEUBEL, Fritz, (1898-1966)
DEWALQUE, Gilles Joseph Gustave, (1828-1905)
DEWEY, Henry, (1876-1965)
DIETZ, Robert Sinclair, (1914-)
DINES, Henry George, (1891-?1964)
DINHAM, Charles Hawker, (1883-1955)
DIXON, Ernest Edward Leslie, (1876-1963)
DOELL, Richard Rayman, (1923-)
DONNELLY, Ignatius, (1831-1901)
DORSEY, George Edwin, (1892-1953)
DOUGLAS, George Vibert, (1892?-1958)
DOWLING, Donaldson Bogart, (1858-1925)
DRĂGHICEANU, Mathei M., (1844-1939)
DRESSER, John Alexander, (1866-1954)
DRYSDALE, Charles Wales, (1885-1917)
DUMBLE, Edwin Theodore, (1852-1927)
DUTTON, Clarence Edward, Major, (1841-1912)
DUTY, Richard Emerson, (1908-1966)
DYER, William Spafford, (1894-1941)
EARDLEY, Armand John, (1901-1972)
EASTWOOD, Tom, (1889?-1970)
EDMUNDS, Francis Hereward, (1893-1960)
EDWARDS, Merwin Guy, (1889-1969)
EGAN, Frederick William, (1836-1901)
EKBLAW, George Elbert, (1895-1972)
ELDRIDGE, George Homans, (1854-1905)
ELLES, Gertrude Lilian, (1872-1960)
ELLS, Robert Wheelock, (1845-1911)
ELLSWORTH, Hardy Vincent, (1889-1952)
EMERSON, Benjamin Kendall, (1843-1932)
EMMONS, William Harvey, (1876-1948)
ERDMANN, Charles Edgar, (1897-1971)
ERDMANN, Edvard, (1840-1923)
ESCHER, Berend George, (-1967)
ESKOLA, Pentti Eelis, (1883-1964)
ETHERIDGE, Robert, (1819-1903)
ETHERIDGE, Robert Jr., (1847-1920)
EVANS, Percy, (1892-1974)
EWING, William Maurice, (1906-1974)
EYLES, Victor Ambrose, (1896-1978)
FAIRBAIRN, Harold Williams, (1906-)
FAIRCHILD, John Gifford, (1882-1965)
FALCONER, John Downie, (1876-1947)
FALLOT, Paul, (1889-1960)
FATH, Arthur Earl, (1887-1976)
FEARNSIDES, William George, (1879-1968)
FERGUSON, John Logan, (1893-1973)
FERMOR, Lewis Leigh, Sir, (1880-1954)
FERRONNIÈRE, Georges, (1875-1922)
FIELD, Richard Montgomery, (1885-1961)
FINCH, Elmer Harrison, (1883-1953)
FISHER, Cassius Asa, (1872-1930)
FLETT, John Smith, Sir, (1869-1947)
FOLEY, Edward J., (1908-1971)

STRUCTURAL GEOLOGY 1900 to present Cont'd.

FOOTE, R. Bruce, (-1912)
FORTAU, René, (1867-1920)
FOSLIE, Steinar, (1887-1951)
FOSTER, Clement Le Neve, Sir, (1841-1904)
FOSTER, Margaret Dorothy, (1895-1970)
FOURMARIER, Paul Fréderic Joseph Lucien Lauren, (1877-1970)
FOURNIER, Eugène, (1871-1941)
FOX-STRANGWAYS, Charles Edward, (1844-1910)
FRAZER, Persifor, (1844-1909)
FRECH, Fritz, (1861-1917)
FREI, Ernst, (1894-1974)
FRIEDEL, Georges, (1865-1933)
FRIEDL, Karl, (1898-1966)
FRIES, Carl, Jr., (1910-1965)
FRITSCH, Wolfgang, (1928-1970)
FRITTS, Crawford Ellsworth, (1927-1972)
FUCHS, Alexander, (1874-1935)
FUCHS, Vivian Ernest, Sir, (1908-)
FURCRON, Aurelius Sydney, (1899-1971)
GAGNEBIN, Élie, (1891-1949)
GALLAGHER, David, (1906-1968)
GARDINER, Charles Irving, (?1868-1940)
GARDNER, James Henry, (1883-1964)
GARDNER, Julia Anna, (1882-1960)
GAULT, Hugh Richard, (1915-1961)
GEIJER, Per Adolf, (1886-1976)
GEIKIE, Archibald, Sir, (1835-1924)
GEIKIE, James, (1839-1915)
GENTIL, Louis, (1868-1925)
GEORGE, William Owsley, (1892-1966)
GERBER, Eduard, (1876-1956)
GIBSON, Walcot, (1864-1941)
GIGNOUX, Maurice Irénée Marie, (1881-1955)
GILBERT, Grove Karl, (1843-1918)
GILL, James Rogers, (1922-1972)
GILLERMAN, Elliot "Bud", (1913-1974)
GILLULY, James, (-)
GINTER, Roy Lamont, (1892-1958)
GLANGEAUD, Louis, (1903-)
GLASER, Theodor, (1894-1922)
GLAUERT, Ludwig, (1879-1963)
GLENN, Leonidas Chalmers, (1871-1951)
GOETEL, Walery, (1889-1972)
GOGUEL, Jean, (1908-)
GOODCHILD, John George, (1844-1906)
GORJANOVIĆ-KRAMBERGER, Dragutin, (1856-1937)
GORMAN, Joseph M., (1910-1967)
GOTTSCHE, Carl Christian, (1855-1909)
GÖTZINGER, Gustav, (1880-1969)
GOULD, Charles Newton, (1868-1949)
GRANT, Ulysses Sherman, (1867-1932)
GRANTHAM, Donald Robert, (1896-1969)
GRATON, Louis Caryl, (1880-1970)
GREEN, John Frederick Norman, (?1873-1949)

STRUCTURAL GEOLOGY 1900 to present Cont'd.

GREEN, Upfield, (1834-1917)
GREENE, Alfred Ernest, (1878-1940)
GREENE, Frank Cook, (1886-1975)
GREENLY, Edward, (1861-1951)
GREGORY, Herbert E., (1869-1952)
GREGORY, John Walter, (1864-1932)
GRIESBACH, Karl Ludolf, (1847-1907)
GRIFFITH, Glenn Rodgers Vickery, (1891-1964)
GROEBER, Paul Friedrich Carl, (1885-1968)
GROTH, Jean, (1899-1916)
GROUT, Frank Fitch, (1880-1958)
GROVES, Arthur William, (1903-1956)
GRZYBOWSKI, Józef, (1869-1922)
GUBLER, Jean, (1902-1940)
GUÉBHARD, Adrien, (1849-1924)
GUILLOU, Robert Barton, (1923-1965)
GUNN, William, (1837-1902)
GUNTER, Herman, (1885-1972)
GÜNZLER-SEIFFERT, Hans, (1883-1957)
GUTZWILLER, Andreas, (1845-1917)
GUZIK, Kazimierz, (1911-1970)
HALL, Arthur Lewis, (1872-1955)
HALL, Christopher Webber, (1845-1911)
HALL, Thomas Clifford Fitzwilliam, (1886-1956)
HALLIMOND, Arthur Francis, (1890-1968)
HAM, William Eugene, (1916-1970)
HANSELL, James Myron, (1902-1970)
HANSON, Ernest Albin, (1899-1973)
HARES, Charles Joseph, (1881-1970)
HARKER, Alfred, (1859-1939)
HARRINGTON, Bernard J., (1848-1907)
HARRINGTON, George Leavitt, (1883-1972)
HARRINGTON, Horacio Jaime, (1910-1973)
HARRIS, Gilbert Dennison, (1864-1952)
HARRISON, John Vernon, (1892-1972)
HARTNAGEL, Chris Andrew, (1874-1962)
HATCH, Frederick Henry, (?1864-1932)
HAUPTMAN, Charles R., (1891-1971)
HAVARD, Charles Gentry, (1921-1950)
HAWLEY, James Edwin, (1897-1965)
HAYES, Albert Orion, (1882-1965)
HAYES, Charles Willard, (1858-1916)
HECTOR, James, Sir, (1834-1907)
HEIM, Albert, (1849-1937)
HEIM, Arnold, (?1882-1965)
HEINZELMANN, Karl, (1911-1966)
HELMHACKER, Rudolf Václav, (1840-1915)
HENDRICKSON, Victor James, (1892-1953)
HENNEN, Ray Vernon, (1875-1958)
HENRY, Gerard, (1890-1943)
HERALD, Frank Edward, (1887-1965)
HERITSCH, Franz, (1882-1945)
HERNÁNDEZ-PACHECO, Francisco, (1899-1976)
HERON, Alexander Macmillan, (1884-1971)

STRUCTURAL GEOLOGY 1900 to present Cont'd.

HEROY, William Bayard, Sr., (1883-1971)
HERPERS, Henry F., Jr., (1915-1952)
HERRICK, Clarence Luther, (1858-1904)
HESS, Frank L., (1871-1955)
HESS, Harry Hammond, (1906-1969)
HEWETT, Donnel Foster, (1881-1971)
HEWITT, Robert Leigh, (1911-1956)
HIBBERT-WARE, Samuel, (1782-1848)
HIBSCH, Josef Emanuel, (1852-1940)
HIGAZY, Riad Abdel-Megid, (1919-1967)
HILL, Frank A., (1858-1915)
HILL, James Bastian, (1861-1927)
HILL, James Wilcott, (1920-1954)
HILL, Robert Thomas, (1858-1941)
HINCH, John de Witt, (?1875-1931)
HINTERLECHNER, Karl, (1874-1932)
HLÂVKA, Karel, (1905-)
HOBBS, William Herbert, (1864-1953)
HODGE, Edwin Thomas, (1887-1970)
HOERNES, Rudolf, (1850-1912)
HOFFMANN, George Christian, (1837-1917)
HOLDEN, Roy Jay, (1870-1945)
HOLLAND, Thomas Henry, Sir, (1868-1947)
HOLLINGWORTH, Sydney Ewart, (1899-1966)
HOLLISTER, Joseph Steffens, (1903-1953)
HOLMES, Arthur, (1890-1965)
HOLMES, T. Vincent, (?1841-1923)
HOLMQUIST, Per Johan, (1866-1946)
HOLMSEN, Gunnar, (1880-1976)
HOLTEDAHL, Olaf, (1885-1975)
HOPPER, Walter Everett, (1886-1971)
HORNE, John, (1848-1928)
HOWE, Ernest, (1875-1932)
HOWELL, Edwin Eugene, (1845-1911)
HOWELL, Henry Hyatt, (1834-1915)
HUDDLE, John Warfield, (1907-1975)
HUDSON, George Henry, (1855-1934)
HUFFMAN, Edward Archibald, (1912-1954)
HUGHES, Thomas McKenny, (1832-1917)
HULL, Edward, (1829-1917)
HULL, Joseph Poyer Deyo, (1889-1967)
HUME, William Fraser, (1867-1949)
HURLEY, Patrick Mason, (1912-)
HUTCHINSON, Lon Lewis, (1877-1947)
HUTCHINSON, William W., (-)
HUTTENLOCHER, Heinrich, (1890-1954)
HUTTON, Colin Osborne, (1910-1971)
HUTTON, Frederick Wollaston, (1836-1905)
IDDINGS, Joseph Paxson, (1857-1920)
IRELAND, Hubert Andrew, (1904-1973)
IRVING, John Duer, (1874-1918)
ISLAMOV, Oskar Islamovich, (1910-1978)
JACOB, Charles Edward, (1914-1970)
JACOB, Charles François Étienne, (1878-1962)

STRUCTURAL GEOLOGY 1900 to present Cont'd.

JACOBS, Elbridge Churchill, (1873-1957)
JACQUET, Fernand, (1908-1937)
JAGGAR, Thomas Augustus, (1871-1953)
JAHN, Jaroslav Jiljí, (1865-1934)
JEHU, Thomas John, (1871-?1943)
JELÍNEK, František, (1911-1966)
JEWETT, John Mark, (1896-1970)
JILLSON, Willard Rouse, (1890-)
JOESTING, Henry Rochambeau, (1903-1965)
JOHANNSEN, Albert, (1871-1962)
JOHNSEN, Arrien, (1877-1934)
JOHNSON, Charles George, (1914-1969)
JOHNSTON, Ashton William, (1910-)
JOHNSTON, William Drumm, Jr., (1899-1972)
JOLEAUD, Léonce, (1880-1938)
JONES, Brynmor, (1902-1941)
JONES, Hubert Cecil, (1881-1963)
JONES, Owen Thomas, (1878-1967)
JONES, William Richard, (1890-1970)
JORDAN, Louise, (1908-1966)
JOSSÉ, Kazimierz, (1892-1942)
JUDSON, Sidney Arthur, (1887-1971)
JUKES-BROWNE, Alfred John, (1851-1914)
JUNG, Jean, (1896-)
JUNNER, Norman Ross, (1891-1973)
KAISER, Friedrich Wilhelm Erich, (1871-1934)
KAISIN, Félix, Sr., (1879-1948)
KALÁŠEK, Jan, (1903-)
KARPINSKY, Alexander Petrovich, (1847-1936)
KATZER, Bedřich, (1861-1925)
KAY, Fred Hall, (1885-1943)
KEITH, Arthur, (1864-1944)
KEMP, James Furman, (1859-1926)
KENNEDY, Horas Tristram, Lieutenant, (1889-1917)
KENNEDY, Luther Eugene, (1891-1966)
KENNEDY, William, (-1926)
KENNERLEY, John Brian, (1937-1976)
KERFORNE, Fernand, (1864-1927)
KETTNER, Radim, (1891-1967)
KEW, William Stephen Webster, (1890-1961)
KEYES, Charles Rollin, (1864-1942)
KILIAN, Charles Constant Wilfrid, (1862-1925)
KILROE, James Robinson, (1848-1927)
KIMBALL, Kent K., (1892-1953)
KINAHAN, George Henry, (1829-1908)
KING, Clarence Rivers, (1842-1901)
KIRBY, Maurice E. "Bud", (1909-1963)
KITSON, Albert Ernest, Sir, (1868-1937)
KLEINPELL, William Darwin, (1898-1959)
KLEMM, Gustav, (1858-1938)
KLINE, Virginia Harriett, (1910-1959)
KLOMPÉ, Theodorus Hendricus Franciscus, (?1904-1963)
KNIGHT, Garold Louis "Jerry", (1894-1968)
KNIGHT, Samuel Howell, (1892-1975)

STRUCTURAL GEOLOGY 1900 to present Cont'd.

KNOPF, Adolph, (1882-1966)
KOBER, Leopold, (1883-1970)
KOCH, Ferdo, (1874-?1936)
KOCH, Lauge, (1892-1964)
KOCKEL, Carl Walter, (1898-1966)
KODYM, Odolen, (1898-1963)
KOLDERUP, Niels-Henrik, (1898-1971)
KORN, Johannes, (1862-1927)
KOSCHMANN, Albert Herbert, (1896-1962)
KOSSMAT, Franz, (1871-1938)
KOUTEK, Jaromír, (1902-)
KRAFFT VON DELLMENSINGEN, Albrecht, (1871-1901)
KRAMER, William Baltser, III, (1896-1963)
KRAUS, Ernst, (1889-1970)
KRECHTER, Vladimir, (?1897-1967)
KREIDLER, William Lynn, (1911-1975)
KREUTZ, Felix Franz Xaver, (1844-1910)
KRISHNAN, Maharajapuram Sitaram, Padma Bhusan, (1898-1970)
KRYNINE, Paul Dimitri, (1902-1964)
KUENEN, Phillip Henry, (1902-1976)
KUMMEL, Henry Barnard, (1867-1945)
KYNASTON, Herbert, (1868-1915)
LACHMANN, Richard, (1885-1916)
LACOSTE, Jean Pierre Victor, (1901-1942)
LACROIX, François Antoine Alfred, (1863-1948)
LAFLAMME, Joseph Clovis Kemner, Monseigneur, (1849-1910)
LAHEE, Frederic Henry, (1884-1968)
LAKE, Philip, (1865-1949)
LAMARE, Pierre, (1894-1967)
LAMBE, Lawrence Morris, (1863-1919)
LAMBERT, Alan, (1837-1927)
LAMMERS, Edward Chauncey Hinman, (1910-1971)
LAMPLUGH, George William, (1859-1926)
LANCASTER, L. Kenneth, (1906-1971)
LANDES, Henry, (1867-1936)
LANDES, Robert William, (1910-1969)
LANE, Alfred Church, (1863-1948)
LANQUINE, Antonin, (-)
LAPPARENT, Albert Auguste de, (1839-1908)
LAPPARENT, Albert Félix de, (1905-1975)
LAPWORTH, Charles, (1842-1920)
LARSEN, Esper Signius, III, (1912-1961)
LAWSON, Andrew Cowper, (1861-1952)
LEANZA, Armando F., (1919-1975)
LEATHEROCK, Otto, (1901-1941)
LEE, Willis Thomas, (1864-1926)
LEECH, John George Chambré, (1890-1966)
LEES, George Martin, (1898-1955)
LEIGHTON, Morris Morgan, (1887-1971)
LEMOINE, Paul, (1878-1940)
LEONARD, Hugh, (1841-1909)
LEPSIUS, Karl Georg Richard, (1851-1915)
LEWIS, G. Edward, (-)
LEWIS, Joseph Volney, (1869-1969)

STRUCTURAL GEOLOGY 1900 to present Cont'd.

LIEBUS, Adalbert, (1876-1945)
LIGHTFOOT, Ben, (1888-1966)
LIMANOWSKI, Mieczyslaw, (1876-1947)
LINDGREN, Waldemar, (1860-1939)
LJUNGNER, Erik, (1892-1954)
LOBJOIT, William Michael, (1932-1967)
LÓCZY, Lajos, (1849-1920)
LOHEST, Max, (1857-1926)
LONSDALE, John Tipton, (1895-1960)
LORY, Pierre, (1866-1956)
LOTZE, Franz, (1903-1971)
LOUGHLIN, Gerald Francis, (1880-1946)
LOUGHRIDGE, Robert Hills, (1843-1917)
LOW, Albert Peter, (1861-1952)
LÖWL, Ferdinand, (1856-1908)
LUBBOCK, John, Lord Avebury, (1834-1913)
LUCIUS, Michel, (1876-1961)
LUGEON, Maurice, (1870-1953)
LUKASZEWICZ, Józef, (1863-1928)
LUKOVIĆ, Milan, (1889-)
LUPTON, Charles Thomas, (1878-1935)
LUTAUD, Léon, (1883-1964)
LYDEKKER, Richard, (1849-1915)
LYELL, Charles, Sir, (1797-1875)
LYMAN, Benjamin Smith, (1835-1920)
MAACK, Reinhard, (1892-1969)
MACBRIDE, Thomas Huston, (1848-1934)
MACDONALD, Simon, (1927-1971)
MACGREGOR, Murray, (1884-1966)
MACHADE E COSTA, Alfredo Augusto d'Olveira, (1870-?1952)
MacHENRY, Alexander, (1843-1919)
MACLAREN, James Malcolm, (1873-1935)
MACPHERSON, José, Don, (1839-1902)
MADIGAN, Cecil Thomas, (1889-1946)
MAGNUSSON, Nils Harald, (1890-1976)
MARGETIČ, Marko, (1891-1955)
MARSHALL, Patrick, (1869-1950)
MARTIN, David, (1842-1918)
MARTIN, Francis Ivan "Shorty", (1896-1952)
MARTIN, George C., (1875-1943)
MASON, Arnold Caverly, (1906-1961)
MATĚJKA, Alois, (1898-)
MATHER, Kirtley Fletcher, (1888-1978)
MATHEWS, Edward Bennett, (1869-1944)
MATSON, George Charlton, (1873-1940)
MATTHES, François Emile, (1874-1948)
MAUFE, Herbert Brantwood, (1879-1946)
MAWSON, Douglas, Sir, (1882-1958)
McCALLEY, Henry, (1852-1904)
McCONNEL, Roger H., (1908-1971)
McCONNELL, Richard George, (1857-1942)
McDONALD, Stanley Matthews, (1915-1950)
McGEE, William John, (1853-1912)
McGEHEE, Julius Rex, (1905-1963)

STRUCTURAL GEOLOGY 1900 to present Cont'd.

McGUINNESS, Charles Lee, (1914-1971)
McHENRY, Alexander, (1843-1919)
McINTYRE, Peter, (1863-1940)
McKAY, Alexander, (1841-1917)
McMANNIS, William J., (1924-1973)
MEAD, Warren Judson, (1883-1960)
MEDLICOTT, Henry Benedict, (1829-1905)
MEINZER, Oscar Edward, (1876-1948)
MEKEL, Johannes Antonius Alphonsus, (1891-1942)
MENDENHALL, Walter Curran, (1871-1957)
MERRITT, Clifford A., (-)
MICHALSKI, Aleksander Octavianovich, (1855-1904)
MICHEL-LÉVY, Auguste, (1844-1911)
MIDDLEMISS, Charles Stewart, (1859-1945)
MILES, Keith Rodney, (1915-1972)
MILLER, Benjamin Le Roy, (1874-1944)
MILLER, Don John, (1919-1961)
MILLER, Wendell Zerbe, (1892-1958)
MILLER, Willard L., (1896-1973)
MILLER, William John, (1880-1965)
MILTHERS, Paul Christian Vilhelm Madsen, (1865-1962)
MIRĂUŢĂ, Orest, (1931-1967)
MISER, Hugh Dinsmore, (1884-1969)
MITCHELL, George Hoole "Mick", (1902-1976)
MOHOROVIČIĆ, Andrya, (1857-1936)
MOJSISOVICS, Johann August Georg Edmund, Edler von Mojsvár, (1839-1907)
MOLENGRAAFF, Gerard Johan Hendrik, (1887-1961)
MOLENGRAAFF, Gustaaf Adolf Frederik, (1860-1942)
MOLYNEUX, Arthur John Charles, (?1866-1921)
MONNETT, Victor Elvert, (1889-1972)
MOORE, Carl Allphin, (1911-1973)
MORAN, Thomas, (1837-1926)
MORET, Léon, (1890-1972)
MORGAN, Conwy Lloyd, (?1852-1936)
MORGAN, Peter (Percy) Gates, (1867-1927)
MORRIS, Frederick Kuhne, (1885-1962)
MORSE, Paul Franklin, (1897-1929)
MOSELEY, John Reed, (1900-1959)
MOURLON, Michel Félix, (1845-1915)
MUIR, John, (1838-1914)
MUIR, John Malcolm, (1885-1938)
MULLER, Siemon William, (1900-1970)
MUNN, Malcolm John, (1874-1949)
MUNTEANI-MURGOCI, Gheorghe, (1872-1925)
MUNTHE, Henrik Vilhelm, (1860-1958)
MURRAY, Albert Nelson, (1894-1961)
MYERS, John Clark, (1894-1970)
MYLIUS, Louis Aubrey, (1890-1965)
NARAMORE, Chester "Chet", (1879-1961)
NAUMANN, Edmund, (1854-1927)
NEALE, Ernest Richard Ward, (1923-)
NETOLICKÝ, Jan, (1934-1963)
NEWLAND, David Hale, (1872-1943)

STRUCTURAL GEOLOGY 1900 to present Cont'd.

NININGER, Robert D., (-)
NOBLE, Levi Fatzinger, (1882-1965)
NOÉ, Emilio, (1881-1940)
NOLAN, Joseph, (1841-1902)
NORDENSKIÖLD, Nils Adolf Erik, (1832-1901)
NORTON, William Harmon, (-)
NORWOOD, Charles Joseph, (1853-1927)
NOVARESE, Vittorio, (1861-1948)
NOWACK, Ernst, (1891-1946)
NOWAK, Jan, (1880-1940)
OATES, Frank, (1890-1945)
OGILVIE, William, (1846-1912)
OGILVIE GORDON, Maria Matilda, Dame, (1864-1939)
OLDHAM, Richard Dixon, (1858-1936)
OLLERENSHAW, Neil, (-)
OPOLSKI, Zdzislaw, (1884-1941)
ORDOÑEZ, Ezequiel, (1867-1950)
ORTON, Edward Francis Baxter, (1829-1899)
OSBERGER, Rudolf A. J., (1924-1972)
OSBORNE, George Davenport, (1899-1955)
OSTADAL, Rudolf, (1897-1970)
OWENS, Eifion Wyn, (1951-1975)
PALMER, Rupert William, (1890-1922)
PARGETER, Reginald Clive, (1919-1972)
PARK, James, (1857-1946)
PASTORE, Franco, (1885-1958)
PAUK, František, (1905-)
PAULCKE, Wilhelm, (1873-1949)
PAVLOVIĆ, Stojan, (1903-)
PEABODY, Harlan William, (1881-1960)
PEACH, Benjamin Neeve, (1842-1926)
PECORA, William Thomas, (1913-1972)
PEGAU, Albert August, (1894-1963)
PENCK, Friedrich Carl Albrecht, (1858-1945)
PENCK, Walther, (1888-1923)
PENNEL, Harry David, (1907-1971)
PENNING, William Henry, (1838-1902)
PEPPERBERG, Leon J., (1883-1937)
PERKINS, George Henry, (1844-1933)
PERRINE, Irving, (1884-1955)
PERRY, Thomas Gregory, (1919-1972)
PETERSEN, Georg, (1898-)
PETHŐ, Gyula, (1848-1902)
PETKOVIĆ, Kosta, (1903-)
PETKOVIĆ, Vladamir K., (1873-1934)
PETRASCHEK, Wilhelm, (1876-1967)
PFANNENSTIEL, Max, (1902-1976)
PICKLESIMER, Harold Greeley, (1916-1952)
PIETZSCH, Kurt, (1884-1964)
PILGRIM, Henry Guy Ellcock, (1875-1943)
PIRSSON, Louis Valentine, (1860-1919)
PLESSMANN, Werner, (1928-1972)
PLUMMER, Frederick Byron, (1886-1947)
PLUMMER, Helen Jeanne (née Skewes), (1891-1951)

STRUCTURAL GEOLOGY 1900 to present Cont'd.

POCOCK, Roy Woodhouse, (1887-1954)
POLANSKI, Jorge, (1892-1975)
POLDERVAART, Arie, (1918-1964)
POLJAK, Josip, (1882-1962)
POLLARD, Ernest Rex, (1922-1966)
POLLARD, William, (1870-1946)
POLWHELE, Thomas Roxburgh, (1831-1909)
POND, Walter Franklin, (1885-1974)
POOR, Russell Spurgeon, (1899-1972)
PORRIO, Cesare, (1865-1940)
POSTEL, Albert Williams, (1909-1966)
POTEET, James Harold, (1896-1965)
POWELL, John Wesley, (1834-1902)
POWERS, Sidney, (1890-1932)
PREISWERK, Heinrich, (1876-1940)
PRESS, Frank, (1924-)
PRETTYMAN, Thomas Mann, (1888-1940)
PRICE, Paul Holland, (1898-)
PRIESTLEY, Raymond Edward, Sir, (1886-1974)
PRINGLE, John, (1877-1948)
PROSSER, Charles Smith, (1860-1916)
PUGH, William John, Sir, (1892-1974)
PUMPELLY, Raphael, (1837-1923)
PUTNAM, William Clement, (1908-1963)
QUIRKE, Terence Thomas, (1886-1947)
RADLEY, Ernest George, (1880-1945)
RAEBURN, Colin, (1894-1970)
RAGGATT, Harold George, Sir, (1900-1968)
RAJNATH, Shri, (1904-1972)
RAMA RAO, Bellu, (1890-1970)
RAMSAY, Wilhelm, (1865-1928)
RANSOME, Frederick Leslie, (1868-1935)
RAVIER, Jean, (1922-)
READ, Herbert Harold, (1889-1970)
READE, Thomas Mellard, (1832-1909)
REECE, Alan W., (1921-1960)
REED, Frederick Richard Cowper, (1869-1946)
REESIDE, John Bernard, Jr., (1889-1958)
RÉVIL, Joseph, (1849-1931)
RICH, John Lyon, (1884-1956)
RICHARDS, James Taylor "Dick", (1892-1953)
RICHARDSON, Charles Henry, (1862-1935)
RICHARDSON, F. D. S. "Ricardo", (-1966)
RICHARDSON, George Burr, (1872-1949)
RICHEY, James Ernest, (1886-1968)
RICHTHOFEN, Paul Wilhelm Ferdinand von, Freiherr, (1833-1905)
RIES, Heinrich, (1871-1951)
RIGGS, Robert Jennings, (1888-1965)
ROARK, Louis, (1890-1950)
ROBBIE, James Andrew, (1910-1977)
ROBERTS, Morgan Evan, (1894-1941)
ROBINSON, G. D., (-)
ROE, Frank Walton "Peter", (?1914-1965)
ROGERS, Arthur William, (1872-1946)

STRUCTURAL GEOLOGY 1900 to present Cont'd.

ROGERS, Gaillard Sherburne, (1889-1919)
ROMAN, Albert Frédéric, (1871-1943)
ROMANES, James, (1886-1942)
ROMER, Eugeniusz, (1871-1954)
ROMINGER, Carl Ludwig, (1820-1907)
RONES, Morris, (1922-1965)
RØRDAM, Kristian, (1860-1939)
ROSS, Clarence Samuel, (1880-1975)
ROUBAULT, Marcel Edouard, (1905-1974)
ROUNDY, Paul Vere, (1884-1937)
ROWE, Richard Burton, (1872-1902)
ROZLOZSNIK, Pál, (1880-1940)
RUBEY, William Walden, (1898-1974)
RUNNER, Joseph James, (1885-1970)
RUSS, Leon Fair, (1886-1953)
RUSSELL, Loris Shano, (1904-)
RUSSELL, Philip G., (1891-1971)
RUSSELL, Richard Joel, (1895-1971)
RUSSELL, William Low, (1897-1976)
RUTKOWSKI, Feliks, (1878-1964)
RUTLEY, Frank, (1842-1904)
RUTTEN, Martin Gerard, (1910-1970)
SAFFORD, James Merrill, (1822-1907)
SALISBURY, Rollin D., (1858-1922)
SALOMON-CALVI, Wilhelm, (1868-1941)
SAMBASIVA IYER, Valiathur S., (-?1943)
SAMPSON, David Noel, (1917-1971)
SARDESON, Frederick William, (1866-1959)
SAUER, Gustav Adolf, (1825-1932)
SCHAFARZIK, Ferenc, (1854-1927)
SCHAFFER, Franz Xavier, (1876-1954)
SCHALCH, Ferdinand, (1848-1918)
SCHALLER, Waldemar Theodore, (1882-1967)
SCHARDT, W. Hans, (1858-1931)
SCHEIBE, Robert, (1859-1923)
SCHEUMANN, Karl Hermann, (1881-1964)
SCHMIDT, Carl, (1862-1923)
SCHMIDT, Robert Eligius, (1902-1973)
SCHNEEGANS, Daniel, (1907-1949)
SCHOTTLER, Wilhelm, (1869-1932)
SCHRADER, François Jean Daniel "Franz", (1844-1924)
SCHRADER, Frank Charles, (1860-1944)
SCHRAMM, Eck Frank, (1883-1967)
SCHRÖDER, Hans Adolf Karl Eckart, (1899-1968)
SCHWARZ, Ernest Hubert Lewis, (1873-1928)
SCHWENNESEN, Alvin Theodore, (1883-1962)
SCHWINNER, Robert Gangolf, (1878-1953)
SCOTT, James Frew, (1902-1956)
SCRUTON, Philip Challacombe, (1922-1962)
SEARS, Julian Ducker, (1891-1970)
SEE, Thomas Jefferson Jackson, (1866-)
SELWYN, Alfred Richard Cecil, (1824-1902)
SHALER, Nathaniel Southgate, (1841-1906)
SHARMAN, George, (1832-1914)

STRUCTURAL GEOLOGY 1900 to present Cont'd.

SHAW, Eugene Wesley, (1881-1935)
SHEARER, Harold Kurtz, (1889-1946)
SHEPARD, Edward Martin, (1854-1934)
SHERLOCK, R. L., (1875-1948)
SHERRILL, Richard Ellis, (1899-1953)
SHORT, Maxwell Naylor, (1889-1952)
SIBLY, Thomas Franklin, Sir, (1883-1948)
SIMMONS, William Charles, (1888-1951)
SINCLAIR, Ewart Gladstone, (1884-1927)
SKOCYLAS-CISZEWSKA, Kamila, (1902-1971)
SMITH, Bernard, (1881-1936)
SMITH, Eugene Allen, (1841-1927)
SMITH, George Otis, (1871-1944)
SMITH, Ned Myron, (1923-1976)
SMIT SIBINGA, Gerard Leonard, (1895-1963)
SMOCK, John Conover, (1842-1925)
SMOLUCHOWSKI, Marian, (-)
SNIDER, Luther Crocker, (1882-1947)
SOBRAL, José María, (-1961)
SÖLCH, Johann, (1883-1951)
SPENCER, Joseph William Winthrop, (1851-1921)
SPENGLER, Erich, (1886-1962)
SPITZ, Albrecht, (1883-1918)
SPROULE, John Campbell, (1905-1970)
STAUB, Rudolf, (1890-1961)
STEBINGER, Eugene, (1883-1951)
STEHR, Raymond Albert, (1903-1967)
STEIGER, George, (1869-1944)
STEJSKAL, Jan, (1902-1972)
STEVENSON, John James, (1841-1924)
STEWART, Irvine E., (1893-1965)
STILLE, Hans, (1876-1966)
ST. JOHN, Orestes Hawley, (1841-1921)
STOČES, Bohuslav, (1890-)
STOCKDALE, Paris Buell, (1896-1962)
STOW, Marcellus Henry, (1902-1957)
STRAHAN, Aubrey, Sir, (1852-1928)
STREIFF, Viktor, (1909-1973)
STUTZER, Otto, (1881-1936)
SUBRAMANIAM, Anantharama Parameshwara, (1921-1973)
SUESS, Eduard, (1831-1914)
SUESS, Franz Eduard, (1867-1941)
SUTTON, Arle Herbert, (1896-1972)
SUTTON, John, (-)
SVOBODA, Josef, (1898-)
SWANN, David H., (1915-1968)
SWARTZ, Joel Howard, (1893-1971)
ŚWIDERSKI, Bohdan, (1892-1943)
ŚWIDZINSKI, Henryk, (1904-1969)
SYMES, Richard Glascott, (1840-1906)
SZÁDECZKY, Gyula, Baron Szádeczky-Kardoss of Szádecsne and Kafdosfalva, (1860-1935)
SZATSKIY, Mikolai Sergiechewich, (1895-1960)
TAFF, Joseph Alexander, (1862-1944)

STRUCTURAL GEOLOGY 1900 to present Cont'd.

TANDY, James Hiram "Spot", (1892-1962)
TANTON, Thomas Leslie, (1890-1971)
TAYLOR, Garvin Lawrence, (1908-1974)
TAYLOR, Griffith, (1881-1963)
TAYLOR, James Carlton, (1924-1976)
TAYLOR, James Haward, (1909-1968)
TCHERNYCHEV, Féodoss Nikolaievich, (1856-1914)
TEALE, Edmund Oswald, Sir, (1874-1971)
TEALL, Jethro Justinian Harris, Sir, (1849-1924)
TEISSEYRE, Henryk, (1903-1975)
TEISSEYRE, Wawrzyniec, (1860-1939)
TELLER, Friedrich, (1852-1913)
TEMPLEMAN, Andrew, (1887-1945)
TERMIER, Henri François Emile, (1897-)
TERMIER, Pierre, (1859-1930)
THIENHAUS, Rolf, (1913-1968)
THIES, Roland K., (1905-1973)
THOMAS, Herbert Henry, (1876-1935)
THOMPSON, David Grosh, (1888-1943)
THOMPSON, Jean Cleveland, (1887-1929)
THOMSON, Joseph Ellis, (1882-1944)
TIDDEMAN, Richard Hill, (1842-1917)
TIDY, Anthony John Lintott, (1928-1969)
TING, Ven Kiang, (?1887-1936)
TIPPER, George Howlett, (1881-1947)
TOIT, Alexander Logie du, (1878-1948)
TÖRÖK, Zoltán, (1893-1963)
TOULA, Franz, (1845-1920)
TOWSE, Donald Frederick, (1924-)
TOZER, Charles Fallows, (1918-1966)
TRECHMANN, Charles Taylor, (1885-1964)
TRIBOLET-HARDY, Frédéric Maurice de, (1852-1929)
TROTTER, Frederick Murray, (1897-1968)
TRÜMPY, Rudolf D., (-)
TSCHIKHACHEV, Peter A., (-)
TWITCHELL, Mayville William, (1866-1927)
TYRRELL, Joseph Burr, (1858-1957)
UDDEN, Johan August, (1859-1932)
UGLOW, William Lawrence, (1884-1926)
UHLIG, Victor Karl, (1857-1911)
ULRICH, Edward Oscar, (1857-1944)
UMBGROVE, Johannes Herman Frederik, (1899-1954)
UNTERMANN, Billie R., (1906-1973)
URBAN, Karel, (1905-)
USSHER, William Augustus Edmond, (1849-1920)
USSING, Niels Viggo, (1864-1911)
VAN HISE, Charles Richard, (1857-1918)
VAN HORN, Frank Robertson, (1872-1933)
VARDABASSO, Silvio, (1891-1966)
VAUGHAN, Thomas Wayland, (1870-1952)
VEATCH, Arthur Clifford, (1878-1938)
VENING MEINESZ, Felix Andries, (1887-1966)
VERRI, Antonio, (1839-1925)
VER WIEBE, Walter August, (1887-1961)

STRUCTURAL GEOLOGY 1900 to present Cont'd.

VESTAL, Franklin Earl, (1884-1969)
VETTERS, Hermann, (1880-1941)
VIENNOT, Pierre, (1891-1931)
VIGILEV, Boris, (1883-1924)
VINCIENNE, Henri, (1898-1965)
WADIA, Darashaw Nosherwan, (1883-1969)
WAGNER, Georg, (1885-1972)
WÄHNER, Franz, (1856-1932)
WALCOTT, Charles Doolittle, (1850-1927)
WALDBAUM, David Robert, (1937-1974)
WALKER, Thomas Leonard, (1867-1942)
WALL, George Parkes, (1832-1912)
WALLACE, Robert Charles, (1881-1955)
WARD, Freeman, (1879-1943)
WARD, Leonard Keith, (1879-1964)
WARING, Gerald Ashley, (1883-1971)
WASHBURNE, Chester Wesley, (1883-1971)
WASSON, Theron, (1887-1970)
WATSON, Thomas Leonard, (1871-1924)
WDOWIARZ, Jan, (1903-1956)
WEAVER, Charles Edwin, (1880-1958)
WEAVER, Paul, (1888-1964)
WEBB, John Benwell, (1901-1975)
WEEKS, Herbert John, (1897-1960)
WEGENER, Alfred Lothar, (1880-1930)
WEIDMAN, Samuel, (1870-1945)
WEIGELT, Johannes, (1890-1948)
WEIGNER, Stanislaw Wiktor, (1886-1935)
WELLER, James Marvin, (1899-1976)
WELLER, Stuart, (1870-1927)
WELLS, Francis Gerrit, (1898-1969)
WELLS, Roger Clark, (1877-1944)
WESLEY, George Rutherford, (1905-1965)
WESTGATE, Lewis Gardner, (1868-1948)
WESTON, Thomas Chesmer, (1832-1910)
WHEELER, Orby Clinton, (1898-1970)
WHITE, Charles Abiathar, (1826-1910)
WHITE, George Willard, (1903-   )
WHITE, Israel Charles, (1848-1927)
WHITE, William Harrison, (1913-1972)
WHITEAVES, John Joseph Frederick, (1835-1909)
WHITEHEAD, Talbot Haes, (1890-1966)
WHITTIER, William Harrison, (1889-1951)
WHYMPER, Edward, (1840-1911)
WIIK, Fredrik Johan, (1839-1909)
WILCKENS, Otto Rudolf, (1876-1943)
WILCOCKSON, William Howson, (1891-1976)
WILDER, Frank Alonzo, (1870-1930)
WILHELM, Oscar, (1897-1975)
WILKINSON, William Donald, (1901-1969)
WILKMAN, Wanold Wrydon, (1872-?1938)
WILLETT, Richard Wright, (1912-1974)
WILLIAMS, Mervyn, (1934-1970)
WILLIS, Bailey, (1857-1949)

STRUCTURAL GEOLOGY 1900 to present Cont'd.

WILSON, Bernard Wright, (1922-1976)
WILSON, Eldred Dewey, (1898-1965)
WILSON, J. S. Grant, (1855-1908)
WILSON, John Tuzo, (1908-)
WILSON, Malcolm Earl, Sr., (1885-1960)
WILSON, Reginald Charles, (1888-1943)
WILSON, Roy Arthur, (1891-1970)
WILSON, Vernon, (1907-1971)
WINCHELL, Horace Vaughn, (1865-1923)
WINCHESTER, Dean Eddy, (1883-1936)
WINDHAUSEN, Anselm, (1882-1932)
WINKLER-HERMADEN, Artur, (1890-1963)
WIŚNIOWSKI, Tadeusz, (1865-1933)
WOOD, Robert Harrison, (1887-1956)
WOOD, Virgil O., (1892-1954)
WOODRUFF, Elmer Grant, (1872-1952)
WOODRUFF, John Grant, (1898-1971)
WOODWARD, Harry Page, (1858-1917)
WOODWARD, Herbert Preston, (1899-1968)
WOODWARD, Horace Bolingbroke, (1848-1914)
WOODWARD, Robert Simpson, (1849-1924)
WOODWORTH, Jay Backus, (1865-1925)
WOOLNOUGH, Walter George, (1876-1958)
WRATHER, William Embry, (1883-1963)
WRIGHT, Frederick Eugene, (1877-1953)
WRIGHT, William Bourke, (1876-1939)
WUNDERLICH, Hans-Georg, (1928-1974)
WYNNE, Arthur Beavor, (1835-1906)
YABE, Hisakatsu, (1878-1969)
YACHEVSKY, Leonard Antonovich, (1858-1916)
YOUNG, Alfred Prentice, (1841-1919)
YOUNG, George Albert, (1898-1947)
YOUNG, John, (1835-1902)
YOUNGQUIST, Walter, (-)
ZANS, Verners Aleksandrs, (1904-1961)
ZAPLETAL, Karel, (1903-1972)
ZAPP, Alfred Dexter, (1916-1962)
ZITTEL, Karl Alfred von, (1839-1904)
ZÖLLICH, Martin Sylvester, (1913-1940)
ZUBER, Stanislaw, (1893-1947)
ZUFFARDI, Pietro, (1885-1915)
ZWIERZYCKI, Józef, (1888-1961)

Time unknown

D'ISENGARD, Luigi, Barone, (-)
MACGREGOR, A. M., (-)
TEGENGREN, Felix, (-)

STRUCTURAL GEOLOGY See also VULCANOLOGY

VERTEBRATE PALAEONTOLOGY

Up to 1800

ALBERTUS MAGNUS, (1193-1280)
BAIN, Andrew Geddes, (1797-1864)
BALD, Robert, (1776-1861)
BERTRAND DE DOUE, Jacques Mathieu, (1776-1862)
BLAINVILLE, Henri Marie Ducrotay de, (1777-1850)
BLUMENBACH, Johann Friedrich, (1752-1840)
BOJANUS, Ludwig Heinrich, (1776-1827)
BREISLAK, Scipione, (1748-1826)
BRONGNIART, Alexandre, (1778-1847)
CAMPER, Pieter (Petrus), (1722-1789)
CIAMPINI, Giovanni Giustino, (1633-1698)
CIMARELLI, Vincenzo Maria, (-1660)
CLIFT, William, (1775-1849)
CUVIER, Jean Léopold Nicolas Frédéric, (1769-1832)
DUVERNOY, George Louis, (1777-1855)
FAUJAS DE SAINT FOND, Barthélémy, (1741-1819)
FISCHER VON WALDHEIM, Johann Gotthelf Friedrich, (1771-1853)
GEOFFROY SAINT-HILAIRE, Étienne, (1772-1844)
HOME, Everard, Sir, (1756-1832)
JEFFERSON, Thomas, (1743-1826)
KIDD, John, (1775-1851)
LAMARCK, Jean-Baptiste Pierre Antoine de Monet, Chevalier de, (1744-1829)
MARUM, Martin van, (1750-1838)
MICHAELIS, Christian Friedrich, (fl. 1766-1780)
MONTI, Giuseppe, (1682-1760)
PEALE, Charles Willson, (1741-1827)
PLATT, Joshua, (1699-1773)
SCHEUCHZER, Johann Gaspar, (1702-1729)
SCHEUCHZER, Johann Jakob, (1672-1733)
SCHEUCHZER, Johann Jakob, (1672-1733)
SCHLOTHEIM, Ernst Friedrich von, Freiherr, (1764-1832)
SERRES, Marcel de, (1783-1862)
STELLER, Georg Wilhelm, (1709-1746)
STENO, Nicolaus, (1631-1687)
SWEDENBORG, Emanuel, (1688-1772)
WARREN, John Collins, (1778-1856)
WISTAR, Caspar, (1761-1818)

1800-1849

AGASSIZ, Louis Jean Rodolphe, (1807-1873)
AICHHORN, Sigmund, (1814-1892)
APLIN, Christopher d'Oyley Hale, (1819-1875)
ARCHIAC, Étienne Jules Adolphe Dexmier de Simon, Vicomte d', (1802-1868)
ATHERSTONE, William Guybon, Honourable, (-1898)
AUSTIN, Charles Edward, (1819-1893)
BAER, Karl Ernst von, (1792-1876)
BAILY, William Hellier, (1819-1888)
BAIN, Andrew Geddes, (1797-1864)
BAIRD, Spencer Fullerton, (1823-1887)
BALD, Robert, (1776-1861)
BALD, William, (-?1857)
BALSAMO-CRIVELLI, Giuseppe, Marchese, (1800-1874)

VERTEBRATE PALAEONTOLOGY 1800-1849 Cont'd.

BARKAS, Thomas Pallister, (1819-1891)
BEHN, Wilhelm Friedrich Georg, (1808-1878)
BENEDEN, Pierre Joseph van, (1809-1893)
BERTRAND DE DOUE, Jacques Mathieu, (1776-1862)
BEYRICH, August Heinrich Ernst, (1815-1896)
BIBRA, Ernst von, Freiherr, (1806-1878)
BLAINVILLE, Henri Marie Ducrotay de, (1777-1850)
BLUMENBACH, Johann Friedrich, (1752-1840)
BOJANUS, Ludwig Heinrich, (1776-1827)
BOLL, Jacob, (1828-1880)
BOUCHER DE CRÈVECOEUR DE PERTHES, Jacques, (1788-1868)
BOURGEOIS, Louis Alexis, Abbé, (1819-1878)
BRANDT, Johann Friedrich von, (1802-1879)
BRAUN, Karl Friedrich Wilhelm, (1800-1864)
BREISLAK, Scipione, (1748-1826)
BRONGNIART, Alexandre, (1778-1847)
BRONN, Heinrich Georg, (1800-1862)
BUCKLAND, William, Reverend, (1784-1856)
BURMEISTER, Karl Hermann Konrad, (1807-1892)
BUSK, George, (1807-1886)
CARTER, Henry John, (1813-1895)
CARTER, James, (1813-1895)
CAUTLEY, Proby T., Colonel Sir, (1802-1871)
CLIFT, William, (1775-1849)
CONYBEARE, William Daniel, Reverend, (1787-1857)
CRAHAY, Jacques Guillaume, (1789-1855)
CURIONI, Giulio, (1796-1878)
CUVIER, Jean Léopold Nicolas Frédéric, (1769-1832)
DAVIES, William, (1814-1891)
DAWSON, John William, Sir, (1820-1899)
DEAN, Bashford, (1867-1928)
DESNOYERS, Jules, (1801-1887)
DE ZIGNO, Achille, Barone, (1813-1888)
DUVERNOY, George Louis, (1777-1855)
EGERTON, Philip de Malpas Grey, Sir, (1806-1881)
EHRLICH, Franz Carl, (1808-1886)
EICHWALD, Eduard Iwanowitsch von, (1795-1876)
ENNISKILLEN, William Willoughby Cole, 3rd Earl of, (1807-1886)
ERLE, C., Reverend, (1790-1870?)
EUDES-DESLONGCHAMPS, Jacques Amand, (1794-1867)
EVANS, John, Sir, (1823-1908)
EWALD, Julius Wilhelm, (1811-1891)
FALCONER, Hugh, (1808-1865)
FAUJAS DE SAINT FOND, Barthélémy, (1741-1819)
FILHOL, Edouard, (1814-1883)
FISCHER-OOSTER, Carl von, (1807-1875)
FISCHER VON WALDHEIM, Johann Gotthelf Friedrich, (1771-1853)
FITCH, Robert, (1802-1895)
FITZINGER, Leopold Joseph, (1802-1884)
FRAAS, Oskar Friedrich von, (1824-1897)
FUHLROTT, Johann Carl, (1804-1877)
GASTALDI, Bartolomeo, (1818-1879)
GAUDIN, Charles Théophile, (1822-1866)
GEINITZ, Hanns Bruno, (1814-1900)

VERTEBRATE PALAEONTOLOGY 1800-1849 Cont'd.

GEOFFROY SAINT-HILAIRE, Étienne, (1772-1844)
GERVAIS, François Louis Paul, (1816-1879)
GIEBEL, Christoph Gottfried Andreas, (1820-1881)
GODMAN, John Davidson, (1794-1830)
GOLDFUSS, Georg August, (1782-1848)
GORDON, George, Reverend, (1801-1893)
GRANT, Robert Edmond, (1793-1874)
GUNN, John, Reverend, (1801-1890)
GUTBIER, Christian August von, (1798-1866)
HAAST, John Francis Julius von, Sir, (1824-1887)
HANCOCK, Albany, (1806-1873)
HASTINGS, Barbara, Marchioness of, (1810-1858)
HAWKINS, Thomas, (1810-1889)
HAYS, Isaac, (1796-1879)
HÉBERT, Edmond, (1812-1890)
HECKEL, Johann Jakob, (1790-1857)
HENSEL, Reinhold Friederich, (1826-1881)
HITCHCOCK, Edward, (1793-1864)
HOCHSTETTER, Ferdinand von, Ritter, (1829-1884)
HOME, Everard, Sir, (1756-1832)
HOSIUS, August, (1825-1896)
HUNTER, Robert, Reverend, (1823-1897)
HUOT, Jean Jacques Nicolas, (1790-1845)
HUXLEY, Thomas Henry, (1825-1895)
JÄGER, Georg Friedrich, (1785-1866)
JEFFERSON, Thomas, (1743-1826)
KAUP, Johann Jakob, (1803-1873)
KEYSERLING, Alexander Andreevich, Count, (1815-1891)
KIDD, John, (1775-1851)
KING, Samuel William, (1821-1868)
KJERULF, Theodor, (1825-1888)
KLAASSEN, Hendericus Martinus, (1828-1910)
KLIPSTEIN, August Wilhelm, (1801-1894)
KNER, Rudolf, (1810-1869)
KORNHUBER, Andreas, (1824-1905)
KRAUSS, Ferdinand, (1812-1890)
LAING, Samuel, (1810-1897)
LAMARCK, Jean-Baptiste Pierre Antoine de Monet, Chevalier de, (1744-1829)
LARTET, Edouard Armand Isidore Hippolyte, (1801-1871)
LEA, Isaac, (1792-1886)
LE HON, Henri Sebastian, (1809-1872)
LEIDY, Joseph, (1823-1891)
LONG, Stephen Harriman, (1784-1864)
LÜTKEN, Christian Fredrik, (1827-1901)
MACKESON, Henry Bean, (1811-1894)
MANTELL, Gideon Algernon, (1790-1852)
MARUM, Martin van, (1750-1838)
MAYER-EYMAR, Karl David Wilhelm, (1826-1907)
McCOY, Frederick, Sir, (1823-1899)
MENEGHINI, Giuseppe Giovanni Antonio, (1811-1889)
MEYER, Christian Erich Hermann von, (1801-1869)
MITCHELL, Hugh, Reverend, (1822-1894)
MORREN, Charles, (1807-1858)

VERTEBRATE PALAEONTOLOGY 1800-1849 Cont'd.

MORTILLET, Gabriel de, (1820-1898)
MUÑIZ, Francisco Javier, (1795-1871)
MURCHISON, Roderick Impey, Sir, (1792-1871)
NEUGEBOREN, Johann Ludwig, (1806-1887)
NEWBERRY, John Strong, (1822-1892)
NODOT, Léonhard, (1802-1859)
NORDMANN, Alexander von, (1803-1866)
NOULET, Jean Baptiste, (1802-1890)
OGILBY, William, (1808-1873)
OMBONI, Giovanni, (1829-1910)
OWEN, Richard, Sir, (1804-1892)
PARISH, Woodbine, Sir, (1796-1882)
PARKER, William Kitchen, (1823-1890)
PEACH, Charles William, (1800-1886)
PEALE, Charles Willson, (1741-1827)
PEALE, Titian Ramsay, (1799-1885)
PENGELLY, William, (1812-1894)
PENTLAND, Joseph Barclay, (1797-1873)
PEREIRA DA COSTA, Francisco Antonio, (1809-1889)
PETERS, Karl, (1825-1881)
PIPER, George Harry, (1819-1897)
PLIENINGER, Wilhelm Heinrich Theodor von, (1795-1879)
POMEL, Nicolas Auguste, (1821-1898)
PRATT, Samuel Peace, (1789-1863)
PRESL, Jan Svatopluk, (1791-1849)
PROBST, Joseph, (1823-1905)
QUATREFAGES DE BRÉAU, Jean Louis Armand de, (1810-1892)
QUEKETT, John Thomas, (1815-1861)
REDFIELD, John Howard, (1815-1895)
REDFIELD, William C., (1789-1857)
REEVE, James, (1833-1920)
RICHTER, Reinhard, (1813-1884)
ROUAULT, Marie, (1813-1881)
ROYLE, John Forbes, (-1857?)
RÜTIMEYER, Karl Ludwig, (1825-1895)
SARS, Michael, (1805-1869)
SARTORIUS VON WALTERSHAUSEN, Wolfgang, (1809-1876)
SAVI, Paolo, (1798-1871)
SCHAAFHAUSEN, Hermann, (1816-1893)
SCHIMPER, Karl Friedrich, (1803-1867)
SCHLEGEL, Hermann, (1804-1884)
SCHLOTHEIM, Ernst Friedrich von, Freiherr, (1764-1832)
SCHMERLING, Philippe Charles, (1791-1836)
SCHMID, Ernst Erhard, (1815-1885)
SCHRENCK, Leopold von, (1826-1894)
SERRES, Marcel de, (1783-1862)
SHARPE, Daniel, (1806-1856)
SISMONDA, Eugenio, (1815-1870)
STEPHENS, William John, (1829-1890)
STOPPANI, Antonio, (1824-1891)
STORMS, Raymond, (1834-1897)
STROBEL, Pellegrino, (1821-1895)
STUR, Dénes, (1827-1893)
THIOLLIÈRE, Victor, (1801-1859)

VERTEBRATE PALAEONTOLOGY 1800-1849 Cont'd.

TROSCHEL, Franz Hermann, (1810-1882)
UBAGHS, Johann Casimir, (1829-1894)
WAGNER, Johann Andreas, (1797-1861)
WARREN, John Collins, (1778-1856)
WINCHELL, Alexander, (1824-1891)
WISTAR, Caspar, (1761-1818)
WOOD, Searles Valentine, (1798-1880)
ZADDACH, Ernst Gustav, (1817-1881)
ZITTEL, Karl Alfred von, (1839-1904)

1850-1899

ABEL, Othenio, (1875-1946)
ADAMS, Andrew Leith, (-1882)
AGASSIZ, Louis Jean Rodolphe, (1807-1873)
AICHHORN, Sigmund, (1814-1892)
AIM, William Lennie, (?1872-1938)
ALLEN, Joel Asaph, (1838-1921)
AMEGHINO, Florentino, (1854-1911)
AMMON, Ludwig von, (1850-1922)
ANDERSON, Charles, (1876-1944)
ANDERSSON, Johann Gunnar, (1874-1960)
ANDREWS, Charles William, (1866-1924)
APLIN, Christopher d'Oyley Hale, (1819-1875)
ARCHIAC, Étienne Jules Adolphe Dexmier de Simon, Vicomte d', (1802-1868)
ATHERSTONE, William Guybon, Honourable, (-1898)
AUSTIN, Charles Edward, (1819-1893)
BAER, Karl Ernst von, (1792-1876)
BAILY, William Hellier, (1819-1888)
BAIN, Thomas Charles John, (1829-1893)
BAIRD, Spencer Fullerton, (1823-1887)
BALD, Robert, (1776-1861)
BALSAMO-CRIVELLI, Giuseppe, Marchese, (1800-1874)
BARBOUR, Erwin Hinckley, (1856-1947)
BARKAS, Thomas Pallister, (1819-1891)
BARON, Richard, Reverend, (1847-1907)
BATE, Dorothea Minola Alice, (1879-1951)
BATESON, William, (1861-1926)
BAUR, Georg Hermann Carl Ludwig, (1859-1893)
BEADNELL, Hugh John Llewellyn, Major, (1874-1944)
BEASLEY, Henry Charles, (1836-1919)
BECKLES, Samuel H., (-1890)
BEHN, Wilhelm Friedrich Georg, (1808-1878)
BELLOTTI, Christoforo, (1823-1919)
BENEDEN, Pierre Joseph van, (1809-1893)
BENSLEY, Benjamin Arthur, (1875-1934)
BERENDT, Gottlieb Michael, (1836-1920)
BERTRAND DE DOUE, Jacques Mathieu, (1776-1862)
BEYRICH, August Heinrich Ernst, (1815-1896)
BIBRA, Ernst von, Freiherr, (1806-1878)
BLANCKENHORN, Max Ludwig Paul, (1861-1947)
BLANFORD, William Thomas, (1832-1905)
BOLL, Jacob, (1828-1880)

VERTEBRATE PALAEONTOLOGY 1850-1899 Cont'd.

BONAPARTE, Charles Lucien Jules Laurent, Prince of Canino, (1803-1857)
BOSNIASKI, Sigismund Grzymala, (1831-)
BOTTI, Ulderigo, (1822-1905)
BOUCHER DE CRÈVECOEUR DE PERTHES, Jacques, (1788-1868)
BOULE, Marcellin, (1861-1942)
BOURGEOIS, Louis Alexis, Abbé, (1819-1878)
BRANDES, Theodor, (-1916)
BRANDT, Alexander von, (1844-1932)
BRANDT, Johann Friedrich von, (1802-1879)
BRANSON, Edwin Bayer, (1877-1950)
BRAUN, Karl Friedrich Wilhelm, (1800-1864)
BREUIL, Henri, Abbé, (1877-1961)
BROILI, Ferdinand, (1874-1946)
BRONN, Heinrich Georg, (1800-1862)
BROOM, Robert, (1866-1951)
BROWN, Barnum, (1873-1963)
BUCKLAND, William, Reverend, (1784-1856)
BURCKHARDT, Carl Rudolf, (1866-1908)
BURMEISTER, Karl Hermann Konrad, (1807-1892)
BUSK, George, (1807-1886)
CALDERON Y ARANA, Salvador, (1851-1911)
ČAPEK, Václav, (1862-1926)
CAPELLINI, Giovanni, (1833-1922)
CARTER, Henry John, (1813-1895)
CARTER, James, (1813-1895)
CASE, Ermine Cowles, (1871-1953)
CASTRO Y SUERO, Manuel Fernandez de, (1825-1895)
CAUTLEY, Proby T., Colonel Sir, (1802-1871)
CHANTRE, Ernest, (1843-)
CHARLESWORTH, Edward, (1813-1893)
CHOFFAT, Paul, (1849-1919)
CLAYPOLE, Edward Waller, (1835-1901)
CODRINGTON, Thomas, (1829-1918)
COLENUTT, George William, (?1862-1944)
COLLETT, Robert, (1842-1913)
COLLOT, Louis Marie François, (1846-1915)
COMMONT, Victor, (1866-1918)
CONYBEARE, William Daniel, Reverend, (1787-1857)
COPE, Edward Drinker, (1840-1897)
CORNER, Frank, (1861-1939)
CORNET, François Léopold, (1834-1887)
COTTEAU, Gustave Honoré, (1818-1894)
CRAHAY, Jacques Guillaume, (1789-1855)
CREDNER, Carl Hermann Georg, (1841-1913)
CROOK, Alja Robinson, (1864-1930)
CURIONI, Giulio, (1796-1878)
DALL, William Healey, (1845-1927)
DAMES, Wilhelm Barnim, (1843-1898)
DAVID, Tannatt William Edgeworth, Sir, (1858-1934)
DAVIES, Henry Nathaniel, (-1920)
DAVIES, William, (1814-1891)
DAVIS, James William, (1846-1893)
DAWKINS, William Boyd, Sir, (1837-1929)

VERTEBRATE PALAEONTOLOGY 1850-1899 Cont'd.

DAWSON, John William, Sir, (1820-1899)
DEAN, Bashford, (1867-1928)
DEICHMÜLLER, Johannes, (1854-)
DELVAUX, Émile, (1837-1901)
DEPÉRET, Charles Jean Julien, (1854-1929)
DESNOYERS, Jules, (1801-1887)
DE ZIGNO, Achille, Barone, (1813-1888)
DIBLEY, George Edward, (1857-1943)
DITMARS, Raymond Lee, (1876-1942)
DOLLO, Louis Antoine Marie Joseph, (1857-1931)
DOUGLASS, Earl, (1862-1931)
DREVERMANN, Fritz, (1875-1932)
DUBOIS, Eugène, (1858-1940)
DUPONT, Édouard François, (1841-1911)
DUVERNOY, George Louis, (1777-1855)
EASTMAN, Charles Rochester, (1868-1918)
EGERTON, Philip de Malpas Grey, Sir, (1806-1881)
EHRLICH, Franz Carl, (1808-1886)
EICHWALD, Eduard Iwanowitsch von, (1795-1876)
ELLIOT SMITH, Grafton, Sir, (1871-1937)
ENNISKILLEN, William Willoughby Cole, 3rd Earl of, (1807-1886)
ERLE, C., Reverend, (1790-1870?)
ETHERIDGE, Robert Jr., (1847-1920)
EUDES-DESLONGCHAMPS, Eugène, (1830-1889)
EUDES-DESLONGCHAMPS, Jacques Amand, (1794-1867)
EVANS, John, Sir, (1823-1908)
EWALD, Julius Wilhelm, (1811-1891)
FALCONER, Hugh, (1808-1865)
FERRONNIÈRE, Georges, (1875-1922)
FERRY, Henri de, (1826-1869)
FIGGINS, Jesse Dade, (1867-1944)
FILHOL, Edouard, (1814-1883)
FILHOL, Henri, (1843-1902)
FISCHER, Paul Henri, (1835-1893)
FISCHER-OOSTER, Carl von, (1807-1875)
FISCHER VON WALDHEIM, Johann Gotthelf Friedrich, (1771-1853)
FITCH, Robert, (1802-1895)
FITZINGER, Leopold Joseph, (1802-1884)
FLOWER, William Henry, Sir, (1831-1899)
FOLLMANN, Otto, (1856-1926)
FONDOUCE, Paul Cazalis de, (1838-1932)
FONTANNES, Charles François, (1839-1886)
FOOT, Frederick J., (-1867)
FORSYTH MAJOR, Charles Immanuel, (1843-1923)
FRAAS, Eberhard, (1862-1915)
FRAAS, Oskar Friedrich von, (1824-1897)
FRAIPONT, Julien, (1857-1910)
FRITSCH, Anton, (1832-1913)
FRITSCH, Karl von, Freiherr, (1838-1906)
FRÜH, Jakob, (1852-)
FUCHS, Theodor, (1842-1925)
FUHLROTT, Johann Carl, (1804-1877)
FÜRBRINGER, Max, (1846-1920)
GAÁL, István, (1877-1957)

VERTEBRATE PALAEONTOLOGY 1850-1899 Cont'd.

GARDINER, Charles Irving, (?1868-1940)
GASTALDI, Bartolomeo, (1818-1879)
GAUDIN, Charles Théophile, (1822-1866)
GEINITZ, Hanns Bruno, (1814-1900)
GERVAIS, François Louis Paul, (1816-1879)
GIDLEY, James Williams, (1866-1931)
GIEBEL, Christoph Gottfried Andreas, (1820-1881)
GILMORE, Charles Whitney, (1874-1945)
GOELDI, Emil August, (1859-1917)
GOODRICH, Edwin Stephen, (1868-1946)
GORDON, George, Reverend, (1801-1893)
GORJANOVIĆ-KRAMBERGER, Dragutin, (1856-1937)
GOSSELET, Jules Auguste, (1832-1916)
GRANDIDIER, Alfred, (1836-1921)
GRANGER, Walter, (1872-1941)
GRANT, Robert Edmond, (1793-1874)
GREGORY, William King, (1876-1970)
GUNN, John, Reverend, (1801-1890)
GÜNTHER, Albert Charles Lewis Gotthilf, (1830-1914)
GUTBIER, Christian August von, (1798-1866)
HAAST, John Francis Julius von, Sir, (1824-1887)
HAECKEL, Ernst Heinrich Philipp August, (1834-1919)
HALAVÁTS, Gyula, (1853-1926)
HAMY, Jules Ernest Théodore, (1842-1908)
HANCOCK, Albany, (1806-1873)
HARDMAN, Edward Townley, (1845-1887)
HARGER, Oscar, (1843-1882)
HARKNESS, Robert, (1816-1878)
HARLÉ, Edouard, (1850-1922)
HARTT, Charles Frederic, (1840-1878)
HASSE, Carl, (1841-1922)
HASTINGS, Barbara, Marchioness of, (1810-1858)
HATCHER, John Bell, (1861-1904)
HAUFF, Bernhard, (1866-1950)
HAUTHAL, Rudolf, (1854-1928)
HAWKINS, Thomas, (1810-1889)
HAY, Oliver Perry, (1846-1930)
HAYS, Isaac, (1796-1879)
HÉBERT, Edmond, (1812-1890)
HECKEL, Johann Jakob, (1790-1857)
HECTOR, James, Sir, (1834-1907)
HELLMAN, Milo, (1872-1947)
HENSEL, Reinhold Friederich, (1826-1881)
HERZ, Otto, (1853-1905)
HILBER, Vincenz, (1853-1931)
HIRSZBERG, Franciszek, (1863-1933)
HITCHCOCK, Edward, (1793-1864)
HOCHSTETTER, Ferdinand von, Ritter, (1829-1884)
HOFMANN, Adolf, (1853-1913)
HOLLAND, William Jacob, (1848-1932)
HOOLEY, Reginald Walter, (1866-1923)
HOSIUS, August, (1825-1896)
HOVEY, Horace Carter, Reverend, (1833-1914)
HOWES, George Bond, (1853-1905)

VERTEBRATE PALAEONTOLOGY 1850-1899 Cont'd.

HUENE, Friedrich von, Freiherr, (1875-1969)
HULKE, John Whitaker, (1830-1895)
HUMBERT, Alois, (1829-1887)
HUNTER, Robert, Reverend, (1823-1897)
HUNTER-SELKIRK, John R. S., (1835-1898)
HUOT, Jean Jacques Nicolas, (1790-1845)
HUTCHINSON, Henry Neville, Reverend, (?1855-1927)
HUXLEY, Thomas Henry, (1825-1895)
ISSEL, Arturo, (1842-1922)
JAECKEL, Otto, (1863-1929)
JÄGER, Georg Friedrich, (1785-1866)
JAMES, W. Warwick, (?1874-1965)
JOHNSON, Walter, (1867-1950)
JOHNSTON, Mary Sophia, (1875-1955)
JORDAN, David Starr, (1851-1931)
KADIČ, Ottokár, (1876-1957)
KAFKA, Josef, (1858-1929)
KARPINSKY, Alexander Petrovich, (1847-1936)
KAUP, Johann Jakob, (1803-1873)
KEITH, Arthur, Sir, (1866-1955)
KENNARD, Alfred Santer, (1870-1948)
KEYSERLING, Alexander Andreevich, Count, (1815-1891)
KIAER, Johan Aschehong, (1869-1931)
KING, Samuel William, (1821-1868)
KITCHIN, Finlay Lorimer, (1870-1934)
KITTL, Ernst Anton Leopold, (1854-1913)
KJERULF, Theodor, (1825-1888)
KLAASSEN, Hendericus Martinus, (1828-1910)
KLAATSCH, Hermann, (1863-1916)
KLIPSTEIN, August Wilhelm, (1801-1894)
KNER, Rudolf, (1810-1869)
KNIGHT, Charles Robert, (1875-1953)
KNIGHT, Wilbur Clinton, (1858-1903)
KNIPE, Henry Robert, (1855-1918)
KOCH, Antal von Bodrog, (1843-1927)
KOKEN, Ernst Friedrich Rudolph Karl von, (1860-1912)
KORNHUBER, Andreas, (1824-1905)
KOVALEVSKY, Vladimir Onufrievich, (1843-1883)
KRAUSS, Ferdinand, (1812-1890)
KREFFT, Gerhard, (1830-1881)
LAING, Samuel, (1810-1897)
LAMBE, Lawrence Morris, (1863-1919)
LANG, Arnold, (1855-1914)
LANKESTER, Edwin Ray, Sir, (1847-1929)
LARTET, Edouard Armand Isidore Hippolyte, (1801-1871)
LAUBE, Gustav C., (1839-1923)
LEA, Isaac, (1792-1886)
LE HON, Henri Sebastian, (1809-1872)
LEIDY, Joseph, (1823-1891)
LEMOINE, Armand Victor, (1837-1897)
LERICHE, Maurice, (1875-1948)
LIEBE, Karl Leopold Theodor, (1828-1894)
LIEBUS, Adalbert, (1876-1945)
LIOY, Paolo, (1834-1911)

VERTEBRATE PALAEONTOLOGY 1850-1899 Cont'd.

LOMAS, Joseph, (1860-1908)
LONG, Stephen Harriman, (1784-1864)
LOOMIS, Frederic Brewster, (1873-1937)
LÖRENTHEY, Imre, (1867-1917)
LORTET, Louis Charles, (1836-1909)
LUBICZ-NIEZABITOWSKI, Edward, (1875-1946)
LULL, Richard Swann, (1867-1957)
LÜTKEN, Christian Fredrik, (1827-1901)
LYDEKKER, Richard, (1849-1915)
MACKESON, Henry Bean, (1811-1894)
MAKOWSKY, Alexander, (1833-1908)
MANTELL, Gideon Algernon, (1790-1852)
MARINONI, Camillo, (1845-1883)
MARSH, Othniel Charles, (1831-1899)
MARTIN, Handel Tongue, (1862-1931)
MARTIN, Henri, (1873?-1936)
MAŠKA, Karel Jaroslov, (1851-1915)
MATTHEW, William Diller, (1871-1930)
MAYER-EYMAR, Karl David Wilhelm, (1826-1907)
McCOY, Frederick, Sir, (1823-1899)
MELI, Romolo, (1852-1920)
MENEGHINI, Giuseppe Giovanni Antonio, (1811-1889)
MENZEL, Karl Gustav Johannes, (1875-1914)
MERCERAT, Alcides, (-)
MERRIAM, John Campbell, (1869-1945)
MEYER, Christian Erich Hermann von, (1801-1869)
MIALL, Louis Compton, (1842-1921)
MICHAEL, Richard, (1869-1925)
MIEG-KROH, Mathieu, (1849-1911)
MIGEOD, Frederick William Hugh, (1872-1952)
MILLER, Loye Holmes, (1874-1970)
MILLER, Paul C., (1876-1945)
MILNE-EDWARDS, Alphonse, (1835-1900)
MITCHELL, Hugh, Reverend, (1822-1894)
MORENO, Francisco Josué Pascasio, (1852-1919)
MORGAN, Edward Delmar, (1840-1909)
MORREN, Charles, (1807-1858)
MORTILLET, Gabriel de, (1820-1898)
MOY-THOMAS, James Alan, (-1944)
MUÑIZ, Francisco Javier, (1795-1871)
MURCHISON, Charles, (1830-1879)
MURCHISON, Roderick Impey, Sir, (1792-1871)
NAUMANN, Edmund, (1854-1927)
NEHRING, Alfred, (1845-1904)
NEUGEBOREN, Johann Ludwig, (1806-1887)
NEWBERRY, John Strong, (1822-1892)
NEWTON, Alfred, (1829-1907)
NEWTON, Edwin Tulley, (1840-1930)
NODOT, Léonhard, (1802-1859)
NOPCSA VON FELSŐ-SZILVÁS, Ferenc, baron, (1877-1933)
NORDMANN, Alexander von, (1803-1866)
NOULET, Jean Baptiste, (1802-1890)
NÜESCH, Jakob, (1845-1915)
OGILBY, William, (1808-1873)

VERTEBRATE PALAEONTOLOGY 1850-1899 Cont'd.

OMBONI, Giovanni, (1829-1910)
OSBORN, Henry Fairfield, (1857-1935)
OWEN, Richard, Sir, (1804-1892)
PANTANELLI, Dante, (1844-1913)
PARISH, Woodbine, Sir, (1796-1882)
PARKER, William Kitchen, (1823-1890)
PARKS, William Arthur, (1868-1936)
PAUW, L. F. de, (-)
PEACH, Charles William, (1800-1886)
PEALE, Titian Ramsay, (1799-1885)
PENGELLY, William, (1812-1894)
PENTLAND, Joseph Barclay, (1797-1873)
PEREIRA DA COSTA, Francisco Antonio, (1809-1889)
PETERS, Karl, (1825-1881)
PFIZENMAYER, Eugen Wilhelm, (1869-)
PHLEPS, Otto, (1868-1928)
PILGRIM, Henry Guy Ellcock, (1875-1943)
PIPER, George Harry, (1819-1897)
PLIENINGER, Wilhelm Heinrich Theodor von, (1795-1879)
POMEL, Nicolas Auguste, (1821-1898)
POMPECKJ, Josef Felix, (1867-1930)
PORTIS, Alessandro, (1853-1931)
POULTON, Edward Bagnall, Sir, (1856-1943)
PRATT, Samuel Peace, (1789-1863)
PRIEM, Fernand, (1857-1919)
PRIMICS, György, (1849-1893)
PROBST, Joseph, (1823-1905)
QUATREFAGES DE BRÉAU, Jean Louis Armand de, (1810-1892)
QUEKETT, John Thomas, (1815-1861)
RAGAZZONI, Giuseppe, (-1898)
RAW, Frank, (?1875-1961)
REDFIELD, John Howard, (1815-1895)
REDFIELD, William C., (1789-1857)
REEVE, James, (1833-1920)
REMELÉ, Adolf Karl, (1839-1915)
REPOSSI, Emilio, (1876-1931)
REYNOLDS, Sidney Hugh, (1867-1949)
RICHTER, Reinhard, (1813-1884)
RISTORI, Giuseppe, (1856-1905)
ROBERTS, George Edward, (1831-1865)
ROGER, Otto, (1841-1915)
ROMAN, Albert Frédéric, (1871-1943)
ROTH, Santiago, (1850-1924)
ROUAULT, Marie, (1813-1881)
ROYLE, John Forbes, (-1857?)
RÜTIMEYER, Karl Ludwig, (1825-1895)
RZEHÅK, Anton, (1855-1923)
SACCO, Federico, (1864-1948)
SARS, Michael, (1805-1869)
SARTORIUS VON WALTERSHAUSEN, Wolfgang, (1809-1876)
SAUVAGE, Henri Émile, (1842-1917)
SAVI, Paolo, (1798-1871)
SCHAAFHAUSEN, Hermann, (1816-1893)
SCHIMPER, Karl Friedrich, (1803-1867)

VERTEBRATE PALAEONTOLOGY 1850-1899 Cont'd.

SCHLEGEL, Hermann, (1804-1884)
SCHLOSSER, Max, (1854-1933)
SCHLUETER, Clemens August, (1835-1906)
SCHMID, Ernst Erhard, (1815-1885)
SCHMIDT, Oskar, (1823-1886)
SCHOETENSACK, Otto, (1850-1912)
SCHÖNFELD, Georg, (1878-1926)
SCHRENCK, Leopold von, (1826-1894)
SCHROEDER, Henry Carl, (1859-1927)
SCHWALBE, Gustav, (1844-1916)
SCHWERTSCHLAGER, Joseph, (1853-1924)
SCOTT, William Berryman, (1858-1947)
SEDGWICK, Adam, (1854-1913)
SEELEY, Harry Govier, (1839-1909)
SEGUENZA, Luigi, (1873-1908)
SELLARDS, Elias Howard, (1875-1961)
SERRES, Marcel de, (1783-1862)
SHARPE, Daniel, (1806-1856)
SHERBORN, Charles Davies, (1861-1942)
SHUFELDT, Robert Wilson, (1850-1934)
SISMONDA, Eugenio, (1815-1870)
SNOW, Francis Huntington, (1840-1908)
SPENCER, William Kingdon, (1878-1955)
STANTON, Timothy William, (1860-1953)
ŞTEFANESCU, Gregoriu, (1838-1911)
ŞTEFĂNESCU, Sabba, (1857-1931)
STEHLIN, Hans Georg, (1870-1942)
STEPHENS, William John, (1829-1890)
STERNBERG, Charles Hazelius, (1850-1943)
ST. JOHN, Orestes Hawley, (1841-1921)
STOLICZKA, Ferdinand, (1838-1874)
STOPPANI, Antonio, (1824-1891)
STORMS, Raymond, (1834-1897)
STORRIE, John, (1844-1901)
STROBEL, Pellegrino, (1821-1895)
STUR, Dénes, (1827-1893)
SUSCHKIN, Peter Petrovich, (1868-1928)
SWINNERTON, Henry Hurd, (1875-1967)
THEVENIN, Armand, (1870-1918)
THIOLLIÈRE, Victor, (1801-1859)
THOMAS, Abram Owen, (1876-1931)
THOMAS, Philippe, (1843-1910)
TOKUNAGA, Shigeyasu, (1874-1940)
TOLL, Eduard von, Baron, (1858-1902)
TRAQUAIR, Ramsay Heatley, (1840-1912)
TROSCHEL, Franz Hermann, (1810-1882)
TROUESSART, Edouard Louis, (1842-1927)
TURTON, William Henry, (1862-1942)
TWELVETREES, William Harper, (1848-1919)
UBAGHS, Johann Casimir, (1829-1894)
VACEK, Michael, (1848-1925)
VAILLANT, Léon Louis, (1834-1914)
VERRI, Antonio, (1839-1925)
VETTER, Benjamin, (1848-1893)

VERTEBRATE PALAEONTOLOGY 1850-1899 Cont'd.

WAGNER, Johann Andreas, (1797-1861)
WARD, John, (1837-1906)
WARREN, John Collins, (1778-1856)
WEIDENREICH, Franz, (1873-1947)
WEISS, Christian Ernst, (1833-1890)
WESTON, Thomas Chesmer, (1832-1910)
WIELAND, George Reber, (1865-1953)
WILLISTON, Samuel Wendell, (1852-1918)
WIMAN, Carl Johan Josef Ernst, (1867-1944)
WINCHELL, Alexander, (1824-1891)
WINCHELL, Newton Horace, (1829-1914)
WINGE, Herluf, (1857-1923)
WINKLER, Tiberius Cornelius, (1822-1897)
WOLDŘICH, Jan Nepomuk, (1834-1906)
WOOD, Searles Valentine, (1798-1880)
WOOD-MASON, James, (1846-1893)
WOODWARD, Arthur Smith, Sir, (1864-1944)
WOODWARD, Henry, (1832-1921)
WORSFORD, Frederick H., (1870-1954)
WORTMAN, Jacob Lawson, (1856-1926)
WRIGHT, Albert Allen, (1846-1905)
ZADDACH, Ernst Gustav, (1817-1881)
ZITTEL, Karl Alfred von, (1839-1904)

1900 to present

ABEL, Othenio, (1875-1946)
ADAMETZ, Lotte, (1879-1966)
AGOGINO, George, (-)
AIM, William Lennie, (?1872-1938)
ALLEN, Joel Asaph, (1838-1921)
AMEGHINO, Florentino, (1854-1911)
AMMON, Ludwig von, (1850-1922)
ANDERSON, Charles, (1876-1944)
ANDERSSON, Johann Gunnar, (1874-1960)
ANDREWS, Charles William, (1866-1924)
ANDREWS, Roy Chapman, (1884-1960)
ARAMBOURG, Camille, (1885-1969)
ASTRE, Gaston, (1896-)
AUGUSTA, Josef, (1903-1968)
BARBOUR, Erwin Hinckley, (1856-1947)
BARBOUR, George Brown, (1891-1977)
BARBOUR, Thomas, (1884-1946)
BARON, Richard, Reverend, (1847-1907)
BATE, Dorothea Minola Alice, (1879-1951)
BATESON, William, (1861-1926)
BAYER, Josef, (1882-1931)
BEADNELL, Hugh John Llewellyn, Major, (1874-1944)
BEASLEY, Henry Charles, (1836-1919)
BELLOTTI, Christoforo, (1823-1919)
BENSLEY, Benjamin Arthur, (1875-1934)
BERENDT, Gottlieb Michael, (1836-1920)
BERNSEN, J. J. A., Pater, (1888-1932)
BIRD, Roland T., (-1978)
BISHOP, Walter William, (1931-1977)

VERTEBRATE PALAEONTOLOGY 1900 to present Cont'd.

BLACK, Davidson, (1884-1934)
BLANC, Alberto Carlo, (1906-1960)
BLANCKENHORN, Max Ludwig Paul, (1861-1947)
BLANFORD, William Thomas, (1832-1905)
BONAPARTE, Charles Lucien Jules Laurent, Prince of Canino, (1803-1857)
BOTTI, Ulderigo, (1822-1905)
BOULE, Marcellin, (1861-1942)
BRADY, Lionel Francis "Major", (1880-1963)
BRANDES, Theodor, (-1916)
BRANDT, Alexander von, (1844-1932)
BRANSON, Carl Colton, (1906-1975)
BRANSON, Edwin Bayer, (1877-1950)
BREUIL, Henri, Abbé, (1877-1961)
BROILI, Ferdinand, (1874-1946)
BROOM, Robert, (1866-1951)
BROTZEN, Fritz, (1902-1968)
BROWN, Barnum, (1873-1963)
BURCKHARDT, Carl Rudolf, (1866-1908)
BUWALDA, John Peter, (1886-1954)
BYSTROV, Alexei Petrovich, (1899-1959)
CALDERON Y ARANA, Salvador, (1851-1911)
ČAPEK, Václav, (1862-1926)
CAPELLINI, Giovanni, (1833-1922)
CASE, Ermine Cowles, (1871-1953)
CASTRO Y SUERO, Manuel Fernandez de, (1825-1895)
CHANTRE, Ernest, (1843-)
CHARLESWORTH, Edward, (1813-1893)
CHOFFAT, Paul, (1849-1919)
CLAYPOLE, Edward Waller, (1835-1901)
CODRINGTON, Thomas, (1829-1918)
COLBERT, Edwin H., (1905-)
COLENUTT, George William, (?1862-1944)
COLLETT, Robert, (1842-1913)
COLLOT, Louis Marie François, (1846-1915)
COMMONT, Victor, (1866-1918)
COOK, Margaret C., (1899-1968)
CORNER, Frank, (1861-1939)
CREDNER, Carl Hermann Georg, (1841-1913)
CROOK, Alja Robinson, (1864-1930)
CURRIE, Philip John, (1949-)
DALL, William Healey, (1845-1927)
DART, Raymond Arthur, (1893-)
DAVID, Tannatt William Edgeworth, Sir, (1858-1934)
DAVIES, Henry Nathaniel, (-1920)
DAWKINS, William Boyd, Sir, (1837-1929)
DEICHMÜLLER, Johannes, (1854-)
DELVAUX, Émile, (1837-1901)
DEPÉRET, Charles Jean Julien, (1854-1929)
DIBLEY, George Edward, (1857-1943)
DIGBY, Bassett, (-)
DITMARS, Raymond Lee, (1876-1942)
DOLGOPOL DE SÁEZ, Mathilde, (1901-1957)
DOLLO, Louis Antoine Marie Joseph, (1857-1931)

VERTEBRATE PALAEONTOLOGY 1900 to present Cont'd.

DOUGLASS, Earl, (1862-1931)
DREVERMANN, Fritz, (1875-1932)
DUBOIS, Eugène, (1858-1940)
DUPONT, Édouard François, (1841-1911)
EASTMAN, Charles Rochester, (1868-1918)
EDINGER, Tilly, (1897-1967)
EFREMOV, Ivan Antonovich, (1907-1972)
ELLIOT SMITH, Grafton, Sir, (1871-1937)
EPPS, Frank James, (?1891-1974)
ETHERIDGE, Robert Jr., (1847-1920)
EVANS, John, Sir, (1823-1908)
FEJÉRVÁRY, Géza Gyula Imre, Baron von Komlóskeresztes, (1894-1932)
FERRONNIÈRE, Georges, (1875-1922)
FIGGINS, Jesse Dade, (1867-1944)
FILHOL, Henri, (1843-1902)
FISCHER, Ernst, (1888-1914)
FOLLMANN, Otto, (1856-1926)
FONDOUCE, Paul Cazalis de, (1838-1932)
FORSTER-COOPER, Clive, (1880-1947)
FORSYTH MAJOR, Charles Immanuel, (1843-1923)
FRAAS, Eberhard, (1862-1915)
FRAIPONT, Julien, (1857-1910)
FRITSCH, Anton, (1832-1913)
FRITSCH, Karl von, Freiherr, (1838-1906)
FRÜH, Jakob, (1852-)
FRY, Willis E., (1926-1973)
FUCHS, Theodor, (1842-1925)
FÜRBRINGER, Max, (1846-1920)
GAÁL, István, (1877-1957)
GARDINER, Charles Irving, (?1868-1940)
GEINITZ, Hanns Bruno, (1814-1900)
GEMMELLARO, Mariano, (1879-1921)
GIDLEY, James Williams, (1866-1931)
GILMORE, Charles Whitney, (1874-1945)
GLAUERT, Ludwig, (1879-1963)
GLORY, André, Abbé, (-)
GOELDI, Emil August, (1859-1917)
GOODRICH, Edwin Stephen, (1868-1946)
GORIZDRO-KULCZYCKA, Zinaida, (1884-1949)
GORJANOVIĆ-KRAMBERGER, Dragutin, (1856-1937)
GOSSELET, Jules Auguste, (1832-1916)
GRANDIDIER, Alfred, (1836-1921)
GRANGER, Walter, (1872-1941)
GREGORY, William King, (1876-1970)
GROMOVA, Vera Isaacovna, (1891-1973)
GROSS, Walter Robert, (1903-1974)
GÜNTHER, Albert Charles Lewis Gotthilf, (1830-1914)
GUT, H. James, (1904-1969)
HAECKEL, Ernst Heinrich Philipp August, (1834-1919)
HALAVÁTS, Gyula, (1853-1926)
HAMY, Jules Ernest Théodore, (1842-1908)
HARBORT, Erich, (1879-1929)
HARLÉ, Edouard, (1850-1922)
HASSE, Carl, (1841-1922)

VERTEBRATE PALAEONTOLOGY 1900 to present Cont'd.

HATCHER, John Bell, (1861-1904)
HAUFF, Bernhard, (1866-1950)
HAUTHAL, Rudolf, (1854-1928)
HAY, Oliver Perry, (1846-1930)
HECTOR, James, Sir, (1834-1907)
HEINTZ, Anatol, (1898-1975)
HELBING, Hermann, (1880-1938)
HELLMAN, Milo, (1872-1947)
HENNIG, Edwin, (1882-)
HERNÁNDEZ-PACHECO, Eduardo, (-1965)
HERZ, Otto, (1853-1905)
HESSE, Curtis Julian, (1905-1945)
HIBBARD, Claude W., (1905-1973)
HIBBERT-WARE, Samuel, (1782-1848)
HILBER, Vincenz, (1853-1931)
HINTON, Martin Alister Campbell, (1883-1961)
HIRSZBERG, Franciszek, (1863-1933)
HOFFSTETTER, Robert, (1908-)
HOFMANN, Adolf, (1853-1913)
HOLLAND, William Jacob, (1848-1932)
HONNOR, Frank Walter, (?1892-1976)
HOOLEY, Reginald Walter, (1866-1923)
HORNER, Jack, (-)
HOVEY, Horace Carter, Reverend, (1833-1914)
HOWARD, Edgar Billings, (1887-1943)
HOWES, George Bond, (1853-1905)
HUENE, Friedrich von, Freiherr, (1875-1969)
HUSSAKOF, Louis, (1881-1965)
HUTCHINSON, Henry Neville, Reverend, (?1855-1927)
ISSEL, Arturo, (1842-1922)
JACKSON, James Frederick, (1896-1966)
JAECKEL, Otto, (1863-1929)
JAMES, W. Warwick, (?1874-1965)
JANENSCH, Werner, (1878-1969)
JOHANSON, Donald C., (-)
JOHNSON, Hope, (-)
JOHNSON, Walter, (1867-1950)
JOHNSTON, Mary Sophia, (1875-1955)
JOLEAUD, Léonce, (1880-1938)
JORDAN, David Starr, (1851-1931)
JURADO, Martín Doello, (1884-1948)
KADIĆ, Ottokár, (1876-1957)
KAFKA, Josef, (1858-1929)
KAISEN, Ove C., (1902-1953)
KARPINSKY, Alexander Petrovich, (1847-1936)
KAY, James LeFoy, (1892-1971)
KEITH, Arthur, Sir, (1866-1955)
KELLOGG, Arthur Remington, (1892-1969)
KENNARD, Alfred Santer, (1870-1948)
KIAER, Johan Aschehong, (1869-1931)
KIELAN-JAWOROWSKA, Zofia, (-)
KITCHIN, Finlay Lorimer, (1870-1934)
KITCHING, James, (-)
KITTL, Ernst Anton Leopold, (1854-1913)

VERTEBRATE PALAEONTOLOGY 1900 to present Cont'd.

KLAASSEN, Hendericus Martinus, (1828-1910)
KLAATSCH, Hermann, (1863-1916)
KLÄHN, Hans, (1884-1933)
KNIGHT, Charles Robert, (1875-1953)
KNIGHT, Samuel Howell, (1892-1975)
KNIGHT, Wilbur Clinton, (1858-1903)
KNIPE, Henry Robert, (1855-1918)
KOCH, Antal von Bodrog, (1843-1927)
KOENIGSWALD, Gustav Heinrich Ralph von, (1902-)
KOKEN, Ernst Friedrich Rudolph Karl von, (1860-1912)
KORNHUBER, Andreas, (1824-1905)
KRAGLIEVICH, Lucas, (1886-1932)
KUTASSY, Endre, (1898-1938)
KYRLE, Georg, (1887-1937)
LAMBE, Lawrence Morris, (1863-1919)
LAMBRECHT, Kálmán, (1889-1936)
LANG, Arnold, (1855-1914)
LANKESTER, Edwin Ray, Sir, (1847-1929)
LAPPARENT, Albert Félix de, (1905-1975)
LAUBE, Gustav C., (1839-1923)
LEAKEY, Louis Seymour Bazett, (1903-1972)
LEAKEY, Mary Douglas (née Nicol), (1913-)
LEAKEY, Richard Erskine, (1944-)
LE GROS CLARK, Wilfrid Edward, Sir, (1895-1971)
LERICHE, Maurice, (1875-1948)
LIEBE, Karl Leopold Theodor, (1828-1894)
LIEBUS, Adalbert, (1876-1945)
LIOY, Paolo, (1834-1911)
LOMAS, Joseph, (1860-1908)
LOOMIS, Frederic Brewster, (1873-1937)
LÖRENTHEY, Imre, (1867-1917)
LORTET, Louis Charles, (1836-1909)
LUBICZ-NIEZABITOWSKI, Edward, (1875-1946)
LULL, Richard Swann, (1867-1957)
LÜTKEN, Christian Fredrik, (1827-1901)
LYDEKKER, Richard, (1849-1915)
MAKOWSKY, Alexander, (1833-1908)
MALDONADO-KOERDELL, Manuel, (1908-1972)
MARSTON, Alvan Theophilus, (1889-1971)
MARTIN, Handel Tongue, (1862-1931)
MARTIN, Henri, (1873?-1936)
MAŠKA, Karel Jaroslov, (1851-1915)
MATTHEW, William Diller, (1871-1930)
MAXSON, John Haviland, (1906-1966)
MAYER-EYMAR, Karl David Wilhelm, (1826-1907)
MEFFERD, Ralph L., (1909-1964)
MEHL, Maurice Goldsmith, (1887-1966)
MELI, Romolo, (1852-1920)
MENZEL, Karl Gustav Johannes, (1875-1914)
MERRIAM, John Campbell, (1869-1945)
MIALL, Louis Compton, (1842-1921)
MICHAEL, Richard, (1869-1925)
MIEG-KROH, Mathieu, (1849-1911)
MIGEOD, Frederick William Hugh, (1872-1952)

VERTEBRATE PALAEONTOLOGY 1900 to present Cont'd.

MILLER, Loye Holmes, (1874-1970)
MILLER, Paul C., (1876-1945)
MILNE-EDWARDS, Alphonse, (1835-1900)
MITZOPOULOS, Maximos K., (1896-1968)
MORELLET, Lucien, (1882-1945)
MORENO, Francisco Josué Pascasio, (1852-1919)
MORGAN, Edward Delmar, (1840-1909)
MOY-THOMAS, James Alan, (-1944)
NAUMANN, Edmund, (1854-1927)
NEHRING, Alfred, (1845-1904)
NEWTON, Alfred, (1829-1907)
NEWTON, Edwin Tulley, (1840-1930)
NOPCSA VON FELSÖ-SZILVÁS, Ferenc, baron, (1877-1933)
NÜESCH, Jakob, (1845-1915)
OAKLEY, Kenneth Page, (1911-)
OBRUCHEV, Dmitri, (1900-1970)
OERTEL, Walter, (1889-1924)
OMBONI, Giovanni, (1829-1910)
ORLOV, Uri Aleksandrovich, (1893-1966)
OSBORN, Henry Fairfield, (1857-1935)
PABST, Marie B., (1909-1963)
PANTANELLI, Dante, (1844-1913)
PARKS, William Arthur, (1868-1936)
PARRINGTON, Francis Rex, (-)
PATRIZI, Saverio, (1902-1957)
PAUW, L. F. de, (-)
PEABODY, Frank Elmer, (1914-1958)
PHLEPS, Otto, (1868-1928)
PILGRIM, Henry Guy Ellcock, (1875-1943)
PIVETEAU, Jean, (1898-)
POMPECKJ, Josef Felix, (1867-1930)
PORTIS, Alessandro, (1853-1931)
POULTON, Edward Bagnall, Sir, (1856-1943)
PRIEM, Fernand, (1857-1919)
PROBST, Joseph, (1823-1905)
PRUVOST, Pierre Eugène Marie Joseph, (1890-1967)
RAKOVEC, Ivan, (1899-)
RAMA RAO, Lakshmeswar, (1896-1974)
RAW, Frank, (?1875-1961)
RECK, Hans, (1886-1937)
REICHEL, Manfred, (1896-)
REMELÉ, Adolf Karl, (1839-1915)
REPOSSI, Emilio, (1876-1931)
REYNOLDS, Sidney Hugh, (1867-1949)
REYNOLDS, Thomas Emmett, (-1972)
RISTORI, Giuseppe, (1856-1905)
ROGER, Otto, (1841-1915)
ROMAN, Albert Frédéric, (1871-1943)
ROMER, Alfred Sherwood, (1894-1973)
ROTH, Santiago, (1850-1924)
RUSSELL, Loris Shano, (1904-)
RUTTEN, Louis Martin Robert, (1884-1946)
RZEHÁK, Anton, (1855-1923)
SACCO, Federico, (1864-1948)

VERTEBRATE PALAEONTOLOGY 1900 to present Cont'd.

SAINTY, James Edward, (?1882-1967)
SAUVAGE, Henri Émile, (1842-1917)
SÄVE-SÖDERBERGH, Gunnar, (1910-1948)
SCHAUB, Samuel, (1882-1962)
SCHLAIKJER, Erich Maren, (1905-1972)
SCHLOSSER, Max, (1854-1933)
SCHLUETER, Clemens August, (1835-1906)
SCHOETENSACK, Otto, (1850-1912)
SCHÖNFELD, Georg, (1878-1926)
SCHROEDER, Henry Carl, (1859-1927)
SCHRUEDER, Antje, (1887-1952)
SCHWALBE, Gustav, (1844-1916)
SCHWERTSCHLAGER, Joseph, (1853-1924)
SCOTT, William Berryman, (1858-1947)
SEDGWICK, Adam, (1854-1913)
SEELEY, Harry Govier, (1839-1909)
SEGUENZA, Luigi, (1873-1908)
SELLARDS, Elias Howard, (1875-1961)
SHERBORN, Charles Davies, (1861-1942)
SHUFELDT, Robert Wilson, (1850-1934)
SHULER, Ellis William, (1881-1954)
SIMPSON, George Gaylord, (1902-)
SMITH, James Leonard Brierley, (1897-)
SNOW, Francis Huntington, (1840-1908)
SORENSEN, Carl Christian, (1893-1965)
SPENCER, William Kingdon, (1878-1955)
ŠPINAR, Zdeněk Vlastimil, (1906-)
STANTON, Timothy William, (1860-1953)
ŞTEFANESCU, Gregoriu, (1838-1911)
ŞTEFĂNESCU, Sabba, (1857-1931)
STEHLIN, Hans Georg, (1870-1942)
STERNBERG, Charles Hazelius, (1850-1943)
STERNBERG, Charles M., (1885-)
STIRTON, Ruben Arthur, (1901-1966)
ST. JOHN, Orestes Hawley, (1841-1921)
STOCK, Chester, (1892-1950)
STORRIE, John, (1844-1901)
STRAW, Stephen Henry, (?1892-1963)
SUSCHKIN, Peter Petrovich, (1868-1928)
SWINNERTON, Henry Hurd, (1875-1967)
TASNÁDI-KUBACSKA, András, (1902-1977)
TEILHARD DE CHARDIN, Pierre, Father, S.J., (1881-1955)
THEVENIN, Armand, (1870-1918)
THOMAS, Abram Owen, (1876-1931)
THOMAS, Philippe, (1843-1910)
THORPE, Malcolm Rutherford, (1891-1958)
TOKUNAGA, Shigeyasu, (1874-1940)
TOLL, Eduard von, Baron, (1858-1902)
TRAQUAIR, Ramsay Heatley, (1840-1912)
TROUESSART, Edouard Louis, (1842-1927)
TROXELL, Edward Leffingwell, (1884-1972)
TURTON, William Henry, (1862-1942)
TWELVETREES, William Harper, (1848-1919)
UNTERMANN, Gerhard Ernest, Jr., (1898-1975)

VERTEBRATE PALAEONTOLOGY 1900 to present Cont'd.

VACEK, Michael, (1848-1925)
VAILLANT, Léon Louis, (1834-1914)
VANDER HOOF, Vertress Lawrence, (1904-1964)
VERRI, Antonio, (1839-1925)
VERTÉS, László, (1914-1968)
VIANNA, António, (1887-1949)
VIRET, Jean, (1894-1970)
VOLLRATH, Paul, (-1929)
WARD, John, (1837-1906)
WATSON, David Meredith Seares, (1886-1973)
WEIDENREICH, Franz, (1873-1947)
WEIGELT, Johannes, (1890-1948)
WEILER, Wilhelm, (1890-1972)
WELLES, Samuel Paul, (-)
WEPFER, Emil, (1882-1930)
WESTON, Thomas Chesmer, (1832-1910)
WETMORE, Alexander, (1886-)
WIELAND, George Reber, (1865-1953)
WILLISTON, Samuel Wendell, (1852-1918)
WIMAN, Carl Johan Josef Ernst, (1867-1944)
WINCHELL, Newton Horace, (1829-1914)
WINGE, Herluf, (1857-1923)
WOLDŘICH, Jan Nepomuk, (1834-1906)
WOODWARD, Arthur Smith, Sir, (1864-1944)
WOODWARD, Henry, (1832-1921)
WORSFORD, Frederick H., (1870-1954)
WORTMAN, Jacob Lawson, (1856-1926)
WRIGHT, Albert Allen, (1846-1905)
ZÁZVORKA, Vlastislav, (1903-)
ZEUNER, Frederick Everard, (1905-1963)
ZITTEL, Karl Alfred von, (1839-1904)

VULCANOLOGY
Up to 1800
ALBERTUS MAGNUS, (1193-1280)
ARDUINO, Giovanni, (1714-1795)
ARISTOTLE, (384-322 B.C.)
AVICENNA, (980-1037)
BERTRAND DE DOUE, Jacques Mathieu, (1776-1862)
BRYDONE, Patrick, (1736-1818)
CASSIUS, Dio Cocceianus, ([155-235] A.D.)
DESMAREST, Nicolas, (1725-1815)
DOLOMIEU, Dieudonné Sylvain Guy Tancrède, called Déodat, Gratet de, (1750-1801)
HAMILTON, William, Sir, (1730-1803)
HUMBOLDT, Friedrich Wilhelm Heinrich Alexander von, Freiherr, (1769-1859)
HUTTON, James, (1726-1797)
KIRCHER, Athanasius, (1601-1680)
LAVOISIER, Antoine Laurent, (1743-1794)
MONTICELLI, Teodoro, Abate, (1757-1845)
MORO, Anton Lazzaro, (1687-1740)
PLINY THE ELDER (CAIUS PLINIUS SECUNDUS), (?20 A.D.-79 A.D.)
RAY, John, (1627-1705)

VULCANOLOGY Up to 1800 Cont'd.

SICULUS, Diodorus, (ca.85-21 B.C.)
SPALLANZANI, Lazzaro, Abate, (1729-1799)
STRABO, (63B.C.-23 A.D.?)
TOWNSON, Robert, (1763-1827)
TROIL, Uno von, (1746-1803)

1800-1849
BARROW, John, Jr., (1808-1898)
BERTRAND DE DOUE, Jacques Mathieu, (1776-1862)
BRYDONE, Patrick, (1736-1818)
BRYSON, Alexander, (1816-1866)
DANA, James Dwight, (1813-1895)
DAUBENY, Charles Giles Brindle, (1795-1867)
DESMAREST, Nicolas, (1725-1815)
DOLOMIEU, Dieudonné Sylvain Guy Tancrède, called Déodat, Gratet de, (1750-1801)
DUBOIS DE MONTPÉREUX, Frédéric, (1798-1850)
ERLE, C., Reverend, (1790-1870?)
FORBES, James David, (1809-1868)
FOUQUÉ, Ferdinand André, (1828-1904)
GEMMELLARO, Carlo, (1787-1866)
GREEN, William Lowthian, (1819-1890)
HAMILTON, William, Sir, (1730-1803)
HENDERSON, Ebenezer, (1784-1858)
HOFF, Karl Ernst Adolf von, (1771-1837)
HOOKER, William Jackson, Sir, (1785-1865)
HUMBOLDT, Friedrich Wilhelm Heinrich Alexander von, Freiherr, (1769-1859)
JOHNSTRUP, Johannes Frederik, (1818-1894)
LECOQ, Henri, (1802-1871)
MALLET, Robert, (1810-1881)
MONTICELLI, Teodoro, Abate, (1757-1845)
PHILLIPS, John, (1800-1874)
ROTH, Justus Ludwig Adolph, (1818-1892)
SAINTE-CLAIRE DEVILLE, Charles, (1814-1876)
SARTORIUS VON WALTERSHAUSEN, Wolfgang, (1809-1876)
SCACCHI, Arcangelo, (1810-1893)
SCHMIDT, Johann Friedrich Julius, (1825-1884)
SCROPE, George Poulett, (né Thomson), (1797-1876)
TOWNSON, Robert, (1763-1827)
TROIL, Uno von, (1746-1803)

1850-1899
ANDERSON, Tempest, (1846-1913)
BARROW, John, Jr., (1808-1898)
BASSANI, Francesco, (1853-1916)
BERGEAT, Alfred, Edmund, (1866-1924)
BERTRAND DE DOUE, Jacques Mathieu, (1776-1862)
BRANCA, Wilhelm von, (1844-1928)
BRYSON, Alexander, (1816-1866)
BUSZ, Karl Heinrich Emil Georg, (1863-1930)
CLELAND, Herdman Fitzgerald, (1869-1935)
DANA, Edward Salisbury, (1849-1935)
DANA, James Dwight, (1813-1895)

VULCANOLOGY 1850-1899 Cont'd.

DAUBENY, Charles Giles Brindle, (1795-1867)
DAVID, Tannatt William Edgeworth, Sir, (1858-1934)
DAY, Arthur Louis, (1869-1960)
DE LORENZO, Giuseppe, (1871-1957)
DE ROSSI, Michele Stefano, (1834-1898)
DOELTER y CISTERLICH, Cornelio August, (1850-1930)
DUTTON, Clarence Edward, Major, (1841-1912)
EAKLE, Arthur Starr, (1862-1931)
ERLE, C., Reverend, (1790-1870?)
FALB, Rudolf, (-)
FLETT, John Smith, Sir, (1869-1947)
FLORES REYES, Teodoro, (1873-1955)
FORBES, James David, (1809-1868)
FOUQUÉ, Ferdinand André, (1828-1904)
FRANCO, Pasquale, (1852-1907)
GALE, Hoyt Stoddard, (1876-1952)
GEMMELLARO, Carlo, (1787-1866)
GEMMELLARO, Gaetano Giorgio, (1832-1904)
GERLAND, Georg Karl Cornelius, (1833-1919)
GREEN, William Lowthian, (1819-1890)
HAGUE, Arnold, (1840-1917)
HEILPRIN, Angelo, (1853-1907)
HEMPEL, Walther Matthias, (1851-1916)
HENDERSON, Ebenezer, (1784-1858)
HOBSON, Bernard, (1860-1933)
HOFF, Karl Ernst Adolf von, (1771-1837)
HOOKER, William Jackson, Sir, (1785-1865)
HOVEY, Edmund Otis, (1862-1924)
HUMBOLDT, Friedrich Wilhelm Heinrich Alexander von, Freiherr, (1769-1859)
JAGGAR, Thomas Augustus, (1871-1953)
JOHNSTON, Mary Sophia, (1875-1955)
JOHNSTON-LAVIS, Henry James, (1856-1914)
JOHNSTRUP, Johannes Frederik, (1818-1894)
JUDD, John Wesley, (1840-1916)
KOTÔ, Bundjirô, (1856-1935)
LARSEN, Esper Signius Jr., (1879-1961)
LASAULX, Arnold von, (1839-1886)
LECOQ, Henri, (1802-1871)
LIBBEY, William, (1855-1927)
LOEWINSSON-LESSING, Franc Julyevich, (1859-1939)
MACHADE E COSTA, Alfredo Augusto d'Olveira, (1870-?1952)
MALLADRA, Alessandro, (1865-1945)
MALLET, Robert, (1810-1881)
MARTIN, George C., (1875-1943)
MATTEUCCI, Raffaele Vittorio, (1862-1909)
MERCALLI, Giuseppe, (1850-1914)
MEUNIER, Étienne Stanislas, (1843-1925)
MEYER, Hans, (1858-1929)
MONTESSUS DE BALLORE, Fernand Jean Baptiste Marie Bernard, Comte de, (1851-1923)
MOORE, Elwood S., (1879-1966)
NASON, Henry Bradford, (1831-1895)
OMORI, Fusakichi, (1868-1923)

VULCANOLOGY 1850-1899 Cont'd.

ORDOÑEZ, Ezequiel, (1867-1950)
O'REILLY, Joseph P., (1829-1905)
OSWALD, Felix, (1866-1958)
PHILLIPS, John, (1800-1874)
PLATANIA, Gaetano, (1867-1938)
ROCKWOOD, Charles Greene, (1843-1913)
RODOLPH, Emil, (1853-1915)
ROTH, Justus Ludwig Adolph, (1818-1892)
RUDOLPH, Emil, (1853-1915)
SABATINI, Venturino, (1879-1922)
SAINTE-CLAIRE DEVILLE, Charles, (1814-1876)
SAPPER, Karl Theodor, (1866-1945)
SARTORIUS VON WALTERSHAUSEN, Wolfgang, (1809-1876)
SCACCHI, Arcangelo, (1810-1893)
SCHMIDT, Johann Friedrich Julius, (1825-1884)
SCROPE, George Poulett, (né Thomson), (1797-1876)
SEEBACH, Karl Albert Ludwig von, (1839-1880)
SEIKEI, Sekiya, (1855-1896)
SILVESTRI, Orazio, (-1890)
STANLEY, William Ford Robinson, (1828-1909)
TANNHÄUSER, Felix, (1874-1924)
THORODDSEN, Thorvaldur, (1855-1921)
VERBEEK, Rogier Diederick Marius, (1845-1926)
WEIDMAN, Samuel, (1870-1945)
WICHMANN, Carl Ernst Arthur, (1851-1927)
WILLIAMS, John Francis, (1862-1891)
WOLF, Theodor, (1841-1924)
YAMASAKI, Naomasa, (1870-1929)

1900 to present

ACLAND, Henry Dyke, (-1936)
ANDERSON, Tempest, (1846-1913)
ASHBY, Donald Frederick, (?1914-1943)
BARTH, Thomas Frederik Weiby, (1899-1971)
BARTRUM, John Arthur, (1885-1949)
BASSANI, Francesco, (1853-1916)
BERGEAT, Alfred, Edmund, (1866-1924)
BRADY, Lionel Francis "Major", (1880-1963)
BRANCA, Wilhelm von, (1844-1928)
BULLERWELL, William, (1916-1977)
BUSZ, Karl Heinrich Emil Georg, (1863-1930)
CLELAND, Herdman Fitzgerald, (1869-1935)
COCKBURN, Alexander Murray, (1901-1959)
COMSTOCK, Theodore Bryant, (1849-1915)
COOMBS, Howard A., (-)
COTTON, Charles Andrew, Sir, (1885-1970)
COTTON, Leo Arthur, (1883-1963)
DANA, Edward Salisbury, (1849-1935)
DAVID, Tannatt William Edgeworth, Sir, (1858-1934)
DAY, Arthur Louis, (1869-1960)
DE LORENZO, Giuseppe, (1871-1957)
DOELTER y CISTERLICH, Cornelio August, (1850-1930)
DUBYANSKY, Viktor Viktorovich, (1880-1925)
DUTTON, Clarence Edward, Major, (1841-1912)

VULCANOLOGY 1900 to present Cont'd.

EAKLE, Arthur Starr, (1862-1931)
EDWARDS, Austin Burton, (1909-1960)
ESCHER, Berend George, (-1967)
FAIRCHILD, John Gifford, (1882-1965)
FISHER, Bernard, (1918-1946)
FLETT, John Smith, Sir, (1869-1947)
FLORES REYES, Teodoro, (1873-1955)
FOSHAG, William Frederick, (1894-1956)
FOUQUÉ, Ferdinand André, (1828-1904)
FRANCO, Pasquale, (1852-1907)
FULLER, Richard Eugene, (1897-1976)
GALE, Hoyt Stoddard, (1876-1952)
GEMMELLARO, Gaetano Giorgio, (1832-1904)
GERLAND, Georg Karl Cornelius, (1833-1919)
GORSHKOV, Georgii Stepanovich, (1921-1975)
HAGUE, Arnold, (1840-1917)
HANNESSON, Pálmi, (1898-1956)
HEILPRIN, Angelo, (1853-1907)
HEMPEL, Walther Matthias, (1851-1916)
HERNÁNDEZ-PACHECO, Francisco, (1899-1976)
HIBBERT-WARE, Samuel, (1782-1848)
HOBSON, Bernard, (1860-1933)
HOVEY, Edmund Otis, (1862-1924)
HUXLEY, Julian Sorell, Sir, (1887-1975)
JAGGAR, Thomas Augustus, (1871-1953)
JOHNSTON, Mary Sophia, (1875-1955)
JOHNSTON-LAVIS, Henry James, (1856-1914)
JUDD, John Wesley, (1840-1916)
KEMMERLING, Georges Laure Louis, (1888-1932)
KOMOROWICZ, Maurycy, (1881-1923)
KOTÔ, Bundjirô, (1856-1935)
KUENEN, Phillip Henry, (1902-1976)
KUNO, Hisashi, (1910-1969)
LARSEN, Esper Signius Jr., (1879-1961)
LOEWINSSON-LESSING, Franc Julyevich, (1859-1939)
MACHADE E COSTA, Alfredo Augusto d'Olveira, (1870-?1952)
MALLADRA, Alessandro, (1865-1945)
MARKHININ, Y., (-)
MARTIN, George C., (1875-1943)
MARWICK, John, (1891-1978)
MATTEUCCI, Raffaele Vittorio, (1862-1909)
MERCALLI, Giuseppe, (1850-1914)
MEUNIER, Étienne Stanislas, (1843-1925)
MEYER, Hans, (1858-1929)
MONTESSUS DE BALLORE, Fernand Jean Baptiste Marie Bernard, Comte de, (1851-1923)
MOORE, Elwood S., (1879-1966)
MULFORD, John W., (1902-1964)
NAGATA, Takesi, (1913-)
OMORI, Fusakichi, (1868-1923)
ORDOÑEZ, Ezequiel, (1867-1950)
O'REILLY, Joseph P., (1829-1905)
OSWALD, Felix, (1866-1958)
PALMER, Harold Schjoeth, (1890-1959)

VULCANOLOGY 1900 to present Cont'd.

PANTÓ, Gábor, (1917-1972)
PASTORE, Franco, (1885-1958)
PLATANIA, Gaetano, (1867-1938)
PRATT, Wallace Everette, (1885-)
PUTNAM, William Clement, (1908-1963)
RECK, Hans, (1886-1937)
ROBINSON, G. D., (-)
ROCKWOOD, Charles Greene, (1843-1913)
RODOLPH, Emil, (1853-1915)
RUDOLPH, Emil, (1853-1915)
SABATINI, Venturino, (1879-1922)
SAPPER, Karl Theodor, (1866-1945)
SCHWARZBACH, Martin, (-)
SELGA, Miguel, Reverend Father, S. J., (1879-1956)
SHAND, Samuel James, (1882-1957)
SIKES, Harold Lesley, (1881-1943)
STANLEY, William Ford Robinson, (1828-1909)
STÜBEL, Alfons, (-?1904)
TANNHÄUSER, Felix, (1874-1924)
TAYLOR, Griffith, (1881-1963)
TAZIEFF, Haroun, (1914-)
THORODDSEN, Thorvaldur, (1855-1921)
TRISTÁN, José Fidel, (-1932)
TRYGGVASON, Tómas, (1907-1965)
UMBGROVE, Johannes Herman Frederik, (1899-1954)
VERBEEK, Rogier Diederick Marius, (1845-1926)
VERHOOGEN, Jean, (-)
WEIDMAN, Samuel, (1870-1945)
WELLS, Francis Gerrit, (1898-1969)
WENTWORTH, Chester Keeler, (1891-1969)
WICHMANN, Carl Ernst Arthur, (1851-1927)
WOLF, Theodor, (1841-1924)
YAMASAKI, Naomasa, (1870-1929)
ZAMBONINI, Ferruccio, (1880-1932)

GEOLOGISTS (Speciality unstated or unknown)

Up to 1800

ANDRADA E SILVA, José Bonifácio de, (1762-1838)
ASCOLI, Cacco d', (1257-1327)
ASHMOLE, Elias, (1617-1692)
ASSMAN, C. G., (-)
AUBUISSON DE VOISINS, Jean François d', (1769-1841)
BALD, Robert, (1776-1861)
BANKS, Sir Joseph, (1743-1820)
BARROW, John, Sir, (1764-1848)
BARTRAM, John, (1699-1777)
BARTRAM, William, (1739-1823)
BASSI, Ferdinando, (1714-1774)
BELGRADO, Jacopo, Padre, (fl.1738)
BERTRAND, Élie, (1712-1797)
BESSON, Henri, (fl.1786)
BLACK, Joseph, (1728-1799)
BRAHE, Tyge (Tycho), (1546-1601)
BROWN, E., (fl.1685)
BUFFON, Georges Louis Leclerc, comte de, (1707-1788)
BURNET, Thomas, (1635-1715)
BUTLER, Samuel, (1774-1839)
CANALI, Luigi, (1759-1841)
CLARKE, Edward Daniel, (1769-1822)
CLERK, John, Jr., Lord Eldin, (1757-1812)
COLLOT, Georges Henri Victor, (1750-1805)
COOPER, Thomas, (1759-1840)
COTTA, Heinrich, (1763-1844)
DAMPIER, William, (1652-1715)
DARBY, William, (1775-1854)
DARWIN, Erasmus, (1731-1802)
DEGLI ALESSANDRI, Alessandro, (1461-1523)
DESCARTES, René du Perrot, (1596-1650)
DRAKE, Daniel, (1785-1852)
DRAYTON, Michael, (1563-1631)
FORSTER, Johann Georg Adam, (1754-1794)
FORTIS, Giovanni Battista Alberto, Abate, (1741-1803)
FOURIER, Jean Baptiste Joseph, (1768-1830)
FUCHS, Leonhart, (1501-1566)
GILBERT, Davies, (1767-1839)
GMELIN, Karl Christian, (1762-1837)
GROSSE, Karl Friedrich August, (1768-1847)
GYLLENBERG, Fredrik, (fl.1750)
JOSSELYN, John, (ca.1608-1675)
KAIN, John H., (1759-1831)
KIDD, John, (1775-1851)
KLUK, Krzysztof, (1739-1796)
LEEUWENHOEK, Antoni van, (1632-1723)
LEHMANN, Johann Christoph, (1675-1739)
LESKE, Nathanael Gottfried, (1751-1786)
LESSER, Friedrich Christian, (1692-1754)
MONBODDO, James Burnett, Lord, (1714-1799)
NEWCOMEN, Thomas, (1663-1739) [PROSPECTORS, DIVINERS AND MINING ENGINEERS]
PALEY, William, Archdeacon, (1743-1805)
PARROT, Georg Friedrich, (1767-1852)

GEOLOGISTS (Speciality unstated or unknown) Up to 1800 Cont'd.

PHILLIPS, Richard, (1778?-1851)
PINI, Ermenegildo, (1739-1825)
PYE SMITH, John, (1774-1851)
PYTHAGORAS, (fl.532 B.C.)
SHERLEY, Thomas, (1638-1678)
SMITH, John, Captain, (1580-1631)
SPILBERGEN, Joris van, (1568-1620)
STIFFT, Christian Ernst, (1780-1855)
TAVERNIER, Jean Baptiste, baron d'Aubonne, (1605-1689)
TILAS, Daniel, (1712-1772)
TOULMIN, George Hoggart, (ca.1750-1817)
URE, Andrew, (1778-1857)
USHER, James, Archbishop, (1581-1656)
VALLTRAVERS, Johann Rodolph von, (1723-)
VANDELLI, Domenico, (1735-1816)
VOLCKMANN, Georg Anton, (1636-1721)
VOLNEY, Constantin François Chasseboeuf, Comte de, (1757-1820)
WELD, Isaac, Jr., (1774-1856)
WHISTON, William, (1666-1753)

1800-1849

ABERT, James W., Sr., (1820-1897)
ADAMS, John Couch, (1819-1892)
ANDRADA E SILVA, José Bonifácio de, (1762-1838)
ARRHENIUS, Johan, (1811-1889)
ASSMAN, C. G., (-)
AUBUISSON DE VOISINS, Jean François d', (1769-1841)
AULD, William, (fl.1810-1812)
BAKER, William, (1787-1853)
BALD, Robert, (1776-1861)
BANKS, Sir Joseph, (1743-1820)
BARROW, John, Sir, (1764-1848)
BARROW, John, Jr., (1808-1898)
BARTRAM, William, (1739-1823)
BARZANÒ, Gaetano, (-?1859)
BELZA, Józef, (1805-1888)
BLACK, Samuel, (1785-1841)
BRADETZKY, Samuel, (-)
BRAYLEY, Edward William, (1802-1870)
BRUIMANN, Vilmos, (1819-?1896)
BUCHANAN, John, (1819-1898)
BUCKLEY, Samuel Botsford, (1809-1884)
BURT, William Austin, (1792-1858)
BUTLER, Samuel, (1774-1839)
CALCARA, Pietro, (1819-1854)
CAMPANI, Giovanni, (1820-1891)
CANALI, Luigi, (1759-1841)
CHRISTY, Henry, (1810-1865)
CLARKE, Edward Daniel, (1769-1822)
CLERK, John, Jr., Lord Eldin, (1757-1812)
COLLOMB, Edouard, (1801-1875)
COLLOT, Georges Henri Victor, (1750-1805)
COOPER, Thomas, (1759-1840)
CORNELIUS, Elias C., (1794-1832)

GEOLOGISTS (Speciality unstated or unknown) 1800-1849 Cont'd.

COSSA, Alfonso, (-)
COTTA, Heinrich, (1763-1844)
ČŽJŽEK, Johann Baptist Anton Karl, (1806-1855)
CZYRNIAŃSKI, Emilian, (1824-)
DALLAS, William Sweetland, (1824-1890)
DARBY, William, (1775-1854)
DA RICO, Nicolò, Conte, (-?1845)
DARWIN, Erasmus, (1731-1802)
DARWIN, Francis Sacheverell, Sir, (1786-1859)
DEFFNER, Carl, (1817-1878)
DIEFFENBACH, Ernst, (1811-1855)
DRAKE, Daniel, (1785-1852)
DUMONT, André Hubert, (1809-1857)
EATON, T. D., (-?1829)
ERMAN, Georg Adolf, (1806-1877)
ESZTERHÁZY, Miklós, Herczeg, (1816-1894)
EVANS, John, (1812-1861)
FITZ-ROY, Robert, Vice-Admiral, (1805-1865)
FLEMING, John, Reverend, (1785-1857)
FORTIS, Giovanni Battista Alberto, Abate, (1741-1803)
FOURIER, Jean Baptiste Joseph, (1768-1830)
FOX, Robert Were, (1789-1877)
GALTON, Francis, Sir, (1822-1911)
GILBERT, Davies, (1767-1839)
GMELIN, Karl Christian, (1762-1837)
GREGORY, Augustus Charles, (1819-1905)
GREGORY, Joshua William, (-1850)
GREY, George, Sir, (1812-1898)
GROSSE, Karl Friedrich August, (1768-1847)
HARE, Robert, (1781-1858)
HIGGINS, Henry Hugh, Reverend, (1814-1893)
HOLLAND, Henry, Sir, (1788-?1873)
HOMAN, Christian Horrebow, (1826-1880)
HONEYMAN, David, Reverend, (1814-1889)
ISBISTER, Alexander Kennedy, (1822-1883)
JAMES, Edwin, (1797-1861)
KAIN, John H., (1759-1831)
KEATING, William Hypolitus, (1799-1840)
KIDD, John, (1775-1851)
KINGSLEY, Charles, Reverend, (1819-1875)
LABĘCKI, Hieronim, (1809-1862)
LE CONTE, Joseph, (1823-1901)
LEWIS, Tayler, (1802-)
LLOYD, J. A., Colonel, (-1856?)
LOWRY, Joseph Wilson, (1803-1879)
LUXMOORE, Edward Bouverie, (1829-1893)
MACGILLIVRAY, William, (1796-1852)
MANTELL, Walter Baldock Durant, Honourable, (1820-1895)
MARCY, Randolph Barnes, (1812-1887)
MARSCHALL, August Friedrich, Count of Burgholtzhausen and Tromsdorf, (1805-1887)
MELION, Josef Vincenc, (1813-1905)
MITCHELL, Elisha, Reverend, (1793-1857)
MIVART, St. George, (1827-1900)

GEOLOGISTS (Speciality unstated or unknown) 1800-1849 Cont'd.

MURRAY, John, III, (1808-1892)
NENDTVICH, Károly, (1811-?1892)
OKEN, Lorenz, (1779-1851)
OLBERS, Elam Willgott, (1816-1895)
O'SULLIVAN, T. H., (-1882)
PAGE, David, (1814-1879)
PALEY, William, Archdeacon, (1743-1805)
PARANDIER, A. N., (1804-1901)
PARKER, Samuel, Reverend, (1779-1866)
PAROLINI, Alberto Cav., (1788-1867)
PARROT, Georg Friedrich, (1767-1852)
PHILLIPS, Richard, (1778?-1851)
PINI, Ermenegildo, (1739-1825)
PRATT, Samuel Peace, (1789-1863)
PRUGBERGER, József, (1818-1889)
PYE SMITH, John, (1774-1851)
RAFINESQUE-SCHMALTZ, Constantine Samuel, (1783-1840)
REQUIEN, Esprit, (-1852)
RICHARDSON, John, Sir, (1787-1865)
RIVIÈRE, Auguste, (1809-1877)
RIVOT, Louis Edouard, (1820-1869)
RÓMER, Flóris, (1815-?1889)
SCHÖNN, Rudolf, (-1889)
SEELY, Henry Martyn, (1828-1917)
SELKIRK, Dunbar James Douglas, Earl of, (1809-1885)
SEUBERT, Moritz August, (1818-1878)
SILLIMAN, Benjamin, Sr., (1779-1864)
SMITH, Augustus, (1804-1872)
SOPWITH, Thomas, (1803-1879)
SPITZWEG, Carl, (1808-1885)
STIFFT, Christian Ernst, (1780-1855)
ST. JOHN, Samuel, (1813-1876)
SYMMES, John Cleves, (1780-1829)
THIELE, Friedrich, (1814-1899)
THIRRIA, Charles Edouard, (1796-1868)
TOULMIN, George Hoggart, (ca.1750-1817)
URE, Andrew, (1778-1857)
VALLTRAVERS, Johann Rodolph von, (1723-)
VANDELLI, Domenico, (1735-1816)
VILLEFAGNE D'INGIHOUL, Hilarion Noel de, Baron, (-?1837)
VOLNEY, Constantin François Chasseboeuf, Comte de, (1757-1820)
VOSKOBOINIKOV, N. A., (-)
WATERHOUSE, John, (1806-1879)
WATERTON, Charles, (1782-1865)
WELD, Isaac, Jr., (1774-1856)
WHITE, Walter, (-)
WILBERFORCE, Samuel, Bishop, (1805-1873)
WOOD, Edward, (1808-1877)
WOODROW, James, (1828-1907)
XÁNTUS, János, (1825-1894)
ZIETEN, Carl Hartwig von, (1785-1846)
ZSIGMONDY, Vilmos, (1821-1888)

GEOLOGISTS (Speciality unstated or unknown) Cont'd.

1850-1899

ABBE, Ernst, (1840-1905)
ABERT, James W., Sr., (1820-1897)
ADAMS, John Couch, (1819-1892)
ADAMS, Mikhail Ivanovich, (-)
AEBERHARDT, Berchtold, (1872-1912)
ALMERA, Jaume, Canon, (1854-1919)
ANDERSSON, Axel R., (1858-1952)
ANDERSSON, Gunnar, (1865-1928)
ANDREWS, W. S., (1847-1929)
ARRHENIUS, Johan, (1811-1889)
BAKER, William, (1787-1853)
BALD, Robert, (1776-1861)
BALDACCI, Luigi, (1850-1927)
BALLOU, William Hosea, (1857-1937)
BANNISTER, Henry Martyn, (-)
BARNES, William Henry, (1845-1935)
BARROW, John, Jr., (1808-1898)
BÄRTLING, Richard, (1878-1936)
BARZANÒ, Gaetano, (-?1859)
BAUER, Grace M., (1878-1965)
BAYER, František, (1854-1936)
BECQUEREL, Henri, (1852-1908)
BEEDE, Joshua William, (1871-1940)
BEHREND, Paul, (-?1905)
BELINFANTE, Leo Baptiste Louis, (1861-1937)
BELT, Thomas, (1832-1878)
BELZA, Józef, (1805-1888)
BERDAU, Feliks, (-?1888)
BÉZIER, T., (-1926)
BIAGI, Giuseppi, (1859-1905)
BICKMORE, Albert Smith, (1839-1914)
BØGGILD, Ove Balthazar, (1872-1956)
BÖLSCHE, Wilhelm, (1843-1893)
BOMBICCI PORTA, Luigi, (1833-1903)
BONOMINI, Celestino, Don, (1866-?1946)
BORGSTRÖM, Leonard Henrik, (1876-?1954)
BOWMAN, Amos, (1839-1894)
BRAYLEY, Edward William, (1802-1870)
BROWNRIDGE, Charles, Colonel, (?1862-1951)
BRUIMANN, Vilmos, (1819-?1896)
BRYAN, William Jennings, (1860-1925)
BUCHANAN, John, (1819-1898)
BUCKLEY, Ernest Robertson, (1872-1912)
BUCKLEY, Samuel Botsford, (1809-1884)
BUOHY, Victor, (1843-1887)
BURT, William Austin, (1792-1858)
BUTLER, Samuel, (1835-1902)
CALCARA, Pietro, (1819-1854)
CAMMIADE, Louis Aimé, (1872-?1951)
CAMPANI, Giovanni, (1820-1891)
CHALUPCZYŃSKI, Adam, (-1887)
CHAPIN, James Henry, (1832-1892)
CHOROSZEWSKI, Wincenty Ignacy Jędrzej, (1845-1901)

GEOLOGISTS (Speciality unstated or unknown) 1850-1899 Cont'd.

CHRISTENSEN, Carl Christian, (1860-1935)
CHRISTY, Henry, (1810-1865)
CLARK, John Willis, (1833-1910)
CLÉMENT, Charles Adolphe Antoine, (1860-1933)
CLERICI, Enrico, (1862-1938)
COLE, Edward Maule, Reverend, (1833-1911)
COLLOMB, Edouard, (1801-1875)
CONTI, Piero Ginori, (1856-1939)
COSSA, Alfonso, (-)
COZZAGLIO, Arturo, (1862-)
CRAMP, Alfred William, (?1873-1950)
CROFTS, William Hastings, (1861-1935)
ČŽJŽEK, Johann Baptist Anton Karl, (1806-1855)
DALLAS, William Sweetland, (1824-1890)
DAMASIO, José Victorio, (-)
DAMPIER, William Cecil, Sir, (1867-1952)
DARBY, William, (1775-1854)
DARROW, Clarence, (1857-1938)
DARWIN, Francis Sacheverell, Sir, (1786-1859)
DAWSON, George Mercer, (1849-1901)
DE AGOSTINI, Giovanni, (-)
DEFFNER, Carl, (1817-1878)
DE GREGORIO, Antonio, Marchese, (1855-1930)
DEMARCHI, Lamberto, (1841-1907)
DIEFFENBACH, Ernst, (1811-1855)
DI MARCHI, Luigi, (1857-1936)
D'OSSAT, Gioacchino De Angelis, (1865-1957)
DRAKE, Daniel, (1785-1852)
DREGER, Julius, (1861-1945)
DRUMMOND, Henry, (1851-1897)
DUBICKI, Józef, (-?1890)
DUMONT, André Hubert, (1809-1857)
DYBOWSKI, Benedykt, (1833-1930)
ELLISON, John, (1859-1936)
ERMAN, Georg Adolf, (1806-1877)
ESZTERHÁZY, Miklós, Herczeg, (1816-1894)
EVANS, John, (1812-1861)
FELLENIUS, Wolmar, (1876-?1957)
FENNELL, Charles William, (?1863-1935)
FENNEMA, Reinder, (1849-?1897)
FENNER, Edward George, (?1873-1941)
FINO, Vincenzo, (1848-1914)
FISKE, John, (1842-1901)
FITZ-ROY, Robert, Vice-Admiral, (1805-1865)
FLEMING, John, Reverend, (1785-1857)
FLODKVIST, Herman, (1875-)
FOURNEL, Henri, (-)
FOX, Robert Were, (1789-1877)
FRIEDLAENDER, Immanuel, (-?1918)
FUCHS, Carl Wilhelm C., (1837-1886)
FULLER, Homer Taylor, (1838-1910)
FUNKQUIST, Herman Per Anton, (1870-1952)
GALLI, Edward Maryan de, (-?1893)
GALTON, Francis, Sir, (1822-1911)

GEOLOGISTS (Speciality unstated or unknown) 1850-1899 Cont'd.

GERBER, Frigyes, (1856-1897)
GERRARD, John, (?1850-1920)
GERTZ, Otto Daniel, (1878-1948)
GILL, David, Sir, (1843-1914)
GIULIETTI, G. M., (-?1928)
GLENN, John, (-)
GLINOJECKI, Jan, (-)
GORDON, Charles Henry, (1857-1934)
GRAD, Charles, (1842-?1905)
GREGORY, Augustus Charles, (1819-1905)
GREGORY, Joshua William, (-1850)
GREY, George, Sir, (1812-1898)
GRIGGS, Robert F., (-)
GÜLDNER, Max, (1872-1947)
GUTMANN, Max von, Ritter, (1857-1930)
HAITINGER, Max, (1868-1946)
HANUSZ, István, (1840-1909)
HARE, Robert, (1781-1858)
HAUAN, Kund Møller, (1840-1898)
HESSELMAN, Oscar August Henrik Wilhelm, (1874-1943)
HICE, Richard Robert, (1865-1925)
HIGGINS, Henry Hugh, Reverend, (1814-1893)
HILLIERS, John K., (-)
HINTON, Henry Arthur, (?1874-1937)
HIRSZBERG, Franciszek, (1863-1933)
HOLBROOK, Levi, (1836-1922)
HOLLAND, Henry, Sir, (1788-?1873)
HOLLANDE, Dieudonné, (1845-1921)
HOLLENDER, Artur, (-?1946)
HOLMES, William Henry, (1846-1933)
HOMAN, Christian Horrebow, (1826-1880)
HONEYMAN, David, Reverend, (1814-1889)
HOTBLACK, John Turner, (1848-1915)
HOWITT, Alfred William, (1830-1908)
HOWORTH, Henry Hoyle, Sir, (1842-1923)
HÜCKEL, Edward, (-?1896)
ISBISTER, Alexander Kennedy, (1822-1883)
JACK, Robert Logan, (1845-1921)
JACKSON, William Henry, (1843-1942)
JAMES, Edwin, (1797-1861)
JENKIN, Henry Charles Fleeming, (1833-1885)
JONES, William Rupert, (1855-1915)
JUNGHUHN, Franz Wilhelm, (1808-1864)
KENNICOTT, Robert, (1835-1866)
KINGSLEY, Charles, Reverend, (1819-1875)
KJELLMARK, Knut, (1866-?1944)
KNIES, Jan, (1860-1925)
KOLTHOFF, Gustaf, (1845-1914)
KOVALEVSKY, Alexander, (1840-1901)
LABĘCKI, Hieronim, (1809-1862)
LAGRELIUS, Axel, (1863-1944)
LEATHERS, Walter S., (1874-)
LE CONTE, Joseph, (1823-1901)
LEINER, Ludwig, (1830-1901)

GEOLOGISTS (Speciality unstated or unknown) 1850-1899 Cont'd.

LEWIS, Tayler, (1802-)
LILJEVALL, Georg Gideon, (1848-1928)
LINDSTRÖM, Axel Fredrik, (1839-1911)
LITTLE, George, (1838-1924)
LLOYD, J. A., Colonel, (-1856?)
LOMNICKI, Marian Alojzy R., (1845-1915)
LOTTI, Bernardino, (1847-1933)
LOWRY, Joseph Wilson, (1803-1879)
LUKASIEWICZ, Ignacy, (-)
LUTHER, Daniel Dana, (1840-1923)
LUXMOORE, Edward Bouverie, (1829-1893)
MacCURDY, George Grant, (1863-1947)
MACGILLIVRAY, William, (1796-1852)
MACINTOSH, Charles, (1839-1922)
MACOUN, John, (1831-1920)
MANTELL, Walter Baldock Durant, Honourable, (1820-1895)
MARCY, Randolph Barnes, (1812-1887)
MARIANI, Ernesto, (-?1939)
MARINELLI, Giovanni, (1846-1900)
MARSCHALL, August Friedrich, Count of Burgholtzhausen and Tromsdorf, (1805-1887)
MARTIN, Daniel Strobel, (1842-1925)
MAUCH, Carl Gottlieb, (1837-1875)
MAUCHER, Wilhelm, (1879-1930)
McLEAN, William, (?1877-1946)
MELION, Josef Vincenc, (1813-1905)
MENEGUZZO, Giovanni, (-)
MICHALSKI, Stanislaw, (1865-1949)
MITCHELL, Elisha, Reverend, (1793-1857)
MIVART, St. George, (1827-1900)
MOORE, Andrew Charles, (1866-1928)
MOORE, Henry Cecil, (1835-1908)
MURDOCH, James Barclay, (1831-1906)
MURRAY, John, III, (1808-1892)
MUSHKETOV, Ivan Vassilievitch, (1850-1902)
NEGRI, Cristoforo, (-)
NENDTVICH, Károly, (1811-?1892)
NESSIG, Wilhelm Robert, (1861-1932)
NIELSEN, Kristian Brünnich, (1872-1942)
NIEVO, Ippolito, (1867-1916)
NIEWIADOMSKI, Wincenty, (-?1892)
NILSSON, Lars Albert, (1860-1906)
OKEN, Lorenz, (1779-1851)
OLBERS, Elam Willgott, (1816-1895)
OSSOWSKI, Godfryd, (-?1897)
O'SULLIVAN, T. H., (-1882)
PAGE, David, (1814-1879)
PARANDIER, A. N., (1804-1901)
PARKER, Samuel, Reverend, (1779-1866)
PAROLINI, Alberto Cav., (1788-1867)
PARROT, Georg Friedrich, (1767-1852)
PATTEN, William, (1861-1932)
PELLOUX, Alberto, (-)
PETERSON, Olof August, (1865-1933)

GEOLOGISTS (Speciality unstated or unknown) 1850-1899 Cont'd.

PETERSSON, Walfrid, (1862-1933)
PETRA (PETANDER), August Herman, (1853-1928)
PHRAGMÉN, Gösta, (-?1944)
PIWOWAR, Adam, (1874-1939)
PLATT, William, (?1867-1943)
POZZI, Giovanni, (-?1891)
PRATT, Samuel Peace, (1789-1863)
PRENTICE, Sydney Clark, (1870-1943)
PRUGBERGER, József, (1818-1889)
PYE SMITH, John, (1774-1851)
RAMA RAU, S. Sethu, Rao Bahadur, (1874-1929)
RAMÈS, Jean-Baptiste, (1832-1894)
RANSON, John, (?1833-1866)
RAU, Karl, (1870-1966)
REHMAN, Antoni, (-)
REINACH, Albert von, Baron, (1842-1905)
REQUIEN, Esprit, (-1852)
RETZIUS, Emilia, (-?1904)
RICHARDSON, John, Sir, (1787-1865)
RICHERT, Johann Gustaf, (1857-1954)
RIVA-PALAZZI, Giovanni, (1838-1913)
RIVIÈRE, Auguste, (1809-1877)
RIVOT, Louis Edouard, (1820-1869)
ROBERTS, Robert Davis, (1851-1911)
ROBINSON, Thomas, (1828-1906)
ROMANES, George John, (1848-1894)
RÓMER, Flóris, (1815-?1889)
ROSENKJAER, H. N., (1851-1907)
ROVERETO, Gaetano, (1870-1952)
SAJÓHELYI, Frigyes, (1848-1940)
SARASIN, Edouard, (1843-1917)
SARASIN, Paul, (1856-1929)
SAWICKI, Ludomir, (-?1928)
SCHAEFFER, Charles Frederic August, (1860-1903)
SCHÖNN, Rudolf, (-1889)
SCHREIBER, Hans, (1859-?1936)
SEAMAN, Elizabeth Cochrane, (1865-1922)
SEELY, Henry Martyn, (1828-1917)
SELKIRK, Dunbar James Douglas, Earl of, (1809-1885)
SERNANDER, Rutger, (1866-1944)
SEUBERT, Moritz August, (1818-1878)
SIEMIRADZKI, Józef, (1858-1933)
SIGMOND, Elek, (1873-1939)
SILLIMAN, Benjamin, Sr., (1779-1864)
SJÖGREN, Sten Anders Hjalmar, (1856-1922)
ŚLOŚARSKI, Antoni, (-?1897)
SMITH, Augustus, (1804-1872)
SOPWITH, Thomas, (1803-1879)
SORET, Charles, (-?1903)
SPIERS, William, Reverend, (-)
SPITZWEG, Carl, (1808-1885)
STIFFT, Christian Ernst, (1780-1855)
ST. JOHN, Samuel, (1813-1876)
SULIMIRSKI, Filip, (-?1885)

GEOLOGISTS (Speciality unstated or unknown) 1850-1899 Cont'd.

SZTOLCMAN, J., (-)
TANNER, Väinö, (1881-1948)
TELLINI, Achille, (1866-1938)
TERNETZ, Carl, (1870-1928)
TERRIGI, Guglielmo, (1831-1892)
THIELE, Friedrich, (1814-1899)
THIRRIA, Charles Edouard, (1796-1868)
TITTONI, Tommaso, (1855-1931)
TONINI, Lorenzo, ([1859]-1926)
TREJDOSIEWICZ, Jan, (-)
TRUSCOTT, Samuel John, (?1870-1950)
TUXEN, Christian Fr. A., (1850-1903)
UGOLINI, Riccardo, (1874-1936)
UHLIG, Carl Ludwig Gustav, (1872-?1938)
URE, Andrew, (1778-1857)
UZIELLI, Gustavo, (1839-1911)
VACCHELLI, Nicola, Generale, (1870-1932)
VERRI, Pietro, (-)
VOIGT, Friedrich Gustav, (1844-1886)
WALLIN, Gustaf, (1871-1940)
WATERHOUSE, John, (1806-1879)
WATERTON, Charles, (1782-1865)
WELD, Isaac, Jr., (1774-1856)
WHITE, Walter, (-)
WILBERFORCE, Samuel, Bishop, (1805-1873)
WILLIAMS, Arthur James, (1877-1954)
WINKLER, Clemens Alexander, (1838-1904)
WIŚNIOWSKI, Sygurd, (-)
WOOD, Edward, (1808-1877)
WOODROW, James, (1828-1907)
WORMS, Henry de, Baron, Lord Pirbright, (1840-1903)
XÁNTUS, János, (1825-1894)
YNGSTRÖM, Lars, (1870-1944)
YOUNG, George William, (1862-1929)
ZAHÁLKA, Čeněk, (1856-1938)
ZÀMARA, Giuseppe, (1837-1917)
ZAPALOWICZ, Hugo, (-)
ZAWISZA, Jan, (-)
ZDZITOWIECKI, Seweryn, (-?1881)
ŽEBROWSKI, Teofil, (-?1887)
ZIELENIEWSKI, Michal, (-?1896)
ZSIGMONDY, Vilmos, (1821-1888)
ŽUKOWSKI, Wladyslaw, (1873-1952)

1900 to present

ADAMS, Mikhail Ivanovich, (-)
AEBERHARDT, Berchtold, (1872-1912)
AIROLDI, Marco, (1900-1937)
ALMERA, Jaume, Canon, (1854-1919)
ANDERSSON, Axel R., (1858-1952)
ANDERSSON, Gunnar, (1865-1928)
ANDREWS, W. S., (1847-1929)
ARMASHEVSKY, Peter Jakovlevich, (-1919)
ARRHENIUS, Olof, (1895-)

GEOLOGISTS (Speciality unstated or unknown) 1900 to present Cont'd.

ASSARSSON, Gunnar, (-1963)
ATHANASIU, Sava, (-?1946)
AZARI, Mario, (-?1932)
BABIČ, Bogdan, (1891-1961)
BACKLUND, Erik, (1892-1938)
BAIN, Samuel McCutcheon, (-)
BALDACCI, Luigi, (1850-1927)
BALLOU, William Hosea, (1857-1937)
BANDL, Erich, (1880-1960)
BARNES, William Henry, (1845-1935)
BÄRTLING, Richard, (1878-1936)
BATTAGLIA, Rafaello, (1896-1958)
BAUER, Grace M., (1878-1965)
BAYER, František, (1854-1936)
BEASLEY, Cyril George, (1901-?1957)
BECQUEREL, Henri, (1852-1908)
BEEDE, Joshua William, (1871-1940)
BEHREND, Paul, (-?1905)
BELINFANTE, Leo Baptiste Louis, (1861-1937)
BENEDICKS, Carl, (-1958)
BERCKHEMER, Fritz, (1890-?1954)
BÉZIER, T., (-1926)
BIAGI, Giuseppi, (1859-1905)
BICKMORE, Albert Smith, (1839-1914)
BJÖRNSSON, Sven, (1905-1950)
BJORULF, Hjalmar, (-?1956)
BLEDSOE, Lucybelle, (1923-1966)
BOAGA, Giovanni, (1902-1961)
BODMAN, Gösta, (-1961)
BØGGILD, Ove Balthazar, (1872-1956)
BOMBICCI PORTA, Luigi, (1833-1903)
BONOMINI, Celestino, Don, (1866-?1946)
BORGSTRÖM, Leonard Henrik, (1876-?1954)
BREDDIN, Hans, (1900-1973)
BRENNER, Thord, (1892-1949)
BROWNRIDGE, Charles, Colonel, (?1862-1951)
BRUET, Edmond, (1890-)
BRYAN, William Jennings, (1860-1925)
BUCKLEY, Ernest Robertson, (1872-1912)
BUMP, James Dye, (1903-1959)
BURDICK, Clifford, (-)
BURTON, Ernest St. John, (?1874-1962)
BUTLER, Samuel, (1835-1902)
CALDENIUS, Carl, (-1961)
CAMMIADE, Louis Aimé, (1872-?1951)
CANNON, Herbert Graham, (1897-1963)
CARPANESE, Tullio, (-?1940)
CEHÅK, Karl, (1903-1957)
CHOROSZEWSKI, Wincenty Ignacy Jędrzej, (1845-1901)
CHRISTENSEN, Carl Christian, (1860-1935)
CLARK, John Willis, (1833-1910)
CLÉMENT, Charles Adolphe Antoine, (1860-1933)
CLERICI, Enrico, (1862-1938)
COLE, Edward Maule, Reverend, (1833-1911)

GEOLOGISTS (Speciality unstated or unknown) 1900 to present Cont'd.

CONTI, Piero Ginori, (1856-1939)
COZZAGLIO, Arturo, (1862-)
CRAMER, Rudolf, (1882-1949)
CRAMP, Alfred William, (?1873-1950)
CREDNER, Wilhelm, (1892-1948)
CROFTS, William Hastings, (1861-1935)
DAMASIO, José Victorio, (-)
DAMPIER, William Cecil, Sir, (1867-1952)
DANYSZ-FLESZAROWA, Regina, (1888-1969)
DARÁNYI, Ignác, (-1925)
DARROW, Clarence, (1857-1938)
DAS-GUPTA, Hemchandra C., (1878-1933)
DAWSON, George Mercer, (1849-1901)
DE ARAUJO FERRAZ, Jorge Palmiro, (1883-1926)
DE GEER, Sten, (1886-1933)
DE GREGORIO, Antonio, Marchese, (1855-1930)
DEMARCHI, Lamberto, (1841-1907)
D'ERASMO, Geremia, (1887-1962)
DE WAAL, Philip Henry Overbeck, Reverend, (-1942)
DI MARCHI, Luigi, (1857-1936)
DODD, C., (-1928)
D'OSSAT, Gioacchino De Angelis, (1865-1957)
DREGER, Julius, (1861-1945)
DRURY, James, (1879-)
DYBOWSKI, Benedykt, (1833-1930)
EDDINGTON, Arthur Stanley, Sir, (1882-1944)
EHRINGHAUS, Arthur, (-)
EKSTRÖM, Gunnar, (-1960)
ELLISON, John, (1859-1936)
ENEROTH, Olof, (-1945)
ERIKSSON, J. V., (1883-1930)
FELLENIUS, Wolmar, (1876-?1957)
FENNELL, Charles William, (?1863-1935)
FENNER, Edward George, (?1873-1941)
FERUGLIO, Egidio, (1897-1954)
FINO, Vincenzo, (1848-1914)
FISHER, Ronald Aylmer, (1890-1962)
FISKE, John, (1842-1901)
FLODKVIST, Gunnar, (-?1946)
FLODKVIST, Herman, (1875-)
FRANCHI, Secondo, (-?1939)
FRANK, Wilhelm, (1891-1914)
FREIRE DE ANDRADE, Carlos Benot, (1893-1956)
FRIEDLAENDER, Immanuel, (-?1918)
FRÖDIN, John, (-1961)
FULLER, Homer Taylor, (1838-1910)
FUNKQUIST, Herman Per Anton, (1870-1952)
FUSSENEGGER, Siegfried, (1894-1960)
GALTON, Francis, Sir, (1822-1911)
GERBER, Frigyes, (1856-1897)
GERRARD, John, (?1850-1920)
GERTZ, Otto Daniel, (1878-1948)
GILL, David, Sir, (1843-1914)
GIULIETTI, G. M., (-?1928)

GEOLOGISTS (Speciality unstated or unknown) 1900 to present Cont'd.

GLEAVES, John T., (-1969)
GORDON, Charles Henry, (1857-1934)
GRAD, Charles, (1842-?1905)
GRANÖ, Johannes Gabriel, (1882-1956)
GREIG, Arthur, (1893-)
GRIGGS, Robert F., (-)
GRILL, Emanuele, (1884-1961)
GÜLDNER, Max, (1872-1947)
GUSTAFSSON, Johan Peter, (-?1943)
GUTMANN, Max von, Ritter, (1857-1930)
GWINNELL, Russell Frost, (1880-1913)
HADDING, Assar, (1886-1962)
HAHN, Emily, (1904-)
HAITINGER, Max, (1868-1946)
HALLGREN, Gunnar, (1907-)
HANIEL, Curt Alfons, (1884-1914)
HANUSZ, István, (1840-1909)
HARRASSOWITZ, Hermann, (-1956)
HEIMANS, E., (-?1937)
HELLAND-HANSEN, Björn, (-1958?)
HESSELMAN, Oscar August Henrik Wilhelm, (1874-1943)
HEWITT, Henry Dixon, (-?1966)
HICE, Richard Robert, (1865-1925)
HINTON, Henry Arthur, (?1874-1937)
HIRSZBERG, Franciszek, (1863-1933)
HÖGBOM, Ivar, (1892-1962)
HOLBROOK, Levi, (1836-1922)
HOLLANDE, Dieudonné, (1845-1921)
HOLLENDER, Artur, (-?1946)
HOLLOWAY, Lionel E., (1898-1965)
HOLMBOE, Jens, (-?1943)
HOLMES, William Henry, (1846-1933)
HOLMSTRÖM, Leonard, (-?1922)
HORN, Hans Lorenz, (1918-1941)
HÖRNER, Nils Gustaf, (1896-1950)
HOTBLACK, John Turner, (1848-1915)
HOWARTH, W. E., (-1956)
HOWITT, Alfred William, (1830-1908)
HOWORTH, Henry Hoyle, Sir, (1842-1923)
HUBRECHT, P. F., (-?1929)
HUTCHINS, Forbes M., (-)
IPPEN, Joseph Anton, (-?1918)
JACK, Robert Logan, (1845-1921)
JACKSON, William Henry, (1843-1942)
JÄGER, Robert, (1890-1915)
JAMOTTE, André, (1903-1951)
JEWELL, Willard B., (-)
JOHANSSON, Simon, (1881-1944)
JONES, William Rupert, (1855-1915)
JUNGHUHN, Franz Wilhelm, (1808-1864)
KANERVA, John, (1882-)
KAUDERN, Walter, (1881-1924)
KJELLMARK, Knut, (1866-?1944)
KLINGER, Fritz Erdmann, (-?1955)

GEOLOGISTS (Speciality unstated or unknown) 1900 to present Cont'd.

KNIES, Jan, (1860-1925)
KOLTHOFF, Gustaf, (1845-1914)
KONTHER, Fritz, (1895-1968)
KOVALEVSKY, Alexander, (1840-1901)
KOWALSKI, Ludwik, (1885-1943)
KREJCI-GRAF, Karl, (-?1956)
KROPACZEK, Boleslaw, (1886-1914)
KRYZ, Ferdinand, (1879-1958)
KULHAY, Gyula, (1910-1945)
KÜMEL, Friedrich, (1908-1955)
LAGRELIUS, Axel, (1863-1944)
LAKE, Horace Alexander, (1887-1957)
LEATHERS, Walter S., (1874-)
LE CONTE, Joseph, (1823-1901)
LEINER, Ludwig, (1830-1901)
LILJEVALL, Georg Gideon, (1848-1928)
LINDSTRÖM, Axel Fredrik, (1839-1911)
LINSBAUER, Alfred, (1943-1969)
LITTLE, George, (1838-1924)
LOMNICKI, Marian Alojzy R., (1845-1915)
LORENZI, Arrigo, (-1953)
LOTTI, Bernardino, (1847-1933)
LOW, Julian W., (1904-1968)
LUTHER, Daniel Dana, (1840-1923)
MacCURDY, George Grant, (1863-1947)
MACINTOSH, Charles, (1839-1922)
MACOUN, James Melville, (-1920)
MACOUN, John, (1831-1920)
MAGISTRETTI, Luigi, (-?1959)
MAUCHER, Wilhelm, (1879-1930)
McLAUGHLIN, R. P., (-)
McLEAN, William, (?1877-1946)
MELION, Josef Vincenc, (1813-1905)
MERRICK, F. Kermit, (1915-1963)
MICHALSKI, Stanislaw, (1865-1949)
MICHELS, Franz-Xaver, (1900-1973)
MIKKOLA, Erkki, (-?1940)
MOORE, Andrew Charles, (1866-1928)
MOORE, Henry Cecil, (1835-1908)
MORANDINI, Giuseppe, (1907-1969)
MUHAMMAD, Wali, (1935-1963)
MURDOCH, James Barclay, (1831-1906)
MUSHKETOV, Ivan Vassilievitch, (1850-1902)
NÄGELE, Julius, (1906-1973)
NASH, Carlton S., (1914-)
NATURSKI, Jan, (-1936)
NESSIG, Wilhelm Robert, (1861-1932)
NIELSEN, Kristian Brünnich, (1872-1942)
NIETHAMMER, Gottlob August, (1882-1915)
NIEVO, Ippolito, (1867-1916)
NILSSON, Lars Albert, (1860-1906)
NORDENSKIÖLD, Karl Erik, (-?1954)
NORIN, Erik, (-)
OLSEN, Gregers Vang, (1929-1960)

GEOLOGISTS (Speciality unstated or unknown) 1900 to present Cont'd.

OSVALD, Hugo, (1892-)
PÅLSSON, Sveinn, (-)
PARANDIER, A. N., (1804-1901)
PATTEISKY, Karl, (-?1967)
PATTEN, William, (1861-1932)
PEIRANO, Abel Antonio, (1896-1969)
PERRIER, Carlo, (1886-1948)
PETERHANS, Emil, (1899-1931)
PETERSON, Olof August, (1865-1933)
PETRA (PETANDER), August Herman, (1853-1928)
PHRAGMÉN, Gösta, (-?1944)
PIWOWAR, Adam, (1874-1939)
PLATT, William, (?1867-1943)
PLEININGER, Felix, (-?1954)
PONIATOWSKI, Stanislaw Augusta, (1884-)
PRENTICE, Sydney Clark, (1870-1943)
PRICHARD, Hesketh Vernon Hesketh, (1876-1922)
PRINCIPI, Paolo, (-)
RABOWSKI, Ferdynand, (1884-1940)
RAMA RAU, S. Sethu, Rao Bahadur, (1874-1929)
RANSON, John, (?1833-1866)
RATHSBURG, Alfred, (-)
RAU, Karl, (1870-1966)
REHMAN, Antoni, (-)
REINACH, Albert von, Baron, (1842-1905)
RETZIUS, Emilia, (-?1904)
RICHERT, Johann Gustaf, (1857-1954)
RIVA-PALAZZI, Giovanni, (1838-1913)
ROBERTS, Robert Davis, (1851-1911)
ROBINSON, Thomas, (1828-1906)
ROSENKJAER, H. N., (1851-1907)
ROVERETO, Gaetano, (1870-1952)
SAJÓHELYI, Frigyes, (1848-1940)
SARASIN, Edouard, (1843-1917)
SARASIN, Paul, (1856-1929)
SAWICKI, Ludomir, (-?1928)
SCHAEFFER, Charles Frederic August, (1860-1903)
SCHARF, Willi, (1896-1971)
SCHMÖGER, Carl August, (-1970?)
SCHREIBER, Hans, (1859-?1936)
SEAMAN, Elizabeth Cochrane, (1865-1922)
SEELY, Henry Martyn, (1828-1917)
SEEMAN, Fritz, (1884-1914)
SERNANDER, Rutger, (1866-1944)
SIDENBLADH, Elias, (1836-1914)
SIEMIRADZKI, Józef, (1858-1933)
SIGMOND, Elek, (1873-1939)
SIMMONS, Eric Warr, Lieutenant, (1893-1915)
SIMONOVIĆ, Radivoj, (-1950)
SJÖGREN, Sten Anders Hjalmar, (1856-1922)
SOLOMONICA, Paul, (1908-1954)
SOWERBY, Arthur de Carle, (1885-1954)
STEFANINI, Giuseppe, (1882-1938)
STEINER-BALZER, Alfred, (1879-1968)

GEOLOGISTS (Speciality unstated or unknown) 1900 to present Cont'd.

STRAKHOV, Nikolai Mikhailovich, (1900-)
TAMMEKANN, August, (1894-1959)
TANNER, Väinö, (1881-1948)
TELLINI, Achille, (1866-1938)
TERNETZ, Carl, (1870-1928)
TITTONI, Tommaso, (1855-1931)
TONINI, Lorenzo, ([1859]-1926)
TRIBUSSON, Boris, (1924-1957)
TRUSCOTT, Samuel John, (?1870-1950)
TUXEN, Christian Fr. A., (1850-1903)
UGOLINI, Riccardo, (1874-1936)
UHLIG, Carl Ludwig Gustav, (1872-?1938)
UZIELLI, Gustavo, (1839-1911)
VACCHELLI, Nicola, Generale, (1870-1932)
VELIKOVSKY, Immanuel, (1895-)
VERRI, Pietro, (-)
WALLÉN, Axel, (1877-1935)
WALLIN, Gustaf, (1871-1940)
WERENSKIOLD, Werner, (-1962)
WEYL, Richard, (-)
WIENHOLZ, Eva, (1930-1969)
WILLIAMS, Arthur James, (1877-1954)
WINKLER, Clemens Alexander, (1838-1904)
WOLF, Key, (1886-1960)
WOODROW, James, (1828-1907)
WORMS, Henry de, Baron, Lord Pirbright, (1840-1903)
WRÅK, Walter Olof, (-?1933)
YNGSTRÖM, Lars, (1870-1944)
YOUNG, George William, (1862-1929)
ZACCAGNA, Domenico, (-?1940)
ZAHÅLKA, Čeněk, (1856-1938)
ZALESKI, Stanislaw, (-?1933)
ZÀMARA, Giuseppe, (1837-1917)
ZAPALOWICZ, Hugo, (-)
ZAPART, Janusz Stanislaw, (1930-1961)
ZENARI, Silvia, (1895-1956)
ŽUKOWSKI, Wladyslaw, (1873-1952)

Time unknown

CAPORIACCO, Ludovico di, (-)
D'AREZZO, Restoro, (-)
DULHUNTY, John, (-)
JURKIEWICZ, Karol, (-)
KARSTEN, Carl Johann, (-)
LASZKIEWICZ, Antoni, (-)
LUBOMESKI, Augusz, (-)
MORGAN, William Fellowes, (-)
PISULEWSKI, Szymon, (-)
SAPALSKI, Jozef, (-)
SEVASTIANOV, D. P., (-)

GEOLOGISTS (Speciality unstated or unknown) Time unknown Cont'd.

SZAFARKIEWICZ, Jozef, (-)
VOITESTI, Ion Popescu, (-)